DRUG METABOLISM HANDBOOK

DRUG METABOLISM HANDBOOK

CONCEPTS AND APPLICATIONS

Edited by

Ala F. Nassar, Paul F. Hollenberg, and JoAnn Scatina

A JOHN WILEY & SONS, INC., PUBLICATION

Library of Congress Cataloging-in-Publication Data

Drug metabolism handbook : concepts and applications / [edited by] Ala F. Nassar,
Paul F. Hollenberg, JoAnn Scatina.
    p. ; cm.
  Includes bibliographical references and index.
  ISBN 978-0-470-11803-0 (cloth)
    1. Drugs–Metabolism–Handbooks, manuals, etc. I. Nassar, Ala F. II. Hollenberg,
Paul F. III. Scatina, JoAnn.
    [DNLM: 1. Pharmaceutical Preparations–metabolism. QV 38 D79366 2009]
  RM301.55.D766 2009
  615′.7—dc22

                                                                            2008035478

Printed in the United States of America

10 9 8 7 6 5 4 3 2 1

■■■■ CONTENTS

v

9. Approaches to Performing Metabolite Elucidation: One Key to Success in Drug Discovery and Development

Ala F. Nassar

10. Structural Modifications of Drug Candidates: How Useful Are They in Improving Metabolic Stability of New Drugs? Part I: Enhancing Metabolic Stability

Ala F. Nassar

11. Structural Modifications of Drug Candidates: How Useful Are They in Improving PK Parameters of New Drugs? Part II: Drug Design Strategies

Ala F. Nassar

PREFACE

Studies on absorption, distribution, metabolism, and elimination-toxicology (ADME-Tox) have progressed over the years to the point where they now play a major role in drug discovery and development. Until the late 1980s, the primary role of drug metabolism groups in the pharmaceutical industry was to provide ADME-Tox information to support the regulatory package.

During the last decade, with the rapid rise in new molecular entities (NMEs) arising from combinatorial chemistry and high-throughput biological screening, an urgent need has arisen for the determination of the ADME properties of these NMEs at very early stages in the drug discovery pipeline in order to facilitate the selection of "ideal" drug candidates for further development. Back integration of key studies into the discovery phase has resulted in earlier identification of potential DM/PK and safety liabilities. This information aids in decision making and in many instances, has been incorporated into criteria for compound advancement into the development phase. Given the need for earlier and more rapid evaluation of a larger number of compounds, drug metabolism scientists have developed and incorporated novel approaches into early drug discovery including "humanized" *in vitro*-based cell systems, sophisticated automation, higher-throughput ADME assays and screens, ultrasensitive analytical technologies, and computational models in order to accelerate the examination of the drug metabolism pathways for their NMEs. The success of this approach is evident, as the number of failures due to DM/PK liabilities has dramatically decreased.

Clearly, there is a growing need for improving and expanding education of students as well as current practitioners involved in investigations in these areas. The needs for continuing education in the rapidly expanding and dynamic area of ADME-Tox studies are not being met at the university level and oftentimes this is being taught in a piecemeal fashion on the job in pharmaceutical industries. The goal of this book is to provide a systematic approach for the education of younger researchers and students at the university level and to improve their knowledge of drug metabolism by presenting in-depth coverage of the drug disposition process, pharmacokinetic drug–drug interactions, theory, and evaluation approaches and improving the decision-making process in structural modification of drug candidates to reduce toxicity.

ADME-Tox experts in the field from both industry and academia have joined forces and offered their time to write this book, introducing students to modern concepts and practices of ADME-Tox. This book provides the basic training in the area of drug metabolism and disposition, including training programs for students and new employees in the pharmaceutical industry. Mastery of the material in this text will allow them to apply state-of-the-art research tools to *in vitro* and *in vivo* metabolism studies and contribute greatly to their abilities to perform pharmaceutical research in support of industrial, academic, and regulatory agency needs. This textbook consists of five parts. Part I provides an introduction to drug metabolism. Part II presents the *in vitro* and *in vivo* technologies. Part III presents an important area of drug–drug interaction. Part IV discusses the toxicity and Part V presents the regulatory perspectives.

The editors and contributing authors greatly appreciate the commitment of the publishers to make this book available to our scientific colleagues in developing countries to enhance their knowledge in the area of ADME-Tox and to help them in furthering their careers in this very important area of research. Finally, we thank our many colleagues worldwide who have contributed to the development of the knowledge and techniques described in this book. We feel very fortunate to be able to participate in an area of scientific pursuit in which cooperation and collaboration between investigators in industry, academic institutions, and regulatory agencies is so strongly encouraged and highly valued.

A.F. NASSAR
P.F. HOLLENBERG
J. SCATINA

■ LIST OF CONTRIBUTORS

Christopher A. Bradfield, PhD, McArdle Lab for Cancer Research, 1400 University Ave., Madison, WI 53706-159; bradfield@oncology.wisc.edu

Chuan Chen, PhD, Arena Pharmaceuticals, Department of Drug Metabolism and Pharmacokinetics, 6510 Nancy Ridge Drive, San Diego, CA 92121

Carl D. Davis, Amgen Inc., Department of Pharmacokinetics and Drug Metabolism, One Amgen Center Drive, Thousand Oaks, CA 91320; carld@amgen.com

Sujal V. Deshmukh, Merck Research Laboratories, 33 Avenue Louis Pasteur, BMB 4-112, Boston, MA 02115; sujal_deshmukh@merck.com

Stephen Ferguson, Cellzdirect Inc., 4301 Emperor Blvd., Durham, NC 27703

Helen Gill, PhD, Product Development Manager, Cyprotex Discovery Ltd., 15 Beech Lane, Macclesfield, Cheshire, SK10 2DR, United Kingdom; h.gill@CYPROTEX.com

Mark P. Grillo, Amgen Inc., Department of Pharmacokinetics and Drug Metabolism, South San Francisco, CA 94080; mark_grillo@msn.com, grillo@amgen.com

Umesh M. Hanumegowda, Pharmaceutical Candidate Optimization, Bristol-Myers Squibb, Wallingford, CT 06492-7660; umesh.hanumegowda@bms.com

Paul F. Hollenberg, PhD, Maurice H. Seevers Collegiate Professor Department of Pharmacology, University of Michigan, 2301 MSRB III, 1150 West Medical Center Drive, Ann Arbor, MI 48109-5632, USA; phollen@umich.edu

Patricia B. Hoyer, PhD, University of Arizona, Department of Physiology, 1501 N. Campbell Ave., #4122, Tucson, AZ 85724-5051; hoyer@u.arizona.edu

Eli G. Hvastkovs, PhD., Asst. Professor, Department of Chemistry, East Carolina University, 300 Science and Technology Bldg., Greenville, NC 27858; hvastkovse@ecu.edu

Aileen F. Keating, PhD, University of Arizona, Department of Physiology, 1501 N. Campbell Ave., Tucson, AZ 85724-5051; akeating@email.arizona.edu

Sean Kim, Bristol Myers Company, 5 Research Parkway, Wallingford, CT 06492

Roberta S. King, University of Rhode Island, 203 Fogarty Hall, 41 Lower College Road, Kingston, RI 02881; rking@uri.edu

Lawrence H. Lash, Department of Pharmacology, Wayne State University School of Medicine, 540 East Canfield Avenue, Detroit, MI 48201; l.h.lash@wayne.edu

Edward LeCluyse, Cellzdirect Inc., 4301 Emperor Blvd., Durham, NC 27703

Louis Leung, Wyeth Research, Department of Biotransformation, 500 Arcola Road, Collegeville, PA 19426; leungl@wyeth.com

Mark N. Milton, PhD, VP, Nonclinical Development, Tempo Pharmaceuticals Inc., 161 First Street, Suite 2A, Cambridge, MA02142; mnmilt@gmail.com

Ala F. Nassar, Vion Pharmaceuticals Inc., 4 Science Park, New Haven, CT 06511; nassaral@aol.com

Aram Oganesian, Principal Research Scientist II, Wyeth Research, Drug Safety and Metabolism, 500 Arcola Road, Collegeville, PA 19426; oganesa@wyeth.com

Melvin Reichman, Director Lankenau Chemical Genomics Center, 100 Lancaster Avenue, Wynnewood, PA 19096; reichmanm@mlhs.org

Dan Rock, Amgen Inc., Pharmacokinetics and Drug Metabolism, AW2-D/3392, 1201 Amgen Court West, Seattle, WA 98119-3105

A. David Rodrigues, Bristol-Myers Squibb, Metabolism and Pharmacokinetics, BMS Research & Development Mail Stop F13-04, P.O. Box 4000, Princeton, NJ 08543-4000; david.rodrigues@bms.com

James F. Rusling, University of Connecticut, Department of Chemistry, U-60, and Department of Pharmacology (Health Center), 55 N. Eagleville Road, Storrs, CT 06269-3060; James.Rusling@uconn.edu

Abu J.M. Sadeque, PhD, Arena Pharmaceuticals, Department of DMPK, 6510 Nancy Ridge Drive, San Diego, CA 92121; asadeque@arenapharm.com

JoAnn Scatina, VP, Department of Drug Safety and Metabolism, Wyeth Research, 500 Arcola Road, Collegeville, PA 19426; ScatinJ@wyeth.com

John B. Schenkman, Professor, Department of Cell Biology, University of Connecticut Health Center, Farmington, CT

Thomas R. Sharp, Pfizer Global Research and Development, Eastern Point Road, Groton, CT 06340; thomas.r.sharp@pfizer.com

Michael Sinz, Bristol-Myers Squibb, Metabolism and Pharmacokinetics, 5 Research Parkway, Wallingford, CT 6492; michael.sinz@bms.com

Aaron L. Vollrath, McArdle Lab for Cancer Research, 1400 University Avenue, Madison, WI 53706; alvollra@facstaff.wisc.edu

Hongbing Wang, University of Maryland, School of Pharmacy, Department of Pharmaceutical Science, 20 Penn Street, HSF II RM 549, Baltimore, MD 21201; hwang@rx.umaryland.edu

Larry C. Wienkers, Amgen Inc., Pharmacokinetics and Drug Metabolism, AW2-D/3392, 1201 Amgen Court West, Seattle, WA 98119-3105; wienkers@amgen.com

Wen Xie, University of Pittsburgh, Department of Pharmaceutical Sciences, 633 Salk Hall, 3501 Terrace Street, Pittsburgh, PA 15261; wex6@pitt.edu

Zheng Yang, Metabolism and Pharmacokinetics, Pharmaceutical Candidate Optimization, Bristol-Myers Squibb, Rt. 206 and Province Line Road, Princeton, NJ 08543; yangz@bms.com

Lilian G. Yengi, Wyeth Pharmaceuticals, Department of Drug Safety and Metabolism, 500 Arcola Road, Collegeville, PA 19426; yengil@wyeth.com

PART I

INTRODUCTION

CHAPTER 1

Historical Perspective

ROBERTA S. KING

1.1 CONTROVERSIES SPANNING PAST, PRESENT, AND FUTURE

Two major issues have been debated throughout the history of drug metabolism, and are still disputed to some degree. One is the name itself, the other is the physiological purpose of "drug" metabolism. In the 1800s and early 1900s, the generally agreed purpose of these reactions was reflected in the most widely used name, *detoxication mechanisms*. However, *detoxication* became widely recognized as a misnomer because not all parent compounds were toxic and not all metabolites were less or nontoxic. A better term was not invented until the 1950s when the term "drug metabolism" was coined. While handy, this term was still not entirely valid, and it needed silent agreement that "drug" be not restricted to medicinal compounds (Bachmann and Bickel, 1985–86). Thus, *xenobiotic metabolism* became popular starting in the 1970s, especially in circles studying carcinogens and environmental compounds. Xenobiotic, by definition, included all compounds foreign to the organism, not just medicinal ones. However, even in the early 1900s many examples were already known of metabolism of endogenous compounds, for example, steroids undergoing glucuronidation. These early examples of endogenous substrates were generally dismissed because they typically occurred at much higher concentrations than normally present, so-called "supraphysiological" concentrations. While none of these three terms could be considered ideal, in 1947 R.T. Williams concluded that the field of detoxication included "… all those metabolic processes not specifically covered by the main streams of fat, carbohydrate and protein intermediary metabolism" (Williams, 1947). Williams went on to explain that, "Detoxication is, in fact, the study of the metabolism of organic compounds other than lipids, carbohydrates, proteins and closely related natural compounds, although the lines of demarcation between these two groups is by no means a sharp one …" (Williams, 1947; Bachmann and Bickel, 1985–86). Thereby, the earliest clear description of the field mainly described what it was not, and later terms were not much more precise.

The physiological purpose of these reactions was also widely debated. Of note is that several of the early theories are still considered at least partially valid. The first theories tried to answer the question of how these transformations were related to

Drug Metabolism Handbook: Concepts and Applications, Edited by Ala F. Nassar, Paul F. Hollenberg, and JoAnn Scatina
Copyright © 2009 by John Wiley & Sons, Inc.

detoxication. For example, in 1917, Berczeller published the first article trying to answer why conjugation would result in detoxication (Berczeller, 1917; Bachmann and Bickel, 1985–86). His theory, later to be disproved, was that conjugations such as glucuronidation and sulfonation led to a change in the surface tension of dilute aqueous solutions. He related this to *in vivo* conditions by indicating that the conjugates, with less surface activity than the parent drug, would be more easily removed from cellular surfaces while the parent drugs would accumulate at the surfaces to toxic concentrations (Bachmann and Bickel, 1985–86). The next hypothesis, proposed in 1922 by Sherwin (Sherwin, 1922), was based on the idea of "chemical defense" against accumulation of foreign compounds. This hypothesis is still considered valid as one role of drug metabolism. Sherwin proposed that the body needed to completely destroy foreign compounds or excrete them in the urine. Thus, for Sherwin, the purpose of oxidation and reduction reactions was to destroy the foreign molecules. If complete destruction was not possible, then conjugation reactions could make the molecule more aqueous soluble and more easily removed through the urine. In 1925, Schüller (Schüller, 1925) proposed a theory also similar to the modern view of the physiological purpose of drug metabolism. Schüller proposed that conjugation reactions led to an increase in the water solubility of compounds leading to a change in the distribution of the compound in the body (Schüller, 1925). The third theory, which resonates somewhat with today's metabolism scientists, was proposed by Quick in 1932 (Quick, 1932). His view was that conjugations resulted in an increase in acidity, converting a weak acid parent into a strong acid metabolite which could be eliminated more easily. One benefit of this hypothesis was that it included in its rationale the acetylation reaction, which was ignored by Sherwin and Schüller. In his 1947 comprehensive review of metabolism (Williams, 1947), R.T. Williams criticized each of these potential roles of metabolism. Yet rather than proposing his own theory, Williams concluded that "interpretations worthy of the status of theories were lacking at that time" (Bachmann and Bickel, 1985–86).

In 2008, the field is described as an "elaborate defense system against foreign compounds and against the accumulation of potentially toxic endogenous molecules" (Meyer, 2007). However, the same author also adds recognition of the modern "concept of molecular links between xenobiotic metabolism and endogenous pathways of sterol, lipid, bile acid and energy homeostasis." To complicate matters, the current generation of genetic technologies has revealed that all of the enzyme families contributing to drug metabolism include both members with selectivity for endogenous compounds and members with selectivity for exogenous compounds, and that homologues are present throughout diverse species from bacteria to human. Indeed, gene knockout studies have shown that multiple members of the "drug"-metabolizing enzyme families are essential to life or reproductive processes, as well as serving as a means to defend against xenobiotics (Sheets, 2007). Thus, in 2008, the field has no more specific name than the popular term "drug metabolism," yet its physiological function has broadened to include endogenous compounds and endogenous regulatory pathways.

1.2 1800s: DISCOVERY OF MAJOR DRUG METABOLISM PATHWAYS (CONTI AND BICKEL, 1977)

Many current students of metabolism are surprised to learn that drug metabolism experiments were first conducted and published more than 180 years ago, first in dogs in 1824 and next in humans in 1841 (Wöhler, 1824; Ure, 1841). While these experiments

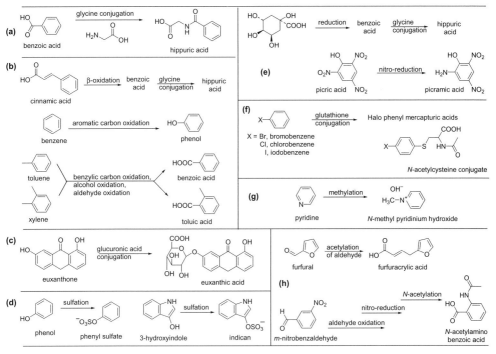

Figure 1.1   Examples of first-observed metabolism reactions. **(a)** glycine conjugation; **(b)** β-oxidation, aromatic carbon oxidation, benzylic carbon oxidation, aldehyde oxidation; **(c)** glucuronide conjugation; **(d)** sulfate conjugation; **(e)** reduction; **(f)** glutathione conjugation; **(g)** methylation; **(h)** acetylation.

were quite rudimentary by today's standards, they laid the foundation for all of metabolism by establishing the paradigm that the body could take up exogenous compounds, perform chemical reactions on them, and then remove them via the urine in a chemically altered form. Also noteworthy is that by 1893, one or more examples of each major metabolism pathway had been published (Table 1.1, Fig. 1.1). Nearly all of this early work was published in German, often by German scientists. This was a time of significant advances in chemistry and medicine in German laboratories. However, the metabolism studies were limited by the difficulty in purifying and proving structure identification of the proposed metabolites. Advances in organic chemistry and analytical methods often directly led to advances in the metabolism studies.

Glycine conjugation has the honor of being the first-described metabolism reaction. In 1841 and 1842, two scientists independently ingested benzoic acid and observed a compound in the urine "in copious amounts," and "without any apparent unhealthy effects" (Ure, 1841; Keller, 1842; Conti and Bickel, 1977). The excreted compound was similar to benzoic acid, but also contained nitrogen. Identification of the compound as hippuric acid, the glycine conjugate of benzoic acid, was made in 1845 by a French scientist named Dessaignes (Dessaignes, 1845).

The benzoic acid studies soon led to the observation of oxidation as a preliminary step to conjugation, by observing that ingestion of cinnamic acid also caused excretion

TABLE 1.1 Early development of the major drug metabolism pathways.[a]

Year	Reaction name	Substrate	Intermediate	Product	Original citation (Conti and Bickel, 1977)
1824, 1841	Glycine conjugation	Benzoic acid	None	Hippuric acid	Wöhler (1824), Ure (1841)
1842,	β-oxidation	Cinnamic acid	Benzoic acid	Hippuric acid	Erdmann and Marchand (1842a, b)
1848,	Aldehyde oxidation	Benzaldehyde	None	Benzoic acid	Wöhler and Frerichs (1848)
1867	Aromatic and benzylic carbon oxidation	Benzene	None	Phenol	Schultzen and Naunyn (1867)
		Toluene	Benzyl alcohol	Benzoic acid (via further oxidation of alcohol to carboxylic acid)	
1844, 1870	Glucuronide conjugation	Mango leaves	None	Euxanthic acid, Euxanthone and oxidized glucose	Erdmann (1844), Baeyer (1870)
1851, 1876	Sulfate conjugation	Benzene	Phenol	Conjugated phenol	Staedeler (1851), Baumann (1876)
		Hydroxyindole		Indole-sulfate	
1863	Reduction	Quinic acid	Benzoic acid	Hippuric acid	Lautemann (1863)
		Picric acid		Picramic acid	Karplus (1893)
1879, 1884	Glutathione conjugation	Bromobenzene	None	Mercapturic acid conjugates	Baumann and Preuss (1879), Jaffe (1879), Baumann (1884)
		Chlorobenzene		Acetylcysteine conjugates	
1887	Methylation	Pyridine acetate	Pyridine	N-methyl pyridinium hydroxide	His (1887)
1887, 1893	Acetylation	Furfural	N-acetylfurfural	Furfuracrylic acid	Jaffe and Cohn (1887), Cohn (1893)
		m-nitrobenzaldehyde		N-acetyl amino benzoic acid	

[a]Structures of substrates and products are shown in Fig. 1.1.

of hippuric acid into the urine (Erdmann and Marchand 1842a, b; Wöhler and Frerichs, 1848). We now call this biotransformation of cinnamic acid to benzoic acid, β-oxidation. Wöhler and Frerichs (1848) also discovered aldehyde oxidation when they found that dogs and cats excreted a "considerable amount" of benzoic acid after treatment with benzaldehyde (oil of bitter almonds) (Conti and Bickel, 1977, p. 11).

Nothing of the mechanism of these transformations was yet understood. During this era, the body was considered simply a chemical reaction container. Indeed, the chemists found many transformations "absolutely puzzling" (Conti and Bickel, 1977, p. 8) because, outside the organism, they could only be reproduced only under very harsh conditions, if at all. For example, oxidation of benzene to phenol had never been accomplished when, in 1867, Schultzen and Naunyn published their very clear determination of phenol in the urine of humans and dogs after ingestion of benzene. Even in the twenty-first century, only rather harsh and non-physiological conditions are known to chemically transform benzene to phenol (March, 1992).

In 1844 Erdmann observed that euxanthic acid isolated from urine of cows fed mango leaves could be hydrolyzed to euxanthone, but it took until 1870 to characterize the sugar moiety as an oxidized form of glucose. Similarly, conjugated phenols were observed in 1851, but it took until 1876 to confirm that the conjugate was a sulfate producing sulfuric acid upon hydrolysis (Staedeler, 1851; Baumann, 1876). Baumann is known as the "father of sulfation" and subsequently showed that many other ingested chemicals were also excreted as sulfates in the urine, often after preliminary oxidation to a phenol (Baumann, 1876). Methylation was first described in 1887 after feeding pyridine acetate to dogs and isolating the *N*-methyl product from the urine (His, 1887). Also in 1887, Jaffe and Cohn first observed a product that appeared to be conjugated with acetic acid (Jaffe and Cohn, 1887), although it took until 1893 for Cohn to confirm *N*-acetylation as a major conjugation reaction (Cohn, 1893).

While biochemists, chemists, physiologists, and pharmacologists contributed to these discoveries, the emphasis was on studying the metabolism of "foreign" compounds. More importantly, the emphasis was the fate of the chemical compound, rather than the organism that transformed it. Indeed, Bachmann and Bickel concluded that before circa 1870, "the whole matter was a biochemical curiosity rather than a physiologically meaningful process" (Bachmann and Bickel, 1985–86, p. 213). Evidence that this view began to change comes from Nencki's thesis of 1870, foreseeing that, "… one will on the one hand be able to establish laws allowing predictions on the fate of new compounds, and on the other hand gain increasing insight into the organism as a 'chemical agent'." (Nencki, 1870) Beginning in about 1876 with Baumann's phenyl sulfate (Baumann, 1876), it was often found that the excreted compounds were much less toxic than their parent compounds. And by 1893, enough evidence existed for introduction of the term "detoxication" in a textbook of physiological chemistry (Neumeister, 1893; Bachmann and Bickel, 1985–86, p. 213). Detoxication still stands as one major physiological role of metabolism, but is no longer recognized as the only role.

1.3 1900–1950s: CONFIRMATION OF MAJOR PATHWAYS AND MECHANISTIC STUDIES

As better techniques for identification of compounds were developed, the major reactions were confirmed and put upon somewhat stronger structural foundations; and in two cases the active cofactors for the reactions were elucidated. For example, Lipmann earned the 1953 Nobel Prize in Medicine, in part, for his 1945 publication of the role of coenzyme A in acetylation of sulfanilamides (Lipmann, 1945). In 1953, Cantoni published evidence for S-adenosyl methionine as the active cofactor for methylation reactions (Cantoni, 1953). The biosynthetic source of glycine was elucidated in 1946 (Shemin, 1946), but it was not yet understood how the conjugation reaction occurred under physiological conditions.

Nencki's (1870) prediction of gaining insight into the organism as a chemical agent was also developed in significant ways. In the 1800s, the blood was often considered the localized source of the transformations, and the emphasis was on animal or human ingestion of a compound and isolation of excretion products from the urine. In the early 1900s, techniques such as hepatectomy and perfusion of livers and kidneys of laboratory animals proved the alternative paradigm of organ-based metabolism. One novel approach (Hemingway, Pryde and Williams, 1934) used serially perfused dog organs (such as liver or spleen) in combination with the kidney to demonstrate that the liver was the main site of glucuronic acid conjugation. In 1936, Potter and Elvehjem described an improved alternative to the commonly used tissue slice and tissue mince methods (Potter and Elvehjem, 1936). Their device used a glass test tube as mortar with a motor-driven blown-glass pestle for nondestructive tissue homogenization. Claude added differential centrifugation to the improved homogenization technique, and made possible subcellular localization of metabolism via isolation of individual tissue organelles (Claude, 1940). He also coined the still-popular term "microsome" (Claude, 1943).

Of note is that during the early twentieth century, most metabolism studies were conducted within biochemistry departments of university medical schools. Also during this time, because of (or in spite of) the World Wars, metabolism research expanded across Europe to England and North America, and by 1950 the United States had replaced Germany as the dominant origin of metabolism publications.

Approximately 100 years after the first published discoveries, "Modern" metabolism science was founded by the Welshman R.T. Williams, who in 1947 published the first text devoted to metabolism entitled *Detoxication Mechanisms: The Metabolism of Drugs and Allied Organic Compounds*. This text was expanded in a second edition in 1959 with a modified title, *The Metabolism and Detoxication of Drugs, Toxic Substances, and Other Organic Compounds*. In these texts, Williams brought systematic organization and clarity to what had previously been broad and disconnected research, "so that working hypotheses (could) be advanced" (Williams, 1947, quoted from Caldwell, 2006). It was Williams who proposed that metabolism occurred through reactions representing two distinct phases leading to the still popular terms, "phase I" and "phase II" metabolism reactions. Another revolutionary, B.B. Brodie of the United States National Institutes of Health, first published in 1948 and "led the field into its modern phase" (Bachmann and Bickel, 1985–86, p. 188). As foreseen by Nencki in 1870 (Nencki, 1870), predictive rules for functional group transformations began to emerge, and the enzymology of the reactions began to be elucidated.

1.4 1950s–1980: MODERN DRUG METABOLISM EMERGES, WITH ENZYMATIC BASIS

Because of advances in analytical technologies and biochemical methods, metabolism studies took off starting in about 1950. For example, partition chromatography improved separations and allowed differentiation of drug versus metabolites. Isotope-tracer methods (mostly 14C and 15N) allowed metabolites of foreign compounds to be detected and quantified at nontoxic doses. Absorption spectrophotometry also improved both quantification and identification of drug versus metabolites. For the first time, the cofactors of the enzymatic reactions were fully elucidated, and the enzymatic biosynthesis pathways of the cofactors were established. For example, the molecular mechanism of glucuronidation utilizing the reactive cofactor, uridine-3′,5′-diphosphate glucuronic acid (UDPGA) was published in 1954 (Dutton and Storey, 1954). Advances in tissue fractionation methods provided the means to prove the enzymatic basis of metabolism. Indeed, Brodie and coworkers published the first review of this emphasis in 1958 entitled, "Enzymatic metabolism of drugs and other foreign compounds" (Brodie, Gillette and La Du, 1958). In the late 1950s and early 1960s, the cytochrome P450 enzyme system was discovered and characterized as the source of microsomal oxidations of several drugs and steroid hormones (Mason, 1957a, b; Klingenberg, 1958; Omura and Sato, 1962; Omura *et al.*, 1965). By the end of this period, the role of cytochrome P450 in drug metabolism was well established and the important modifying factors (inhibition, induction, and polymorphisms) were beginning to be understood (Conney *et al.*, 1980; Netter, 1980; Orrenius, Thor and Jernström, 1980; Ritchie *et al.*, 1980).

1.5 1980–2005: FIELD DRIVEN BY IMPROVED TECHNOLOGIES

Major advances during this 25-year period were driven by improved technologies. The major human cytochrome P450 members were identified, purified, and characterized from human liver tissue starting in 1983 (Wang *et al.*, 1983). Shortly after, development of cloning and heterologous expression techniques led to single isoform preparation with relative ease. Robotics for liquid handling and improvements in fluorescent technologies have driven the development of successful high-throughput screening procedures. Advanced separation systems linked to mass spectrometry or NMR detection systems have driven metabolite structure analysis and quantitation. Advances in viable hepatocyte isolation have improved the ability to characterize inducers of specific enzymes. Other advances include *in silico* modeling of all stages of metabolism including *in vitro* to *in vivo* modeling and computational prediction of human drug metabolism. Improvements in protein crystallization techniques, availability of powerful synchrotron sources, and improvements in 3D-structure generation software have led to the generation of multiple structures of each enzyme family providing an appreciation of structure variability and invariability. More recently, the study of transporters and transport systems has driven a paradigm shift toward the recognition of the important contributions of transport to drug development and drug safety.

1.6 2005+: HIGH TECHNOLOGY

While some would describe the metabolism field as "mature," technological advances still push the field forward in dramatic ways. We are now in an age of high technology in metabolism science, as evidenced by the emphasis in Part II of this text. Indeed, most of the knowledge described in this text would not be possible without the technological advances of the recent 30 years. Much of this modern innovative technology has been driven by the pharmaceutical industry, including highly advanced analytical instrumentation and high-throughput technologies. It has also been influenced by great advances of the "genomics era," through recognition of the genetic basis of variation in drug metabolism among individuals and populations. The future of the "drug" metabolism field will be driven by further technological advances toward the purposes of elucidating and understanding the impact of metabolic "cross talk": the complex three-way interactions among endogenous compounds and their regulatory pathways, exogenous compounds, and disease states.

ACKNOWLEDGMENTS

The author acknowledges three English language reviews of the early metabolism literature as major sources: Bachmann and Bickel (1985–86), Caldwell (2006), and Conti and Bickel (1977). All quotations are as provided in the reviews in English translation. Publications published prior to 1900 were not accessed directly and are listed as cited in Conti and Bickel (1977).

REFERENCES

Bachmann, C., Bickel, M.H. (1985–86). History of drug metabolism: the first half of the 20th century. *Drug Metab. Rev.*, *16* (3), 185–253.

Baeyer, A. (1870). *Ann. Chem. Pharm.*, *155*, 257.

Baumann, E. (1876). Concerning the occurrence of brenzcatechin in the urine. *Pflügers Arch. Physiol.*, *12*, 69.

Baumann, E. (1884). Ueber cystin und cystein. *Z. Physiol. Chem.*, *8*, 299.

Baumann, E., Preusse, C. (1879). *Z. Physiol. Chem.*, *3*, 156.

Berczeller, L. (1917). The excretion of substances foreign to the organism in the urine. *Biochem. Z.*, *84*, 75–79.

Brodie, B.B., Gillette, J.R., La Du, B.N. (1958). Enzymatic metabolism of drugs and other foreign compounds. *Annu. Rev. Biochem.*, *27* (3), 427–454.

Caldwell, J. (2006). Drug metabolism and pharmacogenetics: the British contribution to fields of international significance. *Br. J. Pharmacol.*, *147*, S89–S99.

Cantoni, G.L. (1953). S-Adenosylmethionine; a new intermediate formed enzymically from L-methionine and adenosine triphosphate. *J. Biol. Chem.*, *204*, 403–416.

Claude, A. (1940). Particulate components of normal and tumor cells. *Science*, *91* (2351), 77–78.

Claude, A. (1943). The constitution of protoplasm. *Science*, *97* (2525), 451–456.

Cohn, R. (1893). Concerning the occurrence of acetylated conjugates following the administration of aldehydes. *Z. Physiol. Chem.*, *17*, 274.

Conney, A.H., Buening, M.K., Pantuck, E.J., Pantuck, C.B., Fortner, J.G., Anderson, K.E., Kappas, A. (1980). Regulation of human drug metabolism by dietary factors. *Ciba Found. Symp.*, *76*, 147–167.

Conti, A., Bickel, M.H. (1977). History of drug metabolism: discoveries of the major pathways in the 19th century. *Drug Metab. Rev.*, *6* (1), 1–50.

Dessaignes, V. (1845). *C. R. Acad. Sci.*, *21*, 1224.

Dutton, G.J., Storey, I.D. (1954). Uridine compounds in glucuronic acid metabolism. I. The formation of glucuronides in liver suspensions. *Biochem. J.*, *57* (2), 275–283.

Erdmann, O.L. (1844). *J. Prakt. Chem.*, *33*, 190.

Erdmann, O.L., Marchand, R.F. (1842a). Metabolism of cinnamic acid to hippuric acid in animals. *Ann. Chem. Pharm.*, *44*, 344.

Erdmann, O.L., Marchand, R.F. (1842b). *J. Prakt. Chem.*, *26*, 491.

Hemingway, A., Pryde, J., Williams, R.T. (1934). The biochemistry and physiology of glucuronic acid: the site and mechanism of the formation of conjugated glucuronic acid. *Biochem. J.*, *28* (1), 136–142.

His, W. (1887). On the metabolic products of pyridine. *Arch. Ex. Pathol. Pharmakol.* 22, 253–260.

Jaffe, M. (1879). Ueber die nach einfuhrung von brombenzol und chlorbenzol im organismus entstehenden schwefelhaltigen sauren. *Ber. Deut. Chem. Ges.*, *12*, 1092.

Jaffe, M., Cohn, R. (1887). *Ber. Dtsch. Chem. Ges.*, *20*, 2311.

Karplus, J.P. (1893). *Z. Klin. Med.*, *22*, 210.

Keller, W. (1842). On the conversion of benzoic acid into hippuric acid. *Ann. Chem. Pharm.*, *43*, 108.

Klingenberg, M. (1958). Pigments of rat liver microsomes. *Arch. Biochem. Biophys.*, *75*, 376–386.

Lautemann, E. (1863). Concerning the reduction of quinic acid to benzoic acid and the conversion of the same to hippuric in animals. *Ann. Chem. Pharm.*, *125*, 9.

Lipmann, F. (1945). Acetylation of sulfanilamide by liver homogenates and extracts. *J. Biol. Chem.*, *160*, 173.

March, J. (1992). *Advanced Organic Chemistry*, 4th edn, John Wiley & Sons, Inc., New York, pp. 553–554, 700.

Mason, H.S. (1957a). Mechanisms of oxygen metabolism. *Adv. Enzymol. Relat. Sub. Biochem.*, *19*, 79–233.

Mason, H.S. (1957b). Mechanism of oxygen metabolism. *Science*, *125* (3259), 1185–1188.

Meyer, U.A. (2007). Endo-xenobiotic crosstalk and the regulation of cytochromes P450. *Drug Metab. Rev.*, *39* (2), 639–646.

Nencki, M. (1870). *du Bois-Reymond's Arch. Anat. Physiol.*, 399. As quoted in Conti, A. and Bickel, M.H. (1977). History of drug metabolism: discoveries of the major pathways in the 19th century. *Drug Metab. Rev.*, *6*, 1–50.

Netter, K.J. (1980). Inhibition of oxidative drug metabolism in microsomes. *Pharmacol. Ther.*, *10* (3), 515–535.

Neumeister, R. (1893). *Lehrbuch der physiologischen Chemie mit Berucksichtigung der patholo-gischen Verhaltnisse*, 2, 346, Gustav Fischer, Jena.

Omura, T., Sato, R. (1962). A new cytochrome in liver microsomes. *J. Biol. Chem.*, *237*, 1375–1376.

Omura, T., Sato, R., Cooper, D.Y., Rosenthal, O., Estabrook, R.W. (1965). Function of cytochrome P-450 of microsomes. *Fed. Proc.*, *24* (5), 1181–1189.

Orrenius, S., Thor, H., Jernström, B. (1980). The influence of inducers on drug-metabolizing enzyme activity and on formation of reactive drug metabolites in the liver. *Ciba Found. Symp.*, *76*, 25–42.

Potter, V.R., Elvehjem, C.A. (1936). The effect of selenium on cellular metabolism. The rate of oxygen uptake by living yeast in the presence of sodium selenite. *Biochem. J.*, *30* (2), 189–196.

Quick, A.J. (1932). The relationship between chemical structure and physiological response. II. The conjugation of hydroxy- and methoxy-benzoic acids. *J. Biol. Chem.*, *97*, 403.

Ritchie, J.C., Sloan, T.P., Idle, J.R., Smith, R.L. (1980). Toxicological implications of polymorphic drug metabolism. *Ciba Found. Symp.*, *76*, 219–244.

Schüller, J. (1925). Detoxication combinations within the body. *Arch. Exp. Pathol. Pharmakol.*, *106*, 265.

Schultzen, O., Naunyn, B. (1867). The behavior of benzene-derived hydrocarbons in the animal organism. *du Bois-Reymond's Arch Anat Physiol.*, 1867, 349.

Sheets, J.J. (2007). Ronald Estabrook's early guidance of a postdoctoral fellow concerning the intricacies of steroid metabolism by cytochromes P450. *Drug Metab. Rev.*, *39* (2–3), 281–283.

Shemin, D. (1946). The biological conversion of L-serine to glycine. *J. Biol. Chem.*, *162*, 297–307.

Sherwin, C.P. (1922). The fate of foreign organic compounds in the animal body. *Physiol. Rev.*, *2*, 238–276.

Staedeler, G. (1851). *Ann. Chem. Pharm.*, *77*, 17.

Ure, A. (1841). On gouty concretions; with a new method of treatment. *Pharm. J. Trans.*, *1*, 24.

Wang, P.P., Beaune, P., Kaminsky, L.S., Dannan, G.A., Kadlubar, F.F., Larrey, D., Guengerich, F.P. (1983). Purification and characterization of six cytochrome P-450 isozymes from human liver microsomes. *Biochemistry*, *22*, 5375–5383.

Williams, R.T. (1947). *Detoxication Mechanisms: The Metabolism of Drugs and Allied Organic Compounds*, Chapman and Hall, London.

Williams, R.T. (1959). *The Metabolism and Detoxication of Drugs, Toxic Substances, and Other Organic Compounds*, 2nd edn, Chapman and Hall, London.

Wöhler, F. (1824). *Tiedemann's Z. Physiol.*, *1*, 142.

Wöhler, F., Frerichs, F.T. (1848). Concerning the modifications which particular organic materials undergo in their transition to the urine. *Ann. Chem. Pharm.*, *65*, 335.

CHAPTER 2

Factors Affecting Metabolism

ROBERTA S. KING

Multiple pathways generally compete for metabolism of any particular compound. Thus, the relative amounts of each metabolite formed may be somewhat or even quite different from one species to another, one gender to the other, one age group to another, among otherwise apparently similar individuals, and even from one time point to another in a single individual. Furthermore, the proportion of each metabolite formed *in vivo* may be different than the proportion formed *in vitro*. Major influences causing these differences are availability of relevant enzyme systems, availability of relevant cofactors, presence of modulators, and transport properties of the parent and/or metabolites. Thus, in general, *in vitro* metabolism provides examples of the *possible* biotransformations of a particular agent, whereas *in vivo* metabolism is limited by the specific conditions of availability of enzyme, cofactor, modulators, and transport.

Investigation of the differences in metabolism among species, age groups, genders, polymorphic enzymes, and presence of modulators is applicable to both drug development and drug therapy. In development, knowledge of these differences in metabolism helps to correlate pharmacokinetic (PK), toxicology, and structure–effect studies from animal models to humans. During development and therapy, knowledge of these individual differences in metabolism helps to predict and understand differences in drug response and adverse effects. Because metabolism is an important criterion determining the rate of clearance of most drugs, differences in clearance can cause significant difference in individual dose–response and incidence of adverse effects. This is especially important for drugs with a *narrow therapeutic window*, that is, those having a small difference between minimum plasma concentration needed to reach a therapeutic effect and maximum plasma concentration to prevent adverse effects.

The major influences affecting metabolism can be categorized into genetic, environmental, and physiological factors, although the distinction between the categories is confounded by the fact that environmental and physiological factors can modify genetic susceptibilities. *Environmental factors* are thought of as temporal factors, with varying consequences according to their concentration over time. Environmental factors are usually limited to exogenous "small" molecules such as other drugs, compounds

Drug Metabolism Handbook: Concepts and Applications, Edited by Ala F. Nassar, Paul F. Hollenberg, and JoAnn Scatina
Copyright © 2009 by John Wiley & Sons, Inc.

naturally present in foods, and compounds otherwise absorbed through our environment. The environmental factors may act directly on an enzyme (as inhibitor or activator) or indirectly on the genetic regulatory system (as suppressor or inducer). *Genetic factors* are coded into one's genome. Genetic factors include variations among species and individuals in the regulatory (non-coded) and/or the transcribed (coded) region of each enzyme system. By definition, genetic factors are inherited (with allowances for natural selection of variants) and, therefore, should not change over one's lifetime. *Physiological factors* include those related to age, disease state, and gender (including hormonal cycling). A fourth category of factors influencing metabolism is currently emerging. This category is the interplay or cross talk among disease and endogenous and/or exogenous regulators of drug metabolism (Meyer, 2007).

Modulators such as enzyme inhibitors and enzyme inducers are considered environmental factors affecting metabolism. As described more fully in Chapter 18, *enzyme inhibitors* reduce formation of a particular metabolite, allowing buildup of the parent and/or a shifting of biotransformation toward other possible metabolites. Inhibitors bind directly to the enzyme. Three main types of inhibitors are relevant to drug metabolism. *Competitive* (reversible) inhibitors bind directly at the active site and competitively block access of substrate to the active site. *Noncompetitive* (reversible) inhibitors bind outside the active site not blocking access of the substrate, yet decrease the rate of catalysis. *Irreversible* inhibitors bind covalently to the enzyme, and activity can be regenerated only by biosyntheses of more enzyme. *Enzyme activators* are analogous to noncompetitive inhibitors, but have a stimulating effect on enzyme activity. Cytochrome P450 3A4 is especially susceptible to drug–drug interactions via inhibition and activation, although all cytochrome P450s have known inhibitors. As described more fully in Chapter 19, *enzyme inducers* enhance gene transcription of particular enzyme systems, increasing the amount of that particular enzyme available, thus increasing the proportion of metabolism catalyzed by that particular enzyme. Enzyme inducers bind to one or more transcription factors or promoter elements in the DNA region that regulates expression of the gene. This causes increased transcription of the gene, resulting in more mRNA, and more protein synthesized. The effect of enzyme inducers lasts several days, depending on the degradation rate of the enzyme rather than on clearance of the inducer. *Enzyme suppressors* have the opposite effect from inducers, decreasing transcription and protein expression. Cytochrome P450 3A4 is especially susceptible to drug–drug interactions via the induction/suppression mechanism.

An often overlooked, yet important, factor affecting metabolism is cofactor supply. For each transformation, a specific cofactor is required to supply the source of the reactive transfer moiety. For example, oxidative reactions catalyzed by cytochrome P450 need a ready supply of NAD(P)H. Transferases need a ready supply of the transfer moiety such as uridine-5′-diphosphate glucuronic acid (UDPGA) for glucuronidation or 3′-phosphoadenosine-5′-phosphosulfate (PAPS) for sulfonation. Each cofactor has a limited cellular supply and regeneration system. For example, it is generally accepted that sulfotransferase (SULT) reactions are limited by depletion and slow regeneration of PAPS supply, whereas UDPGA supply for glucuronidation is abundant and readily regenerated. Genetic and environmental factors also modify availability of cofactors via inhibition, activation, induction, or suppression of the enzymes controlling regeneration of each cofactor.

Physiological factors affecting drug metabolism include those related to age, disease state, and gender (including hormonal cycling). Age differences are typically observed in the very young or in the elderly. For example, fetal tissue and newborns lack significant expression of all drug-metabolizing enzymes except CYP3A7 and SULT (Hines, 2007). Other enzyme systems develop over days, weeks, and months. In the elderly, total metabolism is often reduced due to lower hepatic blood flow and diminished activity of nearly all metabolism enzymes. Hepatic insufficiency caused by one or more of a number of diseases also reduces hepatic metabolism. Similarly, renal insufficiency generally reduces renal metabolism. More complicated regulatory mechanisms control the differences in metabolism seen in hormonal cycling and possibly also in endocrine disorders. Evidence is emerging for significant regulatory cross talk among endogenous and exogenous regulators of drug metabolism (Meyer, 2007).

Genetic differences are responsible for many well-known differences in metabolism among different individuals. When a trait has differential expression in >1% of the population, the trait is considered to have a *genetic polymorphism*. These differences may be within coded and/or non-coded regions of the gene for each enzyme member. Of genetic differences affecting drug metabolism, the cytochrome P450 system is considered of greatest importance. Indeed, genetic screens for CYP2D6 and CYP2C19 status (Roche Diagnostics, 2008), and CYP2C9 status (Nanosphere, 2008) have recently been approved by the FDA (US Food and Drug Administration, 2005, 2008a). These tests are currently optional to assist in dosing of, for example, tricyclic antidepressants and antipsychotics (de Leon, Armstrong and Cozza, 2006), β-blockers, antiepileptics, proton pump inhibitors, and warfarin (US Food and Drug Administration, 2008b).

Genetic polymorphisms for three additional drug metabolism systems are currently recognized as clinically relevant: thiopurine methyltransferase (TPMT), *N*-acetyltransferase 2 (NAT2), and UDP-glucuronosyltransferase (UGT1A1). TPMT catalyzes degradation of thiopurine anticancer agents (mercaptopurine, thioguanine). Individuals who are homozygous-deficient for TPMT have little or no enzyme activity and accumulate toxic levels of the corresponding antimetabolites. While only 0.3% of patients are affected (Woodson, Dunnette and Weinshilboum, 1982), avoidance of life-threatening bone marrow suppression requires substantial dose reduction in these patients (Relling *et al.*, 1999). In contrast, up to 50% of patient populations have a NAT2 genetic polymorphism causing "slow" metabolism and increased incidence of adverse effects from isoniazid (Evans, 1989). A common allele for decreased UGT1A1 activity is present in up to 30% of patient populations, and can cause increased adverse effects after irinotecan treatment (Ando *et al.*, 2000). While the adverse effects from the isoniazid and irinotecan treatments are not life-threatening to individuals with low-activity genotypes, lower dosing can impart quality of life benefits while retaining effectiveness of the treatment.

REFERENCES

Ando, Y., Saka, H., Ando, M., Sawa, T., Muro, K., Ueoka, H., Yokoyama, H., Saitoh, S., Shimokata, K., Hasegawa, Y. (2000). Polymorphisms of UDP-glucuronosyltransferase gene and irinotecan toxicity: a pharmacogenetic analysis. *Cancer Res.*, *60* (24), 6921–6926.

Evans, D.A. (1989). *N*-acetyltransferase. *Pharmacol. Ther.*, *42*, 157–234.

de Leon, J., Armstrong, S.C., Cozza, K.L. (2006). Clinical guidelines for psychiatrists for the use of pharmacogenetic testing for CYP450 2D6 and CYP450 2C19. *Psychosomatics*, *47* (1), 75–85.

Hines, R.N. (2007). Ontogeny of human hepatic cytochromes P450. *J. Biochem. Mol. Toxicol.*, *21* (4), 169–175.

Meyer, U.A. (2007). Endo-xenobiotic crosstalk and the regulation of cytochromes P450. *Drug Metab. Rev.*, *39* (2), 639–646.

Nanosphere (2008). *Verigene Warfarin Metabolism Nucleic Acid Test*, http://www.nanosphere.us/ VerigeneWarfarinMetabolismNucleicAcidTest_4472.aspx (accessed 12 February 2008).

Relling, M.V., Hancock, M.L., Rivera, G.K., Sandlund, J.T., Ribeiro, R.J., Krynetski, E.Y., Pui, C.H., Evans, W.E. (1999). Mercaptopurine therapy intolerance and heterozygosity at the thiopurine S-methyltransferase gene locus. *J. Natl. Cancer Inst.*, *91* (23), 2001–2008.

Roche Diagnostics (2008). *Background Information: AmpliChip CYP450*, http://www.roche.com/ med_backgr-ampli.htm (updated 21 January 2008, accessed 12 February 2008).

US Food and Drug Administration (2005). US FDA approved the Roche AmpliChip™ CYP450 Array for the rapid genotyping of both *CYP2C19* and *CYP2D6* variants for diagnostic use. *Fed. Regist.*, *70* (46), 11865–11867.

US Food and Drug Administration (2008a). *FDA News, FDA Clears Genetic Lab Test for Warfarin Sensitivity*, http://www.fda.gov/bbs/topics/NEWS/2007/NEW01701.html (release date: 17 September 2007, accessed 12 February 2008).

US Food and Drug Administration (2008b). *Center for Drug Evaluation and Research, Table of Valid Genomic Biomarkers in the Context of Approved Drug Labels*, http://www.fda.gov/cder/ genomics/genomic_biomarkers_table.htm (date created: 15 September 2006, updated 29 January 2008, accessed 12 February 2008).

Woodson, L.C., Dunnette, J.H., Weinshilboum, R.M. (1982). Pharmacogenetics of human thiopurine methyltransferase: kidney-erythrocyte correlation and immunotitration studies. *J. Pharmacol. Exp. Ther.*, *222* (1), 174–181.

CHAPTER 3

Biotransformations in Drug Metabolism

ROBERTA S. KING

3.1 DRUG METABOLISM IN DRUG DEVELOPMENT AND DRUG THERAPY

The major function of drug metabolism is to facilitate removal of compounds from the organism, thereby preventing unwanted accumulation of foreign compounds or of potentially toxic levels of endogenous compounds. This physiological function of metabolism is quantified using two pharmacokinetic terms, *bioavailability* and *clearance*. Bioavailability indicates the percentage of administered drug that reaches the systemic circulation, and is indicated by percent absorbed minus pre-systemic metabolism. Clearance indicates the rate of removal of the drug from the systemic circulation, and is indicated by the rates of post-systemic metabolism and excretion. Bioavailability and clearance combine to affect the total drug exposure which is usually quantified by *area under the concentration-versus-time curve* (AUC). These relationships are clarified in Fig. 3.1.

In addition to clearance, metabolism can affect the spectrum of consequences of drug action including both the desired therapeutic effects and undesired effects. During the early stages of drug development, metabolism-based strategies allow scientists to improve the pharmacokinetic properties of a lead compound by blocking or inserting sites of facile metabolism. During the clinical studies and post-marketing stages, metabolism data can be used to promote safe use of a drug, including dosing adjustments or warnings about drug–drug interactions. The most common fate of metabolism is the formation of *inactive products*—metabolites having no physiological effect (desired or undesired). In the case of inactive products, the rate of clearance would be equal to the loss of parent drug. If the drug is converted to inactive metabolites pre-systemically or too quickly for the desired action, then analogues can be developed with these sites of metabolism blocked, thereby slowing metabolic clearance. A second possible fate of drug metabolism is the formation of *active metabolites*—those having the desired activity to an extent similar to (or greater than) the parent drug. In the case of active metabolites (called "prodrugs" under certain circumstances), clearance of all active

Drug Metabolism Handbook: Concepts and Applications, Edited by Ala F. Nassar, Paul F. Hollenberg, and JoAnn Scatina
Copyright © 2009 by John Wiley & Sons, Inc.

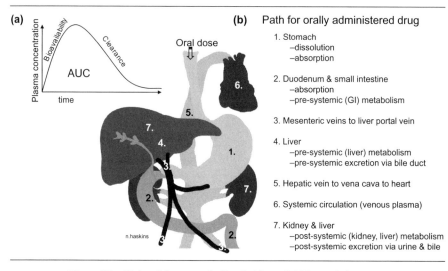

(a)

Plasma concentration

Bioavailability

Clearance

AUC

time

(b) Path for orally administered drug

Oral dose

1. Stomach
   –dissolution
   –absorption

2. Duodenum & small intestine
   –absorption
   –pre-systemic (GI) metabolism

3. Mesenteric veins to liver portal vein

4. Liver
   –pre-systemic (liver) metabolism
   –pre-systemic excretion via bile duct

5. Hepatic vein to vena cava to heart

6. Systemic circulation (venous plasma)

7. Kidney & liver
   –post-systemic (kidney, liver) metabolism
   –post-systemic excretion via urine & bile

n.haskins

Figure 3.1 Role of drug metabolism in bioavailability and clearance.

metabolites needs to be considered. A third possible fate of drug metabolism is the formation of *toxic metabolites*. Toxic metabolites are undesirable, and analogues can be developed with these sites blocked to prevent formation. A fourth possible fate of drug metabolism is the formation of metabolites with physiological action unrelated to the desired mechanism of action. These sites can be blocked, or the metabolite can be considered a lead compound to achieve the unrelated action. For example, an early antihistamine, promethazine, has been successfully used for its anticholinergic (anti-emetic) activity. Table 3.1 summarizes the effective uses of metabolism science in drug development and drug therapy.

The most famous example of the importance of metabolism science to drug therapy and drug development, published in 1990 (Monahan *et al.*, 1990), involved a young woman presenting to the emergency room with a potentially lethal arrhythmia (palpitations, syncope, *torsades de pointes*). After the woman was treated successfully with the standard of care for arrhythmia, the investigation began to shed light on the cause. The patient's drug history showed that 10 days prior to admission, she was prescribed terfenadine (Seldane®—an antihistamine) and cefaclor (Ceclor®—a cephalosporin antibiotic). On the eighth day of terfenadine therapy, the patient began a self-medicated course of ketoconazole (Nizoral®—an azole antifungal) for vaginal candidiasis. She was also taking medroxyprogesterone acetate, although this fact is irrelevant. Existing knowledge led to the prediction that ketoconazole, a known potent cytochrome P450 (CYP) inhibitor, decreased the metabolism-based clearance of terfenadine leading to a high plasma concentration of terfenadine (Fig. 3.2). At these relatively high concentrations, an unrelated effect of terfenadine was observed—inhibition of potassium channels in the heart leading to impaired control of heart rate. For a time, terfenadine was marketed with strict warnings about coadministration of certain drugs. However, instead of giving up on their investment or being satisfied with a risky prescribing scenario, scientists at Hoechst Marion Roussel (currently Sanofi-Aventis) effectively used

TABLE 3.1   Effective uses of metabolism science in drug development and drug therapy.

Development	Therapy
Find therapeutically active metabolites	Major component controlling duration of action
Find toxic or reactive metabolites	
Alter rate of metabolic clearance	Rate of "clearance by metabolism"
Prodrug strategy	Prodrugs, active metabolites, toxic metabolites
Predict potential drug interactions	
Predict individual variation in response	Individual variation in response to drug therapy
Understand similarity of humans to animals for PK, toxicology, structure–effect studies	
	Genetic differences
	Induction
	Inhibition
	Physiological state differences

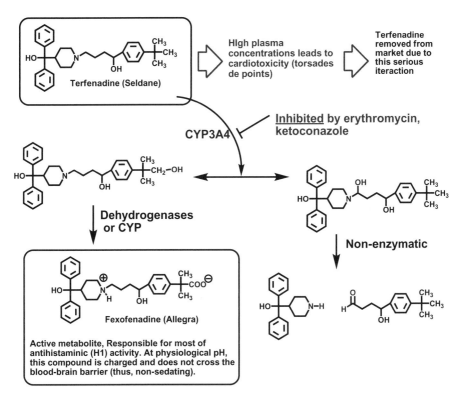

Figure 3.2   Example of effective use of metabolism science: fexofenadine (Allegra) as a superior replacement for terfenadine (Seldane).

metabolism science and developed the highly successful fexofenadine (Allegra) as a superior replacement (Fig. 3.2). This exemplifies an effective case of taking a "lemon" (problematic product) and producing "lemonade" (a highly successful derivative product).

3.2   PREDICTION OF METABOLITE AND RESPONSIBLE ENZYME

As a professor of pharmacy, I often hear the question, "How does one predict exactly what metabolites will be formed?" Well, it is pretty easy to predict the metabolites that *may potentially* form and which enzymes *may be* responsible for their formation. Recently published computational models report 80–90% prediction accuracy for the three major CYP members (Sheridan *et al.*, 2007; Terfloth, Bienfait and Gasteiger, 2007), but it is more difficult to predict in a given individual the exact proportions of each potential metabolite. I also hear the related question, "How does one know whether a particular metabolite will be clinically active or inactive, or whether it will cause an adverse affect?" Again, predictions are somewhat accurate for common functional groups and well-studied groups of drug molecules, yet the predictions are best confirmed experimentally.

A curated and well-maintained source for specific links to original literature for CYP substrates, inhibitors, and inducers is available to the public at http://medicine. iupui.edu/flockhart/table.htm. Two versions of the table are maintained: one relevant for *in vitro* metabolism, and the other relevant for real clinical situations under expected dose ranges. In addition, through the author's work at the University of Rhode Island, a comprehensive publicly available database has been developed which focuses on both phase I and phase II metabolites formed, enzymes responsible, and potential for individual variation. The data represented have been gathered through course projects by students of the College of Pharmacy and currently include many top 100 prescribed drugs (2005 list) and additional drugs used for the treatment of cardiovascular and renal diseases (http://www.uri.edu/pharmacy/departments/bps/faculty/king.html).

Of course, the major goal of the pharmaceutical industry is to make useful medications for human and veterinary use. While humans, agricultural animals, and pets are the major focus of these medications, obvious ethical limitations require much research to be conducted *in silico, in vitro*, or limited to laboratory animal species. Thus, there will always be a need to correlate data from several experimental approaches including comparing *in vitro* studies with *in vivo*, various animal species with humans, among humans of diverse genetic backgrounds, and in the presence of potential modifiers.

As a foundation, one needs to understand two categories of information about metabolism: first, the transformations and, second, the enzymes that catalyze these transformations. If one learns the transformations possible for each type of functional group, then possible metabolites can be predicted for any structure. If one also learns the enzyme systems and cofactors present (or lacking) in each species or tissue type, then the prediction can be narrowed further. Lastly, if one learns the major modifiers of metabolism and which pathways are susceptible to modification, then complete predictions can be made. Neither category is more important than the others, although most texts emphasize only one at the expense of the others, depending on whether they are written from the view of the chemist or of the enzymologist.

3.3   FUNCTIONAL GROUP BIOTRANSFORMATIONS:
PHASE I, PHASE II, CATALYSIS

In vivo, each compound is typically metabolized through a number of parallel and/or sequential reactions, a so-called *metabolism pathway*, and the relative amounts of each metabolite formed may be somewhat or even quite different from one situation to another. The total metabolism of any compound is controlled by the total of all pathways available to metabolize the drug and is modified by the specific conditions of availability of enzyme, cofactor, modulators, and transport. As early as 1947, R.T. Williams proposed that metabolism occurred through reactions representing two distinct phases, "phase I" and "phase II" metabolism reactions (Williams, 1947). These terms are still popular, although we must be aware that the order is not exclusive (phase I not always followed by phase II; phase II not always preceded by phase I). *Phase I* reactions are most commonly described as "functionalization" reactions and include oxidations, reductions, and hydrolyses. These reactions (Fig. 3.3) introduce a new polar functional group to the parent drug (oxidations), modify an existing functional group to be more polar (reductions), or unmask existing polar functional groups (hydrolyses). The most common functional groups exposed or introduced in the phase I reactions are hydroxyl (–OH), amino (–NH–), and carboxylic acid (–COOH).

Phase II reactions are most commonly described as *conjugation* reactions, and include glucuronidation, sulfonation, glycine/glutamine conjugation, acetylation, methylation, and glutathione conjugation (Fig. 3.4). These are substitution-type reactions and they link a new group either to the parent drug or to a phase I metabolite. Some conjugations cause a dramatic increase in the polarity, and thus excretion, of a drug by

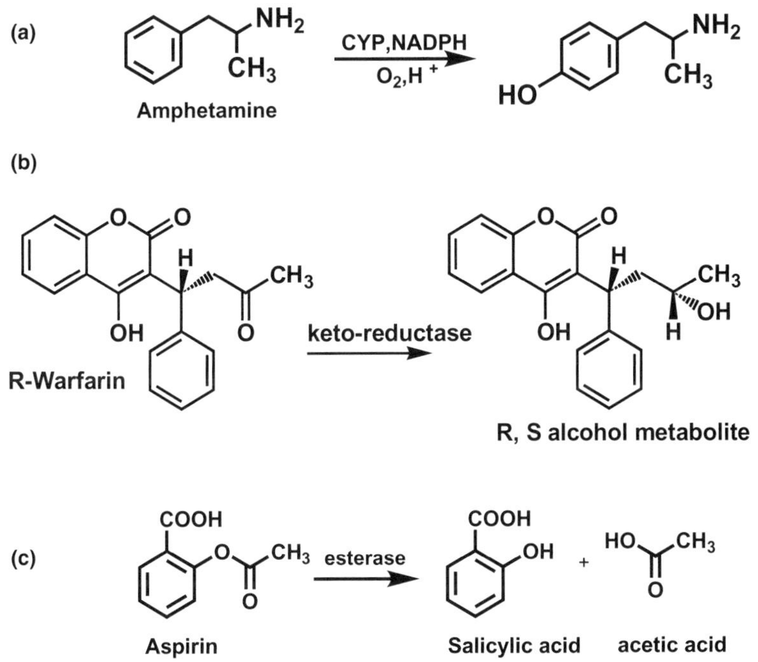

Figure 3.3 Examples of phase I reactions. (**a**) Introduce an –OH by oxidation of a carbon atom (aromatic hydroxylation). (**b**) Modify existing group to an –OH by reduction (ketone reduction). (**c**) Expose an existing –OH by hydrolysis (ester hydrolysis).

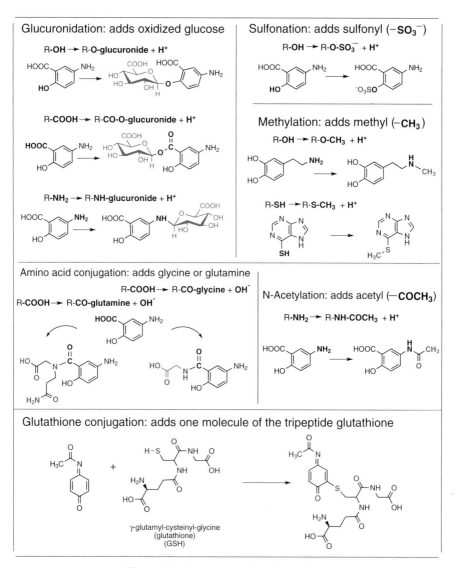

Figure 3.4 Examples of phase II reactions.

adding an ionized functional group: sulfonation, glucuronidation, and amino acid conjugation. Other conjugation reactions are just likely to cause termination of therapeutic activity: methylation and acetylation. One type of conjugation reaction protects against reactive metabolites: glutathione conjugation.

Of note is that all of these biotransformations occur under very mild "physiological" conditions: typically neutral pH, approximately 37°C, aqueous solvent, and atmospheric pressure. Only with enzymes as catalysts can these reactions occur under such mild conditions. Every enzyme acts to increase a rate of reaction by a combination of two components, approximation and catalysis. *Approximation* increases the local concentration of the reactants and facilitates the proper three-dimensional orientation of the reactants. This occurs by non-covalent enzyme–ligand interactions such as hydrogen bonds, complementary charge–charge interactions, and attractive lipophilic forces. *Catalysis* serves to lower the activation energy of the reaction. Most enzyme systems are able to stabilize the transition state (and destabilize the reactants) by non-covalent molecular forces such as hydrogen bonding, charge–charge interactions, and dipole interactions. The oxidation/reduction reactions are catalyzed by assisting the electron transport and in the formation of the reactive oxygen moiety. The rate of metabolism by any one enzyme molecule is controlled by the number of product molecules formed per active site per minute, which is called the "turnover rate" of each active site. The rate of metabolism contributed by any one enzyme member (i.e. CYP3A4) is the turnover rate multiplied by the total number of active sites of that member (dependent on CYP3A4 protein expression). And finally, the rate of total metabolism of an administered drug is controlled by the total of all pathways and enzymes available to metabolize the drug.

3.4 OXIDATIONS AND CYTOCHROME P450

Oxidations comprise the bulk of phase I metabolism reactions, and members of the CYP *superfamily* catalyze the majority of these drug oxidation reactions. Of the nearly 60 human CYP members currently identified (http://drnelson.utmem.edu/CytochromeP450.html), only about 10 contribute significantly to metabolism of marketed drugs (Fig. 3.5). Individual *members* are coded by separate genes, and the members are grouped and named according to amino acid sequence similarity of the gene product. Within each CYP *gene family* (CYP1, CYP2, and CYP3), the members have more than 40% sequence identity. Each family is divided into subfamilies indi-

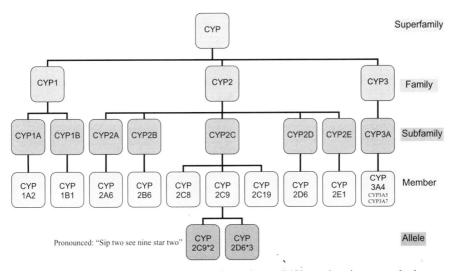

Figure 3.5 Relationships and nomenclature of cytochrome P450 members important for human drug metabolism.

TABLE 3.2 Major P450 reactions classified according to category and mechanism.[a]

Category	Mechanism	Biotransformation

A. sp[3] Carbon hydroxylation	i. Hydrogen atom abstraction
	ii. Oxygen rebound
		1. Benzylic carbon hydroxylation[b]
		2. Allylic carbon hydroxylation
		3. Aliphatic carbon hydroxylation
		  a. end of chain (ω)
		  b. penultimate (ω-1)
		  c. cyclic
		4. Alcohol/aldehyde carbon oxidation

B. sp[2] Epoxidation	i. Hydrogen atom abstraction ii. Oxygen rebound	5. Alkene epoxidation 6. Aromatic hydroxylation[b]
C. Heteroatom release	i. Nonbonded electron abstraction ii. Hydrogen atom abstraction iii. Oxygen rebound iv. Nonenzymatic rearrangement	7. N-dealkylation[b]/deamination 8. O-dealkylation[b] 9. Oxidative cleavage

D. Heteroatom oxygenation	i. Nonbonded electron abstraction
	ii. Nonbonded electron abstraction
	iii. Oxygen rebound
		10. Nitrogen oxidation
		  a. Primary amine
		  b. Secondary amine
		  c. Tertiary amine
		11. Sulfur oxidation
		  a. Thiol (sulfide)
		  b. Sulfoxide

[a]From Guengerich (2001).
[b]Indicates the four most common cytochrome P450 oxidation reactions in human drug metabolism.

cated by a capital letter (CYP2C, CYP2D, CYP3A). Each *subfamily* contains members with more than 55% amino acid sequence identity (CYP2C8, CYP2C9, CYP2C19). That brings us to the member level, and remember that members are defined as being coded by separate genes. Members can be further subdivided into allele or variant categories. *Alleles* are responsible for hereditary variation and are defined as alternative versions of a gene resulting from mutation or duplication/deletion events. Each person has two separately inherited alleles that together make up the individual's genotype for that gene. *Genetic polymorphism* is used to describe genes that contain an alternative allele present in >1% of a given population. For example, CYP2C9 includes a wild-type allele (CYP2C9*1) and several alternative alleles (CYP2C9*2, CYP2C9*3). Although not applicable to the CYP superfamily, some enzyme superfamilies include member genes that encode multiple sequence *variants* (commonly by variable splicing). These variants are indicated by appending "_v#" to the member name, such as SULT2B1_v1 and SULT2B1_v2.

There are at least 11 different common oxidation reactions catalyzed by CYP (and many additional less common reactions) (Table 3.2). How can one enzyme family catalyze so many different reactions? The easiest way to rationalize all these different possible oxidations is by thinking about the CYP catalytic mechanism (Fig. 3.6). The major reactions can be understood using a single reactive oxygen moiety (FeO^{3+}) and only four mechanistic subcategories (Table 3.2): (A) sp[3] carbon hydroxylation, (B) sp[2]

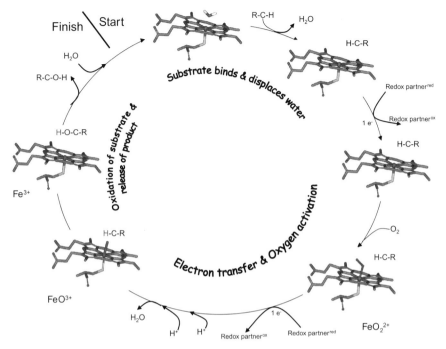

Figure 3.6 P450 catalytic mechanism cycle. First, substrate binds by displacing water molecule from heme iron. Second, electron transfer steps and oxygen activation occur. Third, the substrate is oxidized and the product is released.

carbon epoxidation, (C) heteroatom release, and (D) heteroatom oxygenation. This "unified" view of the major P450 reactions was originally described by Guengerich in 1984 (Guengerich and McDonald, 1984; Guengerich, 2001). *Regioselectivity*, which is defined as the atom or atoms oxidized in a molecule when multiple sites are possible, is partially controlled by the ease of hydrogen atom abstraction and the stability of the resulting carbon radical. Mechanism, however, does not singularly control the regioselectivity. Each CYP superfamily member has a slightly different three-dimensional structure of the substrate-binding pocket, each of which exhibits unique structural preferences for substrate-binding affinity and orientation toward the reactive oxygen. Thus, mechanistic and structural features combine to define the substrate selectivity and regioselectivity of each CYP. These features are described in more detail in Section 3.5 of this chapter.

Because of the combination of mechanistic and structural features controlling CYP oxidations, the probability of forming each potential product is a combination of three criteria including abundance of enzyme isoform catalyst, abundance of each substrate functional group, and relative reactivity of each substrate functional group. Because of its abundance in the liver and GI tract, CYP3A4 catalyzes the largest fraction of human drug metabolism (Fig. 3.7) (Shimada *et al.*, 1994). Thus, the pool of total formed metabolites is enriched with functional groups oxidized by CYP3A4 within substrate structures preferred by the CYP3A4-binding pocket. CYP2D6, in contrast, only makes up only a

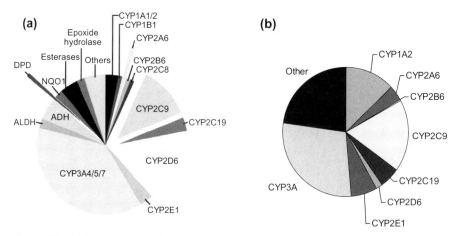

Figure 3.7   **(a)** Proportion of all marketed drugs metabolized by each phase I enzyme. Enzyme polymorphisms associated with changes in drug effects are shown offset from the corresponding pie charts (reprinted with permission from Evans and Relling, 1999). **(b)** Average relative proportions of CYPs present in human liver (Shimada *et al.*, 1994; Emoto *et al.*, 2006).

$$\text{FeO}^{3+} \overset{\text{i}}{\longrightarrow} \left[\text{FeOH}^{3+}\right] \overset{\text{ii}}{\longrightarrow} \text{Fe}^{3+}$$

Figure 3.8   Carbon hydroxylation under the unified mechanism. (i) H• abstraction from an sp³ carbon; and (ii) oxygen rebound (Guengerich, 2001).

very small percentage of the total human liver P450, yet oxidizes approximately 25% of all drug molecules (Fig. 3.7) (Shimada *et al.*, 1994; Evans and Relling, 1999). This contradiction occurs because CYP2D6-preferred substrates are highly represented within the pool of all therapeutic drug molecules. Lastly, certain functional groups are more readily oxidized than others due to the stability of the carbon radical or ease of nonbonded electron abstraction. For example, the pool of CYP-oxidized metabolites is enriched in benzylic alcohols due to the relative stability of benzylic carbon radicals. As a result of these combined structural and mechanistic features, four oxidation reactions are most commonly observed in human drug metabolism: (i) aromatic hydroxylation, (ii) *N*-dealkylation, (iii) *O*-dealkylation, and (iv) benzylic carbon hydroxylation.

3.4.1   CYP Carbon Hydroxylation Reactions

sp³ Carbon hydroxylation by CYP is very common and can be used to exemplify the unified mechanism proposed by Guengerich (Guengerich, 2001). This mechanism considers the heme iron to be in the oxyferryl state (FeO^{3+}) for the oxidation step (Fig. 3.6). As shown in Fig. 3.8, carbon hydroxylation occurs in two steps: (i) H• abstraction

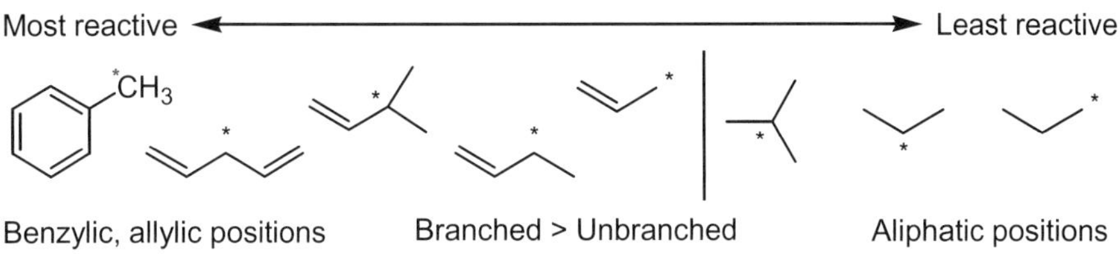

Most reactive ←————————————————————————→ Least reactive

Benzylic, allylic positions   Branched > Unbranched   Aliphatic positions

Figure 3.9 Order of reactivity of sp³ carbons toward cytochrome P450 oxidation. Asterisk (*) indicates position of carbon radical in each type of functional group.

from an sp³ carbon, and (ii) oxygen rebound. Therefore, the oxidized carbon must have a C–H bond. In the absence of enzyme conformational limitations, the most easily oxidized carbon will be one with the most easily abstracted hydrogen atom, although rearrangements may occur to provide the most stable carbon radical (Fig. 3.8).

Multiple sp³ carbon atom types are hydroxylated by CYP, including benzylic, allylic, and aliphatic carbons (Table 3.2). Benzylic positions are the most reactive, followed by allylic, followed by aliphatic (Fig. 3.9). Indeed, benzylic carbon hydroxylation is one of the four most commonly observed CYP reactions. Within each group, branched carbon positions are more reactive than unbranched ones. While relatively unreactive, the terminal methyls of tertiary-butyl and isopropyl groups are commonly oxidized by CYP because of the statistical probability (9x, 6x) of having any one terminal C–H bond oxidized. As discussed in a later section, steric hindrance around the heme FeO^{3+} limits hydroxylations of sterically hindered carbons, which encourages hydroxylation at the terminal (ω) and penultimate (ω-1) positions of aliphatic chains. Occasionally aliphatic 6-membered rings are oxidized to the least sterically hindered product (usually 4-*trans* product). Examples of benzylic and allylic carbon hydroxylation are shown in Fig. 3.10. Examples of aliphatic carbon hydroxylation are shown in Fig. 3.11.

Alcohol and aldehyde oxidation can also be catalyzed by CYP via sp³ carbon hydroxylation (Guengerich, 2001). While these oxidations are often considered products of alcohol dehydrogenase and aldehyde dehydrogenase, respectively, CYP can also contribute to their oxidation. Figure 3.12 shows the unified mechanism for oxidation of ethanol and acetaldehyde by CYP via carbon hydroxylation. Specifically, ethanol and acetaldehyde are known to be very good substrates for CYP2E1.

3.4.2 CYP Epoxidation Reactions

Oxidation of sp² carbons results in formation of an epoxide or arene oxide (Fig. 3.13). In most cases, the epoxides do not survive to excretion, but are hydrolyzed by epoxide hydrolase to a *trans*-diol. Carbamazepine is a well-known exception to this rule: a significant portion of the epoxide is stable enough for renal excretion (Fig. 3.14). In the case of the arene oxide, the return to aromaticity drives a rearrangement reaction which produces the phenol. The complete conversion to the phenol is called aromatic hydroxylation and is one of the four most common oxidations in drug metabolism (Fig. 3.13). In general, the order of preference for aromatic hydroxylation position is *para* > *ortho* >> *meta*. When two substituents are present, the least sterically hindered product is generally favored (Fig. 3.14).

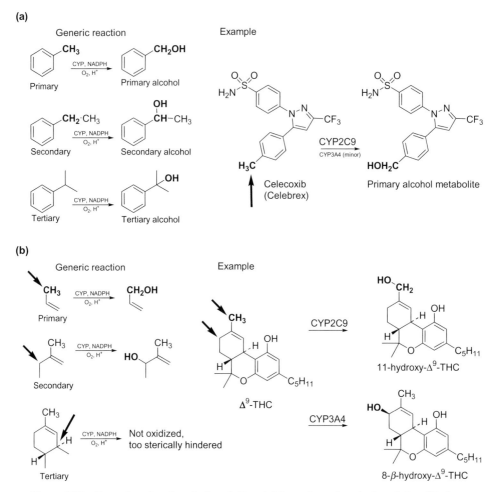

Figure 3.10 Examples of carbon hydroxylation: **(a)** benzylic carbon hydroxylation (Paulson *et al.*, 2000; Tang *et al.*, 2000), and **(b)** allylic carbon hydroxylation (Bornheim *et al.*, 1992; Lemberger 1973).

3.4.3 CYP Heteroatom Release Reactions

The remaining two of the four most commonly observed P450 drug oxidations result in the reactions called *N*-dealkylation and *O*-dealkylation. The P450 enzyme catalyzes oxygen insertion on a C–H which is bonded to an amine nitrogen (or ether oxygen). The resulting carbinolamine (or hemiacetal) is not stable and can spontaneously degrade, releasing the aldehyde/ketone and amine (or alcohol). Thus, the bond between the heteroatom and the carbon is broken. As shown in Fig. 3.15, the reaction on amines can be rationalized using four steps: (i) one-electron transfer, (ii) proton abstraction, (iii) oxygen rebound, and (iv) nonenzymatic rearrangement. In general, the most commonly *N*-dealkylated groups are methyl, ethyl, *n*-propyl, isopropyl, allyl, and

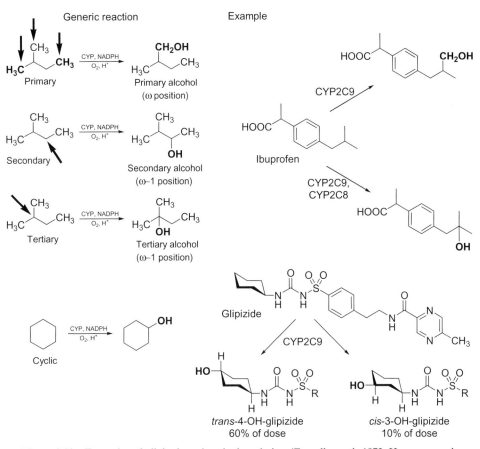

Figure 3.11   Examples of aliphatic carbon hydroxylation (Fuccella *et al.*, 1973; Hamman *et al.*, 1997; Rudy *et al.*, 1991).

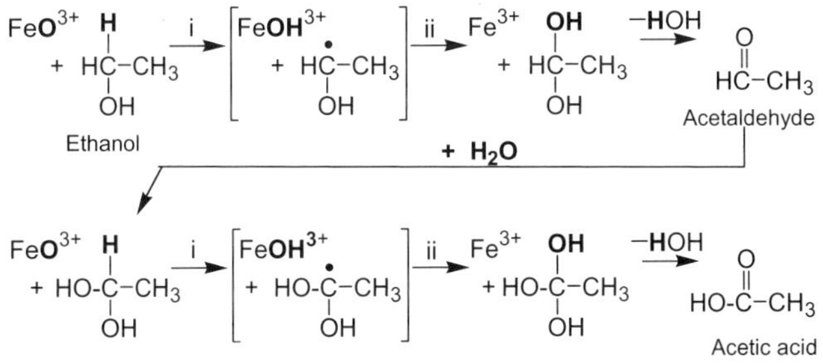

Figure 3.12   Mechanism for oxidation of ethanol and hydrated acetaldehyde by CYP2E1 (Guengerich, 2001).

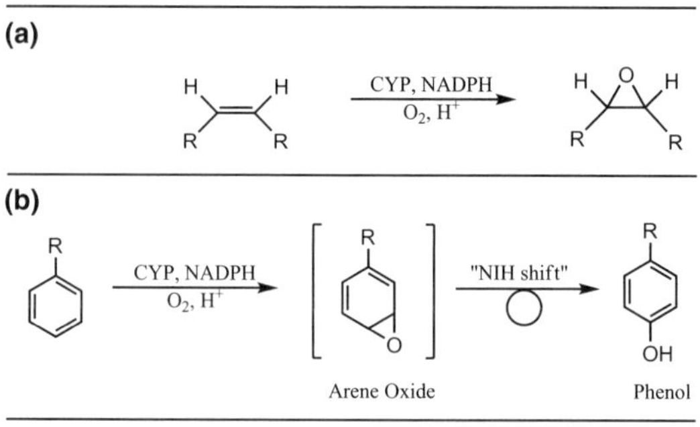
(a)
CYP, NADPH
O2, H+
(b)
CYP, NADPH
O2, H+
"NIH shift"
Arene Oxide
Phenol

Figure 3.13 Oxidation of sp^2 systems. **(a)** Alkenes are oxidized to epoxide. **(b)** Arenes (aromatic) are oxidized to arene oxides; rearrangement of the arene oxide results in aromatic hydroxylation.

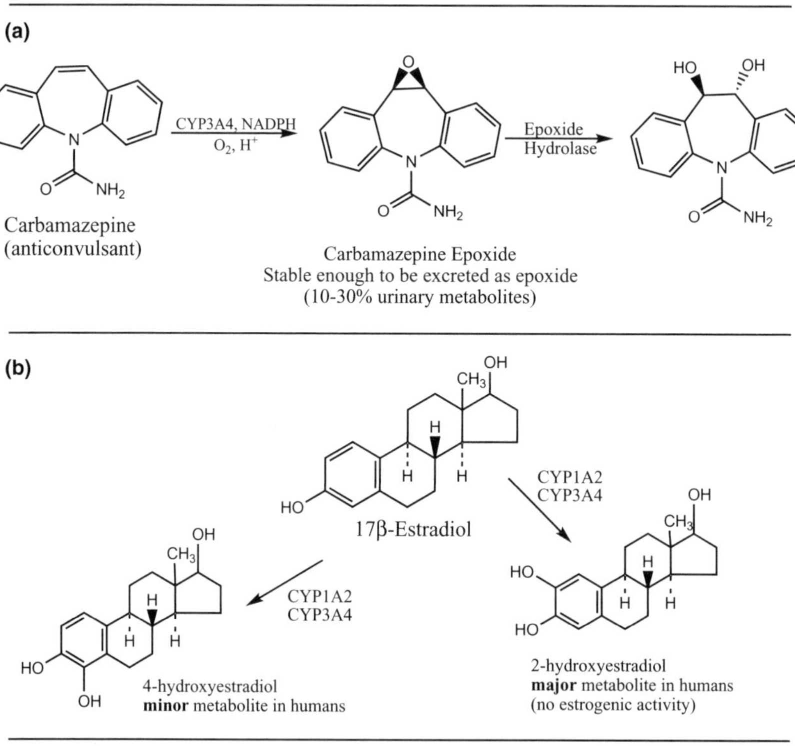
(a)
CYP3A4, NADPH
O2, H+
Epoxide
Hydrolase
Carbamazepine
(anticonvulsant)
Carbamazepine Epoxide
Stable enough to be excreted as epoxide
(10-30% urinary metabolites)
(b)
17β-Estradiol
CYP1A2
CYP3A4
CYP1A2
CYP3A4
4-hydroxyestradiol
minor metabolite in humans
2-hydroxyestradiol
major metabolite in humans
(no estrogenic activity)

Figure 3.14 **(a)** Example of alkene epoxidation and **(b)** aromatic hydroxylation.

benzyl. For amines, tertiary amines are favored for dealkylation over secondary over primary amines. When a primary amine is dealkylated, the reaction is usually called *N*-deamination, rather than *N*-dealkylation, but the mechanism is the same. Examples of *N*-dealkylation/deamination are shown in Fig. 3.17.

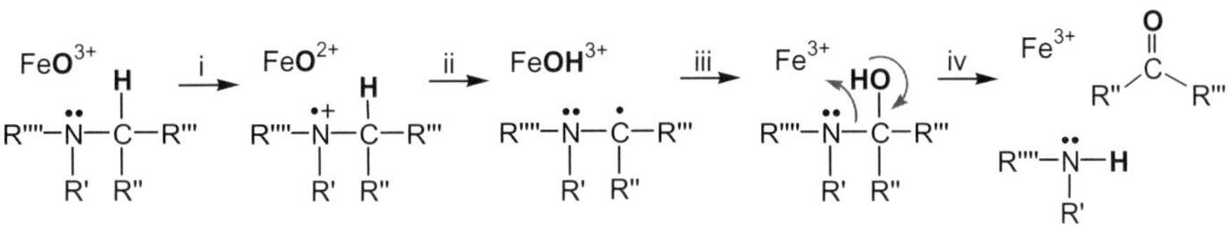

Figure 3.15 Unified P450 mechanism for N-dealkylation of amine nitrogen. Mechanistic steps defined as (i) nonbonded electron transfer, (ii) hydrogen atom abstraction, (iii) oxygen rebound, and (iv) non-enzymatic rearrangement (Guengerich, 2001).

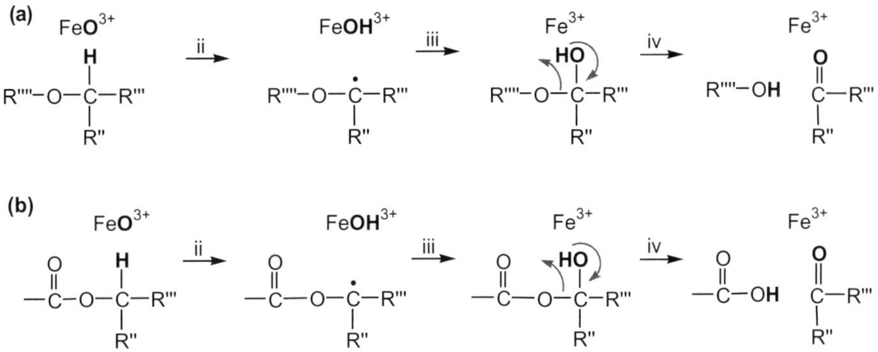

Figure 3.16 Unified P450 mechanism for O-dealkylation of ethers **(a)** and oxidative cleavage of esters **(b)**. Mechanistic steps defined as (ii) hydrogen atom abstraction, (iii) oxygen rebound, and (iv) non-enzymatic rearrangement (Guengerich, 2001).

O-dealkylations generally occur near terminal positions on the molecule, and methoxy groups are especially susceptible (O-demethylation). As shown in Fig. 3.16, the reaction on ethers can be rationalized similarly to N-dealkylation, but starting at step ii: (ii) proton abstraction, (iii) oxygen rebound, and (iv) nonenzymatic rearrangement (Guengerich, 2001). The same mechanism holds for another reaction, oxidative cleavage of esters and amides (Fig. 3.16). *In vitro*, oxidative cleavage can be distinguished from ester hydrolysis by whether a ketone/aldehyde or an alcohol, respectively, accompanies the carboxylic acid product (Guengerich, 1987; Peng et al., 1995), and by whether the reaction is dependent on presence of CYP and its electron-donating system. Oxidative cleavage also occurs on amides, carbamates, and ureas (Peng et al., 1995). Examples of O-dealkylation and oxidative cleavage are shown in Fig. 3.17.

3.4.4 CYP Heteroatom Oxygenation Reactions

The final category of CYP-catalyzed oxidations involves the direct oxidation of heteroatoms (N, S). Originally these reactions were thought to be exclusively catalyzed by flavin-containing monooxygenase (FMO) enzyme system, but more recently the role of CYPs has been recognized. As shown in Fig. 3.18, the amine oxidation can be rationalized using three steps: (i) first one-electron transfer, (ii) second one-electron transfer, and (iii) oxygen rebound. Primary amine nitrogens are enzymatically oxidized

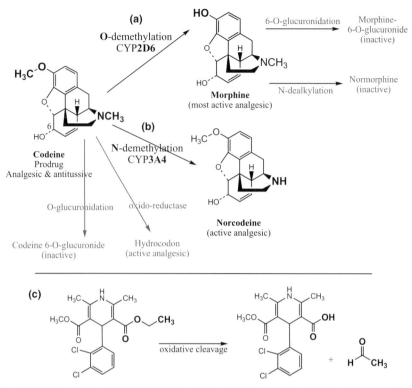

Figure 3.17   Examples of N-dealkylation, O-dealkylation, and oxidative cleavage. **(a)** N-dealkylation and O-dealkylation of codeine. **(b)** Oxidative cleavage of ester functional group in felodipine. Amides, carbamates, and ureas are also cleaved through oxidation.

$$FeO^{3+} \qquad FeO^{2+} \qquad FeO^{+} \qquad Fe^{3+}$$

Figure 3.18   Unified P450 mechanism for N-oxygenation. Mechanistic steps defined as (i) first nonbonded electron transfer, (ii) second nonbonded electron transfer, and (iii) oxygen rebound (Guengerich, 2001).

to the respective *N*-hydroxylamine, with further spontaneous oxidation to the nitroso and nitro form (Fig. 3.19a). Secondary amines are enzymatically oxidized to the respective secondary *N*-hydroxylamine, with further spontaneous oxidation to the nitrone form (Fig. 3.19b). Tertiary amines are enzymatically oxidized to the amine oxide (*N*-oxide) (Fig. 3.19c). CYPs also catalyze oxidation of sulfur atoms as shown in Fig. 3.20. Most sulfoxidation reactions are catalyzed by a single isoform, CYP3A4. Indeed, a recent computational regioselectivity model indicated that CYP3A4-catalyzed sulfur

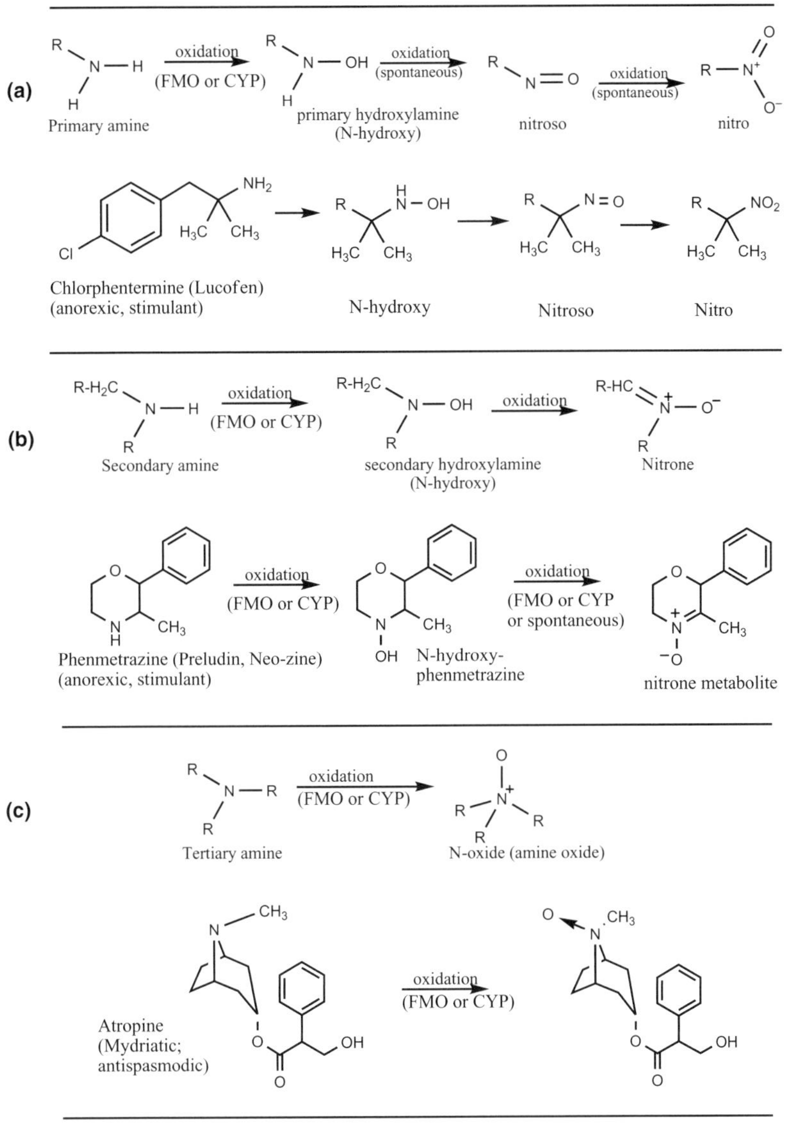

Figure 3.19   Examples of nitrogen oxidation: **(a)** primary amine oxidation, **(b)** secondary amine oxidation, and **(c)** tertiary amine oxidation. Each generic reaction is shown first, followed by a specific example.

oxidation is the major reaction for nearly all drug molecules containing an oxidizable sulfur (Sheridan *et al.*, 2007).

CYP also catalyzes less common reactions that are of interest within the field of drug metabolism including oxidation of 1,4-dihydropyridines, oxidative deformylation, oxidation of a double bond with 1,2-migration, and steroid aromatization (Guengerich,

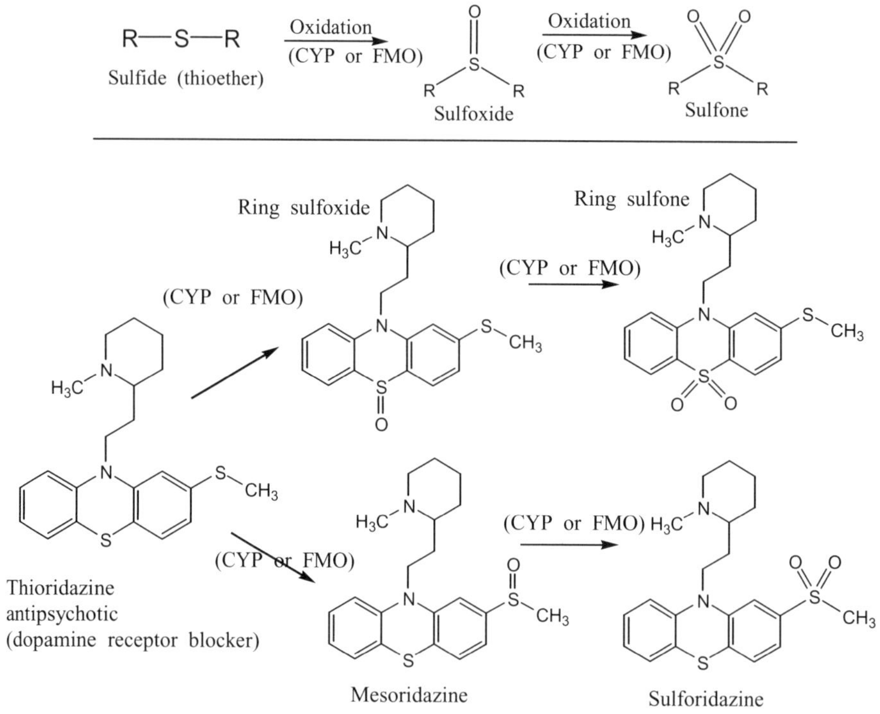

Figure 3.20 Examples of sulfur oxidation.

2001). Examples of these reactions relevant to drug metabolism are shown in Fig. 3.21.

3.5 ENZYMOLOGY AND MODIFIERS OF CYPs

A number of features differentiate the many CYPs, including relative expression levels and 3D protein structure of the substrate-binding cavity and access channel. Relative expression is controlled by tissue localization (GI versus liver versus kidney versus other), age- and gender-related differences, presence of genetic polymorphisms with functional relevance, and susceptibility to induction/suppression. Inhibitors and activators act via direct binding with the enzyme and, thus, their effects are influenced by differences in the 3D protein structure of the substrate-binding cavity and access channel. Substrate selectivity and regioselectivity are also defined by the features of the substrate-binding cavity and access channel.

3.5.1 3D Protein Structure of the Substrate-Binding Cavity and Access Channel

The 3D protein structure of the substrate-binding cavity and access channel is unique for each CYP and, thus, controls the differential binding of ligands. Substrates and

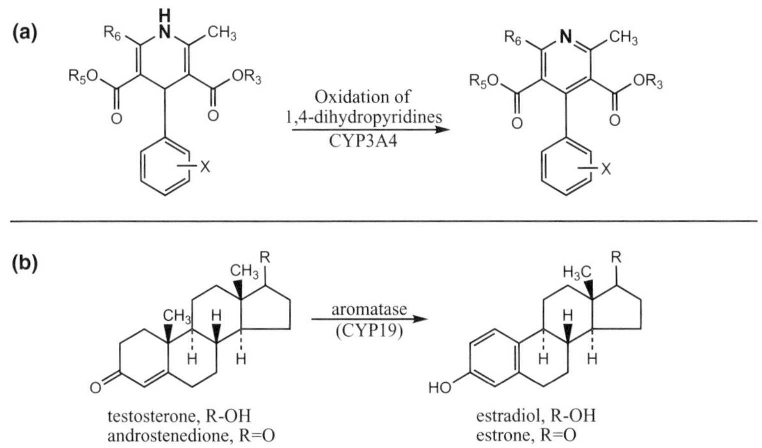

Figure 3.21 Other oxidations relevant to drug metabolism. **(a)** Oxidation of 1,4-dihydropyridines, and **(b)** steroid aromatization.

inhibitors must be able to enter into the binding cavity through the available access channel and have a degree of affinity for the binding cavity. Although the ligand and the protein side chains have some conformational flexibility, studies have shown that this flexibility is limited by atomic and bonding interactions. Thus, each CYP member is limited to a range of ligands that can interact with its binding cavity, and these limitations define the isoform-, substrate-, inhibitor-, and regioselectivity of each P450 member.

3.5.1.1 Isoform Selectivity for Substrates and Inhibitors
Isoform selectivity refers to the ability of a particular enzyme member to accept a certain ligand into the binding pocket, either as a substrate or as an inhibitor. For example, a potential ligand may be excluded from the binding pocket when it is larger than the access channel or the cavity available for ligand binding. In contrast, a potential ligand may be small enough to fit within the binding cavity but lack complementary structural features necessary for appropriate binding affinity with a particular enzyme isoform. By definition, P450 *substrates* are ligands that can bind with suitable affinity and with an oxidizable functional group within catalytic distance of the activated heme oxygen. Note that some substrates can also inhibit the metabolism of other substrates via competitive binding at the active site. Rational inspection of the pool of substrate structures oxidized by each particular CYP member has allowed vague, but fairly reliable, predictions as to which general classes of ligands will be substrates for each major P450 (Lewis, 2003; Totah and Rettie, 2005). The available X-ray crystal structures and homology models of human P450s support these predictions. A more quantitative approach recently confirmed and extended these predictions for CYP3A4, 2D6, and 2C9 (Terfloth, Bienfait and Gasteiger, 2007). Combining all these sources, the authors have prepared a modified decision tree exemplified in Fig. 3.22. In brief, CYP1A2 has a relatively flat yet broad binding pocket and accepts mostly planar molecules such as aromatic rings. CYP2E1 and 2A6 have relatively small binding pockets and accept only rather small substrates. In contrast, CYP3A4 and 2C8 have quite large binding pockets,

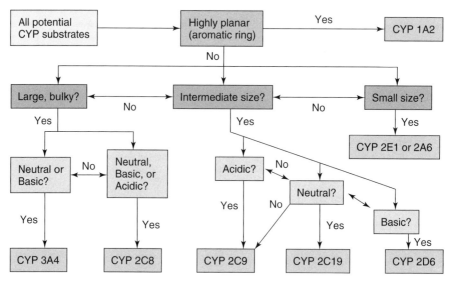

Figure 3.22   Decision tree for prediction of dominant cytochrome P450 isoform based on substrate physicochemical features.

accepting up to rather large bulky molecules. CYP2D6 and 2C19 have preference for basic molecules, and CYP2C9 has a strong preference for molecules with one or more acidic functional groups.

These preferences can be rationalized by observation of significant differences in the relative features of the binding cavity for each major human CYP as defined by X-ray crystal structures and homology models. For example, CYP3A4- and 2C8-binding cavities are rather large (calculated to be 1386 and 1438 Å^3, respectively). CYP2E1- and 2A6-binding cavities are relatively small (2A6 calculated to be 260 Å^3). Other CYP members have intermediate size binding cavities (2D6, 2C9, 2C19), each of which is differentiated through the physicochemical properties of the distinct amino acid side chains lining the access channel and within the binding cavity. For example, CYP2C9 has a strong preference for substrates with acidic functional groups. The structure of CYP2C9 shows a positively charged region at the opening of the access channel, which could attract molecules with acidic functional groups and repel molecules with basic functional groups. While both CYP2C8 and CYP3A4 have rather large binding pockets, visualization of the binding cavities shows that they should each have distinct steric restrictions on ligand binding. In contrast, visualization of CYP2A6-binding cavity rationalizes why this P450 member accepts only rather small ligands.

3.5.1.2   Regioselectivity   The decision tree, however, does not address *regioselectivity*, that is, which atom will be oxidized by a particular CYP member when multiple sites are available. Each CYP member exhibits a unique "metabolite fingerprint" because regioselectivity is influenced not only by relative reactivity of the oxidizable atoms but also by binding affinity and orientation relative to the activated heme oxygen. A recently published computational study confirmed and extended prior

TABLE 3.3 Regioselectivity preferences of CYP3A4, CYP2D6, and CYP2C9.

CYP3A4
1. *N*-dealkylations (*N*-demethylation is most preferred)
2. Sulfur oxidations (whenever available)
3. Allylic and benzylic carbon hydroxylation
4. Aromatic carbon hydroxylation, especially to catechol
5. Oxidizes atoms near end of molecule

CYP2D6
1. Aromatic methoxy (phenol *O*-demethylation)
2. Aromatic methyl oxidation (benzylic carbon oxidation)
3. *O*-demethylation
4. Oxidizes atoms at end or near middle of long molecule

CYP2C9
1. Aromatic methoxy (phenol *O*-demethylation)
2. Aromatic methyl oxidation (benzylic carbon oxidation)
3. Sulfur oxidations
4. *N*-dealkylations
5. Aromatic hydroxylation is not particularly favored, but when it occurs, it results in catechol formation
6. Oxidizes atoms at end or near middle of long molecule

Sheridan's model predicts these observed preferences (Sheridan *et al.*, 2007).

understanding of the structural parameters associated with the regioselectivity of three major CYP members: CYP3A4, 2C9, and 2C19 (Sheridan *et al.*, 2007). Specifically, all three studied CYP members were found to preferentially oxidize non-sterically hindered atoms, as measured by the magnitude of hydrogen atom exposure. Atoms at the exposed ends of molecules were also found to be preferentially oxidized by all three CYP members. CYP2D6 and 2C9 were found to *allow* oxidation of atoms in the middle of a long molecule; presumably because 2D6 and 2C9 are less sterically hindered in the region proximal to the heme and allow more "folding" of the substrate relative to CYP3A4.

The computational model also found that each CYP member exhibited distinct regioselective preferences which result in quite different metabolic fingerprints (Table 3.3). For example, Sheridan's model confirmed the long-standing knowledge that the stability of the intermediate carbon radical is a major influence for regioselectivity of CYP3A4. The model also showed that CYP3A4 prefers to oxidize sulfur atoms whenever they are available in a ligand, and to also oxidize aromatic carbons to form phenols and catechols.

3.5.2 Tissue Localization and Relative Expression Levels

3.5.2.1 CYP3A4 The major sites for drug metabolism are the liver and the mucosa of GI tract, with some metabolism occurring in the kidney. While all drug-metabolizing CYP members are expressed to some extent in the liver, CYP3A4 is the only member that is also highly expressed in the mucosa of the GI tract. Thus, CYP3A4 is responsible for much "first-pass" metabolism. Newborn and fetal livers express very little (or no) CYP3A4 and instead express the related member CYP3A7 (Hines, 2007). Some adults express an additional CYP3A family member, CYP3A5, in the liver and GI mucosa, whereas CYP3A5 appears to be the dominant form in the kidney of most individuals. Because the isoform- and regioselectivity of these related members slightly differ, the relative amounts of 3A4 versus 3A5 versus 3A7 can cause variation in the metabolites formed in each individual. However, at this time, only a few compounds have been characterized for differential metabolism by CYP3A4, 3A5, or 3A7.

In addition to differences in constitutive levels, CYP3A4 expression is widely affected by temporal factors. This is mostly attributable to endogenous and exogenous inducers/suppressors of expression. Exogenous CYP3A4 inducers are well characterized and act mostly through the nuclear transcription factors and response elements. Recently, a few endogenous CYP3A4 expression regulators have been identified such as vitamin D3, growth hormone, and triiodothyronine. Gender, age, and disease-state differences in CYP3A4 expression may also act via endogenous regulators of gene expression.

3.5.2.2 CYP2D6 CYP2D6 constitutes only a small fraction (~2%) of the total enzyme content of typical human liver tissue, yet it contributes to the metabolism of approximately 25% of all marketed drugs. This paradox occurs because CYP2D6 substrates are highly represented among the total pool of drug structures. Because of wide variation in CYP2D6 allele expression due to gene duplication, deletion, or mutation events, individuals may have abolished, decreased, normal, or ultrarapid CYP2D6 activity. The phenotype (CYP2D6 activity) has been separated into four groups: poor metabolizers (PMs), intermediate metabolizers (IMs), extensive metabolizers (EMs), and ultrarapid metabolizers (UMs). The majority (70–90%) of individuals in most studied populations are CYP2D6 EMs (normal activity), 5–30% are IMs (decreased activity), and 0–8% are PMs (abolished activity) (Ingelman-Sundberg *et al.*, 2007). However, gene duplications resulting in much higher prevalence of the ultrarapid phenotype (UM) have been selected for in Northeast Africa (*CYP2D6*2×N* genotype) and Oceania (*CYP2D6*1×N* genotype). A secondary effect modifying CYP2D6 metabolism is that presence of CYP2D6 inhibitors (or competitive substrates) can temporally cause lower 2D6-catalyzed metabolism. Thus, individuals with inherited intermediate expression levels may, in the presence of one or more 2D6 inhibitors, metabolize CYP2D6 substrates more slowly than predicted from the genotype alone. Thus, CYP2D6 is responsible for a large portion of pharmacokinetic variation among individuals. A comprehensive review on CYP2D6 has been recently published (Zanger, Raimundo and Eichelbaum, 2004).

3.5.2.3 CYP2C9 CYP2C9 prefers acidic molecules as substrates and oxidizes such drug categories as nonsteroidal anti-inflammatory drugs, oral antidiabetics, warfarin, antiepileptics, and hypnotics. While CYP2C9 represents ~20% of the average hepatic CYP content, it metabolizes only ~10% of all marketed drugs. CYP2C9 is affected by all three mechanisms of variability (genetic polymorphism, enzyme induction, and enzyme inhibition) and is responsible for individual differences in drug response and side effects. The polymorphic activity of CYP2C9 is due mainly to two common coding variants, *CYP2C9*2* and *CYP2C9*3*. These variant alleles are present in up to 10% of Caucasians, less in Africans, and not at all in Asians (Ingelman-Sundberg *et al.*, 2007). The *CYP2C9*3* allele results in production of protein which has only about 10% of the intrinsic clearance for most substrates versus the wild-type CYP2C9. The effect of the CYP2C9*2 variant varies with substrate, but generally slightly lower catalytically efficiency than the wild-type enzyme.

3.5.2.4 CYP2C19 CYP2C19 represents about 5% of the total P450 content in typical human liver tissue, and metabolizes approximately 3% of all marketed drugs. CYP2C19 is affected by genetic polymorphisms and enzyme inhibition. Individuals carrying two defective *CYP2C19* genes (poor metabolizers) represent up to 5% of Caucasian and African populations, and up to 20% of Asian populations (Ingelman-Sundberg *et al.*, 2007). This poor CYP2C19 metabolizer phenotype has proven beneficial for the treatment of GI disorders with proton pump inhibitors. The poor metabolizers have reduced metabolism of these drugs, with concomitant higher drug plasma levels and increased responsiveness to treatment. The pharmacokinetics of the tricyclic antidepressants amitriptyline and clomipramine, and the selective serotonin reuptake inhibitors (SSRI) sertraline and citalopram, are affected by CYP2C19 genotype/phenotype, but significant clinical effect or risk is as yet unclear.

REFERENCES

Bornheim, L.M., Lasker, J.M., Raucy, J.L. (1992). Human hepatic microsomal metabolism of delta 1-tetrahydrocannabinol. *Drug Metab. Dispos.*, *20* (2), 241–246.

Emoto, C., Murase, S., Iwasaki, K. (2006). Approach to the prediction of the contribution of major cytochrome P450 enzymes to drug metabolism in the early drug-discovery stage. *Xenobiotica*, *36* (8), 67–683.

Evans, W.E., Relling, M.V. (1999). Pharmacogenomics: translating functional genomics into rational therapeutics. *Science*, *286*, 487–491.

Fuccella, L.M., Tamassia, V., Valzelli, G. (1973). Metabolism and kinetics of the hypoglycemic agent glipizide in man: comparison with glibenclamide. *J. Clin. Pharmacol.*, *13*, 68–75.

Guengerich, F.P. (1987). Oxidative cleavage of carboxylic esters by cytochrome P-450. *J. Biol. Chem.*, *262* (18), 8459–8462.

Guengerich, F.P. (2001). Common and uncommon cytochrome P450 reactions related to metabolism and chemical toxicity. *Chem. Res. Toxicol.*, *14* (6), 611–650.

Guengerich, F.P., McDonald, T.L. (1984). Chemical mechanisms of catalysis by cytochromes P-450: a unified view. *Acc. Chem. Res.*, *17*, 9–16.

Hamman, M.A., Thompson, G.A., Hall, S.D. (1997). Regioselective and stereoselective metabolism of ibuprofen by human cytochrome P450 2C. *Biochem. Pharmacol.*, *54* (1), 33–41.

Hines, R.N. (2007). Ontogeny of human hepatic cytochromes P450. *J. Biochem. Mol. Toxicol.*, *21* (4), 169–175.

Ingelman-Sundberg, M., Sim, S.C., Gomez, A., Rodriguez-Antona, C. (2007). Influence of cytochrome P450 polymorphisms on drug therapies: pharmacogenetic, pharmacoepigenetic and clinical aspects. *Pharmacol. Ther.*, *116*, 496–526.

Lemberger, L. (1973). Tetrahydrocannabinol metabolism in man. *Drug Metab. Dispos.*, *1* (1), 461–468.

Lewis, D.F.V. (2003). Essential requirements for substrate binding affinity and selectivity toward human CYP2 family enzymes. *Arch. Biochem. Biophys.*, *409*, 32–44.

Monahan, B.P., Ferguson, C.L., Killeavy, E.S., Lloyd, B.K., Troy, J., Cantilena, L.R., Jr (1990). Torsades de pointes occurring in association with terfenadine use. *JAMA*, *264* (21), 2788–2790.

Paulson, S.K., Hribar, J.D., Liu, N.W., Hajdu, E., Bible, R.H., Jr, Piergies, A., Karim, A. (2000). Metabolism and excretion of [(14)C]celecoxib in healthy male volunteers. *Drug Metab. Dispos.*, *28* (3), 308–314.

Peng, H.-W., Raner, G.M., Vaz, A.D.N., Coon, M.J. (1995). Oxidative cleavage of esters and amides to carbonyl products by cytochrome P450. *Arch. Biochem. Biophys.*, *318* (2), 333–339.

Rudy, A.C., Knight, P.M., Brater, D.C., Hall, S.D. (1991). Stereoselective metabolism of ibuprofen in humans: administration of R-, S- and racemic ibuprofen. *J. Pharmacol. Exp. Ther.*, *259* (3), 1133–1139.

Rudy, A.C., Knight, P.M., Brater, D.C., Hall, S.D. (1991). Stereoselective metabolism of ibuprofen in humans: administration of R-, S- and racemic ibuprofen. *J. Pharmacol. Exp. Ther.*, *259* (3), 1133–1139.

Sheridan, R.P., Korzekwa, K.R., Torres, R.A., Walker, M.J. (2007). Empirical regioselectivity models for human cytochromes P450 3A4, 2D6, and 2C9. *J. Med. Chem.*, *50* (14), 3173–3184.

Shimada, T., Yamazaki, H., Mimura, M., Inui, Y., Guengerich, F.P. (1994). Interindividual variations in human liver cytochrome P-450 enzymes involved in the oxidation of drugs, carcinogens and toxic chemicals: studies with liver microsomes of 30 Japanese and 30 Caucasians. *J. Pharmacol. Exp. Ther.*, *270* (1), 414–423.

Tang, C., Shou, M., Mei, Q., Rushmore, T.H., Rodrigues, A.D. (2000). Major role of human liver microsomal cytochrome P450 2C9 (CYP2C9) in the oxidative metabolism of celecoxib, a novel cyclooxygenase-II inhibitor. *J. Pharmacol. Exp. Ther.*, *293* (2), 453–459.

Terfloth, L., Bienfait, B., Gasteiger, J. (2007). Ligand-based models for the isoform specificity of cytochrome P450 3A4, 2D6, and 2C9 substrates. *J. Chem. Inf. Model.*, *47*, 1688–1701.

Totah, R.A., Rettie, A.E. (2005). Cytochrome P450 2C8: substrates, inhibitors, pharmacogenetics, and clinical relevance. *Clin. Pharmacol. Ther.*, *77* (5), 341–352.

Williams, R.T. (1947). *Detoxication Mechanisms: The Metabolism of Drugs and Allied Organic Compounds*. Chapman and Hall, London.

Zanger, U.M., Raimundo, S., Eichelbaum, M. (2004). Cytochrome P450 2D6: overview and update on pharmacology, genetics, biochemistry. *Naunyn Schmiedebergs Arch. Pharmacol.*, *369* (1), 23–37.

CHAPTER 4

In Vivo Metabolite Kinetics

ZHENG YANG

4.1 INTRODUCTION

Following the administration of a drug, the pharmacokinetics and disposition of the metabolite(s) that are formed can be described and modeled. Such information is important because any metabolite can be pharmacologically active and contribute to the overall efficacy of the drug of interest. For example, atorvastatin (Lipitor®) produces two active hydroxylated metabolites that are equipotent to the parent drug *in vitro* and contribute to the prolonged inhibition of 3-hydroxy-3-methylglutaryl coenzyme A (HMG-CoA) reductase in an *ex vivo* assay (Christians, Jacobsen and Floren, 1998; Siedlik *et al.*, 1999). Another example is clopidogrel (Plavix®) whose pharmacological activity is due to the production of a reactive thiol metabolite that irreversibly blocks the ADP binding and receptor activation of the platelet $P2Y_{12}$ receptor (Savi *et al.*, 2000; Pereillo *et al.*, 2002). Information on active metabolites and their kinetics in the body is essential to comprehending the pharmacokinetic–pharmacodynamic (PK–PD) relationships of drugs and provides important guidance to their use in the clinic. With recent advancement in bioanalytical techniques, many of pharmacologically active metabolites can be identified and quantified in various biological fluids at the early stages of drug discovery. Understanding the pharmacological activity and pharmacokinetics of metabolites may provide useful insights into the PK–PD disconnect exhibited by a new chemical entity during *in vivo* efficacy testing. Moreover, the structural insights from active metabolites offer an excellent opportunity in the design of novel drug molecules (Fura *et al.*, 2004; Fura, 2006).

From a safety point of view, recent discussions on metabolites in safety testing (Baillie *et al.*, 2002; Hastings *et al.*, 2003; Smith and Obach, 2005, 2006; Davis-Bruno and Atrakchi, 2006) and the FDA Guidance on this topic also highlight the importance of understanding metabolite kinetics. In this context, it is imperative to make a distinction between a major circulating metabolite and a major metabolic pathway, and to understand the factors that determine the exposure of a metabolite in circulation and its appearance in excreta (e.g. urine and bile). In addition, knowledge of how metabolites distribute and accumulate in tissues is valuable to biotransformation-based

Drug Metabolism Handbook: Concepts and Applications, Edited by Ala F. Nassar, Paul F. Hollenberg, and JoAnn Scatina
Copyright © 2009 by John Wiley & Sons, Inc.

toxicity (Humphreys and Unger, 2006). Lastly, metabolite concentrations are often monitored as a way to detect alterations in the clearance of a drug in the presence of an inhibitor or inducer (Wilkinson, 2002). Appreciating the relevance of various metabolite-related pharmacokinetic parameters to the change in parent drug disposition is important to assess drug–drug interaction potential of various therapeutic agents. Clearly, in all the aforementioned areas, a thorough understanding of metabolite kinetics as well as factors governing such processes is essential.

In this chapter, the concepts and principles of *in vivo* drug metabolite kinetics will be reviewed and discussed along with examples. The important equations in the chapter are indicated by an asterisk (*). There are several excellent reviews on the subject (Houston, 1982; Pang, 1985). In addition, the book *Clinical Pharmacokinetics: Concepts and Applications* by Rowland and Tozer (1995) is highly recommended for everyone who is interested in exploring this topic further.

4.2 *IN VIVO* METABOLITE KINETIC CONCEPTS AND PRINCIPLES

4.2.1 General Considerations

In this chapter, linear pharmacokinetics is assumed throughout various kinetics models and equation derivations. The metabolite formed is assumed not to convert back to the parent drug; that is, there is no interconversion or reversible metabolism. All doses, drug amounts, and drug concentrations are in molar units in order to avoid conversions between a drug and a metabolite due to a difference in molecular weight.

4.2.2 Formation Clearance versus Elimination Clearance of Metabolite

The kinetics of a metabolite in the systemic circulation after dosing of a parent drug is governed by both its formation and its disposition. As illustrated in Fig. 4.1, the appearance of the metabolite in circulation is a function of enzyme activities in an eliminating organ (e.g. liver) and permeation of the metabolite formed across cellular membranes. Sequential metabolism and excretion into excreta (e.g. bile and urine) affect the concentration of the metabolite formed in the eliminating organ, thus also impacting on the rate and extent of the metabolite appearance in circulation. Once in the systemic circulation, the disposition of the metabolite in the body is the same as that of any other xenobiotics, which includes distribution and elimination processes.

In a simplified scheme, one could envision the metabolite kinetics using the model as shown in Fig. 4.2. Accordingly, the rate of change in the plasma and tissue concentration of a metabolite is given by the following equations:

$$V_d(m)\frac{dC(m)}{dt} = CL_f \cdot C - CL(m) \cdot C(m) + [CL_{d,t}(m) \cdot C_t(m) - CL_d(m) \cdot C(m)] \quad (4.1)$$

$$V_{d,t}(m)\frac{dC_t(m)}{dt} = CL_d(m) \cdot C(m) - CL_{d,t}(m) \cdot C_t(m) \quad (4.2)$$

where C is the parent drug concentration in plasma; $C(m)$ and $C_t(m)$ are the metabolite concentrations in plasma and tissues, respectively; $V_d(m)$ and $V_{d,t}(m)$ are the volumes

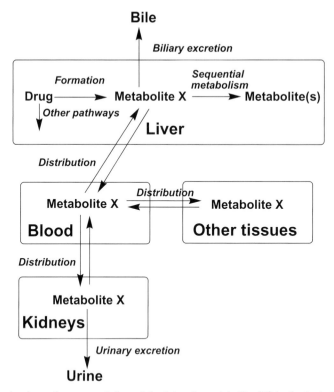

Figure 4.1 A schematic representation of the fate of a metabolite (X) in the body following its generation in the liver.

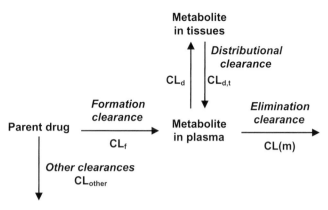

Figure 4.2 A simplified model for metabolite kinetics.

of distribution of the metabolite in plasma and tissues, respectively; CL_f is the clearance of the parent drug that forms the metabolite, also known as the formation clearance of the metabolite; $CL(m)$ is the total body clearance of the metabolite, also known as the metabolite elimination clearance; $CL_d(m)$ and $CL_{d,t}(m)$ are the distributional clearances of the metabolite between plasma and tissues. Integrating Equations 4.1 and 4.2 from time zero to infinity and appreciating the fact that the metabolite concentration in the body at either time zero or infinity is zero, the following equations are obtained:

$$CL_f \cdot AUC - CL(m) \cdot AUC(m) +$$
$$[CL_{d,t}(m) \cdot AUC_t(m) - CL_d(m) \cdot AUC(m)] = 0 \tag{4.3}$$

$$CL_d(m) \cdot AUC(m) - CL_{d,t}(m) \cdot AUC_t(m) = 0 \tag{4.4}$$

where AUC, $AUC(m)$, and $AUC_t(m)$ are the total areas under the concentration–time curve of the parent drug in plasma, metabolite in plasma, and metabolite in tissues, respectively. Substituting Equation 4.4 into Equation 4.3 and rearranging the equation, it follows that:

$$\frac{AUC(m)}{AUC} = \frac{CL_f}{CL(m)} \tag{4.5}$$

where $AUC(m)/AUC$ is the metabolite-to-parent (M/P) plasma AUC ratio. It is important to note that Equation 4.5 is valid only if the AUC is a total area from time zero to infinity. By the same token, the following relationship also exists:

$$\frac{C_{ss}(m)}{C_{ss}} = \frac{CL_f}{CL(m)} \tag{4.6}$$

where $C_{ss}(m)/C_{ss}$ is the M/P steady-state plasma concentration (C_{ss}) ratio. Equations 4.5 and 4.6 can be expanded by substituting CL_f with $f_m \cdot CL$, where CL is the total body clearance of the parent drug and f_m is the fraction of the total body clearance of the parent drug that forms the metabolite. Accordingly, the M/P AUC or C_{ss} ratio can be represented as follows:

$$\frac{AUC(m)}{AUC} = f_m \cdot \frac{CL}{CL(m)} \tag{4.7*}$$

$$\frac{C_{ss}(m)}{C_{ss}} = f_m \cdot \frac{CL}{CL(m)} \tag{4.8*}$$

Equations 4.7 and 4.8 indicate that the M/P AUC or C_{ss} ratio is determined by both the formation and the elimination clearance of the metabolite. In other words, an increase in the M/P ratio can be due to an increase in the formation clearance, decrease in the elimination clearance, or a combination of both.

4.2.3 Major Circulating Metabolite versus Major Metabolic Pathway

A major circulating metabolite is the metabolite that represents a substantial portion (e.g. ≥10%) of drug-related components in circulation. The exact percentage of the cutoff value for defining a major metabolite in safety testing is a subject of debate at the moment (Baillie *et al.*, 2002; Hastings *et al.*, 2003; FDA Guidance for Industry, 2008). However, the exact cutoff value may not be important; rather, considerations should be given to the structure and potential pharmacology and toxicology of a circulating metabolite prior to further safety assessment (Smith and Obach, 2005, 2006). Furthermore, it has been suggested that absolute abundance (i.e. metabolite concentrations in circulation) is more relevant than relative abundance (i.e. percentage) when considering metabolites in safety testing (Smith and Obach, 2005).

A major metabolic pathway is the route that constitutes a major fraction of biotransformation pathways. Although the same metabolite in some cases can be both a major circulating metabolite and the product of a major metabolic pathway, it is important to make a distinction between these two concepts. To appreciate the difference, please consider the following example as shown in Fig. 4.3. Drug X is metabolized to two metabolites (M1 and M2). The CL_f and $CL(M1)$ of M1 are 90 and 1000 mL/min, respectively. The CL_f and $CL(M2)$ of M2 are 10 and 5 mL/min, respectively. Based on Equation 4.5, the M/P AUC ratio of M1 and M2 is calculated to be 0.09 (=90 mL/min/1000 mL/min) and 2 (=10 mL/min/5 mL/min), respectively. The result demonstrates that M2, not M1, is the major circulating metabolite of drug X. However, the fraction of the

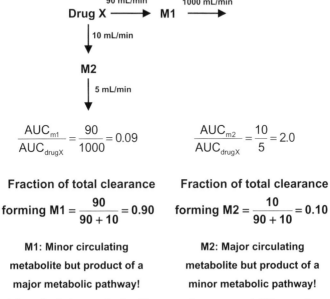

Figure 4.3 A hypothetical example that illustrates the conceptual difference between a major circulating metabolite and the product of a major metabolite of biotransformation.

total clearance of drug X forming M1 is 0.90 (=90 mL/min/(90 mL/min + 10 mL/min)), whereas the fraction of drug X forming M2 is only 0.10. This result indicates that although M1 is a minor metabolite in circulation, it represents the product of the major metabolic pathway of drug X. This hypothetical example illustrates the point that a high M/P AUC ratio for a particular metabolite does not necessarily indicate that the pathway is the major biotransformation route of the parent drug. As illustrated in Fig. 4.1, the fate of a metabolite upon its formation in the liver (i.e. entrance into circulation, biliary secretion, and sequential metabolism) and upon its reaching the systemic circulation would affect the metabolite levels in circulation. Consequently, the M/P AUC or C_{ss} ratio is governed by both the formation and the elimination clearance of the metabolite.

There are many examples in drug discovery and development that contrast the differences between a major circulating metabolite and a product of the major metabolic pathway. For example, muraglitazar underwent glucuronidation as the major elimination pathway in humans (Wang *et al.*, 2006). However, the glucuronide of muraglitazar was only a minor drug-related component in circulation, with an M/P AUC ratio less than 0.05. In another example, the formation of trifluoromethoxy salicylic acid from the substance P receptor antagonist CP-122721 represented only 1.4–3.5% of the dose in human excreta (Colizza, Awad and Kamel, 2007). However, trifluoromethoxy salicylic acid accounted for 29–56% of the total radioactivity in human circulation, whereas the parent drug was only up to 5.6%. The results indicate that trifluoromethoxy salicylic acid is a major circulating metabolite of CP-122721, but its formation pathway is only a minor metabolic pathway in humans. Clearly, understanding these metabolite kinetic concepts is important when considering the safety testing of metabolites observed in human circulation and excreta.

A true determination of the major biotransformation pathway of a drug is to measure the amount (e.g. radioactivity) of a particular metabolite as a fraction of dose in excreta. This can be achieved following the administration of a radiolabeled drug and profiling of excreta (Roffey *et al.*, 2007). The approach, however, can be confounded with sequential metabolism. Therefore, for this approach to be valid, it is important to understand the entire biotransformation pathways.

The pharmacokinetic relationship between a metabolite and a parent drug in excreta can be examined by considering the model scheme shown in Fig. 4.2. CL_{other} and $CL(m)$ are assumed to be the excretory clearance of the drug and metabolite, respectively. Based on mass balance, the following equations exist:

$$Ae + Ae(m) = Dose \tag{4.9}$$

$$Ae = (1 - f_m) \cdot Dose \tag{4.10}$$

$$Ae(m) = f_m \cdot Dose \tag{4.11}*$$

$$\frac{Ae(m)}{Ae} = \frac{CL_f}{CL_e} \tag{4.12}*$$

where Ae and Ae(m) are the total amount of the parent drug and the metabolite recovered in excreta, respectively; CL_e is the excretory clearance of the parent drug into the urine or bile, with a sum of the CL_e and CL_f equal to the total body clearance

(CL) of the parent drug; and dose is the dose of the drug administered. Equations 4.9–4.12 are valid only if the recovery of the parent drug and the metabolite is complete in excreta. The ratio of Ae(m) and Ae, which equals the ratio of the formation and excretory clearances of the parent drug, is sometimes referred to as the metabolic ratio.

Also, it is clear from Equation 4.11 that Ae(m), when normalized with dose, equals the f_m, a fraction that represents the extent of that particular biotransformation pathway to the total elimination processes. It is worth pointing out that the *extent* (f_m) of metabolism cannot be confused with the *rate* (CL_f) of metabolism, although a high CL_f generally leads to a high f_m. This is because the f_m is dependent on not only the rate of metabolism (CL_f) per se but also on the rate of removal from other pathways (CL_{other}). In the cases of tolbutamide and warfarin, they have a very low rate of metabolism in humans (i.e. metabolic clearance $< 1\,mL/min/kg$), but they are extensively metabolized, with an f_m approaching 1.0 (Thummel *et al.*, 2006).

4.2.4 Formation Rate-Limited versus Elimination Rate-Limited Metabolite Kinetics

Similar to the M/P AUC or C_{ss} ratio, the concentration–time profile of a metabolite is also governed by the formation and elimination processes. In this case, however, not only does the clearance play a role, but also the volume of distribution of the parent drug and metabolite is important. This is because both clearance and volume, as independent pharmacokinetic parameters, determine the half-life (or elimination rate constant), which in turn governs the shape of concentration–time profiles. The half-life ($t_{1/2}$) as a function of the elimination rate constant (k_{el}), volume of distribution, and clearance is shown in the following equations:

$$t_{1/2}(m) = \frac{0.693}{k_{el}(m)} = \frac{0.693 \cdot V_d(m)}{CL(m)} \tag{4.13}$$

$$t_{1/2} = \frac{0.693}{k_{el}} = \frac{0.693 \cdot V_d}{CL} \tag{4.14}$$

where $t_{1/2}(m)$ is the half-life of the metabolite when dosed by itself; $t_{1/2}$ is the half-life of the parent drug; $k_{el}(m)$ and k_{el} are the elimination rate constants of the metabolite and parent drug, respectively. Accordingly, when the formation rate constant k_f is much slower than the $k_{el}(m)$ (i.e. the rate-limiting step is the formation of the metabolite), the concentration–time profile of the metabolite is parallel to that of the parent (Fig. 4.4a). In this case, the apparent $t_{1/2}$ of the metabolite when dosed as a parent drug is the same as that of the parent, but the true half-life ($t_{1/2}(m)$) of the metabolite when dosed by itself is much shorter than that of the parent. This case is known as formation rate-limited metabolite kinetics. In some cases, metabolite kinetics can also be elimination rate-limited. The situation arises when the k_f is much faster than the $k_{el}(m)$ (i.e. the rate-limiting step is the elimination of the metabolite). As a result, the concentration–time profile of the metabolite is no longer parallel to that of the parent, with an elimination $t_{1/2}$ of the metabolite longer than that of the parent (Fig. 4.4b). In this case, the $t_{1/2}$ of the metabolite when dosed as a parent drug is the same as the $t_{1/2}(m)$ obtained after dosing the metabolite by itself. The situation is much like the flip-flop kinetics

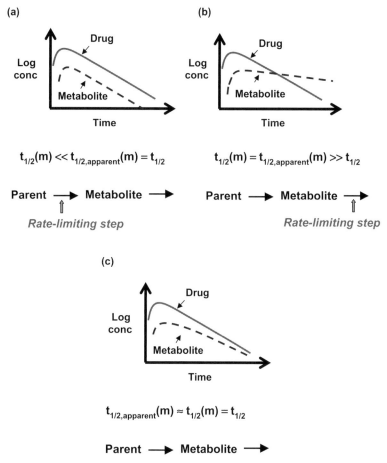

Figure 4.4 Effects of the rate-limiting step in metabolite kinetics on the shape of metabolite curves and metabolite half-lives after a single dose of a parent drug, where $t_{1/2}$ is the terminal half-life of the parent drug; $t_{1/2}(m)$ is the terminal half-life of the metabolite after intravenous dosing of the metabolite itself; and $t_{1/2,apparent}(m)$ is the apparent terminal half-life of the metabolite after dosing of the parent drug. **(a)** Formation rate-limited. **(b)** Elimination rate-limited. **(c)** Neither of steps is rate-limiting.

following oral administration, where the apparent oral $t_{1/2}$ is longer than that after intravenous dosing. The third situation is that the k_f is equal to the $k_{el}(m)$. In this case, neither formation nor elimination process is rate-limiting. The concentration–time profile of the metabolite is nearly parallel to that of the parent (Fig. 4.4c), with an apparent $t_{1/2}$ of the metabolite slightly longer than that of the parent, which is also longer than its true $t_{1/2}(m)$.

Knowledge of whether metabolite kinetics is formation or elimination rate-limited has a direct implication on how a metabolite is cleared from the body, how quickly it reaches steady state, and how much accumulation it may have relative to the parent. This is of great importance in the situations where a metabolite is pharmacologically

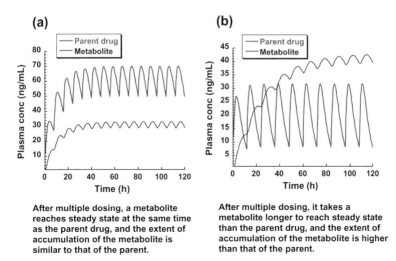

Figure 4.5   Effects of the rate-limiting step in metabolite kinetics on the time to reach steady state and extent of accumulation of a metabolite during multiple dosing of a parent drug. **(a)** Formation rate-limited. **(b)** Elimination rate-limited.

active or toxic. For formation rate-limited metabolite kinetics, the metabolite reaches the steady state at the same time as the parent drug with an extent of accumulation similar to that of the parent (Fig. 4.5a). By the same token, the metabolite is cleared from the body at the same rate as the parent. However, for elimination rate-limited metabolite kinetics, it takes much longer for the metabolite to reach steady state than the parent and the metabolite has a higher accumulation ratio than that of the parent (Fig. 4.5b). Similarly, the metabolite is cleared from the body at a slower rate than that of the parent. As a result, when a metabolite is pharmacologically active or toxic, it may lead to a disconnect between parent drug concentrations and pharmacological response or toxicity. For example, atorvastatin is an HMG CoA reductase inhibitor that has a mean plasma elimination $t_{1/2}$ of ~14 hours in humans (Williams and Feely, 2002). It undergoes the extensive first-pass metabolism in the liver and produces two active hydroxylated metabolites in humans that have equipotent activity to the parent drug *in vitro* (Christians, Jacobsen and Floren, 1998; Siedlik *et al.*, 1999). Using an *ex vivo* assay that measures the inhibition of HMG-CoA reductase in plasma samples, the plasma $t_{1/2}$ of inhibitory activity was estimated to be 20–45 hours (Siedlik *et al.*, 1999; Williams and Feely, 2002), significantly longer than that of the parent drug. The results indicate that the active metabolites of atorvastatin likely exhibit elimination rate-limited kinetics in plasma.

4.2.5   Pre-systemically Formed versus Systemically Formed Metabolite

Pre-systemically formed metabolites are generally referred to as the metabolites produced during the first pass after oral administration, which generally involves metabolism in the gut wall and liver. Systemically formed metabolites are the metabolites that are produced during the general circulation such as the ones after intravenous dosing or subsequent organ passes after reaching the systemic circulation following oral administration. Recognizing the anatomical location of the liver in the body, it is unlikely that hepatic metabolism during the first pass is any different from that during subsequent passes or systematic circulation, provided that no saturation in metabolism occurs. On the other hand, the gut wall likely plays a more pronounced role during the first pass than that during subsequent circulation because of its anatomical location and blood flow to that region. In addition, numerous drug-metabolizing enzymes (e.g. cytochrome P450 enzymes, sulfotransferases, and UDP-glucuronosyltransferases) are expressed in enterocytes and have been shown to metabolize drugs such as midazolam, 17α-ethinylestradiol, terbutaline, and isoproterenol (Thummel and Shen, 2002). Therefore, the gut wall, in addition to the liver, is an important organ in generating pre-systemically formed metabolites.

The M/P plasma AUC ratio, based on Equation 4.7, should be independent of the route of drug administration, provided that the f_m does not change with the dosing route and pharmacokinetics is linear between the routes of administration. However, this may not be the case when a drug is administered orally, where the M/P AUC ratio after oral dosing is often different from that after intravenous dosing. When the M/P AUC ratio after oral dosing is higher than that after intravenous dosing, it indicates that there are additional metabolites formed pre-systemically and these metabolites enter the systemic circulation directly. In this case, it is as if a portion of metabolites is directly dosed and, as a result, the f_m is changed between intravenous and oral routes. One good example is midazolam. The plasma AUC ratio between 1′-hydroxy-midazolam and midazolam in humans after oral administration was ~0.70, whereas the M/P AUC ratio after IV dosing was ~0.12 (Thummel *et al.*, 1996), indicating a substantial amount of 1′-hydroxy-midazolam was produced during the first pass after oral administration and the gut wall could contribute significantly to the metabolite formed pre-systematically. Indeed, in anhepatic patients, the average gut wall extraction ratio after intraduodenal administration was 0.43, significantly higher than that (0.08) after intravenous dosing (Paine *et al.*, 1996). When the M/P AUC ratio after oral dosing is lower than that after intravenous dosing, it suggests that the formation clearance may be saturated during the first-pass metabolism. This is possible as drug concentrations in the gut wall and portal vein during the first pass are generally higher than those after intravenous dosing, which could lead to the saturation of drug-metabolizing enzymes, particularly when a biotransformation reaction is characterized by a low Michaelis–Menten constant (Km) or when the oral dose is high.

In addition to the M/P plasma AUC ratio, pre-systemically formed metabolites could have a significant impact on the shape of the metabolite curve when a metabolite exhibits formation rate-limited kinetics. Using a model scheme originally proposed by Houston (Houston, 1982), the effect of pre-systemically formed metabolites on the shape of the metabolite curve can be examined. When there is no pre-systemically formed metabolite, the metabolite curve is parallel to that of the parent (Fig. 4.6a), as expected from the formation rate-limited kinetics. When 50% of drug dose is converted to the metabolites pre-systemically, the metabolite curve starts to show a biphasic phenomenon, with metabolites in circulation coming from two sources—pre-systemically and systemically formed (Fig. 4.6b). When nearly 90% of drug dose is converted to a metabolite pre-systemically during the first-pass metabolism, the metabolite curve exhibits a classic biphasic shape (Fig. 4.6c). The initial portion of the metabolite curve is largely made up of the pre-systemically formed metabolite, which can peak earlier than that of the parent and have a shorter $t_{1/2}$ than that of the parent depending on the

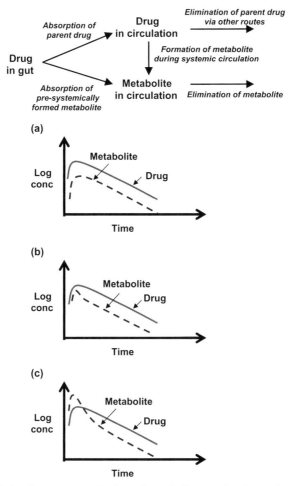

Figure 4.6 Effects of pre-systemically formed metabolites on the shape of metabolite curves following a single oral dose of a parent drug. **(a)** No first-pass extraction. **(b)** First-pass extraction is 50% of dose administered. **(c)** First-pass extraction is 90% of dose administered.

sampling duration. But if studied long enough with good assay sensitivity, the metabolite curve will eventually be parallel to that of the parent as expected from the formation rate-limited metabolite kinetics.

One example to illustrate the impact of pre-systemically formed metabolite on the metabolite curve is shown in Fig. 4.7. In this example, midazolam was administered intravenously and orally, with 1′-hydroxy-midazolam and midazolam measured in plasma (Thummel *et al.*, 1996). After intravenous dosing, the concentration–time curves of 1′-hydroxy-midazolam in both subjects 4 and 8 were parallel to those of the parent drug, indicating formation rate-limited metabolite kinetics. After oral administration,

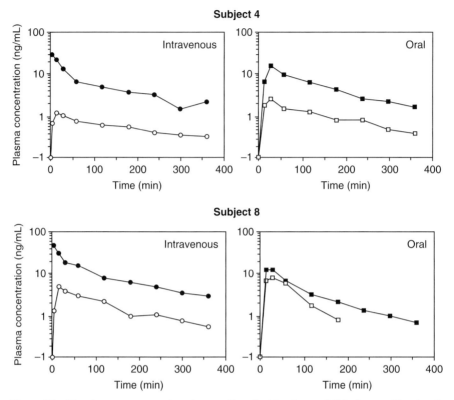

Figure 4.7   The plasma concentration–time profiles of midazolam and 1′-hydroxy-midazolam in two human subjects following intravenous (1 mg) and oral (2 mg) doses of midazolam. Solid symbols correspond to midazolam and open symbols indicate 1′-hydroxy-midazolam (from Thummel *et al.*, 1996. Adapted with permission from Macmillan Publishers Ltd.).

however, the metabolite curve behaved differently between the two subjects. The metabolite curve in subject 4 was still parallel to that of the parent, whereas the metabolite curve in subject 8 was no longer parallel to the parent curve, with an apparent $t_{1/2}$ of the metabolite shorter than that of the parent at the time interval studied. The estimated M/P AUC ratio in subject 4 after intravenous and oral administration was 0.13 and 0.43, respectively. In subject 8, the corresponding M/P AUC ratio was 0.19 and 1.2, respectively. Clearly, the fold difference in the M/P AUC ratio between the two routes in subject 8 was bigger than that in subject 4, indicating more metabolite was produced pre-systemically (possibly in the gut wall) in subject 8 than in subject 4. As a result, the metabolite curve in subject 8 after oral administration was affected significantly. Consistent with this assessment is the fact that the oral bioavailability in subject 4 was in good agreement with the predicted hepatic availability, whereas the observed bioavailability in subject 8 was much lower than the one predicted from the hepatic availability, indicating a significant first-pass metabolism in the gut wall.

It is important to note again that, under linear conditions (i.e. no saturation in liver metabolism), hepatic metabolism is the same between the first pass and systemic cir-

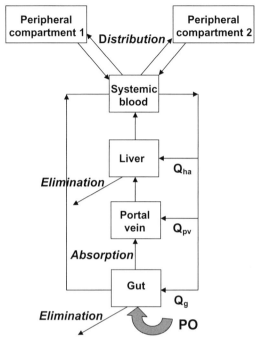

Figure 4.8 A representative hybrid physiologically based pharmacokinetic model that is allowed to differentiate pre-systemic and systemic processes following oral administration of a drug. Q_{pv}: portal vein blood flow; Q_{ha}: hepatic artery blood flow; Q_g: gut blood flow.

culation. However, the gut wall may play a more important role in metabolism during the first pass than that during the systemic circulation following oral administration.

One of the challenges in using compartmental models to describe the metabolite kinetics influenced by pre-systemically formed metabolite is the fact that the conventional models cannot differentiate pre-systemic and systemic processes. To overcome this obstacle, one has to rely on physiologically based pharmacokinetic models (Vossen *et al.*, 2007). However, this also presents a challenge as we often do not have sufficient data in tissues to describe tissue partition coefficients. As a result, a semi-physiologically based pharmacokinetic model (or hybrid model) has been developed (Yamano *et al.*, 2001; Ito *et al.*, 2003; Yang *et al.*, unpublished data), where the gut and liver are taken out from the conventional compartments and set up in a physiologically and anatomically meaningful way linking the blood compartment. A representation of this kind of models is shown in Fig. 4.8. The model is of particular importance to our understanding of metabolite kinetics following oral administration.

4.2.6   *In Situ*-Generated versus Exogenously Dosed Metabolite

An *in situ*-generated metabolite refers to the metabolite formed in an eliminating organ after dosing of the parent drug. An exogenously dosed metabolite is the synthetic standard of a metabolite administered to the body. This is often performed in order to determine the elimination clearance of the metabolite or assess pharmacological activity or toxicity associated with the metabolite. The exogenously dosed metabolite is also known as the preformed metabolite.

From a pharmacokinetic point of view, an *in situ*-generated metabolite may behave differently from an exogenously dosed metabolite (Pang, 1985; Prueksaritanont, Lin and Baillie, 2006). A number of factors contribute to the differences. First of all, metabolites formed, if not always, are more hydrophilic than the parent drug. As a result, the extent of distribution of an exogenously dosed metabolite into tissues may be different from that of *in situ*-generated metabolites. The differences also reside in the ability of the exogenously dosed metabolite in accessing the influx and efflux drug transporters and drug-metabolizing enzymes in eliminating organs for further metabolism and excretion. For an *in situ*-generated metabolite, its metabolism may be affected in the presence of the parent drug and vice versa, which cannot be reproduced by doing the metabolite exogenously. In addition, for formation rate-limited metabolite kinetics, the kinetics of an *in situ*-generated metabolite in the body is governed by the kinetics of the parent drug. Collectively, these kinetics differences between an *in situ*-generated and an exogenously dosed metabolite may translate to the exposure differences in the tissues that are important to the pharmacology and toxicology of the metabolite. Specific examples of this kind are nicely summarized by Prueksaritanont, Lin and Baillie (2006).

4.2.7 Sequential Metabolism

Many times, a primary metabolite is formed followed by sequential metabolism to produce a secondary metabolite. A typical example is hydroxylation followed by glucuronidation. Using a model similar to the one shown in Fig. 4.2, with an addition of the secondary metabolite that is eliminated through excretion into the urine or bile (Fig. 4.9), the following equations can be obtained:

$$\frac{AUC(m1)}{AUC} = f_m \cdot \frac{CL}{CL(m1)} \tag{4.15}*$$

$$\frac{AUC(m2)}{AUC} = f_m \cdot f_m(m1) \cdot \frac{CL}{CL(m2)} \tag{4.16}*$$

where $AUC(m1)$, $AUC(m2)$, and AUC are the AUCs of the primary metabolite, secondary metabolite, and parent drug, respectively; $CL(m1)$, $CL(m2)$, and CL are the elimination clearances of the primary metabolite, secondary metabolite, and parent drug, respectively; f_m is the fraction of the total body clearance of the parent drug forming the primary metabolite; and $f_m(m1)$ is the fraction of the total body clearance of the primary metabolite forming the secondary metabolite. The metabolic ratio in this case equals:

$$\frac{Ae(m1) + Ae(m2)}{Ae} = \frac{CL_f}{CL_e} \tag{4.17}*$$

where Ae, $Ae(m1)$, and $Ae(m2)$ are the amounts of the parent drug, primary metabolite, and secondary metabolite in excreta, respectively. Similarly, the ratio of $Ae(m2)$ to

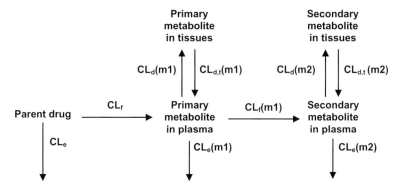

Figure 4.9 A kinetic model for sequential metabolism. CL_e, $CL_e(m1)$, and $CL_e(m2)$ are excretory clearances of the parent drug, primary metabolite, and secondary metabolite, respectively; CL_f and $CL_f(m1)$ are the formation clearance of the primary and secondary metabolites, respectively; $CL_d(m1)$ and $CL_{d,t}(m1)$ are the distributional clearance of the primary metabolite to and from tissues, respectively; $CL_d(m2)$ and $CL_{d,t}(m2)$ are the distributional clearance of the secondary metabolite to and from tissues, respectively.

Ae(m1) equals the ratio of the formation clearance of the secondary metabolite $(CL_f(m1))$ and the excretory clearance of the primary metabolite $(CL_e(m1))$ and is shown in the following equation:

$$\frac{Ae(m2)}{Ae(m1)} = \frac{CL_f(m1)}{CL_e(m1)} \tag{4.18}*$$

Once again, Equations 4.15–4.18 are valid only if the AUC is the total area of the curve from time zero to infinity and the recovery of the parent drug and metabolites is complete in excreta.

For sequential metabolism that involves a common secondary metabolite produced from two parallel, primary metabolic pathways, the kinetic processes are more complex than the scheme shown in Fig. 4.9. Interested readers can refer to the paper by Pang (1995) for detailed kinetic treatment.

4.3 EFFECT OF INHIBITION AND INDUCTION ON METABOLITE KINETICS

In a drug–drug interaction study, metabolite concentrations associated with an affected clearance pathway are often measured. Intuitively, one would think that, when the formation clearance is inhibited, that metabolite concentrations and AUC would decrease, and when the formation clearance is induced, the metabolite concentration and AUC would increase. However, this may not be entirely true based on the metabolite kinetic principles. Considering the model scheme shown in Fig. 4.2, one could rearrange Equation 4.7 and obtain the following equation:

$$AUC(m) = f_m \cdot \frac{CL \cdot AUC}{CL(m)} \tag{4.19}$$

Recalling that drug dose equals CL·AUC, the following equation exists:

$$AUC(m) = f_m \cdot \frac{Dose}{CL(m)} \tag{4.20}*$$

By the same token, the following equation can be obtained at steady state:

$$C_{ss}(m) = f_m \cdot \frac{Dose/\tau}{CL(m)} \tag{4.21}*$$

where τ is the dosing interval. Equations 4.20 and 4.21 indicate that the AUC and C_{ss} of a metabolite are a function of dose, f_m, and elimination clearance of the metabolite. When f_m is ~1.0, meaning that the formation of the metabolite is a predominant elimination pathway of the parent drug, then Equations 4.20 and 4.21 can be simplified to the following equations:

$$AUC(m) = \frac{Dose}{CL(m)} \tag{4.22}$$

$$C_{ss}(m) = \frac{Dose/\tau}{CL(m)} \tag{4.23}$$

In this case, the AUC and $C_{ss}(m)$ of the metabolite would not change regardless of whether its formation clearance is inhibited or induced. Conceptually, this makes sense. When the formation clearance, as a predominant clearance pathway of the parent drug, is inhibited, the parent drug concentration would increase. The rate of formation, however, would not change, as it equals the product of the formation clearance and parent drug concentrations. As a result, AUC and $C_{ss}(m)$ of the metabolite would not change regardless of what happens to its formation clearance. However, the M/P AUC or C_{ss} ratio would change because of the changes in the parent drug concentrations. This is best illustrated through the simulations shown in Fig. 4.10a. In this case, the formation clearance of the metabolite is the total body clearance of the parent drug (i.e. $f_m = 1$). When the formation clearance is inhibited by 50%, the average parent drug concentrations at steady state increase by twofold as expected. On the one hand, the average metabolite concentration at steady state remains the same. There are many clinical drug–drug interactions that exemplify the case cited. For example, in the presence of rifampicin, the plasma AUC of midazolam after oral administration reduced from 35.8 to 3.7 µg/h/mL (Gorski *et al.*, 2003). On the other hand, the average urinary recovery of 1′-hydroxy-midazolam before and after rifampicin treatment was comparable (74% versus 64%), indicating that the extent of formation of 1′-hydroxy-midazolam, the major elimination pathway of midazolam in humans, is similar in the presence and absence of rifampicin.

When the f_m no longer approaches 1, the AUC and $C_{ss}(m)$ of the metabolite would change when the formation clearance of the metabolite is affected. The extent of change is dependent on the change in the f_m. Figure 4.10b shows a compound that has an f_m of 0.8 for the formation of metabolite A. In this case, when the formation clearance of metabolite A is inhibited by 50%, the average steady-state concentrations of the parent drug are increased by 1.7-fold. Notice that the change is not twofold due

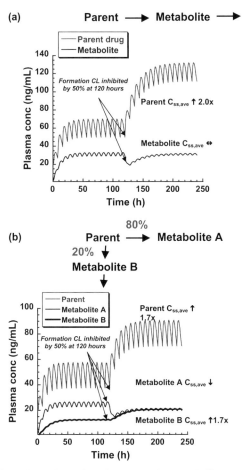

Figure 4.10   Inhibition of the formation clearance of a metabolite on metabolite kinetics at steady state during a multiple-dose regimen. **(a)** $f_m = 1$; **(b)** $f_m = 0.8$.

to a shunting effect to another pathway (metabolite B). Accordingly, the average steady-state concentration of metabolite B is also increased by 1.7-fold. The average steady-state concentration of metabolite A, of which the formation clearance is inhibited, was marginally decreased by 17%. Clearly, the existence of additional pathways may significantly attenuate the extent of inhibition or induction on a particular pathway.

A sensitive way to detect the change in an affected formation clearance pathway is to compare the M/P AUC or C_{ss} ratio in the presence and absence of an inhibitor or inducer. Recall from Equation 4.5 that:

$$\frac{AUC(m)}{AUC} = \frac{CL_f}{CL(m)}$$

Assuming no change in the CL(m) in the presence of an inhibitor or inducer, Equation 4.5 becomes:

$$\left\{\frac{AUC(m)}{AUC}\right\}_{affected} = \frac{CL_{f,affected}}{CL(m)} \tag{4.24}$$

Therefore, the M/P AUC ratio in the presence and absence of an inhibitor or inducer is:

$$\frac{\left\{\dfrac{AUC(m)}{AUC}\right\}_{affected}}{\left\{\dfrac{AUC(m)}{AUC}\right\}} = \frac{CL_{f,affected}}{CL_f} \tag{4.25}*$$

Equation 4.25 indicates that a change in the M/P AUC ratio directly reflects the change in the metabolite formation clearance.

In addition to the M/P plasma AUC ratio, the metabolic ratio (Ae(m)/Ae) in excreta can be used to assess the alteration in the formation clearance of a metabolite. Recall from Equation 4.12 that:

$$\frac{Ae(m)}{Ae} = \frac{CL_f}{CL_e}$$

Assuming no change in the CL_e (renal or biliary clearance) of the parent drug in the presence and absence of an inhibitor or inducer, the following equation exists:

$$\frac{\left\{\dfrac{Ae(m)}{Ae}\right\}_{affected}}{\left\{\dfrac{Ae(m)}{Ae}\right\}} = \frac{CL_{f,affected}}{CL_f} \tag{4.26}*$$

Again, like the M/P AUC ratio, a change in the metabolic ratio directly reflects the change in the metabolite formation clearance.

To appreciate these effects, simulations were conducted to examine the effects of f_m and inhibition of the metabolite formation clearance on the metabolite AUC, parent AUC, and M/P AUC (or metabolic) ratio (Fig. 4.11). It is apparent from the graphs that the higher the inhibition of metabolite formation clearance, the more affected the metabolite AUC. However, the f_m attenuates this effect as discussed earlier. The higher the f_m value, the less affected the metabolite AUC even when the formation clearance is inhibited by ~100%. On the other hand, the higher the f_m, the more affected the parent AUC. Lastly, the change in the M/P AUC ratio or metabolic ratio is inversely related to the extent of the inhibition of the metabolite formation clearance and is independent of the f_m as shown in Equations 4.25 and 4.26. And this change largely results from the alteration in the parent AUC or Ae. As the f_m approaches unity, the metabolite AUC or Ae(m) will be largely unchanged, assuming that the inhibitor or inducer does not affect the elimination clearance of the metabolite.

(a)

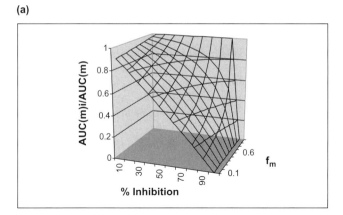

(b)

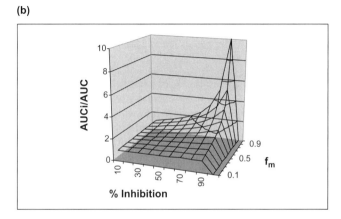

(c)

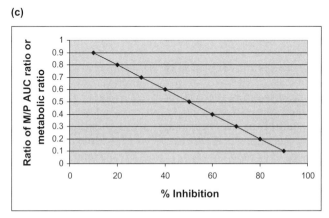

Figure 4.11 Sensitivity of various pharmacokinetic parameters to the inhibition of the metabolite formation clearance as a function of f_m. **(a)** Metabolite AUC. **(b)** Parent AUC. **(c)** M/P AUC ratio or metabolic ratio.

4.4   DETERMINATION OF FORMATION AND ELIMINATION CLEARANCE OF METABOLITE

The elimination clearance (CL(m)) of a metabolite can be determined by intravenous dosing of a synthetic standard of the metabolite, with a caveat that disposition kinetics may be different between an *in situ*-generated metabolite and an exogenously dosed one (Pang, 1985; Prueksaritanont, Lin and Baillie, 2006). Once the CL(m) is determined, the formation clearance (CL_f) of the metabolite can be determined based on the M/P AUC or C_{ss} ratio as shown in Equation 4.5 or Equation 4.6.

However, sometimes, a synthetic standard of the metabolite is not available or not administered intravenously to obtain the CL(m). In this case, one could still make a case to estimate the CL(m) and CL_f from dosing the parent drug alone, provided that a number of assumptions are met and metabolite and parent drug data are available in plasma and excreta. In one scenario, one could estimate the CL(m) from the metabolite data in plasma and excreta. Considering the model scheme shown in Fig. 4.2, it can be shown that:

$$CL(m) = \frac{Ae(m)}{AUC(m)} \qquad (4.27)*$$

In this case, the recovery of a metabolite in excreta does not have to be complete, as long as the collection time interval for excreta is the same as that for the AUC of the metabolite. The key assumptions, however, are that the metabolite produced in the body is entirely available to the systemic circulation, as if it was dosed systemically, and there is no sequential metabolism. Conceivably, the CL(m) can be accurately determined from Equation 4.27 if the metabolite is produced in the liver and/or gut and is eliminated exclusively into the urine unchanged. However, if there is a significant sequential metabolism involved, Equation 4.27 would underestimate the true CL(m), as Ae(m) does not account for the entire amount eliminated. In addition, if a metabolite is produced in the liver but eliminated significantly via biliary secretion, Equation 4.27 would overestimate the true CL(m), as the metabolite in the bile may come directly from the liver. Once the CL(m) is determined, the CL_f of the metabolite can be determined from the M/P AUC or C_{ss} ratio using Equation 4.5 or Equation 4.6.

In another scenario, if the formation of the metabolite represents a predominant pathway of the metabolic clearance of the parent drug, one could estimate the CL_f of the metabolite by subtracting the total body clearance of the parent drug from its excretory clearance. Alternatively, the CL_f can be calculated from the product of the metabolic ratio and the excretory clearance of the parent drug using Equation 4.12, with additional assumptions that the recovery of the parent drug and metabolite is complete and there is no sequential metabolism. The excretory clearance of the parent drug can be calculated as the amount of the parent drug in excreta divided by its plasma AUC estimated at the time interval as the excreta. Once the CL_f is determined, the CL(m) of the metabolite can be determined from the M/P AUC or C_{ss} ratio using Equation 4.5 or Equation 4.6.

The situation could be further complicated if metabolite information is not available in plasma. In this case, with a number of assumptions, one could still estimate the CL_f of the metabolite but not the CL(m). Recalling that drug dose equals

CL·AUC and CL_f equals $f_m·CL$, the following equation exists based on Equation 4.11:

$$CL_f = \frac{Ae(m)}{AUC} \qquad (4.28)*$$

where AUC is the total area under the curve of the parent drug. It is important to note that Equation 4.28 is valid only if the recovery of the metabolite in excreta is complete and there is no sequential metabolism of the metabolite. If the recovery of the metabolite in excreta is incomplete (e.g. true recovery is 50% of the dose versus observed recovery is 25% over a finite period of time), it may lead to an underestimation of the CL_f. In addition, for orally administered drugs, one has to assume that the metabolite is largely formed after the parent drug reaches the systemic circulation (i.e. minimal first-pass metabolism). Violating this assumption leads to an overestimation of the CL_f. Alternatively, as discussed earlier, the CL_f can be calculated from the product of the metabolic ratio and the excretory clearance of the parent drug using Equation 4.12, if both parent and metabolite data are available in excreta. In this case, the assumptions on complete recovery in excreta and no sequential metabolism still apply. In addition, the formation of the metabolite is assumed to represent a predominant metabolic pathway of the parent drug.

For sequential metabolism, the same principles along with the assumptions discussed before can be applied to determine the formation (CL_f) and elimination clearance ($CL(m1)$) of the primary metabolite (model scheme shown in Fig. 4.9). With plasma and excreta data available for the parent drug, primary metabolite, and secondary metabolite, the $CL_f(m1)$, $CL_e(m1)$, and $CL(m1)$ of the primary metabolite are calculated as follows:

$$CL_f(m1) = \frac{Ae(m2)}{AUC(m1)} \qquad (4.29)$$

$$CL_e(m1) = \frac{Ae(m1)}{AUC(m1)} \qquad (4.30)$$

$$CL(m1) = \frac{Ae(m1) + Ae(m2)}{AUC(m1)} \qquad (4.31)*$$

The key assumptions are (i) the primary metabolite produced in the body entirely available in the systemic circulation, (ii) no sequential metabolism of the secondary metabolite, and (iii) complete recovery of metabolites in excreta. As a result, the CL_f calculated from Equation 4.29 always overestimates the true CL_f as the secondary metabolite can form locally in the liver upon the formation of the primary metabolite. The CL_e can also be overestimated using Equation 4.30 if biliary secretion is a part of the excretory clearance pathways of the primary metabolite. Once the $CL(m1)$ is known, the CL_f of the primary metabolite can be determined from the M/P AUC or C_{ss} ratio between the primary metabolite and the parent drug using Equation 4.5 or Equation 4.6. In the case of lack of the primary metabolite data in plasma, one could determine the CL_f but not $CL(m1)$ of the primary metabolite using the following equation:

$$CL_f = \frac{Ae(m1) + Ae(m2)}{AUC} \tag{4.32}$$

In this case, the assumptions of no sequential metabolism of the secondary metabolite and complete recovery of metabolites in excreta also apply. In addition, for orally administered drugs, minimal first-pass metabolism is assumed.

4.5 SUMMARY

In vivo drug metabolite kinetics is a complex process in part because it is governed by both formation and elimination of the metabolite(s) of interest. As a result, the M/P AUC or C_{ss} ratio is a function of both the formation and the elimination clearance of the metabolite. The conceptual difference between a major circulating metabolite and the product of a major metabolic pathway needs to be appreciated in the safety testing of metabolites observed in circulation and excreta. Also, the rate of metabolism is not necessarily related to the extent of metabolism, as a low rate of metabolism can still lead to extensive metabolism if no other pathways exist. The rate-limiting step in the formation and elimination of a metabolite is a key determinant to the pharmacokinetic characteristics of the metabolite, including the shape of curve, rate of removal from the body, time to reach steady state, and extent of accumulation. Furthermore, the extent of metabolites formed pre-systemically in the gut wall/lumen following oral administration of a parent drug could have a significant impact on the metabolite kinetics and the shape of the metabolite curve. In addition, a change in the metabolic ratio or M/P plasma AUC ratio directly reflects the change in the formation clearance of a metabolite, whereas an alteration in the metabolite or parent AUC is also a function of the f_m. Overall, a clear understanding and appreciation of metabolite kinetic principles is essential to assess the pharmacology and toxicology of metabolites and to guide the better use of therapeutic agents in the clinic.

ACKNOWLEDGMENTS

The author would like to thank Drs. Punit Marathe and A. David Rodrigues for their critical review of the book chapter and valuable comments and suggestions.

REFERENCES

Baillie, T.A., Cayen, M.N., Fouda, H., Gerson, R.J., Green, J.D., Grossman, S.J., Klunk, L.J., LeBlanc, B., Perkins, D.G., Shipley, L.A. (2002). Drug metabolites in safety testing. *Toxicol. Appl. Pharmacol.*, *182*, 188–196.

Christians, U., Jacobsen, W., Floren, L.C. (1998). Metabolism and drug interactions of 3-hydroxy-3-methylglutaryl coenzyme A reductase inhibitors in transplant patients: are the statins mechanistically similar? *Pharmacol. Ther.*, *80*, 1–34.

Colizza, K., Awad, M., Kamel, A. (2007). Metabolism, pharmacokinetics, and excretion of the substance P receptor antagonist CP-122,721 in humans: structural characterization of the novel major circulating metabolite 5-trifluoromethoxy salicylic acid by high-performance liquid chromatography-tandem mass spectrometry and NMR spectroscopy. *Drug Metab. Dispos.*, *35*, 884–897.

Davis-Bruno, K.L., Atrakchi, A. (2006). A regulatory perspective on issues and approaches in characterizing human metabolites. *Chem. Res. Toxicol.*, *19*, 1561–1563.

FDA Guidance for Industry: Safety Testing of Drug Metabolite. February 2008. http://www.fda.gov/CDER/GUIDANCE/6897fnl.pdf.

Davis-Bruno, K.L., Atrakchi, A. (2006). A regulatory perspective on issues and approaches in characterizing human metabolites. *Chem. Res. Toxicol.*, *19*, 1561–1563.

Fura, A. (2006). Role of pharmacologically active metabolites in drug discovery and development. *Drug Discov. Today*, *11*, 133–141.

Fura, A., Shu, Y.-Z., Zhu, M., Hanson, R.L., Roongta, V., Humphreys, W.G. (2004). Discovering drugs through biological transformation: role of pharmacologically active metabolites in drug discovery. *J. Med. Chem.*, *47*, 4339–4351.

Gorski, J.C., Vannaprasaht, S., Hamman, M.A., Ambrosius, W.T., Bruce, M.A., Haehner-Daniels, B., Hall, S.D. (2003). The effect of age, sex, and rifampin administration on intestinal and hepatic cytochrome P450 3A activity. *Clin. Pharmacol. Ther.*, *74*, 275–287.

Hastings, K.L., El-Hage, J., Jacobs, A., Leighton, J., Morse, D., Osterberg, R.E. (2003). Letter to the editor on drug metabolites in safety testing. *Toxicol. Appl. Pharmacol.*, *190*, 91–92.

Houston, J.B. (1982). Drug metabolite kinetics. *Pharmacol. Ther.*, *15*, 521–552.

Humphreys, W.G., Unger, S.E. (2006). Safety assessment of drug metabolites: characterization of chemically stable metabolites. *Chem. Res. Toxicol.*, *19*, 1564–1569.

Ito, K., Ogihara, K., Kanamitsu, S., Itoh, T. (2003). Prediction of the in vivo interaction between midazolam and macrolides based on in vitro studies using human liver microsomes. *Drug Metab. Dispos.*, *31*, 945–954.

Paine, M.F., Shen, D.D., Kunze, K.L., Perkins, J.D., Marsh, C.L., McVicar, J.P., Barr, D.M., Gillies, B.S., Thummel, K.E. (1996). First-pass metabolism of midazolam by the human intestine. *Clin. Pharmacol. Ther.*, *60*, 14–24.

Pang, K.S. (1985). A review of metabolite kinetics. *J. Pharmacokinet. Biopharm.*, *13*, 633–662.

Pang, K.S. (1995). Kinetics of sequential metabolism: contribution of parallel, primary metabolic pathways to the formation of a common secondary metabolite. *Drug Metab. Dispos.*, *23*, 166–177.

Pereillo, J.M., Maftouh, M., Andrieu, A., Uzabiaga, M.F., Fedeli, O., Savi, P., Pascal, M., Herbert, J.M., Maffrand, J.P., Picard, C. (2002). Structure and stereochemistry of the active metabolite of clopidogrel. *Drug Metab. Dispos.*, *30*, 1288–1295.

Prueksaritanont, T., Lin, J.H., Baillie, T.A. (2006). Complicating factors in safety testing of drug metabolites: kinetic differences between generated and preformed metabolites. *Toxicol. Appl. Pharmacol.*, *217*, 143–152.

Roffey, S.J., Obach, R.S., Gedge, J.I., Smith, D.A. (2007). What is the objective of the mass balance study? A retrospective analysis of data in animal and human excretion studies employing radiolabeled drugs. *Drug Metab. Rev.*, *39*, 17–43.

Rowland, M., Tozer, T.N. (1995). *Clinical Pharmacokinetics: Concepts and Applications*, 3rd edn, Williams & Wilkins, Baltimore, MD.

Savi, P., Pereillo, J.M., Uzabiaga, M.R., Combalbert, J., Picard, C., Maffrand, J.P., Pascal, M., Herbert, J.M. (2000). Identification and biological activity of the active metabolite of clopidogrel. *Thromb. Haemost.*, *84*, 891–896.

Siedlik, P.H., Olson, S.C., Yang, B.B., Stern, R.H. (1999). Erythromycin coadministration increases plasma atorvastatin concentrations. *J. Clin. Pharmacol.*, *39*, 501–504.

Smith, D.A., Obach, R.S. (2005). Seeing through the mist: abundance versus percentage. Commentary on metabolites in safety testing. *Drug Metab. Dispos.*, *33*, 1409–1417.

Smith, D.A., Obach, R.S. (2006). Metabolites and safety: what are the concerns, and how should we address them? *Chem. Res. Toxicol.*, *19*, 1570–1579.

Thummel, K.E., O'Shea, D., Paine, M.F., Shen, D.D., Kunze, K.L., Perkins, J.D., Wilkinson, G.R. (1996). Oral first-pass elimination of midazolam involves both gastrointestinal and hepatic CYP3A-mediated metabolism. *Clin. Pharmacol. Ther.*, *59*, 491–502.

Thummel, K.E., Shen, D.D. (2002). The role of the gut mucosa in metabolically based drug-drug interactions, in *Drug–Drug Interactions* (ed. A.D. Rodrigues), Marcel Dekker, New York, pp. 359–385.

Thummel, K.E., Shen, D.D., Isoherranen, N., Smith, H.E. (2006). Design and optimization of dosage regimens: pharmacokinetic data, in *Goodman & Gilman's The Pharmacological Basis of Therapeutics*, 11th edn (eds L.L. Brunton, J.S. Lazo, K.L. Parker), McGraw-Hill, New York, pp. 1787–1888.

Vossen, M., Sevestre, M., Niederalt, C., Jang, I.J., Willmann, S., Edginton, A.N. (2007). Dynamically simulating the interaction of midazolam and the CYP3A4 inhibitor itraconazole using individual coupled whole-body physiologically-based pharmacokinetic (WB-PBPK) models. *Theor. Biol. Med. Model.*, *4*, 13. doi:10.1186/1742-4682-4-13.

Wang, L., Zhang, D., Swaminathan, A., Xue, Y., Cheng, P.T., Wu, S., Mosqueda-Garcia, R., Aurang, C., Everett, D.W., Humphreys, W.G. (2006). Glucuronidation as a major metabolic clearance pathway of 14c-labeled muraglitazar in humans: metabolic profiles in subjects with or without bile collection. *Drug Metab. Dispos.*, *34*, 427–439.

Wilkinson, G.R. (2002). In vivo probes for studying induction and inhibition of cytochrome P450 enzymes in humans, in *Drug–Drug Interactions* (ed. A.D. Rodrigues), Marcel Dekker, New York, pp. 439–504.

Williams, D., Feely, J. (2002). Pharmacokinetic-pharmacodynamic drug interactions with HMG-CoA reductase inhibitors. *Clin. Pharmacokinet.*, *41*, 343–370.

Yamano, K., Yamamoto, K., Katashima, M., Kotaki, H., Takedomi, S., Matsuo, H., Ohtani, H., Sawada, Y., Iga, T. (2001). Prediction of midazolam-CYP3A inhibitors interaction in the human liver from in vivo/in vitro absorption, distribution, and metabolism data. *Drug Metab. Dispos.*, *29*, 443–452.

CHAPTER 5

Pharmacogenetics and Pharmacogenomics

LILIAN G. YENGI

5.1 INTRODUCTION: PHARMACOGENETICS AND PHARMACOGENOMICS

The fields of pharmacogenetics, genomics, and drug transporters have profoundly impacted drug metabolism research by providing plausible mechanisms for interindividual variability in drug response and metabolism-related toxicity. They have provided tools with which to understand enzyme regulation, identify factors that affect drug exposure, the potential for drug–drug interactions, and species differences in drug disposition. Knowledge from these fields is being used to form the scientific basis for designing appropriate clinical studies and data interpretation, leading to the development and use of safer and more efficacious drugs.

Pharmacogenetics is the study of the effects of genetic differences between individuals on interindividual responses to medicines (Ginsburg *et al.*, 2005; Eichelbaum, Ingelman-Sundberg and Evans, 2006). It is an old discipline, which has been invigorated by an increased understanding of molecular biology and development of associated technological tools with which to study these mechanisms (Goldstein, Tate and Sisodiya, 2003). In its simplest form, genetics is the scientific study of heredity, dating back to Mendel, who showed that the inheritance pattern of certain traits in pea plants followed simple statistical rules and described a fundamental unit of heredity, which he called allele. While Mendel used *allele* to refer to what we now know as gene, allele is currently used to denote a specific variant of a particular gene. Pharmacogenetics can therefore be viewed as a more specific and narrow subfield of genetics. Though an old discipline, the term having been around since the 1950s has enjoyed a new lease of life in the last 10–20 years, mainly due to advances in our understanding of the impact of genetics on efficacy and safety of drugs, emergence of global tools with which to study multiple genes at any given time, and the completion of the human genome project.

Pharmacogenomics is a more recent term, coined to define a more holistic or global approach, in which the expression levels, regulation, functions, and interactions of

Drug Metabolism Handbook: Concepts and Applications, Edited by Ala F. Nassar, Paul F. Hollenberg, and JoAnn Scatina

multiple genes are simultaneously studied, and their effects on overall variability in drug response determined (Nebert, 1999). It is sometimes used interchangeably with pharmacogenetics, despite these subtle differences. These fields are continually being integrated in various aspects of the life sciences, including drug discovery and development, with the expectations that they would lead to the development of safer drugs that can be tailored to subsets of patients based on their genetic makeup, the so-called personalized medicines (Wolf, Smith and Smith, 2000; Hedgecoe and Martin, 2003). Genome-wide screening is used to identify single nucleotide polymorphisms (SNPs) that co-segregate with certain diseases with a view to using these SNP regions as potential targets for drug development, and global gene expression analysis is used preclinically and clinically to identify surrogate markers of toxicity (toxicogenomics) and/or efficacy (Searfoss *et al.*, 2003). That individuals differ in their response to drug use is not new and there have always been suggestions that a genetic component could be responsible. However, this could not be directly proven due to lack of appropriate technologies. Nevertheless, differences in plasma levels of antidepressants such as nortriptyline and imiprimine were attributed to genetic factors (Hammer and Sjoqvist, 1967; Alexanderson, Evans and Sjoqvist, 1969) and impact of genetic variations in the toxicity of drugs such as the antitubercular drug Isoniazid, were postulated, and then proven as early as the 1950s (Bonicke and Reif, 1953; Evans, Manley and Mc, 1960). The CYP2D6 polymorphism, which resulted in differences in the metabolism of debrisoquine among different individuals, is a more recent and famous example of the impact of genetic variations on exposure and disposition drugs (Mahgoub *et al.*, 1977; Eichelbaum, 1986). Many functional polymorphisms have now been identified in many gene families, with the variant alleles occurring with varying frequencies among different ethnic populations (Garte *et al.*, 2001; Ingelman-Sundberg, Daly and Nebert, 2007). Those polymorphisms with major implications on the development and clinical applications of major pharmaceutical drugs will be discussed later in this chapter.

5.2 PHARMACOGENOMICS OF DRUG-METABOLIZING ENZYMES (DME)

Interest in gene regulation and its potential impact on the efficacy and safety of new drug candidates is steadily increasing. One of the consequences of transcriptional regulation is clinical drug–drug interactions with coadministered drugs and also the observation that some drugs induce the enzymes involved in their metabolism, hence altering their kinetic properties, including those compounds still in development, where exposures could alter following multiple dosing in safety studies. It is therefore important that the effects of new drug compounds on the expression of drug metabolism genes form an integral part of drug metabolism studies, both preclinically and during clinical development. The discovery and characterization of the cytochrome P450s (CYPs) in the 1950s, their solubilization, and elucidation of the mechanisms underlying their regulation are critical milestones in our understanding of the biotransformation of xenobiotics including pharmaceutical drugs. Although it had been known since their characterization that some P450s were inducible, it was knowledge of the interactions between steroid hormones and their targets within the cell that led to the hypothesis that induction of these microsomal enzymes by ligands was via a mechanism similar to that of steroid hormones. This hypothesis was later supported in subsequent experiments that eventually led to the characterization and identification of the aryl hydrocarbon hydroxylase (Ah) receptor as responsible for CYP1A1 induction (Poland, Glover and Kende, 1976; Guenthner and Nebert, 1977). Several nuclear receptors have now been cloned and the downstream genes they regulate have been identified (Bertilsson *et al.*, 1998; Blumberg *et al.*, 1998; Kliewer *et al.*, 1998). These include pregnane X receptor PXR, with CYP3A4, CYP2C, and UGT as the prototypic downstream genes, constitutive androstane receptor (CAR), which regulates genes in the CYP2B family and UGTs, liver X receptor (LXR), the master regulator of cholesterol homeostasis, farnesoid X receptor (FXR), peroxisome proliferator-activated receptor (PPAR), vitamin D receptor (VDR), glucocorticoid receptor (GR), and hepatocyte nuclear receptor 4 (HNF4). These receptors control a complex network of endogenous pathways and often exhibit cross talk between them in relation to their ligands and downstream regulated genes (Kliewer, Lehmann and Willson, 1999; Repa and Mangelsdorf, 2000; Rushmore and Kong, 2002). It is therefore not surprising that regulation of nuclear receptors by drug candidates has the potential not only to impact targeted pathways, but also affect off-target genes leading to unexpected efficacy and/or safety outcomes. The role of nuclear hormone receptors in drug metabolism and drug–drug interactions will be discussed in detail in later chapters of this book.

5.3 POLYMORPHISMS IN GENES ENCODING DME AND IMPLICATIONS FOR DRUG DISCOVERY AND DEVELOPMENT

Following the concept of two-phase elimination of xenobiotics proposed by R.T. Williams, reactions such as oxidation, reduction, and hydrolysis are considered phase I and activating, while conjugation reactions are considered phase II and detoxifying in nature. Although other phase I enzyme systems exist, the CYPs have emerged as the principal enzymes catalyzing most phase I reactions, while phase II reactions are catalyzed by several gene families including uridine diphosphate glucuronosyltransferase (UGT), glutathione transferases (GSTs), *N*-acetyltransferase (NAT), and sulfotransferases (SULTs). Phase I reactions are often the initial stage in converting often highly lipophilic environmental chemicals, which cannot readily be excreted from the body, to more water-soluble, hydrophilic species (Guengerich, 1991b; Goldstein and Faletto, 1993). Phase II enzymes then catalyze the attachment of highly water-soluble moieties to these intermediate products, thus increasing their hydrophilicity. The net result of phase I and phase II reactions is generally to detoxify and eliminate the compound from the body, although they can also result in activation of some compounds to their ultimate toxic and/or carcinogenic end products (Guengerich, 1991a, b; Goldstein and Faletto, 1993). Research over the years has shown that functional polymorphisms in both phase I and II enzymes are remarkably common and some have been associated with disease susceptibility and prognoses, and interindividual variability in drug response and toxicity. The following section will review some of the functional polymorphisms in DME that impact the safety and/or efficacy of drugs.

5.3.1 P450 Polymorphisms

Although it is now common knowledge that many P450s exhibit genetic polymorphisms with implications for drug efficacy and safety, the first direct association between CYP polymorphisms and alterations in the pharmacokinetics of a drug was not made until the late 1970s and 1980s. Many of the variations were first identified by the occurrence of adverse reactions in patients or volunteers taking normal doses of drugs later found to be metabolized primarily by polymorphic CYPs. Despite the fact that more than 50 human CYPs have been identified, only genes in the first three families (CYP1 to 3) appear to be responsible for the metabolic transformation of most drugs and xenobiotics. Hence polymorphisms in genes belonging to these families are likely to be relevant in determining the safety and efficacy of most pharmaceuticals. Indeed many of these genes are functionally polymorphic, with different allelic frequencies among different ethnic populations.

CYP1A1 is one of the few major CYPs involved in the metabolism of xenobiotics that appears to have been conserved from an evolutionary perspective. CYP1A1 is constitutively expressed at very low levels predominantly in extrahepatic tissues but is highly induced in response to several endogenous and exogenous ligands. About 20 single nucleotide polymorphisms have been identified at the CYP1A1 locus, with 10 resulting in amino acid substitutions (Daly *et al.*, 1993; Saito *et al.*, 2003; Ingelman-Sundberg, Daly and Nebert, 2007). However, none of them significantly affect protein function and therefore have very little impact on the metabolism of CYP1A1 drug substrates, although some of the allelic variants have been found to be in linkage disequilibrium with certain diseases including various cancers (Taioli *et al.*, 2003; Vineis *et al.*, 2004). This evolutionary conservation in CYP1A1 sequence suggests a critical endogenous role for CYP1A1, although endogenous substrates for CYP1A1 have not been identified. However, CYP1A1 is regulated by the Ah receptor, which has an important role in cell cycle and it has been hypothesized that CYP1A1 might be an important mediator in some cell types (Ingelman-Sundberg, 2001).

CYP1A2 is expressed almost exclusively in the liver and constitutes approximately 12% of total hepatic CYP content. It is involved in the metabolism of various toxic and carcinogenic compounds and several drugs, including caffeine, clozapine (where CYP1A2 has a major role), fluvoxamine, olanzapine, tacrine, theophylline, and acetaminophen (Murayama *et al.*, 2004; Chetty and Murray, 2007). Although CYP1A2-mediated metabolism is often a minor pathway in the metabolism of many drugs, it may become critical in the absence of the major contributing CYP or when the major isoform is saturated (e.g. in CYP2D6 poor metabolizers). Increased abundance of CYP1A2 due to induction may also result in clinically significant interactions. In addition, constitutive expression of CYP1A2 mRNA levels in human liver samples can vary by as much as 15-fold and the protein by up to 60-fold (Shimada *et al.*, 1994; Hammons *et al.*, 2001). Therefore, CYP1A2 polymorphisms have the potential to alter the disposition of many drugs including those in which it only plays a minor role. Approximately 40 single nucleotide variant alleles have been identified at the CYP1A2 locus, five of which result in alterations to the amino acid sequence of the encoded protein (Ingelman-Sundberg, Daly and Nebert, 2007). The 60-fold interindividual variation detected in the activity of CYP1A2 protein may be the phenotypic manifestation of these polymorphisms (Rasmussen *et al.*, 2002; Pucci *et al.*, 2007). A transition polymorphism (C to A) in the first intron of CYP1A2 has been associated with variations in CYP1A2 inducibility in healthy volunteers who are cigarette smokers (Sachse *et al.*, 1999) and has been postulated to be responsible for tardive dyskinesia (TD), a common and potentially irreversible side effect associated with long-term treatment with antipsychotics (Basile *et al.*, 2000). However, studies in a Turkish population found no association between this polymorphism and schizophrenic patients with TD (Boke *et al.*, 2007). A polymorphism (G to A) in the 5′-flanking region of *CYP1A2* in Japanese subjects has been associated with a significant decrease in CYP1A2-mediated demethylation of caffeine (Nakajima *et al.*, 1999). The implications of this and other CYP1A2 polymorphisms in relation to clinical drug interactions remain to be ascertained.

CYP2D6 was the first human CYP isozyme identified to be polymorphic with clinical drug implication. In the 1970s, it had been known that the metabolism of certain drugs such as debrisoquine and spartein showed interindividual variations as evidenced by different plasma levels of the drugs (Eichelbaum *et al.*, 1979). About a decade later, several scientists were able to show that these differences were due to genetic differences in the enzyme responsible for the metabolism of these drugs. This gene was later identified as CYP2D6, located on the long arm of chromosome 22. Because of this unique place in the history of pharmacogenetics, the CYP2D6 locus has been extensively studied and to date, more than 70 alleles have been identified, with different frequencies occurring in different ethnic populations (Ikenaga *et al.*, 2005; Beverage *et al.*, 2007; Ingelman-Sundberg, Daly and Nebert, 2007). Some of these allelic variants produce nonfunctional proteins, some have reduced function, and others have no effect on phenotype. There are also individuals with multiple alleles resulting in an increase in function. Phenotypically, these are respectively called poor metabolizers (PMs), intermediate metabolizers (IMs), extensive metabolizers (EMs), and ultrarapid metabolizers (UMs). The PM phenotype appears to be more common among Europeans (3–11%), low in African populations, and 1–2% in Asians (Scordo *et al.*, 2004). The UM phenotype is the most common phenotype in Ethiopians, with a reported frequency of up to 39% (Daly *et al.*, 1993; Aklillu *et al.*, 1996). The importance of CYP2D6 polymorphisms in pharmaceutical development is evidenced by the long list of drugs whose efficacy and safety is impacted by CYP2D6 polymorphisms (Meyer *et al.*, 1996; Ingelman-Sundberg, 2001).

The importance of CYP2C8 in the metabolism of several drugs is becoming evident; hence interest in CYP2C8 polymorphisms and impact on drug disposition and clinical drug interactions is increasing. CYP2C8 plays a major role in the metabolism of the thiazolidinediones (glitazones) class of antidiabetic drugs. For instance *in vitro* and *in vivo* studies suggest that pioglitazone, a peroxisome proliferator-activated receptor gamma (PPAR-γ) agonist used in the treatment of type 2 diabetes, is metabolized mainly by CYP2C8, and to a lesser extent by CYP3A4, whereas rosiglitazone is metabolized by CYP2C9 and CYP2C8 (Jaakkola *et al.*, 2006; Scheen, 2007). The antimalarial agent, amodiaquin, is almost exclusively metabolized by CYP2C8, and several variant forms of CYP2C8 with low activity toward amodiaquin clearance have been identified and estimated at 1–4% in African populations (Gil and Gil Berglund, 2007). This obviously raises safety concerns for carriers of these defective alleles, since amodiaquin is poorly tolerated, causing hematologic and hepatic toxicity (Brasseur, 2007). Although a functional allele, CYP2C8*3, coding for the Arg139Lys and Lys399Arg amino acid substitutions has been shown to confer higher *in vivo* metabolic capacity than the wild-type CYP2C8*1 allele, these pharmacokinetic differences were quantitatively moderate and did not impact the overall clearance of the drug (Kirchheiner *et al.*, 2006; Pedersen, Damkier and Brosen, 2006). Nevertheless, developers of drugs that are predominantly metabolized by CYP2C8 should take into consideration the frequency of these polymorphisms in the populations tested, especially if the drug has a narrow safety window.

CYP2D6, CYP2C9, and CYP2C19 polymorphisms are the most clinically relevant and the most widely studied. This is due to the fact that these three isozymes contribute to the metabolism of about 40% of drugs currently on the market (Ingelman-Sundberg,

Oscarson and McLellan, 1999). The CYP2C9 enzyme, the most abundant among human CYP2C isoforms, metabolizes many therapeutically important drugs, including most nonsteroidal anti-inflammatory drugs (NSAIDs), *S*-warfarin, phenytoin, and losartan (Scordo *et al.*, 2004; Zand *et al.*, 2007). The impact of genetic variations at the CYP2C9 and CYP2C19 loci on the metabolism and safety of the anticoagulant, warfarin, and antiepileptic, phenytoin is well documented. About 30 *CYP2C9* alleles have been reported with more than 20 additional SNPs identified (Ingelman-Sundberg, Daly and Nebert, 2007). The CYP2C9*2 variant is the most common detrimental allele of CYP2C9 among Caucasians (8–12%). It is a C430T transition in exon 3, resulting in an Arg144Cys amino acid substitution. The frequency of the CYP2C9*3 allele varies between 3% and 8% among Caucasians and is an Ile359Leu amino acid change due to an A1075C transition in exon 7. The frequency of these two alleles is lower in Asians and black Africans. Patients with CYP2C9*2 and/or CYP2C9*3 variants metabolize warfarin slowly and are more likely to have elevated plasma levels of the drug, potentially leading to adverse events, including hemorrhaging during initiation of warfarin therapy, before dose stabilization (Crespi and Miller, 1997; Gage and Lesko, 2008). Therefore, taking the genetically determined metabolic capacities of CYP2C9 into account during clinical development of new drugs that are substrates for CYP2C9 or during therapy has the potential to improve individual risk/benefit relationships (Rettie and Tai, 2006). Warfarin and drugs with similar risk/benefit parameters in particular will benefit from this information because of the narrow therapeutic index for each patient (Krynetskiy and McDonnell, 2007). Indeed the FDA recently updated the warfarin label, to include information on pharmacogenetic testing and to encourage the use of this information in choosing doses for individuals initiating warfarin therapy (Gage and Lesko, 2008).

CYP2C19 is involved in the metabolism of many drugs including *S*-mephenytoin, diazepam, omeprazole, proguanil, citalopram, *R*-warfarin, and many antidepressants (Herrlin *et al.*, 1998; Zand *et al.*, 2007). More than 20 CYP2C19 alleles have been identified, with some resulting in decreases in the activity of the encoded enzyme. The CYP2C19*2 and CYP2C19*3 alleles have been associated with poor metabolism of CYP2C19 drug substrates, accounting for 50% to 90% of the CYP2C19 poor metabolizer phenotype (Nakamoto *et al.*, 2007; Zand *et al.*, 2007). The frequency of the poor metabolizer phenotype ranges between 2% and 5% in Caucasians, and 18% to 23% in Asians (Ohkubo *et al.*, 2006).

CYP3A4 is the most abundant CYP in human liver and is responsible for the metabolism of more than 50% of the drugs currently on the market (Bertilsson *et al.*, 1998; Siest, Jeannesson and Visvikis-Siest, 2007). Over 30 SNPs have been identified in both the coding and promoter region of CYP3A4, with some variant alleles resulting in variations in CYP3A4 activity, including CYP3A4*4, CYP3A4*5, and CYP3A4*18 (Henningsson *et al.*, 2005; Hu *et al.*, 2005; Ruzilawati, Suhaimi and Gan, 2007). It is interesting that although CYP3A4 catalyzes the largest number of pharmaceutical drugs, the pharmacokinetic properties of none of its drug substrates appear to be notably affected by the genetic mutations that have so far been described (Siest, Jeannesson and Visvikis-Siest, 2007). In addition, the extensive interindividual differences in drug response for drugs metabolized by CYP3A4 do not seem to have any genetic basis. Although the reason for this variation may be multifactorial, it is believed that extensive variation in the levels of CYP3A4 protein in different individuals is likely to play a part. It has been further suggested that polymorphisms in other genes, including those involved in the transcriptional regulation of CYP3A4, may contribute to this variation (Siest *et al.*, 2004; Siest, Jeannesson and Visvikis-Siest, 2007). Furthermore, expression or polymorphism of CYP3A5, the other CYP3A family gene expressed in adults, has also been implicated in CYP3A4 interindividual variability. However, there have been conflicting reports regarding the importance of CYP3A5 and CYP3A7 (fetal CYP3A) expression/polymorphisms in overall CYP3A-mediated enzyme activities (Fukuda *et al.*, 2004; Jin *et al.*, 2005; Wojnowski and Kamdem, 2006).

5.3.2 Phase II DME Polymorphisms

NATs are a phase II gene family, with two human isoforms, NAT-1 and NAT-2. NATs are responsible for the acetylation conjugation of arylamine drugs such as isoniazid, dapsone, and procainamide (Cascorbi, 2006; Yalin *et al.*, 2007). NAT2 is expressed predominantly in the human liver, whereas NAT1 can be detected in a wide variety of different tissues (Cascorbi, 2006). NATs were among the first family of enzymes reported to exhibit pharmacogenetic variations. The role of NAT in the metabolism of the antitubercular drug isoniazid was identified as far back as the 1950s (Bonicke and Reif, 1953; Weber and Hein, 1985; Sim, Westwood and Fullam, 2007) and *N*-acetylation has since been shown to be the cause of interindividual variation in the metabolism and toxicity of isoniazid (Evans, Manley and Mc, 1960; Weber and Hein, 1985). More than 36 SNPs have been identified at the NAT-2 locus (Hein *et al.*, 2007). Some of these alleles (NAT2*5B, *6A, and *7B) and/or combinations of haplotypes are functional, resulting in the slow, intermediate, or rapid acetylator phenotypes (Pistorius *et al.*, 2007). The slow acetylator phenotype occurs at frequencies of up to 95% in some populations (Relling *et al.*, 1992; Ilett *et al.*, 1993; Lin *et al.*, 1993). The serum half-life of isoniazid varies from 0.9 to 1.8 hours in rapid acetylators and from 2.2 to 4.4 hours in slow acetylators (Peloquin *et al.*, 1997). An association has also been reported between SNPs in NAT2 and adverse events caused by sulfasalazine, a first-line treatment for inflammatory bowel disease and a rheumatoid arthritis drug. A study showed that Chinese patients with inflammatory bowel disease who were slow acetylators, experienced more adverse events (36%) following treatment with sulfasalazine compared with patients with the common (wild-type) allele (Chen *et al.*, 2007; Taniguchi *et al.*, 2007). These studies emphasize the importance of NAT2 polymorphisms in the metabolism and/or toxicity of drugs predominantly eliminated via acetylation.

UGTs are a major phase II detoxication system, responsible for the glucuronidation of many pharmacologically active xenobiotic compounds and endogenous substances in humans and have profound effects on the disposition, metabolism, and elimination of many drugs (Burchell *et al.*, 2000, 2005). UGTs are responsible for converting potentially toxic compounds into water-soluble glucuronides that can be easily excreted from the body (Burchell and Coughtrie, 1989). In humans, they exist as a supergene family comprising two families, UGT1 and UGT2, located on chromosomes 2q37 and 4q13, respectively. Human UGT1 family consists of nine members (A1, A3, A4, A5, A6, A7, A8, A9, and A10) with an identical carboxyl terminus but different amino-terminal domains (28). UGT1A genes have five exons and share complete sequence homology from exons 2 to 5, with alternative splicing resulting in sequence differences in exon 1. The UGT2 family is further divided into two subfamilies (UGT2A and UGT2B) with each gene comprising six exons. UGT2A contains two genes (2A1 and 2A2) while UGT2B contains seven genes (2BIX4, 2BIX7, 2BIX10, 2BIX11, 2BIX15, 2BIX17, and 2BIX28) (Miners, McKinnon and Mackenzie, 2002). This remarkably small number of enzymes is responsible for the conjugation of thousands of xenobiotics and plays a critical role in the detoxication of endogenous toxins such as bilirubin and hormones such as the thyroid hormones (Burchell, 2003; Yamanaka *et al.*, 2007).

Several polymorphisms have been reported at UGT loci, the functional significance of which is still not known for some. Perhaps the best-known clinical implications of UGT polymorphisms are Gilbert's syndrome and Crigler–Najjar syndromes I and II, which are nonhemolytic unconjugated hyperbilirubinemias, resulting from mutations in or lack of UGT1A1 activity (Tukey and Strassburg, 2000; Paoluzzi *et al.*, 2004). UGT1A1*28 is a functional UGT promoter polymorphism associated with Gilbert's disease (Monaghan *et al.*, 1996; Strassburg and Manns, 2000), which has also been implicated in toxicity associated with irinotecan therapy. Since UGTs are responsible for direct conjugation of parent drugs and also metabolites resulting from phase I biotransformation, functional polymorphisms of UGTs can affect the rate at which a parent drug is cleared or the accumulation of a metabolite to toxic levels. The formation of reactive intermediates such as the acyl glucuronide can be impacted by UGT polymorphisms. Indeed several allelic variants of UGT proteins have been reported as essential determinants of susceptibility to drug-related toxicities. Irinotecan is an anticancer drug approved for use in patients with metastatic colorectal cancer. It is metabolized to an active metabolite, SN-38, which is glucuronidated by hepatic UGTs. Diarrhea is the major irinotecan-related toxicity, believed to be due to direct enteric injury caused by SN-38, which has been shown to accumulate in mouse intestines after intraperitoneal administration (Araki *et al.*, 1993). Furthermore, studies in humans have shown an inverse relationship between SN-38 glucuronidation rates and the severity of diarrhea in patients treated with increasing doses of irinotecan (Gupta *et al.*, 1994), indicating that glucuronidation of SN-38 protects against irinotecan-induced gastrointestinal toxicity. Therefore, the conversion of SN-38 to SN-38 glucuronide by hepatic UGTs is a critical step in the detoxication of this active metabolite that is presumably responsible for the toxicity of irinotecan. It has been shown that UGT1A1 is the major UGT isoform responsible for SN-38 glucuronidation (Nagar and Blanchard, 2006), hence it is reasonable to presume that polymorphisms in UGT1A1 that reduce the activity of the encoded UGT1A1 protein would compromise an individual's ability to deactivate SN-38. Indeed the FDA recently revised the package insert for irinotecan to warn of this association between UGT1A1*28 and potential irinotecan toxicity (Ando *et al.*, 2007).

Other variant alleles of UGT1A1, UTG1A1*6, and UGT1A1*27 found in the coding regions are known to reduce the activity of the enzyme and have also been postulated to impact the ability of variant individuals to metabolize drugs that undergo UGT1A-mediated conjugation (Ando *et al.*, 2007). For instance, UGT1A1*6, together with UGT2B7 and CYP2D6 variant alleles, was associated with a low ability to glucuronidate carvedilol, an adrenergic beta-blocker, strongly affecting the pharmacokinetics of this drug in Japanese subjects (Ohno *et al.*, 2004).

Functional polymorphisms also appear common in the other UGT genes, including UGT1A4, UGT1A6, UGT1A7, UGT1A8, UGT1A9, UGT1A10, and UGT2B7 (Burchell *et al.*, 2000; Guillemette *et al.*, 2000; Tukey and Strassburg, 2000; Saeki *et al.*, 2005; Thibaudeau *et al.*, 2006). Although some of these have been found to encode proteins with reduced activities toward their substrates, the clinical implications of these variations have not yet been definitively defined. For instance, while UGT1A7 and UGT1A9 are also involved in the metabolism of irinotecan and exhibit allelic variations, only UGT1A7 alleles have been shown to influence SN-38 glucuronidation kinetics (Gagne *et al.*, 2002; Jinno *et al.*, 2003b; Lankisch *et al.*, 2005). Two SNPs in exon

1 of UGT1A10 (G177A and C605T) resulting in amino acid alterations have been shown to encode proteins with reduced activities toward UGT substrates in recombinant systems, suggesting that these substitutions can influence the metabolism of drugs that are primarily eliminated by glucuronidation (Saeki *et al.*, 2002; Jinno *et al.*, 2003a).

SULTs are also a major phase II detoxication enzyme system. These cytosolic enzymes evolved from one supergene superfamily and catalyze the sulfation of many xenobiotics including drugs, and endogenous compounds including steroids, bile acids, thyroid hormones, and neurotransmitters (Weinshilboum *et al.*, 1997). Although SULTs catalyze phase II reactions, which are generally considered to render compounds less functional and more soluble, sulfation is also involved in biosynthetic pathways such as steroid biosynthesis and also serves as the terminal step in the bioactivation of many compounds to mutagens (Hobkirk and Glasier, 1993). UGTs and SULTs often share similar substrate requirements and it appears the relative contribution of each system to the disposition of a common substrate is dictated by the subcellular location of the isozymes, kinetic properties, and the availability of obligatory cofactors (Burchell and Coughtrie, 1997).

At least three SULT families (SULT1, SULT2, and SULT4) exist in humans with 10 functional genes, localized on five different chromosomes, identified to date (Nagata and Yamazoe, 2000; Coughtrie, 2002; Gamage *et al.*, 2006). Genetic polymorphisms have been detected in all human SULT genes and although some allelic variants have been associated with disease onset, epidemiology, or prognosis (Wang *et al.*, 2002; Wu *et al.*, 2003; Glatt and Meinl, 2004), definitive correlations between SULT allelic variants and interindividual variations in drug response have not been made. Perhaps this is not surprising, considering that not many drugs are metabolized exclusively by SULTs and also the fact that SULT and UGTs share common substrates; hence, any deficiencies in SULT catalytic activity is likely compensated for by UGT conjugation. Nevertheless, it has been suggested that SULT1A1 polymorphisms may be important in women who are on adjuvant tamoxifen therapy because SULT1A1 catalyzes the sulfation of a variety of phenolic and estrogenic compounds, including 4-hydroxytamoxifen, the active metabolite of tamoxifen. Indeed it has been shown that patients who are homozygous for SULT1A1*2 have significantly reduced platelet sulfotransferase activity and platelet enzyme activity correlates strongly with SULT protein expression (Raftogianis *et al.*, 1997; Nowell *et al.*, 2002).

GSTs are a multifunctional phase II supergene family of enzymes that catalyze many reactions between glutathione (GSH) and lipophilic compounds with electrophilic centers, including those that are cytotoxic and/or genotoxic. The results of GST-mediated conjugation are often stable covalently bound adducts which are usually no longer toxic and can be excreted from the body. For instance, PAHs found in cigarette smoke produce electrophilic substrates for GSTs (Ketterer *et al.*, 1992; Raunio *et al.*, 1995). Aflatoxin B_1 ($AFBIX_1$) is a potent hepatocarcinogen in many animals and its carcinogenic activity is derived from its activation to AFB_1 exo-8,9-epoxide primarily by CYPs. Several studies have indicated that the extent of GSH conjugation of AFB_1 is a major factor in determining risk of AFB_1 carcinogenicity in different species (Hayes *et al.*, 1991a, b; Guengerich *et al.*, 1998). Hence, amounts of the CYP enzymes involved in epoxide formation, as well as the levels of the detoxification of AFB1 and the 8,9-epoxide by systems such as GSTs and epoxide hydrolase, determine the toxicity of this compound (Hayes *et al.*, 1992). GSTs form an important component of the cellular response to xenobiotics and endogenous chemicals, and also contribute to physiological processes such as steroid hormone synthesis and tyrosine catabolism (Litwack, Ketterer and Arias, 1971; Johansson and Mannervik, 2001).

The GSTs have been grouped into seven classes based on sequence homology (Alpha, Mu, Pi, Theta, Zeta, Omega, and Zigma), with about 17 genes identified in humans (Mannervik, Board and Hayes, 2005). Despite the multiplicity of this super-gene family, polymorphisms have been identified at almost all GST gene loci and appear to play an important role in the etiology of several malignancies and other diseases (Katoh *et al.*, 1996; Yengi *et al.*, 1996; Strange, Lear and Fryer, 1998). Although several lines of evidence suggest that GST polymorphisms may be important in clinical outcomes of cancer therapy, a clear relationship between GST polymorphisms and drug response has not be established, with data from different studies often conflicting (Anderer *et al.*, 2000; Ambrosone *et al.*, 2001; Mossallam, Hamid and Samra, 2006; Lo and Ali-Osman, 2007). In addition, these associations between clinical outcome and GST polymorphisms do not seem to be supported by pharmacological evidence such as changes in pharmacokinetic properties in patients carrying GST mutations (Innocenti and Ratain, 2002). This could be due to the fact that there is some redundancy in the GST superfamily, with multiple isozymes sharing the same substrate requirements. In addition, GSTs appear to play only a minor role in the metabolic transformation of a few drugs (Eaton and Bammler, 1999; Ates *et al.*, 2004; Kuhne *et al.*, 2008), hence the impact of their polymorphisms is likely to be minimal.

5.3.3 Nuclear Receptors and Transporter Polymorphisms

Genetic variations have been identified at gene loci encoding drug transporters and nuclear receptors responsible for the transcriptional regulation of major DMEs. Interest in deciphering nuclear receptor polymorphisms is driven by suggestions that the high degree of variability in human DME expression and/or activity, may result more from polymorphisms in these regulatory proteins than from direct polymorphisms in the individual enzymes (Harper *et al.*, 2002; Bosch *et al.*, 2006; Martinez-Jimenez *et al.*, 2007). As already mentioned, drugs that are metabolized by CYP3A4 exhibit high interindividual variability in their pharmacokinetic parameters, yet none of the polymorphisms identified at the CYP 3A4 locus can account for this variation, suggesting polymorphisms in PXR protein (Garcia-Martin *et al.*, 2002). In fact PXR has been shown to exhibit alternative splicing, with the PXR-2 splice variant lacking a contiguous stretch of 37 amino acids, with an effect on the basal and/or induced expression of PXR-regulated genes (Dotzlaw *et al.*, 1999; Enoru-Eta *et al.*, 2006). In addition, up to 24 SNPs, occurring in different ethnic populations, have been identified (Hustert *et al.*, 2001; Koyano *et al.*, 2002). Hustert *et al.* (2001) reported six missense mutations that result in variant PXR proteins and studied associations between these PXR variants and CYP3A4 transcription in transiently transfected LS174T cells. Although they reported that three variants affected the basal and/or induced expression of the reporter gene constructs containing two copies of the ER6 motif of the proximal promoter region of CYP3A4, they also suggested that further studies are needed to determine whether this correlation between PXR variants and CYP3A4 promoter activity may be used to predict CYP3A4 activity during drug development or treatment. Although several AhR allelic variants with functional implications and strong effects on AhR phenotypes have been documented in laboratory animals (Poland and Glover, 1990), the few polymorphisms of the human AhR do not appear to have strong and obvious consequences on drug metabolism, although a few have been linked to decreases in

CYP1A activity and induction by 2,3,7,8-tetrachlorodibenzo-p-dioxin (TCDD) (Smart and Daly, 2000; Wong, Okey and Harper, 2001; Harper *et al.*, 2002). Several SNPs have been identified at the CAR locus and two variants have been reported functional in reporter gene assays (Ikeda *et al.*, 2003, 2005). Again the clinical implications of these polymorphisms remain to be established. The clinical importance of genetic polymorphisms in drug transporter genes has also been recognized and extensively reviewed (Kim, 2002; Mizuno *et al.*, 2003; Marzolini, Tirona and Kim, 2004).

5.4 APPLICATIONS OF GENETICS AND GENOMICS IN DRUG DISCOVERY AND DEVELOPMENT

5.4.1 Pharmacogenomics Applications

One of the most attractive possibilities of genomics is the ability to use this technology in the early stages of drug discovery, to predict the safety/toxicity profile of new compounds based on gene expression similarities with previous compounds with similar mechanisms of action. The field toxicogenomics evolved from this premise. The potential benefit is that it would reduce or eliminate the need for costly preclinical animal safety studies (Suter, Babiss and Wheeldon, 2004; Luhe *et al.*, 2005). Thus, by developing databases of the gene expression profiles of compounds with known types of toxicities, the toxic liabilities of new compounds could be determined by querying these databases. Companies such as Gene Logic, in partnership with pharmaceutical companies, developed software and databases, which aimed to apply toxicity-based gene expression assessments to lead selection and optimization. Several pharmaceutical companies have now developed proprietary databases that are applied to rank-order compounds based on toxicogenomics data (Waring *et al.*, 2003; Nie *et al.*, 2006).

Pharmacogenomics, specifically toxicogenomics, is being used mechanistically, to determine the underlying molecular basis of toxicological findings. These data can be particularly useful when several mechanisms can result in a similar pathological end points, or when species- and/or gender-specific differences in toxicity are observed. These data are important when extrapolating data from nonclinical species to humans and have the potential to reduce the attrition rate of compounds in the preclinical stages. The ability to identify which genes are changing in response to treatment with a particular compound, can also be used to determine off-target toxicity (Wang and LeCluyse, 2003; Keshava and Caldwell, 2006). Such an approach was used to determine that thyroid hypertrophy and reductions in thyroid hormone levels following treatment with a nonsteroidal progestin receptor agonist, resulted from induction of UGTs which are known to conjugate thyroid hormones (Yengi *et al.*, 2007).

One of the most successful applications of genomics data is to predict the potential for new chemical entities to cause drug–drug interactions with coadministered drugs that may result in loss of efficacy and ultimately toxicity (Ereshefsky and Dugan, 2000; Rodriguez-Novoa *et al.*, 2006). Interest in gene regulation has increased continuously, with many regulatory agencies requesting information on the ability of new compounds to induce DME. In fact, the FDA draft guidance on *in vitro* and *in vivo* drug interaction studies lends credence to the fact that the agency considers induction data integral to their ability to appropriately evaluate new chemical entities for clinical drug–drug interaction liabilities (FDA, 2006). Preclinical gene regulation studies are performed primarily *ex vivo*, in livers from preclinical safety studies. Increasingly, however, studies conducted *in vitro* using systems such as primary hepatocytes and cell-based gene reporter assays are yielding more human relevant data, as the need for species to species extrapolation is eliminated (Smith, 2000; Luo *et al.*, 2002; Vermeir *et al.*, 2005; Enoru-Eta *et al.*, 2006; FDA, 2006; Yengi *et al.*, 2007). In addition to drug–drug interaction liabilities, gene regulation can also increase toxic liabilities of compounds by altering rates of metabolism and/or elimination of toxic species. For drugs whose toxicity results from their metabolism to more active and toxic metabolites, such as acetaminophen, induction or down-regulation of genes involved in formation of the metabolite or elimination can greatly affect the safety profile of the compound (Douidar and Ahmed, 1987; DiPetrillo *et al.*, 2002). Genome-based studies can determine early in the discovery process whether metabolism genes are regulated, enabling the design of more appropriate safety and/or clinical studies. Also, since most DMEs are involved in normal biochemical processes such as steroid hormone biosynthesis and elimination, cholesterol homeostasis, and bilirubin metabolism, regulation of DMEs can lead to undesirable end points (You, 2004; Rifkind, 2006). Some studies have shown that drugs associated with idiosyncratic drug reactions (IDRs) also up-regulate some protective genes *in vivo*, suggesting that gene expression profiling could potentially be used to identify compounds with a potential to cause IDRs (Uetrecht, 2003).

Traditionally, gene regulation studies were conducted by determining enzyme activities; however, mRNA data are increasingly being generated either alone or in combination with enzyme activity data, because of the lack of isozyme-specific probe substrates (Smith, 2000; Yengi *et al.*, 2007). In addition, there is increasing evidence that compounds that cause mechanism-based inhibition (MBI) of DMEs can also act as inducers of these enzymes (McConn and Zhao, 2004; Zhou *et al.*, 2005). Occurring post-transcriptionally, mRNA data can help delineate potential toxicological processes taking place at the molecular level that may not be observed at the level of activity because of continuous inactivation of the protein.

5.4.2   Pharmacogenetic Applications

As has been discussed, the efficacy and/or safety of a drug predominantly metabolized by a polymorphic gene can be greatly impacted by the genotype of the individual taking the drug (Evans and Relling, 1999; Shi, Bleavins and de la Iglesia, 2001). Information on the genotypes of participants in clinical trials not only provides data that could be used to explain potential individual differences in response, but also ensures that those with an altered phenotype are properly monitored. This is particularly important with drugs that have a narrow therapeutic window outside of which severe toxicity or therapeutic failure may occur (Linder, Prough and Valdes, 1997). For drugs predominantly metabolized by a polymorphic DME, most companies routinely genotype subjects during clinical trials; however, timing of these studies varies. While some companies genotype clinical samples as a matter of routine, some take a diagnostic approach, only genotyping if the data indicate the genetic variation has an impact on the pharmacokinetics and/or safety profile of the compound. For drugs metabolized by DMEs for which the impact of allelic variants is still unknown, other approaches such as targeted or genome-wide SNP analysis is used to identify SNPs and/or haplotypes that may co-segregate with any observed variations (Swan *et al.*, 2005; Haga and Ginsburg, 2006).

The idea that polymorphisms as yet unidentified either in target or off-target genes may play a causative role in unexpected adverse (idiosyncratic) drug reactions is increasingly gaining traction. It is currently difficult to accurately predict which drugs will be associated with a significant incidence of IDRs. Formation of reactive metabolites and covalent binding of these intermediates with cellular components such as proteins and DNA has been associated with IDRs and is used as a screening tool to predict IDR liabilities of compounds (Williams *et al.*, 2002; Liebler and Guengerich, 2005). However, since not all reactive metabolites cause IDR and some cause reactions in a species-specific manner, these predictions only provide partial solutions (Gardner *et al.*, 2005; Liebler and Guengerich, 2005). It has been suggested that polymorphisms in DMEs, the proteins to which reactive metabolites covalently bind, receptors, and/or protective genes all have the potential to alter the metabolic clearance of a drug, potentially accounting for individual or species differences in susceptibility to IDR. Use of genomics and genetic technologies to determine gene expression profiles and genotypes therefore has the potential to develop into a more effective screening tool for IDR, as genetically susceptible subsets of the population could be identified (Hosford *et al.*, 2004; Hughes *et al.*, 2004).

The "omics" technologies have been incorporated into traditional drug safety and metabolism studies, to provide information on the efficacy and toxicity of compounds in development and to elucidate the underlying molecular mechanisms responsible. The FDA and other regulatory agencies recognize that in order to reduce the time and costs associated with bringing new drugs to market, drug companies have to be innovative in their approaches, and embrace new technologies and processes that challenge traditional paradigms. To this effect, the Critical Path Initiative was launched by the FDA to encourage collaborations among government regulators, the academic community, and regulated businesses to modernize scientific tools and processes, and harness the potential of bioinformatics to, among others, evaluate and predict the safety, effectiveness, and manufacturability of candidate medical products (FDA, 2004), the ultimate objective being a smarter and more rational approach to the development of safer and more effective drugs.

REFERENCES

Aklillu, E., Persson, I., Bertilsson, L., Johansson, I., Rodrigues, F., Ingelman-Sundberg, M. (1996). Frequent distribution of ultrarapid metabolizers of debrisoquine in an Ethiopian population carrying duplicated and multiduplicated functional cyp2d6 alleles. *J. Pharmacol Exp. Ther.*, *278*, 441–446.

Alexanderson, B., Evans, D.A., Sjoqvist, F. (1969). Steady-state plasma levels of nortriptyline in twins: influence of genetic factors and drug therapy. *Br. Med. J.*, *4*, 764–768.

Ambrosone, C.B., Sweeney, C., Coles, B.F., Thompson, P.A., McClure, G.Y., Korourian, S., Fares, M.Y., Stone, A., Kadlubar, F.F., Hutchins, L.F. (2001). Polymorphisms in glutathione S-transferases (GSTM1 and GSTT1) and survival after treatment for breast cancer. *Cancer Res.*, *61*, 7130–7135.

Anderer, G., Schrappe, M., Brechlin, A.M., Lauten, M., Muti, P., Welte, K., Stanulla, M. (2000). Polymorphisms within glutathione S-transferase genes and initial response to glucocorticoids in childhood acute lymphoblastic leukaemia. *Pharmacogenetics*, *10*, 715–726.

Ando, Y., Fujita, K., Sasaki, Y., Hasegawa, Y. (2007). UGT1AI*6 and UGT1A1*27 for individualized irinotecan chemotherapy. *Curr. Opin. Mol Ther.*, *9*, 258–262.

Araki, E., Ishikawa, M., Iigo, M., Koide, T., Itabashi, M., Hoshi, A. (1993). Relationship between development of diarrhea and the concentration of SN-38, an active metabolite of CPT-11, in the intestine and the blood plasma of athymic mice following intraperitoneal administration of CPT-11. *Jpn J. Cancer Res.*, *84*, 697–702.

Ates, N.A., Tursen, U., Tamer, L., Kanik, A., Derici, E., Ercan, B., Atik, U. (2004). Glutathione S-transferase polymorphisms in patients with drug eruption. *Arch. Dermatol. Res.*, *295*, 429–433.

Basile, V.S., Ozdemir, V., Masellis, M., Walker, M.L., Meltzer, H.Y., Lieberman, J.A., Potkin, S.G., Alva, G., Kalow, W., Macciardi, F.M., Kennedy, J.L. (2000). A functional polymorphism of the cytochrome P450 1A2 (CYP1A2) gene: association with tardive dyskinesia in schizophrenia. *Mol. Psychiatry*, *5*, 410–417.

Bertilsson, G., Heidrich, J., Svensson, K., Asman, M., Jendeberg, L., Sydow-Backman, M., Ohlsson, R., Postlind, H., Blomquist, P., Berkenstam, A. (1998). Identification of a human nuclear receptor defines a new signaling pathway for CYP3A induction. *Proc. Natl. Acad. Sci. U.S.A.*, *95*, 12208–12213.

Beverage, J.N., Sissung, T.M., Sion, A.M., Danesi, R., Figg, W.D. (2007). CYP2D6 polymorphisms and the impact on tamoxifen therapy. *J. Pharm. Sci.*, *96*, 2224–2231.

Blumberg, B., Sabbagh, W. Jr, Juguilon, H., Bolado, J., Jr, van Meter, C.M., Ong, E.S., Evans, R.M. (1998). SXR, a novel steroid and xenobiotic-sensing nuclear receptor. *Genes Dev.*, *12*, 3195–3205.

Boke, O., Gunes, S., Kara, N., Aker, S., Sahin, A.R., Basar, Y., Bagci, H. (2007). Association of serotonin 2A receptor and lack of association of CYP1A2 gene polymorphism with tardive dyskinesia in a Turkish population. *DNA Cell Biol.*, *26*, 527–531.

Bonicke, R., Reif, W. (1953). Enzymatic inactivation of isonicotinic acid hydrizide in human and animal organism. *Naunyn Schmiedebergs Arch. Exp. Pathol. Pharmakol.*, *220*, 321–323.

Bosch, T.M., Deenen, M., Pruntel, R., Smits, P.H., Schellens, J.H., Beijnen, J.H., Meijerman, I. (2006). Screening for polymorphisms in the PXR gene in a Dutch population. *Eur. J. Clin. Pharmacol.*, *62*, 395–399.

Brasseur, P. (2007). Tolerance of amodiaquine. *Med. Trop. (Mars)*, *67*, 288–290.

Burchell, B. (2003). Genetic variation of human UDP-glucuronosyltransferase: implications in disease and drug glucuronidation. *Am. J. Pharmacogenomics*, *3*, 37–52.

Burchell, B., Coughtrie, M.W. (1989). UDP-glucuronosyltransferases. *Pharmacol. Ther.*, *43*, 261–289.

Burchell, B., Coughtrie, M.W. (1997). Genetic and environmental factors associated with variation of human xenobiotic glucuronidation and sulfation. *Environ. Health Perspect.*, *105* (Suppl. 4), 739–747.

Burchell, B., Lockley, D.J., Staines, A., Uesawa, Y., Coughtrie, M.W. (2005). Substrate specificity of human hepatic udp-glucuronosyltransferases. *Methods Enzymol.*, *400*, 46–57.

Burchell, B., Soars, M., Monaghan, G., Cassidy, A., Smith, D., Ethell, B. (2000). Drug-mediated toxicity caused by genetic deficiency of UDP-glucuronosyltransferases. *Toxicol. Lett.*, *112*–113, 333–340.

Cascorbi, I. (2006). Genetic basis of toxic reactions to drugs and chemicals. *Toxicol. Lett.*, *162*, 16–28

Chen, M., Xia, B., Chen, B., Guo, Q., Li, J., Ye, M., Hu, Z. (2007). N-acetyltransferase 2 slow acetylator genotype associated with adverse effects of sulphasalazine in the treatment of inflammatory bowel disease. *Can. J. Gastroenterol.*, *21*, 155–158.

Chetty, M., Murray, M. (2007). CYP-mediated clozapine interactions: how predictable are they? *Curr. Drug Metab.*, *8*, 307–313.

Coughtrie, M.W. (2002). Sulfation through the looking glass—recent advances in sulfotransferase research for the curious. *Pharmacogenomics J.*, *2*, 297–308.

Crespi, C.L., Miller, V.P. (1997). The R144C change in the CYP2C9*2 allele alters interaction of the cytochrome P450 with NADPH:cytochrome P450 oxidoreductase. *Pharmacogenetics*, *7*, 203–210.

Daly, A.K., Cholerton, S., Gregory, W., Idle, J.R. (1993). Metabolic polymorphisms. *Pharmacol. Ther.*, *57*, 129–160.

DiPetrillo, K., Wood, S., Kostrubsky, V., Chatfield, K., Bement, J., Wrighton, S., Jeffery, E., Sinclair, P., Sinclair, J. (2002). Effect of caffeine on acetaminophen hepatotoxicity in cultured hepatocytes treated with ethanol and isopentanol. *Toxicol. Appl. Pharmacol.*, *185*, 91–97.

Dotzlaw, H., Leygue, E., Watson, P., Murphy, L.C. (1999). The human orphan receptor PXR messenger RNA is expressed in both normal and neoplastic breast tissue. *Clin. Cancer Res.*, *5*, 2103–2107.

Douidar, S.M., Ahmed, A.E. (1987). A novel mechanism for the enhancement of acetaminophen hepatotoxicity by phenobarbital. *J. Pharmacol. Exp. Ther.*, *240*, 578–583.

Eaton, D.L., Bammler, T.K. (1999). Concise review of the glutathione S-transferases and their significance to toxicology. *Toxicol. Sci.*, *49*, 156–164.

Eichelbaum, M. (1986). Polymorphic oxidation of debrisoquine and sparteine. *Prog. Clin. Biol. Res.*, *214*, 157–167.

Eichelbaum, M., Ingelman-Sundberg, M., Evans, W.E. (2006). Pharmacogenomics and individualized drug therapy. *Annu. Rev. Med.*, *57*, 119–137.

Eichelbaum, M., Spannbrucker, N., Steincke, B., Dengler, H.J. (1979). Defective N-oxidation of sparteine in man: a new pharmacogenetic defect. *Eur. J. Clin. Pharmacol.*, *16*, 183–187.

Enoru-Eta, J., Yengi, L.G., Kao, J., Scatina, J. (2006). A reporter cellular model for evaluating induction of CYP3A4 by new chemical entities. *Drug Dev. Res.*, *67*, 470–475.

Ereshefsky, L., Dugan, D. (2000). Review of the pharmacokinetics, pharmacogenetics, and drug interaction potential of antidepressants: focus on venlafaxine. *Depress. Anxiety*, *12* (Suppl. 1) 30–44.

Evans, D.A., Manley, K.A., McKusick, K.V. (1960). Genetic control of isoniazid metabolism in man. *Br. Med. J.*, *2*, 485–491.

Evans, W.E., Relling, M.V. (1999). Pharmacogenomics: translating functional genomics into rational therapeutics. *Science*, *286*, 487–491.

FDA (2004). *Critical Path Initiative*, http://www.fda.gov/oc/initiatives/criticalpath/ (accessed 28 October 2008).

FDA (2006). Guidance for Industry. Drug Interaction Studies—Study Design, Data Analysis, and Implications for Dosing and Labeling, Food and Drug Administration, Washington, DC, http://www.fda.gov/cder/guidelines/dft.htm (accessed 28 October 2008).

Fukuda, T., Onishi, S., Fukuen, S., Ikenaga, Y., Ohno, M., Hoshino, K., Matsumoto, K., Maihara, A., Momiyama, K., Ito, T., Fujio, Y., Azuma, J. (2004). CYP3A5 genotype did not impact on nifedipine disposition in healthy volunteers. *Pharmacogenomics J.*, *4*, 34–39.

Gage, B.F., Lesko, L.J. (2008). Pharmacogenetics of warfarin: regulatory, scientific, and clinical issues. *J. Thromb. Thrombolysis*, *25*, 45–51. (Epub 1 October 2007).

Gagne, J.F., Montminy, V., Belanger, P., Journault, K., Gaucher, G., Guillemette, C. (2002). Common human UGT1A polymorphisms and the altered metabolism of irinotecan active metabolite 7-ethyl-10-hydroxycamptothecin (SN-38). *Mol. Pharmacol.*, *62*, 608–617.

Gamage, N., Barnett, A., Hempel, N., Duggleby, R.G., Windmill, K.F., Martin, J.L., McManus, M. E. (2006). Human sulfotransferases and their role in chemical metabolism. *Toxicol. Sci.*, *90*, 5–22.

Garcia-Martin, E., Martinez, C., Pizarro, R.M., Garcia-Gamito, F.J., Gullsten, H., Raunio, H., Agundez, J.A. (2002). CYP3A4 variant alleles in white individuals with low CYP3A4 enzyme activity. *Clin. Pharmacol. Ther.*, *71*, 196–204.

Gardner, I., Popovic, M., Zahid, N., Uetrecht, J.P. (2005). A comparison of the covalent binding of clozapine, procainamide, and vesnarinone to human neutrophils in vitro and rat tissues in vitro and in vivo. *Chem. Res. Toxicol.*, *18*, 1384–1394.

Garte, S., Gaspari, L., Alexandrie, A.K., Ambrosone, C., Autrup, H., Autrup, J.L., Baranova, H., Bathum, L., Benhamou, S., Boffetta, P., Bouchardy, C., Breskvar, K., Brockmoller, J., Cascorbi,

I., Clapper, M.L., Coutelle, C., Daly, A., Dell'Omo, M., Dolzan, V., Dresler, C.M., Fryer, A., Haugen, A., Hein, D.W., Hildesheim, A., Hirvonen, A., Hsieh, L.L., Ingelman-Sundberg, M., Kalina, I., Kang, D., Kihara, M., Kiyohara, C., Kremers, P., Lazarus, P., Le Marchand, L., Lechner, M.C., van Lieshout, E.M., London, S., Manni, J.J., Maugard, C.M., Morita, S., Nazar-Stewart, V., Noda, K., Oda, Y., Parl, F.F., Pastorelli, R., Persson, I., Peters, W.H., Rannug, A., Rebbeck, T., Risch, A., Roelandt, L., Romkes, M., Ryberg, D., Salagovic, J., Schoket, B., Seidegard, J., Shields, P.G., Sim, E., Sinnet, D., Strange, R.C., Stucker, I., Sugimura, H., To-Figueras, J., Vineis, P., Yu, M.C., Taioli, E. (2001). Metabolic gene polymorphism frequencies in control populations. *Cancer Epidemiol. Biomarkers Prev.*, *10*, 1239–1248.

Gil, J.P., Gil Berglund, E. (2007). CYP2C8 and antimalaria drug efficacy. *Pharmacogenomics*, *8*, 187–198.

Ginsburg, G.S., Konstance, R.P., Allsbrook, J.S., Schulman, K.A. (2005). Implications of pharmacogenomics for drug development and clinical practice. *Arch. Intern. Med.*, *165*, 2331–2336.

Glatt, H., Meinl, W. (2004). Pharmacogenetics of soluble sulfotransferases (SULTs). *Naunyn Schmiedebergs Arch. Pharmacol.*, *369*, 55–68.

Goldstein, D.B., Tate, S.K., Sisodiya, S.M. (2003). Pharmacogenetics goes genomic. *Nat. Rev. Genet.*, *4*, 937–947.

Goldstein, J.A., Faletto, M.B. (1993). Advances in mechanisms of activation and deactivation of environmental chemicals. *Environ. Health Perspect.*, *100*, 169–176.

Guengerich, F.P. (1991a). Molecular advances for the cytochrome P-450 superfamily. *Trends Pharmacol. Sci.*, *12*, 281–283.

Guengerich, F.P. (1991b). Reactions and significance of cytochrome P-450 enzymes. *J. Biol. Chem.*, *266*, 10019–10022.

Guengerich, F.P., Johnson, W.W., Shimada, T., Ueng, Y.F., Yamazaki, H., Langouet, S. (1998). Activation and detoxication of aflatoxin B1. *Mutat. Res.*, *402*, 121–128.

Guenthner, T.M., Nebert, D.W. (1977). Cytosolic receptor for aryl hydrocarbon hydroxylase induction by polycyclic aromatic compounds. Evidence for structural and regulatory variants among established cell cultured lines. *J. Biol. Chem.*, *252*, 8981–8989.

Guillemette, C., Ritter, J.K., Auyeung, D.J., Kessler, F.K., Housman, D.E. (2000). Structural heterogeneity at the UDP-glucuronosyltransferase 1 locus: functional consequences of three novel missense mutations in the human UGT1A7 gene. *Pharmacogenetics*, *10*, 629–644.

Gupta, E., Lestingi, T.M., Mick, R., Ramirez, J., Vokes, E.E., Ratain, M.J. (1994). Metabolic fate of irinotecan in humans: correlation of glucuronidation with diarrhea. *Cancer Res.*, *54*, 3723–3725.

Haga, S.B., Ginsburg, G.S. (2006). Prescribing BiDil: is it black and white? *J. Am. Coll. Cardiol.*, *48*, 12–14.

Hammer, W., Sjoqvist, F. (1967). Plasma levels of monomethylated tricyclic antidepressants during treatment with imipramine-like compounds. *Life Sci.*, *6*, 1895–1903.

Hammons, G.J., Yan-Sanders, Y., Jin, B., Blann, E., Kadlubar, F.F., Lyn-Cook, B.D. (2001). Specific site methylation in the 5′-flanking region of CYP1A2 interindividual differences in human livers. *Life Sci.*, *69*, 839–845.

Harper, P.A., Wong, J.Y., Lam, M.S., Okey, A.B. (2002). Polymorphisms in the human AH receptor. *Chem. Biol. Interact.*, *141*, 161–187.

Hayes, J.D., Judah, D.J., McLellan, L.I., Kerr, L.A., Peacock, S.D., Neal, G.E. (1991a). Ethoxyquin-induced resistance to aflatoxin B1 in the rat is associated with the expression of a novel alpha-class glutathione S-transferase subunit, Yc2, which possesses high catalytic activity for aflatoxin B1-8,9-epoxide. *Biochem. J.*, *279* (Pt 2), 385–398.

Hayes, J.D., Judah, D.J., McLellan, L.I., Neal, G.E. (1991b). Contribution of the glutathione S-transferases to the mechanisms of resistance to aflatoxin B1. *Pharmacol. Ther.*, *50*, 443–472.

Hayes, J.D., Judah, D.J., Neal, G.E., Nguyen, T. (1992). Molecular cloning and heterologous expression of a cDNA encoding a mouse glutathione S-transferase Yc subunit possessing high catalytic activity for aflatoxin B1-8,9-epoxide. *Biochem. J.*, *285* (Pt 1), 173–180.

Hedgecoe A., Martin, P. (2003). The drugs don't work: expectations and the shaping of pharmacogenetics. *Soc. Stud. Sci.*, *33*, 327–364.

Hein, D.W., Sim, E., Boukouvala, S., Grant, D.M., Minchin, R.F. (2007). *Consensus Arylamine N-Acetyltransferase (NAT) Gene Nomenclature*, http://www.louisville.edu/medschool/pharmacology/NAT.html (accessed 28 October 2008).

Henningsson, A., Marsh, S., Loos, W.J., Karlsson, M.O., Garsa, A., Mross, K., Mielke, S., Vigano, L., Locatelli, A., Verweij, J., Sparreboom, A., McLeod, H.L. (2005). Association of CYP2C8, CYP3A4, CYP3A5, and ABCB1 polymorphisms with the pharmacokinetics of paclitaxel. *Clin. Cancer Res.*, *11*, 8097–8104.

Herrlin, K., Massele, A.Y., Jande, M., Alm, C., Tybring, G., Abdi, Y.A., Wennerholm, A., Johansson, I., Dahl, M.L., Bertilsson, L., Gustafsson, L.L. (1998). Bantu Tanzanians have a decreased capacity to metabolize omeprazole and mephenytoin in relation to their CYP2C19 genotype. *Clin. Pharmacol. Ther.*, *64*, 391–401.

Hobkirk, R., Glasier, M.A. (1993). Generation of estradiol within the pregnant guinea pig uterine compartment with special reference to the myometrium. *J. Steroid Biochem. Mol. Biol.*, *44*, 291–297.

Hosford, D.A., Lai, E.H., Riley, J.H., Xu, C.F., Danoff, T.M., Roses, A.D. (2004). Pharmacogenetics to predict drug-related adverse events. *Toxicol. Pathol.*, *32* (Suppl. 1), 9–12.

Hu, Y.F., He, J., Chen, G.L., Wang, D., Liu, Z.Q., Zhang, C., Duan, L.F., Zhou, H.H. (2005). CYP3A5*3 and CYP3A4*18 single nucleotide polymorphisms in a Chinese population. *Clin. Chim. Acta*, *353*, 187–192.

Hughes, A.R., Mosteller, M., Bansal, A.T., Davies, K., Haneline, S.A., Lai, E.H., Nangle, K., Scott, T., Spreen, W.R., Warren, L.L., Roses, A.D. (2004). Association of genetic variations in HLA-B region with hypersensitivity to abacavir in some, but not all, populations. *Pharmacogenomics*, *5*, 203–211.

Hustert, E., Zibat, A., Presecan-Siedel, E., Eiselt, R., Mueller, R., Fuss, C., Brehm, I., Brinkmann, U., Eichelbaum, M., Wojnowski, L., Burk, O. (2001). Natural protein variants of pregnane X receptor with altered transactivation activity toward CYP3A4. *Drug Metab. Dispos.*, *29*, 1454–1459.

Ikeda, S., Kurose, K., Jinno, H., Sai, K., Ozawa, S., Hasegawa, R., Komamura, K., Kotake, T., Morishita, H., Kamakura, S., Kitakaze, M., Tomoike, H., Tamura, T., Yamamoto, N., Kunitoh, H., Yamada, Y., Ohe, Y., Shimada, Y., Shirao, K., Kubota, K., Minami, H., Ohtsu, A., Yoshida, T., Saijo, N., Saito, Y., Sawada, J. (2005). Functional analysis of four naturally occurring variants of human constitutive androstane receptor. *Mol. Genet. Metab.*, *86*, 314–319.

Ikeda, S., Kurose, K., Ozawa, S., Sai, K., Hasegawa, R., Komamura, K., Ueno, K., Kamakura, S., Kitakaze, M., Tomoike, H., Nakajima, T., Matsumoto, K., Saito, H., Goto, Y., Kimura, H., Katoh, M., Sugai, K., Minami, N., Shirao, K., Tamura, T., Yamamoto, N., Minami, H., Ohtsu, A., Yoshida, T., Saijo, N., Saito, Y., Sawada, J. (2003). Twenty-six novel single nucleotide polymorphisms and their frequencies of the NR1I3 (CAR) gene in a Japanese population. *Drug Metab. Pharmacokinet.*, *18*, 413–418.

Ikenaga, Y., Fukuda, T., Fukuda, K., Nishida, Y., Naohara, M., Maune, H., Azuma, J. (2005). The frequency of candidate alleles for CYP2D6 genotyping in the Japanese population with an additional respect to the -1584C to G substitution. *Drug Metab. Pharmacokinet.*, *20*, 113–116.

Ilett, K.F., Chiswell, G.M., Spargo, R.M., Platt, E., Minchin, R.F. (1993). Acetylation phenotype and genotype in aboriginal leprosy patients from the north-west region of Western Australia. *Pharmacogenetics*, *3*, 264–269.

Ingelman-Sundberg, M. (2001). Pharmacogenetics: an opportunity for a safer and more efficient pharmacotherapy. *J. Intern. Med.*, *250*, 186–200.

Ingelman-Sundberg, M., Daly, A.K., Nebert, D.W. (2007). *Home Page of the Human Cytochrome P450 (CYP) Allele Nomenclature Committee*, http://www.cypalleles.ki.se/ (accessed 28 October 2008).

Ingelman-Sundberg, M., Oscarson, M., McLellan, R.A. (1999). Polymorphic human cytochrome P450 enzymes: an opportunity for individualized drug treatment. *Trends Pharmacol. Sci.*, *20*, 342–349.

Innocenti, F., Ratain, M.J. (2002). Update on pharmacogenetics in cancer chemotherapy. *Eur. J. Cancer*, *38*, 639–644.

Jaakkola, T., Laitila, J., Neuvonen, P.J., Backman, J.T. (2006). Pioglitazone is metabolised by CYP2C8 and CYP3A4 in vitro: potential for interactions with CYP2C8 inhibitors. *Basic Clin. Pharmacol. Toxicol.*, *99*, 44–51.

Jin, M., Gock, S.B., Jannetto, P.J., Jentzen, J.M., Wong, S.H. (2005). Pharmacogenomics as molecular autopsy for forensic toxicology: genotyping cytochrome P450 3A4*1B and 3A5*3 for 25 fentanyl cases. *J. Anal. Toxicol.*, *29*, 590–598.

Jinno, H., Saeki, M., Tanaka-Kagawa, T., Hanioka, N., Saito, Y., Ozawa, S., Ando, M., Shirao, K., Minami, H., Ohtsu, A., Yoshida, T., Saijo, N., Sawada, J. (2003a). Functional characterization of wild-type and variant (T202I and M59I) human UDP-glucuronosyltransferase 1A10. *Drug Metab. Dispos.*, *31*, 528–532.

Jinno, H., Tanaka-Kagawa, T., Hanioka, N., Saeki, M., Ishida, S., Nishimura, T., Ando, M., Saito, Y., Ozawa, S., Sawada, J. (2003b). Glucuronidation of 7-ethyl-10-hydroxycamptothecin (SN-38), an active metabolite of irinotecan (CPT-11), by human UGT1A1 variants, G71R, P229Q, and Y486D. *Drug Metab. Dispos.*, *31*, 108–113.

Johansson, A.S., Mannervik, B. (2001). Human glutathione transferase A3-3, a highly efficient catalyst of double-bond isomerization in the biosynthetic pathway of steroid hormones. *J. Biol. Chem.*, *276*, 33061–33065.

Katoh, T., Nagata, N., Kuroda, Y., Itoh, H., Kawahara, A., Kuroki, N., Ookuma, R., Bell, D.A. (1996). Glutathione S-transferase M1 (GSTM1) and T1 (GSTT1) genetic polymorphism and susceptibility to gastric and colorectal adenocarcinoma. *Carcinogenesis*, *17*, 1855–1859.

Keshava, N., Caldwell, J.C. (2006). Key issues in the role of peroxisome proliferator-activated receptor agonism and cell signaling in trichloroethylene toxicity. *Environ. Health Perspect.*, *114*, 1464–1470.

Ketterer, B., Harris, J.M., Talaska, G., Meyer, D.J., Pemble, S.E., Taylor, J.B., Lang, N.P., Kadlubar, F.F. (1992). The human glutathione S-transferase supergene family, its polymorphism, and its effects on susceptibility to lung cancer. *Environ. Health Perspect.*, *98*, 87–94.

Kim, R.B. (2002). Pharmacogenetics of CYP enzymes and drug transporters: remarkable recent advances. *Adv. Drug Deliv. Rev.*, *54*, 1241–1242.

Kirchheiner, J., Thomas, S., Bauer, S., Tomalik-Scharte, D., Hering, U., Doroshyenko, O., Jetter, A., Stehle, S., Tsahuridu, M., Meineke, I., Brockmoller, J., Fuhr, U. (2006). Pharmacokinetics and pharmacodynamics of rosiglitazone in relation to CYP2C8 genotype. *Clin. Pharmacol. Ther.*, *80*, 657–667.

Kliewer, S.A., Lehmann, J.M., Willson, T.M. (1999). Orphan nuclear receptors: shifting endocrinology into reverse. *Science*, *284*, 757–760.

Kliewer, S.A., Moore, J.T., Wade, L., Staudinger, J.L., Watson, M.A., Jones, S.A., McKee, D.D., Oliver, B.B., Willson, T.M., Zetterstrom, R.H., Perlmann, T., Lehmann, J.M. (1998). An orphan nuclear receptor activated by pregnanes defines a novel steroid signaling pathway. *Cell*, *92*, 73–82.

Koyano, S., Kurose, K., Ozawa, S., Saeki, M., Nakajima, Y., Hasegawa, R., Komamura, K., Ueno, K., Kamakura, S., Nakajima, T., Saito, H., Kimura, H., Goto, Y., Saitoh, O., Katoh, M., Ohnuma, T., Kawai, M., Sugai, K., Ohtsuki, T., Suzuki, C., Minami, N., Saito, Y., Sawada, J. (2002). Eleven novel single nucleotide polymorphisms in the NR1I2 (PXR) gene, four of which induce non-synonymous amino acid alterations. *Drug Metab. Pharmacokinet.*, *17*, 561–565.

Krynetskiy, E., McDonnell, P. (2007). Building individualized medicine: prevention of adverse reactions to warfarin therapy. *J. Pharmacol. Exp. Ther.*, *322*, 427–434.

Kuhne, A., Sezer, O., Heider, U., Meineke, I., Muhlke, S., Niere, W., Overbeck, T., Hohloch, K., Trumper, L., Brockmoller, J., Kaiser, R. (2008). Population pharmacokinetics of melphalan and glutathione S-transferase polymorphisms in relation to side effects. *Clin. Pharmacol. Ther.*, *83*, 749–759. (Epub 3 October 2007).

Lankisch, T.O., Vogel, A., Eilermann, S., Fiebeler, A., Krone, B., Barut, A., Manns, M.P., Strassburg, C.P. (2005). Identification and characterization of a functional TATA box polymorphism of the UDP glucuronosyltransferase 1A7 gene. *Mol. Pharmacol.*, *67*, 1732–1739.

Liebler, D.C., Guengerich, F.P. (2005). Elucidating mechanisms of drug-induced toxicity. *Nat. Rev. Drug Discov.*, *4*, 410–420.

Lin, H.J., Han, C.Y., Lin, B.K., Hardy, S. (1993). Slow acetylator mutations in the human polymorphic N-acetyltransferase gene in 786 Asians, blacks, Hispanics, and whites: application to metabolic epidemiology. *Am. J. Hum. Genet.*, *52*, 827–834.

Linder, M.W., Prough, R.A., Valdes, R. Jr. (1997). Pharmacogenetics: a laboratory tool for optimizing therapeutic efficiency. *Clin. Chem.*, *43*, 254–266.

Litwack, G., Ketterer, B., Arias, I.M. (1971). Ligandin: a hepatic protein which binds steroids, bilirubin, carcinogens and a number of exogenous organic anions. *Nature*, *234*, 466–467.

Lo, H.W., Ali-Osman, F. (2007). Genetic polymorphism and function of glutathione S-transferases in tumor drug resistance. *Curr. Opin. Pharmacol.*, *7*, 367–374.

Luhe, A., Suter, L., Ruepp, S., Singer, T., Weiser, T., Albertini, S. (2005). Toxicogenomics in the pharmaceutical industry: hollow promises or real benefit? *Mutat. Res.*, *575*, 102–115.

Luo, G., Cunningham, M., Kim, S., Burn, T., Lin, J., Sinz, M., Hamilton, G., Rizzo, C., Jolley, S., Gilbert, D., Downey, A., Mudra, D., Graham, R., Carroll, K., Xie, J., Madan, A., Parkinson, A., Christ, D., Selling, B., LeCluyse, E., Gan, L.S. (2002). CYP3A4 induction by drugs: correlation between a pregnane X receptor reporter gene assay and CYP3A4 expression in human hepatocytes. *Drug Metab. Dispos.*, *30*, 795–804.

McConn, D.J., Zhao, Z. (2004). Integrating in vitro kinetic data from compounds exhibiting induction, reversible inhibition and mechanism-based inactivation: in vitro study design. *Curr. Drug Metab.*, *5*, 141–146.

Mahgoub, A., Idle, J.R., Dring, L.G., Lancaster, R., Smith, R.L. (1977). Polymorphic hydroxylation of Debrisoquine in man. *Lancet*, *2*, 584–586.

Mannervik, B., Board, P.G., Hayes, J.D., Listowsky, I., Pearson, W.R. (2005). Nomenclature for mammalian soluble glutathione transferases. *Methods Enzymol.*, *401*, 1–8.

Martinez-Jimenez, C.P., Jover, R., Donato, M.T., Castell, J.V., Gomez-Lechon, M.J. (2007). Transcriptional regulation and expression of CYP3A4 in hepatocytes. *Curr. Drug Metab.*, *8*, 185–194.

Marzolini, C., Tirona, R.G., Kim, R.B. (2004). Pharmacogenomics of the OATP and OAT families. *Pharmacogenomics*, *5*, 273–282.

Meyer, U.A., Amrein, R., Balant, L.P., Bertilsson, L., Eichelbaum, M., Guentert, T.W., Henauer, S., Jackson, P., Laux, G., Mikkelsen, H., Peck, C., Pollock, B.G., Priest, R., Sjoqvist, F., Delini-Stula, A. (1996). Antidepressants and drug-metabolizing enzymes—expert group report. *Acta Psychiatr. Scand.*, *93*, 71–79.

Miners, J.O., McKinnon, R.A., Mackenzie, P.I. (2002). Genetic polymorphisms of UDP-glucuronosyltransferases and their functional significance. *Toxicology*, *181*–182, 453–456.

Mizuno, N., Niwa, T., Yotsumoto, Y., Sugiyama, Y. (2003). Impact of drug transporter studies on drug discovery and development. *Pharmacol. Rev.*, *55*, 425–461.

Monaghan, G., Ryan, M., Seddon, R., Hume, R., Burchell, B. (1996). Genetic variation in bilirubin UPD-glucuronosyltransferase gene promoter and Gilbert's syndrome. *Lancet*, *347*, 578–581.

Mossallam, G.I., Abdel Hamid, T.M., Samra, M.A. (2006). Glutathione S-transferase GSTM1 and GSTT1 polymorphisms in adult acute myeloid leukemia; its impact on toxicity and response to chemotherapy. *J. Egypt Natl. Canc. Inst.*, *18*, 264–273.

Murayama, N., Soyama, A., Saito, Y., Nakajima, Y., Komamura, K., Ueno, K., Kamakura, S., Kitakaze, M., Kimura, H., Goto, Y., Saitoh, O., Katoh, M., Ohnuma, T., Kawai, M., Sugai, K., Ohtsuki, T., Suzuki, C., Minami, N., Ozawa, S., Sawada, J. (2004). Six novel nonsynonymous CYP1A2 gene polymorphisms: catalytic activities of the naturally occurring variant enzymes. *J. Pharmacol. Exp. Ther.*, *308*, 300–306.

Nagar, S., Blanchard, R.L. (2006). Pharmacogenetics of uridine diphosphoglucuronosyltransferase (UGT) 1A family members and its role in patient response to irinotecan. *Drug Metab. Rev.*, *38*, 393–409.

Nagata, K., Yamazoe, Y. (2000). Pharmacogenetics of sulfotransferase. *Annu. Rev. Pharmacol. Toxicol.*, *40*, 159–176.

Nakajima, M., Yokoi, T., Mizutani, M., Kinoshita, M., Funayama, M., Kamataki, T. (1999). Genetic polymorphism in the 5′-flanking region of human CYP1A2 gene: effect on the CYP1A2 inducibility in humans. *J. Biochem. (Tokyo)*, *125*, 803–808.

Nakamoto, K., Kidd, J.R., Jenison, R.D., Klaassen, C.D., Wan, Y.J., Kidd, K.K., Zhong, X.B. (2007). Genotyping and haplotyping of CYP2C19 functional alleles on thin-film biosensor chips. *Pharmacogenet. Genomics*, *17*, 103–114.

Nebert, D.W. (1999). Pharmacogenetics and pharmacogenomics: why is this relevant to the clinical geneticist? *Clin. Genet.*, *56*, 247–258.

Nie, A.Y., McMillian, M., Parker, J.B., Leone, A., Bryant, S., Yieh, L., Bittner, A., Nelson, J., Carmen, A., Wan, J., Lord, P.G. (2006). Predictive toxicogenomics approaches reveal underlying molecular mechanisms of nongenotoxic carcinogenicity. *Mol. Carcinog.*, *45*, 914–933.

Nowell, S., Sweeney, C., Winters, M., Stone, A., Lang, N.P., Hutchins, L.F., Kadlubar, F.F., Ambrosone, C.B. (2002). Association between sulfotransferase 1A1 genotype and survival of breast cancer patients receiving tamoxifen therapy. *J. Natl. Cancer Inst.*, *94*, 1635–1640.

Ohkubo, Y., Ueta, A., Ando, N., Ito, T., Yamaguchi, S., Mizuno, K., Sumi, S., Maeda, T., Yamazaki, D., Kurono, Y., Fujimoto, S., Togari, H. (2006). Novel mutations in the cytochrome P450 2C19 gene: a pitfall of the PCR-RFLP method for identifying a common mutation. *J. Hum. Genet.*, *51*, 118–123.

Ohno, A., Saito, Y., Hanioka, N., Jinno, H., Saeki, M., Ando, M., Ozawa, S., Sawada, J. (2004). Involvement of human hepatic UGT1A1, UGT2B4, and UGT2B7 in the glucuronidation of carvedilol. *Drug Metab. Dispos.*, *32*, 235–239.

Paoluzzi, L., Singh, A.S., Price, D.K., Danesi, R., Mathijssen, R.H., Verweij, J., Figg, W.D., Sparreboom, A. (2004). Influence of genetic variants in UGT1A1 and UGT1A9 on the in vivo glucuronidation of SN-38. *J. Clin. Pharmacol.*, *44*, 854–860.

Pedersen, R.S., Damkier, P., Brosen, K. (2006). The effects of human CYP2C8 genotype and fluvoxamine on the pharmacokinetics of rosiglitazone in healthy subjects. *Br. J. Clin. Pharmacol.*, *62*, 682–689.

Peloquin, C.A., Jaresko, G.S., Yong, C.L., Keung, A.C., Bulpitt, A.E., Jelliffe, R.W. (1997). Population pharmacokinetic modeling of isoniazid, rifampin, and pyrazinamide. *Antimicrob. Agents Chemother.*, *41*, 2670–2679.

Pistorius, S., Goergens, H., Engel, C., Plaschke, J., Krueger, S., Hoehl, R., Saeger, H.D., Schackert, H.K. (2007). N-Acetyltransferase (NAT) 2 acetylator status and age of tumour onset in patients with sporadic and familial, microsatellite stable (MSS) colorectal cancer. *Int. J. Colorectal Dis.*, *22*, 137–143.

Poland, A., Glover, E. (1990). Characterization and strain distribution pattern of the murine Ah receptor specified by the Ahd and Ahb-3 alleles. *Mol. Pharmacol.*, *38*, 306–312.

Poland, A., Glover, E., Kende, A.S. (1976). Stereospecific, high affinity binding of 2,3,7,8-tetrachlorodibenzo-p-dioxin by hepatic cytosol. Evidence that the binding species is receptor for induction of aryl hydrocarbon hydroxylase. *J. Biol. Chem.*, *251*, 4936–4946.

Pucci, L., Geppetti, A., Maggini, V., Lucchesi, D., Maria Rossi, A., Longo, V. (2007). CYP1A2 F21L and F186L polymorphisms in an Italian population sample. *Drug Metab. Pharmacokinet.*, *22*, 220–222.

Raftogianis, R.B., Wood, T.C., Otterness, D.M., Van Loon, J.A., Weinshilboum, R.M. (1997). Phenol sulfotransferase pharmacogenetics in humans: association of common SULT1A1 alleles with TS PST phenotype. *Biochem. Biophys. Res. Commun.*, *239*, 298–304.

Rasmussen, B.B., Brix, T.H., Kyvik, K.O., Brosen, K. (2002). The interindividual differences in the 3-demthylation of caffeine alias CYP1A2 is determined by both genetic and environmental factors. *Pharmacogenetics*, *12*, 473–478.

Raunio, H., Husgafvel-Pursiainen, K., Anttila, S., Hietanen, E., Hirvonen, A., Pelkonen, O. (1995). Diagnosis of polymorphisms in carcinogen-activating and inactivating enzymes and cancer susceptibility—a review. *Gene*, *159*, 113–121.

Relling, M.V., Lin, J.S., Ayers, G.D., Evans, W.E. (1992). Racial and gender differences in N-acetyltransferase, xanthine oxidase, and CYP1A2 activities. *Clin. Pharmacol. Ther.*, *52*, 643–658.

Repa, J.J., Mangelsdorf, D.J. (2000). The role of orphan nuclear receptors in the regulation of cholesterol homeostasis. *Annu. Rev. Cell Dev. Biol.*, *16*, 459–481.

Rettie, A.E., Tai, G. (2006). The pharmocogenomics of warfarin: closing in on personalized medicine. *Mol. Interv.*, *6*, 223–227.

Rifkind, A.B. (2006). CYP1A in TCDD toxicity and in physiology-with particular reference to CYP dependent arachidonic acid metabolism and other endogenous substrates. *Drug Metab. Rev.*, *38*, 291–335.

Rodriguez-Novoa, S., Barreiro, P., Jimenez-Nacher, I., Soriano, V. (2006). Overview of the pharmacogenetics of HIV therapy. *Pharmacogenomics J.*, *6*, 234–245.

Rushmore, T.H., Kong, A.N. (2002). Pharmacogenomics, regulation and signaling pathways of phase I and II drug metabolizing enzymes. *Curr. Drug Metab.*, *3*, 481–490.

Ruzilawati, A.B., Suhaimi, A.W., Gan, S.H. (2007). Genetic polymorphisms of CYP3A4: CYP3A4*18 allele is found in five healthy Malaysian subjects. *Clin. Chim. Acta*, *383*, 158–162.

Sachse, C., Brockmoller, J., Bauer, S., Roots, I. (1999). Functional significance of a C–>A polymorphism in intron 1 of the cytochrome P450 CYP1A2 gene tested with caffeine. *Br. J. Clin. Pharmacol.*, *47*, 445–449.

Saeki, M., Ozawa, S., Saito, Y., Jinno, H., Hamaguchi, T., Nokihara, H., Shimada, Y., Kunitoh, H., Yamamoto, N., Ohe, Y., Yamada, Y., Shirao, K., Muto, M., Mera, K., Goto, K., Ohmatsu, H., Kubota, K., Niho, S., Kakinuma, R., Minami, H., Ohtsu, A., Yoshida, T., Saijo, N., Sawada, J. (2002). Three novel single nucleotide polymorphisms in UGT1A10. *Drug Metab. Pharmacokinet.*, *17*, 488–490.

Saeki, M., Saito, Y., Jinno, H., Sai, K., Kaniwa, N., Ozawa, S., Komamura, K., Kotake, T., Morishita, H., Kamakura, S., Kitakaze, M., Tomoike, H., Shirao, K., Minami, H., Ohtsu, A., Yoshida, T., Saijo, N., Kamatani, N., Sawada, J. (2005). Genetic polymorphisms of UGT1A6 in a Japanese population. *Drug Metab. Pharmacokinet.*, *20*, 85–90.

Saito, T., Egashira, M., Kiyotani, K., Fujieda, M., Yamazaki, H., Kiyohara, C., Kunitoh, H., Kamataki, T. (2003). Novel nonsynonymous polymorphisms of the CYP1A1 gene in Japanese. *Drug Metab. Pharmacokinet.*, *18*, 218–221.

Scheen, A.J. (2007). Pharmacokinetic interactions with thiazolidinediones. *Clin. Pharmacokinet.*, *46*, 1–12.

Scordo, M.G., Caputi, A.P., D'Arrigo, C., Fava, G., Spina, E. (2004). Allele and genotype frequencies of CYP2C9, CYP2C19 and CYP2D6 in an Italian population. *Pharmacol. Res.*, *50*, 195–200.

Searfoss, G.H., Jordan, W.H., Calligaro, D.O., Galbreath, E.J., Schirtzinger, L.M., Berridge, B.R., Gao, H., Higgins, M.A., May, P.C., Ryan, T.P. (2003). Adipsin, a biomarker of gastrointestinal toxicity mediated by a functional gamma-secretase inhibitor. *J. Biol. Chem.*, *278*, 46107–46116.

Shi, M.M., Bleavins, M.R., de la Iglesia, F.A. (2001). Pharmacogenetic application in drug development and clinical trials. *Drug Metab. Dispos.*, *29*, 591–595.

Shimada, T., Yamazaki, H., Mimura, M., Inui, Y., Guengerich, F.P. (1994). Interindividual variations in human liver cytochrome P-450 enzymes involved in the oxidation of drugs, carcinogens and toxic chemicals: studies with liver microsomes of 30 Japanese and 30 Caucasians. *J. Pharmacol. Exp. Ther.*, *270*, 414–423.

Siest, G., Jeannesson, E., Berrahmoune, H., Maumus, S., Marteau, J.B., Mohr, S., Visvikis, S. (2004). Pharmacogenomics and drug response in cardiovascular disorders. *Pharmacogenomics*, *5*, 779–802.

Siest, G., Jeannesson, E., Visvikis-Siest, S. (2007). Enzymes and pharmacogenetics of cardiovascular drugs. *Clin. Chim. Acta*, *381*, 26–31.

Sim, E., Westwood, I., Fullam, E. (2007). Arylamine N-acetyltransferases. *Expert Opin. Drug Metab. Toxicol.*, *3*, 169–184.

Smart, J., Daly, A.K. (2000). Variation in induced CYP1A1 levels: relationship to CYP1A1, Ah receptor and GSTM1 polymorphisms. *Pharmacogenetics*, *10*, 11–24.

Smith, D.A. (2000). Induction and drug development. *Eur. J. Pharm. Sci.*, *11*, 185–189.

Strange, R.C., Lear, J.T., Fryer, A.A. (1998). Glutathione S-transferase polymorphisms: influence on susceptibility to cancer. *Chem. Biol. Interact.*, *111*–112, 351–364.

Strassburg, C.P., Manns, M.P. (2000). Jaundice, genes and promoters. *J. Hepatol.*, *33*, 476–479.

Suter, L., Babiss, L.E., Wheeldon, E.B. (2004). Toxicogenomics in predictive toxicology in drug development. *Chem. Biol.*, *11*, 161–171.

Swan, G.E., Valdes, A.M., Ring, H.Z., Khroyan, T.V., Jack, L.M., Ton, C.C., Curry, S.J., McAfee, T. (2005). Dopamine receptor DRD2 genotype and smoking cessation outcome following treatment with bupropion SR. *Pharmacogenomics J.*, *5*, 21–29.

Taioli, E., Gaspari, L., Benhamou, S., Boffetta, P., Brockmoller, J., Butkiewicz, D., Cascorbi, I., Clapper, M.L., Dolzan, V., Haugen, A., Hirvonen, A., Husgafvel-Pursiainen, K., Kalina, I., Kremers, P., Le Marchand, L., London, S., Rannug, A., Romkes, M., Schoket, B., Seidegard, J., Strange, R.C., Stucker, I., To-Figueras, J., Garte, S. (2003). Polymorphisms in CYP1A1, GSTM1, GSTT1 and lung cancer below the age of 45 years. *Int. J. Epidemiol.*, *32*, 60–63.

Taniguchi, A., Urano, W., Tanaka, E., Furihata, S., Kamitsuji, S., Inoue, E., Yamanaka, M., Yamanaka, H., Kamatani, N. (2007). Validation of the associations between single nucleotide polymorphisms or haplotypes and responses to disease-modifying antirheumatic drugs in patients with rheumatoid arthritis: a proposal for prospective pharmacogenomic study in clinical practice. *Pharmacogenet. Genomics*, *17*, 383–390.

Thibaudeau, J., Lepine, J., Tojcic, J., Duguay, Y., Pelletier, G., Plante, M., Brisson, J., Tetu, B., Jacob, S., Perusse, L., Belanger, A., Guillemette, C. (2006). Characterization of common UGT1A8, UGT1A9, and UGT2B7 variants with different capacities to inactivate mutagenic 4-hydroxylated metabolites of estradiol and estrone. *Cancer Res.*, *66*, 125–133.

Tukey, R.H., Strassburg, C.P. (2000). Human UDP-glucuronosyltransferases: metabolism, expression, and disease. *Annu. Rev. Pharmacol. Toxicol.*, *40*, 581–616.

Uetrecht, J. (2003). Screening for the potential of a drug candidate to cause idiosyncratic drug reactions. *Drug Discov. Today*, *8*, 832–837.

Vermeir, M., Annaert, P., Mamidi, R.N., Roymans, D., Meuldermans, W., Mannens, G. (2005). Cell-based models to study hepatic drug metabolism and enzyme induction in humans. *Expert Opin. Drug Metab. Toxicol.*, *1*, 75–90.

Vineis, P., Veglia, F., Anttila, S., Benhamou, S., Clapper, M.L., Dolzan, V., Ryberg, D., Hirvonen, A., Kremers, P., Le Marchand, L., Pastorelli, R., Rannug, A., Romkes, M., Schoket, B., Strange, R.C., Garte, S., Taioli, E. (2004). CYP1A1, GSTM1 and GSTT1 polymorphisms and lung cancer: a pooled analysis of gene-gene interactions. *Biomarkers*, *9*, 298–305.

Wang, H., LeCluyse, E.L. (2003). Role of orphan nuclear receptors in the regulation of drug-metabolising enzymes. *Clin. Pharmacokinet.*, *42*, 1331–1357.

Wang, Y., Spitz, M.R., Tsou, A.M., Zhang, K., Makan, N., Wu, X. (2002). Sulfotransferase (SULT) 1A1 polymorphism as a predisposition factor for lung cancer: a case-control analysis. *Lung Cancer*, *35*, 137–142.

Waring, J.F., Cavet, G., Jolly, R.A., McDowell, J., Dai, H., Ciurlionis, R., Zhang, C., Stoughton, R., Lum, P., Ferguson, A., Roberts, C.J., Ulrich, R.G. (2003). Development of a DNA microarray for toxicology based on hepatotoxin-regulated sequences. *EHP Toxicogenomics*, *111*, 53–60.

Weber, W.W., Hein, D.W. (1985). N-acetylation pharmacogenetics. *Pharmacol. Rev.*, *37*, 25–79.

Weinshilboum, R.M., Otterness, D.M., Aksoy, I.A., Wood, T.C., Her, C., Raftogianis, R.B. (1997). Sulfation and sulfotransferases 1: Sulfotransferase molecular biology: cDNAs and genes. *FASEB J.*, *11*, 3–14.

Williams, D.P., Kitteringham, N.R., Naisbitt, D.J., Pirmohamed, M., Smith, D.A., Park, B.K. (2002). Are chemically reactive metabolites responsible for adverse reactions to drugs? *Curr. Drug Metab.*, *3*, 351–366.

Wojnowski, L., Kamdem, L.K. (2006). Clinical implications of CYP3A polymorphisms. *Expert Opin. Drug Metab. Toxicol.*, *2*, 171–182.

Wolf, C.R., Smith, G., Smith, R.L. (2000). Science, medicine, and the future: pharmacogenetics. *BMJ*, *320*, 987–990.

Wong, J.M., Okey, A.B., Harper, P.A. (2001). Human aryl hydrocarbon receptor polymorphisms that result in loss of CYP1A1 induction. *Biochem. Biophys. Res. Commun.*, *288*, 990–996.

Wu, M.T., Wang, Y.T., Ho, C.K., Wu, D.C., Lee, Y.C., Hsu, H.K., Kao, E.L., Lee, J.M. (2003). SULT1A1 polymorphism and esophageal cancer in males. *Int. J. Cancer*, *103*, 101–104.

Yalin, S., Hatungil, R., Tamer, L., Ates, N.A., Dogruer, N., Yildirim, H., Karakas, S., Atik, U. (2007). N-acetyltransferase 2 polymorphism in patients with Diabetes Mellitus. *Cell Biochem. Funct.*, *25*, 407–411.

Yamanaka, H., Nakajima, M., Katoh, M., Yokoi, T. (2007). Glucuronidation of thyroxine in human liver, jejunum, and kidney microsomes. *Drug Metab. Dispos.*, *35*, 1642–1648.

Yengi, L., Inskip, A., Gilford, J., Alldersea, J., Bailey, L., Smith, A., Lear, J.T., Heagerty, A.H., Bowers, B., Hand, P., Hayes, J.D., Jones, P.W., Strange, R.C., Fryer, A.A. (1996). Polymorphism at the glutathione S-transferase locus GSTM3: interactions with cytochrome P450 and glutathione S-transferase genotypes as risk factors for multiple cutaneous basal cell carcinoma. *Cancer Res.*, *56*, 1974–1977.

Yengi, L.G., Xiang, Q., Shen, L., Appavu, C., Kao, J., Scatina, J. (2007). Application of pharmacogenomics in drug discovery and development: correlations between transcriptional modulation and preclinical safety observation. *Drug Metab. Lett.*, *1*, 41–48.

You, L. (2004). Steroid hormone biotransformation and xenobiotic induction of hepatic steroid metabolizing enzymes. *Chem. Biol. Interact.*, *147*, 233–246.

Zand, N., Tajik, N., Moghaddam, A.S., Milanian, I. (2007). Genetic polymorphisms of cytochrome P450 enzymes 2C9 and 2C19 in a healthy Iranian population. *Clin. Exp. Pharmacol. Physiol.*, *34*, 102–105.

Zhou, S., Yung Chan, S., Cher Goh, B., Chan, E., Duan, W., Huang, M., McLeod, H.L. (2005). Mechanism-based inhibition of cytochrome P450 3A4 by therapeutic drugs. *Clin. Pharmacokinet.*, *44*, 279–304.

CHAPTER 6

Introduction to Drug Transporters

LOUIS LEUNG and ARAM OGANESIAN

6.1 INTRODUCTION

The role of transporter proteins in drug and endogenous substance disposition has increasingly gained recognition over the last decade (Mizuno *et al.*, 2003; Beringer and Slaughter, 2005; Anzai, Kanai and Endou, 2006). For example, the role of transporters in biliary excretion was not recognized as late as 1995 and as a result, species differences in biliary excretion were not readily explainable (Lin, 1995). It has now become clear that transporters are responsible both for the uptake and efflux of drugs and other chemicals in various tissues and may be key determinants of the pharmacokinetic characteristics of a drug. Inhibition or induction of different drug transporters in different tissues can lead to clinical drug–drug interactions that may result in adverse effects or lack of efficacy (Bodo *et al.*, 2003; Van Montfoort *et al.*, 2003; Sai, 2005). Inhibition or induction of transporters that are responsible for endogenous substance disposition (e.g. bile acids) can also result in toxicological consequences (Bodo *et al.*, 2003). This review is not intended to be a comprehensive and in-depth review of all transporter-related topics, but rather focus on the role of drug transporters in drug development. There are many excellent reviews on various topics on transporters (Bodo *et al.*, 2003; Mizuno *et al.*, 2003; Van Montfoort *et al.*, 2003; Beringer and Slaughter, 2005; Sai, 2005; Anzai, Kanai and Endou, 2006; Endres *et al.*, 2006; Zhang *et al.*, 2006; Alrefai and Gill, 2007; Lin, 2007; Urquhart, Tirona and Kim, 2007). The purpose of this review is to provide the readers with fundamental knowledge of drug transporters and their roles in drug development relating to drug absorption, distribution, metabolism, excretion and toxicity, clinical drug–drug interactions, and *in vitro*/*in vivo* methodologies.

6.2 TRANSPORTER CLASSIFICATION, LOCALIZATION, AND FUNCTIONS

Table 6.1 is a summary of transporter families and the individual genes expressed. Transporters have been classified as primary, secondary, or tertiary active transporters.

Drug Metabolism Handbook: Concepts and Applications, Edited by Ala F. Nassar, Paul F. Hollenberg, and JoAnn Scatina

TABLE 6.1   Human drug transporter gene families.

Gene family	Gene product (gene symbol)
Multidrug resistance protein/P-glycoprotein	MDR1/P-gp (*ABCB1*)
Bile salt export pump	BSEP (*ABCB11*)
Multidrug resistance-associated protein	MRP1 (*ABCC1*)
	MRP2 (*ABCC2*)
	MRP3 (*ABCC3*)
	MRP4 (*ABCC4*)
Breast cancer resistance protein	BCRP (*ABCG2*)
Sodium taurocholate cotransporting peptide	NTCP (*SLC10A1*)
Oligopeptide transporters	PEPT 1 (*SLC15A1*)
	PEPT 2 (*SLC15A2*)
Organic anion-transporting polypeptides	OATP-A/OATP1A2 (*SLC21A3*)
	OATP-B/OATP2B1 (*SLC21A9*)
	OATP-C/OATP1B1 (*SLC21A6*)
	OATP8/OATP1B3 (*SLC21A8*)
	OATP-D/OATP3A1 (*SLC21A11*)
	OATP-E/OATP4A1 (*SLC21A12*)
	OATP-F/OATP1C1 (*SLC21A14*)
Organic cation transporters	OCT1 (*SLC22A1*)
	OCT2 (*SLC22A2*)
	OCT3 (*SLC22A3*)
Novel organic cation transporters	OCTN1 (*SLC22A4*)
	OCTN2 (*SLC22A5*)
Organic anion transporters	OAT1 (*SLC22A6*)
	OAT2 (*SLC22A7*)
	OAT3 (*SLC22A8*)
	OAT4 (*SLC22A9*)

Primary active transporters require ATP-dependent hydrolysis as the first step in catalysis. Examples are ATP-binding cassette transporters such as multidrug resistance protein (MDR), multidrug resistance-associated protein (MRP), and breast cancer resistance protein (BCRP). Secondary or tertiary active transporters are driven by an exchange or cotransport of intracellular and/or extracellular ions with the substrate. Examples include organic anion transporter (OAT), organic anion-transporting polypeptide (OATP), sodium taurocholate cotransporting peptide (NTCP), organic cation transporter (OCT), novel organic cation transporter (OCTN), and oligopeptide transporter (PEPT). The major tissue localization and examples of common substrates and inhibitors of the different transporter families are summarized in Table 6.2 (see also Fig. 6.1).

6.2.1   ABC Transporters

6.2.1.1   MDR1P-gp (ABCB1)   Of the known transporters, P-glycoprotein (P-gp) is by far the most characterized and understood efflux transporter. P-gp belongs to the ATP-binding cassette transporter superfamily and is encoded by the MDR1 (multidrug resistance) gene (ABCB1) in humans. It is predominantly expressed in tissues including the luminal surface of intestinal epithelia, the renal proximal tubule, the bile canalicular

TABLE 6.2 Major human transporters, tissue localization, substrates, and inhibitors[a]

Transporter (gene)	Transporter family	Major tissue localization	Examples of common substrates	Examples of inhibitors
		ABC transporters		
MDR1/P-gp (*ABCB1*)	Multidrug resistance protein/P-glycoprotein	Intestinal enterocytes, liver, kidney, blood–brain barrier, placenta, adrenal, testes, tumor cells	Digoxin, paclitaxel, tacrolimus, atorvastatin	Cyclosporine, terfenadine, verapamil, ketoconazole, quinidine, ritonavir, elacridar (LY-335979), valspodar (PSC-833), GF-120918
BSEP (*ABCB11*)	Bile salt export pump	Liver	Bile salts: paclitaxel, vinblastine, pravastatin	Cyclosporine, troglitazone
MRP1 (*ABCC1*)	MRP	Liver, intestine, kidney, brain, testis	Substrates overlap among MRP1, MRP2, and MRP3.	Ketoconazole
MRP2 (*ABCC2*)	MRP	Intestine, liver, kidney, brain, placenta	Glutathione, glucuronide, and sulfate conjugates.	Cyclosporine, rifampin, omeprazole
MRP3 (*ABCC3*)	MRP	Intestine, pancreas, placenta, adrenal cortex, liver, kidney, prostate	E17βG, indinavir, cisplatin, etoposide, methotrexate, vincristine	
MRP4 (*ABCC4*)	MRP	Prostate, lung, adrenals, ovary, testis, pancreas, small intestine	Nucleoside analogues and cyclic nucleotides (cGMP and cAMP).	
MRP5 (*ABCC5*)	MRP	Skeletal muscle, heart, and brain	Nucleoside analogues and cyclic nucleotides (cGMP and cAMP).	
BCRP (*ABCG2*)	Breast cancer resistance protein	Placenta, intestine, liver, breast, brain, tumor cells	Broad substrate specificity, partly overlapping with P-gp and MRPs. Doxorubicin, daunorubicin, topotecan, rosuvastatin, sulfasalazine	Elacridar (GF120918), gefitinib

TABLE 6.2 Continued

Transporter (gene)	Transporter family	Major tissue localization	Examples of common substrates	Examples of inhibitors
		Solute carrier transporters		
NTCP (*SLC10A1*)	Sodium taurocholate co-transporting peptide	Liver, pancreas	Bile salts; taurocholate, rosuvastatin	
PEPT1 (*SLC15A1*) PEPT2 (*SLC15A2*)	Oligopeptide transporters	Intestine, kidney	Dipeptides, tripeptides, and peptidomimetics; ampicillin, amoxicillin, captopril, valacyclovir	
OATP-A/OATP1A2 (*SLC21A3*)	OATP	Brain, liver, testis, prostate	Bile salts, ouabain, *N*-methyl quinidine, certain hormones, and/or hormone conjugates	
OATP-B/OATP2B1 (*SLC21A9*)	OATP	Liver, intestine, pancreas, lung, ovary, testes, spleen	Estrone-3-sulfate, DHEAS, pravastatin	
OATP-C/OATP1B1 (*SLC21A6*)	OATP	Liver	Estrone-3-sulfate, DHEAS, taurocholate, rifampin, rosuvastatin, methotrexate, pravastatin, thyroxine	Cyclosporine, rifampin
OATP8/OATP1B3 (*SLC21A8*)	OATP	Liver	Digoxin, pravastatin, methotrexate, rifampin	
OATP-D/OATP3A1 (*SLC21A11*)	OATP	Ubiquitous, strong expression in leukocytes, spleen, tumors	E-3-sulfate, benzylpenicillin	
OATP-E/OATP4A1 (*SLC21A12*)	OATP	Ubiquitous, skeletal muscle, tumors	Taurocholate, E-17β-G	

Transporter	Type	Tissue distribution	Substrates	Inhibitors
OCT1 (*SLC22A1*)	OCT	Liver	Small organic cations; acyclovir, amantadine, desipramine, ganciclovir, quinidine, quinine, metformin	Disopyramide, midazolam, phenformin, phenoxy-benzamine, quinidine, quinine, Ritonavir, verapamil
OCT2 (*SLC22A2*)	OCT	Kidney, brain	Small organic cations; amantadine, cimetidine, memantine	Desipramine, phenoxy-benzamine, quinine
OCT3 (*SLC22A3*)	OCT	Skeletal muscle, liver, brain placenta, kidney, heart	Small organic cations; cimetidine	Desipramine, prazosin, phenoxy-benzamine
OAT1 (*SLC22A6*)	OAT	Kidney, brain	Organic anions; acyclovir, adefovir, methotrexate, zidovudine	Probenecid, cefadroxil, cefamandole, cefazolin
OAT2 (*SLC22A7*)	OAT	Liver, kidney	Organic anions; zidovudine, salicylates	
OAT3 (*SLC22A8*)	OAT	Kidney, brain	Organic anions; methotrexate, zidovudine, cimetidine, E-3-sulfate	Probenecid, cefadroxil, cefamandole, cefazolin
OCTN1 (*SLC22A4*)	Novel organic cation transporters	Kidney, skeletal muscle, prostate, placenta, heart	Small organic cations; quinidine, verapamil	
OCTN2 (*SLC22A5*)	Novel organic cation transporters	Kidney, skeletal muscle, prostate, lung, heart, liver	Small organic cations; quinidine, verapamil	

[a]Adapted and expanded from Xia, Milton and Gan (2007) and Zhang *et al.* (2006).

(a)

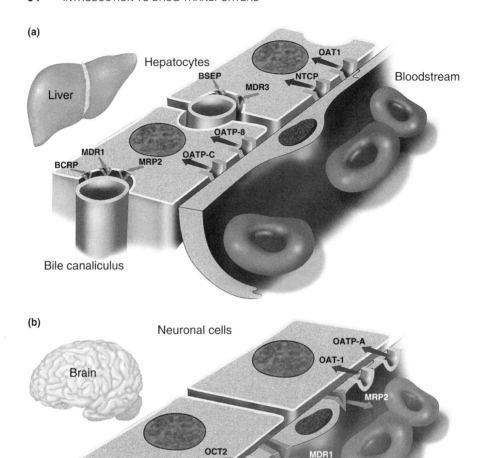

Figure 6.1 Human drug transporters expressed in different tissues: **(a)** liver, **(b)** brain, **(c)** kidney, and **(d)** intestine (from Ho and Kim, 2005).

membrane of hepatocytes, and the blood–brain barrier (BBB). P-gp can play an important role in limiting intestinal drug absorption and brain penetration, and in facilitating renal or biliary excretion of drug substrates. P-gp substrates tend to be organic amphipathic molecules with large structural diversities. Drugs that are P-gp substrates represent diverse therapeutic categories; more comprehensive examples of drugs that are P-gp substrates are summarized in Table 6.3.

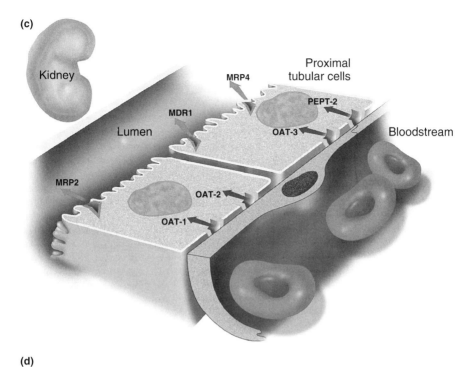

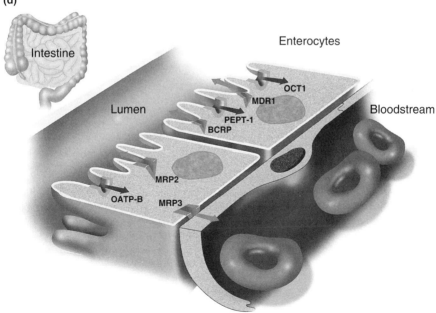

Figure 6.1 *Continued*

TABLE 6.3   Examples of drugs that are P-gp substrates.[a]

Analgesics	Antiepileptic	Antihypertensives	Oncology
Asimadoline	Phenytoin	Losartan	Actinomycin D
[d-penicillamine-	Antifungals	β-Blockers	Bisantrene
2,5]-enkephalin	Itraconazole	Celiprolol	Cisplatin
(DPDPE)	Ketoconazole	Talinolol	Cytarabine
Fentanyl	Antihistamines	Other cardiovascular	Chlorambucil
Methadone	Cetirizine	Atorvastatin	Docetaxel
Morphine	Fexofenadine	Digoxin	Doxorubicin
Antibiotics	Anti-inflamatory	Diltiazem	Daunorubicin
Dactinomycin	Methylprednisolone	Quinidine	Fluorouracil
Erythromycin	Methotrexate	Verapamil	Epirubicin
Gramicidin D	Corticosteroids	Immunosuppressants	Etoposide
Sparfloxacin	Corticosterone	Cyclosporine A (CsA)	Hydroxyurea
Valinomycin	Dexamethasone	FK506	Lapatinib
Antigout agents	Hydrocortisone	Sirolimus	Mitoxantrone
Colchicine	Triamcinolone	Tacrolimus	Octreotide
Antidiarrheal	Diagnostic dyes	HIV protease inhibitors	Paclitaxel
Loperamide	Hoechst 33342	Amprenavir	Tamoxifen
Antiemetics	Rhodamine 123	Indinavir	Teniposide
Domperidone	H2-receptor antagonists	Lopinavir	Topotecan
Ondansetron	Cimetidine	Nelfinavir	Vinblastine
Antidepressant	Ranitidine	Ritonavir	Vincristine
Amitriptyline		Saquinavir	Vinorelbine
Nortriptyline			

[a]Adapted and expanded from Matheny *et al.* (2001) and Endres *et al.* (2006).

6.2.1.2   *BSEP (ABCB11)*

The bile salt export pump (BSEP) is located on the canalicular membrane of hepatocytes, and mediates the transport of bile acids such as taurocholic acid into the bile. Mutations in BSEP have been associated with cholestatic liver disease (Sai, 2005). BSEP has recently been shown to transport non-bile acid substrates such as pravastatin (Hirano *et al.*, 2005) and vinblastine (Akashi, Tanaka and Takikawa, 2006).

6.2.1.3   *MRPs (ABCC)*

The multidrug resistance associated proteins (MRPs) have been shown to be associated with drug resistance in cancer cells. The exact physiological role of MRPs is not known. MRPs are involved in the efflux of substrates or drugs from the liver or kidney into the peripheral blood (e.g. MRP1, MRP3, and MRP6), or from the liver, kidney, and small intestines into the bile, urine, or intestinal lumen (MRP2).

MRP1 (ABCC1) is ubiquitous and is present at the basolateral surface of various tissues including the liver, intestine, brain, kidney, and testis. In the liver and kidney, MRP1 is responsible for the secretion of substrates into peripheral blood. MRP1 substrates are glutathione, glucuronide and sulfate conjugates, and organic anions such as methotrexate and estradiol-17β-D-glucuronide.

MRP2 (ABCC2) is by far the most well-studied and important MRP in human drug disposition and toxicity. MRP2 is present at the apical surface of many tissues and include the liver, kidney, brain, and intestines, and is responsible for biliary and renal excretion and efflux into intestinal lumen of substrates. Substrates for MRP2 overlap with those of MRP1, and include glutathione, glucuronide and sulfate conjugates, and organic anions such as indinavir and cisplatin. A mutation of human MRP2 was shown to lead to hyperbilirubinemia such as the Dubin–Johnson syndrome due to a defect in the biliary excretion of bilirubin glucuronides (Paulusma et al., 1997).

MRP3 (ABCC3) is present at the basolateral surface of various tissues including the liver, kidney, and intestines. Like MRP1, MRP3 is responsible for the secretion of substrates into peripheral blood in the kidneys and liver. Substrates for MRP3 overlap with those of MRP1 and MRP2, and include glutathione, glucuronide and sulfate conjugates, and organic anions such as vincristine and etoposide.

MRP4 (ABCC4) and MRP5 (ABCC5) are present at the apical and basolateral surface, respectively, of many tissues including lungs, intestines, and testes. Substrates for these MRPs are primarily nucleoside analogues and cyclic nucleotides (cGMP and cAMP).

MRP6 (ABCC6) is present at the basolateral surface of the liver and kidneys, and like MRP1 and MRP3, is responsible for the secretion of substrates into peripheral blood in the kidneys and liver. Substrates for MRP6 include cisplatin and daunorubicin.

6.2.1.4 BCRP (ABCG2)

6.2.1.4 BCRP (ABCG2) Breast cancer resistance protein (BCRP) has only one ATP-binding cassette and six putative transmembrane domains (Rocchi et al., 2000), suggesting that BCRP is a half-transporter, which may function as a homo- or heterodimer. BCRP is located on the apical surface of the liver, intestines, and breasts, and exhibits broad substrate specificity partly overlapping with P-gp and MRP substrates, such as topotecan and daunorubicin.

6.2.2 Solute Carrier Transporters

6.2.2.1 PEPT1 (SLC15A1) and PEPT2 (SLC15A2) PEPT1 and PEPT2 are peptide transporters found in the brush border membrane of the intestines and renal cortex, respectively, and are involved in absorption of dipeptides, tripeptides, and peptidomimetic drugs.

6.2.2.2 Organic Anion Transporting Polypeptides Organic anion-transporting polypeptides (OATPs) are a superfamily of uptake transporters with wide tissue distribution including the liver, gut, and BBB. They are responsible for the uptake from blood or lumen of a variety of amphipathic organic anions such as steroid conjugates, bile acids, and drugs.

6.2.2.2.1 OATP1A2 (SLC21A3/SLCO1A2) OATP1A2 (OATP-A), the first OATP to be cloned (from the liver), is expressed in the BBB and was detected in the brain, lungs, liver, biliary epithelial cells, kidneys, and testes (Van Montfoort et al., 2003). OATP1A2 exhibits a broad substrate specificity including hydrophobic organic anions such as bile acid and bilirubin, neutral compounds such as ouabain, and organic cations such as N-methyl quinidine.

6.2.2.2.2 OATP1B1 (SLC21A6/SLCO1B1) OATP1B1 (OATP-C) was the second OATP to be cloned, and is only expressed in the liver responsible for the uptake of substrates across the basolateral membrane of hepatocytes. OATP1B1 seems to prefer organic anions including bilirubin, steroid hormone conjugates, and the drugs methotrexate and pravastatin.

6.2.2.2.3 OATP1B3 (SLC21A8/SLCO1B3) OATP1B3 (OATP-8) is primarily expressed in the liver, and similar to OATP1B1, transports steroid hormones and organic anions such as pravastatin. In addition to ouabain, OATP1B3 also mediates the hepatic uptake of the uncharged cardiac glycoside digoxin. OATP1B3, together with OATPs 1B1 and 2B1, is primarily responsible for the uptake of sodium-independent bile salt, organic anions, and drugs into the liver.

6.2.2.2.4 OATP2B1 (SLC21A9/SLCO2B1) OATP2B1 (OATP-B) is primarily expressed at the basolateral membrane of hepatocytes, and was also detected in the kidney, small intestine, brain, spleen, placenta, lung, heart, and ovary. OATP2B1 has narrow substrate selectivity when compared with members of the OATP1 family, and transports organic anions including estrone-3-sulfate but not bile salts.

6.2.2.2.5 OATP3A1 (SLC21A11/SLCO3A1), OATP4A1 (SLC21A12/SLCO4A1), and OATP1C1 (SLC21A14/SLCO1C1) OATP3A1 (OATP-D) and OATP4A1 (OATP-E) were cloned from human kidney (Tamai *et al.*, 2000) and are ubiquitous in their tissue distribution. OATP3A1 transports estrone-3-sulfate and benzylpenicillin and OATP4A1 transports taurocholate and estradiol-17β-glucuronide. OATP1C1 (OATP-F) was cloned from human brain (Pizzagalli *et al.*, 2002) and its substrate specificity appears to be narrow.

6.2.2.3 Organic Cation Transporters The OCT family of transporters is responsible for the uptake of organic cations into the kidney or liver from blood, and represents the first step in renal or biliary excretions (Koepsell, Lips and Volk, 2007).

6.2.2.3.1 OCT1 (SLC22A1) OCT1 is the first member of the OCT family and was cloned from the human liver (Zhang *et al.*, 1997). OCT1 is primarily found on the sinusoidal membrane of hepatocytes and mediates the uptake of quaternary ammonium compounds such as tetraethylammonium (TEA) and *N*-methylquinine, and drugs including quinidine, quinine, and metformin.

6.2.2.3.2 OCT2 (SLC22A2) OCT2 was cloned from human kidney and is primarily expressed on the basolateral membrane of renal proximal tubules. OCT2 is involved in basolateral uptake of organic cations into proximal tubule cells, the first step in the secretion in proximal tubules of cations and drugs such as the weak base cimetidine. OCT2 is also present in neurons in the brain, and may mediate the uptake of cationic drugs.

6.2.2.3.3 OCT3 (SLC22A3) OCT3 is found in various tissues including the liver, kidney, and brain, and shares substrate selectivity as OCT1 and OCT2. Like OCT2, OCT3 is involved in basolateral uptake of organic cations into proximal tubule cells leading to subsequent secretion into proximal tubules.

6.2.2.4 *Novel Organic Cation Transporters*

6.2.2.4.1 OCTN1 (SLC22A4) OCTN1 is a polyspecific and sodium-independent cation exchanger found in various tissues including kidney, skeletal muscle, placenta, and heart but not in the liver. OCTN1 transports cations and cationic drugs such as TEA, quinidine, verapamil, and the zwitterion carnitine. Localized in the brush border membrane of the renal proximal tubules, OCTN1 mediates the secretion of various organic cations across the luminal membrane.

6.2.2.4.2 OCTN2 (SLC22A5) OCTN2, also known as carnitine transporter, was also found in various tissues including kidney and liver. Like OCTN1, OCTN2 is involved in the secretion of cations and cationic drugs across the luminal membrane in renal proximal tubule cells. OCTN2 mediates the uptake of carnitine across the luminal membrane, and the cotransport was enhanced in the presence of sodium.

6.2.2.5 *Organic Anion Transporters*
The OAT family of proteins is involved in the uptake (OAT1, OAT2, or OAT3) of organic anions into the kidney or liver from the blood (OAT1, OAT2, or OAT3), or efflux into the lumen in the kidney (OAT4). OATs contribute collectively to renal or biliary drug elimination.

6.2.2.5.1 OAT1 (SLC22A6) OAT1 was cloned from human kidney (Race *et al.*, 1999) and is primarily expressed at the basolateral membrane of renal proximal tubule cells and was also found in the brain and placenta. OAT1 is involved in the uptake of organic anions across the basolateral membrane of the kidney proximal tubule cells, such as nonsteroidal anti-inflammatory drugs.

6.2.2.5.2 OAT2 (SLC22A7) OAT2 is expressed at the basolateral membrane of the renal proximal tubule cells and also at the sinusoidal membrane of hepatocytes, and is involved in the transport of organic anions such as salicylates and zidovudine.

6.2.2.5.3 OAT3 (SLC22A8) Like OAT2, OAT3 is expressed at the basolateral membrane of the renal proximal tubule cells, and in addition is found in the brain and skeletal muscle. OAT3 mediates uptake of nonsteroidal inflammatory drugs, cimetidine, antiviral drugs, and estrone-3-sulfate.

6.2.2.5.4 OAT4 (SLC22A9) OAT4 is expressed at the brush border membrane of the renal proximal tubule cells, and mediates the secretion across the luminal membrane of the renal proximal tubule cells of organic anions such as estrone-3-sulfate, methotrexate, and zidovudine.

6.3 CLINICAL DRUG–DRUG INTERACTIONS

Since transporters are present on the apical or basolateral membrane of various tissues, drug–drug interactions via inhibition or induction of a particular transporter can occur at the respective sites leading to altered pharmacokinetics or metabolic disposition of the concomitant drugs. Examples of clinical drug–drug interactions involving drug transporters are summarized in Tables 6.4 and 6.5. The expected effects of a putative

TABLE 6.4 Transporter-mediated drug–drug interactions reported in literature.[a]

Transporter	No. of studies reported	No. of showing no effect	No. with effects eliciting >20% change (PK)
P-gp	251	56	189
OATP	24	5	19
MRP2	13	3	10
MRP1	3	1	2
OAT	8	3	5
OAT1	2	1	1
OATP-C (OATP2; OATP1BIX1)	13	1	6
OATP-A (OATP1A2)	3	0	3
OATP-B (OATP2BIX1)	1	0	1
OCT	5	0	5
OCT2	4	1	3
BCRP	1	0	1
Transporter unidentified	2	0	2

[a]Survey of University of Washington Drug-Drug Interactions (DDI) database conducted in September 2007.

TABLE 6.5 Drug–drug interactions involving P-gp reported in University of Washington DDI database.

Inhibitor (precipitant)	Object ("victim") drug
(R)-verapamil	Doxorubicin (Wilson *et al.*, 1995), talinolol (Schwarz *et al.*, 1999)
Atazanavir	Ritonavir (Boffito *et al.*, 2004, 2006; Von Hentig *et al.*, 2007), saquinavir (Boffito *et al.*, 2004, 2006; Von Hentig *et al.*, 2007)
Azithromycin	Fexofenadine (Gupta *et al.*, 2001), ximelagatran (Dorani *et al.*, 2007)
Biricodar (VX-710)	Paclitaxel (Rowinsky *et al.*, 1998; Seiden *et al.*, 2002)
Carvedilol	Digoxin (Baris *et al.*, 2006)
Cefuroxime	Ximelagatran (Dorani *et al.*, 2007)
Cetirizine	Pilsicainide (Tsuruoka *et al.*, 2006)
Clarithromycin	Colchicine (Rollot *et al.*, 2004), digoxin (Wakasugi *et al.*, 1998; Kurata *et al.*, 2002; Zapater *et al.*, 2002; Rengelshausen *et al.*, 2003; Tanaka *et al.*, 2003) glyburide (glibenclamide) (Lilja *et al.*, 2007)
Clotrimazole	Tacrolimus (Vasquez *et al.*, 2005)
Cremophor EL	Saquinavir (Martin-Facklam *et al.*, 2002)
Cremophor RH40	Digoxin (Tayrouz *et al.*, 2003)
Curcumin	Talinolol (Juan *et al.*, 2007)
Cyclosporine	Atorvastatin (Asberg *et al.*, 2001; Lemahieu *et al.*, 2005) cerivastatin (Mück *et al.*, 1999), docetaxel (Malingré *et al.*, 2001a), doxorubicin (Bartlett *et al.*, 1994; Rushing *et al.*, 1994) etoposide (Lum *et al.*, 2000; Lum *et al.*, 1992) idarubicin (Pea *et al.*, 1999; Smeets *et al.*, 2001) irinotecan (Innocenti *et al.*, 2004), paclitaxel (Meerum Terwogt *et al.*, 1999; Britten *et al.*, 2000; Meerum Terwogt *et al.*, 1998) prednisolone (Potter *et al.*, 2004), saquinavir (Brinkman, Huysmans and Burger, 1998), sirolimus (Zimmerman *et al.*, 2003), sitagliptin (Krishna *et al.*, 2007), vincristine (Bertrand *et al.*, 1992)
Diltiazem	Methylprednisolone (Booker *et al.*, 2002)
Docetaxel	Cyclosporine (Malingré *et al.*, 2001b), vinorelbine (Airoldi *et al.*, 2003)
Elacridar (GF120918)	Doxorubicin (Planting *et al.*, 2005), paclitaxel (Malingré *et al.*, 2001c)

TABLE 6.5 *Continued*

Inhibitor (precipitant)	Object ("victim") drug
Erythromycin	Atorvastatin (Siedlik *et al.*, 1999), fexofenadine (Petri *et al.*, 2006), simvastatin (Kantola, Kivistö and Neuvonen, 1998), talinolol (Schwarz *et al.*, 2000), ximelagatran (Eriksson *et al.*, 2006), digoxin (Eberl *et al.*, 2007)
Gemfibrozil	Loperamide (Niemi *et al.*, 2006)
Grapefruit juice	Dextromethorphan (Di Marco *et al.*, 2002), erythromycin (Kanazawa, Ohkubo and Sugawara, 2001), fexofenadine (Glaeser *et al.*, 2007), talinolol (Schwarz *et al.*, 2005)
Itraconazole	Celiprolol (Lilja *et al.*, 2003), cimetidine (Karyekar *et al.*, 2004), digoxin (Partanen, Jalava and Neuvonen, 1996; Jalava, Partanen and Neuvonen, 1997) fexofenadine (Shon *et al.*, 2005; Shimizu *et al.*, 2006; Shimizu *et al.*, 2006; Uno *et al.*, 2006) loperamide (Niemi *et al.*, 2006), paroxetine (Yasui-Furukori *et al.*, 2007), quinidine (Kaukonen, Olkkola and Neuvonen, 1997)
Ketoconazole	Cyclosporine (Akhlaghi *et al.*, 2001), irinotecan (Kehrer *et al.*, 2002), ritonavir (Khaliq *et al.*, 2000), saquinavir (Khaliq *et al.*, 2000), tacrolimus (Floren *et al.*, 1997; Tuteja *et al.*, 2001)
Lansoprazole	Erythromycin (Frassetto *et al.*, 2007)
Lopinavir and ritonavir	Fexofenadine (Van Heeswijk *et al.*, 2006)
Ms209	Docetaxel (Diéras *et al.*, 2005)
Nelfinavir	Azithromycin (Amsden *et al.*, 2000), cyclosporine (Frassetto *et al.*, 2005)
Oc144-093 (Ont-093)	Docetaxel (Kuppens *et al.*, 2005), paclitaxel (Chi *et al.*, 2005)
Omeprazole	Saquinavir (Winston *et al.*, 2006)
Orange juice	Celiprolol (Lilja, Juntti-Patinen and Neuvonen, 2004), dextromethorphan (Di Marco *et al.*, 2002)
Paclitaxel	Doxorubicin (Moreira *et al.*, 2001), epirubicin (Fogli *et al.*, 2002)
Pilsicainide	Cetirizine (Tsuruoka *et al.*, 2006)
Posaconazole	Cyclosporine (Sansone-Parsons *et al.*, 2007), tacrolimus (Sansone-Parsons *et al.*, 2007)
Probenecid	Fexofenadine (Yasui-Furukori *et al.*, 2005)
Quinidine	Fentanyl (Sadeque *et al.*, 2000; Kharasch *et al.*, 2004) loperamide (Skarke *et al.*, 2003), morphine (Kharasch *et al.*, 2003)
Rifampin	Ezetimibe (Oswald *et al.*, 2006)
Ritonavir	Buprenorphine (McCance-Katz *et al.*, 2006), digoxin (Penzak *et al.*, 2004; Ding *et al.*, 2004) fexofenadine (Van Heeswijk *et al.*, 2006), Indinavir (Haas *et al.*, 2003), loperamide (Tayrouz *et al.*, 2001), prednisone (Penzak *et al.*, 2005), saquinavir (Buss *et al.*, 2001)
Saquinavir	Cyclosporine (Brinkman, Huysmans and Burger, 1998)
St. John's wort extract (*Hypericum perforatum*)	Fexofenadine (Wang *et al.*, 2002)
Surfactant TPGS	Talinolol (Bogman *et al.*, 2005)
Talinolol	Digoxin (Westphal *et al.*, 2000)
Tariquidar (XR9576)	Erythromycin (Kurnik, Wood and Wilkinson, 2006)
Telithromycin	Digoxin (Nenciu, Laberge and Thirion, 2006)
Tenofovir	Saquinavir (Chittick *et al.*, 2006)
Valspodar (PSC 833)	Dexamethasone (Kovarik *et al.*, 1998), digoxin (Kovarik *et al.*, 1999), doxorubicin (Sonneveld *et al.*, 1996; Giaccone *et al.*, 1997; Advani *et al.*, 2001; Minami *et al.*, 2001) etoposide (Boote *et al.*, 1996), paclitaxel (Patnaik *et al.*, 2000; Chico *et al.*, 2001)

TABLE 6.5 *Continued*

Inhibitor (precipitant)	Object ("victim") drug
Verapamil	Doxorubicin (Kerr *et al.*, 1986), fexofenadine (Tannergren *et al.*, 2003; Yasui-Furukori *et al.*, 2005) ranolazine (Lemma *et al.*, 2006), risperidone (Nakagami *et al.*, 2005), simvastatin (Kantola, Kivistö and Neuvonen, 1998)
Zosuquidar (LY335979)	Doxorubicin (Sandler *et al.*, 2004), vinorelbine (Lê *et al.*, 2005)

Inducer	Object ("victim") drug
Anticonvulsants inducers	Irinotecan (Crews *et al.*, 2002)
Antivirals	Paclitaxel (Nannan Panday *et al.*, 1999)
Avasimibe	Digoxin (Sahi *et al.*, 2003)
Clarithromycin	Colchicine (Rollot *et al.*, 2004)
Corticosteroids	Tacrolimus (Anglicheau *et al.*, 2003)
Cyclosporine	Cerivastatin (Mück *et al.*, 1999)
Docetaxel	Cyclosporine (Malingré *et al.*, 2001b)
Efavirenz	Erythromycin (Mouly *et al.*, 2002)
Itraconazole	Cimetidine (Karyekar *et al.*, 2004)
Ketoconazole	Cyclosporine (Akhlaghi *et al.*, 2001)
Lopinavir	Amprenavir (Basso *et al.*, 2002; Kashuba *et al.*, 2005)
Lopinavir and ritonavir	Phenytoin (Lim *et al.*, 2004)
Mefloquine	Ritonavir (Khaliq *et al.*, 2001)
Methylprednisolone	Tacrolimus (Shimada *et al.*, 2002)
Nelfinavir or indinavir	Cyclosporine (Frassetto *et al.*, 2005)
Phenytoin	Lopinavir (Lim *et al.*, 2004), ritonavir (Lim *et al.*, 2004)
Pilsicainide	Cetirizine (Tsuruoka *et al.*, 2006)
Posaconazole	Cyclosporine (Sansone-Parsons *et al.*, 2007)
Prednisone	Etoposide (Kishi *et al.*, 2004)
Red wine	Cyclosporine (Tsunoda *et al.*, 2001)
Rifampin	(R)-Talinolol (Zschiesche *et al.*, 2002), (S)-talinolol (Zschiesche *et al.*, 2002), carvedilol (Giessmann *et al.*, 2004), celecoxib (Jayasagar *et al.*, 2003), celiprolol (Lilja, Niemi and Neuvonen, 2004), dicloxacillin (Putnam *et al.*, 2005), digoxin (Greiner *et al.*, 1999; Drescher *et al.*, 2003; Larsen *et al.*, 2007) erythromycin (Paine *et al.*, 2002a), ezetimibe (Oswald *et al.*, 2006), fexofenadine (Hamman *et al.*, 2001), indinavir (Justesen *et al.*, 2004), linezolid (Egle *et al.*, 2005), morphine (Fromm *et al.*, 1997), tacrolimus (Hebert *et al.*, 1999), talinolol (Westphal *et al.*, 2000)
Saquinavir	Cyclosporine (Brinkman, Huysmans and Burger, 1998)
St. John's wort extract (*Hypericum perforatum*)	(R)-Verapamil (Tannergren *et al.*, 2004), (S)-verapamil (Tannergren *et al.*, 2004), amitriptyline (Johne *et al.*, 2002), cyclosporine (Barone *et al.*, 2000, 2001; Breidenbach *et al.*, 2000; Mai *et al.*, 2000; Ruschitzka *et al.*, 2000; Bauer *et al.*, 2003; Dresser *et al.*, 2003) digoxin (Johne *et al.*, 1999; Dürr *et al.*, 2000) fexofenadine (Xie *et al.*, 2005), simvastatin (Sugimoto *et al.*, 2001), tacrolimus (Mai *et al.*, 2003), talinolol (Schwarz *et al.*, 2007)
Tenofovir	Atazanavir (Taburet *et al.*, 2004)
Tipranavir	Loperamide (Mukwaya *et al.*, 2005)

TABLE 6.6   The expected effect of a putative inhibitor on the systemic exposure (AUC) of a drug that is a substrate for efflux or uptake transporters in the intestine, kidney, liver, or brain.

Tissue	Transporter	Membranelocalization	Drug substrate transport direction	Change in drug substrate AUC by inhibitors
Intestine	MDR1 (P-gp)	Brush border	Efflux into lumen	↑
	MRP2	Brush border	Efflux into lumen	↑
	MRP3	Basolateral	Efflux into blood	↓
	BCRP	Brush border	Efflux into lumen	↑
	OATP2B1	Brush border	Uptake from lumen	↓
	PEPT1/2	Brush border	Uptake from lumen	↓
Kidney	MDR1 (P-gp)	Brush border	Efflux into lumen	↑
	MRP2	Brush border	Efflux into lumen	↑
	MRP4	Brush border	Efflux into lumen	↑
	OCTN1/2	Brush border	Efflux into lumen	↑
	OAT4	Brush border	Efflux into lumen	↑
	OAT1/2/3	Basolateral	Uptake from blood	↑
	OCT2	Basolateral	Uptake from blood	↑
Liver	MDR1 (P-gp)	Canalicular	Efflux into bile	↑
	MRP2	Canalicular	Efflux into bile	↑
	BSEP	Canalicular	Efflux into bile	↑
	BCRP	Canalicular	Efflux into bile	↑
	MRP1/3	Sinusoidal	Efflux into blood	↓
	OATP1A2/1B1/ 1B3/2B1	Sinusoidal	Uptake from blood	↑
	OCT1	Sinusoidal	Uptake from blood	↑
	OAT2	Sinusoidal	Uptake from blood	↑
Brain	MDR1 (P-gp)	Luminal	Efflux into luminal blood	↑[a]
	MRP1/4/5	Luminal	Efflux into luminal blood	↑[a]
	BCRP	Luminal	Efflux into luminal blood	↑[a]
	OATP1A2	ND	ND	ND

[a]Drug concentration in brain.
ND, not determined.

inhibitor on the systemic exposure (area under the curve, AUC) of a drug that is a substrate of efflux or uptake transporters in the intestine, kidney, liver, or brain are summarized in Table 6.6.

6.3.1   P-Glycoprotein

Since P-gp is present on the apical membrane of various tissues including the intestinal epithelia, renal proximal tubules, hepatocytes, and BBB, drug–drug interactions via inhibition or induction of P-gp can thus occur at the respective sites leading to altered pharmacokinetics or metabolic disposition of the concomitant drugs. Classic examples are P-gp mediated drug–drug interactions involving the prototype P-gp substrate digoxin, and concomitant drugs that are P-gp inhibitors such as quinidine, talinolol, clarithromycin, itraconazole, and atorvastatin. Since digoxin can be administered intravenously or orally, the effect of P-gp inhibitors on digoxin pharmacokinetics can be attributed to effect on systemic clearance (biliary or renal) or absorption (intestinal). Inhibition of P-gp-mediated biliary or renal clearance, or inhibition of intestinal P-gp-mediated efflux, was thought to result in decreased systemic clearance or increased oral absorption, leading to elevated systemic exposure (AUCs) of digoxin. For drugs with narrow therapeutic windows such as digoxin, anticancer agents, and immunosuppressants, adverse effects resulting from P-gp-related drug interactions are particularly problematic. Although the role of P-gp in limiting brain penetration of drugs has been shown in animal models using mdr1a (–/–) mice (Seelig, Gottschlich and Devant, 1994; Schinkel *et al.*, 1996), there are few examples where clinical drug–drug interactions were demonstrated at the BBB. Using radiolabeled verapamil as the P-gp substrate and cyclosporine A (CsA) as the P-gp inhibitor, the brain:blood AUC ratio of radioactivity in humans was shown to increase by about 88% at 2.8-μM pseudo-steady-state CsA blood concentration (Sasongko *et al.*, 2005). The authors concluded that the extent of P-gp-based drug interactions at the human BBB at therapeutic CsA blood concentrations is not likely to be very pronounced.

6.3.2 BCRP

Like P-gp, drug–drug interactions involving inhibition of BCRP could occur at various organ sites including the liver, kidney, and small intestine due to the expression of BCRP in these organs. GF120918 is a potent inhibitor of BCRP and P-gp, and increases the oral bioavailability and plasma AUC of topotecan by two- to threefold (Kruijtzer *et al.*, 2002). Since topotecan is a substrate for BCRP and is only a weak-to-moderate substrate of P-gp, the effect of GF120918 on the pharmacokinetics of topotecan was thought to be due to inhibition of BCRP-mediated intestinal efflux leading to increased oral drug absorption.

6.3.3 OATP

Organic anion-transporting polypeptides (OATPs) are a superfamily of uptake transporters with wide tissue distribution including the liver, gut, and BBB. They are responsible for the transport of a variety of amphipathic organic anions such as steroid conjugates, bile acids, and drugs. OATP1B1 has been implicated as a factor in the interactions between fruit juices and fexofenadine (Dresser *et al.*, 2002), in which the systemic exposure of orally administered fexofenadine is reduced by approximately 30% when taken with grapefruit, apple, or orange juice. The mechanism appears to be the inhibition of OATP1B1-mediated intestinal absorption of fexofenadine by fruit juices and this has been demonstrated using cell lines expressing OATPs (Cvetkovic *et al.*, 1999). Hepatic drug–drug interactions that might involve OATP1B1 have also been reported. Cyclosporine A, an OATP1B1 inhibitor, increased the plasma AUC of cerivastatin by almost fourfold (Mück *et al.*, 1999). Gemfibrozil, an inhibitor of OATP1B1 and also of MRP2 (Sasaki *et al.*, 2002), increased the plasma AUC of HMG-CoA reductase inhibitors including simvastatin, lovastatin, and cerivastatin by at least twofold (Backman *et al.*, 2000; Kyrklund *et al.*, 2001; Backman *et al.*, 2002). Genetic polymorphisms in OATP1A2 expression has been shown, which appeared to be ethnicity-dependent (Lee *et al.*, 2005) and may contribute to interindividual variability in drug disposition.

6.3.4 OAT

Since OATs are responsible for the renal excretion of a wide variety of anionic drugs, inhibition of OATs may result in decreased clearance and increased AUCs of drugs, the elimination of which is primarily due to renal clearance. For example, the OAT1 inhibitor probenecid (Hosoyamada *et al.*, 1999) reduced the renal excretion of benzyl-penicillin, acyclovir, cepahalosporins, and cidofovir (Roberts *et al.*, 1981; Overbosch *et al.*, 1988; Tsuji *et al.*, 1990; Cundy, 1999), presumably by inhibiting the active tubular secretion of these drugs. Since OATs are also present in the liver and brain, there remains the possibility for drug–drug interactions to occur by reducing hepatic or brain drug uptake via inhibition of OATs.

6.3.5 OCTs

Like OATs, OCTs are responsible for the renal excretion of a wide variety of cationic drugs; inhibition of OCTs therefore may decrease the clearance and increase AUC of drugs where renal drug clearance is primarily responsible for drug elimination. Examples of drug–drug interactions involving OCTs are the increase in AUC and decrease in renal clearance of procainamide and metformin (Somogyi, McLean and Heinzow, 1983; Somogyi *et al.*, 1987) when coadministered with cimetidine, itself a substrate for active renal excretion (Drayer *et al.*, 1982). Metformin, a biguanide antihyperglicemic agent used for the treatment of type 2 diabetes mellitus, was taken up into the liver by OCT1. Genetic variation in OCT1 affects the response to metformin presumably due to variability in liver drug concentration achieved (Reitman and Schadt, 2007; Shu *et al.*, 2007).

6.3.6 MRP and BSEP

Instead of adverse effects that may result from transporter-mediated drug–drug interactions, MRP2 and BSEP have been implicated in drug and endogenous substrate-induced toxicity. Identification of ligands that are substrates or inhibitors of MRP2 or BSEP may help to reduce the possibility of drug-induced toxicity. MRP2 is involved in the biliary excretion of anion drugs and their conjugates at the bile canalicular membrane. It is believed that the toxic effects of methotrexate in the intestines result from the active excretion of methotrexate into the bile by MRP2, with subsequent accumulation in the intestine leading to the observed toxicity (Masuda *et al.*, 1997). The excretion of the reactive glucuronides of diclofenac by MRP2 and its accumulation in bile is also thought to contribute to the toxic effects diclofenac has on bile canalicular membranes (Seitz and Boelsterli, 1998). Inhibition of BSEP has been implicated in drug-induced cholestasis caused by intracellular accumulation of toxic bile salts (taurocholate). Inhibition of BSEP-mediated transport of bile salt is also suggested as a mechanism responsible for cholestasis caused by cyclosporine and estradiol-17β-glucuronides (Stieger *et al.*, 2000). Other examples of transporter involvement in toxicity include troglitazone, a diabetic agent that was withdrawn from the market due to liver toxicity. It is believed the toxicity was via a cholestatic mechanism, since troglitazone and its sulfate conjugate have been shown to inhibit BSEP-mediated taurocholate transport (Funk *et al.*, 2001).

6.4 POLYMORPHISMS AND REGULATION OF DRUG TRANSPORTERS

Polymorphisms in various drug transporters have been identified including MDR1, MRP2, OATPs, OATs, and OCTs (Kerb, Hoffmeyer and Brinkmann, 2001; Kim *et al.*, 2001; Suzuki and Sugiyama, 2002; Tirona and Kim, 2002; Ieiri, Takane and Otsubo, 2004; Marzolini *et al.*, 2004; Ho and Kim, 2005). Unlike drug metabolism enzymes where polymorphisms have well-documented effects on drug pharmacokinetics and disposition, the clinically relevant effects of drug transporter polymorphisms are not as common and remain to be firmly established. An example of variation of drug response due to drug transporter genetic variation was suggested for OCT1, where reduced function polymorphisms of OCT1 in individuals were thought to affect the response to metformin, a biguanide for the treatment of type 2 diabetes and the hepatic uptake of which is mediated by OCT1 (Shu *et al.*, 2007). Drug transporters like cytochrome P450 (CYP) enzymes are subject to modulations via nuclear receptors. The biology and functions of nuclear receptors are discussed in great detail in separate chapters and will only be briefly described herein. Nuclear receptors that are known to be involved in the regulation of various drug transporters include pregnane X receptor (PXR), constitutive androstane receptor (CAR), glucocorticoid receptor (GR), hepatocyte nuclear factor 4α (HNF4α), farnesoid X receptor (FXR) (Lee *et al.*, 2006), and vitamin D receptor (VDR). Examples of drug transporters that are regulated by nuclear receptors, the nuclear receptors involved, and their ligands are summarized in Table 6.7.

6.4.1 P-Glycoprotein (P-gp)

P-gp was shown to be regulated by both PXR and CAR. Coadministration of PXR ligands, such as rifampin, induces P-gp protein expression and decreased digoxin plasma

TABLE 6.7 Regulation of drug transporters: nuclear receptors and ligands.[a]

Target gene	Nuclear receptor	Ligands	Effect on target
MDR1	PXR	Flucloxacillin, artemisinin, rifampin,	Up
	CAR	hyperforin	Up
MRP2	PXR	Rifampin, hyperforin, phenobarbital,	Up
	CAR	chenodeoxycholic acid	Up
	FXR		Up
MRP4	CAR	Phenobarbital	Up
BSEP	FXR	Cholic acid, chenodeoxycholic acid,	Up
		lithocholic acid	Down
BCRP	PPARγ	Rosiglitazone	Up
		GW9662	Down
NTCP	GR	Dexamethasone	Up
OATP8	FXR	Chenodeoxycholic acid	Up
OATP-A	PXR		Up

[a]Adapted from Urquhart, Tirona and Kim (2007).

levels (Greiner *et al.*, 1999; Geick, Eichelbaum and Burk, 2001). Administration of St. John's wort extract also decreased digoxin plasma levels and increased duodenal P-gp expression (Dürr *et al.*, 2000), presumably via activation of PXR. Antimalarial artemisinin drugs were shown to induce P-gp expression by activation of PXR and CAR (Burk *et al.*, 2005a, b). Certain P-gp substrates such as cyclosporine, amitriptyline, and verapamil were shown to have elevated plasma levels during inflammation (Pheterson *et al.*, 1987; Chen *et al.*, 1994; Mayo *et al.*, 2000). Reduced transporter expression, as a result of decreased expression of PXR in the liver, was thought to be a possible mechanism leading to increased systemic exposure of P-gp substrates during an inflammation response.

6.4.2 BCRP

The BCRP was recently shown to be regulated by the nuclear receptor peroxisome proliferators-activated receptor gamma (PPARγ) in human dendritic cells (Szatmari *et al.*, 2006), as evident by increased BCRP mRNA and protein expression following treatment with the PPARγ agonist rosiglitazone.

6.4.3 MRP2

MRP2 has been shown to be regulated by PXR, CAR, and FXR. Rifampicin and hyperforin (PXR ligands), phenobarbital (a CAR activator), and chenodeoxycholic acid (a FXR ligand) increased the expression of MRP2 (Kast *et al.*, 2002). Drug–drug interactions involving induction of MRP2 though remain to be demonstrated.

6.4.4 BSEP

The BSEP is regulated by FXR. BSEP expression was shown to be up-regulated by the FXR ligand chenodeoxycholic acid and decreased by the FXR antagonist lithocholic acid (Ananthanarayanan *et al.*, 2001; Yu *et al.*, 2002). FXR, together with its target genes, is essential in maintaining normal liver function. For example, individuals affected with progressive familial intrahepatic cholestasis type I, which was caused by mutations in BSEP, were shown to have decreased expression of FXR (Chen *et al.*, 2002; Alvarez *et al.*, 2004).

6.4.5 OATP

OATP1A2 was suggested to be a target gene for PXR in breast carcinoma due to a strong correlation between OATP1A2 and PXR expression (Miki *et al.*, 2006). The FXR ligand chenodeoxycholic acid was shown to up-regulate OATP1B3 in a human hepatoma cell line (Korjamo *et al.*, 2005).

6.5 *IN VITRO* METHODS IN EVALUATION OF DRUG TRANSPORTERS

In vitro and *in vivo* methods to assess the involvement of drug transporters in drug disposition and interactions have been the topic of detailed discussions in several recent reviews (Lin, 2007; Xia, Milton and Gan, 2007). Several common *in vitro* approaches to assess transporter involvement are discussed here. A variety of *in vitro* and *in vivo* models are available to assess the role of different transporters in drug disposition. Among *in vitro* systems, the colon carcinoma cell line (Caco-2) is one of the most widely used models for predicting intestinal drug absorption (Polli *et al.*, 2001; Korjamo *et al.*, 2005). Caco-2 cells express a variety of drug transporters including MDR1/P-gp and MRP2, and can be used to assess drug–drug interactions involving multiple transporters. However, the presence of multiple transporters can also be a limitation because it can be difficult to determine the involvement of a particular transporter, unless there are specific inhibitors and/or substrates available. As a result, Caco-2 cells are used mainly to evaluate P-gp interactions in a monolayer-based format, with selective inhibitors or substrates of P-gp. Various factors such as culture time, passage number, and culture conditions can affect P-gp expression; hence, these conditions need to be optimized and standardized to ensure data reproducibility. Other cell lines include canine kidney cell line (MDCK). Their advantage over Caco-2 cells is shorter culture time and the fact that they can be specifically transfected to overexpress human P-gp, thus increasing sensitivity and specificity (Polli *et al.*, 2001). Sandwich-cultured hepatocytes represent a potentially useful *in vitro model* to study drug transporters (Annaert *et al.*, 2001; Annaert and Brouwer, 2005). Biliary excretion in long-term sandwich-cultured rat hepatocytes has been shown to correlate well with *in vivo* biliary excretion, though evaluation of *in vitro–in vivo* correlation in humans remains difficult. Sandwich-cultured hepatocytes may also allow the study of the biliary excretion of metabolites, provided the relevant metabolic activities are maintained.

Efforts in cloning and transgenic technologies have also been put toward understanding transporter biology and their importance in drug discovery and development. Single-transfected cells that stably or transiently express individual transporters are used to determine whether a compound is a substrate or inhibitor of particular transporters (Sasaki *et al.*, 2004; Lee *et al.*, 2005). Double-transfected cells or monolayers that stably express human or rat uptake and efflux transporters have been developed (Sasaki *et al.*, 2004) to better understand the synergistic role of uptake and efflux transporters under *in vivo* conditions. Although these expression systems have the advantages of allowing the qualitative identification of specific drug transporters involved in a drug's disposition, quantitative prediction of *in vitro–in vivo* correlations remains to be established before these systems can gain widespread applications.

In addition to cell-based systems, canalicular membrane vesicles (CMVs) are used to assess transport of substances and drugs into the bile (Ishizuka *et al.*, 1999; Shilling *et al.*, 2006). CMVs have been shown to provide a good prediction of transporter-mediated biliary drug clearance. CMVs express many of the efflux transporters found on the canalicular membrane of hepatocytes, including P-gp, MRP2, and BSEP. CMVs can be prepared from livers of various species using similar procedures and therefore allow for the simultaneous assessment of species differences or similarities. One limitation of CMVs, however, is variability in quality between different livers (human and nonhuman primates), and strict adherence to protocols to ensure reproducible expression of the various transporters is critical. Sinusoidal membrane vesicles (SMVs) are prepared from the sinusoidal (basolateral) membrane of hepatocytes and are used to evaluate transporters expressed on the basolateral membranes of hepatocytes (Mizuno *et al.*, 2003). They are, however, not as widely used as more validated systems. Scientists from the FDA have published their perspective on using *in vitro* methods to assess whether a drug candidate is a substrate or inhibitor of efflux transporters such as P-gp, MRP2, or BCRP (Zhang *et al.*, 2006). The preferred methods were monolayer methods such as Caco-2 or MDCK cells expressing the transporter of interest, while *in vitro* methods such as the ATPase assays (Ramachandra *et al.*, 1996; Taguchi *et al.*, 1997) were deemed not as informative.

Like CYP enzymes, the inhibition of drug transporters can result in changes in drug clearance and in drug exposure, leading to either adverse effects or inadequate pharmacological activity (Mizuno *et al.*, 2003; Beringer and Slaughter, 2005). Since P-gp is the most well-understood drug transporter with respect to its role in clinical drug–drug interactions, considerable efforts have been put into identifying and characterizing P-gp substrates or inhibitors during drug discovery and development, using *in vitro* models such as bidirectional transport employing monolayers (e.g. Caco-2 and MDCK cells), uptake/efflux of fluorescent or radiolabeled probes (e.g. Calcein-AM and rhodamine-123), or simulation of ATPase activity using membrane vesicles derived from tissues or cells expressing P-gp (Polli *et al.*, 2001). Bidirectional transport using monolayers expressing P-gp is considered the most reliable method for identifying P-gp substrates or inhibitors, since the method has the advantages of allowing the direct measurement of efflux across the cell barrier and evaluation of P-gp involvement. Similar to studies involving CYPs, criteria can be set up to identify whether a drug is a substrate (e.g. efflux ratios >2) or an inhibitor (e.g. IC_{50} values in inhibiting digoxin efflux) of P-gp. However, unlike CYPs, the extrapolation of *in vitro* P-gp inhibition data to clinical situations has not been well established, in part due to the wide tissue distribution of P-gp (i.e. intestinal lumen, liver, kidney, and BBB) and that relevant inhibitor concentration that needs to be considered (Bjornsson *et al.*, 2003).

Drug transporters have been recognized to play an active role in renal and biliary excretion of many xenobiotics and endogenous substances. Since *in vivo* studies of renal or biliary excretion may not be always readily conducted during the drug discovery phase, and realizing that species differences exist in renal or biliary drug clearance, *in vitro* approaches in assessing renal or biliary clearance have been proposed, and may provide guidance on the significance of these pathways in drug clearance. Together with physicochemical properties such as molecular weight, solubility, and permeability, the identification of transporter involvement (particularly for efflux transporters P-gp and MRP2) greatly enhances our capability to predict compounds that are renally or biliary excreted. Considerable efforts continue to be exerted during the drug discovery phase to identify and select drug candidates that are not (or poor) P-gp substrates in an attempt to improve oral absorption, and for CNS-targeted drugs, to improve brain penetration. The absence of P-gp-mediated efflux, together with high passive permeability, can aid in the selection of CNS drugs (Mahar Doan *et al.*, 2002).

6.6 TRANSPORTERS–DRUG-METABOLIZING ENZYMES INTERPLAY

Due to the overlapping substrate and inhibitor selectivity for CYP3A and P-gp, it has become increasingly recognized that drug–drug interactions may also be mediated by this transporter–enzyme interplay. Wu and Benet (2005) recently proposed a revised Biopharmaceutics Classification System (BCS) that may be useful in predicting overall drug disposition, including routes of elimination and effects of efflux and uptake transporters on oral drug absorption, and the clinical significance of transporter–enzyme interplay on oral bioavailability and drug–drug interactions. Transporter–drug-metabolizing enzyme interplay, primarily involving P-gp-CYP3A4 in the intestine, is thought to limit the oral absorption of certain drugs. Though *in vitro* systems, such as cell monolayers expressing CYP3A4 and P-gp, to assess the role of transporter–enzyme interplay on drug absorption have been used (Paine *et al.*, 2002a, b; Cummins *et al.*, 2004), the quantitative extrapolation of *in vitro* data to *in vivo* situations remains to be established.

It has been suggested (Wu and Benet, 2005) that drug–drug interactions as a result of transporter–enzyme interplay might be more significant for drugs with low solubility and extensive metabolism (class 2 drugs). An example of a drug where transporter–enzyme interplay may play an important role in its metabolic disposition is sirolimus, a macrolide antibiotic used as an immunosuppressive agent to prevent allograft rejection in kidney transplantation. Sirolimus was metabolized by CYP3A and a majority of the administered dose was excreted into the bile (Leung *et al.*, 2006). Preclinical *in vitro* studies, using models including liver microsomes and Caco-2 cell monolayers expressing CYP3A, indicated that sirolimus was susceptible to macrolide ring opening via hydrolysis/dehydration, and cytochrome P450 CYP3A-mediated oxidation and P-gp-mediated efflux (Paine *et al.*, 2002b). *In vitro* results therefore suggested that sirolimus would be subject to CYP3A- and non-CYP-mediated metabolic clearance, as well as P-gp-mediated biliary excretion. Moreover, its oral bioavailability could be limited by intestinal and hepatic CYP3A-catalyzed first-pass effect and intestinal P-gp-mediated efflux. Inhibition of P-gp in a CYP3A4-transfected Caco-2 cellular system in the apical to basolateral direction decreased the extraction ratio of sirolimus, suggesting that inhibition of intestinal drug efflux can lead to decreased intestinal metabolism of the concomitant drug due to decreased access of the drug to the enzyme by reducing recycling (Paine *et al.*, 2002b; Cummins *et al.*, 2004). In contrast, inhibition of P-gp in the basolateral to apical direction increased the extraction ratio of sirolimus, suggesting that inhibition of hepatic drug efflux can increase the extent of hepatic metabolism of the concomitant drug due to increased access of the drug to the active enzyme. An example of a drug where lack of proper integration of drug–drug interaction data can lead to drastic consequence is mibefradil, a calcium channel blocker used for hypertension, which was withdrawn from the market in 1998 due to serious drug interactions with concomitant medications that are CYP3A or P-gp substrates. Continued research is needed in this area to further demonstrate the clinical significance of transporter–enzyme interplay, and to assess approaches to predict likely clinical outcome from *in vitro* methodologies for drugs that are inhibitors of transporters but not of CYP enzymes.

6.7   OUTLOOK

Despite advances made in the drug transporter field in the last decade, further understanding and improved tools in certain areas are needed. An area of continued importance is understanding and prediction of clinical drug–drug interactions involving concomitant substrates and inhibitors of drug transporters. Due to the ubiquitous nature of drug transporters and possible sites of substrate–inhibitor interaction including absorption (intestinal lumen), distribution (BBB, cell uptake), and elimination (liver, kidney), challenges continue with respect to the usage of the most appropriate *in vivo* and *in vitro* models and parameters in the prediction of clinical outcome. Although parameters such *in vitro* K_i and plasma inhibitor concentrations have been proposed to predict drug transporter-mediated drug–drug interactions (Zhang *et al.*, 2006; Zhang *et al.*, 2008), the appropriateness of the model remains unclear due to the complex nature of drug transporter kinetics that do not necessarily follow simple

Michaelis–Menten kinetics. There remains limited published information on species differences in drug transporters, though differences in transporter activity, such as P-gp, MRP2, OATPs, and OATs, were observed between human and other animal species (Lin and Yamazaki, 2003; Mizuno *et al.*, 2003; Van Montfoort *et al.*, 2003; Shilling *et al.*, 2006). Therefore, due to observed or potential species differences in various drug transporter activities, the use of animal models to extrapolate to human situations should be performed with caution.

The study of the regulation of drug transporters has made tremendous advances in the past decade, and will continue to be an area of active research. Recognition of the role of nuclear receptors in drug transporter expression, and their associated polymorphism, has provided insight into the molecular mechanism for transporter-mediated drug–drug and drug–endogenous substrate interactions and drug toxicity, and provided the opportunity of individualized treatment based on interindividual variability in drug disposition and response. Inflammation-mediated changes in drug disposition have been noted, and might be related to decreased mRNA and protein expressions of drug transporters such as P-gp, MRPs, or OATPs during periods of inflammation (Petrovic, Teng and Piquette-Miller, 2007). Future research will need to provide further understanding of the molecular mechanisms involved in the regulation of drug transporters.

The emerging importance of drug transporters, notably P-gp, in drug disposition and drug–drug interactions has recently prompted scientists from the FDA to publish their scientific perspective on the subject matter (Zhang *et al.*, 2006). It was suggested that our current knowledge allowed the use of *in vitro* models, such as a bidirectional transport assay, to evaluate whether a drug candidate is a substrate or inhibitor of P-gp. Positive outcomes from the *in vitro* studies may warrant confirmation in clinical drug interaction studies employing a P-gp inhibitor (if the drug candidate is a potential P-gp substrate) or a P-gp substrate such as digoxin (if the drug candidate is a potential P-gp inhibitor). Due to current limited knowledge or availability of standardized methods, routine *in vitro* evaluation of P-gp induction or other transporter-based interactions was not recommended at this time. Therefore, additional knowledge and established methods for various transporters should continue to be an active area of research.

REFERENCES

Advani, R., Fisher, G.A., Lum, B.L., Hausdorff, J., Halsey, J., Litchman, M., Sikic, B.I. (2001). A phase I trial of doxorubicin, paclitaxel, and valspodar (PSC 833), a modulator of multidrug resistance. *Clin. Cancer Res.*, 7, 1221–1229.

Airoldi, M., Cattel, L., Marchionatti, S., Recalenda, V., Pedani, F., Tagini, V., Bumma, C., Beatrice, F., Succo, G., Maria Gabriele, A. (2003). Docetaxel and vinorelbine in recurrent head and neck cancer: pharmacokinetic and clinical results. *Am. J. Clin. Oncol.*, 26, 378–381.

Akashi, M., Tanaka, A., Takikawa, H. (2006). Effect of cyclosporin A on the biliary excretion of cholephilic compounds in rats. *Hepatol. Res.*, 34, 193–198.

Akhlaghi, F., Keogh, A.M., McLachlan, A.J., Kaan, A. (2001). Pharmacokinetics of cyclosporine in heart transplant recipients receiving metabolic inhibitors. *J. Heart. Lung. Transplant.*, 20, 431–438.

Alrefai, W.A., Gill, R.K. (2007). Bile acid transporters: structure, function, regulation and pathophysiological implications. *Pharm. Res.*, 24, 1803–1823.

Alvarez, L., Jara, P., Sánchez-Sabaté, E., Hierro, L., Larrauri, J., Díaz, M.C., Camarena, C., De la Vega, A., Frauca, E., López-Collazo, E., Lapunzina, P. (2004). Reduced hepatic expression of farnesoid X receptor in hereditary cholestasis associated to mutation in ATP8B1. *Hum. Mol. Genet.*, *13*, 2451–2460.

Amsden, G.W., Nafziger, A.N., Foulds, G., Cabelus, L.J. (2000). A study of the pharmacokinetics of azithromycin and nelfinavir when coadministered in healthy volunteers. *J. Clin. Pharmacol.*, *40*, 1522–1527.

Ananthanarayanan, M., Balasubramanian, N., Makishima, M., Mangelsdorf, D.J., Suchy, F.J. (2001). Human bile salt export pump promoter is transactivated by the farnesoid X receptor/bile acid receptor. *J. Biol. Chem.*, *276*, 28857–28865.

Anglicheau, D., Flamant, M., Schlageter, M.H., Martinez, F., Cassinat, B., Beaune, P., Legendre, C., Thervet, E. (2003). Pharmacokinetic interaction between corticosteroids and tacrolimus after renal transplantation. *Nephrol. Dial. Transplant.*, *18*, 2409–2414.

Annaert, P.P., Brouwer, K.L. (2005). Assessment of drug interactions in hepatobiliary transport using rhodamine 123 in sandwich-cultured rat hepatocytes. *Drug Metab. Dispos.*, *33*, 388–394.

Annaert, P.P., Turncliff, R.Z., Booth, C.L., Thakker, D.R., Brouwer, K.L. (2001). P-glycoprotein-mediated in vitro biliary excretion in sandwich-cultured rat hepatocytes. *Drug Metab. Dispos.*, *29*, 1277–1283.

Anzai, N., Kanai, Y., Endou, H. (2006). Organic anion transporter family: current knowledge. *J. Pharmacol. Sci.*, *100*, 411–426.

Asberg, A., Hartmann, A., Fjeldså, E., Bergan, S., Holdaas, H. (2001). Bilateral pharmacokinetic interaction between cyclosporine A and atorvastatin in renal transplant recipients. *Am. J. Transplant.*, *1*, 382–386.

Backman, J.T., Kyrklund, C., Kivistö, K.T., Wang, J.S., Neuvonen, P.J. (2000). Plasma concentrations of active simvastatin acid are increased by gemfibrozil. *Clin. Pharmacol. Ther.*, *68*, 122–129.

Backman, J.T., Kyrklund, C., Neuvonen, M., Neuvonen, P.J. (2002). Gemfibrozil greatly increases plasma concentrations of cerivastatin. *Clin. Pharmacol. Ther.*, *72*, 685–691.

Baris, N., Kalkan, S., Güneri, S., Bozdemir, V., Guven, H. (2006). Influence of carvedilol on serum digoxin levels in heart failure: is there any gender difference? *Eur. J. Clin. Pharmacol.*, *62*, 535–538.

Barone, G.W., Gurley, B.J., Ketel, B.L., Lightfoot, M.L., Abul-Ezz, S.R. (2000). Drug interaction between St. John's wort and cyclosporine. *Ann. Pharmacother.*, *34*, 1013–1016.

Barone, G.W., Gurley, B.J., Ketel, B.L., Abul-Ezz, S.R. (2001). Herbal supplements: a potential for drug interactions in transplant recipients. *Transplantation*, *71*, 239–241.

Bartlett, N.L., Lum, B.L., Fisher, G.A., Brophy, N.A., Ehsan, M.N., Halsey, J., Sikic, B.I. (1994). Phase I trial of doxorubicin with cyclosporine as a modulator of multidrug resistance. *J. Clin. Oncol.*, *2*, 835–842.

Basso, S., Solas, C., Quinson, A.M., Ravaux, I., Poizot-Martin, I., Bacconier, J., Durand, A., Lacarelle, B. (2002). Pharmacokinetic interaction between lopinavir/r and amprenavir in salvage therapy. *J. Acquir. Immune. Defic. Syndr.*, *31*, 115–117.

Bauer, S., Störmer, E., Johne, A., Krüger, H., Budde, K., Neumayer, H.H., Roots, I., Mai, I. (2003). Alterations in cyclosporin A pharmacokinetics and metabolism during treatment with St John's wort in renal transplant patients. *Br. J. Clin. Pharmacol.*, *55*, 203–211.

Beringer, P.M., Slaughter, R.L. (2005). Transporters and their impact on drug disposition. *Ann. Pharmacother.*, *39*, 1097–1108.

Bertrand, Y., Capdeville, R., Balduck, N., Philippe, N. (1992). Cyclosporin A used to reverse drug resistance increases vincristine neurotoxicity. *Am. J. Hematol.*, *40*, 158–159.

Bjornsson, T.D., Callaghan, J.T., Einolf, H.J., Fischer, V., Gan, L., Grimm, S., Kao, J., King, S.P., Miwa, G., Ni, L., Kumar, G., McLeod, J., Obach, R.S., Roberts, S., Roe, A., Shah, A., Snikeris,

F., Sullivan, J.T., Tweedie, D., Vega, J.M., Walsh, J., Wrighton, S.A. (2003). The conduct of in vitro and in vivo drug–drug interaction studies: a Pharmaceutical Research and Manufacturers of America (PhRMA) perspective. *Drug Metab. Dispos.*, *31*, 815–832.

Bodo, A., Bakos, E., Szeri, F., Varadi, A., Sarkadi, B. (2003). The role of multidrug transporters in drug availability, metabolism and toxicity. *Toxicol. Lett.*, *140*–141, 133–143.

Boffito, M., Kurowski, M., Kruse, G., Hill, A., Benzie, A.A., Nelson, M.R., Moyle, G.J., Gazzard, B.G., Pozniak, A.L. (2004). Atazanavir enhances saquinavir hard-gel concentrations in a ritonavir-boosted once-daily regimen. *AIDS*, *18*, 1291–1297.

Boffito, M., Maitland, D., Dickinson, L., Back, D., Hill, A., Fletcher, C., Moyle, G., Nelson, M., Gazzard, B., Pozniak, A. (2006). Pharmacokinetics of saquinavir hard-gel/ritonavir and atazanavir when combined once daily in HIV Type 1-infected individuals administered different atazanavir doses. *AIDS Res. Hum. Retroviruses*, *22*, 749–756.

Bogman, K., Zysset, Y., Degen, L., Hopfgartner, G., Gutmann, H., Alsenz, J., Drewe, J. (2005). P-glycoprotein and surfactants: effect on intestinal talinolol absorption. *Clin. Pharmacol. Ther.*, *77*, 24–32.

Booker, B.M., Magee, M.H., Blum, R.A., Lates, C.D., Jusko, W.J. (2002). Pharmacokinetic and pharmacodynamic interactions between diltiazem and methylprednisolone in healthy volunteers. *Clin. Pharmacol. Ther.*, *72*, 370–382.

Boote, D.J., Dennis, I.F., Twentyman, P.R., Osborne, R.J., Laburte, C., Hensel, S., Smyth, J.F., Brampton, M.H., Bleehen, N.M. (1996). Phase I study of etoposide with SDZ PSC 833 as a modulator of multidrug resistance in patients with cancer. *J. Clin. Oncol.*, *14*, 610–618.

Breidenbach, T., Kliem, V., Burg, M., Radermacher, J., Hoffmann, M.W., Klempnauer, J. (2000). Profound drop of cyclosporin A whole blood trough levels caused by St. John's wort (Hypericum perforatum). *Transplantation*, *69*, 2229–2230.

Brinkman, K., Huysmans, F., Burger, D.M. (1998). Pharmacokinetic interaction between saquinavir and cyclosporine. *Ann. Intern. Med.*, *129*, 914–915.

Britten, C.D., Baker, S.D., Denis, L.J., Johnson, T., Drengler, R., Siu, L.L., Duchin, K., Kuhn, J., Rowinsky, E.K. (2000). Oral paclitaxel and concurrent cyclosporin A: targeting clinically relevant systemic exposure to paclitaxel. *Clin. Cancer Res.*, *6*, 3459–3468.

Burk, O., Arnold, K.A., Geick, A., Tegude, H., Eichelbaum, M. (2005a). A role for constitutive androstane receptor in the regulation of human intestinal MDR1 expression. *Biol. Chem.*, *386*, 503–513.

Burk, O., Arnold, K.A., Nussler, A.K., Schaeffeler, E., Efimova, E., Avery, B.A., Avery, M.A., Fromm, M.F., Eichelbaum, M. (2005b). Antimalarial artemisinin drugs induce cytochrome P450 and MDR1 expression by activation of xenosensors pregnane X receptor and constitutive androstane receptor. *Mol. Pharmacol.*, *67*, 1954–1965.

Buss, N., Snell, P., Bock, J., Hsu, A., Jorga, K. (2001). Saquinavir and ritonavir pharmacokinetics following combined ritonavir and saquinavir (soft gelatin capsules) administration. *Br. J. Clin. Pharmacol.*, *52*, 255–264.

Chen, H.L., Chang, P.S., Hsu, H.C., Ni, Y.H., Hsu, H.Y., Lee, J.H., Jeng, Y.M., Shau, W.Y., Chang, M.H. (2002). FIC1 and BSEP defects in Taiwanese patients with chronic intrahepatic cholestasis with low gamma-glutamyltranspeptidase levels. *J. Pediatr.*, *140*, 119–124.

Chen, Y.L., Le Vraux, V., Leneveu, A., Dreyfus, F., Stheneur, A., Florentin, I., De Sousa, M., Giroud, J.P., Flouvat, B., Chauvelot-Moachon, L. (1994). Acute-phase response, interleukin-6, and alteration of cyclosporine pharmacokinetics. *Clin. Pharmacol. Ther.*, *55*, 649–660.

Chi, K.N., Chia, S.K., Dixon, R., Newman, M.J., Wacher, V.J., Sikic, B., Gelmon, K.A. (2005). A phase I pharmacokinetic study of the P-glycoprotein inhibitor, ONT-093, in combination with paclitaxel in patients with advanced cancer. *Invest. New Drugs*, *23*, 311–315.

Chico, I., Kang, M.H., Bergan, R., Abraham, J., Bakke, S., Meadows, B., Rutt, A., Robey, R., Choyke, P., Merino, M., Goldspiel, B., Smith, T., Steinberg, S., Figg, W.D., Fojo, T., Bates, S. (2001). Phase

I study of infusional paclitaxel in combination with the P-glycoprotein antagonist PSC 833. *J. Clin. Oncol.*, 19, 832–842.

Chittick, G.E., Zong, J., Blum, M.R., Sorbel, J.J., Begley, J.A., Adda, N., Kearney, B.P. (2006). Pharmacokinetics of tenofovir disoproxil fumarate and ritonavir-boosted saquinavir mesylate administered alone or in combination at steady state. *Antimicrob. Agents. Chemother.*, 50, 1304–1310.

Crews, K.R., Stewart, C.F., Jones-Wallace, D., Thompson, S.J., Houghton, P.J., Heideman, R.L., Fouladi, M., Bowers, D.C., Chintagumpala, M.M., Gajjar, A. (2002). Altered irinotecan pharmacokinetics in pediatric high-grade glioma patients receiving enzyme-inducing anticonvulsant therapy. *Clin. Cancer Res.*, 8, 2202–2209.

Cummins, C.L., Jacobsen, W., Christians, U., Benet, L.Z. (2004). CYP3A4-transfected Caco-2 cells as a tool for understanding biochemical absorption barriers: studies with sirolimus and midazolam. *J. Pharmacol. Exp. Ther.*, 308, 143–155.

Cundy, K.C. (1999). Clinical pharmacokinetics of the antiviral nucleotide analogues cidofovir and adefovir. *Clin. Pharmacokinet.*, 36, 127–143.

Cvetkovic, M., Leake, B., Fromm, M.F., Wilkinson, G.R., Kim, R.B. (1999). OATP and P-glycoprotein transporters mediate the cellular uptake and excretion of fexofenadine. *Drug Metab. Dispos.*, 27, 866–871.

Di Marco, M.P., Edwards, D.J., Wainer, I.W., Ducharme, M.P. (2002). The effect of grapefruit juice and seville orange juice on the pharmacokinetics of dextromethorphan: the role of gut CYP3A and P-glycoprotein. *Life Sci.*, 71, 1149–1160.

Diéras, V., Bonneterre, J., Laurence, V., Degardin, M., Pierga, J.Y., Bonneterre, M.E., Marreaud, S., Lacombe, D., Fumoleau, P. (2005). Phase I combining a P-glycoprotein inhibitor, MS209, in combination with docetaxel in patients with advanced malignancies. *Clin. Cancer Res.*, 11, 6256–6260.

Ding, R., Tayrouz, Y., Riedel, K.D., Burhenne, J., Weiss, J., Mikus, G., Haefeli, W.E. (2004). Substantial pharmacokinetic interaction between digoxin and ritonavir in healthy volunteers. *Clin. Pharmacol. Ther.*, 76, 73–84.

Dorani, H., Schützer, K.M., Sarich, T.C., Wall, U., Logren, U., Ohlsson, L., Eriksson, U.G. (2007). Pharmacokinetics and pharmacodynamics of the oral direct thrombin inhibitor ximelagatran co-administered with different classes of antibiotics in healthy volunteers. *Eur. J. Clin. Pharmacol.*, 63, 571–581.

Drayer, D.E., Romankiewicz, J., Lorenzo, B., Reidenberg, M.M. (1982). Age and renal clearance of cimetidine. *Clin. Pharmacol. Ther.*, 31, 45–50.

Drescher, S., Glaeser, H., Mürdter, T., Hitzl, M., Eichelbaum, M., Fromm, M.F. (2003). P-glycoprotein-mediated intestinal and biliary digoxin transport in humans. *Clin. Pharmacol. Ther.*, 73, 223–231.

Dresser, G.K., Bailey, D.G., Leake, B.F., Schwarz, U.I., Dawson, P.A., Freeman, D.J., Kim, R.B. (2002). Fruit juices inhibit organic anion transporting polypeptide-mediated drug uptake to decrease the oral availability of fexofenadine. *Clin. Pharmacol. Ther.*, 71, 11–20.

Dresser, G.K., Schwarz, U.I., Wilkinson, G.R., Kim, R.B. (2003). Coordinate induction of both cytochrome P4503A and MDR1 by St John's wort in healthy subjects. *Clin. Pharmacol. Ther.*, 73, 41–50.

Dürr, D., Stieger, B., Kullak-Ublick, G.A., Rentsch, K.M., Steinert, H.C., Meier, P.J., St, Fattinger, K. (2000). John's Wort induces intestinal P-glycoprotein/MDR1 and intestinal and hepatic CYP3A4. *Clin. Pharmacol. Ther.*, 68, 598–604.

Eberl, S., Renner, B., Neubert, A., Reisig, M., Bachmakov, I., König, J., Dörje, F., Mürdter, T.E., Ackermann, A., Dormann, H., Gassmann, K.G., Hahn, E.G., Zierhut, S., Brune, K., Fromm, M.F. (2007). Role of p-glycoprotein inhibition for drug interactions : evidence from in vitro and pharmacoepidemiological studies. *Clin. Pharmacokinet.*, 46, 1039–1049.

Egle, H., Trittler, R., Kümmerer, K., Lemmen, S.W. (2005). Linezolid and rifampin: drug interaction contrary to expectations? *Clin. Pharmacol. Ther.*, 77, 451–453.

Endres, C.J., Hsiao, P., Chung, F.S., Unadkat, J.D. (2006). The role of transporters in drug interactions. *Eur. J. Pharm. Sci.*, 27, 501–517.

Eriksson, U.G., Dorani, H., Karlsson, J., Fritsch, H., Hoffmann, K.J., Olsson, L., Sarich, T.C., Wall, U., Schützer, K.M. (2006). Influence of erythromycin on the pharmacokinetics of ximelagatran may involve inhibition of P-glycoprotein-mediated excretion. *Drug Metab. Dispos.*, 34, 775–782.

Floren, L.C., Bekersky, I., Benet, L.Z., Mekki, Q., Dressler, D., Lee, J.W., Roberts, J.P., Hebert, M.F. (1997). Tacrolimus oral bioavailability doubles with coadministration of ketoconazole. *Clin. Pharmacol. Ther.*, 62, 41–49.

Fogli, S., Danesi, R., Gennari, A., Donati, S., Conte, P.F., Del Tacca, M. (2002). Gemcitabine, epirubicin and paclitaxel: pharmacokinetic and pharmacodynamic interactions in advanced breast cancer. *Ann. Oncol.*, 13, 919–927.

Frassetto, L., Baluom, M., Jacobsen, W., Christians, U., Roland, M.E., Stock, P.G., Carlson, L., Benet, L.Z. (2005). Cyclosporine pharmacokinetics and dosing modifications in human immunodeficiency virus-infected liver and kidney transplant recipients. *Transplantation*, 80, 13–17.

Frassetto, L.A., Poon, S., Tsourounis, C., Valera, C., Benet, L.Z. (2007). Effects of uptake and efflux transporter inhibition on erythromycin breath test results. *Clin. Pharmacol. Ther.*, 81, 828–832.

Fromm, M.F., Eckhardt, K., Li, S., Schänzle, G., Hofmann, U., Mikus, G., Eichelbaum, M. (1997). Loss of analgesic effect of morphine due to coadministration of rifampin. *Pain*, 72, 261–267.

Funk, C., Pantze, M., Jehle, L., Ponelle, C., Scheuermann, G., Lazendic, M., Gasser, R. (2001). Troglitazone-induced intrahepatic cholestasis by an interference with the hepatobiliary export of bile acids in male and female rats. Correlation with the gender difference in troglitazone sulfate formation and the inhibition of the canalicular bile salt export pump (Bsep) by troglitazone and troglitazone sulfate. *Toxicology*, 167, 83–98.

Geick, A., Eichelbaum, M., Burk, O. (2001). Nuclear receptor response elements mediate induction of intestinal MDR1 by rifampin. *J. Biol. Chem.*, 276, 14581–14587.

Giaccone, G., Linn, S.C., Welink, J., Catimel, G., Stieltjes, H., van der Vijgh, W.J., Eeltink, C., Vermorken, J.B., Pinedo, H.M. (1997). A dose-finding and pharmacokinetic study of reversal of multidrug resistance with SDZ PSC 833 in combination with doxorubicin in patients with solid tumors. *Clin. Cancer Res.*, 3, 2005–2015.

Giessmann, T., Modess, C., Hecker, U., Zschiesche, M., Dazert, P., Kunert-Keil, C., Warzok, R., Engel, G., Weitschies, W., Cascorbi, I., Kroemer, H.K., Siegmund, W. (2004). CYP2D6 genotype and induction of intestinal drug transporters by rifampin predict presystemic clearance of carvedilol in healthy subjects. *Clin. Pharmacol. Ther.*, 75, 213–222.

Glaeser, H., Bailey, D.G., Dresser, G.K., Gregor, J.C., Schwarz, U.I., McGrath, J.S., Jolicoeur, E., Lee, W., Leake, B.F., Tirona, R.G., Kim, R.B. (2007). Intestinal drug transporter expression and the impact of grapefruit juice in humans. *Clin. Pharmacol. Ther.*, 81, 362–370.

Greiner, B., Eichelbaum, M., Fritz, P., Kreichgauer, H.P., von Richter, O., Zundler, J., Kroemer, H.K. (1999). The role of intestinal P-glycoprotein in the interaction of digoxin and rifampin. *J. Clin. Invest.*, 104, 147–153.

Gupta, S., Banfield, C., Kantesaria, B., Marino, M., Clement, R., Affrime, M., Batra, V. (2001). Pharmacokinetic and safety profile of desloratadine and fexofenadine when coadministered with azithromycin: a randomized, placebo-controlled, parallel-group study. *Clin. Ther.*, 23, 451–466.

Haas, D.W., Johnson, B., Nicotera, J., Bailey, V.L., Harris, V.L., Bowles, F.B., Raffanti, S., Schranz, J., Finn, T.S., Saah, A.J., Stone, J. (2003). Effects of ritonavir on indinavir pharmacokinetics in cerebrospinal fluid and plasma. *Antimicrob. Agents. Chemother.*, *47*, 2131–2137.

Hamman, M.A., Bruce, M.A., Haehner-Daniels, B.D., Hall, S.D. (2001). The effect of rifampin administration on the disposition of fexofenadine. *Clin. Pharmacol. Ther.*, *69*, 114–121.

Hebert, M.F., Fisher, R.M., Marsh, C.L., Dressler, D., Bekersky, I. (1999). Effects of rifampin on tacrolimus pharmacokinetics in healthy volunteers. *J. Clin. Pharmacol.*, *39*, 91–96.

Hirano, M., Maeda, K., Hayashi, H., Kusuhara, H., Sugiyama, Y. (2005). Bile salt export pump (BSEP/ABCB11) can transport a nonbile acid substrate, pravastatin. *J. Pharmacol. Exp. Ther.*, *314*, 876–882.

Ho, R.H., Kim, R.B. (2005). Transporters and drug therapy: implications for drug disposition and disease. *Clin. Pharmacol. Ther.*, *78*, 260–277.

Hosoyamada, M., Sekine, T., Kanai, Y., Endou, H. (1999). Molecular cloning and functional expression of a multispecific organic anion transporter from human kidney. *Am. J. Physiol.*, *276*, F122–128.

Ieiri, I., Takane, H., Otsubo, K. (2004). The MDR1(ABCB1) gene polymorphism and its clinical implications. *Clin. Pharmacokinet.*, *43*, 553–576.

Innocenti, F., Undevia, S.D., Ramírez, J., Mani, S., Schilsky, R.L., Vogelzang, N.J., Prado, M., Ratain, M.J. (2004). A phase I trial of pharmacologic modulation of irinotecan with cyclosporine and phenobarbital. *Clin. Pharmacol. Ther.*, *76*, 490–502.

Ishizuka, H., Konno, K., Shiina, T., Naganuma, H., Nishimura, K., Ito, K., Suzuki, H., Sugiyama, Y. (1999). Species differences in the transport activity for organic anions across the bile canalicular membrane. *J. Pharmacol. Exp. Ther.*, *290*, 1324–1330.

Jalava, K.M., Partanen, J., Neuvonen, P.J. (1997). Itraconazole decreases renal clearance of digoxin. *Ther. Drug Monit.*, *19*, 609–613.

Jayasagar, G., Krishna Kumar, M., Chandrasekhar, K., Madhusudan Rao, Y. (2003). Influence of rifampicin pretreatment on the pharmacokinetics of celecoxib in healthy male volunteers. *Drug Metab. Drug Interact.*, *19*, 287–295.

Johne, A., Brockmöller, J., Bauer, S., Maurer, A., Langheinrich, M., Roots, I. (1999). Pharmacokinetic interaction of digoxin with an herbal extract from St John's wort (Hypericum perforatum) *Clin. Pharmacol. Ther.*, *66*, 338–345.

Johne, A., Schmider, J., Brockmöller, J., Stadelmann, A.M., Störmer, E., Bauer, S., Scholler, G., Langheinrich, M., Roots, I. (2002). Decreased plasma levels of amitriptyline and its metabolites on comedication with an extract from St. John's wort (Hypericum perforatum). *J. Clin. Psychopharmacol.*, *22*, 46–54.

Juan, H., Terhaag, B., Cong, Z., Bi-Kui, Z., Rong-Hua, Z., Feng, W., Fen-Li, S., Juan, S., Jing, T., Wen-Xing, P. (2007). Unexpected effect of concomitantly administered curcumin on the pharmacokinetics of talinolol in healthy Chinese volunteers. *Eur. J. Clin. Pharmacol.*, *63*, 663–668.

Justesen, U.S., Andersen, A.B., Klitgaard, N.A., Brøsen, K., Gerstoft, J., Pedersen, C. (2004). Pharmacokinetic interaction between rifampin and the combination of indinavir and low-dose ritonavir in HIV-infected patients. *Clin. Infect. Dis.*, *38*, 426–429.

Kanazawa, S., Ohkubo, T., Sugawara, K. (2001). The effects of grapefruit juice on the pharmacokinetics of erythromycin. *Eur. J. Clin. Pharmacol.*, *56*, 799–803.

Kantola, T., Kivistö, K.T., Neuvonen, P.J. (1998). Erythromycin and verapamil considerably increase serum simvastatin and simvastatin acid concentrations. *Clin. Pharmacol. Ther.*, *64*, 177–182.

Karyekar, C.S., Eddington, N.D., Briglia, A., Gubbins, P.O., Dowling, T.C. (2004). Renal interaction between itraconazole and cimetidine. *J. Clin. Pharmacol.*, *44*, 919–927.

Kashuba, A.D., Tierney, C., Downey, G.F., Acosta, E.P., Vergis, E.N., Klingman, K., Mellors, J.W., Eshleman, S.H., Scott, T.R., Collier, A.C. (2005). Combining fosamprenavir with lopinavir/ritonavir substantially reduces amprenavir and lopinavir exposure: ACTG protocol A5143 results. *AIDS*, *19*, 145–512.

Kast, H.R., Goodwin, B., Tarr, P.T., Jones, S.A., Anisfeld, A.M., Stoltz, C.M., Tontonoz, P., Kliewer, S., Willson, T.M., Edwards, P.A. (2002). Regulation of multidrug resistance-associated protein 2 (ABCC2) by the nuclear receptors pregnane X receptor, farnesoid X-activated receptor, and constitutive androstane receptor. *J. Biol. Chem.*, *277*, 2908–2915.

Kaukonen, K.M., Olkkola, K.T., Neuvonen, P.J. (1997). Itraconazole increases plasma concentrations of quinidine. *Clin. Pharmacol. Ther.*, *62*, 510–517.

Kehrer, D.F., Mathijssen, R.H., Verweij, J., de Bruijn, P., Sparreboom, A. (2002). Modulation of irinotecan metabolism by ketoconazole. *J. Clin. Oncol.*, *20*, 3122–3129.

Kerb, R., Hoffmeyer, S., Brinkmann, U. (2001). ABC drug transporters: hereditary polymorphisms and pharmacological impact in MDR1, MRP1 and MRP2. *Pharmacogenetics*, *2*, 51–64.

Kerr, D.J., Graham, J., Cummings, J., Morrison, J.G., Thompson, G.G., Brodie, M.J., Kaye, S.B. (1986). The effect of verapamil on the pharmacokinetics of adriamycin. *Cancer Chemother. Pharmacol.*, *18*, 239–242.

Khaliq, Y., Gallicano, K., Venance, S., Kravcik, S., Cameron, D.W. (2000). Effect of ketoconazole on ritonavir and saquinavir concentrations in plasma and cerebrospinal fluid from patients infected with human immunodeficiency virus. *Clin. Pharmacol. Ther.*, *68*, 637–646.

Khaliq, Y., Gallicano, K., Tisdale, C., Carignan, G., Cooper, C., McCarthy, A. (2001). Pharmacokinetic interaction between mefloquine and ritonavir in healthy volunteers. *Br. J. Clin. Pharmacol.*, *51*, 591–600.

Kharasch, E.D., Hoffer, C., Whittington, D., Sheffels, P. (2003). Role of P-glycoprotein in the intestinal absorption and clinical effects of morphine. *Clin. Pharmacol. Ther.*, *74*, 543–554.

Kharasch, E.D., Hoffer, C., Altuntas, T.G., Whittington, D. (2004). Quinidine as a probe for the role of p-glycoprotein in the intestinal absorption and clinical effects of fentanyl. *J. Clin. Pharmacol.*, *44*, 224–233.

Kim, R.B., Leake, B.F., Choo, E.F., Dresser, G.K., Kubba, S.V., Schwarz, U.I., Taylor, A., Xie, H.G., McKinsey, J., Zhou, S., Lan, L.B., Schuetz, J.D., Schuetz, E.G., Wilkinson, G.R. (2001). Identification of functionally variant MDR1 alleles among European American and African Americans. *Clin. Pharmacol. Ther.*, *70*, 189–199.

Kishi, S., Yang, W., Boureau, B., Morand, S., Das, S., Chen, P., Cook, E.H., Rosner, G.L., Schuetz, E., Pui, C.H., Relling, M.V. (2004). Effects of prednisone and genetic polymorphisms on etoposide disposition in children with acute lymphoblastic leukemia. *Blood*, *103*, 67–72.

Koepsell, H., Lips, K., Volk, C. (2007). Polyspecific organic cation transporters: structure, function, physiological roles, and biopharmaceutical implications. *Pharm. Res.*, *24*, 1227–1251.

Korjamo, T., Honkakoski, P., Toppinen, M.R., Niva, S., Reinisalo, M., Palmgren, J.J., Monkkonen, J. (2005). Absorption properties and P-glycoprotein activity of modified Caco-2 cell lines. *Eur. J. Pharm. Sci.*, *26*, 266–279.

Kovarik, J.M., Purba, H.S., Pongowski, M., Gerbeau, C., Humbert, H., Mueller, E.A. (1998). Pharmacokinetics of dexamethasone and valspodar, a P-glycoprotein (mdr1) modulator: implications for coadministration. *Pharmacotherapy*, *18*, 1230–1236.

Kovarik, J.M., Rigaudy, L., Guerret, M., Gerbeau, C., Rost, K.L. (1999). Longitudinal assessment of a P-glycoprotein-mediated drug interaction of valspodar on digoxin. *Clin. Pharmacol. Ther.*, *66*, 391–400.

Krishna, R., Bergman, A., Larson, P., Cote, J., Lasseter, K., Dilzer, S., Wang, A., Zeng, W., Chen, L., Wagner, J., Herman, G. (2007). Effect of a single cyclosporine dose on the single-dose pharmacokinetics of sitagliptin (MK-0431)., a dipeptidyl peptidase-4 inhibitor, in healthy male subjects. *J. Clin. Pharmacol.*, *47*, 165–174.

Kruijtzer, C.M., Beijnen, J.H., Rosing, H., ten Bokkel Huinink, W.W., Schot, M., Jewell, R.C., Paul, E.M., Schellens, J.H. (2002). Increased oral bioavailability of topotecan in combination with the breast cancer resistance protein and P-glycoprotein inhibitor GF120918. *J. Clin. Oncol.*, *20*, 2943–2950.

Kuppens, I.E., Bosch, T.M., van Maanen, M.J., Rosing, H., Fitzpatrick, A., Beijnen, J.H., Schellens, J.H. (2005). Oral bioavailability of docetaxel in combination with OC144-093 (ONT-093). *Cancer Chemother. Pharmacol.*, *55*, 72–78.

Kurata, Y., Ieiri, I., Kimura, M., Morita, T., Irie, S., Urae, A., Ohdo, S., Ohtani, H., Sawada, Y., Higuchi, S., Otsubo, K. (2002). Role of human MDR1 gene polymorphism in bioavailability and interaction of digoxin, a substrate of P-glycoprotein. *Clin. Pharmacol. Ther.*, *72*, 209–219.

Kurnik, D., Wood, A.J., Wilkinson, G.R. (2006). The erythromycin breath test reflects P-glycoprotein function independently of cytochrome P450 3A activity. *Clin. Pharmacol. Ther.*, *80*, 228–234.

Kyrklund, C., Backman, J.T., Kivistö, K.T., Neuvonen, M., Laitila, J., Neuvonen, P.J. (2001). Plasma concentrations of active lovastatin acid are markedly increased by gemfibrozil but not by bezafibrate. *Clin. Pharmacol. Ther.*, *69*, 340–345.

Larsen, U.L., Hyldahl Olesen, L., Guldborg Nyvold, C., Eriksen, J., Jakobsen, P., Østergaard, M., Autrup, H., Andersen, V. (2007) Human intestinal P-glycoprotein activity estimated by the model substrate digoxin. *Scand. J. Clin. Lab. Invest.*, *67*, 123–134.

Lê, L.H., Moore, M.J., Siu, L.L., Oza, A.M., MacLean, M., Fisher, B., Chaudhary, A., de Alwis, D. P., Slapak, C., Seymour, L. (2005). Phase I study of the multidrug resistance inhibitor zosuquidar administered in combination with vinorelbine in patients with advanced solid tumours. *Cancer Chemother. Pharmacol.*, *56*, 154–160.

Lee, F.Y., Lee, H., Hubbert, M.L., Edwards, P.A., Zhang, Y. (2006). FXR, a multipurpose nuclear receptor. *Trends. Biochem. Sci.*, *31*, 572–580.

Lee, W., Glaeser, H., Smith, L.H., Roberts, R.L., Moeckel, G.W., Gervasini, G., Leake, B.F., Kim, R.B. (2005). Polymorphisms in human organic anion-transporting polypeptide 1A2 (OATP1A2): implications for altered drug disposition and central nervous system drug entry. *J. Biol. Chem.*, *280*, 9610–9617.

Lemahieu, W.P., Hermann, M., Asberg, A., Verbeke, K., Holdaas, H., Vanrenterghem, Y., Maes, B.D. (2005). Combined therapy with atorvastatin and calcineurin inhibitors: no interactions with tacrolimus. *Am. J. Transplant.*, *5*, 2236–2243.

Lemma, G.L., Wang, Z., Hamman, M.A., Zaheer, N.A., Gorski, J.C., Hall, S.D. (2006). The effect of short- and long-term administration of verapamil on the disposition of cytochrome P450 3A and P-glycoprotein substrates. *Clin. Pharmacol. Ther.*, *79*, 218–230.

Leung, L.Y., Lim, H.K., Abell, M.W., Zimmerman, J.J. (2006). Pharmacokinetics and metabolic disposition of sirolimus in healthy male volunteers after a single oral dose. *Ther. Drug Monit.*, *28*, 51–61.

Lilja, J.J., Backman, J.T., Laitila, J., Luurila, H., Neuvonen, P.J. (2003). Itraconazole increases but grapefruit juice greatly decreases plasma concentrations of celiprolol. *Clin. Pharmacol. Ther.*, *73*, 192–198.

Lilja, J.J., Juntti-Patinen, L., Neuvonen, P.J. (2004). Orange juice substantially reduces the bioavailability of the beta-adrenergic-blocking agent celiprolol. *Clin. Pharmacol. Ther.*, *75*, 184–190.

Lilja, J.J., Niemi, M., Neuvonen, P.J. (2004). Rifampicin reduces plasma concentrations of celiprolol. *Eur. J. Clin. Pharmacol.*, *59*, 819–824.

Lilja, J.J., Niemi, M., Fredrikson, H., Neuvonen, P.J. (2007). Effects of clarithromycin and grapefruit juice on the pharmacokinetics of glibenclamide. *Br. J. Clin. Pharmacol.*, *63*, 732–740.

Lim, M.L., Min, S.S., Eron, J.J., Bertz, R.J., Robinson, M., Gaedigk, A., Kashuba, A.D. (2004). Coadministration of lopinavir/ritonavir and phenytoin results in two-way drug interaction through cytochrome P-450 induction. *J. Acquir. Immune. Defic. Syndr.*, *36*, 1034–1040.

Lin, J.H. (1995). Species similarities and differences in pharmacokinetics. *Drug Metab. Dispos.*, *23*, 1008–1021.

Lin, J.H. (2007). Transporter-mediated drug interactions: clinical implications and in vitro assessment. *Expert Opin. Drug Metab. Toxicol.*, *3*, 81–92.

Lin, J.H., Yamazaki, M. (2003). Role of P-glycoprotein in pharmacokinetics: clinical implications. *Clin. Pharmacokinet.*, *42*, 59–98.

Lum, B.L., Kaubisch, S., Fisher, G.A., Brown, B.W., Sikic, B.I. (2000). Effect of high-dose cyclosporine on etoposide pharmacodynamics in a trial to reverse P-glycoprotein (MDR1 gene) mediated drug resistance. *Cancer Chemother. Pharmacol.*, *5*, 305–311.

Lum, B.L., Kaubisch, S., Yahanda, A.M., Adler, K.M., Jew, L., Ehsan, M.N., Brophy, N.A., Halsey, J., Gosland, M.P., Sikic, B.I. (1992). Alteration of etoposide pharmacokinetics and pharmacodynamics by cyclosporine in a phase I trial to modulate multidrug resistance. *J. Clin. Oncol.*, *10*, 1635–1642.

McCance-Katz, E.F., Moody, D.E., Smith, P.F., Morse, G.D., Friedland, G., Pade, P., Baker, J., Alvanzo, A., Jatlow, P., Rainey, P.M. (2006). Interactions between buprenorphine and antiretrovirals. II. The protease inhibitors nelfinavir, lopinavir/ritonavir, and ritonavir. *Clin. Infect. Dis.*, *43* (Suppl 4), S235–S246.

Mahar Doan, K.M., Humphreys, J.E., Webster, L.O., Wring, S.A., Shampine, L.J., Serabjit-Singh, C.J., Adkison, K.K., Polli, J.W. (2002). Passive permeability and P-glycoprotein-mediated efflux differentiate central nervous system (CNS) and non-CNS marketed drugs. *J. Pharmacol. Exp. Ther.*, *303*, 1029–1037.

Mai, I., Krüger, H., Budde, K., Johne, A., Brockmöller, J., Neumayer, H.H., Roots, I. (2000). Hazardous pharmacokinetic interaction of Saint John's wort (Hypericum perforatum) with the immunosuppressant cyclosporin. *Int. J. Clin. Pharmacol. Ther.*, *38*, 500–502.

Mai, I., Störmer, E., Bauer, S., Krüger, H., Budde, K., Roots, I. (2003). Impact of St John's wort treatment on the pharmacokinetics of tacrolimus and mycophenolic acid in renal transplant patients. *Nephrol. Dial. Transplant.*, *18*, 819–822.

Malingré, M.M., Richel, D.J., Beijnen, J.H., Rosing, H., Koopman, F.J., Ten Bokkel Huinink, W.W., Schot, M.E., Schellens, J.H. (2001a). Coadministration of cyclosporine strongly enhances the oral bioavailability of docetaxel. *J. Clin. Oncol.*, *19*, 1160–1166.

Malingré, M.M., Ten Bokkel Huinink, W.W., Mackay, M., Schellens, J.H., Beijnen, J.H. (2001b). Pharmacokinetics of oral cyclosporin A when co-administered to enhance the absorption of orally administered docetaxel. *Eur. J. Clin. Pharmacol.*, *57*, 305–307.

Malingré, M.M., Beijnen, J.H., Rosing, H., Koopman, F.J., Jewell, R.C., Paul, E.M., Ten Bokkel Huinink, W.W., Schellens, J.H. (2001c). Co-administration of GF120918 significantly increases the systemic exposure to oral paclitaxel in cancer patients. *Br. J. Cancer*, *84*, 42–47.

Martin-Facklam, M., Burhenne, J., Ding, R., Fricker, R., Mikus, G., Walter-Sack, I., Haefeli, W.E. (2002). Dose-dependent increase of saquinavir bioavailability by the pharmaceutic aid cremophor EL. *Br. J. Clin. Pharmacol.*, *53*, 576–581.

Marzolini, C., Paus, E., Buclin, T., Kim, R.B. (2004). Polymorphisms in human MDR1 (P-glycoprotein): recent advances and clinical relevance. *Clin. Pharmacol. Ther.*, *75*, 13–33.

Masuda, M., I'Izuka, Y., Yamazaki, M., Nishigaki, R., Kato, Y., Ni'inuma, K., Suzuki, H., Sugiyama, Y. (1997). Methotrexate is excreted into the bile by canalicular multispecific organic anion transporter in rats. *Cancer Res.*, *57*, 3506–3510.

Matheny, C.J., Lamb, M.W., Brouwer, K.R., Pollack, G.M. (2001). Pharmacokinetic and pharmacodynamic implications of P-glycoprotein modulation. *Pharmacotherapy*, *21*, 778–796.

Mayo, P.R., Skeith, K., Russell, A.S., Jamali, F. (2000). Decreased dromotropic response to verapamil despite pronounced increased drug concentration in rheumatoid arthritis. *Br. J. Clin. Pharmacol.*, *50*, 605–613.

Meerum Terwogt, J.M., Beijnen, J.H., ten Bokkel Huinink, W.W., Rosing, H., Schellens, J.H. (1998). Co-administration of cyclosporin enables oral therapy with paclitaxel. *Lancet, 352* (9124), 285.

Meerum Terwogt, J.M., Malingré, M.M., Beijnen, J.H., ten Bokkel Huinink, W.W., Rosing, H., Koopman, F.J., van Tellingen, O., Swart, M., Schellens, J.H. (1999). Coadministration of oral cyclosporin A enables oral therapy with paclitaxel. *Clin. Cancer Res., 5*, 3379–3384.

Miki, Y., Suzuki, T., Kitada, K., Yabuki, N., Shibuya, R., Moriya, T., Ishida, T., Ohuchi, N., Blumberg, B., Sasano, H. (2006). Expression of the steroid and xenobiotic receptor and its possible target gene, organic anion transporting polypeptide-A, in human breast carcinoma. *Cancer Res., 66*, 535–542.

Minami, H., Ohtsu, T., Fujii, H., Igarashi, T., Itoh, K., Uchiyama-Kokubu, N., Aizawa, T., Watanabe, T., Uda, Y., Tanigawara, Y., Sasaki, Y. (2001). Phase I study of intravenous PSC-833 and doxorubicin: reversal of multidrug resistance. *Jpn. J. Cancer Res., 92*, 220–230.

Mizuno, N., Niwa, T., Yotsumoto, Y., Sugiyama, Y. (2003). Impact of drug transporter studies on drug discovery and development. *Pharmacol. Rev., 55*, 425–461.

Moreira, A., Lobato, R., Morais, J., Silva, S., Ribeiro, J., Figueira, A., Vale, D., Sousa, C., Araújo, F., Fernandes, A., Oliveira, J., Passos-Coelho, J.L. (2001). Influence of the interval between the administration of doxorubicin and paclitaxel on the pharmacokinetics of these drugs in patients with locally advanced breast cancer. *Cancer Chemother. Pharmacol., 48*, 333–337.

Mouly, S., Lown, K.S., Kornhauser, D., Joseph, J.L., Fiske, W.D., Benedek, I.H., Watkins, P.B. (2002). Hepatic but not intestinal CYP3A4 displays dose-dependent induction by efavirenz in humans. *Clin. Pharmacol. Ther., 72*, 1–9.

Mück, W., Mai, I., Fritsche, L., Ochmann, K., Rohde, G., Unger, S., Johne, A., Bauer, S., Budde, K., Roots, I., Neumayer, H.H., Kuhlmann, J. (1999). Increase in cerivastatin systemic exposure after single and multiple dosing in cyclosporine-treated kidney transplant recipients. *Clin. Pharmacol. Ther., 65*, 251–261.

Mukwaya, G., MacGregor, T., Hoelscher, D., Heming, T., Legg, D., Kavanaugh, K., Johnson, P., Sabo, J.P., McCallister, S. (2005). Interaction of ritonavir-boosted tipranavir with loperamide does not result in loperamide-associated neurologic side effects in healthy volunteers. *Antimicrob. Agents. Chemother., 49*, 4903–4910.

Nakagami, T., Yasui-Furukori, N., Saito, M., Tateishi, T., Kaneo, S. (2005). Effect of verapamil on pharmacokinetics and pharmacodynamics of risperidone: in vivo evidence of involvement of P-glycoprotein in risperidone disposition. *Clin. Pharmacol. Ther., 78*, 43–51.

Nannan Panday, V.R., Hoetelmans, R.M., van Heeswijk, R.P., Meenhorst, P.L., Inghels, M., Mulder, J.W., Beijnen, J.H. (1999). Paclitaxel in the treatment of human immunodeficiency virus 1-associated Kaposi's sarcoma—drug–drug interactions with protease inhibitors and a nonnucleoside reverse transcriptase inhibitor: a case report study. *Cancer Chemother. Pharmacol., 43*, 516–519.

Nenciu, L.M., Laberge, P., Thirion, D.J. (2006). Telithromycin-induced digoxin toxicity and electrocardiographic changes. *Pharmacotherapy, 26*, 872–876.

Niemi, M., Tornio, A., Pasanen, M.K., Fredrikson, H., Neuvonen, P.J., Backman, J.T. (2006). Eur Itraconazole, gemfibrozil and their combination markedly raise the plasma concentrations of loperamide. *J. Clin. Pharmacol., 62*, 463–472.

Oswald, S., Giessmann, T., Luetjohann, D., Wegner, D., Rosskopf, D., Weitschies, W., Siegmund, W. (2006). Disposition and sterol-lowering effect of ezetimibe are influenced by single-dose coadministration of rifampin, an inhibitor of multidrug transport proteins. *Clin. Pharmacol. Ther., 80*, 477–485.

Oswald, S., Haenisch, S., Fricke, C., Sudhop, T., Remmler, C., Giessmann, T., Jedlitschky, G., Adam, U., Dazert, E., Warzok, R., Wacke, W., Cascorbi, I., Kroemer, H.K., Weitschies, W., von Bergmann, K., Siegmund, W. (2006). Intestinal expression of P-glycoprotein (ABCB1), multidrug resistance associated protein 2 (ABCC2), and uridine diphosphate-glucuronosyltransferase

1A1 predicts the disposition and modulates the effects of the cholesterol absorption inhibitor ezetimibe in humans. *Clin. Pharmacol. Ther.*, *79*, 206–217.

Overbosch, D., Van Gulpen, C., Hermans, J., Mattie, H. (1988). The effect of probenecid on the renal tubular excretion of benzylpenicillin. *Br. J. Clin. Pharmacol.*, *25*, 51–58.

Paine, M.F., Wagner, D.A., Hoffmaster, K.A., Watkins, P.B. (2002a). Cytochrome P450 3A4 and P-glycoprotein mediate the interaction between an oral erythromycin breath test and rifampin. *Clin. Pharmacol. Ther.*, *72*, 524–535.

Paine, M.F., Leung, L.Y., Lim, H.K., Liao, K., Oganesian, A., Zhang, M.Y., Thummel, K.E., Watkins, P.B. (2002b). Identification of a novel route of extraction of sirolimus in human small intestine: roles of metabolism and secretion. *J. Pharmacol. Exp. Ther.*, *301*, 174–186.

Partanen, J., Jalava, K.M., Neuvonen, P.J. (1996). Itraconazole increases serum digoxin concentration. *Pharmacol. Toxicol.*, *79*, 274–276.

Patnaik, A., Warner, E., Michael, M., Egorin, M.J., Moore, M.J., Siu, L.L., Fracasso, P.M., Rivkin, S., Kerr, I., Litchman, M., Oza, A.M. (2000). Phase I dose-finding and pharmacokinetic study of paclitaxel and carboplatin with oral valspodar in patients with advanced solid tumors. *J. Clin. Oncol.*, *18*, 3677–3689.

Paulusma, C.C., Kool, M., Bosma, P.J., Scheffer, G.L., ter Borg, F., Scheper, R.J., Tytgat, G.N., Borst, P., Baas, F., Oude Elferink, R.P. (1997). A mutation in the human canalicular multispecific organic anion transporter gene causes the Dubin-Johnson syndrome. *Hepatology*, *25*, 1539–1542.

Pea, F., Damiani, D., Michieli, M., Ermacora, A., Baraldo, M., Russo, D., Fanin, R., Baccarani, M., Furlanut, M. (1999). Multidrug resistance modulation in vivo: the effect of cyclosporin A alone or with dexverapamil on idarubicin pharmacokinetics in acute leukemia. *Eur. J. Clin. Pharmacol.*, *55*, 361–368.

Penzak, S.R., Shen, J.M., Alfaro, R.M., Remaley, A.T., Natarajan, V., Falloon, J. (2004). Ritonavir decreases the nonrenal clearance of digoxin in healthy volunteers with known MDR1 genotypes. *Ther. Drug Monit.*, *26*, 322–330.

Penzak, S.R., Formentini, E., Alfaro, R.M., Long, M., Natarajan, V., Kovacs, J. (2005). Prednisolone pharmacokinetics in the presence and absence of ritonavir after oral prednisone administration to healthy volunteers. *J. Acquir. Immune. Defic. Syndr.*, *40*, 573–580.

Petri, N., Borga, O., Nyberg, L., Hedeland, M., Bondesson, U., Lennernas, H. (2006). Effect of erythromycin on the absorption of fexofenadine in the jejunum, ileum and colon determined using local intubation in healthy volunteers. *Int. J. Clin. Pharmacol. Ther.*, *44*, 71–79.

Petrovic, V., Teng, S., Piquette-Miller, M. (2007). Regulation of drug transporters during infection and inflammation. *Mol. Interv.*, *7*, 99–111.

Pheterson, A.D., Miller, L., Fox, C.F., Estroff, T.W., Sweeney, D.R. (1987). Multifocal neurological impairment caused by infection-induced rise in blood lithium and amitriptyline. *Int. J. Psychiatry Med. 1986.*, *16*, 257–262.

Pizzagalli, F., Hagenbuch, B., Stieger, B., Klenk, U., Folkers, G., Meier, P.J. (2002). Identification of a novel human organic anion transporting polypeptide as a high affinity thyroxine transporter. *Mol. Endocrinol.*, *16*, 2283–2296.

Planting, A.S., Sonneveld, P., van der Gaast, A., Sparreboom, A., van der Burg, M.E., Luyten, G. P., de Leeuw, K., de Boer-Dennert, M., Wissel, P.S., Jewell, R.C., Paul, E.M., Purvis, N.B.Jr, , Verweij, J. (2005). A phase I and pharmacologic study of the MDR converter GF120918 in combination with doxorubicin in patients with advanced solid tumors. *Cancer Chemother. Pharmacol.*, *55*, 91–99.

Polli, J.W., Wring, S.A., Humphreys, J.E., Huang, L., Morgan, J.B., Webster, L.O., Serabjit-Singh, C.S. (2001). Rational use of in vitro P-glycoprotein assays in drug discovery. *J. Pharmacol. Exp. Ther.*, *299*, 620–628.

Potter, J.M., McWhinney, B.C., Sampson, L., Hickman, P.E. (2004). Area-under-the-curve monitoring of prednisolone for dose optimization in a stable renal transplant population. *Ther. Drug Monit.*, *26*, 408–414.

Putnam, W.S., Woo, J.M., Huang, Y., Benet, L.Z. (2005). Effect of the MDR1 C3435T variant and P-glycoprotein induction on dicloxacillin pharmacokinetics. *J. Clin. Pharmacol.*, *45*, 411–421.

Race, J.E., Grassl, S.M., Williams, W.J., Holtzman, E.J. (1999). Molecular cloning and characterization of two novel human renal organic anion transporters (hOAT1 and hOAT3) *Biochem. Biophys. Res. Commun.*, *255*, 508–514.

Ramachandra, M., Ambudkar, S.V., Gottesman, M.M., Pastan, I., Hrycyna, C.A. (1996). Functional characterization of a glycine 185-to-valine substitution in human P-glycoprotein by using a vaccinia-based transient expression system. *Mol. Biol. Cell*, *7*, 1485–1498.

Reitman, M.L., Schadt, E.E. (2007). Pharmacogenetics of metformin response: a step in the path toward personalized medicine. *J. Clin. Invest.*, *117*, 1226–1229.

Rengelshausen, J., Göggelmann, C., Burhenne, J., Riedel, K.D., Ludwig, J., Weiss, J., Mikus, G., Walter-Sack, I., Haefeli, W.E. (2003). Contribution of increased oral bioavailability and reduced nonglomerular renal clearance of digoxin to the digoxin-clarithromycin interaction. *Br. J. Clin. Pharmacol.*, *56*, 32–38.

Roberts, D.H., Kendall, M.J., Jack, D.B., Welling, P.G. (1981). Pharmacokinetics of cephradine given intravenously with and without probenecid. *Br. J. Clin. Pharmacol.*, *11*, 561–564.

Rocchi, E., Khodjakov, A., Volk, E.L., Yang, C.H., Litman, T., Bates, S.E., Schneider, E. (2000). The product of the ABC half-transporter gene ABCG2 (BCRP/MXR/ABCP) is expressed in the plasma membrane. *Biochem. Biophys. Res. Commun.*, *271*, 42–46.

Rollot, F., Pajot, O., Chauvelot-Moachon, L., Nazal, E.M., Kélaïdi, C., Blanche, P. (2004). Acute colchicine intoxication during clarithromycin administration. *Ann. Pharmacother.*, *38*, 2074–2077.

Rowinsky, E.K., Smith, L., Wang, Y.M., Chaturvedi, P., Villalona, M., Campbell, E., Aylesworth, C., Eckhardt, S.G., Hammond, L., Kraynak, M., Drengler, R., Stephenson, J. Jr, Harding, M.W., Von Hoff, D.D. (1998). Phase I and pharmacokinetic study of paclitaxel in combination with biricodar, a novel agent that reverses multidrug resistance conferred by overexpression of both MDR1 and MRP. *J. Clin. Oncol.*, *16*, 2964–2976.

Ruschitzka, F., Meier, P.J., Turina, M., Lüscher, T.F., Noll, G. (2000). Acute heart transplant rejection due to Saint John's wort. *Lancet*, *355* (9203), 548–549.

Rushing, D.A., Raber, S.R., Rodvold, K.A., Piscitelli, S.C., Plank, G.S., Tewksbury, D.A. (1994). The effects of cyclosporine on the pharmacokinetics of doxorubicin in patients with small cell lung cancer. *Cancer*, *74*, 834–841.

Sadeque, A.J., Wandel, C., He, H., Shah, S., Wood, A.J. (2000). Increased drug delivery to the brain by P-glycoprotein inhibition. *Clin. Pharmacol. Ther.*, *68*, 231–237.

Sahi, J., Milad, M.A., Zheng, X., Rose, K.A., Wang, H., Stilgenbauer, L., Gilbert, D., Jolley, S., Stern, R.H., LeCluyse, E.L. (2003). Avasimibe induces CYP3A4 and multiple drug resistance protein 1 gene expression through activation of the pregnane X receptor. *J. Pharmacol. Exp. Ther.*, *306*, 1027–1034.

Sai, Y. (2005). Biochemical and molecular pharmacological aspects of transporters as determinants of drug disposition. *Drug Metab. Pharmacokinet.*, *20*, 91–99.

Sandler, A., Gordon, M., De Alwis, D.P., Pouliquen, I., Green, L., Marder, P., Chaudhary, A., Fife, K., Battiato, L., Sweeney, C., Jordan, C., Burgess, M., Slapak, C.A.A. (2004). Phase I trial of a potent P-glycoprotein inhibitor, zosuquidar trihydrochloride (LY335979)., administered intravenously in combination with doxorubicin in patients with advanced malignancy. *Clin. Cancer Res.*, *10*, 3265–3272.

Sansone-Parsons, A., Krishna, G., Martinho, M., Kantesaria, B., Gelone, S., Mant, T.G. (2007). Effect of oral posaconazole on the pharmacokinetics of cyclosporine and tacrolimus. *Pharmacotherapy*, *27*, 825–834.

Sasaki, M., Suzuki, H., Ito, K., Abe, T., Sugiyama, Y. (2002). Transcellular transport of organic anions across a double-transfected Madin-Darby canine kidney II cell monolayer expressing both human organic anion-transporting polypeptide (OATP2/SLC21A6) and multidrug resistance-associated protein 2 (MRP2/ABCC2). *J. Biol. Chem.*, *277*, 6497–6503.

Sasaki, M., Suzuki, H., Aoki, J., Ito, K., Meier, P.J., Sugiyama, Y. (2004). Prediction of in vivo biliary clearance from the in vitro transcellular transport of organic anions across a double-transfected Madin-Darby canine kidney II monolayer expressing both rat organic anion transporting polypeptide 4 and multidrug resistance associated protein 2. *Mol. Pharmacol.*, *66*, 450–459.

Sasongko, L., Link, J.M., Muzi, M., Mankoff, D.A., Yang, X., Collier, A.C., Shoner, S.C., and Unadkat, J.D. (2005). Imaging P-glycoprotein transport activity at the human blood-brain barrier with positron emission tomography. *Clin. Pharmacol. Ther.*, *77*, 503–514.

Schinkel, A.H., Wagenaar, E., Mol, C.A., van Deemter, L. (1996). P-glycoprotein in the blood-brain barrier of mice influences the brain penetration and pharmacological activity of many drugs. *J. Clin. Invest.*, *97*, 2517–2524.

Schwarz, U.I., Gramatté, T., Krappweis, J., Berndt, A., Oertel, R., von Richter, O., Kirch, W. (1999). Unexpected effect of verapamil on oral bioavailability of the beta-blocker talinolol in humans. *Clin. Pharmacol. Ther.*, *65*, 283–290.

Schwarz, U.I., Gramatté, T., Krappweis, J., Oertel, R., Kirch, W. (2000). P-glycoprotein inhibitor erythromycin increases oral bioavailability of talinolol in humans. *Int. J. Clin. Pharmacol. Ther.*, *38*, 161–167.

Schwarz, U.I., Seemann, D., Oertel, R., Miehlke, S., Kuhlisch, E., Fromm, M.F., Kim, R.B., Bailey, D.G., Kirch, W. (2005). Grapefruit juice ingestion significantly reduces talinolol bioavailability. *Clin. Pharmacol. Ther.*, *77*, 291–301.

Schwarz, U.I., Hanso, H., Oertel, R., Miehlke, S., Kuhlisch, E., Glaeser, H., Hitzl, M., Dresser, G. K., Kim, R.B., Kirch, W. (2007). Induction of intestinal P-glycoprotein by St John's wort reduces the oral bioavailability of talinolol. *Clin. Pharmacol. Ther.*, *81*, 669–768.

Seelig, A., Gottschlich, R., Devant, R.M. (1994). A method to determine the ability of drugs to diffuse through the blood-brain barrier. *Proc. Natl. Acad. Sci. U. S. A.*, *91*, 68–72.

Seiden, M.V., Swenerton, K.D., Matulonis, U., Campos, S., Rose, P., Batist, G., Ette, E., Garg, V., Fuller, A., Harding, M.W., Charpentier, D. (2002). A phase II study of the MDR inhibitor biricodar (INCEL, VX-710) and paclitaxel in women with advanced ovarian cancer refractory to paclitaxel therapy. *Gynecol. Oncol.*, *86*, 302–310.

Seitz, S., Boelsterli, U.A. (1998). Diclofenac acyl glucuronide, a major biliary metabolite, is directly involved in small intestinal injury in rats. *Gastroenterology*, *115*, 1476–1482.

Shilling, A.D., Azam, F., Kao, J., Leung, L. (2006). Use of canalicular membrane vesicles (CMVs) from rats, dogs, monkeys and humans to assess drug transport across the canalicular membrane. *J. Pharmacol. Toxicol. Methods*, *53*, 186–197.

Shimada, T., Terada, A., Yokogawa, K., Kaneko, H., Nomura, M., Kaji, K., Kaneko, S., Kobayashi, K., Miyamoto, K. (2002). Lowered blood concentration of tacrolimus and its recovery with changes in expression of CYP3A and P-glycoprotein after high-dose steroid therapy. *Transplantation*, *74*, 1419–1424.

Shimizu, M., Uno, T., Sugawara, K., Tateishi, T. (2006). Effects of itraconazole and diltiazem on the pharmacokinetics of fexofenadine, a substrate of P-glycoprotein. *Br. J. Clin. Pharmacol.*, *61*, 538–544.

Shon, J.H., Yoon, Y.R., Hong, W.S., Nguyen, P.M., Lee, S.S., Choi, Y.G., Cha, I.J., Shin, J.G. (2005). Effect of itraconazole on the pharmacokinetics and pharmacodynamics of fexofenadine in relation to the MDR1 genetic polymorphism. *Clin. Pharmacol. Ther.*, *78*, 191–201.

Shu, Y., Sheardown, S.A., Brown, C., Owen, R.P., Zhang, S., Castro, R.A., Ianculescu, A.G., Yue, L., Burchard, E.G., Brett, C.M., Giacomini, K.M. (2007). Effect of genetic variation in the organic cation transporter 1 (OCT1) on metformin action. *J. Clin. Invest.*, *117*, 1422–1431.

Siedlik, P.H., Olson, S.C., Yang, B.B., Stern, R.H. (1999). Erythromycin coadministration increases plasma atorvastatin concentrations. *J. Clin. Pharmacol.*, *39*, 501–504.

Skarke, C., Jarrar, M., Schmidt, H., Kauert, G., Langer, M., Geisslinger, G., Lötsch, J. (2003). Effects of ABCB1 (multidrug resistance transporter) gene mutations on disposition and central nervous effects of loperamide in healthy volunteers. *Pharmacogenetics*, *13*, 651–660.

Smeets, M., Raymakers, R., Muus, P., Vierwinden, G., Linssen, P., Masereeuw, R., de Witte, T. (2001). Cyclosporin increases cellular idarubicin and idarubicinol concentrations in relapsed or refractory AML mainly due to reduced systemic clearance. *Leukemia*, *15*, 80–88.

Somogyi, A., McLean, A., Heinzow, B. (1983). Cimetidine-procainamide pharmacokinetic interaction in man: evidence of competition for tubular secretion of basic drugs. *Eur. J. Clin. Pharmacol.*, *25*, 339–345.

Somogyi, A., Stockley, C., Keal, J., Rolan, P., Bochner, F. (1987). Reduction of metformin renal tubular secretion by cimetidine in man. *Br. J. Clin. Pharmacol.*, *23*, 545–551.

Sonneveld, P., Marie, J.P., Huisman, C., Vekhoff, A., Schoester, M., Faussat, A.M., van Kapel, J., Groenewegen, A., Charnick, S., Zittoun, R., Löwenberg, B. (1996). Reversal of multidrug resistance by SDZ PSC 833, combined with VAD (vincristine, doxorubicin, dexamethasone) in refractory multiple myeloma. A phase I study. *Leukemia*, *10*, 1741–1750.

Stieger, B., Fattinger, K., Madon, J., Kullak-Ublick, G.A., Meier, P.J. (2000). Drug- and estrogen-induced cholestasis through inhibition of the hepatocellular bile salt export pump (Bsep) of rat liver. *Gastroenterology*, *118*, 422–430.

Sugimoto, K., Ohmori, M., Tsuruoka, S., Nishiki, K., Kawaguchi, A., Harada, K., Arakawa, M., Sakamoto, K., Masada, M., Miyamori, I., Fujimura, A. (2001). Different effects of St John's wort on the pharmacokinetics of simvastatin and pravastatin. *Clin. Pharmacol. Ther.*, *70*, 518–524.

Suzuki, H., Sugiyama, Y. (2002). Single nucleotide polymorphisms in multidrug resistance associated protein 2 (MRP2/ABCC2): its impact on drug disposition. *Adv. Drug Deliv. Rev.*, *54*, 1311–1331.

Szatmari, I., Vámosi, G., Brazda, P., Balint, B.L., Benko, S., Széles, L., Jeney, V., Ozvegy-Laczka, C., Szántó, A., Barta, E., Balla, J., Sarkadi, B., Nagy, L. (2006). Peroxisome proliferator-activated receptor gamma-regulated ABCG2 expression confers cytoprotection to human dendritic cells. *J. Biol. Chem.*, *281* (33), 23812–23823.

Taburet, A.M., Piketty, C., Chazallon, C., Vincent, I., Gérard, L., Calvez, V., Clavel, F., Aboulke, J.P., Girard, P.M. (2004). Interactions between atazanavir-ritonavir and tenofovir in heavily pretreated human immunodeficiency virus-infected patients. *Antimicrob. Agents. Chemother.*, *48*, 2091–2096.

Taguchi, Y., Yoshida, A., Takada, Y., Komano, T., Ueda, K. (1997). Anti-cancer drugs and glutathione stimulate vanadate-induced trapping of nucleotide in multidrug resistance-associated protein (MRP) *FEBS Lett.*, *401*, 11–14.

Tamai, I., Nezu, J., Uchino, H., Sai, Y., Oku, A., Shimane, M., Tsuji, A. (2000). Molecular identification and characterization of novel members of the human organic anion transporter (OATP) family. *Biochem. Biophys. Res. Commun.*, *273*, 251–260.

Tanaka, H., Matsumoto, K., Ueno, K., Kodama, M., Yoneda, K., Katayama, Y., Miyatake, K. (2003). Effect of clarithromycin on steady-state digoxin concentrations. *Ann. Pharmacother.*, *37*, 178–181.

Tannergren, C., Petri, N., Knutson, L., Hedeland, M., Bondesson, U., Lennernäs, H. (2003). Multiple transport mechanisms involved in the intestinal absorption and first-pass extraction of fexofenadine. *Clin. Pharmacol. Ther.*, *74*, 423–436.

Tannergren, C., Engman, H., Knutson, L., Hedeland, M., Bondesson, U., St, Lennernäs, H. (2004). John's wort decreases the bioavailability of R- and S-verapamil through induction of the first-pass metabolism. *Clin. Pharmacol. Ther.*, *75*, 298–309.

Tayrouz, Y., Ding, R., Burhenne, J., Riedel, K.D., Weiss, J., Hoppe-Tichy, T., Haefeli, W.E., Mikus, G. (2003). Pharmacokinetic and pharmaceutic interaction between digoxin and Cremophor RH40. *Clin. Pharmacol. Ther.*, *73*, 397–405.

Tayrouz, Y., Ganssmann, B., Ding, R., Klingmann, A., Aderjan, R., Burhenne, J., Haefeli, W.E., Mikus, G. (2001). Ritonavir increases loperamide plasma concentrations without evidence for P-glycoprotein involvement. *Clin. Pharmacol. Ther.*, *70*, 405–414.

Tirona, R.G., Kim, R.B. (2002). Pharmacogenomics of organic anion-transporting polypeptides (OATP). *Adv. Drug Deliv. Rev.*, *54*, 1343–1352.

Tsuji, A., Terasaki, T., Tamai, I., Takeda, K. (1990). In vivo evidence for carrier-mediated uptake of beta-lactam antibiotics through organic anion transport systems in rat kidney and liver. *J. Pharmacol. Exp. Ther.*, *253*, 315–320.

Tsunoda, S.M., Harris, R.Z., Christians, U., Velez, R.L., Freeman, R.B., Benet, L.Z., Warshaw, A. (2001). Red wine decreases cyclosporine bioavailability. *Clin. Pharmacol. Ther.*, *70*, 462–467.

Tsuruoka, S., Ioka, T., Wakaumi, M., Sakamoto, K., Ookami, H., Fujimura, A. (2006). Severe arrhythmia as a result of the interaction of cetirizine and pilsicainide in a patient with renal insufficiency: first case presentation showing competition for excretion via renal multidrug resistance protein 1 and organic cation transporter 2. *Clin. Pharmacol. Ther.*, *79*, 389–396.

Tuteja, S., Alloway, R.R., Johnson, J.A., Gaber, A.O. (2001). The effect of gut metabolism on tacrolimus bioavailability in renal transplant recipients. *Transplantation*, *71*, 1303–1307.

Uno, T., Shimizu, M., Sugawara, K., Tateishi, T. (2006). Lack of dose-dependent effects of itraconazole on the pharmacokinetic interaction with fexofenadine. *Drug Metab. Dispos.*, *34*, 1875–1879.

Urquhart, B.L., Tirona, R.G., Kim, R.B. (2007). Nuclear receptors and the regulation of drug-metabolizing enzymes and drug transporters: implications for interindividual variability in response to drugs. *J. Clin. Pharmacol.*, *47*, 566–578.

Van Heeswijk, R.P., Bourbeau, M., Campbell, P., Seguin, I., Chauhan, B.M., Foster, B.C., Cameron, D.W. (2006). Time-dependent interaction between lopinavir/ritonavir and fexofenadine. *J. Clin. Pharmacol.*, *46*, 758–767.

Van Montfoort, J.E., Hagenbuch, B., Groothuis, G.M., Koepsell, H., Meier, P.J., Meijer, D.K. (2003). Drug uptake systems in liver and kidney. *Curr. Drug Metab.*, *4*, 185–211.

Vasquez, E.M., Shin, G.P., Sifontis, N., Benedetti, E. (2005). Concomitant clotrimazole therapy more than doubles the relative oral bioavailability of tacrolimus. *Ther. Drug Monit.*, *27*, 587–591.

Von Hentig, N., Müller, A., Rottmann, C., Wolf, T., Lutz, T., Klauke, S., Kurowski, M., Oertel, B., Dauer, B., Harder, S., Staszewski, S. (2007). Pharmacokinetics of saquinavir, atazanavir, and ritonavir in a twice-daily boosted double-protease inhibitor regimen. *Antimicrob. Agents. Chemother.*, *51*, 1431–1439.

Wakasugi, H., Yano, I., Ito, T., Hashida, T., Futami, T., Nohara, R., Sasayama, S., Inui, K. (1998). Effect of clarithromycin on renal excretion of digoxin: interaction with P-glycoprotein. *Clin. Pharmacol. Ther.*, *64*, 123–128.

Wang, Z., Hamman, M.A., Huang, S.M., Lesko, L.J., Hall, S.D. (2002). Effect of St John's wort on the pharmacokinetics of fexofenadine. *Clin. Pharmacol. Ther.*, *71*, 414–420.

Westphal, K., Weinbrenner, A., Giessmann, T., Stuhr, M., Franke, G., Zschiesche, M., Oertel, R., Terhaag, B., Kroemer, H.K., Siegmund, W. (2000). Oral bioavailability of digoxin is enhanced by talinolol: evidence for involvement of intestinal P-glycoprotein. *Clin. Pharmacol. Ther.*, *68*, 6–12.

Westphal, K., Weinbrenner, A., Zschiesche, M., Franke, G., Knoke, M., Oertel, R., Fritz, P., von Richter, O., Warzok, R., Hachenberg, T., Kauffmann, H.M., Schrenk, D., Terhaag, B., Kroemer, H.K., Siegmund, W. (2000). Induction of P-glycoprotein by rifampin increases intestinal secretion of talinolol in human beings: a new type of drug/drug interaction. *Clin. Pharmacol. Ther.* *68*, 345–355.

Wilson, W.H., Jamis-Dow, C., Bryant, G., Balis, F.M., Klecker, R.W., Bates, S.E., Chabner, B.A., Steinberg, S.M., Kohler, D.R., Wittes, R.E. (1995). Phase I and pharmacokinetic study of the multidrug resistance modulator dexverapamil with EPOCH chemotherapy. *J. Clin. Oncol.*, *13*, 1985–1994.

Winston, A., Back, D., Fletcher, C., Robinson, L., Unsworth, J., Tolowinska, I., Schutz, M., Pozniak, A.L., Gazzard, B., Boffito, M. (2006). Effect of omeprazole on the pharmacokinetics of saquinavir-500 mg formulation with ritonavir in healthy male and female volunteers. *AIDS*, *20*, 1401–1406.

Wu, C.Y., Benet, L.Z. (2005). Predicting drug disposition via application of BCS: transport/absorption/elimination interplay and development of a biopharmaceutics drug disposition classification system. *Pharm. Res.*, *22*, 11–23.

Xia, C.Q., Milton, M.N., Gan, L.S. (2007). Evaluation of drug-transporter interactions using in vitro and in vivo models. *Curr. Drug Metabolism.*, *8*, 341–363.

Xie, R., Tan, L.H., Polasek, E.C., Hong, C., Teillol-Foo, M., Gordi, T., Sharma, A., Nickens, D.J., Arakawa, T., Knuth, D.W., Antal, E.J. (2005). CYP3A and P-glycoprotein activity induction with St. John's Wort in healthy volunteers from 6 ethnic populations. *J. Clin. Pharmacol.*, *45*, 352–356.

Yasui-Furukori, N., Uno, T., Sugawara, K., Tateishi, T. (2005). Different effects of three transporting inhibitors, verapamil, cimetidine, and probenecid, on fexofenadine pharmacokinetics. *Clin. Pharmacol. Ther.*, *77*, 17–23.

Yasui-Furukori, N., Saito, M., Niioka, T., Inoue, Y., Sato, Y., Kaneko, S. (2007). Effect of itraconazole on pharmacokinetics of paroxetine: the role of gut transporters. *Ther. Drug Monit.*, *29*, 45–48.

Yu, J., Lo, J.L., Huang, L., Zhao, A., Metzger, E., Adams, A., Meinke, P.T., Wright, S.D., Cui, J. (2002). Lithocholic acid decreases expression of bile salt export pump through farnesoid X receptor antagonist activity. *J. Biol. Chem.*, *277*, 31441–31447.

Zapater, P., Reus, S., Tello, A., Torrús, D., Pérez-Mateo, M., Horga, J.F. (2002). A prospective study of the clarithromycin-digoxin interaction in elderly patients. *J. Antimicrob. Chemother.*, *50*, 601–606.

Zhang, L., Dresser, M.J., Gray, A.T., Yost, S.C., Terashita, S., Giacomini, K.M. (1997). Cloning and functional expression of a human liver organic cation transporter. *Mol. Pharmacol.*, *51*, 913–921.

Zhang, L., Strong, J.M., Qiu, W., Lesko, L.J., Huang, S.M. (2006). Scientific perspectives on drug transporters and their role in drug interactions. *Mol. Pharm.*, *3*, 62–69.

Zhang, L., Zhang, Y.D., Strong, J.M., Reynolds, K.S., Huang, S.M. (2008). A regulatory viewpoint on transporter-based drug interactions. *Xenobiotica*, *38*, 709–724.

Zimmerman, J.J., Harper, D., Getsy, J., Jusko, W.J. (2003). Pharmacokinetic interactions between sirolimus and microemulsion cyclosporine when orally administered jointly and 4 hours apart in healthy volunteers. *J. Clin. Pharmacol.*, *43*, 1168–1176.

Zschiesche, M., Lemma, G.L., Klebingat, K.J., Franke, G., Terhaag, B., Hoffmann, A., Gramatté, T., Kroemer, H.K., Siegmund, W. (2002). Stereoselective disposition of talinolol in man. *J. Pharm. Sci.*, *91*, 303–311.

PART II

TECHNOLOGIES FOR *IN VITRO* AND *IN VIVO* STUDIES

CHAPTER 7

Automated Drug Screening for ADMET Properties

MELVIN REICHMAN and HELEN GILL

7.1 INTRODUCTION

The pharmacological basis of therapeutics is comprised of two overarching scientific disciplines: pharmacodynamics and pharmacokinetics. It is often summarily stated: "pharmacodynamics is the study of what a drug does to the body; whereas pharmacokinetics is the study of what the body does to a drug." The latter studies the *a*bsorption, *d*istribution, *m*etabolism, *e*xcretion, and *t*oxicology of drugs (ADMET). The survival and growth of the pharmaceutical industry relies on favorable, early clinical trial outcomes. Poor outcomes are commonly due to poor ADMET properties of new chemical entities (NCEs). The difficulty in optimizing ADMET properties of any NCE is universally recognized as an important factor responsible for the increasing cost of bringing a new drug to market. On the other hand, ADMET sciences in drug development can "rescue" drugs that otherwise would fail. While the list of marketed drug withdrawals continues to increase, it appears that significant progress has been made by earlier detection of ADMET issues through automated bioassays amenable to higher throughput screening paradigms. Higher throughput ADMET for drug lead qualification is deployed more commonly in the industry, as well as earlier rather than later preclinical development. There are data suggesting that the paradigm shift of deploying ADMET earlier in the R&D continuum has led to a marked decrease in drug failures. This article traces the development of automation in experimental ADMET to provide an overview of the assays that have been miniaturized to the microtiter plate format and automated on common or specialized laboratory workstations to enable higher throughput ADMET screening.

7.2 BACKGROUND

The term high-throughput screening (HTS) began regularly appearing in the scientific literature in the early 1990s (Harris *et al.*, 1991; Burch, 1993; and see Fig. 7.1). During

Drug Metabolism Handbook: Concepts and Applications, Edited by Ala F. Nassar, Paul F. Hollenberg, and JoAnn Scatina

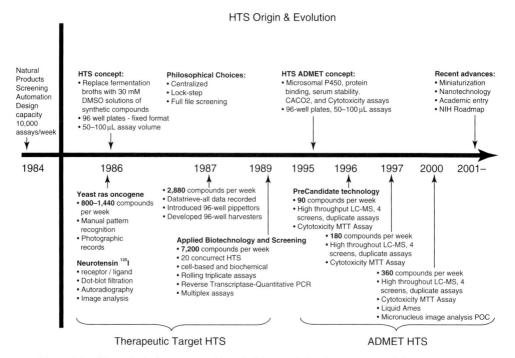

Figure 7.1   Chronological sequence of key decisions and developments pertaining to the origin and evolution of HTS to ADMET (Pereira and Williams, 2007).

the 1990s, HTS methodologies became *the* intense focus throughout the entire pharmaceutical industry. In the 1990s, a new era of technology-driven R&D rapidly emerged that focused on clever bioassays and laboratory automation. Prior to this paradigm shift, the focus in pharmaceutical R&D was on better understanding and refining the traditional principles of pharmacology that helped develop first-in-class antimicrobials, cardiovascular, gastrointestinal, psychoactive, and other drugs (see Sneader, 2005).

At the start of the 1990s, animal models were still used for primary screening and lead optimization assays. Potency and efficacy directly relate to bioavailability in animal models; therefore, when a compound showed activity without overt side effects *in vivo*, then that compound would more likely progress on a preclinical development track than a compound that had only been tested *in vitro*. Before the era of HTS, *in vivo* data were complemented by functional assays in smooth muscles and receptor binding assays. The data from this "troika" of assay types, namely, (1) biochemical, (2) *in vitro* functional, and (3) efficacy *in vivo*, were typically generated in parallel. Such data served medicinal chemists extraordinarily well in the 1970s through around 1990, during which time occurred the greatest number of successful first-in-class drug approvals in history (Ng, 2004). The pharmaceutical industry during this period of productivity was viewed as "the perfect sector."

The productivity of pharmaceutical R&D in the 1970–1990s was due to many factors. The most important was that ADMET properties of new leads were considered

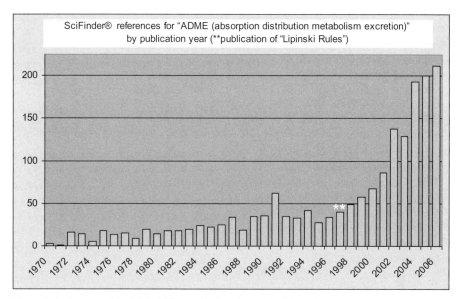

Figure 7.2   Impact of the Lipinski Rules. Christopher Lipinski published two papers in 1997–1998 summarizing the properties of "drug-likeness," or guidelines to determine if a chemical compound with a certain pharmacological or biological activity has properties that would make it a likely orally active drug in humans, based mainly on the observation that most medication drugs are relatively small (<500 MW) and lipophilic (but log P < 5) molecules. Figure graciously provided by Wolfgang Sauer of Merck Serono S.A.

at the very start of pharmaceutical R&D continuum. Primary screens were run *in vivo*, and experienced medicinal chemists and pharmacologists knew there were particular requirements for oral bioavailability, such as that compounds should have molecular weights of <500 and log P values <5 (Buxton, 2006). Nonetheless, today's emphasis on more rigorous ADMET profiling earlier in the drug development process emerged almost immediately following Lipinski's seminal publication (see Fig. 7.2). The "Lipinski Rule of Five" came to form the modern foundation of the field of cheminformatics. Cheminformatics was seen as particularly useful as a *guideline* for what *not* to synthesize.

Medicinal chemists now understand that acquiring data on potency, efficacy, and selectivity of NCEs must be coupled with ADMET for best understanding structure–activity relationships (SARs) and for chemical leads to progress to clinical development (Fig. 7.3). To generate ADMET data in parallel for dozens of therapeutic targets on the hundreds of compounds each month that are synthesized in a large pharma, manual assay methods are simply not practical. For example, if the target under investigation is a kinase, such as C-MET (Gentile, Trusolino and Comoglio, 2008; Goldstein, Gray and Zarrinkar, 2008), over 100 kinase cross-reactivity assays may be performed on newly synthesized compounds to define on-target versus off-target cross-reactivity profiles. While the problem is less intense in ADMET than for kinase inhibitor profiling, either scale of compound profiling *cannot* be achieved manually. Hence, there has been

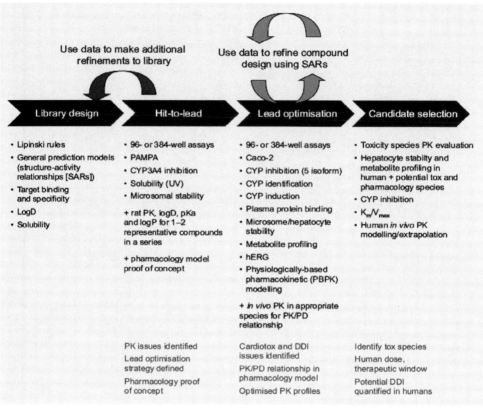

Figure 7.3 A graphical representation of the pharmaceutical R&D continuum with respect to ADMET profiling (adapted from McGee, 2005).

intense interest starting in around 2000 in the application of laboratory automation for ADMET assays *in vitro*.

7.3 MODERN LABORATORY AUTOMATION: ORIGINS IN ADMET

Starting in the early 1980s, scientists running *in vitro* screening assays began enjoying the benefits of commercially available laboratory automation. Hamilton Inc. (Switzerland), Tecan Inc. (Switzerland), and Beckman Inc. (USA) developed automated liquid-handling workstations that exist to this day as modernized "upgrades" to the original releases. These workstations could solvate and serially dilute compound dozens or even hundreds of samples, and then add the required reagents to execute bioassays, which were run only manually then. By that time, the microtiter plate was in broad use, which facilitated the development of standardized laboratory workstations with a compact footprint that could generate hundreds of data points per day. The Beckman Biomek 1000 (Fig. 7.4a) was the first "fully automated" assay workstation and possibly the most popular in the history of pharmaceutical R&D. In fact, given its size, its cost, its simple

(a)

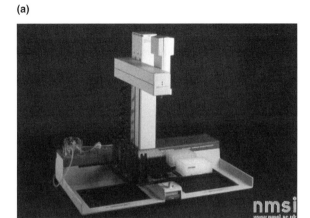

(b)

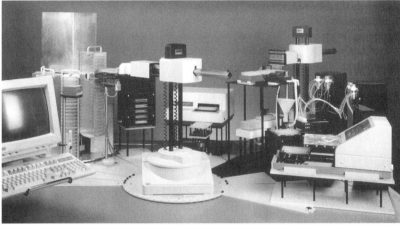

Figure 7.4   **(a)** Biomek 1000 robotic workstation, 1991. The Biomek 1000 was the first instrument designed to allow scientists to perform drug screening on one integrated device. Its features included a graphical user interface anyone could use, a full complement of pipetting heads spanning from 1 μL to 1 mL, a plate washer, and a spectrophotometer (http://www.ingenious.org. uk/See/Scienceandtechnology/Biologyandbiotechnology/?target=SeeLarge&ObjectID= {4DF77B0C-5B7B-CE96-3A8A-29E944E550D3}&viewby=images). **(b)** The Zymate System. Zymark International Inc. (a company founded by former executives of Waters, the leader in analytical instrumentation in the 1980s) introduced the radial table layout known as "pi" system. Their system was the first designed with "plug and play" modules that could be swapped in and out of the system in an extraordinary flexible manner. The Zymate was the first commercially available, fully automated, "standardized" screening platform that enabled 24/7 laboratory operations. The first adapters of the Zymate were analytical chemists who ran protocols to extract and measure drug and drug metabolites from clinical plasma samples (reproduced with permission of Zymark Corporation).

graphical user interface, ability to perform single- and multiple-channel pipetting, plate washing, and plate reading (with the aid of a unique absorbance tool for visible light), as well as other powerful capabilities (simple tool-changing ability, line-level programming, data logging, etc.), the Biomek 1000, arguably, remains unequaled as an automated bioassay workstation. Furthermore, the Biomek 1000 was sufficiently compact to fit in a standard laminar flow hood for preparing and running cultured cell assays.

Zymark advanced the concept of laboratory robotics to new "enablement" with the release of the first fully automated screening platforms equipped with a robotic arm capable of serving a large work-cell around 6 feet in diameter (Fig. 7.4b). Frank Zenie (then President and CEO of Waters, Inc., USA) and other senior executives from the Waters Corporation, the leading analytical instrumentation company at that time, founded Zymark Inc. in 1981. The Zymate systems from Zymark could be deployed for automating virtually any common laboratory assays, including solid-phase extraction, liquid chromatography–mass spectrometry (LC-MS) analysis, and other analytical tasks that were the staples of bioavailability assessment in preclinical development. By 1984, Zymark had validated their Zymate systems for the analysis of drugs in biological fluids. Their systems could run unattended around the clock (on good days) and had a significant impact for enabling high-throughput drug metabolism studies at pharmaceutical companies worldwide (see Fig. 7.4b and Banno and Takahashi, 1991).

7.4 COMPUTATIONAL CHEMISTRY 101: THE LIPINSKI RULES

Physicochemical information is important in drug discovery (Kearns, 2001, see Fig. 7.3) as follows:

- To rank order compounds for drug-like properties
- For rapid evaluation of the synthetic feasibility of compounds for HTS
- To aid in risk assessment of problems for ADMET, formulation, stability, or process
- For monitoring synthetic optimization
- Help objectively define drug advancement criteria to meet established discovery team goals.

Choosing the right balance of ADMET and pharmacokinetic properties is difficult because the rules that determine success remain, ultimately, not precisely defined. An excellent case in point is represented by the heavily cited "Lipinski Rule of Five" (see Fig. 7.2, Lipinski *et al.*, 1997; Lipinski, 2000), which originates from the considerable amount of data gathered by the industry on properties that maximize an *oral* drug candidate's probability of surviving development. The "Lipinski Rules" have been received enthusiastically by industry and academia because the rules provide a simple-minded framework (often, there lies genius in simple-minded frameworks) for defining the chemistry space for oral bioavailability, insofar as we understand this limited chemistry space, and recognizing that the "rules" are only around 75% "effective." Such rules have been applied almost universally to the design and selection of compounds for lead discovery. The Lipinski Rules state that *poor oral bioavailability due to poor absorption or permeation is more likely when*:

- There are more than five H-bond donors expressed as sum of OH and NH.
- The molecular weight is over 500.
- The c log P is over 5.
- There are more than 10 (2×5) H-bond acceptors.

There is also danger in applying "rules" on ADMET too early in the R&D continuum; there is danger in applying them too late. Instead, the SR properties can be used for rank ordering or grouping compounds into high, medium, and low categories that alert the project team that there *may* be bioavailability issues with compounds that violate the Lipinski principles.

Most might agree, however, that the most important of the Lipinski and other rules that guide on oral bioavailability of compounds are for MW (<around 500) and c log P (generally <5), the latter of which may be assessed from its measured (i.e. "observed") log P or log D value.

7.5 AUTOMATED METHODS FOR DETERMINING LOG P/LOG D

The partition coefficient is the ratio of concentrations of un-ionized compound between two immiscible solutions (see Wikipedia under "Partition Coefficient"). To measure the partition coefficient of ionizable solutes, the pH of the aqueous phase is adjusted such that the predominant form of the compound is un-ionized. The logarithm of the ratio of the concentrations of the un-ionized solute in the solvents is called log P.

$$\log P_{oct/wat} = \log\left(\frac{[solute]_{octanol}}{[solute]_{water}^{un-ionized}}\right)$$

The distribution coefficient is the ratio of the sum of the concentrations of all forms of the compound (ionized plus un-ionized) in each of the two phases. For measurements of distribution coefficient, the pH of the aqueous phase is buffered to a specific value such that the pH is not significantly perturbed by the introduction of the compound. The logarithm of the ratio of the sum of concentrations of the solute's various forms in one solvent, to the sum of the concentrations of its forms in the other solvent is called log D:

$$\log D_{oct/wat} = \log\left(\frac{[solute]_{octanol}}{[solute]_{water}^{ionized} + [solute]_{water}^{un-ionized}}\right)$$

The log D is pH-dependent; hence, one must specify the pH at which the log D was measured. Of particular interest is the log D at pH = 7.4 (the physiological pH of blood serum). For un-ionizable compounds, log P = log D at any pH. Generally, Wikipedia provides an excellent summary of log P and log D, and the reader is referred there for additional hyperlinked references.

"Automation" of log P determination by high performance liquid chromatography (HPLC) inferred from retention times on hydrophobic (C-18) columns was first published in 1978 (Unger, Cook and Hollenberg, 1978). There are provisos, however, that always accompany the majority of expedient (*in vitro*) assay methods. Rapid HPLC methods work best for predicting accurate partition coefficients *within a related structural series of compound*; however, HPLC alone without physical partitioning coefficients of at least some members of the series of compounds under investigation is subject to potential error. Therefore, the classical "shake-flask" method is still used, in combination with HPLC with UV (and MS) detection to quantitatively measure the amount of compound distributed between the organic and aqueous phases. The importance of drug partition coefficients is that they often relate directly to blood–brain barrier (BBB) permeability and predict oral bioavailability to a reasonable degree.

Another approach to automated partition coefficient determination is a hydrophobicity scale based on the migration index from microemulsion electrokinetic chromatography (MEEKC) of anionic solutes. The MEEKC migration index has been automated and commercially available as a laboratory workstation termed the "pKa PRO system" (Fig. 7.5), which enables parallel, accurate, separation-based log P measurements that are insensitive to impurities and consume minimal amounts of sample. Several published studies suggest that the instrument provides more reliable correlations between log P and the classic octanol:water partitioning method than reverse-phase liquid chromatography.

Basic compounds should have a pKa value less than 10 to minimize charge effects. The use of MEEKC provides a validated method with good correlation to the traditional shake-flask technique (Neubert *et al.*, 1999). The method requires only microgram quantities of compound and is separation-based, allowing for detection and separation of potential impurities. The device is based on parallel, 96-channel microelectrophoresis. Published studies have noted that it is also possible to evaluate the pKa values of compounds in a high-throughput manner using the same instrument platform, by simply exchanging capillary arrays and buffer solutions.

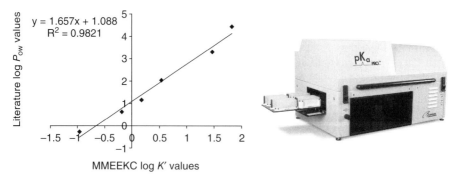

Figure 7.5 The pKa PRO system equipped with the log P application package provides automated measurements of neutral and basic compounds over the log P range of −0.5 to 5.0 with a throughput of up to 46 samples/hour (reproduced with permission from Advanced Analytical Technologies, Inc., http://www.aati-us.com/systems/pKa.html).

7.6 DRUG SOLUBILITY

7.6.1 Background

In order for an orally administered compound to act, it must dissolve in the aqueous environment of the gastrointestinal tract and within the cytoplasm when the target is intracellular. The partition coefficient of a compound is directly related to its aqueous solubility—all else equal within a series. The more lipophilic a compound is, the lower is its aqueous solubility. Poor solubility can limit the absorption of compounds from the gastrointestinal tract, resulting in reduced oral bioavailability. It may also necessitate novel formulation strategies and hence increase cost and delays. Moreover, compound solubility can affect other *in vitro* assays. Poor aqueous solubility is the largest physicochemical problem hindering oral drug activity—hence on completion of compound synthesis, the determination of aqueous solubility is a priority.

Solubility data may prove critical in rationalizing disparate results in other bioassays. For example, a compound must be soluble in the assay buffer used to measure activity. Poorly soluble compounds within a series are more likely to have variable solubility in different buffer systems used in common bioassays, which can confound the SARs that are critical for effective lead optimization. In fact, some have shown that determination of the effective concentration of compounds in bioassays reveals marked errors in potency measurements due to marked underestimation of the actual concentration of compound that is in solution in the bioassay, particularly when a single concentration is used to assess activity by extrapolated IC_{50} values (Popa-Burke *et al.*, 2004).

Determination of compound solubility is, therefore, quite important at every stage of the pharmaceutical R&D continuum including:

1. *The earliest stages of discovery.* Poorly soluble compounds can precipitate in the initial screen and lead to "false negatives," or lead to variable data due to minor changes in assay conditions (including room temperature, minute changes in the pH of a buffer, etc.).

2. *Lead optimization stage.* This, along with animal pharmacology, is the most expensive component of preclinical drug development. The greatest variety of bioassays are deployed at this stage to assess cross-reactivity and mechanism of action of nascent drug leads. The standard operating procedures (SOPs) for every bioassay typically call for specific (i.e. nonuniform) buffer recipes, ions, and cofactors—and sometimes even detergents—that surely affect pH, ionic strength, and potential ability to raise (or lower) the critical micelle concentration (CMC) of compounds in aqueous buffers.

3. *Preclinical development.* Compound solubility may be critical to clinical efficacy. One focus for formulation chemists is to optimize compound solubility, which may be accomplished by various physical formulations. For example, progesterone is rather insoluble as an oil and has very poor bioavailability (<1%); however, micronization may increase solubility (>5%), and therefore markedly enhance efficacy and potency (Fitzpatrick and Good, 2000).

7.6.2 Automated Methods Used to Determine Drug Solubility

Compound solubility is usually determined by either kinetic or thermodynamic methods. Kinetic assessments, which start from a solution of dissolved compound, are now common in an early discovery setting. Often a turbidimetric end point is used for the kinetic solubility measurements. Thermodynamic assessments, where a saturated solution of a compound in equilibrium is prepared from solid, are considered to be the gold standard and take into consideration the crystal lattice of the compound, and therefore, the dissolution rate of the compound. Throughput is generally lower than kinetic measurements and larger quantities of analytical grade of compound are ideally used for these studies. For this reason, thermodynamic assays are usually performed in early development (rather than in the earlier discovery stages) to confirm earlier kinetic solubility results, to rule out potential artifacts, and to generate quality solubility data with crystalline material to support the best selection of potential clinical candidates. The traditional shake-flask method is still frequently used in the drug development phases for measuring thermodynamic solubility. The commercial "pSol Gemini" instrument (pION, Woburn, MA, USA) uses a pH-based approach, which is approved by the FDA to derive a complete solubility–pH profile. This allows experimental validation of prediction calculations for solubility, as well as $c \log P$ and $c \log D$. Less than 100 micrograms of compound is required, making the method suitable for early stages of discovery where compounds can be in very limited supply, although throughput is limited to handfuls of compound per day (Avdeef, 2007; see also http://www.pion-inc.com/papers.htm).

A number of automated methods based on kinetic solubility have been developed, which have dramatically increased the throughput of solubility assessment. A brief overview of the methods can be found in Kearns (2001). The MultiScreen Solubility filter plate (Millipore, MA, USA) and Screening Method provides an automation-compatible, high-throughput means to estimate the aqueous solubility of hundreds of compounds per day. Using a single-point calibration, the screening ratio is easily and quickly derived, and compound solubility is readily approximated. Multiple samples, each requiring approximately 200 nanomoles (~100 µg) per result, can be run in parallel. The method allows for the analysis of approximately 45 compounds (duplicate determinations) per plate with the capability of completing four or more plates in a standard 8-hour day (see Fig. 7.6a and b).

Methods that use turbidimetric detection techniques are now becoming more popular as they bypass the need for HPLC analysis. Turbidimetry measures the onset of precipitation by either absorbance or nephelometry (light scattering) and is often used as the end point for kinetic solubility determination. These techniques are amenable to high-throughput format with rapid data generation and require only small amounts of test compound, usually from readily available dimethyl sulfoxide (DMSO) stock solutions (Bevan and Lloyd, 2000). An additional advantage of the turbidimetric method is that it is easily temperature controlled and incubations can be performed and monitored at physiological temperature.

7.6.3 Summary: Measuring Drug Solubility

Figure 7.7 depicts a simple approach to automate determination of aqueous drug solubility. The following provisos must be considered in establishing and then validly managing a high-throughput ADMET laboratory for determining drug solubility:

- Use the turbidimetric and equilibrium methods for discovery and development projects, respectively.

(a)

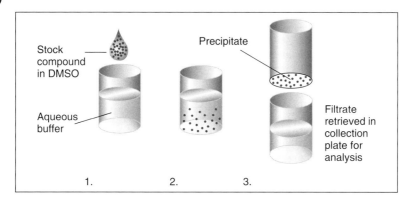

(b)

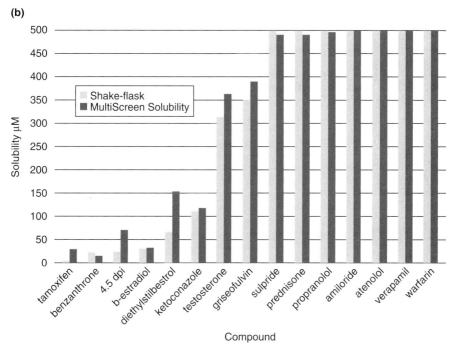

Figure 7.6   Graphical depiction of basis for the MultiScreen solubility filter plate method (a) and its comparison with the shake-flask method (b). The filtrate retrieved in the collection plate may be analyzed by HPLC with ELSD (evaporative light scattering detection) (against a standard curve or known concentration of test compound, or by bioassay). (http://www.millipore.com/catalogue/module.do?id=C8875).

• Measure onset of precipitation by absorbance change or nephelometry.
• Rapid, high-throughput plate-based method which uses
  small amounts of compound in dimethylsulfoxide (DMSO) solution.
• See how compound solubility changes throughout the
  GI tract by performing the assay at a number of pH values.
  Ionisable compounds will show pH dependant solubility.
• Used primarily in drug discovery.

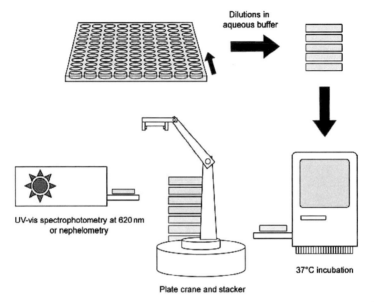

Figure 7.7 An example of drug solubility assay automation using a turbidimetric method (from www.cyprotex.com).

• Assess solubility using different pH levels for ionizable groups.
• Assess solubility data first and then use this as a guide to interpreting results from *in vitro* experiments.
• Use surrogate data for ranking purposes rather than as "rejection filters."
• Use solubility data to aid design of new chemical libraries.
• Use solubility data as a guide for formulation strategies and dosage.
• If the therapeutic indication is an area of high unmet medical need (cancer), oral bioavailability would not be an appropriate parameter for early "Go-No-Go" decision making.
• It may be generally shortsighted to reject a compound based solely on solubility data. Route of administration and therapeutic plasma concentrations must also be considered, particularly in areas of desperate unmet medical need, such as oncology and even diabetic complications.
• Optimizing physical form (formulation) can have a significant impact on solubility.

It is notable that both Figures 7.6b and 7.8 are in reasonable agreement between automated high-throughput, microtiter-based kinetic methods and classical shake-flask methods. Figure 7.9 provides data demonstrating that it can be difficult to effectively mix reagents in wells of a microtiter plate uniformly. Others have shown that there can be considerable variability between different lots of the same compound that are simply distributed on multiple microtiter plates, and that the potency of a compound in a given assay can be off by more than threefold because of solubility issues (Popa-Burke *et al.*, 2004). In summary, great care and attention to details are obligatory for successful automation of pharmacokinetics assays run in parallel, and from whose data compounds will either be "killed" or advance. In the field of ADMET we may say, therefore, that "the executioner lies in the details."

7.7 DRUG PERMEABILITY

7.7.1 Why Is Permeability Important?

Drugs with low membrane permeability (i.e. low lipophilicity) are generally absorbed slowly from solution in the stomach and small intestine, despite that they may readily

no.	compound	solubility ($\mu g/mL$)			comment[a]
		from solid by HPLC	from DMSO soln by HPLC	from nephelometry	
1	acetazolamide	>500	>111	>111	✓
2	allopurinol	337	>68	>68	✓
3	bendroflumethiazide	12	33	13	✓
4	benzocaine	>500	>83	>83	✓
5	benzthiazide	11	20	27	✓
6	betamethasone	63	>196	>196	✓
7	butamben	125	85	100	✓
8	butylparaben	139	88	97	✓
9	chlorpropamide	>500	>138	>138	✓
10	clofazimine	<1	30	1	red soln, insoluble by eye
11	flurbiprofen	>500	>122	>122	✓
12	gemfibrozil	>500	>125	>125	✓
13	hydrocortisone	258	175	>181	✓
14	hydroflumethiazide	187	>166	>166	✓
15	iodipamide	>500	>57	57	✓
16	nitrofurazone	163	>99	>99	✓
17	oxyphenbutazone	>500	<1	40	✓ with DMSO soln by HPLC result
18	phenacemide	182	>89	>89	✓
19	phenylbutazone	320	>154	77	
20	prednisone	98	>179	89	✓
21	propylparaben	273	89	>89	✓
22	tolazamide	350	>174	90	
23	trimethoprim	>500	>145	145	✓
24	tyrosine	no result	<1	91	✓
25	hydroquinine	>500	138	120	✓
26	morin hydrate	192	>151	nd[b]	red soln
27	phenyl salicylate	16	<1	>107	
28	theobromine	>500	>90	90	✓
29	2-hydroxy-3-isopropyl-6-methylbenzoic acid	>500	>97	>97	✓
30	progesterone	6	15	20	✓

[a] ✓ indicates acceptable agreement.

Figure 7.8 A comparison of different methods to measure drug solubility. Results of the nephelometric, kinetic, and equilibrium solubility methods (Kearns, 2001).

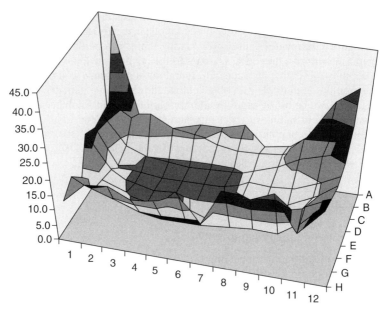

Figure 7.9 An example of the importance of small details in the assay significantly influencing results. Stirring efficiency varies considerably from well to well on a 96-well plate when shaken on an orbital shaker. The shaker has room for four plates. The A12-corner faced the center of the shaking tray. The data are instructive of the principle that for successful conversion of classical ADMET assays to high-throughput automated assays, small details are critical (from http://pion-inc.com/images/PermPosterCPH01.pdf).

dissolve in aqueous media. The ability to assess the rate and extent of absorption across the intestinal tract is critical if a small organic molecule is to be developed as an orally available drug; and generally, this is the objective across the entire pharmaceutical industry. Understanding whether a compound is likely to enter the brain and central nervous system (CNS) via the BBB may also be important for any drug, not only psychoactive medicines. If the compound in question is a potential CNS therapeutic, crossing the BBB is a prerequisite, of course. Conversely, BBB penetration might cause unwanted off-target side effects for those drugs not intended for the CNS. Measuring drug permeability through appropriate monolayer model systems is a popular surrogate for measurement of intestinal permeability and oral bioavailability.

7.7.2 What Factors Determine Permeability?

Permeability—or drug absorption—may be simple (as in the case of ethanol, which is permeable through the gastrointestinal tract by passive diffusion) or the result of a complex composite of different physicochemical properties required for optimal oral absorption, as well as certain biological processes that may actively transport—or eliminate—drugs from the bloodstream. There are three pharmacokinetic processes that govern permeability: (a) passive diffusion, (b) active transport via drug transporters (requiring ATP), and (c) facilitated transport (not requiring ATP).

As a general rule, compounds that are very hydrophilic do not cross cell membranes readily; conversely, compounds that are very lipophilic tend to readily cross cell membranes. In general, medicinal chemists tend to focus on optimizing permeability first, because solubility may be improved through formulation at a later stage. For example, a micronized (fine pulverization) drug formulation can increase its rate of dissolution by increasing its surface area by a factor of 10 or greater. Conversely, pegylation can decrease solubility (rate of release) by a factor of 10 or more (Boyd, 2008). *It is not as easy to optimize drug absorption by formulation protocols as it is to enhance (or decrease) solubility.*

7.7.3 Methods Used for Investigating Permeability

Drug permeability cannot be accurately predicted by physicochemical factors alone because there are many drug transport pathways that are driven by protein transporters that recognize their ligands in ways that cannot be simply predicted. Consequently, a number of validated, cell-monolayer systems *in vitro* have been validated, including the human colon adenocarcinoma cell line, Caco-2 and the Madin Darby canine kidney (MDCK) cell line, to help predict intestinal permeability. These assays are commonly employed during early discovery, especially in lead prioritization for the chemotypes discovered by HTS, as well as for lead optimization to guide the medicinal chemist for enhancing intrinsic potency and efficacy (pharmacodynamics) in parallel with bioavailability (pharmacokinetics). Because of the complexity involved in measuring active transport, most lead optimization efforts focus on passive permeation to better predict the oral bioavailability of drugs. Caco-2 and MDCK (transfected and non-transfected) cell cultures are widely used *in vitro* methods for estimating both passive and active permeability. Although advances have been made in throughput and consistency with the introduction of drug transport plates amenable to automation, there is still a requirement for cell culture and the associated time delay in initiating studies and risks of infection (of cell cultures by environmental microorganisms). This has led to the development and adoption of alternative techniques.

A newer *in vitro* model, known as the parallel artificial membrane permeability assay (PAMPA), has been developed (Figs. 7.10 and 7.12). This approach ranks compounds on their passive diffusion rates alone. PAMPA is increasingly used as the first-line permeability screen during lead profiling. Both methods have been adopted for use in assessing the ability of compounds to cross the BBB (Fig. 7.11).

The functional concept depicted by Fig. 7.10 has been incorporated into the microtiter plate format by Millipore and termed the MultiScreen PAMPA System (also see Fig. 7.6), which is an important basis for fully automated methods to assess permeability. PAMPA data may be used to bin compounds into high- and low-permeability categories. Furthermore, close examination of BBB and high-resolution PAMPA examples led to the presumption that PAMPA can be considered a sophisticated method of measuring lipophilicity (as measured by octanol/water partitioning; see Fig. 7.11).

7.7.4 Summary

Figure 7.12 depicts automation of the PAMPA assay and Fig. 7.13 displays some reference compounds that may be used to "bin" compounds based on their permeability and solubility classes. Recently, Galinis-Luciani, Nguyen and Yazdanian (2007) have published a critical assessment of the PAMPA method (see Fig. 7.11). The authors

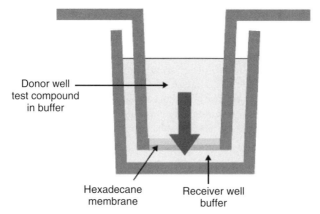

Figure 7.10 The basis for the PAMPA assay depicted graphically. Compound is added to the donor well and the amount of compound transferred across an artificial bilayer into the receiver well is measured by HPLC (from www.cyprotex.com).

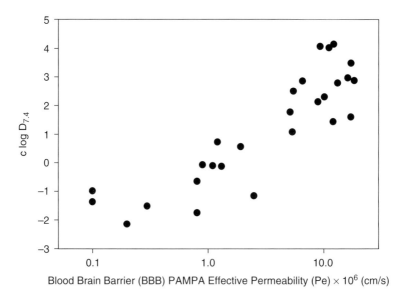

Figure 7.11 Comparison of calculated log D7.4 values versus blood–brain barrier PAMPA effective permeability (from Galinis-Luciani, Nguyen and Yazdanian, 2007). On the one hand, Fig. 7.11 may be interpreted as support for the PAMPA assay as a predictor of blood–brain barrier permeability. On the other hand, the data imply that the PAMPA assay may not yield better information than calculated log D values.

Figure 7.12   An example of modern automation of the PAMPA Evolution 96 permeability assay. Compound and plate preparation were accomplished with the Biomek FX Laboratory Automation Workstation. The figure depicts a fully automated turnkey application for accomplishing the PAMPA assay. (Courtesy of Beckman Coulter.)

Drug	Permeability class	Solubility class
Carbamazepine	High	Low
Ketoprofen	High	Low
Verapamil	High	Low
Metoprolol	High	High
Propranolol	High	High
Atenolol	Low	High
Ranitidine	Low	High

Figure 7.13   Reference compounds that may be used to "bin" test compounds (new chemical entities) into high-versus-low solubility and permeability classes.

concluded that PAMPA data collected at a single pH permitted compounds to be binned as being predicted to have high and low permeability, but the same outcome could be achieved using calculated log D7.4 values. Their conclusion is that the PAMPA assay has little, if any, advantage over mechanistic permeability assays such as the Caco-2 or MDCK cell models. The cellular models provide additional information about active transport. Upon examination of the balance of quality of data and quantity of time, the PAMPA assay may be a very limited asset to a drug discovery effort these authors conclude. Others conclude that the success of using PAMPA in drug discovery depends on careful data interpretation, use of optimal assay conditions, implementation and integration strategies, and education of users with respect to expectations and proper application of the data (Avdeef *et al.*, 2007).

7.8 DRUG TRANSPORTERS IN INTESTINAL ABSORPTION

Understanding the role of transporters in pharmacokinetic processes has become a key objective over recent years. Transporters play an important function in renal clearance, biliary clearance, hepatic uptake, intestinal absorption, and BBB permeability, as well as cancer cell resistance. For a more detailed description of transporters and their roles in drug absorption and elimination, see Chapter 6. Both influx transporters, which facilitate the entry of drugs into cells, and efflux transporters, which limit the entry of drugs or enhance the removal of drugs from the cells, are crucial in determining concentrations of drug in the body and within various cells and tissues (see Fig. 7.14). As is the case with cytochrome P450 (CYP) enzymes, transporter-mediated drug interactions have been reported in both humans and animals. The substrate specificity of the individual transporters often overlaps and therefore the evidence for transporter interactions is often less conclusive. Many inhibitors and inducers can simultaneously affect both drug transport and CYP enzymes. The lack of specific probe substrates and inhibitors for the individual transporters has hindered our development of simple *in vitro* assays.

Our knowledge is growing of the transporters present in the various tissues of the body. In enterocytes, a number of different transporters exist which are involved in both drug uptake and drug efflux, as illustrated in Fig. 7.14. Although many transporters have been identified, P-glycoprotein (P-gp) is the most extensively studied and many of the reported transporter-mediated clinical interactions focus around P-gp. These include clinical drug–drug interactions (DDIs) between the P-gp substrate digoxin and other drugs such as erythromycin, quinidine, verapamil, talinolol, and clarithromycin (Doering, 1979; Leahey *et al.*, 1979; Bigger and Leahey 1982; Klein *et al.*, 1982; Belz *et al.*, 1983; Maxwell, Gilmour-White and Hall, 1989; Wakasugi *et al.*, 1998; Verschraagen *et al.*, 1999; Westphal *et al.*, 2000). The tools for evaluating substrates and inhibitors of P-gp are much advanced compared with the other transporters. The most popular *in vitro* methods for investigating P-gp are Caco-2 cells and either the MDCKII cell line or the LLC cell line transfected with the human *MDR1* gene. These cells can be grown as polarized monolayer and the bidirectional transport of the drug through the monolayer can easily be investigated. The Millipore MultiScreen Permeability plates have been successfully used for full automation of transport (permeability) assays.

A number of higher-throughput methods also exist. For example, ATP is the energy source for moving substrates by the ABC transporters; therefore, generation of inorganic phosphate from ATP hydrolysis can easily be measured using a simple colorimetric reaction. This assay is useful for distinguishing between substrates and inhibitors as well as rapidly estimating the kinetics of the drug transporter interaction (Garrigues *et al.*, 2002; Glavinas *et al.*, 2007).

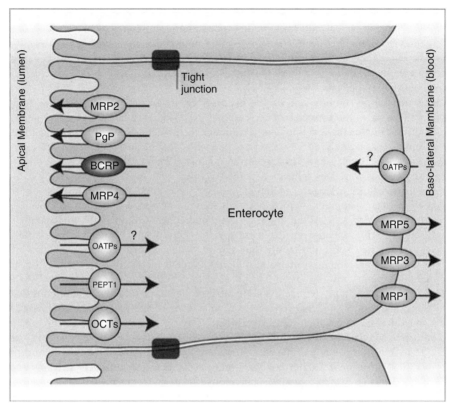

Figure 7.14 Localization of efflux and uptake transporters in the enterocyte. It is important to note that there is still dispute over the presence of OATP transporters in the enterocyte and, for this reason, these have been marked with question marks in this figure (with permission from SOLVO Biotechnology, Budapest, Hungary).

7.9 DRUG METABOLISM

7.9.1 Hepatic Intrinsic Clearance

The ideal scenario is that a drug has an optimal half-life for once-a-day dosing with no toxic effects from drug accumulation. As the majority of drug metabolism occurs in the liver, the most popular *in vitro* models employed for evaluating intrinsic clearance in a discovery setting are either human or animal-derived hepatic microsomes, hepatic S9, or hepatocytes. Typically, microsomes are used as a first line screen, and then the most favorable compounds are then assessed in the presence of hepatocytes. The data generated from both these models are used to select compounds for further development.

7.9.1.1 Liver Microsomes Subcellular fractions such as liver microsomes are one of the most commonly used *in vitro* models of hepatic clearance in drug discovery.

Microsomes are easy to prepare and can be stored for long periods of time. They are easily adaptable to 384-well formats that enable large numbers of compounds to be screened rapidly and cost-effectively. A pool of human donors is typically assessed to reduce the influence of interindividual variability. A wide range of different species is commercially available and therefore the assay is useful in assessing interspecies differences in metabolism.

For metabolic stability studies, the microsomes are incubated with the test compound at 37 °C in the presence of the cofactor, NADPH, which initiates the reaction. Solvent terminates the reaction and precipitates the protein. Following centrifugation, the supernatant is analyzed by HPLC coupled with MS. The substrate concentration is typically set between 1 and 5 μM to ensure adequate analytical sensitivity yet to limit solubility issues or risk of saturation of the enzymes.

In discovery, knowledge of metabolic pathways is limited and, therefore, substrate depletion over a number of time points is now an accepted approach to calculate intrinsic clearance. Due to the low test compound concentration in the final samples, saturation of the response on the mass spectrometer is unlikely and it is generally considered unnecessary to quantify the samples using a standard curve.

Although microsomes contain the full complement of phase I enzymes, cytosolic metabolizing enzymes are absent and phase II metabolism may be overlooked. As a result, microsomes tend to underestimate intrinsic clearance for those compounds that undergo metabolism by these phase II enzymes. One of the most common phase II enzymes, uridine glucuronyl transferase (UGT) is present in microsomes. However, in contrast to CYPs and the flavin-containing monooxygenases, the active site of the UGTs resides in the lumen of the endoplasmic reticulum (ER), and the ER membrane provides a diffusional barrier for substrates, cofactors, and products (Meech and Mackenzie, 1997). In a microsomal incubation, disruption of this barrier is required to remove the latency of UGTs. Traditionally, detergents were added to the incubation; however, the optimal levels of detergent often varied for different substrates which circumvented this as a routine early screen (Lett *et al.*, 1992). More recently, the use of the antibiotic fungal peptide alamethicin has been proposed as a universal substrate-independent *in vitro* method to study glucuronidation activity in human liver microsomes (Fisher *et al.*, 2000). Alamethicin forms well-defined pores in the membrane and allows access to the enzyme without affecting gross membrane structure or general intrinsic enzyme catalytic activity. Supplementation of the microsomal incubations with alamethicin and the cofactors, NADPH and UDGPA, allows both CYP-mediated and UGT-mediated metabolism to be evaluated as an early stage screen. Extending the range of enzymes investigated is likely to improve the prediction of hepatic clearance from microsomal stability data.

The data analysis system at Cyprotex enables issues such as retention time shifts, peaks in blank samples, internal standard fluctuations, atypical profile fluctuations, inconsistencies in the minus NADPH data, and control compound failures to be flagged with warning messages (Fig. 7.15). These warnings can be automatically integrated into comments on the final spreadsheets. As well as dramatically improving efficiency, this type of system acts as an additional checking mechanism and so reduces the chances of error in the data. Checks performed by scientists are still deemed necessary to maintain the quality of the data. However, as these systems become more sophisticated, it is entirely feasible that checking systems relying entirely on machine interpretation will eventually become more commonplace, generating higher quality and less subjective data than manual human intervention.

Auto check warning	
Condition failed	**Warning**
FALSE	Interfering peak
FALSE	Compound detection problem
FALSE	Minus NADPH control low
FALSE	Minus NADPH control high
FALSE	Low response
FALSE	Low internal standard response
FALSE	Using all five time points
FALSE	Variable profile
FALSE	Retention time shifts
FALSE	Test compound not detectable
TRUE	Control compound out of range

Full warning message	Control compound out of range

Accept	FALSE
Reject	TRUE

Figure 7.15   An example of automated checking procedures performed at Cyprotex in the Cloe Screen Microsomal Stability assay to assure QC/QA in automated analytical methods.

7.9.1.2 *Supplemented S9 Fraction*

S9 fraction consists of both cytosolic and microsomal enzymes. As with microsomes, this model is amenable to HTS, and as such is rapid and cost-effective. Despite containing the full complement of hepatic enzymes, the S9 fraction is not energetically competent and requires cofactor supplementation to enable phase II metabolism. The activity of the S9 tends to be lower than that of microsomes and as such higher protein concentrations are required in the incubations, which can lead to issues with extensive binding to the S9 protein. This model tends to be less popular than human liver microsomes as a first line screen.

7.9.1.3 *Hepatocytes*

Hepatocytes have obvious advantages over subcellular fractions in that they are whole cell systems that are energetically competent with active transporters. This avoids issues with latency and circumvents the need for cofactor supplementation. Due to these reasons, the predictive power of hepatocytes tends to be considered superior compared with microsomes. With improvements in cryopreservation techniques, practical issues associated with isolation of the hepatocytes are reduced, and intrinsic clearance assessment using cryopreserved hepatocytes has increased noticeably over the past five years (Hewitt *et al.*, 2007). Despite this, the use of hepatocytes as a model for intrinsic clearance still remains a second line screen for many companies, mainly a consequence of increased running costs. Hepatocytes tend to be the system of choice for applications such as metabolite profiling and induction studies which are performed later and are less cost-sensitive.

7.9.2 Approaches for Interpretation of Intrinsic Clearance Data

Often, classification bands are used to categorize compounds into low, medium, or high clearance. These classification bands are calculated assuming an extraction ratio (the

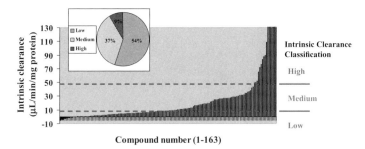

Figure 7.16 Classification of microsomal intrinsic clearance for a set of 163 FDA-approved drugs. Drugs were screened in the Cloe Screen Microsomal Stability assay at Cyprotex. From the set of drugs screened, over 90% of the FDA-approved drugs were classified as low or medium intrinsic clearance in the microsomal stability assay. Only 9% of the drugs were rapidly cleared.

TABLE 7.1 Classification bands for categorizing compounds into high, medium, and low clearance.

	Intrinsic clearance (µL/min/mg protein)	
Clearance category[a]	Human liver microsomes	Human hepatocytes
Low	≤8.6 µL/min/mg protein	≤3.5 µL/min/10⁶ cells
Medium	Between 8.6 and 47 µL/min/mg protein	Between 3.5 and 19 µL/min/10⁶ cells
High	≥47 µL/min/mg protein	≥19 µL/min/10⁶ cells

[a]Assumes:
- Low clearance equivalent to an extraction ratio of less than or equal to 0.3 and high clearance equivalent to an extraction ratio greater than or equal to 0.7 (hepatic clearance = extraction ratio × hepatic blood flow).
- Conversion of hepatic clearance to intrinsic clearance uses the equation: $CL_{int} = CL_H/(fu(1 - E))$ and assumes fu = 1.
- Scaling factors to convert from mL/min/70 kg to µL/min/mg protein assumes 1800 g liver/70 kg man and 40 mg protein/g liver (microsomes) or 99×10^6 cells/g liver (hepatocytes).

fraction of drug that is eliminated from the blood by an organ) of 0.3 and 0.7 for the low and high boundaries, respectively, as defined in Wilkinson and Shand (1975). Figure 7.16 suggests that the large majority of FDA-approved drugs (>90%) have low or medium clearance classification in the microsomal stability assay (Table 7.1).

The caveat with using the classification approach is that only microsomal clearance is taken into account for the prediction of hepatic clearance. Therefore, a common approach is to predict hepatic clearance from the intrinsic clearance by incorporating hepatic blood flow, plasma protein binding, and microsomal binding.

7.9.3 Drug–Drug Interactions

DDIs are a significant safety concern as substantial changes in blood and tissue concentrations of drug or metabolite can occur. These fluctuating exposure levels can alter the safety and efficacy profile of a drug and/or its metabolites, especially for drugs with a narrow therapeutic range. DDIs remain a major regulatory hurdle, which can lead to early termination of development, refusal of approval, prescribing restrictions, or withdrawal of drugs from the market. Therefore, the potential for drug interactions should be evaluated early. These studies should be of sufficient quality as the outcome ultimately influences the design of clinical trials. Metabolic drug interaction studies focus on the CYP family as these enzymes play a key role in the metabolic clearance of drugs. The availability of human tissue is a valuable tool in the *in vitro* assessment of drug interactions.

7.9.3.1 Cytochrome P450 Inhibition The two most popular *in vitro* systems for assessment of CYP inhibition are human liver microsomes and recombinant CYP enzymes. These systems investigate the potential of the test compound to inhibit the metabolism of a probe substrate by the main CYP isoforms. The probe substrate selected should be predominantly metabolized by a single CYP enzyme and should have a simple metabolic scheme preferably with no sequential metabolism. Initial evaluation of time and protein linearity as well as enzyme kinetic studies should be performed during the setup of these screening assays. For early discovery screening, an IC_{50} (concentration at which the test compound causes 50% inhibition of the probe substrate metabolism) is deemed sufficient for ranking and selection purposes. IC_{50} studies should be performed using a probe substrate concentration at or below its Michaelis–Menten constant (K_m) with a minimum of six inhibitor concentrations and under linear conditions with respect to time and protein. Typically, compounds are categorized into the following classification bands (Krippendorff *et al.*, 2007):

- Potent inhibition $IC_{50} < 1\,\mu M$
- Moderate inhibition IC_{50} between 1 and $10\,\mu M$
- No or weak inhibition $IC_{50} > 10\,\mu M$.

If IC_{50} studies indicate that an investigational drug does not inhibit the main CYP isoforms, then corresponding *in vivo* inhibition-based interaction studies of the investigational drug and concomitant medications eliminated by these pathways are typically not required.

For later stage studies where decisions are made regarding clinical studies and where inhibition is observed in the IC_{50} screen, a more detailed K_i determination is considered necessary. The K_i (or inhibition constant) is the affinity with which the inhibitor binds to the enzyme. K_i determination has several advantages over IC_{50} determination in that it is independent on type and concentration of substrate and incubation conditions and, as such, is more reproducible between laboratories. Data are useful in identifying the type of inhibition (i.e. competitive, noncompetitive, uncompetitive, or mixed) and Eadie Hofstee plots tend to be the preferred method for differentiating between the various types of reversible inhibition.

Probe substrates, which form fluorescent products, have been introduced as an alternative to the traditional probe substrates. These substrates are common in an early discovery setting as fluorescent plate-based analysis is faster than LC-MS analysis and so throughput can be enhanced. However, there are several disadvantages in using fluoroprobes for P450 inhibition. Firstly, most of these substrates are not P450 isoform-specific, and so they have to be screened using individually expressed enzyme. The absence of other P450 enzymes will eliminate competing pathways of metabolism. Secondly, clinical assessment of these substrates is difficult in humans and therefore understanding their relevance to the *in vivo* situation is limited. Thirdly, there is a poor correlation of inhibitory potential between different fluoroprobes (Stresser *et al.*, 2000). Finally, if the inhibitor or its metabolite fluoresces, then interference in the assay can occur, leading to false results. As a consequence, regulatory authorities have not accepted fluorescent probes, and the majority of companies still currently employ the traditional probe substrate methods, which utilize LC-MS/MS. One successful approach to reduce analytical time in these traditional methods has been to combine supernatants from multiple samples into a single "cassette," following termination of the individual reactions. Coupled with rapid gradient ultra-performance liquid chromatography methods, this enables analysis times to be dramatically reduced and enables HTS of the traditional methods to be a more realistic proposition.

7.9.3.2 *Mechanism-Based (Irreversible) CYP Inhibition*   Mechanism-based inhibition, also known as "suicide inhibition" or "time-dependent inhibition," is an irreversible interaction where a covalent bond is formed between a reactive metabolite and the active site of the enzyme, destroying the enzyme's activity. The consequences of mechanism-based inhibition are considered to be more serious than reversible inhibition because the inactivated enzyme must be resynthesized before activity is restored. In addition, the irreversible inactivation usually implies the formation of a covalent bond between the metabolite and the enzyme, which can lead to hapten formation and in some cases, can trigger an autoimmune response. The regulatory authorities are now aware that this is a significant problem and are proposing that NCEs are screened for the potential to cause mechanism-based inhibition. Clinically important mechanism-based inhibitors of CYP include antibacterials (e.g. clarithromycin, erythromycin, and isoniazid), anticancer agents (e.g. tamoxifen and irinotecan), anti-HIV agents (e.g. ritonavir and delavirdine), antihypertensives (e.g. dihydralazine, verapamil, and diltiazem), sex steroids and their receptor modulators (e.g. gestodene and raloxifene), and several herbal constituents (e.g. bergamottin and glabridin) (Zhou *et al.*, 2005).

The design of the protocol for mechanism inhibition involves pre-incubating the test compound at a single concentration with NADPH and 10 times the final concentration of human liver microsomes for 30 minutes. If the test compound is a mechanism-based inhibitor, then it will be activated in the presence of the NADPH and bind irreversibly to the active site of the enzyme. Following the pre-incubation stage, an aliquot of the incubation is diluted 10-fold with buffer containing the CYP isoform-specific substrate and then a second incubation occurs. Mechanism-based inhibitors are not dependent on the probe substrate concentration; therefore, it is not necessary to perform the studies at the K_m of the substrate. By diluting the test compound 10-fold for the final incubation with the marker substrate, any potential reversible inhibition is minimized. Several control incubations are included in addition to the test well where the pre-incubation is performed in the absence of NADPH, and absence of test compound. Details of all incubations are illustrated in Fig. 7.17. The percentage inhibition following pre-incubation is calculated for the test compound using the following equation (the peak area ratio is the peak area of the probe substrate metabolite on LC-MS/MS divided by the peak area of the internal standard on LC-MS/MS):

$$\%\text{Inhibition following pre-incubation} = \left(1 - \left(\frac{\text{Peak area ratio in presence of NADPH}}{\text{Peak area ratio in absence of NADPH}}\right)\right) \times 100$$

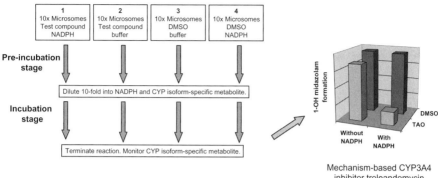

Figure 7.17 Overview of mechanism-based inhibition protocol. Pre-incubation 1: This pre-incubation consists of test compound, NADPH, and 10-fold the concentration of microsomes. If the test compound is a mechanism-based inhibitor, then it should be activated and bind irreversibly to the active site. Pre-incubation 2: This pre-incubation consists of test compound, buffer, and 10-fold the concentration of microsomes. As no NADPH is present in the pre-incubation, no metabolic activation of the test compound by CYP should occur in the pre-incubation step; therefore, mechanism-based inhibition should not occur. Pre-incubation 3: This pre-incubation is one of the vehicle control incubations where the test compound is replaced with DMSO, and buffer replaces the NADPH. It consists of DMSO, buffer, and 10-fold the concentration of microsomes. As the test compound and NADPH are absent, no reversible or mechanism-based inhibition of the CYP-specific isoform should occur. Pre-incubation 4: This pre-incubation is one of the vehicle control incubations where DMSO replaces the test compound. It consists of DMSO, NADPH, and 10-fold the concentration of microsomes. As the test compound is absent, no reversible or mechanism-based inhibition of the CYP-specific isoform should occur.

If the compound is identified to be a mechanism-based inhibitor, then more detailed studies can be performed at multiple test compound concentrations and multiple pre-incubation times to calculate the rate of inactivation (K_{inact}) and the inhibition constant (K_I).

Compared with reversible inhibition, mechanism-based inhibition more frequently causes pharmacokinetic–pharmacodynamic DDIs, as the inactivated enzyme has to be replaced by newly synthesized CYP protein. The resultant drug interactions may lead to adverse drug effects, including some fatal events. For example, when mechanism-based inhibitors of CYP3A4 are coadministered with terfenadine, cisapride, or astemizole, torsades de pointes (a life-threatening ventricular arrhythmia associated with QT prolongation) may occur (Zhou *et al.*, 2005). This method is amenable to HTS and due to the clinical implications of such an interaction, the industry is now addressing this issue at a much earlier stage of drug discovery.

7.9.3.3 *CYP Reaction Phenotyping*

Drug-metabolizing enzyme identification studies, or reaction phenotyping studies, are a set of experiments that identify the specific enzymes (usually CYP enzymes) responsible for metabolism of a drug. *In vitro* identification of drug-metabolizing CYP enzymes helps predict the potential for *in vivo* DDIs and the impact of polymorphic enzyme activity on drug disposition. Furthermore, if suitable *in vitro* studies at therapeutic concentrations indicate that the main CYP isoforms do not metabolize an investigational drug, then clinical studies to evaluate the effect of CYP inhibitors/inducers on the elimination of the investigational drug will not be required. The methods for assessing reaction phenotyping include recombinant enzymes, potential of selective P450 inhibitors to inhibit test compound, antibody inhibition studies, and/or correlation analyses with individual donor activity. All these techniques can be adapted to automated plate-based systems.

7.9.3.4 *CYP Induction* Induction of CYP can either lead to a reduction in therapeutic effect as a result of increased metabolic clearance, or adverse effects as a consequence of increased formation of an active metabolite. The regulatory authorities currently recommend that hepatocytes be used for the purpose of assessing P450 induction of NCEs. Traditionally, this has been performed by incubating the potential inducers with freshly isolated cultured human hepatocytes prior to quantification of CYP-specific probe substrate metabolism. However, it is becoming more popular to assess mRNA levels and/or protein levels in parallel with activity. This can be helpful when both inhibition and induction are occurring simultaneously. More recently, several studies have also reported that cryopreserved hepatocytes can be used for CYP induction studies (Roymans *et al.*, 2005), and this has become more widely accepted due to the obviously practical advantages.

Due to the fact that the regulatory authorities currently suggest a minimum of three donors be investigated in these studies, it can be expensive to perform these studies, and therefore higher-throughput methods are attractive. One such method that is growing in popularity is the aromatic hydrocarbon receptor (AhR) and pregnane X receptor (PXR) reporter gene assay. However, there is still question as to the *in vivo* relevance of these alternative approaches and the data tend to be used as supporting evidence alongside other more acceptable methods.

7.9.3.5 *Reactive Metabolites* Idiosyncratic drug toxicity is a major concern for the pharmaceutical industry, as often the toxicity is not detected until after launch of the drug. A high proportion of drugs that cause idiosyncratic toxicity have the capability of forming reactive metabolites. Walgren, Mitchell and Thompson (2005) found that from a set of 21 drugs that have either been withdrawn from the US market due to hepatotoxicity or have a black box warning for hepatotoxicity, there was evidence for the formation of reactive metabolites in five out of six drugs that were withdrawn, and 8 out of 15 drugs that had black box warnings.

Reactive metabolites are capable of covalently binding to macromolecules by nucleophilic substitution which is thought in some circumstances to instigate toxic effects. The approach by Merck and several other companies has been to impose a panel of screening assays that can detect reactive metabolite formation and the risk can then be minimized by the appropriate structural modification during the lead optimization stage (Evans *et al.*, 2004). The use of chemical trapping agents such as reduced glutathione (Yan *et al.*, 2005; Masubuchi, Makino and Murayama, 2007) or glutathione ethyl ester (Soglia *et al.*, 2004) has become a popular early screen for detecting reactive metabolites. The experiments are usually performed in the presence of liver microsomes and stable adducts formed can be detected by LC-MS/MS. These studies are useful in providing indirect information on the structure of the reactive species. One disadvantage to this screen is that no single small molecule serves as universal surrogate and screening through a panel of trapping agents may be needed.

Covalent binding to proteins is another recognized method for evaluating reactive metabolite formation. As radiolabeled compound is required for these studies, their

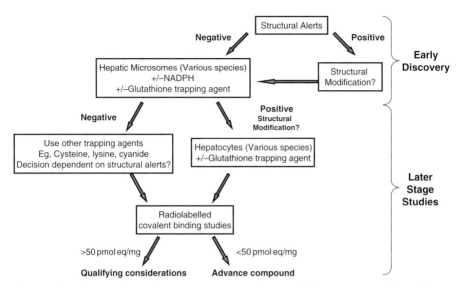

Figure 7.18 Suggested screening strategy for reactive metabolite assessment (adapted from Evans *et al.*, 2004).

suitability as early screening tools is restricted. Covalent binding studies can either be performed *in vitro*, by investigating binding to liver microsomes, or *in vivo*, by dosing to rats and measuring adduct formation in the liver or plasma proteins. Merck have proposed a cutoff point for the acceptable level of covalent binding of 50 pmol drug equivalent/mg of protein (Evans *et al.*, 2004) (Fig. 7.18). Compounds that are lower than this value are advanced further. Compounds that exceed this level have to pass a number of other qualifying considerations before being considered for further development. These considerations include dose, availability of alternative therapies, whether the condition is life threatening, and whether structural modification is possible.

CYP-mediated bioactivation of drugs can lead to intermediates that are so reactive that they bind irreversibly to the active site of the enzyme resulting in loss of enzyme activity. This phenomenon is known as mechanism or time-dependent inhibition, which was discussed earlier in this chapter. Identifying if a compound is a mechanism-based inhibitor is also an important screening tool in the detection of reactive metabolites. Although reactive metabolite formation is an obvious risk factor, not all compounds identified in the screening tools described result in toxicity. Therefore, the approach adopted by Merck seems to be logical. Risks should be minimized at the early lead optimization stage by appropriate structural modification, and qualifying considerations should also be taken into account before deciding to stop further development of a compound.

7.9.4 The Potential Pitfalls of Extrapolating From *In Vitro* to *In Vivo* in Drug Clearance

Extrapolating from *in vitro* to *in vivo* can be problematic and error prone. The design and types of studies should be planned carefully before embarking on drug metabolism screening. Some of the potential pitfalls are described in the next section.

7.9.4.1 Nonspecific Binding   Nonspecific binding reduces the concentration of free drug available to either be metabolized by an enzyme or inhibit an enzyme. Nonspecific binding either to the incubation vessel or to the incubation constituents (e.g. microsomes) can alter the concentration of free drug. There are now several studies where knowledge of the extent of microsomal binding has led to a better understanding of the relationship between *in vitro* metabolism and *in vivo* pharmacokinetics. As well as studies investigating improvements in clearance prediction (Obach, 1997, 1999; Carlile *et al.*, 1999), it has been demonstrated that nonspecific microsomal binding can account for underestimation of inhibitor potency (i.e. overestimation of IC_{50} or K_i values) when dealing with lipophilic basic drugs (Tran *et al.*, 2002; Margolis and Obach, 2003), with the potential implication being an underestimation of risk from DDIs. The risks of not considering nonspecific microsomal binding in estimating inhibitory potency may be greater when dealing with mechanism-based inhibitors, due to the relatively high microsomal concentrations that are typically used during pre-incubation of the inhibitor in these experiments.

7.9.4.2 Interindividual Variability   Interindividual variability can exist both *in vivo* and in the *in vitro* metabolizing systems. The *in vivo* clearance can be influenced dramatically by CYP450 induction (e.g. smoking can induce CYP1A2), polymorphisms in metabolism, gender, age, and physiological conditions such as stress. For *in vitro* studies, pooled microsomes or pooled hepatocytes can reduce the problems associated with interindividual variability. However, this will mask the true extent of variability likely to take place *in vivo*.

7.9.4.3 Metabolism in Extrahepatic Tissue   *In vitro* hepatic metabolizing systems only address clearance in a single organ. In reality, although the liver is the major site of drug metabolism, many other tissues are capable of metabolizing drugs. If the extrahepatic metabolism is significant, the *in vitro* intrinsic clearance is likely to underpredict the *in vivo* clearance.

7.9.4.4 Renal and Biliary Clearance   As well as metabolic clearance, renal and biliary clearances contribute to total *in vivo* clearance. Presently, no routine widely accepted *in vitro* assays exist for biliary or renal clearance, which are appropriate for drug discovery, and as such the reliable prediction of total *in vivo* clearance is a primary unmet need in early ADMET/PK. In the future as a greater understanding of the biology of the processes is gained, it is possible that passive permeability data and transporter data can be used in combination with physiologically based PK modeling techniques to predict these processes.

7.9.4.5 Equilibrium between Blood and Hepatocytes   It is assumed that there is rapid equilibrium of drugs between the blood and hepatocytes. When the intrinsic clearance in hepatocytes is much greater than the efflux clearance from hepatocytes to blood, the *in vivo* intrinsic clearance is rate-limited by the influx process from blood to hepatocytes. Such an incorrect assumption of rapid equilibrium can be one of the reasons why *in vivo* intrinsic clearance is less than *in vitro* intrinsic clearance (Ito *et al.*, 1998).

7.9.5 Transporters in Hepatocytes

The subject of drug metabolism is made even more complex by the role of transporters in the distribution of drug into the hepatocyte and in the biliary excretion of the drug (Fig. 7.19). Interest in this area is growing rapidly. So far, the majority of studies have focused on P-gp, an efflux transporter present in the gastrointestinal tract, brain, kidney, and liver. More recently, interest has grown in other transporter families such as OATP (organic anion transporting polypeptides) and MRP (multidrug resistance associated proteins). These families of transporters among others play a major role in the efflux and uptake processes in the hepatocyte. A number of transporters also exist on the bile canalicular membrane and are involved in hepatobiliary excretion of drugs.

If a compound is a substrate of any of these transporters, the concentration of compound within the cell and in turn, at the site of the drug-metabolizing enzymes, may be influenced. Furthermore, if the compound is an inhibitor of a transporter, then this could affect the metabolism and elimination of a competing drug resulting in DDIs. Presently, it would appear that the specificity of drug transporters is less well characterized. The impact of this is twofold. Firstly, it may indicate the greater likelihood that multiple transporters may be involved in the membrane transport of a single drug, and so the contribution of the individual transporters to the overall net membrane transport needs to be taken into consideration when extrapolating from systems expressing a single transporter protein. Secondly, the absence of specific probe substrates and selective inhibitors may hinder the identification of the mechanisms of drug transport. Presently, the link between drug transporters and the prediction of pharmacokinetics is limited. By gaining a better insight into the value of *in vitro* transporter assays, it will

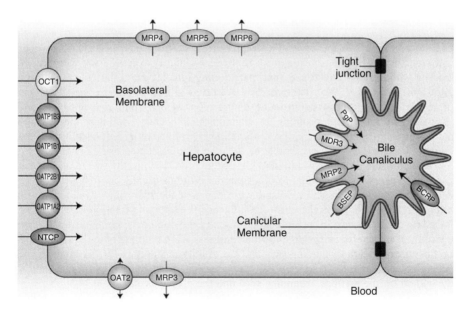

Figure 7.19   Localization of efflux and uptake transporters in the hepatocyte (with permission from SOLVO Biotechnology, Budapest, Hungary).

enable *in vitro* screening methods to be more widely accepted and integrated into the panel of routine assays currently being assessed for ADMET. Incorporating the transporter information will enable plasma and tissue levels to be more accurately modeled, leading to an enhancement in the overall prediction of pharmacokinetics.

7.9.6   The Future of Drug Metabolism

As new experimental and analytical technologies become available on the market, both the speed and quantity of data generation will increase. For example, chip technology is now being introduced for recombinant CYP screening (Gilardi *et al.*, 2002; Lee *et al.*, 2005). Advances in the understanding of the complexity of drug metabolism and its interplay with other biological processes such as drug transporters will enable the accuracy of clearance prediction to increase. As this happens, there will be a greater reliance on pharmacokinetic modeling techniques enabling experimental data on individual parameters to be used in combination to predict the pharmacokinetics. More complex experimental models such as bioreactors are also being developed, which are closer to the physiology of the intact organ (Allen, Khetani and Bhatia, 2005; Sivaraman *et al.*, 2005; De Bartolo *et al.*, 2006). These systems can offer flow conditions and long-term culture of human hepatocytes. Recreation of the *in vivo* microarchitecture appears to retain liver-specific function more effectively (Hoffmaster, 2007). However, at present, throughput is restricted with these systems. Yet by combining assessment of metabolism with transporter information and potential toxicity evaluation in a more physiologically relevant system, it may be a smarter way of predicting drug disposition.

7.10   PLASMA PROTEIN BINDING

Distribution refers to the reversible transfer of drug from one location to another within the body. Delivery of drug to tissues by blood, ability to cross tissue membranes, binding within blood and tissues, and partitioning into fat can all affect distribution (Rowland and Tozer, 1995). The propensity of drugs to bind to plasma proteins is one of several key factors in determining distribution. A drug that is extensively and strongly bound to plasma proteins has limited access to sites of action in the cell, and metabolism and elimination may be slower. Plasma protein binding is typically a reversible interaction; however, irreversible or covalent binding can occasionally occur.

Human serum albumin and α_1-acid glycoprotein (AAG) are the two main binding proteins in plasma with binding to other plasma proteins usually occurring to a much lesser extent. Generally, acidic drugs have a higher tendency to bind to albumin and basic drugs to AAG. Several factors can influence the unbound fraction in the plasma including drug concentration, binding affinity, and the number of binding sites. At low concentrations (i.e. below the dissociation constant), the fraction unbound is dependent on the concentration of binding sites and the affinity of the drug for the binding site. At high concentrations (i.e. above the dissociation constant), the fraction unbound is dependent on the number of binding sites and the drug concentration.

The levels of the individual proteins in plasma can fluctuate dramatically depending on age, disease, and pregnancy. Therefore, identifying which plasma protein the drug binds to and to what extent can be useful in understanding distribution in certain patient groups.

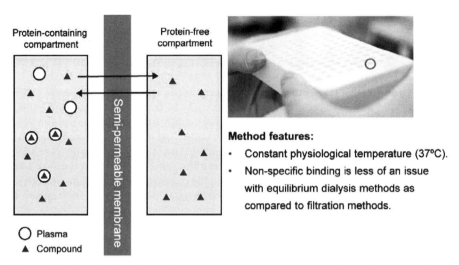

Method features:
- Constant physiological temperature (37°C).
- Non-specific binding is less of an issue with equilibrium dialysis methods as compared to filtration methods.

○ Plasma
▲ Compound

Figure 7.20 Equilibrium dialysis is one of the most common methods used for determining plasma protein binding. With the availability of 96-well equilibrium dialysis plates, throughput is now improved dramatically over the traditional Dianorm systems. (From www.cyprotex.com.)

7.10.1 Methods for Determining Protein Binding

7.10.1.1 Conventional Methods   Although several higher-throughput methods (often based on investigating binding to a single plasma protein such as HSA) have been developed, none appear to offer the quality associated with equilibrium dialysis, and as such this remains the most popular method in industry to date (Fig. 7.20). The apparatus consists of two compartments containing buffer and plasma respectively separated by a semipermeable membrane that allows drug but not protein to pass across. Equilibrium dialysis has many advantages in that it can be used with whole plasma under a temperature-controlled environment. The apparatus is usually made from Teflon that reduces effect of binding. The traditional, and quite laborious, protocol for equilibrium dialysis using Dianorm cells has now largely been replaced with plate-based and automation friendly versions. Coupled with advances in LC-MS/MS, this has improved throughput dramatically.

Ultrafiltration is another popular technique for determining plasma protein binding. It is more rapid than the equilibrium dialysis method and filter plates are now commercially available. However, there are several perceived drawbacks to this method that include issues with binding to the filters and possible changes in binding as the concentration of drug in the plasma increases during centrifugation.

7.10.1.2 Alternative Methods   Advances in chromatographic technology have led to the development of highly automated rapid methods for determining binding to HSA. These methods use HPLC columns with immobilized protein attached. A paper by Valko *et al.* (2003) illustrates that data generated using a column-based method correlate relatively well with literature plasma protein binding measurements, and the method appears to be more accurate for highly bound compounds compared with ultrafiltration methods. Even compounds with 99.2% and 99.3% can be distinguished and have retention times that differ by several minutes. An earlier paper by Talbert *et al.* (2002) describes how kinetic rate constants and equilibrium constants can be simultaneously determined using these immobilized protein columns by monitoring the retention time and peak width.

HSA immobilized on beads, microparticles, or sensor chips (Sjöholm *et al.*, 1979; Ahmad *et al.*, 2003; Schuhmacher *et al.*, 2004) can also be used to assess protein binding. The TRANSIL technology developed by Sovicell GmbH (formerly Nimbus Biotechnology) consists of either human serum albumin or AAG covalently immobilized on an inert surface. The beads are designed to minimize interactions with drug molecules, and thus reduce any nonspecific binding effects. The protein is randomly oriented on the bead surface, and no loss of function due to the covalent immobilization procedure is observed. The beads are separated from the free fraction at the end of the incubation period by filtration or centrifugation.

7.11 QC/QA CONSIDERATIONS

This review of automation in modern ADMET has provided a brief survey of the variety of assays *in vitro* that are amenable to automation and that are deployed by all modern pharma to assess the pharmacokinetic profiles of nascent drug leads. Some pharma have established highly organized and fully automated ADMET profiling laboratories. There are now contract research organizations (CROs), such as Cyprotex, NovaScreen, PanLabs, CEREP, and others, which have made available panels of semi- and fully automated ADMET assays that are run in parallel from the same stock solutions of compounds. When one set of compounds is submitted to a centralized lab for such cross-profiling, the results from one assay to another are more likely comparable to one another without major confounds that could arise due to, for example, varying compound solubility as a result of the pH varying across the different assays.

Apart from the intrinsic assay issues, there are "extrinsic" issues that should be considered whenever comparing data from different labs in different assays. As shown in Fig. 7.9, small nuances in SOPs can have large impacts on variables such as mixing efficiency in an assay. Some labs may control better than others the potential "edge effects" due to more rapid evaporation of outer wells of a microtiter plate than those toward the middle of the plate, which may cause precipitation of compounds due to changes in drug concentration, or pH, or ionic strength. For these reasons, results should always be compared with a reference compound, as has been mentioned earlier (as in Fig. 7.13).

Furthermore, there is always an inherent presumption that the designated structure for a compound powder in a vial is correct, and that the presumed structure is in fact the entity that elicits the primary response in the primary screening assay for determining potency and efficacy (pharmacodynamics), as well as that the same structure mediates the responses from ADMET cross-profiling assays—particularly those that may involve toxicity. This, however, is an assumption until proven, and proving this assumption is not always straightforward.

For example, the standard acceptance criteria for compound purity vary from site to site, and even within a site. Within a site, acceptance criteria for registered (accepted into the corporate compound library) may be as follows; however, the purity targets could vary ±10% between sites:

1. A compound synthesized in-house by discrete organic synthetic methods (i.e. handmade compounds by chemists in-house or by contract research) may have the requirement of being >95% pure as assessed by LC-MS and with structure properly assigned based on NMR.
2. Compounds made by combinatorial synthesis may require only 90% purity by only LC-MS.
3. Compound purchased for HTS may require only 80% purity, and it is not clear whether the analytical data are valid.

All these compound types from the different sources, which may range in age from weeks to years, and which have been stored under various controlled or uncontrolled (ambient) conditions, are found in a "typical" HTS compound library. The "same" compound may exist in both solution and powder forms, and as different stock concentrations and multiple lots, respectively. Different lots of the "same" compound solution may have been freeze-thawed several times and the DMSO solvent may contain varying amounts of water, which not only is known to affect compound solubility (Cheng *et al.*, 2003) but may also affect compound stability.

It should quickly become apparent to the reader that without rigorous compound (lot) management, comparison of data between different assays run at different times may be problematic, irrespective of whether the assays are pharmacodynamic or pharmacokinetic in nature. Figure 7.15 depicts an approach that may be adopted to assist QC/QA processes using automated rules to mimic checking procedures typically employed by a scientist.

7.12   SUMMARY

There is an overall drive in the pharmaceutical industry to improve efficiency and reduce operating costs. Sophisticated instruments are now widely available for liquid handling and sample analysis, which enable rapid screening of large numbers of compounds. As a consequence, the bottleneck has now shifted to the stage of data and information handling which is traditionally a resource-intensive process. Intelligent workflow techniques play a major role in addressing these challenges. By automating these processes, it allows more efficient use of internal resources, enables knowledge to be captured from individuals and shared, and reduces subjective human intervention. Ultimately this leads to less errors, better consistency, reduced timelines, and lower costs. It appears that a greater reliance on automated *in vitro* methods earlier in the pharmaceutical R&D continuum is having an impact on improving the failure rate of new drugs (see Fig. 7.21).

Over the next decade, the pharmaceutical industry will continue to radically change its approach to address the lack of efficiency in its current working practice. Implementation of *in silico* techniques, quicker and better screening techniques, and smarter use of experimental data will hopefully move a step forward in achieving this goal. The move toward identifying efficacious, safer drugs with optimal pharmacokinetic parameters at an early stage will continue, enabling efficiencies both in terms of time and money, and ultimately allowing more effective drugs to reach the market quicker.

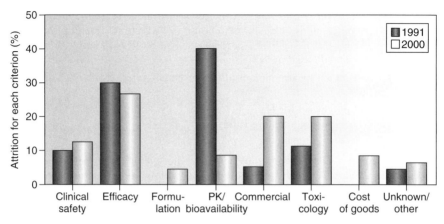

Figure 7.21   Drug attrition in drug development in 1991 and 2000 (from Kola and Landis, 2004). The text discusses the reasons for the recent progress in ADMET due to better and earlier filtering from the development queue—both experimentally and computationally—those compounds that display poor bioavailability.

Smarter decision making and interpretation of the data will be required with ADMET properties considered to be whole rather than individual properties. Integrating multiple ADMET properties with physiochemical data to predict whole body pharmacokinetics using physiologically based pharmacokinetic modeling will continue to grow in popularity as models become more accurate (Lüpfert and Reichel, 2005; De Buck *et al.*, 2007).

REFERENCES

Ahmad, A., Ramakrishnan, A., McLean, M.A., Breau, A.P. (2003). Use of surface plasmon resonance biosensor technology as a possible alternative to detect differences in binding of enantiomeric drug compounds to immobilized albumins. *Biosens. Bioelectron.*, *18* (4), 399–404.

Allen, J.W., Khetani, S.R., Bhatia, S.N. (2005). *In vitro* zonation and toxicity in a hepatocyte bioreactor. *Toxicol. Sci.*, *84* (1), 110–119.

Avdeef, A. (2007). Solubility of sparingly-soluble drugs. *Adv. Drug Deliv. Rev.*, *59* (7), 568–590.

Avdeef, A., Bendels, S., Di, L., Faller, B., Kansy, M., Sugano, K., Yamauchi, Y. (2007). PAMPA–critical factors for better predictions of absorption. *J. Pharm. Sci.*, *96* (11), 2893–2909.

Banno, K., Takahashi, R. (1991). Laboratory automation system for the analysis of drugs in biological fluids. *Anal. Sci.*, *7* (3), 511–513.

Belz, G.G., Doering, W., Munkes, R. *et al.* (1983). Interaction between digoxin and calcium antagonists and antiarrhythmic drugs. *Clin. Pharmacol. Ther.*, *33*, 410–417.

Bevan, C.D., Lloyd, R.S. (2000). A high-throughput screening method for the determination of aqueous drug solubility using laser nephelometry in microtiter plates. *Anal. Chem.*, *72* (8), 1781–1787.

Bigger, J.T., Jr, Leahey, E.B., Jr (1982). Quinidine and digoxin: an important interaction. *Drugs*, *24*, 229–239.

Boyd, B.J. (2008). Past and future evolution in colloidal drug delivery systems. *Expert Opin. Drug Deliv.*, *5* (1), 69–85.

Burch, R.M. (1993). Mass ligand-binding screening strategies for identification of leads for new drug discovery. *NIDA Res. Monogr.*, *134*, 37–45.

Buxton, I. (2006). Pharmacokinetics and pharmacodynamics: The dynamics of drug absorption, distribution, action, and elimination. In Brunton, L. *et al.* (eds.), *Goodman & Gilman's The Pharmacological Basis of Therapeutics*, 11th ed. McGraw-Hill, pp. 1–137.

Buxton, I. (2006). Pharmacokinetics and pharmacodynamics: The dynamics of drug absorption, distribution, action, and elimination In . *NIDA Res. Monogr.*, *134*, 37–45.

Carlile, D.J., Hakooz, N., Bayliss, M.K., Houston, J.B. (1999). Microsomal prediction of *in vivo* clearance of CYP2C9 substrates in humans. *Br. J. Clin. Pharmacol.*, *47*, 625–635.

Cheng, X., Hochlowski, J., Tang, H., Hepp, D., Beckner, C., Kantor, S., Schmitt, R. (2003). Studies on repository compound stability in DMSO under various conditions. *J. Biomol. Screen.*, *8* (3), 292–304.

De Bartolo, L., Salerno, S., Morelli, S., Giorno, L., Rende, M., Memoli, B., Procino, A., Andreucci, V.E., Bader, A., Drioli, E. (2006). Long-term maintenance of human hepatocytes in oxygen-permeable membrane bioreactor. *Biomaterials*, *27* (27), 4794–4803.

De Buck, S.S., Sinha, V.K., Fenu, L.A., Nijsen, M.J., Mackie, C.E., Gilissen, R.A.H.J. (2007). Prediction of human pharmacokinetics using physiologically based modelling: a retrospective analysis of 26 clinically tested drugs. *Drug Metab. Dispos.*, *35* (10), 1766–1780.

Doering, W. (1979). Quinidine-digoxin interaction: pharmacokinetics, underlying mechanism and clinical implications. *N. Engl. J. Med.*, *301*, 400–404.

Evans, D.C., Watt, A.P., Nicoll-Griffith, D.A., Baillie, T.A. (2004). Drug-protein adducts: an industry perspective on minimizing the potential for drug bioactivation in drug discovery and development. *Chem. Res. Toxicol.*, *17* (1), 3–16.

Fisher, M.B., Campanale, K., Ackermann, B.L., VandenBranden, M., Wrighton, S.A. (2000). *In vitro* glucuronidation using human liver microsomes and the pore-forming peptide alamethicin. *Drug Metab. Dispos.*, *28* (5), 560–566.

Fitzpatrick, L.A., Good, A. (2000). Micronized progesterone: clinical indications and comparison with current treatments. *Fertil. Steril.*, *73* (3), 654–754.

Galinis-Luciani, D., Nguyen, L., Yazdanian, M. (2007). Is PAMPA a useful tool for discovery? *J. Pharm. Sci.*, *96* (11), 2886–2892.

Garrigues, A., Nugier, J., Orlowski, S., Ezan, E. (2002). A high-throughput screening microplate test for the interaction of drugs with P-glycoprotein. *Anal. Biochem.*, *305* (1), 106–114.

Gentile, A., Trusolino, L., Comoglio, P.M. (2008). The Met tyrosine kinase receptor in development and cancer. *Cancer Metastasis Rev.*, *27* (1), 85–94.

Gilardi, G., Meharenna, Y.T., Tsotsou, G.E., Sadeghi, S.J., Fairhead, M., Giannini, S. (2002). Molecular Lego: design of molecular assemblies of P450 enzymes for nanobiotechnology. *Biosens. Bioelectron.*, *17* (1–2), 133–145.

Glavinas, H., Kis, E., Pál, A., Kovács, R., Jani, M., Vági, E., Molnár, E., Bánsághi, S., Kele, Z., Janáky, T., Báthori, G., Von Richter, O., Koomen, G.J., Krajcsi, P. (2007). ABCG2 (breast cancer resistance protein/mitoxantrone resistance-associated protein) ATPase assay: a useful tool to detect drug-transporter interactions. *Drug Metab. Dispos.*, *35* (9), 1533–1542.

Goldstein, D.M., Gray, N.S., Zarrinkar, P.P. (2008). High-throughput kinase profiling as a platform for drug discovery. *Nat. Rev. Drug Discov.*, *7* (5), 391–397.

Harris, S.R., Garlick, R.K., Miller, J.J., Jr, Harney, H.N., Monroe, P.J. (1991). Complement C5a receptor assay for high throughput screening. *J. Recept. Res.*, *11* (1–4), 115–128.

Hewitt, N.J., Gómez Lechón, M.J., Houston, J.B., Hallifax, D., Brown, H.S., Maurel, P., Kenna, J. G., Gustaysson, L., Lohmann, C., Skonberg, C., Guillouzo, A., Tuschl, G., Li, A.P., LeCluyse, E. (2007). Primary hepatocytes: current understanding of the regulation of metabolic enzymes and transporter proteins, and pharmaceutical practice for the use of hepatocytes in metabolism, enzyme induction, transporter, clearance, and hepatotoxicity studies. *Drug Metab. Rev.*, *39*, 159–234.

Hoffmaster, K.A. (2007). *Presentation: AAPS Workshop on Drug Transporters in ADME: from the Bench to the Bedside, March 5–7.*

Ito, K., Iwatsubo, T., Kanamitsu, S., Nakajima, Y., Sugiyama, Y. (1998). Quantitative prediction of *in vivo* drug clearance and drug interactions from *in vitro* data on metabolism together with binding and transport. *Annu. Rev. Pharmacol. Toxicol.*, *38*, 461–499.

Kearns, D.H. (2001). High throughput physicochemical profiling for drug discovery. *J. Pharm. Sci.*, *90* (11), 1838–1858.

Klein, H.O., Lang, R., Weiss, E., Di Segni, E., Libhaber, C., Guerrero, J., Kaplinsky, E. (1982). The influence of verapamil on serum digoxin concentration. *Circulation*, *65*, 998–1003.

Kola, I., Landis, J. (2004). Can the pharmaceutical industry reduce attrition rates? *Nat. Rev. Drug Discov.*, *3* (8), 711–771.

Krippendorff, B.F., Lienau, P., Reichel, A., Huisinga, W. (2007). Optimizing classification of drug-drug interaction potential for CYP450 isoenzyme inhibition assays in early drug discovery. *J. Biomol. Screen.*, *12* (1), 92–99.

Leahey, E.B., Jr, Reiffel, J.A., Heissenbuttel, R.H. *et al.* (1979). Enhanced cardiac effect of digoxin during quinidine treatment. *Arch. Intern. Med.*, *139*, 519–521.

Lee, M.Y., Park, C.B., Dordick, J.S., Clark, D.S. (2005). Metabolizing enzyme toxicology assay chip (MetaChip) for high-throughput microscale toxicity analyses. *Proc. Natl. Acad. Sci. U.S.A.*, *102* (4), 983–987.

Lett, E., Kriszt, W., de Sandro, V., Ducrotov, G., Richert, L. (1992). Optimal detergent activation of rat liver microsomal UDP-glucuronosyl transferases toward morphine and 1-naphthol: contribution to induction and latency studies. *Biochem. Pharmacol.*, *43* (7), 1649–1653.

Lipinski, C.A. (2000). Drug-like properties and the causes of poor solubility and poor permeability. *Pharmacol. Toxicol. Met.*, *44*, 235–249.

Lipinski, C.A., Lombardo, F., Dominy, B.W., Feeney, P.J. (1997). Experimental and computational approaches to estimate solubility and permeability in drug discovery and development settings. *Adv. Drug. Del. Rev.*, *23*, 3–25.

Lüpfert, C., Reichel, A. (2005). Development and application of physiologically based pharmacokinetic-modeling tools to support drug discovery. *Chem. Biodivers.*, *2* (11), 1462–1486.

McGee, P. (2005). Tracking drug safety from the ground up. *Drug Discov. Devel.*, 20–28.

Margolis, J.M., Obach, R.S. (2003). Impact of nonspecific binding to microsomes and phospholipid on the inhibition of cytochrome P4502D6: implications for relating *in vitro* inhibition data to *in vivo* drug interactions. *Drug Metab. Dispos.*, *31* (5), 606–611.

Masubuchi, N., Makino, C., Murayama, N. (2007). Prediction of *in vivo* potential for metabolic activation of drugs into chemically reactive intermediate: correlation of *in vitro* and *in vivo* generation of reactive intermediates and *in vitro* glutathione conjugate formation in rats and humans. *Chem. Res. Toxicol.*, *20* (3), 455–464.

Maxwell, D.L., Gilmour-White, S.K., Hall, M.R. (1989). Digoxin toxicity due to interaction of digoxin with erythromycin. *BMJ*, *298*, 572.

Meech, R., Mackenzie, P.I. (1997). Structure and function of uridine diphosphate glucuronosyl-transferases. *Clin. Exp. Pharmacol. Physiol.*, *24* (12), 907–915.

Neubert, R.H.H., Schwarz, M.A., Mrestani, Y., Plätzer, M., Raith, K. (1999). Affinity capillary electrophoresis in pharmaceutics. *Pharm. Res.*, *16*, 1663–1673.

Ng, R. (2004). *Drugs: From Discovery to Approval*, John Wiley & Sons, Inc.

Obach, R.S. (1997). Nonspecific binding to microsomes: impact on scale-up of in vitro intrinsic clearance to hepatic clearance as assessed through examination of warfarin, imipramine, and propranolol. *Drug Metab. Dispos.*, *25* (12), 1359–1369.

Obach, R.S. (1999). Prediction of human clearance of twenty-nine drugs from hepatic microsomal intrinsic clearance data: an examination of in vitro half-life approach and nonspecific binding to microsomes. *Drug Metab. Dispos.*, *27* (11), 1350–1359.

Pereira, D.A., Williams, J.A. (2007). Origin and evolution of high throughput screening. *Br. J. Pharmacol.*, *152*, 53–61.

Popa-Burke, I.G., Issakova, O., Arroway, J.D., Bernasconi, P., Chen, M., Coudurier, L., Galasinski, S., Jadhav, A.P., Janzen, W.P., Lagasca, D., Liu, D., Lewis, R.S., Mohney, R.P., Sepetov, N., Sparkman, D.A., Hodge, C.N. (2004). Streamlined system for purifying and quantifying a diverse library of compounds and the effect of compound concentration measurements on the accurate interpretation of biological assay results. *Anal. Chem.*, *76* (24), 7278–7287.

Rowland, M., Tozer, T. (1995). *Clinical Pharmacokinetics, Concepts and Applications*, 3rd edn, Lippincott Williams & Wilkins.

Roymans, D., Annaert, P., Van Houdt, J., Weygers, A., Noukens, J., Sensenhauser, C., Silva, J., Van Looveren, C., Hendrickx, J., Mannens, G., Meuldermans, W. (2005). Expression and induction potential of cytochromes P450 in human cryopreserved hepatocytes. *Drug Metab. Dispos.*, *33* (7), 1004–1016.

Schuhmacher, J., Kohlsdorfer, C., Bühner, K., Brandenburger, T., Kruk, R. (2004). High-throughput determination of the free fraction of drugs strongly bound to plasma proteins. *J. Pharm. Sci.*, *93* (4), 816–830.

Sivaraman, A., Leach, J.K., Townsend, S., Iida, T., Hogan, B.J., Stolz, D.B., Fry, R., Samson, L.D., Tannenbaum, S.R., Griffith, L.G. (2005). A microscale *in vitro* physiological model of the liver: predictive screens for drug metabolism and enzyme induction. *Curr. Drug Metab.*, *6* (6), 569–591.

Sjöholm, I., Ekman, B., Kober, A., Ljungstedt-Påhlman, I., Seiving, B., Sjödin, T. (1979). Binding of drugs to human serum albumin: XI. The specificity of three binding sites as studied with albumin immobilized in microparticles. *Mol. Pharmacol.*, *16*, 767–777.

Sneader, W. (2005). *Drug Discovery: A History*, John Wiley & Sons, Inc.

Soglia, J.R., Harriman, S.P., Zhao, S., Barberia, J., Cole, M.J., Boyd, J.G., Contillio, L.G. (2004). The development of a higher throughput reactive intermediate screening assay incorporating micro-bore liquid chromatography-micro-electrospray ionization-tandem mass spectrometry and glutathione ethyl ester as an in vitro conjugating agent. *J. Pharm. Biomed. Anal.*, *36*, 105–116.

Stresser, D.M., Blanchard, A.P., Turner, S.D., Erve, J.C., Dandeneau, A.A., Miller, V.P., Crespi, C. L. (2000). Substrate dependent modulation of CYP3A4 catalytic activity:analysis of 27 test compounds with four fluorometric substrates. *Drug Metab. Dispos.*, *28*, 1440–1448.

Talbert, A.M., Tranter, G.E., Holmes, E., Francis, P.L. (2002). Determination of drug-plasma protein binding kinetics and equilibria by chromatographic profiling: exemplification of the method using L-tryptophan and albumin. *Anal. Chem.*, *74*, 446–452.

Tran, T.H., Von Moltke, L.L., Venkatakrishnan, K., Granda, B.W., Gibbs, M.A., Obach, R.S., Harmatz, J.S., Greenblatt, D.J. (2002). Microsomal protein concentration modifies the apparent inhibitory potency of CYP3A inhibitors. *Drug Metab. Dispos.*, *30* (12), 1441–1445.

Unger, S.H., Cook, J.R., Hollenberg, J.S. (1978). Simple procedure for determining octanol–aqueous partition, distribution, and ionization coefficients by reversed-phase high-pressure liquid chromatography. *J. Pharm. Sci.*, *67* (10), 1364–1367.

Valko, K., Nunhuck, S., Bevan, C., Abraham, M.H., Reynolds, D.P. (2003). Fast gradient HPLC method to determine compounds binding to human serum albumin. Relationships with octanol/water and immobilized artificial membrane lipophilicity. *J. Pharm. Sci.*, *92* (11), 2236–2248.

Verschraagen, M., Koks, C.H., Schellens, J.H., Beijnen, J.H. (1999). P-glycoprotein system as a determinant of drug interactions: the case of digoxin-verapamil. *Pharmacol. Res.*, *40*, 301–306.

Wakasugi, H., Yano, I., Ito, T., Hashida, T., Futami, T., Nohara, R., Sasayama, S., Inui, K. (1998). Effect of clarithromycin on renal excretion of digoxin: interaction with P-glycoprotein. *Clin. Pharmacol. Ther.*, *64*, 123–128.

Walgren, J.L., Mitchell, M.D., Thompson, D.C. (2005). Role of metabolism in drug-induced idiosyncratic hepatotoxicity. *Crit. Rev. Toxicol.*, *35*, 325–361.

Westphal, K., Weinbrenner, A., Giessmann, T., Stuhr, M., Franke, G., Zschiesche, M., Oertel, R., Terhaag, B., Kroemer, H.K., Siegmund, W. (2000). Oral bioavailability of digoxin is enhanced by talinolol: evidence for involvement of intestinal P-glycoprotein. *Clin. Pharmacol. Ther.*, *68*, 6–12.

Wilkinson, G.R., Shand, D.G. (1975). Commentary: a physiological approach to hepatic drug clearance. *Clin. Pharm. Ther.*, *18*, 377–390.

Yan, Z., Maher, N., Torres, R., Caldwell, G.W., Huebert, N. (2005). Rapid detection and characterization of minor reactive metabolites using stable-isotope trapping in combination with tandem mass spectrometry. *Rapid Commun. Mass Spectrom.*, *19* (22), 3322–3330.

Zhou, S., Yung Chan, S., Cher Goh, B., Chan, E., Duan, W., Huang, M., McLeod, H.L. (2005). Mechanism-based inhibition of cytochrome P450 3A4 by therapeutic drugs. *Clin. Pharmacokinet.*, *44* (3), 279–304.

CHAPTER 8

Mass Spectrometry

THOMAS R. SHARP

8.1 INTRODUCTION

"It is a fundamental tenet of chemistry that the structural formula of any compound contains coded within it all that compound's chemical, physical and biological properties." (Katritzky, Karelson and Lobanov, 1997)

If we accept this statement as philosophical truth, the determination of the structure of organic molecules is of paramount interest. Determining the structure will permit determining all other properties of the molecule. Therefore, methods for determining structure are important. Mass spectrometry has become one of the premier methods for determining organic structure, and is generally the first line of experimentation done (if not the only experimentation necessary) for determining the structure of organic molecules. Mass spectrometry, coupled especially with liquid chromatography (normal or reverse phase), is a foundation experimental technique in any structure elucidation effort, and in drug metabolism in particular. The combination of high-resolution chromatography, for handling complex mixtures, and the structural specificity and sensitivity for detecting low-level components of mass spectrometry—both in low relative abundance in a mixture and low absolute amounts of analyte—is unsurpassable for tackling such problems. Sometimes it is sufficient to solve the problem. Always it is necessary to have done the experiments to provide solid foundation for subsequent required work—by isolation, NMR, synthesis, and so on—to finally arrive at the answer.

"Why learn anything about spectral interpretation when the computer can do all the work? The answer to this question is simple, as most conscientious users quickly realize. The library search often does not provide a realistic answer or (worse) may provide an answer that looks correct but is not. Even software programs that profess to 'interpret' unknown spectra can only provide probable answers. After that, you are left to your own devices." (Smith, 2004)

Drug Metabolism Handbook: Concepts and Applications, Edited by Ala F. Nassar, Paul F. Hollenberg, and JoAnn Scatina

Much of the literature focuses on the instrumentation and techniques for how to generate mass spectral data. There is a substantial body of older literature dealing with interpretation of mass spectra and cataloging of fragmentation patterns. This older body of literature is primarily concerned, however, with electron ionization (EI) mass spectra. Even the more recently published textbooks and treatises, discussing the benefits of the newer soft ionization techniques,[1] do not discuss in much depth how to interpret mass spectra generated by these techniques. The statement that soft ionization techniques generate protonated molecular ions is often all that is seen. Little else is offered. The primary purpose in composing this chapter is to discuss mass spectra generated by the soft ionization methods—electrospray, in the main—and phenomena that often appear in these spectra, and to understand what these phenomena mean.

We will discuss instrumentation briefly here insofar as to provide a reference point for our subsequent discussions of data, and to provide entry into the vast amount of literature on instrumentation and instrumental advancements. Starting with spectral interpretation, we will describe the EI literature for completeness, so that the reader has a connection with this literature. Some of the numerological and interpretational rules are directly transferable from the EI literature to the interpretation of soft ionization spectra. Examining EI-induced fragmentation patterns is instructive to a degree, especially when considering the higher-energy regimes invoked when generating fragmentation of soft ionization-produced molecular ions using tandem mass spectrometry.

8.2 A BRIEF HISTORY

> "… If there is a mortal sin…, it is the teaching of science without connecting to the history of science…. The lack of historical perspective was dramatically demonstrated… by the student who asked if Galileo and Einstein ever met (Klose, 1987)."

Mass spectrometry, as a distinct discipline, is approximately 100 years old. Grayson (2002) has compiled an excellent history covering all aspects of the field of mass spectrometry for the fiftieth anniversary of the annual meetings of the American Society for Mass Spectrometry. We will make a few arbitrarily selected historical comments here. The mass spectrometry community has made significant efforts in the last 20 years to record and preserve the history of the field directly from some of the early workers in the field before that very personal perspective is lost. Much of this accumulated history has been published. Some oral histories with surviving pioneers in the field have been recorded and deposited at the Chemical Heritage Foundation.[2]

Most practitioners of the precise art and subtle science of mass spectrometry acknowledge the field to have originated with the work of J.J. Thomson and associates (Griffiths, 1997), published in 1910–1912, using the parabola mass spectrograph. Seminal

---

[1] The soft ionization methods, which will be discussed later, most often produce a molecular ion in which a charge-carrying species is attached to the neutral molecule. Typically, an H^+ is the attaching species. Many structural classes of compounds, however, show the strong tendency to scavenge and attach monovalent cations, such as Na^+.

[2] Chemical Heritage Foundation, 315 Chestnut Street, Philadelphia, Pennsylvania 19106. http://www.chemheritage.org.

discoveries that he and his coworkers made include the fact that the elements could be polyisotopic, by discovering the isotopes of neon. Thomson was awarded the 1906 Nobel Prize in physics (http://nobelprize.org/physics/laureates/1906/index.html) for his work on "investigations on the conduction of electricity by gases," work not directly involved with mass spectrometry, but which helped lay the foundations for it. Thomson's direct involvement in early mass spectrometry ended (Svec, 1985) with the publication of the parabola mass spectrograph work in 1913. Francis Aston (Squires, 1998; Downard, 2007a, b), a student of Thomson's, continued the work, is attributed as coining the term "mass spectrometry," and advanced the field significantly. Aston was awarded the 1922 Nobel Prize in chemistry (http://nobelprize.org/chemistry/laureates/1922/index.html) for his discoveries of a large number of the stable isotopes of the elements using mass spectrometry. Aston and A.J. Demster continued to determine the isotopic composition of the elements until finishing in 1935 with the publication of the isotopic composition of platinum and iridium. Aston is attributed as stating, in a presentation given not long after this 1935 publication, that "mass spectroscopy had served its purpose and would die away as a field for research" (Svec, 1985).

An unfortunate but seminal feat that mass spectrometry accomplished was the isolation of sufficient quantities of the fissile isotope of uranium, ^{235}U, to produce the first uranium-based fission atomic bombs during World War II. Numerous instrumental and other innovations derived from the efforts in the Manhattan Project. Physical separation of the uranium isotopes was accomplished using a large production array of mass spectrometers. Each individual machine was called a calutron (Nier, 1989; Yergey and Yergey, 1997; Settle, 2002). Gaseous diffusion was used to provide initial enrichment of feedstock (in the form of uranium hexafluoride) to the calutrons. Final enrichment was done mass spectrometrically. Gaseous diffusion did not replace mass spectrometric enrichment until after World War II had ended. The name "calutron" was coined by Ernest Lawrence. The physical principles of magnetic separation were based on Lawrence and other's work building and using cyclotrons for subatomic particle acceleration in laboratories at the University of California—Berkeley (now the Lawrence-Berkeley Laboratories).

The first industrial interest—in particular, chemical applications—of this heretofore physics laboratory oddity started in the petroleum industry in the 1950s (Meyerson, 1986; Quayle, 1987). Application of mass spectrometry to the structure determination of organic compounds, and quantitation of components of mixtures started with the faith that the petroleum industry placed in mass spectrometry during and after World War II. Mass spectrometry seemed an ideal technique to address the detection and determination of gases and volatile components of petroleum distillates from refinery streams. This faith and interest is directly responsible for creation of the first commercially available mass spectrometer systems, produced by Consolidated Engineering Corporation (CEC) in the United States (Meyerson, 1986) and Metropolitan-Vickers (later to become Associated Electrical Industries, AEI) in the United Kingdom (Quayle, 1987). An outpost of mass spectrometry investigation also grew up in Australia, influenced initially by encouragement and a visit from Aston (Morrison, 1991; Downard and de Laeter, 2005).

The importance of mass spectrometry in planetary and interplanetary research to date has been summarized by Neir (1985) and Grayson (2002). Mass spectrometers have been key pieces of instrumentation to ride atmospheric and interplanetary probes. Being well adapted to analysis of gaseous samples, they are ideal for studying the atmospheres of the earth and other planets, recording information on the chemical composition of those atmospheres. The Viking lander missions to Mars illustrate this application well by the examination of the composition of the Martian atmosphere (McElroy, Yung and Nier, 1976) and the search for signs of life on Mars (Biemann *et al.*, 1976).

Forensic application of mass spectrometry has extended into the arena of professional sports in addition to applications in crime detection and resolution. Recognition of gas chromatography–mass spectrometry (GC-MS) as an important tool in this area was made in 1967. The 1968 Winter Olympics saw the first time in which contestants were monitored for a few stimulants and narcotics (Grayson, 2002). Application to the horse racing industry followed a number of years later. The efforts continue to this day in sports monitoring and horse racing. The drug designers have become more sophisticated, and efforts to discover what they create must keep pace. An example of a relatively recent new performance-enhancing drug was reported by Catlin and associates (Catlin *et al.*, 2004).

Expansion of the utility of mass spectrometry to large and in particular biologically relevant molecules has been a direct result of the desorption ionization techniques—field desorption, fast atom bombardment (FAB), matrix-assisted laser desorption—and the atmospheric pressure ionization (API) techniques—atmospheric pressure chemical ionization (APCI) and electrospray (ESI). The latter are particularly important in achieving the interface between condensed-phase chromatographic techniques—HPLC and supercritical fluid chromatography—and the mass spectrometer. The importance of electrospray ionization, in particular, has been recognized by the 2002 awarding of the Nobel Prize (Fenn, 2003) for chemistry to John Fenn and Koichi Tanaka. Fenn and Tanaka, along with Kurt Wüthrich,[3] were recognized for their development of methods for identification and structure analysis of biological molecules.

8.3 GENERAL BACKGROUND IN MASS SPECTROMETRY

Understanding the powers that various chromatography–mass spectrometry techniques have presupposes that one understands the information that mass spectrometry in its various forms and nuances can provide. We will cursorily survey the field of instrumentation to provide a minimum reference here, with entry points into the more extensive detailed literature. The detailed focus will be on the interpretation of data and experiments.

Discussions of some of the basic mass spectrometric principles in the following will reinforce previous awareness. Should the reader's curiosity stray beyond these bounds, we invite consultation with the literature, and provide entry points to those topics in the following. Recent reasonably comprehensive textbooks on the general topic of mass spectrometry have been published by de Hoffmann and Stroobant (2002), Gross (2004), and Watson and Sparkman (2007). A treatise on the general subject of mass spectrometry, slightly more up-to-date than the textbooks, has been published by

---

[3] http://nobelprize.org/chemistry/laureates/2002/index.html. Fenn's and Tanaka's work applied electrospray to the mass spectrometric examination of large molecules. Wüthrich's contributions were in application of NMR to this area.

Burinsky (2006). A description of the state of the art on combined chromatography–mass spectrometry (including a discussion on LC-NMR) has been published by Sharp and Marquez (2006). Volume 101 (2001), issue 2 of *Chemical Reviews* covers many of the current active areas of development in mass spectrometry.

There is no one miracle technique which will solve all problems. Specialized techniques address specialized problems. It is one's responsibility as a good scientist to know the breadth of applicability of a technique, what it is good for—and what it is NOT good for—and to use what is needed when it is needed. Do not fall into the trap—if one's favorite tool is a hammer, then everything looks like a nail—of attempting to solve all problems with one's favorite technique.

8.3.1 The Mass Spectrometry Literature

The mass spectrometry research literature is scattered through many journals and publications. *The Journal of the American Chemical Society* and *Analytical Chemistry* have been consistent contributors. In recent years, several journals specializing in fundamentals and applications of mass spectrometry have come and gone. *The International Journal of Mass Spectrometry and Ion Processes* began publication in 1968 and has gone through an evolution in its title, being formerly known as *The International Journal of Mass Spectrometry and Ion Physics*. The title has been further simplified to *The International Journal of Mass Spectrometry*. *Organic Mass Spectrometry* started in 1968, specializing in organic chemical applications of mass spectrometry. *Biomedical Mass Spectrometry* (later changing to *Biomedical and Environmental Mass Spectrometry*, then to *Biological Mass Spectrometry*) started in 1986 as a companion journal specializing in the biomedical and environmental application areas, as the title implies. These two journals combined in 1995 to become *The Journal of Mass Spectrometry*, currently being published. *Mass Spectrometry Reviews* started in 1982. *Rapid Communications in Mass Spectrometry* started in 1987 as a means of providing (as the title implies) rapid dissemination of new results. *The Journal of the American Society for Mass Spectrometry* started in 1990 as the official archive of the American Society for Mass Spectrometry. *European Journal of Mass Spectrometry* is another specialty journal, published since 1995, containing fundamental and applied work. Specific application areas are also located in the specialty journals for those particular areas.

Two primary mass spectrometry conferences occur on a regular basis. The International Mass Spectrometry Conference is held every three years, and is typically held in a European city (Hamming and Foster, 1972; Quayle, 1987). The proceedings of this conference, in the form of extended abstracts of presented papers, are often published in book form as the series, titled *Advances in Mass Spectrometry*. Proceedings have been published occasionally as a special volume of the *International Journal of Mass Spectrometry and Ion Processes*. The annual conference of the American Society for Mass Spectrometry rotates among numerous cities on the North American continent (Hamming and Foster, 1972). The Society originated as the American Society for Testing and Materials Committee E-14 on mass spectrometry, starting with an annual meeting in 1953. The Society was formally organized in 1969. Extended abstracts of papers presented at this meeting are also collected and distributed, but do not constitute official publications, as they are distributed only to members of the Society and to conference attendees.

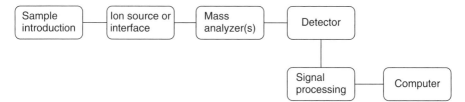

Figure 8.1   Modular diagram of a mass spectrometer system (from Burinsky, 2006).

8.4   INSTRUMENTATION

Figure 8.1 shows a very simplified (and simple-minded) modular diagram of a modern mass spectrometer system. We will make brief commentary on these various parts. Our main focus is to discus what to do with the data after one detects the ions—interpretation!

Mass spectrometry is an inherently gas-phase (vacuum[4]) experiment. The low-pressure gas-phase nature, however, should immediately suggest incompatibilities. Facile sample introduction into an instrument suggests compromising the vacuum chamber. Developing practical, efficient methods and mechanisms for accommodating the transition from samples at atmospheric pressure and in condensed phase (solids or liquids) into a vacuum system—into the mass spectrometer—has been an ongoing effort. Once into the instrument, ion sources and ionization methods convert molecules to charged particles, if they are not already charged.

A molecular ion or fragment ion requires a net electrostatic charge in order to be influenced by and steered by an external magnetic and/or electric field. Once charged particles are produced, a number of different mass filters or mass analyzers are available to achieve sorting of particles—molecular ions, fragments, clusters, and so on—according to their mass-to-charge ratio. This is an opportune time to emphasize that the mass spectrometer does not measure mass, but measures mass-to-charge ratio. Commonly, and until recent times, most data collected by mass spectrometers related to particles that contained a single excess electrostatic charge. The mass-to-charge ratio of a particle reduced to the "mass" of the particle divided by the +1 or −1 electrostatic charge. Although multiply charged ions had been reported in the literature, starting with some of Thomson's original observations (Kiser, 1965), they were uncommon until electrospray was developed for direct observation of macromolecules.

Detection[5] is accomplished most typically by using an electron multiplier, which works essentially the same as a photomultiplier. The ion beam impinges on the first dynode of a multiplier stack. Variations in recent years have replaced the nude electron multiplier with a phosphorescent dynode—the ion beam strikes the dynode and emits a pulse of light, which is then detected and amplified by a photomultiplier. This technique is reminiscent of the original detection mechanism used by Thomson and others, a variation on a phosphorescent television screen picture tube. Microchannel

---

[4] An entire technological field of vacuum science and techniques exists in order to achieve the necessary vacuum and pumping speed, which we shall ignore at present.

[5] Detectors, like vacuum systems and technology, are another extensive science in and of themselves, which we will simple take for granted here.

plate detectors and other variations on this theme have been introduced, some to implement two-dimensional detection. This property recalls the Mattauch–Herzog classic double focusing mass analyzer geometry, which focused a high-resolution ion beam on a focal plane for, among other things, photographic plate detection of mass spectra.

The "vacuum zone" of a mass spectrometer is classically considered to include the ion source, mass analyzer, and detector components of Fig. 8.1. Sample introduction mechanisms allow being able to efficiently bridge the gap between atmosphere and vacuum. Direct insertion probe locks for EI and FAB ionization required complex valving systems. Much of the technological effort expended in developing chromatography–mass spectrometry interfaces dealt with the condensed-phase and high-pressure to low-pressure transition, along with disposing of excess solvents and carrier gases. A significant shift occurred with the perfection of the API sources, now used with electrospray and several other methods. The vacuum chamber boundary moved from between the ion source and the sample introduction boxes of Fig. 8.1 to between the ion source and mass analyzer boxes. Sample introduction, desolvation, and ionization now occur at atmospheric pressure. Introduction of a representative plume of vapors into the mass spectrometer for mass analysis is now done through a small sampling cone strategically placed in the atmospheric ionization chamber.

8.5 INLET SYSTEMS: GETTING SAMPLES IN

8.5.1 Direct Insertion Probe Techniques

Distillation of samples placed on a direct insertion probe, inserted into the mass spectrometer and placed in close proximity to the ion source, was at one time the most common and most frequently used mechanism for conducting mass spectral analysis. The heat generated by the ion source—typically an EI or CI ion source—caused components to volatilize out of the sample. If a pure compound, heating to above the melting point of the compound produced vapors, which were ionized and mass-analyzed. If a mixture, components would distill off the probe according to their melting points and volatilities. It is a crude but effective evaluation of homogeneity of a sample as well as providing identity information about the components. Independently heated probes made the experiment all the more reliable and useful, as rates of heating could be controlled. Samples that would thermally decompose during the distillation could be addressed by using a direct exposure probe—one in which the sample was placed on an electrically heated wire filament, and the filament heated ballistically to volatilize sample very quickly. Running the race between intact volatilization and thermal decomposition often produced valuable structural information. FAB ionization was also used primarily as a direct insertion probe technique. Sample, dispersed in a matrix, was inserted on a probe into proximity with the ion source, and bombarded by a beam of fast atoms (or ions, in the case of liquid secondary ion mass spectrometry or SIMS).

Replacement of a direct insertion probe with the effluent of a gas chromatographic column constitutes the GC-MS experiment, described later. FAB was adapted to serving liquid chromatographic effluents by the flow-FAB experiment, also described later. Once being the standard sample introduction method, instruments capable of conducting direct insertion probe experiments are becoming rare.

8.5.2   Liquid and Gas Septum Inlet

Gases and thermally stable liquids can be introduced into the mass spectrometer by injecting a sample through a septum into an evacuated, heated expansion bulb. The resulting vapor can be leaked into the source of the mass spectrometer, and the spectrum recorded. The caution necessary here is decomposition of the analyte by contact with hot walls of the expansion bulb and inlet tubing. All-glass inlet systems, in which the sample only contacts glass walls, are available, but this does not always solve these kinds of problems. For gases, one must use a gas-tight syringe or other types of transfer device. Introduction of samples by this mechanism presupposes EI or CI is being used. Such inlet systems are also becoming rare in commercial instruments.

8.5.3   Gas Chromatography–Mass Spectrometry

GC-MS is a mature technique. Gohlke and McLafferty (1993) reviewed events in the early history of its development. The first successful experiments using this technique were conducted in late 1955, coupled to a time-of-flight (TOF) mass analyzer. Other analyzers available at that time could not scan fast enough to be able to record an adequate full-scan mass spectrum during the elution of a gas chromatographic peak, even though gas chromatography was being done at that time on packed columns, with very wide chromatographic peaks. Improvements in the scanning speed of magnetic sector analyzers, and the advent of the quadrupole mass filter, supplanted the original TOF analyzers until recent years when TOF technology improved.

Both gas chromatography and mass spectrometry are gas-phase experiments. However, coupling a gas chromatographic effluent to a mass spectrometer requires dealing with the problems associated with making the transition from a high-pressure gas-phase experiment to a low-pressure gas-phase experiment. Two older books on GC-MS by McFadden (1973) and Message (1984) describe the early days in the maturation of the GC-MS technique. The problems of coping with relatively high flows of carrier gas, in the case of packed column chromatography, and of enrichment of the analyte in a large excess of carrier gas have been dealt with straightforwardly.

Several types of carrier gas separators were developed to accommodate packed column gas chromatography—to accommodate the high gas flows and the necessity to enrich analytes. Most assume that helium is used as the carrier gas, and rely on the small molecular size of helium in relation to the larger size of analytes of interest. Typical packed column gas flows accommodated were 30 mL/min (or more) of helium carrier gas, flowing into the separator, and 3 to 5 mL/min flowing on into the mass spectrometer.

Because of the substantially lower total gas flow through a fused silica capillary gas chromatography column, a separator is no longer needed. With flexible fused silica columns, the column end can be inserted directly into the mass spectrometer. Typical carrier gas flows through capillary columns of 1 to 2 mL/min can be readily accommodated by the pumping systems on modern mass spectrometers. The end of the capillary column can be positioned in close proximity to the ion source of the mass spectrometer, if not actually inserted into it. The sensitivity of detection is thus maximized, since this positioning delivers the entire sample directly into the ion source. Separator technology is not entirely obsolete, however. Megabore fused silica columns are often operated with carrier gas flows approaching those of packed columns. The higher flows need to

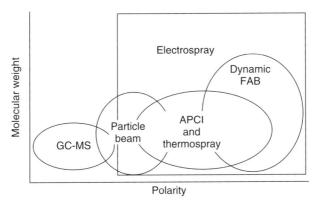

Figure 8.2 Applicability of various chromatography–mass spectrometry techniques to the molecular domain.

be processed through a jet separator, with the requisite reduction in total gas flow and analyte enrichment.

Not all ionization methods are available for use in GC-MS experiments. Because the GC experiment is a gas-phase experiment, only those ionization methods that operate in the gas phase are appropriate. Standard EI and CI methods are applicable. Recent years have seen a reintroduction of field ionization coupled to GC-MS experiments. The literature suggests spectra of quality comparable to that generated by conventional EI.

The gas-phase limitation imposed by the very nature of the GC-MS experiment limits its applicability to chemical space. The Venn diagram in Fig. 8.2 illustrates an approximate mapping of applicability of the combined chromatography–mass spectrometry techniques discussed here. Using only relative molecular mass and polarity as two dimensions that can map chemical space, the figure indicates that the applicability of GC-MS is limited to those relatively small, thermally stable, nonpolar compounds that can be readily volatized (or made to volatilize by derivatization (Knapp, 1979) to pass through a gas chromatographic column.

8.5.4 Liquid Chromatography–Mass Spectrometry (LC-MS)

Combined LC-MS is a rapidly maturing technique. Peter Arpino's cartoon (Fig. 8.3) captures well the essence of the difficulties associated with coupling the liquid chromatographic experiment with the mass spectrometric experiment. The historical perspective presented in the following is intended to illustrate the difficulties in achieving a useful LC-MS interface, and the variety of devices and the ingenuity of the people attempting, and eventually succeeding, to make it happen. The LC-MS experiment has only become an effective experiment since the early to mid-1990s.

Attempts to couple condensed-phase chromatographic methods with mass spectrometry start with the work reported by Victor Tal'Rose[6] (Baldwin, Burlingame and Nikolaev, 2004) and colleagues in 1968 (Tal'Rose et al., 1968). By referring to

---

[6] Victor Tal'Rose recently passed away.

Figure 8.3 Combining liquid chromatography with mass spectrometry involves a difficult court-ship (from Arpino, 1982).

condensed-phase chromatographic methods here, we primarily imply HPLC. Attempts, with varying degrees of success and acceptance, have been reported to interface other condensed-phase separation methods, such as thin-layer chromatography (Wilson, 1999; Busch, 2004a) and supercritical fluid chromatography (Randall and Wahrhaftig, 1981; Smith and Udseth, 1983). Yergey *et al.* (1990) summarized the progress on development of effective LC-MS interfaces, and compiled a review and comprehensive bibliography of interface development and applications up to 1990. They do not cover electrospray, as the explosion of efforts and applications of this technique only started in this time frame.

Unlike combined GC-MS, LC-MS presents a significant incompatibility between the condensed-phase liquid chromatographic experiment and the gas-phase mass spectro-metric experiment. The pumping system of the mass spectrometer must be capable of handling the gas load introduced by the chromatographic experiment. Typical gas loads generated by a gas chromatographic experiment are 3 to 5 mL/min for packed columns (after jet separation!) and 0.5 to 1 mL/min for capillary columns. HPLC mobile phase must be vaporized. Typical mobile phase flow is 1 to 3 mL/min. Hexane, converted to gas phase, produces 180 to 540 mL/min of vapor. Chloroform produces 280 to 840 mL/min. Methanol produces 350 to 1650 mL/min. Water produces 1250 to 3720 mL/min. The design and engineering of an interface to connect a liquid chromatographic experi-ment with a mass spectrometric experiment must accommodate these large gas flows—approximately 1000× that of the gas chromatographic experiment—and still permit both the chromatographic experiment and the mass spectrometric experiment to work properly.

8.5.4.1 Direct Liquid Introduction (DLI) DLI is the simplest and most straightforward approach. It was first attempted and reported by Tal'Rose *et al.* (1968). Baldwin and McLafferty (1973) reported their attempts, utilizing CI. Chapman *et al.* (1983) reported their experiences with interfacing to a magnetic sector instrument. A small portion of chromatographic effluent (10 to 20 μL/min) is sprayed into the ion source through a pinhole leak into the mass spectrometer. CI is most frequently used, with the residual solvent molecules in the spray acting as the CI reagent.

8.5.4.2 Atmospheric Pressure Chemical Ionization (APCI) Rather than leaking a portion of a chromatographic effluent into the vacuum system of the mass spectrometer, the effluent is introduced via a heated nebulizer into an API source. The source is outside the vacuum chamber, operating at or above atmospheric pressure. Ionization occurs in a corona discharge, struck in the plume of effluent sprayed from the heated probe. The solvent vapors provide a source of protonating reagent for the proton transfer CI reactions in the vapor. The corona discharge region is sampled through a sampling orifice by the mass spectrometer. The discharge was first struck by proximity with the ionizing radiation from a ^{63}Nickel radioactive needle, and later by a needle with an applied high voltage (Carroll *et al.*, 1975). The ionization is soft, as the majority of compounds produce abundant protonated molecular ions.

8.5.4.3 Continuous Flow FAB Interfaces FAB ionization revolutionized the application of mass spectrometry to large, thermally fragile, nonvolatile molecules. This desorption technique was described by Barber and colleagues in 1981 (Barber *et al.*, 1981; Surman and Vickerman, 1981). Dependence of the technique on a dispersing matrix and its inherent incompatibility with flowing sample introduction into the mass spectrometer prevented its immediate application to chromatographic separations. However, a mobile phase adulterated with 10% (by volume) of glycerol, the classical FAB matrix, was infused through a glass frit, the frit being the FAB target. Low flow chromatographic effluent (circa 2 μL/min) could be analyzed by FAB ionization (Ito *et al.*, 1986). Caprioli *et al.* (Caprioli, Fan and Cottrell, 1986) showed that the frit was not necessary, effluent being introduced directly onto the target through a hole in the center. The fine art of obtaining useful results from either of these approaches involved balancing the rate of evaporation of effluent from the FAB target with replenishment of fluid by chromatographic flow. Adjusting solvent composition, glycerol content in the mobile phase and externally applied heat to the target were some of the operational variables.

8.5.4.4 The Moving Belt Interface Scott *et al.* (1974) and McFadden, Schwartz and Bradford (1976) first described this mechanical interface in 1974 and 1976. A diagram of a commercialized moving belt interface is shown in Fig. 8.4. The interface first consisted of a spool of wire, which was unrolled off one spool and onto another. As the wire was wound from spool to spool, the effluent from a liquid chromatographic separation was applied to the wire. As the wire was transported through the ion source of a mass spectrometer, it was heated to desorb analytes that had been applied to it. The desorbed materials were ionized and mass-analyzed in the mass spectrometer. Later, a recycling continuous loop of wire, or a belt of an appropriate material, was used, with added mechanisms for "cleaning" the belt to avoid chromatographic confusion. The moving belt is a versatile interface, offering applicability in EI, CI, and FAB (Stroh *et al.*, 1985) ionization modes. However, it is mechanically the most complex, both in

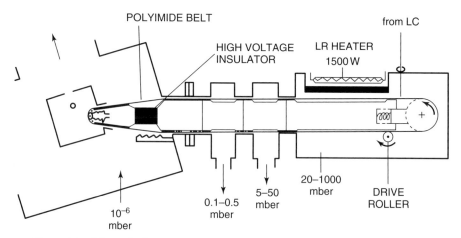

Figure 8.4 Schematic diagram of a moving belt interface. Courtesy of V.G. Analytical.

operation and optimization of the interface, and in its installation into a mass spectrometer.

8.5.4.5 Thermospray The term "thermospray" refers to both a vaporization LC-MS interface device and a soft ionization method. Vestal and coworkers (Blakley, Carmody and Vestal, 1980) developed thermospray, beginning in 1976. Plasmaspray is a variation on the thermospray concept, implemented by V.G. Instruments on their line of magnetic sector and quadrupole instruments. The technique and device were designed as an alternate means of rapidly vaporizing solvent and nonvolatile analytes by heating the tip of a stainless steel tube through which the LC effluent flowed. The resulting spray of vapor and droplets was directed against a heated probe to complete droplet vaporization. An electron beam from a hot filament bombarded this vapor, producing CI, and the vapor was sampled with a sampling orifice into the mass spectrometer. The accidental observation that a filament and electron beam was not needed was made when, unknown to the operators, the filament had burned out and as many ions were being produced and detected without the filament as with the filament. Mass spectra produced by thermospray are soft ionization spectra. The molecular species is typically a protonated molecular ion or an adduct—often an NH_4^+ cationated species, because thermospray works best from ammoniated mobile phases. Thoughts on the mechanism(s) involved in this process were reviewed by Vestal (1983). The book by Yergey *et al.* (1990) contains an extensive treatise on this technique. Empirical observations show that ionization is quite efficient when the LC separations are carried out in a polar solvent, which includes ionic solutes such as ammonium acetate. Separations carried out in a nonpolar solvent, without salts, require an assisting electron beam from a filament, if the application works at all.

8.5.4.6 MAGIC The MAGIC acronym stands for "monodisperse aerosol generation interface for chromatography." This LC-MS interfacing technique was first described by Willoughby and Browner (1984), and evolved into the particle beam interface. Various instrument manufacturers implemented it under a variety of names. It contained no moving parts. Volatile solvents work best. Small portions of split aqueous flows could also be accommodated. Desolvation was done by evaporation and impinging with a "nebulizing gas." Only sufficient external heat was supplied to counteract the cooling caused by expansion and evaporation of the solvent. Reduction of net flow into the mass spectrometer is done by using the Ryhage jet separator technique (developed for packed column GC-MS). This interface presents the possibility of conducting EI on samples introduced into the mass spectrometer, EI spectra being the "richest" in fragmentation and structural information. EI spectra could be obtained of sufficiently high quality to permit spectral library searching against, for example, the NIST library. The particle beam technique occupies a niche in applications space. Compounds that can produce fragmentation-rich EI mass spectra (e.g. via a probe distillation experiment), but which will not pass well through a gas chromatographic column, work in liquid chromatographic experiments interfaced to the mass spectrometer through a particle beam interface.

Semi-volatile and nonvolatile compounds of interest in environmental sciences (Bellar, Behymer and Budde, 1990), such as the larger polycyclic aromatic hydrocarbons and pesticides (Miles, Doerge and Bajic, 1992), have been effectively addressed. It is theoretically possible to introduce reagent gases into the ion source of the mass spectrometer and do CI on materials introduced through a particle beam interface. Extrel Corporation also developed and commercialized an ion source that included a fast atom source and a target for doing FAB ionization. The mobile phase was adulterated with a low percentage of glycerol. The low percentage, in general, did not interfere significantly with the chromatography. The glycerol acted as a continuously renewing matrix from which to do FAB ionization, the glycerol being sprayed onto the FAB target along with the analyte.

8.5.4.7 *Electrospray* The term "electrospray" refers both to the interface design and the ionization mechanism, as do the "thermospray" and "atmospheric pressure chemical ionization" terms. Abbe Jean-Antoine Nollet's 1750 experiments on electrostatic spraying of liquids[7] constitute the first experiments reported with this technique (Fig. 8.5[8]). Nollet demonstrated a number of ways in which electrostatically charged spraying devices could effectively spray fluids. John Fenn's (Yamashita and Fenn, 1984) more recent elaboration of the electrospray phenomenon has led to a revolution in the application of mass spectrometry, first to biologically relevant large molecules, and subsequently in the application to polar molecules in general.

8.6 IONIZATION METHODS: MAKING IONS

To perform mass spectrometry, one must make ions from neutral molecules. Ionization methods have advanced from the classic EI, through CI (Harrison, 1992), field desorption (Wood, 1982), FAB (Barber *et al.*, 1981; Surman and Vickerman, 1981), and thermospray (Vestal, 1983) to the API techniques currently favored. EI is classic, but is restricted in its applicability to thermally stable, volatilizable compounds. Field

---

[7] This interesting historical footnote was brought to our attention by Dr. Thomas Covey, MDS Sciex Instrument Corporation.
[8] Information available on the history of chemical instrumentation web site maintained by Prof. Euginii Katz, Department of Chemistry, The Hebrew University of Jerusalem. See http://chem.ch.huji.ac.il/~eugeniik/history/nollet.htm.

Figure 8.5 Abbe Jean-Antoine Nollet's experiments with the electrostatic spraying of liquids.

desorption was always a specialized niche technique applicable to some larger compounds. FAB enjoyed a meteoric rise in use first reported in 1981; but, has all but disappeared now, being replaced by the API techniques: atmospheric pressure chemical ionization (APCI) (Carroll *et al.*, 1975) and electrospray (ESI) (Yamashita and Fenn, 1984). Matrix-assisted laser desorption ionization (MALDI) (Hillenkamp and Karas, 1990) has shown significant utility for characterizing larger proteins, approximately 100 kDa and larger.

The evolution of the ionization techniques has progressed from the hard EI technique, which produces extensive fragmentation, to the soft ionization API techniques, which produce in general only molecular ions.[9] In achieving the goal of being able to

[9] Hardness and softness of ionization refer to the energetics of forming an ion from a neutral molecule, and the extent of spontaneous fragmentation resulting from the ionization process. Hard ionization techniques induce extensive fragmentation and have a high probability of completely destroying the molecular ion. Soft ionization techniques induce minimal or no fragmentation. The ideal soft ionization technique would produce a molecular ion species and nothing else. This is not to imply that substantial amounts of energy need to be deposited in samples in order to induce evaporation, desorption, and/or ionization. See the discussion by Hillenkamp and Karas (1990).

ionize larger, more polar, less thermally stable molecules, and in optimizing for production of molecular ions, much of the structure-indicating fragmentation information has been lost. MS-MS experiments restore structural information content. They select the molecular ion produced by the soft ionization technique in the first mass analyzer of a tandem instrument and activate it by collision with a target gas. The product ion mass spectrum of fragments resulting from collision-induced decomposition (CID) of the parent ion in a later mass analyzer is recorded. The effect of inducing fragmentation by collisional activation can also be achieved through collision with solvent molecules in the API ion source. Various implementations of this procedure are called up-front CID, cone voltage fragmentation, and "poor man's MS-MS." Results obtained this way are comparable to true MS-MS product ion spectra. The strict interpretation of the results in terms of precursor–product ion relationships is lost, but this limitation does not negate the utility of the method when a true tandem instrument is not available.

Discovering and developing new ionization methods is one of the many topics that punctuate the history of mass spectrometry. All of the mass analyzers rely on the ability to use external force fields—magnetic fields, electric fields, radio frequency fields—to steer electrostatically charged particles through space. Methods to convert neutral molecules to charged species while preserving structural information are a perpetual topic of interest and development. Electric discharge in a gas produces ions, but is highly energetic, producing primarily atomic species. It was the ionization method used in Thomson's early parabola mass spectrograph studies that utilized electric discharge, and in the early evaluations of isotopic composition of the elements. EI—a beam of electrons interacting with neutral molecule vapors—is the classical ionization method, producing radical cation molecular ions (when the molecule is stable enough to withstand dislodging of a valence electron) and extensive fragmentation. Development of softer ionization methods is driven by the attempt to produce a molecular ion species without destroying structural integrity of the molecule. CI using a variety of different reagent gases, field ionization (FI), field desorption ionization, FAB, APCI, electrospray (ESI), and MALDI are the more well-known and prominent ionization methods. The order in which they are listed is an approximate order from hardness to softness. This list is also not comprehensive. Vestal (2001) reviews these and a number of other minor specialty ionization methods.

Which method should one use for a particular application? The first factor to consider is the type of experiment to be used. Not all of these methods are compatible with combined chromatography–mass spectrometry experiments. Figure 8.2 gives a first impression of where, within the universe of chemical compounds, various techniques can be applied. The nature of GC-MS limits use to EI and CI, with some FI, and to compounds that are sufficiently volatile and thermally stable to pass successfully through a gas chromatography column. Attempts have been made to utilize nearly all of the major ionization methods for LC-MS, with varying degrees of success. Figure 8.2 again tries to capture the circumstances. FAB and MALDI are desorption ionization methods, and present particular engineering difficulties over how to present sample while preserving chromatographic separation. Workable solutions have been developed for FAB. While the general nature of pharmaceutically relevant compounds makes electrospray and APCI nearly ideal for this application area, they do not work well for all compounds. In short, there is no universally applicable ionization method.

8.6.1   Electron Ionization

Electron ionization (also known as electron impact) is the classical ionization method. Vapors of the sample are bombarded with a beam of high-energy electrons (typically at 70 eV), produced by an electrically heated filament. During interaction of an energetic electron from the ionizing beam with an analyte molecule, an outer shell electron is displaced from the neutral molecule, producing a positively charged radical cation. Even with a low-energy transfer efficiency, a large amount of energy is still deposited into the molecule. Substantial fragmentation of the molecule is expected. The timescale of interaction of the ionizing electron with the neutral molecule is very short with respect to any vibrational events within the molecule. As the energy starts to redistribute into the various vibrational modes of the molecule, vibrations increase until a bond is broken and a fragment produced.

Organic mass spectrometry under EI is typically performed at 70 eV. This energy is chosen because the generalized shape of the ionization efficiency curve for most gases and organic molecules plateaus at approximately 60 to 80 eV. Standard libraries of EI spectra are recorded at 70 eV, as the mass spectrum of most compounds by EI at 70 eV is very reproducible from instrument to instrument and from time to time. The EI source is very reliable and easy to operate. Low eV observations, however, can be useful. Less energy is transferred to the molecule during ionization, and less fragmentation occurs. For particularly fragile molecules, low eV observations can often help to identify the molecular ion.

8.6.2   Chemical Ionization

This ionization technique, first christened as chemical ionization, was described by Munson and Field (1966). The first observation of the CH_5^+ species in a high-pressure ionization source, however, was reported by Tal'Rose and Lyubimova (1952). Field summarized the early work (Field, 1968) and published a historical retrospective (Field, 1990). A reagent gas is bombarded by electrons to form a reagent ion in the gas phase. The sample of interest is introduced into the mixture of neutral reagent gas and reagent gas ion. Sample molecules are ionized by ion–molecule reaction with the reagent ion, typically by proton transfer, often producing an abundant protonated molecular ion, $[M+H]^+$. The attraction of CI is that compounds, which yield a very low abundance molecular ion, if at all, by EI, will yield molecular mass information, in the form of an $[M+H]^+$ under CI. The most commonly used reagent gases used for CI mass spectrometry are methane, isobutane, and ammonia. Methane was used in the original observations. Isobutane often produced less fragmentation and/or more abundant $[M+H]^+$. Ammonia presents selectivity in ionizing compounds which have proton affinities higher than ammonia, but not ionizing compounds with proton affinities lower than ammonia. A number of exotic reagent gases have been explored for even more selective CI experiments. Negative ion chemical ionization is also a useful variation on this experiment, applicable to highly electronegative molecules. Alex Harrison's compendial book summarizes the fundamental and applied work done to explore this technique (Harrison, 1992).

8.6.3   Field Ionization and Field Desorption Ionization

Field desorption ionization was first described by Beckey (1969), comparing and contrasting to field ionization. Neutral molecules "introduced" into an electric field with a very large field gradient will ionize. Thermal evaporation of a substance and introduction into the field gradient characterizes field ionization. Field desorption includes evaporation of target analyte from an activated emitter, upon which a fine brush-like structure of micro-needles had been grown. The activated emitter is dipped in a solution of analyte. After introducing into the vacuum chamber of the mass spectrometer, solvent is allowed to evaporate. The emitter is then heated by electric current until the analyte desorbs, producing (when successful) a spectrum of abundant protonated molecular ion. The technique was first demonstrated for monosaccharides, these and larger polysaccharides being notoriously difficult compounds mass spectrometrically. Mechanisms of how this technique works have been investigated (Holland, Soltmann and Sweeley, 1976). Wood (1982) catalogs a list of successful applications of field desorption. The technique, however, tends to be very dependent on the technique of the operator and the skill with which the emitters are prepared (Schulten and Beckey, 1972; Lehmann and Schulten, 1978). For this reason, the technique was not widely used, and has fallen into disfavor, FAB (see further discussion) being the preferred replacement.

8.6.4 Plasma Desorption Mass Spectrometry

Plasma desorption is a particle-induced desorption technique described in 1974 by Torgerson et al. (Torgerson, Skowronski and Macfarlane, 1974). A sample is deposited on a foil target and placed in the proximity of a ^{252}Californium radioactive source in a TOF mass spectrometer. Upon fission, a ^{252}CF nucleus splits into two high-energy fission fragments, which are ejected in opposite directions. One fragment triggers a clock, while the other bombards the sample foil, causing desorption and ionization of target analyte molecules on the film. The desorbed ions are then accelerated and mass-analyzed in the TOF mass spectrometer. It is a soft ionization technique which produced information on large, nonvolatile molecules. It preceded FAB as one of the few methods at the time for directly observing peptides, proteins, and other biologically important molecules. Reviews describe techniques and successful applications (Macfarlane, 1983; Sundqvist and Macfarlane, 1985; Macfarlane et al., 1994). This technique, while having significant impact during its height, was subsequently replaced by FAB.

8.6.5 Fast Atom Bombardment and Secondary Ion Mass Spectrometry

FAB was first described in 1981 by Barber et al. (1981) and by Surman and Vickerman (1981) in a pair of back-to-back published papers. It is a desorption ionization technique, an offshoot of SIMS (Honig, 1985), in which the surface of a liquid matrix containing the target analyte molecule is bombarded with high-energy neutral atoms, rather than with fast ions as in SIMS. Both positive and negative ions are produced by the same process. The yield of negative ions is generally at least an order of magnitude less than that of positive ions. The technique makes large, nonvolatile, and thermally labile molecules accessible for study by mass spectrometry. Field desorption addressed these same classes of molecules, but FAB proved to be a much simpler and more reliable technique.

Sample was dispersed in a drop of matrix on the end of a direct insertion probe. Selection of a matrix compatible with the chemistry of the target analytes became an art form. While glycerol, the originally used matrix, was relatively general purpose, a number of other widely used matrices, such as *o*-nitrobenzyl alcohol (Meili and Seibl, 1984; Sweetman and Blair, 1988) or a eutectic mixture of dithiothreitol and dithioerythritol (Witten *et al.*, 1986), are prevalent in the literature. FAB mass spectra are mixture spectra, features in the spectrum being from either the target molecule or the matrix. Features derived from commonly used matrices are cataloged by Costello (1991).

8.6.6 Matrix-Assisted Laser Desorption Ionization

Hillenkamp and Karas first described this technique in 1985 (Karas *et al.*, 1987) and reviewed early progress (Hillenkamp and Karas, 1990). Laser desorption of target analytes from surfaces had been exploited extensively, but was restricted to certain classes of compounds, in particular those of no more than approximately 1000 daltons. Matching wavelength of irradiating light from a laser with the absorption maxima of target analytes imposed further technical difficulties. By contrast, dispersing a target analyte (or mixture of analytes) in a matrix of a light-absorbing small molecule matrix, and irradiating with a wavelength at which the matrix absorbed, effected absorption and energy transfer into the target analytes without extensive fragmentation of the analytes and without the necessity of tuning the irradiating wavelength to the absorption maximum of the target. A number of other advantages also presented. The molecular mass boundary of laser desorption could be extended to intact proteins. In the process of examining this application, a number of small aromatic molecules which absorbed in the 270-nm region were examined for utility as matrices (Beavis and Chait, 1989). Attempts to find the magic matrix, choosing or tailoring the matrix to the application, resulted in a significant amount of research activity, similar to searching for that special FAB matrix. Today, MALDI competes effectively with electrospray (see next section) as the ionization method of choice, in particular for large proteins and other macromolecules.

8.6.7 Electrospray

Malcolm Dole (Dole *et al.*, 1968) described his group's attempts to measure the relative molecular masses of oligomers of synthetic polymers based on electrospray ionization in 1968. Fenn's group elaborated on the Dole observations, and built the electrospray ion source shown in Fig. 8.6 (Yamashita and Fenn, 1984). In essence, liquid sample containing molecules of interest is flowed into the source through a small tube, on which a high electrostatic voltage is placed. The liquid is sprayed from the tube tip, forming a Taylor cone and subsequently a stream of droplets, each electrostatically charged. A countercurrent flow of dry nitrogen gas opposes the spray, causing evaporation of solvent. As the droplets become smaller because of solvent loss, they reach the Rayleigh instability limit, where coulomb charge repulsion overcomes solvent surface tensions and the droplets fragment further. Repetition of this process ultimately produces solvent-free solute ions, which are subsequently swept into the sampling orifice of the mass spectrometer and detected.

The detailed mechanism of electrospray is still an active and lively topic of discussion at current scientific conferences. Numerous research laboratories are actively probing the subject. Volume 35, issue 7 (2000), of the *Journal of Mass Spectrometry* published a series of five papers (Amad *et al.*, 2000; Cole, 2000; Gamero-Castaño and Fernandez de la Mora, 2000; Kebarle, 2000; Van Berkel, 2000) from research groups

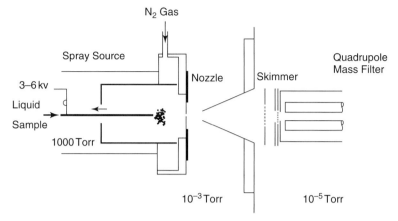

Figure 8.6   Electrospray ion source (from Yamashita and Fenn, 1984).

working in this area, followed by a lengthy printed discussion (Fernandes de la Mora *et al.*, 2000) in the subsequent issue.

Two models have been repeatedly discussed in the literature—ion evaporation, attributed to Iribarne and Thomson (1976), and coulomb repulsion, attributed to Dole. Both models rely on the Rayleigh instability of charged droplets to become smaller and smaller. They differ in that coulomb repulsion relies on the continuous repeating of the process until one is left with a single isolated desolvated ion. Ion evaporation invokes the idea that, before a droplet becomes small enough to contain a single solute molecule, the electrostatic field strength at the droplet surface becomes sufficiently intense to eject a surface ion from the droplet into the ambient gas. Kebarle reviewed the data available in 1991 (Blades, Ikonomou and Kebarle, 1991), and updates and interprets (Kebarle, 2000) the summed results to propose that the data indicate that the mechanisms constitute a continuum, and that elements of both interplay, depending upon physical properties such as hydrophobicity or hydrophilicity, most notably the size of the molecule under consideration—small ions are produced by ion evaporation, large ions are produced by coulomb repulsion—and other environmental factors such as ion concentration in solution. Vestal (2001) gives an additional concise summary of the current thoughts on how electrospray works, but more importantly summarizes a number of design variations of electrospray ion sources extant on modern instruments.

8.6.8   Atmospheric Pressure Chemical Ionization

APCI is similar to electrospray in that ionization occurs at atmospheric pressure. The effluent of, for example, an HPLC column is passed through a heated nebulizer and sprayed into the source chamber. Ionization occurs in a corona discharge, struck in the plume of effluent. The solvent vapors provide a source of protonating reagent for the proton transfer CI reactions in the vapor. The corona discharge region of the spray plume is sampled through a sampling orifice by the mass spectrometer. When this experiment was first reported, the discharge was struck by proximity with the ionizing radiation of a radioactive nickel needle (Carroll *et al.*, 1975). The radioactive needle

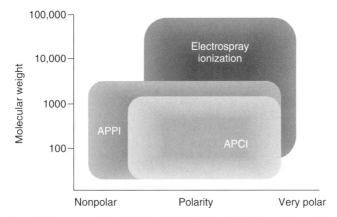

Figure 8.7 Applicable range of molecules ionizable by APPI (from Syage, Short and Cai, 2008).

was later replaced by a needle with an applied high voltage. APCI ionization is soft, as the majority of compounds produce abundant protonated molecular ions. It is, however, somewhat harsher than electrospray, and is often used for compounds that do not have readily protonatable sites in the molecule, and require higher energies to make ions. Regarding the "quality" of the spectra, some comparisons have been made to FAB, using extent of fragmentation as a measure of quality.

8.6.9 Atmospheric Pressure Photoionization (APPI)

APPI is a variation on the electrospray and APCI techniques. Instead of a corona discharge or a high-voltage potential to effect ionization, a high-intensity ultraviolet lamp is used to effect photoionization of analytes in a chromatographic effluent stream (Robb, Covey and Bruins, 2000). Photoionization detection combined with gas chromatography has been studied and available since the 1970s. Photoionization, coupled with addition of dopants such as toluene or methanol into the chromatographic effluent to control the photochemistry occurring in the sprayed effluent, has been demonstrated to present advantages of sensitivity and extending the range of applicable types of compounds. Developers claim to be able to address a group of nonpolar analytes which are not accessible by either electrospray or APCI (Fig. 8.7). Study of the mechanism of APPI (Syage, 2004), where the range of applications was discussed, has been published (Syage, Short and Cai, 2008).

8.6.10 DART and DESI

Two new ionization methods to appear recently are desorption electrospray ionization (DESI) and direct analysis in real time (DART). They are the first in a new area of ionization mechanisms being explored which are collectively referred to as open-air ionization sources. Recall the earlier discussed progression from forming ions in an evacuated chamber to forming ions in an atmospheric pressure chamber. Both DART and DESI form ions in an "open atmosphere" and sample the resulting plume through an entrance cone into the mass spectrometer.

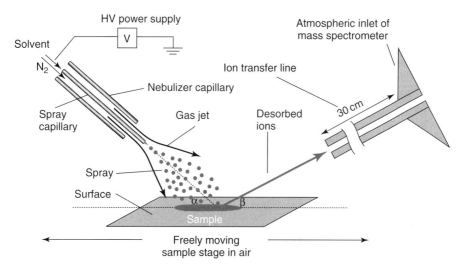

Figure 8.8 Desorption electrospray ionization source (from Takáts *et al.*, 2004).

DESI has been described in detail (Takáts *et al.*, 2004, 2005). Figure 8.8 shows the configuration of the source and sampling arrangement. A spray of solvent, generated by an electrospray spraying device, is impinged on a surface. This beam of impinging particles consists of gas-phase solvent ions, ionic clusters, and charged microdroplets. These charged projectiles act similar to the bombarding particles (ions, atoms, photons) in the desorption ionization methods, chemically sputtering target analyte molecules off the surface. The nearby mass spectrometer entrance orifice introduces a sampling of this collection of sputtered chemical species for mass analysis. The process has been likened to a "power-washing" process.[10] In a very short time, a number of successful applications have developed using this technique, from analyzing thin-layer chromatographic separations to doing two-dimensional imaging of surfaces and of thin sections of biological tissues.

The applicability of DESI to the two-dimensional mapping of a surface resurrects significant interest in the combination of mass spectrometry and thin-layer chromatography. Reports have periodically appeared in the literature on developing devices and techniques for accomplishing this coupling, such as enclosing an entire x-y-translatable stage (such as implied by Fig. 8.8, and actually used by Takáts *et al.* 2004) within a vacuum chamber and adding a layer of matrix for FAB ionization (Busch, 1995). A relatively recent review (Wilson, 1999) summarizes those efforts. The innovation comes here in that the entire x-y-translatable stage apparatus can be placed in the open air.

The DART ionization source (Cody, Laramée and Durst, 2005) uses an electric discharge to produce a plasma of electrons, ions, and excited states of a gas, typically nitrogen, which is then formed into a high-velocity gas stream. This gas stream can be directed toward a liquid or solid sample, or it can interact with vapor-phase samples. As with the DESI source, the angular and distance orientation of the ionizing gas stream toward the sample surface and the mass spectrometer entrance influences the

---

[10] The power-washing analogy is attributed to Dr. Gary van Berkel (2007).

quantitative aspects of DART sampling. However, regarding efficient operation of the technique, and for qualitative applications, these orientation parameters are not critical. This noncritical nature makes sample preparation and presentation for study by this technique very easy. Therein lies the attraction, especially when identity of substances is the goal of the experiment and absolute amounts or percentage compositions of mixtures are not necessary results of the analysis.

8.7 MASS ANALYSIS: SORTING BY SIZE

8.7.1 Time-of-Flight Analyzer

The TOF mass analyzer is the simplest type of mass analyzer. Ions in a beam produced by an ion source have a "constant" kinetic energy, but have velocities that differ according to the mass of the particle. Recalling that kinetic energy is expressed as $0.5\,mv^2$, particles of higher mass will have lower velocities and will take longer to travel along a tube of fixed length than lighter particles with higher velocities. By measuring the elapsed time of travel of particles at a detector, a fixed distance from the ion source is correlated to the mass of the particles.

8.7.2 Quadrupole and Triple Quadrupole

The quadrupole mass filter has several properties that make it very useful. It can be scanned very fast over a wide mass range. It has very high transmission efficiency. A high percentage of the ions of a selected mass which are introduced into the filter are transmitted. The ratio of the DC voltage and the radio frequency voltage applied to the quadrupole determines the mass of the particle transmitted. If the DC voltage is turned off, the quadrupole becomes a high-efficiency ion pipe, transmitting all masses. The quadrupole can operate efficiently at higher pressures and much lower voltages than magnetic sector instruments (see further discussion). These several advantages make it a good mass spectrometer to couple to a gas chromatograph—especially a capillary chromatograph which produces very narrow chromatographic peaks. This also presents a significant advantage in coupling to ultrahigh-pressure liquid chromatographs, which also produce very narrow chromatographic peaks. Quadrupoles can be very small, making them ideal for benchtop laboratory instrument applications. The primary disadvantage of this device is that, operated under typical conditions, it is a low-mass resolution device.

Campana has presented an excellent discussion of quadrupolar mass filter theory (Campana, 1980). The quadrupole arrangement is not the only type of mass filter that works according to the basic principles of this device. A number of n-pole mass filters have been described, including a monopole. A hexapole is commonly used in several commercial instruments. In a three-dimensional "quadrupole," the ion trap has been developed.

The triple quadrupole mass analyzer was first assembled by Yost, Enke and Morrison as a device for conducting photoionization experiments. It is a tandem mass spectrometric arrangement where three quadrupole mass filters are connected such that the exiting ion beam of one quadrupole is introduced as the source of ions for the next quadrupole in series. The first quadrupole was used to select a specific precursor ion from a heterogeneous beam of ions and introduce it into the second quadrupole. The second quadrupole was used in its RF-only ion pipe mode to contain and constrain the selected ions. These selected ions were then irradiated with light in order to induce photochemical reactions. The products were then introduced into the third quadrupole and mass-analyzed to identify the photochemical products. After conducting these experiments, the originators and many others recognized that this configuration could also be used, by conducting collisional activation with a collision gas in the central RF-only quadrupole, to do definitive structural and quantitative experiments. The triple quadrupole configuration is probably now the most common tandem configuration, and perhaps the most common quadrupole configuration in use in modern chemical laboratories.

8.7.3 Ion Trap

Wolfgang Paul shared the 1989 Nobel Prize in physics (http://nobelprize.org/nobel_prizes/physics/laureates/1989/) for his work on the ion trap. The trap is a three-dimensional quadrupole, which can be used in many of the same capacities as the triple quadrupole. Additionally, tandem mass spectrometric experiments with multiple stages of collisional activation and decomposition beyond the single-stage MS-MS experiments of the triple quadrupole can be conducted.

8.7.4 Magnetic Sector

The magnetic sector mass analyzer is the stereotypical mass spectrometer. When one mentions a mass spectrometer, this image is conjured. Single-focusing instruments have been commercially built. Double-focusing magnetic machines combine a magnetic momentum analyzer sector with an energy-focusing electric sector. A number of double-focusing sector geometries have been developed, including the forward and reverse Nier–Johnson designs, the Mattauch–Herzog focal plane design, and numerous others. The Nier–Johnson is the most commonly seen and commonly commercially manufactured geometry. Variations of this geometry have been used in producing multi-sector tandem mass spectrometers. The sector instrument has largely become obsolete in recent years. Its expense, complexity, and difficulty of operation have caused its demise in favor of quadrupole instruments, TOF instruments, and ion traps. Improvements in the reflectron time-of-flight (rTOF) mass analyzer have even challenged the one remaining niche that the sector instruments, and the equally technologically intense Fourier transform mass spectrometers (FT-MS) occupy, namely the high-resolution accurate mass measurement arena.

8.7.5 Ion Cyclotron Resonance Fourier Transform Mass Analyzer

The FT-MS uses the principles of the ion cyclotron resonance (ICR) experiment to make mass measurements. The mass-to-charge ratio of particles undergoing cyclotron resonance in a strong magnetic field correlates with the period of their orbiting in the magnetic field. This frequency domain is transformed into a mass-to-charge domain by use of the mathematical Fourier transform. Many portions of the technology used in FT-MS originated from Fourier transform NMR. A recent review describes this and related high-resolution techniques (Marshall and Hendrickson, 2008). The two strengths FT-MS presents are the potentially extremely high resolution the instrument can generate for making accurate mass measurements, and its tandem mass spectrometry capabilities. The price for this performance is a large, expensive, and relatively complex instrument.

8.7.6 Tandem Mass Spectrometry (MS-MS)

Coupling two or more mass analyzers in series produces a tandem mass spectrometer (an MS-MS) (McLafferty, 1983), the best known of which is the triple quadrupole. A number of tandem sector geometries have been built, some of which have been commercially available (Crow, Tomer and Gross, 1983). The historical first tandem sector instruments were by Futrell and Miller (1966) and by McLafferty (1981). Hybrid geometries, combining a magnetic sector instrument with a quadrupole, an ion trap, or a TOF analyzer, have also been built and characterized and are described in the literature. The ion traps, triple quadrupoles and, more recently, the tandem quadrupole–TOF instruments are the more abundant and user-friendly instruments. The quadrupole and the sector geometries provide the ability to perform tandem-in-space MS-MS experiments. The ion traps and FT-MS instruments permit performing tandem-in-time MS-MS, permitting MS^n (mass spectrometry raised to the nth degree, multiple stages of collisional activation and mass analysis) experiments. Multiple MS-MS stages in space would require a mass analyzer for each MS-MS stage, rapidly making tandem-in-space experiments impractical for an n-value greater than 2.

8.8 DETECTION: SEEING WHAT HAS BEEN GENERATED

8.8.1 Presentation of Mass Spectra

Wide variation is seen in the absolute intensity of signals in a mass spectrum. To compensate for this variation, a mass spectrum of a compound or a mixture is typically presented as a "stick" or histogram plot. The most intense peak in the spectrum is used to normalize the other peaks in the spectrum, is referred to as the base peak, and is assigned to a 100% relative abundance. The intensities of the other peaks in the spectrum are expressed as percentages of the base peak, and plotted on the y-axis of a spectral plot. Rather than presenting an actual peak shape (in reality, approximately Gaussian if one examines the actual raw data), the abundance of an "ion" in a mass spectrum is represented as a single line. The graphical representation generated by a computer data system is often annotated with an "absolute signal intensity" corresponding to the intensity of the base peak so that absolute intensity information is not lost. This number is usually in "counts," a count being the signal intensity interval detected by the analog-to-digital conversion electronics. Different instruments will have different intervals, so that the actual meaning of the count value, in millivolts of signal for example, will have to be traced to the specific instrument. The number is useful in comparing intensities of different spectra.

The x-axis is the customary "mass" scale, more correctly noted as the "m/z" scale. Numerically, the values correspond to daltons (or atomic mass units) of mass. Most often, it is assumed to be an integral scale, peaks being centered on the whole number integers. On mass spectrometers with sufficient resolving power, and with "mass" assignments made with sufficient precision, positioning of peaks at other than integral mass values is correct, and constitutes useful information. Casual practice typically uses whole numbers for the masses of the atoms, while the only atom for which a true integer mass is correct is ^{12}C, upon which the International Union of Pure and Applied Chemists (IUPAC) atomic mass unit is assigned. While the deviation from integral relative molecular mass is small for molecules with relative molecular masses of approximately 300 daltons or less, and is most often neglected, the significance of nonintegral masses becomes more important as the molecular mass increases above 300 daltons. Large numbers of hydrogen atoms in the molecule accumulate excess mass. Sixty-five hydrogen atoms will add approximately 0.5 dalton above the integer to the relative molecular mass.

In 1991, Cooks and Rockwood (1991) championed a movement to clarify an ambiguity in referencing the typically presented x-axis in plots of mass spectra. The values along the x-axis have been casually referred to masses, and the axis the mass axis. Fortunately, the axis is generally labeled in print as the "m/z" axis. However, the frequent appearance in recent years of mass spectra, where ions are multiply charged or negatively charged, introduces semantic and algebraic inconsistencies. While most mass spectra concern ions that carry a single positive electrostatic charge, the m/z axis reduces arithmetically to be equivalent to the "mass" axis; no longer so with multiply charged ions, as in proteins. Negative ion spectra are typically presented in the same orientation as positive ions spectra, with the x-axis of the plot starting at zero on the lower left corner and with increasing absolute value of m/z proceeding toward the right, leading to an ambiguity which could lead to misinterpretation. Proposing the Thomson as a unit of measure (comparable to the Dalton), and incorporating the sign (positive or negative) from the charge state of the ion would remedy these ambiguities. Although a reasonable proposal, the suggestion has not yet caught hold.

8.8.2 Data Systems and Signal Processing: Making Sense of It

The ease of processing and displaying mass spectral data that we enjoy today is often taken for granted. Meyerson (1986) provides an enlightening view into the history of using "computer" systems to help with mass spectral data reduction and interpretation. Achieving an effective combination of computing power to control and reduce data from a combined GC-MS system warranted a full paper published in *Analytical Chemistry* in 1970 (Reynolds *et al.*, 1970). In the abstract, the authors claim spectral information to have been "acquired [and] made available to the chemist within minutes in an on-line graphic system." In the next year, Hertz *et al.* (Hertz, Hites and Biemann, 1971) reported success with the progenitor of the library search for identification of compounds by their mass spectra.

8.8.3 The Chromatography–Mass Spectrometry Data Model

In order to compensate for possible separatory shortcomings of the chromatography, a number of data-processing algorithms have been developed to make use of the fact that repetitive scan GC-MS data are really a three-dimensional matrix of data. The x-axis describes the time dependence of the separation. The y-axis describes the mass spectrum of the GC effluent at any "instant" in time, and the z-axis describes the intensity of the signal. Imagine then a cube, with the lower left corner of the front face assigned to be the origin of our data space. Along the bottom edge of the front face is the time axis (the x-axis). The left edge of the front face is the intensity axis (the y-axis), and the axis from front to back is the m/z axis (the z-axis). In general, this model applies to any repetitive scan chromatography–spectroscopy experiment.

The total ion current (TIC) chromatogram is a summation of all the mass intensities for each individual mass spectrum, plotted as a function of time. The TIC chromatogram can be visualized as a projection of all the intensities of the mass spectra onto the front surface of the data cube. (Strictly speaking, the projection of data onto the front surface of a transparent cube would actually be the base peak intensity (BPI) chromatogram, in which the BPI of each spectrum is plotted as a function of time. Summing the intensities of all peaks in each spectrum, and plotting the sum as a function of time, is the TIC chromatogram.) Slicing the cube in a plane parallel to the front face and at a particular point on the m/z axis would expose a selected ion chromatogram of the time-dependent intensity of that particular m/z value. Slicing the cube orthogonal to the front face at any particular point in time would present on the exposed surface a mass spectrum of the GC effluent at that particular point in time.

8.8.4 Full (Repetitive) Scan Data Collection versus Selected Ion Recording

Full-scan experiments set the mass spectrometer in a cycle of repeatedly scanning a selected range of m/z values. The experiment is done primarily for structure elucidation, and the richness of structural information the fragmentation (induced or spontaneous) provides. It is relatively insensitive. Observing a chromatographic peak relies on sufficient accumulated signal in the mass spectrum of a component to be discernable above background. If background in a given experiment is high, components can be easily missed. Selected ion recording (SIR or SIM) is used for quantitation. The experiment shows greater sensitivity because one maximizes the time that one is looking at the signal that tells one about what one wants to know, and discards everything else. Consider that, in a repetitive scan experiment, a finite amount of time is required for the mass spectrometer to scan across a range of m/z values, typically 1000 daltons. One would envision spending equal amounts of time observing each of those nominal 1000 channels. Only a few (or even one) of those channels contain signals of interest—the molecular ion and/or fragments indicating the compound of interest. The remaining 99.9% of the time, one is looking elsewhere. By performing an SIR experiment, one is watching the information channel(s) containing the signal of interest nearly 100% of the time. One would expect to see sensitivity increases approaching three orders of magnitude. The price, of course, is that the maximum information is being measured on the targeted signal, but all other information about the composition of the original sample is being discarded.

8.9 INTERPRETATION OF MASS SPECTRAL DATA

8.9.1 Mass Spectral Fragmentation Compendia and Libraries

The concept of the organic functional group—that a particular organic functional group will behave in essentially the same way, regardless of other functional groups attached to it—holds true in the mass spectrometer, too. Meyerson (1986) gives insight into the beginnings of the revolutionary concept that mass spectral data correlate to structure of organic compounds.

Compilations of mass spectral fragmentation data for organic compounds are useful sources for help when interpreting mass spectra. Two of the better-known libraries are the NIST/EPA/MSDC mass spectral database, maintained by the United States National Institute of Standards and Technology, and the Wiley-McLafferty database, vended by John Wiley & Sons Publishing Company. These two differ in that the NIST collection contains only one spectrum representing a given compound. The Wiley-McLafferty collection contains a larger number of spectra, but includes multiple spectra for some compounds, intending to capture information on variability of spectra from observation to observation and instrument to instrument. There are a number of smaller and more specialized collections scattered in the literature. The Thermodynamics Research Center at Texas A&M University still collects compounds and compiles data, which eventually are incorporated into the NIST and Wiley-McLafferty libraries.

In addition to the searchable library compilations, several compendial books on the EI fragmentation behavior of compounds have been published (Budzikiewicz, Djerassi and Williams, 1967; McLafferty and Venkataraghavan, 1982; Porter, 1985). They are dated, but nevertheless effectively capture the collective fragmentation information prior to their publication. Unfortunately, all of these information sources discuss EI spectra. EI fragmentation rules can be of limited assistance in interpreting MS-MS product ion spectra. One small collection of CI information has been published (Harrison, 1992). Fragmentation behavior of molecules under the soft ionization conditions—CI, FAB, API—still remains largely scattered throughout the chemical literature.

Compilation of spectral catalogs for the soft ionization techniques—CI, FAB, API—has not occurred, for at least two reasons. First, the exact nature of a soft ionization spectrum is much more dependent upon instrument and sampling conditions. Second, for many compounds analyzed by the soft ionization techniques, the spectrum consists principally of a molecular ion. A compilation of soft ionization spectra, in its ideal form, would be a compilation of molecular masses. Movements have surfaced several times over recent years to promote a compilation of MS-MS product ion spectra. These too are dependent upon exact operating conditions. An acceptable standard set of conditions has eluded everyone's efforts to make such a project happen.

All of these aspects for extracting structural information from mass spectral data come into play in some way when one is presented with a structure elucidation problem; however, this is not the complete set of tools and techniques. Finding that one obscure but revealing literature reference, flashes of insight in the middle of the night, snippets from one's collective personal and professional experiences, interactions with others practicing the art, are all ill-definable but nevertheless important contributors to this process. One can apply a degree of systematic practice, but structure elucidation from spectral (and other) data is still largely an empirical method, and still requires significant amounts of time simply staring at the data.

8.9.2 Molecular Weight versus Relative Molecular Mass

"Molecular weight" is an incorrect scientific term! The numerical value being discussed, rather, is more correctly known as a relative molecular mass. It is a mass, measuring a quantity of matter, not the influence of an external gravitational field on that matter. It is a relative mass, not an absolute mass measured in grams, because it is expressed relative to the standard atomic mass of 12.0000 atomic mass units (daltons), defined in reference to the ^{12}C isotope of carbon. It is a relative molecular mass because it is a sum of the relative atomic masses of the atoms contained within the molecule. The distinction between the incorrect term "atomic weight" and the correct term "relative atomic mass" follows the same logic, only applied to the atomic masses of the elements on the periodic table (Busch, 2004b).

The atomic mass unit is given the name "Dalton." The standard is defined as being 1/12 of the mass of an atom of the most abundant naturally occurring isotope of carbon, ^{12}C. The ^{12}C standard was adopted by the IUPAC in 1960 (Grayson, 2002). The previous standard had been assigned to be 1/16 of the atomic mass of oxygen. Two scales developed around this ambiguously defined oxygen standard, depending on whether one used the chemically determined atomic mass of oxygen, or the mass of the most abundant stable isotope, ^{16}O. The chemically determined mass incorporated the natural abundances of the minor stable oxygen isotopes, ^{17}O (0.04%) and ^{18}O (0.2%). The other was based upon mass measurement specifically of ^{16}O. The two scales are offset by 0.0044 daltons. Accurate masses reported prior to 1960 are referenced to the oxygen standard. Care should be used when comparing these numbers with more modern measurements. Note should be made to which of the two oxygen standards the older measurements are referenced. The extensive table of accurate mass values, calculated and published in Beynon's book (Beynon, 1960), is based upon the oxygen standard. If using those values, mass differences in that table should be correct, regardless of the original standard. Absolute masses should be corrected for the current definition.

8.9.3   Monoisotopic versus Chemical (Average) Relative Molecular Mass

The relative molecular mass that is measured by the mass spectrometer is numerically different from the chemical (average) molecular mass used to calculate molar concentrations or reaction stoichiometries. For the elements commonly found in organic molecules, the numbers are close and in many contexts interchangeable. However, when one discusses a mass spectrum, and an accurate mass measurement made by a mass spectrometer, the differences are noticeable. For a molecule that contains atoms of polyisotopic elements such as chlorine or bromine, or when the mass of the molecule gets large, the numbers become even more significantly different. One must exercise care when discussing molecular masses and be explicit about which kind of molecular mass number one is discussing. Two examples are calculated in Table 8.1. The normal hydrocarbon hexatriacontane, $C_{36}H_{74}$, has a nominal relative molecular mass of 506 daltons, an exact monoisotopic molecular mass of 506.5791 daltons, and a chemical

TABLE 8.1   Example calculations.

Element	No. of atoms	Nominal atomic mass	Mass	Chemical atomic mass	Mass	Accurate atomic mass	Mass
Hexatriacontane							
C	36	12	432	12.011	432.396	12.0000	432.0000
H	74	1	74	1.008	74.592	1.0078	74.5772
r.m.m.			506		506.988		506.5772
Chlorinated compound							
C	14	12	168	12.011	168.154	12.0000	168.0000
H	12	1	12	1.008	12.096	1.0078	12.0936
N	1	14	14	14.007	14.007	14.0031	14.0031
O	1	16	16	15.999	15.999	15.9949	15.9949
F	1	19	19	18.998	18.998	18.9984	18.9984
^{35}Cl	2	35	70	35.450	70.900	34.9689	69.9378
r.m.m.			299		300.154		299.0278

molecular mass of 506.98 daltons. Note that the difference between the nominal molecular mass and the monoisotopic molecular mass of hexatriacontane derives from the large number of hydrogen atoms in the molecule. Hydrogen is a mass-sufficient element, implying that its monoisotopic accurate mass (1.007825 daltons) is larger than its nominal (integral) atomic mass of 1 dalton. The cumulative excess mass of this large number of hydrogen atoms accounts for the 0.58 dalton above the nominal molecular mass.[11] All three molecular masses for hexatriacontane are correct, when used in their proper contexts, but are substantially different from each other and can generate errors when improperly used.

The second example in Table 8.1 includes several additional elements. Nitrogen (^{14}N) is mass-sufficient, while oxygen (^{16}O), fluorine (^{19}F), and chlorine (^{35}Cl in the table) are mass-deficient—having monoisotopic accurate masses smaller than their nominal (integral) atomic masses. The mass sufficiencies and deficiencies make small contributions to the discrepancies in the three kinds of molecular masses. The largest contributor, however, is the fact that chlorine is polyisotopic, with two naturally occurring isotopes of significant abundance, differing by 2 daltons in mass. The nominal and accurate mass calculations in Table 8.1 use only the mass for ^{35}Cl. The chemical molecular mass calculation uses the periodic table mass, which is a weighted average of the masses and abundances of the ^{35}Cl and ^{37}Cl isotopes. See the discussion of polyisotopic elements and isotope patterns in the next section.

8.9.4 Polyisotopic Elements and Isotope Patterns

Thomson's discovery and characterization of the isotopes of neon (Griffiths, 1997), and subsequent work by Francis Aston[12] (Griffiths, 1997) (a student of Thomson's) and others, measuring the accurate masses of the isotopes, elaborated the isotopic diversity of the periodic table. Elements that have more than one naturally occurring isotope are called polyisotopic elements. The most familiar polyisotopic element to organic chemists is carbon, with two common stable isotopes—^{12}C (98.9% natural abundance) and ^{13}C (1.1% natural abundance)[13] (see Table 8.2). Next are chlorine and bromine, each with two very abundant isotopes separated by 2 daltons. Although hydrogen, oxygen and nitrogen have well-known naturally occurring stable isotopes, they are of low natural abundance. In practice, hydrogen is treated as monoisotopic in small molecules, where the number of hydrogen atoms is relatively small. Deuterium is only 0.015% naturally abundant. The stable isotopes of oxygen and nitrogen are also of sufficiently low abundance—^{17}O, 0.038%; ^{18}O, 0.2%; ^{15}N, 0.37%—that, unless a molecule contains significant numbers of these heteroatoms, these elements do not contribute significantly to an isotope pattern. Fluorine and phosphorus are monoisotopic.[14] Sulfur is also a common polyisotopic element appearing in organic molecules. DeBievre and Barnes (1985) report the natural abundances of the isotopes of the elements. Abundances and accurate masses of the isotopes can also be found in the *Handbook*

---

[11] Sixty-four hydrogen atoms cumulatively contribute 0.5 dalton of additional mass above the integral mass.

[12] Aston was awarded the Nobel Prize in chemistry in 1922 (http://nobelprize.org/chemistry/laureates/1922/index.html).

[13] ^{14}C is a naturally occurring isotope, but is a radioactive (unstable) isotope. Its natural abundance is low enough not to be of concern in mass spectrometry, unless one makes special efforts, for example, using accelerator mass spectrometry.

[14] Naturally occurring stable isotopes! We are not considering ^{18}F and ^{32}P here.

of Chemistry and Physics . Masses and abundances of selected isotopes are given in Table 8.2.

In conventional mass spectra, isotope patterns deriving from the presence of poly-isotopic elements are striking and quite noticeable. An excellent example is the ESI spectrum shown in Fig. 8.9 for a compound with a $C_{14}H_{12}NOFCl_2$ empirical formula. The m/z 300 [M+H]$^+$ and the m/z 269 neutral loss fragment (resulting from neutral

TABLE 8.2 Accurate masses and natural abundances of selected isotopes.

Isotope	Accurate mass	Natural abundance
^{12}C	12.0000	98.9
^{13}C	13.00336	1.1
^{1}H	1.007825	99.99
^{2}H	2.0140	0.01
^{14}N	14.00307	99.6
^{15}N	15.0001	0.4
^{16}O	15.99491	99.8
^{17}O	16.9991	0.04
^{18}O	17.99916	0.2
^{19}F	18.9984	100.0
^{23}Na	22.9898	100.0
^{31}P	30.9738	100.0
^{32}S	31.9721	95.0
^{34}S	33.9679	4.2
^{35}Cl	34.9689	75.8
^{37}Cl	36.9659	24.2
^{79}Br	78.9184	50.7
^{81}Br	80.9163	49.3

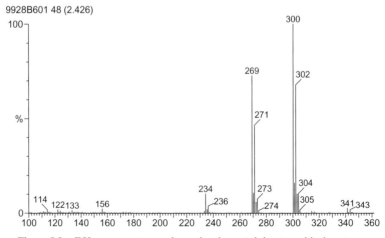

Figure 8.9 ESI mass spectrum of a molecule containing two chlorine atoms.

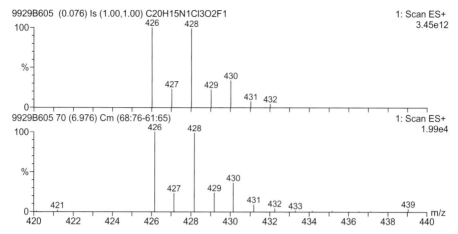

Figure 8.10 Comparison of the isotope pattern for the observed $[M+H]^+$ (bottom panel) with the predicted model (top panel) for $C_{20}H_{15}NO_2FCl_3$.

loss of methylamine) clearly show Cl_2 isotope patterns which match well with simulations.[15] The m/z 234 fragment results from further loss of one of the chlorine atoms, illustrating a clear Cl_1 isotope pattern, and the utility of the isotope patterns in assigning structure.

The close match of simulation (Fig. 8.10, upper panel) and observation (Fig. 8.10, lower panel) of the $[M+H]^+$ of another chlorine-containing compound, along with some chemical considerations of the history of the sample, permitted assigning the structure of the unknown compound based solely upon the isotope pattern.

The most extensively polyisotopic element is tin, with 10 stable isotopes ranging in abundance from 0.4% to 33%, over a 12-dalton range. Figure 8.11 illustrates a dramatic example of an isotope pattern for monooctyl tin ethylhexylthioglycolate (Structure 8.1). The upper panel of Fig. 8.11 expands the cationated molecular ion region, $[M+Na]^+$, for this compound, taken from an LC-MS experiment conducted using APCI. Which species is the molecular ion? The answer is "all of them." The empirical formula for all of these species is $C_{38}H_{74}O_6S_3Sn_1Na$. While the chemical relative molecular mass for this compound is correctly calculated from the empirical formula to be 864.9 daltons (based upon periodic table atomic masses), the mass spectrometer reveals these 10 species, all of the same elemental composition but differing by isotopic content, all representing the molecule under discussion. The 38 carbon atoms and the three sulfur atoms contribute to this complex isotope pattern, but the primary contributor is the single tin atom.

Revisiting the question, "Which is the molecular ion species?" and calculating using isotopic masses, the measured values still bear further examination. For the most commonly encountered elements in organic molecules, the lowest mass isotope is the most abundant—for example, carbon, oxygen, sulfur, and chlorine. The calculated molecular mass using lowest-mass-isotope masses in general corresponds to the most abundant

[15] The well-done mass spectrometry data systems all have isotope pattern modeling programs built in. A number of other stand-alone simulation programs are also available. For example, see http://www.alchemistmatt.com/.

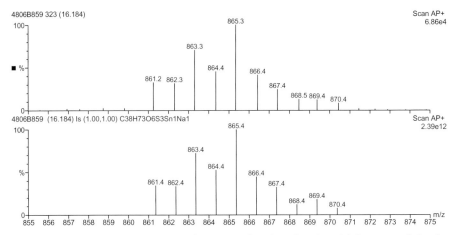

Figure 8.11   Observed isotope pattern (upper panel) and predicted model (lower panel) for the [M+Na]$^+$ of dioctyl tin EHTG, taken from a positive ion APCI LC-MS experiment.

Structure 8.1   Monooctyl tin ethylhexyl thioglycolate. $C_{38}H_{74}O_6S_3Sn$.

species observed. Using ^{112}Sn, the species corresponding to this molecular ion should appear in Fig. 8.11 at m/z 857. Any signal at m/z 857 in the spectrum is insignificant. ^{112}Sn is only 1% abundant. The more appropriately designated molecular ion species here is the one at m/z 865, containing ^{120}Sn (the most abundant isotope of tin, at 33% natural abundance). The lower panel of Fig. 8.11 compares a theoretical prediction of the isotope pattern of this compound with the experimental observation. For complex isotope patterns such as these, one can (and should) simulate the isotope pattern with isotope pattern modeling software, available in the better mass spectrometer data systems, and in several data system-independent programs. When discussing molecular mass results, especially mass spectrometric results, for such polyisotope-containing molecules, one should specify how one is calculating results.

Further, the 865.4-dalton assignment for this species in Fig. 8.11 and the position at which the peak is plotted, are correct and not a drafting artifact. The only isotope whose mass is truly an integer is ^{12}C, and only because it is arbitrarily assigned to be 12.0000 daltons. All other isotopic masses are either mass-sufficient or mass-deficient. The

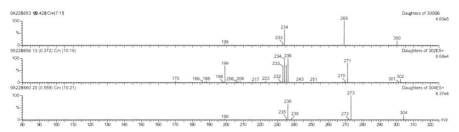

Figure 8.12   MS-MS product ion spectra of the m/z 300 (top panel), m/z 302 (middle panel), and m/z 304 (bottom panel) [M+H]⁺.

relative molecular mass of the m/z 865 species truly is 865.36 daltons. The sufficiency results from the accumulated effect of the 74 hydrogen atoms,[16] offset by the deficiency contributed by all of the other mass-deficient elements (excluding carbon) in the empirical formula.

The consideration of molecular mass is further complicated by the existence of a number of significantly polyisotopic elements. Given this complication, determination (and reporting) of the correct molecular mass is controversial. Polyisotopic elements immediately divide practitioners into two camps. One camp says the molecular mass is that value calculated using the lowest mass isotope of each element in the molecule. The other camp says to calculate using the mass of the most abundant isotope of each element. In most instances, these two rules give rise to the same number. The most abundant isotopes of carbon, hydrogen, oxygen, nitrogen, sulfur, phosphorus, and all of the halogens are indeed the lowest mass isotope. Boron-containing molecules or organometallic compounds containing a polyisotopic transition metal or heavy metal may more appropriately deserve application of the second rule (Sharp *et al.*, 1986; Kowalski, Sharp and Stang, 1987). Applying either of these rules, however, becomes even more complicated when a molecule contains, for example, three or more chlorine atoms, two or more bromine atoms, or more than one polyisotopic metal atom (such as tin).

The interpretation of isotope patterns changes when applied to MS-MS product ion spectra. Consider the product ion spectrum, in the top panel of Fig. 8.12, of the same m/z 300 [M+H]⁺ discussed earlier. The experiment was performed on a triple quadrupole instrument by selecting the m/z 300 species with the first analyzer under unit resolution conditions. The m/z 300 species was collisionally activated in the second quadrupole collision cell. The product ion spectrum resulting from CID was recorded in the third quadrupole mass analyzer. Recall that the empirical formula of this molecular ion is $C_{14}H_{13}NOFCl_2$. Why have the very prominent isotope patterns disappeared? Because one has selected the m/z 300 species as the parent, there is no longer any consideration of polyisotopy. This parent is a monoisotopic parent—namely $C_{14}H_{13}NOF^{35}Cl_2$. Methylamine neutral loss produces a single species—m/z 269. Further fragmentation, sequentially losing chlorine atoms, produces single species—m/z 234 and m/z 199—because the only chlorine isotope that can be lost from this selected m/z 300 parent is a ^{35}Cl atom.

---

[16] Sixty-four hydrogen atoms cumulatively contribute 0.5 dalton of additional mass above the integral mass.

Why does an isotope pattern reappear in the product ion spectrum of the m/z 302 parent (Fig. 8.12, middle panel)? The elemental composition of the parent is now $C_{14}H_{13}NOF^{35}Cl_1{}^{37}Cl_1$. The m/z 271 species shows no polyisotopic signature because its formation results from loss of CH_3NH_2 from the parent to give $C_{13}H_8OF^{35}Cl_1{}^{37}Cl_1$— only one possible elemental composition. Loss of the first chlorine atom, however, now permits producing two possible elemental compositions—$C_{13}H_8OF^{35}Cl_1$ and $C_{13}H_8OF^{37}Cl_1$ at m/z 234 and m/z 236. They are of equal abundance because of the equal probability of losing either a ^{35}Cl or a ^{37}Cl atom in the fragmentation process. The fact that the natural abundances of the chlorine isotopes are approximately 3 to 1 is irrelevant here. Selecting the m/z 302 species as the parent sets the isotope ratio of ^{35}Cl to ^{37}Cl in this experiment to be 1 to 1. The isotope pattern disappears on losing the second chlorine to give the m/z 199 species, since the source of polyisotopy is now completely gone.

All evidence of isotope patterns in the product ion spectrum of the m/z 304 parent (Fig. 8.12, bottom panel) disappears because the selected parent is monoisotopic—$C_{14}H_{13}NOF^{37}Cl_2$. Justification for the absence of isotope patterns is identical to that of interpretation of the m/z 300 product ion spectrum. The detailed explanation of this excellent example serves to illustrate the care one needs to exercise in interpreting isotope patterns (or the lack thereof) in MS-MS spectra. At the same time, it demonstrates the structure-indicating power of careful interpretation.

With the recent development of the combined quadrupole TOF tandem instrument geometry, MS-MS product ion spectra are starting to appear in the literature with isotope patterns reflecting the natural abundance of the involved elements. Considering this discussion, how does this occur? Note that our selection of parent ions for CID in the experiments described earlier was done under unit resolution conditions. Only species with a single (although nominal) mass are permitted to pass into the collision cell—thus the argument of above, justifying the selection of a specific isotope content. With a tandem instrument composed of a quadrupole and a TOF mass analyzer, the resolution of the first mass analyzer can be set to substantially less than unit resolution, such that the entire isotope pattern can be selected as the "parent." This approach preserves the isotope pattern information while providing MS-MS product ion information. This new consideration must be added while interpreting isotope patterns in MS-MS spectra—namely to identify on what kind of instrument the experiment is being done, and how the instrument is set up.

8.9.5   High-Resolution Measurements versus Accurate Mass Measurements

The accurate mass measurement technique is being actively rediscovered because of recent improvements in instrumentation. John Beynon (Beynon, 1954, 1960) established the validity of using high-precision mass measurements (to at least four decimal places of precision) to determine elemental compositions of ions with high certainty, capitalizing on the nonintegral nature of the precise atomic masses of the isotopes. Accurate mass measurements were until recently the nearly exclusive province of magnetic sector instruments. Making such measurements presented difficulties, and required careful attention to detail. rTOF instruments and FT-MS instruments have challenged magnetic sector dominance of this area and facilitated a somewhat easier route to making these measurements. Accurate mass measurement became accepted in the synthetic organic chemistry community as an appropriate measure of the "proof" of a structure. Acceptance criteria for including accurate mass measurements in the suite of data defending structure assignment of a newly synthesized compound are given in the instructions for authors of the synthetic and other journals,[17] even though Clayton Heathcock (Heathcock, 1990) (then editor-in-chief of the *Journal of Organic Chemistry*) acknowledged that combustion analysis, for which accurate mass measurements are often used as a substitute, is really a measure of purity and not of identity. Accurate mass measurements are a measure of identity.

Biemann's discussion (Biemann, 1990) defines the circumstances and limitations very carefully, where one should and should not use these kinds of measurements, and illustrates with examples applied to the elucidation of fragmentation pathways (the least demanding application), evidence supporting the confirmation of structural assignments and the determination of the structure of natural products (the most demanding application). Busch (1994) discusses the utility of an accurate mass measurement in confirming the identification of a phthalate contaminant as an example. Gilliam *et al.* (Gilliam, Landis and Occolowitz, 1983, 1984) report on procedures and practices for making accurate mass measurements in a production type of environment, and the utility the measurements provide in a pharmaceutical discovery setting.

Improvements in FT-MS instrumentation have permitted a re-evaluation of the value of accurate mass measurements to the analysis of very complex mixtures. Alan Marshall's group's work, for example, on vegetable oils (Wu, Rodgers and Marshall, 2004) and on crude and refined petroleum fractions (Marshall and Rodgers, 2004) has resurrected concepts that were developed in the early 1950s when mass spectrometry was being vigorously applied in the petroleum industry. The very high resolving power accessible on a high magnetic field FT-MS instrument emphasizes the potential elemental composition heterogeneity that can exist at a single mass in these complex mixtures, and demonstrates how much structural and compositional information can be extracted from a single mass spectrum. The mass deficiency of a single oxygen or sulfur atom imparts enough difference to distinguish compositions containing these elements from the more abundant, and more readily expected hydrocarbon compositions in a diesel oil sample (Fig. 8.13). A veritable forest of compositions appears in the negative ion ESI spectrum of olive oil, due to a large number of oxygen atoms present in the compositions (Fig. 8.14).

A little recognized systematic error in the calculation of accurate masses of, for example, small radical cation molecular ions (as in EI) or protonated molecular ions (as seen in the soft ionization methods) is the fact that the electron has a small but finite mass. The accurate masses of radical cations, in which a valence electron has been removed, of anions that have been created by capture of an electron, and of protonated species produced by soft ionization processes, should take into consideration this small mass difference (Mamer and Lesimple, 2004). For example, there is a small difference between the relative atomic mass of a neutral hydrogen atom and a proton. The accepted accurate mass of $^1H^0$ is 1.007825 daltons. The accurate mass of $^1H^+$ is 1.007276 daltons. To be completely correct, expected accurate masses of protonated molecular ions, $[M+H]^+$, produced by electrospray should be calculated using the mass of one H^+, rather than all of neutral hydrogen atoms. Mamer and Lesimple do acknowledge, however, that, for large molecules, the error is of little consequence.

Notwithstanding the efforts of groups attempting to precisely define acceptable practices and requirements (Gross, 1994), confusion still exists on the differences

---

[17] Please consult, for example, recent instructions for authors in the *Journal of Organic Chemistry* or the *Journal of the American Society for Mass Spectrometry*.

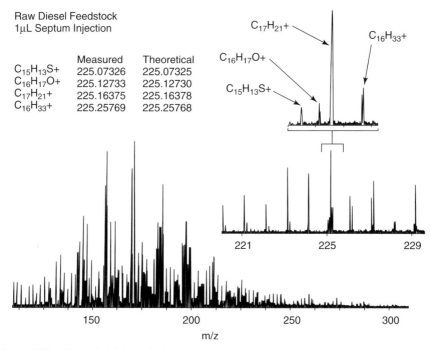

Raw Diesel Feedstock
1μL Septum Injection

	Measured	Theoretical
$C_{15}H_{13}S+$	225.07326	225.07325
$C_{16}H_{17}O+$	225.12733	225.12730
$C_{17}H_{21}+$	225.16375	225.16378
$C_{16}H_{33}+$	225.25769	225.25768

Figure 8.13 Example high-resolution experiment on complex mixtures, showing the multiple elemental compositions possible at a single nominal mass. Four elemental compositions at m/z 225 in the positive ion electrospray mass spectrum of a diesel oil feedstock (from Wu, Rodgers and Marshall, 2004).

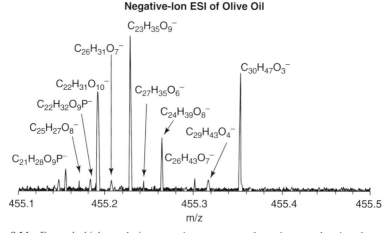

Negative-Ion ESI of Olive Oil

Figure 8.14 Example high-resolution experiment on complex mixtures, showing the multiple elemental compositions possible at a single nominal mass. Multiple elemental compositions at m/z 455 in the negative ion electrospray spectrum of olive oil (from Marshall and Rodgers, 2004).

between definition of the terms "high resolution" and "accurate mass" (Kondrat, 1999). High resolution, as in any context, implies the ability to make confident and reliable measurements of small differences. In mass spectrometry, high resolution implies the ability to measure the small differences deriving from different elemental compositions that have the same nominal mass (see the examples in Figs 8.13 and 8.14). Measurement can be made at sufficiently high-enough resolution to see the heterogeneity of compositions at a given nominal mass without necessarily being able to assign the precise accurate masses of those compositions. The partial mass spectra shown in Fig. 8.13 truly are high-resolution spectra. Conversely, accurate mass measurement refers to the ability to confidently assign a precise relative molecular mass (customarily four decimal places or more) to a signal. Accurate mass measurements can be made at low resolution, given a stable, reproducible scan by the instrument, appropriate accurate mass reference standards, and the confidence that one is observing a mass spectral peak composed of a single elemental composition (Tyler, Clayton and Green, 1996). The spectra in Fig. 8.13 are additionally accurate mass-assigned spectra because of the care taken in making the mass assignments. Indeed, Beynon's original demonstration (Beynon, 1954) of the utility of calculating elemental compositions from accurate mass measurements was done on a single-focusing magnetic sector instrument at a resolving power of only 250, while they were building an instrument intending to achieve a 2500 resolving power to enable them to make better measurements. Note these low resolving powers, in contrast to the 5000 to 15,000 resolving powers used by Gilliam *et al.* (Gilliam, Landis and Occolowitz, 1983, 1984) and the 350,000 resolving power used by Marshall *et al.* (Marshall and Rodgers, 2004) and Wu *et al.* (Wu, Rodgers and Marshall, 2004).

The instrument manufacturers often imply that accurate mass measurements (and therefore elemental composition determinations) are a single measurement solution to many structural problems. This is an oversimplification. A recent report, stemming from a metabolomics application, presents data that high mass measurements, even at sub-1 ppm accuracy, are insufficient to "identify" metabolites (Kind and Fiehn, 2006). Isotopic abundance pattern analysis is a necessary additional constraint to eliminate false elemental composition candidates.

8.9.6 Unimolecular Decompositions versus Granddaughter[18] Ions

The theoretical concept behind MS-MS involves unimolecular decomposition to establish ion-genetic relationships between fragments and their parent structures. The difficulty of interpreting a conventional mass spectrum consisting of more than one component is determining which fragment belongs to which parent. Another problem occurs when a fragment gives rise to another fragment via a sequence of fragmentations. This dilemma is tremendously improved by having a chromatographic separation on the front of the experiment, giving some assurance that one is looking at the mass

---

[18] Some individuals have taken offense at the usage of the originally coined term "daughter ion" in describing the ion-genetic relationships resulting from a collision-induced decomposition experiment. The more palatable term "product ion" has come into use as an alternative. We have attempted to refer to product ions in our discussions. Notwithstanding, the term "granddaughter" conveys a certain meaning about the skipping-of-generations relationship between products formed by two or more sequential chemical reactions from the original parent. No comparable companion term to "product ion" has yet arisen which conveys this meaning. For this reason, we still use the term "granddaughter ion" here in this limited context.

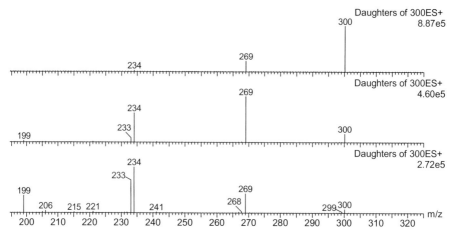

Figure 8.15   MS-MS product ion spectra of the m/z 300 [M+H]⁺ at 10 volts (top panel), 20 volts (middle panel), and 30 volts (bottom panel) interface cone voltages.

spectrum of a single compound, and that all fragments are ultimately derived from one parent. This is not the case when one still has co-eluting components. MS-MS experiments seek to eliminate this uncertainty. These experiments assume an "infinitely thin" collision zone and single collisions of the parent with a collision gas, depositing only sufficient energy to induce one unimolecular decomposition reaction. These assumptions are not sustained when utilizing a physically large quadrupole collision cell. Multiple collisions at high energy are possible. Product ions can also undergo collisional activation before they leave the collision cell. Granddaughter and great-granddaughter ions can be formed and detected in the MS-MS spectrum. How does one tell unimolecular decomposition from multistep decomposition? Fragmentations requiring the breaking of multiple bonds at multiple sites in the molecule can indicate multistep decomposition.

Examination of the collision energy dependence of the product ion spectrum is a functional test. To illustrate, the product ion spectra of the m/z 300 [M+H]⁺ of a molecule of empirical formula $C_{14}H_{12}NOFCl_2$, at three different collision energies, are shown in Fig. 8.15. This is the same molecule on which the isotope pattern discussion in Section 8.9.4 is based. The top panel, at low collision energy, shows significant survival of the [M+H]⁺ and loss of a 31-dalton species, corresponding to neutral loss of methylamine— a sensible fragmentation, as the original molecule contains an aliphatic N-methylamine side chain. Only just appearing is an m/z 234 fragment, corresponding to further loss of one of the chlorine atoms. That the methylamine and chlorine moieties are distant from each other in the original structure implies that these are two separate events, and that the m/z 269 species is losing the chlorine. If chlorine were to be lost directly from the parent ion, an m/z 265 species would be present, and it is not.

Increasing the collision energy, as illustrated in the two lower panels of Fig. 8.15 further destroys the [M+H]⁺ and induces more extensive fragmentation of the m/z 269 species. It increases the relative abundance of the first chlorine loss, and induces a further loss of the second chlorine atom, producing the m/z 199 species. The collision energy dependence on the abundance and appearance of the product ions indicates

Structure 8.2 Lecithin general structure (phosphatidylcholine) for a, b = 7, dipalmitoylphosphatidylcholine. $C_{40}H_{81}NO_8P$; r.m.m. 734.5700.

that these species are the result of a sequence of fragmentations, and helps one correlate back to the original structure of the molecule.

8.9.7 When Is a Molecular Ion Not an [M+H]⁺?

A molecular ion species should not always be automatically assumed to be a protonated species, one mass unit heavier than the neutral molecular species. Cationic species, such as quaternary ammonium compounds, will appear in the mass spectrum at the mass of the cationic species. A biologically relevant example is the family of phosphatidylcholines (also known as the lecithins), represented by Structure 8.2. Mass spectra of these compounds, recorded under appropriate conditions, show M^+ species, not $[M+H]^+$. The molecular ion species for the dipalmitoyl phosphatidylcholine species, whose empirical formula and relative molecular mass are given with Structure 8.2, would appear at m/z 734.6. Note also the 0.6 dalton of excess mass above the integral value because of the large number of mass-sufficient hydrogen atoms in this molecule.

8.9.8 Deuterium Exchange Experiments

Substantial amounts of work appear in the literature in recent years on the study of backbone amide hydrogen–deuterium exchange studies of protein folding, based upon the pH dependence of the rates of H-D exchange of amide protons (Bai *et al.*, 1993). The rate minimum for this exchange is between pH 2 and pH 3 at 0 °C. Half-lives for exchange can be greater than 1 hour at these conditions for unprotected amides. This slowness of exchange permits conducting exchange experiments on intact proteins, then cleaving those proteins with acid-active peptidases and determining the regions of the original protein accessible to exchange. These constitute solution state probes of tertiary and higher structure of proteins. Review papers by Englander and colleagues (Krishna *et al.*, 2004; Englander, 2006) describe this important technique.

The utility of deuterium exchange, however, is not limited to large molecules. Identifying and confirming the structures of small molecules can benefit considerably by determining the numbers (and nature) of exchangeable hydrogen atoms in the molecule. Olsen *et al.* (2000) report on the general utility of deuterium exchange LC-MS experiments for structure elucidation and impurity identification. However, one should not forget the importance of the chemical nature of the exchanging hydrogen. Exchange is not instantaneous and universal for exchangeable hydrogen atoms. Such

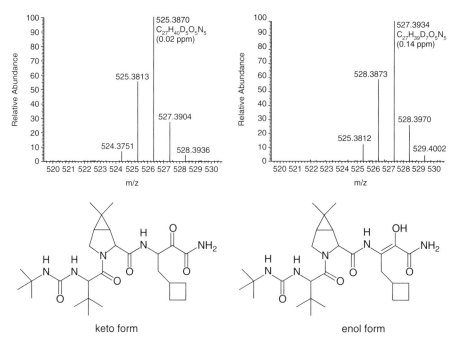

Figure 8.16 Differentiation of keto (left structure and mass spectrum) and enol (right structure and mass spectrum) by hydrogen–deuterium exchange (from Chen *et al.*, 2007).

a serendipitous observation in the identification of a process-related impurity (Alsante *et al.*, 2004) in a pharmaceutical drug candidate exemplifies the benefit of an inventory of exchangeable hydrogen atoms in an impurity identification.

An example where deuterium exchange played an important role to imply a keto-enol tautomerization in a pharmaceutical drug substance has been reported (Chen *et al.*, 2007). An impurity exhibiting the same relative molecular mass as the drug substance was observed. Conducting a deuterium exchange LC-MS experiment showed that the drug substance contained five exchangeable hydrogen atoms, while the impurity contained six exchangeable hydrogens, the structural explanation being that the two molecular forms were keto and enol tautomers. The enol tautomer contained an additional hydroxyl group with a readily exchangeable hydrogen (Fig. 8.16).

The DART ionization source permits performing even more facile hydrogen–deuterium exchange experiments. Investigations into the contamination of pet foods by melamine are reported by Vail *et al.* (Vail, Jones and Sparkman, 2007). The hydrogen–deuterium exchange was conducted simply by placing a source of D_2O vapors in proximity with the ion source and the sample. Exchange was effected in the gas phase between the melamine molecules and the D_2O (see Fig. 8.17). The six exchangeable hydrogen atoms in the melamine molecule plus the additional deuterium to form the $[M+D]^+$ are accounted for in the spectrum of the exchanged compound. The exchanged spectrum in Fig. 8.17 points out a concern. Is the deuterium incorporation chemically specific for exchangeable hydrogen atoms or nonspecific? If exchange were to be demonstrated to stop at seven and only seven hydrogens, then exchange is chemically specific. If the molecule could be shown to incorporate more than seven

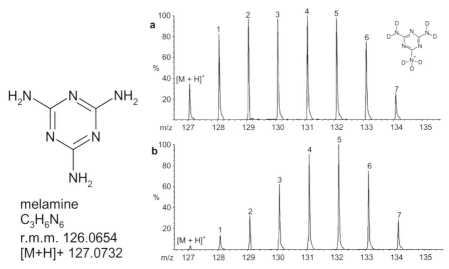

H₂N — N — NH₂
(melamine structure)

melamine
$C_3H_6N_6$
r.m.m. 126.0654
[M+H]+ 127.0732

Figure 8.17 Hydrogen-deuterium exchange of melamine in a DART ion source. **(a)** Standard melamine in the presence of deuterium oxide. **(b)** Melamine detected in contaminated pet food in the presence of deuterium oxide. From Vail *et al.* 2007

Structure 8.3 Anidulafungin. $C_{58}H_{73}N_7O_{17}$; r.m.m. 1139.5063.

deuterium atoms, then exchange would not have the specific chemical implication implied here. While not shown here, communications with the authors and with R. Cody (R.B.Cody,2007,personal communications) indicate that inducing more complete hydrogen–deuterium exchange in melamine and in other compounds used to test this phenomenon indeed shows that exchange is specific and consistent with the chemical nature of the hydrogen atoms involved.

Anidulafungin (Structure 8.3) (Cappelletty and Eiselstein-McKitrick, 2007) is a semisynthetic antifungal compound, derived from a natural product. The molecular ion region of the positive ion electrospray mass spectrum of this compound is shown in Fig. 8.18 (Z. Zhang and R. Morris, 2007, personal communications). The expected monoisotopic [M+H]⁺ is observed at m/z 1140.9. (The mass calibration in Fig. 8.18 is

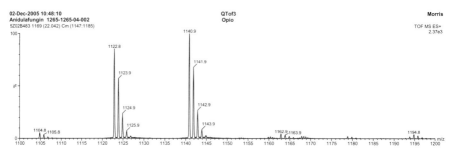

Figure 8.18 Positive ion electrospray mass spectrum of the molecular ion region of anidulafungin.

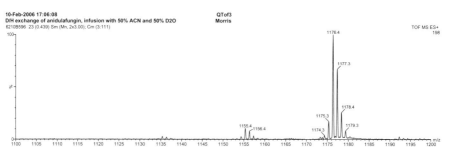

Figure 8.19 Deuterium exchanged molecular ion region of anidulafungin.

off by ~0.4 daltons!) A small sodiated molecular ion, $[M+Na]^+$, is observed at m/z 1162.9, resulting from this molecule's scavenging Na^+ from the instrument system. The species at m/z 1122.8 and m/z 1104.8 result from mass spectrometrically induced dehydrations of the molecule, of which there are ample opportunities in this molecule. Isotopic abundances are consistent with expectations for this empirical formula.

Evidence supporting the correctness of the anidulafungin structure is available by hydrogen–deuterium exchange experiments. The 14 hydrogen atoms explicitly shown in Structure 8.3—a phenolic hydrogen, five amide hydrogen atoms, and eight aliphatic alcohol hydrogens—exchange in an aqueous environment. By dissolving anidulafungin in D_2O, the mass spectrum of this thoroughly exchanged material is shown in Fig. 8.19. The m/z 1176.4 species is the $[M+Na]^+$. The isotope pattern suggests that deuterium exchange is not 100%, because of the significant abundances at m/z 1175.3 and m/z 1174.3. However, the mass of the $[M+Na]^+$ has moved up 14 daltons, corresponding to deuterium exchange of the 14 exchangeable hydrogens. The m/z 1155.4 species is the $[M+H]^+$ with exchanged hydrogens. Recall that the reason it has moved up 15 daltons is that the charge carrier, an H^+, has also been replaced by a D^+ under the exchange conditions—in actual fact, an $[M+D]^+$.

8.9.9 Cluster Ions and Adduct Ions

The quest for an ionization method that guarantees a molecular ion for all compounds has driven the development of softer and softer methods. While approaching this goal,

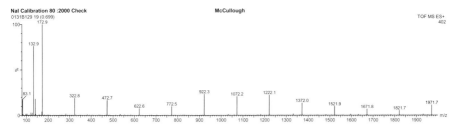

Figure 8.20 Sodium iodide cluster spectrum by electrospray.

a wealth of new exceptions to the rule has been observed. Cluster ions and adduct ions form in the matrix (for FAB ionization) or in the spray of the API methods depending on the nature of the compound, the matrix, solvent, or extraneous salts present in the sample or in the instrument. While complicating the interpretation of such results and requiring the appropriate care in interpretation, the presence of these species can be useful (Fig. 8.20).

The ubiquitous nature of the sodium cation often leads to Na^+ cationization and the presence of $[M+Na]^+$ in the mass spectrum in addition to, or in place of, the $[M+H]^+$. Efforts to either remove or add Na^+ from the sample result in changes in the relative abundances of the $[M+H]^+$ and $[M+Na]^+$. Addition will often cause replacement of other labile protons with additional sodium atoms. Addition of lithium or potassium can replace the Na, with the appropriate mass changes, to confirm the identity of the Na adducts. Ammonium cationization will often occur from LC-MS mobile phases containing an ammonium buffer salt. Acetonitrile solvent adducts will also often form.

An example of a molecule that exhibits an extreme dependence on solvent conditions is shown in the partial spectra of Fig. 8.21. The molecule has a 404-dalton relative molecular mass. The material is produced as the sodium salt. Samples of the exact same material have been dissolved in and introduced into the electrospray ion source using three different solvents. The m/z 405 species is the $[M+H]^+$. The m/z 422 species is an $[M+NH_4]^+$. The m/z 427 species is $[M+Na]$. The m/z 468 species corresponds to an $[M+Na+actronitrile]^+$. Examining any of these three spectra in isolation, one cannot easily tell which is the real $[M+H]^+$. However, once one knows that a particular class of molecule—such as this one—behaves in this manner, one can use the presence of the cluster ions as additional supporting evidence for assigning the molecular mass of the compound. This phenomenon adds uncertainty about assuming the behavior of unknowns. If one can safely assume that an unknown impurity or degradant is sufficiently structurally similar to the parent molecule to behave in the same manner, then the adduct formation helps to confirm molecular mass assignments. Looking for 17 daltons (NH_4^+), 22 daltons (Na^+), 38 daltons (K^+), and 41 daltons (acetonitrile) mass differences, or various combinations of these numbers, will provide some level of comfort in assuming adduction behavior.

HPLC columns and instruments that have been used for numerous projects and been exposed to numerous mobile phases will often be heavily "contaminated" with extraneous buffer and other cations and anions. Some organic molecule structures show a propensity to scavenge such stray ions from the instrument and column. From where does the Na^+ in an $[M+Na]^+$ derive if neither the sample nor the mobile phase expressly contains Na^+? Stray Na^+ or other ions are ubiquitous. In negative ion detec-

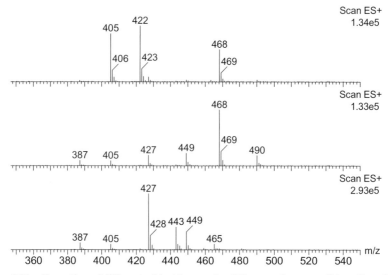

Figure 8.21   Formation of different adduct ions under different solvent conditions. Samples of a 404-dalton molecule dissolved in and introduced in 1:1 acetonitrile:20 mM aqueous ammonium acetate (top panel), 1:1 acetonitrile:water (middle panel), and 1:1 methanol:water (bottom panel).

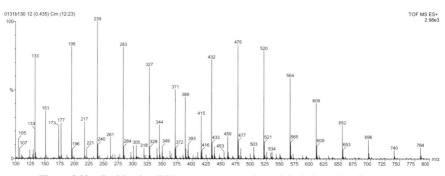

Figure 8.22   Positive ion ESI mass spectrum of a poly(ethylene glycol) mixture.

tion mode, the presence of $[M-H+114]^+$ adducts derives from adduction with trifluoro-acetic acid, an often-used mobile phase adulterant to produce low-pH mobile phases. Residual trifluoroacetic acid often abounds in mass spectrometers.

8.9.10 Polymer Patterns

Another striking feature that can be seen in the overall appearance of a mass spectrum is a repeating series of peaks at a regular interval—a "polymer pattern." It is almost as striking as an isotope pattern. Such a pattern is illustrated by the spectrum shown in Fig. 8.22, an ESI mass spectrum of a mixture of poly(ethylene glycol) molecules—PEG (Structure 8.4). Various PEG mixtures and similar polymers are frequently used as

Structure 8.4   Poly(ethylene glycol). $(C_2H_4O)_nH_2O$; r.m.m. $44N + 18$.

Structure 8.5   Octoxynol "Triton X". $C_8H_{17} C_6H_4 [C_2H_4O]_nOH$; r.m.m. $206 + [44]_n$.

mass calibrants and accurate mass reference standards. The spectrum shown is of a mixture of two commercially available PEG preparations of 200- and 400-dalton average molecular masses.[19] The constant interval between the repeating peaks is 44 daltons, the mass of the PEG repeating unit. The m/z 239 peak is the $[M+H]^+$ of a PEG molecule with five repeating units—PEG$_5$. The m/z 261 peak is the $[M+Na]^+$ of PEG$_5$. The m/z 476 species is the $[M+NH_4]^+$ of PEG$_{10}$. The sample solution was prepared by dissolving the PEG in aqueous ammonium acetate, providing a source of ammonium cation. This spectrum illustrates the interesting phenomenon of Na$^+$ cationization, and of preferential NH_4^+ cationization by the larger PEG polymers. Recognizing these features, the complicated appearing spectrum can be easily interpreted. One can visually pick out the components of each series, and estimate the average molecular masses of the original PEG starting materials. An obvious PEG polymer pattern was key in identifying a drug-excipient interaction in a formulation containing PEG (Fig. 8.24). The drug substance, a carboxylic acid, formed a series of esters with the terminal alcohol group of the PEG molecules. The mass spectrum clearly showed this pattern of peaks, separated by 44 daltons, with the masses offset from those of PEG by the mass of the drug substance.

A positive ion electrospray mass spectrum of a sample of octoxynol (1,1,3,3-tetramethyl)butylphenyl (poly)ethylene glycol, Triton X-100, Structure 8.5) (Octoxynol. Merck Index, 12th edn. Entry # 6858) is shown in Fig. 8.23. The characteristic polymer pattern is prominent. The peaks in the spectrum are a series of ammoniated molecular ions, $[M+NH_4]^+$. This surfactant molecule is efficient at scavenging ammonium cation from the instrument system.

Table 8.3 calculates the average relative molecular mass for the triton sample recorded in Fig. 8.23. The number contribution is the product of the chain length and the abundance of a chain length (simple peak height from the spectrum). The average chain length, 9.5 units, is the ratio of the sum of number contributions divided by the sum of contributions (proportional to the number of molecules in the sample). The average molecular mass of the polymer sample, 644 daltons, can be calculated from the average chain length. Visual inspection of the spectrum, suggesting a center of the distribution between m/z 620 and m/z 664, is consistent with the results of the calculation. The distribution width can be used as a measure of polymer "polydispersity"—in this case, a chain length of 9.5 units ±2.7.

---

[19] Note here that the term "average molecular mass" applied to a polymer implies yet another nuance in the usage of the term "molecular mass."

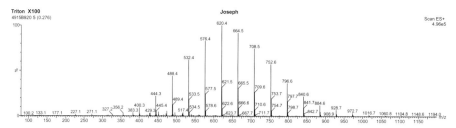

Figure 8.23   Electrospray mass spectrum of octoxynol (Triton X).

TABLE 8.3 Calculation of average molecular mass and polydispersity for Triton X-100.

Length	Mass	Empirical formula	Abundance	Number contribution
0	224.2	$C_{14}H_{26}NO_1$		0
1	268.2	$C_{16}H_{30}NO_2$		0
2	312.3	$C_{18}H_{34}NO_3$		0
3	356.3	$C_{20}H_{38}NO_4$	8	24
4	400.3	$C_{22}H_{42}NO_5$	18	72
5	444.3	$C_{24}H_{46}NO_6$	39	195
6	488.4	$C_{26}H_{50}NO_7$	78	468
7	532.4	$C_{28}H_{54}NO_8$	109	763
8	576.4	$C_{30}H_{58}NO_9$	145	1,160
9	620.4	$C_{32}H_{62}NO_{10}$	175	1,575
10	664.5	$C_{34}H_{66}NO_{11}$	161	1,610
11	708.5	$C_{36}H_{70}NO_{12}$	131	1,441
12	752.5	$C_{38}H_{74}NO_{13}$	102	1,224
13	796.5	$C_{40}H_{78}NO_{14}$	64	832
14	840.6	$C_{42}H_{82}NO_{15}$	38	532
15	884.6	$C_{44}H_{86}NO_{16}$	22	330
16	928.6	$C_{46}H_{90}NO_{17}$	13	208
17	972.6	$C_{48}H_{94}NO_{18}$	7	119
18	1,016.7	$C_{50}H_{98}NO_{19}$	3	54
19	1,060.7	$C_{52}H_{102}NO_{20}$	1	19
20	1,104.7	$C_{54}H_{106}NO_{21}$		0
			$\Sigma = 1{,}114$	$\Sigma = 10{,}626$
			Average length	9.5
			Average mol. mass	643.7

Such average molecular mass determinations are analogous to two figures of merit used in characterizing polymer samples, namely the number-average relative molecular mass, $\overline{M}_n$, and the weight-average relative molecular mass, $\overline{M}_w$.[20] $\overline{M}_n$, indicates the average chain length in a polymer sample, and is classically measured by determination of colligative properties (Yau, Kirkland and Bly, 1979), such as freezing point depression or vapor-phase osmometry. $\overline{M}_w$ correlates with physical properties of larger

---

[20] Note that we are using the correct terminology here, rather than the more customary and historically used number-average and weight-average molecular weights.

polymers. Literature on determination of $\overline{M}_n$ and $\overline{M}_w$ show that FAB, electrospray, and MALDI as very appropriate techniques for direct determination. Limitations presented by MALDI are discussed by Byrd and McEwen (2000).

The calculation in Table 8.3, derived from the spectrum in Fig. 8.23, is directly related to $\overline{M}_n$ and is a much more facile and direct measure of $\overline{M}_n$. $\overline{M}_w$ can also be calculated directly from the data in Table 8.3, and is 665.6 daltons for this sample. Polydispersity, as defined in polymer chemistry, is the ratio of $\overline{M}_w/\overline{M}_n$, calculated here to be 1.03. Response factors of individual species of a polymeric molecule in the mass spectrometer can vary with the chain length of the molecule (R.E. Borjas, 2007, personal communications). For some types of polymer, lower mass members of the series will, for example, have a higher molar ion yield than will higher mass members of the series. The distribution observed in the mass spectrum will therefore be biased in favor of shorter chain length. Not all polymers, however, exhibit this phenomenon. To a first approximation, however, and for comparative purposes, the measurements are useful and much easier to do than, for example, the classical freezing point depression.

Why examine polymers and polymer patterns in the mass spectrometer here? A number of polymeric materials are used regularly as excipients in pharmaceutical formulations—PEG, various surfactants such as Triton X-100, and others. The opportunity then presents for a drug substance to interact chemically with the polymeric material, which will then appear in analyses of the formulation as an impurity or a degradant. Polymeric materials, which are not UV-active, and not detectable by a typical UV-visible detector in an LC-MS run, will become visible when a UV-active drug substance attaches to a polymer molecule and labels it with a chromophore. Such polymeric materials will often elute in an HPLC chromatogram as broad chromatographic peaks detected by a UV-visible detector, the mass spectrometer TIC, or both. The mass spectrum under the chromatographic peak will show a polymer pattern of a polymeric excipient, perchance offset in mass by attachment of the drug substance molecule or a portion of it. Carboxylic acid drug substances have been observed to esterify with PEG. An example of interaction with cetirizine (the acid portion of Structure 8.6) with PEG to form a family of esters in a formulation is illustrated in the FAB mass spectrum of these compounds in Fig. 8.24.

8.9.11 Multiply Charged Ions

ESI-generated multiply charged ions are most frequently discussed in the examination of peptides and intact proteins. The general observation is that peptides need to be on

Structure 8.6   Cetirizine PEG esters. $C_{21}H_{25}N_2O_3Cl[C_2H_4O]_n$ (monoisotopic); r.m.m. $388 + [44]_n$.

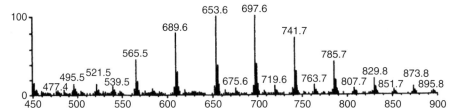

Figure 8.24 Fast atom bombardment spectrum of cetirizine poly(ethylene glycol) ester mixture.

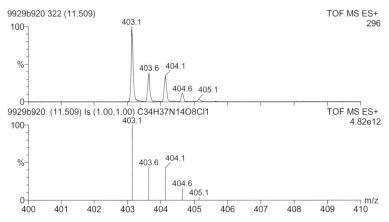

Figure 8.25 Comparison of the isotope pattern of the doubly charged molecular ion region (top panel) and the isotope pattern modeled (bottom panel) from the $C_{34}H_{37}N_{14}O_8Cl$ elemental composition of the m/z 403 doubly charged molecular ion.

the order of 1500–2000 daltons or larger in order to form abundant multiply charged ions (i.e. doubly charged ions). Although doubly charged molecular ions have been reported in the EI mass spectra of polycyclic aromatic hydrocarbons (Myerson and Vander Haar, 1962; Sharp *et al.*, 1988), mention of multiply charged ions in the context of classic mass spectra of small molecules is rare. In our work on impurities and degradants of small pharmaceutical molecules, however, we are starting to see an increasing frequency of doubly charged molecular ions in electrospray mass spectra—molecules with relative molecular masses as low as 500 to 600 daltons. The factors that determine whether or not and to what extent a molecule will form doubly charged molecular ions are still under investigation (Gaskell, 1997). How does one tell if one has a doubly charged species? The spacing between the isotope peaks indicates multiple charging.

ESI mass spectra of compounds often contain van der Waals clusters of the molecule. Numerologically, these species appear at two times the relative molecular mass of the molecule plus one. Consider the example spectrum shown in Fig. 8.25. The m/z 403 species could be assigned to be the [M+H]⁺ of the compound in question, indicating a relative molecular mass of 402 daltons. It was indeed the most abundant peak in the

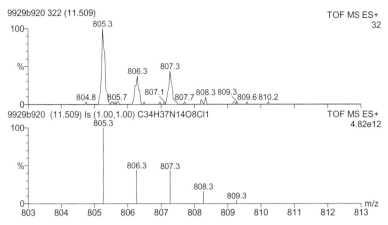

Figure 8.26   Comparison of the isotope pattern of the singly charged molecular ion region (top panel) and the isotope pattern modeled (bottom panel) from the $C_{34}H_{37}N_{14}O_8Cl$ elemental composition of the m/z 805 singly charged molecular ion.

spectrum. A proton-bound dimer would correspond to a $[2M+H]^+$ peak at m/z 805, consistent with the expanded spectral portion shown in Fig. 8.26. The relative abundance of the m/z 805 species is approximately 10% that of the m/z 403 species, consistent with typical relative abundance observations of $[M+H]^+$ and $[2M+H]^+$. The convention would be to assign the $[M+H]^+$ to be the largest peak in the spectrum. The fact that a 402-dalton molecular mass exhibited an illogical mass difference between this and the drug substance molecule under investigation prompted re-examination of the m/z 403 region and the putative assignment of this species as the $[M+H]^+$.

Close examination reveals that the peaks comprising the isotope pattern of the m/z 403 species in Fig. 8.25 appear at 0.5-dalton intervals. The nature of the instrument on which this spectrum was obtained, and the confidence in mass assignments, indicate that these mass assignments were indeed correct, and that this spectrum represents a doubly charged ion. Recall that the mass spectrometer does not measure mass but mass-to-charge ratio. Typically, for small molecules, ions carry only a single electrostatic charge, and the m/z value reduces to the mass. The interval between isotope peaks in a pattern such as this should be 1 dalton. The 0.5-dalton spacing indicates that the m/z 403 species is really a $[M+2H]^{2+}$. The m/z 805 species is the true $[M+H]^+$, and the molecular mass is really 804 daltons. This reassigned molecular mass and a consideration of the synthetic chemistry of the drug substance being studied were sufficient to propose a reasonable identity of the impurity.

8.9.12   Calculating the Charge State and Chemical Molecular Mass of a Protein

With the increasing application of ESI to the examination of intact proteins (illustrated by the example of Fig. 8.27), one observes a pattern outwardly resembling a polymer pattern. This feature is not in fact a polymer pattern but a multiply charged ion envelope of a single chemical species. The spectra in Figs. 8.22 to 8.24 represent the $[M+H]^+$ of collections of molecules of different sizes—the spectrum in Fig. 8.27 that of

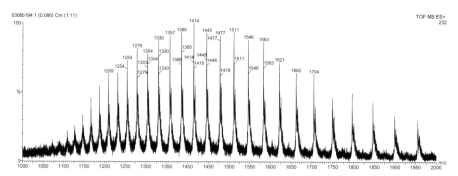

Figure 8.27 Positive ion ESI mass spectrum of bovine serum albumin. The average molecular mass observed here for this protein is 66,424 daltons. There are two additional higher mass proteins in this sample.

a single species. The protein used here is bovine serum albumin, a single molecule with a chemical molecular mass calculated from this spectrum of 66,424 daltons. The definitive way to differentiate a polymer pattern from a multiply charged ion series of a protein is by evaluating the spacing of the peaks. A polymer pattern will show peaks spaced at a constant mass interval across the entire series. A multiply charged ion series will show a steadily increasing mass spacing as one traverses the pattern from low mass to high mass.

It is useful to point out some of the differences in interpretation and thought process needed to interpret mass spectra of proteins. The shift from monoisotopic molecular masses in small molecules to chemical molecular masses in peptides and proteins is necessary when the chemical entities exceed 1500 to 2000 daltons. Mass calibration procedures and philosophies of instrument operation change. The concept and application of chemical molecular mass accuracy changes from 0.1 to 0.2 dalton accuracy for nominal monoisotopic measurements to 0.01–0.02% for proteins (acceptable 0.01% accuracy for the bovine serum albumin example used here is ±7 daltons). In short, the mass spectrometry of proteins and other intact biopolymers deserves an entire monograph of its own. We will include some discussions here, however, to introduce the area.

The first useful fact to extract from the spectrum of a protein is the relative molecular mass of the protein. Mass spectrometer data systems from the major manufacturers incorporate programs and algorithms for deconvoluting these complex mass spectra into spectra from which the relative molecular masses can be directly read. Some programs require significant input and intervention in order to give useful results. Others are very automatic. How does one spot-check the results of such programs, rather than relying on blind faith? The procedure for hand calculating chemical molecular masses is discussed elsewhere (de Hoffmann and Stroobant, 2002, pp. 37–39), and will be briefly derived here. Example calculations will illustrate the procedure.

A peak in the mass spectrum will have an assigned "mass" value, m, determined by the mass spectrometer data system. It will have an unknown charge state, q, the number of electrostatic charges carried by it. If we represent the chemical molecular mass of a protein by M, we can relate M, m, and q by Equation 8.1:

$$M = mq \tag{8.1}$$

Each peak in the multiply charged envelope will have this same relationship, differing by the numerical values of the assigned mass and the charge state. For two adjacent peaks in the multiply charged envelope, their electrostatic charges, q, will differ by one charge. Expressing relationships for the two peaks as Equations 8.2 and 8.3, with the charge relationship as Equation 8.4:

$$M = m_1 q_1 \tag{8.2}$$

$$M = m_2 q_2 \tag{8.3}$$

$$q_2 = q_1 + 1 \tag{8.4}$$

Setting the expressions for M in terms of m_1 and m_2 equal, substituting for q_2, and rearranging gives Equation 8.5:

$$m_1/m_2 - 1 = 1/q_1 \tag{8.5}$$

Note that Equations 8.4 and 8.5 are written assuming one is using adjacent peaks in the multiply charged envelope that differ by only one electrostatic charge unit. One could easily use any pair of peaks in the envelope, adjusting the equations for the difference in charge state. For example, for peaks separated by three charge states, Equation 8.5 changes to $3/q_1$. Using adjacent peaks is easiest. Substituting the appropriate numerical values into Equation 8.6, one can calculate q_1 for the corresponding m_1. Once one charge state is known, the rest are assumed, each differing by 1. With the charge states in hand, M of the protein can be calculated from any of the sets of numbers, according to Equation 8.6. Subtracting q corrects for the charge carriers, namely the H^+, that have been attached to the protein. As a check of the result, M should be calculated for different peaks in the spectrum to verify that the charge states are assigned correctly:

$$M = (mq) - q \tag{8.6}$$

Figure 8.28 is the positive ion electrospray mass spectrum of insulin. All three prominent species in the spectrum are molecular ions. The pattern is a multiply charged envelope because the differences between the m/z values decrease from high mass to

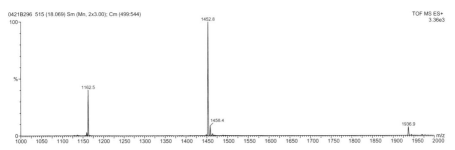

Figure 8.28   Electrospray mass spectrum of insulin.

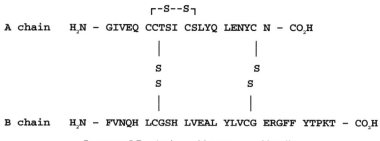

Structure 8.7   Amino acid sequence of insulin.

low mass. This feature differentiates this pattern from a polymer pattern (discussed earlier).

The amino acid sequences of the two polypeptide chains of insulin are shown in Structure 8.7 (using the one letter coding for the amino acids). The disulfide bridges interconnecting the two chains are also shown. The A chain contains 21 amino acids (average molecular mass 2381.7 daltons), with an intramolecular disulfide bridge between cysteine residues A_6 and A_{11}. The B chain contains 30 amino acid residues (average molecular mass 3430.0 daltons). Two disulfide bridges connect the A and B chains, connecting between cysteine residues A_7 and B_7 and between A_{20} and B_{19}. This composite structure gives a chemical molecular mass of 5807.7 daltons.

We will apply this calculation to the data presented in Fig. 8.28 for insulin. We first arbitrarily choose the m/z 1452.8 peak as m_1, with its corresponding unknown charge state q_1. Since the electrostatic charge increases with decreasing m/z, the next higher charge state species, with $q_2 = q_1 + 1$, corresponds to the m/z 1162.5 peak. Substituting and calculating:

$$1/q_1 = (1452.8/1162.5) - 1 \tag{8.7}$$

$$1/q_1 = 0.2497 \tag{8.8}$$

$$q_1 = 4.004 \tag{8.9}$$

Rounding to the nearest integer, q_1 is calculated to be 4, and the m/z 1452.8 species therefore corresponds to the +4 charge state. The m/z 1162.5 species corresponds to the +5 charge state. The m/z 1936.9 species corresponds to the +3 charge state. The chemical molecular mass, M, is calculated using the numbers for the +4 charge state as follows:

$$M = (1452.8)(4) - 4 \tag{8.10}$$

$$M = 5807.2 \text{ daltons} \tag{8.11}$$

Calculating M from the +5 and +3 charge states gives, respectively, 5807.5 and 5807.7 daltons. One could conduct a statistical evaluation of the results of several combinations of numbers to derive a precision for the molecular mass determination. The data system programs that deconvolute these protein spectra permit such an evaluation.

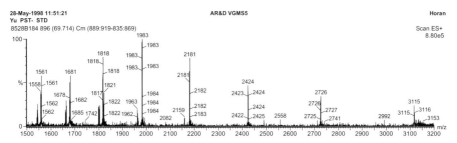

Figure 8.29 Multiply charged ESP mass spectrum of pST.

AFPAMPLSSL FANAVLRAQH LHQLAADTYK EFERAYIPEG QRYSIQNAQA

AFCFSETIPA PTGKDEAQQR SDVELLRFSL LLIQSWLGPV QFLSRVFTNS

LVFGTSDRVY EKLKDLEEGI QALMRELEDG SPRAGQILKQ TYDKFDTNLR

SDDALLKNYG LLSCFKKDLH KAETYLRVMK CRRFVESSCA F

Structure 8.8 Amino acid sequence of porcine somatotropin (pST).

Note the species in Fig. 8.28 at m/z 1458.4. It differs from the m/z 1452.8 species by 5.6 m/z units. Noting that the m/z 1452.8 species is the +4 charge state of insulin, and assuming that this species is of the same charge state, the difference is $5.6 \times 4 = 22.4$ daltons. This mass increment corresponds to a Na^+ cationated molecular ion of insulin, in which one of the H^+ charge carriers is replaced by Na^+. The molecular mass of insulin from this species calculates to 5807.6 daltons. Note here that resolving power of the mass analyzer becomes important in making mass assignments and in being able to differentiate between species with small mass differences. These species are 22 daltons different in mass, but because they carry 4 electrostatic charges, they are only separated by 5.6 m/z units.

Revisiting the spectrum of bovine serum albumin in Fig. 8.27, calculations indicate that the m/z 1477 peak corresponds to a +45 charge state, and the chemical molecular mass of bovine serum albumin, from this spectrum, is approximately 66,420 daltons. The 66,424-dalton value cited earlier was generated by a data system deconvolution program, taking into account the entire spectrum and the decimal places of the assigned m/z values not shown in the figure.

Porcine somatotropin (pST) is another protein we will examine. A spectrum of the protein is shown in Fig. 8.29. Its amino acid sequence is given in Structure 8.8. An average molecular mass of this protein is calculated, using the m/z 2181 species (charge state +10) in Fig. 8.29 to be 21,800 daltons. From the sequence, the chemical molecular mass calculates to 21,802 daltons (fully reduced). Note that the state of oxidation of the two known disulfide links in the native protein structure would cause the molecular mass to vary by 4 daltons. The best figure of merit in the literature for accuracy in measuring the average molecular mass of a protein is ±0.01%. This precision corresponds to ±2 daltons for this protein. To achieve this, one must take significant care in making the measurement. A more typical routine figure of merit is ±0.02% (±4 daltons).

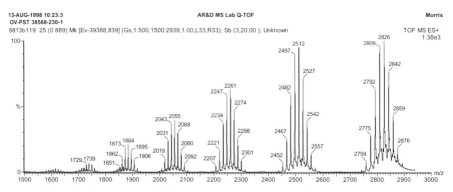

Scheme 8.1 Reversible Schiff base formation between *o*-vanillin and a lysine residue.

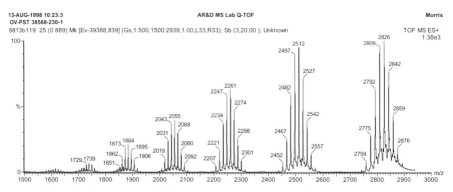

Figure 8.30 Multiply charged ESP mass spectrum of ortho-vanillylated porcine somatotropin.

The question arises about whether mass spectrometry can evaluate the state of reduction/oxidation of this protein by the measured molecular mass determination alone. The difference between fully oxidized pST and fully reduced pST is within the range of imprecision of the experimental measurements. The state of oxidation cannot be reliably measured by determining the average molecular mass.

Experimentation aimed at providing stabilization of this protein involved derivatization of the ε-amino groups of lysine residues in the protein by Schiff base formation with aromatic aldehydes, specifically here *o*-vanillin. Reaction according to Scheme 8.1 introduces an additional 134 daltons for each site on the protein that is derivatized. Assessment of average degree of substitution of pST can be done by direct mass spectrometric observation (Sharp *et al.*, 2006).

The multiply charged mass spectrum in Fig. 8.30 illustrates a polymer pattern superimposed on top of a multiply charged envelope. Using the functional test of spacing between the peaks, we see that, for each of the distributions, mass spacing between peaks is constant, indicating a polymer pattern. The spacings correspond to 134 daltons adjusted by the charge state of the particular pattern, the mass of an *o*-vanillyl reside coupled as a Schiff base to an ε-amino group. The centers of the distributions, however, decrease in mass interval with decreasing m/z, indicating a multiply charged envelope. Any of these distributions can be used to calculate an average degree of substitution for this preparation, according to similar procedures illustrated for calculating average molecular masses of polymers described earlier.

Most of the mass spectrometer computer data systems—especially those targeted toward making measurements of biological molecules—contain one or more

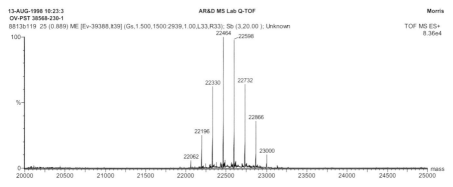

Figure 8.31 MaxEnt transformed singly charged mass spectrum of OV-pST. This spectrum is derived from the one shown in Fig. 8.30.

components for performing the transformation calculation, illustrated before for insulin, to obtain the chemical molecular masses of the macromolecules. If deconvolution of the spectrum in Fig. 8.30 is performed, the spectrum in Fig. 8.31 is obtained. This spectrum represents the distribution of derivatized protein molecules in the sample as if they were singly charged species. The annotations at the tops of the peaks are the chemical molecular masses of the species. The peaks are separated by 134 daltons, the mass of the *o*-vanillyl residue. Calculating the average degree of substitution of this protein sample from the transformed spectrum should give the same value as calculation from the untransformed spectrum of Fig. 8.30. Results are slightly different—from 5.6 for the transformed spectrum to 5.8 for the +8 charge state envelope of Fig. 8.30. Depending upon the needs of the analysis, these slight differences could be acceptable. The quantitative nature of transformation of multiply charged spectra to singly charged spectra have been discussed (Reinhold and Reinhold, 1992).

REFERENCES

Alsante, K.M., Boutros, P., Couturier, M.A. *et al.* (2004). Pharmaceutical impurity identification: a case study using a multidisciplinary approach. *Journal of Pharmaceutical Sciences*, *93*, 2296–2309.

Amad, M.H., Cech, N.B., Jackson, G.S., Enke, C.G. (2000). Importance of gas-phase proton affinities in determining the electrospray ionization response for analytes and solvents. *Journal of Mass Spectrometry*, *35*, 784–789.

Arpino, P.J. (1982). On-line liquid chromatography/mass spectrometry? An odd couple. *Trends in Analytical Chemistry*, *1*, 154.

Bai, Y., Milne, J.S., Mayne, L., Englander, S.W. (1993). Primary structure effects on peptide group hydrogen exchange. *Proteins: Structure Function and Genetics*, *17*, 75–86.

Baldwin, M.A., Burlingame, A.L., Nikolaev, E. (2004). Victor L. Talroze: 1922–2004. *Journal of the American Society for Mass Spectrometry*, *15*, 1517–1519.

Baldwin, M.A., McLafferty, F.W. (1973). Liquid chromatography-mass spectrometry interface. I: the direct introduction of liquid solutions into a chemical ionization mass spectrometer. *Organic Mass Spectrometry*, *7*, 1111–1112.

Barber, M., Bordoli, R.S., Sedgwick, R.D., Tyler, A.N. (1981). Fast atom bombardment of solids (F.A.B): a new ion source for mass spectrometry. *Chemical Communications*, 325–326.

Beavis, R.C., Chait, B.T. (1989). Factors affecting the ultraviolet laser desorption of proteins. *Rapid Communications in Mass Spectrometry*, *3*, 233–237.

Beckey, H.D. (1969). Field desorption mass spectrometry: a technique for the study of thermally unstable substances of low volatility. *International Journal of Mass Spectrometry and Ion Physics*, *2*, 500–503.

Bellar, T.A., Behymer, T.D., Budde, W.L. (1990). Investigation of enhanced ion abundances from a carrier process in high-performance liquid chromatography particle beam mass spectrometry. *Journal of the American Society for Mass Spectrometry*, *1*, 92–98.

Beynon, J.H. (1954). Qualitative analysis of organic compounds by mass spectrometry. *Nature*, *174*, 735–737.

Beynon, J.H. (1960). *Mass Spectrometry and its Applications to Organic Chemistry*, Elsevier Publishers, Amsterdam.

Biemann, K. (1990). Utility of exact mass measurements. *Methods in Enzymology*, *193*, 295–305.

Biemann, K., Oro, J., Toulmin, P.III, , Orgel, L.E., Nier, A.O., Anderson, D.M., Simmonds, P.G., Flory, D., Diaz, A.V., Rushneck, D.R., Biller, J.A. (1976). Search for organic and volatile inorganic compounds in two surface samples from the Chryse Planitia region of Mars. *Science*, *194*, 72–76.

Blades, A.T., Ikonomou, M.G., Kebarle, P. (1991). Mechanism of electrospray mass spectrometry. Electrospray as an electrochemical cell. *Analytical Chemistry*, *63*, 2109–2114.

Blakley, C.R., Carmody, J.C., Vestal, M.L. (1980). Liquid chromatograph-mass spectrometer for analysis of nonvolatile samples. *Analytical Chemistry*, *52*, 1636–1641.

Budzikiewicz, H., Djerassi, C., Williams, D.H. (1967). *Mass Spectrometry of Organic Compounds*, Holden-Day, Inc., San Francisco, CA.

Burinsky, D.J. (2006). Mass spectrometry, in *Comprehensive Analytical Chemistry. Volume 47: Modern Instrumental Analysis* (eds S. Ahuja, N. Jespersen), Elsevier, Amsterdam, pp. 319–396.

Busch, K.L. (1994). Using exact mass measurements. *Spectroscopy*, *9* (7), 21–22.

Busch, K.L. (1995). Mass spectrometric detectors for samples separated by planar electrophoresis. *Journal of Chromatography A*, *692*, 275–290.

Busch, K.L. (2004a). Planar separations and mass spectrometric detection. *Journal of Planar Chromatography*, *17*, 398–403.

Busch, K.L. (2004b). Masses in mass spectrometry: balancing the analytical scales. *Spectroscopy*, *19* (11), 32–34.

Byrd, H.C.M., McEwen, C.N. (2000). The limitations of MALDI-TOF mass spectrometry in the analysis of wide polydisperse polymers. *Analytical Chemistry*, *72*, 4568–4576.

Campana, J.E. (1980). Elementary theory of the quadrupole mass filter. *International Journal of Mass Spectrometry and Ion Physics*, *33*, 101–117.

Cappelletty, D., Eiselstein-McKitrick, K. (2007). The echinocandins. *Pharmacotherapy*, *27*, 369–388.

Caprioli, R.M., Fan, T., Cottrell, J.S. (1986). Continuous-flow sample prove for fast atom bombardment mass spectrometry. *Analytical Chemistry*, *58*, 2949–2953.

Carroll, D.I., Dzidic, I., Stillwell, R.N., Haegele, K.D., Horning, E.C. (1975). Atmospheric pressure ionization mass spectrometry: corona discharge ion source for use in liquid chromatograph-mass spectrometer-computer analytical system. *Analytical Chemistry*, *47*, 2369–2373.

Catlin, D.H., Sekera, M.H., Ahrens, B.D., Starcevic, B., Chang, Y.-C., Hatton, C.K. (2004). Tetrahydrogestrinone: discovery, synthesis, and detection in urine. *Rapid Communications in Mass Spectrometry*, *18*, 1245–1249.

Chapman, J.R., Harden, E.H., Evans, S., Moore, L.E. (1983). LC-MS interfacing in sector mass spectrometers. *International Journal of Mass Spectrometry and Ion Physics*, *46*, 201–204.

Chen, G., Khusid, A., Daaro, I., Irish, P., Pramanik, B.N. (2007). Structural identification of trace level enol tautomer impurity by on-line hydrogen/deuterium exchange HR-LC/MS in a LTQ-Orbitrap hybrid mass spectrometer. *Journal of Mass Spectrometry*, *42*, 967–970.

Cody, R.B., Laramée, J.A., Durst, H.D. (2005). Versatile new ion source for the analysis of materials in open air under ambient conditions. *Analytical Chemistry*, *77*, 2297–2302.

Cole, R.B. (2000). Some tenets pertaining to electrospray ionization mass spectrometry. *Journal of Mass Spectrometry*, *35*, 763–772.

Cooks, R.G., Rockwood, A.L. (1991). The "Thomson." A suggested unit for mass spectroscopists. *Rapid Communications in Mass Spectrometry*, *5*, 93.

Costello, C.E. (1991). Appendix 3, Mass spectra of matrix materials. *Methods in Enzymology*, *193*, 875–882.

Crow, F.W., Tomer, K.B., Gross, M.L. (1983). Mass resolution in mass spectrometry-mass spectrometry. *Mass Spectrometry Reviews*, *2*, 47–76.

DeBievre, P., Barnes, I.L. (1985). Table of the isotopic composition of the elements as determined by mass spectrometry. *International Journal of Mass Spectrometry and Ion Processes*, *65*, 211–230.

de Hoffmann, E., Stroobant, V. (2002). *Mass Spectrometry: Principles and Applications*, 2nd edn, John Wiley & Sons, Ltd, Chichester, UK.

Dole, M., Mack, L.L., Hines, R.L., Mobley, R.C., Ferguson, L.D., Alice, M.B. (1968). Molecular beams of macroions. *Journal of Chemical Physics*, *49*, 2240–2249.

Downard, K.M. (2007a). Francis William Aston: the man behind the mass spectrograph. *European Journal of Mass Spectrometry*, *13*, 177–190.

Downard, K.M. (2007b). Cavendish's crocodile and dark horse: the lives of Rutherford and Aston in parallel. *Mass Spectrometry Reviews*, *26*, 713–723.

Downard, K.M., de Laeter, J.R. (2005). A history of mass spectrometry in Australia. *Journal of Mass Spectrometry*, *40*, 1123–1139.

Englander, S.W. (2006). Hydrogen exchange and mass spectrometry: a historical perspective. *Journal of the American Society for Mass Spectrometry*, *17*, 1481–1489.

Fenn, J.B. (2003). Electrospray wings for molecular elephants (Nobel Lecture). *Angewandte Chemie (International)*, *42*, 3871–3894.

Fernandes de la Mora, J. Van Berkel, G.J., Enke, C.G., Cole, R.B., Martinez-Sanchez, M., Fenn, J.B. (2000). Electrochemical processes in electrospray ionization mass spectrometry. *Journal of Mass Spectrometry*, *35*, 939–952..

Field, F.H. (1968). Chemical ionization mass spectrometry. *Accounts of Chemical Research*, *1*, 42–49.

Field, F.H. (1990). The early days of chemical ionization: a reminiscence. *Journal of the American Society for Mass Spectrometry*, *1*, 277–283.

Futrell, J.H., Miller, C.D. (1966). Tandem mass spectrometer for study of ion-molecule reactions. *Review of Scientific Instrumentation*, *37*, 1521–1526.

Gamero-Castaño, M., Fernandez de la Mora, J. (2000). Kinetics of small ion evaporation from the charge and mass distribution of multiply charged clusters in electrospray. *Journal of Mass Spectrometry*, *35*, 790–803.

Gaskell, S.J. (1997). Electrospray: principles and practice. *Journal of Mass Spectrometry*, *32*, 677–688.

Gilliam, J.M., Landis, P.W., Occolowitz, J.L. (1983). Accurate mass measurement in fast atom bombardment mass spectrometry. *Analytical Chemistry*, *55*, 1531–1533.

Gilliam, J.M., Landis, P.W., Occolowitz, J.L. (1984). On-line accurate mass measurement in fast atom bombardment mass spectrometry. *Analytical Chemistry*, 56, 2285–2288.

Gohlke, R.S., McLafferty, F.W. (1993). Early gas chromatography/mass spectrometry. *Journal of the American Society for Mass Spectrometry*, 4, 367–371.

Grayson, M.A. (2002). *Measuring Mass: From Positive Rays to Proteins*, Chemical Heritage Press, Philadelphia, p. 149.

Griffiths, I.W. (1997). J.J. Thomson—the centenary of his discovery of the electron and of his invention of mass spectrometry. *Rapid Communications in Mass Spectrometry*, 11, 2–16.

Gross, J.H. (2004). *Mass Spectrometry: A Textbook*, Springer-Verlag, Berlin, p. 518.

Gross, M.L. (1994). Accurate masses for structure confirmation. *Journal of the American Society for Mass Spectrometry*, 5, 57.

Hamming, M.C., Foster, N.G. (1972). *Interpretation of Mass Spectra of Organic Compounds*, Academic Press, New York.

Harrison, A.G. (1992). *Chemical Ionization Mass Spectrometry*, 2nd edn, CRC Press, Boca Raton, FL, p. 224.

Heathcock, C.H. (1990). Editorial. *Journal of Organic Chemistry*, 55, 8A.

Hertz, H.S., Hites, R.A., Biemann, K. (1971). Identification of mass spectra by computer-searching a file of known spectra. *Analytical Chemistry*, 43, 681–691.

Hillenkamp, F., Karas, M. (1990). *Methods in Enzymology*, 193, 280–294.

Holland, J.F., Soltmann, B., Sweeley, C.C. (1976). A model for ionization mechanisms in field desorption mass spectrometry. *Biomedical Mass Spectrometry*, 3, 340–345.

Honig, R.E. (1985). The development of secondary ion mass spectrometry (SIMS): a retrospective. *International Journal of Mass Spectrometry and Ion Processes*, 66, 31–54.

Iribarne, J.V., Thomson, B.A. (1976). On the evaporation of small ions from charged droplets. *Journal of Chemical Physics*, 64, 2287–2294.

Ito, Y., Takeuchi, T., Ishii, D., Goto, M., Mizuno, T. (1986). Direct coupling of micro high performance liquid chromatography with fast atom bombardment mass spectrometry. II: application to gradient elution of bile acids. *Journal of Chromatography*, 385, 201–209.

Karas, M., Bachmann, D., Bahr, U., Hillenkamp, F. (1987). Matrix-assisted ultraviolet laser desorption of non-volatile compounds. *International Journal of Mass Spectrometry and Ion Processes*, 78, 53–68.

Katritzky, A.R., Karelson, M., Lobanov, V.S. (1997). SPR as a means of predicting and understanding chemical and physical properties in terms of structure. *Pure and Applied Chemistry*, 69, 245–248.

Kebarle, P. (2000). A brief overview of the present status of the mechanisms involved in electrospray mass spectrometry. *Journal of Mass Spectrometry*, 35, 804–817.

Kind, T., Fiehn, O. (2006). Metabolomic database annotations via query of elemental compositions: mass accuracy is insufficient even at less than 1 ppm. *BMC Bioinformatics*, 7, 234–246.

Kiser, R.W. (1965). *Introduction to Mass Spectrometry and Its Applications*, Prentice-Hall, Inc., Englewood Cliffs, NJ.

Klose, R.T. (1987). The Joys of Science. *Newsweek*, 26 October 1987.

Knapp, D.R. (1979). *Handbook of Analytical Derivatization Reactions*, John Wiley & Sons, Inc., New York, p. 741.

Kondrat, R. (1999). High resolution mass spectrometry. *Journal of the American Society for Mass Spectrometry*, 10, 661.

Kowalski, M.H., Sharp, T.R., Stang, P.J. (1987). *Organic Mass Spectrometry*, 22, 642–643.

Krishna, M.M.G., Hoang, L., Lin, Y., Englander, S.W. (2004). Hydrogen exchange methods to study protein folding. *Methods*, 34, 51–64.

Lehmann, W.D., Schulten, H.-R. (1978). Quantitative field desorption mass spectrometry: V-discussion of methodology and examples of applications. *Biomedical Mass Spectrometry*, 5, 208–214.

McElroy, M.B., Yung, Y.L., Nier, A.O. (1976). Isotopic composition of nitrogen: implications for the past history of Mars' atmosphere. *Science*, *194*, 70–72.

McFadden, W. (1973). *Techniques of Combined Gas Chromatography-Mass Spectrometry: Applications in Organic Analysis*, John Wiley & Sons, Inc., New York.

McFadden, W.H., Schwartz, H.L., Bradford, D.C. (1976). Direct analysis of liquid chromatographic effluents. *Journal of Chromatography*, *122*, 389–396.

Macfarlane, R.D. (1983). Californium-252 plasma desorption mass spectrometry. large molecules, software, and the essence of time. *Analytical Chemistry*, *55*, 1247A–1264A.

Macfarlane, R.D., Hu, Z.H., Song, S., Pittenauer, E., Schmid, E.R., Allmaier, G., Metzger, J.O., Tusznski, W. (1994). ^{252}Cf-plasma desorption mass spectrometry II—a perspective of new directions. *Biology Mass Spectrometry*, *23*, 117–130.

McLafferty, F.W. (1981). Tandem mass spectrometry. *Science*, *214*, 280–287.

McLafferty, F.W. (1983). *Tandem Mass Spectrometry*, John Wiley & Sons, Inc., New York.

McLafferty, F.W., Venkataraghavan, R. (1982). *Mass Spectral Correlations*, 2nd edn, American Chemical Society Advances in Chemistry Series #40, Washington, DC, p. 124.

Mamer, O.A., Lesimple, A. (2004). Letter to the editor. *Journal of the American Society for Mass Spectrometry*, *15*, 626.

Marshall, A.G., and Hendrickson, C.L. (2008). High resolution mass spectrometers. *Annual Review of Analytical Chemistry*, *1*, 579–599.

Marshall, A.G., Rodgers, R.P. (2004). Petroleomics: the next grand challenge for chemical analysis. *Accounts of Chemical Research*, *37*, 53–59.

Meili, J., Seibl, J. (1984). A new versatile matrix for fast atom bombardment analysis. *Organic Mass Spectrometry*, *19*, 581–582.

Message, G.M. (1984). *Practical Aspects of Gas Chromatography/Mass Spectrometry*, John Wiley & Sons, Inc., New York.

Meyerson, S. (1986). Reminiscences of the early days of mass spectrometry in the petroleum industry. *Organic Mass Spectrometry*, *21*, 197–208.

Miles, C.J., Doerge, D.R., Bajic, S. (1992). Particle beam liquid chromatography-mass spectrometry of national pesticide survey analytes. *Archives of Environmental Contamination and Toxicology*, *22*, 247–251.

Morrison, J.D. (1991). Personal reminiscences of forty years of mass spectrometry in Australia. *Organic Mass Spectrometry*, *26*, 183–194.

Munson, M.S.B., Field, F.H. (1966). Chemical ionization mass spectrometry I: general introduction. *Journal of the American Chemical Society*, *88*, 2621–2630.

Myerson, S., Vander Haar, R.W. (1962). Multiply charged ions in mass spectra. *Journal of Chemical Physics*, *37*, 2458–2462.

Nier, A.O. (1985). Mass spectrometry in planetary research. *International Journal of Mass Spectrometry and Ion Processes 66*, 55, 73.

Nier, A.O. (1989). Some reminiscences of mass spectrometry and the Manhattan Project. *Journal of Chemical Education*, *66*, 385–388.

Octoxynol. Merck Index, 12th edn. Entry # 6858.

Olsen, M.A., Cummings, P.G., Kennedy-Gabb, S., Wagner, B.M., Nicol, G.R., Munson, B. (2000). The use of deuterium oxide as a mobile phase for structural elucidation by HPLC/UV/ESI/MS. *Analytical Chemistry*, *72*, 5070–5078.

Porter, Q.N. (1985). *Mass Spectrometry of Heterocyclic Compounds*, 2nd edn, John Wiley & Sons, Inc., New York.

Quayle, A. (1987). Recollections of mass spectrometry of the fifties in a UK Petroleum Laboratory. *Organic Mass Spectrometry*, *22*, 569–585.

Randall, L.G., Wahrhaftig, A.L. (1981). Direct coupling of a dense (supercritical) gas chromatograph to a mass spectrometer using a supersonic molecular beam interface. *Review of Scientific Instrumentation*, *52*, 1283–1295.

Reinhold, B.B., Reinhold, V.N. (1992). Electrosprayionization mass spectrometry: deconvolution by an entropy-based algorithm. *Journal of the American Society for Mass Spectrometry*, *3*, 207–215.

Reynolds, W.E., Bacon, V.A., Bridges, J.C., Coburn, T.C., Halpern, B., Lederberg, J., Levinthal, E.C., Steed, E., Tucker, R.B. (1970). A computer operated mass spectrometer system. *Analytical Chemistry*, *42*, 1122–1129.

Robb, D.B., Covey, T.R., Bruins, A.P. (2000). Atmospheric pressure photoionization: an ionization method for liquid chromatography-mass spectrometry. *Analytical Chemistry*, *72*, 3653–3659.

Schulten, H.R., Beckey, H.D. (1972). Field desorption mass spectrometry with high temperature activated emitters. *Organic Mass Spectrometry*, *6*, 885–985.

Scott, R.P.W., Scott, C.G., Munroe, M., Hess, J., Jr (1974). Interface for on-line liquid chromatography-mass spectroscopy analysis. *Journal of Chromatography*, *99*, 395–405.

Settle, F.A. (2002). Analytical chemistry and the Manhattan Project. *Analytical Chemistry*, *74*, 36A–43A.

Sharp, T.R., Marquez, B.L. (2006). Hyphenated methods, in *Comprehensive Analytical Chemistry. Volume 47: Modern Instrumental Analysis, Amsterdam* (eds S. Ahuja and N. Jespersen), Elsevier, Amsterdam, pp. 691–754.

Sharp, T.R., Smith, D.E., Heah, P.C., Buhro, W.E., Crocco, G.L., Bodner, G.S., Fernandez, J.M., Gladysz, J.A. (1986). *Advances in Mass Spectrometry*, *10B*, 1359–1360.

Sharp, T.R., Lee, H., Ferguson, A., Marsh, K.N., Harvey, R.G. (1988). 36th Annual Conference on Mass Spectrometry and Allied Topics.

Sharp, T.R., Horan, G.J., Morris, R., Pezzullo, L.H., Stroh, J.G. (2006). Method for determining the average degree of substitution of *o*-vanillin derivatized porcine somatotropin. *Journal of Pharmaceutical and Biomedical Analysis*, *40*, 185–189.

Smith, R.D., Udseth, H.R. (1983). Mass spectrometry with direct supercritical fluid injection. *Analytical Chemistry*, *55*, 2266–2272.

Smith, R.M. (2004). *Understanding Mass Spectra*, 2nd edn, John Wiley & Sons, Inc., New York, Preface, pg xi.

Squires, G. (1998). Francis Aston and the mass spectrograph. *Journal of the Chemical Society, Dalton Transactions*, 3893–3899.

Stroh, J.G., Cook, J.C., Milberg, R.M., Brayton, L., Kihara, T., Huang, Z., Rinehart, K.L.Jr, , Lewis, I.A.S. (1985). On-line liquid chromatography fast atom bombardment mass spectrometry. *Analytical Chemistry*, *57*, 985–991.

Sundqvist, B., Macfarlane, R.D. (1985). ^{252}Cf-Plasma desorption mass spectrometry. *Mass Spectrometry Reviews*, *4*, 421–460.

Surman, D.J., Vickerman, J.C. (1981). Fast atom bombardment quadrupole mass spectrometry. *Chemical Communications*, 324–325.

Svec, H.J. (1985). Mass spectroscopy—ways and means. A historical prospectus. *International Journal of Mass Spectrometry and Ion Processes*, *66*, 3–29.

Sweetman, B.J., Blair, I.A. (1988). 3-nitrobenzyl alcohol has wide applicability as a matrix for FAB MS. *Biomedical and Environmental Mass Spectrometry*, *17*, 337–340.

Syage, J.A. (2004). Mechanism of [M$^+$H]+ formation in photoionization mass spectrometry. *Journal of the American Society for Mass Spectrometry*, *15*, 1521–1533.

Syage, J.A., Short, L.C., Cai, S.-S. (2008). Atmospheric pressure photoionization—the second source for LC-MS? *LCGC North America*, *26* (3), 286–296.

Takáts, Z., Wiseman, J.M., Gologan, B., Cooks, R.G. (2004). Mass spectrometry sampling under ambient conditions with desorption electrospray ionization. *Science, 306,* 471–473.

Takáts, Z., Wiseman, J.M., Gologan, B., Cooks, R.G. (2005). Ambient mass spectrometry using desorption electrospray ionization (DESI): instrumentation, mechanisms and applications in forensics, chemistry, and biology. *Journal of Mass Spectrometry 40,* 1261–1275.

Tal'Rose, V.L., Lyubimova, A.K. (1952). *Doklady Akadamea Nauk SSSR, 86,* 909.

Tal'Rose, V.L., Karpov, G.V., Grdetskii, I.G., Skurat, V.E. (1968). Capillary system for the introduction of liquid mixtures into an analytical mass spectrometer. *Russian Journal of Physical Chemistry, 42,* 1658–1664.

Torgerson, D.F., Skowronski, R.P., Macfarlane, R.D. (1974). New approach to the mass spectrometry of non-volatile compounds. *Biochemical and Biophysical Research Communications, 60,* 616–621.

Tyler, A.N., Clayton, E., Green, B.N. (1996). Exact mass measurement of polar organic molecules at low resolution using electrospray ionization and a quadrupole mass spectrometer. *Analytical Chemistry, 68,* 3561–3569.

Vail, T.M., Jones, P.R., Sparkman, O.D. (2007). Rapid and unambiguous identification of melamine in contaminated pet food based on mass spectrometry with four degrees of confirmation. *Journal of Analytical Toxicology, 31,* 304–312.

Van Berkel, G.J. (2000). Electrolytic deposition of metals on to the high-voltage contact in an electrospray emitter: implications for gas-phase ion formation. *Journal of Mass Spectrometry, 35,* 773–783.

Vestal, M.L. (1983). Ionization techniques for nonvolatile molecules. *Mass Spectrometry Reviews, 2,* 447–480.

Vestal, M.L. (2001). Methods of ion generation. *Chemical Reviews, 101,* 361–375.

Watson, J.T., Sparkman, O.D. (2007). *Introduction to Mass Spectrometry: Instrumentation, Applications and Strategies for Data Interpretation,* 4th ed. John Wiley & Sons, Chichester, U.K.

Willoughby, R.C., Browner, R.F. (1984). Monodisperse aerosol generation interface for combining liquid chromatography with mass spectrometry. *Analytical Chemistry, 56,* 2625–2631.

Wilson, I.D. (1999). The state of the art in thin-layer chromatography-mass spectrometry: a critical appraisal. *Journal of Chromatography A, 856,* 429–442.

Wilson, I.D. (1999). The state of the art in thin-layer chromatography–mass spectrometry: a critical appraisal. *Journal of Chromatography A, 856,* 429–442.

Witten, J.L., Schaffer, M.H., O'Shea, M., Cook, J.C., Hemling, M.E., Rinehart, K.L. (1986). Structure of two cockroach neuropeptides assigned by fast atom bombardment mass spectrometry. *Biochemical and Biophysical Research Communications, 124,* 350–358.

Wood, G.W. (1982). Field desorption ionization: applications. *Mass Spectrometry Reviews, 1,* 63–102.

Wu, Z., Rodgers, R.P., Marshall, A.G. (2004). Characterization of vegetable oils: detailed compositional fingerprints derived from electrospray ionization Fourier transform ion cyclotron resonance mass spectrometry. *Journal of Agricultural and Food Chemistry, 52,* 5322–5328.

Yamashita, M., Fenn, J.B. (1984). Electrospray ion source. Another variation on the free-jet theme. *Journal of Physical Chemistry, 88,* 4452–4459.

Yau, W.W., Kirkland, J.J., Bly, D.D. (1979). *Modern Size-Exclusion Liquid Chromatography,* John Wiley & Sons, Inc., New York, pp. 5–13.

Yergey, A.L., Edmonds, C.G., Lewis, I.A.S., Vestal, M.L. (1990). *Liquid Chromatography/Mass Spectrometry: Techniques and Applications,* Plenum Publishers, New York, p. 306.

Yergey, A.L., Yergey, A.K. (1997). Preparative scale mass spectrometry: a brief history of the Calutron. *Journal of the American Society for Mass Spectrometry, 8,* 943–953.

CHAPTER 9

Approaches to Performing Metabolite Elucidation: One Key to Success in Drug Discovery and Development

ALA F. NASSAR

Absorption, distribution, metabolism, excretion, and toxicology (ADMET) studies are widely used in drug discovery and development to help obtain the optimal balance of properties necessary to convert lead compounds into drugs that are safe and effective for human use. Drug discovery efforts have been aimed at identifying and addressing metabolism issues at the earliest possible stage, by developing and applying innovative liquid chromatography–mass spectrometry (LC-MS)-based techniques and instrumentation, which are both faster and more accurate than prior techniques. Such new approaches are demonstrating considerable potential to improve the overall safety profile of drug candidates throughout the drug discovery and development process. These emerging techniques streamline and accelerate the process by eliminating potentially harmful candidates earlier and improving the safety of new drugs. In the area of drug metabolism, for example, revolutionary changes have been achieved by the combination of LC-MS with innovative instrumentation such as triple quadrupoles, ion traps, Orbitrap, and time-of-flight mass spectrometry. In turn, most ADMET studies have come to rely on LC-MS for the analysis of an ever-increasing workload of potential candidates. This chapter provides a discussion on the importance of LC-MS in supporting metabolic activation testing, metabolite characterization, and radiolabeled-drug testing.

9.1 INTRODUCTION

Mass spectrometry (MS) and nuclear magnetic resonance (NMR) are critical to the success of such ADMET studies. NMR spectroscopic techniques are most often used to confirm and elucidate metabolite identification in drug metabolism studies. Liquid

Drug Metabolism Handbook: Concepts and Applications, Edited by Ala F. Nassar, Paul F. Hollenberg, and JoAnn Scatina

chromatography (LC)-NMR is a good choice for these studies, but MS has advantages over NMR with respect to sensitivity, smaller sample size, and greater speed compared with NMR. LC-MS-NMR has become a commercially available technique and is used in the late discovery stages to confirm and characterize metabolites. LC-MS is an analytical technique that still shows room for development; already, significant improvements have been made in sensitivity and resolution. It is probably the most powerful technique currently available for pharmaceutical analysis, and has significantly accelerated the drug discovery and development process. LC-MS has become the dominant technique for performing almost all of the analyses involved in ADMET studies, and is likely to remain the principal tool for such studies.

The major aim of LC-MS is the application of its analytical power to create straightforward, sensitive, fast, and reliable data. Improved hardware and software for LC-MS have led to greater sensitivity, greater ease-of-use, and improved post-analysis of data (Kubinyi, 1977; Bakke *et al.*, 1995; Lesko *et al.*, 2000; Meyboom, Lindquist and Egberts, 2000; Thompson, 2000; Greene, 2002; Lasser *et al.*, 2002; Taylor *et al.*, 2002; Tiller and Romanyshyn, 2002; Yang *et al.*, 2002; Jemal *et al.*, 2003; Kostiainen *et al.*, 2003; Lin *et al.*, 2003; Nassar, 2003; Nassar and Adam, 2003; Plumb *et al.*, 2003; Roberts, 2003; Kassel, 2004; Nassar and Lopez-Anaya, 2004; Nassar and Talaat, 2004; Balimane *et al.*, 2005; Castro-Perez *et al.*, 2005a; Johnson and Plumb, 2005; Leclercq *et al.*, 2005). Techniques such as electron impact, chemical ionization, atmospheric pressure chemical ionization (APCI), fast atom bombardment, thermospray, gas chromatography–mass spectrometry (GC-MS), and electrospray ionization (ESI) are used in ADMET studies. Metabolite characterization ion trap (MS^n—this refers to multistage MS/MS experiments where n is the number of product ion stages (progeny ions)) and quadrupole time-of-flight (QTOF)-MS are widely used; MS^n because it allows the relatively rapid construction of fragmentation maps and a higher degree of specificity than with other methods. QTOF-MS has advantages for metabolite identification over MS^n or triple quadrupole MS, including fast mass spectral acquisition speed with high full-scan sensitivity, enhanced mass resolution, and accurate mass measurement capabilities which allow for the determination of elemental composition. Exact mass measurement is a valuable tool for solving structure elucidation problems by helping to confirm elemental composition, and is also invaluable in eliminating false positives and determining nontrivial metabolites. Accurate mass measurements are routine experiments performed by modern mass spectrometers. The accuracy of a measurement refers to the degree of conformity of a measured quantity to its actual true value. Accurate mass measurements may not be sufficient if mixtures are analyzed or if the isotopic fine structure of compounds needs to be evaluated. In such cases, high resolving power is needed; this can be easily achieved by using Orbitrap or Fourier-transform ion cyclotron resonance (FTICR)-MS. Also, chromatography or comprehensive chromatography (GC-GC, LC-LC), or multidimensional chromatography (LC-GC) can help to enhance the efficiency.

Data-dependent scans on QTOF or MS^n provide a highly useful tool. Increasing sample complexity, sample volume restrictions, and throughput requirements necessitate that the maximum amount of useful information is extracted from a single experiment. Data-directed analysis (DDA) or data-dependent scans enable intelligent MS and MS/MS acquisitions to be performed automatically on multiple co-eluting components. DDA is able to make intelligent decisions about which ions to select for MS/MS using its inherent high resolution, exact mass measurement capability, and full mass range. One shortcoming, however, is that due to limited sensitivity, DDA often overlooks minor metabolites that could be toxic or active in nature. MS/MS or MS^n fragmentation data contain a tremendous variety of information. Unfortunately, this wealth of information is contained within a mosaic of rearrangements, fragmentations, ion physics, and gas-phase chemistry reactions, which we do not yet fully understand. MS/MS fragmentation data cannot solve all our problems without an expert scientist to analyze and interpret this information, and use it to identify and characterize metabolites. While this analysis does not provide certain, "yes-or-no" answers, it does aid in directing the researcher as to which further studies will be needed. Herein, one demonstrably successful analytical strategy for metabolite characterization and identifying reactive metabolites by LC-MS is presented.

9.2 CRITERIA FOR LC-MS METHODS

The growing realization of the importance of ADMET properties early in the drug discovery process has led to a dramatic increase in the numbers of compounds requiring screening. Errors, as well as false positives and/or false negatives, will always be a risk factor with any screening procedure in the early stages of drug discovery because of the resulting necessity for speed. While discovery scientists seek to improve their turnaround time, it is vital to maintain or even improve the quality of chromatographic resolution of the metabolites produced in screening efforts. The high-throughput methods needed to achieve the required speed would not be possible without the innovative enabling technologies of computing, automation, new sample preparation technologies, and highly sensitive and selective detection systems. To ensure the integrity of the data, the screening procedure should meet criteria such as relevance, effectiveness, speed, robustness, accuracy, and reproducibility (White, 2000; Nassar *et al.*, 2006).

9.3 MATRICES EFFECT

The analysis of compounds in complex biological matrices, such as blood, plasma, bile, urine, and feces samples, is probably the largest application of LC-MS/MS. LC-MS has become the principal technique used in quantitative bioanalysis due to a combination of factors, such as cost, ease, and speed of performing selected reaction monitoring (SRM), and a wide dynamic range and good sensitivity due to favorable signal-to-noise ratio (S/N) (Jemal and Bergum, 1992; Jemal, 2000; Nassar *et al.*, 2001; Nassar, Varshney and Getek, 2001; Zhang and Henion, 2001; Jemal and Ouyang, 2003; Guevremont, 2004; Castro-Perez *et al.*, 2005b; Kapron *et al.*, 2005). It is interesting to note that it was initially thought that the invention of MS/MS would enable the analysis of matrices without separation but suppression effects have prevented this. The presence of endogenous components in the matrix can suppress the analyte response, probably by competition for ESI droplet surface and hence ionization. These effects can cause differences in response between samples in matrices and standards, leading to difficulties in quantitative analysis and compound identification. This emphasizes the importance of the chromatographic step in the analysis, where good separation can reduce or eliminate these effects. Because of the high degree of selectivity routinely provided by SRM, bioanalytical method development time for quantitative determinations of one or several analytes has been reduced to a few days or less. Although the SRM approach demonstrates excellent sensitivity, selectivity, and efficiency, one drawback is that it fails to produce the qualitative information required to support the recognition and structural elucidation of metabolites that could be present in the samples. Metabolite identification and structure elucidation requires development of a method that is separate and distinct from the quantitation of the known components present, which requires additional time, effort, and expertise. During spectral scanning, the duty cycle of a quadrupole mass spectrometer is such that only a small fraction of the total time is spent monitoring any one ion. To obtain an optimum S/N, a quadrupole analyzer must allow a limited number of selected ions to pass. Because most ions are filtered out, much of the qualitative information is lost. A potential strategy to deal with this situation is the use of LC-TOF-MS to generate data that will simultaneously provide qualitative and quantitative information about drug candidate metabolism and disposition (Jemal and Bergum, 1992; Jemal, 2000; Zhang and Henion, 2001; Castro-Perez *et al.*, 2005b).

9.4 METABOLITE CHARACTERIZATION

Metabolite identification is crucial to the drug discovery and development process, as it can be used to investigate the phase I metabolites likely to be formed *in vivo*, the differences between species in drug metabolism, the major circulating metabolites of an administered drug and the phase I and phase II pathways of metabolism, and pharmacologically active or toxic metabolites, and also to help determine the effects of metabolizing enzyme inhibition and/or induction. The ability to produce this information early in the discovery phase is becoming increasingly important as a basis for judging whether a drug candidate merits further development. Table 9.1 shows a list of a wide variety of phase I and phase II biotransformations, together with the mass changes from the parent drug (modified from Mortishire *et al.*, 2005). This table could also be used retroactively to find the potential reactions responsible for certain metabolites.

Metabolite identification enables early detection of potential metabolic liabilities or issues, provides a metabolism perspective to guide synthesis efforts with the aim of either blocking or enhancing metabolism so as to optimize the pharmacokinetic and safety profiles of newly synthesized drug candidates, and assists the prediction of the metabolic pathway(s) of potential drug candidates for development (Taylor *et al.*, 2002; Kantharaj *et al.*, 2003; Nassar, Kamel and Clarimont, 2004a, b). The author's experience has demonstrated that the best combination for accomplishing rapid and accurate metabolite identification involves a robotic system for sample preparation and *in silico* software to predict and find possible metabolites as well as to predict hypothetical metabolite chemical structure (Obach, 1997, 1999; Greene, 2002). This can be combined with the use of LC-MS to determine exact mass measurements (accurate mass) for sample analysis and LC-MS-NMR, and online hydrogen–deuterium (HD) exchange for further metabolite structure confirmation and elucidation. In the following sections, the relative merits of current and potential strategies for dealing with metabolite characterization in various stages of drug discovery and development are examined and illustrated with examples. Techniques, tools, and approaches are suggested for each of these stages, as summarized in Table 9.2.

TABLE 9.1   Common biotransformation reactions.

Metabolic reaction	Monoisotopic mass change
Oxidative debromination	−61.9156
tert-Butyl dealkylation	−56.0624
Hydrolysis of nitrate esters	−44.9851
Decarboxylation	−43.9898
Isopropyl dealkylation	−42.0468
Propyl ketone to acid	−40.0675
tert-Butyl to alcohol	−40.0675
Reductive dechlorination	−33.9611
Hydroxymethylene loss	−30.0106
Nitro reduction	−29.9742
Propyl ether to acid	−28.0675
Deethylation	−28.0312
Decarboxylation	−27.9949
Ethyl ketone to acid	−26.0519
Isopropyl to alcohol	−26.0519
Alcohols dehydration	−18.0105
Dehydration of oximes	−18.0105
Reductive defluorination	−17.9906
Oxidative dechlorination	−17.9662
Sulfoxide to thioether	−15.9949
Thioureas to ureas	−15.9772
Ethyl ether to acid	−14.0519
Demethylation	−14.0157
tert-Butyl to acid	−12.0726
Methyl ketone to acid	−12.0363
Ethyl to alcohol	−12.0363
Two sequential desaturation	−4.0314
Hydroxylation and dehydration	−2.0157
Primary alcohols to aldehyde	−2.0157
Secondary alcohols to ketone	−2.0157
Desaturation	−2.0157
1,4-Dihydropyridines to pyridines	−2.0157
Oxidative defluorination	−1.9957
Oxidative deamination to ketone	−1.0316
Demethylation and methylene to ketone	−0.0365
2-Ethoxyl to acid	−0.0363
Oxidative deamination to alcohol	0.984
Isopropyl to acid	1.943
Demethylation and hydroxylation	1.9792
Ketone to alcohol	2.0157
Methylene to ketone	13.9792
Hydroxylation and desaturation	13.9792
Alkene to epoxide	13.9792
(O, N, S) methylation	14.0157
Ethyl to carboxylic acid	15.9586
Hydroxylation	15.9949
Secondary amine to hydroxylamine	15.9949
Tertiary amine to N-oxide	15.9949
Thioether to sulfoxide, sulfoxide to sulfone	15.9949
Aromatic ring to arene oxide	15.9949

TABLE 9.1 *Continued*

Metabolic reaction	Monoisotopic mass change
Demethylation and two hydroxylation	17.9741
Hydration, hydrolysis (internal)	18.0106
Hydrolysis of aromatic nitriles	18.0106
Hydroxylation and ketone formation	29.9741
Quinone formation	29.9741
Demethylation to carboxylic acid	29.9742
Hydroxylation and methylation	30.0105
Two hydroxylation	31.9898
Thioether to sulfone	31.9898
Alkenes to dihydrodiol	34.0054
Acetylation	42.0106
Three hydroxylation	47.9847
Aromatic thiols to sulfonic acids	47.9847
Glycine conjugation	57.0215
Sulfate conjugation	79.9568
Hydroxylation and sulfation	95.9517
Cysteine conjugation	103.0092
Taurine conjugation	107.0041
S-cysteine conjugation	119.0041
Decarboxylation and glucuronidation	148.0372
N-acetylcysteine conjugation	161.0147
Glucuronide conjugation	176.0321
Two sulfate conjugation	191.9035
Hydroxylation + glucuronide	192.027
GSH conjugation	289.0732
Desaturation + S-GSH conjugation	303.0525
S-GSH conjugation	305.0682
Epoxidation + S-GSH conjugation	321.0631
Two glucuronide conjugation	352.0642

Below is a general list of tools that can be used to identify and characterize metabolites. It is not necessary to use all of these tools for each analysis; for example, LC-NMR can be saved for use in the late stage of drug discovery.

- *In silico*
- Full-scan LC-MS positive and negative modes
- Multistages MS (MS^n) positive and negative modes
- N-rule
- Ring double-bond equivalent (RDBE)
- Isotopic patterns
- Exact mass measurements
- HD experiments
- Stable isotope
- LC-MS-NMR

9.4.1 Use of LC-MS to Identify and Characterize Metabolites

LC-MS is used with various strategies in drug discovery for identifying compounds and/or their metabolites. One area in which the technique is employed is in confirming

TABLE 9.2   Techniques that help in the search for, and identification of, metabolites in drug metabolism.

Stage of compound	Technique	Comments
Very early stage of drug discovery	*In silico* as well as *in vitro* screening	Useful during synthesis to help select/eliminate compounds
Ranking compounds for drug discovery	*In silico* as well as *in vitro* screening	Useful during synthesis efforts to adjust metabolism
	LC-MS and LC-MS/MS techniques	Helps in identifying of simple/ major metabolites, for example, dealkylations, and conjugations such as glucuronide
		Predicts likely metabolites formed *in vivo*
Selected candidate in late drug discovery	*In silico* as well as *in vitro* screening	Aids in determination of metabolic differences between species
	LC-MS and LC-MS/MS techniques	Used to identify potential pharmacologically active or pharmacologically toxic metabolites
	Online HD exchange, ion trap, Q-TOF and Orbitrap	
Nominate compound for clinical trials	Same as selected candidate in the late drug discovery. Also, LC-MS-NMR and LC-Radioactive-MS	Used to determine the percentage of metabolite formed *in vitro* or *in vivo*
		Aids synthesis of metabolites for toxicology testing
		Aids in comparison of human pathways
		Aids identification of drug–drug interactions

Reprinted from Nassar and Lopez-Anaya, 2004, with permission from Thomson Scientific.

the structure of a known compound. A second area of application, and most relevant to this discussion, is the identification of unknown metabolites of drug candidates.

9.5   STRATEGIES FOR IDENTIFYING UNKNOWN METABOLITES

When using full-scan mass spectra to search for metabolites, one should search for the most intense ion with singly charged ions, or doubly charged ions at 0.5 mass-to-charge ratio (m/z), adducts, multiply charged ions, and/or dimers in the full-scan spectrum. According to the nitrogen rule, if a compound contains either no nitrogen atoms or an even number, its molecular ion will be at an even mass number, while an odd number of nitrogen atoms will produce an odd mass number for the molecular ion. This rule, applied to the molecular ion, determines whether the unknown agent has an even or odd number of nitrogen atoms. A check of the isotopic peak of the molecular ions serves to confirm patterns, being aware of interferences from other ions. Lastly, by applying the RDBE rule, one can determine or confirm the total number of rings plus double bonds for a compound containing carbon, hydrogen, nitrogen, oxygen, and elements with the same valences as these. The value of DBE can be calculated according to the following equation:

$$\text{RDBE} = C - \tfrac{1}{2}[H + F + Cl + Br + I] + \tfrac{1}{2}[N + P] + 1$$

Next, one would perform accurate mass measurement to confirm/determine possible molecular formulae, and then compare the product ion MS/MS or MS^n spectra of the parent with the metabolites. After having identified molecular ions for possible metabolites, MS/MS or MS^n should be performed, and comparisons made between the product ion MS/MS or MS^n spectra of the parent and metabolites. Given the structure of the parent drug, with its corresponding fragmentation, elucidation of metabolite structure is greatly facilitated. The specific fragment ion that creates a shift in the m/z value leads to identification of the site of the modification on the molecule.

Once drug candidates reach the late drug discovery/candidate selection phase, LC-MS/MS, QTOF, Orbitrap (high-resolution and exact mass measurement), ion trap (MS^n), and HD exchange come into prominence in order to determine metabolic differences between species and identify potential pharmacologically active or toxic metabolites (Table 9.2). Exact mass analysis serves to confirm the elemental composition of metabolites. The exact mass (accurate mass) shift between a drug candidate and its metabolites can be used to predict the elemental composition of those metabolites. Knowledge of the molecular formulae of unknown metabolites is one of the tools that can be used to identify the metabolites and then propose their structure using additional MS data and tools.

9.5.1 Online HD-LC-MS and Derivatization

HD exchange occurs in solution in the presence of exchangeable (labile) hydrogen atoms in a molecule. When this method is combined with LC-MS, it has the advantage of providing an easy estimation of the number of labile hydrogen atoms in such groups as -OH, -SH, -NH, $-NH_2$, and -COOH (Nassar, 2003). This number is useful in comparing metabolite structure with that of the parent drug to determine the presence or absence of these groups. HD exchange experiments are also valuable for structural elucidation and interpretation of MS/MS fragmentation processes (Nassar, 2003). A method for metabolite identification in drug discovery and development utilizing online HD exchange and a tandem QTOF mass spectrometer coupled with LC (LC-QTOF-MS) recently has been developed (Nassar, 2003). This method apparently works very well for the identification of metabolites produced by dehydrogenation, oxidation, glucuronidation, and dealkylation. It provides discrimination between *N*- or *S*-oxide formation and monohydroxylation; also, it becomes easy to identify conjugations such as quaternary amine glucuronide versus primary or secondary glucuronides using this technique. The generic method is simple, easy, fast, sensitive, robust, and reliable, as well as achieving enhanced throughput; these factors facilitate rapid characterization of metabolites *in vitro* or *in vivo*. This time saving, combined with the particular benefits of QTOF or Orbitrap, such as higher resolution, makes this technique a valuable tool for structure elucidation. This method also works well with low-resolution instruments. Derivatization can be useful to stabilize unstable metabolites, improve chromatographic properties for highly polar compounds, and reduce volatility for volatile metabolites. Furthermore, it can be useful for characterizing chirality and site of metabolism. Many methods of derivatization have been reported that can be used to identify most functional groups. These techniques have the limitation of time-consuming sample preparation. Derivatization of metabolites in a sample demands either isolation of a specific metabolite before derivatization, or derivatization of the entire sample, which

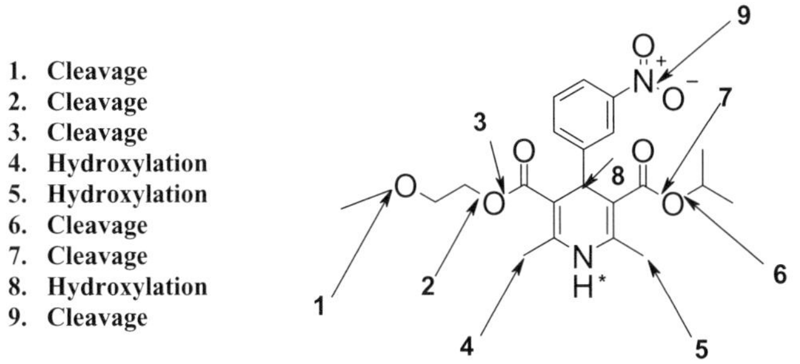

1. Cleavage
2. Cleavage
3. Cleavage
4. Hydroxylation
5. Hydroxylation
6. Cleavage
7. Cleavage
8. Hydroxylation
9. Cleavage

Figure 9.1 The chemical structure of nimodipine with the labile H exchange as indicated by the asterisk. The arrows indicate the positions of the possible metabolites as predicted using Pallas software.

TABLE 9.3 Exchange of labile hydrogens in nimodipine and metabolites formed *in vitro*.

| Compound | Labile hydrogen | Mass $[M_H+H]^+$, m/z | Mass $[M_D+D]^+$, m/z | |
			Predicted	Measured
Nimodipine	1	419	421	421
M-1	1	359	361	361
M-2	2	435	438	438
M-3	1	403	405	405
M-4	2	405	408	408
M-5	0	417	418	418

M_H = the molecular weight in H_2O; M_D = the molecular weight in D_2O.

will create more complex ions in MS. The speed and reduction in preparation time and effort involved with HD-LC-MS, and the fact that LC-MS-NMR has become more widely available and sensitive, have pushed derivatization methods out of favor.

As an example of how this process works, five metabolites of nimodipine formed in human liver microsomes; all were identified and characterized by LC-QTOF (Nassar, 2003). Figure 9.1 shows the chemical structure of nimodipine with the labile H exchange as indicated by asterisks; the arrows indicate the positions of the possible metabolites as predicted using Pallas software. Table 9.3 shows the exchange of labile hydrogens in nimodipine and its metabolites formed *in vitro*. The labile hydrogens ranged from none to two, which provides a significant means of identification, particularly with dehydrogenation metabolites, suggesting that the dehydrogenation took place on the pyridine moiety, resulting in a loss of one labile hydrogen. For example, in Fig. 9.2a, the full-scan mass spectrum of M-3 revealed a protonated molecular ion $[M_H+H]^+$ at m/z 403, 16 amu lower than nimodipine, suggesting that this metabolite was a cleaved product. The product ion spectrum of $[M_H+H]^+$ at m/z 403 showed the fragment ions of m/z 361, 343, 317, and 301. When H_2O was replaced with D_2O in the mobile phase, the full-scan mass spectrum of M-3 revealed a molecular ion $[M_D+D]^+$ at m/z 405 (Fig. 9.2b), 2 mass units higher than $[M_H+H]^+$ M-3, indicating the presence of one exchangeable hydrogen atom

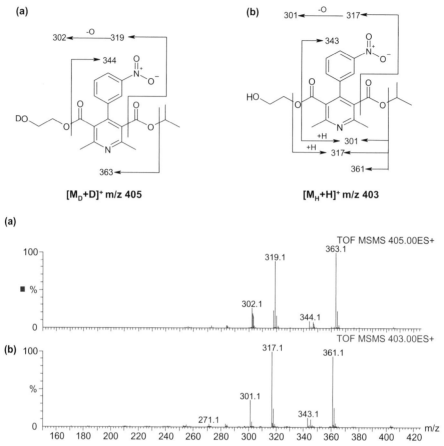

Figure 9.2 QTOF MS/MS spectra of metabolite M-3 in **(a)** D$_2$O and **(b)** H$_2$O. Reprinted from Nassar, 2003, with permission from Preston Publications.

in M-3. The product ion spectrum of m/z 405 showed the fragment ions of m/z 363, 344, 319, and 302. Nimodipine has one exchangeable hydrogen atom, with the cleavage producing another exchangeable hydrogen atom. This eliminates the possibility that M-3 is due to direct cleavage of nimodipine. It is possible that M-3 can be formed from the dehydrogenation and cleavage of nimodipine. The proposed TOF-MS/MS fragmentation for M-3 in D$_2$O and H$_2$O is shown in Fig. 9.2.

The proposed metabolic pathways of nimodipine in hepatic microsomal incubations in H$_2$O and D$_2$O are shown in Figs 9.3 and 9.4. Nimodipine is metabolized by means of dehydrogenation, demethylation of the methoxy group, cleavage of the ester groups by hydrolysis or oxidation, and hydroxylation of methyl groups. HD exchange provided significant information for all five metabolite identifications. One advantage of the HD exchange method is that, with LC-MS/MS, it offers an easy estimation of the number of labile hydrogen atoms in such groups as -SH, -OH, -NH, -NH$_2$, and -COOH. This number is useful in comparing metabolite structure with that of the parent drug to determine the presence or absence of these groups. HD exchange experiments have

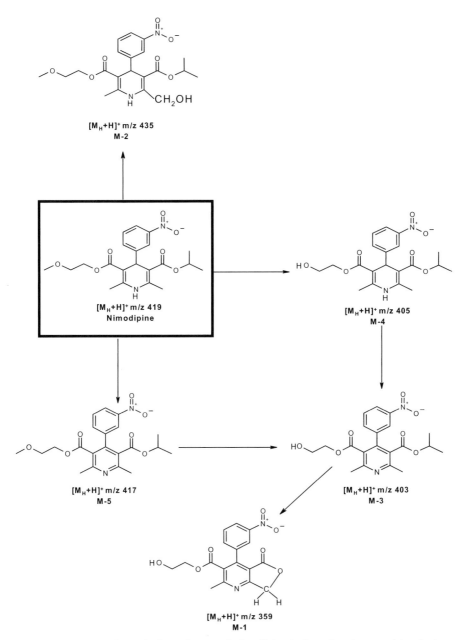

Figure 9.3 Proposed metabolic pathways of nimodipine in hepatic microsomal incubations in H$_2$O. Reprinted from Nassar, 2003, with permission from Preston Publications.

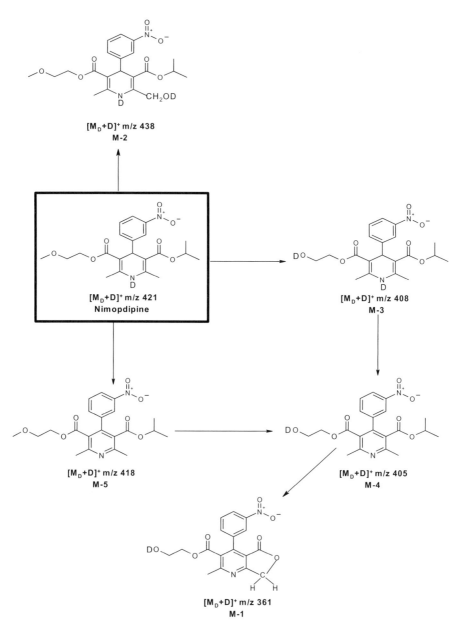

Figure 9.4 Proposed metabolic pathways of nimodipine in hepatic microsomal incubations in D$_2$O. Reprinted from Nassar, 2003, with permission from Preston Publications.

facilitated structural elucidation and interpretation of fragmentation processes as well. These results show that this method should be particularly desirable for identification of metabolites produced by dehydrogenation, oxidation, and dealkylation.

9.5.2   LC-MS-NMR

For determination of the structure of metabolites, NMR spectroscopy has become one of the most commonly used techniques. The sensitivity of NMR has increased significantly with the introduction of modern high-field-strength (500, 600, and 800 MHz) NMR spectrometers. For the purposes of drug metabolism, the most commonly accessible nuclei are ^1H, ^{19}F, ^{13}C, ^{31}P, and ^{15}N. When using NMR, sample concentrations must be between 1 μg and 10 mg, with analysis times between 5 minutes and 1 hour. Several operational modes are available for LC-MS-NMR such as isocratic or gradient elution, continuous flow, stop flow, time-sliced stop flow, peak collection into capillary loops for post-chromatographic analysis, and automatic detection of chromatographic peaks with triggered NMR acquisition. Given considerations for sample concentration and chromatographic resolution, any of these techniques is available.

The online HD-LC-MS method has the ability to rapidly provide characterization of metabolites, such as those formed by dealkylation, those from *in vitro* or *in vivo* samples, and can distinguish between *N*- or *S*-oxide versus monohydroxyl and quaternary amine versus primary or secondary glucuronide. Despite these advantages, NMR is still necessary for identifying the regiochemistry of aromatic oxidation, determining the site of aliphatic oxidation where fragmentation pathways are unavailable or inconclusive, and locating groups such as OH, epoxide, and sulfate by comparing NMR spectra of the parent with the metabolite. When seeking to identify unknown metabolites, NMR and MS are clearly complementary methods, and confirmation of a definite metabolite structure often requires data from both methods. There are strong benefits from the high quality and dual nature of information gained from a single run when using LC-MS-NMR; as the instrumentation and techniques involved undergo continued evolution, efficiency and sensitivity in terms of the chromatographic properties are bringing LC-MS-NMR well on the way to becoming a mainstream approach.

9.5.3   Stable Isotope Labeling

To improve the reliability of metabolite identification by full-scan MS, stable isotopic labeling (^2H, ^{13}C, ^{15}N, ^{18}O, ^{34}S) has been used. Also, the use of stable isotope-labeled drugs allows safe experiments in humans, which provides great advantages over the use of radiolabels. In this type of study, a mixture of known amounts of labeled and non-labeled drug is used for the metabolic experiments and analyzed by LC-MS. The identification of a metabolite should have the following criteria: (i) two peaks with identical shapes and retention times must be recorded in the ion chromatograms; (ii) the mass difference of the two peaks must be the same as the mass difference between the labeled and the unlabeled parent drug; and (iii) the relative abundance ratio of the peaks must be the same as the concentration ratio of the labeled and the unlabeled parent drug. Labeling may not be necessary in those cases where the compound includes one or more Cl or Br atoms, which will show abundant "mass + 2" isotopes with known abundance ratios in the mass spectra. Also, the metabolite standards must be synthesized, and in some cases chemical synthesis of metabolites may be difficult. Alternatively, enzymatic synthesis of metabolites has been found to be an easy and rapid way to produce a few milligrams of desired metabolites for structure confirmation.

9.6 "ALL-IN-ONE" RADIOACTIVITY DETECTOR, STOP FLOW, AND DYNAMIC FLOW FOR METABOLITE IDENTIFICATION

Many of the studies of drug metabolism such as absorption, bioavailability, distribution, biotransformation, excretion, metabolite identification, and other pharmacokinetic research use radiolabeled drugs, with the radioactive isotopes ^{14}C or tritium (^{3}H) most commonly used for the labeling of a given drug. Radioactive labeling is a good fit with HPLC separation, as it allows high-resolution, quantitative detection of unknown metabolites and real-time monitoring by connecting the HPLC-radioactivity detector outlet to MS. The use of these detector interfaces aids the generation of data for structural elucidation of metabolites and biotransformation pathways for an administered drug. The detection sensitivity of both radioactivity and MS is the key to successfully identify and characterize metabolites (Nassar, Bjorge and Lee, 2003; Nassar *et al.*, 2004).

Regulatory policy dictates that exposure to administered radioactivity be held as low as possible in most studies, which demands much greater sensitivity of the radioactivity detector to be able to detect metabolites. There has been great progress in the emerging science of detecting trace amounts of radiolabeled or non-radiolabeled drugs and their metabolites. The availability of these technologies should have a dramatic impact on drug discovery and development for metabolite profiling studies. It has been reported that a microplate scintillation counter combined with capillary LC can be used to enhance sensitivity by eluent fractionation and subsequent offline counting. The limitations with this method are that the sample must be completely dry before counting, that any volatile compounds are likely to be lost, and that there is the potential for apolar compounds to adsorb on the surface of the plate. In addition, this technique has the limitations of time-consuming sample preparation, high analysis costs, and the inability to elucidate metabolite structure.

A new development for metabolite identification in drug discovery and development is a detection method which couples online stop flow and dynamic flow with a radioactivity detector and mass spectrometer. Figure 9.5 shows the hardware schematic diagram of the v.ARC system. This system is simple and requires no custom-made

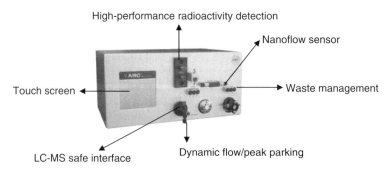

Figure 9.5 Hardware schematic diagram of the LC-ARC system. Courtesy of Aim Research Company, Hochessin, DE.

hardware. The v.ARC system uses ARC's specially designed cells, and is operated under the ARC Data System which controls the entire radio-LC system, including LC and the radioactivity detector. When interfaced with LC-MS, the v.ARC system enhances detection of low-level radioactive peaks while increasing the flexibility and productivity of MS testing. This system has the major benefit of enhancing the sensitivity of radioisotope measurement for metabolite identification in drug metabolism studies. Another advantage to this system is the easy interface with the mass spectrometer, which allows acquisition of mass spectrometric data online. This system dramatically improves the sensitivity for ^{14}C peaks by up to 10-fold over conventional flow-through detection methods, and eliminates the need for a fraction collector and time-consuming sample preparation. In addition, the system produces accurate column recovery, quantification of low-level radioactivity, and consistently high resolution throughout the run. These factors give the combination of radioactivity detector, stop flow, and dynamic flow and mass spectrometry great potential as a powerful tool for improving the sensitivity of radioisotope measurement in metabolite identification studies, which also highlights the impressive progress that has been made in the technology of radioisotope counting in drug metabolism.

When online dynamic flow coupled with a radioactivity detector and mass spectrometer is used, the total run time remains similar to conventional radio-LC. A technique has been developed to collect the peaks online and then infuse them to the MS, which allows acquisition of a higher order of multistage fragmentation for both major and minor metabolites. The multistage MS fragmentation pattern obtained for the metabolites enabled determination of the sites of metabolism.

9.6.1 Online Peak Collection and Multistage Mass Experiments

It is important to note that this system can also be used to collect the peaks of interest online (peak parking), then infuse them for extended periods of time at flow rates as low as $1\,\mu L/min$ while maintaining the column pressure. The peaks of interest can be triggered by radioactivity signals, UV signals, or specified retention time, allowing analysts to use direct infusion with a low flow rate, with only one run and a small sample size. The peak parking feature allows sustained analytical signal input (infusion fashion), which allows any or all of the following experiments to be done as desired: optimizing mass spectrometric condition; tuning for any individual metabolite, tuning for both positive and negative polarities, and optimizing collision energy. Also, it allows the operator to conduct multistage mass experiments, automatically acquire ion mass data, perform MS^n, compare isotopic mass spectra, and perform neutral loss tests. This method provides the capability of identifying the structures of unknown metabolites or impurities, again requiring a limited sample amount and a single run. This is a significant improvement over an offline fraction collector, which may lose volatile compounds during the fraction collection process. Because this method retains these compounds for analysis, it greatly expands the ability to characterize and identify metabolites of a given compound, which in turn is of significant benefit to analysts.

For purposes of evaluating this system, *in vitro* human liver microsomal (HLM) incubations were performed with only [^{14}C]-dextromethorphan (Fig. 9.6), a semisynthetic narcotic cough-suppressing ingredient in a variety of over-the-counter cold and cough medications. The online separation and identification of dextromethorphan metabolites did not require intensive sample preparation, concentration or fraction collection. Following incubation of [^{14}C]-dextromethorphan with human liver microsomes

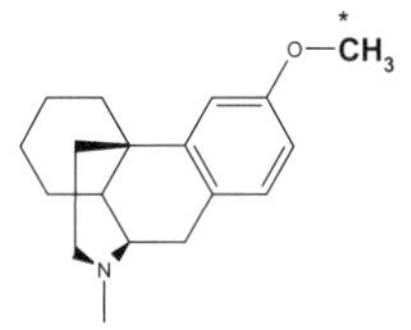

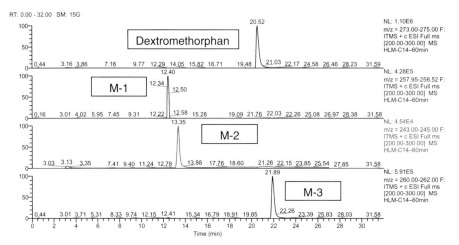

Figure 9.6   Chemical structure of [^{14}C]dextromethorphan, the test compound used for this study (* = ^{14}C).

Figure 9.7   HPLC-MS chromatograms of [^{14}C]dextromethorphan following incubation with pooled human liver microsomes for 60 minutes in the presence of NADPH showing M-1, M-2, M-3, and non-metabolized dextromethorphan. Reprinted from Nassar and Lee, 2007, with permission from Preston Publications.

for 60 minutes in the presence of nicotinamide adenine dinucleotide phosphate (NADPH), three metabolites were detected—M-1, M-2, and M-3—along with the parent drug. The retention times of dextromethorphan and these metabolites were between 12 and 23 minutes with excellent separation efficiency. Metabolites M-1, M-2, and M-3 have retention times of 12.4, 13.3, and 22.9 minutes, respectively. Figure 9.7 shows the LC-MS chromatograms of dextromethorphan following incubation with HLM for 60 minutes in the presence of NADPH. The LC-radio chromatogram shows that M-1 and M-2 did not have radioactive peaks, while M-3 and the parent dextromethorphan did. The full-scan mass spectra for M-1, M-2, and M-3 revealed protonated molecular ions [M+H]$^+$ at m/z 258, 244, and 260, suggesting that M-1 lost one methyl that contained C14, M-2 lost two methyl groups (one of which contained C14), and M-3 lost one methyl group.

The structures of dextromethorphan and metabolites were elucidated by LTQ (linear trap quadrupole)-MS/MS analysis. These peaks were collected online and then infused at flow rate 10 μL/min; MS^2, MS^3, and MS^4 were performed with sufficient time to examine the MS data and decide on the next fragment ions to be used for metabolite characterizations. Separate runs were performed for MS^2, MS^3, and MS^4, each without collecting the peaks; the results were similar. Figure 9.8 represents LC-MS spectra of

(a)

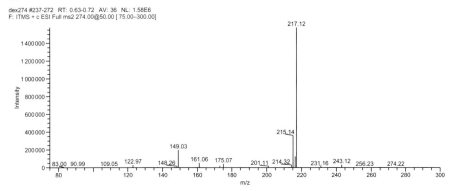

(b)

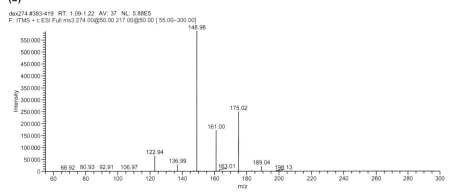

(c)

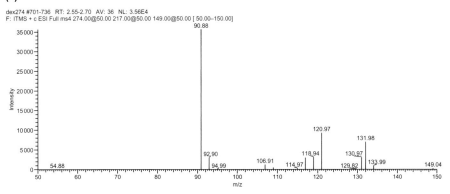

Figure 9.8 Representative LC-MS spectra of $[M+H]^+$ m/z 274, M-1: **(a)** MS^2, **(b)** MS^3, and **(c)** MS^4 following incubation of $[^{14}C]$dextromethorphan with human liver microsomes. Reprinted from Nassar and Lee, 2007, with permission from Preston Publications.

non-metabolized [^{14}C]dextromethorphan [M+H]$^+$ at m/z 274: (a) MS^2, (b) MS^3, and (c) MS^4. The fragment ions are m/z 217, 215, 201, 175, 149, 123, 121, and 91. Comparison of the fragmentation ions for M-1 with the parent drug fragmentation suggests that *N*-dealkylation took place on the dextromethorphan molecule. Figure 9.9 shows proposed metabolic pathways of dextromethorphan in hepatic microsomal incubations. Mass spectrometric analysis showed the presence of dextromethorphan metabolites formed by *N*- and *O*-dealkylation, correlating with previously published results.

One of the major benefits of this method is that it is up to 10 times more sensitive in detecting ^{14}C peaks than commercially available flow-through radioactivity detectors. This method enhances the resolution of radiochromatograms and is able to measure volatile metabolites. Another advantage to this system is the easy interface with the mass spectrometer, which allows acquisition of mass spectrometric data online. The method gives accurate column recovery and quantification of low-level radioactivity and high resolution throughout the run. An important safety benefit is that by using this method, injection size has been reduced, thereby decreasing potential exposure to

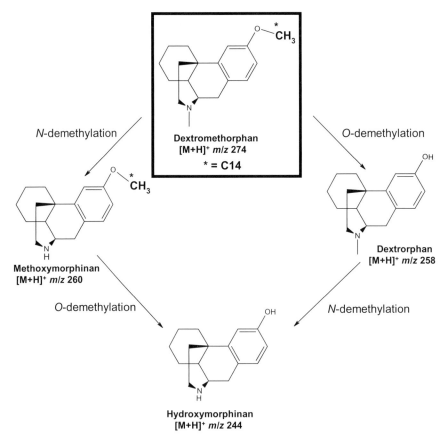

Figure 9.9 Proposed metabolic pathways of dextromethorphan in human hepatic microsomal incubations based on LC-ARC-MS/MS data. Reprinted from Nassar and Lee, 2007, with permission from Preston Publications.

radioactivity and reducing the amount of radioactive wastes. This smaller injection size, as well as the capability to perform many tests on a single run, also reduces both time and expenses significantly. Furthermore, it is easier because it reduces manual operations. This study showed that impressive progress that has been made in the technology of radioisotope counting and metabolite characterization in drug metabolism using LC-ARC. The overall results suggest that the combination of LC-ARC dynamic flow with radioactivity detection and mass spectrometry has great potential as a powerful tool for radioisotope measurement in metabolite identification studies during drug discovery and development.

9.6.2   Metabolic Activation Studies by Mass Spectrometry

Adverse drug reactions can be classified into type A (predictable) and type B (unpredictable or idiosyncratic) reactions. Type A reactions are dose-dependent and predictable based on the pharmacology of the drug; it is possible to reverse the reactions by reducing the dosage or, if necessary, discontinuing the drug altogether. Type B reactions cannot be predicted on the basis of the pharmacology of the drug. Type B reactions include anaphylaxis, blood dyscrasias, hepatoxicity, and skin reactions, and may result from a combination of genetic predisposition and environmental factors. Such unacceptable consequences can result from the occurrence of idiosyncratic drug reactions during late-stage clinical trials or even after a drug has been released. It is this uncertainty that has driven researchers to develop improved means for predicting the potential for a drug to cause such reactions. Currently, it appears that idiosyncratic drug reactions may be due to reactive metabolites. A typical cause of toxicity is the formation of electrophilic reactive metabolites which bind to nucleophilic groups present in vital cellular proteins and/or nucleic acids. Significant evidence from previously published reports indicates that chemically reactive metabolites may be responsible for serious forms of toxicity, including cellular necrosis, mutagenesis, carcinogenesis, teratogenesis, hypersensitivity, and blood disorders.

9.7   STRATEGIES TO SCREEN FOR REACTIVE METABOLITES

LC-MS may be used to screen for reactive electrophiles generated from *in vitro* experiments in several tissues (e.g. liver microsomes) following reaction with glutathione (GSH) (Nassar and Lopez-Anaya, 2004; Nassar, Kamel and Clarimont, 2004a). There are several types of reactive metabolites (epoxides, arene oxides, quinones, quinone imines, quinone methides, iminoquinone methides, nitroso derivatives, nitrenium ions, nitro reduction products, nitro radicals, iminium ions, Michael acceptors, *S*-oxides, aliphatic aldehydes, or hydrolysis/acetylation products) that can be trapped in the presence of either GSH or an equimolar mixture of *N*-acetyl-cysteine and *N*-acetyl-lysine. This method is relatively simple and rapid, and has been implemented for high-throughput screening to identify reactive metabolites generated by bioactivation with phase I and phase II enzymes. Since thousands of compounds must be screened in a typical pharmaceutical industry setting, it is almost impossible, and highly expensive, to have all of them radiolabeled. Using radiolabeled reagents such as GSH and potassium cyanide is easier and more cost-effective for obtaining an accurate measurement of adduct formation. The reactive intermediates may be formed in other subcellular fractions such as liver S9, mitochondria, and cytosol (e.g. valproic acid forms reactive metabolites in subcellular mitochondria), and thus, screening for reactive intermediates in these fractions will minimize false negatives. As an example of such a screening procedure, the author compared *in vitro* cellular and subcellular models for identifying drug candidates with the potential to produce reactive metabolites. For purposes of evaluating these models, six known compounds currently in the clinic but with dose limitations due to hepatotoxicity were selected. Labetalol, acetaminophen, niacin, iproniazid, 8-hydroxyquinoline, and isoniazid were incubated separately with human liver mitochondria, S9, microsomes, cytosol, and hepatocytes. LC-QTOF-MS methods were used to elucidate metabolic profiles of selected compounds. The phase II metabolites were in good agreement with previously published results, although without supplementation of reagents in subcellular fractions, phase II metabolites were not detected. Hepatocytes also formed GSH conjugates but the relative abundance was lower, indicating that the best model to identify potential reactive metabolites may involve the subcellular fractions. The results from incubation of acetaminophen and labetalol, respectively, in subcellular fractions are described elsewhere. The results demonstrated that for acetaminophen, when reactive intermediates are produced, they can be trapped by GSH; while in the case of labetalol, no reactive intermediates were produced and subsequently no GSH adducts were detected.

9.8   SUMMARY

There is no doubt as to the value of LC-MS in drug metabolism, as it has proved to be a sensitive and specific technique, which gives it important roles in ADMET studies. This chapter has demonstrated that LC-MS can provide a rapid, accurate, and relatively easy-to-use approach for metabolism studies without compromising the quality of the assays. Although LC-MS has improved the efficiency of metabolite analysis, it is still a daunting challenge to identify all of the metabolites generated by a particular drug candidate. Despite the high resolution MS provides, neutral-loss and precursor ion scans may miss some metabolites with unexpected fragmentation. Also, the current MS techniques may not provide the sensitivity required to detect trace quantities of metabolites in complex biological matrices. It is the opinion of the author that future implementation plans will apply in the following areas:

1. Modern MS instrumentations can greatly assist the analyst in providing high-quality data in a more rapid and time-efficient manner. Techniques employing QTOF, Orbitrap, micro-, and capillary separation techniques with nanospray MS are emerging and gaining wider use. As the sensitivity and resolution of MS continue to improve, the technique will further enhance metabolite characterization. For example, another recent innovation, the hybrid two-dimensional quadrupole ion trap–Fourier transform ion cyclotron resonance mass spectrometer, offers good potential for metabolite identification with high sensitivity and resolution.

2. Good progress has been made and will continue in the automation of sample preparation and data handling.

3. Advances in column technology and LC-MS have made chiral separation routine; more assays now have direct coupling of various chiral columns with an atmospheric pressure ion source without compromising the detector sensitivity or the LC resolving power.

4. It is not always possible to determine the site of metabolic reaction in a drug molecule with MS. LC-MS-NMR provides unambiguous structural characterization of metabolites. Although the sensitivity of NMR is currently not sufficient for the analysis of metabolites in trace quantities, continuous improvements are being made.

5. Online HD-LC-MS has become more widely used, because it is much faster and requires less preparation than NMR for metabolite confirmation, and further progress is anticipated.

6. An important consideration is that any screening process for idiosyncratic drug reactions will produce false results, both positive and negative, and a goal must be to reduce or eliminate these, which will in turn lead to improved drug safety. More studies are required to improve our knowledge of, and hence the techniques used for, screening with the use of radiolabeled or stable isotope reagents, which would give an accurate measurement of the amount of reactive intermediate. Reactive intermediate formation should be investigated in subcellular fractions, which will aid in reducing false results.

REFERENCES

Bakke, O.M., Manocchia, M., de Abajo, F., Kaitin, K.I., Lasagna, L. (1995). Drug safety discontinuations in the United Kingdom, the United States and Spain from 1974 through 1993: a regulatory perspective. *Clin. Pharmacol. Ther.*, *58* (1), 108–117.

Balimane, P.V., Pace, E., Chong, S., Zhu, M., Jemal, M., Pelt, CK (2005). A novel high-throughput automated chip-based nanoelectrospray tandem mass spectrometric method for PAMPA sample analysis. *J. Pharm. Biomed. Anal.*, *39* (1–2), 8–16.

Castro-Perez, J., Plumb, R., Granger, J.H., Beattie, I., Joncour, K., Wright, A. (2005a). Increasing throughput and information content for *in vitro* drug metabolism experiments using ultra-performance liquid chromatography coupled to a quadrupole time-of-flight mass spectrometer. *Rapid Commun. Mass Spectrom.*, *19* (6), 843–848.

Castro-Perez, J., Plumb, R., Liang, L., Yang, E. (2005b). A high-throughput liquid chromatography/tandem mass spectrometry method for screening glutathione conjugates using exact mass neutral loss acquisition. *Rapid Commun. Mass Spectrom.*, *19* (6), 798–804.

Greene, N. (2002). Computer systems for the prediction of toxicity: an update. *Adv. Drug Deliv. Rev.*, *54* (3), 417–431.

Guevremont, R. (2004). High-field asymmetric waveform ion mobility spectrometry: a new tool for mass spectrometry. *J. Chromatogr. A*, *1058* (1–2), 3–19.

Jemal, M. (2000). High-throughput quantitative bioanalysis by LC/MS/MS. *Biomed. Chromatogr.*, *14* (6), 422–429.

Jemal, M., Bergum, J. (1992). Effect of the amount of internal standard on the precision of an analytical method. *J. Clin. Pharmacol.*, *32* (7), 676–677.

Jemal, M., Ouyang, Z. (2003). Enhanced resolution triple-quadrupole mass spectrometry for fast quantitative bioanalysis using liquid chromatography/tandem mass spectrometry: investigations of parameters that affect ruggedness. *Rapid Commun. Mass Spectrom.*, *17* (1), 24–38.

Jemal, M., Ouyang, Z., Zhao, W., Zhu, M., Wu, W.W. (2003). A strategy for metabolite identification using triple-quadrupole mass spectrometry with enhanced resolution and accurate mass capability. *Rapid Commun. Mass Spectrom.*, *17* (24), 2732–2740.

Johnson, K.A., Plumb, R. (2005). Investigating the human metabolism of acetaminophen using UPLC and exact mass oa-TOF MS. *J. Pharm. Biomed. Anal.*, *39* (3–4), 805–810.

Kantharaj, E., Tuytelaars, A., Proost, P.E., Ongel, Z., Van Assouw, H.P., Gilissen, R.A. (2003). Simultaneous measurement of drug metabolic stability and identification of metabolites using ion-trap mass spectrometry. *Rapid Commun. Mass Spectrom.*, *17* (23), 2661–2668.

Kapron, J.T., Jemal, M., Duncan, G., Kolakowski, B., Purves, R. (2005). Removal of metabolite interference during liquid chromatography/tandem mass spectrometry using high-field asymmetric waveform ion mobility spectrometry. *Rapid Commun. Mass Spectrom.*, *19* (14), 1979–1983.

Kassel, D.B. (2004). Applications of high-throughput ADME in drug discovery. *Curr. Opin. Chem. Biol.*, *8* (3), 339–345.

Kostiainen, R., Kotiaho, T., Kuuranne, T., Auriola, S. (2003). Liquid chromatography/atmospheric pressure ionization-mass spectrometry in drug metabolism studies. *J. Mass Spectrom.*, *38* (4), 357–372.

Kubinyi, H. (1977). Quantitative structure-activity relationships. 7. The bilinear model, a new model for nonlinear dependence of biological activity on hydrophobic character. *J. Med. Chem.*, *20* (5), 625–629.

Lasser, K.E., Allen, P.D., Woolhandler, S.J., Himmelstein, D.U., Wolfe, S.M., Bor, D.H. (2002). Timing of new black box warnings and withdrawals for prescription medications. *J. Am. Med. Assoc.*, *287* (17), 2215–2220.

Leclercq, L., Delatour, C., Hoes, I., Brunelle, F., Labrique, X., Castro-Perez, J. (2005). Use of a five-channel multiplexed electrospray quadrupole time-of-flight hybrid mass spectrometer for metabolite identification. *Rapid Commun. Mass Spectrom.*, *19* (12), 1611–1618.

Lesko, L.J., Rowland, M., Peck, C.C., Blaschke, T.F. (2000). Optimizing the science of drug development: opportunities for better candidate selection and accelerated evaluation in humans. *Pharm. Res.*, *17* (11), 1335–1344.

Lin, J., Sahakian, D.C., de Morais, S.M., Xu, J.J., Polzer, R.J., Winter, S.M. (2003). The role of absorption, distribution, metabolism, excretion and toxicity in drug discovery. *Curr. Top. Med. Chem.*, *3* (10), 1125–1154.

Meyboom, R.H., Lindquist, M., Egberts, A.C. (2000). An ABC of drug-related problems. *Drug Saf.*, *22* (6), 415–423.

Mortishire, R.J., O'Connor, D., Castro-Perez, J.M., Kirby, J. (2005). Accelerated throughput metabolic route screening in early drug discovery using high-resolution liquid chromatography/quadrupole time-of-flight mass spectrometry and automated data analysis. *Rapid Commun. Mass Spectrom.*, *19*, 2659–2670.

Nassar, A.E. (2003). Online hydrogen-deuterium exchange and a tandem quadrupole time-of-flight mass spectrometer coupled with liquid chromatography for metabolite identification in drug metabolism. *J. Chromatogr. Sci.*, *41* (8), 398–404.

Nassar, A.E., Adam, P. (2003). Metabolite characterization in drug discovery utilizing robotic liquid-handling, quadrupole time-of-flight mass spectrometry and *in silico* prediction. *Curr. Drug Metab.*, *4* (4), 259–271.

Nassar, A.E., Lee, D.Y. (2007). Novel approach to performing metabolite identification in drug metabolism. *J. Chromatogr. Sci.*, *45* (3), 113–119.

Nassar, A.E., Lopez-Anaya, A. (2004). Strategies for dealing with reactive intermediates in drug discovery and development. *Curr. Opin. Drug Discov. Devel.*, *7* (1), 126–136.

Nassar, A.E., Bjorge, S.M., Lee, D.Y. (2003). On-line liquid chromatography-accurate radioisotope counting coupled with a radioactivity detector and mass spectrometer for metabolite identification in drug discovery and development. *Anal. Chem.*, *75* (4), 785–790.

Nassar, A.E., Kamel, A.M., Clarimont, C. (2004a). Improving the decision-making process in structural modification of drug candidates: reducing toxicity. *Drug Discov. Today*, *9* (24), 1055–1064.

Nassar, A.E., Kamel, A.M., Clarimont, C. (2004b). Improving the decision-making process in the structural modification of drug candidates: enhancing metabolic stability. *Drug Discov. Today*, *9* (23), 1020–1028.

Nassar, A.E., Talaat, R. (2004). Strategies for dealing with metabolite elucidation in drug discovery and development. *Drug Discov. Today*, *9* (7), 317–327.

Nassar, A.E., Parmentier, Y., Martinet, M., Lee, D.Y. (2004). Liquid chromatography-accurate radioisotope counting and microplate scintillation counter technologies in drug metabolism studies. *J. Chromatogr. Sci.*, *42* (7), 348–353.

Nassar, A.E., Varshney, N., Getek, T., Cheng, L. (2001). Quantitative analysis of hydrocortisone in human urine using a high-performance liquid chromatographic-tandem mass spectrometric-atmospheric-pressure chemical ionization method. *J. Chromatogr. Sci.*, *39* (2), 59–64.

Nassar, A.E., Talaat, R.E., Kamel, A.M. (2006). The impact of recent innovations in the use of liquid chromatography-mass spectrometry in support of drug metabolism studies: are we all the way there yet? *Curr. Opin. Drug Discov. Devel.*, *9* (1), 61–74.

Obach, R.S. (1997). Nonspecific binding to microsomes: impact on scale-up of *in vitro* intrinsic clearance to hepatic clearance as assessed through examination of warfarin, imipramine, and propranolol. *Drug Metab. Dispos.*, *25* (12), 1359–1369.

Obach, R.S. (1999). Prediction of human clearance of twenty-nine drugs from hepatic microsomal intrinsic clearance data: an examination of *in vitro* half-life approach and nonspecific binding to microsomes. *Drug Metab. Dispos.*, *27* (11), 1350–1359.

Plumb, R.S., Stumpf, C.L., Granger, J.H., Castro-Perez, J., Haselden, J.N., Dear, G.J. (2003). Use of liquid chromatography/time-of-flight mass spectrometry and multivariate statistical analysis shows promise for the detection of drug metabolites in biological fluids. *Rapid Commun. Mass Spectrom.*, *17* (23), 2632–2638.

Roberts, S.A. (2003). Drug metabolism and pharmacokinetics in drug discovery. *Curr. Opin. Drug Discov. Devel.*, *6* (1), 66–80.

Taylor, E.W., Jia, W., Bush, M., Dollinger, G.D. (2002). Accelerating the drug optimization process: identification, structure elucidation, and quantification of *in vivo* metabolites using stable isotopes with LC/MS[n] and the chemiluminescent nitrogen detector. *Anal. Chem.*, *74* (13), 3232–3238.

Thompson, T.N. (2000). Early ADME in support of drug discovery: the role of metabolic stability studies. *Curr. Drug Metab.*, *1* (3), 215–241.

Tiller, P.R., Romanyshyn, L.A. (2002). Liquid chromatography/tandem mass spectrometric quantification with metabolite screening as a strategy to enhance the early drug discovery process. *Rapid Commun. Mass Spectrom.*, *16* (12), 1225–1231.

White, R.E. (2000). High-throughput screening in drug metabolism and pharmacokinetic support of drug discovery. *Ann. Rev. Pharmacol. Toxicol.*, *40*, 133–157.

Yang, L., Amad, M., Winnik, W.M., Schoen, A.E., Schweingruber, H., Mylchreest, I., Rudewicz, P.J. (2002). Investigation of an enhanced resolution triple quadrupole mass spectrometer for high-throughput liquid chromatography/tandem mass spectrometry assays. *Rapid Commun. Mass Spectrom.*, *16* (21), 2060–2066.

Zhang, H., Henion, J. (2001). Comparison between liquid chromatography-time of-flight mass spectrometry and selected reaction monitoring liquid chromatography-mass spectrometry for quantitative determination of idoxifene in human plasma. *J. Chromatogr. B Biomed. Sci. Appl.*, *757* (1), 151–159.

CHAPTER 10

Structural Modifications of Drug Candidates: How Useful Are They in Improving Metabolic Stability of New Drugs? Part I: Enhancing Metabolic Stability

ALA F. NASSAR

The rule of three, which attempts to define the activity–exposure–toxicity triangle, presents the single most difficult challenge in the design and advancement of drug candidates to the development stage. Absorption, distribution, metabolism, and excretion (ADME) studies are widely used in drug discovery to optimize this balance of properties necessary to convert lead compounds into drugs that are both safe and yet effective for human patients. Metabolite characterization has become one of the main drivers of the drug discovery process, helping to optimize ADME properties and increase the success rate for drugs. The study of structural modifications produces valuable information which, throughout the process of drug discovery, keeps on giving direction and focus in the effort to improve and balance the relationships between activity, exposure, and toxicity for drug candidates. In this chapter, the strategies for the decision-making process in structural modification of drug candidates are addressed to improve metabolic stability. Several examples are included to show how metabolic stability has influenced and guided drug design.

10.1 BACKGROUND

During the process of metabolism, the molecular structure of a drug is changed from one that is absorbed (lipophilic, or capable of crossing the membrane lipid core of membranes) to one that can be readily eliminated from the body (incapable of crossing

Drug Metabolism Handbook: Concepts and Applications, Edited by Ala F. Nassar, Paul F. Hollenberg, and JoAnn Scatina
Copyright © 2009 by John Wiley & Sons, Inc.

the lipid core of membranes, or hydrophilic). If lipophilic drugs are not metabolized, they will remain in the body for longer than intended, and their cumulative biological effects will eventually cause harm. Thus, the formation of water-soluble metabolites not only enhances drug elimination but also leads to compounds that are generally pharmacologically inactive and relatively nontoxic. Phase I reactions include oxidative, reductive, and hydrolytic biotransformations. These reactions introduce a polar functional group (e.g. OH, COOH, NH_2, and SH) into the drug molecule and usually result in only a small increase in hydrophilicity. This can occur through direct introduction of the functional group (e.g. aromatic and aliphatic hydroxylation) or by modifying existing functionalities (e.g. reduction of ketone and aldehydes to alcohols; oxidation of alcohols to acids; hydrolysis of ester and amides to yield COOH, NH_2, and OH groups; reduction of azo and nitro compounds to give NH_2 moieties; oxidative *N*-, *O*-, and *S*-dealkylation to give NH_2, OH, and SH groups). Phase II reactions form water-soluble conjugated products by reaction with polar and ionizable endogenous compounds such as glucuronic acid, sulfate, glycine, and other amino acids to the functional groups of phase I metabolites. Most phase II biotransformation reactions result in a large increase in drug hydrophilicity, thus they greatly promote the excretion of foreign chemicals. Conjugated metabolites are readily excreted in the urine or bile depending on the molecular weight and are generally pharmacologically inactive and nontoxic. Other phase II pathways, such as methylation and acetylation, serve to terminate biological activity, while glutathione conjugation serves to protect the body against chemically reactive compounds. The opposite may sometimes be true: drugs can form metabolically activated or reactive electrophilic species in a process known as bioactivation. Drugs that already have existing functional groups, such as OH, COOH, and NH, are often directly conjugated by reactions with phase II enzymes.

10.2 INTRODUCTION

There is no doubt as to the value of predicting the ADME/TOX of compounds as early as possible in the drug discovery process. Pharmaceutical companies are coming under increasing pressure to more precisely prove long-term safety, and to provide more data on the safety of their products. Failure to successfully address these concerns causes companies to withdraw drugs from the market, not only those that present human health consequences but also negative economic and public relations impact on the pharmaceutical industry. Because of human health safety concerns, a number of drugs have been withdrawn from the market. Failures in the drug discovery process account for a substantial part of the cost of drug development because many compounds are discarded after substantial investment owing to unforeseen toxicological effects.

The drug discovery process in the pharmaceutical industry has undergone a dramatic change in the last decade due to advances in molecular biology, high-throughput pharmacological screens, and combinatorial synthesis (Kubinyi *et al.*, 1977; Bakke *et al.*, 1995; Pirmohamed *et al.*, 1998; Richard *et al.*, 1998; Lesko *et al.*, 2000; Meyboom *et al.*, 2000; Greene *et al.*, 2002; Lasser *et al.*, 2002; Taylor *et al.*, 2002; Tiller *et al.*, 2002; Uetrecht *et al.*, 2002; Kantharaj *et al.*, 2003; Kostiainen *et al.*, 2003; Nassar *et al.*, 2003a, b, 2004a, b; Plumb *et al.*, 2003; Watt, 2003). The high cost, long development time, and high failure rate in bringing drugs to market have been important factors behind this change. The major reasons for this low success rate include poor pharmacological activity, low bioavailability, or high toxicity. Thus, the design of drugs with optimal potency, pharmacokinetic properties, and less toxicity is challenging given the opposing requirements for absorption and metabolism. During this process, high metabolic liability needs to be avoided which usually leads to poor bioavailability and high clearance as well as the formation of active or toxic metabolites, which also have an impact on the pharmacological and toxicological outcomes. One approach to this dilemma would be to thoroughly explore the structure requirements for *in vitro* pharmacological activity through combinatorial synthesis and identify the key structural features that are essential for activity for each chemical series, and then attempt to improve the metabolism, PK and toxicity properties through structure modifications to the regions of the molecule that have little or no impact on the desired activity. Thus, one can maintain an optimal level of potency while introducing structural features that will improve the metabolism and toxicity characteristics.

10.3   SIGNIFICANCE OF METABOLITE CHARACTERIZATION AND STRUCTURE MODIFICATION

Optimization of the PK properties of candidates is one of the important considerations during the drug discovery process. The rate and sites of metabolism of new chemical entities by drug-metabolizing enzymes are amenable to modulation by appropriate structural changes. Until recently, this optimization has been achieved largely by empirical methods and trial and error. Now, metabolite identification enables early identification of potential metabolic liabilities or issues, provides a metabolism perspective to guide synthesis efforts with the aim of either blocking or enhancing metabolism so as to optimize the pharmacokinetic and safety profiles of newly synthesized drug candidates, and assists the prediction of the metabolic pathways of potential drug candidates. This is crucial to the drug discovery process, as it can be used to investigate the phase I metabolites likely to be formed *in vivo*, differences between species in drug metabolism, the major circulating metabolites of an administered drug, phase I and phase II pathways of metabolism, pharmacologically active or toxic metabolites, and to help determine the effects of metabolizing enzyme inhibition and/or induction. The ability to produce this information early in the discovery phase has become increasingly important as a basis for judging whether a drug candidate merits further development.

Herein, the strategies for the decision-making process in structural modification of drug candidates are discussed to enhance the metabolic stability (Part I). Also, some of the chemical substructures that form reactive metabolites and are involved in toxicities in humans are discussed. Several examples from literature are discussed to show how metabolic stability studies have influenced drug design strategies, and helped to facilitate improvements.

10.4   PART I: ENHANCE METABOLIC STABILITY

10.4.1   *In Vitro* Studies

Various *in vitro* methods are available to evaluate metabolic stability. Among the most popular and widely utilized systems are liver microsomes. Microsomes retain activity of key enzymes involved in drug metabolism which reside in the smooth endoplasmic reticulum, such as cytochrome P450s (CYPs), flavin monooxygenases, and glucuronosyltransferases. Isolated hepatocytes offer a valuable whole cell *in vitro* model, and retain a broader spectrum of enzymatic activities, including not only reticular systems, but cytosolic and mitochondrial enzymes as well. Because of a rapid loss of hepatocyte-specific functions, it has been possible to generate useful data only with short-term hepatocyte incubations or cultures. However, significant strides have been made toward enhancing the viability of both cryopreserved and cultured hepatocytes. Liver slices, like hepatocytes, retain a wide array of enzyme activities, and are used to assess metabolic stability. Furthermore, both hepatocytes and liver slices are capable of assessing enzyme induction *in vitro*. The choice of which system to employ in a drug discovery screening program is dependent on numerous factors such as information about a particular chemical series and availability of tissue. For example, if a pharmacological probe has been shown to induce CYP activity, it may be preferable to use liver slices rather than microsomes for metabolic screening in support of this program. Cross-comparison of metabolic turnover rates in various tissues from various species can be quite beneficial. In addition to providing information on potential rates and routes of metabolism, interspecies comparisons may help in choosing species to be used in efficacy and toxicology models. The use of *in vivo* methods to assess metabolism has advantages and disadvantages. On the plus side, data are being generated within a living species, allowing a full early PK profile for the compound to be assessed. However, the methods are generally of low throughput, are time-consuming, and can suffer from marked species differences. In this respect, *in vitro* studies can be used in conjunction with *in vivo* experiments in order to select the animal model with metabolism most similar to human.

10.5   METABOLIC STABILITY AND INTRINSIC METABOLIC CLEARANCE

Using the concept of intrinsic metabolic clearance (CL_{int}), it has been demonstrated that *in vitro* metabolism rates for a selected set of model substrates correlated well with hepatic extraction ratios determined from isolated perfused rat livers (Rane *et al.*, 1977). Thus, when measured under appropriate conditions, the intrinsic hepatic metabolic clearance (CL_{int}) of human and preclinical species can be predicted by measuring enzyme kinetics parameters: Km (Michaelis–Menten constant) and V_{max} (maximum velocity) ($CL_{int} = V_{max} / Km$) (Segel *et al.*, 1993). However, it is time-consuming to measure Km and V_{max}. In drug discovery, one can use a single substrate concentration (<<Km) to determine the parent drug disappearance half-life ($t_{1/2}$). The intrinsic hepatic metabolic clearance (CL_{int}) can be calculated based on scaling factors from microsomes as well as hepatocytes (Obach *et al.*, 1999a). Then, the hepatic metabolic clearance (CL_h) and hepatic extraction ratio (E_h) can be calculated (Obach *et al.*, 1999b). Integration of metabolic stability studies with the other high-throughput pharmacology screens and other ADME screens such as inhibition screening is essential. Solving a metabolic stability problem may not necessarily lead to a compound with an overall improvement in activity or even PK properties. For example, compounds with improved metabolic stability may be developed only to find out that absorption becomes a problem. Reduction in metabolic clearance may be accompanied by an increase in renal or biliary clearance of the parent drug. It is also possible that improvement of *in vitro* intrinsic clearance (CL_{int}) may result from self-inhibition of one or more drug-metabolizing enzymes. This apparent improvement is only realized during the *in vitro* screening process whereas *in vivo*, the compound may exhibit saturable metabolism, nonlinear PK, or drug–drug interactions. Thus, it is advisable that *in vitro* metabolic stability data be integrated with inhibition screening.

10.6   ADVANTAGES OF ENHANCING METABOLIC STABILITY

Several advantages to be gained from enhancing metabolic stability have been reported (Ariens *et al.*, 1982):

- Increased bioavailability and longer half-life, which in turn should allow lower and less frequent dosing, thus promoting better patient compliance.
- Better congruence between dose and plasma concentration, thus reducing or even eliminating the need for expensive therapeutic monitoring.
- Reduction in metabolic turnover rates from different species which, in turn, may permit better extrapolation of animal data to humans.
- Lower patient-to-patient and intra-patient variability in drug levels, since this is largely based on differences in drug metabolic capacity.
- Diminishing the number and significance of active metabolites and thus lessening the need for further studies on drug metabolites in both animals and man.

10.7   STRATEGIES TO ENHANCE METABOLIC STABILITY

To enhance microsomal stability, *in vitro* metabolism studies can be performed to confirm formation of metabolites as well as to provide quantitative analysis of major metabolites. After identifying moieties that contribute to activity (the pharmacophore) and other moieties necessary for activity, several modifications can be used to enhance metabolic stability. In general, metabolism can be reduced by incorporation of stable functions (blocking groups) at metabolically vulnerable sites. Substrate structure–activity relationships of metabolizing enzymes have to be accommodated within the structure–activity relationships of the actual pharmacological target. However, when it is appropriate to enhance metabolic stability in the molecular design, the following strategies have been used (Testa *et al.*, 1990; Humphrey and Smith, 1992):

- Deactivating aromatic rings toward oxidation by substituting them with strongly electron-withdrawing groups (e.g. CF_3, SO_2NH_2, SO_3-).
- Introducing an *N*-t-butyl group to prevent *N*-dealkylation.
- Replacing a labile ester linkage with an amide group.
- Constraining the molecule in a conformation which is unfavorable to the metabolic pathway, more generally, protecting the labile moiety by steric shielding.
- The phenolic function has consistently been shown to be rapidly glucuronidated. Thus, avoidance of this moiety in a sterically unhindered position is advised in any compound intended for oral use.
- Avoidance of other conjugation reactions as primary clearance pathways, would also be advised in the design stage in any drug destined for oral usage.
- Sometimes the best strategy is to anticipate a likely route of metabolism and prepare the expected metabolite if it has adequate intrinsic activity. For example, often *N*-oxides are just as active as the parent amine, but will not undergo further *N*-oxidation.

Examples from literature for these strategies to enhance metabolic stability in the molecular design are summarized in Table 10.1 (Genin *et al.*, 1996; Bouska *et al.*, 1997;

TABLE 10.1   Enhancement of metabolic stability through structural modification.

Category	Approach/ strategy	Enzyme, pathway	Lead compound
I. Lipophilicity and metabolic stability	Reduce log P and log D	NA	
		NA	$EC50 = 0.078\,mM$, ClogP = 2.07, C7 h = 0.012 μM
	Introduce isosteric atoms or polar functional group	NA	$Ki = 1\,nM$, $AUC_{0\text{-}6h} = 922\,ng/mL\,h$
		NA	
		NA	
		NA	
II. Metabolically labile groups	Remove or block the vulnerable site of metabolism	Benzylic oxidation	$Ki = 66\,nM$, $AUC_{0\text{-}6h} = 40\,ng/mL\,h$

Optimized compound	Experimental model	Therapeutic Class	Reference
 EC50 = 0.058 μM, ClogP = 0.18, C7h = 0.057 μM	Orally dosed monkey	3C protease inhibitor	Dragovich et al. (2003)
 EC50 = 0.047 μM, ClogP = 0.66, C7h = 0.896 μM	Orally dosed monkey	3C protease inhibitor	Dragovich et al. (2003)
 Ki = 5 nM, AUC$_{0-6h}$ = 1872 ng/mL h	Orally dosed rat	CCR5 antagonists	Tagat et al. (2001)
 Ki = 5 nM, AUC$_{0-6h}$ = 2543 ng/mL h	Orally dosed rat	CCR5 antagonists	Tagat et al. (2001)
 Ki = 2.3 nM, AUC$_{0-6h}$ = 3,905 ng/mL h	Orally dosed rat	CCR5 antagonists	Tagat et al. (2001)
 Ki = 7 nM, AUC$_{0-6h}$ = 9243 ng/mL h	Orally dosed rat	CCR5 antagonists	Tagat et al. (2001)
 Ki = 2 nM, AUC$_{0-6h}$ = 1400 ng/mL h	Orally dosed rat	CCR5 antagonists	Palani et al. (2002)

TABLE 10.1 *Continued*

Category	Approach/ strategy	Enzyme, pathway	Lead compound

Allylic oxidation

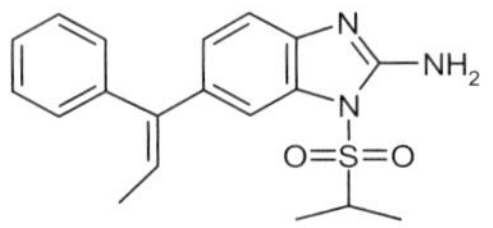

IC$_{50}$ = 0.06 µg/mL, C$_{max}$ = 14–140 ng/mL

Phenyl oxidation

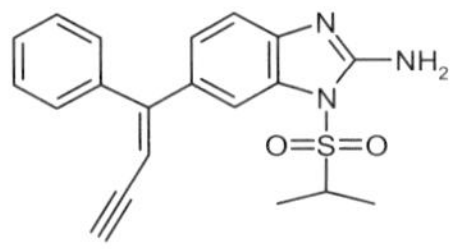

IC$_{50}$ = 0.02 µg/mL, %F = 9

N-oxidation

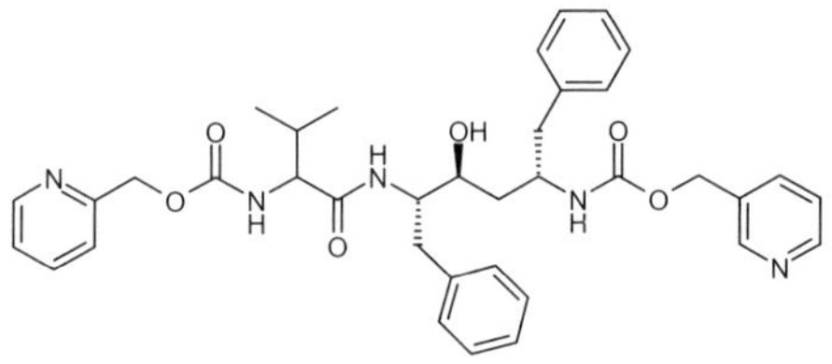

AUC = 1.98 µg·h/mL, %F = 26

N-demethylation

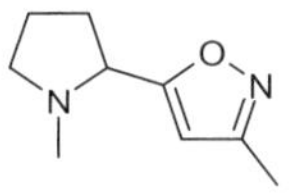

t$_{1/2}$ = 3 h, %F = 1.2

N-dealkylation

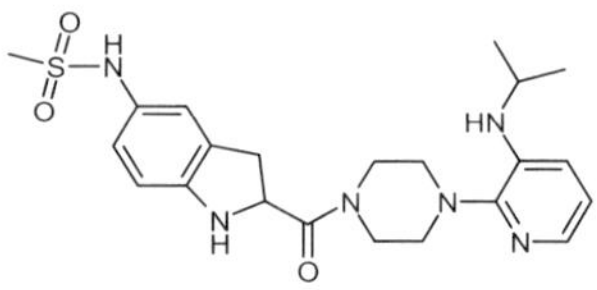

t$_{1/2}$ = 10.8 min

Ester hydrolysis

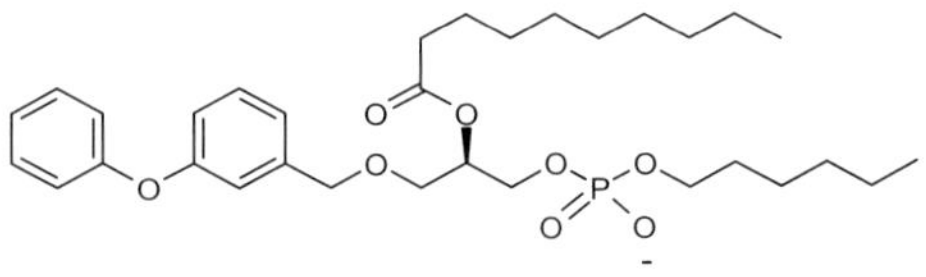

t$_{1/2}$ = 33 min, C$_{max}$ = 465 ng/mL, %F = 4

Optimized compound	Experimental model	Therapeutic Class	Reference
 Ki = 2.1 nM, AUC_{0-6h} = 6500 ng/mL h	Orally dosed rat	CCR5 antagonists	Palani et al. (2002)
 IC_{50} = 0.02 µg/mL, C_{max} = 70–300 ng/mL	Orally dosed monkey	Vinylacetylene antivirals	Victor et al. (1997)
 IC_{50} = 0.04 µg/mL, %F = 23	Orally dosed monkey	Vinylacetylene antivirals	Victor et al. (1997)
 AUC = 4.24 µg·h/mL, %F = 47	Orally dosed rat	HIV protease inhibitors	Kempf et al. (1998)
 $t_{1/2}$ = 24 h, %F = 61.5	Dog liver slices	NAChR ligands	Lin et al. (1997)
 $t_{1/2}$ = 46.8 min	Rat liver microsomes	BHAP reverse transcriptase inhibitors	Genin et al. (1996)
 $t_{1/2}$ = 39 min, C_{max} = 3261 ng/mL, %F = 90	Rat blood and plasma, liver microsomes and homogenate	Phospholipase A inhibitors	Blanchard et al. (1998)

TABLE 10.1 *Continued*

Category	Approach/strategy	Enzyme, pathway	Lead compound
		Amide hydrolysis	$K_i = 0.2\,nM$, 40% and >60% degradation in human cytosole and microsomes, respectively
		Glucuronidation (effect of linker)	UDPGA rate (nmol/min/mg protein) = 0.19, $t_{1/2} = 4.7\,h$
		Glucuronidation (effect of template)	UDPGA rate (nmol/min/mg protein) = 0.05, $t_{1/2} = 5.5\,h$
		Glucuronidation (effect of stereochemistry)	UDPGA rate (nmol/min/mg protein) = 0.02, $t_{1/2} = 7.7\,h$

Note: NA = not applicable; C_{max} = maximum plasma concentration; K_i = inhibition constant; F = bioavailability. Reprinted from Nassar *et al.* 2004, with permission from Elsevier.

Lin *et al.*, 1997; Victor *et al.*, 1997; Blanchard *et al.*, 1998; Kempf *et al.*, 1998; Zhuang *et al.*, 1998; Tagat *et al.*, 2001; Palani *et al.*, 2002; Dragovich *et al.*, 2003). One strategy for improving the metabolic stability of a compound is to reduce the overall lipophilicity (log P, log D) of the structure. This is due to the fact that the metabolizing enzymes generally have a lipophilic binding site and hence more readily accept lipophilic molecules. As exemplified in a Pfizer study of the human rhinovirus 3C protease inhibitor (Dragovich *et al.*, 2003), the lead compound, which exhibited poor oral bioavailability *in vivo* in the monkey, was subjected to PK optimization with the benzyl group being identified as a position for reducing lipophilicity without compromising activity. Introduction of substituents with reduced lipophilicity led to the propargyl analog 2 and the ethyl analog 3, both of which had reduced ClogPs when compared with the parent. These compounds demonstrated improved oral exposure in the monkey, highlighting a simple yet effective way to improve the metabolic profile of a compound. This method is not always successful however, as lipophilic groups are usually involved in binding to the biological target.

An alternative approach toward lowering the lipophilicity of a lead is to introduce isosteric atoms or functional groups into the molecule that impart increased polarity. This could be in the form of a heteroatom (e.g. a nitrogen introduced into a benzene

Optimized compound	Experimental model	Therapeutic Class	Reference
 Ki = 0.069 nM, 10% and <5% degradation in human liver cytosole and microsomes, respectively	Human liver cytosole and microsomes	$5HT_{1A}$ receptor ligand	Zhuang *et al.* (1998)
 UDPGA rate (nmol/min/mg protein) = 0.05, $t_{1/2}$ = 5.5 h	Monkey liver microsomes and plasma	5-LO inhibitors	Bouska *et al.* (1997)
 UDPGA rate (nmol/min/mg protein) = 0.012, $t_{1/2}$ = 14.5 h	Monkey liver microsomes and plasma	5-LO inhibitors	Bouska *et al.* (1997)
 UDPGA rate (nmol/min/mg protein) = 0.01, $t_{1/2}$ = 8.7 h	Monkey liver microsomes and plasma	5-LO inhibitors	Bouska *et al.* (1997)

ring to form the more polar pyridine) or instigating functional group transformations toward more polar functionalities (e.g. conversion of a ketone to the corresponding carboxamide). This strategy was utilized in a Schering-Plough program to enhance the pharmacokinetics of a CCR5 antagonist lead in their HIV-1 inhibitor program (Tagat *et al.*, 2001). The benzamide of the lead compound was identified as a position for reducing lipophilicity, and changes were made to alter the substitution pattern of the ring. The more polar anthranilamide and salicylamide gave significant improvements in the rat oral blood levels, with the area under the curve (AUC) increasing substantially. However, the best results were achieved by introducing a heteroatom into the phenyl ring, with the more polar nicotinamide showing satisfactory oral blood levels in the rat. During this study, the authors noticed a single metabolite of nicotinamide rapidly forming in the plasma samples—they concluded this was due to *N*-oxidation. Thus, the pyridine *N*-oxide was prepared and found to be an excellent compound both in terms of potency and oral bioavailability in rat, dog, and monkey.

A much more elegant approach toward improving metabolic stability is to remove or block the vulnerable site of metabolism. For example, identification of sites potentially labile toward oxidation (e.g. benzylic or allylic positions) can be addressed by blocking the site of metabolism (introduction of a halogen onto the carbon atom involved, replacing the benzylic CH$_2$ with an isostere such as O). Using this approach, workers at Schering-Plough conducted a metabolism-driven optimization of a second-generation lead from the aforementioned CCR5 antagonist program (Palani *et al.*, 2002). The lead compound exhibited good potency against the receptor, but showed very poor oral bioavailability in the rat. The benzylic position was identified as a metabolically labile site, and a variety of isosteres were introduced including O, CHOH, and C O among others. Eventually, the methoxime was identified as a potent, metabolically stable derivative and formed the next-generation lead compound.

The work of Victor *et al.* provides another example of blocking sites of potential oxidative metabolism to improve metabolic stability (Victor *et al.*, 1997). Replacing a methyl group vulnerable to allylic oxidation with acetylene resulted in up to fivefold improvement in blood levels of monkeys. A further improvement of about twofold in oral bioavailabilty in the monkey was achieved when an acetylenyl moiety replaced the methyl and the *p*-fluoro substitution was made on the phenyl ring. It may be that the electron-withdrawing character of fluorine deactivates the aromatic ring toward metabolic oxidation. Or perhaps the somewhat lipophilic character of fluorine coupled with its ability to hydrogen bond provides a unique mixture of physical properties which assist in the absorption process.

Kempf and coworkers employed the principle of bioisosterism in their campaign to synthesize the potent HIV protease inhibitor ritonovir (Kempf *et al.*, 1998). Pyridyl *N*-oxidation was a limiting route of metabolism in one of their early lead compounds. The authors embarked on a systematic search for other heterocyclic groups which would decrease the rate of metabolism while at the same time maintaining adequate solubility to ensure absorption as well as inhibitory activity. Ultimately, pyridyl groups were replaced with metabolism-resistant thiazole groups resulting in analog (ritonovir), which exhibited a nearly twofold improvement in bioavailability while maintaining potent inhibition of HIV protease.

Lin and colleagues (Lin *et al.*, 1997) observed up to about a 50-fold improvement in potency at the nicotinic acetylcholine receptor when the *N*-desmethyl analog was used instead of *N*-methyl analog, which was vulnerable to *N*-demethylation.

Genin and colleagues (Blanchard *et al.*, 1998) realized a significant improvement of about fourfold when they replaced a metabolically vulnerable 3-isopropylamino substituent on the lead compound with an ethoxy substituent. While one might reason that this substituent would also be rapidly metabolized, in this case it proved to be stable, at least when compared with the lead compound.

Functional groups can be open to other types of metabolism other than oxidation; it is well documented that amidases and esterases are present in the liver and can hydrolyze amides and esters, respectively. Additionally, phase II metabolism where polar functionality (e.g. glucuronides and sulfates) is imparted onto a molecule in order to make it more water-soluble and hence easier to clear, can be targeted as a method for improving overall metabolic stability.

Replacing a labile ester linkage with an amide group is known to impart added stability to esterase activity. In Blanchard's study (Blanchard *et al.*, 1998), this can be illustrated by observing the 22-fold improvement in bioavailability in amide when compared with its corresponding ester.

During a metabolism-driven optimization of a series of new arylpiperazine benzamido derivatives as potential ligands for 5-HT1A receptors, Zhuang *et al.* (1998) attributed the low brain uptake in the human subject to rapid metabolism (amid hydrolysis), whereas in other species (rats and monkeys), the metabolic pathway (cleavage of the amid bond by liver microsomes) appears to be much slower. The authors reported that the cyclized amide derivatives provided better than twofold improvement in amide stability compared with the open-chain lead compound.

An extensive body of work on *N*-hydroxyurea inhibitors of 5-lipogenase related to zileuton provides an excellent opportunity to examine structure–activity relationships with respect to conjugation with glucuronic acid (Bouska *et al.*, 1997). It was discovered early on in this investigation that glucuronidation of the *N*-hydroxyl moiety was an activity-limiting step, so it became of pivotal importance to this program to impart metabolic stability toward this conjugation. The *N*-hydroxyurea compounds were divided into three areas for structural modifications: the template, the linking group, and the pharmacophore. The *N*-hydroxyurea was identified in zileuton as an optimal pharmacophore for potency and selectivity; therefore, it was not modified in the search for metabolically stable compounds. The template and link components were the major focus of medicinal chemistry efforts to optimize *in vivo* duration in the cynomolgus monkey. This group systematically modified first the linkage group joining the template with the *N*-hydroxyurea pharmacophore and then the benzthiophene template. Each compound was screened for stability to glucuronidation of the hydroxyurea. Of the linkage moieties, compounds with acetylene linkage groups were generally found to have lower uridine diphosphate glucuronic acid (UDPGA) rates than any of the links tested. This lower UDPGA rate resulted in the longest *in vivo* duration in monkey. Apparently, the rigid conformation required by the acetylenic group disrupts binding in the active site of UDPGA and diminishes conjugation.

Many variations of the template group were also tested in the UDPGA assay. Compounds containing the benzthiophene template found in zileuton demonstrated more rapid rates of glucuronidation. Conversely, biaryl templates such as phenoxyphenyl, phenoxyfuran, and phenoxythiophene demonstrated reduced rates of glucuronidation. These different metabolic stability results may be attributed to changes in lipophilicity and/or remote electronic effects.

Finally, the importance of the stereochemistry of the carbon center immediately adjacent to the hydroxyurea was investigated. *N*-hydroxyurea compounds containing a methyl branch on the linking group had already shown a longer *in vivo* duration than their methylene analogs, which encouraged further work with this methyl branch series. The addition of the methyl group introduced a chiral center and resulted in a compound with separable *R*- and *S*-enantiomers. Approximately twofold difference in the glucuronidation rate, depending on the configuration of the adjacent carbon center, was reported.

10.8 CONCLUSIONS

We have discussed strategies and tools to enhance metabolic stability to enable researchers in the drug companies to design drugs that are safer and more robust for humans. Structural information on metabolites helps in enhancing as well as streamlining the process of developing new drug candidates, which in turn has great value in several important aspects of drug discovery and development. By improving our ability to identify both helpful and harmful metabolites, suggestions for structural modifications will optimize the likelihood that other compounds in the series are more successful. *In silico* and *in vitro* techniques are available to screen compounds for key ADME characteristics, which, when applied within a rational strategy, can make a major contribution to the design and selection of successful drug candidates. Structural modifications to solve a metabolic stability problem may not necessarily lead to a compound with an overall improvement in PK properties. Solving metabolic stability problems at one site could result in the increase in the rate of metabolism at another site, a phenomenon known as metabolic switching. Further, reduction in hepatic clearance may lead to increased renal or biliary clearance of a parent drug or inhibition of one or more drug-metabolizing enzymes. Therefore, it is advisable that *in vitro* metabolic stability data be integrated with other ADME screening.

REFERENCES

Ariens, E.J. *et al.* (1982). *Strategy in Drug Research*, Elsevier Scientific Publishing Company, Amsterdam, pp. 165–178.

Bakke, O.M. *et al.* (1995). Drug safety discontinuations in the United Kingdom, the United States and Spain from 1974 through 1993: a regulatory perspective. *Clin. Pharmacol. Ther.*, *58* (1), 108–117.

Blanchard, S.G. *et al.* (1998). Discovery of bioavailable inhibitors of secretory phospholipase A2. *Pharm. Biotechnol.*, *11*, 445–463.

Bouska, J.J. *et al.* (1997). Improving the in vivo duration of 5-lipoxygenase inhibitors: application of an in vitro glucuronosyltransferase assay. *Drug Metab. Dispos.* 25 (9), 1032–1038.

Dragovich, P. *et al.* (2003). Structure-based design, synthesis, and biological evaluation of irreversible human rhinovirus 3C protease inhibitors. 8. Pharmacological optimization of orally bioavailable 2-pyridone-containing peptidomimetics. *J. Med. Chem.*, *46* (21), 4572–4585.

Genin, M. *et al.* (1996). Synthesis and bioactivity of novel Bis(heteroaryl)piperazine reverse transcriptase inhibitors: structure-activity relationships and increased metabolic stability of novel substituted pyridine analogs. *J. Med. Chem.*, *39* (26), 5267–5275.

Greene, N. *et al.* (2002). Computer systems for the prediction of toxicity: an update. *Adv. Drug Deliv. Rev.*, *54* (3), 417–431.

Humphrey, M.J., Smith, D.A. (1992). Role of metabolism and pharmacokinetic studies in the discovery of new drugs—present and future perspectives. *Xenobiotica*, *22* (7), 743–755.

Kantharaj, E. *et al.* (2003). Simultaneous measurement of drug metabolic stability and identification of metabolites using ion-trap mass spectrometry. *Rapid Commun. Mass Spectrom.*, *17* (23), 2661–2668.

Kempf, D. *et al.* (1998). Discovery of ritonavir, a potent inhibitor of HIV protease with high oral bioavailability and clinical efficacy. *J. Med. Chem.*, *41* (4), 602–617.

Kostiainen, R. *et al.* (2003). Liquid chromatography/atmospheric pressure ionization-mass spectrometry in drug metabolism studies. *J. Mass Spectrom.*, *38* (4), 357–372.

Kubinyi, H. *et al.* (1977). Quantitative structure-activity relationships. 7. The bilinear model, a new model for nonlinear dependence of biological activity on hydrophobic character. *J. Med. Chem.*, *20*, 625–629.

Lasser, K.E. *et al.* (2002). Timing of new black box warnings and withdrawals for prescription medications. *J. Am. Med. Assoc.*, *287* (17), 2215–2220.

Lesko, L.L. *et al.* (2000). Optimizing the science of drug development: opportunities for better candidate selection and accelerated evaluation in humans. *Pharm. Res.*, *14*, 1335–1343.

Lin, N.H. *et al.* (1997). Structure-activity studies on 2-methyl-3-(2(S)-pyrrolidinylmethoxy) pyridine (ABT-089): an orally bioavailable 3-pyridyl ether nicotinic acetylcholine receptor ligand with cognition-enhancing properties. *J. Med. Chem.*, *40* (3), 385–390.

Meyboom, R.H.B. *et al.* (2000). An ABC of drug-related problems. *Drug Saf.*, *22* (6), 415–423.

Nassar, A.-E.F. *et al.* (2003a). Metabolite characterization in drug discovery utilizing robotic liquid-handling, quadrupole time-of-flight mass spectrometry and *in-silico* prediction. *Curr. Drug Metab.*, *4*, 259–271.

Nassar, A.-E.F. *et al.* (2003b). Online hydrogen-deuterium exchange and a tandem quadrupole time-of-flight mass spectrometer coupled with liquid chromatography for metabolite identification in drug metabolism. *J. Chromatog. Sci.*, *41*, 398–404.

Nassar, A.F. *et al.* (2004a). Strategies for dealing with reactive intermediates in drug discovery and development. *Curr. Opin. Drug Discov. Devel.*, *7*, 126–136.

Nassar, A.-E.F. *et al.* (2004b). Strategies for dealing with metabolite elucidation in drug discovery and development. *Drug Discov. Today*, *9*, 317–327.

Nassar, A.E., Kamel, A.M., Clarimont, C. (2004). Improving the decision-making process in the structural modification of drug candidates: enhancing metabolic stability, *Drug Discov. Today*, *9*, (23), 1020–1028.

Obach, R.S. *et al.* (1999a). Prediction of human clearance of twenty-nine drugs from hepatic microsomal intrinsic clearance data: an examination of in vitro half-life approach and nonspecific binding to microsomes. *Drug Metab. Dispos.*, *27* (11), 1350–1359.

Obach, R.S. *et al.* (1999b). Nonspecific binding to microsomes: impact on scale-up of in vitro intrinsic clearance to hepatic clearance as assessed through examination of warfarin, imipramine, and propranolol. *Drug Metab. Dispos.*, *25* (12), 1359–1369.

Palani, A. *et al.* (2002). Synthesis, SAR, and biological evaluation of oximino-piperidino-piperidine amides. 1. Orally bioavailable CCR5 receptor antagonists with potent anti-HIV activity. *J. Med. Chem.*, *45* (14), 3143–3160.

Pirmohamed, M. *et al.* (1998). Adverse drug reactions. *Br. Med. J.*, *316* (7140), 1295–1298.

Plumb, R.S. *et al.* (2003). Use of liquid chromatography/time-of-flight mass spectrometry and multivariate statistical analysis shows promise for the detection of drug metabolites in biological fluids. *Rapid Commun. Mass Spectrom.*, *17* (23), 2632–2638.

Rane, A. *et al.* (1977). Prediction of hepatic extraction ratio from in vitro measurement of intrinsic clearance. *J. Pharmacol. Exp. Ther.*, *200* (2), 420–424.

Richard, A.M. *et al.* (1998). Structure-based methods for predicting mutagenicity and carcinogenicity: are we there yet?. *Mutat. Res.*, *400* (1–2), 493–507.

Segel, I.H. *et al.* (1993). Chapter two: kinetics of unireactant enzymes, in *Enzyme Kinetics*, John Wiley & Sons, Inc., New York, pp. 18–89.

Tagat, J.R. *et al.* (2001). Piperazine-based CCR5 antagonists as HIV-1 inhibitors. II. Discovery of 1-[(2,4-dimethyl-3-pyridinyl)carbonyl]-4- methyl-4-[3(S)-methyl-4-[1(S)-[4-(trifluoromethyl)phenyl]ethyl]-1-piperazinyl]- piperidine N1-oxide (Sch-350634), an orally bioavailable, potent CCR5 antagonist. *J. Med. Chem.*, *44* (21), 3343–3346.

Taylor, E.W. *et al.* (2002). Accelerating the drug optimization process: identification, structure elucidation, and quantification of in vivo metabolites using stable isotopes with LC/MSn and the chemiluminescent nitrogen detector. *Anal. Chem.*, *74* (13), 3232–3238.

Testa, B. *et al.* (1990). Drug metabolism and pharmacokinetics: implications for drug design. *Acta Pharm. Jugosl.* *40* (3), 315–350.

Tiller, P.R. *et al.* (2002). Liquid chromatography/tandem mass spectrometric quantification with metabolite screening as a strategy to enhance the early drug discovery process. *Rapid Commun. Mass Spectrom.*, *16*, 1225–1231.

Uetrecht, J.P. *et al.* (2002). Screening for the potential of a drug candidate to cause idiosyncratic drug reactions. *Drug Discov. Today*, *5*, 832–837.

Victor, F. *et al.* (1997). Synthesis, antiviral activity, and biological properties of vinylacetylene analogs of enviroxime. *J. Med. Chem.*, *40* (10), 1511–1518.

Watt, A.P. (2003). Metabolite identification in drug discovery. *Curr. Opin. Drug Discov. Devel.*, *6* (1), 57–65.

Zhuang, Z.P. *et al.* (1998). Isoindol-1-one analogues of 4-(2′-methoxyphenyl)-1-[2′-[N-(2″-pyridyl)-p-iodobenzamido]ethyl]pipera zine (p-MPPI) as 5-HT1A receptor ligands. *J. Med. Chem.*, *41* (2), 157–166.

CHAPTER 11

Structural Modifications of Drug Candidates: How Useful Are They in Improving PK Parameters of New Drugs? Part II: Drug Design Strategies

ALA F. NASSAR

11.1 INTRODUCTION

One key to successful drug design and development is the process of finding the right combination of diverse properties such as activity, toxicity, exposure, and so on. It is very important to first determine, and then optimize, the exposure–activity–toxicity relationships for drug candidates, and thus their suitability for advancement to development. Therefore, the concern of the drug metabolism scientist is to optimize plasma half-life, drug/metabolic clearance, metabolic stability, and the ratio of metabolic to renal clearance. The effort to achieve these positives must be balanced by the need to minimize or eliminate potential difficulties or dangers such as gut/hepatic-first-pass metabolism, inhibition/induction of drug-metabolizing enzymes by metabolites, biologically active metabolites, metabolism by polymorphically expressed drug-metabolizing enzymes, and formation of reactive metabolites. The optimal result will be a safer drug that undergoes predictable metabolic inactivation or even undergoes no metabolism. Among the approaches available to the drug design team, as they seek to meet these goals, are active metabolites, prodrugs, and hard and soft drugs; these will be discussed with some examples from recent case studies.

11.2 ACTIVE METABOLITES

Some drugs produce metabolites that are pharmacologically active; in some cases, these metabolites surpass their parent drugs as candidates for development. Active metabolites are often indicated by an elevated level of pharmacological effect, which is revealed through pharmacokinetic tests on the parent drug candidate. These metabolites often are subject to phase II metabolism and thus have better safety profiles. Table 11.1

Drug Metabolism Handbook: Concepts and Applications, Edited by Ala F. Nassar,
Paul F. Hollenberg, and JoAnn Scatina
Copyright © 2009 by John Wiley & Sons, Inc.

TABLE 11.1   Metabolism-driven optimization of PK properties by structure modification.

Category	Pre-optimized compound
Active metabolites	

Morphine AUC 0–12 = 614 ng·h/mL

Clarithromycin $t_{1/2}$ = 3.7 h day 1, $t_{1/2}$ = 5.1 h day 7

Prodrug

%F = 17 (rats), %F = 4 (monkeys)

Ampicillin, mean peak serum level = 3.7 µg/mL, %F = 47–49

First-pass effect

Propranolol AUC 0–6 = 132 ng/mL·h

Naltrexone %F = 1.1

Optimized compound	Optimized compound	Reference
Morphine-6-glucuronide, AUC 0–12 = 1672 ng·h/mL		Koren *et al.* (2004)
14-hydroxy-clarithromycin $t_{1/2}$ = 6.2 h day 1, $t_{1/2}$ = 10.1 h day 7		Lode *et al.* (2002)
%F = 30 (rats), %F = 23 (monkeys)		Stearns *et al.* (2002)
Pivampicillin, mean, peak serum level = 7.1 µg/mL, %F = 82–89	Bacampicillin, mean, peak serum level = 8.3 µg/mL, %F = 87–95	Sjövall, Magni and Bergan (1978)
Hemisuccinate ester of propranolol AUC 0–6 = 1075 ng/mL·h		Hasegawa *et al.* (1978)
Anthranilate %F = 49.2	Acetylsalicylate %F = 31.0	Hussain *et al.* (1987)

TABLE 11.1 *Continued*

Category	Pre-optimized compound
Half-life	5-LO inhibitors $t_{1/2}$ = 5.5 h Thrombin $t_{1/2}$ = 3.8 h
PO and IV	5 pyrrolidine ABT-418 $t_{1/2}$ = 0.2 h BMS 187308, AUC 0–inf. (μM·h) = 343 (iv) and 166 (po) at 100 μmol/kg

contains the examples of 14-hydroxy-clarithromycin and morphine 6-glucuronide; they are more potent and less toxic than the parent drugs clarithromycin and morphine, respectively (Dabernat *et al.*, 1991; Bergeron, Bernier and L'Ecuyer, 1992; Burkhardt *et al.*, 2002; Kopecky *et al.*, 2004). Phase I metabolism of clarithromycin produces 14-hydroxy-clarithromycin. Clarithromycin is an important antibacterial drug in the treatment of community-acquired respiratory tract infections. In a double-blind, randomized, two-period crossover study, the pharmacokinetics properties of clarithromycin were determined after single and multiple doses in 12 healthy male volunteers. On day 1, the half-life was 3.7 hours for clarithromycin and 6.2 hours for its metabolite 14-hydroxy-clarithromycin. On day 7, the half-life was longer than on day 1 (5.1 hours for clarithromycin and 10.1 hours for 14-hydroxy-clarithromycin). Also, the *in vitro* activity of clarithromycin and its main metabolite 14-hydroxy clarithromycin against *Haemophilus influenzae* was evaluated. The 14-hydroxy metabolite was more active than the parent compound against *Haemophilus influenzae*. Microdilution broth minimum inhibitory concentrations (MICs) and minimum bactericidal concentrations (MBCs) of both clarithromycin and 14-hydroxy-clarithromycin were determined. For clarithromycin, the MIC50 was 4 mg/L and the MIC90 was 8 mg/L, while the hydroxy metabolite was two- to fourfold more active with an MIC50 and MIC90 of 2 mg/L. As a second example of working with an active metabolite, morphine 6-glucuronide is formed by phase II metabolism of morphine. Morphine 6-glucuronide is more potent than morphine itself without side effects. Morphine-6-glucuronide is a major metabolite of morphine with potent analgesic actions. The PK of morphine and its active metabolite

Optimized compound	Optimized compound	Reference
$t_{1/2} = 16\,h$		Bouska et al. (1997)
$t_{1/2} = 6.6\,h$	$t_{1/2} = 9.7\,h$	Burgey et al. (2003)
ABT-089 $t_{1/2} = 1.6\,h$		Lin et al. (1997)
BMS 193884, AUC 0–inf. ($\mu M \cdot h$) = 840 (iv) at 100 $\mu mol/kg$		Humphreys et al. (2003)

morphine-6-glucuronide were studied in humans, giving AUC_{0-12h} results of 614.3 and 1672.3 ng·h/mL for morphine and morphine 6-glucuronide, respectively. Thus, it becomes clear that morphine 6-glucuronide is a highly potent metabolite, and there is strong evidence that it is important to the overall effect of morphine.

11.3 PRODRUGS

The prodrug approach has been widely used in drug design. Prodrugs were discovered when it was demonstrated that the antibacterial agent protosil was active *in vivo* only when it was metabolized to the actual drug sulfanilamide. Prodrugs most commonly use either oxidative or reductive activation; for example, protosil is activated by reduction of its azo linkage to the amine sulfa drug. Another example of a prodrug is the antipyretic agent phenacetin, which provides its activity after conversion to acetaminophen. Carbamazepine is an anticonvulsant drug that is the metabolic precursor of the active agent carbamazepine 10,11-oxide. Minoxidil is a potent vasodilator that also induces hypertrichosis of facial and body hair (Buhl *et al.*, 1990). Research data show that sulfation is a critical step for the hair growth effects of minoxidil and that it is the sulfated metabolite that directly stimulates hair follicles.

As an example of the effort to improve PK properties through structure modification, the oral bioavailability of a 3-pyridyl thiazole benzenesulfonamide adrenergic receptor agonist (Table 11.1) was investigated in rats, dogs, and monkeys (Stearns *et al.*, 2002). It was shown that the limited bioavailability of this compound was due to poor oral absorption in rats and monkeys (%F = 17 and 4 for rats and monkeys, respectively). A slight modification to the structure of this compound resulted in a significant improvement to its oral bioavailability. The linkage to the pyridine moiety was changed from the 3- to the 2-position so that the pyridyl-nitrogen atom was positioned to the hydrogen bond with the ethanolamine hydroxyl group; this minimized intermolecular interactions that may limit the oral absorption of this compound class. Bioavailability in rats and monkeys improved to %F = 30 and 23 for rats and monkeys, respectively.

The PK evaluation of various structural modifications of ampicillin is another example of how the prodrug strategy can improve PK properties (see Table 11.1). Pivampicillin, talampicillin, and bacampicillin are prodrugs of ampicillin, each of which is produced from the esterification of the polar carboxylate group to form lipophilic, enzymatically labile esters (Sjövall, Magni and Bergan, 1978; Ehrnebo, Nilsson and Boreus, 1979; Ensink *et al.*, 1996). These prodrugs result in nearly total absorption, whereas that of ampicillin is <50%. A randomized crossover study on 11 healthy volunteers to determine the pharmacokinetics of bacampicillin, ampicillin, and pivampicillin showed that the relative bioavailability of bacampicillin (%F = 87–95) and pivampicillin (%F = 82–89) was comparable, whereas ampicillin was less than two-thirds that of the others (%F = 47–49). Additionally, the mean of the individual peak concentrations in serum was 8.3 μg/mL for bacampicillin, 7.1 μg/mL for pivampicillin, and 3.7 μg/mL for ampicillin.

11.4 HARD AND SOFT DRUGS

Hard drugs have been defined as drugs that are biologically active and non-metabolizable *in vivo*. Soft drugs are defined as those that produce predictable and controllable *in vivo* metabolism to form nontoxic products after they have achieved their therapeutic role. For example, quaternary ammonium compounds, such as benzalkonium chloride, are hard antibacterial agents. They are toxic to humans and animals, and their chemical stability (non-metabolizable) makes them unsuitable for general environmental sanitation. Soft analogs of such hard antibacterial agents have been developed, and are less toxic. Although the soft analogs have been shown to possess antibacterial activity in *in vitro* studies, a problem for researchers is that it is likely that their *in vivo* activity will be hampered by their chemical instability. When considering toxicology and pharmacology, a safer drug is one that undergoes predictable metabolic inactivation or even undergoes no metabolism (Schuster *et al.*, 1999; Fromigue, Lagneaux and Body, 2000; Bos *et al.*, 2002).

11.4.1 Hard Drugs

The hard drug approach is very attractive to drug design teams because the problem of toxicity due to reactive or active metabolites is eliminated, and the pharmacokinetics also are simplified because the drugs are excreted primarily through either the bile or kidney. When excretion of a drug occurs mainly through the kidney, differences in its elimination between humans and animals will depend mainly on the renal function, which is a strong predictor of PK profiles. Examples of hard drugs include biophosphonates and certain ACE inhibitors. Because they are not metabolized in animals or humans, and are only eliminated through renal excretion, biophosphonates are very safe with no significant systemic toxicity. Similarly, ACE inhibitors undergo very limited metabolism and are exclusively excreted by the kidney. It is worth noting that biophosphonates and carboxyalkyldipeptide ACE inhibitors were not intentionally designed as hard drugs; rather, their hard qualities are the result of structural improvement.

11.4.2 Soft Drugs

Soft drugs are those that produce predictable *in vivo* metabolism which forms nontoxic products after achieving their therapeutic role. The soft drug approach also aims to produce safer drugs with an increased therapeutic index by integrating metabolism considerations into the drug design process (Bodor and Kaminski, 1980; Bodor, 1984; Bodor and Buchwald, 2000, 2004; Huang *et al.*, 2003). Because most oxidative reactions of drugs are produced by hepatic cytochrome P450 enzyme systems that are often affected by age, sex, disease, and environmental factors, biotransformation becomes more complex and PK is variable. P450 oxidative reactions have the potential to form reactive intermediates and active metabolites which can produce toxicity. The distribution and clearance of such metabolites may be different from those of the parent drug. The design of soft drugs aims to avoid oxidative metabolism as much as possible and to use hydrolytic enzymes to achieve predictability and control of drug metabolism. There are a number of drugs currently on the market, such as esmolol, remifentanil, or loteprednol etabonate, which are products of the soft drug design approach. Other promising drug candidates are currently under investigation in a variety of fields including possible soft antimicrobials, anticholinergics, corticosteroids, beta-blockers, analgetics, ACE inhibitors, antiarrhythmics, and others. An example of this concept in drug design is the isosteric soft analog of the hard antifungal drug cetylpyridinium chloride. While cetylpyridinium is quite effective against both gram-positive and gram-negative bacteria, it is also quite toxic ($LD_{50} = 108\,mg/kg$). The soft analog has a metabolically soft spot (the ester group) built into the structure to replace the ethylene group for detoxification as well as 3-methyl imidazole derivative to replace the pyridine moiety. The soft nature of the new antimicrobial is low toxicity with $LD_{50} = 4110\,mg/kg$.

11.5 PK PARAMETERS

To increase the chances of success for a drug's development, it is essential that a drug candidate has good bioavailability and a desirable half-life, while avoiding problems such as poor absorption and extensive first-pass metabolism. Therefore, an accurate measurement of the pharmacokinetic parameters and a good understanding of the factors that affect the pharmacokinetics will guide drug design. In the following sections, the strategies for structural modification of drug candidates are discussed to improve PK properties such as first-pass effect, half-life, and intravenous versus oral absorption.

11.5.1 First-Pass Effect/Prodrug

Table 11.1 shows three examples of how the first-pass effect can be minimized through structure modification. The first example, propranolol, shows the potential usefulness of the prodrug approach when a highly metabolized drug is protected from first-pass elimination in order to improve bioavailability. Oral dosage of propranolol produces a low bioavailability and a wide variation from patient to patient when compared with intravenous administration; this difference is attributed to first-pass elimination of the drug. Hemisuccinate ester of propranolol was selected as a potential prodrug with the hypothesis that propranolol hemisuccinate ester administration would avoid glucuronide formation during absorption and subsequently be released in the blood by hydrolysis. Following 80 mg po of propranolol hydrochloride and an equivalent dose of its hemisuccinate ester derivative, the AUC_{0-6h} was 1075 and 132 ng/mL·h for hemisuccinate ester and propranolol, respectively (Garceau, Davis and Hasegawa, 1978; Shameem, Imai and Otagiri, 1993; Imai *et al.*, 2003).

A similar approach was successfully employed to enhance the bioavailability of naltrexone [17-(cyclopropylmethyl)-4,5 alpha-epoxy-3,14-dihydroxymorphinan-6-one]. Naltrexone is used in the treatment of opioid addiction and absorbs well from the gastrointestinal tract. However, its systemic bioavailability is very poor, as it undergoes extensive first-pass metabolism when administered orally. Employing the prodrug strategy, a number of prodrug esters on the 3-hydroxyl group were prepared: the anthranilate, acetylsalicylate, benzoate, and pivalate. The oral bioavailability of these prodrugs was determined in dogs. The anthranilate and acetylsalicylate compounds exhibited the greatest enhancement of naltrexone bioavailability (45 and 28 times greater than naltrexone, respectively). No correlation was found between the rates of plasma hydrolysis and bioavailability. Naltrexone-3-acetylsalicylate hydrolyzed in human and dog plasma, with a fast deacetylation step to naltrexone salicylate followed by a slower hydrolysis step to naltrexone (Hussain *et al.*, 1987; Reuning *et al.*, 1989).

An extensive body of work on N-hydroxyurea inhibitors of 5-lipoxygenase related to zileuton provides an excellent opportunity to examine structure-activity relationships (SARs) with respect to conjugation with glucuronic acid (Bouska *et al.*, 1997). In zileuton, the N-hydroxyurea moiety was identified as an optimal pharmacophore for potency and selectivity; therefore, it was not modified in the search for metabolically stable compounds. Medicinal chemistry efforts to optimize *in vivo* duration in the cynomolgus monkey were focused on the template and link components. In general, compounds with acetylene linker groups were found to have lower uridine 5-diphosphoglucuronic acid (UDPGA) rates than any of the links tested. Biaryl templates such as phenoxyphenyl, phenoxyfuran, and phenylthiophene also demonstrated reduced rates of glucuronidation and longer *in vivo* duration in monkeys. Finally, the importance of the stereochemistry of the carbon center immediately adjacent to N-hydroxyurea was investigated. It had already been shown that N-hydroxyurea compounds containing a methyl branch on the linking group had longer *in vivo* durations than their methylene analogs, which led to additional research on this methyl branch series. By adding the methyl group, a chiral center was created, and resulted in a compound with separable R- and S-enantiomers. The glucuronidation rate showed an approximately twofold difference, depending on the configuration of the adjacent carbon center. The result was the identification of a clinical candidate, ABT-761, which demonstrated the longest *in vivo* duration ($t_{1/2}$ = 16 hours) compared with that of zileuton ($t_{1/2}$ = 0.4 hours).

11.5.2 Half-life

Most drugs are administered as a fixed dose given at regular intervals to achieve therapeutic objectives. The plasma half-life is a measure of the duration of drug action. Thus, the half-life of drugs in plasma is one of the most important factors that determine the selection of a dosage regimen. Shorter half-life means more frequent dosing, which tends to reduce the likelihood that the patient will maintain the regimen. The two factors affecting half-life are volume of distribution and elimination clearance. An increase in the first or a reduction of the second will improve the drug's half-life. Modification of the chemical structure is one viable approach to slowing a drug's clearance.

A metabolism-based approach to the optimization of the 3-(2-phenethylamino)-6-methylpyrazinone acetamide template for thrombin is an example of these efforts, which resulted in the identification of several potent thrombin inhibitors with high levels of oral bioavailability and long plasma half-lives. Thrombin is vital to hemostasis because it mediates conversion of fibrinogen to fibrin and activates platelets. By developing the aminopyrazinone acetamide thrombin-inhibitor template, an efficacious, orally bioavailable pre-optimized compound with plasma half-life of 3.8 hours was identified. This compound was found to metabolize through oxidation at the benzylic positions and phase II conjugation of the amino group. These structural areas of metabolism represented major means of clearance of the drug. Their elimination enabled generation of compounds with improved pharmacokinetics. Both the chloro ($t_{1/2}$, 6.6 hours) and cyano ($t_{1/2}$, 9.7 hours) modifications imparted significantly improved half-lives versus the pre-optimized compound (Burgey *et al.*, 2003).

As a second example, structural modification of ABT-418 was required to achieve further improvements in the margin of safety and to identify compounds with favorable oral bioavailability. ABT-418, an analog of (*S*)-nicotine in which the pyridine ring is replaced by the 3-methyl-5-isoxazole moiety, has been shown to possess cognitive-enhancing and anxiolytic-like activities in animal models with an improved safety profile compared with that of nicotine (Lin *et al.*, 1997). One shortcoming of ABT-418 was its very poor bioavailability (%F = 1.2), with a short plasma half-life ($t_{1/2}$ = 0.21 hour). Research on structural modification led to the identification of ABT-089, 2-methyl-3-(2(*S*)-pyrrolidinylmethoxy)pyridine, with a vastly improved oral bioavailability (%F = 61.5) with $t_{1/2}$ = 1.6 hours.

11.5.3 Oral Absorption and Intravenous Dose

The oral bioavailability of a drug is defined as the fraction of an oral dose of the drug that reaches the systemic circulation. An important consideration for orally administered drugs is that the entire blood supply of the upper gastrointestinal tract passes through the liver before reaching the systematic circulation. This means that the drug may be metabolized by the gut wall and/or the liver during the first passage of drug absorption. A drug with high metabolic clearance is always subject to an extensive first-pass effect, thus reducing its bioavailability. By contrast, administering a drug intravenously ensures that all of it enters the blood. Rapid injection will promptly achieve elevated concentrations of the drug in the blood; infusion at a controlled rate produces a constant concentration which can be maintained for any desired length of time. The lipophilicity of a drug is important to its membrane permeability as well as metabolic activity. Higher lipophilicity for a drug generally results in higher permeability and greater clearance, and thereby higher first-pass metabolism.

Bristol-Myers Squibb Pharmaceutical Research Group (Humphrey and Smith, 1992; Humphreys *et al.*, 2003) present a good example of working with structure–metabolism relationships to produce favorable pharmacokinetic properties. They identified an initial lead (2′-amino-*N*-(3,4-dimethyl-5-isoxazolyl)-40-(2-methylpropyl)[1,10-biphenyl]-2-sulfonamide) for endothelin (ET) receptor antagonists. However, testing in pre-clinical animal species and human *in vitro* systems revealed the compound was extensively metabolized due to oxidative biotransformation. The site of metabolism of the candidate was determined and structure–activity and structure–metabolism studies were aimed at optimizing this structural class by finding more metabolically stable analogs that maintained potency. This resulted in the identification of an analog (*N*-(3,4-dimethyl-5-isoxazolyl)-40-(2-oxazolyl)[1,10-biphenyl]-2-sulfonamide) with improved *in vitro* properties; further studies revealed that the new compound had improved pharmacokinetic properties (Table 11.1).

11.6 PK ANALYSIS

Pharmacokinetic analysis is important to drug discovery as it provides insight into how new drug candidates may be absorbed and excreted by the human body (Lesko *et al.*, 2000; Bapiro *et al.*, 2001; Nassar and Adams, 2003; Nassar, Kamel and Clarimont, 2004a, b). Pharmacokinetic studies in early drug discovery involve dosing an animal with the compound of interest and measuring the drug concentration in biological fluids (e.g. plasma) as a function of time. This yields valuable information such as the time required to reach peak concentration and the half-life of the drug. Other parameters such as volume of distribution, clearance, and bioavailability can also be determined from these experiments. This important information is valuable to the decision-making process because it gives medicinal chemists targets for optimization. With many candidates to screen, pharmacokinetic studies are routine but demand a significant amount of time and resources. Sample preparation is a highly important aspect of these experiments, and robotic-based automation is commonly used. The biological fluid is then analyzed by liquid chromatography–mass spectrometry (LC-MS) directly after online or off-line solid-phase extraction, liquid–liquid extraction, or protein precipitation. By reducing the number of pharmacokinetic samples through pooling at certain time points or across all time points, and by the use of an abbreviated standard curve, the analysis is streamlined. The new technology of ultra-performance liquid chromatography (UPLC) can offer significant advantages in resolution, speed, and sensitivity for analytical determinations, which are all important factors for the demands of high-throughput screening in drug discovery. UPLC is more beneficial when coupled with mass spectrometers capable of high-speed acquisitions. This approach allows attainment of more information about a lead candidate while maintaining a rapid analytical turnaround time. Chromatographic separations with analysis times of around 15 seconds can be envisioned; however, such separations will require high-speed MS systems. Also, by using micro and capillary separation techniques with nanospray MS, significant additional sensitivity can be gained.

11.7 CONCLUSIONS

The clinical success of a drug candidate depends upon desirable ADME/TOX properties. An accurate measurement of the pharmacokinetic parameters and a good understanding of the factors that affect the pharmacokinetics will help to guide drug design. By seeking a balance based upon the rule of three (exposure–activity–toxicity), and taking pharmaceutical research beyond the initial candidate or parent drug, drug metabolism scientists have opened up a whole new front in the effort to develop safer and more efficacious drugs. New approaches, evolving technologies, and the wisdom gained from examination of previous difficulties and failures have aided the drug development team to focus their efforts. It is the opinion of the author that improvement in the drug discovery process will be facilitated most by learning how and when to use each of these tools and approaches to design drugs to achieve the maximum benefit, and by increasing cooperation and dialogue among the disciplines involved in the entire process. The author believes that the following suggestions should be considered when developing a new drug:

1. Considering the risks and costs of new drug development, it is critical to eliminate high-risk compounds as early as possible. Fast and reliable *in silico* screens are needed to filter out problematic molecules at the earliest stages of discovery.

2. The metabolism of new drugs should be studied *in vitro* before the initiation of any clinical studies. Early information on *in vitro* metabolic processes in humans, such as the identification of the enzymes responsible for drug metabolism and sources of potential enzyme polymorphism, can be useful in the design of clinical studies, particularly those that examine drug–drug interactions. High metabolic ability usually leads to poor bioavailability and high clearance, and formation of active or toxic metabolites will have an impact on the pharmacological and toxicological outcomes.

3. Appropriate structural modifications to the drug candidate help to optimize the metabolic ability, drug–drug interactions with coadministered drugs due to inhibition and/or induction of drug metabolism pathways, and the rate and sites of metabolism of new chemical entities by drug-metabolizing enzymes.

4. Good *in vitro* activity does not correlate to good *in vivo* activity unless a drug has a good bioavailability and a desirable duration of action. These qualities are essential to successful development.

5. To avoid exclusion of good compounds, the selection of animal species and the experimental design of studies are important in providing a reliable prediction of drug absorption and elimination in humans. A good compound could be excluded on the basis of results from an inappropriate animal species or poor experimental design.

6. Drug candidates should have little or none of the following: gut/hepatic-first-pass metabolism, inhibition/induction of drug-metabolizing enzymes, biologically active metabolites, metabolism by polymorphically expressed drug-metabolizing enzymes, and formation of reactive metabolites. Also, it is important to have the most desirable plasma half-life and ratio of metabolic to renal clearance.

7. One of the most important keys to successful drug design and development is finding the right combination of multiple properties such as activity, toxicity, and exposure. To design good drug candidates, the rule of three should be applied regardless of the therapeutic index.

REFERENCES

Bapiro, T.E., Egnell, A.C., Hasler, J.A., Masimirembwa, C.M. (2001). Application of higher throughput screening (HTS) inhibition assays to evaluate the interaction of antiparasitic drugs with cytochrome P450s. *Drug Metab. Dispos.*, *29*, 30–35.

Bergeron, M.G., Bernier, M., L'Ecuyer, J. (1992). In vitro activity of clarithromycin and its 14-hydroxy-metabolite against 203 strains of *Haemophilus influenzae*. *Infection*, *20*, 164–167.

Bodor, N. (1984). Soft drugs: principles and methods for the design of safe drugs. *Med. Res. Rev.*, *4*, 449–469.

Bodor, N., Buchwald, P. (2000). Soft drug design: general principles and recent applications. *Med. Res. Rev.*, *20*, 58–101.

Bodor, N., Buchwald, P. (2004). Designing safer (soft) drugs by avoiding the formation of toxic and oxidative metabolites. *Mol. Biotechnol.*, *26*, 123–132.

Bodor, N., Kaminski, J.J. (1980). Soft drugs. 2. Soft alkylating compounds as potential antitumor agents. *J. Med. Chem.*, *23*, 566–569.

Bos, H., Henning, R.H., De Boer, E., Tiebosch, A.T., De Jong, P.E., De Zeeuw, D., Navis, G. (2002). Addition of AT1 blocker fails to overcome resistance to ACE inhibition in adriamycin nephrosis. *Kidney Int.*, *61*, 473–480.

Bouska, J.J., Bell, R.L., Goodfellow, C.L., Stewart, A.O., Brooks, C.D., Carter, G.W. (1997). Improving the in vivo duration of 5-lipoxygenase inhibitors: application of an in vitro glucuronosyltransferase assay. *Drug Metab. Dispos.*, *25*, 1032–1038.

Buhl, A.E., Waldon, D.J., Baker, C.A., Johnson, G.A. (1990). Minoxidil sulfate is the active metabolite that stimulates hair follicles. *J. Invest. Dermatol.*, *95*, 553–557.

Burgey, C.S., Robinson, K., Lyle, T.A., Sanderson, P.E.J., Lewis, S.D., Lucas, B.J., Krueger, J.A., Singh, R., Miller-Stein, C., White, R.B., Wong, B., Lyle, E.A., Williams, P.D., Coburn, C.A., Dorsey, B.D., Barrow, J.C., Stranieri, M.T.; Holahan, M.A., Sitko, G.R., Cook, J.J., McMasters, D.R., McDonough, C.M., Sanders, W.M., Wallace, A.A., Clayton, F.C., Bohn, D., Leonard, Y. M., Detwiler, T.J., Lynch, J.J., Yan, Y., Chen, Z., Kuo, L., Gardell, S.J., Shafer, J.A., Vacca, J.P. (2003). Metabolism-directed optimization of 3-aminopyrazinone acetamide thrombin inhibitors. Development of an orally bioavailable series containing P1 and P3 pyridines. *J. Med. Chem.*, *46*, 461–473.

Burkhardt, O., Borner, K., Stass, H., Beyer, G., Allewelt, M., Nord, C., Lode, H. (2002). Single- and multiple-dose pharmacokinetics of oral moxifloxacin and clarithromycin, and concentrations in serum, saliva and faeces. *Scand. J. Infect. Dis.*, *34*, 898–903.

Dabernat, H., Delmas, C., Seguy, M., Fourtillan, J.B., Girault, J., Lareng, M.B. (1991). The activity of clarithromycin and its 14-hydroxy metabolite against Haemophilus influenzae, determined by in-vitro and serum bactericidal tests. *J. Antimicrob. Chemother.*, *27*, 19–30.

Ehrnebo, M., Nilsson, S.O., Boreus, L.O. (1979). Pharmacokinetics of ampicillin and its prodrugs bacampicillin and pivampicillin in man. *J. Pharmacokinet. Biopharm.*, *7*, 429–451.

Ensink, J.M., Vulto, A.G., van Miert, A.S., Tukker, J.J., Winkel, M.B., Fluitman, M.A. (1996). Oral bioavailability and in vitro stability of pivampicillin, bacampicillin, talampicillin, and ampicillin in horses. *Am. J. Vet. Res.*, *57*, 1021–1024.

Fromigue, O., Lagneaux, L., Body, J.J. (2000). Bisphosphonates induce breast cancer cell death in vitro. *J. Bone Miner. Res.*, *15*, 2211–2221.

Garceau, Y., Davis, I., Hasegawa, J. (1978). Plasma propranolol levels in beagle dogs after administration of propranolol hemisuccinate ester. *J. Pharm. Sci.*, *67*, 1360–1363.

Huang, F., Browne, C.E., Wu, W.M., Juhasz, A., Ji, F., Bodor, N. (2003). Design, pharmacokinetic, and pharmacodynamic evaluation of a new class of soft anticholinergics. *Pharm. Res.*, *20*, 1681–1689.

Humphrey, M.J., Smith, D.A. (1992). Role of metabolism and pharmacokinetic studies in the discovery of new drugs—present and future perspectives. *Xenobiotica*, *22*, 743–755.

Humphreys, W.G., Obermeier, M.T., Barrish, J.C., Chong, S., Marino, A.M., Murugesan, N., Wang-Iverson, D., Morrison, R.A. (2003). Application of structure-metabolism relationships in the identification of a selective endothelin A antagonist, BMS-193884, with favourable pharmacokinetic properties. *Xenobiotica*, *33*, 1109–1123.

Hussain, M.A., Koval, C.A., Myers, M.J., Shami, E.G., Shefter, E. (1987). Improvement of the oral bioavailability of naltrexone in dogs: a prodrug approach. *J. Pharm. Sci.*, *76*, 356–358.

Imai, T., Yoshigae, Y., Hosokawa, M., Chiba, K., Otagiri, M. (2003). Evidence for the involvement of a pulmonary first-pass effect via carboxylesterase in the disposition of a propranolol ester derivative after intravenous administration. *J. Pharmacol. Exp. Ther.*, *307*, 1234–1242.

Kopecky, E.A., Jacobson, S., Joshi, P., Koren, G. (2004). Systemic exposure to morphine and the risk of acute chest syndrome in sickle cell disease. *Clin. Pharmacol. Ther.*, *75*, 140–146.

Lesko, L.J., Rowland, M., Peck, C.C., Blaschke, T.F. (2000). Optimizing the science of drug development: opportunities for better candidate selection and accelerated evaluation in humans. *Pharm. Res.*, *17*, 1335–1343.

Lin, N.H., Gunn, D.E., Ryther, K.B., Garvey, D.S., Donnelly-Roberts, D.L., Decker, M.W., Brioni, J.D., Buckley, M.J., Rodrigues, A.D., Marsh, K.G., Anderson, D.J., Buccafusco, J.J., Prendergast, M.A., Sullivan, J.P., Williams, M., Arneric, S. P., Holladay, M.W. (1997). Structure-activity studies on 2-methyl-3-(2(S)-pyrrolidinylmethoxy) pyridine (ABT-089): an orally bioavailable 3-pyridyl ether nicotinic acetylcholine receptor ligand with cognition-enhancing properties. *J. Med. Chem.*, *31*, 385–390.

Nassar, A.E., Adams, P.E. (2003). Metabolite characterization in drug discovery utilizing robotic liquid-handling, quadrupole time-of-flight mass spectrometry and *in silico* prediction. *Curr. Drug Metab.*, *4*, 259–271.

Nassar, A.E., Kamel, A.M., Clarimont, C. (2004a). Improving the decision-making process in structural modification of drug candidates: reducing toxicity. *Drug Discov. Today*, *24*, 1055–1064.

Nassar, A.E., Kamel, A.M., Clarimont, C. (2004b). Improving the decision-making process in the structural modification of drug candidates: enhancing metabolic stability. *Drug Discov. Today*, *23*, 1020–1028.

Reuning, R.H., Ashcraft, S.B., Wiley, J.N., Morrison, B.E. (1989). Disposition and pharmacokinetics of naltrexone after intravenous and oral administration in rhesus monkeys. *Drug Metab. Dispos.*, *17*, 583–589.

Schuster, C., Reinhart, W.H., Hartmann, K., Kuhn, M. (1999). Angioedema induced by ACE inhibitors and angiotensin II-receptor antagonists: analysis of 98 cases. *Schweiz. Med. Wochenschr.*, *129*, 362–369.

Shameem, M., Imai, T., Otagiri, M. (1993). An in-vitro and in-vivo correlative approach to the evaluation of ester prodrugs to improve oral delivery of propranolol. *J. Pharm. Pharmacol*, *45*, 246–452.

Sjövall, J., Magni, L., Bergan, T. (1978). Pharmacokinetics of bacampicillin compared with those of ampicillin, pivampicillin, and amoxycillin. *Antimicrob. Agents Chemother.*, *13*, 90–96.

Stearns, R.A., Miller, R.R., Tang, W., Kwei, G.Y., Tang, F.S., Mathvink, R.J., Naylor, E.M., Chitty, D., Colandrea, V.J., Weber, A.E., Colletti, A.E., Strauss, J.R., Keohane, C.A., Feeney, W.P., Iliff, S.A., Chiu, S.H. (2002). The pharmacokinetics of a thiazole benzenesulfonamide beta 3-adrenergic receptor agonist and its analogs in rats, dogs, and monkeys: improving oral bioavailability. *Drug Metab. Dispos.*, *30*, 771–777.

CHAPTER 12

Minimizing the Potential for Drug Bioactivation of Drug Candidates to Success in Clinical Development

ALA F. NASSAR

12.1 INTRODUCTION

Toxicity problems, especially those that may only occur under unusual or idiosyncratic conditions during the late stages of drug development, are one of the most devastating surprises for pharmaceutical companies. Variations in human drug-metabolizing enzymes can produce subtle evidence of potential toxicity, or none at all, during pre-clinical safety studies. Such problems are also unlikely to show up in all but the largest clinical trials, but if the side effects are serious, can result in product withdrawal. There are indications that some substructures found in drugs can form reactive metabolites that are involved in toxicities in humans. These substructures include arylacetic and arylpropionic acids, aryl hydroxamic acids, oximes, anilines, anilides, hydrazines, hydrazides, hydantoins, quinones, quinone methides, nitroaromatics, heteroaromatics, halogenated hydrocarbons, some halogenated aromatics, chemical groups that can be oxidized to acroleins, and medium-chain fatty acids. Reactive metabolites are unstable, and are intermediates to more stable metabolites. Table 12.1 shows several examples of drugs that undergo metabolic activation and cause adverse reaction in humans, which have been withdrawn from the market or restricted in use with toxicity warnings. Clearly, a drug candidate that does, or may, metabolize to such substructures would increase the risk of failure or withdrawal.

12.1.1 Importance of Reactive Intermediates in Drug Discovery and Development

One key to success in clinical development is to minimize reactive intermediates in drug metabolism. There is a growing consensus that idiosyncratic drug reactions (IDRs) have enormous consequences for patients and the pharmaceutical industry. It is estimated that IDRs account for ~5% of all hospital admissions and occur in 10% to 20%

Drug Metabolism Handbook: Concepts and Applications, Edited by Ala F. Nassar, Paul F. Hollenberg, and JoAnn Scatina
Copyright © 2009 by John Wiley & Sons, Inc.

TABLE 12.1 Examples of chemical structures activating to produce toxic metabolites.

Chemical class	Bio-transformation	Toxic metabolite	Compound Name	Clinical use	Biological effects	Ref
Quinone	Oxidation	Quinone-type	Tacrine	Alzheimer's disease	Hepatic toxicity	Madden et al. (1993)
			Troglitazone	Treat Type II diabetes	Hepatic toxicity	Kassahun et al. (2001; Tettey et al. 2001)
			Minocycline	Antibiotics	Hepatic toxicity Lupus-like syndrome	Shapiro et al. (1997)
			Acetaminophen	Analgesic agent	Hepatic toxicity	Isaacson et al. (1999)
			Aminosalicylic acid	Inflammatory bowel disease	Lupus-like syndrome Pancreatic toxicity Hepatic toxicity Renal toxicity	Liu et al. (1995)
			Amodiaquine	Treat malaria	Hepatic toxicity Agranulocytosis	Ruscoe et al. (1995)
			Phenytoin	Anticonvulsant	Drug-induced hypersensitivity Teratogenicity	Munns et al. (1997)
			Carbamazepine	Anticonvulsant	Teratogenicity	Ju et al. (1999)
			Vesnarinone	Phosphodiesterase inhibitor	Agranulocytosis	Uetrecht et al. (1994)
			Prinomide	Antiinflammatory	Agranulocytosis	Parrish et al. (1997)
			Estrogens	NSAID	Breast cancer Uterine cancer	Mohsin et al. (2004)
			Tamoxifen	NSAID	Endometrial cancer	Fan et al. (2001)
			Fluperlapine	Antipsychotic agent	Agranulocytosis	Mann et al. (1987)

Chemical class	Mechanism	Reactive metabolite	Drug	Therapeutic use	Toxicity	Reference
Aryl nitro	Reduction	Nitroso	Tolcapone	Parkinson's disease	Liver toxicity	Borges et al. (2000)
			Chloramphenicol	Antibiotic	Aplastic anemia, Bone marrow toxicity	Yunis et al. (1980)
Nitrogen-containing aromatic	Oxidation	Nitrenium ion	Dantrolene	Muscle relaxant	Liver toxicity	Utili et al. (1977)
		Free radical	Nimesulide	COX 2 inhibitors	Liver toxicity	Merlani et al. (2001)
			Clozapine	Antipsychotic agent	Agranulocytosis, Liver toxicity, Myocarditis	Idanpaan-Heikkila et al. (1977; Haack et al. 2003)
			Aminopyrine	Painkiller	Agranulocytosis, CNS toxicity	Uetrecht et al. (1995)
Aryl amines	Oxidation to hydroxylamine	Nitroso	Dipyrone	Painkiller	Agranulocytosis	Sabbaga et al. (1993)
			Sulfamethoxazole	Antibacterial agent	Hepatotoxicity, Agranulocytosis, Lupus-like syndrome, Skin rashes	Cribb et al. (1992)
			Dapsone	Antiparasitic	Agranulocytosis, Flu-like syndrome, Hemolytic anemia, Methemoglobinemia, Lupus-erythematosis	Coleman et al. (1994)
			Procainamide	Cardiac antiarrhythmic	Agranulocytosis, Fever	Woosley et al. (1978)
			Nomifensine	Antidepressant	Hemolytic anemia, Allergic reactions	Salama et al. (1983)
			Sulfasalazine	Ulcerative colitis	Abnormal liver function, Decreased blood counts, Allergic reactions, Skin rashes, Fever	Senturk et al. (1997)
			Aminoglutethimide	Breast cancer	Agranulocytosis, Thrombocytopenia, Liver toxicity	Stuart-Harris et al. (1984)

TABLE 12.1 *Continued*

Chemical class	Bio-transformation	Toxic metabolite	Compound		Biological effects	Ref
			Name	Clinical use		
Michael Acceptors	Hydrolysis Oxidation	Aldehyde Co-A conjugate	Felbamate	Anticonvulsant	Aplastic anemia Liver toxicity	Thompson *et al.* (1996; Dieckhaus *et al.* 2002)
			Terbinafine	Antifungal agent	Bone marrow toxicity Liver toxicity Skin rashes	Gupta *et al.* (1998)
			Valproic acid	Anticonvulsant	Liver toxicity	Grillo *et al.* (2001)
			Mianserin	Antidepressant	Agranulocytosis	Inman *et al.* (1988)
			Leflunomide	Inflammatory arthritis	Liver toxicity Agranulocytosis	Jardine (2002)
Carboxylic acids	Glucuronidation	Acyl glucuronides	Diclofenac	NSAID	Liver toxicity Agranulocytosis	Ware *et al.* (1998)
			Zomepirac	NSAID	Liver toxicity	Bailey *et al.* (1996)
			Ibufenac	NSAID	Liver toxicity	Prescott *et al.* (1986)
			Bromfenac	NSAID	Liver toxicity	(Skjodt *et al.*, 1999; Moses *et al.*, 1999)
			Benoxaprofen	NSAID	Liver toxicity	Halsey *et al.* (1982)
			Indomethacine	NSAID	Bone marrow toxicity	Godessart *et al.* (1999)

Reprinted from Nassar, Kamel, and Clairmont, 2004, with permission from Elsevier.

of hospital inpatients (Bakke *et al.*, 1995; Pirmohamed *et al.*, 1998; Richard *et al.*, 1998; Meyboom *et al.*, 2000; Greene *et al.*, 2002; Lasser *et al.*, 2002; Tiller *et al.*, 2002; Uetrecht *et al.*, 2002). Of the new prescription drugs approved in the United States during the period between 1975 and 2000, 10.2% acquired a new "black box" warning or had to be withdrawn from the market because of IDRs (Nassar *et al.*, 2004b). Idiosyncratic drug toxicity is generally believed to be a phenomenon that cannot be readily evaluated experimentally because it is a rare event (usually <1 in 5000 cases), making clinical trials impractical since an extremely large patient population would be required. The current data suggest that IDRs are human-specific events that may not be detectable with experimental animals currently used. IDRs may have a complex pathology and can lead to a large number of symptoms, ranging from nonspecific rashes to specific organ damage, such as agranulocytosis or cholestasis. Immunological reactions can affect several organs including the liver, kidney, or lung, or they may have systemic effects. Hepatotoxicity is the most common type of idiosyncratic reaction. The time to onset of liver reactions varies from several days to almost one year, suggesting that the rare cases of liver injury may be caused by a metabolic idiosyncrasy (Pirmohamed *et al.*, 1998). Immune system responses may account for between 3% and 25% of all IDRs. For example, the antimalarial drug amodiaquine produces life-threatening agranulocytosis and hepatotoxicity in approximately 1 in 2000 patients during prophylactic administration. The unexpected occurrence of IDRs during late clinical trials has led to severe restrictions on use or failure to launch, and in the case of launched drugs, even withdrawal from the market. During clinical development, toxicity of drug candidates accounts for a significant portion of attrition (~20%) (Uetrecht *et al.*, 2002). Some examples of drugs withdrawn from the market due to IDRs are troglitazone, practolol, benoxaprofen, ticrynafen, zomepirac, and nomifensine. While all pharmaceutical companies consider IDRs to be a significant issue and are making efforts to predict them, the elusive nature and mechanisms of IDRs hinder the development of a clear and universal approach (Mehta *et al.*, 1982; Mattocks *et al.*, 1990; Li *et al.*, 2002; Matzinger *et al.*, 2002; Nassar *et al.*, 2004b).

This chapter discusses the recent efforts to overcome or assess the problem of reactive metabolites in drug discovery and development, which might help to evaluate the safety profile of new drug candidates for IDRs during the preclinical phase. Recently developed tools for identifying and determining the potential toxicity of reactive metabolites are discussed. The use of these tools is evaluated at several stages of drug discovery and development, along with their potential for improving drug safety. Also, some of the chemical substructures that form reactive metabolites and are involved in toxicities in humans are discussed.

12.2 IDIOSYNCRATIC DRUG TOXICITY AND MOLECULAR MECHANISMS

12.2.1 Adverse Drug Reactions

From a clinical perspective, adverse drug reactions (ADRs) can be classified as type A (predictable) and type B (unpredictable or idiosyncratic) reactions. Type A reactions, which account for ~80% of all ADRs, are dose-dependent and predictable based on the pharmacology of the drug. Type B, or IDRs, cannot be predicted on the basis of the pharmacology of the drug and lack simple dose dependency. Type B reactions, which include anaphylaxis, blood dyscrasias, hepatotoxicity, and skin reactions, may result from a combination of genetic predisposition and environmental factors, and are reviewed elsewhere (Uetrecht *et al.*, 2002).

12.2.2   Types of Reactive Metabolites

There is significant evidence to indicate that some substructures found in drugs can form reactive metabolites that are involved in toxicities in humans. These substructures include arylacetic and arylpropionic acids, anilines, anilides, hydrazines, hydantoins, quinones, quinone methides, nitroaromatics, heteroaromatics, halogenated hydrocarbons, some halogenated aromatics, chemical groups that can be oxidized to acroleins, and medium-chain fatty acids. The thiazolidinedione ring, a relatively new substructure, can form reactive metabolites that may cause hepatotoxicity (Pirmohamed *et al.*, 1998). Reactive metabolites are unstable, and are intermediates to more stable metabolites. The different types of reactive metabolites and bioactivation have been thoroughly reviewed elsewhere (Pirmohamed *et al.*, 1998). For the purpose of this review, it is sufficient to categorize the three major types of metabolites as electrophiles, polarized double bonds, and free radicals. Previously published reports have provided significant evidence that chemically reactive metabolites may be responsible for serious forms of toxicity, including cellular necrosis, mutagenesis, carcinogenesis, teratogenesis, hypersensitivity, and blood disorders (Fig. 12.1).

12.2.3   Hypotheses for Idiosyncratic Reactions

Several hypotheses have been proposed to explain potential mechanisms for IDRs (Kassahun *et al.*, 1993; Tang *et al.*, 1996, 1997; Chen *et al.*, 2001; Burkhart *et al.*, 2002;

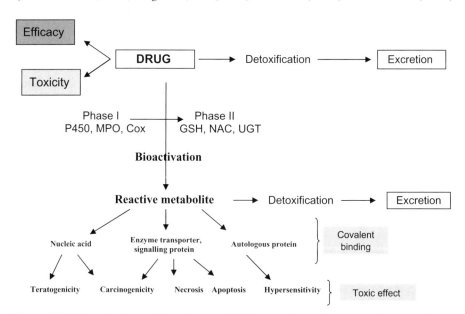

Figure 12.1   Reactive intermediate theory and idiosyncratic reactions/toxic effects. Reprinted from Nassar and Lopez-Anaya, 2004, with permission from Thomson Reuters (Scientific) Limited. Copyright 2004, © The Thomson Corporation.

Judson *et al.*, 2003a, b; Steghens *et al.*, 2003; Evans, 2004; Soglia *et al.*, 2004). These include the hapten, the P-I, and the danger hypotheses. Alternatively, the theory of tolerance toward an immune reaction may explain why the majority of individuals are not susceptible to IDRs.

The hapten hypothesis proposes that modification of an endogenous protein by a reactive metabolite or directly by a reactive parent drug generates a "foreign" protein that, in some cases, leads to an immune-mediated adverse reaction. Immune-mediated mechanisms have been proposed for several drugs including halothane, phenytoins, sulfonamides, and tienilic acid. Failure to down-regulate harmful immune responses due to a "foreign" protein may cause IDRs in susceptible individuals as described in the danger hypothesis (Chen *et al.*, 2001).

The P-I hypothesis provides a new explanation for the occurrence of IDRs in the absence of a reactive intermediate. It is proposed that T cells do not necessarily bind to protein, but may reversibly bind to major histocompatibility complex/T cell receptors (MHC/TCRs). There is a lack of evidence linking such direct interactions with immune system responses (Soglia *et al.*, 2004).

The danger hypothesis proposes that, rather than the "foreignness" of a reactive metabolite triggering the immune system, cell damage produced by reactive metabolites results in a "danger signal." If the reactive metabolites are not toxic enough to cause cell injury, or if the target cells are resistant to the stress induced by reactive metabolites, protein adducts will have a lower probability of escaping the target cells and inducing a specific immune response. This could be one of the reasons why not all drugs that form reactive metabolites cause IDRs, and not all people are susceptible to these reactions (Evans, 2004).

Currently, the risk assessment of a given toxic effect in humans is usually based on a safety margin of the compound, which is the ratio of the no observable adverse effect level (NOAEL) in the most sensitive species and the expected therapeutic dose in man. The NOAEL is typically derived from a rodent and non-rodent species. Clinical and pathological measurements are the gold standard for identification of organ toxicity in animals. The nonclinical studies are conducted in healthy animals, and the risk assessment is extrapolated to healthy and health-compromised patients. This traditional approach to risk assessment has performed reasonably well in predicting most ADRs; however, it has poor predictability for IDRs (Judson *et al.*, 2003a). Therefore, strategies need to be developed or adapted from current drug safety assessment methods to consider reactive intermediates in the overall assessment of the potential risk profile during drug discovery and development.

12.2.3.1 Considerations   To illustrate how to reduce or eliminate reactive metabolites, consider drug-induced adverse reactions, particularly IDRs (referred to as type B reactions). Such reactions are a major issue because, given current techniques and approaches, they often go undetected until late in the process, or even after the drug has been released onto the market, and the consequences are disastrous for all concerned. IDRs mediated through a reactive metabolite may be associated with several mechanisms (Mehta *et al.*, 1982; Mattocks *et al.*, 1990; Li *et al.*, 2002; Matzinger *et al.*, 2002). Currently there is no general approach or "one-size-fits-all" screen that addresses idiosyncratic reactions, because of the difficulties in understanding the mechanisms of these reactions and in accurately predicting clinical results. Some of the strategies under consideration to improve our understanding of these mechanisms and to develop safer drugs include:

1. Avoidance or flagging of chemical functional groups that are known to cause toxicity during drug design (e.g. aromatic and hydroxy amines, phenols, epoxides, acyl halides, acyl glucuronides, thiopenes, furans, fatty acid-like compounds, hydroxylated metabolites, and quinines).
2. Development of suitable *in vitro* and *in vivo* systems to elucidate the role of short-lived, potentially toxic metabolites in the pathogenesis of idiosyncratic toxicity.
3. Identification of chemical functional groups that are associated with low or no toxicity, and development of more metabolically stable drugs to potentially avoid metabolic interactions.

IDRs mediated through a reactive metabolite may be associated with several mechanisms, and it is unlikely that one test or screen will accurately predict all such mechanisms. Although there is no evidence that safer drugs will be identified by addressing the following questions in the drug discovery/development stages, it is considered that drugs without these potential liabilities may have a better safety profile:

1. Does the candidate have the potential to form reactive metabolites based on chemical "structure-alerts"?
2. Does the compound form reactive metabolites in liver, blood, or skin tissue or cells?
3. Is the binding of the drug >50 pmol/mg of microsomal protein?
4. Does the candidate form reactive intermediates that are able to "travel" and react covalently with other tissues?
5. What proteins are affected and what are the effects of the modified proteins?
6. Which genes are affected? Could these affected genes generate a potential IDR?
7. What cellular functions are affected by the reactive intermediate?
8. Bioaccumulation in liver?
9. Glutathione depletion?
10. Drug–drug interactions?
11. Ames and/or micronucleus positive responses?

It remains to be determined whether the answers to these questions will allow the pharmaceutical industry to reduce or eliminate IDRs. Since the major organs involved in IDRs are the liver, skin, and bone marrow, preclinical approaches should consider these target organs to study the effects of reactive intermediates. Thus, a sound strategy should evaluate three major elements: reactive intermediate characterization, covalent binding, and biological impact and/or function.

12.3   KEY TOOLS AND STRATEGIES TO IMPROVE DRUG SAFETY

The exact mechanism(s) for IDRs are unknown, and because many drugs form reactive metabolites, screening processes may eliminate potentially good candidates (false negatives). Current safety screening methods with animal models do not accurately predict IDRs, according to the Multinational Pharmaceutical Company Survey and the outcome

TABLE 12.2 **Selected tools to screen for reactive metabolites, covalent binding and biological effects.**

Stage	Tool	Comment
Early drug discovery	*In silico* methods/drug design/data mining.	Prior to drug synthesis.
Available compounds for screening	Microsomal/CYP time-dependent inactivation. Reactive intermediate screens in human tissues.	*In vitro* studies in cells, subcellular fractions or purified enzymes.
Late drug discovery/ candidate selection	Covalent binding. Oxidative stress. Hapten characterization. Reactive intermediate comparison (human versus preclinical species).	Cellular/subcellular/ enzymatic or preliminary *in vivo* studies. Is the binding >50 pmol/mg microsomal protein?
Preclinical development	Animal models/biomarkers. Proteonomics/metabolomics/genomics. Reactive intermediate exposure in animal models. Prediction of species differences in exposure of reactive intermediates.	Available animal models have limited prediction rates of IDRs, but new tools may improve predictability.
Clinical	Biomarkers/proteonomics/genomics/ metabolomics. Reactive intermediate exposure when considered relevant.	Hapten identification. mRNA. Immunotoxicity. Reactive metabolites (in blood, skin and liver).

Reprinted from Nassar and Lopez-Anaya, 2004, with permission from Thomson Reuters (Scientific) Limited. Copyright 2004, © The Thomson Corporation.

of the International Life Sciences Institute Workshop (Judson *et al.*, 2003a). Many tools are in the exploratory stage, encompassing computational approaches and experimental assays to predict the formation of reactive metabolites *in vivo* and their role in IDRs (Nassar *et al.*, 2004b). For example, during *in vitro* testing, if significant protein binding is found, an attempt is made to discover the chemical basis for the binding so that it can be designed out of the structure. Another potential technique is to trap reactive intermediates by incubating them in human liver microsomes. To evaluate the biological effects of reactive intermediates initially, methods for cytotoxicity testing including tests for cell viability, membrane integrity, protein synthesis, DNA synthesis, glutathione (GSH) level, apoptosis, free radical production, lipid peroxidation, enzyme inhibition, and other enzymatic activities due to oxidative stress are being developed for early drug discovery. Results suggest that these types of screens can be used to differentiate known toxicants from relatively safe drugs. Table 12.2 presents some of the current tools with the advantages and disadvantages summarized and related to the problems facing those in drug discovery and development. The study also attempts to synthesize these tools and strategies into an alternative strategy.

12.3.1 *In Silico* Screens as Filtering Process

A promising new trend in early drug discovery is the collaboration between medicinal and computational chemists and drug metabolism scientists seeking to identify and eliminate potential toxic chemical groups. Structure–activity relationships are well known as a means to identify several toxic end points. One way to improve screening for such potentially toxic compounds would be to increase the use of *in silico* testing methods at earlier stages in drug discovery (Richard *et al.*, 1998; Greene *et al.*, 2002; Judson *et al.*, 2003a, b). Several programs such as TOPKAT, CASE/MULTI-CASE, DEREK, HazardExpert, and OncoLogic are commercially available for the prediction of mutagenicity and carcinogenicity based on chemical structure. These available commercial systems for mutagenicity and/or carcinogenicity prediction differ in their specifics, yet most fall into two major categories. One is automated approaches that rely on the use of statistics for extracting correlations between structure and activity. The others are knowledge-based expert systems that rely on a set of programmed rules distilled from available knowledge and human expert judgment. The advantages of *in silico* techniques include time and money savings, reduced use of laboratory animals, and the ability to rapidly screen large numbers of structures even before synthesis occurs. This high capacity improves prioritization in early discovery for toxicology testing and highlighting toxophores for easy identification. Because it may be difficult to avoid some of these functionalities in the design of new compounds, in some cases their presence should be considered a "structural alert" for the drug metabolism scientist. In turn, studies can be conducted at an earlier stage to determine whether the compound in question undergoes metabolism at the site to generate a potentially toxic intermediate. However, *in silico* methods still cannot completely replace conventional toxicity testing. These systems are not designed/validated to screen for potential IDRs but could be useful as a first step in eliminating compounds with other potential toxic effects.

12.3.2 *In Vitro* Assays

Most *in vitro* assays are amenable to high-throughput screening (HTS), and their use is preferred in early drug discovery. To try to identify reactive metabolites *in vitro*, human tissue, cells, or purified/recombinant enzymes have been used under similar conditions to those in typical drug metabolism studies. Since liver, blood, and skin are the most common targets for the majority of IDRs, these systems have been considered. The human liver and blood systems are the most metabolically active in forming reactive intermediates, while the skin has a lower metabolic activity. Although these *in vitro* assays can identify reactive metabolites, it has been difficult to extrapolate the results to *in vivo* settings. Many drugs have reactive intermediates and *in vitro* assays may result in a significant number of false negatives for IDRs. Metabolic models of cytotoxicity and methods to assess cell death have been reviewed elsewhere (Li *et al.*, 2002; Matzinger *et al.*, 2002). These methods of testing cell viability can be useful to evaluate potential toxic mechanisms. For hepatocellular necrosis, hepatocyte culture and crude cytotoxicity or cell death end points (i.e. enzyme leakage and dye exclusion) can be considered toxicologically relevant. Hepatocyte incubation/cultures have received the most attention (Tiller *et al.*, 2002), while fresh hepatocytes are of limited use. Cryopreserved human hepatocytes are commercially available and technological improvements in preservation will facilitate studies to further understand the mechanisms of IDRs.

12.3.3 Reactive Intermediates: Phase I and Phase II

Phase I and/or phase II enzymes can be involved in drug bioactivation. Several phase I enzymes are recognized as important in forming reactive intermediates such as cytochrome P450s (CYPs), myeloperoxidases (MPOs), cyclooxygenases (COXs), aldehyde oxidase, and flavin-containing monooxygenases (FMOs) (Pritsos *et al.*, 1985; Kassahun *et al.*, 1993; Liu *et al.*, 1994; Munns *et al.*, 1997; Timmins *et al.*, 1999; Cuttle *et al.*, 2000). In theory, any enzyme is capable of forming reactive intermediates, and the involvement of CYPs has received most attention. Metabolism by FMOs has a lower potential for the formation of reactive metabolites (except for certain *S* or *N* groups) because substrates do not bind directly to the enzyme and electrons are not transferred, avoiding formation of free radicals. In general, phase II metabolism, involving enzymes such as glucuronyl transferases, sulfatase, and GSH transferases, are less frequently recognized as being involved in reactive intermediate formation, despite being responsible for the formation of many reactive intermediates. The study of acyl glucuronides has attracted much interest, as their reactivity is strongly related to covalent binding and potential toxicity.

12.3.3.1 *Trapping Reactive Intermediates* Reactive electrophile screenings generated from *in vitro* experiments in several tissues (i.e. liver microsomes) have been designed to react with GSH and allow for subsequent analysis of the GSH adducts by liquid chromatography–mass spectrometry (LC-MS) (Chen *et al.*, 2001). Several types of reactive metabolites (epoxide, arene oxide, quinone, quinone imine, quinone methide, iminoquinone methide, nitroso, nitrenium ion, nitro reduction, nitro radical, iminium ion, free radical, *S*-oxidation, Michael acceptor, *S*-oxide, aliphatic aldehyde, or hydrolysis/acetylation) can be trapped in the presence of either GSH or an equimolar mixture of *N*-acetyl-cysteine and *N*-acetyl-lysine. This method is relatively simple and rapid, and has been implemented for HTS to identify reactive metabolites generated by bioactivation with phase I and phase II enzymes (Kassahun *et al.*, 1993; Tang *et al.*, 1996, 1997; Chen *et al.*, 2001; Soglia *et al.*, 2004). A preliminary evaluation of this method was conducted using 20 commercially available compounds with known toxicological profiles in liver microsomes at substrate concentrations of 10 μM (Chen *et al.*, 2001). The results indicated that the method is unlikely to produce false-negative response because relatively safe compounds did not generate GSH conjugates, while 8 of the 10 compounds that are known to generate reactive metabolites resulted in positive responses. The two compounds that are known to generate reactive metabolites, but did not produce positive responses in the current assay were valproic acid and phenytoin. The reason for this was that the formation of the valproic acid reactive metabolite (2,4-dieneVPA) requires P450-oxidation and β-oxidation catalyzed by a mitochondrial coenzyme A-dependent process rather than microsomal enzymes (Kassahun *et al.*, 1993; Tang *et al.*, 1996, 1997; Soglia *et al.*, 2004). The reactive metabolite of phenytoin may be a free radical instead of an epoxide which cannot form a stable GSH conjugate (Liu *et al.*, 1994; Cuttle *et al.*, 2000). This method was further evaluated using 43 compounds as positive controls and 16 compounds as negative controls. The results indicated that 40 of 43 compounds tested gave positive results as expected from the literature. Those producing false negatives were felbamate, trimethoprim, and sulfamethoxazole. The potential reason for this was a relatively low *in vitro* concentration used in the assay (100 μM). On the other hand, all 16 of the compounds used as negative controls were found to be negative as expected, indicating that this method may be useful in determining the potential for a compound to form reactive intermediates. Although feasible for HTS, the strategy has limited applications to trap the most stable GSH adducts. Those unstable GSH adducts can spontaneously regenerate GSH and the reactive metabolite. A highly reactive metabolite can react with an active site residue of the enzyme that forms it and thus may not be trapped by GSH. If the reactive metabolite is a free radical, it will most likely abstract a hydrogen atom from GSH rather than react with it. Alternative trapping reagents may be used for more reactive intermediates (Timmins *et al.*, 1999).

Since thousands of compounds must be screened in a typical pharmaceutical industry setting, it is almost impossible, and very expensive, to have all of them radiolabeled. Using radiolabeled reagents such as GSH and KCN is easier and more cost-effective to obtain an accurate measurement of adduct formation (Nassar *et al.*, 2004a). The reactive intermediates may be formed in other subcellular fractions such as liver S9, mitochondria, and cytosol, and thus, screening for reactive intermediates in these fractions will minimize false negatives. As an example, valproic acid forms reactive metabolites in subcellular mitochondria, as explained earlier. We have compared *in vitro* cellular and subcellular models for identifying drug candidates with the potential to produce reactive metabolites. For purposes of evaluating these models, six known compounds currently in the clinic with dose limitations due to hepatotoxicity were selected. Labetalol, acetaminophen, niacin, iproniazid, 8-hydroxyquinoline, and isoniazid were incubated separately with human liver mitochondria, S9, microsomes, cytosol, and hepatocytes. Liquid chromatography hybrid quadrupole time of flight mass spectrometry (LC-QTOF-MS) methods were used to elucidate metabolic profiles of selected compounds. The phase II metabolites were in good agreement with previously published results. Without supplementation of reagents in subcellular fractions, phase II metabolites were not detected. Hepatocytes also formed GSH conjugates but the relative abundance was lower, indicating that the best model to identify potential reactive metabolites may involve the subcelluar fractions (Nassar *et al.*, 2004a). Using the results for acetaminophen, for example, when reactive intermediates are produced, they can be trapped by GSH, while in the case of labetalol, no reactive intermediates were produced and subsequently no GSH adducts were detected. Studies should be done to improve our knowledge of, and hence techniques used for, screening with the use of radiolabeled reagents, which would give an accurate measurement of the amount of reactive intermediates (Nassar *et al.*, 2004a). Also, we should investigate reactive intermediate formation in subcellular fractions, which will aid in reducing false results (Nassar *et al.*, 2004a).

12.3.3.2 *Trapping Free Radicals* Techniques for trapping free radicals are well established. Free radicals resulting from reduction and oxidation can be trapped with spin-trapping reagents (Pritsos *et al.*, 1985; Timmins *et al.*, 1999; Sanders *et al.*, 2000). For example, a method for trapping free radicals was used on hydrazine analogs and their derivatives which are responsible for hemolytic and hepatotoxic events, presumably via the alkyl or aryl free radicals that oxidize essential cysteinyl residues in proteins or covalently react with biomacromolecules (Gorrod, Whittesea and Lam, 1991). Another example is the bioactivation of phenytoin to a free radical species, proposed to be mediated by COX-1, which can be trapped with the spin trapping agent α-phenyl-*N*-t-butylnitrone (PNB) *in vitro* in embryo cell cultures (Liu *et al.*, 1994; Munns *et al.*, 1997; Timmins *et al.*, 1999; Cuttle *et al.*, 2000). The techniques to trap free radicals need further evaluation to develop high-throughput methodology and establish their role in IDRs.

12.3.3.3 *Trapping Iminium Ion* Trapping iminium ions with cyanide is also a well-known technique (Gorrod, Whittesea and Lam, 1991). The technique can be implemented as a screen for the detection of reactive iminium intermediates by trapping them with radiolabeled cyanide. With (S)-nicotine as the reference compound, the extent of radiolabeled cyanide incorporated into the test compounds can be quantified. If compounds have higher cyanide incorporation than (S)-nicotine, it is considered indicative of the formation of iminium intermediates. However, the iminium ions can also form GSH adducts depending on their relative stability. For example, U-89843 incubated in liver microsomes formed N-acetylcysteinyl and GSH adducts in NADPH-supplemented rat liver microsomes supporting a bioactivation pathway potentially involving an iminium intermediate (Zhao *et al.*, 1996a, b). The reactive metabolites of both DMP 406 and mianserin reacted with a range of nucleophiles, but in many cases the reaction was reversible. The best nucleophile for trapping these reactive metabolites was cyanide (Gorrod, Whittesea and Lam, 1991). These results suggest the potential need for a combination of trapping methods to evaluate the formation of reactive intermediates.

Taken together, the results suggest the need for a combination of trapping methods to evaluate the formation of reactive intermediates.

12.4 PEROXIDASES

Peroxidases are present in polymorphonuclear granulocyte (PMN) and play a significant role in the bioactivation of many drugs in humans. Several *in vitro* systems have been used to try to mimic bioactivation processes *in vivo* (i.e. myeloperoxidase [MPO]/ H_2O_2/Cl_2, HOCl, chloroperoxidase, and horseradish peroxidase [HRP]/H_2O_2). A preliminary evaluation of the nucleophile-activated system was conducted to assess its ability to predict drug-induced agranulocytosis (Richard *et al.*, 1998). In this study, DMP-406 (a clozapine analog), which caused agranulocytosis in dogs, was used as the testing agent, with clozapine and mianserin as positive controls, and olanzapine as a negative control. Clozapine is thought to cause agranulocytosis via a reactive nitrenium ion metabolite produced by neutrophils. It has been demonstrated that the products of clozapine oxidation by HRP/H_2O_2 induced apoptosis in neutrophils at therapeutic concentrations (Colburn, 2003). The major reactive intermediate of DMP-406 is an imine ion in activated neutrophils and is similar to the reactive intermediate responsible for mianserin-induced agranulocytosis. DMP-406 did not increase apoptosis at concentrations <50 µM, while mianserin increased apoptosis at a concentration 10 µM above its therapeutic concentration, indicating that this assay lacks predictability (resulting in false negatives) for both drugs. Olanzapine increased apoptosis at the same concentration as clozapine (1 µM), but this concentration is above the therapeutic concentration of olanzapine, consistent with *in vivo* results. There was no increase in apoptosis with any drug in the absence of HRP/H_2O_2, which forms the reactive intermediates of these drugs. These preliminary results indicate that this assay may be unable to reliably predict the ability of different types of drugs to cause agranulocytosis and that different drugs may induce agranulocytosis by different mechanisms. Drugs that form nitrenium ion intermediates might be better predicted by this system.

12.5 ACYL GLUCURONIDATION AND *S*-ACYL-COA THIOESTERS

Phase II biotransformation is generally considered to be a bioinactivation pathway, but some drugs become bioactivated during this phase (Madden *et al.*, 1993; Takeyama *et al.*, 1993; Liu *et al.*, 1995; Shapiro *et al.*, 1997; Cziraky *et al.*, 1998; Isaacson *et al.*, 1999; Kassahun *et al.*, 2001; Tettey *et al.*, 2001; Nassar *et al.*, 2004a; Schonbeck *et al.*, 2004).

Since acyl glucuronide metabolites of some carboxylic acids might cause fatal IDRs in some patients (Takeyama *et al.*, 1993), much attention has been focused on the formation of reactive intermediates through glucuronidation and the methods to screen these conjugates to select for safer drugs (Takeyama *et al.*, 1993; Cziraky *et al.*, 1998; Schonbeck *et al.*, 2004). Some acyl glucuronides are reactive intermediates, which bind covalently to protein by mechanisms that may or may not result in the cleavage of the glucuronic acid moiety. Carboxylic acids can be metabolized by alternative pathways: (i) by glucuronide conjugation to form *O*-acyl glucuronides and (ii) by CoA conjugation to form *S*-acyl-CoA thioesters. Both pathways result in the formation of potentially reactive, electrophilic intermediates, due to the reactivity of the carbonyl carbon. The electrophilic carbon center can react with nucleophilic targets on macromolecules, forming covalent adducts with biomacromolecules, for example, proteins. Three major factors have been suggested to determine the reactivity of acyl glucuronides *in vivo*. The first is the relative stability of a given acyl glucuronide in aqueous buffer at pH 7.4/37°C. Some compounds with short, first-order half-lives (<0.5 hour) result in a relatively high concentration of drug–protein adducts in human tissues and are associated with a high risk for causing IDRs in patients (Takeyama *et al.*, 1993; Liu *et al.*, 1995; Ruscoe *et al.*, 1995). In contrast, acyl glucuronides with longer half-lives (>5 hours) have rarely been reported to cause IDRs. Second is the degree of substitution at the α-carboxyl carbon of the aglycone (an increase in degree of substitution leads to a decrease in acyl glucuronide reactivity). The third factor is the overall drug exposure (including the up-concentration of the acyl glucuronides) in particular tissue compartments. A recent example of apparent success using this screening strategy is the assessment of telmisartan 1-*O*-acyl glucuronide, the principal metabolite of telmisartan in humans. Telmisartan 1-*O*-acyl glucuronide exhibited a long, first-order degradation half-life of 26 hours compared with the short half-life of diclofenac (0.5 hour), suggesting telmisartan has a low potential for covalent binding (Cziraky *et al.*, 1998). The current literature indicates a low incidence/absence of ADRs for telmisartan, supporting the potential use of this screening approach.

S-acyl-CoA thioesters have received significant attention as reactive intermediates involved in IDRs (Takeyama *et al.*, 1993). They can be formed with carboxylic acid drugs, and the resultant thioesters can be much more reactive than the acyl glucuronides due to the nature of the thioester bond. Similarly to acyl glucuronides, the reactivity of synthetic *S*-acyl-CoA thioesters at pH 7.4/37°C may be an important indicator of their potential to covalently bind to biomolecules. The covalent binding of seven structurally different carboxylic acid drugs for GSH was compared at pH 7.4/37°C. The results indicated that hydrolysis rates might be a good predictor for GSH conjugation, and similarly to acyl glucuronides, substitution at the α-carbon to the thioester bond affects reactivity (ISSX 2003). An HTS assay could be easily implemented to assess reactivity or stability of *S*-acyl-CoA thioesters, but additional studies are necessary to further evaluate the role of this metabolic pathway in IDRs.

12.6 COVALENT BINDING

Several methods are available to detect or quantify covalent binding of drugs and their metabolites to macromolecules, including radiochemical and immunological methods, and protein analysis (LC-MS, matrix-assisted laser desorption ionization time-of-flight (MALDI-TOF), and quadrupole time-of-flight (Q-TOF)). A large number of studies indicate that there is a good correlation between drug-induced toxicities and protein covalent binding (Takeyama *et al.*, 1993; Liu *et al.*, 1995; Ruscoe *et al.*, 1995). The alkylation of certain proteins has been suggested to lead to formation of neoantigens that may trigger immunomediated hepatotoxicity. Immunochemical techniques have been used to identify the protein adducts of hepatotoxic compounds. In an effort to understand the significance of *in vitro* covalent binding, studies have been carried out in animals, where GSH adduct and/or protein adducts were detected in biological fluids (i.e. in samples of bile) (Munns *et al.*, 1997; Ju *et al.*, 1999). Subsequent experiments with radiolabeled compound can reveal more detail on the propensity of the reactive intermediate to covalently bind cellular proteins *in vitro* and *in vivo*. From a quantitative viewpoint, it has been suggested that as a rule of thumb, compounds with potential toxic profiles have a binding affinity of 50 pmol/mg microsomal protein; however, in order to quantify the covalent binding, radiolabeled drug is required (ISSX 2002, 2003). An example of the use of this approach in drug discovery has been presented recently using tritiated analogs to rank compounds with lower covalent binding in microsomal protein (Munns *et al.*, 1997). Another example of this approach has been discussed using four radiolabeled compounds including imipramine and diclophenac, which exhibited >50 pmol/mg suggesting potential risk for documented IDRs. However, additional studies may be necessary to further evaluate the clinical relevance of the 50 pmol/mg value since only liver microsome proteins have been considered. There are many other enzymes that can also form reactive metabolites in other cellular fractions (cytosol or mitochondria), blood, lung, and so on.

The question remains as to what is an acceptable level of covalent binding to proteins/macromolecules in the liver (or other organs) for a drug candidate to be taken into development. Another important consideration is that covalent binding might be an important detoxification pathway for some reactive intermediates, and it should therefore not always be perceived as a negative attribute for a drug in development. Without a sound understanding of the potential implications/mechanisms of covalent binding, a safe and effective drug could be eliminated from further development.

12.6.1.1 *Reactive Oxygen Species (ROS)*

It has been proposed that some IDRs may be the result of oxidative stress (increases in the intracellular levels of ROS) (Uetrecht *et al.*, 1989; Takeyama *et al.*, 1993; Masubuchi *et al.*, 2000, 2002). Generally, ROS are generated as by-products from cellular metabolism, primarily in the mitochondria. When the cellular production of ROS exceeds the antioxidant capacity of the cell, cellular macromolecules such as lipids, proteins, and DNA can be damaged. To prevent oxidative stress under normal physiological conditions, these free radicals are neutralized by an elaborate antioxidant defense system consisting of enzymes (e.g. catalase, superoxide dismutase, and GSH peroxidase), and numerous nonenzymatic antioxidants (e.g. vitamins A, E and C, GSH, ubiquinone, and flavonoids) (Nikulina *et al.*, 2000; Yin *et al.*, 2001; Miyamoto *et al.*, 2003). GSH-associated metabolism is a major mechanism for cellular protection against agents or reactive intermediates. It has been reported that several mechanisms could contribute to cell death associated with oxidative stress, which can be summarized in three steps (O'Donovan *et al.*, 2000). The first stage is the recognition of stress by sensitive protein(s) (e.g. the depletion of GSH), mitochondrial damage, inactivation of critical cellular functions, activation of transcription factors, defense gene expression, protein expression, protein function, and release of pro-inflammatory cytokinase. Stage 2 is the subsequent activation of cellular defenses through phase II metabolism enzymes (e.g. uridine diphosphate-glucuronyltransferase (UDP-GT)), GSH-related enzymes (e.g. glutathione S-transferase (GST)), heat shock proteins (e.g. Hsp72), antioxidants, and cell cycle inhibitors. At stage 3, the cells of tolerant individuals are able to dynamically protect themselves from continued stress, but the cells of susceptible individuals may not be able to do so, leading to premature apoptotic death. Several strategies of antioxidative defense include transition metals inactivated by chelating proteins (e.g. ferritin), and ROS reduced enzymatically (e.g. by the glutathione peroxidase) or nonenzymatically by antioxidants (e.g. by vitamin E, vitamin C, and glutathione).

12.7 MECHANISTIC STUDIES

Screening for cytotoxicity, covalent binding, and oxidative stress may help to characterize reactive intermediates. However, a greater understanding of the role of the reactive metabolite will allow for better decisions to be made in the final assessment of quality candidates. The apparent lack of predictability of IDRs by animal models may be due to (i) differences in drug metabolism (and, therefore, reactive metabolites being formed) between species, (ii) species differences in immune-mediated IDRs, and (iii) predisposing factors in humans that do not usually exist in animal models (e.g. alcohol, coadministration of drugs, or disease). Some of the following mechanistic studies have furthered our understanding of the safety profile of a drug candidate with reactive intermediates.

12.7.1 Covalent Binding to Albumin/Plasma Proteins

In general, covalently modified proteins can be repaired or degraded; if not, they may impair important cellular functions, which could be directly pathogenic (Parrish *et al.*, 1997). Several acyl glucuronides have been demonstrated to covalently bind albumin/plasma proteins *in vitro* and *in vivo* (human volunteers). Studies of covalent binding to albumin can be performed with or without radiolabeled compounds. To facilitate these studies, radiolabeled material is preferred and can be useful to identify plasma proteins. Synthetic standards of intermediate metabolites (i.e. acyl glucuronides) are required to assess covalent binding to these proteins. Although non-radiolabeled material can be used to identify albumin covalent binding, quantitation via this approach is difficult. The purpose of these *in vitro* or *in vivo* studies is to characterize the disposition of reactive intermediates in plasma; these studies cannot answer the question of whether plasma proteins can function as protectants of reactive intermediate toxicity or are pathogenic after modification. For some acyl glucuronides, current, indirect evidence suggests a link between covalent plasma binding of tissues and potential for IDRs.

12.7.2 Time-Dependent Inhibition

CYP time-dependent inhibition studies can potentially identify reactive intermediates when they are covalently bound to metabolizing enzymes, as this interaction decreases the activity of the latter. CYP time-dependent inhibition assays are widely used and implemented in a high-throughput manner during early drug discovery as part of the candidate selection criteria. However, these data may be used only as preliminary information since some reactive intermediates do not exhibit time dependency. For example, acetaminophen (APAP) is not a CYP inhibitor but its reactive intermediate N-acetyl-4-benzoquinone imine (NAPQI) is formed by CYP2E1. The stability of the reactive intermediate may be an important factor that determines the effect on time dependency. A highly reactive intermediate may react with the metabolizing enzyme, rendering it unable to distribute to other cell compartments or tissues, while a reactive intermediate with moderate reactivity may diffuse into additional cell compartments. For example, APAP reactive metabolite appears to distribute from the endothelium reticulum into the mitochondria where it reacts with critical proteins. On the other hand, tienilic acid induces hepatotoxicity, and patients form anti-LKM2 (liver kidney microsomes type 2) antibodies in the serum which recognizes CYP2C9. Tienilic acid covalently binds to CYP2C9, the major CYP metabolizing enzyme, and inhibits its activity in a time-dependent manner (Mohsin *et al.*, 2004). The time inhibition assay may record reactive intermediates with moderate reactivity as false negatives, and may be limited to the activity of the enzyme that is being measured.

12.7.3 Antioxidants/Trapping Reagents

In vivo and *in vitro* studies have been conducted with supplementation/depletion of antioxidants to elucidate toxicity mechanisms. The results of these studies indicate that antioxidants can prevent covalent binding of several types of reactive intermediates to biomolecules and they can also prevent oxidative stress generated by non-covalent interactions. For example, addition of antioxidants such as GSH or ascorbic acid resulted in significant prevention of CYP1A2-dependent cytotoxicity and protein-reactive metabolite formation of tacrine. In another *in vitro* example during early discovery, the amount of reactive intermediate covalently bound to microsomal protein was reduced on addition of GSH to the microsomal incubation, as determined using radio-labeled compounds (Munns *et al.*, 1997). Another interesting approach is the use of the spin-trap agent PNB *in vivo*. This trapping agent was used with thalidomide, which initiates embryonic DNA oxidation and teratogenicity in rabbits, both of which were abolished by pre-treatment with PNB.

Several potential biomarkers of oxidative stress can be monitored to evaluate the potential toxic effects of drugs (Uetrecht *et al.*, 1994; Masubuchi *et al.*, 2000). Drugs that undergo redox cycling or form free radicals can generate toxic effects through oxidative stress without forming covalent adducts with biomolecules. This suggests that certain groups of compounds can be screened out in the drug design. The overall data in this area of research suggest that more study is required to select the appropriate biomarker(s) of oxidative stress that can aid in the assessment of the safety profile of a new potential drug candidate. The potential advantage of an oxidative stress biomarker is that it can be monitored during *in vitro* and/or *in vivo* studies.

12.7.3.1 *Biological Markers (Biomarkers)* Biomarkers can be measured and quantified, providing useful information for a wide range of clinical and preclinical uses (Colburn, 2003; Ilyin, Belkowski and Plata-Salaman, 2004; Lesko and Atkinson, 2001). Some potential examples of biological parameters which can be measured include: concentration of specific enzyme(s) and/or specific hormones; specific gene phenotype distribution in a population; presence of biological substances which are useful as indicators for health and physiology related assessments such as disease risk, psychiatric disorders, environmental exposure and its effects; disease diagnosis; metabolic processes; substance abuse; pregnancy; cell line development; and epidemiologic studies.

These and other parameters can be used to identify a toxic effect in an individual organism and can be used to extrapolate between species. Biomarkers can serve to confirm diagnoses, monitor treatment effects or disease progression, and predict clinical results.

Biomarkers are clearly indicated as having important roles in drug development for a number of situations, such as their ability to provide a rational basis for selection of lead compounds, as a help in determining the ability to work toward qualification and use of a biomarker as a surrogate end point. Changes in a biomarker can provide useful indicators for pathophysiology, which in turn are important in identifying a suitable therapeutic target. For example, the association of elevated serum cholesterol levels with an increased incidence of coronary heart disease provides an underlying rationale for developing drugs that lower cholesterol by inhibiting 3-hydroxy-3-methylglutaryl coenzyme A reductase. Thus, total cholesterol is a good example of a clinical biomarker that has been qualified for use as a surrogate end point (Schonbeck *et al.*, 2004; Cziraky *et al.*, 1998).

Biomarkers are also important to the preclinical assessment of the potential benefits and harmful effects of a new drug candidate. Screening tests in animals using biomarkers, such as reduction of blood pressure, provide important demonstration that a compound is likely to produce the intended therapeutic activity in patients. By measuring blood levels during adverse events, such as seizures, in animal toxicology studies may help guide the design of dose escalation studies in humans and serve as a surrogate for preventing or reducing the likelihood of similar adverse events in humans. Biomarkers for potential toxicity play an equally important role. For example, a drug found to prolong the QT interval in animals may warn of potential cardiovascular risk in subsequent clinical studies. Also, biomarkers could be glutathione conjugates and/or glutathionylated or oxidized proteins detectable both in vitro and in vivo. Biomarkers of oxidative stress involving gene/protein expression at least require a whole cell system. It is highly unlikely that a single compound/gene/protein/function will be an effective biomarker, however combinations of these may prove more successful. Several potential biomarkers of oxidative stress can be monitored to evaluate the potential toxic effects of drugs. Drugs that undergo redox cycling or form free radicals can generate toxic effects through oxidative stress without forming covalent adducts with biomolecules. This suggests that certain groups of compounds can be screened out in the drug design phase. The potential advantage of an oxidative stress biomarker is that it can be monitored during in vitro and/or in vivo studies. Pharmacokinetic-pharmacodynamic studies using biomarkers may be particularly useful; for example, one such study showed good correlation between the hypotensive effects of an antiarrhythmic drug in dogs and humans. One shortcoming is that most biomarkers, used singly, are unlikely to capture all the effects of a drug, and thereby fulfill the most stringent criterion for a surrogate end point, although by using several of them in combination, it is more likely, and desirable, to produce evidence which is consistent enough to point in a particular direction.

12.8 PRECLINICAL DEVELOPMENT

The main goal in preclinical development is the extrapolation of efficacy and safety data from animal and *in vitro* models to humans (Fan *et al.*, 2001). This approach should be considered when a metabolic pathway is suspected to play a major role in this extrapolation. One example of the usefulness of this technique is illustrated by tamoxifen, which results in the development of hepatic tumors in rats, although these tumors were not detected in mice in the rodent carcinogenicity assay (Fan *et al.*, 2001). The proposed metabolic pathway responsible for tumorigenesis involves sequential bioactivation of tamoxifen via α-hydroxylation and *O*-sulfonation, and the resultant reactive metabolite reacts with DNA. *In vitro* bridging studies with hepatic subcellular fractions, which formed part of the human risk assessment, demonstrated that rat microsomes were 3-fold more active than human microsomes in forming α-hydroxytamoxifen, and 5-fold more active than human hepatic cytosol with regard to *O*-sulfonation (bioactivation). In contrast, the rate of *O*-glucuronidation (bioinactivation) in human hepatic microsomes was at least 100-fold greater than for rat hepatic microsomes. The dose of tamoxifen required to induce tumors in rats was 40 mg/kg, which is ~150-fold greater than the therapeutic dose in man (0.3 mg/kg). This data suggests a safety margin of 150,000 for risk of tumors, which is consistent with the clinical experience of the drug to date, and is consistent with the mouse data. This example, along with others, illustrates the need for a better understanding of the connection between exposure of reactive intermediates and their reactivity in the assessment of a safety profile. However, a reliable animal model to predict IDRs is needed to be able to justify the use of metabolic profiles in the safety profile assessment. Currently, species differences in metabolic pathways may be considered in the attempt to bridge safety data between animals and humans. Additional knowledge of biomarkers may also help improve the development of this approach.

12.9 CLINICAL DEVELOPMENT: STRATEGY

Drugs dosed at ≤10 mg/day appear to have a low incidence of IDRs, while low potent drugs (β-lactam antibiotics, sulfonamides, phenobarbital, phenytoin, carbamazepine, tricyclic antidepressants, and non-steroidal anti-inflammatory drugs administered at unusually high doses) administered at doses of >10 mg/day may exhibit a higher incidence of IDRs. It has been suggested that more potent drugs should be developed to reduce the risk of IDRs. For example, olanzapine forms a reactive nitrenium ion similar to that formed by clozapine, which is considered to be responsible for clozapine-induced agranulocytosis. Olanzapine, however, is not associated with a significant incidence of agranulocytosis. A critical difference between the two drugs is that of the dose delivered: clozapine is administered at a dose of several hundred mg/day while olanzapine is administered at a maximum dose of 10 mg/day. Another example is troglitazone, which, when administered at 200 to 600 mg/day, resulted in serious hepatic injury and was withdrawn from the market by the FDA in 2000. Rosiglitazone, an analog of troglitazone, is administered at 4 to 8 mg/day and is not hepatotoxic (Mann *et al.*, 1987). The *in vitro* induction of CYP enzymes by rosiglitazone and troglitazone suggests that other thiazolidinediones may have the potential to cause clinically significant drug interactions if administered at sufficiently high doses (Borges *et al.*, 2000).

In an effort to characterize the spectrum of IDRs more accurately, the following issues may be considered during early-to-late clinical phases:

1. Study of high-risk patients to identify pharmacokinetic and pharmacodynamic factors that influence susceptibility to drug toxicity.

2. Development of computer-based schemes to monitor for adverse reactions and adverse events in primary and secondary care.

3. Encouragement to report ADRs to regulatory agencies.

4. Identification of risk factors for different types of drug toxicity using pharmaco-epidemiological approaches.

5. Identification of multigenic predisposing factors to permit the prediction of individual susceptibility.

12.10 CONCLUSION AND FUTURE POSSIBILITIES

During the past decade, there has been an enormous increase in our understanding of how cells and organisms respond to the generation of metabolites, which are chemically reactive. An important consideration is that any screening process will produce false positive and false negative results, and our goal must be to reduce or eliminate these, which in turn will improve drug safety.

There are two major inter-related points of emphasis in the effort to understand and manage IDRs. Firstly, research should focus on the process or mechanism by which IDRs occur. Unfortunately, due to their relative rarity and unpredictability, these have proven difficult to determine with any certainty. Secondly, further studies are required to improve our knowledge in and hence techniques used for screening, and this will be aided by continued improvement in cooperation and dialogue between pharmaceutical companies and academia. For example, greater advances could be gained if researchers involved in this field shared data to correlate human toxicity with animal toxicity or functional assays. Future screens could focus on biomarkers for oxidative stress. Since there is a general inconsistency in the correlation of toxicity with covalent binding, further validation of *in vitro* covalent binding with regard to *in vivo* toxicity is required. In our opinion, future implementation plans should focus on: (i) correlating covalent binding in different *in vitro* systems (animal models and human); (ii) defining biomarkers for oxidative stress; (iii) correlating covalent binding *in vitro* with findings in animals; (iv) continued improvement of databases of genomics/proteomics; and (v) extrapolating data to humans. It is worth stating that drug safety should be further investigated by pharmaceutical companies to conduct post-marketing studies, which are required by the Food and Drug Administration when a safety question arises during the pre-approval period.

REFERENCES

Bailey, M.J. *et al.* (1996). Chemical and immunochemical comparison of protein adduct formation of four carboxylate drugs in rat liver and plasma. *Chem. Res. Toxicol.*, 9, 659–666.

Bakke, O.M. *et al.* (1995). Drug safety discontinuations in the United Kingdom, the United States and Spain from 1974 through 1993: a regulatory perspective. *Clin. Pharmacol. Ther.*, 58 (1), 108–117.

Borges, N. *et al.* (2000). Tolcapone-related liver dysfunction: implications for use in Parkinson's disease therapy. *Drug Saf.*, 26 (11), 743–747.

Burkhart, C. *et al.* (2002). Non-covalent presentation of sulfamethoxazole to human CD4+ T cells is independent of distinct human leucocyte antigen-bound peptides. *Clin. Exp. Allergy*, 32 (11), 1635–1643.

Chen, W.G. *et al.* (2001). Reactive metabolite screen for reducing candidate attrition in drug discovery, in *Biological Reactive Intermediates VI. Chemical and Biological Mechanisms in Susceptibility to and Prevention of Environmental Diseases* (eds P.M. Dansetter, R. Snyder, M. Delaforge, G.G. Gibson, H. Geim, D.J. Jollow, T.J. Monks and I.G. Sipes), Kluwer Academic/Plenum Press, New York, pp. 521–524.

Colburn, W.A. (2003). Biomarkers in drug discovery and development: from target identification through drug marketing. *J. Clin. Pharmacol.*, *43* (4), 329–341.

Coleman, M.D. *et al.* (1994). Reduction of dapsone hydroxylamine to dapsone during methaemoglobin formation in human erythrocytes in vitro. IV: Implications for the development of agranulocytosis. *Biochem. Pharmacol.*, *48* (7), 1349–1354.

Cribb, A.E. *et al.* (1992). Sulfamethoxazole is metabolized to the hydroxylamine in humans. *Clin. Pharmacol. Ther.*, *51*, 522–526.

Cuttle, L. *et al.* (2000). Phenytoin metabolism by human cytochrome P450: involvement of P450 3A and 2C forms in secondary metabolism and drug-protein adduct formation. *Drug Metab. Dispos.*, *28* (8), 945–950.

Cziraky, M. *et al.* (1998). Clinical positioning of HMG-CoA reductase inhibitors in lipid management protocols. *Pharmacoeconomics*, *14* (Suppl. 3), 29–38.

Dieckhaus, C.M. *et al.* (2002). Mechanisms of idiosyncratic drug reactions: the case of felbamate. *Chem. Biol. Interact.*, *142* (1–2), 99–117.

Evans, D.C. (2004). Drug-protein adducts: an industry perspective on minimizing the potential for drug bioactivation in drug discovery and development. *Chem. Res. Toxicol.*, *17* (1), 3–16.

Fan, P.W. *et al.* (2001). Bioactivation of tamoxifen to metabolite E quinone methide: reaction with glutathione and DNA. *Drug Metab. Dispos.*, *29*, 891–896.

Godessart, N. *et al.* (1999). Role of COX-2 inhibition on the formation and healing of gastric ulcers induced by indomethacin in the rat. *Adv. Exp. Med. Biol.*, *469*, 157–163.

Gorrod, J.W., Whittesea, C.M.C., Lam, S.P. (1991). Trapping of reactive intermediates by incorporation of 14C-sodium cyanide during microsomal oxidation. *Adv. Exp. Med. Biol.*, *283*, 657–664.

Greene, N. *et al.* (2002). Computer systems for the prediction of toxicity: an update. *Adv. Drug Deliv. Rev.*, *54* (3), 417–431.

Grillo, M.P. *et al.* (2001). Effect of alpha-fluorination of valproic acid on valproyl-S-acyl-CoA formation in vivo in rats. *Drug Metab. Dispos.*, *29*, 1210–1215.

Gupta, A.K. *et al.* (1998). Severe neutropenia associated with oral terbinafine therapy. *J. Am. Acad. Dermatol.*, *38*, 765–767.

Haack, M.J. *et al.* (2003). Toxic rise of clozapine plasma concentrations in relation to inflammation. *Eur. Neuropsychopharmacol.*, *13* (5), 381–385.

Halsey, J.P. *et al.* (1982). Benoxaprofen: side-effect profile in 300 patients. *BMJ*, *284*, 1365–1368.

Idanpaan-Heikkila, J. *et al.* (1977). Agranulocytosis during treatment with chlozapine. *Eur. J. Clin. Pharmacol.*, *11*, 193–198.

Ilyin, S.E., Belkowski, S.M., Plata-Salaman, C.R. (2004). Biomarker discovery and validation: technologies and integrative approaches. *Trends Biotechnol.*, *22* (8), 411–416.

Inman, W.H. *et al.* (1988). Blood disorders and suicide in patients taking mianserin or amitriptyline. *Lancet*, *2* (8602), 90–92.

Isaacson, J. *et al.* (1999). Index of suspicion. Case 3. Acetaminophen overdose. *Pediatr. Rev.*, *20*, 309–310, 312–313.

Jardine, D.L. (2002). Hodgkin's disease following methotrexate therapy for rheumatoid arthritis. *N. Z. Med. J.*, *115* (1156), 293–294.

Ju, C. *et al.* (1999). Detection of 2-hydroxyiminostilbene in the urine of patients taking carbamazepine and its oxidation to a reactive iminoquinone intermediate. *J. Pharmacol. Exp. Ther.*, *288*, 51–56.

Judson, P.N. *et al.* (2003a). A comprehensive approach to argumentation. *J. Chem. Inf. Comput. Sci.*, *43*, 1356–1363.

Judson, P.N. *et al.* (2003b). Using argumentation for absolute reasoning about the potential toxicity of chemicals. *J. Chem. Inf. Comput. Sci.*, *43*, 1364–1370.

Kassahun, K. *et al.* (1993). In vivo formation of the thiol conjugates of reactive metabolites of 4-ene VPA and its analog 4-pentenoic acid. *Drug Metab. Dispos.*, *21* (6), 1098–1106.

Kassahun, K. *et al.* (2001). Studies on the metabolism of troglitazone to reactive intermediates in vitro and in vivo. Evidence for novel biotransformation pathways involving quinone methide formation and thiazolidinedione ring scission. *Chem. Res. Toxicol.*, *14*, 62–70.

Lasser, K.E. *et al.* (2002). Timing of new black box warnings and withdrawals for prescription medications. *JAMA*, *287* (17), 2215–2220.

Lesko, L.J., Atkinson, A.J., Jr (2001). Use of biomarkers and surrogate endpoints in drug development and regulatory decision making: criteria, validation, strategies. *Annu. Rev. Pharmacol. Toxicol.*, *41*, 347–366.

Li, A.P. *et al.* (2002). A review of the common properties of drugs with idiosyncratic hepatotoxicity and the "multiple determinant hypothesis" for the manifestation of idiosyncratic drug toxicity. *Chem. Biol. Interact.*, *142* (1–2), 7–23.

Liu, L. *et al.* (1994). In vivo phenytoin-initiated oxidative damage to proteins and lipids in murine maternal hepatic and embryonic tissue organelles: potential molecular targets of chemical teratogenesis. *Toxicol. Appl. Pharmacol.*, *125* (2), 247–255.

Liu, Z.C. *et al.* (1995). Oxidation of 5-aminosalicylic acid by hypochlorous acid to a reactive iminoquinone. Possible role in the treatment of inflammatory bowel diseases. *Drug Metab. Dispos.*, *23*, 246–250.

Madden, S. *et al.* (1993). An investigation into the formation of stable, protein-reactive and cytotoxic metabolites from tacrine in vitro. Studies with human and rat liver microsomes. *Biochem. Pharmacol.*, *46*, 13–20.

Mann, K. *et al.* (1987). Differential effects of a new dibenzo-epine neuroleptic compared with haloperidol. Results of an open and crossover study. *Pharmacopsychiatry*, *20*, 155–159.

Masubuchi, Y. *et al.* (2000). Possible mechanism of hepatocyte injury induced by diphenylamine and its structurally related nonsteroidal anti-inflammatory drugs. *J. Pharmacol. Exp. Ther.*, *292* (3), 982–987.

Masubuchi, Y. *et al.* (2002). Role of mitochondrial permeability transition in diclofenac-induced hepatocyte injury in rats. *Hepatology*, *35* (3), 544–551.

Mattocks, A.R. *et al.* (1990). Trapping of short-lived electrophilic metabolites of pyrrolizidine alkaloids escaping from perfused rat liver. *Toxicol. Lett.*, *54* (1), 93–99.

Matzinger, P. *et al.* (2002). The danger model: a renewed sense of self. *Science*, *296* (5566), 301–305.

Mehta, J.R. *et al.* (1982). Trapping of DNA-reactive metabolites of therapeutic or carcinogenic agents by carbon-14-labeled synthetic polynucleotides. *Cancer Res.*, *42* (8), 2996–2999.

Merlani, G. *et al.* (2001). Fatal hepatoxicity secondary to nimesulide. *Eur. J. Clin. Pharmacol.*, *57*, 321–326.

Meyboom, R.H.B. *et al.* (2000). An ABC of drug-related problems. *Drug Saf.*, *22* (6), 415–423.

Miyamoto, Y. *et al.* (2003). Oxidative stress caused by inactivation of glutathione peroxidase and adaptive responses. *Biol. Chem.*, *384* (4), 567–574.

Mohsin, S.K. *et al.* (2004). Progesterone receptor by immunohistochemistry and clinical outcome in breast cancer: a validation study. *Mod. Pathol.*, *17*, 1545–1554.

Moses, P.L. *et al.* (1999). Severe hepatotoxicity associated with bromfenac sodium. *Am. J. Gastroenterol.*, *94* (5), 1393–1396.

Munns, A.J. *et al.* (1997). Bioactivation of phenytoin by human cytochrome P450: characterization of the mechanism and targets of covalent adduct formation. *Chem. Res. Toxicol.*, *10* (9), 1049–1058.

Nassar, A.E. *et al.* (2004a). Detecting and minimizing reactive intermediates in R&D. *Curr. Drug Discov.*, July, 20–25.

Nassar, A.F. *et al.* (2004b). Strategies for dealing with reactive intermediates in drug discovery and development. *Curr. Opin. Drug Discov. Dev.*, *7*, 126–136.

Nassar, A.E. Kamel, A.M., Clarimont, C. (2004). Improving the decision-making process in structural modification of drug candidates: reducing toxicity. *Drug Discov. Today*, *9* (24), 1055–1064.

Nassar, A.E., Lopez-Anaya, A. (2004). Stagegies for dealing with reactive intermediates in drug discovery and development. *Curr. Opin. Drug Discov. Devel.*, *7* (1), 126–136.

Nikulina, M.A. *et al.* (2000). Glutathione depletion inhibits IL-1 beta-stimulated nitric oxide production by reducing inducible nitric oxide synthase gene expression. *Cytokine*, *12* (9), 1391–1394.

O'Donovan, D.J. *et al.* (2000). Mitochondrial glutathione and oxidative stress: implications for pulmonary oxygen toxicity in premature infants. *Mol. Genet. Metab.*, *71* (1–2), 352–358.

Parrish, D.D. *et al.* (1997). Activation of CGS 12094 (prinomide metabolite) to 1,4-benzoquinone by myeloperoxidase: implications for human idiosyncratic agranulocytosis. *Fundam. Appl. Toxicol.*, *35*, 197–204.

Pirmohamed, M. *et al.* (1998). Adverse drug reactions. *BMJ*, *316* (7140), 1295–1298.

Prescott, L.F. *et al.* (1986). Effects of non-narcotic analgesics on the liver. *Drugs*, *32* (4), 129–147.

Pritsos, C.A., Constantinides, P.P., Tritton, T.R., Heimbrook, D.C., Sartorelli, A.C. (1985). Use of high-performance liquid chromatography to detect hydroxyl and superoxide radicals generated from mitomycin C. *Anal. Biochem.*, *150* (2), 294–299.

Richard, A.M. *et al.* (1998). Structure-based methods for predicting mutagenicity and carcinogenicity: are we there yet? *Mutat. Res.*, *400* (1–2), 493–507.

Ruscoe, J.E. *et al.* (1995). The effect of chemical substitution on the metabolic activation, metabolic detoxication, and pharmacological activity of amodiaquine in the mouse. *J. Pharmacol. Exp. Ther.*, *273*, 393–404.

Sabbaga, J. *et al.* (1993). Acute agranulocytosis after prolonged high-dose usage of intravenous dipyrone—a different mechanism of dipyrone toxicity? *Ann. Hematol.*, *66* (3), 153–155.

Salama, A. *et al.* (1983). The role of metabolite-specific antibodies in nomifensine-dependent immune hemolytic anemia. *N. Engl. J. Med.*, *313*, 469–474.

Sanders, S.P., Bassett, D.J., Harrison, S.J., Pearse, D., Zweier, J.L., Becker, P.M. (2000). Measurements of free radicals in isolated, ischemic lungs and lung mitochondria. *Lung*, *178* (2), 105–118.

Schonbeck, U. *et al.* (2004). Inflammation, immunity, and HMG-CoA reductase inhibitors: statins as antiinflammatory agents? *Circulation*, *109* (21 Suppl. 1), II18–II26.

Senturk, T. *et al.* (1997). Seizures and hepatotoxicity following sulphasalazine administration. *Rheumatol. Int.*, *17* (2), 75–77.

Shapiro, L.E. *et al.* (1997). Comparative safety of tetracycline, minocycline, and doxycycline. *Arch. Dermatol.*, *133*, 1224–1230.

Skjodt, N.M. *et al.* (1999). Clinical pharmacokinetics and pharmacodynamics of bromfenac. *Clin. Pharmacokinet.*, *36* (6), 399–408.

Soglia, J.R. *et al.* (2004). The development of a higher throughput reactive intermediate screening assay incorporating micro-bore liquid chromatography-micro-electrospray ionization-tandem mass spectrometry and glutathione ethyl ester as an in vitro conjugating agent. *J. Pharm. Biomed. Anal.*, *36* (1), 105–116.

Steghens, J.P. *et al.* (2003). Fast liquid chromatography-mass spectrometry glutathione measurement in whole blood: micromolar GSSG is a sample preparation artifact. *J. Chromatogr. B Analyt. Technol. Biomed. Life Sci.*, *798* (2), 343–349.

Stuart-Harris, R.C. *et al.* (1984). Aminoglutethimide in the treatment of advanced breast cancer. *Cancer Treat. Rep.*, *11*, 189–204.

Takeyama, N. *et al.* (1993). Oxidative damage to mitochondria is mediated by the Ca(2+)-dependent inner-membrane permeability transition. *Biochem. J.*, *294* (Pt 3), 719–725.

Tang, W. *et al.* (1996). Characterization of thiol-conjugated metabolites of 2-propylpent-4-enoic acid (4-ene VPA), a toxic metabolite of valproic acid, by electrospray tandem mass spectrometry. *J. Mass Spectrom.*, *8*, 926–936.

Tang, W. *et al.* (1997). A comparative investigation of 2-propyl-4-pentenoic acid (4-ene VPA) and its alpha-fluorinated analogue: phase II metabolism and pharmacokinetics. *Drug Metab. Dispos.*, *25* (2), 219–227.

Tettey, J.N. *et al.* (2001). Enzyme-induction dependent bioactivation of troglitazone and troglitazone quinone in vivo. *Chem. Res. Toxicol.*, *14* (8), 965–974.

Thompson, C.D. *et al.* (1996). Synthesis and in vitro reactivity of 3-carbamoyl-2-phenylpropionaldehyde and 2-phenylpropenal: putative reactive metabolites of felbamate. *Chem. Res. Toxicol.*, *9*, 1225–1229.

Tiller, P.R. *et al.* (2002). Liquid chromatography/tandem mass spectrometric quantification with metabolite screening as a strategy to enhance the early drug discovery process. *Rapid Commun. Mass Spectrom.*, *16*, 1225–1231.

Timmins, G.S., Liu, K.J., Bechara, E.J., Kotake, Y., Swartz, H.M. (1999). Trapping of free radicals with direct in vivo EPR detection: a comparison of 5,5-dimethyl-1-pyrroline-N-oxide and 5-diethoxyphosphoryl-5-methyl-1-pyrroline-N-oxide as spin traps for HO* and SO4. *Free Radic. Biol. Med.*, *27* (3–4), 329–333.

Uetrecht, J.P. *et al.* (1989). Idiosyncratic drug reactions: possible role of reactive metabolites generated by leukocytes. *Pharm. Res.*, *6* (4), 265–273.

Uetrecht, J.P. *et al.* (1994). Metabolism of vesnarinone by activated neutrophils: implications for vesnarinone-induced agranulocytosis. *J. Pharmacol. Exp. Ther.*, *270*, 865–872.

Uetrecht, J.P. *et al.* (1995). Oxidation of aminopyrine by hypochlorite to a reactive dication: possible implications for aminopyrine-induced agranulocytosis. *Chem. Res. Toxicol.*, *8* (2), 226–233.

Uetrecht, J.P. *et al.* (2002). Screening for the potential of a drug candidate to cause idiosyncratic drug reactions. *Drug Discov. Today*, *5*, 832–837.

Utili, R. *et al.* (1977). Dantrolene-associated hepatic injury. Incidence and character. *J. Gastroenterol.*, *72*, 610–616.

Ware, J.A. *et al.* (1998). Immunochemical detection and identification of protein adducts of diclofenac in the small intestine of rats: possible role in allergic reactions. *Chem. Res. Toxicol.*, *1998* (11), 164–171.

Woosley, R.L. *et al.* (1978). Effect of acetylator phenotype on the rate at which procainamide induces antinuclear antibodies and the lupus syndrome. *N. Engl. J. Med.*, *298*, 1157–1159.

Yin, J.H. *et al.* (2001). Inducible nitric oxide synthase neutralizes carbamoylating potential of 1,3-bis(2-chloroethyl)-1-nitrosourea in c6 glioma cells. *J. Pharmacol. Exp. Ther.*, *297* (1), 308–315.

Yunis, A.A. *et al.* (1980). Chloramphenicol toxicity: pathogenetic mechanisms and the role of the p-NO2 in aplastic anemia. *Clin. Toxicol.*, *17* (3), 359–373.

Zhao, Z. *et al.* (1996a). Bioactivation of 6,7-dimethyl-2,4-di-1-pyrrolidinyl-7H-pyrrolo[2,3-d] pyrimidine (U-89843) to reactive intermediates that bind covalently to macromolecules and produce genotoxicity. *Chem. Res. Toxicol.*, *9* (8), 1230–1239.

Zhao, Z. *et al.* (1996b). In vitro and in vivo biotransformation of 6,7-dimethyl-2,4-di-1-pyrrolidinyl-7H-pyrrolo[2,3-D]pyrimidine (U-89843) in the rat. *Drug Metab. Dispos.*, *24* (2), 187–198.

CHAPTER 13

Screening for Reactive Metabolites Using Genotoxicity Arrays and Enzyme/DNA Biocolloids

JAMES F. RUSLING, ELI G. HVASTKOVS, and JOHN B. SCHENKMAN

13.1 INTRODUCTION

Therapeutic drugs and chemicals used in our bodies or in our environment must be guaranteed safe to the people who are exposed to them. Extensive procedures for screening and predicting toxicity have been developed in the pharmaceutical industry. Nevertheless, drugs that are toxic to some subset of the population are not always identified by these procedures. About 30% of drug development failures are linked to toxicity issues, and unfortunately some of these do not come to light until clinical trials or even after the drug is introduced to the market. In addition, drug costs correlate with drug development failures (Caldwell and Yan, 2006). For these reasons, predicting drug toxicity at the earliest stages of development has become a critical goal (Nasser, Kamel and Clarimont, 2004).

A wide variety of strategies have been proposed for early toxicity prediction, including *in silico* methods along with a range of *in vitro* and *in vivo* biological approaches (Nasser, Kamel and Clarimont, 2004; Mayne, Ku and Kennedy, 2006; Kramer, Sagartz and Morris, 2007). Established methods use microsomes, cell cultures, or animal models and tend toward utilization of biochemical end points that are the result of complex responses to the drug (Kramer, Sagartz and Morris, 2007). These methods are typically combined into a panel of methodologies that in many cases provide a reasonably good prediction of human *in vivo* toxicity (Mayne, Ku and Kennedy, 2006). Nonetheless, unpredicted or idiosyncratic drug toxicity can result from interindividual variations in human biochemistry and resulting drug behavior in specific individuals that may be impossible to predict from batteries of toxicity tests and sometimes even from human clinical trials limited to subsets of the population that will eventually use the drug.

Certainly existing toxicity testing and prediction methods are important, viable, and useful. However, there is an unfilled niche for simple, cheap, high-throughput,

Drug Metabolism Handbook: Concepts and Applications, Edited by Ala F. Nassar,
Paul F. Hollenberg, and JoAnn Scatina
Copyright © 2009 by John Wiley & Sons, Inc.

biochemically based screening assays that can be arranged into biosensor or biosensor array formats that are currently emerging (Stoll *et al.*, 2004; Rusling, Hvastkovs and Schenkman, 2007). Inexpensive toxicity devices of this sort could be used at very early stages of drug development and contribute important information about toxic drug candidates to aid in screening decisions. While the development of such sensing devices is in its infancy, this approach has the potential to help decrease drug development expenditures and, ultimately, costs of drugs to the public. In this chapter, we discuss fabrication and measurement principles for such devices, as well as recent examples of such approaches.

Reactions of DNA with drug molecules or their enzyme-generated metabolites can produce covalently linked nucleobase adducts which may initiate cancer (Jacoby, 1980; Friedberg, 2003; Scharer, 2003). These adducts most often occur on guanines and adenines in DNA, and serve as good biomarkers for cancer risk (Warren and Shields, 1997; Phillips *et al.*, 2000). They are also convenient biomarkers for detecting reactive metabolites and predicting drug toxicity (Tarun and Rusling, 2005a; Rusling, Hvastkovs and Schenkman, 2007). The reactive metabolites also modify proteins and other biomolecules, but DNA adducts are readily measured by a number of modern bioanalytical techniques including liquid chromatography–mass spectrometry (LC-MS). "Bioactivation" is the term used to denote generation of reactive metabolites by cytochrome P450s (cyt P450) and other metabolic enzymes. The process of forming reactive metabolites and causing DNA damage is usually called genotoxicity. Many substrates yield DNA-reactive metabolites, including styrene, benzo[*a*]pyrene (B[*a*]P), nitrosamines, napthylamines, and tamoxifen and other chemotherapeutic agents (Bond, 1989; Cavalieri *et al.*, 1990; Pauwels *et al.*, 1996; Umemoto *et al.*, 2001; Wang *et al.*, 2001). These reactive intermediates can also damage proteins and other biomolecules. In general, there is much more published information about the chemistry of reactive intermediates from nondrug molecules such as pollutants, since information for drug candidates that never came to market may be considered proprietary.

Cyt P450s are metabolic iron-heme enzymes (P—Fe) that catalyze the transfer of oxygen atoms to organic substrates (Schenkman *et al.*, 1993; Lippard and Berg, 1994; Ortiz de Montellano, 2005). Figure 13.1 depicts an accepted mechanistic model for cyt P450 enzyme catalysis that can lead to bioactivation. The resting state of the catalytic iron heme has water bound to the distal side of P—Fe(III) **(1)**. This water freely exchanges with its environment and does not participate in oxygen transfer reactions (Lippard and Berg, 1994). This form of the enzyme binds the substrate RH and eliminates the distal water **(2)**. Substrate is not bound directly to the iron heme of P—Fe(III), but sits above it in a hydrophobic pocket within the protein. Next, **2** is reduced by one electron to the P—Fe(II) state by an NADPH-dependent reductase to give **3**. P—Fe(II) **3** then binds dioxygen at the distal site to form a ferrous-dioxygen or ferric superoxy complex **(4)** (Guengerich, Bell and Okazaki, 1995; Guengerich, 2001; Ortiz de Montellano and De Voss, 2002). This P—Fe(II)—O_2 complex is converted to a P—Fe(III)—OOH complex via a one-electron reduction to give **5**, then protonated to **6**. The electron to generate **5** comes from an NADPH-dependent reductase in most cases. P—Fe(III)—OOH **(6)** may also be generated by addition of hydrogen peroxide or organic peroxides in a reversible process called the peroxide shunt. Protonation of **6** and dioxygen bond cleavage leads to formation of the highly reactive heme-iron(IV)-oxo radical cation {(P•+)Fe(IV)-oxo, **7**}, which transfers oxygen to the bound substrate to generate the product (ROH). The product dissociates from the enzyme and the distal site is reoccupied by water for another catalytic cycle. In addition, exposure of ferrous form **3** to

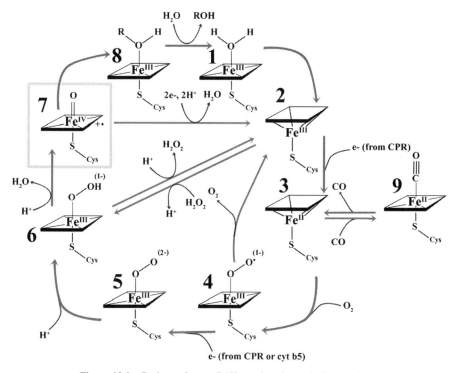

Figure 13.1   Pathway for cyt P450-catalyzed metabolic reactions.

carbon monoxide produces the P—Fe(II)—CO complex **9** that absorbs light at ~450 nm and gives cyt P450s their characteristic name.

A simple example of bioactivation involves the conversion of styrene to styrene oxide by cyt P450 enzymes. In this case, styrene oxide is the only enzyme-generated metabolite. We use this example to show how the natural bioactivation process can be mimicked to produce and detect DNA adducts.

DNA damage can be used as an end point for formation of the reactive metabolite styrene oxide. In Fig. 13.2, we see that if styrene is converted to styrene oxide by a cyt P450 enzyme in the presence of DNA, DNA will be damaged, mainly by forming covalent adducts with guanine bases (Tarun and Rusling, 2005a; Rusling, Hvastkovs and Schenkman, 2007). Further, we show how thin films of metabolic enzymes and DNA grown on arrays or colloidal particles can facilitate detection of these damage events, rapidly and nonspecifically by novel array technology, or more specifically by capillary LC-MS/MS (capLC-MS).

In the next section, we describe how these concepts can be utilized to make electrochemiluminescent arrays for drug screening, specifically addressing formation of reactive intermediates, genotoxicity, and enzyme inhibition. In Section 13.3, we show how the same concepts can be used for the rapid, sensitive identification and quantitation of major reactive intermediates and their nucleobase adducts by capLC-MS. In Section 13.4, we survey alternative emerging technologies including optical protein microarrays that can be applied to drug toxicity screening.

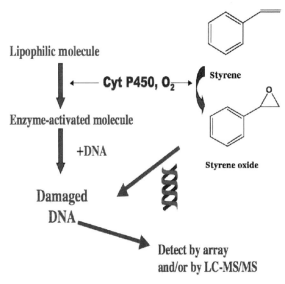

Figure 13.2 Illustration of scheme used for toxicity screening from the formation of reactive intermediates using DNA damage as a measured end point.

13.2 ELECTROCHEMICAL AND ELECTROCHEMILUMINESCENT ARRAYS

A major research effort in our laboratory has been directed toward developing biosensor arrays for genotoxicity screening (Rusling, 2004, 2005; Rusling *et al.*, 2007; Rusling, Hvastkovs and Schenkman, 2007). These arrays all utilize a versatile layer-by-layer film fabrication technique (Lvov, 2000, 2001; Ariga, Hill and Ji, 2007; Zhang, Chen and Zhang, 2007) to grow 20- to 40-nm-thick films containing metabolic enzymes and DNA (Fig. 13.3). These films facilitate the two-step process illustrated in Fig. 13.2. That is, the enzyme reaction produces metabolites that may or may not damage DNA, and the array measurement step detects the relative degree of DNA damage.

The enzyme reaction is the first step of array operation. It was established long ago that metabolic specificity is a property of the terminal oxidase, cyt P450 (Rahimtula *et al.*, 1978; Werringloer, Kawano, Estabrook, 1980; White, Sligar, Coon, 1980; Adams and Adams, 1992). Thus, cyt P450 enzymes in the films can be activated by small concentrations (≤ 1 mM) of hydrogen peroxide, in the reverse of the well-known hydrogen peroxide shunt (see Fig. 13.1), to give identical products as when the natural oxidoreductase system is used. A few reports documented cases of altered product distributions when cyt P450s were activated using peroxides as opposed to the natural electron donor/reductase cycle (Bichara *et al.*, 1996; Kupfer *et al.*, 2001; Ortiz de Montellano, 2005). However, in all cases examined in our laboratory, hydrogen peroxide activation gives the same metabolites as natural cyt P450 activation by NADPH and cyt P450 reductase. Thus, peroxide activation is appropriate and convenient for use in these sensors. In addition, we shall see that these arrays can also be made using liver micro-

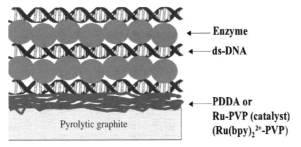

Figure 13.3 Idealized representation of enzyme/DNA films that can be made on any solid surface. The enzyme reaction produces metabolites in close proximity to DNA. In arrays, resulting damage to DNA from the metabolites is detected by voltammetry or electrochemiluminescence.

$$ClRu^{II}\text{-}PVP \leftrightarrow ClRu^{III}\text{-}PVP + e^- \text{ (at electrode)} \qquad (13.1)$$

$$ClRu^{III}\text{-}PVP + G \to ClRu^{II}\text{-}PVP + G^{\bullet} + H^+ \qquad (13.2)$$

Scheme 13.1 Catalytic electrochemical pathway for detection of DNA damage using ClRu-PVP.

somes as the source of metabolic enzymes, and activated by NADPH and cyt P450 reductase as in living systems.

The enzymes in the "nanoreactor spots" on the arrays synthesize reactive metabolites near large concentrations of DNA in each film spot. The rate of DNA damage from metabolite–nucleobase adduct formation is a measure of relative genotoxicity, and can be detected by voltammetric (Wang *et al.*, 2005) or electrochemiluminescent (Hvastkovs *et al.*, 2007) measurement steps.

Arrays for genotoxicity screening arose from of our research on single electrode biosensors that established the optimized parameters and operating conditions (Rusling, 2004, 2005; Rusling, Hvastkovs and Schenkman, 2007). The enzyme reaction is the first step of sensor operation. We ascertained that pH ~5.5 was optimum for DNA damage (Rusling, 2004, 2005), and that enzyme/polyion films 20–40 nm thick eliminate performance limitations from mass transport of reactant entering the film and product leaving (Munge *et al.*, 2003). The second step in the assay is detection of DNA damage, and we found that electrochemical detection with very good S/N can be done with square wave voltammetry (SWV) for DNA oxidation using a soluble electrochemical catalyst, $Ru(bpy)_3^{2+}$ (Zhou *et al.*, 2003; Wang *et al.*, 2005). Alternatively, a catalytic Ru-polyvinylpyridine polymer [$Ru(bpy)_2^{2+}$Cl-PVP or ClRu-PVP) can be incorporated within the film for "reagentless" sensing (Mugweru, Yang and Rusling, 2004). The signal in the measurement step after the enzyme reaction results predominantly from catalytic electrochemical oxidation of guanines in DNA (Scheme 13.1). Here, the catalyst active sites have better access to guanines in the partly unfolded, damaged DNA than in intact ds-DNA. Improved access of catalyst to the guanines provides faster rates of the reaction (Eq. 13.2) in damaged DNA than in undamaged ds-DNA and consequently larger signals (Rusling, 2004, 2005; Rusling *et al.*, 2007).

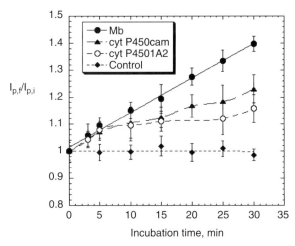

Figure 13.4 Influence of incubation time with 50-μM benzo[*a*]pyrene and 1-mM H₂O₂ on the final/initial peak current ratios from SWV of PDDA/DNA/(enzyme/DNA)₂ films in an eight-electrode array. Control is PDDA/DNA/(Mb/DNA)₂ film in 50-μM benzo[*a*]pyrene alone (four replicates for Mb films, three each for cyt P450 films). Amounts of proteins in the films in nmol cm⁻² were 0.26 for Mb, 0.060 for cyt P450cam, and 0.054 for cyt P540 1A2. Adapted with permission from Wang *et al.* (2005). Copyright 2005 American Chemical Society.

Prototype genotoxicity screening sensors in the format described combined enzyme bioactivation leading to potential DNA damage, mimicking events in the human liver. Direct measurement of formation rates of altered nucleobases by LC-MS from similar films that had been reacted with damage agents and hydrolyzed confirmed that the sensors actually detect DNA damage (Tarun and Rusling, 2005a). Sensor and LC-MS results also correlated well with animal genotoxicity estimated by TDL₀, that is, the lowest dose producing carcinogenicity, in mice and rats (Tarun and Rusling, 2005a; Yang, Wang and Rusling, 2005).

Inclusion of several cyt P450s in electrochemical arrays facilitates analysis of the relative ability of the enzymes to generate genotoxic intermediates. Our first such array had eight electrodes and included cyt P450cam, human cyt P450 1A2, and myoglobin (Mb) (Wang *et al.*, 2005). Mb is a good control in these arrays as it catalyzes some of the reactions of cyt P450s, but usually much less efficiently. In an array with cyt P450s, it should be the least active enzyme or have no activity at all. B[*a*]P was used as a test substrate for these arrays because of its well-understood metabolism resulting in several DNA-reactive metabolites from cyt P450-catalyzed oxidations (Rogan *et al.*, 1993; Todorovic *et al.*, 1997; Neilson, 1998). DNA damage was detected by increases in SWV peak ratios using soluble $Ru(bpy)_3^{2+}$ as the catalyst. Figure 13.4 shows the influence of enzyme incubation time on final/initial sensor peak ratio from such an array. Initial slopes of these plots divided by the amount of each enzyme in the array spots give relative turnover rates in {(nmol enzyme)⁻¹ min⁻¹)}, which were 3.0 for cyt P450cam, 3.5 for cyt P450 1A2, and 0.9 for Mb. Cyt P450cam and cyt P450 1A2 showed threefold higher activity for bioactivation of B[*a*]P for DNA damage than the control enzyme Mb.

The ability to detect metabolite-based DNA damage simultaneously for several enzymes represents a promising approach to identify and characterize genotoxicity pathways. However, the electrochemical arrays described require individual, electronically addressed electrodes and reproducible microelectronic chip manufacture. While this is certainly possible, an instrumentally more convenient approach is based on optical detection of electrochemiluminescence (ECL) from damaged DNA using a similar catalytic detection scheme. This approach requires no special electronics, and detection can been done with a simple charge-coupled device (CCD) camera.

Using ECL, application of a voltage to a polymer in a simple array can produce light. In the absence of DNA, when oxidation catalyst $Ru(bpy)_3^{2+}$ is oxidized to $Ru(bpy)_3^{3+}$, ECL has been generated by using a sacrificial reductant, often tripropylamine or oxalate. This produces a photoexcited $[Ru(bpy)_3^{2+}]^*$ by a pathway involving reaction of a radical form of the reductant with electrochemically generated $Ru(bpy)_3^{3+}$ (Ege, Becker and Bard, 1984; Kenten et al., 1991; Xu et al., 1994). Alternatively, $Ru(bpy)_3^+$ is formed by reduction of $Ru(bpy)_3^{2+}$ by the radical, followed by combination of Ru^I and Ru^{III} complexes to give $[Ru(bpy)_3^{2+}]^*$ (Rubinstein and Bard, 1980), which emits light at 610 nm. $Ru(bpy)_3^{2+}$-labeled DNA using a sacrificial reductant to produce ECL has been developed as the basis of a very sensitive method to detect oligonucleotide hybridization (Leland and Powell, 1990; Blackburn et al., 1991; Xu and Bard, 1995).

The polymer $[Ru(bpy)_2(PVP)_{10}]^{2+}$ (Ru-PVP, PVP = polyvinylpyridine) shown in Fig. 13.5 can be used for "reagentless" DNA sensors since it can be electrochemically activated to react with DNA in films and generate ECL directly (Dennany, Forster and Rusling, 2003). That is, DNA itself is the sacrificial reductant, removing the need for an external sacrificial reductant in DNA detection. ECL can be measured from a single sensor by positioning an optical fiber under an electrode coated with a Ru-PVP/enzyme/DNA film, and directing the light to a photomultiplier tube. This type of sensor allows the simultaneous recording of voltammetric and ECL signals (Dennany, Forster and Rusling, 2003; So et al., 2007). While these sensors are excellent for exploratory studies, high-throughput analysis requires an array format.

The chemistry of the detection process is summarized in Scheme 13.2. ECL is generated from guanine moieties present on DNA strands in the films (Dennany, Forster and Rusling, 2003). Electrochemical oxidation generates Ru^{III}-PVP, and guanine radicals are initially formed (Eq. 13.3) by catalytic oxidation of guanines (Eq. 13.4). These guanine radicals can react with the metallopolymer (Eq. 13.5) to produce electronically excited $Ru^{II}*$ sites in the film (Eq. 13.6). Alternatively, Ru^I-PVP can be formed from reaction of Ru^{II}-PVP with $G^•$, and can then react with Ru^{III}-PVP to produce the excited $Ru^{II}*$ state to give ECL. As in purely electrochemical detection of DNA damage, catalyst active sites gain better access to guanines in partly unfolded, damaged DNA than in intact ds-DNA, thus increasing the rate of the catalytic process and consequently the output of ECL light.

ECL is well suited to arrays without individually addressable electrodes. An ECL array can be spotted directly on a conductive plate, for example, a single pyrolytic graphite (PG) block electrode. We have developed convenient, prototype high-throughput toxicity screening arrays based on multiple spots of DNA, enzyme, and the Ru-PVP polymer (Fig. 13.5). Up to 50 individual spots, each an LbL film containing DNA, enzyme, and Ru-PVP, can be manually micropipetted onto the 1×1 in. array plate. We are also evaluating automated spotting devices that can be used to deposit hundreds of spots on similar arrays.

As in the single probe sensors, the enzyme reaction is run first on this array, which then is washed and placed into an electrochemical cell housed in a dark box with a CCD camera (Fig. 13.6) (Hvastkovs et al., 2007). Upon application of 1.25 V versus

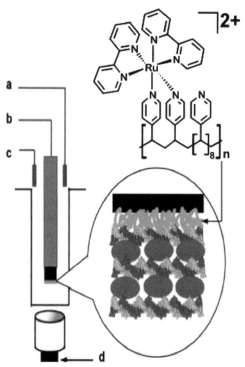

Figure 13.5 Sensor system for simultaneous ECL and voltammetry detection. The reference electrode **(a)**, working electrode **(b)**, and Pt counter electrode **(c)** are located in a glass cell with an extended cylindrical glass base. A fiber optic cable **(d)** is positioned on the outside of the cell directly under the sensor surface and leads to a photomultiplier tube. Ru-PVP (structure shown), DNA (double helix), and enzyme (ovals) layers forming the active sensor film are shown on the right. Only one layer of RuPVP is shown on the sensor, although several can be used to enhance S/N.

$$\text{Ru}^{II}\text{-PVP} \leftrightarrow \text{Ru}^{III}\text{-PVP} + e^- \text{ (at electrode)} \tag{13.3}$$

$$\text{Ru}^{III}\text{-PVP} + G \rightarrow \text{Ru}^{II}\text{-PVP} + G^{\bullet} + H^+ \tag{13.4}$$

$$G^{\bullet} + \text{Ru}^{III}\text{-PVP} \rightarrow G_{2ox} + \text{Ru}^{II*}\text{-PVP} \tag{13.5}$$

$$\text{Ru}^{II*}\text{-PVP} \rightarrow \text{Ru}^{II}\text{-PVP} + h\nu(610 \text{ nm}) \tag{13.6}$$

Scheme 13.2 Electrocatalytic oxidation of guanine employing ECL generating Ru-PVP polymer.

SCE, the electrochemical oxidation of Ru-PVP initiates the measurement chemistry in Scheme 13.2 and generates light from each spot, which is measured with the CCD camera over an integration time of ~20 seconds. As in all catalytic DNA detection schemes, larger signals are obtained from damaged DNA because of better accessibility of the guanines to the catalyst as the ds-DNA unravels.

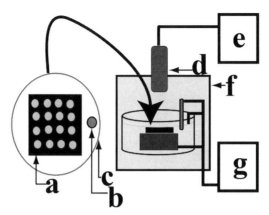

Figure 13.6   ECL arrays for toxicity screening: **(a)** array, **(b)** reference electrode, **(c)** counter electrode, **(d)** CCD camera, **(e)** computer, **(f)** dark box housing, and **(g)** potentiostat to apply voltage for ECL generation. Adapted with permission from Hvastkovs *et al.* (2007). Copyright 2007 American Chemical Society.

ECL arrays can be configured to measure the time course of reactions catalyzed by a single enzyme, or can contain a collection of enzymes for simultaneous comparison of reactive intermediate formation kinetics. Results can be reorganized and presented in a number of ways by computer software. Figure 13.7 shows an ECL array with 49 RuPVP/DNA/enzyme spots designed to study rates of DNA damage for the bioactivation of B[*a*]P by the human enzyme cyt P450 1B1. Individual array spots were subjected to 0.5-mM H_2O_2 and 100-μM B[*a*]P for various times marked on the figure. We can see that light intensity increased with the time of the enzyme reaction. Control spots were exposed to B[*a*]P alone or H_2O_2 alone, and showed little change in ECL. These controls are necessary to confirm that H_2O_2 activates the enzymes but does not damage DNA. Figure 13.7b illustrates a useful way of expressing the array data, as a plot of the final/initial ECL ratio.

Figure 13.8 illustrates oxidation of B[*a*]P with five enzymes in a single array (Hvastkovs *et al.*, 2007). The data have been rearranged by computer software so that representative spots for a single enzyme lie in a specific row. Relative rates of DNA damage were estimated simultaneously in this way for these five enzymes in ~1 minute of enzyme reaction time and 20 seconds of array development time. The slopes of the linear ECL increases correlate with DNA damage rates, and are compared in the graph on the right of Fig. 13.8. Bioactivation producing DNA damage taken as the initial slopes of these graphs was in the order cyt P4501B1 > cyt P4501A2 > cyt P450cam > cyt P4502E1 > Mb, the same as the order of relative metabolic activity of these enzymes toward B[*a*]P (Hvastkovs *et al.*, 2007). The slope of the linear ECL increase estimates the relative activity of different enzymes to produce reactive metabolites.

We have also developed and tested ECL arrays for *N*-nitrosamines (Krishnan *et al.*, 2007). Nitrosamines have been reported to be carcinogens in more than 30 species (Hecht, 1998). Cyt P450s, in particular CYP2E1, bioactivate *N*-nitrosamines to reactive intermediates, mainly by α-hydroxylation. *N*-Nitrosopyrrolidine (NPYR) (Preussmann and Stewart, 1984) was used as the test compound. Adducts form from reaction of DNA with NPYR metabolites (Scheme 13.3) (Wang *et al.*, 2001).

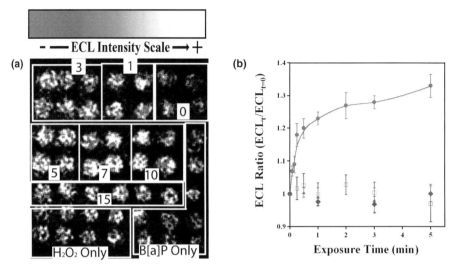

Figure 13.7 **(a)** CCD image of ECL array with 49 individual RuPVP/DNA/enzyme spots all containing cyt P450 1B1. Boxes denote spots that were exposed to 0.5-mM H₂O₂ + 100-µM B[*a*]P for the labeled times (min). Controls (bottom) were exposed to B[*a*]P or H₂O₂ only for increasing amounts of time from 1 to 7 minutes as viewed from right to left (not marked for clarity). **(b)** ECL ratio plot demonstrating the increase in ECL intensity versus time of enzyme reaction. Controls show ECL response versus time exposed to B[*a*]P only (squares), H₂O₂ only (diamonds), and 0.5-mM H₂O₂ + 100-µM B[*a*]P + 30µM of inhibitor αNF (triangles). Adapted with permission from Hvastkovs *et al.* (2007). Copyright 2007 American Chemical Society.

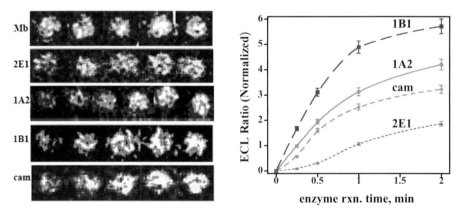

Figure 13.8 ECL array results for enzyme reactions with 100-µM benzo[*a*]pyrene + 0.5-mM H₂O₂ **(a)** reconstructed images for reaction times of 0, 1, 3, 5, and 7 minutes for cyt P450 enzymes and myoglobin on the same array. CCD images are recolorized with the brighter spots showing more DNA damage. **(b)** ECL initial/final ratios normalized for the amount of enzyme in each spot. Adapted with permission from Hvastkovs *et al.* (2007). Copyright 2007 American Chemical Society.

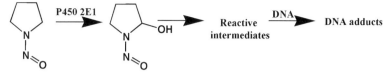

Scheme 13.3   Reaction pathway for the cyt P450 mediated bioactivation of NPYR.

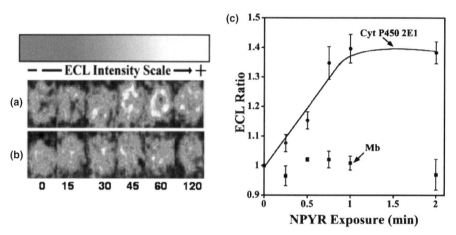

Figure 13.9   ECL array data after exposure to NPYR: **(a)** digitally reconstructed image demonstrating CCD-captured ECL emitted from RuPVP/DNA/cyt P450 2E1 array and **(b)** RuPVP/DNA/Mb array exposed to 150-μM NPYR + 1-mM H_2O_2 for the denoted amounts of time (s). **(c)** Ratio plot demonstrating the ECL signal increase from films containing cyt P450 2E1 or Mb. Adapted with permission from Krishnan *et al.* (2007). Copyright 2007 Royal Society of Chemistry.

Reconstructed ECL arrays (Krishnan *et al.*, 2007) show the ECL response when the DNA/cyt P450 2E1 and DNA/Mb films are exposed to increasing times of damage solution (Fig. 13.9a,b). Each spot contains enzyme, DNA, and RuPVP. The increase in light intensity with time of the Cyt P450 2E1 reaction indicates increased DNA damage via generation of a reactive intermediate. No change in ECL occurred for RuPVP/DNA/Mb spots, showing that Mb does not bioactivate NPYR. Figure 13.9c shows the ECL data in the ratio plot format, and is another way to present the result that cyt P450 2E1 bioactivates NPYR, but Mb does not.

The future vision for this ECL toxicity arrays includes screening for the formation of reactive intermediates (Krishnan *et al.*, 2007; Rusling, Hvastkovs and Schenkman, 2007; So *et al.*, 2007), identifying the enzymes producing them (Wang *et al.*, 2005; Hvastkovs *et al.*, 2007), and applications in drug–drug interactions involving enzyme inhibition (Mugweru and Rusling, 2006). For this type of array, we will need also to utilize phase II and multienzyme metabolic reaction in the arrays. This development is currently under way in our laboratories.

We can also use liver microsomes as enzyme sources in the arrays. Figure 13.10 shows a digitally reconstructed image of an array in which each spot contains DNA,

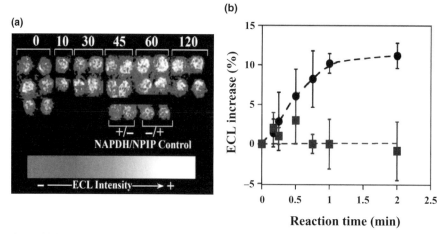

Figure 13.10 Data from ECL arrays using rat liver microsomes (RLMs) as the enzyme source. **(a)** Reconstructed image showing the results of an array exposed to NADPH and 1-mM NPIP for increasing times (in seconds at top). Spot had film architecture (DNA/RuPVP)$_2$DNA(RuPVP/RLM/DNA)$_3$. Controls (bottom) were exposed to only NPIP in buffer or NADPH alone. **(b)** ECL percentage change versus reaction time (●) demonstrating from array runs. Controls (■) correspond to spots incubated in NPIP (no NADPH) under the same conditions.

rat liver microsomes (RLMs), and RuPVP. The spots were exposed to an NADPH-generating system containing 1-mM *N*-nitrosopiperidine (NPIP) for increasing amounts of time at 37 °C. NPIP is a known genotoxic nitrosamine compound that induces esophageal and liver tumors after exposure (Young-Scaime *et al.*, 1995). The use of liver microsomes allows the cyt P450 enzymes in the array to be activated by the natural *in vivo* route (Scheme 13.1) through electron transfer from NADPH via with cyt P450 reductase (CPR) (Ortiz de Montellano, 2005). NPIP is metabolically activated by the cyt P450 enzymes present in the RLM and diffuses through the films to react with nucleophilic DNA. Figure 13.10 shows the increase in ECL intensity with reaction time. Figure 13.10 also shows control experiments where spots were incubated with only NPIP in buffer (no NADPH) or the NADPH system alone. Controls show negligible ECL increases. The relative turnover for NPIP was found by dividing the initial slope of the ECL plot by the amount of RLM present in the film estimated by a quartz crystal microbalance (QCM). For this example, turnover rate was 660 min^{-1} (mg RLM)$^{-1}$.

13.3 DNA/ENZYME BIOCOLLOIDS FOR LC-MS TOXICITY SCREENING

Progress in techniques like electrospray ionization (ESI) and matrix-assisted laser desorption ionization (MALDI) has made MS unprecedented in popularity for bio-analysis. ESI has facilitated coupling of capillary liquid chromatography (capLC) to MS (Vanhoutte *et al.*, 1997; Appruzzese and Vouros, 1999; Gangl, Turesky and Vouros, 2001), minimizing the need to isolate and purify DNA adduct samples and facilitating small sample size and ultrahigh sensitivity in detection of nucleobase adducts (Tarun

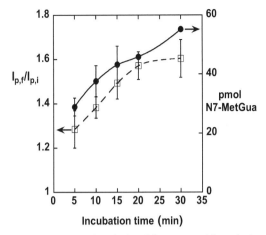

Figure 13.11   Comparisons of electrochemical toxicity sensor with nucleobase adduct formation rate. Peak current ratio $I_{p,f}/I_{p,i}$ (final/initial) for sensors consisting of $(PDDA/DNA)_2$ films on graphite electrode-based ($\square$), and pmol N7-methylguanine found by LC-MS ($\bullet$) of hydrolyzed DNA from similar films on carbon cloth. Both assays after incubation of DNA films with 2-mM MMS. Adapted with permission from Tarun and Rusling (2005b). Copyright 2005 American Chemical Society.

and Rusling, 2005a). LC-MS methods are among the most specific and versatile methods to detect DNA adducts, and can provide chemical structures and rates of formation of specific nucleobase adducts.

In the early days of developing toxicity sensors, we used LC-MS to measure DNA adducts to confirm that the sensors actually detect DNA damage (Zhou *et al.*, 2003; Tarun and Rusling, 2005b). We measured rates of altered nucleobases by LC-MS using DNA films on carbon cloth that had been reacted with damage agents and hydrolyzed (Yang, Wang and Rusling, 2005; Tarun and Rusling, 2006). In an early example, response of electrochemical toxicity sensors gave good correlations with the rate of formation of N-7-methylguanine found by LC-MS after incubation with methyl methane sulfonate (MMS) (Fig. 13.11) and epoxides. Such correlations provided direct evidence that the slopes of sensor response versus enzyme reaction time measure relative rates of DNA damage. Sensor and LC-MS results also correlated well with animal genotoxicity estimated by carcinogenicity index TDL_o in rodents (Tarun and Rusling, 2005b, 2006).

Preliminary results showed that films of enzymes and DNA similar to those used in the arrays could provide nucleobase adducts for LC-MS detection after running the enzyme reaction. These early films for LC-MS were made on carbon cloth, but we later found that increased reaction rates resulting in more sensitive adduct detection could be effected when the films were made on 500-nm-diameter silica colloids. We shall see in the succeeding discussion that these enzyme/DNA biocolloids combined with neutral hydrolysis of the DNA, and capLC-MS with online sample preconcentration provides a very rapid and sensitive method for measuring and identifying reactive metabolites that may be involved in genotoxicity. Together, ECL arrays and capLC-MS analyses provide complementary tools for toxicity studies. The ECL arrays are rapid, inexpensive, high-throughput screening tools, while the capLC-MS approach is a bit more

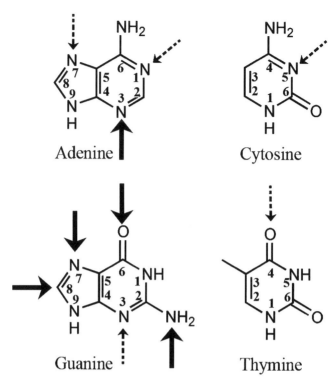

Scheme 13.4 Sites of DNA alkylation subject to attach by reactive drug metabolites. Arrows show major (bold, solid arrow) and minor (broken arrow) alkylation sites.

time-consuming and costly, but provides a wealth of information on reactive intermediates, their nucleobase adducts, and the rates of formation of these products.

Before we describe specific DNA/enzyme biocolloids, we digress to summarize basic DNA chemistry necessary for understanding their operation. Many DNA adducts are formed by alkylating agents (Marnett and Burcham, 1993), and nitrogen and oxygen atoms on nucleobases are active sites (Scheme 13.4). Reactive metabolites are often good alkylating agents or electrophiles and react at one of these nucleophilic sites on the bases (Scheme 13.5). In practice, guanine is the most reactive base. An example of an N7-guanine adduct is shown in Scheme 13.6.

Scheme 13.5 also illustrates the useful sample preparation method of neutral thermal hydrolysis, which selectively ejects damaged DNA nucleobases. Alkylation at N7-guanine and N3-adenine makes the N-glycosidic bonds thermally labile for G and A derivatized at these positions (Zamenhof and Arikawa, 1966; Zoltewicz and Clark, 1972). Thus, simple heating of damaged DNA produces a sample enriched in N7-guanine- and N3-adenine adducts, with a much smaller fraction of native nucleobases than would be obtained by acid or enzymatic hydrolysis (Jacoby, 1980; Tarun and Rusling, 2005b). For this reason, neutral thermal hydrolysis is an excellent sample preparation method for LC-MS.

The use of nanoparticles for biocatalysis was pioneered by Lvov and Caruso (Caruso et al., 1998; Caruso and Möhwald, 1999; Caruso and Schüler, 2000; Lvov and Caruso,

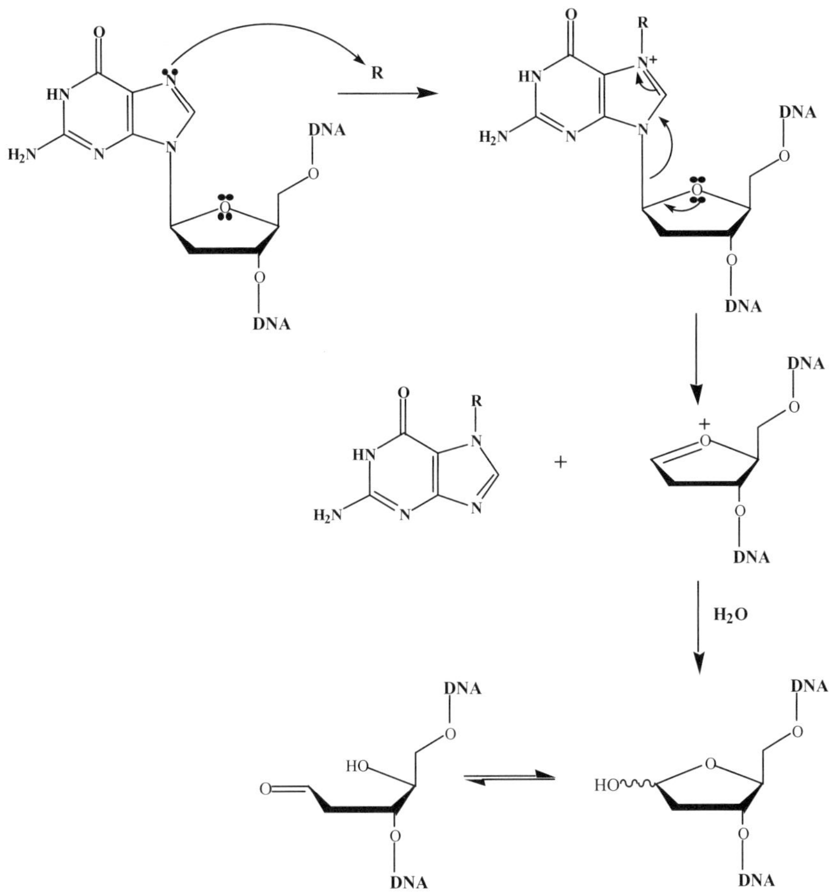

Scheme 13.5 Mechanism of *N*7-alkylation of guanine and subsequent depurination of the damaged nucleobase that can be effected by neutral thermal hydrolysis. R = electrophilic reactive metabolite.

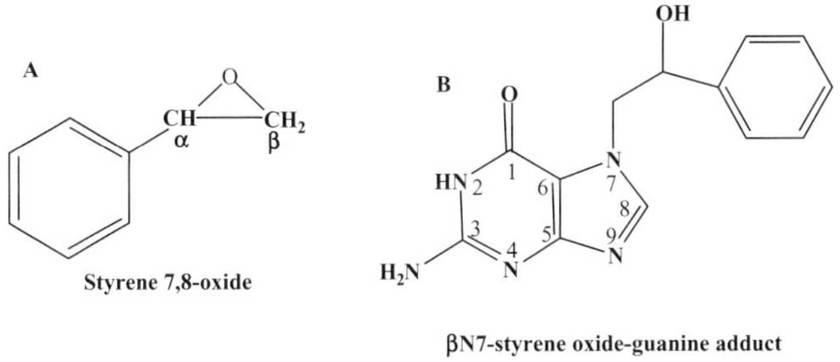

Styrene 7,8-oxide

βN7-styrene oxide-guanine adduct

Scheme 13.6 An example of an *N*7-guanine adduct formed from reaction of the styrene metabolite styrene 7,8-oxide with guanine moieties in DNA.

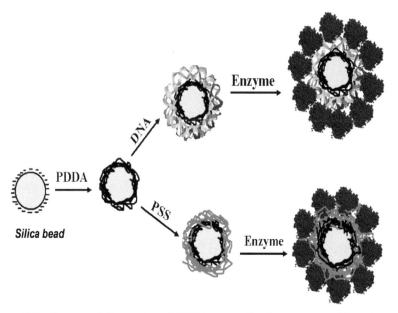

Scheme 13.7 Conceptual illustration of DNA/enzyme film formation on silica microbeads. Layer-by-layer electrostatic assembly is employed to immobilize positively charged PDDA on the negatively charged microbeads, followed by sequential adsorption of negatively charged poly(styrene sulfonate) (PSS) or DNA, and a positively charged enzyme layer.

2001; Fang *et al.*, 2002; Shutava *et al.*, 2004). They used alternate electrostatic layer-by-layer (LbL) film assembly (Lvov, 2000, 2001; Ariga, Hill and Ji, 2007; Zhang, Chen and Zhang, 2007), the same method used for our sensor arrays, to form active enzyme films of glucose oxidase, horseradish peroxidase, and urease on spherical nanoparticles. Enzyme and DNA/enzyme *biocolloids* for reactive intermediate screening are illustrated in Scheme 13.7. Their tiny size offers a very large active surface area and the ability to employ tiny solution volumes to conserve enzyme and obtain more product per unit time for subsequent determination of major and minor products of both the enzyme and DNA reactions.

To demonstrate the biocolloid approach, we show results for styrene oxide and 4-(methylnitrosoamino)-1-(3-pyridyl)-butanone (NNK) oxidized using human cyt P450 2E1 (CYP 2E1) (Bajrami *et al.* , submitted). Cyt P450 2E1 is a prime catalyst in the metabolism of styrene to styrene oxide (Scheme 13.6) *in vivo* (Tanaka, Terada and Misawa, 2000). NNK is an *N*-nitroso procarcinogen in tobacco smoke that has been implicated in cancer risk (Scheme 13.8) (Hecht, 2003; Wang *et al.*, 2003). Styrene and NNK were oxidized with cyt P450 2E1 biocolloids activated by 1 mM H_2O_2. Styrene forms styrene oxide exclusively from the enzyme reaction, while metabolites of NNK undergo hydrolysis to 4-hydroxy-1-(3-pyridyl)-butanone (HPB) (Wang *et al.*, 2003) (Scheme 13.8). Figure 13.12 shows styrene oxide and HPB production as detected by GC for styrene oxide and LC-MS for HPB. The rate of production of both metabolites was initially similar, but over 5 minutes the rate of styrene oxide production was larger than that of HPB.

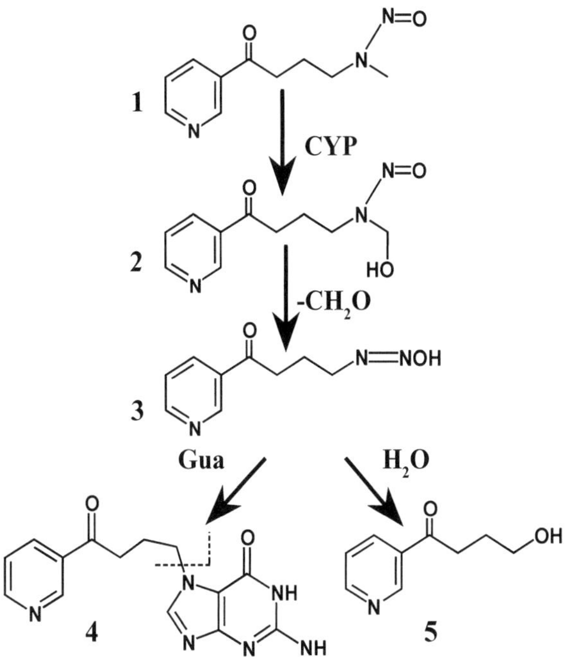

Scheme 13.8 Pathway for cyt P450 (CYP)-catalyzed bioactivation of NNK **(1)** (Wang *et al.*, 2003), which undergoes α-hydroxylation to unstable intermediate **(2)** that loses formaldehyde to form 4-oxo-4-(3-pyridyl)-1-butanediazohydroxide **(3)**. Reaction of **3** with guanine produces several adducts including pyridyloxobutylation at *N*7 **(4**, 7-[4-oxo-4-(3-pyridyl)but-1-yl]Gua). Hydrolysis of **3** produces 4-hydroxy-1-(3-pyridyl)-1-butanone (HPB, **5)**. The dashed line in **4** represents fragmentation in MS.

Figure 13.13 shows data for DNA adducts obtained from NNK metabolites made using DNA/cyt P450 2E1 biocolloids. Figure 13.13a is the capLC/MS-MS chromatogram and 13b is the SRM spectrum showing the product m/z peaks after reaction and neutral thermal hydrolysis. The m/z 299 → 152 is consistent with the previously reported pyridyloxobutylation of guanine via attachment at the most likely *N*7 location in Scheme 13.8 (Wang *et al.*, 2003), which shows an *N*7-pyridyloxobutylated guanine that would produce m/z 152 (Gua) upon fragmentation.

The multiple peaks in Fig. 13.13a are probably due to several positional isomers (Wang *et al.*, 2003), since monitoring at this m/z also detects adducts with linkages at positions other than *N*7 of guanine. The m/z 299 peaks for the hydrolysate are consistent with the formation of pyridyloxobutyl-guanine from the enzyme-catalyzed oxidation of NNK.

In this case, HPB is not the DNA-reactive metabolite. The precursor to HPB is a short-lived species that reacts with DNA (**3**, Scheme 13.8). Figure 13.13c shows that the integrated area of this peak increases linearly similar to HPB in Fig. 13.12. Thus, rate of formation of the specific metabolite produced directly mirrors the rate of formation of the major DNA adducts in the films.

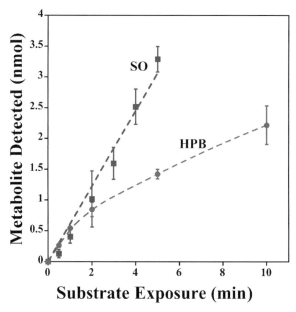

Figure 13.12 Influence of reaction time on amount of styrene oxide (SO, ■) from 1% styrene and HPB (●) from 100-mM NNK found from reaction with PSS/cyt P450 2E1 colloids activated by 1-mM H_2O_2 in pH 5.5 buffer. Adapted with permission from Bajrami *et al.* (2008). Copyright 2007 American Chemical Society.

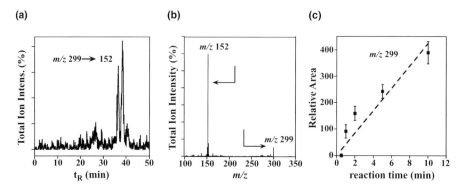

Figure 13.13 CapLC/MS-MS results for NNK reaction catalyzed by DNA/cyt P450 2E1 microbeads in pH 5.5 buffer. **(a)** Chromatogram after a 15-minute reaction. **(b)** Mass spectrum analyzed by SRM (m/z 299 → 152) of same sample as in panel a. © Influence of reaction time on total integrated area of the peaks in panel a at t_R = 37. Adapted with permission from Bajrami *et al.* (2008). Copyright 2007 American Chemical Society.

These results show that DNA/enzyme biocolloids can be used to measure formation rates and confirm structures of both metabolites and DNA adducts. This technique can be used for substrates that do not produce DNA adducts as well as those that do, and can thus be used in other chemical and drug discovery studies such as enzyme inhibition. Finally, we have recently made biocolloids that employ liver microsomes and microsomes genetically engineered to contain a single cyt P450 enzyme and its reductase, so that tedious purification of individual enzymes can be avoided. These microsome biocolloids contain cyt P450 reductase and can be activated with NADPH, exactly as in conventional biocatalysis with microsomes (Krishnan *et al.*, in preparation).

13.4 ALTERNATIVE ARRAYS AND OTHER NOVEL APPROACHES

The distillation of vast numbers of potential lead drug candidates involves initial screening of prospective hit molecules for possible toxicity that might be missed in later *in vivo* tests. Such *in vitro* screens include those for genotoxicity, hERG channel block, drug–drug interactions, and metabolite-mediated toxicity. In addition, these assays often include a standard cytotoxicity (cell destruction) assay (Kramer, Sagartz and Morris, 2007). These assays must be able to detect potential toxicity from thousands of potential candidates in a high-throughput and cost-efficient manner, which is not always realized. In addition, as significant monetary losses in pharmaceutical development occur upon late stage discovery of toxicity, preliminary *in vitro* screens must selectively and sensitively identify toxic candidates while keeping those deemed acceptable in the development pipeline (Nicholson *et al.*, 2002).

13.4.1 Traditional Cell-Based Toxicity Detection

Typically, initial *in vitro* toxicity assays are designed to monitor degrees of cytotoxicity arising from exposure to a compound of interest. Cytotoxicity is typically induced via some form of DNA damage, or genotoxicity (Kramer, Sagartz and Morris, 2007). The bacterial reverse mutation mutagenicity assay (Ames test) (Ames, Mcann and Yamasaki, 1975), the SOS/*umu* test (Oda *et al.*, 1985; Liebler and Guengerich, 2005), the micronucleus test (MNT) (Galloway *et al.*, 1994), the chromosomal aberration (CA) test, the mouse lymphoma assay (MLA) (Kramer, Sagartz and Morris, 2007), and the Comet assay (Singh *et al.*, 1998; Sasaki *et al.*, 2000) for chromosome breakage are currently employed as indirect genotoxicity screens for potential drug candidates and environmental contaminants. The most well known is the Ames test that assays the ability of a compound to induce frame or base pair shifts that facilitate the growth of a bacterial strain in the absence of certain amino acids (Ames, Mcann and Yamasaki, 1975). These assays predominantly measure cell growth, protein expression, or, in the case of the Comet assay, the electrophoresis gel tail length (Singh *et al.*, 1998), to gauge relative genotoxicity. The desirable quality of these tests is that predefined thresholds provide *yes/no* answers to facilitate developmental decisions. In cases generating a "yes" toxicity answer, the tested compound may be shelved or redesigned unless a certain amount of genotoxicity is tolerated, for instance in cancer drug development (Kramer, Sagartz and Morris, 2007). Despite their ubiquitous and required use, generally these tests have low specificity providing negative output for certain classes of compounds, such as aneugens (aneuploidy inducing), or give false positives upon exposure to non-genotoxic compounds (Hastwell *et al.*, 2006).

High-throughput modifications to these mutagenicity and clastogenicity (chromosome breakage) tests have been reported, typically employing microtiter plate formats (Kramer, Sagartz and Morris, 2007). For instance, the use of such a platform imparts a bit of high-throughput nature to the electrophoresis-based Comet (Kiskinis, Suter and Hartmann, 2002) and bacterial mutagenesis Ames II assays (Flukiger-Isler *et al.*, 2004).

Laser scanning cytometry allowed for more rapid analysis of stained erythrocytes in the MNT assay (Styles *et al.*, 2001). A similar detection approach was employed using bone marrow cells in an *in vitro* approach to the MNT assay (Shuga *et al.*, 2007). However, newer genotoxicity screens designed to improve upon the traditional Ames assays have yet to attract significant mainstream industrial attention due to high throughput or specificity limitations (Van Gompel *et al.*, 2005).

Employing a fusion plasmid coupling genes for green fluorescent protein (GFP) to growth arrest and DNA damage (GADD) protein, Hastwell *et al.* demonstrated higher specificity and comparable sensitivity in determining a host of genotoxic agents compared with the Ames test (Hastwell *et al.*, 2006). In addition, the GADD-GFP assay was able to determine the action of aneugens, which the Ames assay cannot. These benefits were hypothetically due to the eukaryotic nature of the assay and the importance of the p53-binding motif designed into the GADD protein. Due to the removal of DNA repair mechanisms in the bacterial strains for Ames assays, some relatively benign xenobiotics produced characteristic false-positive outputs. The GADD-GFP system accounted for these false positives presumably due the use of repair-oriented p53 competent cells in the system (Hastwell *et al.*, 2006).

13.4.2 Metabolite Screening

Another important aspect for *in vitro* toxicity screening is the consideration that many compounds administered *in vivo* are toxic upon metabolism (bioactivation) (Oda *et al.*, 1985; Kramer, Sagartz and Morris, 2007). A common drawback that exists in the typical mutagenicity/toxicity screens is that addition of exogenous liver homogenate (S9 fraction) is necessary to account for reactive metabolite genotoxicity. Indeed, throughput in the aforementioned GPDD-GFP test was hampered by this modification (Hastwell *et al.*, 2006). Although the addition of a liver fraction introduces a variety of enzymes involved in general metabolism, no information is garnered regarding enzyme specificity or mechanism. In an attempt to ascertain such information in toxicity screening, purified or microsomal preparations of cyt P450 enzymes in conjunction with cell lines have been incorporated. Arrays of cyt P450 3A4 and 2D6 sol-gel immobilized in contacting a MCF7 breast cancer cell monolayer were used to determine metabolite efficacy in promoting cytotoxicity. Scheme 13.9 demonstrates the metabolite enzyme toxicity analysis chip (MetaChip) technology, which involves suspending cyt P450 enzymes in a sol-gel mixture and clamping the sol-gel to a monolayer of cells. Cytoxan and Tegafur were metabolized by the immobilized cyt P450s, the respective reactive metabolites diffused into the proximal cancer cells, and cell viability was measured by fluorescence after cell staining (Lee *et al.*, 2005b).

Overexpression of cyt P450 3A4 in supersomes and liver cells has also been exploited to screen for metabolite toxicity. The toxicity of several drugs with known adverse drug reactions (ADRs) toward the HepG2 cell line was reported with cell viability measured using MTT colorimetric and ATP chemiluminescent assays. Ketoconozole inhibition elucidated the action of the particular P450 isoform in the metabolism of these drugs (Vignati *et al.*, 2005). Idiosyncratic drug reactions (IDRs) were studied employing a proteomics approach where a mixture of drugs causing hepatotoxicity *in vivo* induced intersecting extracellular protein biomarkers from the cells (Gao *et al.*, 2004). Drug–protein interactions can be studied by hepatocyte or microsomal incubation of a scrutinized compound and radioactive or LC-MS analysis of adducted proteins (Evans *et al.*, 2004). Further metabolite analysis can be done employing LC-MS methods after

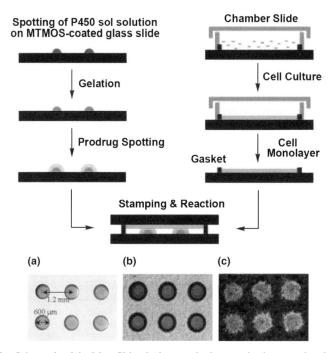

Scheme 13.9   Schematic of the MetaChip platform and microscopic photographs of sol-gel spots. Shown are 30-nL P450 sol-gel spots (**a**), 30-nL sol-gel spots with 60 nL of prodrug solution after being stamped by MCF7 cell monolayer (**b**), and MCF7 cell monolayer after removal from sol-gel array and staining (**c**). Adapted with permission from Lee *et al.* (2005b). Copyright 2005 National Academy of Sciences.

adduction or "trapping" with nucleophilic compounds, such as glutathione (Liebler and Guengerich, 2005; Park *et al.*, 2005; Caldwell and Yan, 2006).

Beyond detection of toxicity at the cellular level, several *in vitro* tests have been designed to predict histological (tissue) level toxicity. Inter-organ interactions were modeled using six different organ and tissue cell types placed in the same well of a specially designed microtiter plate connected with a substrate containing common fluid (Li, Bode and Sakai, 2004). *In vitro* modeling of the blood–brain barrier (BBB) has been accomplished employing brain endothelial and glial cells attached in multi-well plates fitted with a porous membrane modeling the capillary barrier. Transfer between the two wells was monitored and can be used to estimate pharmaceutical transfer between capillaries and the brain (Cecchelli *et al.*, 1999). Embryonic stem cells have been used as a screen for potential terotogenic compounds. The embryonic stem cell test (EST) has the ability to accurately determine strong embryotoxic materials. However, this assay needs to be refined to include maternal metabolism enzymes to preclude false positives (McNeish, 2004).

High-content screening (HCS) is a method that measures several cytotoxicity parameters simultaneously. HepG2 hepatocyte cells in 96-well plates were incubated for three days and exposed to several fluorescent dyes to analyze calcium content, mitochondrial membrane potential, DNA content, and cell number. This system was able to predict the hepatotoxicity of 90% of varying degree liver toxins. Of those, 98% are known to cause some form of *in vivo* toxicity demonstrating the predictive ability of the HCS system (O'Brien *et al.*, 2006).

In order to more accurately model the action of cancerous cell growths, breast cancer cells were encapsulated in alginate poly-L-lysine alginate (ALA) microcapsules and exposed to varying chemotherapeutic drugs. Compared with monolayer culture assays, the studied drugs exhibited altered toxicity profiles in the 3-D microcapsule assay, which demonstrated the complex nature of actual tumor cells and elucidated some serious drawbacks to monolayer cell-based assay formats (Zhang *et al.*, 2005).

13.4.3   -Omics Technologies

Genome sequence alone does not predict disease phenotypes, giving rise to a significant number of approaches designed to analyze gene expression to elucidate such phenotypic outcomes. These fields are termed the -omics fields: genomics (genetic complement), transcriptomics (gene expression), proteomics (cell signaling and protein synthesis), metabolomics (cellular metabolite flux), and metabonomics (endogenous small molecule flux) (Nicholson *et al.*, 2002).

The use of DNA microarrays in transcriptomics or toxicogenomics promises to be a powerful tool in the elucidation of DNA damage mechanism via drug and environmental toxin exposure. In such experiments, selected animals or cell lines are exposed to a compound of interest and after sufficient time, RNA is isolated from the exposed tissue or cells. The RNA is amplified, fluorescently labeled, and allowed to hybridize to array-immobilized DNA probes (Butte, 2002). Additionally, the RNA can undergo reverse-transcription producing the original DNA sequence which can then be analyzed. The overall goal is to determine the regulation of a set of genes due to physiological changes from drug exposure. From a genotoxic standpoint, damage induced by a drug would elicit a damage response from a host of genes that will be detected in the microarray (Ulrich and Friend, 2002). In this fashion, a selection of direct genotoxic agents was discerned from indirect genotoxic agents (i.e. mode of action not directly on DNA itself) based on differences in gene expression profiles after exposure to mouse lymphoma cells (Hu *et al.*, 2004). Gene expression analysis of rat liver cells exposed to a series of marketed thiazolidinedione diabetes drugs was performed to elucidate the IDR mechanism that caused recall of one of the drugs. This method identified several expressed genes involved in DNA damage response, cyt P450 induction, and cell regulatory cycles demonstrating that the IDR response was multifaceted in nature. Also, the DNA expression procedure was able to more sensitively determine IDRs than *in vivo* tests (Vansant *et al.*, 2006).

Microarrays have been employed to measure the damaging effects of electron dosing at 3 eV. Fluorescent DNA microarrays demonstrated that 300 electrons per dT 25-mer oligomer was sufficient to preclude hybridization (Solomun and Sturm, 2007). Overall, despite the promise to elucidate genotoxic mechanism, initial microarray studies have shown less sensitivity than traditional genotoxic assays, such as the MNT assay (Newton, Aardema and Aubrecht, 2004). In addition, questions exist about reproducibility and interpretation, but standardization of protocols can improve interlaboratory findings (Irizarry *et al.*, 2005; Owens, 2005).

Proteomic microarrays have been developed to study translation products in a cellular context. From a toxicity standpoint, the interest lies in the charting of expression levels, emergence of new protein biomarkers, alteration of translated products, or post-translation modifications (Stoll *et al.*, 2004). For instance, microarrays of capture antibodies have been employed to ascertain the changes in protein expression in cancer cells upon radiation treatment (Sreekumar *et al.*, 2001). Sandwich immunoassay arrays have been developed to quantify several different cellular proteins at femtomolar levels (Schweitzer *et al.*, 2000). Limitations in capture antibody development and high-throughput issues will need to be improved before proteomic microarray applications become more widespread (Stoll *et al.*, 2004).

Metabonomics attempts to study the time flux of relative concentrations of endogenous small molecule components of biofluids and tissues (Lindon *et al.*, 2003). In this fashion, a complete end-point analysis due to genetic expression, protein translation, or metabolite exposure of these samples can be discerned and multiple biomarkers of drug exposure can be elucidated. The benefits of this application are numerous and include a global profiling tool at relatively low per sample expense compared with DNA microarray analysis (Griffin and Bollard, 2004). Magic angle resonance (MAR) ^{1}H NMR allows for histological screening on solid tissue samples after drug exposure. This technique requires little sample workup and does not destroy the sample so that it can be used in multiple analyses (Nicholson *et al.*, 2002; Lindon *et al.*, 2003). Coupled with LC-MS, vast amounts of metabolite and other proteomic biomarker data due to drug exposure can be discerned. An example of metabonomic toxicology was presented by Nicholson *et al.* with the rapid determination of fatty acid metabolism impairment by a failed pharmaceutical candidate. The metabonomic profile of the drug was consistent with several lines of *in vivo* rodent and primate experiments (Mortishire-Smith *et al.*, 2004). Despite high promise, drawbacks of this approach are the throughput and sensitivity of the NMR, the complexity of the data analysis, and initial start-up expenses in acquiring equipment with sufficient frequency resolution. Overall, due to the expense and high-throughput limitations, the -omics fields will likely be used extensively in detailed studies of further developed drug candidates to elucidate findings from preliminary *in vitro* genotoxicity screens (Newton, Aardema and Aubrecht, 2004).

13.4.4 Emerging Toxicity Detection Technologies

In addition to some of the aforementioned cell-based cytotoxicity assays, other novel means of detecting xenobiotic toxicity have also been reported. Cell morphology changes eliciting a transduction signal alteration have shown promise. Scheme 13.10 shows the experimental setup described by Schwartz *et al.* using scattered light intensity changes to monitor real-time cellular viability. Rat hepatocytes were immobilized on a porous silica photonic crystal and upon exposure to cytotoxic agents cadmium chloride or acetaminophen, the intensity of scattered light at 640 nm increased demonstrating the alteration in the cellular structure due to the xenobiotic action (Schwartz *et al.*, 2006).

In a similar vein, microelectrode-containing microtiter plates have been employed to measure the impedance change between the electrode and the ambient upon exposure to cytotoxic agents. Cells adhered to the electrode surface were exposed to various cytotoxic agents that induce morphological changes. Cytotoxicity affects cell viability, cell number, and electrode adherence producing a resistance change from electrode to the surrounding buffer that correlates with rates of xenobiotic-induced cytotoxicity. The results obtained in this fashion charted well with standard colorimetric detection (Xiao and Luong, 2005; Xing *et al.*, 2005; Zhu *et al.*, 2006).

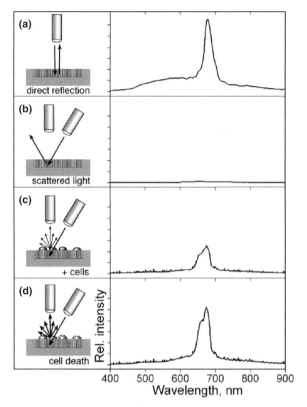

Scheme 13.10 Optical design used to monitor biological events on a polystyrene-filled porous Si photonic crystal, with representative spectra. **(a)** To measure the specular reflection from the photonic crystal, illumination and observation share the same optical path, along the axis normal to the chip surface, leading to a sharp reflectivity peak, as shown on the right of the diagram. **(b)** To measure scattered (diffuse) reflectivity from the photonic crystal, the light source is incident from an off-normal position. The specular reflection from the photonic crystal is no longer observed. **(c)** Placing cells on the surface of the chip introduces scattering centers that direct some of the light into the detection optics. **(d)** Changes in cell morphology alter scattering efficiency, which are detected as changes in the intensity of the spectral peak. Adapted with permission from Schwartz *et al.* (2006). Copyright 2006 American Chemical Society.

Similar electronic microsensors have been employed to monitor certain areas of toxicity upon exposure to exogenous agents. Metabolic oxygen consumption changes and extracellular acidities of tumor cells exposed to drugs were measured concurrently using a multi-parametric electronic sensor. Differences in measured parameters allowed discernment of the mechanism of drug action on the cells (Montrescu *et al.*, 2005). Electrochemical measurement of K^+-induced dopamine release from a microarray of cells demonstrated that exposure to nomifensine, a dopamine transporter inhibitor, can provide a more sensitive toxicity screen than standard staining and counting viable cells (Cui *et al.*, 2006). Also using a microelectrode array format, the decrease in action potentials of cardiac cells was used to estimate the toxicity of pesticides (Natarajan *et al.*, 2006).

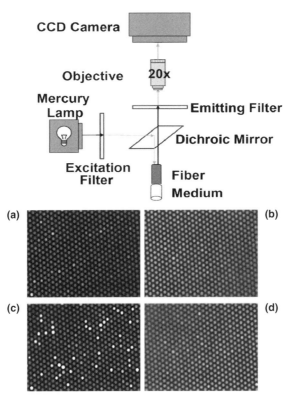

Scheme 13.11 Single-cell array biosensor platform setup and fluorescent images from a small portion of two imaging fiber-based arrays. The left panel shows an array where cells were incubated with medium containing $5\,\mu g\,mL^{-1}$ mitomycin C. Fluorescent signals were acquired at **(a)** t = 0 minute and **(b)** t = 90 minutes. The right panel shows the control array where cells were incubated with only medium and fluorescent signals were taken at **(c)** t = 0 minute and **(d)** t = 90 minutes. The bright spots indicate a fluorescence increase, and the cells exposed to MMC were much brighter than the control. Adapted with permission from Kuang, Biran and Walt (2004). Copyright 2004 American Chemical Society.

Other optical detection strategies to monitor toxicity have also been described. Surface plasmon resonance (SPR) has been used to monitor adsorption processes modeling serum-binding behavior of drugs on surfaces (Cooper, 2002). Alveolar epithelial cells anchored to a fiber optic were employed to assay the cytotoxicity of inhalation hazards employing fiber evanescent wave spectroscopy (FEWS). Wavelengths in the mid-IR range were monitored and compared with a MTT colorimetric assay to ascertain damage by various airborne toxic agents (Riley *et al.*, 2006).

Fiber optic arrays to measure genotoxicity were developed by Walt *et al.* employing *Escherichia coli* carrying a *recA:gfp* fusion reporter plasmid. A single bacterium was loaded into a properly sized microwell on the tip of each fiber optic (shown in Scheme 13.11). Bundling of the individual fibers created a high-density biosensor array. Upon exposure to genotoxic medium, fluorescence from expressed GFP was captured by a distally located CCD camera (Kuang, Biran and Walt, 2004). Likewise, compatible *recA:gfp* and *katG:lux* plasmids inserted into *E. coli* were described to detect and distinguish genotoxic from oxidative cell damage based on increases in fluorescence and bioluminescence, respectively (Mitchell and Gu, 2004a, b). Extending this methodology, *E. coli* containing several different promoters fused with *lux* genes were placed in well plates or constructed chip arrays. Toxicity was assessed from bioluminescence increases detected by a CCD camera and was dependent on reporter gene and employed damage agent (Lee *et al.*, 2005a).

Improvements to the "lab on a chip" concept for biological monitoring continue to be reported. Several unique microfluidic and cell trapping approaches have been described with potential toxicity detection applications (Dittrich and Manz, 2006). *In vitro* protein expression in manufactured oil and water droplets ("biocontainers") was detected fluorescently using a microfluidic approach. Genetic and protein expression components were mixed simultaneously with the oil and water. Expression occurred along a perpendicular reaction strip after mixing, and detection at certain positions along the strip allowed for time-dependent monitoring of protein expression (Dittrich, Jahnz and Schwille, 2005). Also, improvements in such microfluidic devices allow for single-cell immobilization and analysis of the cellular actions upon exposure to an exogenous medium (Dittrich and Manz, 2006). A novel concept in the microfluidics area involves simultaneously monitoring of multiple cell types using a four-compartment flow cell microdevice to mimic *in vivo* metabolism processes. As a test, naphthalene was metabolized in the liver compartment by cyt P450-containing hepatocyte cells followed by flow dispersion into the other compartments of the cell. Depletion of glutathione in the lung compartment denoted reactive metabolite formation (Viravaidya, Sin and Shuler, 2004).

Overall, numerous clever approaches have been explored to measure toxicity of new drug candidates and environmental xenobiotics with varying optical and electrochemical transduction strategies. In order to gain wider acceptance, these platforms must be able to assess chemical toxicity from a wide range of mechanisms in a cost-efficient manner. They must be able to provide more sensitive information, and even chemical structural information, if they are to eventually be used as replacements or in conjunction with traditional toxicity assays. Based on the continual financial driving force to discover toxic pharmaceutical candidates at early stages, we envision that more novel *in vitro* toxicity assays will be developed and employed as initial toxicity screens followed by a more thorough analysis using -*omics* analyses.

13.5 SUMMARY AND FUTURE OUTLOOK

In this chapter, we described new array-based methodology utilizing enzyme/DNA films to screen for reactive drug metabolites based on a DNA-damage end point measured by ECL. Similar enzyme/DNA films on nanoparticles coupled with LC-MS to obtain structure, and rate information for formation of nucleobase adducts and reactive metabolites was also described. Relative rates of DNA damage from both of these biochemical assays can be used to predict relative genotoxicity and the tendency to form reactive metabolites for new drugs. These methods are complementary since the ECL arrays provide rapid screening for reactive metabolites, while the LC-MS approach provides valuable structural and rate of formation information about metabolites and possible nucleobase adducts. Both approaches are adaptable to enzyme inhibition studies. These technologies are currently being developed in higher-throughput versions taking advantage of modern robotics and biomolecule printing devices that handle sub-nanoliter volumes.

This chapter also surveyed recent developments utilizing other types of array and nanoscience-based technologies for drug toxicity assessment. The future appears bright for incorporation of some of these technologies in high-throughput drug toxicity screening. Many of these approaches could be used alongside or in place of certain conventional toxicity tests.

We feel that the new approaches described herein will prove to be valuable in drug toxicity testing along with other reliable tests and *in silico* assessments. We agree with many pharmaceutical scientists that the future lies in developing predictive batteries of tests upon which to base reliable informed decisions about toxicity at early stages in the development life of a drug (Nasser, Kamel and Clarimont, 2004; Caldwell and Yan, 2006; Mayne, Ku and Kennedy, 2006; Kramer, Sagartz and Morris, 2007). Indeed, this is already happening. We feel that it would be advantageous to evaluate newer, faster, more biochemical-based approaches to the toxicity testing toolbox on an experimental basis, and that perhaps drug companies should develop small groups of bioanalytical scientists charged with this specific task. After a period of evaluation, the new technologies could be adopted if they prove valuable, or discarded if they do not. The hope and dream is that such an approach would establish reliable new high-throughput tests in the toxicity toolbox that would eventually lower the number of drug failures late in the development process, and thus lower drug development costs, as well as drug costs to patients.

REFERENCES

Adams, C., Adams, P.A. (1992). A comparative study of pH/activity profiles for the anaerobic H_2O_2 and alkyl hydroperoxide supported N demethylation of N methylaniline catalyzed by alkaline haematin and microsomal cytochrome P-450. *J. Inorg. Biochem.*, *45*, 47–52.

Ames, B.N., Mcann, J., Yamasaki, E. (1975). Methods for detecting carcinogens and mutagens with the Salmonella/mammalian-microsome mutagenicity test. *Mutat. Res.*, *31*, 347–364.

Appruzzese, W.A., Vouros, P. (1999). Analysis of DNA adducts by capillary methods coupled to mass spectrometry: a perspective. *J. Chromatogr. B*, *794*, 97–108.

Ariga, K., Hill, J.P., Ji, Q. (2007). Layer by layer assembly as a versatile bottom-up nanofabrication technique for exploratory research and realistic application. *Phys. Chem. Chem. Phys.*, *9*, 2319–2340.

Bajrami, B., Hvastkovs, E.G., Jensen, G., Schenkman, J.B., Rusling, J.F. (2008). Enzyme-DNA biocolloids for DNA adduct and reactive metabolite detection by chromatography-mass spectrometry. *Anal. Chem.*, *80*, (4) 922–932.

Bichara, N., Ching, M.S., Blake, C.L., Ghabrial, H., Smallwood, R.A. (1996). Propranolol hydroxylation and N-desisopropylation by cytochrome P4502D6: studies using the yeast-expressed enzyme and NADPH/O2 and cumene hydroperoxide-supported reactions. *Drug Metab. Dispos.*, *24*, 112–118.

Blackburn, G.F., Shah, H.P., Kenten, J.H., Leland, J., Kamin, R.A., Link, J., Petermann, J., Powell, M.J., Shah, A., Talley, D.B., Tyagi, S.K., Wilkins, E., Wu, T.-G., Massey, R.J. (1991). Electrochemiluminescence detection for development of immunoassays and DNA probe assays for clinical diagnostics. *Clin. Chem.*, *37*, 1534–1539.

Bond, J.A. (1989). Review of the toxicology of styrene. *CRC Crit. Rev. Toxicol.*, *19*, 227–249.

Butte, A. (2002). The use and analysis of microarray data. *Nat. Rev. Drug Discov.*, *1*, 951–960.

Caldwell, G.W., Yan, L. (2006). Screening for reactive intermediates and toxicity assessment in drug discovery. *Curr. Opin. Drug Discov. Devel.*, *9*, 47–50.

Caruso, F., Lichtenfeld, H., Giersig, M., Möhwald, H. (1998). Electrostatic self-assembly of silica nanoparticle-polyelectrolyte multilayers on polystyrene latex particles. *J. Am. Chem. Soc.*, *120*, 8523–8524.

Caruso, F., Möhwald, H. (1999). Protein multilayer formation on colloids through a stepwise self-assembly technique. *J. Am. Chem. Soc.*, *121*, 6039–6046.

Caruso, F., Schüler, C. (2000). Enzyme multilayers on colloid particles: assembly, stability, and enzymatic activity. *Langmuir*, *16*, 9595–9603.

Cavalieri, E.L., Rogan, E.G., Devaneshan, P.D., Cremonesi, P., Cerny, R.L., Gross, M.L., Bodell, W.J. (1990). Binding of benzo[a]pyrene to DNA by cytochrome P 450 catalyzed one-electron oxidation in rat liver microsomes and nuclei. *Biochemistry*, *29*, 4820–4827.

Cecchelli, R., Dehouck, B., Fenart, L., Buee-Scherrer, V., Duhem, C., Lundquist, S., Rentfel, M., Torpier, G., Dehouck, M.P. (1999). In vitro model for evaluating drug transport across the blood-brain barrier. *Adv. Drug Deliv. Rev.*, *36*, 165–178.

Cooper, M.A. (2002). Optical biosensors in drug discovery. *Nat. Rev. Drug Discov.*, *1*, 515–528.

Cui, H., Ye, J., Chen, Y., Chong, S., Sheu, F. (2006). Microelectrode array biochip: tool for in vitro drug screening based on the detection of a drug effect on dopamine release from PC12 cells. *Anal. Chem.*, *78*, 6347–6355.

Dennany, L., Forster, R.J., Rusling, J.F. (2003). Simultaneous direct electrochemiluminescence and catalytic voltammetry detection of DNA in ultrathin films. *J. Am. Chem. Soc.*, *125*, 5213–5218.

Dittrich, P.S., Jahnz, M., Schwille, P. (2005). A new embedded process for compartmentalized cell-free protein expression and on-line detection in microfluidic devices. *Chembiochem*, *6*, 811–814.

Dittrich, P.S., Manz, A. (2006). Lab on a chip: microfluidics in drug discovery. *Nat. Rev. Drug Dev.*, *5*, 210–218.

Ege, D., Becker, W.G., Bard, A.J. (1984). Electrogenerated chemiluminescent determination of tris(2,2′-bipyridine)ruthenium ion (Ru(bpy)32+) at low levels. *Anal. Chem.*, *56*, 2413–2417.

Evans, D.C., Watt, A.P., Nicoll-Griffith, D.A., Baillie, T.A. (2004). Drug-protein adducts: an industry perspective on minimizing the potential for drug bioactivation in drug discovery and development. *Chem. Res. Toxicol.*, *17*, 3–16.

Fang, M., Grant, P.S., McShane, M.J., Sukhorukov, G.B., Golub, V.O., Lvov, Y. (2002). Magnetic bio/nanoreactor with multilayer shells of glucose oxidase and inorganic nanoparticles. *Langmuir*, *18*, 6336–6344.

Flukiger-Isler, S., Baumeister, M., Braun, K., Gervais, V., Hasler-Nguyen, N., Reimann, R., Van Gompel, J., Wunderlich, H.-G., Engelhardt, G. (2004). Assessment of the performance of the Ames II assay: a collaborative study with 19 coded compounds. *Mutat. Res.*, *558*, 181–197.

Friedberg, E.C. (2003). DNA damage and repair. *Nature*, *421*, 436–440.

Galloway, S.M., Aardema, M.J., Ishidate, M., Jr, Ivett, J.L., Kirkland, D.J., Morita, T., Mosesso, P., Sofuni, T. (1994). Report from working group on in vitro tests for chromosomal aberrations. *Mutat. Res.*, *312*, 241–261.

Gangl, E.T., Turesky, R.J., Vouros, P. (2001). Detection of in vivo formed DNA adducts at the part-per-billion level by capillary liquid chromatography/microelectrospray mass spectrometry. *Anal. Chem.*, *73*, 2397–2404.

Gao, J., Garulacan, L., Storm, S., Hefta, S., Opiteck, G., Lin, J., Moulin, F., Dambach, D. (2004). Identification of in vitro protein biomarkers of idiosyncratic liver toxicity. *Toxicol. In Vitro*, *18*, 533–541.

Griffin, J.L., Bollard, M.E. (2004). Metabonomics: its potential as a tool in toxicology for safety assessment and data integration. *Curr. Drug Metab.*, 5, 389–398.

Guengerich, F.P. (2001). Common and uncommon cytochrome P450 reactions related to metabolism and chemical toxicity. *Chem. Res. Toxicol.*, 14, 611–650.

Guengerich, F.P., Bell, L.C., Okazaki, O. (1995). Interpretations of cytochrome P450 mechanisms from kinetic studies. *Biochemie*, 77, 573–580.

Hastwell, P.W., Chai, L.-L., Roberts, K.J., Webster, T.W., Harvey, J.S., Rees, R.W., Walmsley, R.M. (2006). High specificity and high sensitivity genotoxicity assessment in a human cell line: validation of the GreenScreen HC GADD45α-GFP genotoxicity assay. *Mutat. Res.*, 607, 160–175.

Hecht, S.S. (1998). Biochemistry, biology, and carcinogenicity of tobacco specific N-nitrosamines. *Chem. Res. Toxicol.*, 11, 559–603.

Hecht, S.S. (2003). Tobacco carcinogens, their biomarkers and tobacco-induced cancer. *Nat. Rev. Cancer*, 10, 733–744.

Hu, T., Gibson, D.P., Carr, G.J., Torontali, S.M., Tiesman, J.P., Chaney, J.G., Aardema, M.J. (2004). Identification of a gene expression profile that discriminates indirect-acting genotoxins from direct-acting genotoxins. *Mutat. Res.*, 549, 5–27.

Hvastkovs, E.G., So, M., Krishnan, S., Bajrami, B., Tarun, M., Jansson, I., Schenkman, J.B., Rusling, J.F. (2007). Electrochemiluminescent arrays for cytochrome P450-activated genotoxicity screening. DNA damage from benzo[a]pyrene metabolites. *Anal. Chem.*, 79, 1897–1906.

Irizarry, R.A., Warren, D., Spencer, F., Kim, I.F., Biswal, S., Frank, B.C., Gabrielson, E., Garcia, J.G.N., Geoghegan, J., Germino, G., Griffin, C., Hilmer, S.C., Hoffman, E., Jedlicka, A.E., Kawasaki, E., Martinez-Murillo, F., Morsburger, L., Lee, H., Petersen, D., Quackenbush, J., Scott, A., Wilson, M., Yang, Y., Ye, Q.S., Yu, W. (2005). Multiple-laboratory comparison of microarray platforms. *Nat. Methods*, 2, 345–349.

Jacoby, W.B. (ed.) (1980). *Enzymatic Basis of Detoxification*, Vols I and II, Academic, New York.

Kenten, J.H., Casedei, J., Link, J., Lupold, S., Willey, J., Powell, M., Rees, A., Massey, R.J. (1991). Rapid electrochemiluminescence assays of polymerase chain reaction products. *Clin. Chem.*, 37, 1626.

Kiskinis, E., Suter, W., Hartmann, A. (2002). High throughput Comet assay using 96-well plates. *Mutagenesis*, 17, 37–43.

Kramer, J.A., Sagartz, J.E., Morris, D.L. (2007). The application of discovery toxicology and pathology towards the design of safer pharmaceutical lead candidates. *Nat. Rev. Drug Discov.*, 6, 636–649.

Krishnan, S., Hvastkovs, E.G., Bajrami, B., Choudhary, D., Schenkman, J.B., Rusling, J.F. (2008). Synergistic metabolic toxicity screening using microsome/DNA electrochemiluminescent arrays and nanoreactors. *Anal. Chem.*, 80, (14) 5279–5285.

Krishnan, S., Hvastkovs, E.G., Bajrami, B., Jansson, I., Schenkman, J.B., Rusling, J.F. (2007). Genotoxicity screening for N-nitroso compounds. Electrochemical and electrochemiluminescent detection of human enzyme-generated DNA damage from N-nitrosopyrrolidine. *Chem. Commun.*, 1713–1715.

Kuang, L., Biran, I., Walt, D. (2004). Living bacterial cell array for genotoxin monitoring. *Anal. Chem.*, 76, 2902–2909.

Kupfer, R., Liu, S.Y., Allentoff, A.J., Thompson, J.A. (2001). Comparisons of hydroperoxide isomerase and monooxygenase activities of cytochrome P450 for conversions of allylic hydroperoxides and alcohols to epoxyalcohols and diols: probing substrate reorientation in the active site. *Biochemistry*, 40, 11490–11501.

Lee, J., Mitchell, R., Kim, B., Cullen, D., Gu, M. (2005a). A cell array biosensor for environmental toxicity analysis. *Biosens. Bioelectron.*, 21, 500–507.

Lee, M., Park, C., Dordick, J., Clark, D. (2005b). Metabolizing enzyme toxicology assay chip (MetaChip) for high throughput microscale toxicity analyses. *Proc. Natl. Acad. Sci. U.S.A.*, *102*, 983–987.

Leland, J.K., Powell, M.J. (1990). Electrogenerated chemiluminescence: an oxidative-reduction type ECL reaction sequence using tripropyl amine. *J. Electrochem. Soc.*, *137*, 3127–3131.

Li, A., Bode, C., Sakai, Y. (2004). A novel in vitro system, the integrated discrete multiple organ cell culture (IdMOC) system, for the evaluation of human drug toxicity: comparative cytotoxicity of tamoxifen towards normal human cells from five major organs and MCF-7 adenocarcinoma breast cells. *Chem. Biol. Interact.*, *150*, 129–136.

Liebler, D.C., Guengerich, F.P. (2005). Elucidating the mechanisms of drug-induced toxicity. *Nat. Rev. Drug Dev.*, *4*, 410–420.

Lindon, J.C., Nicholson, J.K., Holmes, E., Antti, H., Bollard, M.E., Keun, H., Beckonert, O., Ebbels, T.M., Reily, M.D., Robertson, D.R., Stevens, G.J., Luke, P., Breau, A.P., Cantor, G.H., Bible, R.H., Niederhauser, U., Senn, H., Schlotterbeck, G., Sideelmann, U.G., Laursen, S.M., Tymiak, A., Car, B.D., Lehman-McKeeman, L., Colet, J.-M., Loukaci, A., Thomas, C. (2003). Contemporary issues in toxicology: the role of metabonomics in toxicology and its evaluation by the COMET project. *Toxicol. Appl. Pharm.*, *187*, 137–146.

Lippard, S.J., Berg, J.M. (1994). *Principles of Bioorganic Chemistry*, University Science Books, Mill Valley, CA.

Lvov, Y. (2000). *Protein Architecture: Interfacing Molecular Assemblies and Immobilization Biotechnology* (eds Y. Lvov and H. Möhwald), Marcel Dekker, New York, pp. 125–167.

Lvov, Y. (2001). *Handbook of Surfaces and Interfaces of Materials*, Vol. 3. Nanostructured Materials, Micelles and Colloids (ed. R.W. Nalwa), Academic Press, San Diego, pp. 170–189.

Lvov, Y., Caruso, F. (2001). Biocolloids with ordered urease multilayer shells as enzymatic reactors. *Anal. Chem.*, *73*, 4212–4217.

McNeish, J. (2004). Embryonic stem cells in drug discovery. *Nat. Rev. Drug Dev.*, *3*, 70–80.

Marnett, L.J., Burcham, P.C. (1993). Endogenous DNA adducts: potential and paradox. *Chem. Res. Toxicol.*, *6*, 771–785.

Mayne, J.T., Ku, W.W., Kennedy, S.P. (2006). Informed toxicity assessment in drug discovery: systems-based toxicology. *Curr. Opin. Drug Discov. Devel.*, *9*, 75–83.

Mitchell, R., Gu, M. (2004a). An Escherichia coli biosensor capable of detecting both genotoxic and oxidative damage. *Appl. Microbiol. Biotechnol.*, *64*, 46–52.

Mitchell, R., Gu, M. (2004b). Construction and characterization of novel dual stress-responsive bacterial biosensors. *Biosens. Bioelectron.*, *19*, 977–985.

Montrescu, E.R., Otto, A.M., Brischwein, M., Zahler, S., Wolf, B. (2005). Dynamic analysis of metabolite effects of chloroacetaldehyde and cytochalasin B on tumor cells using bioelectronic sensor chips. *J. Cancer Res. Clin. Oncol.*, *131*, 683–691.

Mortishire-Smith, R.J., Skiles, G.L., Lawrence, J.W., Spence, S., Nicholls, A.W., Johnson, B.A., Nicholson, J.K. (2004). Use of metabonomics to identify impaired fatty acid metabolism as the mechanism of drug induced toxicity. *Chem. Res. Toxicol.*, *17*, 165–173.

Mugweru, A., Rusling, J.F. (2006). Studies of DNA damage inhibition by dietary antioxidants using metallopolyion/DNA sensors. *Electroanalysis*, *18*, 327–332.

Mugweru, A., Yang, J., Rusling, J.F. (2004). Comparison of hemoglobin and myoglobin for in-situ metabolite generation in chemical toxicity sensors using a metallopolymer catalyst for DNA damage detection. *Electroanalysis*, *16*, 1132–1138.

Munge, B., Estavillo, C., Schenkman, J.B., Rusling, J.F. (2003). Optimizing electrochemical and peroxide-driven oxidation of styrene with ultrathin polyion films containing cytochrome P450cam and myoglobin. *Chembiochem*, *4*, 82–89.

Nasser, A.E.F., Kamel, A.M., Clarimont, C. (2004). Improving the decision making process in structural modification of drug candidates: reducing toxicity. *Drug Discov. Today*, *9*, 1055–1064.

Natarajan, A., Molnar, P., Sieverdes, K., Jamshidi, A., Hickman, J. (2006). Microelectrode array readings of cardiac action potentials as a high throughput method to evaluate pesticide toxicity. *Toxicol. In Vitro*, *20*, 375–381.

Neilson, A.H. (ed.) (1998). *PAHs and Related Compounds*, Springer, Berlin.

Newton, R.K., Aardema, M., Aubrecht, J. (2004). The utility of DNA microarrays for characterizing genotoxicity. *Environ. Health Pers.*, *112*, 420–422.

Nicholson, J.K., Connelly, J., Lindon, J.C., Holmes, E. (2002). Metabonomics: a platform for studying drug toxicity and gene function. *Nat. Rev. Drug Discov.*, *1*, 156–161.

O'Brien, P., Irwin, W., Diaz, D., Cofield, H., Krejsa, C., Slaughter, M., Gao, B., Kaludercic, N., Angeline, A., Bernardi, P., Brain, P., Hougham, C. (2006). High concordance of drug-induced human hepatotoxicity with in vitro cytotoxicity measured in a novel cell based model using high content screening. *Arch. Toxicol.*, *80*, 580–604.

Oda, Y., Nakamura, S., Oki, I., Kato, T., Shinagawa, H. (1985). Evaluation of a new system (umu-test) for the detection of environmental mutagens and carcinogens. *Mutat. Res.*, *147*, 212–229.

Ortiz de Montellano, P.R. (2005). *Cytochrome P-450: Structure, Mechanism, and Biochemistry*, 3rd edn, Kluwer Academic/Plenum, New York.

Ortiz de Montellano, P.R., De Voss, J.J. (2002). Oxidizing species in the mechanism of cytochrome P450. *Nat. Prod. Rep.*, *19*, 477–493.

Owens, J. (2005). Do microarrays match up? *Nat. Rev. Drug Discov.*, *4*, 459.

Park, K., Williams, D.P., Naisbitt, D.J., Kitteringham, N.R., Pirmohamed, M. (2005). Investigation of toxic metabolites during drug development. *Toxicol. Appl. Pharmacol.*, *207*, S425–S434.

Pauwels, W., Vodiceka, P., Servi, M., Plna, K., Veulemans, H., Hemminki, K. (1996). Adduct formation on DNA and haemoglobin in mice intraperitoneally administered with styrene. *Carcinogenisis*, *17*, 2673–2680.

Phillips, D.H., Farmer, P.B., Beland, F.A., Nath, R.G., Poirier, M.C., Reddy, M.V., Turtletaub, K.W. (2000). Methods of DNA adduct determination and their application to testing compounds for genotoxicity. *Environ. Mol. Mutagen.*, *35*, 222–233.

Preussmann, R., Stewart, B.W. (1984). N-Nitroso carcinogens, in *Chemical Carcinogenesis* (ed. C.E. Searle), ACS monograph 182, American Chemical Society, Washington, DC, pp. 643–828.

Rahimtula, A.D., O'Brien, P.J., Seifried, H.E., Jerina, D.M. (1978). The mechanism of action of cytochrome P-450. Occurrence of the "NIH shift" during hydroperoxide-dependent aromatic hydroxylations. *Eur. J. Biochem.*, *89*, 133–141.

Riley, M.R., Lucas, P., Le Coq, D., Juncker, C., Boesewetter, D.E., Collier, J.L., DeRosa, D.M., Katterman, M.E., Boussard-Pledel, C., Bureau, B. (2006). Lung cell fiber evanescent wave spectroscopic biosensing of inhalation health hazards. *Biotechnol. Bioeng.*, *95*, 599–612.

Rogan, E.G., Devanesan, P.D., Ramakrishna, N.V.S., Higgenbotham, S., Padvavathi, N.S., Chapman, K., Cavalieri, E.L., Jeong, H., Jankowiak, R., Small, G.J. (1993). Identification and quantitation of benzo[a]pyrene-DNA adducts formed in mouse skin. *Chem. Res. Toxicol.*, *6*, 356–363.

Rubinstein, I., Bard, A.J. (1980). Polymer films on electrodes. 4. Nafion-coated electrodes and electrogenerated chemiluminescence of surface-attached tris(2,2'-bipyridine)ruthenium(2+). *J. Am. Chem. Soc.*, *102*, 6642–6644.

Rusling, J.F. (2004). Sensors for toxicity of chemicals and oxidative stress based on electrochemical catalytic DNA oxidation. *Biosens. Bioelectron.*, *20*, 1022–1028.

Rusling, J.F. (2005). *Electrochemistry of Nucleic Acids and Proteins* (eds E. Palecek, F. Scheller and J. Wang), Elsevier, pp. 433–450.

Rusling, J.F., Hvastkovs, E.G., Hull, D.O., Schenkman, J.B. (2007). Biochemical applications of ultrathin films of enzymes, polyions and DNA (feature article). *Chem. Commun.*, DOI: 10.1039/b709121b.

Rusling, J.F., Hvastkovs, E.G., Schenkman, J.B. (2007). Toxicity screening using biosensors that measure DNA damage. *Curr. Opin. Drug Discov. Devel.*, *10*, 67–73.

Sasaki, Y.F., Sekihashi, K., Izumiyama, F., Nishidate, E., Saga, A., Ishida, K., Truda, S. (2000). The Comet assay with multiple mouse organs: comparison of Comet assay results and carcinogenicity with 208 chemicals selected from the IARC monographs and U.S. NTP Carcinogenicity Database. *Crit. Rev. Toxicol.*, *30*, 629–799.

Scharer, O.D. (2003). Chemistry and biology of DNA repair. *Angew. Chem. Int. Ed. Engl.*, *42*, 2946–2974.

Schenkman, J.B., Greim, H. (eds) (1993). *Cytochrome P450*, Springer-Verlag, Berlin.

Schwartz, M.P., Derfus, A.M., Alvarez, S.D., Bhatia, S.N., Sailor, M.J. (2006). The smart petri dish: a nanostructured photonic crystal for real-time monitoring of living cells. *Langmuir*, *22*, 7084–7090.

Schweitzer, B., Wiltshire, S., Lambert, J., O'Malley, S., Kukanskis, K., Zhu, Z., Kingsmore, S.F., Lizardi, P.M., Ward, D.C. (2000). Immunoassays with rolling circle DNA amplification: a versatile platform for ultrasensitive antigen determination. *Proc. Natl. Acad. Sci. U.S.A.*, *97*, 10113–10119.

Shuga, J., Zhang, J., Samson, L.D., Lodish, H.F., Griffith, L.G. (2007). In vitro erythropoiesis from bone marrow-derived progenitors provides a physiological assay for toxic and mutagenic compounds. *Proc. Natl. Acad. Sci. U.S.A.*, *104*, 8737–8742.

Shutava, T., Zheng, Z., John, V., Lvov, Y. (2004). Microcapsule modification with peroxidase-catalyzed phenol polymerization. *Biomacromolecules*, *5*, 914–921.

Singh, N.P., McCoy, M.T., Tice, R.R., Schneider, E.L. (1998). A simple technique for quantitation of low levels of DNA damage in individual cells. *Exp. Cell. Res.*, *175*, 184–191.

So, M., Hvastkovs, E.G., Schenkman, J.B., Rusling, J.F. (2007). Electrochemiluminescent/voltammetric toxicity screening sensor using enzyme-generated DNA damage. *Biosens. Bioelectron.*, *23* (4), 492–498.

Solomun, T., Sturm, H. (2007). Bringing electrons and microarray technology together. *J. Phys. Chem. B*, *111*, 10636–10638.

Sreekumar, A., Nyati, M.K., Varambally, S., Barrette, T.R., Ghosh, D., Lawrence, T.S., Chinnaiyan, A.M. (2001). Profiling of cancer cells using protein microarrays: discovery of novel radiation-regulated proteins. *Cancer Res.*, *61*, 7585–7593.

Stoll, D., Bachmann, J., Templin, M.F., Joos, T.O. (2004). Microarray technology: an increasing variety of screening tools for proteomic research. *Drug Discov. Today: Targets*, *3*, 24–31.

Styles, J.A., Clark, H., Festing, M.F.W., Rew, D.A. (2001). Automation of mouse micronucleus genotoxicity assay by laser scanning cytometry. *Cytometry*, *44*, 153–155.

Tanaka, E., Terada, M., Misawa, S. (2000). Cytochrome P450 2E1: its clinical and toxicological role. *J. Clin. Pharm. Ther.*, *3*, 165–175.

Tarun, M., Rusling, J.F. (2005a). Measuring DNA nucleobase adducts using neutral hydrolysis and liquid chromatography-mass spectrometry. *Crit. Rev. Eukaryot. Gene Expr.*, *15*, 295–315.

Tarun, M., Rusling, J.F. (2005b). Quantitative measurement of DNA adducts using neutral hydrolysis and LC-MS. Validation of genotoxicity sensors. *Anal. Chem.*, *77*, 2056–2062.

Tarun, M., Rusling, J.F. (2006). Genotoxicity screening using biocatalyst-DNA films and capillary LC-MS/MS. *Anal. Chem.*, *78*, 624–627.

Todorovic, R., Ariese, F., Devenesan, P., Jankowiak, R., Small, G.J., Rogan, E.G., Cavalieri, E.L. (1997). Determination of benzo[a]pyrene- and 7,12-dimethylbenz[a]anthracene-DNA adducts formed in rat mammary glands. *Chem. Res. Toxicol.*, *10*, 941–947.

Ulrich, R., Friend, S.H. (2002). Toxicogenomics and drug discovery: will new technologies help us produce better drugs. *Nat. Rev. Drug Discov.*, *1*, 84–88.

Umemoto, A., Komaki, K., Monden, Y., Suwa, M., Kanno, Y., Kitagawa, M., Suzuki, M., Lin, C.-X., Ueyama, Y., Momen, M.A., Ravindernath, A., Shibutani, S. (2001). Identification and quantification of tamoxifen-DNA adducts in the liver of rats and mice. *Chem. Res. Toxicol.*, *14*, 1006–1013.

Van Gompel, J., Woestenborghs, F., Beerens, D., Mackie, C., Cahill, P.A., Knight, A.W., Billinton, N., Tweats, D.J., Walmsley, R.M. (2005). An assessment of the utility of the yeast GreenScreen assay in pharmaceutical screening. *Mutagenesis*, *20*, 449–454.

Vanhoutte, K., Dongen, W., Hoes, I., Lemiere, F., Esmans, E.L., Van Onckelen, H., Van den Eckhout, E., Soest, R.E.J., Hudson, A.J. (1997). Development of a nanoscale liquid chromatography/electrospray mass spectrometry methodology for the detection and quantification of DNA adducts. *Anal. Chem.*, *69*, 3161–3168.

Vansant, G., Pezzoli, P., Saiz, R., Birch, A., Duffy, C., Ferre, F., Monforte, J. (2006). Gene expression analysis of troglitazone reveals its impact on multiple pathways in cell culture: a case for in vitro platforms combined with gene expression analysis for early (idiosyncratic) toxicity screening. *Int. J. Toxicol.*, *25*, 85–94.

Vignati, L., Turlizzi, E., Monaci, S., de Grossi, P., Kanter, P., Monshouwe, M. (2005). An in vitro approach to detect metabolite toxicity due to CYP3A4-dependent bioactivation of xenobiotics. *Toxicology*, *216*, 154–167.

Viravaidya, K., Sin, A., Shuler, M. (2004). Development of a microscale cell culture analog to probe naphthalene toxicity. *Biotechnol. Prog.*, *20*, 316–323.

Wang, B., Jansson, I., Schenkman, J.B., Rusling, J.F. (2005). Evaluating enzymes that generate genotoxic benzo[a]pyrene metabolites using sensor arrays. *Anal. Chem.*, *77*, 1361–1367.

Wang, M., Cheng, G., Sturla, S.J., Shi, Y., McIntee, E.J., Villalta, P.W., Upadhyaya, P., Hecht, S.S. (2003). Identification of adducts formed by pyridyloxobutylation of deoxyguanosine and DNA by 4-(acetoxymethylnitrosamino)-1-(3-pyridyl)-1-butanone, a chemically activated form of tobacco specific carcinogens. *Chem. Res. Toxicol.*, *16*, 616–626.

Wang, M., McIntee, E.J., Shi, Y., Cheng, G., Upadhyaya, P., Villalta, P.W., Hecht, S.S. (2001). Reactions of acetoxy-N-nitrosopyrrolidine with deoxyguanosine and DNA. *Chem. Res. Toxicol.*, *14*, 1435–1445.

Warren, A.J., Shields, P.G. (1997). Molecular epidemiology: carcinogen-DNA adducts and genetic susceptibility. *Proc. Soc. Exp. Biol. Med.*, *216*, 172–180.

Werringloer, J., Kawano, S., Estabrook, R.W. (1980). Comparison of the interaction of cumene hydroperoxide and hydrogen peroxide with liver microsomal cytochromes, in *Microsomes, Drug Oxidations, and Chemical Carcinogenesis* (eds M.C. Coon, A.H. Conney, R.W. Estabrook and H.V. Gelboin), Academic Press, New York, pp. 403–406.

White, R.E., Sligar, S.G., Coon, M.W. (1980). Evidence for a homolytic mechanism of peroxide oxygen-oxygen bond cleavage during substrate hydroxylation by cytochrome P-450. *J. Biol. Chem.*, *255*, 11108–11111.

Xiao, C., Luong, H.T. (2005). Assessment of cytotoxicity by emerging impedance spectroscopy. *Toxicol. Appl. Pharm.*, *206*, 102–112.

Xing, J.Z., Zhu, L., Jackson, J., Gabos, S., Sun, X.-J., Wang, X.-b., Xu, X. (2005). Dynamic monitoring of cytotoxicity on microelectrode sensors. *Chem. Res. Toxicol.*, *18*, 154–161.

Xu, X.-H., Bard, A.J. (1995). Immobilization and hybridization of DNA on an aluminum(III) alkanebisphosphonate thin film with electrogenerated chemiluminescent detection. *J. Am. Chem. Soc.*, *117*, 2627–2631.

Xu, X.-H., Yang, H.C., Mallouk, T.E., Bard, A.J. (1994). Immobilization of DNA on an aluminum(III) alkanebisphosphonate thin film with electrogenerated chemiluminescent detection. *J. Am. Chem. Soc.*, *116*, 8386–8387.

Yang, J., Wang, B., Rusling, J.F. (2005). Genotoxicity sensor response correlated with DNA nucleo-base damage rates. *Mol. Biosyst.*, *1*, 251–259.

Young-Scaime, R., Wang, M., Chung, F.-L., Hecht, S.S. (1995). Reactions of acetoxy-N-nitrosopyrrolidine and acetoxy-N-nitrosopiperidine with deoxyguanosine: formation of N2-tetrahydrofuranyl and N2-tetrahydropyranyl adducts. *Chem. Res. Toxicol.*, *8*, 607–616.

Zamenhof, S., Arikawa, S. (1966). Depurination and alkylation of DNAs of different base compositions. *Mol. Pharmacol.*, *2*, 570–573.

Zhang, X., Chen, H., Zhang, H. (2007). Layer-by-layer assembly: from conventional to unconventional methods. *Chem. Commun.*, 1395–1405.

Zhang, X., Wang, W., Yu, W., Xie, Y., Zhang, X., Zhang, Y., Ma, X. (2005). Development of an in vitro multicellular tumor spheroid model using microencapsulation and its application in anticancer drug screening and testing. *Biotechnol. Prog.*, *21*, 1289–1296.

Zhou, L., Yang, J., Estavillo, C., Stuart, J.D., Schenkman, J.B., Rusling, J.F. (2003). Toxicity screening by electrochemical detection of DNA damage by metabolites generated *in-situ* in ultrathin DNA-enzyme films. *J. Am. Chem. Soc.*, *125*, 1431–1436.

Zhu, J., Wang, X., Yu, W., Abassi, Y.A. (2006). Dynamic and label-free monitoring of natural killer cell cytotoxic activity using electronic cell sensor arrays. *J. Immunol. Methods*, *309*, 25–33.

Zoltewicz, J.A., Clark, D.F. (1972). Kinetics and mechanism of the hydrolysis of guanosine and 7-methylguanosine nucleosides in perchloric acid. *J. Org. Chem.*, *37*, 1193–1197.

PART III

DRUG INTERACTIONS

CHAPTER 14

Enzyme Inhibition

PAUL F. HOLLENBERG

14.1 INTRODUCTION

Inhibition of the catalytic activities of the drug-metabolizing enzymes has received considerable attention from clinical pharmacologists and pharmaceutical companies in the past 20 years, because of the fact that several widely used drugs or drugs in the late stages of development were shown to cause life-threatening adverse affects when given to patients taking other drugs due to their ability to inhibit the metabolism of those drugs. The inhibition of the metabolism of one drug by another drug taken concurrently by the patient has now been shown to be a very important cause of adverse drug reactions (ADRs). The potential for drugs to interact with one another by competing for metabolism by the same drug-metabolizing enzyme continues to be a significant health concern and, if anything, will probably become of greater concern in the coming years. ADRs are suggested to be responsible for up to 10% of all hospital admissions (Kohler *et al.*, 2000). In addition, drug–drug interactions are estimated to be implicated in up to 20% of all ADRs (Levy *et al.*, 1980). The probability that ADRs associated with drug–drug interaction increase as the number of drugs administered concomitantly increases is generally accepted as true, and the occurrence of ADRs is believed to increase very rapidly as the number of drugs given simultaneously to a patient increases. When the activity of a given drug-metabolizing enzyme is modified by inhibition, this modification may not only change the pharmacokinetics of the drug whose metabolism is inhibited by leading to an increase in the levels of the parent drug in the patient but, where more than one enzyme may be responsible for the metabolism of a drug and the different enzymes may result in the formation of different products, it may result in a significant change in the profile of the metabolites formed. If any of the different products formed have significantly enhanced or decreased biological and/or toxicological properties, this may lead to significantly altered biological or toxicological activity. Since many of the drug-metabolizing enzymes may be viewed as being relatively non-specific (i.e. very promiscuous) in their substrate specificity, it is not surprising that two or more drugs can compete for metabolism by the same enzyme resulting in inhibition of the metabolism of one or both of the drugs. This can lead to enhanced metabolism

Drug Metabolism Handbook: Concepts and Applications, Edited by Ala F. Nassar, Paul F. Hollenberg, and JoAnn Scatina

of one of the drugs by a second enzyme that shows specificity for forming another product from the drugs whose metabolism is inhibited. In the situation where two or more drugs are competing for metabolism by the same enzyme, the effective K_I (the inhibitor concentration that results in half maximal inhibition) for the inhibition of the metabolism of one drug by the other may be comparable to the K_M (the substrate concentration that gives half maximal velocity) for its metabolism. If the peak plasma levels in normal clinical usage approximate the K_M (or K_I) for the first drug, this can result in significant inhibition of the metabolism of the second drug. In addition to inhibition, the induction of the drug-metabolizing enzymes also is an important factor in many of the drug interactions documented in the literature (Lin and Lu, 1998). This is discussed in a subsequent chapter.

In addition to the potential for drug–drug interactions leading to ADRs, the pharmaceutical industry is also very interested in enzyme inhibition as a basis for developing drugs that target specific enzymes of interest. For example, some of the cytochrome P450s as well as monoamine oxidase are targets for drug development because the products of their normal reactions may be of importance under certain conditions. A good example is the yeast P450 51 (lanosterol protein α-demethylase), which is a target for many common fungicides currently on the market. Aromatase (P450 19A1), which oxidizes the androgens androstandione and testosterone to form the estrogens estrone and 17β-estradiol, respectively, is also a target for inhibition in cancer therapy.

The development of technologies that can be used to study the interactions of drugs with individual human P450 enzymes *in vitro* by using either the purified recombinant P450 enzymes in reconstituted systems or enzyme selective substrates in human microsomal preparations or freshly isolated primary hepatocytes, all of which are now readily available commercially, has led to major breakthroughs in studying drug–drug interactions *in vitro*. The development of these *in vitro* technologies has made evaluation of drug–drug interactions an integral part of the drug development process such that *in vitro* drug–drug interaction studies are now routinely performed by pharmaceutical companies. However, the *in vitro–in vivo* correlations for drug–drug interactions have been much more problematic. Details of the development of these methodologies are covered in a subsequent chapter.

Competition between two or more different drugs for metabolism by the same enzyme resulting in the inhibition of the metabolism of one or more of the competing drugs may result in unexpected elevations in the plasma concentration(s) that can ultimately result in a variety of adverse effects. Depending on the final plasma concentrations of the inhibited drugs, this can ultimately result in a variety of minor as well as serious adverse effects and, in some cases, has resulted in fatalities. The onset of inhibition in patients is usually rapid following a single administration of the inhibitory compound. However, in some cases, as will be discussed later, the inhibition is not observed immediately.

Several very prominent drugs have been withdrawn from the market and some drugs in late-stage development have ultimately been stopped due to their potential for ADRs resulting from inhibition of the drug-metabolizing enzymes. Drug–drug interactions responsible for the limited utility of a given drug or even its withdrawal from the market generally result in significant financial losses to the pharmaceutical companies involved. Therefore, pharmaceutical companies are continuing to develop and utilize a variety of new approaches for the prediction of possible drug–drug interactions resulting from inhibition of the drug-metabolizing enzymes. Our knowledge of the structures, catalytic mechanisms, and regulation of the drug-metabolizing enzymes has increased greatly during the last decade due to a number of major advances in molecular biology and biochemical approaches to the study of enzymes. Although the mechanistic aspects of the inhibition of the drug-metabolizing enzymes *in vitro* are much better understood than previously, the accurate prediction of the occurrence and consequences of drug–drug interactions due to inhibition continue to be an important unsolved challenge. This chapter will focus on approaches used to investigate the inhibition of the drug-metabolizing enzymes. The theoretical underpinnings for understanding the various mechanisms by which inhibition may occur as well as experimental design and the interpretation of inhibition data will be discussed. Particular attention is paid to the cytochrome P450 enzymes since many of the clinically relevant pharmacokinetic drug interactions are in one way or another related to the P450s. In addition, the principles used in the study of the P450s and for understanding the mechanisms of their inhibition are applicable, in general, to all of the other drug-metabolizing enzymes. Thus, although the focus will be on P450s, the general principles described here are applicable to all drug-metabolizing enzymes.

14.2 MECHANISMS OF ENZYME INHIBITION

The inhibitors of the drug-metabolizing enzymes can be divided into three general categories that differ in their mechanisms. They are (i) compounds that bind reversibly in the active site or an effector site of the enzyme, (ii) compounds that form quasi-irreversible complexes with a prosthetic group of the enzyme in question, and (iii) compounds that bind irreversibly to the prosthetic group of the protein or to the apoprotein itself (Guengerich, 1999). The most common cause of drug–drug interactions is generally believed to be reversible inhibition. Reversible inhibition includes competitive, noncompetitive, uncompetitive, product inhibition, inhibition by transition-state analogs, and inhibition by slow, tight-binding inhibitors. Reversible inhibition may occur rapidly following a single dose of the inhibitory drug. It is transient and, following elimination of the inhibitor from the body, the normal metabolic functions of the enzymes will be regained. As opposed to the quasi-irreversible and irreversible inhibitors that exhibit both dose-dependent and time-dependent inhibition of catalytic activity, reversible inhibitors generally only exhibit dose-dependent inhibition of activity.

14.3 COMPETITIVE INHIBITION

Competitive inhibition occurs when the inhibitor binds reversibly to the enzyme and prevents the binding of the substrate to the catalytically active site of the enzyme. This is generally believed to result from the fact that the inhibitor shares some degree of structural similarity with the drug whose metabolism is being inhibited. The inhibitor itself may or may not be metabolized by the enzymes that it inhibits. Depending on the substrate specificity of the drug-metabolizing enzyme in question, some of which exhibit relatively little apparent specificity, the structural similarities between the substrate whose metabolism is inhibited and the competitive inhibitor may or may not be apparent. Competition of the inhibitor with the drug whose metabolism is inhibited for occupancy of the active site of the drug-metabolizing enzyme, may involve simple competition for binding to a lipophilic or hydrophilic domain in the active site, or it may involve the formation of hydrogen bonds or ionic bonds with specific amino acid residues in the active site (Ortiz de Montellano and Correia, 2005). In some cases, a competitive inhibitor may bind reversibly to a site other than the active site and block

$$
\begin{array}{ccccccccc}
& k_1 & & k_2 & & k_3 & & k_4 & \\
E + S & \rightarrow & ES + I & \rightarrow & ES^* & \rightarrow & EP & \rightarrow & E + P \\
& \leftarrow & & \leftarrow & & \leftarrow & & \leftarrow & \\
& k_{-1} & & k_{-2} & & k_{-3} & & k_{-4} & \\
\end{array}
$$

$$
\begin{array}{ccc}
& k_{1'} & \\
E + I & \rightarrow & EI \\
& \leftarrow & \\
& k_{-1'} & \\
\end{array}
$$

Scheme   14.1

$$
v = \frac{V_{max}\,[S]}{K_M\,(1 + [I]/K_I) + [S]}
$$

Scheme   14.2

substrate binding to the active site. Competitive inhibition is commonly observed when two different substrates of the same drug-metabolizing enzyme are present in the *in vitro* system under study or are given at the same time to the patient. The classic enzyme kinetic concept for this process is depicted in Scheme 14.1.

The intermediate depicted by ES* may or may not be present in any appreciable concentration. The kinetic constants k_1, $k_{1'}$, and k_{-4} may be lumped together and termed "k_{on}" and k_{-1}, $k_{-1'}$, and k_4 may be lumped together as the "k_{off}." For classic competitive inhibition, the steady-state apparent K_M value for the reaction of the substrate under study will increase by the factor $(1+[I]/K_I)$ in the presence of the inhibitor while the V_{max} does not change. The inhibition constant, K_I, for a given compound can be estimated from steady-state kinetic studies with increasing concentrations of the competitive inhibitor since the apparent K_M value and the K_I are related by the expression in Scheme 14.2, where the apparent K_M is equal to $K_M\,(1+[I]/K_I)$.

The most commonly used method for characterizing inhibition data is to systematically vary the concentrations of both the substrate and the inhibitor. Initial estimates for the kinetic parameters can be obtained from data for control studies run in the absence of inhibitor and from a double-reciprocal plot of the inhibition data. These analyses provide preliminary estimates of V_{max} (the maximal velocity at saturating substrate concentrations), K_M, and K_I. Once an estimate of the K_M for the reaction is known and competitive inhibition is assumed initially, the K_I can be determined by varying the concentration of the inhibitor at a single-substrate concentration. Data obtained by varying the concentration of the inhibitor at several different single-substrate concentrations can be used to provide information regarding the K_M and the K_I as well as the type of inhibition. When both substrate and inhibitor concentrations are varied, the data can be fit to equations for not only competitive inhibition but also noncompetitive, uncompetitive, and mixed type and the fits can be compared. In general, the most prevalent type of inhibition is competitive with the partial mixed type of inhibition being the second most prevalent.

Competitive inhibition can be identified by a common intersection point on the ordinate in either an Eadie–Hofstee (V versus V/S) or Lineweaver–Burk (1/V versus 1/S) plot. An alternative method involves using the Dixon plot, where 1/V is plotted versus the concentration of the substrate, S (Bronson *et al.*, 1995). Since the weighting of the data points is a problem for all of the linear transforms of kinetic data

(Cornish-Bowden, 1979), various nonlinear regression programs have been developed and are readily available for analysis of data obtained from kinetic studies on various drug-metabolizing enzymes (EZ-Fit, GraFit, KaleidaGraph, SigmaPlot, etc.).

In principle, for studies involving reversible inhibition, the K_I is equivalent to the binding constant for the inhibitor to the enzyme (K_D), whereas the K_M usually is not equivalent to the binding constant but is composed of a variety of kinetic terms. However, when the inhibitor is also a substrate for the enzyme it inhibits and the competitive inhibition is due to the fact that the two substrates are competing with each other for metabolism, then the K_I is not equal to the actual binding constant (K_D) of the inhibitor.

The *in vitro* assays are most reliable when used with either purified enzymes in the appropriate reconstituted system or with microsomal or cytosolic preparations, in which there is only one enzyme responsible for metabolizing the probe substrate to the product that is measured, and the enzyme under study is the only one responsible for binding the compound being investigated as the inhibitor. Since the concentrations of marker substrates and inhibitors added to the growth medium of cells in culture do not necessarily reflect the intracellular concentrations experienced by the drug-metabolizing enzymes, studies aimed at determination of the type of inhibition are difficult to perform with freshly isolated hepatocytes, and the interpretation of data obtained from these studies with respect to the calculation of meaningful kinetic constants and the type of inhibition is suspect.

When using microsomal preparations or purified reconstituted enzyme systems *in vitro*, competitive inhibition is observed relatively frequently in these types of studies. It is generally presumed that this is also the case *in vivo* and approaches have been developed for characterizing inhibitory interactions *in vivo* by analyzing various pharmacokinetic parameters (Black *et al.*, 1996). Since many of the most important drug-metabolizing enzymes are relatively nonspecific and have a variety of different drugs as substrates, competition of several drugs for metabolism by a specific drug-metabolizing enzyme is a common occurrence leading to drug–drug interactions in patients who are simultaneously administered a number of different drugs.

14.4   NONCOMPETITIVE INHIBITION

Some drugs can inhibit the drug-metabolizing enzymes even when they are not metabolized by the enzyme that they affect. A simple noncompetitive inhibitor binds to a site on the enzyme that is distinct from that of the substrate and thus, it is not metabolized by that enzyme and does not alter substrate binding and vice versa. The substrate and the inhibitor bind reversibly and independently. Although the bound inhibitor has the effect of apparently decreasing the amount of enzyme present (decrease in V_{max} with no change in K_M), it does not inactivate the enzyme. In this case, the enzyme–substrate–inhibitor complex is formed; however, it is unable to function catalytically. This type of inhibition, although discussed routinely in any presentation of enzyme kinetics, is relatively uncommon and is not routinely encountered in studies of the drug-metabolizing enzymes. This type of behavior might be observed in studies with microsomes or expressed enzymes when some of the enzyme involved in metabolism is inactivated during the course of the study, for example, by the formation of reactive intermediates which inactivate the protein or by the presence of a tight-binding inhibitor, and can be differentiated from these possibilities by approaches described later.

14.5 UNCOMPETITIVE INHIBITION

Similar to noncompetitive inhibition, uncompetitive inhibition is readily defined but is rarely observed in drug metabolism studies done *in vitro*. In uncompetitive inhibition, the inhibitor does not bind to the free enzyme but, in fact, only binds to the enzyme that already has substrate bound and prevents subsequent metabolism. This favors the equilibrium shown in Scheme 14.1 to form the ES complex and in this case leads to the formation of a nonproductive enzyme–substrate–inhibitor complex. Therefore, both the V_{max} and the K_M decrease proportionately so that the ratio of the V_{max} to K_M remains constant. In a Lineweaver–Burk plot of the data, parallel lines would be observed for plots of 1/V versus 1/S at different concentrations of the inhibitor.

The easiest way to estimate the inhibition constants for these classical mechanisms involves determining the initial rates of metabolism at various substrate concentrations in the absence and in the presence of a single concentration of the inhibitor under study. From each of the double-reciprocal plots, the values V_{max} and K_M are determined graphically or by linear regression. Subsequently, the calculations may be performed using the relationship between K_M in the presence of the inhibitor and K_I from Scheme 14.2 (competitive inhibition), Scheme 14.3 (noncompetitive inhibition), and Scheme 14.4 (uncompetitive inhibition).

14.6 PRODUCT INHIBITION

Due to the relative lack of specificity of many of the drug-metabolizing enzymes and their ability to metabolize a substrate on several different parts of the parent molecule to give different metabolites, it is possible that a product of an initial reaction catalyzed by a drug-metabolizing enzyme may bind to the enzyme again in the same or a different orientation and inhibit the initial reaction for the metabolism of another substrate molecule. Thus, the product may have physical characteristics that are not significantly different from that of the initial substrate and it could thereby bind competitively and inhibit metabolism of the parent compound either by serving as a substrate for a second reaction or by just binding in the substrate binding site. For example, cytochrome P450 2E1 hydroxylates benzene to give the phenol and then hydroxylates it once again to form the hydroquinone (in this case, the phenol competes with the benzene for metabolism and vice versa). It is possible in some cases that the first product may have a greater affinity for binding to the active site than the parent drug.

$$v = \frac{V_{max}\,[S]}{[S]\,(1 + [I]/K_I) + K_M}$$

Scheme 14.3

$$v = \frac{V_{max}\,[S]}{([S] + K_M)\,(1 + [I]/K_I)}$$

Scheme 14.4

14.7 TRANSITION-STATE ANALOGS

Transition-state analogs resemble the transition state for the enzymatic reaction catalyzed by the enzyme. That is to say, they resemble the transient complex formed within the catalytic cycle that has the maximum free energy. These compounds are tight-binding noncovalently bound inactivators in which the k_{on} rate is rapid and the k_{off} is slow. The inactivation of the enzyme occurs rapidly and, under typical assay conditions, there is no time dependence. In principle, the enzyme activity can be restored by removal of the inhibitor by a variety of approaches including gel filtration, centrifugal separation, or dialysis. In this case, the kinetics are characterized by a decrease in the V_{max} but no alteration in the K_M. Transition-state inhibitors have been developed to target a variety of different clinically important enzymes since the transition state tends to be uniquely characteristic of one kind of enzyme reaction, whereas substrates may be shared by two or more different enzymes. So far, transition-state analogs have been of relatively minor importance in the study or the targeting of drug-metabolizing enzymes.

14.8 SLOW, TIGHT-BINDING INHIBITORS

These compounds are characterized by the fact that they exhibit relatively slow k_{on} rates but have even slower k_{off} rates. Thus, they have very high affinities for the target enzyme. In these cases, the binding may be noncovalent or possibly even covalent, but due to the slowness of the binding, the decrease in the activity of the enzyme may be time-dependent and can easily be mistaken for mechanism-based inactivation. However, in this case, gel filtration, centrifugal separation, or dialysis may be used to restore enzyme activity, even though the process may be slow. Although rapid formation of a complex between the inhibitor and the enzyme may occur initially, the tight binding of the inhibitor to the enzyme resulting in inhibition may require a conformational change of the enzyme as a result of the initial binding, a displacement of a water molecule in the enzyme active site, a change in the protonation state of an amino acid residue in the active site, or reversible formation of a covalent bond (Silverman, 1995).

14.9 MECHANISM-BASED INACTIVATORS

Specific covalent binding of reactive intermediates formed by the drug-metabolizing enzymes within the active sites of the enzymes that initially form them is oftentimes referred to as "mechanism-based," "catalysis-dependent," or "suicide" inactivation. A broad definition of mechanism-based or suicide inactivators would include any compound that is activated via the normal catalytic mechanism of the enzyme that it subsequently inactivates. However, we will restrict the definition to that of Silverman (1995), who defines it as "an unreactive compound the structure of which resembles that of either the substrate or the product of the target enzyme and that undergoes a catalytic transformation by the enzyme to a species that, before release from the active site, inactivates the enzyme." Key to this concept is the requirement that the inactivation occurs without the release of the reaction product from the enzyme active site. This definition rules out other types of inactivators such as affinity labels, transition-state analogs, and slow, tight-binding inhibitors.

Although mechanism-based inactivation is generally thought to be a relatively unusual occurrence for enzyme-catalyzed reactions, it has been observed more frequently than might be expected with the cytochrome P450s. Presumably, this results from the increased reactivity of many of the oxygenated intermediates formed by these enzymes during the course of the oxygenation reactions. Thus, mechanism-based inactivation has been encountered with P450s as well as other drug-metabolizing enzymes with a variety of different drugs. The utility of mechanism-based inactivators in the design of new drugs has attracted significant interest recently since, in principle, these compounds could be designed so that they would only inhibit the target enzyme. They are also of significant interest because of their potential for use in elucidating mechanisms catalyzed by the drug-metabolizing enzymes. In addition, a great deal of effort has been put into the development of mechanism-based inactivators as diagnostic inhibitors of the various drug-metabolizing enzymes in order to identify which enzymes may be involved in catalyzing a particular reaction. A primary focus of many of these studies has been aimed at identifying which form(s) of the various cytochrome P450s are involved in catalyzing a particular reaction in microsomal preparations. Following catalytic formation of the reactive intermediate, these compounds result in the formation of covalent bonds to amino acid residues or prosthetic groups in the enzyme active site that cannot be broken to regenerate the catalytically active enzyme. Although the onset of inhibition by these compounds either *in vivo* or *in vitro* may occur more slowly than that observed with reversible or quasi-irreversible inhibitors, the final effect of the mechanism-based inactivators is generally much more profound than observed with other inhibitors since the inhibition of drug metabolism can only be reversed *in vivo* by the synthesis of new, catalytically active enzymes. Due to the formation of a covalent bond, gel filtration, dialysis, or separation by centrifugation is not effective in regenerating the catalytic activity.

The relevant kinetic scheme for mechanism-based inactivation is shown in Scheme 14.5. In this scheme, MBI is the mechanism-based inactivator, E-MBI is the initial Michaelis–Menten enzyme–substrate complex, E-MBI* is the reactive intermediate complexed to the enzyme, E-MBI** is the inactivated enzyme, and P is the product that leaves the enzyme. The derivation and the kinetic implications of the various kinetic constants have been described previously by Silverman (1995). Parameters that are routinely determined to describe mechanism-based inactivators include k_{inact}, K_I, and the partition ratio. The k_{inact} and the K_I are defined in a manner analogous to the k and K_M in standard enzyme-catalyzed reactions. The partition ratio, a concept introduced by Walsh (Wang and Walsh, 1978), is determined from the ratio of k_3/k_4 and is a measure of the number of latent inactivator molecules metabolized and released as product for each inactivator molecule metabolized to give a form that reacts with the active enzyme to inactivate it. It can be thought of as the number of cycles of metabolism that can be traversed by the enzyme, on average, before it is inactivated by the mechanism-based inactivator. The partition ratio can also be considered to be a measure of the efficiency of the mechanism-based inactivator and is dependent on a number of

$$
\begin{array}{ccccccc}
& k_1 & & k_2 & & k_4 & \\
E + MBI & \rightarrow & E \cdot MBI & \rightarrow & E \cdot MBI^* & \rightarrow & EMBI^{**} \\
& \leftarrow & & & \downarrow k_3 & & \\
& k_{-1} & & & E + P & &
\end{array}
$$

Scheme   14.5

factors including the reactivity of the reactive intermediate, the proximity of the appropriate target molecule(s) (nucleophile, electrophile, or radical) in the active site to the site of formation of the reactive intermediate, the rate of diffusion of the reactive intermediate from the active site of the enzyme, and the presence in the active site of the enzyme of other molecules which may react with the reactive intermediate and render it harmless (e.g. water, or low-molecular-weight nucleophiles or electrophiles). It is not dependent on the concentration of the protein or inactivator. Partition ratios ranging from almost zero to several thousand have been reported for various mechanism-based inactivators. The most efficient mechanism-based inactivator would have a partition ratio of zero in that every reactive intermediate formed would inactivate the enzyme. In this case, in a standard assay where metabolite formation is measured, the conclusion would be that the inactivator is not a substrate for the enzyme since there would be no measurable product formation.

As pointed out by Ortiz de Montellano and Correia (2005), mechanism-based inactivators may exhibit much better specificity for their target enzymes than reversible inhibitors since (i) the initial step(s) for the binding of the inhibitor to the specific enzyme must satisfy all of the constraints imposed on reversible inhibitors; (ii) the inhibitor must also be acceptable as a substrate for the enzyme due to the requirement that it must undergo catalytic activation to form a reactive species; and (iii) the reactive species formed as a result of metabolism by the target enzyme results in an irreversible modification of that enzyme which permanently removes it from the pool of active enzymes.

A variety of different criteria are routinely used to assess whether a substrate is a mechanism-based inactivator of an enzyme (Silverman, 1995; Kent, Jushchyshyn and Hollenberg, 2001; Ortiz de Montellano and Correia, 2005). The criteria include:

1. The loss of enzyme activity must exhibit first-order kinetics. Thus, a plot of the logarithm of the percent enzyme activity remaining versus time should give a straight line. The slope of the straight line gives the k_{app}, the rate constant for the inactivation reaction at the concentration of the inactivator used.

2. The inactivation should exhibit saturation kinetics with respect to the concentration of the inactivator. The plot of k_{app} versus [I] should be hyperbolic and a linear transformation of the data (e.g. a plot of $1/k_{app}$ versus $1/[I]$) can be used to calculate K_I, the concentration of the inactivator which gives the half-maximal rate of inactivation and the k_{inact}, the rate of inactivation at saturating concentrations of inactivator. The half-life, $t_{1/2}$, can be calculated from the relationship $t_{1/2} = 0.693/k_{inact}$.

3. The inactivation requires that all of the typical cofactors be present and that metabolism is occurring. It should be possible to demonstrate metabolism of the inactivator to a reactive intermediate which can then lead to the formation of product or to inactivation.

4. In the presence of one of its typical substrates, the enzyme should be protected from inactivation and thus show a slower rate of inactivation.

5. The inactivation should be irreversible. It should not be possible to show a regain of activity upon dialysis, gel filtration, or centrifugation since the inhibitor should be covalently bound to the apoprotein or to a prosthetic group of the enzyme.

6. Following inactivation, there should be a 1:1 stoichiometry of inactivator bound to enzyme molecule inactivated.

7. The presence of exogenous "scavenger" nucleophiles should not cause a decrease in the rate of inactivation or the stoichiometry of the reaction.

8. The inactivation should not exhibit any lag time. Some substrates of drug-metabolizing enzymes are converted to reactive intermediates that may leave the enzyme active site and form covalent adducts with proteins as well as with DNA or RNA. Therefore, it is necessary to differentiate between specific inactivation by mechanism-based inactivators and nonspecific inactivation which may result from extensive modification of the protein by reactive intermediates released from the active site which bind nonspecifically to the protein and eventually will lead to inactivation. Oftentimes, careful analysis of the time course for the reaction is required in order to differentiate between the two, since the latter case often exhibits some lag in the inactivation as the level of modification of the enzyme must increase to a certain point before significant inactivation is observed.

Under optimal conditions, all of these criteria will be readily observable for mechanism-based inactivator of a drug-metabolizing enzyme. However, not all of these tests may be applicable in every situation.

Experiments with mechanism-based inactivators do not need to be restricted to purified enzymes. Initial studies may be done with relatively crude enzyme preparations (e.g. cytosolic fractions or microsomes) using probe substrates demonstrated to be specific for the individual enzymes under investigation in order to monitor inactivation. Radioactive labeling studies can also be performed with such preparations followed by SDS-PAGE (sodium dodecyl sulfate-polyacrylamide gel electrophoresis) or HPLC (high performance liquid chromatography) separation of the enzymes of interest to examine the specificity and stoichiometry of the labeling.

14.10   INHIBITORS THAT ARE METABOLIZED TO REACTIVE PRODUCTS THAT COVALENTLY ATTACH TO THE ENZYME

This group of inhibitors may oftentimes be grouped with the mechanism-based inactivators previously discussed. Some of the same criteria may apply, such as saturation kinetics with respect to the concentration of the inactivator, the requirement for the presence of all of the typical cofactors and that metabolism is occurring, protection from inactivation by normal substrates, and the irreversible nature of the inactivation. However, the stoichiometry for the inactivation may be much greater than one, the presence of exogenous "scavengers" such as glutathione may affect both the rate of inactivation and the stoichiometry and the inactivation reaction may exhibit a lag time. Thus, the kinetics may not show first-order loss of activity with respect to time. However, it is important to realize that deviations from pseudo-first-order kinetics may not be easy to detect.

14.11   SUBSTRATE INHIBITION

Substrate inhibition kinetics are observed when a substrate causes a decrease in the rate of product formation as its concentration increases above a certain level. Although the exact mechanism for this type of inhibition with drug-metabolizing enzymes is not well understood, it has been observed in some cases, particularly with the cytochrome

P450s (Shou *et al.*, 2001). In classical enzymology, this behavior is oftentimes seen with enzymes that function using a ping-pong mechanism. For cytochrome P450s, this has been explained by proposing a two-site model (Shou *et al.*, 2001). In this model, the substrate can bind to the active site to yield a productive complex and substrate can also bind to an inhibitory site which is assumed to be nonproductive. When both sites are occupied with substrates, the turnover leading to product formation is no longer as rapid as that seen with the initial binding of the substrate to the productive site. In this model, the K_S (the binding constant) for substrate binding to the catalytic site is generally smaller than the K_I for the binding of the substrate to the inhibitory site, indicating that the substrate has a binding affinity to the catalytically active site greater than that for binding to the inhibitory site.

14.12   PARTIAL INHIBITION

Partial inhibition is observed when the enzyme is saturated with inhibitor but the inhibition is not complete. This phenomenon has been primarily observed in P450-catalyzed reactions and corresponds to a mixed-type inhibition. The cause of the partial inhibition is the formation of a Michaelis–Menten enzyme–substrate complex which also has the inhibitor bound and yet it is still catalytically active. The kinetics of this reaction can best be understood in terms of the substrate binding in its pre-ferred binding site in the active site, and in terms of the inhibitor binding either in the active site, to a site where it does not necessarily compete with the substrate, or outside of the active site, in some way affecting the catalytic activity. Partial inhibi-tion may result either from steric hindrance that hampers optimal binding of the sub-strate to its binding site in the active site or from a conformational change of the enzyme.

14.13   INHIBITION OF CYTOCHROME P450 ENZYMES

As discussed already, enzyme inhibition is a very important issue in drug discovery and development as well as in the clinic. The inhibitory effects of one drug on the disposi-tion, efficacy, or toxicity of another drug have major implications for the continued development and use of that entity. In many cases, drug–drug interactions can result in alterations in the pharmacokinetics of one or both agents. Although enzyme inhibi-tion can affect all of the various drug-metabolizing enzymes through a variety of mechanisms, the most common drug interactions generally occur through the cyto-chrome P450 system since they play such an important role in the metabolism of many drugs. Therefore, the rest of this chapter will focus on various aspects of the inhibition of these enzymes and we will now address some things that are more specific to the P450s; however, in many cases, the topics discussed are applicable to other drug-metabolizing enzymes although maybe not as widely as those areas discussed previously.

Generally, the metabolism of a drug by a cytochrome P450 represents the rate-limiting step in its metabolism; therefore, inhibition of the P450s is recognized by the FDA and other regulatory agencies as an important cause of drug–drug interactions (US Food and Drug Administration, Center for Drug Evaluation and Research, 1997). The catalytic cycle for drug metabolism by the cytochrome P450s provides for a number of potential points at which inhibition of metabolism may occur. The primary steps in the catalytic cycle for cytochrome P450-catalyzed reactions are (i) binding of the substrate; (ii) one-electron reduction of the ferric (Fe^{+3}) enzyme to the ferrous (Fe^{+2}) form by the NADPH-cytochrome P450 reductase; (iii) binding of the molecular oxygen by the ferrous (Fe^{+2}) iron; (iv) transfer of the second electron from the reductase to the ferrous-oxy-substrate complex with the subsequent release of one molecule of water and the formation of an activated oxygen intermediate; (v) insertion of the activated oxygen into the substrate to form the oxygenated product; and (vi) release of the oxygenated product from the enzyme leading to the regeneration of the native ferric (Fe^{+3}) form of the enzyme that can now bind another molecule of drug and undergo another catalytic cycle (White and Coon, 1980). Three of the preceding steps: (i) binding of the substrate, (ii) binding of the molecular oxygen to the ferrous (Fe^{+2}) enzyme, and (iii) the catalytic step in which the activated oxygen is transferred from the heme iron to the enzyme-bound substrate, appear to be particularly susceptible to inhibition (Ortiz de Montellano and Correia, 2005). Another potential target for inhibition of cytochrome P450-catalyzed drug metabolism is the transfer of electrons from the P450 reductase to the cytochrome P450, which occurs at two points in the catalytic cycle (after the addition of the substrate and after the binding of molecular oxygen to the ferrous P450). In general, inhibitors of these two steps interfere with the transfer of electrons from the reductase to the P450 by accepting electrons directly from the reductase and therefore they do not cause inhibition by interaction directly with the P450 protein. Thus, they are relatively nonspecific with respect to the cytochrome P450 forms that they inhibit and have the potential to inhibit all P450-catalyzed drug metabolism (Hollenberg, 2002).

14.14   REVERSIBLE INHIBITORS

The cytochrome P450s can be inhibited by all of the mechanisms already described in this chapter. The inhibitors of the P450s can be divided into three general categories that differ in their mechanisms: (i) inhibitors that bind reversibly to the enzyme; (ii) inhibitors that form quasi-irreversible complexes with either the ferric (Fe^{+3}) or ferrous (Fe^{+2}) iron of the heme prosthetic group; and (iii) inhibitors that bind irreversibly to the apoprotein or to the prosthetic heme, or that cause covalent binding of the prosthetic heme or a degradation product of the heme to the apoprotein. In general, reversible inhibitors interfere with the catalytic cycle prior to the formation of the activated oxygen intermediate and they may be competitive, noncompetitive, uncompetitive, product, or transition-state inhibitors. Those inhibitors that act during or subsequent to the formation of the activated oxygen intermediate are generally either quasi-irreversible or irreversible inhibitors.

As already indicated, some reversible inhibitors of the P450s may act by coordinating with the iron in the prosthetic heme atom. Coordination of a strong ligand to the ferric (Fe^{+3}) heme iron shifts the iron from the high- to the low-spin form giving rise to the "type II" difference spectrum (Schenkman, Sligar and Cinti, 1981). This change in the spin state occurs concomitantly with a change in the redox potential of the P450 that makes the reduction by the P450 reductase more difficult (Guengerich, 1983). Conversely, binding of a substrate to the active site of the P450 often shifts the heme iron to a more high-spin form causing a change in the redox potential which facilitates reduction. Thus, the inhibition of P450s by ligands to the heme iron which cause a shift in the spin state to a more low-spin form is a result not only of the occupation of the sixth coordination site of the heme iron by the ligand, but also the change in the reduction potential of the iron.

Inhibitors that bind to both the iron in the heme and to a hydrophobic binding site in the P450 active site are generally much more effective inhibitors than those that utilize only one of those interactions (Ortiz de Montellano and Correia, 2005). A number of nitrogen-containing aliphatic and aromatic compounds exhibit these properties and are relatively potent reversible inhibitors. Some of the most widely used compounds in this class are derivatives of imidazole and pyridine.

14.15 QUASI-IRREVERSIBLE INHIBITORS

Another category of inhibitors includes those compounds that inhibit by forming quasi-irreversible complexes with the iron atom in the heme prosthetic group. In general, these compounds require initial activation by the enzyme resulting in the formation of transient intermediates that can then coordinate very tightly to the prosthetic heme in the P450 active site leading to inhibition. Although these linkages are not covalent, the coordination is generally so tight that the inhibitory complexes can only be broken down to release the native catalytically active enzyme under special experimental conditions. Several different classes of compounds have been shown to be capable of forming these metabolic intermediates that coordinate so effectively with the P450 that they are converted to a catalytically nonfunctional state. These classes include compounds containing a dioxymethylene function and nitrogen-containing compounds including acyl hydrazines, 1,1-disubstituted hydrazines, and a variety of alkyl amines that can be oxidized to nitroso compounds by the P450s. Aryl and alkyl methylenedioxy compounds are oxidized by the P450s to yield species that can coordinate very tightly to the prosthetic heme iron, leading to the formation of metabolite intermediate complexes (MI complexes) that are extremely stable. The requirement for an initial oxidation by the P450 has been unequivocally demonstrated (Franklin, 1971). The resulting ferrous MI complex is extremely stable as demonstrated by the fact that it can be isolated intact from rats treated with isosafrole. The ferric MI complex is much less stable than the ferrous complex and incubation with various lipophilic molecules can result in displacement of the inhibitor from the active site, leading to regeneration of the native catalytically active enzyme (Elcombe *et al.*, 1975). The chemical nature of the side-chain substituent on the methylenedioxy compound plays an important role in determining the stability of the MI complex. Both the lipophilicity and the size of the alkyl side chain are important since alkyl chains of one to three carbon atoms lead to the formation of relatively unstable complexes while side chains containing larger alkyl groups result in the formation of stable MI complexes (Murray, Hetnarski and Wilkinson, 1985).

A second class of compounds that can form MI complexes with the P450 heme iron is the relatively large class composed of aromatic and alkyl amines. Included in this class are a number of clinically important antibiotics such as troleandomycin and erythromycin (Mansuy *et al.*, 1977). Oxidation of these amines results in the formation of intermediates that can coordinate very tightly to the ferrous heme resulting in optical spectra exhibiting an absorbance maxima in the range of 445–455 nm. Following oxidation, primary amines are capable of forming these types of MI complexes although secondary or tertiary amines are not able to do this. However, secondary and tertiary amines may also form MI complexes following conversion to primary amines by N-dealkylation. The primary amines initially undergo hydroxylation followed by a two-electron oxidation of the resulting hydroxylamine to give the C-nitroso group, which currently is thought to be responsible for chelation with the iron to form the MI complex (Mansuy *et al.*, 1977). Although the formation of MI complexes might be considered to fall into the category of mechanism-based inactivation, a better description would probably be transformation of the substrates by the enzyme(s) to slow, tight-binding inhibitors since the linkages are not covalent and oxidation by an oxidant such as Fe $(CN)_6^{3-}$ will oxidize the ferrous complex and release the ligand, thereby regenerating the native ferric enzyme which is now fully competent to carry out metabolism.

14.16 MECHANISM-BASED INACTIVATORS

Over the past decade, considerable interest has developed with respect to mechanism-based inactivators of the cytochrome P450s. Three general classes of mechanism-based inactivators of the cytochrome P450s have been identified. These classes consist of compounds that (i) bind covalently to the P450 apoprotein, (ii) bind irreversibly to the prosthetic heme group, and (iii) cause covalent binding of the heme prosthetic group or its degradation product to the apoprotein (Osawa and Pohl, 1987). These modifications, when they occur *in vivo*, appear to greatly accelerate the degradation of the inactivated protein. It is now clear that many mechanism-based inactivators may inactivate a given P450 by more than one mechanism and may inactivate different forms of P450s by different mechanisms. Thus, the predominant way in which a mechanism-based inactivator inactivates a P450 may be determined by a number of factors including the identity of the form of P450 being studied. So far, the factors that determine how a specific mechanism-based inactivator will modify a particular form of P450 are not well understood.

A wide variety of different types of compounds have been found to be excellent mechanism-based inactivators of the cytochrome P450s. These include (i) acetylenes and terminal aryl and alkyl olefins such as 2-ethynylnaphthalene, 5-phenyl-1-pentyne, 9-ethynylphenanthrene, 10-dodecynoic acid, 17α-ethynylestradiole, gestodene, mifepristone (RU486), and secobarbital; (ii) a variety of different organosulfur compounds such as carbon disulfide, disulfiram, cimetidine, thiophenes, mercaptosteroids, isothiocyanates, thioureas, parathion, dialkylsulfides, and diethyldithiocarbamate; (iii) furanocoumarins such as 8-methoxypsoralen, bergamottin, and L-754,394 (a Merck compound synthesized as a potential inhibitor of the HIV protease); (iv) 1-aminobenzotriazole and its N-aralkylated derivatives; and (v) halogenated compounds such as chloramphenicol and various N-monosubstituted dichloroacetamides (Kent, Jushchyshyn and Hollenberg, 2001; Ortiz de Montellano and Correia, 2005).

The marked lack of substrate specificity of many of the P450s involved in drug metabolism, as evidenced by the differences in the sizes and the chemical properties of their substrates as well as the number of different substrates that they can metabolize, speaks to the apparent flexibility of the active sites of these enzymes. This apparent promiscuity of the P450s offers the possibility of designing a multitude of mechanism-based inactivators having a variety of different potentially reactive moieties. Thus, it would seem to be a relatively straightforward exercise to design mechanism-based inactivators that will inhibit a desired P450. However, the problem in designing mechanism-based inactivators for P450s lies not in designing one that will inactivate a P450 but in designing one that will be specific for a single form of P450.

The development of reversible, quasi-irreversible, and irreversible inhibitors of the P450s involved in the metabolism of drugs, other xenobiotics, and endogenous substrates such as steroid hormones, and our understanding of the mechanisms of action of these various types of inhibitors have increased remarkably over the past decade and have provided important insights that can be used to develop highly selective isozyme-specific inhibitors of these P450s. The development of highly selective inhibitors of P450s is of great importance not only for studies aimed at probing the three-dimensional structures of the proteins, their mechanisms of action, and the biological roles of the various P450s, but also because of their potential as modulators of cytochrome P450 activity that can be used as therapeutic agents in a fashion similar to the way that inhibitors of the P450s involved in steroid metabolism have been used to treat endocrine disorders and also as anticancer agents. In addition, since different P450s play major roles in the metabolic activation and detoxication of a variety of different chemical carcinogens as well as other toxins, the development of inhibitors that can be used to selectively inactivate these enzymes is of great interest. These types of inhibitors have the potential to shift the balance between various pathways for the metabolism of a chemical carcinogen or toxin such that those pathways involved in metabolic activation will be minimized, whereas those pathways leading to detoxication will be enhanced. Thus, these agents would be of great value in improving the quality of human life. In addition, a better understanding of inhibition mechanisms and selectivity may lead to more efficient screening protocols for drug–drug interactions in the pharmaceutical industry.

REFERENCES

Black, D.J., Kunze, K.L., Wienkers, L.C., Gidal, B.E., Seaton, T.L., McDonnell, N.D., Evans, J.S., Bauwens, J.E., Trager, W.F. (1996). Warfarin-fluconazole II. A metabolically based drug interaction: in vivo studies. *Drug Metab. Dispos.*, *24*, 422.

Bronson, D.D., Daniels, D.M., Dixon, J.T., Redick, C.C., Haaland, P.D. (1995). Virtual kinetics: using statistical experimental design for rapid analysis of enzyme inhibitor mechanisms. *Biochem. Pharmacol.*, *50*, 823.

Cornish-Bowden, A. (1979). *Fundamentals of Enzyme Kinetics*, Butterworths, London.

Elcombe, C.R., Bridges, J.W., Gray, T.J.B., Nimmo-Smith, R.H., Netter, K.J. (1975). Studies on the interaction of safrole with rat hepatic microsomes. *Biochem. Pharmacol.*, *24*, 1427.

Franklin, M.R. (1971). The enzymic formation of a methylenedioxyphenyl derivative exhibiting an isocyanide-like spectrum with reduced cytochrome P-450 in hepatic microsomes. *Xenobiotica*, *1*, 581.

Guengerich, F.P. (1983). Oxidation-reduction properties of rat liver cytochromes P450 and NADPH-cytochrome P450 reductase related to catalysis in reconstituted systems. *Biochemistry*, *22*, 2811.

Guengerich, F.P. (1999). Inhibition of drug metabolizing enzymes: molecular and biochemical aspects, in *Handbook of Drug Metabolism* (ed. T.F. Woolf), Marcel Dekker, New York, p. 203.

He, K., Iyer, K., Hayes, R.N., Sing, M.W., Woolf, T.F., Hollenberg, P.F. (1998). Inactivation of P450 3A4 by bergamottin, a component of grapefruit juice. *Chem. Res. Toxicol.*, *11*, 252.

Hollenberg, P.F. (2002). Characteristics and common properties of inhibitors, inducers and activators of CYP enzymes. *Drug Metab. Rev.*, *34*, 17.

Kent, U.M., Jushchyshyn, M.I., Hollenberg, P.F. (2001). Mechanism-based inactivators as probes of cytochrome p450 structure and function. *Current Drug Metab.*, *2*, 215.

Kohler, G.I., Bode-Boger, S.M., Busse, R., Hoopman, M., Welete, T., Boger, R. (2000). Drug-drug interactions in medical patients: effects of in-hospital treatment and relation to multiple drug use. *Int. J. Clin. Pharmacol. Ther.*, *38*, 504.

Levy, M., Kewitz, H., Altwein, W., Hillebrand, J., Eliakim, M. (1980). Hospital admissions due to adverse drug reactions: a comparative study from Jerusalem and Berlin. *Eur. J. Clin. Pharmacol.*, *17*, 25.

Lin, J.H., Lu, A.Y.H. (1998). Inhibition and induction of cytochrome P450 and the clinical implications. *Clin. Pharmacokinet.*, *35*, 361.

Mansuy, D., Gans, P., Chottard, J.-C., Bartoli, J.-F. (1977). Nitrosoalkanes as Fe(II) ligands in the 455-nm-absorbing cytochrome P-450 complexes formed from nitroalkanes in reducing conditions. *Eur. J. Biochem.*, *76*, 607.

Murray, M., Hetnarski, K., Wilkinson, C.F. (1985). Selective inhibitory interactions of alkoxy-methylene-dioxybenzenes towards mono-oxygenase activity in rat hepatic microsomes. *Xenobiotica*, *15*, 369.

Ortiz de Montellano P.R. and Correia, M.A. (2005). Inhibition of cytochrome P450 enzymes, in *Cytochrome P450. Structure, Mechanism and Structure*, 3rd edn (ed. P.R. Ortiz de Montellano), Kluwer Academic/Plenum Press, New York, p. 247.

Osawa, Y., Pohl, L.R. (1987). Covalent binding of the prosthetic heme to protein: a potential mechanism for the suicide inactivation or activation of hemoproteins. *Chem. Res. Toxicol.*, *2*, 131.

Schenkman, J.B., Sligar, S.G., Cinti, D.L. (1981). Substrate interactions with cytochrome P-450. *Pharmacol. Ther.*, *12*, 43.

Shou, M., Lin, Y., Lu, P., Tang, C., Mei, Q., Cui, D., Tang, W., Ngui, J.S., Lin, C.C., Singh, R., Wong, B.K., Yergey, J.A., Lin, J.H., Pearson, P.G., Baillie, T.A., Rodrigues, A.D., Rushmore, T.H. (2001). Enzyme kinetics of cytochrome P450-mediated reactions. *Current Drug Metab.*, *2*, 17.

Silverman, R.B. (1995). Mechanism-based enzyme inactivators. *Methods Enzymol.*, *249*, 240.

US Food and Drug Administration, Center for Drug Evaluation and Research (CDER) (1997). *Guidance for Industry: Drug Metabolism/Drug Interaction Studies in the Drug Development Process: Studies In Vitro*, www.fda.gov/cder/guidance/clin3.pdf (accessed April 1997).

Wang, E., Walsh, C. (1978). Suicide substrates for the alanine racemase of *Escherichia coli* B. *Biochemistry*, *17*, 1313.

White, R.E., Coon, M.J. (1980). Oxygen activation by cytochrome P450. *Annu. Rev. Biochem.*, *49*, 315.

CHAPTER 15

Evaluating and Predicting Human Cytochrome P450 Enzyme Induction

MICHAEL SINZ, SEAN KIM, STEPHEN FERGUSON, and EDWARD LECLUYSE

15.1 INTRODUCTION

Cytochrome P450 (CYP) isoforms are the predominate enzymes involved in the biotransformation and elimination of xenobiotics. Since their identification and characterization in 1962, they have become one of the most published enzyme systems in pharmaceutical research (Omura and Sato, 1962). Given their prevalence for metabolizing and eliminating drugs, interactions or events that manipulate the concentration or function of CYP enzymes have great impact on drug disposition, toxicity, and pharmacology. The most common alterations to drug-metabolizing enzyme function are due to (i) genetic polymorphisms, (ii) enzyme inhibition, and (iii) enzyme induction. The focus of this chapter will be on CYP enzyme induction, including methods to evaluate, predict, and quantitate drug interactions due to enzyme induction. Enzyme induction is the action of creating more enzymes than is normally present in a biological system and can be manifested by increased gene transcription or decreased protein or mRNA degradation. As early as 1954, the first report of enzyme induction appeared in a manuscript by Brown *et al.*, who described the enzyme-inducing effects of various food diets when given to rodents and the resulting increased enzymatic effects of *N*-demethylation in liver homogenates (Brown, Miller and Miller, 1954). This was followed some time later by the first review on microsomal enzyme induction in 1967 (Conney, 1967). Interestingly, some of the original nomenclature used to describe CYP enzymes was based on enzyme-inducing agents employed to increase CYP-isoform levels making them easier to characterize. For example, BNF-B was used to identify a CYP1A enzyme and PB-4/5 for a CYP2B enzyme, after treatment with the inducing agents β-naphthoflavone and phenobarbital (PB), respectively.

The increase in enzyme activity caused by a drug is reflected in an increased hepatic clearance of drugs metabolized by the induced enzyme. Various terms are used to describe the drug that causes enzyme induction (perpetrator or inducer) and the drug

Drug Metabolism Handbook: Concepts and Applications, Edited by Ala F. Nassar,
Paul F. Hollenberg, and JoAnn Scatina
Copyright © 2009 by John Wiley & Sons, Inc.

affected by the induced enzyme (victim or substrate). Pharmacokinetically, the drugs affected by enzyme induction generally demonstrate reduced AUC (area under the curve), C_{max} (maximum concentration), and half-life as a reflection of increased clearance. For example, coadministration of the antidiabetic drug rosiglitazone (a CYP2C8 substrate) with the inducing agent rifampicin leads to 65%, 31%, and 62% decreases in AUC, C_{max}, and half-life of rosiglitazone, respectively (Park *et al.*, 2004). A more dramatic example is illustrated by the interaction between midazolam (a CYP3A4 substrate) and St. John's wort (an herbal antidepressant and CYP3A4 inducer), where the AUC and C_{max} decreased by 79% and 65%, respectively. However, the half-life of midazolam decreased only by 9%, which is anticipated for a high-clearance drug such as midazolam (Mueller *et al.*, 2006). These pharmacokinetic changes have significant pharmacological consequences because the increased metabolism (elimination) reduces the duration and pharmacological effect of a coadministered drug. For example, the induction of CYP3A4 results in reduced ethinylestradiol levels that can lead to unexpected pregnancies and reduced cyclosporine concentrations that can lead to organ transplant rejection (both drugs are predominately metabolized by CYP3A4). Drug–drug interactions are known to occur via enzyme induction for several CYP enzymes, most notably CYP3A4. In general, the decrease in AUC observed ranges from 15% to 98% depending on the potency of the inducing agent as well as the fraction metabolized and overall elimination of a drug (victim) via the induced pathway.

In addition to the aforementioned scenarios where the perpetrator drug causes the induction and a second victim drug is affected, there exists a situation where a single drug acts as both perpetrator and victim, referred to as "autoinduction." Carbamazepine is an anticonvulsant that causes CYP3A4 enzyme induction in epileptic patients and this induction increases the clearance of carbamazepine itself because carbamazepine is also a CYP3A4 substrate (Bertilsson *et al.*, 1980). The antimalarial drug artemisinin also exhibits this effect by inducing CYP2B6 which is significantly involved in artemisinin elimination (Simonsson *et al.*, 2003). Autoinduction is a phenomenon that can occur anytime a drug induces an enzyme that is also predominately involved in its own metabolic clearance. Therefore, when understanding the overall liability of an enzyme inducer, a complete understanding of the major mechanisms of elimination and the enzymes involved in elimination is important to assess potential autoinduction.

Drug–drug interactions are occurring with increasing frequency due to the rise in multiple prescriptions taken by individual patients. Recently it was estimated that 32 million Americans take three or more prescriptions daily (American Heart Association, www.americanheart.org). Hence, the potential for drug–drug interactions is also increasing and the need to develop compounds lacking the ability to induce drug-metabolizing enzymes has become a significant concern while developing new drugs. Most pharmaceutical companies have implemented several layers of screening and characterization to assess the potential liability of enzyme induction. In early drug discovery stages, large numbers of compounds are screened by means of high-throughput assays to either eliminate compounds with enzyme induction potential or reduce the liability by selecting compounds with lower potencies for the inductive effect. During preclinical development, characterization in more complex human systems helps to further define the induction potential of a compound. Ultimately, the effect or lack of effect of an enzyme inducer is evaluated in human volunteers/patients with well-characterized CYP probe substrates (as indicated in Table 15.1) where the actual magnitude of pharmacokinetic change is measured.

TABLE 15.1 *In vivo* probe substrates and enzyme inducers used in clinical drug interaction studies (oral administration).[a]

CYP	*In vivo* probe substrates	Inducers
CYP1A	Theophylline, caffeine	Smokers versus nonsmokers
CYP2B6	Efavirenz	Rifampicin
CYP2C8	Rosiglitazone, repaglinide	Rifampicin
CYP2C9	Warfarin, tolbutamide	Rifampicin
CYP2C19	Omeprazole, esoprazole, lansoprazole, pantoprazole	Rifampicin
CYP2D6	Desipramine, dexamethasone, atomoxetine	None identified
CYP2E1	Chlorzoxazone	Ethanol
CYP3A4	Midazolam	Rifampicin, carbamazepine

[a]From FDA (2006).

15.2 RECEPTOR-BASED *IN VITRO* BIOASSAYS

For many drug-metabolizing enzymes, transcriptional activation and subsequent induction are mediated by nuclear hormone receptors that function as transcription factors, such as CAR (constitutive androstane receptor), PXR (pregnane X receptor or SXR-steroid X receptor), and PPAR (peroxisome proliferator-activated receptor) (Honkakoski, Sueyoshi and Negishi, 2003; Wang and LeCluyse, 2003; Qatanani *et al.*, 2005; Tirona and Kim, 2005). One notable exception is AhR (aromatic hydrocarbon receptor) which regulates the expression of CYP1A enzymes (Mandal, 2005). AhR is not a member of the nuclear receptor (NR) family, but belongs to the basic helix-loop-helix PER-ARNT-SIM (PAS) transcription factor family.

The identification of receptors that regulate CYP/transporter expression has led to the development of several high-throughput *in vitro* methods such as receptor binding and transactivation assays which assess the interaction between receptors and potential ligands (Vignati *et al.*, 2004; Zhu *et al.*, 2004). These assays are widely employed throughout drug discovery and development to identify potential inducers of CYP and transporter enzymes (Stanley *et al.*, 2006). In practice, these assays are utilized, in a parallel or sequential manner, along with lower-throughput but more definitive cell-based assays. As described in the preceding chapter, the human pregnane X receptor (hPXR, SXR, NR1I2) is a nuclear hormone receptor/transcription factor principally responsible for the induction of CYP3A4, as well as the expression of CYP2B6, CYP2C8/9, Pgp, MRP-2, and OATP-2 (Kliewer, Goodwin and Willson, 2002). PXR-derived assays are the most common high-throughput assays due to the simplicity of ligand-based activation and the importance of target genes in drug interactions. As such, the following discussion will mainly focus on PXR-based bioassays.

Ligand-binding assays generally consist of expressed receptors incubated with test compounds and a tight-binding radiolabeled ligand. Competition (displacement) of the radiolabeled ligand with the test compound is measured and an IC_{50} (inhibitor concentration that causes 50% inhibition) is determined. Cell-based PXR transactivation-reporter assays employ (i) an expression vector containing some form of the CYP3A4 5'-promoter region coupled to a reporter gene (typically luciferase) and (ii) an hPXR expression vector. Due to increased expression of luciferase in response to ligand–receptor interaction, the induction response can be measured as an increased production of luminescent product and an EC_{50} (concentration at which 50% efficacy is found) value determined from the concentration–response curve. Typically these assays are augmented by co-transfecting an expression vector for human PXR to enhance the magnitude of induction response because most immortalized cell lines generally do not significantly express PXR (Goodwin, Hodgson and Liddle, 1999; El-Sankary et al., 2001). It has been shown that the choice of host cell has an impact on the magnitude of induction response. For example, rifampicin treatment of HepG2 cells induces luciferase activity, while the same treatment in COS7 or NIH3T3 cell lines results in minimal responses (Goodwin, Hodgson and Liddle, 1999). In addition, it does not appear that a maximal PXR response requires a cell of hepatic origin. For example, the African Green Monkey kidney cell line, CV-1, is widely used in PXR reporter assays. However, because the ligand–PXR interaction is a complex process requiring multiple co-regulators, the most appropriate host cell lines presumably have all the regulatory partners necessary to mediate an induction response.

In general, there is a good correlation between EC_{50} and IC_{50} values in PXR trans-activation and binding assays, respectively (Zhu et al., 2004). However, there are examples where significant binding to the PXR receptor does not lead to transactivation or enzyme induction. These unique situations can be caused by PXR antagonists or ligands which bind, but do not elicit the appropriate displacement of corepressors or recruitment of co-activators. Such is the example of docetaxel and paclitaxel, two structurally similar anticancer agents, in which both drugs bind to PXR; however, only paclitaxel binding results in significant gene activation and enzyme induction (Harmsen et al., 2007). Ectesinascidin 743 (ET-743) is an experimental anticancer drug that can also inhibit PXR transactivation at nanomolar concentrations via PXR antagonism (Synold, Dussault and Forman, 2001). With the increasing number of compounds being tested simultaneously in PXR receptor binding and reporter assays, more PXR antagonists (characterized by a strong PXR binding but no or weak PXR transactivation) are expected to be identified. However, it is not yet clear what the clinical impact that PXR antagonists will have on the expression of PXR target genes in humans.

The change in hPXR transactivation-reporter activities has also been shown to correlate with the changes in CYP enzyme activities measured from primary human hepatocytes, which are widely accepted as the standard in vitro assessment of enzyme induction (Luo et al., 2002). Presently, the PXR transactivation assay is the preferred method to screen new chemical entities over the PXR binding assay due to the improved correlation to human drug–drug interactions, fewer false positives, and no radiolabeled reagents. While many reporter assays utilize transiently transfected systems, the variation in transfection efficiency and the resulting induction responses from one transfection to another is not ideal in an industrial paradigm of high-throughput screening. By introducing an antibiotic resistance gene as a selection marker, one can establish a stably transfected cell such as the DPX-2 cell line to minimize run-to-run variations (Lemaire, de Sousa and Rahmani, 2004; Trubetskoy et al., 2005). Another approach is to cryopreserve a single large batch of transfected cells and thus ensure a more consistent induction response from one experiment to another. The bulk cryopreservation method does not appear to affect the functionality of the cells as the induction responses (EC_{50} and maximum response) observed between cryopreserved and fresh cells were found to be similar (Zhu et al., 2007).

While PXR-mediated gene transcription is primarily initiated by direct ligand binding, several alternative mechanisms of gene induction have been identified. CAR is constitutively activated and predominately sequestered in the cytoplasm by accessory proteins, such as HSP90 and CAR cytoplasmic retention protein (CCRP), in order to regulate transcriptional activity. Translocation of CAR to the nucleus can be initiated by direct agonist binding to the receptor but also through a partially elucidated ligand-independent mechanism involving kinases. Phenobarbital is an example of a drug that does not bind to CAR yet causes nuclear translocation and transcriptional activation of a target gene, CYP2B6 (Qatanani *et al.*, 2005). Similarly, the inactive AhR resides in the cytoplasm and can be activated upon ligand binding or via protein tyrosine kinases. Omeprazole-mediated induction of CYP1A has been shown to proceed through this kinase-mediated pathway although this premise has come under some debate (Hu *et al.*, 2007). Induction of CYP2E1 is notably different from other CYP enzymes in that no receptor is known to be involved. CYP2E1 inducers, such as ethanol, increase the stability of mRNA and/or protein and thus result in increased enzyme activity due to diminished degradation pathways (Gonzalez, 2007).

The development of CAR-based assays has proven to be more difficult than PXR assays. CAR transactivation assays frequently encounter a high basal CAR activity and CAR often spontaneously translocates to the nucleus when overexpressed in cell lines (Chang and Waxman, 2006). These experimental challenges hamper the identification of CAR activators and only a few compounds have been identified to date. Also due to dual mechanisms of CAR activation, direct and indirect, screening for CAR modulators (potential CYP2B6 inducers) involves a nuclear translocation assay to complement a transactivation assay. The CAR nuclear translocation assays are frequently conducted in primary hepatocytes due to the spontaneous translocation of CAR in cell lines as described earlier. The translocation of hCAR from cytoplasm to nucleus can be monitored and measured either by confocal microscopy using an expressed CAR protein tagged with fluorescent proteins such as GFP or EYFP or by measuring the ratio of CAR proteins between cytoplasm and nucleus using Western blot analysis (Faucette *et al.*, 2007). Recently, a human splice variant of CAR was identified and termed "CAR3." A subsequent analysis showed that CAR3 has ~80% lower basal activity compared with wild-type CAR (CAR1) when expressed in COS-1 cells (Auerbach *et al.*, 2005; Faucette *et al.*, 2007). The CAR3-based transactivation assay can more easily identify potential CAR activators by lowering the basal activity that can otherwise mask a potential ligand-activated transactivation response.

It should be noted that the ligand binding domains (LBDs) of many nuclear receptors are different between various animal species and human (Wang and LeCluyse, 2003). For example, rodent and human PXR only share ~76% LBD sequence homology. Therefore, *in vitro* bioassays utilizing animal PXR are generally not predictive of the potential induction effect in humans. Nonetheless, enzyme induction in animals can complicate preclinical safety assessments in which these animal species are utilized (Worboys and Carlile, 2001). Frequently, enzyme induction (especially autoinduction) is found in multiple-dose animal pharmacokinetic or toxicokinetic studies, and the use of animal *in vitro* assays can aid in the investigation of induction mechanisms as well as species differences.

While nuclear hormone receptor-based assays provide reliable high-throughput screening opportunities, one disadvantage is that each assay only predicts one particular receptor-mediated process at a time. In some cases, multiple nuclear hormone receptors can work in concert and provide a cumulative effect on target genes (Wang and LeCluyse, 2003). This nuclear receptor cross talk with target genes is best evaluated in more complex cell-based models to be described in subsequent sections.

15.3 IMMORTALIZED HEPATOCYTES

Primary human hepatocytes are commonly used in drug metabolism-related studies as well as in the assessment of hepatotoxicants (Soars *et al.*, 2007). However, their limited supply, rapid decline in expression of key drug-metabolizing enzymes, and significant donor-to-donor variation have fostered research toward the generation of human hepatocyte-like cells that can provide a continuous supply of cells while maintaining stable expression of necessary enzymes and transporters. There are currently three different approaches being taken to generate cell lines with these characteristics: refinement of hepatocarcinoma-derived cells into an appropriate phenotype, immortalization of human primary hepatocytes, and differentiation of stem cells into hepatocyte-like cells (to be discussed in Section 15.14).

Immortalized cells are defined as cells that can grow and divide indefinitely under optimal culture conditions (Hahn, 2002). Immortalization of cells can occur naturally as exemplified by cells of tumor origin; however, a considerable effort is being made to convert primary cells into non-tumorigenic immortalized cells (Sinz and Kim, 2006). There are many different approaches used to immortalize primary hepatocytes but the most common methods of immortalization are (i) overexpression of SV40 large T antigen (viral oncogene) or (ii) expression of telomerase reverse transcriptase (TERT). While viral transformation is relatively simpler, the cells produced through this method tend to be genetically unstable and readily lose their phenotypic characteristics. Combined, both SV40 T antigen and TERT-mediated transformations are predicted to produce genetically more stable cells (Cascio, 2001). In all types of cells, the goal is to either find or produce a cell line that maintains the phenotypic characteristics of the parental tissue (hepatocytes) and thus provide an unlimited and stable supply of cells for ADMET (absorption, distribution, metabolism, elimination, toxicity) research. The goal of this section is to review the most widely studied immortalized cells used in the assessment of CYP induction as well as the methodologies that are employed.

15.4 HepG2 AND HepG2-DERIVED CELL LINES

In the absence of a suitable alternative to primary hepatocytes, several lines of hepatocarcinoma cells have been used in the study of drug metabolism. Among them, the HepG2 cell line (established in 1979 from a hepatoma tissue) is the most well characterized and frequently used. While retaining some liver-specific functions, HepG2 cells have low expression and activity of CYP enzymes, as well as phase II enzymes (Vermeir *et al.*, 2005). Mostly fetal isoforms of CYPs, such as CYP1A1 and CYP3A7, are expressed in high levels and the expression of other phase I and II enzymes are much lower than those in primary hepatocytes (Wilkening, Stahl and Bader, 2003). The same observation was made in recent comprehensive reports on the expression analysis of phase I and II enzymes in HepG2 cells (Westerink and Schoonen, 2007a, b). Compared with the expression levels in cryopreserved human hepatocytes, HepG2 cells express much lower levels of all major CYPs (CYP1A1, -1A2, -2A6, -2B6, -2C8, 2C9, -2C19, -2D6, -2E1, and -3A4) as well as phase II enzymes (UGT1A1 and -1A6). However, the transcript levels of several isozymes of sulfotransferases (SULTs), glutathione-*S*-transferases (GSTs), *N*-acetyltransferases (NATs), as well as epoxide hydrolase (EPHX1) were similar to those in primary human hepatocytes in these studies.

In these studies, the transcript levels of several CYPs including CYP3A4 and -2B6 as well as phase II enzymes such as EPHX1, SULTs, and GTSs were increased in response to model inducers. The most significant induction was observed with CYP1A1/2 RNA expression and activity by known AhR agonists (BNF, 3MC, TCDD, indirubin, and indigo). This observation is consistent with previous reports of CYP1A1/2 induction by other xenobiotics including but not limited to polyaromatic hydrocarbons, omeprazole, and quinine (Bapiro *et al.*, 2002; Iwanari *et al.*, 2002; Baliharova *et al.*, 2003). The induction of CYP3A4 in HepG2 cells was also reported with *o, p'*-DDT (main isomer of DDT) which induced both activity and mRNA levels of CYP3A4 via an activation of PXR (Medina-Diaz and Elizondo, 2005; Medina-Diaz *et al.*, 2007). Others reported the induction of CYP3A4 by PB and dexamethasone (Hewitt and Hewitt, 2004) as well as induction of UGT1A1 by the dietary flavonoid, chrysin (Walle *et al.*, 2000; Smith *et al.*, 2005). However, due to the low basal expression of enzymes and minimal induction responses, HepG2 cells are not considered an appropriate *in vitro* model to study induction of CYPs.

Recently, HepG2 cells have been recognized for their heterogeneity in genetic characteristics in that several different cell subtypes can coexist in a cell population. Hewitt *et al.* reported that even the source of the cells can have an impact on basal activities in HepG2 cells (Hewitt and Hewitt, 2004). By enriching only the type of cells that have better induction responses through means of altered nutritional requirements, several HepG2 subtypes have been established. Amphioxus ACTIVTox cells are a highly selected subclone of HepG2 cells developed by Amphioxus Cell Technologies (Stem Cell Innovations, Houston, TX). The ACTIVTox cells have been shown to be useful in assessing the induction of both CYP1A and CYP3A4 as well as toxicity of new chemical entities (www.stemcellinnovations.com). The WGA cell line is also a highly differentiated subclone of HepG2 cells enriched by culturing the cells in a defined medium that lacks arginine (Rencurel *et al.*, 2005). Unlike the parental HepG2 cells or other immortalized cells, the WGA cells were shown to be responsive to PB-induced CYP2B6 induction, which holds promise as a cell line to study CAR-mediated CYP regulatory mechanisms.

15.5   HBG-BC2

The human hepatoma cell line, HBG, was isolated from the resected tumor of a male patient (Le Jossic *et al.*, 1996). When seeded at confluency, these cells undergo a differentiation process, resulting in expression of liver-specific functions, as well as drug-metabolizing enzymes. These differentiated cells are known as BC2 cells (O'Connor *et al.*, 2005). Gomez-Lechon *et al.* performed an extensive analysis of the biotransformation properties of BC2 cells by measuring basal and inducible expression of phase I and II enzymes (Gomez-Lechon *et al.*, 2001). The analysis revealed low but measurable activities of CYP1A1/2, -2A6, -2B6, -2C9, and -3A4. Upon treatment of the cells with appropriate inducers, the activity of CYP1A1/2 was greatly increased while no or marginal induction was observed for CYP2B6. Even though the authors observed induction of CYP3A4 in response to 100-μM dexamethasone, no induction of CYP3A4 activity (testosterone 6β-hydroxylase) was found with either rifampicin or phenobarbital suggesting that BC2 cells may not be a good model to assess induction potential of suspected CYP3A inducers. Thus, recent studies using BC2 cells have focused more on their use in generating metabolites or assessing toxicities mediated by biotransformation (Fabre *et al.*, 2003; O'Connor *et al.*, 2005).

15.6 HEPATOCARCINOMA-DERIVED CELL LINE: HepaRG

The HepaRG cell line was isolated from a resected liver tumor of a female patient suffering from hepatitis C-related hepatocellular carcinoma (HCC) (Gripon *et al.*, 2002). When the cells are cultured at confluency for several weeks in the presence of 2% DMSO and hydrocortisone, a highly differentiated hepatocyte-like cell line develops. Subsequent examination revealed that these cells express several liver-specific markers such as albumin, transferrin, CYP2E1, CYP3A4, aldoase B and GSTα, and retain a susceptibility to hepatitis B virus infection which was previously validated only in primary hepatocytes. A comprehensive expression analysis performed by Aninat *et al.* (2006) showed that RNA expressions of major nuclear hormone receptors (PXR, AhR, and PPARα) were similar between primary human hepatocyte and differentiated HepaRG cells (aided by the culture of cells in DMSO). Interestingly, the expression of CAR was high compared with that of HepG2 cells when HepaRG cells were cultured in the presence of DMSO, and was 20–30% of that in primary human hepatocytes. Previously, none of the immortalized cell lines have reported substantial expression of CAR and thus the HepaRG cell line may be of interest in studying CAR-mediated induction. In this study, some of CYP enzymes also required the addition of DMSO (2%) to exhibit similar expression levels as in primary hepatocytes. For example, mRNA levels of CYP2B6, -2E1, and -3A4 were markedly increased in the presence of DMSO while CYP1A2, -2C9, -2D6, and other liver-specific markers were relatively unchanged. The enzyme activities of CYP1A, -2C9, -2E1, and -3A4 were also highly enhanced by the addition of DMSO. However, in the presence of DMSO, HepaRG cells were refractory to induction by several prototypical inducers and the investigators postulated that the DMSO treatment induced the expression of CYPs to the maximum level that can be achieved. This hypothesis was supported by the observation that prototypical inducers increased enzyme activities of CYP3A4, -2C9, and -1A by approximately 10-, 2-, and 63-fold, respectively, in HepaRG cells grown without DMSO. Le Vee *et al.* also examined the expression of hepatic transporters in HepaRG cells (Le Vee *et al.*, 2006). The transcript levels of OCT-1, OATP-C, NTCP, OATP-8, BSEP, and MRP-2 were approximately 10–55% of those found in primary human hepatocytes, while the expressions of OATP-B, MDR-1, and MRP-3 were higher in HepaRG than in primary human hepatocytes. The measurement of uptake transporter activities showed that OCT-1, OATP/OAT2, and NTCP activities were detectable but lower than in primary human hepatocytes. However, the activities of efflux transporters such as MDR-1 (Pgp) and MRP were similar or even higher than those in primary human hepatocytes. The treatment of HepaRG cells with known inducers of transporters such as rifampicin, phenobarbital, and chenodeoxycholic acid significantly increased the mRNA expression of MDR-1, MRP2, and BSEP, respectively. Overall, the HepaRG cell line could represent a reliable surrogate *in vitro* model to primary human hepatocytes as an unlimited source of hepatocyte-like cells while retaining high expression as well as inducibility of drug-metabolizing enzymes and transporters (Guillouzo *et al.*, 2007).

15.7 CELLS OF INTESTINAL ORIGIN: LS180 AND Caco-2

The PXR is highly expressed in the liver, as well as in the intestine, and is known to regulate the expression of intestinal CYP3A4 and MDR-1. The induction of CYP3A4 and MDR-1 in the intestine can negatively impact the absorption of orally delivered drugs and thus a suitable cell line of intestinal origin has been sought after to study the regulation of these genes. LS180 and Caco-2 cells are human colon carcinoma cell lines routinely used and well characterized in studying intestinal absorption of drugs and regulation of CYP3A4 and MDR-1. In LS180 cells, CYP3A4 and MDR-1 have been shown to be responsive to induction stimuli by reserpine, rifampicin, PB, and verapamil (Schuetz, Beck and Schuetz, 1996). The rank order for inducers of MDR-1 was similar to that of CYP3A4 inducers in this study, suggesting a parallel regulation of these genes in LS180 cells. While inducers of MDR-1 and CYP3A4 demonstrate similar changes in LS180 cells, similar responses were not observed for Caco-2 cells (Li *et al.*, 2003). MDR-1 induction was often prominent even in the absence of CYP3A4 induction which could be due to a lower expression of PXR in Caco-2 cells (Li *et al.*, 2003; Brandin *et al.*, 2007).

In addition, induction of CYP1A2 was also reported when LS180 cells were exposed to a prototypical CYP1A inducer, omeprazole (Brandin *et al.*, 2007). A variety of herbal and food supplements were shown to induce the expression of CYP1A2, -3A4, and MDR-1 in this study suggesting herbal/food–drug interactions can be studied in the LS180 cell line. Interestingly, vitamin supplements appear to play an important role in the regulation of drug-metabolizing enzymes and transporter in these intestinal cells. Tocotrienols, major components of vitamin E, induced expression of MDR-1 and UGT1A1 but not CYP3A4 in LS180 cells (Zhou *et al.*, 2004). This was in contrast to primary hepatocytes where only the induction of CYP3A4 by tocotrienols was observed. Subsequently it was shown that the expression of nuclear receptor corepressor (NCoR) which negatively regulates PXR activity was higher in LS180 cells than in primary human hepatocytes. By attenuating the expression of NCoR, the ability of tocotrienols to induce CYP3A4 in LS180 cells was restored. The addition of supra-physiological concentrations of 1α, 25-dihydroxyvitamin D3 (1α, 25-$(OH)_2$-D_3) increased the otherwise low expression of CYP3A4 in Caco-2 and TC7 cells (Schmiedlin-Ren *et al.*, 1997; Engman *et al.*, 2001). Further induction of CYP3A4 by known PXR ligands was not demonstrated in 1α, 25-$(OH)_2$-D_3-treated Caco-2 cells, suggesting that PXR is not involved in vitamin D_3-mediated CYP3A4 induction (Schmiedlin-Ren *et al.*, 2001). Overall, LS180 cells appear to be a suitable model to study the regulation of CYP3A4 and MDR-1 while Caco-2 cells are better suited for drug absorption studies (Aiba *et al.*, 2005).

15.8 IMMORTALIZED HUMAN HEPATOCYTE CELL LINES: Fa2N-4

Developed by researchers at Multicell Technologies, the Fa2N-4 cell line is widely used for the assessment of CYP induction. This cell line is derived from primary hepatocytes (12-year-old female donor) through a clonal selection process following immortalization using the SV40 large T antigen. While most immortalized human hepatocyte cell lines have lost basal expression as well as inducibility of CYPs, Fa2N-4 cells retain normal hepatocellular morphology as well as expression and inducibility of CYPs and transporters. Using known inducers of CYP enzymes, Mills *et al.* reported concentration-dependent increases in both transcript levels and enzyme activities of CYP3A4, -2C9, and -1A2 (Mills *et al.*, 2004). These changes were comparable to the induction responses observed in primary human hepatocytes. In addition, UGT1A and MDR-1 were also induced by rifampicin treatment in this cell line. However, it is generally believed that Fa2N-4 cells have no or very low expression of CAR and thus induction of CYP2B6 cannot be evaluated in this cell line (Hariparsad *et al.*, 2008).

The procedure for assessing enzyme induction in Fa2N-4 cells is well defined. While it is similar to the protocol using primary hepatocytes, there are some subtle differences. Typically, Fa2N-4 cells are propagated on a collagen substratum in a proprietary medium. Following trypsinization, cells are plated in a wide variety of formats (6-, 12-, 24-, and 96-well). After one to two days of adaptation, the cells are treated once daily for two to three consecutive days with test articles and control compounds (negative/ positive). Whereas the cells are immortalized, they proliferate throughout this three- to five-day process. Hence, some consistency must be maintained to control the cell number/density from experiment to experiment to decrease variability of the induction response. The induction experiment can start either by subculturing actively proliferat- ing cells or by thawing and plating cryopreserved Fa2N-4 cells. The latter is more widely used as it eliminates the need for maintaining cells in culture and also possible negative effect of multiple passages on the overall induction response.

Fa2N4 have several attributes that make them amendable to the high-throughput environment of drug discovery. Fa2N-4 cells appear to maintain their induction response even in higher-density well formats, such as the 96-well plate format. Ripp *et al.* have developed a high-throughput induction assay using 96-well plates and produced detailed concentration–response curves for known inducers of CYP3A4 (Ripp *et al.*, 2006). In addition, Youdim *et al.* reported the use of 96-well plate incubations of Fa2N-4 cells with a cocktail of probe substrates (Youdim *et al.*, 2007).

15.9 OTHER IMMORTALIZED HEPATOCYTE CELL LINES

The expansion of immortalized hepatocyte cell lines as well as development of more sophisticated immortalization techniques is expected to produce proliferating yet func- tional hepatocytes that can be used in many areas of ADMET research. While the Fa2N-4 cell line is commonly used, there are several other immortalized hepatocyte cell lines currently under evaluation. NKNT-3 is a reversibly immortalized human hepato- cyte cell line that was generated by retroviral transfer of the SV40 large T antigen that can be subsequently excised by Cre/Lox site-specific recombination (Kobayashi *et al.*, 2000). NKNT-3 cells express CYP2C9 and CYP3A4, which can be further enhanced by coculturing NKNT-3 cells with another immortalized human hepatocyte stellate cell line, TWNT-1 (Watanabe *et al.*, 2003). YOCK-13 is also a reversibly immortalized, insulin-secreting human hepatocyte cell line generated using human TERT as the immortalizing gene. YOCK-13 cells express markers of hepatocyte differentiation that include albumin, asialoglycoprotein receptor, bilirubin-uridine diphosphate glucronosyl- transferase, CYP-associated enzyme 3A4, glutamine synthetase, and GST-π (Okitsu *et al.*, 2004). OSUM-29 is an immortalized human fetal hepatocyte cell line which has demonstrated expression of CYP1A1, CYP1A2, and CYP3A4 when cultured in a 3-D radial-flow bioreactor (Akiyama *et al.*, 2004). Finally, the HepZ cell line was immortal- ized using antisense RNA against Rb and P53 genes, which control cell-cycle regulation and have been shown to exhibit PB-inducible CYP activities (Werner *et al.*, 1999).

15.10 PRIMARY HUMAN HEPATOCYTE CULTURE SYSTEMS

The use of cultured primary human hepatocytes has become the "gold standard" *in vitro* system for assessing the potential for chemicals to induce human CYP expres- sion (LeCluyse *et al.*, 2005). Enzyme induction data from *in vitro* experiments with primary human hepatocytes are known to correlate well with clinical observations when pharmacologically relevant concentrations of test article are used (LeCluyse *et al.*, 2000). When hepatocytes are cultured using appropriate conditions that facilitate liver-like cell morphology and expression of liver-specific proteins (e.g. albumin), CYP enzymes are effectively induced *in vitro* analogous to the *in vivo* situation in terms of the magnitude and specificity of CYP induction (LeCluyse *et al.*, 2000; Runge *et al.*, 2000). Several reviews on the application of primary human hepatocytes in assessing and predicting human drug interactions due to enzyme induction have been recently published (Dickins, 2004; Hewitt, de Kanter and LeCluyse, 2007; Soars *et al.*, 2007).

The process of inducing CYP expression is a complex series of responses (e.g. receptor binding, receptor translocation, receptor activation, recruitment of co-activators and corepressors, receptor dimerization, receptor binding to target DNA, chromatin reorganization, stabilization of mRNA, and stabilization of protein). As a result, a metabolically competent system capable of metabolizing drugs in a liver-like manner that contains human gene targets, human receptors, human co-activators, human protein stabilization machinery, and human enzymes with analogous properties to human liver (e.g. mechanism-based inhibition) is optimal to most effectively model the human inducibility by xenobiotics and their hepatic metabolites. Primary human hepatocyte culture systems have been shown to effectively model these responses and are recognized by the FDA as an effective tool for assessing xenobiotic induction potential (FDA, 2006).

Species differences in induction response can occur at multiple levels of the process. Most importantly, many of the species differences in induction response arise due to unique interactions between xenobiotics and nuclear receptors (transcription factors), making prediction of human induction responses difficult (Wang and LeCluyse, 2003). For example, rifampicin is a potent and effective inducer of human CYP3A4 but not rat CYP3A (Silva, Day and Nicoll-Griffith, 1999; LeCluyse *et al.*, 2000; Zhu *et al.*, 2000). This is effectively modeled in cultures of rat and human hepatocytes and when performing *in vivo* induction studies in rats and humans. Similar species differences are also known for omeprazole where the drug has been shown to be a more effective inducer of human CYP1A2 than of rat CYP1A (Shih *et al.*, 1999; LeCluyse *et al.*, 2000). Also, the EC_{50} of omeprazole induction of human CYP1A2 is much larger than the exposure in patients treated with omeprazole for gastric ulcers; thus, it rarely causes significant clinical drug interactions due to induction because the concentration of omeprazole required to induce CYP1A2 *in vitro* generally exceeds the concentrations achieved *in vivo* (Daujat *et al.*, 1992; Dilger, Zheng and Klotz, 1999; LeCluyse *et al.*, 2000). Omeprazole is a drug that illustrates the need to assess enzyme induction in an appropriate species and at a therapeutic drug concentration in order to obtain the correct induction response/prediction. Phenobarbital and other anticonvulsants (e.g. phenytoin) are known to induce human CYP2B, CYP2C, and CYP3A enzymes. In human patients, phenytoin has been reported to induce the metabolism of multiple drugs including cyclophosphamide, thiotepa, and tacrolimus (de Jonge *et al.*, 2005; Wada *et al.*, 2007). However, clear species differences exist for the induction of CYPs by anticonvulsants. For example, induction of the murine CYP2Cs by phenobarbital and phenytoin has been reported to be mediated through CAR, rather than PXR, as the major determinant for murine Cyp2c29 and Cyp2c37 induction (Jackson *et al.*, 2004, 2006). This is a clear distinction from the human CYP2Cs which appear to be regulated by both PXR and CAR (Chen *et al.*, 2003, 2004; Ferguson *et al.*, 2005).

Comparison of immortalized liver cell systems with primary human hepatocyte cultures reveals differences in their respective abilities to proliferate. This distinction may be important because the cell-signaling pathways that cause a liver cell to become immortalized may influence the pathways involved in the regulation of CYP expression. As previously discussed, the expression level of CYPs and transcription factors are much lower in immortalized cell systems compared with primary hepatocytes (Hewitt, de Kanter and LeCluyse, 2007). Therefore, to comprehensively assess the induction (and/or suppression) of multiple pharmacologically important genes, an immortalized cell system may be limited.

The CAR is thought to be essential for phenobarbital-type induction and appears to play a role in the regulation of CYP2B6, CYP2C9, CYP2C8, and UGT1A1. Therefore, assessing induction potential of new chemical entities in a CAR responsive model system is important to evaluate the full potential of new chemical entities to induce CYP expression. Primary human hepatocyte systems and immortalized cell systems have both been shown to effectively model AhR and PXR receptor pathways as discussed earlier. However, CAR's unique constitutive activity coupled with its spontaneous translocation to nuclei in immortalized cell systems makes it difficult to model in immortalized cell lines (Kawamoto et al., 1999). Primary human hepatocyte culture systems effectively model CAR activation via both translocation and ligand activation, and presently are the most effective model system to assess human CAR activation (Wang et al., 2004; Faucette et al., 2007).

Immortalized cell systems are often deficient in terms of baseline enzymatic activity but can model induced expression and enzymatic activity of certain pathways and enzymatic target genes (see Section 15.3). However, due to the poor metabolism of compounds in these systems, it is difficult to evaluate the effects of metabolites on these induction pathways. Primary human hepatocyte systems are known to be metabolically active and are useful in studying the metabolic stability of chemicals (Jacobson et al., 2007). This, coupled with their demonstrated ability to effectively model multiple induction pathways and other pharmacologically important events such as mechanism-based inhibition, highlights the utility of this model system.

A limitation of primary hepatocyte culture systems has been tissue availability. While fresh human hepatocytes are the standard for evaluating *in vitro* induction of CYP enzymes, attachable cryopreserved hepatocytes can also be used and data using cryopreserved cells are accepted by the FDA. The drug-metabolizing enzymes remain inducible after cryopreservation, and due to the significant variation in activities of drug-metabolizing enzymes between individual human livers, cryopreserved cells generate results essentially indistinguishable from freshly isolated cells (Schehrer, Regan and Westendorf, 2000; Hewitt et al., 2007). mRNA, protein expression, and activities of CYP1A2, 2B6, 2C9, 2E1, and 3A4 in cryopreserved hepatocytes have been shown to be inducible by many common inducers (Roymans et al., 2005), as well as the activity of multiple uridine diphosphate-glucuronosyl-transferases, carboxylesterases, and sulfotransferases (Reinach et al., 1999; Nishimura et al., 2004). The advantage of cryopreserved cells is that experiments can be routinely conducted, as the experiments are not dependent on the sporadic availability of fresh hepatocytes.

Most research currently uses a "sandwich" culture model, where isolated hepatocytes are plated on simple collagen (or other extracellular matrix) and sandwiched between the collagen basement membrane and an extracellular matrix overlay such as Matrigel (LeCluyse et al., 2000, 2005; Hamilton et al., 2001). This environment appears optimal for maintenance of hepatocyte cultures in liver-like cell morphology and over multiple days in culture often forms liver-like bile canaliculi (Bi, Kazolias and Duignan, 2006; Marion, Leslie and Brouwer, 2007). This culture system affords the opportunity to maintain mature hepatocytes in a liver-like environment capable of assessing the cumulative effects of metabolism, induction, inhibition (time-dependent), and metabolite effects over multiple days, thus replicating the more steady-state-like effects on CYP activity and gene expression.

Data should be collected from at least three individual donors, treated with the test compound, vehicle, and positive controls (known inducers, Table 15.2) for two to three days. Drug treatment begins 48 hours after cell plating once the cells have had an opportunity to reform cell–cell contacts. Typically, two to three days of drug treatment are necessary for RNA expression and three days for enzyme activity assessments. Each hepatocyte donor preparation is deemed acceptable if the recommended positive control activity demonstrates a greater than twofold increase in enzyme activity of the probe substrate. A minimum of three test compound concentrations, based on the expected human plasma drug levels are suggested, one of which should be an order of magnitude greater than this concentration. In the absence of knowledge of human plasma levels, concentrations ranging over at least two orders of magnitude should be studied.

TABLE 15.2   Common CYP inducers and CYP probe substrates used in *in vitro* studies.

CYP Enzyme	Inducer	Substrate/probe
1A2	**Omeprazole, β-naphthoflavone, 3-methylcholantrhrene,** lansoprazole	**Phenacetin,** theophylline, caffeine, acetaminophen, tacrine
2A6	**Dexamethasone,** pyrazole	**Coumarin, nicotine,** butadiene
2B6	**Phenobarbital,** phenytoin	**Efavirenz, buproprion,** propofol, *S*-mephenytoin
2C8	**Rifampicin,** phenobarbital	**Taxol,** repaglinide, rosiglitazone, amodiaquine
2C9	**Rifampicin,** phenobarbital	*S*-**warfarin, tolbutamide, diclofenac,** phenytoin, flurbiprofen
2C19	**Rifampicin**	*S*-**mephenytoin,** fluoxetine, omeprazole, esoprazole, lansoprazole, pantoprazole, citalopram, diazepam, hexobarbital, imipramine, proguanil, propanolol
2D6	None identified	**Dextromethorphan, bufuralol,** codeine, ethylmorphine, desipramine, atomoxetine, nicotine, debrisoquine
2E1	Ethanol, isoniazid, acetone	**Chlorzoxazone,** aniline, *p*-nitrophenol, lauric acid, acetaminophen, caffeine, dapsone, enflurane, theophylline
3A4/3A5	**Rifampicin,** rifabutin, Rifapentine, sulfinpyrazone, carbamazepine, dexamethasone, phenytoin, troleandomycin, troglitazone, taxol	**Midazolam, testosterone,** dextromethorphan, triazolam, terfenadine, buspirone, felodipine, lovastatin, eletriptan, sildenafil, simvastatin, triazolam, acetaminophen, carbamezapine, cyclosporin, digitoxin, diazepam, erythromycin, fluoxetine, nifedipine, quinidine, saquinavir, cortisol, terfenadine, verapamil

Drugs highlighted in bold are preferred inducers and substrates.
Compiled from Tucker, Houston and Huang (2001) and FDA (2006).

The draft FDA guidance on drug interaction studies recommends analysis of catalytic activity of the major drug-metabolizing enzymes CYP1A2 and CYP3A4 in freshly isolated or attachable cryopreserved hepatocyte cultures. Since co-induction of CYP2C, CYP2B, and P-gp occurs with CYP3A, negative results *in vitro* with CYP3A may also eliminate the need to address interactions eliminated by these CYPs and P-gp. However, when utilizing data on only CYP1A1/2 and CYP3A4, it may be possible to underpredict clinical induction. This can be due to cross talk between nuclear receptors, for example, PXR and CAR (Faucette *et al.*, 2006). When evaluating only CYP1A and CYP3A and not CYP2B6, induction via CAR could be misrepresented. Enzymatic analysis can be conducted in microsomes prepared from hepatocytes or *in situ*, using media-based assays and CYP-specific probe substrates, where the metabolite formed can be attributed to one major CYP enzyme. The enzyme activity measurement is thought to be more reflective of actual enzyme changes (pharmacokinetic changes) that occur in patients. Additional methods to quantitate *in vitro* enzyme induction are CYP protein concentration (by Western immunoblotting) and mRNA expression (by reverse transcriptase–polymerase chain reaction, RT-PCR). RNA expression is particularly useful when both enzyme induction and inhibition are occurring with the same test article, such as with troleandomycin and DPC681 (Luo *et al.*, 2003; McGinnity *et al.*, 2006). An advantage of primary human hepatocyte culture systems is their ability to model metabolism, inhibition, and induction (Luo *et al.*, 2002). We have examined the effects of rifampicin and phenobarbital (prototypical PXR and CAR activators), ketoconazole (a potent competitive CYP3A4 inhibitor), and several metabolism-dependent inhibitors including erythromycin, troleandomycin, and ritonavir on CYP2B6 and CYP3A4 mRNA, protein content, and enzymatic activity over multiple time points. Generally, clear distinctions were observed between early time points (where inhibition would be predicted to be prominent) and longer time points where contributions from both inhibition and induction would be operative. These studies highlight the power of simultaneously probing induction and inhibition in a single, metabolically competent cell system. Whereas measurement of enzyme activities from incubations is considered the most reliable measure of induction potential, most often, both RNA expression and enzyme activity are measured to better understand the overall effects of a test article.

A relatively new area of research has recently emerged in the area of down-regulation of CYP expression (also called suppression). Suppression is generally thought to be the opposite of induction; however, the mechanisms controlling induction of CYPs may not be the most prominent factors for basal CYP expression in various tissues. For example, in human liver, factors such as GR and HNF4α have been shown to be important for basal expression, but appear to have supportive roles for classical CYP induction (Pascussi *et al.*, 2001). Primary human hepatocyte culture systems have also been shown to model CYP suppression. The most intuitive example is that of interleukins on CYP gene expression (Abdel-Razzak *et al.*, 1995; Chen *et al.*, 1995a, b). It has been known historically that inflammation results in decreased metabolic capacity in humans, and these studies identified specific inflammatory factors. Additional research is needed to probe the mechanisms by which this occurs at the transcriptional level. Another recent example of CYP suppression was with colchicine where Dvorak and coworkers demonstrated suppression of several CYPs including CYP2C9 and CYP3A4 (Dvorak *et al.*, 2003). The mechanism of this suppression appears to be down-regulation of the glucocorticoid receptor which is important for the basal expression of CYPs on multiple levels including down-regulation of nuclear receptor expression and the possible impact on chromatin histone deacetylation (Dvorak *et al.*, 2003). These examples highlight the need to understand chemical interactions that may affect the basal, as well as the induced levels of CYPs, and primary human hepatocyte culture systems have the advantage of modeling these processes. Additional modifications to culture systems such as coculture (e.g. with Kupfer cells) will provide even better models of the human liver and the effects of chemicals to perturb inflammatory pathways in the liver.

Numerous studies have been reported using primary hepatocyte culture systems to assess induction of a variety of gene targets from both phase I and II enzymes, transporters, and genes involved in endogenous processes, such as gluconeogenesis (Raucy *et al.*, 2002, 2004; Raucy, 2003; Kodama *et al.*, 2004). In summary, primary cultures of human hepatocytes are an effective tool to assess the potential for human induction by drugs and drug candidates due to their more liver-like origin, cell morphology, metabolic competence, gene expression profile, and cell-signaling pathways. Technologies continue to emerge to better utilize these systems to probe human hepatic metabolism and drug interactions with new promising drugs.

15.11 INTERPRETATION AND QUANTITATIVE PREDICTION OF ENZYME INDUCTION

The common end points for assessment of enzyme induction are RNA expression (measured by real-time RT-PCR), protein level (measured by ELISA or Western blotting), and enzyme activity (measured by quantitating probe drug turnover or metabolite formation). Enzyme activity is the actual measure of increased enzyme capacity to turn over a substrate (which is the basis of a drug–drug interaction) and most accepted by regulatory agencies. Table 15.2 lists the most common *in vitro* positive control inducers and probe substrates for each CYP enzyme. However, quantitation of RNA expression and protein level often provide a more direct measurement of actual gene activation or increased enzyme concentration when the test article is also a potent enzyme inhibitor. Complications arise when interpreting enzyme activity measurements when a compound both induces and inhibits an enzyme. For example, ritonavir and troleandomycin are both potent inhibitors and inducers of CYP3A4 and result in increased CYP3A4 gene transcription, but generally exhibit normal or reduced enzyme activity (Luo *et al.*, 2002). Also, DPC681 is an inducer and metabolism-based inhibitor of CYP3A4 and an *in vitro* enzyme activity assay revealed no change in enzyme activity (Luo *et al.*, 2003). Additional *in vitro* studies in human hepatocytes indicated increased CYP3A4 protein concentrations and clinical studies revealed autoinduction when patients were dosed with DPC681 for several days. These examples illustrate the complicated nature of predicting drug interactions in patients when competing enzyme inhibition and induction occur *in vitro* and *in vivo*. In most cases, measuring RNA expression and enzyme activity is advised to take advantage of both methods (transcription and enzyme activity). Western blotting analysis is used less frequently in the industrial setting due to additional time and resources necessary to perform these analyses.

While *in vitro* receptor- or cell-based assays yield important qualitative assessments (mild, moderate, strong inducer) of enzyme induction, a quantitative assessment of how the induction will impact the exposure of a "victim" drug and resulting drug–drug interactions remains elusive. The data from *in vitro* experiments can take several forms, such as fold increase over a vehicle-treated control incubation, percent increase of a positive control, or EC_{50} value. A recently published draft FDA guidance for industry on drug interactions proposed two numerical end points to be used in the prediction: percent increase of a positive control and EC_{50} (FDA, 2006). The guidance indicates that a drug that produces a change (enzyme activity) equal to or greater than 40% of the positive control is a potential inducer and requires further *in vivo* evaluation. As for the EC_{50}, the guidance suggests that this parameter can be used to compare the potency of different compounds. While simple rank ordering of EC_{50} values and comparisons to known CYP3A4 inducers, such as rifampicin, can provide useful information on potency, more accurate predictions of drug interactions incorporate clinically relevant drug concentrations, such as total or free drug concentrations (portal vein, C_{max}, or steady state concentration), after a therapeutic dose of drug (Ripp *et al.*, 2006; Sinz *et al.*, 2006; Hewitt, de Kanter and LeCluyse, 2007). Percent activation at a therapeutic drug concentration (C_{max}) and rank ordering (EC_{50}) have been employed when evaluating PXR transactivation data (Sinz *et al.*, 2006). Using this approach, it was shown that compounds causing >40% transactivation at a therapeutic drug concentration are predicted to "likely" have drug interactions, compounds with percent transactivation between 15% and 40% "may" elicit drug interactions, and those with percent transactivation <15% are "not" anticipated to cause drug interactions (Sinz *et al.*, 2006). For example, hyperforin has an $EC_{50} = 0.04\,\mu M$ in the hPXR trans- activation assay, an approximate efficacious plasma concentration of $0.3\,\mu M$, and a percent transactivation of 80% at the efficacious concentration in the hPXR transacti- vation assay. Pioglitazone has an $EC_{50} = 31\,\mu M$ in the hPXR transactivation assay, an approximate efficacious plasma concentration of $3.8\,\mu M$, and a percent transactivation of only 10% at the efficacious concentration. Contrasting these parameters between the two drugs, one would predict that hyperforin will cause CYP3A4 drug interactions while pioglitazone will not, similar to results observed in actual clinical drug interaction studies (Sinz *et al.*, 2006).

Combining values and obtaining a ratio of E_{max}/EC_{50} (E_{max} is maximum effect) can be predictive in assessing the likelihood of a drug interaction much like the approach used for drug inhibition assessments (I/K_i). More complex calculations include multiple parameters, such as EC_{50}, E_{max}, and drug concentration, which have been used to assess induction effects for multiple induction models (transactivation, immortalized hepato- cytes, and primary hepatocytes). The fundamental equation used in these models is the same and can be described as follows:

$$E = \frac{C \times E_{max}}{C + EC_{50}}$$

where E is the inductive effect and C is the drug concentration. However, there is considerable discussion of which drug concentration to use in these calculations. Kato *et al.* showed that the steady-state concentration of unbound drug is useful in predicting clinical induction from human hepatocyte assays (Kato *et al.*, 2005) and Ripp *et al.* demonstrated that, using the Fa2N-4 cell line, unbound therapeutic plasma concentra- tion of drug ($C_{max}(fu)$) can be used to determine a relative induction score (RIS) (Ripp *et al.*, 2006):

$$RIS = \frac{E_{max} \times C_{max}(fu)}{EC_{50} + C_{max}(fu)}$$

More recently, Hewitt *et al.* proposed that the NOEL (no observed effect level) be added to the equation to improve the predictive power (Hewitt, de Kanter and LeCluyse, 2007). The following equation describes a model that combines these parameters (where NOEL is defined as "the highest concentration at which no induction response is observed"):

$$\text{Induction risk factor} = \frac{E_{max} \times C_{max}(fu)}{EC_{50} + C_{max}(fu)} \times \frac{NOEL}{C_{max}(fu)}$$

Some difficulties exist when using cell-based systems, such as effects of protein binding, partial agonists, cytotoxicity, and insolubility of test compounds. While the effect of protein binding on enzyme induction in *in vitro* or *in vivo* systems is not clearly defined, the use of protein binding (i.e. $(C_{max}(fu))$) in some situations appears to be more predictive. However, the use of total drug concentration represents a more conservative approach which defines a maximal induction effect. Certainly more studies on the *in vitro* and *in vivo* relationship between protein binding and intracellular receptor-mediated effects will be required in the future.

Another caveat in applying these quantitative *in vitro* approaches, such as hPXR transactivation or cell-based induction assays, are concentration-response plots that reveal partial agonists or bell-shaped curves which make EC_{50} and E_{max} determinations difficult. In these situations, it may be difficult to establish a concentration–response relationship at concentrations above a certain threshold. In the future, a better understanding of molecular mechanisms of induction as well as advancement in tools to study the interaction will enable more sophisticated quantitative prediction of clinical enzyme induction. Cytotoxicity assessments should be conducted with all cell-based systems to ensure that a negative response is not due to cell death. Often this can be as simple as visualizing the cells under a microscope (cell morphology) or more sophisticated measures of cell membrane permeability or cell death. An understanding of drug solubility in cell culture media can also help avoid false-negative results when compounds are highly insoluble. Both cytotoxicity and insolubility often display bell-shaped response curves when inducers are evaluated, making it difficult to assess the full induction potential of such compounds.

Finally, the aforementioned methods work well for rank ordering of these compounds and predicting the magnitude of a drug interaction for compounds that are highly metabolized and metabolized exclusively by the individual CYP enzyme being induced (e.g. midazolam). However, for victim drugs whose elimination includes other pathways (i.e. renal or biliary) and/or whose metabolism is through multiple CYP enzymes, these methods can overestimate the change in exposure of a victim drug (Rowland and Tozer, 1988; Zhang *et al.*, 2007). Hence, a full understanding of elimination pathways (metabolic and non-metabolic) of victim drugs is necessary to predict exposure changes due to drug–drug interactions *in vivo*.

15.12 CLINICAL DRUG INTERACTION STUDIES

When a drug is predicted, based on *in vitro* data, to induce CYP enzymes in humans, then a clinical drug–drug interaction study is necessary to assess the magnitude of pharmacokinetic changes resulting from the induction. The purpose of these drug interaction studies is to determine if an interaction is significant enough to warrant discontinuation, dosage adjustment, or therapeutic drug monitoring of the drug itself or a coadministered drug. In some instances, enzyme induction is not anticipated and a drug–drug interaction study may still be conducted in order to include labeling information indicating "no potential for drug interactions due to enzyme induction" (i.e. safety or marketing advantages). The majority of drug interaction clinical trials are related to CYP3A4, although other enzymes such as CYP2C9 or CYP1A2, as well as transporters such as MDR1, have been studied (Malmstrom *et al.*, 1998; Sahi *et al.*, 2003; Shadle *et al.*, 2004). Probe substrates recommended by the FDA for various CYP enzymes are listed in Table 15.1 (FDA, 2006). The general design of a CYP3A4 drug–drug interaction study includes 12–20 healthy volunteers (or sometimes patients) including males and females between the ages of 18 and 60. Patients are often required to abstain from medications, herbal supplements, or other dietary components, such as grapefruit juice, which may affect CYP3A4 enzyme levels during the study. Women who are included in such studies are often of non-childbearing age or required to use alternative methods of birth control other than oral contraceptives which may be affected during the trial.

In the case of CYP3A4 enzyme induction studies, midazolam either po or IV is used as a probe substrate to indirectly assess CYP3A4 enzyme changes by measuring pharmacokinetic changes of midazolam before and after drug treatment. Oral doses of midazolam capture the CYP3A4 induction effects from both the liver and intestine while an IV dose of midazolam measures the liver effect alone (Ma *et al.*, 2006a, b). Oral doses of midazolam range between 2 and 5 mg and IV doses are generally ~2 mg, while plasma sample collection periods for po and IV are typically 0.5–8 hours and 0.25–6 hours, respectively. Midazolam is dosed just prior to drug treatment and once again following several days of continuous drug treatment. The dose and duration of drug administration (perpetrator-inducer) between the two doses of midazolam is based on the expected use for drug treatment. For both perpetrator and victim, the dosages should be employed that maximize the likelihood of finding an interaction; hence, the highest approved or predicted dose of each drug should be used in such studies. Typical treatment durations are approximately seven days, as illustrated in Fig. 15.1, during which several samples are taken to measure perpetrator drug levels. Also, a third or multiple midazolam doses may be given several days after the last dose of drug in order to assess the de-induction phase and determine when patients have restabilized to basal CYP3A4 levels. Depending on the half-life of the compound, the

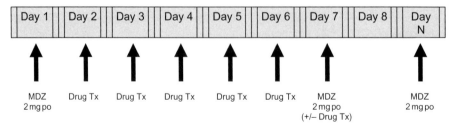

Figure 15.1 A typical CYP3A4 clinical drug–drug interaction design. The probe drug, midazolam (MDZ), is dosed before and after several days of drug treatment (Drug Tx). MDZ may also be administered sometime after the last drug treatment (Day N) to assess the de-induction phase.

time needed to return to normal CYP3A4 levels may vary. For example, the CYP3A4 de-induction phase of pleconaril was studied for a length of 34 days due to the 180-hour half-life of the drug, and it was determined that enzyme levels returned to normal sometime between days 6 and 13 after the last dose of pleconaril (Ma *et al.*, 2006a). Most often, a drug with a reasonable half-life will allow enzyme levels to return to pretreatment levels within three to four days.

Typical pharmacokinetic parameters that are compared between the two doses of midazolam include AUC and C_{max}. The least squares geometric mean ratios with 90% confidence intervals for midazolam AUC and C_{max} are calculated and a drug interaction is established if these values fall outside of the predefined and generally accepted bioequivalence range of 0.80 and 1.25. Relevant *in vitro* or *in vivo* drug interaction information describing the drug's inducing effects on substrates is presented under Drug–Drug Interactions of the Clinical Pharmacology section of the label. Often, additional information describing the interaction and steps necessary to avoid interactions will be presented in the Precautions, Warnings/Precautions, Dosage and Administration, and/or Contraindication sections as necessary (FDA, 1999).

15.13 INDUSTRIAL AND REGULATORY PERSPECTIVES

In an industrial setting, there is a balance of resources dedicated toward first screening compounds for the potential to be enzyme inducers and second, greater characterization of positive hits from screening assays. Most often, the earlier screening assays are high throughput and have the ability to accommodate large numbers of compounds in a short period of time. These characteristics, coupled with a low cost per compound, allow such assays to be placed in early drug discovery. Most often, these high-throughput assays are PXR binding or transactivation assays in 96- or 384-well formats. Although excellent screening and sometimes predictive of human induction potential, assays based on receptors and transcription factors are considered supportive evidence for a compound's induction potential by the FDA as they do not directly reflect the actual enzyme activity (FDA, 2006; Sinz *et al.*, 2006).

Gaining in popularity are the immortalized hepatocyte models which afford many of the advantages of nuclear receptor-based assays (availability and speed) and often incorporate more than one receptor-mediated process in a single system (e.g. Fa2N-4 cell line incorporates CYP1A and 3A induction processes). In many ways, these are similar to primary human hepatocyte experiments and accepted by the FDA as models to assess enzyme induction. However, caution should be noted when using immortalized hepatocyte models, in that they do not incorporate all potential nuclear hormone/transcription factor-mediated CYP induction pathways. Even when CYP1A and CYP3A4 are shown to properly induce with known positive controls, this implies that AhR- and PXR-specific pathways are functioning properly, but not necessarily all possible pathways. For example, CYP3A4 enzyme induction, which can be mediated by both PXR and CAR, may demonstrate an attenuated CYP3A4 induction response in cell-based models that only incorporate PXR and not both PXR and CAR (e.g. Fa2N-4) (Hariparsad *et al.*, 2008). Nevertheless, immortalized hepatocytes are an acceptable model to screen and/or characterize the enzyme induction potential of drug candidates. For the most part, immortalized hepatocytes can be used in conjunction with primary hepatocytes to better characterize the enzyme induction potential but not in lieu of primary human hepatocytes.

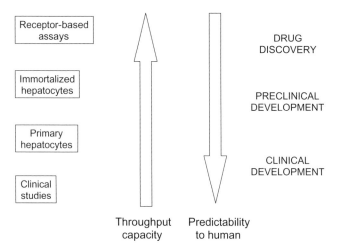

Figure 15.2   Illustration of *in vitro* models and clinical studies used to predict or assess the enzyme induction potential of a drug candidate.

Figure 15.2 describes the variety of assays, throughput capacity, predictability to human drug–drug interactions, and approximate time frame that each assay may be employed during drug discovery/development. Although Fig. 15.2 incorporates all possible *in vitro* human model systems, not all companies will employ the entire suite of assays. In the very least, nearly all companies will eventually incorporate primary human hepatocytes as an induction model to assess PXR- and AhR-mediated enzyme induction (CYP3A4/2C/2B6 and CYP1A, respectively), as this is the standard procedure to assess CYP induction prior to clinical drug–drug interaction studies.

As described here and in the previous chapter, there exist significant species differences in ligand affinities for various nuclear hormone receptors or transcription factors. Hence, most animal models do not properly recapitulate the effects observed in humans, and *in vivo* animal models (mice, rats, dogs) are generally not applicable in predicting enzyme induction potential in humans. Enzyme induction in animals is, however, problematic when assessing the toxicological properties of a new drug candidate. Autoinduction in animals can affect drug exposure to the extent that high drug exposures necessary to assess toxicity cannot be achieved. This poses a serious problem when attempting to achieve adequate exposure–safety margins in the absence of increasing drug exposures with increasing doses of drug.

15.14   FUTURE MODELS

There are multiple new models of enzyme induction being developed with varying degrees of success and validation. All appear promising; however, additional evaluation and a greater understanding of how best to employ these models are still necessary.

In silico simulations and structure–activity relationships (SARs) are useful tools to help eliminate potentially problematic chemotypes early in drug development. Virtual screening can be conducted on large numbers of compounds, for example, chemical library screening. In addition, these models may be useful in defining structural features or structural modifications necessary to eliminate or attenuate the affinity of a ligand to PXR. Successful induction models generally combine nuclear receptor crystal structures with *in silico* structural biology (Handschin, Podvinec and Meyer, 2003). However, merely docking the compound into the ligand-binding domain may not be sufficient in capturing a complete picture of transcriptional events. For example, the large ligand-binding domain of hPXR accommodates ligands of all sizes. Larger ligands can alter the protein structure after binding and smaller molecules have been shown to bind at several sites within the LBD (Handschin, Podvinec and Meyer, 2003). Another complicating factor in ligand-binding pharmacophore models are adjacent sites on the nuclear receptor (allosteric sites) which can be involved in conformational changes that affect ligand or DNA binding. For example, Ekins *et al.* reported that agonists and antagonists bind to distinct regions on PXR and that the effect of ketoconazole on CYP3A transcription was mediated through binding to the AF-2 region of PXR and not the LBD (Ekins *et al.*, 2007; Wang *et al.*, 2007).

"Stem cells" are defined as "omnipotent progenitor cells that retain the ability to replicate indefinitely *in vitro* as undifferentiated cells and under appropriate conditions, differentiate into specialized cell types, such as hepatocytes, neurons and cardiomyocytes" (Pouton and Haynes, 2005; Vermeir *et al.*, 2005). Stem cells can be derived from the inner cell mass of a blastocyst (embryonic stem cell, ESC) or can be obtained from a variety of tissues, such as bone marrow, cord blood, liver, and adipose tissue (adult stem cell, ASC) (McNeish, 2004; Pouton and Haynes, 2007).

Petersen *et al.* demonstrated that bone marrow stem cells can be a potential source of hepatic oval cells, which can further differentiate into two types of liver epithelia cells: ductular cells and hepatocytes (Petersen *et al.*, 1999). Subsequently, it was found that human and murine hematopoietic stem cells in bone marrow can differentiate into hepatocyte-like cells (Alison *et al.*, 2000; Lagasse *et al.*, 2000), opening the possibility of generating functional hepatocytes from non-hepatic adult stem cells. In general, human stem cells express low levels of CYPs but higher levels of GST (Shao *et al.*, 2007). Under the presence of fibroblast growth factor (FGF) and hepatocyte growth factor (HGF), they differentiate into functional hepatocyte-like cells that express high levels of PB-inducible CYP activity and some selected liver-specific functions, such as secretion of urea and albumin and LDL uptake (Schwartz *et al.*, 2002; Lee *et al.*, 2004).

Embryonic stem cells provide another potential source of hepatocytes. The most comprehensive analysis of expression and induction of drug-metabolizing enzymes in human embryonic stem cells was reported by Ek *et al.* (2007). Hepatocyte-like cells derived from hESC under the influence of bFGF expressed CYP1A2, CYP3A4/7, and low levels of CYP1A1 and CYP2C proteins. When treated with a cocktail of known CYP inducers, the protein expressions of CYP1A2 and CYP3A4 were also increased, suggesting the inducibility of these genes is retained. However, the activities of CYP1A1 and CYP3A4 were lower than those in primary hepatocytes (3% and 6%, respectively). An examination of CYP expression in hepatocytes derived from murine ESCs also demonstrated expression of Cyp2b10, Cyp2c29, Cyp2d9, Cyp3a11, and Cyp7a1 as well as PB-inducible CYP activities (Tsutsui *et al.*, 2006). While stem cells are not widely used in ADMET research today (Sinz and Kim, 2006), advancements in stem cell technology and characterization will greatly expand their application in drug metabolism and safety assessment.

Humanized nuclear receptor transgenic mice and knockout mice can be useful in assessing *in vivo* enzyme induction or mechanisms of enzyme induction as described in the previous chapter. A newer humanized model, known as the chimeric mouse, involves transplanting human hepatocytes into a mouse liver, thus providing a useful *in vivo* model capable of assessing enzyme regulation by drug candidates. Humanized chimeric mice were developed by injecting human hepatocytes into SCID mice (80% of the hepatocytes in the liver were of human origin) (Tateno *et al.*, 2004). CYP1A1/2 and CYP3A4 mRNA, protein content, and enzyme activity have been shown to be induced in chimeric mice treated with known human inducers and in hepatocytes isolated from humanized chimeric mice (Katoh *et al.*, 2005a, b; Nishimura *et al.*, 2005). These *in vivo* models (transgenic or chimeric) provide an advantage over previous *in vitro* models, as the systems are dynamic, with drug absorption, distribution, metabolism, and elimination occurring naturally (albeit a humanized mouse). Drug candidates can be administered over several days at doses/exposures equivalent to those expected in humans to enable improved assessments of human induction potential.

Recently, there is an increased effort to validate nonhuman primates (rhesus or cynomolgus monkey) as an animal model for CYP3A induction. The high sequence homology (~96%) between human and monkey PXR is presumed to result in similar induction profiles (Prueksaritanont *et al.*, 2006; Nishimura *et al.*, 2007). CYP1A, -2B, and -3A in the cynomolgus monkey were found to respond to known human inducers, such as β-naphthoflavone, phenobarbital, and dexamethasone, respectively, and CYP3A64 responded to rifampicin induction in the rhesus monkey (Bullock *et al.*, 1995; Prueksaritanont *et al.*, 2006). A more recent *in vitro* report by Nishimura *et al.* indicated that cynomolgus monkey hepatocytes responded to CYP1A and -3A inducers; however, only CYP1A1 responded in the appropriate manner to omeprazole treatment and CYP1A2 did not (Nishimura *et al.*, 2007). This indicates some unknowns still exist due either to the enzyme inducer, omeprazole, or the mechanism(s) of CYP1A2 induction between monkeys and humans. However, if successfully validated, monkeys could provide a useful *in vivo* model to assess human induction at therapeutically relevant concentrations, thus providing a more quantitative prediction of enzyme induction.

15.15 CONCLUSIONS

One of the biggest challenges in predicting human drug–drug interactions from *in vitro* experiments is the lack of understanding between *in vitro* drug concentrations and drug concentration at the site of action in the liver. This is complicated by several factors such as protein binding, nonspecific binding *in vitro*, the use of total versus free (unbound) efficacious plasma concentrations, or the use of systemic versus portal vein drug concentrations, and compounds with extremely low or high liver-to-plasma ratios. The two most common approaches in all human *in vitro* models are (i) total efficacious systemic C_{max} or C_{ss} concentration or (ii) unbound drug concentration at the portal vein (Ripp *et al.*, 2006; Sinz *et al.*, 2006; Hewitt, de Kanter and LeCluyse, 2007). The former approach does not account for protein binding effects on drug disposition, yet is considered a conservative estimate that uses a high (total) drug concentration after the first-pass effect (metabolism/elimination). The latter approach does account for protein binding on drug disposition and reflects the maximum drug absorbed and presented to the liver.

Equally challenging is the extrapolation from *in vitro* induction studies when the compound being evaluated is both an inhibitor and inducer of a CYP enzyme. If the compound is an inhibitor, enzyme induction can be overlooked when only enzyme activity is assessed. Due to the common nature of CYP inhibition, conducting both activity and mRNA analysis is encouraged. Even if a complete understanding of the separate *in vitro* inhibition and induction potentials of a compound is assessed, the combined effects *in vivo* are difficult to predict (e.g. DPC681, Luo *et al.*, 2003).

In summary, the last decade has seen a significant progression forward in our understanding and the predictability of enzyme induction, both *in vitro* and *in vivo*. The use of nuclear hormone receptor and hepatocyte-based cellular assays, combined with our improved understanding of the mechanisms leading to enzyme and transporter induction, has allowed us to predict and understand enzyme induction to a much greater degree than in the past.

ACKNOWLEDGMENTS

The authors wish to thank Drs Jasminder Sahi, Kenneth Santone, and Gillian Wallace for their contributions and comments.

REFERENCES

Abdel-Razzak, Z., Corcos, L., Fautrel, A., Guillouzo, A. (1995). Interleukin-1 beta antagonizes phenobarbital induction of several major cytochromes P450 in adult rat hepatocytes in primary culture. *FEBS Lett.*, *366* (2–3), 159–164.

Aiba, T., Susa, M., Fukumori, S., Hashimoto, Y. (2005). The effects of culture conditions on CYP3A4 and MDR1 mRNA induction by 1alpha,25-dihydroxyvitamin D(3) in human intestinal cell lines, Caco-2 and LS180. *Drug Metab. Pharmacokinet.*, *20* (4), 268–274.

Akiyama, I., Tomiyama, K., Sakaguchi, M. *et al.* (2004). Expression of CYP3A4 by an immortalized human hepatocyte line in a three-dimensional culture using a radial-flow bioreactor. *Int. J. Mol. Med.*, *14* (4), 663–668.

Alison, M.R., Poulsom, R., Jeffery, R. *et al.* (2000). Hepatocytes from non-hepatic adult stem cells. *Nature*, *406* (6793), 257.

Aninat, C., Piton, A., Glaise, D. *et al.* (2006). Expression of cytochromes P450, conjugating enzymes and nuclear receptors in human hepatoma HepaRG cells. *Drug Metab. Dispos.*, *34* (1), 75–83.

Auerbach, S.S., Stoner, M.A., Su, S., Omiecinski, C.J. (2005). Retinoid X receptor-alpha-dependent transactivation by a naturally occurring structural variant of human constitutive androstane receptor (NR1I3). *Mol. Pharmacol.*, *68* (5), 1239–1253.

Baliharova, V., Skalova, L., Maas, R.F., De Vrieze, G., Bull, S., Fink-Gremmels, J. (2003). The effects of benzimidazole anthelmintics on P4501A in rat hepatocytes and HepG2 cells. *Res. Vet. Sci.*, *75* (1), 61–69.

Bapiro, T.E., Andersson, T.B., Otter, C., Hasler, J.A., Masimirembwa, C.M. (2002). Cytochrome P450 1A1/2 induction by antiparasitic drugs: dose-dependent increase in ethoxyresorufin O-deethylase activity and mRNA caused by quinine, primaquine and albendazole in HepG2 cells. *Eur. J. Clin. Pharmacol.*, *58* (8), 537–542.

Bertilsson, L., Hojer, B., Tybring, G., Osterloh, J., Rane, A. (1980). Autoinduction of carbamazepine metabolism in children examined by a stable isotope technique. *Clin. Pharmacol. Ther.*, *27* (1), 83–88.

Bi, Y.A., Kazolias, D., Duignan, D.B. (2006). Use of cryopreserved human hepatocytes in sandwich culture to measure hepatobiliary transport. *Drug Metab. Dispos.*, *34* (9), 1658–1665.

Brandin, H., Viitanen, E., Myrberg, O., Arvidsson, A.K. (2007). Effects of herbal medicinal products and food supplements on induction of CYP1A2, CYP3A4 and MDR1 in the human colon carcinoma cell line LS180. *Phytother. Res.*, *21* (3), 239–244.

Brown, R.R., Miller, J.A., Miller, E.C. (1954). The metabolism of methylated aminoazo dyes. IV. Dietary factors enhancing demethylation in vitro. *J. Biol. Chem.*, *209* (1), 211–222.

Bullock, P., Pearce, R., Draper, A. *et al.* (1995). Induction of liver microsomal cytochrome P450 in cynomolgus monkeys. *Drug. Metab. Dispos.*, *23* (7), 736–748.

Cascio, S.M. (2001). Novel strategies for immortalization of human hepatocytes. *Artif. Organs*, *25* (7), 529–538.

Chang, T.K., Waxman, D.J. (2006). Synthetic drugs and natural products as modulators of constitutive androstane receptor (CAR) and pregnane X receptor (PXR). *Drug Metab. Rev.*, *38* (1–2), 51–73.

Chen, J., Nikolova-Karakashian, M., Merrill, A.H., Jr, Morgan, E.T. (1995a). Regulation of cytochrome P450 2C11 (CYP2C11) gene expression by interleukin-1, sphingomyelin hydrolysis, and ceramides in rat hepatocytes. *J. Biol. Chem.*, *270* (42), 25233–25238.

Chen, J.Q., Strom, A., Gustafsson, J.A., Morgan, E.T. (1995b). Suppression of the constitutive expression of cytochrome P-450 2C11 by cytokines and interferons in primary cultures of rat hepatocytes: comparison with induction of acute-phase genes and demonstration that CYP2C11 promoter sequences are involved in the suppressive response to interleukins 1 and 6. *Mol. Pharmacol.*, *47* (5), 940–947.

Chen, Y., Ferguson, S.S., Negishi, M., Goldstein, J.A. (2003). Identification of constitutive androstane receptor and glucocorticoid receptor binding sites in the CYP2C19 promoter. *Mol. Pharmacol.*, *64* (2), 316–324.

Chen, Y., Ferguson, S.S., Negishi, M., Goldstein, J.A. (2004). Induction of human CYP2C9 by rifampicin, hyperforin, and phenobarbital is mediated by the pregnane X receptor. *J. Pharmacol. Exp. Ther.*, *308* (2), 495–501.

Conney, A.H. (1967). Pharmacological implications of microsomal enzyme induction. *Pharmacol. Rev.*, *19* (3), 317–366.

Daujat, M., Peryt, B., Lesca, P., Fourtanier, G., Domergue, J., Maurel, P. (1992). Omeprazole, an inducer of human CYP1A1 and 1A2, is not a ligand for the Ah receptor. *Biochem. Biophys. Res. Commun.*, *188* (2), 820–825.

Dickins, M. (2004). Induction of cytochromes P450. *Curr. Top. Med. Chem.*, *4* (16), 1745–1766.

Dilger, K., Zheng, Z., Klotz, U. (1999). Lack of drug interaction between omeprazole, lansoprazole, pantoprazole and theophylline. *Br. J. Clin. Pharmacol.*, *48* (3), 438–444.

Dvorak, Z., Modriansky, M., Pichard-Garcia, L. *et al.* (2003). Colchicine down-regulates cytochrome P450 2B6, 2C8, 2C9, and 3A4 in human hepatocytes by affecting their glucocorticoid receptor-mediated regulation. *Mol. Pharmacol.*, *64* (1), 160–169.

Ek, M., Soderdahl, T., Kuppers-Munther, B. *et al.* (2007). Expression of drug metabolizing enzymes in hepatocyte-like cells derived from human embryonic stem cells. *Biochem. Pharmacol.*, *74* (3), 496–503.

Ekins, S., Chang, C., Mani, S. *et al.* (2007). Human pregnane x receptor antagonists and agonists define molecular requirements for different binding sites. *Mol. Pharmacol.*, *72* (3), 592–603.

El-Sankary, W., Gibson, G.G., Ayrton, A., Plant, N. (2001). Use of a reporter gene assay to predict and rank the potency and efficacy of CYP3A4 inducers. *Drug Metab. Dispos.*, *29* (11), 1499–1504.

Engman, H.A., Lennernas, H., Taipalensuu, J., Otter, C., Leidvik, B., Artursson, P. (2001). CYP3A4, CYP3A5, and MDR1 in human small and large intestinal cell lines suitable for drug transport studies. *J. Pharm. Sci.*, *90* (11), 1736–1751.

Fabre, N., Arrivet, E., Trancard, J. *et al.* (2003). A new hepatoma cell line for toxicity testing at repeated doses. *Cell Biol. Toxicol.*, *19* (2), 71–82.

Faucette, S.R., Sueyoshi, T., Smith, C.M., Negishi, M., LeCluyse, E.L., Wang, H. (2006). Differential regulation of hepatic CYP2B6 and CYP3A4 genes by constitutive androstane receptor but not pregnane X receptor. *J. Pharmacol. Exp. Ther.*, *317* (3), 1200–1209.

Faucette, S.R., Zhang, T.C., Moore, R. *et al.* (2007). Relative activation of human pregnane X receptor versus constitutive androstane receptor defines distinct classes of CYP2B6 and CYP3A4 inducers. *J. Pharmacol. Exp. Ther.*, *320* (1), 72–80.

FDA (1999). Guidance for industry in vivo drug metabolism/drug interaction studies—study, design, data analysis, and recommendations for dosing and labelling.

FDA (2006). FDA Draft Guideline: drug interaction studies—study design, data analysis, and implications for dosing and labeling.

Ferguson, S.S., Chen, Y., LeCluyse, E.L., Negishi, M., Goldstein, JA. (2005). Human CYP2C8 is transcriptionally regulated by the nuclear receptors constitutive androstane receptor, pregnane X receptor, glucocorticoid receptor, and hepatic nuclear factor 4alpha. *Mol. Pharmacol.*, *68* (3), 747–757.

Gomez-Lechon, M.J., Donato, T., Jover, R. *et al.* (2001). Expression and induction of a large set of drug-metabolizing enzymes by the highly differentiated human hepatoma cell line BC2. *Eur. J. Biochem.*, *268* (5), 1448–1459.

Gonzalez, F.J. (2007). The 2006 Bernard B. Brodie Award Lecture. Cyp2e1. *Drug Metab. Dispos.*, *35* (1), 1–8.

Goodwin, B., Hodgson, E., Liddle, C. (1999). The orphan human pregnane X receptor mediates the transcriptional activation of CYP3A4 by rifampicin through a distal enhancer module. *Mol. Pharmacol.*, *56* (6), 1329–1339.

Gripon, P., Rumin, S., Urban, S. *et al.* (2002). Infection of a human hepatoma cell line by hepatitis B virus. *Proc. Natl. Acad. Sci. U.S.A.*, *99* (24), 15655–15660.

Guillouzo, A., Corlu, A., Aninat, C., Glaise, D., Morel, F., Guguen-Guillouzo, C. (2007). The human hepatoma HepaRG cells: a highly differentiated model for studies of liver metabolism and toxicity of xenobiotics. *Chem Biol. Interact.*, *168* (1), 66–73.

Hahn, W.C. (2002). Immortalization and transformation of human cells. *Mol. Cells*, *13* (3), 351–361.

Hamilton, G.A., Jolley, S.L., Gilbert, D., Coon, D.J., Barros, S., LeCluyse, E.L. (2001). Regulation of cell morphology and cytochrome P450 expression in human hepatocytes by extracellular matrix and cell-cell interactions. *Cell Tissue Res.*, *306* (1), 85–99.

Handschin, C., Podvinec, M., Meyer, U.A. (2003). In silico approaches, and in vitro and in vivo experiments to predict induction of drug metabolism. *Drug News Perspect.*, *16* (7), 423–434.

Hariparsad, N., Carr, B., Evers, R., Chu, X. (2008). Comparison of immortalized Fa2N-4 cells and human hepatocytes as in vitro models for cytochrome P450 induction. *Drug Metab. Dispos.*, *36* (6), 1046–1055.

Harmsen, S., Meijerman, I., Beijnen, J.H., Schellens, J.H. (2007). The role of nuclear receptors in pharmacokinetic drug-drug interactions in oncology. *Cancer Treat. Rev.*, *33* (4), 369–380.

Hewitt, N.J., Hewitt, P. (2004). Phase I and II enzyme characterization of two sources of HepG2 cell lines. *Xenobiotica*, *34* (3), 243–256.

Hewitt, N.J., de Kanter, R., LeCluyse, E. (2007). Induction of drug metabolizing enzymes: a survey of in vitro methodologies and interpretations used in the pharmaceutical industry—do they comply with FDA recommendations? *Chem. Biol. Interact.*, *168* (1), 51–65.

Hewitt, N.J., Lechon, M.J., Houston, J.B. *et al.* (2007). Primary hepatocytes: current understanding of the regulation of metabolic enzymes and transporter proteins, and pharmaceutical practice for the use of hepatocytes in metabolism, enzyme induction, transporter, clearance, and hepatotoxicity studies. *Drug Metab. Rev.*, *39* (1), 159–234.

Honkakoski, P., Sueyoshi, T., Negishi, M. (2003). Drug-activated nuclear receptors CAR and PXR. *Ann. Med.*, *35* (3), 172–182.

Hu, W., Sorrentino, C., Denison, M.S., Kolaja, K., Fielden, M.R. (2007). Induction of cyp1a1 is a nonspecific biomarker of aryl hydrocarbon receptor activation: results of large scale screening of pharmaceuticals and toxicants in vivo and in vitro. *Mol. Pharmacol.*, *71* (6), 1475–1486.

Iwanari, M., Nakajima, M., Kizu, R., Hayakawa, K., Yokoi, T. (2002). Induction of CYP1A1, CYP1A2, and CYP1B1 mRNAs by nitropolycyclic aromatic hydrocarbons in various human tissue-derived cells: chemical-, cytochrome P450 isoform-, and cell-specific differences. *Arch. Toxicol.*, *76* (5–6), 287–298.

Jackson, J.P., Ferguson, S.S., Moore, R., Negishi, M., Goldstein, J.A. (2004). The constitutive active/androstane receptor regulates phenytoin induction of Cyp2c29. *Mol. Pharmacol.*, *65* (6), 1397–1404.

Jackson, J.P., Ferguson, S.S., Negishi, M., Goldstein, J.A. (2006). Phenytoin induction of the cyp2c37 gene is mediated by the constitutive androstane receptor. *Drug Metab. Dispos.*, *34* (12), 2003–2010.

Jacobson, L., Middleton, B., Holmgren, J. *et al.* (2007). An optimized automated assay for determination of metabolic stability using hepatocytes: assay validation, variance component analysis, and in vivo relevance. *Assay Drug Dev. Technol.*, *5* (3), 403–415.

de Jonge, M.E., Huitema, A.D., van Dam, S.M., Beijnen, J.H., Rodenhuis, S. (2005). Significant induction of cyclophosphamide and thiotepa metabolism by phenytoin. *Cancer Chemother. Pharmacol.*, *55* (5), 507–510.

Kato, M., Chiba, K., Horikawa, M., Sugiyama, Y. (2005). The quantitative prediction of in vivo enzyme-induction caused by drug exposure from in vitro information on human hepatocytes. *Drug Metab. Pharmacokinet.*, *20* (4), 236–243.

Katoh, M., Matsui, T., Nakajima, M. *et al.* (2005a). In vivo induction of human cytochrome P450 enzymes expressed in chimeric mice with humanized liver. *Drug Metab. Dispos.*, *33* (6), 754–763.

Katoh, M., Watanabe, M., Tabata, T. *et al.* (2005b). In vivo induction of human cytochrome P450 3A4 by rifabutin in chimeric mice with humanized liver. *Xenobiotica*, *35* (9), 863–875.

Kawamoto, T., Sueyoshi, T., Zelko, I., Moore, R., Washburn, K., Negishi, M. (1999). Phenobarbital-responsive nuclear translocation of the receptor CAR in induction of the CYP2B gene. *Mol. Cell. Biol.*, *19* (9), 6318–6322.

Kliewer, S.A., Goodwin, B., Willson, T.M. (2002). The nuclear pregnane X receptor: a key regulator of xenobiotic metabolism. *Endocr. Rev.*, *23* (5), 687–702.

Kobayashi, N., Fujiwara, T., Westerman, K.A. *et al.* (2000). Prevention of acute liver failure in rats with reversibly immortalized human hepatocytes. *Science*, *287* (5456), 1258–1262.

Kodama, S., Koike, C., Negishi, M., Yamamoto, Y. (2004). Nuclear receptors CAR and PXR cross talk with FOXO1 to regulate genes that encode drug-metabolizing and gluconeogenic enzymes. *Mol. Cell. Biol.*, *24* (18), 7931–7940.

Lagasse, E., Connors, H., Al-Dhalimy, M. *et al.* (2000). Purified hematopoietic stem cells can differentiate into hepatocytes in vivo. *Nat. Med.*, *6* (11), 1229–1234.

Le Jossic, C., Glaise, D., Corcos, L. *et al.* (1996). trans-Acting factors, detoxication enzymes and hepatitis B virus replication in a novel set of human hepatoma cell lines. *Eur. J. Biochem.*, *238* (2), 400–409.

Le Vee, M., Jigorel, E., Glaise, D., Gripon, P., Guguen-Guillouzo, C., Fardel, O. (2006). Functional expression of sinusoidal and canalicular hepatic drug transporters in the differentiated human hepatoma HepaRG cell line. *Eur. J. Pharm. Sci.*, *28* (1–2), 109–117.

LeCluyse, E.L., Alexandre, E., Hamilton, G.A. *et al.* (2005). Isolation and culture of primary human hepatocytes. *Methods Mol. Biol.*, *290*, 207–229.

LeCluyse, E.L., Madan, A., Hamilton, G., Carroll, K., DeHaan, R., Parkinson, A. (2000). Expression and regulation of cytochrome P450 enzymes in primary cultures of human hepatocytes. *J. Biochem. Mol. Toxicol.*, *14* (4), 177–188.

Lee, K.D., Kuo, T.K., Whang-Peng, J. *et al.* (2004). In vitro hepatic differentiation of human mesenchymal stem cells. *Hepatology*, *40* (6), 1275–1284.

Lemaire, G., de Sousa, G., Rahmani, R.A. (2004). PXR reporter gene assay in a stable cell culture system: CYP3A4 and CYP2B6 induction by pesticides. *Biochem. Pharmacol.*, *68* (12), 2347–2358.

Li, Q., Sai, Y., Kato, Y., Tamai, I., Tsuji, A. (2003). Influence of drugs and nutrients on transporter gene expression levels in Caco-2 and LS180 intestinal epithelial cell lines. *Pharm. Res.*, *20* (8), 1119–1124.

Luo, G., Cunningham, M., Kim, S. *et al.* (2002). CYP3A4 induction by drugs: correlation between a pregnane X receptor reporter gene assay and CYP3A4 expression in human hepatocytes. *Drug Metab. Dispos.*, *30* (7), 795–804.

Luo, G., Lin, J., Fiske, W.D. *et al.* (2003). Concurrent induction and mechanism-based inactivation of CYP3A4 by an L-valinamide derivative. *Drug Metab. Dispos.*, *31* (9), 1170–1175.

Ma, J.D., Nafziger, A.N., Rhodes, G., Liu, S., Bertino, J.S., Jr (2006a). Duration of pleconaril effect on cytochrome P450 3A activity in healthy adults using the oral biomarker midazolam. *Drug Metab. Dispos.*, *34* (5), 783–785.

Ma, J.D., Nafziger, A.N., Rhodes, G., Liu, S., Gartung, A.M., Bertino, J.S., Jr (2006b). The effect of oral pleconaril on hepatic cytochrome P450 3A activity in healthy adults using intravenous midazolam as a probe. *J. Clin. Pharmacol.*, *46* (1), 103–108.

McGinnity, D.F., Berry, A.J., Kenny, J.R., Grime, K., Riley, R.J. (2006). Evaluation of time-dependent cytochrome P450 inhibition using cultured human hepatocytes. *Drug Metab. Dispos.*, *34* (8), 1291–1300.

McNeish, J. (2004). Embryonic stem cells in drug discovery. *Nat. Rev. Drug Discov.*, *3* (1), 70–80.

Malmstrom, K., Schwartz, J., Reiss, T.F. *et al.* (1998). Effect of montelukast on single-dose theophylline pharmacokinetics. *Am. J. Ther.*, *5* (3), 189–195.

Mandal, P.K. (2005). Dioxin: a review of its environmental effects and its aryl hydrocarbon receptor biology. *J. Comp. Physiol. [B]*, *175* (4), 221–230.

Marion, T.L., Leslie, E.M., Brouwer, K.L. (2007). Use of sandwich-cultured hepatocytes to evaluate impaired bile acid transport as a mechanism of drug-induced hepatotoxicity. *Mol. Pharm.*, *4* (6), 911–918.

Medina-Diaz, I.M., Arteaga-Illan, G., de Leon, M.B. *et al.* (2007). Pregnane X receptor-dependent induction of the CYP3A4 gene by o,p'-1,1,1,-trichloro-2,2-bis (p-chlorophenyl)ethane. *Drug Metab. Dispos.*, *35* (1), 95–102.

Medina-Diaz, I.M., Elizondo, G. (2005). Transcriptional induction of CYP3A4 by o,p'-DDT in HepG2 cells. *Toxicol. Lett.*, *157* (1), 41–47.

Mills, J.B., Rose, K.A., Sadagopan, N., Sahi, J., de Morais, S.M. (2004). Induction of drug metabolism enzymes and MDR1 using a novel human hepatocyte cell line. *J. Pharmacol. Exp. Ther.*, *309* (1), 303–309.

Mueller, S.C., Majcher-Peszynska, J., Uehleke, B. *et al.* (2006). The extent of induction of CYP3A by St. John's wort varies among products and is linked to hyperforin dose. *Eur J. Clin. Pharmacol.*, *62* (1), 29–36.

Nishimura, M., Imai, T., Morioka, Y., Kuribayashi, S., Kamataki, T., Naito, S. (2004). Effects of NO-1886 (Ibrolipim), a lipoprotein lipase-promoting agent, on gene induction of cytochrome P450s, carboxylesterases, and sulfotransferases in primary cultures of human hepatocytes. *Drug Metab. Pharmacokinet.*, *19* (6), 422–429.

Nishimura, M., Koeda, A., Suganuma, Y. *et al.* (2007). Comparison of inducibility of CYP1A and CYP3A mRNAs by prototypical inducers in primary cultures of human, cynomolgus monkey, and rat hepatocytes. *Drug Metab. Pharmacokinet.*, *22* (3), 178–186.

Nishimura, M., Yokoi, T., Tateno, C. *et al.* (2005). Induction of human CYP1A2 and CYP3A4 in primary culture of hepatocytes from chimeric mice with humanized liver. *Drug Metab. Pharmacokinet.*, *20* (2), 121–126.

O'Connor, J.E., Martinez, A., Castell, J.V., Gomez-Lechon, M.J. (2005). Multiparametric characterization by flow cytometry of flow-sorted subpopulations of a human hepatoma cell line useful for drug research. *Cytometry A*, *63* (1), 48–58.

Okitsu, T., Kobayashi, N., Jun, H.S. *et al.* (2004). Transplantation of reversibly immortalized insulin-secreting human hepatocytes controls diabetes in pancreatectomized pigs. *Diabetes*, *53* (1), 105–112.

Omura, T., Sato, R. (1962). A new cytochrome in liver microsomes. *J. Biol. Chem.*, *237*, 1375–1376.

Park, J.Y., Kim, K.A., Kang, M.H., Kim, S.L., Shin, J.G. (2004). Effect of rifampin on the pharmacokinetics of rosiglitazone in healthy subjects. *Clin. Pharmacol. Ther.*, *75* (3), 157–162.

Pascussi, J.M., Drocourt, L., Gerbal-Chaloin, S., Fabre, J.M., Maurel, P., Vilarem, M.J. (2001). Dual effect of dexamethasone on CYP3A4 gene expression in human hepatocytes. Sequential role of glucocorticoid receptor and pregnane X receptor. *Eur. J. Biochem.*, *268* (24), 6346–6358.

Petersen, B.E., Bowen, W.C., Patrene, K.D. *et al.* (1999). Bone marrow as a potential source of hepatic oval cells. *Science*, *284* (5417), 1168–1170.

Pouton, C.W., Haynes, J.M. (2005). Pharmaceutical aspplications of embryonic stem cells. *Adv. Drug Deliv. Rev.*, *57* (13), 1918–1934.

Pouton, C.W., Haynes, J.M. (2007). Embryonic stem cells as a source of models for drug discovery. *Nat. Rev. Drug Discov.*, *6* (8), 605–616.

Prueksaritanont, T., Kuo, Y., Tang, C. *et al.* (2006). In vitro and in vivo CYP3A64 induction and inhibition studies in rhesus monkeys: a preclinical approach for CYP3A-mediated drug interaction studies. *Drug Metab. Dispos.*, *34* (9), 1546–1555.

Qatanani, M., Moore, DD. (2005). CAR, the continuously advancing receptor, in drug metabolism and disease. *Curr. Drug Metab.*, *6* (4), 329–339.

Raucy, J.L. (2003). Regulation of CYP3A4 expression in human hepatocytes by pharmaceuticals and natural products. *Drug Metab. Dispos.*, *31* (5), 533–539.

Raucy, J.L., Lasker, J., Ozaki, K., Zoleta, V. (2004). Regulation of CYP2E1 by ethanol and palmitic acid and CYP4A11 by clofibrate in primary cultures of human hepatocytes. *Toxicol. Sci.*, *79* (2), 233–241.

Raucy, J.L., Mueller, L., Duan, K., Allen, S.W., Strom, S., Lasker, J.M. (2002). Expression and induction of CYP2C P450 enzymes in primary cultures of human hepatocytes. *J. Pharmacol. Exp. Ther.*, *302* (2), 475–482.

Reinach, B., de Sousa, G., Dostert, P., Ings, R., Gugenheim, J., Rahmani, R. (1999). Comparative effects of rifabutin and rifampicin on cytochromes P450 and UDP-glucuronosyl-transferases expression in fresh and cryopreserved human hepatocytes. *Chem. Biol. Interact.*, *121* (1), 37–48.

Rencurel, F., Stenhouse, A., Hawley, S.A. *et al.* (2005). AMP-activated protein kinase mediates phenobarbital induction of CYP2B gene expression in hepatocytes and a newly derived human hepatoma cell line. *J. Biol. Chem.*, *280* (6), 4367–4373.

Ripp, S.L., Mills, J.B., Fahmi, O.A. *et al.* (2006). Use of immortalized human hepatocytes to predict the magnitude of clinical drug-drug interactions caused by CYP3A4 induction. *Drug Metab. Dispos.*, *34* (10), 1742–1748.

Rowland, M., Tozer, T.N. (1988). *Clinical Pharmacokinetics: Concepts and Applications*, Lea & Febiger, Philadelphia.

Roymans, D., Annaert, P., Van Houdt, J. *et al.* (2005). Expression and induction potential of cytochromes P450 in human cryopreserved hepatocytes. *Drug Metab. Dispos.*, *33* (7), 1004–1016.

Runge, D., Michalopoulos, G.K., Strom, S.C., Runge, D.M. (2000). Recent advances in human hepatocyte culture systems. *Biochem. Biophys. Res. Commun.*, *274* (1), 1–3.

Sahi, J., Milad, M.A., Zheng, X. *et al.* (2003). Avasimibe induces CYP3A4 and multiple drug resistance protein 1 gene expression through activation of the pregnane X receptor. *J. Pharmacol. Exp. Ther.*, *306* (3), 1027–1034.

Schehrer, L., Regan, J.D., Westendorf, J. (2000). UDS induction by an array of standard carcinogens in human and rodent hepatocytes: effect of cryopreservation. *Toxicology*, *147* (3), 177–191.

Schmiedlin-Ren, P., Thummel, K.E., Fisher, J.M., Paine, M.F., Lown, K.S., Watkins, P.B. (1997). Expression of enzymatically active CYP3A4 by Caco-2 cells grown on extracellular matrix-coated permeable supports in the presence of 1alpha,25-dihydroxyvitamin D3. *Mol. Pharmacol.*, *51* (5), 741–754.

Schmiedlin-Ren, P., Thummel, K.E., Fisher, J.M., Paine, M.F., Watkins, P.B. (2001). Induction of CYP3A4 by 1 alpha,25-dihydroxyvitamin D3 is human cell line-specific and is unlikely to involve pregnane X receptor. *Drug Metab. Dispos.*, *29* (11), 1446–1453.

Schuetz, E.G., Beck, W.T., Schuetz, J.D. (1996). Modulators and substrates of P-glycoprotein and cytochrome P4503A coordinately up-regulate these proteins in human colon carcinoma cells. *Mol. Pharmacol.*, *49* (2), 311–318.

Schwartz, R.E., Reyes, M., Koodie, L. *et al.* (2002). Multipotent adult progenitor cells from bone marrow differentiate into functional hepatocyte-like cells. *J. Clin. Invest.*, *109* (10), 1291–1302.

Shadle, C.R., Lee, Y., Majumdar, A.K. *et al.* (2004). Evaluation of potential inductive effects of aprepitant on cytochrome P450 3A4 and 2C9 activity. *J. Clin. Pharmacol.*, *44* (3), 215–223.

Shao, J., Stapleton, P.L., Lin, Y.S., Gallagher, E.P. (2007). Cytochrome p450 and glutathione S-transferase mRNA expression in human fetal liver hematopoietic stem cells. *Drug Metab. Dispos.*, *35* (1), 168–175.

Shih, H., Pickwell, G.V., Guenette, D.K., Bilir, B., Quattrochi, L.C. (1999). Species differences in hepatocyte induction of CYP1A1 and CYP1A2 by omeprazole. *Hum. Exp. Toxicol.*, *18* (2), 95–105.

Silva, J.M., Day, S.H., Nicoll-Griffith, D.A. (1999). Induction of cytochrome-P450 in cryopreserved rat and human hepatocytes. *Chem. Biol. Interact.*, *121* (1), 49–63.

Simonsson, U.S., Jansson, B., Hai, T.N., Huong, D.X., Tybring, G., Ashton, M. (2003). Artemisinin autoinduction is caused by involvement of cytochrome P450 2B6 but not 2C9. *Clin. Pharmacol. Ther.*, *74* (1), 32–43.

Sinz, M.W., Kim, S. (2006). Stem cells, immortalized cells and primary cells in ADMET assays. *Drug Discov. Today Technol.*, *3* (1), 79–85.

Sinz, M.W., Kim, S., Zhu, Z. *et al.* (2006). Evaluation of 170 xenobiotics as transactivators of human pregnane X receptor (hPXR) and correlation to known CYP3A4 drug interactions. *Curr. Drug Metab.*, *7* (4), 375–388.

Smith, C.M., Graham, R.A., Krol, W.L. *et al.* (2005). Differential UGT1A1 induction by chrysin in primary human hepatocytes and HepG2 Cells. *J. Pharmacol. Exp. Ther.*, *315* (3), 1256–1264.

Soars, M.G., McGinnity, D.F., Grime, K., Riley, R.J. (2007). The pivotal role of hepatocytes in drug discovery. *Chem. Biol. Interact.*, *168* (1), 2–15.

Stanley, L.A., Horsburgh, B.C., Ross, J., Scheer, N., Wolf, C.R. (2006). PXR and CAR nuclear receptors which play a pivotal role in drug disposition and chemical toxicity. *Drug Metab. Rev.*, *38* (3), 515–597.

Synold, T.W., Dussault, I., Forman, B.M. (2001). The orphan nuclear receptor SXR coordinately regulates drug metabolism and efflux. *Nat. Med.*, *7* (5), 584–590.

Tateno, C., Yoshizane, Y., Saito, N. *et al.* (2004). Near completely humanized liver in mice shows human-type metabolic responses to drugs. *Am. J. Pathol.*, *165* (3), 901–912.

Tirona, R.G., Kim, R.B. (2005). Nuclear receptors and drug disposition gene regulation. *J. Pharm. Sci.*, *94* (6), 1169–1186.

Trubetskoy, O., Marks, B., Zielinski, T., Yueh, M.F., Raucy, J. (2005). A simultaneous assessment of CYP3A4 metabolism and induction in the DPX-2 cell line. *Aaps. J.*, *7* (1), E6–13.

Tsutsui, M., Ogawa, S., Inada, Y. *et al.* (2006). Characterization of cytochrome P450 expression in murine embryonic stem cell-derived hepatic tissue system. *Drug Metab. Dispos.*, *34* (4), 696–701.

Tucker, G.T., Houston, J.B., Huang, S.M. (2001). Optimizing drug development: strategies to assess drug metabolism/transporter interaction potential—toward a consensus. *Pharm. Res.*, *18* (8), 1071–1080.

Vermeir, M., Annaert, P., Mamidi, R.N., Roymans, D., Meuldermans, W., Mannens, G. (2005). Cell-based models to study hepatic drug metabolism and enzyme induction in humans. *Expert Opin. Drug Metab. Toxicol.*, *1* (1), 75–90.

Vignati, L.A., Bogni, A., Grossi, P., Monshouwer, M. (2004). A human and mouse pregnane X receptor reporter gene assay in combination with cytotoxicity measurements as a tool to evaluate species-specific CYP3A induction. *Toxicology*, *199* (1), 23–33.

Wada, K., Takada, M., Ueda, T. *et al.* (2007). Drug interactions between tacrolimus and phenytoin in Japanese heart transplant recipients: 2 case reports. *Int. J. Clin. Pharmacol. Ther.*, *45* (9), 524–528.

Walle, T., Otake, Y., Galijatovic, A., Ritter, J.K., Walle, U.K. (2000). Induction of UDP-glucurono-syltransferase UGT1A1 by the flavonoid chrysin in the human hepatoma cell line hep G2. *Drug Metab. Dispos.*, *28* (9), 1077–1082.

Wang, H., Faucette, S., Moore, R., Sueyoshi, T., Negishi, M., LeCluyse, E. (2004). Human constitutive androstane receptor mediates induction of CYP2B6 gene expression by phenytoin. *J. Biol. Chem.*, *279* (28), 29295–29301.

Wang, H., Huang, H., Li, H. *et al.* (2007). Activated pregnenolone X-receptor is a target for ketoconazole and its analogs. *Clin. Cancer Res.*, *13* (8), 2488–2495.

Wang, H., LeCluyse, E.L. (2003). Role of orphan nuclear receptors in the regulation of drug-metabolising enzymes. *Clin. Pharmacokinet.*, *42* (15), 1331–1357.

Watanabe, T., Shibata, N., Westerman, K.A. *et al.* (2003). Establishment of immortalized human hepatic stellate scavenger cells to develop bioartificial livers. *Transplantation*, *75* (11), 1873–1880.

Werner, A., Duvar, S., Muthing, J. *et al.* (1999). Cultivation and characterization of a new immortalized human hepatocyte cell line, HepZ, for use in an artificial liver support system. *Ann. N. Y. Acad. Sci.*, *875*, 364–368.

Westerink, W.M., Schoonen, W.G. (2007a). Phase II enzyme levels in HepG2 cells and cryopreserved primary human hepatocytes and their induction in HepG2 cells. *Toxicol. In Vitro*, *21* (8), 1592–1602.

Westerink, W.M., Schoonen, W.G. (2007b). Cytochrome P450 enzyme levels in HepG2 cells and cryopreserved primary human hepatocytes and their induction in HepG2 cells. *Toxicol. In Vitro*, *21* (8), 1581–1591.

Wilkening, S., Stahl, F., Bader, A. (2003). Comparison of primary human hepatocytes and hepatoma cell line Hepg2 with regard to their biotransformation properties. *Drug Metab. Dispos.*, *31* (8), 1035–1042.

Worboys, P.D., Carlile, D.J. (2001). Implications and consequences of enzyme induction on preclinical and clinical drug development. *Xenobiotica*, *31* (8–9), 539–556.

Youdim, K.A., Tyman, C.A., Jones, B.C., Hyland, R. (2007). Induction of cytochrome P450: assessment in an immortalized human hepatocyte cell line (Fa2N4) using a novel higher throughput cocktail assay. *Drug Metab. Dispos.*, *35* (2), 275–282.

Zhang, H., Davis, C.D., Sinz, M.W., Rodrigues, A.D. (2007). Cytochrome P450 reaction-phenotyping: an industrial perspective. *Expert Opin. Drug Metab. Toxicol.*, *3* (5), 667–687.

Zhou, C., Tabb, M.M., Sadatrafiei, A., Grun, F., Blumberg, B. (2004). Tocotrienols activate the steroid and xenobiotic receptor, SXR, and selectively regulate expression of its target genes. *Drug Metab. Dispos.*, *32* (10), 1075–1082.

Zhu, W., Song, L., Zhang, H., Matoney, L., LeCluyse, E., Yan, B. (2000). Dexamethasone differentially regulates expression of carboxylesterase genes in humans and rats. *Drug Metab. Dispos.*, *28* (2), 186–191.

Zhu, Z., Kim, S., Chen, T. *et al.* (2004). Correlation of high-throughput pregnane X receptor (PXR) transactivation and binding assays. *J. Biomol. Screen*, *9* (6), 533–540.

Zhu, Z., Puglisi, J., Connors, D. *et al.* (2007). Use of cryopreserved transiently transfected cells in high-throughput pregnane X receptor transactivation assay. *J. Biomol. Screen*, *12* (2), 248–254.

CHAPTER 16

An Introduction to Metabolic Reaction Phenotyping

CARL D. DAVIS and A. DAVID RODRIGUES

16.1 INTRODUCTION

The role of drug-metabolizing enzymes (DMEs) and metabolic clearance in the overall determination of a drug's oral bioavailability and pharmacokinetics (PK) is well established and fully incorporated into the process of discovery, design, and development of new and better therapeutic agents. Given the very many different types, classes, and members of enzyme families known to metabolize drugs and other xenobiotics, the task of optimizing metabolic stability can appear somewhat daunting. Fortunately, certain enzymes feature more predominantly than others, and drug optimization often begins and ends with them.

In this chapter we will describe a few of the major enzyme families, the experimental and kinetic methods used to identify their role in the overall metabolic clearance of a drug (so-called "reaction phenotyping" or "enzyme mapping") and illustrate some concerns related to the contribution and interindividual variation of some notable enzymes. An attempt has been made to provide a useful and comprehensive outline of what is requisite for drug discovery, but an introductory text can only go so far. So for greater depth of analysis and discussion, the reader is referred to the many reviews published on bioanalysis, enzyme kinetics, substrates and biotransformation, drug–drug interactions (DDIs), adverse drug reactions, and *in silico* methods and pharmacogenetics, a few of which are listed at the end of this chapter.

16.2 DRUG METABOLISM

The membrane-bound hemoproteins that belong to the superfamily of cytochrome P450s (CYPs), by virtue of expression, number, diversity, and substrate promiscuity, dominate most other enzymes in their contribution to drug clearance. Members of three CYP subfamilies in particular (CYP1, CYP2, and CYP3) are responsible for the

Drug Metabolism Handbook: Concepts and Applications, Edited by Ala F. Nassar, Paul F. Hollenberg, and JoAnn Scatina

metabolic clearance of most drugs and xenobiotics (Rendic and DiCarlo, 1997; Guengerich, 2003; Williams *et al.*, 2004). The flavin-containing monooxygenases (FMOs), which tend to mediate *N*- and *S*-oxidation reactions, and the uridine 5′-diphosphate glucuronosyltransferases (UGTs) that catalyze drug conjugation with uridine 5′-diphosphate glucuronic acid (UDPGA), are two other important classes of enzymes that are also found co-localized with CYPs in the endoplasmic reticulum. In addition to these membrane-bound forms, there are significant families of cytosolic DMEs, such as *N*-acetyl transferases (NATs), sulfotransferases (SULTs), glutathione transferases (GSTs), alcohol dehydrogenases (ADHs), and aldehyde dehydrogenases (ALDHs). Almost all biotransformation and metabolic clearance pathways encountered in drug metabolism involve one or more of these DMEs—most of them involve CYPs, UGTs, and/or FMOs—and although extrahepatic metabolism can be very important, especially on a case-by-case basis, the highest levels of many important enzymes are found in the liver and this organ dominates the metabolic clearance of most drugs.

DMEs have evolved in part to regulate some endogenous reactions (including steroid synthesis and bilirubin clearance) and also, if not primarily, to protect living organisms from exposure to environmental toxins (Guengerich, 1989). Orally ingested xenobiotics such as drugs, environmental chemicals, toxins, and so on, follow a common fate. If soluble and permeable, the drug in question is absorbed from the intestine and may undergo metabolism in the enterocytes forming the microvilli of the duodenum and proximal jejunum. The fraction that is absorbed and not metabolized in the gut subsequently passes into the hepatic portal vein where it enters the highly perfused liver, where, in many cases, it undergoes metabolism by DMEs such as CYPs. This combination of absorption and metabolism (gut and liver) is known as the "first-pass effect" and determines the oral bioavailability of a drug (i.e. the amount of the dose that reaches the systemic circulation and is available for effect on target tissues).

In general, the function of metabolic clearance is to convert a lipophilic molecule into a more polar and more aqueous soluble metabolite that can then be eliminated in the bile and/or urine. The liver possesses the highest levels of the major and most commonly involved DMEs, and subject to dose and saturation kinetics, it is usually the single most important organ of metabolic clearance. The liver thus serves, on one hand, as the *de facto* organ of detoxification, thus protecting us from an accidental or deliberate exposure to poisons and toxins; on the other hand, the liver acts as an effective barrier to achieving efficacious exposure from the oral administration of reasonable dosages and practicable dose regimens. Balancing these properties often determines the safety margins of a drug—achieving exposure for efficacy while not causing significant adverse reactions.

Most other organs and tissues in humans and other mammals are known to express DMEs that can play a significant or even a major role in metabolic clearance. For instance, the lungs receive 100% of the cardiac output, and the nasal and pulmonary epithelia express DMEs that can serve to protect against systemic exposure as well as inhaled toxins, including solvents and volatile agents such as naphthalene (Buckpitt *et al.*, 2002; Castell, Donato and Gómez-Lechón, 2005). Keratinocytes, epithelial cells, and the major cell type in the epidermis, are also known to express DMEs that can protect against permeable agents absorbed via skin contact (Swanson, 2004). The liver and extrahepatic DMEs can lead to the formation of more reactive and/or more toxic metabolites that can cause toxicity locally or *ex situ*, which can become a concern in drug development and ultimately determine the manifestation and prevalence of any ensuing toxicity in humans. For a more detailed discussion on this aspect of drug metabolism, the reader is referred to Chapter 21, which deals with the specific roles and complexities involved in detoxification and bioactivation reactions that can ultimately define the safety profile of a drug.

The expression, localization, and substrate selectivity of the DMEs are key factors essential to understanding and predicting tissue-specific effects (be they benign or malignant) and individual susceptibility to adverse drug reactions. The focus of this chapter will deal specifically with how one (i) measures the rate of metabolism, (ii) identifies the enzymes involved, (iii) extrapolates from *in vitro* to *in vivo* clearance, and (iv) considers metabolic clearance when predicting the potential for DDIs that affect a drug. In addition, examples are provided in order to illustrate why reaction phenotyping is important in the risk evaluation and safety profile of a new drug. In drug discovery, these studies usually provide *a priori* predictions of human metabolism without any human clinical data for the potential drug candidate. Therefore, it is often necessary to draw parallels between results obtained with the test drug and those from drugs in the clinic possessing similar characteristics.

16.3 OBJECTIVES OF REACTION PHENOTYPING IN DRUG DISCOVERY

Rarely, if ever, does a compound offer a perfect profile of efficacy, safety, and ease of development. Therefore, the primary goal in drug discovery is to find the compound that offers the best balance of properties with an acceptable risk–benefit ratio and thus minimize, rather than negate, the chance of failure in development. Before advancing a drug into clinical trials, it is desirable to know (i) the intrinsic clearance of a drug and from that the predicted *in vivo* clearance in humans and (ii) the enzymes involved in its metabolism (reaction phenotype). These are useful since knowing its *in vitro* clearance in human and animal assays helps one to use *in vivo* animal data to help predict a drug's human pharmacokinetics (including half-life and oral bioavailability). At the same time, knowing the enzyme(s) involved in the biotransformation reaction(s) allows one to predict the consequences of individual variability in metabolism and the risks associated with coadministration of drugs that alter the function of such enzyme(s).

These days mention is often made of a drug acting as a "perpetrator" of DDIs, with the drug itself causing offense by acting as an inhibitor or inducer. In reaction phenotyping, arguably one is tasked with the identification of the risks associated with the drug as a potential "victim" of DDIs. As will be discussed, the higher the metabolic clearance, and the fewer the number of enzymes involved in that clearance, the greater is the chance of a significant and potentially dangerous interaction occurring. Such an interaction manifests itself as an increase in the plasma C_{max} (maximal plasma concentration) or AUC (area under the plasma concentration–time curve) of the victim drug with a twofold or greater increase usually considered significant.

16.4 METABOLIC STABILITY SCREENING

The simplest way to minimize the risks associated with drug metabolism is to find a drug that is not cleared by metabolism (assuming that a prolonged half-life is not inherently a concern). Today's pharmaceutical R&D efforts incorporate metabolic screening in the drug discovery process, with the goal of finding and developing drugs without significant metabolism-mediated clearance (Parkinson, 1996; Rodrigues, 1999; Venkatakrishnan, Von Moltke and Greenblatt, 2001; Bjornsson *et al.*, 2003; Williams *et al.*, 2003a; Zhang *et al.*, 2007). This does not guarantee good or even sufficient oral bioavailability—solubility and permeability are two other, often critical, factors that must be considered—however, and all else being equal, a drug with greater metabolic stability will have higher oral exposure and lower interindividual variability. The key advantage to screening metabolic stability *in vitro* is that it allows one to evaluate many more compounds than is possible by *in vivo* screening in preclinical animal models. Hepatic DMEs most frequently involved in drug metabolism, *in vitro* stability, and clearance assays and screens are designed accordingly. For example, liver microsomes (prepared from liver tissue) and hepatocytes (fresh and/or cryopreserved) are most often used to measure and compare metabolic stability. With liver and hepatocytes available from human donors and preclinical animal models, *in vitro* assays in conjunction with allometric-scaling methods can help to bridge *in vitro* and *in vivo* studies to support predictions of PK and metabolism in humans, as well as identify biotransformation reactions that appear specific to humans and/or animals.

Liver microsomes provide a rich source of CYPs, FMOs, and UGTs, are available from human and various animal species, relatively cheap to prepare, simple to use, and can be stored frozen for relatively long periods. Fresh hepatocytes, in contrast, require more handling than liver microsomes yet offer the significant advantage of possessing a full complement of microsomal, cytosolic, and mitochondrial hepatic DMEs. Progress has been made to improve cryopreservation and thawing methods, which has increased the ease of handling of hepatocytes (Gómez-Lechón *et al.*, 2006). Thawing can result in the loss of metabolic activity; therefore, cryopreserved cells tend not to be used to make a definitive prediction of *in vivo* clearance with freshly isolated hepatocytes used instead. Cryopreserved cells are useful, however, as a preliminary screen of metabolic stability and/or as a bioreactor to generate metabolites for identification. Hepatocytes also allow one to look directly at hepatotoxicity associated with the drug and/or its metabolite(s) and, potentially, evaluate directly the role of a specific CYP in that toxicity (Mennes *et al.*, 1994; Li *et al.*, 1999; Hewitt *et al.*, 2002).

Liver microsomes and hepatocytes require different buffers for optimal activity of DMEs, and there are a few other differences; however, in general the *in vitro* methodologies for predicting metabolic clearance in these systems are similar. Often, the rate of loss of test compound over a single predetermined time period is used to rank order metabolic stability. This is very useful in identifying the worst performing compounds, yet it does lack precision and accuracy in predicting properties such as the *in vivo* clearance of a drug. Other methods are needed to scale properties from *in vitro* to *in vivo* and are employed as a compound advances through the screening process.

Two matters must be considered when assessing results from these kinds of screens: (i) the rate of metabolism in liver microsomes does not *per se* indicate which enzyme or enzymes are involved in the reaction; and (ii) measuring the rate of loss of compound does not specify what metabolite or metabolites are formed.

16.5   DRUG-METABOLIZING ENZYMES

The role of enzymes in drug metabolism has long been known and the last 40 years in particular have produced a wealth of data describing and elaborating on many of these enzymes and provided a mechanistic insight into the pharmacokinetics observed in animals and patients. There are very many DMEs expressed in humans and animals with considerable variety in their substrate affinity and expression between tissues and between species. Fortunately, a few enzymes are encountered more frequently than others and we will focus on three such enzyme families: the CYPs, the FMOs, and the UGTs. Understanding how these work, and knowing how to measure their activity and predict their contribution to metabolic clearance, will be relevant to most small molecule drugs.

16.5.1 Cytochrome P450s (CYPs)

The CYPs [E.C. 1.14.14.1] arguably are the single most important family of enzymes involved in regulating exposure to pharmacologically active, toxic, or potentially toxic xenobiotics. These enzymes possess a common ancestor dating back over a billion years and exhibit highly conserved regions in their amino acid sequence across different species including humans. The CYPs are membrane-bound hemoproteins that are found in the smooth endoplasmic reticulum and can be prepared as microsomal fractions for *in vitro* use. CYPs are known to have a structurally diverse array of substrates that usually undergo metabolism via oxidation. The CYP-catalyzed monooxygenase reaction is the most commonly encountered and can be described simply as:

$$RH + O_2 + 2H^+ + 2e^{-1} \rightarrow ROH + H_2O$$

Substrate oxidation proceeds via redox cycling between the ferric to ferrous states of the heme protoporphyrin group in the active site of the enzyme, with the NAD(P)H oxidoreductase and cytochrome b5 accessory flavoproteins involved in the electron transfer from nicotinamide adenine dinucleotide phosphate (NADPH) to the heme and the hemo-substrate complex, respectively (Fig. 16.1). Ultimately, this redox process catalyzes the incorporation of oxygen from molecular oxygen producing water and a metabolite that is usually more polar, more water-soluble, and more easily eliminated in the bile or urine.

In most cases, CYP-mediated metabolic clearance is an elimination process and thus serves as a potential mechanism of detoxification via removal of a potentially hazardous xenobiotic. There are instances, however, in which the transition intermediates formed in the CYP catalytic cycle can prove to be highly reactive. If they react with the heme (diltiazem, tamoxifen or troleandomycin, and CYP3A4) or apoprotein (dihydralazine and CYP1A2; tamoxifen and CYP2B6; tienilic acid and CYP2C9), mechanism-based inactivation can occur that results in irreversible inactivation of the CYP (covalent binding to the heme will completely inhibit the CYP reaction, whereas alkylation of the amino acid can cause a loss of activity but not necessarily complete inactivation).

Mechanism-based inactivation can exacerbate and prolong the risk of DDIs and toxicity from coadministered drugs with effects overcome only once a new enzyme has been synthesized to replace that inactivated. Or if the reactive species is sufficiently stable to escape the active site and covalently bind with nucleophilic macromolecules such as proteins, RNA, or DNA, it may culminate directly in adverse drug reactions and toxicity. (For a comprehensive review of mechanisms and methods related to mechanism-based inactivation, see Venkatakrishnan and Obach, 2007 and references contained therein.)

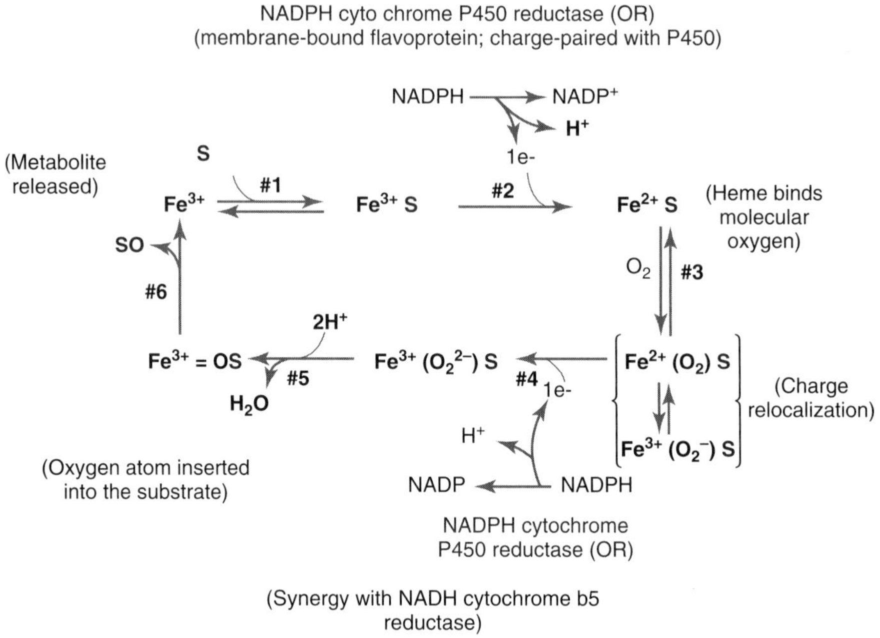

Figure 16.1 Cytochrome P450 (CYP)-catalyzed oxidation and redox cycling. Binding of the substrate to the low-spin ferric heme iron of the P450 (step #1) initiates the transfer of an electron from the cytochrome P450 oxidoreductase (CYPOR) accessory protein to form a substrate complex with a ferrous heme iron (step #2). This can bind with molecular oxygen (step #3), followed by charge relocalization presumed to form a more stable ferric complex that undergoes a second reduction step with CYPOR, with or without involvement from the cytochrome b5 accessory protein (step #4), to form water and an $(Fe-O)^{3+}$ complex or O_2^- superoxide anion, with subsequent incorporation of an oxygen atom into the substrate (step #5). In the last step, the oxidized product (metabolite) is released returning the active site to its original low-spin ferric heme (step #6).

CYPs are divided into subfamilies based on their shared sequence in protein homology, with >40% sequence similarity designating the family and >55% shared sequence homology designating the subfamily (the family is represented by an Arabic numeral, the subfamily by a capital letter, and then another Arabic numeral designates the individual gene member, e.g. CYP3A4). Over 50 different CYPs have been identified in humans, with CYP1A2, 2A6, 2B6, 2C8, 2C9, 2C19, 2D6, 2E1, 2J2, 3A4, 3A5, 4F2, 4F3a, 4F3b, and 4F12 responsible for the metabolic clearance of most drugs. In fact, historically CYP1A2, 2C8, 2C9, 2C19, 2D6, and 3A4 are the most important CYPs in this regard and are the focus of most reaction-phenotyping efforts, at least in the early stages. (The other enzymes are often assessed via a process of elimination.)

The same nomenclature is used for CYPs from human and from other species. Several CYPs are well conserved across species and share sufficient identity that they have the same name (CYP1A1, 1A2, 1B1, and 2E1); however, although these CYPs are generally similar in sequence and function, it is important to note that they are not identical to the human forms. Even a small difference in amino acid sequence can affect substrate affinity and the ability of the enzyme to metabolize a substrate and/or the regiospecificity in the site of the molecule metabolized. (For a complete and updated repository of the classified CYPs, see Home Page of the Human Cytochrome P450 (CYP) Allele Nomenclature Committee at http://www.cypalleles.ki.se/.)

As a family, the CYPs unite to catalyze the oxidation and metabolic clearance of a highly diverse collection of lipophilic substrates that are otherwise not readily eliminated in urine or bile. This lack of specialization distinguishes the CYPs from more efficient enzymes like carbonic anhydrase and provides a safeguard against the tremendous diversity and risk present in nature's chemistry. Individual CYPs, however, tend to be more selective in their substrate characteristics and can be distinguished somewhat by their preferred substrate properties, as shown in Table 16.1. CYP1A1 is often involved in the metabolism of aryl hydrocarbons and although not constitutively expressed in liver tissue, its mRNA is highly inducible with translation to the protein also reported (Drahushuk *et al.*, 1998). CYP1A1 and CYP1A2 show an overlap in their substrate affinity; however, CYP1A2 preferentially metabolizes polycyclic aromatic hydrocarbons and their amine analogues. Both CYP1A1 and 1A2 are involved in the metabolism of procarcinogens to carcinogens; however, CYP1A2 has been shown to be the primary enzyme involved in the metabolism of drugs such as phenacetin and theophylline. CYP2A6 shows a preference for smaller, neutral, less lipophilic molecules such as ketones or nitrosamines. CYP2C9 and CYP2C19 are able to metabolize acidic molecules, with CYP2C9 preferring acidic drugs that have a reactive site six to eight carbon lengths removed from an H-bonding residue (e.g. diclofenac). In contrast CYP2D6, constitutive in liver, is better able to metabolize basic drugs, usually with the reactive site six to eight carbons removed from a basic nitrogen atom (e.g. fluoxetine, nortriptyline, dextromethorphan, and bufuralol). CYP2E1 is highly inducible and often effective in metabolizing small molecules (e.g. ethanol, acetaminophen, halothane, and para-nitrophenol), and CYP3A4 has an active site big enough to accommodate multiple substrates including relatively large and/or very lipophilic compounds (e.g.

TABLE 16.1   Human cytochrome P450s and their substrate selectivity.

CYP	Substrate	Example substrates
CYP1A1, 1A2	Planar lipophilic; neutral or basic molecules with log P 1.39–6.35 (CYP1A1); log P 0.08–3.61 (CYP1A2)	Phenacetin, dapsone, tacrine, ramelteon
CYP2C9	Medium-sized acidic molecules with 1–2 hydrogen bond acceptors; log P 0.89–5.18	Diclofenac, warfarin, tolbutamide
CYP2C19	Medium-sized basic molecules with 2–3 hydrogen bond donor/acceptor atoms; log P 1.49–4.42	(S)-mephenytoin, omeprazole
CYP2D6	Medium-sized basic molecules with a protonatable nitrogen 5–7 Å from the site of metabolism; log P 0.75–5.04	Fluoxetine, dextromethorphan, nortriptyline, bufuralol
CYP2E1	Small, neutral molecules (structurally diverse); log P −1.35–3.63	Para-nitrophenol, ethanol, chlorzoxazone
CYP3A4	Large, lipophilic molecules (structurally diverse); log P 0.97–7.54	Testosterone, cyclosporin A, nifedipine, midazolam, simvastatin

Adapted from Lewis and Dickins (2002) and Lewis, Jacobs and Dickins (2004).

simvastatin, testosterone, cyclosporine A, and midazolam), a combination of properties that has led to its involvement in a majority of CYP-mediated drug reactions and significant consequences that result from DDIs (Guengerich *et al.*, 2002; Lewis and Dickins, 2002).

Substrate and protein conformation, intra- and intermolecular H-bonding, and electronic and steric constraints all influence the propensity for metabolism and make it difficult to model or predict *in silico* how a substrate will behave *a priori*. Despite this complexity, advances have been made with *in silico* models, especially with regard to modeling CYP inhibition governed by binding properties and thus simpler than predicting the site or rate of metabolism (Stouch *et al.*, 2003; O'Brien and de Groot, 2005). Nonetheless, when beginning reaction-phenotyping studies, it is always worth considering the structure and physicochemical properties of the molecule and how this relates to the preferences known for the CYPs described.

A fundamental concern that drives the need to perform reaction-phenotyping studies is that if a new drug is metabolized to a significant degree (>40% of metabolic clearance), then it is at risk of displaying markedly different exposure when coadministered with an inhibitor of that same enzyme. For instance, terfenadine (the original prodrug of fexofenadine or Allegra) is a classic example of a high-metabolic-clearance drug in humans. Terfenadine was shown to undergo extensive first-pass metabolism, with its clearance mediated predominantly by CYP3A4 (Garteiz *et al.*, 1982; Yun, Okerholm and Guengerich, 1993). Unfortunately, terfenadine can also bind to the hERG (human *ether-a-go-go*-related gene) potassium channel receptor and cause QT prolongation and cardiovascular toxicity including potentially fatal arrhythmias (Woosley *et al.*, 1993). When terfenadine was dosed with CYP3A4 inhibitors such as ketoconazole, itraconazole, and grapefruit juice, there was an increase in its exposure and that of its metabolites. This was associated with an increase in prolongation of QT effects, linked to fatal cardiac arrhythmias in patients (Honig *et al.*, 1993a, b, 1996). As a consequence of such poor clinical control, terfenadine was soon withdrawn from the market and now serves as a worst-case example of what can go wrong insofar as a narrow therapeutic index and too specific (single CYP) reaction phenotype is concerned.

16.5.1.1 *Human CYP Polymorphisms and Individual Variability*   Humans show significant interindividual variability in their total and relative expression of DMEs such as CYPs, UGTs, FMOs, SULTs, GSTs, and NATs. This variability extends to the relative exposure seen with drugs and their metabolites in the patient population. By knowing what enzymes are involved in the metabolism of a drug, it is possible to make a prediction of the likely variability in its exposure.

Variability in DMEs can involve several factors, some of which are easier to characterize than others and some are compound- or drug-specific. One source of interindividual variability in humans stems from enzyme induction that can be elicited by environmental influences, such as pollutants or lifestyle and dietary factors such as cigarette smoking and eating cooked meat, which are both known to induce CYP1A1, 1A2, among others; the popular herbal supplements like St. John's wort (*Hypericum perforatum* L.) are also known to induce CYP3A (Markowitz, 2003). Variability in drug metabolism and exposure can also stem from powerful CYP inhibitors found in natural products, for example, furanocoumarins related to and including bergamottin in grapefruit juice (Girennavar *et al.* 2006).

Another important contributor to individual variation in CYP activity is genetics, with heterozygous and homozygous mutants in CYP expression and/or function often resulting in polymorphisms. A polymorphism is defined as an allelic variant that is expressed with ≥1% frequency in the human population. CYP polymorphic phenotypes usually describe bimodal or multimodal distribution in the pharmacokinetics of a drug given at an equivalent dose to healthy humans. Based on their phenotype, patients can be classified as poor metabolizers (PMs), intermediate metabolizers (IMs), extensive metabolizers (EMs), and in the instance of CYP2D6, ultrarapid metabolizers (UMs). At the genetic level, there are numerous allelic variants leading to various mutations that can affect the ability of that individual to metabolize a particular substrate or class of substrates. This is evident with ethnic differences seen in the frequency and expression of CYP polymorphisms. For instance, CYP2D6 slow metabolizers comprise ~7% of Caucasian and only ~0.8% of Asian populations, whereas CYP2C19 slow metabolizers comprise ~2% of Caucasians and ~19% of Asians. This is enough to merit a distinctive shift in focus when evaluating the safety and development of CYP2D6 or CYP2C19 drugs in these different patient populations and markets. CYP3A4 is highly variable within and between humans yet is generally not considered to be polymorphic in terms of its phenotype (i.e. it is present and functionally active in most people). In contrast, its closest family member, CYP3A5, is found in only 25% of Caucasians and only 75% of Africans and people of African descent (Kuehl et al. 2001; Mouly, 2005). An overlapping substrate affinity of CYP3A4 and CYP3A5 likely contributes significantly to the interindividual variability in drug exposure. Since CYP3A5 is known to be expressed in the intestine and other extrahepatic tissues, its contribution to metabolic clearance could be underestimated if only human liver microsomal systems are used to predict metabolic clearance.

In Table 16.2 are listed the frequencies of a few known polymorphisms in Caucasians versus Asians reported for CYP2D6, CYP2C19, and also CYP2A6, 2C9, and 3A4. It should be noted that these represent only a very few number of the genetic polymorphisms already identified for DMEs. Numerous mutant forms have been identified that have greater, lesser, or even no discernible effect on CYP activity, including the possibility of multiple gene copies that can lead to markedly higher CYP protein expression. One example of which is the ultrarapid CYP2D6 metabolizers found at relatively high frequencies in some African populations (~29%) and Southern Europeans (~10%). In contrast to null polymorphisms and slow metabolizer phenotype, a CYP2D6 substrate will show much higher clearance and thus much lower exposure of drug in patients possessing multiple copies of the CYP2D6 gene and potentially much higher exposure to metabolite(s) formed from that drug, for example, the antidepressant nortriptylline (Dalén et al., 1998).

Although there exist some very useful technologies that facilitate genotyping of patients (e.g. the Roche Diagnostics P450 AmpliChip), there can be considerable complexity involved in translating a particular genotype to a predicted phenotype (Jain, 2005; Li et al., 2006). Advances are expected that ultimately could allow individualized dose regimens to become economically feasible. Looking to the future, this will provide more therapeutic opportunities for R&D with both new drugs and also the opportunity to revisit the use of older drugs that were withdrawn due to a poor safety profile in the average population. Meanwhile, the reaction phenotype of a drug when coupled with its predicted or known safety margins does allow one to define the risk involved in administering to various populations. For instance, a new drug eliminated primarily by a polymorphic metabolic clearance with a narrow safety margin is unlikely ever to be developed beyond testing in humans under strictly controlled clinical conditions.

TABLE 16.2 Common polymorphisms of human cytochrome P450s (CYP) in Asian and Caucasian populations.

CYP		Poor metabolizer frequency (%)	Major variant	Allele frequency (%)	Mutation effect	References
CYP2A6	Asian	2–4	2A6*4	11–18	Gene deletion; nonenzyme	Ariyoshi et al. (2000)
	Caucasian	~0	2A6*2	4	Gene deletion	Kwon et al. (2001); Oscarson et al. (1999a, b)
CYP2C9	Asian	0.04	2C9*3	2.1	Stop codon; inactive	Takanashi et al. (2000)
	Caucasian	1	2C9*2	8–13	Reduced affinity	Aithal et al. (2000)
			2C9*3	7–9	Altered substrate specificity	
CYP2C19	Asian	19	2C19*2	32	Splicing defect	Kaneko et al. (1999a, b)
			2C19*3	6–10	Stop codon; inactive	
	Caucasian	2	2C19*2	13	Inactive	Griese et al. (2001)
CYP2D6	Asian	0.8	2D6*5	6.2	Gene deletion; no enzyme	Hiroi, Imaoka and Funae (1998)
			2D6*10	33–50	Unstable (intermediate metabolizer)	Ramamoorthy, Tyndale and Sellers (2001); Shimada et al. (2001)
	Caucasian	7	2D6*4	12–21	Inactive	Sachse et al. (1997)
CYP3A4	Asian	1	3A4*6	1	Frameshift	Hsieh et al. (2001)
	Caucasian	~0	0	0	—	Ozdemir et al. (2000)

Adapted from Mizutani, 2003, with references therein.

16.5.2 Flavin-Containing Monooxygenases

FMOs [E.C. 1.14.13.8] are found in species from bacteria to mammals. They are identified by several classes: external monooxygenases are the most common form and depend on NADPH or nicotinamide adenine dinucleotide (NADH) to reduce the flavin (flavin mononucleotide, FMN, or flavin adenine dinucleotide, FAD) to the reactive form that can bind molecular oxygen; in contrast, the rarely encountered internal monooxygenases (e.g. lactate monooxygenase) can use the substrate itself to reduce the flavin. In humans, genes for ~57 CYPs have been reported (plus approximately five pseudogenes), whereas to date, only five functional human FMOs are known (FMO1-5) with six pseudogenes. Of these enzymes, FMO1 has the broadest substrate specificity; FMO2 is the narrowest because of its restricted active site; FMO3 has the highest functional activity and also a high expression; FMO4 shows the lowest expression—with its low activity this is considered insignificant with respect to drug metabolism (with possible exceptions); and FMO5 has significant expression, albeit with relatively low activity and very narrow substrate selectivity.

FMOs show distinctive and potentially characteristic differences in their relative expression in human tissues: FMO1 is the major FMO expressed in fetal liver and adult human kidney and intestine; FMO2 is the major FMO expressed in human lung, albeit inactive in Caucasians, Asians, and the majority of African-Americans (attributed to the FMO2*2 allele associated with expression of a nonfunctional truncated protein encoded by a nonsense mutation g.23238C > T (Q472X)); and FMO3 is regarded generally as the FMO most highly expressed in the liver (Dolphin et al., 1998; Cashman 2000; Whetstine, 2000; Krueger and Williams 2005). Some studies have, however, shown that FMO5 can be expressed at higher levels in human liver (3.5 to 34 pmol/mg microsomal protein for FMO5 versus 12.5 to 117 pmol/mg microsomal protein for FMO3) (Overby et al., 1997). This could reflect how well the tissues were handled in the preparation of liver microsomes (Cashman and Zhang 2006). In general, FMO3 and FMO5 are the forms most likely to be involved in FMO-mediated drug metabolism.

With liver FMO3 measured at immunoquantified levels on average ~60% of those observed for CYP3A4 (Overby, Carver and Philpot, 1997) and considering its relatively broad substrate selectivity, FMO3 has become the most significant form involved in FMO-mediated drug metabolism. When microsome preparations are properly handled, FMO activity can demonstrate a k_{cat} at 30–60 per minute, which is considerably greater than that typically seen with CYPs (e.g. k_{cat} of 1–20 per minute) (Krueger and Williams, 2005). It is thus conceivable that even with CYP3A4 substrates, FMO3 may constitute the primary metabolic clearance pathway and potentially mitigate CYP3A4 DDIs. With FMOs and CYPs are both NADPH-dependent enzymes, and thus both active in the liver microsomal assays used typically for metabolic stability studies. The two families of enzymes can even compete to form the same metabolite(s) as reaction products (albeit with FMOs being more specific than CYPs and tending to favor S- and N-oxidation reactions). This is a consideration when performing reaction-phenotyping studies and evaluating CYP versus FMO reactions.

One important characteristic of FMOs is that they are thermally unstable when compared with CYPs, which is useful for reaction-phenotyping studies. Conversely, however, if liver microsomes are not properly handled, FMO activity can be lost in vitro, leading to a prediction that underestimates its contribution to the total metabolic clearance.

16.5.2.1 *Human FMO Polymorphisms and Individual Variability* No inducers have been found for FMOs and DDIs appear limited in effect since few drugs are

metabolized primarily by FMOs (itopride is an example of a drug metabolized primarily by FMOs and not by CYPs) (Mushiroda *et al.*, 2000). In general, the FMOs appear to behave as high-capacity enzymes with a low affinity for substrates relative to CYPs. To date, 34, 57, 40, 30, and 40 genomic allelic variants have been found for FMO1, 2, 3, 4, and 5, respectively. In addition to various wild-type forms are variants that are inactive or show decreased or functional activity (Cashman, 2004). The most well-known polymorphism is associated with FMO3, and Trimethylaminuria (TMAU). FMO3 exclusively catalyzes the *N*-oxidation and clearance of trimethylamine (TMA), a product of choline metabolism (Cashman *et al.*, 1997; Lang *et al.*, 1998). A rare subpopulation of humans (0.1% to 1% incidence in Caucasians) is deficient in FMO3 and displays a slow metabolizer phenotype status in their clearance of TMA. As TMA accumulates in excreta such as breath, sweat, and urine, it produces a characteristic "fish odor" syndrome (Arseculeratne *et al.*, 2007).

Polymorphisms in FMO2 are also known, with a 1414C allele expressed in 13% of African-Americans coding for a full-length active form of FMO2, rather than the 1414T allele that has a premature stop codon producing a truncated and inactive form of FMO2 in human lung microsomes from Caucasians and Asians (Dolphin *et al.*, 1998; Whetstine, 2000). In contrast, 26% of African-Americans and 4.5% of Hispanics express an active form of FMO2 in their lung (Cashman, 2000). Since the polymorphic and active form of FMO2 can metabolize the formation of reactive and toxic sulfenic acid metabolites of thioureas, it is plausible that individual and ethnic differences exist in the susceptibility and incidence of pulmonary toxicity (Henderson *et al.*, 2004).

In kidney microsomes, the relative expression of detected FMOs was found to be FMO1 > FMO5 > FMO3. Although an FMO1 polymorphism has not been identified, levels of this flavoprotein appeared to be greater in kidney microsomes from African-Americans than from Caucasians (Krause, Lash and Elfarra, 2003).

16.5.3 Uridine 5′-Diphosphate Glucuronosytransferases

The UGTs [E.C. 2.4.1.17] are a third class of membrane-bound enzymes expressed in the liver and intestine and often play an important role in the first-pass effect. Like CYPs and FMOs, UGTs also utilize a nucleotide cofactor, in this case, UDPGA. Unlike CYPs and FMOs, UGTs effect metabolism not by oxygenation of the substrate but by *N*- and *O*-conjugation of the substrate by group transfer of glucuronic acid (supplied by UDPGA). This typically generates a polar glucuronide conjugate that is amenable to biliary and/or renal elimination. With phenols and carboxylic acids as the typical substrates for UGTs, relative to CYPs there are few drugs that are directly eliminated by this class of enzymes. However, many oxidized products are formed from CYP-mediated reactions that can serve as substrates for conjugation and UGTs are often involved in regulating the exposure of such metabolites.

In addition to their role in drug clearance, the UGT-mediated formation of acyl glucuronides and ether conjugates has been linked to reactivity and toxicity (e.g. diclofenac) (Grillo *et al.*, 2003a, b). This can be a concern in the development of a drug and careful consideration should be given to evaluating the risks involved. This is discussed in detail in Chapter 21.

16.5.3.1 Human UGT Polymorphisms and Individual Variability UGTs are classified by primary amino acid sequence into families and subfamilies. In humans, to

date, two gene families have been identified with a growing number of members that are currently grouped into three subfamilies (UGT1A, UGT2A, and UGT2B). Table 16.3 shows the major UGTs and their reported expression in liver, intestine, kidney, and/or nasal/pulmonary epithelia.

Gene mutations in the coding region and/or promoter of the UGT1A1 gene lead to decreased activity including its ability to conjugate the endogenous substrate bilirubin. This results in an observed polymorphism of UGT1A1 activity in humans, with interindividual variation in the levels of unconjugated bilirubin. A profound (90%) reduction in bilirubin clearance is seen in patients suffering from the rare Crigler–Najjar syndrome, whereas a moderate (40%) decrease is reported on average for patients suffering from the more commonplace Gilbert syndrome, which is seen in 10% of the US population and associated with the UGT1A1*28 homozygous form (Tukey and Strassburg, 2000). Inhibition of UGT1A1 can present symptomatically like Crigler–

TABLE 16.3 **Tissue distribution and substrates of major human uridine 5′-diphosphate glucuronyltransferases (UGTs).**

UGT	Liver	Intestine	Kidney	Nasal/pulmonary	Substrates
1A1	√	√			Bilirubin; 3-OH-estradiol (major)
1A3	√	√			Hexafluoro-1α, 25-dihydroxyvitamin D3; naproxen; 3-OH-estradiol
1A4	√	√			Trifluoroperazine; amitriptyline; cyclobenzaprine; lamotrigine; olanzapine
1A6	√	√			Serotonin; 1-naphthol
1A7		√			7-ethyl-10-hydroxycamptothecin (SN38)
1A8		√			Propofol; 3-OH-estradiol
1A9	√		√		Propofol (major)
1A10		√			3-OH-estradiol
2A1				√	Scopoletin
2B4	√	√			Hyodeoxycholic acid (bile acid)
2B7	√	√	√		Zidovudine; morphine-6-glucuronidation; clofibric acid; diclofenac; naloxone; valproic acid; denopamine (substrate affinity encompasses UGT2B4)
2B10	√	√			Nicotine, cotinine
2B11	√				12-HETE; 15-HETE; 13-HODE
2B15	√	√			S-Oxazepam
2B17	√				Testosterone; alizarin (fastest rate), also eugenol, scopoletin, galagin
2B28	√				None identified

12-HETE: 12-hydroxyeicosatetraenoic acid; 15-HETE: 15-hydroxyeicosatetraenoic acid; 13-HODE: 13-hydroxyoctadecadienoic acid (also substrates for UGT2B10) (Fournel-Gigleux *et al.*, 1989; Jedlitschky *et al.*, 1999; King *et al.*, 2000; Tukey and Strassburg, 2000; Fisher *et al.*, 2001; Lévesque *et al.*, 2001; Gagné *et al.*, 2002; Turgeon *et al.*, 2003a, b; Williams *et al.*, 2004; Court, 2005; Kaji and Kume, 2005; Jakobsson *et al.*, 2006; Miners *et al.*, 2006; Barre *et al.*, 2007; Chen *et al.*, 2007; Kaivosaari *et al.*, 2007).

Najjar or Gilbert syndrome, and in UGT1A1-deficient individuals, could well present as profound hyperbilirubinemia. Irinotecan, an ester camptothecin derivative used to treat metastatic colon cancer, is metabolized by carboxyesterases to form SN-38, a carboxylic acid and active and toxic form of the drug. SN-38 is subsequently cleared by UGT1A1-mediated conjugation and examples exist that show increased incidence and severity of neutropenia and diarrhea in UGT1A1-deficient patients (Toffoli and Cecchin, 2004; Toffoli *et al.*, 2006).

16.6 ASSAY OPTIMIZATION FOR METABOLIC CLEARANCE AND REACTION-PHENOTYPING STUDIES

The first step in reaction phenotyping is to optimize metabolic clearance and predict total metabolic clearance *in vivo*. In general, this is done by measuring the rate of change of a therapeutically relevant substrate concentration determined over an appropriate time course. To detail all mechanisms of metabolic clearance and assay protocols is beyond the scope of this chapter. Therefore, the present discussion will focus on the most commonly encountered DME systems, specifically (i) how to optimize CYP, FMO, and/or UGT activity in liver microsomes and (ii) the tools available to identify the specific enzyme or enzymes involved and the importance of integrating the data. Beyond the necessity for routine and standardized metabolic stability screening, drug metabolism is best approached on a case-by-case basis and it is important to note that the major enzymes involved in the metabolism of a new drug may not be the commonplace DMEs described herein. In addition to obvious structural characteristics—aromatic amines tend to be substrates for *N*-acetyltransferases, monoamines substrates for monoamine oxidases, epoxides substrates of epoxide hydrolases, and so on—*in vivo* excretion balance and biotransformation studies are extremely helpful in identifying metabolic reactions that are not covered by standardized assays. Often the clearest indication of an "atypical" metabolic clearance mechanism is seen when a compound is found to be metabolically stable in liver microsomes and yet is rapidly and/or extensively metabolized *in vivo* (i.e. relatively little unchanged parent compound is measured in excreta). In such circumstances, one must be prepared to take a different approach and consider using other assay conditions, subfractions such as cytosol or mitochondria, and non-hepatic tissues such as gut, skin, nasal mucosa, kidney microsomes, with experimental protocols adapted accordingly. Sequential pathways of metabolism can be encountered that involve several of these enzymes: microsomal, cytosolic, hepatic, as well as extrahepatic (e.g. carbamazepine). In these cases, radiolabeled and stable-label substrates are very useful in understanding the mechanisms involved.

16.6.1   Optimizing CYP Activity in Human Liver Microsomes

Livers procured from animals or human organ donors can be perfused to isolate hepatocytes or homogenized and centrifuged to isolate liver mitochondrial, S9, microsomal, and/or cytosol subfractions. Liver microsomes are the cheapest and easiest system to use and the most commonly used enzyme system for *in vitro* drug metabolism studies. The rate and extent of turnover of test compound are measured in incubations supplemented with an excess of NADPH as a cofactor (directly or as part of a regenerating system). Buffers like potassium phosphate or Tris are used to control the pH, with reactions normally performed at pH 7.4 and 37 °C to simulate physiological conditions. Magnesium chloride is often included in the incubation medium to modify the ionic strength and sometimes this can affect the turnover of substrate. Test compounds usually are added in solution, with the solvent often a mix of dimethylsulfoxide (DMSO) and acetonitrile or methanol and aqueous buffer. It is important to keep the total volume of organic solvent as low as possible since DMSO, acetonitrile, and methanol are each known to differentially inhibit CYP activity (Chauret, Gauthier and Nicoll-Griffith, 1998; Hickman *et al.*, 1998). If not taken into account, the quantification and prediction of a particular CYP's role in the drug's metabolism can be incorrectly estimated.

Human liver microsomes (HLMs) are usually prepared from a pool of donors (typically from N = 12 to 20), with male and female donors included. This helps account for the known heterogeneity in relative expression and activity that exists in the human population. If the individuals are chosen correctly, the metabolic clearance in the HLM pool will represent an "average" but fully functioning human liver (i.e. one without any null polymorphisms and with basal activity of the inducible CYP forms). Animals tend to be more homogeneous in their enzyme expression and liver microsomes are often prepared from the liver of a single untreated animal (or a pooled group if animals have been pretreated with an inducer or potential inducer). Variability does exist even in animals, and may also need to be taken into account; for instance, CYP2D15 polymorphism affects the metabolism and pharmacokinetics of celecoxib in dogs (Paulson *et al.*, 1999). More commonly encountered are gender-specific differences in CYP expression; for example, CYP2C11 is a male-specific enzyme in the rat (Czerniak, 2001). These animal polymorphisms and animal-specific DMEs may show up in the safety and tolerability studies conducted in support of a regulatory filing submitted prior to human clinical trials. Such findings may necessitate additional mechanistic studies to show their clinical relevance.

16.6.2 *In Vitro* Tools for Identifying Cytochrome P450s

Of the human DMEs, the CYPs are the most mature and advanced in terms of reaction-phenotyping tools. In fact, *in vitro* procedures have become increasingly standardized. However, it is important to remember that the different *in vitro* data sets have to be consistent in order to enable an integrated *in vitro* CYP reaction phenotype (Fig. 16.2), sound decision making, and successful *in vitro–in vivo* correlations (Zhang *et al.*, 2007).

16.6.2.1 Selective Inhibitors Possibly the most useful and direct way of identifying what CYP metabolizes a drug is to co-incubate it in the absence and presence of a known selective inhibitor of that enzyme. In this instance, the activity is measured in the absence (solvent alone) and presence of the inhibitor. The "% inhibition" or fractional loss of activity (versus solvent alone) represents the fraction (or %) of the metabolism catalyzed by the enzyme in question:

$$\% \text{ Inhibition} = \frac{\text{vo} - \text{vi}}{\text{vo}} * 100$$

$$\text{or fractional loss of activity for each CYPi} = \frac{\text{vo} - \text{vi}}{\text{vo}}$$

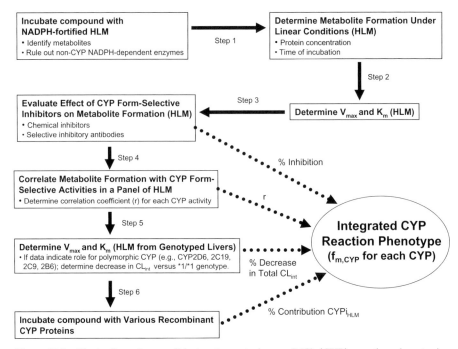

Figure 16.2   Illustration of a possible *in vitro* cytochrome P450 (CYP) reaction-phenotyping strategy. Adapted from Zhang *et al.* (2007), where V_{max} is the maximal rate of reaction; K_m is the Michaelis constant; HLM is human liver microsomes; f_m is the fraction metabolized; CYP is cytochrome P450s, CL_{int} is intrinsic clearance; and r is the correlation coefficient.

where vo is the disappearance rate constant or velocity of reaction and vi is the disappearance rate constant in the presence of an inhibitor; CYPi represents a specific CYP or CYP-mediated reaction.

The fraction metabolized (f_m) is an important factor in understanding the consequences of a coadministered inhibitor:

$$\text{Total Clearance} = \text{Total Clearance} * (f_{\text{non-metabolic}} + f_m)$$
$$1 = f_{\text{non-metabolic}} + f_m$$

where $f_{\text{non-metabolic}}$ is the fraction cleared by mechanisms other than metabolic clearance (e.g. renal and/or biliary elimination); f_m is the fraction metabolized—measured as the fraction of dose recovered as metabolites in excreta.

In a situation where all of the clearance is governed by metabolism, then $f_m = 1$. In reaction phenotyping, this concept is extended further to define $f_{m,CYP}$, which is the fraction metabolized by CYPs, and further still to $f_{m,CYPi}$, which describes the fraction metabolized by a specific CYP (the sum of these individual reactions should account for the total $f_{m,CYP}$). For example, if $f_{m,CYP3A4} = 1$, then the entire CYP-mediated clearance is by CYP3A4. If in this case f_m also accounts for most of the *in vivo* clearance ($f_m \sim 1$), then CYP3A4 determines the *in vivo* clearance and coadministered CYP3A4-selective inhibitors of this enzyme could have a great effect on total exposure, especially relative to a drug with a low f_m or one with an equivalent f_m that results from multiple CYP-mediated reactions, where each $f_{m,CYPi}$ is <0.5. Identifying the fraction metabolized by a specific CYP is the goal of reaction phenotyping. The product of f_m and $f_{m,CYP}$ determines the overall contribution of that CYP to clearance:

$$Clearance_{CYPi} = Total\ Clearance * f_m * f_{m,CYPi}$$

Knowing this for a drug candidate allows one to identify some of the risks associated with DDIs, polymorphisms, and interindividual variability in the target population.

Given that CYPs are distinguished in part by their promiscuity, there are relatively few drugs and chemicals that are uniquely selective. A group of tool compounds has been identified as most appropriate for the purpose of *in vitro* reaction phenotyping. Table 16.4 lists the substrates and inhibitors identified by their selectivity for the major CYPs and that come recommended in the FDA draft guidelines for *in vitro* characterization of drug candidates (FDA, 2006). In the table are listed the preferred choice of substrate and inhibitor and acceptable alternative compounds. (The FDA guidelines are subject to review and a change to the preferred list of substrates and inhibitors is very possible. It is advised, if practicable, to perform studies using the most current recommendations available.) It should be noted that with CYP3A reactions, in particular, it is common to use multiple selective yet structurally dissimilar probe substrates or inhibitors to account for some of the more complex kinetics associated with this enzyme (Kenworthy *et al.*, 1999). The same also appears to be true for CYP2C9, with a panel of CYP2C9 inhibitors differentiating in their potency toward inhibition of (*S*)-flurbiprofen, (*S*)-warfarin, phenytoin, tolbutamide, and diclofenac (Kumar *et al.*, 2006). The heteroactivation of CYP2C9 substrates by dapsone also supports the kinetic similarity with CYP3A4 and suggests a "two-site" binding model is appropriate for CYP2C9 (Hutzler, Hauer and Tracy, 2001). This is further supported by the discovery of a second binding site for warfarin, based on studies of a crystal structure of CYP2C9, and the apparent capacity of a relatively large active site ($\sim470\,\text{Å}^3$) that can accommodate the simultaneous binding of multiple ligands (Williams *et al.*, 2003b).

If a single CYP metabolizes a drug, then coadministered inhibitors of that CYP should demonstrate a profound effect on the rate and extent of metabolism. This is most easily discerned by measuring the rate of formation of a specific metabolite and, for substrates with shorter half-lives, depletion-kinetic approaches also work quite well. Table 16.5 lists the most common CYP-selective chemical inhibitors used and the recommended concentrations for reaction phenotyping (based on the literature reported maximal inhibition of various CYP-specific biotransformation reactions). For a substrate metabolized by multiple CYPs, a single selective inhibitor will affect only one component of the turnover and several inhibitors are needed to characterize the compound (the individual effects are summed to account for the total clearance). In reality, the more CYPs that are involved in the metabolic clearance of a compound, the harder it becomes to identify the contribution of each enzyme to the total clearance. Fortunately, this is also the preferred profile of a drug, and the same issue applies *in vivo*—consequently, CYP DDIs become unlikely to profoundly alter its clearance or AUC and render it a so-called "victim."

16.6.2.2 *CYP Inhibitory Antibodies*   CYP inhibitory antibodies are an alternative to chemical inhibitors and yield similar data (e.g. % inhibition for each CYP form).

TABLE 16.4 Cytochrome P450 (CYP) inhibitors, substrates, and inducers (United States FDA recommended *in vitro* tools).

CYP	Inhibitor Preferred	Inhibitor Acceptable	Substrate Preferred	Substrate Acceptable	Inducer Preferred	Inducer Acceptable
1A2	Furafylline	α-Naphthoflavone	Phenacetin-O-deethylation	7-Ethoxyresorufin-O-deethylation Theophylline-*N*-demethylation Caffeine-3-*N*-demethylation Tacrine 1-hydroxylation	Omeprazole β-Naphthoflavone 3-Methylcholanthrene	Lansoprazole
2A6	Tranylcypromine	Pilocarpine	Coumarin-7-hydroxylation Nicotine C-oxidation		Dexamethasone	Pyrazole
2B6	Methoxsalen	Tryptamine 3-Isopropenyl-3-methyl diamantane 2-Isopropenyl-2-methyl adamantane Sertraline, phencyclidine Triethylenethiophosphoramide (thioTEPA) Clopidogrel, ticlopidine	Efavirenz hydroxylase Bupropion-hydroxylation	Propofol hydroxylation S-mephenytoin-*N*-demethylation	Phenobarbital	Phenytoin
2C8	Montelukast Quercetin	Trimethoprim, gemfibrozil Rosiglitazone, pioglitazone	Taxol 6-hydroxylation	Amodiaquine *N*-deethylation Rosiglitazone para-hydroxylation	Rifampin	Phenobarbital
2C9	Sulfaphenazole	Fluconazole Fluvoxamine Fluoxetine	Tolbutamide methyl-hydroxylation S-warfarin 7-hydroxylation Diclofenac 4'-hydroxylation	Flurbiprofen 4'-hydroxylation Phenytoin-4-hydroxylation	Rifampin	Phenobarbital
2C19		Ticlopidine Nootkatone	S-mephenytoin 4'-hydroxylation	Omeprazole 5-hydroxylation Fluoxetine O-dealkylation	Rifampin	
2D6	Quinidine		(±)-bufuralol 1'-hydroxylation Dextromethorphan O-demethylation	Debrisoquine 4-hydroxylation	None identified	
3A4/5	Ketoconazole Itraconazole	Azamulin Troleandomycin Verapamil	Midazolam 1'-hydroxylation Testosterone 6 β-hydroxylation	Erythromycin *N*-demethylation Dextromethorphan *N*-demethylation Triazolam 4-hydroxylation Terfenadine C-hydroxylation	Rifampin	Phenobarbital Phenytoin, taxol Rifapentine Troglitazone Dexamethasone

Adapted from FDA (2006).

TABLE 16.5 Common chemical inhibitors and recommended concentrations for cytochrome P450 (CYP) reaction phenotyping *in vitro*.

CYP form(s)	Inhibitor	Inhibitor concentrations (µM)
CYP1A2	Furafylline	$10–30^{a}$
	α-Naphthoflavone	1
CYP2A6	Methoxsalen	1
CYP2B6	ThioTEPA	50
CYP2C8	Montelukast	0.1^{b}
CYP2C9	Sulfaphenazole	10
CYP2C19	Benzylnirvanol	1
CYP2D6	Quinidine	<2
CYP2D6 and CYP3A4/5		10
CYP2E1	Diethyldithiocarbamate	50
CYP3A4/5 and others	Ketoconazole	1 and 10
CYP3A4/5	Troleandomycin	50^{a}

[a]Inhibition potency increases with preincubation in the presence of NADPH (time-dependent inhibitors).
[b]Conditions may be subject to the concentration of human liver microsomal protein used in the assay.
Data from Muralidharan *et al.* (1991); Kunze and Trager (1993); Chang, Gonzalez and Waxman (1994); Baldwin *et al.* (1995); Bourrié *et al.* (1996); Eagling, Tjia and Back (1998); Yamazaki and Shimada (1998); Zhang *et al.* (2001); Suzuki *et al.* (2002); Cai *et al.* (2004); Turpeinen *et al.* (2004); Walsky *et al.* (2005).

Unlike chemical inhibitors that tend to work competitively, inhibitory antibodies work noncompetitively by modulating the conformation of the CYP hemoprotein complex. This can simplify the kinetic optimization of substrate inhibition otherwise based on K_{m} and K_{i} properties of substrate and inhibitor, respectively. At the National Institutes of Health (NIH), Gelboin and Krausz have isolated and characterized specific CYP inhibitory monoclonal antibodies for the major human DMEs and shown ascites and purified MAbs to selectively inhibit up to 80–90% of human liver microsomal activity (Gelboin and Krausz, 2006). They even identified an MAb selective for CYP2C9*2 allele that does not significantly inhibit the CYP2C9*1 wild-type form, despite just one amino acid difference in their sequence (this level of specificity should be taken into account, especially when using liver microsomes from individual donors). MAbs are potentially most valuable when the available chemical inhibitors lack the desired selectivity; with CYP2A6 and 2B6, especially, inhibitory MAbs may be the only way to provide a reasonably definitive assessment of their contribution to total clearance.

16.6.2.3 Correlation Analysis A more labor-intensive but occasionally useful approach to reaction phenotyping is to measure the rate of metabolism and/or intrinsic clearance in a panel of HLMs prepared from individual donors (usually a minimum of $N \geq 12$ is recommended). This exploits the relationship between rate of metabolism and total amount of active enzyme present in the liver microsomes (see Section 16.8). The observed interindividual variability of the new drug is then correlated with the metabolism of known CYP-specific substrates using the same human liver donors. For metabolic clearance or specific metabolic pathways governed by a single CYP, a positive correlation should exist between the archetypal probe compound and the new drug. Benetton *et al.* (2007), Ghosal *et al.* (2007), Lee *et al.* (2007a, b, 2008), Obach, Margolis and Logman (2007), and Yoon *et al.* (2007) offer recent examples of correlation analyses coupled with other reaction-phenotyping methods to identify the major enzymes involved in the metabolic clearance of selegiline, vicroviroc, eupatilin, mirodenafil, CP-122,721, and benidipine, respectively. The substrates listed in Table 16.4 are commercially available and considered to have the most favorable specificity for a particular CYP; these are best suited for most correlation analyses.

A significant disadvantage to this method is the possibility of covariates and the difficulty in discerning multiple pathways of metabolic clearance, especially difficult if *in vitro* metabolism is relatively slow. As such, correlation analysis is not used as a sole means of identifying the reaction phenotype but as another tool to help identify those CYPs without ideal selective inhibitors. It can also help to illustrate quite clearly the potential for interindividual variability in metabolic clearance and possible effect on safety margins for a drug.

16.6.2.4 *Recombinant Enzymes*   Very selective tools are made for drug metabolism studies by transfecting a human cDNA for a particular CYP into a vector and its transfer into a rapidly growing and easily cultured cell line, such as lymphoblastoid cells, *Escherichia coli*, or insect cells. The culture is then grown to produce a batch of cells from which microsomes are made that possess enzyme activity enriched with a single human CYP or other enzyme (CYP, FMO, and UGT recombinant enzymes are commercially available). To secure functional CYP activity, these systems require co-expression with NADPH oxidoreductase and, for some substrates, co-expression also with cytochrome b5. A panel of recombinant CYP enzymes (e.g. CYP1A2, 2A6, 2B6, 2C8, 2C9, 2C19, 2E1, 3A4, and 3A5, or a subset thereof) is incubated with the test compound, using buffer and conditions similar to those described for liver microsomal assays. By measuring the turnover of compound in the individual CYP microsomes, the potential of each enzyme to metabolize the substrate can be quite easily determined and provides useful data to help predict its phenotype in humans. Recombinant and reconstituted enzyme systems are especially useful in showing the potential role in drug metabolism for a known polymorphic enzyme (e.g. CYP2C19, CYP2D6, and UGT1A1). This can help devise a strategy to evaluate and define any related issues. Enzyme kinetics and *in vitro* to *in vivo* extrapolation from liver microsomal or recombinant enzyme data are covered in a later section.

16.6.3   Optimizing FMO Activity in Human Liver Microsomes

Conditions that are optimal for liver microsomal CYP activity usually enable FMO reactions to take place (although the true optimum for the flavoproteins is at a higher non-physiological pH of 9–10). FMOs, especially in the absence of NADPH, are less thermally stable than CYPs and livers need to be handled carefully and microsomes prepared under controlled and proper conditions to maintain full activity. Without this, there is a risk of underestimating the contribution of FMOs to metabolic clearance. This thermal instability also provides a useful tool to aid in the reaction phenotyping of gross FMO activity in HLMs by substituting as a nonselective *in vitro* inhibitor (Cashman, 2005).

Reaction phenotyping of FMO activity tends to follow on from CYP studies, that is, if a significant contribution of NADPH-dependent metabolic clearance is not accounted for by CYPs, then evaluating a contribution from FMO-mediated clearance becomes an obvious next step. This is done by using a short (e.g. five minutes) preincubation of HLM at 45 °C in the absence and presence of saturating levels of NADPH. Without NADPH, the FMOs are thermally inactivated while CYP activity remains viable, while NADPH provides a protective effect and acts as a control (it is recommended that one also compare activity with the same HLM without the heat treatment, i.e. the CYP and FMO are both constitutively active). Another approach to distinguish FMO versus CYP involvement is to use 1-benzylimidazole as a nonselective CYP inhibitor (Ziegler, 1988; Grothusen et al., 1996).

If these experiments demonstrate FMO to be a major contributor to metabolic clearance in vitro, then the next step is to identify the specific enzyme form(s) involved. The reported characteristics for different FMOs allow one to take a rational approach in identifying the specific FMO with a process of elimination based on anecdotal evidence. Although FMO5 has the highest expression in the human liver, it tends to have much lower activity and much narrower substrate selectivity than FMO3. Despite possible exceptions in which FMO5 cannot be discounted, in most cases the activity measured in HLM will reflect primarily FMO3-mediated metabolism. As with CYPs, recombinant FMO enzymes are available that allow one to easily demonstrate functional activity and possibly selectivity. If turnover is seen with FMO3 and FMO5, then, if necessary, a correlation analysis can be performed using a panel of HLMs from a range of donors characterized for FMO3 and/or FMO5 activity. Since most drugs metabolized by human FMOs are substrates of FMO3, reaction phenotyping usually starts with substrates relatively selective for this form (used either to correlate with interindividual variability in rate of metabolism and/or as potential competitive inhibitors of the test drug). TMA, (S)-nicotine, cimetidine, clozapine, and ranitidine are typical substrates for FMO3 and lend themselves to such correlation analyses (TMA: Lang et al., 1998; (S)-nicotine: Cashman et al., 1992; cimetidine: Cashman et al., 1993; clozapine: Fang, 2000; ranitidine: Kang et al., 2000). As mentioned previously, covariates are always a risk when it comes to correlation analyses and the potential contribution from other FMOs, and also from CYPs, should be considered before concluding a major role for FMO in metabolic clearance.

Substrates and metabolic reactions that appear to be relatively selective for the individual FMOs are shown in Table 16.6. Reaction phenotyping is made difficult by a paucity of truly selective probe compounds and an even worse dearth of FMO form-selective inhibitors. One alternative approach is to perform a correlation analysis of activity using tissues known to have a relatively greater expression of a particular FMO (Table 16.7). For instance, kidney microsomes can be used as an enriched source of FMO1 and FMO2, lung microsomes from a panel of African-Americans with the 1414C allele provide a source of active FMO2, and microsomes from the small intestine are enriched with FMO5 activity (Cashman and Zhang, 2006). Obtaining these tissues

TABLE 16.6   Known substrates for the major human flavin-containing monooxygenases (FMOs).

FMO	Substrates
FMO1	Xanomeline, S-oxidation of the S-methyl metabolite of disulfiram[a]
FMO2	Thioureas, thioethers
FMO3	Trimethylamine, N'-oxidation of (S)-nicotine, clozapine-N-oxidation, cimetidine-S-oxide formation, ranitidine-N-oxidation
FMO4	Unstable form with no selective substrate identified
FMO5	Thioethers with proximal carboxylic acid[b]

[a]Cashman and Zhang (2006).
[b]Ohmi et al. (2003).
References contained within Cashman (2000)—unless indicated otherwise.

TABLE 16.7   Relative expression of flavin-containing monooxygenases (FMOs) in various adult human tissues.

Adult tissue	Copies per ng RNA				
	FMO1	FMO2	FMO3	FMO4	FMO5
Liver	96	987	23,087	4,882	26,540
Kidney	6,198	4,683	531	2,510	1,628
Small Intestine	523	929	74	403	2,586
Lung	596	115,896	2,224	738	2,274
Brain	3.1	141	11	20	57

Data normalized to HPRT (hypoxanthine phosphoribosyl transferase) and expressed as copies per ng RNA (adapted from Cashman and Zhang, 2006).

is very difficult and expensive and care must be taken when handling the tissues to conserve the FMO activity. Except in the most unusual circumstances, it is unlikely that such reductive measures will be taken in drug discovery.

16.6.4   Optimizing UGT Activity in Human Liver Microsomes

The *in vitro* to *in vivo* extrapolation of UGT activity is notoriously difficult to do and optimizing *in vitro* liver microsomal conditions for metabolism is normally done on a compound-specific basis. This does not lend itself readily to quantitative screening, at least not beyond a "yes" or "no" conclusion about whether or not a compound is a substrate for UGT enzymes. Usually the bias of UGT phenotyping is to identify possible substrates that may act competitively with others and result in DDIs (UGT1A1 and UGT2B7 being the main concern). With this class of enzymes, often it is necessary to correlate *in vitro* studies in liver microsomes or hepatocytes with *in vivo* data, and use that to decide on its value insofar as discriminating compounds for the structure-activity relationship (SAR) and oral exposure. With UGTs expressed on the luminal side of the endoplasmic reticulum, liver microsomes need to be "activated" to obtain optimal activity. The pore-forming agent alamethicin has become commonly used for this purpose (Fulceri *et al.*, 1994; Fisher *et al.*, 2000). Other methods include using detergents like Brij58 or mechanical techniques like sonication to disrupt the membranes and enable the substrates to access the UGTs (Vanstapel and Blanckaert, 1988; Shepherd *et al.*, 1989; Coughtrie *et al.*, 1991; Visser *et al.*, 1993; Soars *et al.*, 2001; Soars, Ring and Wrighton, 2003). UGTs utilize UDPGA as a requisite cofactor, and usually this is supplemented *in vitro* at high concentrations (10 mM). With UDPGA and NADPH both quite soluble in aqueous buffers, they can be added together allowing both UGT and CYP reactions to be measured simultaneously.

UGT-mediated clearance and metabolism are best predicted using hepatocytes rather than liver microsomes (Miners *et al.*, 2006). The advantage to hepatocytes is that they possess the full complement of native liver DMEs, which includes CYPs and UGTs. They also provide a more "natural" environment than liver microsomes, and appear not to have a significant issue with the release of UGT inhibitory unsaturated fatty acids that can affect the liver microsomal metabolism of lamotrigine, zidovudine, and 4-methylumbelliferone (Rowland *et al.*, 2006, 2007; Uchaipichat *et al.*, 2006b). Hepatocytes do suffer the disadvantage of being harder to obtain and use than liver microsomes, with cell viability and yield important additional factors to consider. The use of cryopreserved hepatocytes rather than freshly isolated cells helps to circumvent some of these problems.

There are very few UGT-selective inhibitors identified to date, nor are there many UGT isozyme-specific probes available for correlation analysis. Therefore, reaction phenotyping of UGTs is presently much harder to do than for CYPs. Probably the most useful tool available is the recombinant UGT enzymes. Unfortunately, the UGTs are not very well characterized with respect to their absolute abundance in the liver or other tissues, and scaling from recombinant or other *in vitro* data to human *in vivo* metabolic clearance is not yet possible. Correlation analysis with substrates and reactions that are relatively selective is probably the next best approach to take in reaction phenotyping a new drug. Kaji and Kume (2005), Tachibana *et al.* (2005), Yamanaka *et al.* (2005), Chen *et al.* (2007), Katoh *et al.* (2007), Mano, Usui and Kamimura (2007a), and Yu *et al.* (2007) provide a few examples of correlation analyses as used in the reaction phenotyping of afloqualone, trans-3′-hydroxycotinine, fluoroquinolones (levofloxacin, grepafloxacin, moxifloxacin, and sitafloxacin), racemic flurbiprofen, tranilast, and mitiglinide, respectively.

Table 16.3 lists the major human UGTs and substrates with reactions known to be mediated by those enzymes and their known expression in several major tissues. Correlation analysis can be done with these substrates; however, since many are metabolized by multiple UGTs and other enzymes (e.g. CYPs), several such studies may be needed using several different reference compounds. Although absolute human levels of UGTs are not yet available, there are tissue-specific differences in the relative expression of UGTs that may help to pinpoint a role for certain UGTs in the conjugative clearance of a new drug (e.g. UGT1A8 in small intestine and UGT1A9 in kidney; Table 16.3).

The current lack of good *in vitro* tools, coupled with the complex and compound-specific kinetics that are often encountered with glucuronidation reactions, makes it difficult to identify with confidence the contribution of a single UGT to the total clearance of a drug. At this stage, often the goal in drug discovery is to (i) identify any potential risk of UGT1A1 inhibition and issues associated with hyperbilirubinemia and (ii) select a compound that is metabolized by multiple isoforms (UGT and CYP). More work is needed to support quantitative UGT reaction phenotyping, and one should proceed cautiously if developing a UGT substrate with a relatively narrow therapeutic index. This is true also if a potentially reactive acyl glucuronide metabolite is formed, for example, the acylglucuronide metabolite of diclofenac and its putative role in immunotoxicity (Ebner *et al.*, 1999; Grillo *et al.*, 2003a; Naisbitt *et al.*, 2007).

16.7 *IN VITRO* TO *IN VIVO* EXTRAPOLATION OF METABOLIC CLEARANCE

16.7.1 *In Vitro* Intrinsic Clearance

By characterizing a drug's properties *in vitro* and *in vivo* using animal systems and available *in vitro* human systems, it becomes possible to predict *a priori* the *in vivo* properties of a drug in humans. Clearance (CL) and volume of distribution (Vd) are key properties in deciding the exposure and duration of a drug *in vivo* with half-life derived from both ($t_{1/2}$ = Vd/CL). Therefore, in addition to identifying the enzymes involved in the metabolism of a drug, one also needs to predict the total metabolic clearance and contribution of individual metabolic clearance reactions to the overall clearance. This is done simply using a rate or metabolic half-life of metabolism to calculate the intrinsic clearance of a drug that in turn is scaled to predict its *in vivo* clearance in the animal or human. In general, if the *in vitro* assays correctly predict the *in vivo* clearance measured in animal models, then there is some confidence that the same will hold true for the prediction in humans.

Simple liver microsomal assays are very useful for differentiating compounds with metabolic clearance issues. As discussed previously, metabolic stability screens are often used as a first tier assay, with a single fixed protein concentration used and loss of test compound measured at a single time point relative to the starting concentration (e.g. $C_{10\ minutes}$ relative to C_0). These assays differentiate compounds with rapid turnover from those with no measurable turnover and thus easily identify those compounds with metabolic stability issues. However, in reaction phenotyping and *in vitro* to *in vivo* extrapolation, intrinsic clearance is a more useful measure and accuracy of prediction becomes extremely important. Intrinsic clearance or CL_{int} can be defined as the proportionality constant between the concentration of substrate at the enzyme (C_E) and its rate of metabolism in the absence of physiological constraints:

$$Rate(v) = CL_{int} \times C_E$$

For a single substrate-binding reaction, the chemical reaction can be described as:

$$\mathbf{E + S} \underset{k_{-1}}{\overset{k_1}{\rightleftharpoons}} \mathbf{ES} \leftrightarrow \mathbf{EP} \underset{k_{-2}}{\overset{k_2}{\rightleftharpoons}} \mathbf{P + E}$$

Further simplification is possible if this reaction demonstrates Briggs–Haldane kinetics: (i) a quick rate of substrate, S, binding to the enzyme, E; (ii) a relatively slow reaction step in the formation of EP; (iii) a rapid rate of release of product, P (the latter defined by k_2); and (iv) the reaction is irreversible. These are reasonable assumptions for most reactions and allow the chemical reaction scheme to be reduced to a pseudo-first-order reaction:

$$\mathbf{E + S} \underset{k_{-1}}{\overset{k_1}{\rightleftharpoons}} \mathbf{ES} \overset{k_2}{\longrightarrow} \mathbf{P + E}$$

where $k_2 = k_{cat}$, catalytic rate constant or turnover number.

The enzyme kinetics involved in a typical bimolecular reaction can be daunting but fortunately can also be greatly simplified—subject to various assumptions and the correct use of appropriate assay conditions—while still providing a reasonably close approximation. For most CYP-mediated reactions, it is assumed that the substrate and enzyme are in equilibrium and in quasi-steady-state conditions, with the enzyme-substrate binding complex (ES) forming and dissociating at an equal rate (i.e. d[ES]/dt ~ 0), which is not true initially but becomes a valid assumption upon mixing of substrate and enzyme.

When these assumptions are true, the kinetics for the reaction can be simply described using the classical Michaelis–Menten equation:

$$v = \frac{V_{max} * [S]}{K_m + [S]}$$

where v = rate or −dS/dt; V_{max} = the maximal rate of reaction; [S] = substrate concentration and K_m = Michaelis constant, the concentration at which the rate of reaction is half of its maximum.

If a substrate is metabolized by multiple pathways, then an additive approach is used to calculate the kinetics of metabolism:

$$v = \frac{V_{max1}*[S]}{K_{m1}+[S]} + \frac{V_{max2}*[S]}{K_{m2}+[S]} + \frac{V_{maxn}*[S]}{K_{mn}+[S]}$$

Michaelis–Menten kinetics are usually determined by nonlinear regression analysis of substrate concentration ([S]) versus the rate of reaction (v), which in the classical case results in a hyperbolic curve (Fig. 16.3a). The data can also be transformed to make linear plots. The Eadie–Hoftsee transformation plots v/[S] against v, the slope of which provides the K_m and the intercept on the y-axis provides V_{max}. This transformation is especially useful in identifying multiple pathways in the metabolism of a compound. Figure 16.3b illustrates the Eadie–Hofstee transformation of a hypothetical drug that is metabolized by a single enzyme (or multiple enzymes with identical properties), with a straight-line plot generated. Figure 16.3c illustrates an Eadie–Hofstee transformation of a hypothetical drug that is now metabolized by two distinct enzymes, with a biphasic profile generated as a result of the low- and high-affinity enzymes.

More complicated kinetic models account for homo- and hetero-cooperativity, where multiple molecules of the same or different substrate, respectively, bind to the same molecule or complex of enzyme and modulate the catalytic function of the enzyme. This can result in sigmoidal kinetics, often represented in a simplified form using the Hill equation:

$$v = \frac{V_{max}*[S]^{\gamma}}{S_{50}^{\gamma}+[S]^{\gamma}}$$

where γ is the Hill coefficient: if $\gamma = 1$, then the reaction obeys classical Michaelis–Menten kinetics.

These equations all possess the mathematical form that represents the hyperbolic nature of CYP-mediated reactions. This means that the rate of metabolism or turnover increases as the substrate concentration increases, with the rate eventually limited by the amount of enzyme present and its innate catalytic capacity (known as the enzyme's k_{cat} or turnover number). This leads to an observed V_{max} in the reaction ($V_{max} = k_{cat}*$amount of active enzyme present, Ai). By classical Michaelis–Menten kinetics, the V_{max} occurs at twice the K_m of that drug for that enzyme, and when [S] $\ll$ K_m, v/[S] = V_{max}/K_m, which is the apparent intrinsic clearance. Therefore, by measuring the apparent *in vitro* V_{max} and K_m for each metabolic reaction involved, one obtains a definitive kinetic analysis for the drug—in addition to being able to calculate the intrinsic clearance, which helps to predict the overall metabolic clearance, one can use the V_{max} and K_m to predict nonlinearity in the drug's pharmacokinetics and predict the effect of enzyme induction or inhibition on drug clearance.

The typical and most definitive method used to measure K_m and V_{max} is to incubate a range of substrate concentrates and measure the rate of metabolism at each concentration under linear kinetic conditions. This means that substrate depletion is minimized during the course of incubation, with total turnover ideally <8% (practically <20% is often used because of the constraints imposed by analytical methods). In practice, linear conditions are identified for enzyme content and time for the lowest concentration of substrate and then used to measure rate of metabolism of a range of

(a)

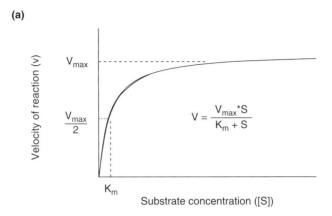

(b)

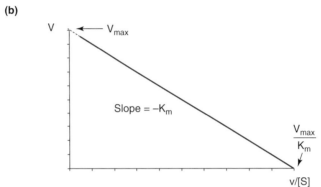

(c)

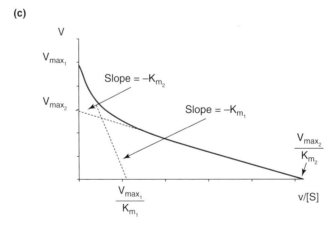

Figure 16.3   Michaelis–Menten kinetics and Eadie–Hofstee transformations: **(a)** Michaelis–Menten kinetics—single enzyme; **(b)** Eadie–Hofstee transformation—single-enzyme reaction; **(c)** Eadie–Hofstee transformation—two-enzyme reaction; v = reaction rate or velocity; V_{max} = maximal rate of reaction; [S] = substrate concentration; K_m = Michaelis constant.

substrate concentrations (e.g. 1 to 1000 µM, subject to the sensitivity limits of the assay). A typical profile is shown in Fig. 16.3a. The K_m and V_{max} are calculated using nonlinear regression analysis of the concentration versus rate data (various commercial software are available for this purpose).

Measuring the rate of formation of metabolites requires the use of either radiolabeled substrate, with resolution achieved by liquid chromatography and/or by quantification relative to calibration curves prepared from reference standards. Radiolabel and metabolite standards are not always available, especially in the early drug discovery stage. In which case, substrate-depletion kinetics are normally used to estimate intrinsic clearance. This is achieved easily enough by measuring the half-life of substrate turnover in the liver microsomal incubation and applying some further simplification to the equations involved:

$$V = \frac{-d[S]}{dt} = \frac{V_{max}*[S]}{K_m + [S]}$$

This equation transposes to:

$$V_{max}*dt = \frac{-d[S]*(K_m + [S])}{[S]}$$

Over one half-life ($t_{1/2}$), $[S] = 0.5*[S]_0$ and the equation can be reduced to:

$$\frac{V_{max}*t_{1/2}}{K_m} = Ln2 + \frac{0.5*[S]_0}{K_m}$$

For this to be valid, the $[S]_0 \ll K_m$, in which case:

$$\frac{0.5*[S]_0}{K_m} \ll Ln2$$

Usually a low concentration of substrate is used (typically ~1 µM) and assumed to be below the K_m with the experimental data inspected to confirm the assumption. If valid, the equation reduces further to:

$$\frac{V_{max}*t_{1/2}}{K_m} = Ln2$$

which is easily rearranged to provide an apparent intrinsic clearance (CL'_{int}):

$$\frac{V_{max}}{K_m} = \frac{Ln2}{t_{1/2}} = CL'_{int}$$

where $t_{1/2}$ for substrate depletion is obtained by linear regression of the log percentage $[S]_0$ remaining v time to obtain a value for the slope (k, or elimination rate constant) (Fig. 16.4), since:

$$t_{1/2} = -Ln2/k$$

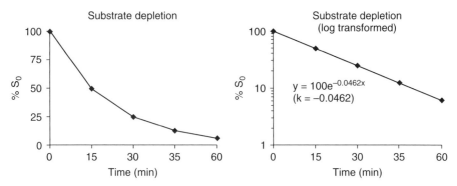

Figure 16.4 Illustration of a substrate-depletion plot for $t_{1/2}$ measurement and calculation of intrinsic clearance. S_0 = initial substrate concentration.

To properly balance the units the full equation assumed is:

$$CL'_{int} = \frac{Ln2}{t_{1/2}(min)} * \frac{volume\ incubation\ (mL)}{microsomal\ protein\ (mg)}$$

Units for CL'_{int} are usually expressed as mL/min/mg microsomal protein.

The substrate-depletion approach is convenient and useful in discovery, yet provides only a gross measure of metabolic clearance and represents the summed processes of all enzymes involved at the substrate concentration used. The advantage to Michaelis–Menten kinetics is that the role and potential for saturation of individual metabolic pathways can be determined and predictions made across a range of relevant drug concentrations and scenarios.

16.7.2 Scaling *In Vitro* Intrinsic Clearance to Predicted *In Vivo* Intrinsic Clearance

Liver microsomal data are scaled by applying an empirical factor to adjust CYP activity in liver microsomes to an equivalent of gram liver weight. Houston, at the University of Manchester, UK, reported that a numeric factor of "×45" gave a product of liver microsomal activity that improved the prediction of *in vivo* metabolic clearance, that is, 45 mg of liver microsomal protein contains the equivalent functional CYP activity of 1 g of liver (also referred to as mg microsomal protein per gram liver or MPPGL) (Houston, 1994). A similar concept is applied to hepatocytes and termed "hepatocellularity," which assumes approximately 3 million cells as equivalent to 1 mg of liver microsomal protein or approximately 120 million cells is equivalent in functional CYP activity to 1 g of liver. Recently, Barter *et al.* performed a meta-analysis that showed 95% confidence intervals for the geometric mean of 27–32 MPPGL and 74–131 million cells per g liver for hepatocellularity (Barter *et al.*, 2007).

One more factor to consider with respect to scaling of CL'_{int} is liver weight. Values for liver weight are available for most species including mouse, rat, monkey, dog, and human (2, 10, 160, 320, and 1500 g, respectively) (Davies and Morris, 1993). Clearance is often represented per kg of body weight rather than per animal (i.e. per liver) and for conversion, the generalized body weights assumed for mouse, rat, monkey, dog, and human are usually 0.025, 0.25, 5, 10, and 70 kg, respectively (Davies and Morris, 1993).

Putting all of this together, if Michaelis–Menten kinetics are measured for each metabolic pathway, one can scale *in vitro* CL_{int} to an *in vivo* hepatic intrinsic clearance ($CL_{H,int}$) using:

$$CL_{H,int} = \frac{V_{max}}{K_m} * 45\,(MPPGL) * \frac{1500\,g\ \text{human liver}}{70\,kg\ \text{human body weight}}$$

If a substrate-depletion approach is taken, the apparent intrinsic clearance from the sum of all metabolic processes active at the substrate concentration tested will scale using:

$$CL_{H,int} = \frac{Ln2}{t_{1/2}(min)} * \frac{\text{volume incubation (mL)}}{\text{microsomal protein (mg)}} * 45\,(MPPGL) * \frac{1500\,g\ \text{human liver}}{70\,kg\ \text{human body weight}}$$

Liver weight and body weight are substituted accordingly to similarly scale animal liver microsomal metabolic clearance data and correlate with the measured *in vivo* clearance. (If a positive correlation is not observed between *in vitro* and *in vivo* animal data, then additional mechanisms may be involved, e.g. non-CYP, non-microsomal, and/or extrahepatic metabolism.)

16.8 *IN VITRO* TO *IN VIVO* EXTRAPOLATION AND PHYSIOLOGICAL MODELS OF HEPATIC CLEARANCE

Intrinsic clearance represents the capacity of a particular enzyme to metabolize a specific substrate in the absence of physiological limitations. The amount of circulating drug presented to the liver enzymes and cleared from the blood depends on the rate of hepatic blood flow (Q_H). There are several models of hepatic blood flow available, with each assuming that hepatic clearance (CL_H) is perfusion rate-limited with only unbound drug entering the hepatocyte and available for metabolism and/or excretion (Wilkinson and Shand, 1975).

Possibly the simplest and most commonly used approach is the venous equilibrium or "well-stirred" model. This model assumes a rapidly distributed and uniform concentration of drug in the whole liver (i.e. no concentration gradient across the liver or cells):

$$CL_H = \frac{Q_H * (f_{ub} * CL_{H,int})}{Q_H + (f_{ub} * CL_{H,int})}$$

where f_{ub} = fraction of drug unbound in blood. This can be reduced further to provide bioavailability (F_H):

$$F_H = \frac{Q_H}{Q_H + (f_{ub} * CL_{H,int})}$$

Other models that can be used include the sinusoidal or "parallel tube" model that assumes a concentration gradient across the liver:

$$CL_H = Q_H - Q_H * \exp(-f_{ub} * CL_{H,int}/Q_H)$$

There is also a dispersion blood flow model that accommodates a radial distribution of drug from blood through the liver:

$$CL_H = Q_H * \frac{1-4a}{(1+a)^2 * \exp[(a-1)/2Dn] - (1-a)^2 * \exp[-(a+1)/2Dn]}$$

where a = $(1 = 4Rn*Dn)^{1/2}$; Dn = 0.17 (dispersion number); Rn = $f_{ub}*CL_{H,int}/Q_H$ (efficiency number).

High and low hepatic extraction drugs show markedly different characteristics with respect to hepatic blood flow. Using the well-stirred model as an example and assuming no significant protein binding (i.e. f_{ub} = 1):

$$CL_H = \frac{Q_H * CL_{H,int}}{Q_H + CL_{H,int}}$$

With a high-clearance drug, the *in vitro* clearance greatly exceeds the flow of blood through the liver (i.e. $CL_{H,int} \gg Q_H$), in which case the equation reduces to:

$$CL_H = Q_H$$

Consequently high-clearance drugs are hepatic blood flow limited, which means that increased expression of the enzymes involved in metabolism and/or an increase in the free fraction from protein displacement will alter pharmacokinetics and pharmacodynamics less than changes in hemodynamics such as blood flow.

In contrast, if the hepatic blood flow greatly exceeds the metabolic clearance (i.e. $Q_H \gg CL_{int}$), then the equation reduces to:

$$CL_H = CL_{H,int}$$

CYP inducers, for instance, can profoundly increase intrinsic clearance and thus increase hepatic clearance and the dose required to achieve exposure for efficacy—if induction is sufficient to turn clearance from low to high, then hepatic blood flow will become the rate-limiting factor. Likewise, coadministered CYP inhibitors can alter the phenotype of a drug effectively converting high clearance to a low clearance, thus profoundly affecting the drug's AUC and effects *in vivo*. Inducers and inhibitors can thus change the phenotype in a patient resulting in a "phenocopy" that mimics a rapid metabolizer or slow metabolizer, respectively.

To predict the effect of metabolic clearance on oral bioavailability:

$$F = Fa * (1 - CL_{non\text{-}renal}/Q_H)$$

where F = bioavailability; Fa = fraction of dose absorbed; $CL_{non\text{-}renal}$ = any pathway not mediated by renal clearance, for example, metabolic, biliary, or chemical instability.

If metabolic clearance governs elimination, then this equation becomes:

$$F = Fa * (1 - CL_H/Q_H)$$

For a more detailed review of these methods, enzyme kinetics and the assumptions made, and examples of how they can be applied to the prediction of *in vivo* clearance, see Segel (1975), Wilkinson and Shand (1975), Houston (1994), Obach *et al.* (1997), and Obach (1999).

16.8.1 cDNA-Expressed Human Recombinant Enzymes

cDNA-transfected recombinant enzyme sources are another useful tool for drug reaction phenotyping. These are made by transfecting a lymphoblastoid, bacterial, or insect cell source with negligible basal CYP or other DME activity, with a vector containing cDNA for a particular human enzyme (recombinant CYPs, FMOs, and UGTs are all commercially available). The cDNA for human cytochrome P450 oxidoreductase (CYPOR) is also included as an essential redox partner to support CYP activity, with or without the cDNA for human cytochrome b5 (cyt b5). The transfected cells are grown in culture and used to prepare microsomes enriched with the specific recombinant enzyme, and used to investigate the role of a specific enzyme in the potential clearance and biotransformation of a drug.

The kinetics associated with CYP-mediated reactions are quite complex and not all substrates necessarily behave the same way. CYPOR and cyt b5 are accessory proteins and electron donors involved in the CYP catalytic cycle and should be included when using artificial constructs such as recombinant or reconstituted enzyme systems. CYPOR is essential for all CYP-mediated biotransformation reactions (see Fig. 16.1). Cyt b5 in contrast is more substrate-specific in its role and although usually involved in stimulating CYP-mediated reactions (Venkatakrishnan *et al.*, 2000), cyt b5 can also inhibit CYP-catalyzed reactions (Gorsky and Coon, 1986). It was also shown that apo-protein of cyt b5 was as effective as holoprotein in stimulating the reactions of several CYPs, with a role postulated for divalent cations in influencing the ability of substrates and cyt b5 to bind with CYP (Yamazaki *et al.*, 1995, 1996). Occasionally for these reasons (e.g. optimization of metabolic stability), assays will benefit from further evaluation of buffers, salts, and effect of redox partner concentration on drug biotransformation and turnover, and whether this improves the accuracy of prediction of *in vivo* clearance from *in vitro* data.

The same kinetic analyses that are done with liver microsomes can also be done with recombinant enzymes, for example, K_m and V_{max} determinations performed and/or a substrate-depletion kinetic approach as discussed previously. Theoretically, the recombinant enzyme should have the same K_m as that of the native liver microsomal form, assuming no significant differences in protein binding and free fraction of test substrate. However, to use recombinant enzyme systems for *in vitro* to *in vivo* extrapolation of metabolic clearance, one needs to account for the relative expression of those enzymes in the human liver and other tissues of concern. For reference, Table 16.8 lists the range and mean values reported for most of the major CYPs measured in microsomes prepared from human liver, small intestine, and skin. There is a significant degree of variability evident in the values reported by the laboratories referenced, which presumably reflects the variability in the donors themselves, as well as differences in the methods used to prepare tissue microsomes and quantify the expression of holoprotein.

TABLE 16.8 Immunoquantified expression of different cytochrome P450 (CYP) forms in the human liver, intestine, and skin microsomal fractions.

CYP	Liver[a]				Intestine[b]		Skin[c]	
	Range (pmol/mg-protein)		Range R.A. (%)		Range (pmol/mg-protein)	Mean R.A. (%)	Mean (pmol/mg-protein)	R.A. (%)
	Min	Max	Min	Max				
CYP1A1	19	67	7.5	13	3.6–7.7	7.4	3.6	16
CYP1A2	14	68	5.5	13	ND	ND		
CYP2A6					BLD	BLD	2.0	9
CYP1B1							0.035	0.2
CYP2B6	1	45	0.4	8.4	BLD	BLD		
CYP2C8	12	64	4.5	12	ND	ND		
CYP2C9	50	96	20	18	2.9–27	11		
CYP2C19	8	20	3.1	3.7	<0.6–3.9	1.3		
Total CYP2C	**60**	**64**	**24**	**12**				
CYP2D6	5	11	2	2.1	<0.2–3.1	0.66		
CYP2E1	22	52	8.6	9.8	BLD	BLD	11	49
CYP2J2					<0.2–3.1	1.2		
CYP3A4	37	108	15	20	8.8–150	57 (variable)		
CYP3A5	1	117	0.4	22	4.9–25	21 (variable)	5.6	25
Total CYP3A	**96**	**262**	**38**	**49**	**Highly variable**			
Total pmol CYP/mg	**255**	**534**						

[a] The ranges shown are based on the mean values reported, based on data from Shimada et al. (1994), Rodrigues (1999), Snawder and Lipscomb (2000), Venkatakrishnan et al. (2000), Rowland-Yeo et al. (2003), and Paine et al. (2006).

[b] Paine et al. (2006).

[c] Bergström et al. (2007).

R.A. = relative abundance; BLD = below limits of detection; ND = not detected.

These values can be used to scale recombinant enzyme activity using the methods described as follows. The recombinant enzymes are incubated at a known concentration of CYP which provides a rate of reaction (typically pmoles metabolized/min/pmol CYPi). Michaelis–Menten kinetics can be measured for each individual enzyme (CYPi) and biotransformation pathway, and summed to give a total rate of metabolism at a specific concentration of substrate:

$$V_{total} = \sum_{i=1} \frac{V_{maxi} * S}{K_{mi} + S}$$

Using rates or apparent intrinsic clearance, the nominal absolute expression per mg of human microsomal protein can be factored to provide a microsomal equivalent that, if all the CYPs are functionally active and expressed at levels corresponding to those published (Table 16.8), will sum to give a total intrinsic clearance that is theoretically equivalent to liver microsomal intrinsic clearance:

$$V_{HLM} = \sum_{i=1}^{n} A_i * V_{rCYPi}$$

where V_{HLM} = predicted rate of metabolism in HLM; V_{rCYPi} = rate of metabolism in CYPi; A_i = abundance of CYPi.

16.8.1.1 *Relative Abundance or Total Normalized Rate (TNR)*   The simplest approach to scaling rate of metabolism in recombinant human enzymes to an *in vivo* equivalent is to assume that the rate of metabolism by each molecule of recombinant CYP incubated can be scaled by its expression in a human liver (as listed in Table 16.8). This relative abundance (Nakajima *et al.*, 1999; Soars *et al.*, 2003) or total normalized rate (TNR) (Rodrigues, 1999) assumes in effect that the recombinant enzyme is identical in activity to the native enzyme. If correct, the relative contribution of a specific enzyme to the overall metabolism in HLMs can be calculated quite easily:

$$\% \text{ Contribution CYPi}_{HLM} = \frac{Ai * V_{rCYPi}}{V_{HLM}} * 100$$

This can also be expressed as the fraction metabolized by a specific CYP ($f_{m.CYP}$):

$$f_{m.CYP} = \frac{Ai * V_{rCYPi}}{V_{HLM}}$$

which assumes that all of the metabolism is governed by CYPs expressed in the human liver microsomes.

16.8.1.2 *Relative Activity Factors (RAFs)*   In reality, the k_{cat} of the recombinant enzyme can differ markedly from that of the native enzyme form. This was recognized early on in the development and use of recombinant enzymes for drug metabolism reactions and a relative activity factor (RAF) was proposed to make allowance for this disconnect (Crespi, 1995).

The RAF is determined for each individual enzyme by measuring the kinetics (e.g. V_{max}, intrinsic clearance) of a prototypic substrate, that is, one known to be selective for the enzyme of interest. The substrate, for example midazolam, is incubated with HLM and, in this case, also with the recombinant human CYP3A4 (rCYP3A4). Under equivalent conditions for CYP content and unbound concentrations of substrate, and assuming no difference in substrate affinity (e.g. K_m), rCYP3A4 should demonstrate the same substrate-dependent rate of metabolism as CYP3A4 in HLMs. If there is a difference between them in relative activity, it is presumed to result from a difference between the recombinant and native forms in the turnover number, necessitating a correction factor for relative activity (the "RAF"):

$$RAF_{CYPi} = \frac{\text{Rate of index reaction CYPi in HLM}}{\text{Rate of index reaction in rCYPi}}$$

This assumes the K_m of midazolam for rCYP3A4 is the same as liver CYP3A4, and that no correction for free fraction is needed.

It is useful when using a specific batch of recombinant enzymes and liver microsomes for reaction phenotyping to assume that any difference in the turnover kinetics of a prototypic substrate will apply equally to all other substrates of that CYP. Consequently, one need only measure the kinetics of a suitable reference compound once per batch of CYPi to obtain a RAF value that can be applied to scale and extrapolate the metabolic clearance of any other drug or compound that is incubated in the same batch of rCYPi.

When the relative abundances of the CYPs involved are applied, the sum of individual reactions in recombinant enzymes should equal that measured in HLMs:

$$V_{HLM} = \sum_{i=1}^{n} Ai * (V_{CYPi} * RAF_{CYPi})$$

When substrates are relatively stable (i.e. showing slow rates of turnover), Michaelis–Menten kinetics of the rate of formation of metabolite(s) is usually the best and possibly only way to get an accurate prediction of clearance (as opposed to measuring the loss of parent substrate). However, if the compound is rapidly and extensively metabolized, then intrinsic clearance can be determined instead using substrate-depletion kinetics ($S < K_m$).

The RAF adjusted rate of reaction by a specific CYP in HLMs is calculated as follows:

$$V_{HLM.CYPi} = A_i * (V_{CYPi} * RAF_{CYPi})$$

The contribution of that CYP to the overall rate of metabolism in HLMs is then calculated:

$$\% \text{ Contribution of CYPi} = \frac{V_{HLM.CYPi}}{V_{HLM}} * 100$$

The relationship between relative abundance, RAF, and the fraction metabolized by a specific CYP can then be represented by:

$$f_{m.CYPi} = \frac{Ai * (V_{CYPi} * RAF_{CYPi})_i}{V_{HLM}}$$

In practicality, the relative abundance or TNR approach is very useful, especially in early drug discovery and in the preliminary stages of reaction phenotyping. For single high-clearance pathways, it is often all that is needed, often corroborated easily by CYP-selective inhibition studies. However, if multiple enzymes show significant turnover of test substrate, and when differentiation and accuracy of prediction assumes greater significance (e.g. a potential CYP polymorphic clearance and a relatively narrow safety margin), the more laborious RAF method becomes the preferred and definitive approach.

To illustrate how these approaches work, Fig. 16.5 shows four hypothetical scenarios that can be observed *in vitro* with recombinant CYPs. In the first example (Fig. 16.5a), with "compound A," all of the rCYPs show the same extent and rate of turnover. This compound will therefore scale by the relative abundances of those enzymes in human tissues (assuming no kinetic correction is needed for k_{cat} and/or K_m). Using the relative abundance of CYPs reported in the liver (Shimada *et al.*, 1994; Lasker *et al.*, 1998), the small intestine (Paine *et al.*, 2006), and skin (Bergström *et al.*, 2007), this hypothetical compound is predicted to have the reaction phenotype shown in the bar plots. In the liver, CYP3A is predicted to be the major CYP, with significant contributions to hepatic metabolism expected from 2C9, 1A2, and 2E1. Although CYP3A4 is again predicted to be the major contributor to metabolism in the intestine, the differential expression of CYPs in this tissue predicts a significant role for CYP3A5, 2C9 and 1A1, and liver and intestinal CYP DDIs should be limited in effect. In contrast, in the skin, CYP2E1 is predicted to be the major CYP, with contributions from CYP3A5, 1A1, and 1B1 to the metabolism of compound A in keratinocytes.

In the second example (Fig. 16.5b), which is the most commonly encountered scenario, especially for large lipophilic molecules, the greatest turnover of "compound B" was measured in CYP3A4. In this particular example, compound B is simulated to have 20% turnover in the CYPs. With compound B, CYP3A4 is predicted to dominate its metabolic clearance in liver and intestine; however, with negligible CYP3A4 expression in the keratinocyte, its reaction phenotype in the skin is predicted similar to that for compound A. A drug with the characteristics of compound B is likely to be a potential victim of CYP3A4 DDIs.

In the third example (Fig. 16.5c), "compound C" is most rapidly metabolized by the recombinant CYP2D6 enzyme. In this example, the other CYP enzymes again show significant turnover and the contribution of CYP2D6 depends on its relative expression. Since CYP2D6 protein is expressed in liver microsomes at only about 2% of the total CYP content, whereas CYP3A4 is more like 30% of the total, only a minor role for CYP2D6 is predicted in the metabolic clearance of this hypothetical drug.

In the fourth simulated example (Fig. 16.5d), recombinant human CYP1A1 most efficiently metabolized "compound D." CYP1A1 is not constitutively expressed in human liver; therefore, the hepatic clearance cannot be scaled accurately to predict *in vivo* clearance. CYP1A1 in the gut and skin is likely to have a major role in extrahepatic metabolism, which could prove significant, for instance, if reactive or toxic and/or active metabolite(s) are formed. With CYP1A1 known to be highly inducible in the liver, the lung and other tissues, and assuming that metabolic clearance accounts for most, if not all, of the total clearance, it is reasonable to expect considerable interindividual variability in a human population dosed with a compound possessing a phenotype such as this. If drug safety could be affected by such a pronounced interindividual variability, then additional studies may be warranted.

(a)

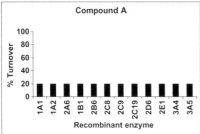

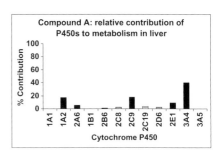

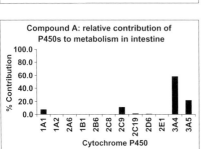

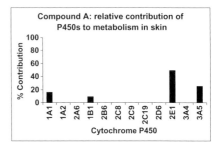

(b)

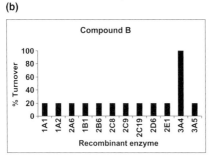

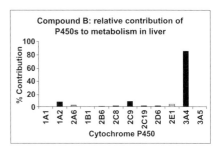

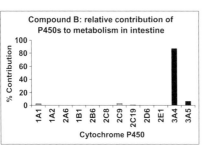

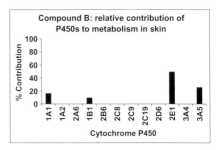

Figure 16.5 **(a)** Scaling of human recombinant cytochrome P450 (CYP) turnover using relative abundance in the liver, intestine and skin. Hypothetical representation: a theoretical compound that is metabolized to an equal extent in human recombinant enzymes. The relative expression of those enzymes varies between liver, intestine, and skin and affects the relative contribution of each enzyme to the total metabolism in that tissue. **(b)** Scaling of human recombinant CYP turnover using relative abundance in the liver, intestine and skin. Hypothetical representation: a theoretical compound that is metabolized rapidly and extensively in human recombinant CYP3A4 and to an equal extent in the other human recombinant enzymes. The relative expression of those enzymes varies between liver, intestine, and skin and affects the relative contribution of each enzyme to the total metabolism in that tissue.

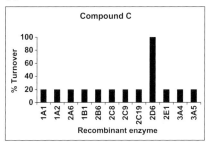

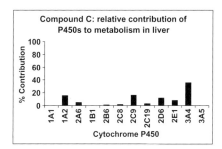

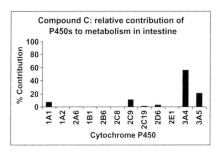

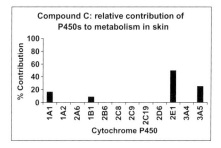

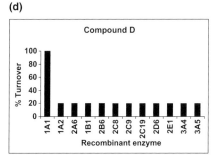

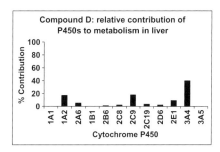

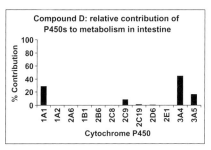

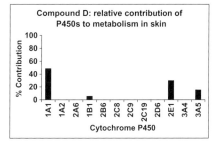

Figure 16.5 **(c)** Scaling of human recombinant CYP turnover using relative abundance in the liver, intestine, and skin. Hypothetical representation: a theoretical compound that is metabolized rapidly and extensively in human recombinant CYP2D6 and to an equal extent in the other human recombinant enzymes. The relative expression of those enzymes varies between the liver, intestine, and skin and affects the relative contribution of each enzyme to the total metabolism in that tissue. **(d)** Scaling of human recombinant CYP turnover using relative abundance in the liver, intestine, and skin. Hypothetical representation: a theoretical compound that is metabolized rapidly and extensively in human recombinant CYP1A1 and to an equal extent in the other human recombinant enzymes. The relative expression of those enzymes varies between the liver, intestine, and skin and affects the relative contribution of each enzyme to the total metabolism in that tissue.

The example of "compound D" also illustrates some issues that apply to DMEs—including hepatic, extrahepatic, CYP, and non-CYP enzymes (e.g. UGTs)—that are not as well defined in their relative expression and thus not so readily scaled to predict their role in the *in vivo* metabolism. It is clear that relatively minor enzymes can scale quite differently between different human tissues. For most orally administered drugs, this is unlikely to be relevant. However, if different routes of administration are part of the drug's life cycle management, then tissue-specific metabolism could well become a concern, especially if these enzymes mediate the formation of CYP-specific metabolites and/or demonstrate tissue-specific toxicity.

These hypothetical scenarios assume that the recombinant and native forms are essentially identical with the same intrinsic clearance *in vitro*. If that is not the case, a RAF is needed to correct the predicted rate. Figure 16.6 uses the example of compound C from Fig. 16.5c to illustrate the potential effect of RAFs on the predicted relative contribution of CYP2D6 to the total hepatic metabolic clearance. In Fig. 16.6a, compound C showed 20% turnover in each of the recombinant CYPs except for CYP2D6 where it is 100%; however, in this case, an index substrate in a separate theoretical incubation showed that the rate of metabolism of CYP2D6 in liver microsomes was much greater than that with an equivalent concentration of the same batch of recombinant CYP2D6. In this case, a TNR approach will be wrong and a correction factor must be used. In this example, a RAF of ×10 was applied and CYP2D6 is now predicted to be the major contributor to metabolic clearance in the liver. In this scenario, CYP2D6 polymorphisms could become a major issue in the development of this compound as a drug. In Fig. 16.6b, the same compound is used to illustrate the opposite scenario in which the index reaction for CYP2D6 showed hypothetically much higher rates of metabolism in the same batch of recombinant CYP2D6. A correction factor is now needed to avoid overestimating the contribution of this enzyme to the total metabolic clearance. Applying a RAF of 0.1 for the recombinant enzyme results in CYP2D6 predicted to have no significant role in the metabolic clearance of compound C.

These are relatively simple but realistic illustrations that serve to demonstrate the potential complexity involved in reaction phenotyping. If the kinetics are not performed or applied correctly, there is a real possibility in making a wrong prediction. Rather than rely on any single method, instead it is recommended that multiple assays be used for reaction phenotyping, and HLM CYP inhibition studies and/or correlation analysis should be used to corroborate recombinant CYP assays, or vice versa, or help identify any anomalies. These hypothetical examples also serve to illustrate that differences may exist locally in the metabolism of a drug, which could prove particularly significant if the route of administration is not oral (e.g. transdermal, intranasal, and inhalation). This can prove especially significant if metabolites are formed that are specific to the enzymes expressed locally to the site of administration.

In Table 16.9 are shown the results of an integrated reaction phenotype for etoricoxib, ABT-761, zileuton, naproxen, clarithromycin, and celecoxib (see Zhang *et al.*, 2007 and references therein). In each case, a specific reaction was monitored using a panel of HLM from at least 10 donors, HLM incubated with and without selective CYP inhibitors, and results from incubations with recombinant human CYPs scaled using a total normalized rate to determine a relative contribution of each to the total hepatic clearance. Knowing the relative fraction metabolized (f_m) and the contribution from each CYP ($f_{m,CYP}$) to the metabolism of that drug, one can predict the potential effect of a coadministered inhibitor on its relative exposure.

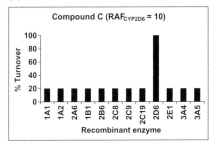

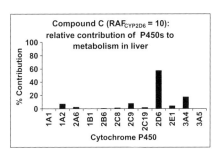

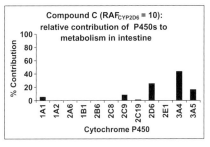

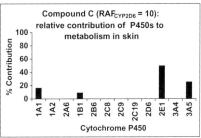

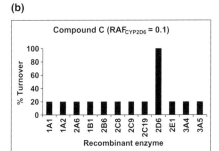

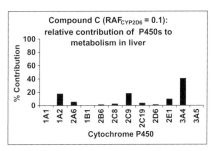

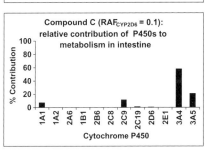

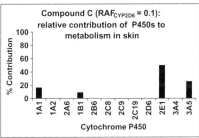

Figure 16.6 **(a)** Effect of RAFs on the scaling of human recombinant cytochrome P450 (CYP) turnover. Hypothetical representation: a theoretical compound that is metabolized rapidly and extensively in human recombinant CYP2D6 and to an equal extent in the other human recombinant enzymes. A RAF correction is needed since the relative activity of CYP2D6 was greater in the native form than in the recombinant form. The relative expression of those enzymes varies between the liver, intestine, and skin and affects the relative contribution of each enzyme to the total metabolism in that tissue. **(b)** Effect of RAFs on the scaling of human recombinant CYP turnover. Hypothetical representation: a theoretical compound that is metabolized rapidly and extensively in human recombinant CYP2D6 and to an equal extent in the other human recombinant enzymes. A RAF correction is needed since the relative activity of CYP2D6 was lower in the native form than in the recombinant form. The relative expression of those enzymes varies between the liver, intestine, and skin and affects the relative contribution of each enzyme to the total metabolism in that tissue.

TABLE 16.9 **Examples of an integrated *in vitro* cytochrome P450 (CYP) reaction phenotype: major CYP(s) shown for each reaction.**

Compound	Reaction	CYP form(s)	Human liver microsomes		% Total normalized rate (rCYPs)[c]
			Correlation coefficient $(r^2)^a$	% Inhibition[b]	
Etoricoxib	Hydroxylation	3A4	0.64	65	58
		1A2	<0.40	5.0	2.0
		2C9	<0.40	10	6.0
		2D6	<0.40	10	26
ABT-761	Hydroxylation	3A4	0.67	55	66
		2C9	0.40	20	19
Zileuton	Sulfoxidation	3A4	0.67	60	95
	Hydroxylation	1A2	0.58	45	57
Naproxen	O-demethylation	2C9	0.67	50	56
		1A2	0.46	55	33
Clarithromycin	N-demethylation	3A4	0.81	>95	100
	Hydroxylation	3A4	0.85	>95	100
Celecoxib	Hydroxylation	2C9	0.92	80	86
		3A4	0.55	20	6.0

[a]Reaction rates were measured in a panel of human liver microsomes from at least 10 different organ donors. Rates were correlated with various CYP form-selective activities and a correlation matrix setup.
[b]Reaction rates were measured in the presence of human liver microsomes co-incubated with solvent or CYP form-selective inhibitor. Inhibition expressed as % loss of activity in the presence of solvent alone.
[c]Reaction rate (turnover) was determined after incubation with a panel of individual expressed (recombinant) CYP proteins (rCYPs). Turnover for each CYP form was then normalized as described in Rodrigues (1999). Adapted from Zhang *et al.*, 2007 and references therein.

16.8.2 Significance of Metabolic Clearance and Reaction Phenotype

There are numerous examples in which drug metabolism is implicated in toxicity through either impaired clearance in the case of polymorphic mechanisms or increased exposure in the case of reactive or toxic metabolites. Knowing the enzymes involved can help understand the mechanism of toxicity, which can optimize the evaluation of a prospective backup compound, and identify individual susceptibility to such toxicity, which can optimize the design of a clinical study with screening and monitoring of the patient population. For instance, drugs metabolized primarily by CYP2D6, CYP2C19, and CYP2C9 are likely to be subject to significant interindividual variability in their exposure and effects as a result of known genetic polymorphisms that affect the relative expression of these enzymes in humans. Nonetheless, the drugs can still be very useful, for example, fluoxetine (CYP2D6), omeprazole (CYP2C19), and warfarin (CYP2C9); however, close attention must be paid to the safety margins and the feasibility of using genotyped/phenotyped patients in the clinical evaluation and even the selection criteria used to enroll patients into a clinical study (e.g. the CYP2C19 slow metabolizer phenotype is much more prevalent in Asians than in Caucasians) (Goldstein *et al.*, 1997; Shimoda *et al.*, 1999; Yokono *et al.*, 2001; Mizutani, 2003; Niu, Luo and Hao 2004; Bertilsson, 2007).

In general, the consequences of higher metabolic clearance are tied to the potential effect of coadministered inhibitors on the total exposure of the new drug. Figure 16.7

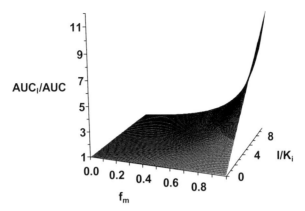

Figure 16.7 Simulation of fraction metabolized, potency of inhibition, and predicted effect on exposure: f_m = fraction metabolized; AUC = area under the curve (with and without inhibitor); I = concentration of inhibitor; and K_i = inhibitor constant.

illustrates the relationship between fraction metabolized (f_m) and the effect of inhibition (I/K_i) on the total exposure of the drug (i.e. the ratio of area under the curve, with and without the inhibitor—AUC_I/AUC). Drugs that are not cleared significantly by metabolism (e.g. $f_m < 0.3$) will not suffer much when coadministered with an inhibitor of that enzyme, regardless of how potent the inhibitor is or the concentration at which it is given or attains. With a low-metabolic-clearance drug, its AUC will be very similar, if not identical, with or without the inhibitor (i.e. $AUC_I/AUC = 1$). In contrast, a drug that is cleared entirely by metabolism ($f_m = 1$) will be markedly affected by inhibition of its clearance, with a several-fold increase in AUC possible.

The relationship between the fraction of drug metabolized and the exposure and potency of a coadministered inhibitor and the predicted effect on the drug AUC of its exposure can be calculated using the following formula:

$$\frac{AUC_i}{AUC} = \frac{1}{\dfrac{f_m}{1+[I]/K_i} + (1 - f_m)}$$

where f_m = fraction metabolized; I = concentration of inhibitor, for example plasma; K_i = inhibition constant; AUC_i = area under the curve with the inhibitor; AUC = area under the curve without the inhibitor.

To be even more specific, the contribution of a particular CYP to the total metabolic clearance can be indicated by $f_m \cdot f_{mCYP}$, where f_{mCYP} is the fraction metabolized by each CYP form, which substitutes into the equation to give:

$$\frac{AUC_i}{AUC} = \frac{1}{\dfrac{f_m \cdot f_{mCYP}}{1+[I]/K_i} + (1 - f_m \cdot f_{mCYP})}$$

If there is only one enzyme involved, then $f_m = f_m \cdot f_{mCYP}$; however, if more enzymes are involved, then selective inhibition is only going to block part of the clearance pathway and the total effect on AUC will be less. Consequently, a drug with a narrow therapeutic index that is metabolically cleared by several CYPs is considered, all else being equal, safer than a drug cleared principally by only one CYP. It is possible to determine f_m after the completion of a radiolabeled study, at which point the dose has been recovered and excreta profiled for parent drug, phase I (oxidative) and phase II (conjugates) metabolism (Zhang *et al.*, 2007).

The effect of inhibitors is compounded further for drugs that are metabolized significantly by enzymes in both the intestine and liver (e.g. CYP3A4 and 3A5). In addition to its effect on hepatic clearance, the inhibitor will now significantly affect the amount of drug available for absorption from the intestine. The subsequent change in AUC can be predicted using the following equation:

$$\frac{AUC_i}{AUC} = \frac{F_{Gi}}{F_G} * \frac{1}{\dfrac{f_m}{1+[I]/K_i} + (1-f_m)}$$

where F_{Gi} = fraction unmetabolized available from the gut in the presence of inhibitor; F_G = fraction unmetabolized and available from the gut without an inhibitor.

In Fig. 16.8 are simulations for F_{Gi}/F_G ranging from 1 (i.e. no effect on gut metabolism) to 10 (i.e. a 10-fold increase in the amount of drug absorbed). The effect on AUC can be exponential if the inhibitor also knocks out the hepatic clearance and f_m is high (e.g. $f_m > 0.7$).

By knowing the enzymes involved in metabolic clearance and their relative contribution, one can simulate the potential effect of inhibition on exposure and safety. The relative expression of CYPs differs from one tissue to another and this can be another major consideration when predicting oral bioavailability and/or the effects of DDIs. For example, a drug that has a high metabolic clearance and is selective for CYP3A4/5 is very likely to demonstrate a DDI given the relative expression of CYP3A in the gut and the liver (e.g. the CYP3A inhibitors itraconazole produced a greater than 10-fold increase observed in the C_{max} and AUC of simvastatin, a known selective CYP3A4 substrate with high first-pass metabolism) (Neuvonen, Kantola and Kivistö, 1998).

The importance of this factor is also illustrated in examples of polymorphic expression that results in reduced f_m*f_{mCYP}, which can lower the oral clearance and alter the susceptibility to interactions from CYP-selective inhibitors. For instance, the CYP2C9*3 allelic variant can result in a lower metabolic clearance of CYP2C9 substrates such as flurbiprofen (Kumar *et al.*, 2006). This is demonstrated *in vivo*, with a recent study reporting a gene-dependent effect on the apparent oral clearance of flurbiprofen in healthy volunteers. Specifically, the CYP2C9 inhibitor, fluvoxamine (400 mg oral dose), showed only a 1.1- and 1.4-fold change in the AUC of flurbiprofen (50 mg oral dose) in the two subjects identified as having the CYP2C9*3/*3 genotype. In comparison, a 2.4-fold change in the median AUC of flurbiprofen was measured in subjects with the CYP2C9*1*3 genotype (n = 8), and a threefold change in subjects with the CYP2C9*1*1 wild-type form (n = 11). Although the subject numbers are admittedly small for the CYP2C9*3*3 group, the authors concluded that, by analysis of fractional clearances, the degree of drug interaction observed was determined by genotype-related differences in the fraction metabolized by CYP2C9 (Kumar *et al.*, 2008).

Further illustration of the role of fraction metabolized in the clearance and AUC is provided by the significance of DDIs elicited via induction of clearance. Induction of

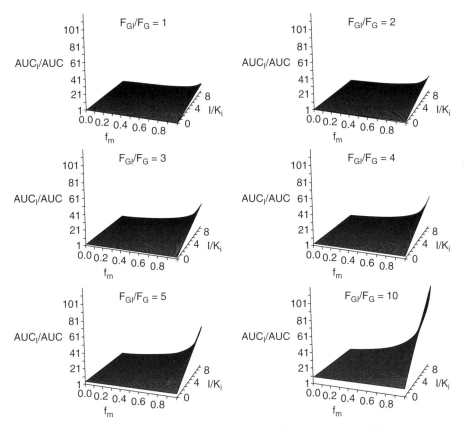

Figure 16.8 Simulation of gut first pass, fraction metabolized, potency of inhibition, and predicted effect on exposure: f_m = fraction metabolized; AUC = area under the curve (with and without inhibitor); I = concentration of inhibitor; and K_i = inhibitor constant; F_G = fraction unmetabolized and available from the gut (with and without inhibitor).

CYPs, for instance, effectively can lead to an increase in the f_m to the point that metabolic clearance can become the major contributor to the total clearance of a drug. In such circumstances, drugs can become markedly susceptible to the effects of coadministered inhibitors with the result that profound changes in exposure become possible.

16.8.3 Advanced Reaction Phenotyping

The methods described earlier can be applied to each individual metabolic pathway involved and can be used to predict the relative contribution of each enzyme in the formation of a specific metabolite from a particular concentration of drug. The antidepressant mirtazapine was studied by Störmer *et al.* (2000) and Störmer, von Moltke and Greenblatt (2000), with the 8-hydroxylation, *N*-desmethylation, and *N*-oxidation pathways evaluated. Table 16.10 summarizes the results of these studies and shows how the contribution of each CYP differs subject to the metabolite formed. In these studies, Störmer *et al.* showed that the formation of the 8-hydroxy metabolite was mediated by CYP2D6, 1A2, 3A4, and 2C9, with a pronounced trend to move away from CYP2D6

TABLE 16.10 Examples of *in vitro* cytochrome P450 (CYP) reaction phenotyping of the major biotransformation pathways involved in the metabolism of mirtazapine.

	Mirtazapine		
	% Predicted contribution		
Mirtazapine 8-hydroxylation	2.5 µM	25 µM	250 µM
CYP1A2	30	50	55
CYP2C8	—	—	—
CYP2C9	—	—	10
CYP2D6	65	40	15
CYP3A4	5	10	20
Mirtazapine *N*-desmethylation			
CYP1A2	45	20	10
CYP2C8	<5	5	15
CYP2C9	—	—	—
CYP2D6	—	—	—
CYP3A4	50	65	75
Mirtazapine *N*-oxide			
CYP1A2	85	45	15
CYP2C8	—	—	—
CYP2C9	—	—	—
CYP2D6	—	—	—
CYP3A4	15	55	85
Biotransformation pathways			
Mirtazapine 8-hydroxylation	55	45	30
Mirtazapine *N*-desmethylation	45	50	55
Mirtazapine *N*-oxidation	<5	5	15

Adapted from Störmer *et al.* (2000) and Störmer, von Moltke and Greenblatt (2000).

at higher concentrations *in vitro*. The formation of the *N*-desmethyl metabolite of mirtazapine was mediated by CYP3A4, 1A2, and 2C8, with CYP3A4 becoming dominant at higher concentrations *in vitro*. The third major metabolite measured, the *N*-oxide, was mediated by CYP3A4 and 1A2, with a move again toward CYP3A4 dominating at higher concentrations *in vitro*. These studies also demonstrated how the relative formation of the metabolites can change, subject to the concentration presented to the metabolizing enzymes; in this case, *N*-desmethylation of mirtazapine became the major pathway at 250 µM. This serves as an excellent example of how reaction-phenotyping methods can be applied to predict *in vivo* biotransformation and kinetics, and illustrates how the biotransformation of a drug depends on the concentration at which it is assessed.

It is recognized that although *in vitro* studies are very useful, especially in the discovery and early development phases, they lack the dynamic environment and complexity offered by real organs and tissues. If, for example, a novel drug candidate has demonstrated a relatively narrow safety margin in preclinical studies, then it follows that variability in its exposure becomes a significant factor in how well the drug will be tolerated and/or efficacious in patients. However, if the exposure can be reasonably well controlled, and the drug has the potential to address a significant unmet medical need, then reaction phenotyping *in vivo* can be useful in helping to predict or define the likely variability in exposure and risks to the target or general population. For instance, the oral bioavailability of a drug in humans can be measured with and without coadministration of inhibitors such as ketoconazole, itraconazole, or grapefruit juice, which allows one to determine under controlled clinical conditions the real potential for CYP3A4-mediated interactions. Another approach that can be taken is to genotype patients for known variants in CYPs, such as CYP2C9 or CYP2D6, to identify any genetic predisposition toward a high or low clearance of the drug.

Reaction phenotyping extends beyond the DMEs and tissues discussed in this chapter. When adverse effects are seen in preclinical studies, knowing the enzymes involved and their relative contribution to the observed toxicity can prove insightful and help to optimize a chemical series and find a safer molecule. Occasionally, especially when a specific metabolite or fragment is involved, it may even be possible to use reaction-phenotyping studies to define the relevance and risk to patients of a species- or gender-specific toxicity observed in animal models. (Chapter 21 addresses these concepts and application in more detail.)

16.9 SUMMARY

Optimizing the metabolic stability of a new chemotype in drug discovery can improve oral bioavailability and help to identify a drug candidate that will achieve efficacious exposure in the target patient population. In drug discovery, the goal is to identify the compound with the best balance of properties and develop the one that offers the least risk going into clinical trials. By identifying the specific enzymes involved in a drug's metabolic clearance, one can identify the risk of polymorphic variability and the risk of high-magnitude DDIs.

The methods discussed can be applied systematically in drug discovery from a higher-throughput early stage, in which the goal is to eliminate the compounds with the highest possible risk in development, through more predictive methods that enable forecasting of clinical outcomes. Irrespective of the stage of development, however, one has to leverage existing tools (e.g. selective inhibitors and recombinant enzymes) and acquire high-quality *in vitro* CYP reaction phenotype data. Such data are essential when attempting to obtain a fully integrated reaction phenotype, enable decision making, submit data for regulatory review prior to first-in-man, and when prioritizing clinical drug interactions studies, human radiolabel studies, or PK studies with CYP genotyped (phenotyped) subjects.

At the present time, the reaction-phenotyping "tool kit" for other DMEs (e.g. SULTs, UGTs, and FMOs) is poorly developed and additional progress is needed. In most cases, reaction phenotype has been rudimentary and based largely on turnover with recombinant enzyme preparations (Schrag *et al.*, 2004; Bowalgaha *et al.*, 2005; Court, 2005; Picard *et al.*, 2005; Uchaipichat *et al.*, 2006a; Mano, Usui and Kamimura, 2007a, b). As a result, there is a need for selective substrates, immuno-reagents, and chemical inhibitors, so that reaction-phenotyping studies can be performed with native microsomes (e.g. FMOs and UGTs) and cytosol (e.g. SULTs). With the advent of newer approaches (e.g. small interfering RNA or "siRNA"), it may be possible to reaction-phenotype multiple DMEs using primary hepatocytes or other cell-based systems (Chen *et al.*, 2006; Yu, 2007).

Many other (non-CYP) enzymes are expressed polymorphically and are also present in the gut and liver. Therefore, the interplay between these enzymes and CYPs and the potential consequences for inter-subject variability, drug interactions, and oral bioavailability should always be considered (Nishimura and Naito, 2006).

ACKNOWLEDGMENT

The authors wish to thank Dr. David M. Stresser of BD Biosciences and Dr. Yang Xu of Amgen for their helpful critique of the manuscript.

REFERENCES

Aithal, G.P., Day, C.P., Leathart, J.B., Daly, A.K. (2000). Relationship of polymorphism in cyp2c9 to genetic susceptibility to diclofenac-induced hepatitis. *Pharmacogenetics*, *10*, 511–518.

Ariyoshi, N., Takahashi, Y., Miyamoto, M., Umetsu, Y., Daigo, S., Tateishi, T., Kobayashi, S., Mizorogi, Y., Loriot, M.A., Stücker, I., Beaune, P., Kinoshita, M., Kamataki, T. (2000). Structural characterization of a new variant of the CYP2A6 gene (CYP2A6*1B) apparently diagnosed as heterozygotes of CYP2A6*1A and CYP2A6*4C. *Pharmacogenetics*, *10*, 687–693.

Arseculeratne, G., Wong, A.K., Goudie, D.R., Ferguson, J. (2007). Trimethylaminuria (fish-odor syndrome): a case report. *Arch. Dermatol.*, *143*, 81–84.

Baldwin, S.J., Bloomer, J.C., Smith, G.J., Ayrton, A.D., Clarke, S.E., Chenery, R.J. (1995). Ketoconazole and sulphaphenazole as the respective selective inhibitors of P4503A and 2C9. *Xenobiotica*, *25*, 261–270.

Barre, L., Fournel-Gigleux, S., Finel, M., Netter, P., Magdalou, J., Ouzzine, M. (2007). Substrate specificity of the human UDP-glucuronosyltransferase UGT2B4 and UGT2B7. Identification of a critical aromatic amino acid residue at position 33. *FEBS J.*, *274*, 1256–1264.

Barter, Z.E., Bayliss, M.K., Beaune, P.H., Boobis, A.R., Carlile, D.J., Edwards, R.J., Houston, J.B., Lake, B.G., Lipscomb, J.C., Pelkonen, O.R., Tucker, G.T., Rostami-Hodjegan, A. (2007). Scaling factors for the extrapolation of in vivo metabolic drug clearance from in vitro data: reaching a consensus on values of human microsomal protein and hepatocellularity per gram of liver. *Curr. Drug. Metab.*, *8*, 33–45.

Benetton, S.A., Fang, C., Yang, Y.O., Alok, R., Year, M., Lin, C.C., Yeh, L.T. (2007). P450 phenotyping of the metabolism of selegiline to desmethylselegiline and methamphetamine. *Drug Metab. Pharmacokinet.*, *22*, 78–87.

Bergström, M.A., Ott, H., Carlsson, A., Neis, M., Zwadlo-Klarwasser, G., Jonsson, C.A., Merk, H.F., Karlberg, A.T., Baron, J.M. (2007). A skin-like cytochrome P450 cocktail activates prohaptens to contact allergenic metabolites. *J. Invest. Dermatol.*, *127*, 1145–1153.

Bertilsson, L. (2007). Metabolism of antidepressant and neuroleptic drugs by cytochrome p450s: clinical and interethnic aspects. *Clin. Pharmacol. Ther.*, *82*, 606–609.

Bjornsson, T.D., Callaghan, J.T., Einolf, H.J., Fischer, V., Gan, L., Grimm, S., Kao, J., King, S.P., Miwa, G., Ni, L., Kumar, G., McLeod, J., Obach, R.S., Roberts, S., Roe, A., Shah, A., Snikeris, F., Sullivan, J.T., Tweedie, D., Vega, J.M., Walsh, J., Wrighton S.A., Pharmaceutical Research and Manufacturers of America (PhRMA), Drug Metabolism/Clinical Pharmacology Technical Working Group, FDA Center for Drug Evaluation and Research (CDER) (2003). The conduct of in vitro and in vivo drug-drug interaction studies: a Pharmaceutical Research and Manufacturers of America (PhRMA) perspective. *Drug Metab. Dispos.*, *31*, 815–832.

Bourrié, M., Meunier, V., Berger, Y., Fabre, G. (1996). Cytochrome P450 isoform inhibitors as a tool for the investigation of metabolic reactions catalyzed by human liver microsomes. *J. Pharmacol. Exp. Ther.*, *277*, 321–332.

Bowalgaha, K., Elliot, D.J., Mackenzie, P.I., Knights, K.M., Swedmark, S., Miners, J.O. (2005). S-Naproxen and desmethylnaproxen glucuronidation by human liver microsomes and recombinant human UDP-glucuronosyltransferases (UGT): role of UGT2B7 in the elimination of naproxen. *Br. J. Clin. Pharmacol.*, *60*, 423–433.

Buckpitt, A., Boland, B., Isbell, M., Morin, D., Shultz, M., Baldwin, R., Chan, K., Karlsson, A., Lin, C., Taff, A., West, J., Fanucchi, M., Van Winkle, L., Plopper, C. (2002). Naphthalene-induced respiratory tract toxicity: metabolic mechanisms of toxicity. *Drug Metab. Rev.*, *34*, 791–820.

Cai, X., Wang, R.W., Edom, R.W., Evans, D.C., Shou, M., Rodrigues, A.D., Liu, W., Dean, D.C., Baillie, T.A. (2004). Validation of (-)-N-3-benzyl-phenobarbital as a selective inhibitor of CYP2C19 in human liver microsomes. *Drug Metab. Dispos.*, *32*, 584–586.

Cashman, J.R. (2000). Human flavin-containing monooxygenase: substrate specificity and role in drug metabolism. *Curr. Drug Metab.*, *1*, 181–191.

Cashman, J.R. (2004). The implications of polymorphisms in mammalian flavin-containing monooxygenases in drug discovery and development. *Drug Discov. Today*, *9*, 574–581.

Cashman, J.R. (2005). Some distinctions between flavin-containing and cytochrome P450 monooxygenases. *Biochem. Biophys. Res. Commun.*, *338*, 599–604.

Cashman, J.R., Bi, Y.A., Lin, J., Youil, R., Knight, M., Forrest, S., Treacy, E. (1997). Human flavin-containing monooxygenase form 3: cDNA expression of the enzymes containing amino acid substitutions observed in individuals with trimethylaminuria. *Chem. Res. Toxicol.*, *10*, 837–841.

Cashman, J.R., Park, S.B., Yang, Z.C., Washington, C.B., Gomez, D.Y., Giacomini, K.M., Brett, C.M. (1993). Chemical, enzymatic, and human enantioselective S-oxygenation of cimetidine. *Drug Metab. Dispos.*, *21*, 587–597.

Cashman, J.R., Park, S.B., Yang, Z.C., Wrighton, S.A., Jacob, P., III, Benowitz, N.L. (1992). Metabolism of nicotine by human liver microsomes: stereoselective formation of trans-nicotine N'-oxide. *Chem. Res. Toxicol.*, *5*, 639–646.

Cashman, J.R., Zhang, J. (2006). Human flavin-containing monooxygenases. *Annu Rev. Pharmacol. Toxicol.*, *46*, 65–100.

Castell, J.V., Donato, M.T., Gómez-Lechón, M.J. (2005). Metabolism and bioactivation of toxicants in the lung. The in vitro cellular approach. *Exp. Toxicol. Pathol.*, *57*, 189–204.

Chang, T.K., Gonzalez, F.J., Waxman, D.J. (1994). Evaluation of triacetyloleandomycin, alpha-naphthoflavone and diethyldithiocarbamate as selective chemical probes for inhibition of human cytochromes P450. *Arch. Biochem. Biophys.*, *311*, 437–442.

Chauret, N., Gauthier, A., Nicoll-Griffith, D.A. (1998). Effect of common organic solvents on in vitro cytochrome P450-mediated metabolic activities in human liver microsomes. *Drug Metab. Dispos.*, *26*, 1–4.

Chen, G., Blevins-Primeau, A.S., Dellinger, R.W., Muscat, J.E., Lazarus, P. (2007). Glucuronidation of nicotine and cotinine by UGT2B10: loss of function by the UGT2B10 Codon 67 (Asp > Tyr) polymorphism. *Cancer Res.*, *67*, 9024–9029

Chen, J., Yang, X., Huang, M., Hu, Z., He, M., Duan, W., Chan, E., Sheu, F., Chen, X., Zhou, S. (2006). Small interfering RNA-mediated silencing of cytochrome P450 3A4 gene. *Drug Metab. Dispos.*, *34*, 1650–1657.

Coughtrie, M.W., Blair, J.N., Hume, R., Burchell, A. (1991). Improved preparation of hepatic microsomes for in vitro diagnosis of inherited disorders of the glucose-6-phosphatase system. *Clin. Chem.*, *37*, 739–742.

Court, M.H. (2005). Isoform-selective probe substrates for in vitro studies of human UDP-glucuronosyltransferases. *Methods Enzymol.*, *400*, 104–116.

Crespi, C.L. (1995). Xenobiotic-metabolizing human cells as tools for pharmacological and toxicological research. *Adv. Drug Res.*, *26*, 179–235.

Czerniak, R. (2001). Gender-based differences in pharmacokinetics in laboratory animal models. *Int. J. Toxicol.*, *20*, 161–163.

Dalén, P., Dahl, M.L., Bernal Ruiz, M.L., Nordin, J., Bertilsson, L. (1998). 10-Hydroxylation of nortriptyline in white persons with 0, 1, 2, 3, and 13 functional CYP2D6 genes. *Clin. Pharmacol. Ther.*, *63*, 444–452.

Davies, B., Morris, T. (1993). Physiological parameters in laboratory animals and humans. *Pharm. Res.*, *10*, 1093–1095.

Dolphin C.T., Beckett D.J., Janmohamed A., Cullingford T.E., Smith R.L., Shephard E.A., Phillips, I.R. (1998). The flavin-containing monooxygenase 2 gene (FMO2) of humans, but not of other primates, encodes a truncated, nonfunctional protein. *J. Biol. Chem.*, *273*, 30599–30607.

Drahushuk, A.T., McGarrigle, B.P., Larsen, K.E., Stegeman, J.J., Olson, J.R. (1998). Detection of CYP1A1 protein in human liver and induction by TCDD in precision-cut liver slices incubated in dynamic organ culture. *Carcinogenesis*, *19*, 1361–1368.

Eagling, V.A., Tjia, J.F., Back, D.J. (1998). Differential selectivity of cytochrome P450 inhibitors against probe substrates in human and rat liver microsomes. *Br. J. Clin. Pharmacol.*, *45*, 107–114.

Ebner, T., Heinzel, G., Prox, A., Beschke, K., Wachsmuth, H. (1999). Disposition and chemical stability of telmisartan 1-O-acylglucuronide. *Drug Metab. Dispos.*, *27*, 1143–1149.

Fang, J. (2000). Metabolism of clozapine by rat brain: the role of flavin-containing monooxygenase (FMO) and cytochrome P450 enzymes. *Eur. J. Drug Metab. Pharmacokinet.*, *25*, 109–114.

FDA (2006). *FDA Guidelines for In Vitro Characterization of Drug candidates*, http://www.fda. gov/cder/guidance/6695dft.htm (accessed 31 October 2008).

Fisher, M.B., Campanale, K., Ackermann, B.L., VandenBranden, M., Wrighton, S.A. (2000). In vitro glucuronidation using human liver microsomes and the pore-forming peptide alamethicin. *Drug Metab. Dispos.*, *28*, 560–566.

Fisher, M.B., Paine, M.F., Strelevitz, T.J., Wrighton, S.A. (2001). The role of hepatic and extrahepatic UDP-glucuronosyltransferases in human drug metabolism. *Drug Metab. Rev.*, *33*, 273–297.

Fournel-Gigleux, S., Jackson, M.R., Wooster, R., Burchell, B. (1989). Expression of a human liver cDNA encoding a UDP-glucuronosyltransferase catalysing the glucuronidation of hyodeoxycholic acid in cell culture. *FEBS Lett.*, *243*, 119–122.

Fulceri, R., Bánhegyi, G., Gamberucci, A., Giunti, R., Mandl, J., Benedetti, A. (1994). Evidence for the intraluminal positioning of p-nitrophenol UDP-glucuronosyltransferase activity in rat liver microsomal vesicles. *Arch. Biochem. Biophys.*, *309*, 43–46.

Gagné, J.F., Montminy, V., Belanger, P., Journault, K., Gaucher, G., Guillemette, C. (2002). Common human UGT1A polymorphisms and the altered metabolism of irinotecan active metabolite 7-ethyl-10-hydroxycamptothecin (SN-38). *Mol. Pharmacol.*, *62*, 608–617.

Garteiz, D.A., Hook, R.H., Walker, B.J., Okerholm, R.A. (1982). Pharmacokinetics and biotransformation studies of terfenadine in man. *Arzneimittelforschung*, *32*, 1185–1190.

Gelboin, H.V., Krausz, K. (2006). Monoclonal antibodies and multifunctional cytochrome P450: drug metabolism as paradigm. *J. Clin. Pharmacol.*, *46*, 353–372.

Ghosal, A., Ramanathan, R., Yuan, Y., Hapangama, N., Chowdhury, S.K., Kishnani, N.S., Alton, K.B. (2007). Identification of human liver cytochrome P450 enzymes involved in biotransformation of vicriviroc, a CCR5 receptor antagonist. *Drug Metab. Dispos.*, *35*, 2186–2195.

Girennavar, B., Poulose, S.M., Jayaprakasha, G.K., Bhat, N.G., Patil, B.S. (2006). Furocoumarins from grapefruit juice and their effect on human CYP 3A4 and CYP 1B1 isoenzymes. *Bioorg. Med. Chem.*, *14*, 2606–2612.

Goldstein, J.A., Ishizaki, T., Chiba, K., de Morais, S.M., Bell, D., Krahn, P.M., Evans, D.A. (1997). Frequencies of the defective CYP2C19 alleles responsible for the mephenytoin poor metabolizer phenotype in various Oriental, Caucasian, Saudi Arabian and American black populations. *Pharmacogenetics*, *7*, 59–64.

Gómez-Lechón, M.J., Lahoz, A., Jiménez, N., Vicente Castell, J., Donato, M.T. (2006). Cryopreservation of rat, dog and human hepatocytes: influence of preculture and cryoprotectants on recovery, cytochrome P450 activities and induction upon thawing. *Xenobiotica*, *36*, 457–472.

Gorsky, L.D., Coon, M.J. (1986). Effects of conditions for reconstitution with cytochrome b5 on the formation of products in cytochrome P-450-catalyzed reactions. *Drug Metab. Dispos.*, *14*, 89–96.

Griese, E.U., Ilett, K.F., Kitteringham, N.R., Eichelbaum, M., Powell, H., Spargo, R.M., LeSouef, P.N., Musk, A.W., Minchin, R.F. (2001). Allele and genotype frequencies of polymorphic cytochromes P4502D6, 2C19 and 2E1 in aborigines from western Australia. *Pharmacogenetics*, *11*, 69–76.

Grillo, M.P., Hua, F., Knutson, C.G., Ware, J.A., Li, C. (2003a). Mechanistic studies on the bioactivation of diclofenac: identification of diclofenac-S-acyl-glutathione in vitro in incubations with rat and human hepatocytes. *Chem. Res. Toxicol.*, *16*, 1410–1417.

Grillo, M.P., Knutson, C.G., Sanders, P.E., Waldon, D.J., Hua, F., Ware, J.A. (2003b). Studies on the chemical reactivity of diclofenac acyl glucuronide with glutathione: identification of diclofenac-S-acyl-glutathione in rat bile. *Drug Metab. Dispos.*, *31*, 1327–1336.

Grothusen, A., Hardt, J., Bräutigam, L., Lang, D., Böcker, R. (1996). A convenient method to discriminate between cytochrome P450 enzymes and flavin-containing monooxygenases in human liver microsomes. *Arch. Toxicol.*, *71*, 64–71.

Guengerich, F.P. (1989). Characterization of human microsomal cytochrome P-450 enzymes. *Annu. Rev. Pharmacol. Toxicol.*, *29*, 241–264.

Guengerich, F.P. (2003). Cytochromes P450, drugs, and diseases. *Mol. Interv.*, *3*, 194–204.

Guengerich, F.P., Miller, G.P., Hanna, I.H., Martin, M.V., Léger, S., Black, C., Chauret, N., Silva, J.M., Trimble, L.A., Yergey, J.A., Nicoll-Griffith, D.A. (2002). Diversity in the oxidation of substrates by cytochrome P450 2D6: lack of an obligatory role of aspartate 301-substrate electrostatic bonding. *Biochemistry*, *41*, 11025–11034.

Henderson, M.C., Krueger, S.K., Stevens, J.F., Williams, D.E. (2004). Human flavin-containing monooxygenase form 2 S-oxygenation: sulfenic acid formation from thioureas and oxidation of glutathione. *Chem. Res. Toxicol.*, *17*, 633–640.

Hewitt, N.J., Lloyd, S., Hayden, M., Butler, R., Sakai, Y., Springer, R., Fackett, A., Li, A.P. (2002). Correlation between troglitazone cytotoxicity and drug metabolic enzyme activities in cryopreserved human hepatocytes. *Chem. Biol. Interact.*, *142*, 73–82.

Hickman, D., Wang, J.P., Wang, Y., Unadkat, J.D. (1998). Evaluation of the selectivity of In vitro probes and suitability of organic solvents for the measurement of human cytochrome P450 monooxygenase activities. *Drug Metab. Dispos.*, *26*, 207–215.

Hiroi, T., Imaoka, S., Funae, Y. (1998). Dopamine formation from tyramine by CYP2D6. *Biochem. Biophys. Res. Commun.*, *249*, 838–843.

Honig, P.K., Wortham, D.C., Lazarev, A., Cantilena, L.R. (1996). Grapefruit juice alters the systemic bioavailability and cardiac repolarization of terfenadine in poor metabolizers of terfenadine. *J. Clin. Pharmacol.*, *36*, 345–351.

Honig, P.K., Wortham, D.C., Hull, R., Zamani, K., Smith, J.E., Cantilena, L.R. (1993a). Itraconazole affects single-dose terfenadine pharmacokinetics and cardiac repolarization pharmacodynamics. *J. Clin. Pharmacol.*, *33*, 1201–1206.

Honig, P.K., Wortham, D.C., Zamani, K., Conner, D.P., Mullin, J.C., Cantilena, L.R. (1993b). Terfenadine-ketoconazole interaction. Pharmacokinetic and electrocardiographic consequences. *JAMA*, *269*, 1513–1518.

Houston, J.B. (1994). Utility of in vitro drug metabolism data in predicting in vivo metabolic clearance. *Biochem. Pharmacol.*, *47*, 1469–1479.

Hsieh, K.P., Lin, Y.Y., Cheng, C.L., Lai, M.L., Lin, M.S., Siest, J.P., Huang, J.D. (2001). Novel mutations of CYP3A4 in Chinese. *Drug Metab. Dispos.*, *29*, 268–273.

Hutzler, J.M., Hauer, M.J., Tracy, T.S. (2001). Dapsone activation of CYP2C9-mediated metabolism: evidence for activation of multiple substrates and a two-site model. *Drug Metab. Dispos.*, *29*, 1029–1034.

Jain, K.K. (2005). Applications of AmpliChip CYP450. *Mol. Diagn.*, *9*, 119–127.

Jakobsson, J., Ekström, L., Inotsume, N., Garle, M., Lorentzon, M., Ohlsson, C., Roh, H.K., Carlström, K., Rane, A. (2006). Large differences in testosterone excretion in Korean and Swedish men are strongly associated with a UDP-glucuronosyl transferase 2B17 polymorphism. *J. Clin. Endocrinol. Metab.*, *91*, 687–693.

Jedlitschky, G., Cassidy, A.J., Sales, M., Pratt, N., Burchell, B. (1999). Cloning and characterization of a novel human olfactory UDP-glucuronosyltransferase. *Biochem. J.*, *340*, 837–843.

Kaivosaari, S., Toivonen, P., Hesse, L.M., Koskinen, M., Court, M.H., Finel, M. (2007). Nicotine glucuronidation and the human UDP-glucuronosyltransferase UGT2B10. *Mol. Pharmacol.*, *72*, 761–768.

Kaji, H., Kume, T. (2005). Regioselective glucuronidation of denopamine: marked species differences and identification of human udp-glucuronosyltransferase isoform. *Drug Metab. Dispos.*, *33*, 403–412.

Kaneko, A., Bergqvist, Y., Taleo, G., Kobayakawa, T., Ishizaki, T., Björkman, A. (1999a). Proguanil disposition and toxicity in malaria patients from Vanuatu with high frequencies of CYP2C19 mutations. *Pharmacogenetics*, *9*, 317–326.

Kaneko, A., Lum, J.K., Yaviong, L., Takahashi, N., Ishizaki, T., Bertilsson, L., Kobayakawa, T., Björkman, A. (1999b). High and variable frequencies of CYP2C19 mutations: medical consequences of poor drug metabolism in Vanuatu and other Pacific islands. *Pharmacogenetics*, *9*, 581–590.

Kang, J.H., Chung, W.G., Lee, K.H., Park, C.S., Kang, J.S., Shin, I.C., Roh, H.K., Dong, M.S., Baek, H.M., Cha, Y.N. (2000). Phenotypes of flavin-containing monooxygenase activity determined by ranitidine N-oxidation are positively correlated with genotypes of linked FM03 gene mutations in a Korean population. *Pharmacogenetics*, *10*, 67–78.

Katoh, M., Matsui, T., Yokoi, T. (2007). Glucuronidation of antiallergic drug, Tranilast: identification of human UDP-glucuronosyltransferase isoforms and effect of its phase I metabolite. *Drug Metab. Dispos.*, *35*, 583–589.

Kenworthy, K.E., Bloomer, J.C., Clarke, S.E., Houston, J.B. (1999). CYP3A4 drug interactions: correlation of 10 in vitro probe substrates. *Br. J. Clin. Pharmacol.*, *48*, 716–727.

King, C.D., Rios, G.R., Green, M.D., Tephly, T.R. (2000). UDP-glucuronosyltransferases. *Curr. Drug Metab.*, *1*, 143–161.

Krause, R.J., Lash, L.H., Elfarra, A.A. (2003). Human kidney flavin-containing monooxygenases and their potential roles in cysteine s-conjugate metabolism and nephrotoxicity. *J. Pharmacol. Exp. Ther.*, *304*, 185–191.

Krueger, S.K., Williams, D.E. (2005). Mammalian flavin-containing monooxygenases: structure/function, genetic polymorphisms and role in drug metabolism. *Pharmacol. Ther.*, *106*, 357–387.

Kuehl, P., Zhang, J., Lin, Y., Lamba, J., Assem, M., Schuetz, J., Watkins, P.B., Daly, A., Wrighton, S.A., Hall, S.D., Maurel, P., Relling, M., Brimer, C., Yasuda, K., Venkataramanan, R., Strom, S., Thummel, K., Boguski, M.S., Schuetz, E. (2001). Sequence diversity in CYP3A promoters and characterization of the genetic basis of polymorphic CYP3A5 expression. *Nat. Genet.*, *27*, 383–391.

Kumar, V., Brundage, R.C., Oetting, W.S., Leppik, I.E., Tracy, T.S. (2008). Differential genotype dependent inhibition of CYP2C9 in humans. *Drug Metab Dispos*, *36*, 1242–1248. (Epub ahead of print).

Kumar, V., Wahlstrom, J.L., Rock, D.A., Warren, C.J., Gorman, L.A., Tracy, T.S. (2006). CYP2C9 inhibition: impact of probe selection and pharmacogenetics on in vitro inhibition profiles. *Drug Metab. Dispos.*, *34*, 1966–1975.

Kunze, K.L., Trager, W.F. (1993). Isoform-selective mechanism-based inhibition of human cytochrome P450 1A2 by furafylline. *Chem. Res. Toxicol.*, *6*, 649–656.

Kwon, J.T., Nakajima, M., Chai, S., Yom, Y.K., Kim, H.K., Yamazaki, H., Sohn, D.R., Yamamoto, T., Kuroiwa, Y., Yokoi, T. (2001). Nicotine metabolism and CYP2A6 allele frequencies in Koreans. *Pharmacogenetics*, *11*, 317–323.

Lang, D.H., Yeung, C.K., Peter, R.M., Ibarra, C., Gasser, R., Itagaki, K., Philpot, R.M., Rettie, A.E. (1998). Isoform specificity of trimethylamine N-oxygenation by human flavin-containing monooxygenase (FMO) and P450 enzymes: selective catalysis by FMO3. *Biochem. Pharmacol.*, *56*, 1005–1012.

Lasker, J.M., Wester, M.R., Aramsombatdee, E., Raucy, J.L. (1998). Characterization of CYP2C19 and CYP2C9 from human liver: respective roles in microsomal tolbutamide, S-mephenytoin, and omeprazole hydroxylations. *Arch. Biochem. Biophys.*, *353*, 16–28.

Lee, H.S., Ji, H.Y., Park, E.J., Kim, S.Y. (2007a). In vitro metabolism of eupatilin by multiple cytochrome P450 and UDP-glucuronosyltransferase enzymes. *Xenobiotica*, *37*, 803–817.

Lee, S., Hwang, H.J., Kim, J.M., Chung, C.S., Kim, J.H. (2007b). CYP2C19 polymorphism in Korean patients on warfarin therapy. *Arch. Pharm. Res.*, *30*, 344–349.

Lee, H.S., Park, E.J., Ji, H.Y., Kim, S.Y., Im, G.J., Lee, S.M., Jang, I.J. (2008). Identification of cytochrome P450 enzymes responsible for N-dealkylation of a new oral erectogenic, mirodenafil. *Xenobiotica*, *38*, 21–33.

Lewis, D.F., Dickins, M. (2002). Substrate SARs in human P450s. *Drug Discov. Today*, *7*, 918–925.

Lewis, D.F., Jacobs, M.N., Dickins, M. (2004). Compound lipophilicity for substrate binding to human P450s in drug metabolism. *Drug Discov. Today*, *9*, 530–537.

Li, A.P., Lu, C., Brent, J.A., Pham, C., Fackett, A., Ruegg, C.E., Silber, P.M. (1999). Cryopreserved human hepatocytes: characterization of drug-metabolizing enzyme activities and applications in higher throughput screening assays for hepatotoxicity, metabolic stability, and drug-drug interaction potential. *Chem. Biol. Interact.*, *121*, 17–35.

Li, L., Pan, R.M., Porter, T.D., Jensen, N.S., Silber, P., Russo, G., Tine, J.A., Heim, J., Ring, B., Wedlund, P.J. (2006). New cytochrome P450 2D6*56 allele identified by genotype/phenotype analysis of cryopreserved human hepatocytes. *Drug Metab. Dispos.*, *34*, 1411–1416.

Lévesque, E., Turgeon, D., Carrier, J.S., Montminy, V., Beaulieu, M., Bélanger, A. (2001). Isolation and characterization of the UGT2B28 cDNA encoding a novel human steroid conjugating UDP-glucuronosyltransferase. *Biochemistry*, *40*, 3869–3881.

Mano, Y., Usui, T., Kamimura, H. (2007a). Predominant contribution of UDP-glucuronosyltransferase 2B7 in the glucuronidation of racemic flurbiprofen in the human liver. *Drug Metab. Dispos.*, *35*, 1182–1187.

Mano, Y., Usui, T., Kamimura, H. (2007b). The UDP-Glucuronosyltransferase 2B7 isozyme is responsible for gemfibrozil glucuronidation in the human liver. *Drug Metab. Dispos.*, *35*, 2040–2044.

Markowitz, J.S., Donovan, J.L., DeVane, C.L., Taylor, R.M., Ruan, Y., Wang, J.S., Chavin, K.D. (2003). Effect of St John's wort on drug metabolism by induction of cytochrome P450 3A4 enzyme. *JAMA*, *290*, 1500–1504.

Mennes, W.C., van Holsteijn, C.W., van Iersel, A.A., Yap, S.H., Noordhoek, J., Blaauboer, B.J. (1994). Interindividual variation in biotransformation and cytotoxicity of bromobenzene as determined in primary hepatocyte cultures derived from monkey and human liver. *Hum. Exp. Toxicol.*, *13*, 415–421.

Miners, J.O., Knights, K.M., Houston, J.B., Mackenzie, P.I. (2006). In vitro-in vivo correlation for drugs and other compounds eliminated by glucuronidation in humans: pitfalls and promises. *Biochem. Pharmacol.*, *71*, 1531–1539.

Mizutani, T. (2003). PM frequencies of major CYPs in Asians and. *Caucasians. Drug Metab. Rev.*, *35*, 99–106.

Mouly, S.J., Matheny, C., Paine, M.F., Smith, G., Lamba, J., Lamba, V., Pusek, S.N., Schuetz, E.G., Stewart, P.W., Watkins, P.B. (2005). Variation in oral clearance of saquinavir is predicted by CYP3A5*1 genotype but not by enterocyte content of cytochrome P450 3A5. *Clin. Pharmacol. Ther.*, *78*, 605–618.

Muralidharan, G., Hawes, E.M., McKay, G., Korchinski, E.D., Midha, K.K. (1991). Quinidine but not quinine inhibits in man the oxidative metabolic routes of methoxyphenamine which involve debrisoquine 4-hydroxylase. *Eur. J. Clin. Pharmacol.*, *41*, 471–474.

Mushiroda, T., Douya, R., Takahara, E., Nagata, O. (2000). The involvement of flavin-containing monooxygenase but not CYP3A4 in metabolism of itopride hydrochloride, a gastroprokinetic agent: comparison with cisapride and mosapride citrate. *Drug Metab. Dispos.*, *28*, 1231–1237.

Naisbitt, D.J., Sanderson, L.S., Meng, X., Stachulski, A.V., Clarke, S.E., Park, B.K. (2007). Investigation of the immunogenicity of diclofenac and diclofenac metabolites. *Toxicol. Lett.*, *168*, 45–50.

Nakajima, M., Nakamura, S., Tokudome, S., Shimada, N., Yamazaki, H., Yokoi, T. (1999). Azelastine N-demethylation by cytochrome P-450 (CYP)3A4, CYP2D6, and CYP1A2 in human liver microsomes: evaluation of approach to predict the contribution of multiple CYPs. *Drug Metab. Dispos.*, *27*, 1381–1391.

Neuvonen, P.J., Kantola, T., Kivistö, K.T. (1998). Simvastatin but not pravastatin is very susceptible to interaction with the CYP3A4 inhibitor itraconazole. *Clin. Pharmacol. Ther.*, *63*, 332–341.

Nishimura, M., Naito, S. (2006). Tissue-specific mRNA expression profiles of human phase I metabolizing enzymes except for cytochrome P450 and phase II metabolizing enzymes. *Drug Metab. Pharmacokinet.*, *21*, 357–374.

Niu, C.Y., Luo, J.Y., Hao, Z.M. (2004). Genetic polymorphism analysis of cytochrome P4502C19 in Chinese Uigur and Han populations. *Chin. J. Dig. Dis.*, *5*, 76–80.

O'Brien, S.E., de Groot, M.J. (2005). Greater than the sum of its parts: combining models for useful ADMET prediction. *J. Med. Chem.*, *48*, 1287–1291.

Obach, R.S. (1999). Prediction of human clearance of twenty-nine drugs from hepatic microsomal intrinsic clearance data: an examination of in vitro half-life approach and nonspecific binding to microsomes. *Drug Metab. Dispos.*, *27*, 1350–1359.

Obach, R.S., Baxter, J.G., Liston, T.E., Silber, B.M., Jones, B.C., MacIntyre, F., Rance, D.J., Wastall, P. (1997). The prediction of human pharmacokinetic parameters from preclinical and in vitro metabolism data. *J. Pharmacol. Exp. Ther.*, *283*, 46–58.

Obach, R.S., Margolis, J.M., Logman, M.J. (2007). In vitro metabolism of CP-122,721 ((2S,3S)-2-phenyl-3-[(5-trifluoromethoxy-2-methoxy)benzylamino]piperidine), a non-peptide antagonist of the substance P receptor. *Drug Metab. Pharmacokinet.*, *22*, 336–349.

Ohmi, N., Yoshida, H., Endo, H., Hasegawa, M., Akimoto, M., Higuchi, S. (2003). S-oxidation of S-methyl-esonarimod by flavin-containing monooxygenases in human liver microsomes. *Xenobiotica*, *33*, 1221–1231.

Oscarson, M., McLellan, R.A., Gullstén, H., Agúndez, J.A., Benítez, J., Rautio, A., Raunio, H., Pelkonen, O., Ingelman-Sundberg, M. (1999a). Identification and characterisation of novel polymorphisms in the CYP2A locus: implications for nicotine metabolism. *FEBS Lett.*, *460*, 321–327.

Oscarson, M., McLellan, R.A., Gullstén, H., Yue, Q.Y., Lang, M.A., Bernal, M.L., Sinues, B., Hirvonen, A., Raunio, H., Pelkonen, O., Ingelman-Sundberg, M. (1999b). Characterisation and PCR-based detection of a CYP2A6 gene deletion found at a high frequency in a Chinese population. *FEBS Lett.*, *448*, 105–110.

Overby, L.H., Carver, G.C., Philpot, R.M. (1997). Quantitation and kinetic properties of hepatic microsomal and recombinant flavin-containing monooxygenases 3 and 5 from humans. *Chem. Biol. Interact.*, *106*, 29–45.

Ozdemir, V., Kalow, W., Tang, B.K., Paterson, A.D., Walker, S.E., Endrenyi, L., Kashuba, A.D. (2000). Evaluation of the genetic component of variability in CYP3A4 activity: a repeated drug administration method. *Pharmacogenetics*, *10*, 373–388.

Paine, M.F., Hart, H.L., Ludington, S.S., Haining, R.L., Rettie, A.E., Zeldin, D.C. (2006). The human intestinal cytochrome P450 "pie". *Drug Metab. Dispos.*, *34*, 880–886.

Parkinson, A. (1996). An overview of current cytochrome P450 technology for assessing the safety and efficacy of new materials. *Toxicol. Pathol.*, *24*, 48–57.

Paulson, S.K., Engel, L., Reitz, B., Bolten, S., Burton, E.G., Maziasz, T.J., Yan, B., Schoenhard, G.L. (1999). Evidence for polymorphism in the canine metabolism of the cyclooxygenase 2 inhibitor, celecoxib. *Drug Metab. Dispos.*, *27*, 1133–1142.

Picard, N., Ratanasavanh, D., Prémaud, A., Le Meur, Y., Marquet, P. (2005). Identification of the UDP-glucuronosyltransferase isoforms involved in mycophenolic acid phase II metabolism. *Drug Metab. Dispos.*, *33* (1), 139–146.

Ramamoorthy, Y., Tyndale, R.F., Sellers, E.M. (2001). Cytochrome P450 2D6.1 and cytochrome P450 2D6.10 differ in catalytic activity for multiple substrates. *Pharmacogenetics*, *11*, 477–487.

Rendic, S., Di Carlo, F.J. (1997). Human cytochrome P450 enzymes: a status report summarizing their reactions, substrates, inducers, and inhibitors. *Drug Metab. Rev.*, *29*, 413–580.

Rodrigues, A.D. (1999). Integrated cytochrome P450 reaction phenotyping: attempting to bridge the gap between cDNA-expressed cytochromes P450 and native human liver microsomes. *Biochem. Pharmacol.*, *57*, 465–480.

Rowland, A., Elliot, D.J., Williams, J.A., Mackenzie, P.I., Dickinson, R.G., Miners, J.O. (2006). In vitro characterization of lamotrigine N2-glucuronidation and the lamotrigine-valproic acid interaction. *Drug. Metab. Dispos.*, *34*, 1055–1062.

Rowland, A., Gaganis, P., Elliot, D.J., Mackenzie, P.I., Knights, K.M., Miners, J.O. (2007). Binding of inhibitory fatty acids is responsible for the enhancement of UDP-glucuronosyltransferase 2B7 activity by albumin: implications for in vitro-in vivo extrapolation. *J. Pharmacol. Exp. Ther.*, *321*, 137–147.

Rowland-Yeo, K., Rostami-Hodjegan, A., Tucker, G.T., British Pharmacological Society Winter Meeting University of London (2003). Abundance of cytochrome P450 in human liver: a meta-analysis. *Br. J. Clin. Pharmacol.*, *57* (5), 687.

Sachse, C., Brockmöller, J., Bauer, S., Roots, I. (1997). Cytochrome P450 2D6 variants in a Caucasian population: allele frequencies and phenotypic consequences. *Am. J. Hum. Genet.*, *60*, 284–295.

Schrag, M.L., Cui, D., Rushmore, T.H., Shou, M., Ma, B., Rodrigues, A.D. (2004). Sulfotransferase 1E1 is a low km isoform mediating the 3-O-sulfation of ethinyl estradiol. *Drug Metab. Dispos.*, *32*, 1299–1303.

Segel, I. (1975). Kinetics of unireactant enzymes, chapter 2 in *Enzyme Kinetics. Behavior Analysis of Rapid Equilibrium and Steady-State Systems* (ed. I. Segel), John Wiley & Sons, Inc., New York, pp. 54–55.

Shepherd, S.R., Baird, S.J., Hallinan, T., Burchell, B. (1989). An investigation of the transverse topology of bilirubin UDP-glucuronosyltransferase in rat hepatic endoplasmic reticulum. *Biochem. J.*, *259*, 617–620.

Shimada, T., Tsumura, F., Yamazaki, H., Guengerich, F.P., Inoue, K. (2001). Characterization of (+/−)-bufuralol hydroxylation activities in liver microsomes of Japanese and Caucasian subjects genotyped for CYP2D6. *Pharmacogenetics*, *11*, 143–156.

Shimada, T., Yamazaki, H., Mimura, M., Inui, Y., Guengerich, F.P. (1994). Interindividual variations in human liver cytochrome P-450 enzymes involved in the oxidation of drugs, carcinogens and toxic chemicals: studies with liver microsomes of 30 Japanese and 30 Caucasians. *J. Pharmacol. Exp. Ther.*, *270*, 414–423.

Shimoda, K., Jerling, M., Böttiger, Y., Yasuda, S., Morita, S., Bertilsson, L. (1999). Pronounced differences in the disposition of clomipramine between Japanese and Swedish patients. *J Clin. Psychopharmacol.*, *19*, 393–400.

Snawder, J.E., Lipscomb, J.C. (2000). Interindividual variance of cytochrome P450 forms in human hepatic microsomes: correlation of individual forms with xenobiotic metabolism and implications in risk assessment. *Regul. Toxicol. Pharmacol.*, *32*, 200–209.

Soars, M.G., Gelboin, H.V., Krausz, K.W., Riley, R.J. (2003). A comparison of relative abundance, activity factor and inhibitory monoclonal antibody approaches in the characterization of human CYP enzymology. *Br. J. Clin. Pharmacol.*, *55*, 175–181.

Soars, M.G., Riley, R.J., Findlay, K.A., Coffey, M.J., Burchell, B. (2001). Evidence for significant differences in microsomal drug glucuronidation by canine and human liver and kidney. *Drug Metab. Dispos.*, *29*, 121–126.

Soars, M.G., Ring, B.J., Wrighton, S.A. (2003). The effect of incubation conditions on the enzyme kinetics of udp-glucuronosyltransferases. *Drug Metab. Dispos.*, *31*, 762–767.

Störmer, E., von Moltke, L.L., Greenblatt, D.J. (2000). Scaling drug biotransformation data from cDNA-expressed cytochrome P-450 to human liver: a comparison of relative activity factors and human liver abundance in studies of mirtazapine metabolism. *J. Pharmacol. Exp. Ther.*, *295*, 793–801.

Störmer, E., von Moltke, L.L., Shader, R.I., Greenblatt, D.J. (2000). Metabolism of the antidepressant mirtazapine in vitro: contribution of cytochromes P-450 1A2, 2D6, and 3A4. *Drug Metab. Dispos.*, *28*, 1168–1175.

Stouch, T.R., Kenyon, J.R., Johnson, S.R., Chen, X.Q., Doweyko, A., Li, Y. (2003). In silico ADME/Tox: why models fail. *J. Comput. Aided. Mol. Des.*, *17*, 83–92.

Suzuki, H., Kneller, M.B., Haining, R.L., Trager, W.F., Rettie, A.E. (2002). (+)-N-3-Benzyl-nirvanol and (−)-N-3-benzyl-phenobarbital: new potent and selective in vitro inhibitors of CYP2C19. *Drug Metab. Dispos.*, *30*, 235–239.

Swanson, H.I. (2004). Cytochrome P450 expression in human keratinocytes: an aryl hydrocarbon receptor perspective. *Chem. Biol. Interact.*, *149*, 69–79.

Tachibana, M., Tanaka, M., Masubuchi, Y., Horie, T. (2005). Acyl glucuronidation of fluroquinolone antibiotics by the UDP-glucuronosyltransferase 1A subfamily in human liver miscrosomes. *Drug Metab. Dispos.*, *33*, 803–811.

Takanashi, K., Tainaka, H., Kobayashi, K., Yasumori, T., Hosakawa, M., Chiba, K. (2000). CYP2C9 Ile359 and Leu359 variants: enzyme kinetic study with seven substrates. *Pharmacogenetics*, *10*, 95–104.

Toffoli, G., Cecchin, E. (2004). Uridine diphosphoglucuronosyl transferase and methylenetetrahydrofolate reductase polymorphisms as genomic predictors of toxicity and response to irinotecan-, antifolate- and fluoropyrimidine-based chemotherapy. *J. Chemother.*, *4*, 31–35.

Toffoli, G., Cecchin, E., Corona, G., Russo, A., Buonadonna, A., D'Andrea, M., Pasetto, L.M., Pessa, S., Errante, D., De Pangher, V., Giusto, M., Medici, M., Gaion, F., Sandri, P., Galligioni, E., Bonura, S., Boccalon, M., Biason, P., Frustaci, S. (2006). The role of UGT1A1*28 polymorphism in the pharmacodynamics and pharmacokinetics of irinotecan in patients with metastatic colorectal cancer. *J. Clin. Oncol.*, *24*, 3061–3068.

Tukey, R.H., Strassburg, C.P. (2000). Human UDP-glucuronosyltransferases: metabolism, expression, and disease. *Annu. Rev. Pharmacol. Toxicol.*, *40*, 581–616.

Turgeon, D., Carrier, J.S., Chouinard, S., Bélanger, A. (2003a). Glucuronidation activity of the UGT2B17 enzyme toward xenobiotics. *Drug Metab. Dispos.*, *31*, 670–676.

Turgeon, D., Chouinard, S., Belanger, P., Picard, S., Labbe, J.F., Borgeat, P., Belanger, A. (2003b). Glucuronidation of arachidonic and linoleic acid metabolites by human UDP-glucuronosyltransferases. *J. Lipid. Res.*, *44*, 1182–1191.

Turpeinen, M., Nieminen, R., Juntunen, T., Taavitsainen, P., Raunio, H., Pelkonen, O. (2004). Selective inhibition of CYP2B6-catalyzed bupropion hydroxylation in human liver microsomes in vitro. *Drug Metab. Dispos.*, *32*, 626–631.

Uchaipichat, V., Mackenzie, P.I., Elliot, D.J., Miners, J.O. (2006a). Selectivity of substrate (trifluoperazine) and inhibitor (amitriptyline, androsterone, canrenoic acid, hecogenin, phenylbutazone, quinidine, quinine, and sulfinpyrazone) "probes" for human udp-glucuronosyltransferases. *Drug Metab. Dispos.*, *34*, 449–456.

Uchaipichat, V., Winner, L.K., Mackenzie, P.I., Elliot, D.J., Williams, J.A., Miners, J.O. (2006b). Quantitative prediction of in vivo inhibitory interactions involving glucuronidated drugs from in vitro data: the effect of fluconazole on zidovudine glucuronidation. *Br. J. Clin. Pharmacol.*, *61*, 427–439.

Vanstapel, F., Blanckaert, N. (1988). Topology and regulation of bilirubin UDP-glucuronyltransferase in sealed native microsomes from rat liver. *Arch. Biochem. Biophys.*, *263*, 216–225.

Venkatakrishnan, K., Obach, R.S. (2007). Drug-drug interactions via mechanism-based cytochrome P450 inactivation: points to consider for risk assessment from in vitro data and clinical pharmacologic evaluation. *Curr. Drug Metab.*, *8*, 449–462.

Venkatakrishnan, K., von Moltke, L.L., Court, M.H., Harmatz, J.S., Crespi, C.L., Greenblatt, D.J. (2000). Comparison between cytochrome P450 (CYP) content and relative activity approaches to scaling from cDNA-expressed CYPs to human liver microsomes: ratios of accessory proteins as sources of discrepancies between the approaches. *Drug Metab. Dispos.*, *28*, 1493–1504.

Venkatakrishnan, K., von Moltke, L.L., Greenblatt, D.J. (2001). Human drug metabolism and the cytochromes P450: application and relevance of in vitro models. *J. Clin. Pharmacol.*, *41*, 1149–1179.

Visser, T.J., Kaptein, E., van Toor, H., van Raaij, J.A., van den Berg, K.J., Joe, C.T., van Engelen, J.G., Brouwer, A. (1993). Glucuronidation of thyroid hormone in rat liver: effects of in vivo treatment with microsomal enzyme inducers and in vitro assay conditions. *Endocrinology*, *133*, 2177–2186.

Walsky, R.L., Obach, R.S., Gaman, E.A., Gleeson, J.P., Proctor, W.R. (2005). Selective inhibition of human cytochrome P4502C8 by montelukast. *Drug Metab. Dispos.*, *33*, 413–418.

Whetstine, J.R., Yueh, M.F., McCarver, D.G., Williams, D.E., Park, C.S., Kang, J.H., Cha, Y.N., Dolphin, C.T., Shephard, E.A., Phillips, I.R., Hines, R.N. (2000). Ethnic differences in human flavin-containing monooxygenase 2 (FMO2) polymorphisms: detection of expressed protein in African-Americans. *Toxicol. Appl. Pharmacol.*, *168*, 216–224.

Wilkinson, G.R., Shand, D.G. (1975). Commentary: a physiological approach to hepatic drug clearance. *Clin. Pharmacol. Ther.*, *18*, 377–390.

Williams, J.A., Hurst, S.I., Bauman, J., Jones, B.C., Hyland, R., Gibbs, J.P., Obach, R.S., Ball, S.E. (2003a). Reaction phenotyping in drug discovery: moving forward with confidence? *Curr. Drug. Metab.*, *4*, 527–534.

Williams, P.A., Cosme, J., Ward, A., Angove, H.C., Matak Vinkovi , D., Jhoti, H. (2003b). Crystal structure of human cytochrome P450 2C9 with bound warfarin. *Nature, 424* (*6947*), 464–468.

Williams, J.A., Hyland, R., Jones, B.C., Smith, D.A., Hurst, S., Goosen, T.C., Peterkin, V., Koup, J.R., Ball, S.E. (2004). Drug-drug interactions for UDP-glucuronosyltransferase substrates: a pharmacokinetic explanation for typically observed low exposure (AUCi/AUC) ratios. *Drug Metab. Dispos.*, *32*, 1201–1208.

Woosley, R.L., Chen, Y., Freiman, J.P., Gillis, R.A. (1993). Mechanism of the cardiotoxic actions of terfenadine. *JAMA*, *269*, 1532–1536.

Yamanaka, H., Nakajima, M., Katoh, A., Tamura, O., Ishibashi, H., Yokoi, T. (2005). Trans-3′-hydroxycotinine O- and N-glucuronidations in human liver microsomes. *Drug Metab. Dispos.*, *33*, 23–30.

Yamazaki, H., Johnson, W.W., Ueng, Y.F., Shimada, T., Guengerich, F.P. (1996). Lack of electron transfer from cytochrome b5 in stimulation of catalytic activities of cytochrome P450 3A4. Characterization of a reconstituted cytochrome P450 3A4/NADPH-cytochrome P450 reductase system and studies with apo-cytochrome b5. *J. Biol. Chem.*, *271*, 27438–27444.

Yamazaki, H., Shimada, T. (1998). Comparative studies of in vitro inhibition of cytochrome P450 3A4-dependent testosterone 6beta-hydroxylation by roxithromycin and its metabolites, troleandomycin, and erythromycin. *Drug Metab. Dispos.*, *26*, 1053–1057.

Yamazaki, H., Ueng, Y.F., Shimada, T., Guengerich, F.P. (1995). Roles of divalent metal ions in oxidations catalyzed by recombinant cytochrome P450 3A4 and replacement of NADPH–cytochrome P450 reductase with other flavoproteins, ferredoxin, and oxygen surrogates. *Biochemistry*, *34*, 8380–8389.

Yokono, A., Morita, S., Someya, T., Hirokane, G., Okawa, M., Shimoda, K. (2001). The effect of CYP2C19 and CYP2D6 genotypes on the metabolism of clomipramine in Japanese psychiatric patients. *J. Clin. Psychopharmacol.*, *21*, 549–555.

Yoon, Y.J., Kim, K.B., Kim, H., Seo, K.A., Kim, H.S., Cha, I.J., Kim, E.Y., Liu, K.H., Shin, J.G. (2007). Characterization of benidipine and its enantiomers' metabolism by human liver cytochrome P450 enzymes. *Drug Metab. Dispos.*, *35*, 1518–1524.

Yu, A. (2007). Small interfering RNA in drug metabolism and transport. *Curr. Drug Metab.*, *8*, 700–708.

Yu, L., Lu, S., Lin, Y., Zeng, S. (2007). Carboxyl-glucuronidation of mitiglinide by human UDP-glucuronosyltransferases. *Biochem. Pharmacol.*, *73*, 1842–1851.

Yun, C.H., Okerholm, R.A., Guengerich, F.P. (1993). Oxidation of the antihistaminic drug terfenadine in human liver microsomes. Role of cytochrome P-450 3A(4) in N-dealkylation and C-hydroxylation. *Drug Metab. Dispos.*, *21*, 403–409.

Zhang, H., Davis, C.D., Sinz, M.W., Rodrigues, A.D. (2007). Cytochrome P450 reaction-phenotyping: an industrial perspective. *Expert Opin. Drug Metab. Toxicol.*, *3*, 667–687.

Zhang, W., Kilicarslan, T., Tyndale, R.F., Sellers, E.M. (2001). Evaluation of methoxsalen, tranylcypromine, and tryptamine as specific and selective CYP2A6 inhibitors in vitro. *Drug Metab. Dispos.*, *29*, 897–902.

Ziegler, D.M. (1988). Flavin-containing monooxygenases: catalytic mechanism and substrate specificities. *Drug Metab. Rev.*, *19*, 1–32.

FURTHER READING

Bjornsson, T.D., Callaghan, J.T., Einolf, H.J., Fischer, V., Gan, L., Grimm, S., Kao, J., King, S.P., Miwa, G., Ni, L., Kumar G., McLeod, J., Obach, S.R., Roberts, S., Roe, A., Shah, A., Snikeris, F., Sullivan, J.T., Tweedie, D., Vega, J.M., Walsh, J., Wrighton, S.A., Pharmaceutical Research and Manufacturers of America Drug Metabolism/Clinical Pharmacology Technical Working Groups. (2003). The conduct of in vitro and in vivo drug-drug interaction studies: a PhRMA perspective. *J. Clin. Pharmacol.*, *43*, 443–469. Review.

Emoto, C., Iwasaki, K. (2007). Approach to predict the contribution of cytochrome P450 enzymes to drug metabolism in the early drug-discovery stage: the effect of the expression of cytochrome b(5) with recombinant P450 enzymes. *Xenobiotica*, *37*, 986–999.

Emoto, C., Murase, S., Iwasaki, K. (2006). Approach to the prediction of the contribution of major cytochrome P450 enzymes to drug metabolism in the early drug-discovery stage. *Xenobiotica*, *36*, 671–683.

Ito, K., Hallifax, D., Obach R.S., Houston J.B. (2005). Impact of parallel pathways of drug elimination and multiple cytochrome P450 involvement on drug-drug interactions: CYP2D6 paradigm. *Drug Metab. Dispos.*, *33*, 837–844.

Ito, K., Iwatsubo, T., Kanamitsu, S., Ueda, K., Suzuki, H., Sugiyama, Y. (1998). Prediction of pharmacokinetic alterations caused by drug-drug interactions: metabolic interaction in the liver. *Pharmacol. Rev.*, *50*, 387–412.

Lu, A.Y., Wang, R.W., Lin, J.H. (2003). Cytochrome P450 in vitro reaction phenotyping: a re-evaluation of approaches used for P450 isoform identification. *Drug Metab. Dispos.*, *31*, 345–350. Review.

Madan, A., Usuki, E., Burton, L., Ogilvie, B., Parkinson, A. (2002). In vitro approaches for studying the inhibition of drug-metabolizing enzymes and identifying the drug-metabolizing enzymes responsible for the metabolism of drugs, in *Drug-Drug Interactions* (ed. A.D. Rodrigues), Marcel Dekker, New York, pp. 217–294.

Obach, R.S., Walsky, R.L., Venkatakrishnan, K., Gaman, E.A., Houston, J.B., Tremaine, L.M. (2006). The utility of in vitro cytochrome P450 inhibition data in the prediction of drug-drug interactions. *J. Pharmacol. Exp. Ther.*, *316*, 336–348.

Proctor, N.J., Tucker, G.T., Rostami-Hodjegan, A. (2004). Predicting drug clearance from recombinantly expressed CYPs: intersystem extrapolation factors. *Xenobiotica*, *34*, 151–178.

Rodrigues, A.D. (1999). Applications of heterologous expressed and purified human drug-metabolizing enzymes: an industrial perspective, in *Handbook of Drug Metabolism* (ed. T.F. Woolf), Marcel Dekker, New York, pp. 279–320.

Rodrigues, A.D. (2005). Impact of CYP2C9 genotype on pharmacokinetics: are all cyclooxygenase inhibitors the same? *Drug Metab. Dispos.* *33*, 1567–1575.

Rodrigues, A.D., Rushmore, T. (2002). Cytochrome P450 pharmacogenetics in drug development: in vitro studies and clinical consequences. *Curr. Drug Metab.*, *3*, 289–309.

Shou, M., Lu, T., Krausz, K.W., Sai, Y., Yang, T., Korzekwa, K.R., Gonzalez, F.J., Gelboin H.V. (2000). Use of inhibitory monoclonal antibodies to assess the contribution of cytochromes P450 to human drug metabolism. *Eur. J. Pharmacol.*, *394*, 199–209.

Tang, W., Wang, R.W., Lu, A.Y. (2005). Utility of recombinant cytochrome p450 enzymes: a drug metabolism perspective. *Curr. Drug Metab.*, *6*, 503–517. Review.

Thummel, K.E., Shen, D.D. (2002). The role of the gut mucosa in metabolically based drug-drug interactions, in *Drug-Drug Interactions* (ed. A.D. Rodrigues), Marcel Dekker, New York, pp. 359–385.

Venkatakrishnan, K., von Moltke, L.L., Greenblatt, D.J. (2001). Application of the relative activity factor approach in scaling from heterologously expressed cytochromes p450 to human liver microsomes: studies on amitriptyline as a model substrate. *J. Pharmacol. Exp. Ther.*, *297*, 326–337.

CHAPTER 17

Nuclear Receptor-Mediated Gene Regulation in Drug Metabolism

HONGBING WANG and WEN XIE

17.1 INTRODUCTION

17.1.1 General Features of Nuclear Receptors

The mammalian nuclear receptor (NR) superfamily contains nearly 50 structurally related members, including both receptors for which ligands are known and "orphan receptors" for which there are, as yet, no known ligands (Laudet *et al.*, 1992; Giguere, 1999). Regulation of gene expression at the transcriptional level by NRs has an important role in both cellular development and the body's defensive systems, including drug metabolism and clearance (Kastner and Chambon, 1995; Mangelsdorf *et al.*, 1995; Honkakoski and Negishi, 2000; Moore *et al.*, 2000a). Members of the nuclear receptor family share several structural features. Two of the modulator domains are common structural motifs shared by most of the nuclear receptors (Fig. 17.1a). The highly conserved DNA-binding domain (DBD) that links the receptor to specific promoter regions of its target genes is termed hormone response elements (HREs) or xenobiotic response elements (XREs). As the most conserved domain of NRs, DBD can recognize the response elements that contain one or two consensus core half-sites related to the hexamer ACAACA (steroid receptors) or AGGTCA (estrogen receptors and other nuclear receptors) (Cairns *et al.*, 1991; Lee *et al.*, 1993). The HREs or XREs usually contain two NR half-sites of the consensus hexameric sequences, organized as inverted, everted, or direct repeats with a 3- to 6-base-pair spacing (Fig. 17.1b). The carboxy (C)-terminal portion of the receptor contains a less conserved ligand-binding domain (LBD), which interacts directly with hormones and xenobiotics. Remarkable variations in the shape and size of LBDs provide the structural basis for fitting a wide range of ligands with diverse chemical structures. It appears that the receptors that bind with high affinity to a specific ligand, such as estrogen receptor (ER) and thyroid hormone receptor (TR), have relatively small ligand-binding pockets, whereas the xenobiotic NRs, such as pregnane X receptor (PXR) and the peroxisome proliferator-activated receptors (PPARs), contain much larger and more flexible ligand-binding cavities that

Drug Metabolism Handbook: Concepts and Applications, Edited by Ala F. Nassar, Paul F. Hollenberg, and JoAnn Scatina
Copyright © 2009 by John Wiley & Sons, Inc.

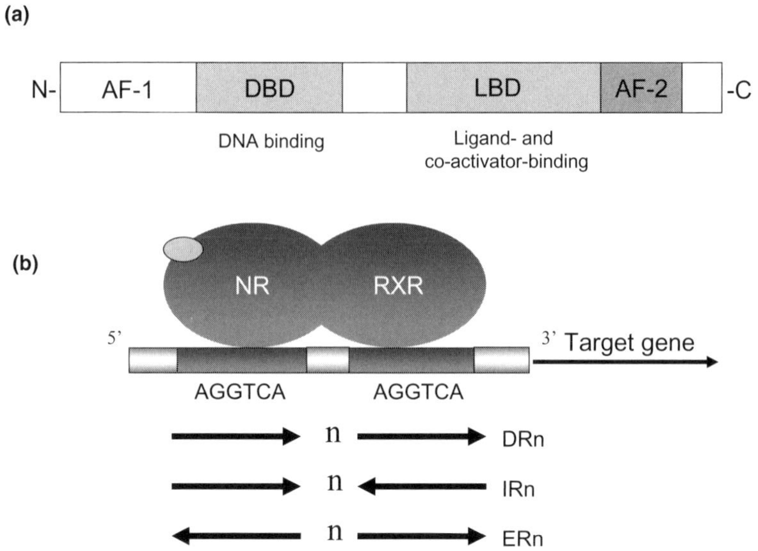

Figure 17.1 Structure and functional domains of nuclear receptors. **(a)** Basal structure of nuclear receptors. AF-1 and DBD are at the amino (N)-terminal, and AF-2 and LBD are at the carboxy (C)-terminal. **(b)** Nuclear receptors can recognize and bind to AGGTCA half-sites spaced by several base pairs as direct repeats separate by n base pairs (DRn), inverted repeats separated by n base pairs (IRn), or everted repeats separated by n base pairs (ERn).

can accommodate the apparently more promiscuous ligands (Nolte *et al.*, 1998; Watkins *et al.*, 2001). Ligand binding induces significant conformational changes in the folding of the LBD, and leads to the recruitment or replacement of transcriptional cofactors, such as co-activators (steroid receptor co-activators, SRCs) and corepressors (nuclear corepressor, NCoR; silencing mediator for retinoid and thyroid hormone receptors, SMRT) (Freedman, 1999; Lanz *et al.*, 1999).

17.1.2 Nuclear Receptors and Drug-Metabolizing Enzymes

An important requirement for physiological homeostasis is the detoxification and removal of endogenous hormones and xenobiotic compounds with biological activities. Much of the detoxification and elimination is carried out by the phase I cytochrome P450 (CYP) enzymes, phase II conjugation enzymes, and "phase III" drug transporters. The products of phase I metabolism are generally more polar and readily excreted than the parent compounds. The phase I enzyme-mediated functionalization also often enables xenobiotics to become better substrates for the phase II conjugating enzymes. The phase II metabolism involves conjugation of hydrophilic moieties, such as sugar, glutathione, and sulfate, to further increase polarity and water solubility of xenobiotics, therefore promoting their excretion and elimination. In addition to phase I and II enzymes, equally important is a group of transporter proteins expressed in various tissues, such as the liver, intestine, brain, and kidney, which modulate the absorption, distribution, and excretion of many drugs.

As our understanding of nuclear receptors grows, it has become evident that nuclear receptors play important roles in the transcriptional regulation of these drug-metabolizing enzymes (DMEs) (Honkakoski and Negishi, 2000; Synold, Dussault and Forman, 2001; Wang and LeCluyse, 2003). The induction profiles of the major DMEs are remarkably linked to the activation of several orphan nuclear receptors. For instance, compounds such as rifampicin (RIF) and phenobarbital (PB) are PXR activators and inducers of CYP3A, CYP2B, UGT1A1, CYP2C9, and multidrug resistance protein 1(MDR1), while 1,4-bis[2-(3,5-dichloropyridyloxy)]benzene (TCPOBOP) and 6-(4-chlorophenyl:imidazo[2,1-b]thiazole-5carbaldehyde O-(3,4-dichlorobenzyl)oxime (CITCO) induce CYP2B, CYP3A, and UGT1A1 through the activation of constitutive androstane receptor (CAR) (Kliewer *et al.*, 1998; Geick, Eichelbaum and Burk, 2001; Ferguson *et al.*, 2002; Maglich *et al.*, 2003). This association of orphan nuclear receptors with particular DMEs not only demonstrates DME specificity, but also defines species-specific differences in the DME induction responses. Therefore, orphan nuclear receptor binding and activation assays have been utilized as useful tools for screening drug candidates *in vitro* for their potentials to cause drug–drug interactions (DDIs) via enzyme induction.

In the following section, we focus on recent advances in our knowledge and understanding of the regulation of DMEs by nuclear receptors. Although it has become increasingly evident that not only PXR and CAR but many other NRs may also be important in regulating the expression of drug disposition genes either directly, or by modulating the activities of PXR and CAR, specific emphasis has been placed on PXR and CAR because of their broad and overlapping xenobiotic specificity and their central role in regulating the expression of multiple DMEs and drug transporters. In addition, the role of a non-NR transcriptional factor, aryl hydrocarbon receptor (AhR), in xenobiotic regulation will also be briefly discussed.

17.2 PREGNANE X RECEPTOR

17.2.1 Cloning and Initial Characterization of PXR

In 1998, several groups published the cloning and initial characterization of the orphan nuclear receptor PXR. Kliewer and colleagues identified mouse PXR on the basis of its sequence homology with other NRs. It was named PXR based on its activation by various natural and synthetic pregnanes (Kliewer *et al.*, 1998). The cloning of human PXR (hPXR) was first reported by Blumberg and colleagues, and this group (led by Dr. Ronald M. Evans at the Salk Institute) initially named this receptor "steroid and xenobiotic receptor (SXR)" based on its activation by steroids and xenobiotics (Blumberg *et al.*, 1998). hPXR was also cloned by Bertilsson and colleagues and was alternately referred to as the pregnane-activated receptor (PAR) (Bertilsson *et al.*, 1998).

PXR shares a common NR modular structure with a conserved *N*-terminal DBD and a *C*-terminal LBD. The DBD contains two zinc-finger motifs, which mediate interaction with specific DNA sequences known as HREs. The LBD, in addition to determining ligand-binding specificity, contains a ligand-inducible transactivation function (AF2) and a motif that directs binding to the common heterodimerization partner retinoid X receptor (RXR). Although DBDs of the mammalian PXRs are highly conserved, sharing more than 95% amino acid identity, the LBDs of PXRs are much more divergent across species than those of other NRs. For example, the human and rat PXR share only 76% amino acid identity in their LBD, whereas most human and rodent NR orthologs share more than 90% amino acid identity (Maglich *et al.*, 2001). The LBD sequence divergence of PXR between species is believed to be responsible for the species-specific response to ligands/drugs.

PXR is highly expressed in the liver and small intestine, where most DMEs are also highly expressed and induced. In rodents, lower levels of PXR mRNA have also been detected in the kidney, stomach, lung, uterus, ovary, and placenta. PXR can be activated by xenobiotics (including many clinical drugs) and endobiotics (such as bile acids and some hormones, including glucocorticoids, anti-glucocorticoids, and estrogens), which underlie the significance of PXR in both xenobiotic and endobiotic responses. PXR orthologs also exhibit interesting and striking species-dependent ligand specificity. For example, the anti-glucocorticoid pregnenolone-16α-carbonitrile (PCN) is an effective activator of mouse and rat PXRs, but has little effect on human PXR. In contrast, the antibiotic RIF can activate the human and rabbit PXRs, but not the mouse and rat PXRs. Detailed elaboration on the PXR ligand species specificity and its biological significance will be discussed later.

Like many other orphan nuclear receptors, activation of PXR by ligand binding leads to formation of PXR/RXR heterodimer and transcriptional activation of target genes by binding to specific PXR response elements (PXREs) in target gene promoters. The classical PXRE is DR-3 which is composed of two copies of direct repeat of the consensus AG(G/T)TCA NR binding motif and separated by three nucleotides. Other types of PXREs, including DR-4, DR-5, ER-6 (everted repeat spaced by 6-bp), ER-8, and IR-0 (inverted repeat without a spacing nucleotide), have also been shown to be bound by the PXR-RXR heterodimers (Kliewer, Goodwin and Willson, 2002; Sonoda *et al.*, 2003; Saini *et al.*, 2004; Xie *et al.*, 2004).

17.2.2   PXR in Phase I CYP Enzyme Regulation

The oxidative CYP enzymes catalyze the metabolic conversion of xenobiotics to polar derivatives that are more readily eliminated. Among the 78 mouse and 65 human CYP enzymes itemized within the current build of the UniGene sequence data bank, a significant number of these enzymes are known to be induced by various environmental pollutants, pharmaceuticals, or steroid metabolites. A more current list of CYPs is available at http://drnelson.utmem.edu/CytochromeP450.html. Among CYP enzymes, the CYP3A isozymes are of particular medical significance since they are involved in the metabolism of more than 50% of clinical drugs as well as nutraceuticals and herbal medicines (Maurel *et al.*, 1996). In 1998, PXR was isolated as a candidate xenobiotic receptor (xenosensor) postulated to regulate CYP3A gene expression (Bertilsson *et al.*, 1998; Blumberg *et al.*, 1998; Kliewer *et al.*, 1998; Lehmann *et al.*, 1998). PXR regulates the expression of both human and rodent CYP3A genes by directly binding to the XREs ER-6 and DR-3 localized in the promoter regions of the human and rodent CYP3A genes, respectively. Both "loss-of-function" gene knockout and "gain-of-function" transgenic mouse studies have provided convincing genetic and pharmacological evidence to support the role of PXR in CYP3A-mediated xenobiotic responses. Targeted disruption of the mouse PXR locus abolished the CYP3A xenobiotic response to prototypic inducers such as PCN and dexamethasone (Xie *et al.*, 2000; Staudinger *et al.*, 2001). In contrast, hepatic expression of an activated form of hPXR in transgenic mice resulted in sustained induction of CYP3A enzymes and enhanced protection against xenobiotic toxicants, such as zoxazolamine and tribromoethanol (Xie *et al.*, 2000).

In addition to CYP3A isozyme, PXR has also been shown to regulate other CYP enzymes, including CYP2B6 (Honkakoski *et al.*, 1998; Goodwin *et al.*, 2001; Dussault *et al.*, 2003; Wang and Negishi, 2003), CYP2B9 (Dvorak *et al.*, 2003), CYP2C8 (Ferguson *et al.*, 2005), and CYP2C9 (Dvorak *et al.*, 2003).

17.2.3 PXR in Phase II Enzyme Regulation

The phase II conjugating enzymes include broad specificity transferases, such as UDP-glucuronosyltransferase (UGT), glutathione *S*-transferase (GST), and sulfotransferases (SULT) (Hayes and Pulford, 1995; Salinas and Wong, 1999; Nagata and Yamazoe, 2000; Tukey and Strassburg, 2000). The phase II conjugation has a dual role in drug metabolism. First, conjugation reactions not only terminate the ability of electrophiles to react with DNA and proteins, they also prevent nucleophiles from interacting with receptor proteins. At the same time, conjugation increases the water solubility of the compounds, which in turn, promotes renal and biliary excretion (Sheweita, 2000). Therefore, the phase II reactions play a critical role in detoxifying both exogenous and endogenous chemicals.

17.2.3.1 PXR in UGT Regulation Glucuronidation, a major metabolic pathway for many endo- and xenobiotics, is catalyzed by enzymes belonging to the family of membrane-bound UGTs. Using UDP-glucuronic acid as a sugar donor, UGTs catalyze the transfer of glucuronic acid to a variety of substrates and thus convert small lipophilic molecules to water-soluble glucuronides. UGT-mediated glucuronidation functions as the principle means to eliminate steroid, heme metabolites, environmental toxin, and drugs from the body (Mackenzie *et al.*, 1997, 2003; Radominska-Pandya *et al.*, 1999; Tukey and Strassburg, 2000). Among UGTs, UGT1A1 is one of the most characterized UGT isoform. UGT1A1 is the critical enzyme responsible for the detoxification of bilirubin. Deficiency in the expression and/or activity of UGT1A1 in patients can lead to severe accumulation of unconjugated bilirubin, medically termed "hyperbilirubinemia," a hallmark of the Crigler–Najjar (CN) syndrome (Tukey and Strassburg, 2000).

Our own study showed that UGT1A1 is under the positive control of PXR. UGT1A1 mRNA and protein expression was up-regulated in transgenic mice that express the activated hPXR (VP-hPXR) and in RIF-treated "humanized" hPXR transgenic mice (Xie *et al.*, 2003). The microsomal glucuronidation activity toward β-estradiol (a UGT1A1 substrate), thyroid hormones, corticosterone, and xenobiotics (such as 4-nitrophenol and 4-OH-PhIP), was also increased in the VP-hPXR transgenic mice. Chen and colleagues showed that PCN, a rodent-specific PXR agonist, could induce UGT expression and increase UGT enzymatic activity in wild-type mice and this induction was abolished in PXR-null mice (Chen, Staudinger and Klaassen, 2003). The identification of a DR3-like PXR responsive element (<u>GGTTCA</u>TAA<u>AGGGTA</u>) in the human UGT1A1 promoter further established UGT1A1 as a direct target gene of PXR. In addition to UGT1A1, the expression of UGT1A9, but not UGT1A2 and 2B5, was also increased by PXR activation, suggesting that the UGT induction by PXR is isoform-specific (Chen, Staudinger and Klaassen, 2003).

In addition to PXR, the expression of UGT1A1 and several other UGT isoforms have also been reported to be regulated by several other nuclear receptors, including CAR (Sugatani *et al.*, 2001; Huang *et al.*, 2003; Xie *et al.*, 2003) and PPARα (Barbier *et al.*, 2003; Ikeda *et al.*, 2005).

17.2.3.2 *PXR in SULT Regulation*   The cytosolic sulfotransferases (SULTs) are another important family of phase II enzymes that catalyze the conjugation of nucleophilic compounds. SULTs catalyze the transfer of a sulfonyl group from a sulfate donor 3′-phosphoadenosine 5′-phosphosulfate (PAPS) to hydroxyl or amino groups of acceptor molecules, forming sulfate or sulfamate conjugates. SULTs are specific to sulfonate small lipophilic molecules, such as steroids, bioamines, and therapeutic drugs. It was believed that, although SULTs and UGTs have similar substrate spectrum, SULTs have higher affinities but lower turnover rates compared with UGTs (Morris and Pang, 1987). Thus, UGT activities are likely to play a more significant role when substrates are abundant due to the higher turnover rate of UGTs, whereas SULTs may play a leading role in cases where concentrations of the substrate are lower. SULT-mediated sulfonation is known to play a significant role in the homeostasis of sex hormones that are present in low concentrations in plasma (Qian *et al.*, 2001).

The expression of rodent hepatic Sult2a9, also called Sult2a1 or dehydroepiandrosterone sulfotransferase (DHEA SULT or STD), is subjected to transcriptional regulation by PXR and an IR-0 element located in the 5′-flanking region of rodent Sult2a gene is required for its activation by PXR (Sonoda *et al.*, 2003). The expression of Sult2a9/2a1/STD has also been shown to be regulated by FXR (Song *et al.*, 2001), CAR (Saini *et al.*, 2004), and LXR (Uppal *et al.*, 2007). Interestingly, all four receptors share the same IR-0 response element to regulate the expression of this SULT isoform. The hierarchy and relative contribution of individual nuclear receptors in Sult2a regulation remain to be determined.

17.2.3.3 *PXR in GST Regulation*   GSTs are soluble homo- and heterodimeric enzymes that use reduced glutathione in conjugation and reduction reactions. GSTs play an important role in the conjugation of electrified chemicals. Since chemical carcinogens are often highly nucleophilic, they also represent potential substrates for GST. For this reason, GSTs are believed to play an important role to protect cells from genotoxic compounds. The cytosolic GST isozymes of rodents and humans can be grouped into several classes, such as Alpha, Mu, Pi, Theta, Omega, and Zeta, based on their amino acid sequences, immunological properties, and substrate specificities (Falkner *et al.*, 2001; Townsend and Tew, 2003). The Alpha, Mu, and Pi classes are the most abundantly expressed GSTs.

GST regulation by PXR was first hinted by several studies of general profiling of gene expression (Maglich *et al.*, 2002; Rosenfeld *et al.*, 2003). It was also reported that GSTA2 expression was induced by PXR and an IR-6 response element was responsible for this transcriptional regulation (Falkner *et al.*, 2001). A more systemic and comparative analysis of GST regulation by PXR was reported in transgenic mice that bear the expression of activated PXR in the liver and intestine (Gong *et al.*, 2006). In this study, it was shown that the expressions of GST Alpha, Pi, and Mu classes are all under the control of PXR. Interestingly, PXR-mediated GST regulation exhibited clear isoform-, tissue-, and gender-specificity: (i) GST Mu was the only isoform that was up-regulated by PXR in both liver and intestine in both sexes; (ii) GST Alpha was induced in the small intestine, but not in the liver; and (iii) PXR had an opposite effect on hepatic GST Pi expression, inducing this class in females, but suppressing it in males.

Paradoxically, although the overall GST expression and activity were increased, activation of PXR sensitized the response to the oxidative xenotoxicant paraquat *in vivo* and in cultured cancer cells. Moreover, heightened paraquat sensitivity in transgenic mice was female-specific. Whether the PXR-mediated, gender-specific GST regulation accounts for the intact paraquat sensitivity in transgenic males remains to be determined. Nevertheless, the regulation of GSTs by PXR suggests that this regulatory pathway may be relevant to carcinogenesis by sensitizing normal and cancerous tissues to oxidative cellular damage (Gong *et al.*, 2006).

17.2.4 PXR in Drug Transporter Regulation

In addition to regulating phase I CYP enzymes and phase II conjugating enzymes, PXR also participates in the regulation of drug transporters. Drug transporters are responsible for uptake and efflux of endogenous and exogenous chemicals, including many clinical prescribed drugs, thus the expression and activity of transporters affect drug efflux and clearance (Klaassen and Slitt, 2005). Organic anion-transporting polypeptides (OATPS) are a family of major uptake drug transporters in the liver. OATPs are localized to the basolateral membrane of hepatocytes and transport compounds, including organic anions, organic cations, and neutral compounds, from blood into hepatocytes. The OATP family includes 9 human, 13 rat, and 15 mouse members (Hagenbuch and Meier, 2003). PXR ligand PCN administration induced Oatp2 mRNA expression in the livers of wild-type mice, but not PXR-null mice (Staudinger *et al.*, 2003). A DR-3 type PXR response element has been identified in the rat Oatp2 promoter (Guo *et al.*, 2002).

Multidrug resistance-associated protein (Mrps) are efflux transporters for structurally diverse amphipathic chemicals and organic anions. The liver-enriched Mrp2 is localized to the canalicular membrane of hepatocytes and is responsible for the hepatobiliary excretion of amphipathic anions (Kast *et al.*, 2002; Kruh and Belinsky, 2003). Natural mutations in MRP2 cause Dubin–Johnson syndrome/hyperbilirubinemia II, a disorder characterized by impaired transfer of anionic conjugates into the bile. Induction of Mrp2 by PXR ligand PCN and dexamethasone was observed in primary hepatocytes isolated from the livers of wild-type mice, but not from PXR-null mice, implicating a role of PXR in the regulation of mouse Oatp2 (Kast *et al.*, 2002). An ER-8 type of NR response element was identified in the promoter of the rat Mrp2 gene that confers the induction of Mrp2 by PXR (Kast *et al.*, 2002). Mrp3 was also reported to be induced by PXR (Teng, Jekerle and Piquette-Miller, 2003). In the intestine, PXR has been shown to stimulate the expression of Mdr1, which encodes an ATP-dependent efflux pump that transports a wide variety of xenobiotics, including many widely used prescription drugs. In humans, activation of PXR in the intestine may decrease intestinal drug absorption by increasing the expression of *Mdr1*. A DR-4 type PXR response element was identified in *Mdr1* gene promoter (Geick, Eichelbaum and Burk, 2001).

17.2.5 Implications of PXR-Mediated Gene Regulation in Drug Metabolism

The implications of PXR-mediated gene regulation in drug metabolism and drug interaction have been recognized since the initial cloning of this receptor. Our body encounters numerous xenobiotic chemicals, including prescription drugs, over-the-counter medications, and herbal medicines. Many of them, especially when accumulated in excess, may exert toxic effects through various mechanisms. The process of drug metabolism is known to largely depend upon a concerted action of phase I and II enzymes, as well as drug transporters. Expressed mainly in the liver and intestine, these enzymes and transporters are capable of recognizing an amazing diversity of xenobiotics to promote their clearance.

The regulation of DMEs by PXR is involved in clinical DDIs, in which one drug accelerates the metabolism of a second medicine and may change or cause adverse results. Because CYP enzymes can recognize a large spectrum of pharmaceutical substrates, a CYP gene-inducing drug is potentially capable of affecting the metabolism and clearance of any co-consumed drugs. As mentioned earlier, the identification of RIF as a potent hPXR agonist has provided an explanation why this antibiotic drug is prone to DDIs. In another example, St. John's wort (SJW), a popular herbal remedy for mild depression, has been reported to trigger severe adverse interactions with several clinical drugs, such as oral contraceptives, the HIV protease inhibitor indinavir, and the immunosuppressant cyclosporin. Such DDIs are likely results of activation of PXR and consequent induction of CYP3A by SJW and the subsequent increased metabolism and/or decreased bioavailability of co-metabolized drugs. In the case of birth control pills, the use of SJW enhances drug clearance, increasing contraceptive failure and thus the birth of "miracle babies." The identification of SJW as a PXR agonist offered a plausible explanation for the propensity of SJW to cause DDIs (Moore et al., 2000b).

Several traditional Chinese medicines (TCMs) have also been implicated in DDIs. TCMs are essential components of alternative medicines. One clinical concern of herbal product use is the effect of herbal products on the metabolism of coadministered drugs. We showed that two TCM herbs Wu Wei Zi (*Schisandra chinensis Baill*) and Gan Cao (*Glycyrrhiza uralensis Fisch*), can activate PXR and induce the expression of several DMEs and transporters, including CYP3A and 2C9 and the multidrug resistance-associated protein 2 (MRP2) in reporter gene assay and in primary hepatocyte cultures (Mu et al., 2006). The anticoagulant warfarin is known to be metabolized by CYP2C9 in humans (Goldstein, 2001). As expected, administration of Wu Wei Zi and Gan Cao extracts in rats resulted in an increased metabolism of coadministered warfarin, reinforcing concerns involving the safe use of herbal medicines and other nutraceuticals to avoid PXR-mediated DDIs (Mu et al., 2006).

Having known the potential of PXR-activating agents in causing DDIs, it is also important to emphasize that PXR activation alone may not be sufficient to predict the propensity of DDIs. Sinz and colleagues recently published the evaluation of 170 xenobiotics in an hPXR transactivation assay and compared these results with known clinical DDIs. Of the 170 xenobiotics tested, 54% of them demonstrated some level of hPXR transactivation. However, by taking into consideration cell culture conditions (solubility, cytotoxicity, appropriate drug concentration in media), as well as *in vivo* pharmacokinetics (therapeutic plasma concentration or C_{max}, distribution, route of administration, dosing regimen, liver exposure, potential to inhibit CYP3A4), the risk potential of CYP3A4 enzyme induction for most compounds reduced dramatically. By employing this overall interpretation strategy, the final percentage of compounds predicted to significantly induce CYP3A4 reduced to 5%, all of which are known to cause DDIs in the clinic (Sinz et al., 2006).

17.2.6 Endobiotic Function of PXR

Even though PXR was initially identified as a "xenobiotic receptor," emerging evidence has pointed to an equally important role of PXR as an "endobiotic receptor" that responds to a wide array of endogenous chemicals (endobiotics), such as bile acids and their intermediates, as well as certain steroid hormones. Another mechanism for PXR's involvement in endobiotic homeostasis is that many endobiotics, such as bile acids, bilirubin, and adrenal steroids, are also substrates of PXR target enzymes and transporters. Activation of PXR by endogenous or xenobiotic ligands has implications in several important physiological and pathological conditions. For this reason, there have been extensive discussions on whether or not PXR can be explored as a therapeutic target (Xie *et al.*, 2004; Gong *et al.*, 2005).

17.2.6.1 *PXR in Bile Acid Detoxification and Cholestasis*

One family of endogenous PXR ligands identified shortly after the cloning of PXR is bile acid. Bile acids are catabolic end products of cholesterol metabolism. They are physiologically important in the formation of bile and solubilizing biliary lipids and promoting their absorption. However, excessive bile acids are potentially toxic. For example, the secondary bile acid lithocholic acid (LCA) has been shown to cause cholestasis in experimental animals and has long been suspected of doing the same in humans. As an average human releases 600 mL of bile a day, the potential for disrupting bile flow (cholestasis) and the resultant accumulation of toxic by-products is significant. Therefore, excess bile acid should be efficiently eliminated to avoid the toxic effect.

PXR has been demonstrated to acts as an LCA sensor and plays an essential role in detoxification of cholestatic bile acids (Staudinger *et al.*, 2001; Xie *et al.*, 2001). Studies in different animal models showed that activation of PXR protected against severe liver damage induced by LCA. Pretreatment of wild-type mice, but not of the PXR-null mice, with PCN reduced the toxic effects of LCA. Moreover, genetic activation of PXR by expressing the activated PXR in the liver of transgenic mice was sufficient to confer resistance to the hepatotoxicity of LCA. The cholestatic preventive effect of PXR was initially reasoned to be due to the activation of CYP3A, an important CYP enzyme responsible for bile acid hydroxylation (Staudinger *et al.*, 2001; Xie *et al.*, 2001). Subsequent identification of SULT2A, a bile acid detoxifying hydroxysteroid sulfotransferase, as a PXR target gene suggested that additional PXR target genes may have also contributed to the phenotype (Sonoda *et al.*, 2002). Several follow-up studies, including those using mice with individual or combined loss of PXR and CAR, have suggested that PXR-responsive bile acid transporter regulation may also play a role in preventing cholestasis (Zhang *et al.*, 2004; Stedman *et al.*, 2005; Uppal *et al.*, 2005). Since activation of PXR was sufficient to prevent cholestasis, it has been suggested that PXR agonists may prove useful in the treatment of human cholestatic liver disease, a notion that has been supported by several clinical observations. Both RIF and SJW have been empirically used to treat cholestatic liver diseases (Kliewer, Goodwin and Willson, 2002). The relief from cholestasis-associated pruritus and amelioration of cholestasis by RIF was associated with increased 6α-hydroxylation of bile acids, which in turn facilitates glucuronidation by the UGTs at the 6α-hydroxy position. Both RIF and SJW are potent agonists of hPXR and both CYP3A and UGT are PXR target genes, suggesting the anti-cholestatic effects of RIF and SJW are mediated by PXR.

17.2.6.2 *PXR in Bilirubin Detoxification and Clearance*

Bilirubin is the catabolic by-product of heme proteins, such as β-globin and CYP enzymes. Accumulation of bilirubin in the blood is potentially hepatotoxic and neurotoxic. For example, an insufficiency in expression of UGT1A1, a key enzyme for the conjugation of bilirubin, in the Crigler–Najjar syndrome and Gilbert's diseases results in severe hyperbilirubinemia. Deficiency of MRP2, a drug transporter responsible for the hepatic excretion of conjugated bilirubin, leads to Dubin–Johnson syndrome, characterized by the accumulation of glucuronidated bilirubin. PXR has been shown to induce the expression of multiple key components in the clearance pathway, including UGT1A1, OATP2, GSTA1 and 2, and MRP2. OATP2 facilitates bilirubin uptake from blood into hepatocytes (Klaassen and Slitt, 2005). GSTA1 and 2 reduce bilirubin back efflux from hepatocytes into blood. MRP2 promotes the canalicular efflux of conjugated bilirubin. Consistent with the pattern of gene regulation, activation of PXR in transgenic mice has been shown to prevent experimental hyperbilirubinemia (Xie *et al.*, 2003).

17.2.6.3 *PXR in Adrenal Steroid Homeostasis and Drug–Hormone Interactions*

PXR plays an important endobiotic role in adrenal steroid homeostasis. Our recent study showed that genetic (by using the VP-hPXR transgene) and pharmacological (by using the hPXR ligand rifampicin) activation of hPXR in mice markedly increased plasma concentrations of corticosterone and aldosterone, the respective primary glucocorticoid and mineralocorticoid in rodents. The increased levels of corticosterone and aldosterone were associated with activation of adrenal steroidogenic enzymes, including *CYP11a1, CYP11b1, CYP11b2*, and *3β-Hsd*. The PXR-activating transgenic mice also exhibited hypertrophy of the adrenal cortex, loss of glucocorticoid circadian rhythm, and lack of glucocorticoid responses to psychogenic stress (Zhai *et al.*, 2007).

Interestingly, the VP-hPXR transgenic mice had normal pituitary secretion of adrenocorticotropic hormone (ACTH) and the corticosterone suppressing effect of dexamethasone (DEX) was intact, suggesting a functional hypothalamus–pituitary–adrenal (HPA) axis despite a severe disruption of adrenal steroid homeostasis. The ACTH-independent hypercortisolism in the PXR-activating transgenic mice is reminiscent of the pseudo-Cushing's syndrome in patients, the clinical hallmark of which is the normal DEX suppression despite a high circulating level of glucocorticoid. Pseudo-Cushing's syndrome is most seen in alcoholic, depressed, or obese subjects. It is of interest to know whether or not these susceptible patients are associated with increased expression and/or activity of PXR. The glucocorticoid effect appeared to be PXR-specific, as the activation of CAR in transgenic mice had little effect. We propose that PXR is a potential endocrine-disrupting factor that may have broad implications in steroid homeostasis and drug–hormone interactions (Zhai *et al.*, 2007).

17.2.6.4 *PXR in Lipid Metabolism*

PXR has also recently been shown to play an endobiotic role by impacting lipid homeostasis (Zhou *et al.*, 2006). Expression of an activated PXR in the livers of transgenic mice resulted in an increased hepatic deposit of triglycerides. This PXR-mediated lipid accumulation was independent of the activation of the lipogenic transcriptional factor sterol regulatory element-binding protein 1c (SREBP-1c) and its primary lipogenic target enzymes, including fatty acid synthase (FAS) and acetyl CoA carboxylase 1 (ACC-1). Instead, the lipid accumulation in transgenic mice was associated with an increased expression of the free fatty acid transporter CD36 and several accessory lipogenic enzymes, such as stearoyl CoA desaturase-1 (SCD-1) and long chain free fatty acid elongase (FAE). Studies using transgenic and knockout mice showed that PXR is both necessary and sufficient for *Cd36* activation. Promoter analyses revealed a DR-3 type of PXR response element in the mouse *Cd36* gene promoter, establishing *Cd36* as a direct transcriptional target of PXR. The hepatic lipid accumulation and *Cd36* induction was also seen in the hPXR "humanized" mice treated with the hPXR agonist rifampicin. The activation of PXR was also associated with an inhibition of pro-β-oxidative genes, such as peroxisome proliferators-activated receptor α (PPARα) and thiolase, and an up-regulation of PPARγ, a positive regulator of CD36. The cross-regulation of CD36 by PXR and PPARγ suggests that this fatty acid transporter may function as a common target of orphan nuclear receptors in their regulation of lipid homeostasis.

17.2.7 Species Specificity of PXR and the Creation of "Humanized" Mice

17.2.7.1 Challenges for Rodents as Drug Metabolism Models Primary hepatocytes are a common *in vitro* system to evaluate metabolism, toxicity, and enzyme induction (Tucker *et al.*, 2001; Weaver, 2001; Weaver *et al.*, 2001). Human hepatocytes are considered the most relevant system to evaluate or predict human metabolism or effects of a new drug. A significant disadvantage of this system is the lack of routine availability of good quality human liver tissues or cells. Other model systems that may provide more consistent access include immortalized hepatocytes or humanized animal models (Xie *et al.*, 2000; Mills *et al.*, 2004). Over the years, it has been perceived that rodent models have limited utility in predicting drug-related human effects due to significant species differences in DMEs, transporters, and nuclear hormone receptors. For example, PCN is an effective CYP3A inducer in rodents but not in humans, and rifampicin induces CYP3A in humans but not in rats (Kocarek *et al.*, 1995). These findings have been attributed to the species differences in the effect of several drugs on CYP3A expression mediated by PXR (Kocarek *et al.*, 1995; Jones *et al.*, 2000). These differences across species demand the development of humanized animal models to evaluate the potential effect of a chemical in humans using an animal model.

17.2.7.2 Species Specificity of the Rodent and Human PXR Both hPXR/SXR and mPXR are highly expressed in the liver and small intestine and share many functional properties, in particular, the regulation of *CYP3A* genes. However, as discussed earlier, these two orthologs are pharmacologically distinct in that strong activators of one receptor are often poor activators of the other. This species-specific ligand profile is reflected by the sequence divergence in the LBDs of the mouse and human receptors. The crystal structure of the hPXR LBD has been solved (Watkins *et al.*, 2001). The hydrophobic ligand-binding cavity of hPXR is composed of a large, smooth surface containing only a small number of polar residues, suggesting that it is not necessary for activators to conform to a restricted orientation. Based on site-directed mutational analysis, the position and nature of these polar residues were found to be crucial for establishing the precise pharmacological activation profile of hPXR (Watkins *et al.*, 2001). Indeed, conversion of four amino acids of mouse PXR that correspond to the hPXR-specific activator SR12813-interacting residues in hPXR produces a hybrid mouse–human PXR that was no longer activated by PCN and was only weakly activated by SR12813 in reporter assays (Watkins *et al.*, 2001). The structural and pharmacological differences between hPXR and mPXR and that of other species might reflect the difference in the diets of rodents and primates and the evolutionary need to respond to a different set of ingested nutrients and xenobiotics.

17.2.7.3 Creation and Characterization of the hPXR "Humanized" Mice Based on the fact that the species origin of the receptor is the determining factor for the ligand specificity between species, we have created transgenic mouse models to determine whether the human receptor is sufficient to establish a human response profile (Xie *et al.*, 2000) (Fig. 17.2). First, hPXR transgenic mice were generated by expressing the hPXR in the mouse liver. The liver-specific expression of the transgene was

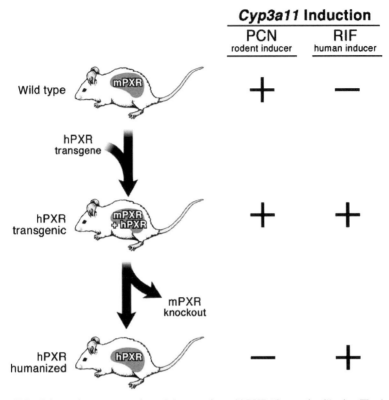

Figure 17.2 Schematic representation of the creation of hPXR "humanized" mice. The humanization was achieved in the liver only when the liver-specific albumin promoter was used to direct the transgene expression, or in both the liver and intestine when the fatty acid-binding protein promoter was used. PCN, pregnenolone-16α-carbonitrile; RIF, rifampicin. "+" and "–" mean induction and lack of induction, respectively.

accomplished by using the mouse albumin promoter. Because the resulting mice harbor both mPXR and hPXR in their livers, the transgenic mice exhibited a chimeric or combined CYP3A response to both the rodent-specific inducer PCN and the human-specific inducer rifampicin. These results imply that mice only expressing hPXR could be fully humanized for the xenobiotic response. These animals were created by breeding the hPXR transgene into the mPXR knockout background. In contrast to the null mice that are devoid of CYP3A induction by steroids, replacement of mPXR with transgenic hPXR restores xenobiotic regulation with a humanized response profile. These mice readily responded to human inducers, such as rifampicin, in the equivalent range of the standard oral dosing regimen in humans (300–600 mg per 70 kg man) and exhibited similar pharmacokinetics of CYP3A regulation (Kolars *et al.*, 1992; Xie *et al.*, 2000). A "fully" human profile of CYP3A inducibility is obtained in the mPXR null/ hPXR transgenic mice. Therefore, these experiments provide compelling evidence that PXR functions as a species-specific xenosensor mediating the adaptive hepatic response. This is also one of the rare examples in which replacing a single transcriptional regulator enables conversion of species-specific gene regulation.

The original "humanized" mice bear the expression hPXR exclusively in the liver (Xie *et al.*, 2000). Since both the DMEs and xenobiotic receptors are also highly expressed in the intestinal tracts, it is conceivable that mouse models with the humanized receptors expressed in both the liver and intestine would represent a more complete humanized mouse model. The liver and intestine dual humanization has been achieved by using the fatty acid-binding protein (FABP) gene promoter that targets the expression of hPXR transgene to both the liver and intestine (Zhou *et al.*, 2006). An alternative strategy is to "knock-in" hPXR in the mouse locus. This would not only direct expression of hPXR in both liver and intestine, but also normalize expression levels and tissue patterns to the endogenous gene.

17.2.7.4 *Significance of Humanized Mice in Drug Metabolism Studies and Drug Development* The creation of mouse models with humanized xenobiotic response may aid pharmaceutical development by predicting potential DDIs (Moore and Kliewer, 2000; Kliewer, Goodwin and Willson, 2002; Sonoda *et al.*, 2003; Xie *et al.*, 2004). For decades, rodent models have been standard components in the assessment of potential toxicity for the development of candidate human drugs. However, their reliability as predictors of the human xenobiotic response is limited due to the species specificity of the xenobiotic response. To date, there has been no reliable system outside of humans to directly and quantitatively assess the DDIs. Primary cultures of human hepatocytes are valuable. However, since the hepatocytes are from individual patients, the utility of human hepatocytes is compromised by interindividual variability, limited tissue resources, and high cost. The humanized mice exhibited a "humanized" hepatic xenobiotic response profile, readily responding to the human-specific inducer RIF in a concentration range equivalent to the standard oral dosing regimen in humans (Xie *et al.*, 2000). The creation of these mice represents a major step toward generating a humanized rodent toxicological model that is continuously renewable and completely standardized. In addition, a PXR-mediated and mechanism-based transfection and reporter gene system has also been shown to be an effective *in vitro* approach to screen for drugs that may be precocious hPXR activators. While the *in vitro* screen is fast, the availability of hPXR "humanized" mice offers a unique screening tool to evaluate DDIs *in vivo*. The humanized mouse models represent important steps in the development of safer human drugs.

Another benefit of the hPXR humanized mice is its utility as a pharmacological model to dissect the function of PXR *in vivo*. The PXR ligand effect on gene expression and/or pathophysiological outcome may require long-term drug treatment. Although PCN is a potent rodent PXR agonist, the wild-type mice cannot tolerate a long-term treatment of PCN due to its toxicity. In this case, the use of humanized mice become necessary since mice can tolerate chronic treatment of RIF, the hPXR agonist. We have successfully used the humanized mice in our recent studies of PXR effect on lipogenesis and adrenal steroid homeostasis (Zhou *et al.*, 2006; Zhai *et al.*, 2007).

17.3 CONSTITUTIVE ANDROSTANE/ACTIVATED RECEPTOR (CAR)

17.3.1 Identification of CAR as the Regulator of CYP2B Genes

Originally termed as MB67 or CARα, the orphan nuclear receptor CAR (NR1I3) was isolated by screening a human liver library with degenerate oligonucleotide probes in

1994 (Baes *et al.*, 1994). The name CAR was initially defined as constitutively activated receptor, because it forms a heterodimer with RXR that binds to retinoic acid response elements (RAREs) and transactivates target genes in the absence of ligands in trans-fection assays (Baes *et al.*, 1994; Choi *et al.*, 1997). Using human (h) CAR cDNA as probe, the mouse counterpart was isolated subsequently, and was designated CARß (Choi *et al.*, 1997). Recently, two androstane metabolites, androstanol (5α-androstan-3α-ol) and androstenol (5α-androstan-16-en-3α-ol), were found to be endogenous antagonists of mouse CAR, giving an alternative name of CAR as constitutive andro-stane receptor. As discussed in the following contents, a breakthrough linkage of CAR with CYP2B genes by Negishi and colleagues has triggered a wealth of sub-sequent studies focusing on the role of CAR on multiple DME and transporter gene expression (Sueyoshi *et al.*, 1999; Xie and Evans, 2001; Ferguson *et al.*, 2002; Wang and Negishi, 2003).

Xenobiotic induction of CYP2B genes in different species has been realized for a long time. Through a series of elegant studies, a specific sequence termed PB-responsive element (PBRE) or PB-responsive unit (PBRU), or later delineated as PB-responsive enhancer module (PBREM), has been identified as the key sequences that are respon-sible for xenobiotic induction of CYP2B genes (Trottier *et al.*, 1995; Park, Li and Kemper, 1996; Honkakoski and Negishi, 1997; Honkakoski *et al.*, 1998). However, substantial progress in our understanding of PBREM activation by PB was obtained upon identification of CAR as the predominant regulator of PBREM activation. Using cell-based transfection assays, a series of known nuclear receptors, including RXR, CAR, liver X receptor (LXR), TR, hepatocyte nuclear factor (HNF4), and chicken ovalbumin upstream promoter-transcription factor (COUP-TF), were screened for their ability to transactivate mouse PBREM in HepG2 and HEK293 cells (Honkakoski *et al.*, 1998). The results revealed that only liver-enriched mouse (m) CAR was able to activate PBREM-driven reporter gene expression. In HepG2 cells transfected with mCAR, co-transfected PBREM was activated in the absence of PB, consistent with constitutive activation of retinoic acid responses by CAR in the absence of retinoic acid (Honkakoski and Negishi, 1998; Sueyoshi *et al.*, 1999). Co-transfection of RXR with mCAR in HepG2 cells resulted in synergistic augmentation of PBREM reporter activity, indicating that CAR activates PBREM by forming a heterodimer with RXR. Electrophoretic mobility shift assays (EMSAs) confirmed that CAR bound to the NR1 and NR2 motifs of the PBREM in the presence but not in the absence of RXR (Honkakoski *et al.*, 1998). The protein that bound to the PBREM in liver of PB-treated mice was purified using NR1-affinity chromatography and identified as mCAR using binding assays and immunoblotting analysis (Honkakoski *et al.*, 1998). Subsequently, a group of structurally diverse CYP2B inducers was shown to activate the PBREM reporter gene via mCAR. A HepG2 stable cell line designated as g2car-3, which con-stitutively expresses mCAR, activates transfected PBREM as well as the endogenous *CYP2B6* gene (Sueyoshi *et al.*, 1999; Sugatani *et al.*, 2001). In addition, treatment of g2car-3 cells with mCAR activators such as TCPOBOP and PB could effectively reac-tivate PBREM activity and *CYP2B6* gene expression initially repressed by androstanol pretreatment. Definite evidence establishing the role of CAR in the regulation of PB induction of the *CYP2B* genes was obtained from experiments using CAR-knockout mice (Wei *et al.*, 2000; Ueda *et al.*, 2002). Two independent mouse lines with targeted ablation of the CAR gene were generated by replacing a segment of the CAR gene that includes part of the DBD, with the coding region of b-galactosidase. As expected, CAR expression was absent in CAR–/– mice, while no overt phenotype was observed in either male or female mice (Wei *et al.*, 2000). Interestingly, induction of CYP2b10 mRNA by PB and TCPOBOP was abolished completely in the CAR–/– homozygotes.

17.3.2   CAR in Other DME Regulation

17.3.2.1   CAR in CYP2Cs Regulation   The cytochrome P450 subfamily CYP2C enzyme is responsible for the metabolism of approximately 20% of therapeutic drugs and many endogenous compounds in humans (Goldstein, 2001). Induction of human CYP2Cs can result in drug tolerance as well as DDIs. Recently, a series of studies by Goldstein and coworkers demonstrated that xenobiotic induction of CYP2Cs is collaboratively mediated by several NRs, including CAR, PXR, GR, and HNF4α (Ferguson *et al.*, 2002; Chen *et al.*, 2003, 2004). In primary cultured human hepatocytes, CYP2C9 is inducible by xenobiotics including PB, RIF, and dexamethasone (Gerbal-Chaloin *et al.*, 2001; Rae *et al.*, 2001; Raucy *et al.*, 2002). A role for CAR in CYP2C9 regulation was suggested by evidence that both constitutive and drug inducible CYP2C9 mRNA expression was elevated in HepG2 cell lines stably transfected with mCAR or hCAR, in the absence or presence of TCPOBOP or PB, respectively (Ferguson *et al.*, 2002). By analyzing the CYP2C9 promoter region, two CAR-responsive elements, a DR5 located from −2900 to −2841 bp, and a DR4 located between −1822 and −1783 bp, were discovered independently (Ferguson *et al.*, 2002; Gerbal-Chaloin *et al.*, 2002). Gel shift assays demonstrated that both DRs bind CAR/RXR heterodimers. Furthermore, transfection experiments in HepG2 cells revealed that constitutively activated CAR increased reporter gene expression in vectors containing either of the DR4 or DR5 promoter elements (Ferguson *et al.*, 2002).

CYP2C8 is the most strongly inducible member of the CYP2C subfamily in human hepatocyte cultures. Examination of the CYP2C8 promoter region revealed that a distal PXR/CAR-binding site located at −8806 bp of the transcriptional start site, confers inducibility of CYP2C8 via the PXR agonist rifampicin and the CAR agonist CITCO (Ferguson *et al.*, 2005). Further analysis of the CYP2C8 basal promoter region revealed several putative binding motifs, including TATA-box, HNF3, CCAAT enhancer-binding protein, HNF4α, and GATA-binding protein, indicating many other potential nuclear factors may be involved in controlling the transcription of CYP2C8. Another important member of the human CYP2C subfamily, CYP2C19, metabolizes a number of clinically important drugs such as omeprazole, diazepam, propranolol, and *S*-mephenytoin. Chen *et al.* (2003) reported that transcriptional regulation of CYP19 gene expression is mediated by CAR, PXR, and GR through the consensus binding sites (CAR-RE; −1891/−1876 bp) and (GRE; −1750/−1736 bp), respectively. In addition, a functional CAR-RE has been localized and characterized in the promoter region of mouse cyp2c37 gene (Jackson *et al.*, 2006). Intriguingly, this CAR-RE only responds to CAR activation but not to PXR. Overall, these data suggest that CAR has a promiscuous role in the regulation of multiple CYP2C gene expression.

17.3.2.2   CAR Regulation of UGT1A1 Gene   For decades, PB has been used in the treatment of hyperbilirubinemia since this compound induces the expression of UGT1A1 enzyme which involves the glucuronidation of elevated bilirubin (Kopecky, Schwarz and Schwenzel, 1975; Schambach and Menzel, 1975; Ishii *et al.*, 1994). The molecular mechanism of PB induction of UGT1A1 remained elusive until a 290-bp distal enhancer module, located from −3483 to −3194 bp of the UGT1A1 promoter, was identified as CAR response element by Sugatani *et al.* (2001). Consisting of three putative nuclear receptor palindromes, this enhancer module can be activated in transfection assays by human and mouse CAR in HepG2 cells, and mouse primary hepatocytes. Site-directed mutagenesis of this module abrogated the response. Furthermore, in a separate study, Huang *et al.* (2003) demonstrated that CAR activation increases hepatic expression of several relevant genes involved in bilirubin clearance, including UGT1A1, MRP2, SLC21A6, GSTA1, and GSTA2, in wild-type but not in CAR-null mice. A similar activation profile was observed also in a line of transgenic mice expressing hCAR but not mCAR in the liver. These results suggest that both human and mouse CAR can respond to elevated levels of bilirubin by regulating simultaneously the expression of UGT1A1, MRP2, SLC21A6, GSTA1, and GSTA2. It is of note that within this 290-bp distal enhancer module in the promoter of UGT1A1, several other xenobiotic response elements have also been recognized subsequently, including a PXRE, two GREs, and an AhR-specific XRE (Xie *et al.*, 2003; Yueh *et al.*, 2003; Kuno, Togawa and Mizutani, 2007).

17.3.2.3 *CAR Regulation of Drug Transporters*

Drug transporters play critical roles in drug absorption and clearance. Several efflux transporters, as well as members of the OATP uptake transporter family, have been identified as transcriptional target genes of the xenobiotic nuclear receptors PXR, CAR, PPAR, and FXR (Dussault *et al.*, 2001; Geick, Eichelbaum and Burk, 2001; Synold, Dussault and Forman, 2001; Guo *et al.*, 2002; Guo, Johnson and Klaassen, 2002; Kast *et al.*, 2002). Multidrug resistance-associated protein 2 (MRP2 or ABCC2) is involved in the transport of organic anions, bile salts, glutathione, and xenobiotics, such as the anticancer drugs cisplatin, anthacyclines, vinca alkaloids, and methotrexate (Konig *et al.*, 1999). Treatment with the PXR-selective activator PCN and the mCAR-specific ligand TCPOBOP resulted in MRP2 induction at the mRNA level. By analyzing the 5′-flanking region of the MRP2 gene, Kast *et al.* isolated an unusual 26-bp sequence, located at 440-bp upstream of the MRP2 transcription initiation site, that contained an everted repeat of AGTTCA spaced by eight nucleotides (ER-8) (Kast *et al.*, 2002). This ER-8 motif was shown to bind both CAR and PXR, and was activated by CAR and PXR ligands in cell-based reporter assays. These findings demonstrated the versatility in the binding abilities of PXR and CAR, and provided the first evidence that CAR and PXR can recognize response element (ER-8) in addition to the previously identified ER-6 element. Interestingly, PB induction of MRP2 was greater in PXR-null mice compared with wild-type animals. Given the fact that PB induction of CYP2B and CYP3A expression was unaffected in PXR-null mice but abolished in CAR-null mice, it is reasonable to speculate that mPXR may interact negatively with PB and that shared PXR/CAR-target genes may be induced by PB more efficaciously in mice lacking PXR due to reduced ligand-binding competition.

Conflicting data have been presented in the literature regarding the contribution of CAR in the regulation of MRP3, another ATP-dependent efflux transporter. Recently, Staudinger *et al.* showed that selective activation of mCAR by TCPOBOP mediates the inducible expression of Oatp2 and Mrp3 (Staudinger *et al.*, 2003). These results were similar to those obtained in PXR-null mice, where PB induction of both Oatp2 and Mrp3 was enhanced. Another report also demonstrated by real-time RT-PCR that both CAR and PXR play critical roles in regulating mouse Mrp3 gene expression (Maglich *et al.*, 2002). In contrast, hCAR appears not to be involved in the regulation of human MRP3 gene expression. Both hCAR- and mCAR-expressing

HepG2 cells have been used to elucidate the role of CAR in regulating induction of human CYP2B6 and UGT1A1 by PB-like inducers (Kawamoto *et al.*, 1999; Sueyoshi *et al.*, 1999; Sugatani *et al.*, 2001). Because of low endogenous CAR expression in HepG2 cells, neither CYP2B6 nor UGT1A1 is inducible by PB. However, higher endogenous CYP2B6 and UGT1A1 expression was observed in HepG2 cells stably transfected with mCAR due to the constitutive activation of CAR in these cells (Sueyoshi *et al.*, 1999). Different from the prototypical CAR target genes, such as CYP2B6 and UGT1A1, human *MRP3* gene expression was induced to a similar extent by PB, but not TCPOBOP in HepG2 and g2car-3 cells, suggesting that PB induction of human MRP3 is mediated by transcription factors other than CAR (Kiuchi *et al.*, 1998; Xiong *et al.*, 2002). Furthermore, sequence analysis of the 5′-flanking region of human MRP3 did not reveal any consensus binding sites for either CAR or PXR, although binding sites were identified for other transcription factors such as Sp1, AP1, AP2, AP3, N-myc, CCAAT/enhancer-binding protein, hepatic nuclear factor-5, and nuclear factor-kB (Fromm *et al.*, 1999; Takada, Suzuki and Sugiyama, 2000). Further study is required to determine if these factors are important in the regulation of *MRP3* gene expression by PB and other compounds. In addition, discrepancies between the results from *in vivo* and *in vitro* studies of the regulation of other transporters should be taken into consideration. For instance, induction of Mdr1 by PCN was observed *in vitro* but not *in vivo* (Salphati and Benet, 1998; Jones *et al.*, 2000). Mrp2 was induced by PB in primary cultured rat hepatocytes but not in rats *in vivo* (Ogawa *et al.*, 2000). These reported *in vitro* and *in vivo* differences suggest that some regulatory mechanisms implicated by *in vitro* methods may not be physiologically relevant under *in vivo* conditions.

17.3.3 Mechanism of CAR Activation

Consistent with its designated name, CAR is constitutively activated in all the immortalized cell lines without xenobiotic stimulation, making the investigation of CAR activation much more challenging. In contrast, CAR activation in primary cultured hepatocytes and intact liver *in vivo* is inducer-dependent, and CAR is localized in the cytoplasm of liver cells and translocated into the nucleus only after exposure to xenobiotic inducers (Kawamoto *et al.*, 1999; Sueyoshi *et al.*, 1999). One of the essential features that are critical for regulating xenobiotic-induced CAR activation involves the nuclear accumulation of this receptor. A number of studies showed that PB and TCPOBOP treatments were associated with decreased CAR expression in cytoplasmic fraction compared with nuclear extracts; this finding was substantiated by the increased binding of a CAR/RXR heterodimer to NR1 in liver nuclear extracts (Kawamoto *et al.*, 1999). These results indicate that nuclear translocation of CAR might be the first activation step in response to PB-type inducers. As a result, the mechanism by which CAR is retained in the cytoplasm has now become a major focus of research interest. Inhibition of nuclear translocation by antagonists such as androstanes might be a possible mechanism for explaining cytoplasmic retention of CAR in primary cells and intact liver. However, physiological serum concentrations of androstane metabolites are several magnitudes lower than concentrations required for repression of CAR activation of the PBREM and induction of *CYP2B* genes in HepG2 cells (Dufort *et al.*, 2001). Moreover, treatment of mouse primary hepatocytes with high concentrations of androstenol could not inhibit PB-driven nuclear accumulation of CAR (Kawamoto *et al.*, 1999). Nuclear translocation is common among steroid receptors such as GR, progesterone receptor (PR), vitamin D receptor (VDR), and PPAR (Picard and Yamamoto, 1987; Guiochon-Mantel *et al.*, 1991; Walker, Htun and Hager, 1999), which are dissociated from their cytoplasmic complexes with chaperone proteins such as heat shock proteins, immunophilins, or p23 protein. Upon direct binding to their agonists, these receptors are released from their complexes and translocate into the nucleus (Defranco *et al.*, 1995; Pratt, Silverstein and Galigniana, 1999). However, it seems that direct ligand binding is not necessary for a drug to stimulate CAR nuclear transloca-tion. Although PB induces nuclear translocation of both mouse and human CAR, it does not bind to either in various *in vitro* assays (Moore *et al.*, 2000a; Tzameli *et al.*, 2000). Similarly, TCPOBOP treatment resulted in mouse and human CAR accumula-tion in the nucleus even though hCAR could neither bind nor be activated by TCPOBOP (Tzameli *et al.*, 2000). Advances in our understanding of the CAR translocation mecha-nism include the discovery that okadaic acid (OA), an inhibitor of protein phosphatase 2A, was able to inhibit PB- and TCPOBOP-driven CAR translocation and CYP2b10 induction in mouse primary hepatocytes (Sidhu and Omiecinski, 1997; Honkakoski and Negishi, 1998; Kawamoto *et al.*, 1999). Nonetheless, subsequent studies have demon-strated that translocation of CAR is necessary but not sufficient to activate this recep-tor. For instance, calcium/calmodulin-dependent kinase (CaMK) inhibitors, KN-93 and KN-62 could efficiently inhibit PB-mediated CYP2B induction without affecting CAR nuclear accumulation (Negishi, 2000). Activation of nuclear receptors is generally associated with co-regulators such as SRCs. Recently, we observed that the known hCAR deactivator clotrimazole could efficiently translocate EYFP-hCAR into nucleus of mouse hepatocytes *in vivo*, but dissociates SRC-1 from the nuclear localized CAR/RXR heterodimer (unpublished data). Taken together, these observations suggest that tight control of an otherwise constitutively active receptor is achieved by cytosolic sequestration *in vivo* and primary hepatocytes through complicated yet unknown mechanisms.

17.3.4 Identification of CAR Activator

One of the major difficulties for investigating drug-induced CAR activation *in vitro* is the high constitutive activation of CAR in immortalized cell lines, due to the spontane-ous accumulation of the receptor in the nucleus in the absence of inducer (Wang and Negishi, 2003). Because of difficulties in evaluation of hCAR activation, the contribu-tion of this receptor to DDIs has remained ambiguous. Mouse CAR-specific target genes and their distinct physiological roles were characterized by utilizing TCPOBOP as a specific mCAR agonist. However, previously identified hCAR ligands such as clotrimazole and 5b-pregnane-3, 20-dione were shown also to activate hPXR. Later, CITCO was identified as a selective hCAR agonist (Maglich *et al.*, 2003). This com-pound was associated with an EC_{50} of 25 nM for hCAR activation of a CYP3A4-XREM reporter construct in transfected CV-1 cells (Maglich *et al.*, 2003). Furthermore, green fluorescent protein (GFP)-CAR fusion protein was accumulated in the nucleus of rat primary hepatocytes after CITCO treatment. CITCO was also used to delineate hCAR target genes in primary cultures of human hepatocyte. The induction of eight DMEs was investigated in three separate sets of human hepatocytes treated with CITCO or hPXR ligand RIF. As expected, CYP2B6 mRNA displayed the most potent induction by 100nM of CITCO. In contrast, CYP3A4 mRNA was more robustly induced by rifampicin than by CITCO. Using TCPOBOP in mice and CITCO in humans, specific CAR target genes were also compared between rodents and humans. Results demonstrated that, in general, human and mouse CAR share a number of similar target genes, including mouse Cyp2b10 and human CYP2B6, the transporter genes mouse Mdr1 and human MDR1, and the phase II sulfotransferase genes mouse SULTN and human SULT1A1. However, to date, there is only limited publication addressing CITCO as a specific hCAR activator; more critical investigation is required to evaluate the specificity of hCAR activation, particularly in *in vitro* cell-based transfection assays by using different cell lines.

Recently, several groups have identified alternative splicing variants of wild-type hCAR with altered functional activities (Auerbach *et al.*, 2003; Arnold, Eichelbaum and Burk, 2004; Jinno *et al.*, 2004; Ikeda *et al.*, 2005). One of these variants, hCAR3, exhibited significantly lower basal activity than wild-type hCAR and was activated extensively by the known hCAR activator CITCO in a cell-based reporter gene assay (Auerbach *et al.*, 2005). In a recent study, we have evaluated the activation profile of hCAR3 more extensively by testing known hCAR activators as well as 16 other CYP3A4 and/or CYP2B inducers (Faucette *et al.*, 2007). Notably, all the known hCAR activators, including the direct activator CITCO, artemisinin and the indirect activators PB and PHN, enhanced CYP2B6 reporter gene expression via hCAR3 in HepG2 cells, suggesting hCAR3 could be activated through both ligand-dependent and -independent mechanisms. Overall, in the absence of other suitable *in vitro* models and based on similar chemical sensitivities to wild-type hCAR, hCAR3 cell-based reporter gene assays may provide a useful tool for *in vitro* screening of hCAR activation in a relatively high-throughput format.

17.3.5   Species Differences in Activation of CAR

Although significant progress on the molecular mechanisms underlying CAR regulation of DME gene expression has been achieved in rodents, information pertaining to hCAR regulation of human DMEs is relatively limited. Two apparent issues that remain to be resolved are whether hCAR is regulated in the same fashion as the rodent CAR and whether CAR is the predominate regulator of CYP2B6 in human liver. At this point, the conclusions regarding the role of CAR in the regulation of CYP2B or other genes are based largely on rodent CAR, primarily mCAR (Hani *et al.*, 1996; Choi *et al.*, 1997). Although hCAR exhibits some common characteristics with its rodent counterparts, such as undergoing nuclear translocation after PB treatment and binding to the PBREM, there are distinct differences between rodent and hCAR. For example, TCPOBOP, the most potent mCAR ligand identified to date, cannot bind or activate either rat or hCAR. All known mCAR inhibitors, such as androstenol, progesterone, and androgens, and CaMK inhibitors do not inhibit hCAR activation. Overall, current evidence suggests that there are clear species-specific differences in CYP2B induction and CAR activation, which severely hamper the extrapolation of animal data to humans. For these reasons, regulation of hCAR by drugs and other xenobiotics has become a more complex and urgent issue that has yet to be solved.

Generation of transgenic PXR-null and CAR-null mice, especially the humanized PXR and CAR mice model, has made it possible to specifically address the effects of these receptors on DME regulation *in vivo*. In CAR-null mice, the strong induction of CYP2b10 gene expression by PB and TCPOBOP was totally absent (Wei *et al.*, 2000). Similarly, in obese Zucker rats, which express extremely low levels of CAR but have normal levels of PXR expression, PB only moderately induces both CYP2B and

CYP3A, whereas CYP3A was significantly induced by the PXR-specific activator PCN (Zelko and Negishi, 2000). Given that xenobiotic induction of DMEs shows striking species specificity due to the species-dependent biochemical properties of nuclear receptors, a CAR humanized mouse line has also been generated (Zhang *et al.*, 2002). This line specifically expresses hCAR in the mCAR-null mice liver driven by the potent albumin promoter. Using this model, opposite responses of mCAR and hCAR to a widely used antiemetic, meclizine, have been established (Huang *et al.*, 2004). Although meclizine treatment augmented the expression of CAR target genes in wild-type mice, both hCAR transactivation and the PB-induced expression of these CAR target genes were inhibited by meclizine treatment in primary hepatocytes isolated from the liver of humanized CAR mice. Furthermore, meclizine decreased the susceptibility of APAP-induced hepatotoxicity in humanized CAR mice (Huang *et al.*, 2004).

17.3.6 Cross talk between CAR, PXR, and Other Nuclear Receptors

Although PXR and CAR have been implicated as the primary regulator of CYP3A and CYP2B, respectively, accumulated evidence has suggested cross talk between these two NRs through recognizing and binding to each other's target gene response elements. Significant overlap exists in the inducers of CYP2B and CYP3A genes in rodent and human primary hepatocytes, such as clotrimazole, RU486, RIF, phenytoin, and PB. Moreover, both receptors can bind to several of the same compounds, including androstanol, clotrimazole, estradiol, and TCPOBOP with different affinities. The current model of cross talk between PXR and CAR is derived primarily from rodent data and commonly depicts symmetrical cross-regulation of CYP2B and CYP3A induction by PXR and CAR, respectively. However, because of documented species differences in nuclear receptor signaling pathways, it is speculative to extrapolate the extents of cross-regulation of *CYP2B* and *CYP3A* genes by rodent PXR and CAR to the corresponding human receptors. The selectivities of hPXR and hCAR for their cross-regulatory genes remain unclear because of the lack of comparisons of CYP2B6 and CYP3A4 promoter binding and functional activation between hPXR and hCAR, as well as the previous unavailability of selective hCAR activators. Recently, we have evaluated the symmetry of hPXR and hCAR cross talk by comparing the selectivities of these receptors for CYP2B6 and CYP3A4 (Faucette *et al.*, 2006). Human hepatocyte studies revealed nonselective induction of both CYP2B6 and CYP3A4 by hPXR activation but marked preferential induction of CYP2B6 by selective hCAR activation. Gel shift assays demonstrated that hPXR exhibited strong and relatively equal binding to all functional response elements in both *CYP2B6* and *CYP3A4* genes, whereas hCAR displayed significantly weak binding to the CYP3A4 proximal ER-6 motif. In cell-based transfection assays, hCAR displayed greater activation of CYP2B6 reporter gene expression compared with CYP3A4 with constructs containing both proximal and distal regulatory elements. Furthermore, in agreement with binding observations, transfection assays using promoter constructs containing repeats of CYP2B6 DR4 and CYP3A4 ER-6 motifs revealed an even greater difference in reporter activation by hCAR. In contrast, hPXR activation resulted in less discernible differences between CYP2B6 and CYP3A4 reporter gene expression. These results lead to a novel model of asymmetrical cross-regulation of these genes that differs from the presumed rodent model (Fig. 17.3). The greater selectivity of hCAR activation resulted in preferred induction of CYP2B6 over CYP3A4, whereas the lower selectivity of hPXR activation conferred efficacious induction of both target genes.

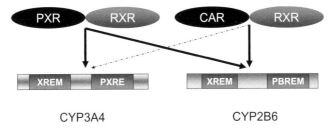

CYP3A4   CYP2B6

Figure 17.3 Asymmetrical model of cross talk between PXR and CAR in the regulation of *CYP2B6* and *CYP3A4* genes. Different from previously presumed rodent model, human PXR and CAR demonstrate an asymmetrical cross-regulation of CYP2B6 and CYP3A4 genes. This model is based on results generated from Faucette *et al.* (2006).

In addition to regulating phase I enzymes, CAR and PXR also cooperate to regulate the expression of phase II enzymes and drug transporters. PXR and CAR have been shown to induce the expression of UGT1A1, OATP2, GSTA1 and 2, all of which are important for bilirubin detoxification. The sharing of bilirubin detoxifying target genes may have accounted for a similar resistance to experimental hyperbilirubinemia in mice with PXR or CAR genetically or pharmacologically activated (Sugatani *et al.*, 2001; Huang *et al.*, 2003; Xie *et al.*, 2003). In another example of receptor cross talk, MRP2, a transporter mediating the efflux of several conjugated compounds across the apical membrane of the hepatocytes into the bile canaliculi, is regulated by FXR, PXR, and CAR (Kast *et al.*, 2002). An ER-8 (everted repeat separated by eight nucleotides) type of NR response element was identified in the promoter of the rat *Mrp2* gene. This element is capable of binding to and conferring PXR, CAR, and FXR responsiveness on a heterologous thymidine kinase promoter (Kast *et al.*, 2002). This is reminiscent of the co-regulation of the rodent Sult2a9/2a1 by FXR, PXR, CAR, and LXR, in which an IR-0 was the shared response element (Song *et al.*, 2001; Sonoda *et al.*, 2002; Saini *et al.*, 2004; Uppal *et al.*, 2007).

Another interesting example of functional interplay between xenobiotic receptors was revealed in the characterization of PXR-null mice in bilirubin clearance (Uppal *et al.*, 2005). Having known that activation of PXR is sufficient to detoxify bilirubin, it was surprising to find that the PXR-null mice also showed increased bilirubin clearance. We proposed that, when both PXR and CAR are present, the ligand-free PXR suppresses the constitutive activity of CAR, maintaining a basal capacity of bilirubin detoxification. The increased bilirubin clearance in PXR-null mice was likely the result of CAR derepression as a consequence of loss of PXR. This notion was supported by the observation that the pattern of enzyme and transporter regulation in the PXR-null mice was remarkably similar to that of transgenic mice expressing activated CAR (Uppal *et al.*, 2005).

In a more recent example, we showed that PXR, LXR, and PPARγ cooperate to regulate the free fatty acid transporter CD36 and to promote fatty acid-mediated hepatic lipogenesis. In this case, three distinct NR response elements were found for each of the three receptors and they are clustered in 500-bp sequences in the mouse *Cd36* gene promoter (J. Zhou and W. Xie, unpublished results). Consistent with the co-regulation of *Cd36*, activation of both PXR and LXR had an additive effect to induce hepatic triglyceride content (J. Zhou and W. Xie, unpublished results).

17.4   ARYL HYDROCARBON RECEPTOR (AhR)

It is noteworthy that transcriptional factors other than NRs also play critical roles in the control of xenobiotic-induced DME expression. One such non-NR transcriptional factor is the AhR, a member of the basic helix-loop-helix PER-ARNT-SIM (PAS) transcription factor family (Burbach, Poland and Bradfield, 1992; Safe, 2001). Similar to several NR family members, AhR expressed predominantly in the cytoplasm as a complex with Hsp90 and several other chaperones without the stimulation of xenobiotic ligands. Once it bonds to its ligands such as TCDD and 3-methylcholanthrene (3-MC), the receptor translocates into the nucleus and forms a heterodimer with its partner, AhR nuclear translocator (ARNT) (Safe, 2001). The newly formed AhR/ARNT heterodimer will bind to the XRE that is located in the promoter region of its target genes.

Like PXR and CAR, activation of AhR has been reported to simultaneously induce both phase I and II enzymes (Rushmore and Pickett, 1990; Ma, 2001) The mammalian CYP1A1, CYP1A2, and CYP1B1 genes are AR target genes identified initially (Hoffman *et al.*, 1991; Gu, Hogenesch and Bradfield, 2000). More recently, it was found that AhR is also implicated in the regulation of UGTs, including UGT1A1 and 1A6 (Emi, Ikushiro and Iyanagi, 1996; Munzel *et al.*, 1998; Klinge *et al.*, 1999; Shelby *et al.*, 2003; Yueh *et al.*, 2003). AhR-binding XREs have been identified in the human and rodent UGT1A1 gene promoter (Yueh *et al.*, 2003). An XRE was reported for the rat *Ugt1a6* gene and found to be both necessary and sufficient to mediate *Ugt1a6* transactivation by AhR (Shelby *et al.*, 2003). Interestingly, the PXR-, CAR-, and AhR-responsive elements are closely localized in the UGT1A1 gene promoter (Zhou, Zhang and Xie, 2005). It remains to be determined whether there are coordination or cross talk between AhR and CAR/PXR in xenobiotic regulation. The existing literature does support cross talk between AhR and nuclear receptor signaling. For example, AhR has been shown to interact with a number of NRs, such as the estrogen receptor α (ERα), COUP-TFI, and estrogen-related receptor α (ERRα). Moreover, the AhR/Arnt heterodimers can utilize certain NR response elements such as the naturally occurring estrogen response elements (Klinge *et al.*, 1999).

17.5   CLOSING REMARKS AND PERSPECTIVES

In the past decade, significant progress has been made in our understanding of the comprehensive role of NRs in the regulation of DMEs and drug transporter system. It has become clear that PXR and CAR can function as master xenosensors to regulate the expression of majority of the DMEs and transporters. Another important concept that has emerged from recent studies is that the interactions between NRs and their target sequences are multifaceted and much more complex than initially estimated. Many of the NRs share marked sequence homology in their DBD, suggesting that there are significant overlaps in their ability to bind to common response elements in a diverse set of target genes. This cross talk between NRs is believed to be the molecular basis for the fail-safe xenobiotic regulatory networks that facilitate host protection.

It also appears that the NR-controlled xenobiotic regulation is a double-edged sword. One of the remaining outstanding challenges is whether the biological action of PXR and CAR makes them drug targets for the treatment of human diseases, such as hyperbilirubinemia, obesity, as well as bile acid-associated cholestasis and colon cancers. Because of the importance of the DMEs in the biotransformation of marketed and future drugs, DDIs involving the induction and inhibition of CYPs are of great clinical importance. However, the potential DDI alone should not exclude NRs as therapeutic target, and this notion is supported by the facts that many of the xenobiotic receptor ligands, including RIF and PB, have been successfully used as clinical drugs.

Our understanding of the NR biology has further expanded our ability to predict the potential pharmacological and toxicological properties in the early stage of drug development. A series of NR-dependent assays has been developed and well evaluated, including *in vitro* cell-based reporter experiments in a high-throughput manner, *in vitro* ligand-binding assays, and transgenic animal models, especially the "humanized" PXR and CAR mice. It is anticipated that these novel methods may lead to a more rational and molecular-based approach to developing drugs with enhanced therapeutic efficacy and improved safety profiles.

ACKNOWLEDGMENTS

The original results from our labs that are described in this chapter were generated with the support of NIH grants ES012479, ES014626, CA107011 (to W.X.) and DK061652 (to H.W.).

REFERENCES

Arnold, K.A., Eichelbaum, M., Burk, O. (2004). Alternative splicing affects the function and tissue-specific expression of the human constitutive androstane receptor. *Nucl. Recept.*, *2* (1), 1.

Auerbach, S.S. *et al.* (2003). Alternatively spliced isoforms of the human constitutive androstane receptor. *Nucleic Acids Res.*, *31* (12), 3194–3207.

Auerbach, S.S. *et al.* (2005). Retinoid X receptor-alpha-dependent transactivation by a naturally occurring structural variant of human constitutive androstane receptor (NR1I3). *Mol. Pharmacol.*, *68* (5), 1239–1253.

Baes, M. *et al.* (1994). A new orphan member of the nuclear hormone receptor superfamily that interacts with a subset of retinoic acid response elements. *Mol. Cell Biol.*, *14* (3), 1544–1551.

Barbier, O. *et al.* (2003). The UDP-glucuronosyltransferase 1A9 enzyme is a peroxisome proliferator-activated receptor alpha and gamma target gene. *J. Biol. Chem.*, *278* (16), 13975–13983.

Bertilsson, G. *et al.* (1998). Identification of a human nuclear receptor defines a new signaling pathway for CYP3A induction. *Proc. Natl. Acad. Sci. U.S.A.*, *95* (21), 12208–12213.

Blumberg, B. *et al.* (1998). SXR, a novel steroid and xenobiotic-sensing nuclear receptor. *Genes Dev.*, *12* (20), 3195–3205.

Burbach, K.M., Poland, A., Bradfield, C.A. (1992). Cloning of the Ah-receptor cDNA reveals a distinctive ligand-activated transcription factor. *Proc. Natl. Acad. Sci. U.S.A.*, *89* (17), 8185–8189.

Cairns, W. *et al.* (1991). Assembly of a glucocorticoid receptor complex prior to DNA binding enhances its specific interaction with a glucocorticoid response element. *J. Biol. Chem.*, *266* (17), 11221–11226.

Chen, C., Staudinger, J.L., Klaassen, C.D. (2003). Nuclear receptor, pregnane X receptor, is required for induction of UDP-glucuronosyltranferases in mouse liver by pregnenolone-16 alpha-carbonitrile. *Drug Metab. Dispos.*, *31* (7), 908–915.

Chen, Y. *et al.* (2003). Identification of constitutive androstane receptor and glucocorticoid receptor binding sites in the CYP2C19 promoter. *Mol. Pharmacol.*, *64* (2), 316–324.

Chen, Y. *et al.* (2004). Induction of human CYP2C9 by rifampicin, hyperforin, and phenobarbital is mediated by the pregnane X receptor. *J. Pharmacol. Exp. Ther.*, *308* (2), 495–501.

Choi, H.S. *et al.* (1997). Differential transactivation by two isoforms of the orphan nuclear hormone receptor CAR. *J. Biol. Chem.*, *272* (38), 23565–23571.

Defranco, D.B. *et al.* (1995). Nucleocytoplasmic shuttling of steroid receptors. *Vitam. Horm.*, *51*, 315–338.

Dufort, I. *et al.* (2001). Comparative biosynthetic pathway of androstenol and androgens. *J. Steroid Biochem. Mol. Biol.*, *77* (4–5), 223–227.

Dussault, I. *et al.* (2001). Peptide mimetic HIV protease inhibitors are ligands for the orphan receptor SXR. *J. Biol. Chem.*, *276* (36), 33309–33312.

Dussault, I. *et al.* (2003). Identification of an endogenous ligand that activates pregnane X receptor-mediated sterol clearance. *Proc. Natl. Acad. Sci. U.S.A.*, *100* (3), 833–838.

Dvorak, Z. *et al.* (2003). Colchicine down-regulates cytochrome P450 2B6, 2C8, 2C9, and 3A4 in human hepatocytes by affecting their glucocorticoid receptor-mediated regulation. *Mol. Pharmacol.*, *64* (1), 160–169.

Emi, Y., Ikushiro, S., Iyanagi, T. (1996). Xenobiotic responsive element-mediated transcriptional activation in the UDP-glucuronosyltransferase family 1 gene complex. *J. Biol. Chem.*, *271* (7), 3952–3958.

Falkner, K.C. *et al.* (2001). Regulation of the rat glutathione S-transferase A2 gene by glucocorticoids: involvement of both the glucocorticoid and pregnane X receptors. *Mol. Pharmacol.*, *60* (3), 611–619.

Faucette, S.R. *et al.* (2006). Differential regulation of hepatic CYP2B6 and CYP3A4 genes by constitutive androstane receptor but not pregnane X receptor. *J. Pharmacol. Exp. Ther.*, *317* (3), 1200–1209.

Faucette, S.R. *et al.* (2007). Relative activation of human pregnane X receptor versus constitutive androstane receptor defines distinct classes of CYP2B6 and CYP3A4 inducers. *J. Pharmacol. Exp. Ther.*, *320* (1), 72–80.

Ferguson, S.S. *et al.* (2002). Regulation of human CYP2C9 by the constitutive androstane receptor: discovery of a new distal binding site. *Mol. Pharmacol.*, *62* (3), 737–746.

Ferguson, S.S. *et al.* (2005). Human CYP2C8 is transcriptionally regulated by the nuclear receptors constitutive androstane receptor, pregnane X receptor, glucocorticoid receptor, and hepatic nuclear factor 4alpha. *Mol. Pharmacol.*, *68* (3), 747–757.

Freedman, L.P. (1999). Increasing the complexity of coactivation in nuclear receptor signaling. *Cell*, *97* (1), 5–8.

Fromm, M.F. *et al.* (1999). Human MRP3 transporter: identification of the 5′-flanking region, genomic organization and alternative splice variants. *Biochim. Biophys. Acta*, *1415* (2), 369–374.

Geick, A., Eichelbaum, M., Burk, O. (2001). Nuclear receptor response elements mediate induction of intestinal MDR1 by rifampin. *J. Biol. Chem.*, *276* (18), 14581–14587.

Gerbal-Chaloin, S. *et al.* (2001). Induction of CYP2C genes in human hepatocytes in primary culture. *Drug Metab. Dispos.*, *29* (3), 242–251.

Gerbal-Chaloin, S. *et al.* (2002). Transcriptional regulation of CYP2C9 gene. Role of glucocorticoid receptor and constitutive androstane receptor. *J. Biol. Chem.*, *277* (1), 209–217.

Giguere, V. (1999). Orphan nuclear receptors: from gene to function. *Endocr. Rev.*, *20* (5), 689–725.

Goldstein, J.A. (2001). Clinical relevance of genetic polymorphisms in the human CYP2C subfamily. *Br. J. Clin. Pharmacol.*, *52* (4), 349–355.

Gong, H. *et al.* (2005). Animal models of xenobiotic receptors in drug metabolism and diseases. *Methods Enzymol.*, *400*, 598–618.

Gong, H. *et al.* (2006). Orphan nuclear receptor pregnane X receptor sensitizes oxidative stress responses in transgenic mice and cancerous cells. *Mol. Endocrinol.*, *20* (2), 279–290.

Goodwin, B. *et al.* (2001). Regulation of the human CYP2B6 gene by the nuclear pregnane X receptor. *Mol. Pharmacol.*, *60* (3), 427–431.

Gu, Y.Z., Hogenesch, J.B., Bradfield, C.A. (2000). The PAS superfamily: sensors of environmental and developmental signals. *Annu. Rev. Pharmacol. Toxicol.*, *40*, 519–561.

Guiochon-Mantel, A. *et al.* (1991). Nucleocytoplasmic shuttling of the progesterone receptor. *EMBO J.*, *10* (12), 3851–3859.

Guo, G.L. *et al.* (2002). Induction of rat organic anion transporting polypeptide 2 by pregnenolone-16alpha-carbonitrile is via interaction with pregnane X receptor. *Mol. Pharmacol.*, *61* (4), 832–839.

Guo, G.L., Johnson, D.R., Klaassen, C.D. (2002). Postnatal expression and induction by pregnenolone-16alpha-carbonitrile of the organic anion-transporting polypeptide 2 in rat liver. *Drug Metab. Dispos.*, *30* (3), 283–288.

Hagenbuch, B., Meier, P.J. (2003). The superfamily of organic anion transporting polypeptides. *Biochim. Biophys. Acta*, *1609* (1), 1–18.

Hani, E-H. *et al.* (1996). Indication for genetic linkage of the phosphoenolpyruvate carboxykinase (PCK1) gene region on chromosome 20q to non-insulin-dependent diabetes mellitus. *Diabetes Metab.*, *22* (6), 451–454.

Hayes, J.D., Pulford, D.J. (1995). The glutathione S-transferase supergene family: regulation of GST and the contribution of the isoenzymes to cancer chemoprotection and drug resistance. *Crit. Rev. Biochem. Mol. Biol.*, *30* (6), 445–600.

Hoffman, E.C. *et al.* (1991). Cloning of a factor required for activity of the Ah (dioxin) receptor. *Science*, *252* (5008), 954–958.

Honkakoski, P. *et al.* (1998). The nuclear orphan receptor CAR-retinoid X receptor heterodimer activates the phenobarbital-responsive enhancer module of the CYP2B gene. *Mol. Cell Biol.*, *18* (10), 5652–5658.

Honkakoski, P., Negishi, M. (1997). Characterization of a phenobarbital-responsive enhancer module in mouse P450 Cyp2b10 gene. *J. Biol. Chem.*, *272* (23), 14943–14949.

Honkakoski, P., Negishi, M. (1998). Protein serine/threonine phosphatase inhibitors suppress phenobarbital-induced Cyp2b10 gene transcription in mouse primary hepatocytes. *Biochem. J.*, *330* (Pt 2), 889–895.

Honkakoski, P., Negishi, M. (2000). Regulation of cytochrome P450 (CYP) genes by nuclear receptors. *Biochem. J.*, *347* (Pt 2), 321–337.

Huang, W. *et al.* (2003). Induction of bilirubin clearance by the constitutive androstane receptor (CAR). *Proc. Natl. Acad. Sci. U.S.A.*, *100* (7), 4156–4161.

Huang, W. *et al.* (2004). Meclizine is an agonist ligand for mouse constitutive androstane receptor (CAR) and an inverse agonist for human CAR. *Mol. Endocrinol.*, *18* (10), 2402–2408.

Ikeda, S. *et al.* (2005). Functional analysis of four naturally occurring variants of human constitutive androstane receptor. *Mol. Genet. Metab.*, *86* (1–2), 314–319.

Ishii, Y. *et al.* (1994). Purification of a phenobarbital-inducible morphine UDP-glucuronyltransferase isoform, absent from Gunn rat liver. *Arch. Biochem. Biophys.*, *315* (2), 345–351.

Jackson, J.P. *et al.* (2006). Phenytoin induction of the cyp2c37 gene is mediated by the constitutive androstane receptor. *Drug Metab. Dispos.*, *34* (12), 2003–2010.

Jinno, H. *et al.* (2004). Identification of novel alternative splice variants of human constitutive androstane receptor and characterization of their expression in the liver. *Mol. Pharmacol.*, *65* (3), 496–502.

Jones, S.A. *et al.* (2000). The pregnane X receptor: a promiscuous xenobiotic receptor that has diverged during evolution. *Mol. Endocrinol.*, *14* (1), 27–39.

Kast, H.R. *et al.* (2002). Regulation of multidrug resistance-associated protein 2 (ABCC2) by the nuclear receptors pregnane X receptor, farnesoid X-activated receptor, and constitutive androstane receptor. *J. Biol. Chem.*, *277* (4), 2908–2915.

Kastner, P.M.M., Chambon, P. (1995). Nonsteroid nuclear receptors: what are genetic studies telling us about their role in real life? *Cell*, *83* (6), 859–869.

Kawamoto, T. *et al.* (1999). Phenobarbital-responsive nuclear translocation of the receptor CAR in induction of the CYP2B gene. *Mol. Cell Biol.*, *19* (9), 6318–6322.

Kiuchi, Y. *et al.* (1998). cDNA cloning and inducible expression of human multidrug resistance associated protein 3 (MRP3). *FEBS Lett.*, *433* (1–2), 149–152.

Klaassen, C.D., Slitt, A.L. (2005). Regulation of hepatic transporters by xenobiotic receptors. *Curr. Drug Metab.*, *6* (4), 309–328.

Kliewer, S.A. *et al.* (1998). An orphan nuclear receptor activated by pregnanes defines a novel steroid signaling pathway. *Cell*, *92* (1), 73–82.

Kliewer, S.A., Goodwin, B., Willson, T.M. (2002). The nuclear pregnane X receptor: a key regulator of xenobiotic metabolism. *Endocr. Rev.*, *23* (5), 687–702.

Klinge, C.M. *et al.* (1999). The aryl hydrocarbon receptor (AHR)/AHR nuclear translocator (ARNT) heterodimer interacts with naturally occurring estrogen response elements. *Mol. Cell Endocrinol.*, *157* (1–2), 105–119.

Kocarek, T.A. *et al.* (1995). Comparative analysis of cytochrome P4503A induction in primary cultures of rat, rabbit, and human hepatocytes. *Drug Metab. Dispos.*, *23* (3), 415–421.

Kolars, J.C. *et al.* (1992). Identification of rifampin-inducible P450IIIA4 (CYP3A4) in human small bowel enterocytes. *J. Clin. Invest.*, *90* (5), 1871–1878.

Konig, J. *et al.* (1999). Conjugate export pumps of the multidrug resistance protein (MRP) family: localization, substrate specificity, and MRP2-mediated drug resistance. *Biochim. Biophys. Acta*, *1461* (2), 377–394.

Kopecky, P., Schwarz, I., Schwenzel, W. (1975). Antepartal phenobarbital therapy for the improvement of fetal bilirubin conjugation. *Arch. Gynakol.*, *219* (1–4), 455–457.

Kruh, G.D., Belinsky, M.G. (2003). The MRP family of drug efflux pumps. *Oncogene*, *22* (47), 7537–7552.

Kuno, T., Togawa, H., Mizutani, T. (2007). Induction of human UGT1A1 by a complex of dexamethasone-GR dependent on proximal site and independent of PBREM. *Mol. Biol. Rep.*, *35* (3), 361–367.

Lanz, R.B. *et al.* (1999). A steroid receptor coactivator, SRA, functions as an RNA and is present in an SRC-1 complex. *Cell*, *97* (1), 17–27.

Laudet, V. *et al.* (1992). Evolution of the nuclear receptor gene superfamily. *EMBO J.*, *11* (3), 1003–1013.

Lee, M.S. *et al.* (1993). Structure of the retinoid X receptor alpha DNA binding domain: a helix required for homodimeric DNA binding, *Science*, *260* (5111), 1117–1121.

Lehmann, J.M. *et al.* (1998). The human orphan nuclear receptor PXR is activated by compounds that regulate CYP3A4 gene expression and cause drug interactions. *J. Clin. Invest.*, *102* (5), 1016–1023.

Ma, Q. (2001). Induction of CYP1A1. The AhR/DRE paradigm: transcription, receptor regulation, and expanding biological roles. *Curr. Drug Metab.*, *2* (2), 149–164.

Mackenzie, P.I. *et al.* (1997). The UDP glycosyltransferase gene superfamily: recommended nomenclature update based on evolutionary divergence. *Pharmacogenetics*, *7* (4), 255–269.

Mackenzie, P.I. *et al.* (2003). Regulation of UDP glucuronosyltransferase genes. *Curr. Drug Metab.*, *4* (3), 249–257.

Maglich, J.M. *et al.* (2001). Comparison of complete nuclear receptor sets from the human, Cae-norhabditis elegans and Drosophila genomes. *Genome. Biol.*, *2* (8), RESEARCH0029.1–7.

Maglich, J.M. *et al.* (2002). Nuclear pregnane x receptor and constitutive androstane receptor regulate overlapping but distinct sets of genes involved in xenobiotic detoxification. *Mol. Pharmacol.*, *62* (3), 638–646.

Maglich, J.M. *et al.* (2003). Identification of a novel human constitutive androstane receptor (CAR) agonist and its use in the identification of CAR target genes. *J. Biol. Chem.*, *278* (19), 17277–17283.

Mangelsdorf, D.J. *et al.* (1995). The nuclear receptor superfamily: the second decade. *Cell*, *83* (6), 835–839.

Maurel, D. *et al.* (1996). Effects of acute tilt from orthostatic to head-down antiorthostatic restraint and of sustained restraint on the intra-cerebroventricular pressure in rats. *Brain Res.*, *736* (1–2), 165–173.

Mills, J.B. *et al.* (2004). Induction of drug metabolism enzymes and MDR1 using a novel human hepatocyte cell line. *J. Pharmacol. Exp. Ther.*, *309* (1), 303–309.

Moore, J.T., Kliewer, S.A. (2000). Use of the nuclear receptor PXR to predict drug interactions. *Toxicology*, *153* (1–3), 1–10.

Moore, L.B. *et al.* (2000a). Orphan nuclear receptors constitutive androstane receptor and preg-nane X receptor share xenobiotic and steroid ligands. *J. Biol. Chem.*, *275* (20), 15122–15127.

Moore, L.B. *et al.* (2000b). St. John's wort induces hepatic drug metabolism through activation of the pregnane X receptor. *Proc. Natl. Acad. Sci. U.S.A.*, *97* (13), 7500–7502.

Morris, M.E., Pang, K.S. (1987). Competition between two enzymes for substrate removal in liver: modulating effects due to substrate recruitment of hepatocyte activity. *J. Pharmacokinet. Bio-pharm.*, *15* (5), 473–496.

Mu, Y. *et al.* (2006). Traditional Chinese medicines Wu Wei Zi (Schisandra chinensis Baill) and Gan Cao (Glycyrrhiza uralensis Fisch) activate pregnane X receptor and increase warfarin clearance in rats. *J. Pharmacol. Exp. Ther.*, *316* (3), 1369–1377.

Munzel, P.A. *et al.* (1998). Aryl hydrocarbon receptor-inducible or constitutive expression of human UDP glucuronosyltransferase UGT1A6. *Arch. Biochem. Biophys.*, *350* (1), 72–78.

Nagata, K., Yamazoe, Y. (2000). Pharmacogenetics of sulfotransferase. *Annu. Rev. Pharmacol. Toxicol.*, *40*, 159–176.

Negishi, M. (2000). Nuclear receptor CAR as a phenobarbital induction signal of CYP2B gene [abstract]. *FASEB J.*, *14*, 1306.

Nolte, R.T. *et al.* (1998). Ligand binding and co-activator assembly of the peroxisome proliferator-activated receptor-gamma. *Nature*, *395* (6698), 137–143.

Ogawa, K. *et al.* (2000). Characterization of inducible nature of MRP3 in rat liver. *Am. J. Physiol. Gastrointest. Liver Physiol.*, *278* (3), G438–446.

Park, Y., Li, H., Kemper, B. (1996). Phenobarbital induction mediated by a distal CYP2B2 sequence in rat liver transiently transfected in situ. *J. Biol. Chem.*, *271* (39), 23725–23728.

Picard, D., Yamamoto, K.R. (1987). Two signals mediate hormone-dependent nuclear localization of the glucocorticoid receptor. *Embo J.*, *6* (11), 3333–3340.

Pratt, W.B., Silverstein, A.M., Galigniana, M.D. (1999). A model for the cytoplasmic trafficking of signalling proteins involving the hsp90-binding immunophilins and p50cdc37. *Cell Signal.*, *11* (12), 839–851.

Qian, Y.M. *et al.* (2001). Targeted disruption of the mouse estrogen sulfotransferase gene reveals a role of estrogen metabolism in intracrine and paracrine estrogen regulation. *Endocrinology*, *142* (12), 5342–5350.

Radominska-Pandya, A. *et al.* (1999). Structural and functional studies of UDP-glucuronosyltransferases. *Drug Metab. Rev.*, *31* (4), 817–899.

Rae, J.M. *et al.* (2001). Rifampin is a selective, pleiotropic inducer of drug metabolism genes in human hepatocytes: studies with cDNA and oligonucleotide expression arrays. *J. Pharmacol. Exp. Ther.*, *299* (3), 849–857.

Raucy, J.L. *et al.* (2002). Expression and induction of CYP2C P450 enzymes in primary cultures of human hepatocytes. *J. Pharmacol. Exp. Ther.*, *302* (2), 475–482.

Rosenfeld, J.M. *et al.* (2003). Genetic profiling defines the xenobiotic gene network controlled by the nuclear receptor pregnane X receptor. *Mol. Endocrinol.*, *17* (7), 1268–1282.

Rushmore, T.H., Pickett, C.B. (1990). Transcriptional regulation of the rat glutathione S-transferase Ya subunit gene. Characterization of a xenobiotic-responsive element controlling inducible expression by phenolic antioxidants. *J. Biol. Chem.*, *265* (24), 14648–14653.

Safe, S. (2001). Molecular biology of the Ah receptor and its role in carcinogenesis. *Toxicol. Lett.*, *120* (1–3), 1–7.

Saini, S.P. *et al.* (2004). A novel constitutive androstane receptor-mediated and CYP3A-independent pathway of bile acid detoxification. *Mol. Pharmacol.*, *65* (2), 292–300.

Salinas, A.E., Wong, M.G. (1999). Glutathione S-transferases—a review. *Curr. Med. Chem.*, *6* (4), 279–309.

Salphati, L., Benet, L.Z. (1998). Modulation of P-glycoprotein expression by cytochrome P450 3A inducers in male and female rat livers. *Biochem. Pharmacol.*, *55* (4), 387–395.

Schambach, K., Menzel, K. (1975). Clinical experiences using phenobarbital in the prevention of hyperbilirubinemia in mature newborn infants in a controlled study. *Z. Arztl Fortbild (Jena)*, *69* (10), 535–537.

Shelby, M.K. *et al.* (2003). Tissue mRNA expression of the rat UDP-glucuronosyltransferase gene family. *Drug Metab. Dispos.*, *31* (3), 326–333.

Sheweita, S.A. (2000). Drug-metabolizing enzymes: mechanisms and functions. *Curr. Drug Metab.*, *1* (2), 107–132.

Sidhu, J.S., Omiecinski, C.J. (1997). An okadaic acid-sensitive pathway involved in the phenobarbital-mediated induction of CYP2B gene expression in primary rat hepatocyte cultures. *J. Pharmacol. Exp. Ther.*, *282* (2), 1122–1129.

Sinz, M. *et al.* (2006). Evaluation of 170 xenobiotics as transactivators of human pregnane X receptor (hPXR) and correlation to known CYP3A4 drug interactions. *Curr. Drug Metab.*, *7* (4), 375–388.

Song, C.S. *et al.* (2001). Dehydroepiandrosterone sulfotransferase gene induction by bile acid activated farnesoid X receptor. *J. Biol. Chem.*, *276* (45), 42549–42556.

Sonoda, J. *et al.* (2002). Regulation of a xenobiotic sulfonation cascade by nuclear pregnane X receptor (PXR). *Proc. Natl. Acad. Sci. U.S.A.*, *99* (21), 13801–13806.

Sonoda, J. *et al.* (2003). A nuclear receptor-mediated xenobiotic response and its implication in drug metabolism and host protection. *Curr. Drug Metab.*, *4* (1), 59–72.

Staudinger, J.L. *et al.* (2001). The nuclear receptor PXR is a lithocholic acid sensor that protects against liver toxicity. *Proc. Natl. Acad. Sci. U.S.A.*, *98* (6), 3369–3374.

Staudinger, J.L. *et al.* (2003). Regulation of drug transporter gene expression by nuclear receptors. *Drug Metab. Dispos.*, *31* (5), 523–527.

Stedman, C.A. *et al.* (2005). Nuclear receptors constitutive androstane receptor and pregnane X receptor ameliorate cholestatic liver injury. *Proc. Natl. Acad. Sci. U.S.A.*, *102* (6), 2063–2068.

Sueyoshi, T. *et al.* (1999). The repressed nuclear receptor CAR responds to phenobarbital in activating the human CYP2B6 gene. *J. Biol. Chem.*, *274* (10), 6043–6046.

Sugatani, J. *et al.* (2001). The phenobarbital response enhancer module in the human bilirubin UDP- glucuronosyltransferase UGT1A1 gene and regulation by the nuclear receptor CAR. *Hepatology*, *33* (5), 1232–1238.

Synold, T.W., Dussault, I., Forman, B.M. (2001). The orphan nuclear receptor SXR coordinately regulates drug metabolism and efflux. *Nat. Med.*, 7 (5), 584–590.

Takada, T., Suzuki, H., Sugiyama, Y. (2000). Characterization of 5′-flanking region of human MRP3. *Biochem. Biophys. Res. Commun.*, 270 (3), 728–732.

Teng, S., Jekerle, V., Piquette-Miller, M. (2003). Induction of ABCC3 (MRP3) by pregnane X receptor activators. *Drug Metab. Dispos.*, 31 (11), 1296–1299.

Townsend, D.M., Tew, K.D. (2003). The role of glutathione-S-transferase in anti-cancer drug resistance. *Oncogene*, 22 (47), 7369–7375.

Trottier, E. *et al.* (1995). Localization of a phenobarbital-responsive element (PBRE) in the 5′-flanking region of the rat CYP2B2 gene. *Gene*, 158 (2), 263–268.

Tucker, C.M. *et al.* (2001). Self-regulation predictors of medication adherence among ethnically different pediatric patients with renal transplants. *J. Pediatr. Psychol.*, 26 (8), 455–464.

Tukey, R.H., Strassburg, C.P. (2000). Human UDP-glucuronosyltransferases metabolism, expression, and disease. *Annu. Rev. Pharmacol. Toxicol.*, 40, 581–616.

Tzameli, I. *et al.* (2000). The xenobiotic compound 1,4-bis[2-(3,5-dichloropyridyloxy)]benzene is an agonist ligand for the nuclear receptor CAR. *Mol. Cell Biol.*, 20 (9), 2951–2958.

Ueda, A. *et al.* (2002). Diverse roles of the nuclear orphan receptor CAR in regulating hepatic genes in response to phenobarbital. *Mol. Pharmacol.*, 61 (1), 1–6.

Uppal, H. *et al.* (2005). Combined loss of orphan receptors PXR and CAR heightens sensitivity to toxic bile acids in mice. *Hepatology*, 41 (1), 168–176.

Uppal, H. *et al.* (2007). Activation of LXRs prevents bile acid toxicity and cholestasis in female mice. *Hepatology*, 45 (2), 422–432.

Walker, D., Htun, H., Hager, G.L. (1999). Using inducible vectors to study intracellular trafficking of GFP- tagged steroid/nuclear receptors in living cells. *Methods*, 19 (3), 386–393.

Wang, H., LeCluyse, E.L. (2003). Role of orphan nuclear receptors in the regulation of drug-metabolising enzymes. *Clin. Pharmacokinet.*, 42 (15), 1331–1357.

Wang, H., Negishi, M. (2003). Transcriptional regulation of cytochrome p450 2B genes by nuclear receptors. *Curr. Drug Metab.*, 4 (6), 515–525.

Watkins, R.E. *et al.* (2001). The human nuclear xenobiotic receptor PXR: structural determinants of directed promiscuity. *Science*, 292 (5525), 2329–2333.

Weaver, R.J. (2001). Assessment of drug-drug interactions: concepts and approaches. *Xenobiotica*, 31 (8–9), 499–538.

Weaver, S.A. *et al.* (2001). Regulatory role of phosphatidylinositol 3-kinase on TNF-alpha-induced cyclooxygenase 2 expression in colonic epithelial cells. *Gastroenterology*, 120 (5), 1117–1127.

Wei, P. *et al.* (2000). The nuclear receptor CAR mediates specific xenobiotic induction of drug metabolism. *Nature*, 407 (6806), 920–923.

Xie, W., Evans, R.M. (2001). Orphan nuclear receptors: the exotics of xenobiotics. *J. Biol. Chem.*, 276 (41), 37739–37742.

Xie, W. *et al.* (2000). Humanized xenobiotic response in mice expressing nuclear receptor SXR. *Nature*, 406 (6794), 435–439.

Xie, W. *et al.* (2001). An essential role for nuclear receptors SXR/PXR in detoxification of cholestatic bile acids. *Proc. Natl. Acad. Sci. U.S.A.*, 98 (6), 3375–3380.

Xie, W. *et al.* (2003). Control of steroid, heme, and carcinogen metabolism by nuclear pregnane X receptor and constitutive androstane receptor. *Proc. Natl. Acad. Sci. U.S.A.*, 100 (7), 4150–4155.

Xie, W. *et al.* (2004). Orphan nuclear receptor-mediated xenobiotic regulation in drug metabolism. *Drug Discov. Today*, 9 (10), 442–449.

Xiong, H. *et al.* (2002). Role of constitutive androstane receptor in the in vivo induction of Mrp3 and CYP2B1/2 by phenobarbital. *Drug Metab. Dispos.*, 30 (8), 918–923.

Yueh, M.F. *et al.* (2003). Involvement of the xenobiotic response element (XRE) in Ah receptor-mediated induction of human UDP-glucuronosyltransferase 1A1. *J. Biol. Chem.*, *278* (17), 15001–15006.

Zelko, I., Negishi, M. (2000). Phenobarbital-elicited activation of nuclear receptor CAR in induction of cytochrome P450 genes. *Biochem. Biophys. Res. Commun.*, *277* (1), 1–6.

Zhai, Y. *et al.* (2007). Activation of pregnane X receptor disrupts glucocorticoid and mineralocorticoid homeostasis. *Mol. Endocrinol.*, *21* (1), 138–147.

Zhang, J. *et al.* (2002). Modulation of acetaminophen-induced hepatotoxicity by the xenobiotic receptor CAR. *Science*, *298* (5592), 422–424.

Zhang, J. *et al.* (2004). The constitutive androstane receptor and pregnane X receptor function coordinately to prevent bile acid-induced hepatotoxicity. *J. Biol. Chem.*, *279* (47), 49517–49522.

Zhou, J. *et al.* (2006). A novel pregnane X receptor-mediated and sterol regulatory element-binding protein-independent lipogenic pathway. *J. Biol. Chem.*, *281* (21), 15013–15020.

Zhou, J., Zhang, J., Xie, W. (2005). Xenobiotic nuclear receptor-mediated regulation of UDP-glucuronosyl-transferases. *Curr. Drug Metab.*, *6* (4), 289–298.

CHAPTER 18

Characterization of Cytochrome P450 Mechanism-Based Inhibition

DAN ROCK and LARRY C. WIENKERS

18.1 INTRODUCTION

Oxidative metabolism mediated by the superfamily of hepatic enzymes known as the cytochrome P450s represents a significant clearance pathway for many drugs prescribed today (Wienkers and Heath, 2005). Of the possible hepatic cytochrome P450 enzymes identified to date, P450 1A2, P450 2C8, P450 2C9, P450 2C19, P450 2D6, and P450 3A4/5 are responsible for greater than 80% of the known oxidative drug metabolism reactions (Fig. 18.1). In light of the small number of P450 enzymes responsible for the metabolism of a large number of drugs, it is not surprising that a significant portion of drugs possess overlapping P450 enzyme specificities (Guengerich, 1997). A consequence of this disproportionate ratio between the numbers of enzymes and substrates is that for many multidrug therapies, it is common for a patient to receive two or more drugs that interact with the same P450 enzyme. Under these polytherapeutic conditions, a patient is significantly at risk for incurring a P450-based, drug–drug interaction (DDI) (Black *et al.*, 1996; Friedman *et al.*, 1999). There are many clinically important mechanism-based P450 inhibitors that span several therapeutic areas including *antiarrhythmics* [e.g. amiodarone (Polasek *et al.*, 2004)], *antibacterials* [e.g. clarithromycin (Ito *et al.*, 2003) and troleandomycin (Yamazaki and Shimada, 1998)], *antidepressants* [e.g. fluoxetine (Murray and Murray, 2003) and paroxetine (Hara *et al.*, 2005)], *anti-HIV agents* [e.g. ritonavir (von Moltke *et al.*, 2000) and delavirdine (Voorman *et al.*, 1998)], *antihypertensives* [e.g. diltiazem (Jones *et al.*, 1999), and verapamil (Wang, Jones and Hall, 2005)], *nonsteroidal anti-inflammatory* drugs (NSAIDs) [suprofen (O'Donnell *et al.*, 2003) and zileuton (Lu *et al.*, 2003)], *steroids/receptor modulators* [e.g. gestodene (Guengerich, 1990), raloxifene (Chen *et al.*, 2002; Baer, Wienkers and Rock, 2007)], and methylenedioxymethamphetamine (MDMA) (Van *et al.*, 2007), and *oncology drugs* [e.g. tamoxifen (Zhao *et al.*, 2002) and irinotecan (Hanioka *et al.*, 2002)]. In addition, P450 mechanism-based inhibitors (MBIs) can be found in environmental sources such as *illicit drugs* [e.g. phecyclidine (PCP) (Jushchyshyn *et al.*, 2005)] and *various dietary*

Drug Metabolism Handbook: Concepts and Applications, Edited by Ala F. Nassar, Paul F. Hollenberg, and JoAnn Scatina
Copyright © 2009 by John Wiley & Sons, Inc.

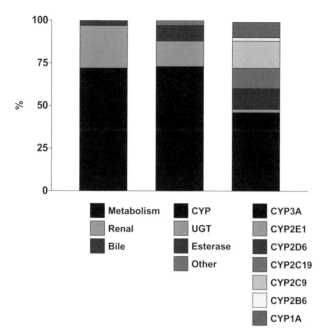

Figure 18.1 A graphical breakdown of the various pathways that contribute to the clearance of the top 200 most prescribed drugs in 2002.

constituents [e.g. bergamottin (He *et al.*, 1998), resveratrol (Chan and Delucchi, 2000), and 8-methoxypsoralen (Koenigs and Trager, 1998a)].

The magnitude of a clinically observed DDI is primarily dependent upon two features. The number of relevant clearance pathways associated with the victim drug's clearance. Restated, the magnitude of change in the systemic concentration of a drug (area under the curve, AUC) is dependent upon the number of elimination pathways associated with clearance (Bertz and Granneman, 1997; Michalets, 1998). For example, for drugs that are principally cleared by P450 3A enzymes, coadministration of potent MBIs of P450 3A4 can yield large changes in drug exposure (Thummel and Wilkinson, 1998). Secondly, how well the increase in victim drug concentration is tolerated by the system. For compounds with a narrow therapeutic index, an increase in drug exposure, through enzyme inhibition, may precipitate untoward side effects, including some fatal events (Fig. 18.2). A particularly tragic example of this situation occurred, when potent P450 3A4 inhibitors were coadministered with terfenadine (a P450 3A4 substrate) (Rodrigues *et al.*, 1995), where the subsequent increase in terfenadine exposure precipitated *torsades de pointes* (a life-threatening ventricular arrhythmia associated with QT prolongation) in many patients (Monahan *et al.*, 1990; Honig *et al.*, 1992). Because of this example and others similar to it, global regulatory agencies now require a complete understanding of the potential for drug inhibition of P450 enzymes for every new molecular entity (Davit *et al.*, 1999; Bjornsson *et al.*, 2003).

As described in Chapter 14, inhibition of cytochrome P450 enzymes can be broadly classified into two categories: reversible (e.g. competitive and noncompetitive) inhibition and mechanism-based (e.g. quasi-irreversible and irreversible) inhibition (Fig. 18.3). Interestingly, when compared with reversible inhibitors of P450 enzymes, MBIs of P450 enzymes are more frequently associated with pharmacokinetic-based

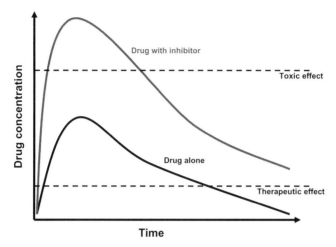

Figure 18.2 A simple cartoon depicting an imaginary time–concentration profile that describes the relationship between drug exposure and toxicity.

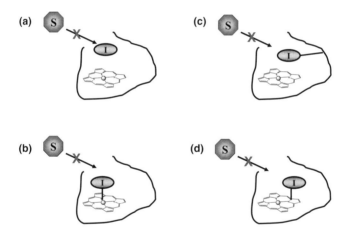

Figure 18.3 A simple schematic describing the possible mechanisms by which an inhibitor (I) affects cytochrome P450 substrates (S) metabolism: **(a)** reversible inhibition, **(b)** quasi-irreversible inhibition, **(c)** irreversible inhibition (apoprotein adduction), and **(d)** irreversible inhibition (heme adduction).

DDIs (Kanamitsu *et al.*, 2000; Ghanbari *et al.*, 2006). This observation is primarily a consequence that the inactivated P450 enzyme has to be replaced by newly synthesized P450 protein before the system is capable of metabolizing the victim drug (Bartkowski *et al.*, 1989). However, predicting the clinical significance of a DDI involving P450 enzyme inactivation is difficult, as the outcomes of the interactions are highly dependent upon a number of factors including (i) the particular P450 enzyme involved (i.e. the zero-order synthesis rate of new or replacement enzyme), (ii) the mechanism of P450 enzyme inactivation (i.e. the inhibitors inhibitory constant (K_I), k_{inact}, and partition ratio), and (iii) the disposition of the inhibited or victim drug (i.e. the number of pathways associated with drug elimination) (Zhou *et al.*, 2005; Kalgutkar, Obach and Maurer, 2007; Riley, Grime and Weaver, 2007).

Cytochrome P450-mediated oxidative metabolism of xenobiotics has been described for over 50 years (Axelrod, 1955; Brodie, Gillette and La Du, 1958), where the enzymatic basis for these reactions was first linked to a unique carbon monoxide-binding pigment found in rat liver microsomes (Klingenberg, 1958). Omura and Sato identified additional properties of the microsomal system in 1962 and coined the name P450 (first used with the hyphen, "P-450") for "pigment 450" because of the absorbance maximum at 450 nm when carbon monoxide was bound to the enzyme in the reduced state (Omura and Sato, 1962). This observation was rapidly followed by experimental evidence, which began to accumulate, for the presence of more than one form of microsomal cytochrome P450, followed by extensive efforts by a number of laboratories to purify P450 enzymes and catalog the oxidative reactions associated with the enzymes (Lu and Coon, 1968). To accommodate the expanding number and diversity of P450 enzymes, a systematic classification based upon protein sequence was developed (Nebert *et al.*, 1987). This system allowed P450 enzymes to be grouped into families, subfamilies, and single enzymes. In this system, the protein sequences within a given gene family are at least 40% identical (e.g. P450 2), the sequences within a given subfamily are >55% identical (e.g. P450 2D and P450 2C), with the last number designating the particular P450 enzyme (e.g. P450 2D6) (Nelson *et al.*, 1996; Nelson, 1999). As stated earlier, in humans, realistically the most important P450s from the point of view of drug metabolism are P450 1A2, P450 2C9, P450 2C19, P450 2D6, and P450 3A4/5 (Wienkers and Heath, 2005).

Prior to developing an in-depth discussion on cytochrome P450 MBIs, it may be worth reviewing the general concepts associated with enzyme catalysis and classical mechanism-based inhibition. In simple terms, in order for a chemical reaction to occur, reactant molecules must possess sufficient energy (activation energy) to cross a potential energy barrier (Fersht, 1974). In this instance, the lower the potential energy barrier to reaction, the more reactants have sufficient energy and in turn the faster the reaction will occur. The high-energy state of the reactants is called the transition state. To illustrate the concept, consider a bond-breaking reaction in which the transition state may be one where the reacting bond, although not completely broken, is vibrating at a high enough frequency that the transition state intermediate is as likely to split apart as to reform (Wolfenden and Snider, 2001). As depicted in Fig. 18.4, the forming reactants or products result in the loss of energy from the transition state. All enzyme reactions (even reactions that precipitate enzyme destruction) function by forming a transition state, with the reactants, of lower free energy than would be found in the uncatalyzed reaction.

There are several means by which enzymes are able to decrease activation energy. The most important of these involves the enzyme initially binding the substrate, in a fashion often described as "lock and key" orientation, which allows the molecule to exist in close proximity to the catalytic groups found within the active enzyme complex (Koshland and Neet, 1968). When the substrate is bound to the enzyme, the enzyme changes conformation and forces the substrate into a strained or distorted structure that mimics the transition state. In this conformation, the substrate is forced into a reactive state, due to the loss of the substrate's translational and rotational entropy, toward the total activation energy.

The traditional view of "lock and key" does not adequately describe the relationship between substrate and enzyme for cytochrome P450 oxidative reactions. As mentioned earlier, the number of chemotypes metabolized by P450 enzymes is quite large while the number of enzymes that carry out these reactions is exceptionally small. This is due

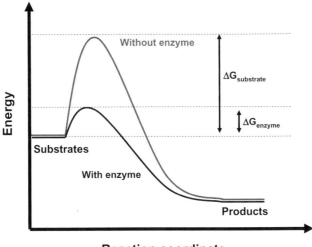

Reaction coordinate

Figure 18.4 A schematic diagram showing the free energy profile of the course of an enzyme-catalyzed reaction. The catalyzed reaction pathway (blue line) goes through the transition state, with standard free energy of activation ΔG_{enzyme}, whereas the uncatalyzed reaction (red line) goes through the transition state with standard free energy of activation $\Delta G_{substrate}$. It should be noted that the energetics of catalysis only relate the lowering of the standard free energy of activation from $\Delta G_{substrate}$ to ΔG_{enzyme} and does not effect the overall free energy change (i.e. the difference between the initial and final states).

to two unique attributes associated with the P450 enzymes. First, the P450s all possess large lipophilic active sites, which are able to accommodate large molecules such as cyclosporine (Fabre *et al.*, 1988). In fact, taken as a whole, P450 enzymes appear to take on a similar structural fold, while in some cases having less than 20% sequence identity and recognizing structurally different substrates (Hasemann *et al.*, 1995). Second, the nature of the P450 oxidative reaction toward nonactivated hydrocarbons is extraordinary efficient. To place the P450 reaction in perspective, oxidations of hydrocarbons of this nature would typically require extremely high temperature and would be nonspecific. At the center of P450 catalysis is the iron protoporphyrin IX (heme) with a thiolate of the conserved cysteine residue as the fifth ligand (White and Coon, 1982).

The sequence of events that make up the P450 catalytic cycle is shown in Fig. 18.5. The simplified P450 cycle begins with heme iron in the ferric state. In step (i), the substrate (R-H) binds to the enzyme, somewhere near the distal region of the heme group and disrupts the water lattice within the enzyme active site (Sligar, 1976). The loss of water elicits a change in the heme iron spin state (from low-spin to high-spin) (Poulos, 1996). Step (ii) involves the transfer of an electron from NADPH via the accessory flavoprotein NADPH-P450 reductase, with the electron flow going from the reductase prosthetic group FAD to FMN to the P450 enzyme (Vermilion and Coon, 1978). The enzyme, now in the ferrous state can bind oxygen (iii) (Estabrook *et al.*, 1971). The resulting iron–oxygen complex is unstable and can generate ferric iron and superoxide anion, or a second electron may enter the system in step (iv). This second electron may come from NADPH-P450 reductase or, in some cases, from cytochrome b_5 (Hildebrandt and Estabrook, 1971). In step (v), a proton is added and the O–O bond

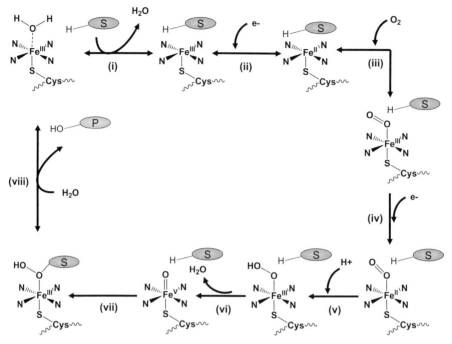

Figure 18.5 The catalytic cycle for cytochrome P450 enzyme oxidation reactions.

is then cleaved in step (vi) (Karuzina and Archakov, 1994), yielding H_2O and an entity shown in the scheme as FeO3+. In step (vii), the existing electron-deficient heme complex abstracts either a hydrogen atom or an electron from the bound substrate (Groves and McClusky, 1978). The subsequent intermediate collapses and the metabolite (R-OH) dissociates from the enzyme active site (viii).

Unlike many of the enzymes described earlier in this chapter, the rate-limiting step in cytochrome P450 enzyme catalysis is the activation of molecular oxygen leading to a highly reactive species (Trager, 1982). Based upon the results of several mechanistic studies, it appears that the electronic characteristics of the iron and oxygen species dictate the rates of substrate oxidation for cytochrome P450 enzymes (Fig. 18.6). As a consequence, the tendency for oxidation of a certain functional group generally follows the relative stability of the radicals that are formed (e.g. *N*-dealkylation > *O*-dealkylation > 2° carbon oxidation > 1° carbon oxidation) (Jones, Shou and Korzekwa, 1996).

While given that these two attributes (reactivity of heme and promiscuous active site) of the P450 enzymes suggest an almost limitless possibility of drug metabolism reactions, the oxygenation of substrates by these enzymes is not completely indiscriminate. For example, the nature of binding and/or oxidation of many P450 substrates can be both region- and stereochemically specific. In the case of warfarin, the anticoagulant medication, it exists as a racemic mixture. Both enantiomers share affinities toward P450 2C9 of less than 10 µM (~4 µM for (*S*)-warfarin and 8 µM for (*R*)-warfarin) (Kunze *et al.*, 1991). However, while both stereoisomers share relative equal affinities toward P450 2C9, only (*S*)-warfarin is metabolized by the enzyme (Rettie *et al.*, 1992). In a second example, quinidine is a pharmaceutical agent that acts as a class I antiarrhyth-

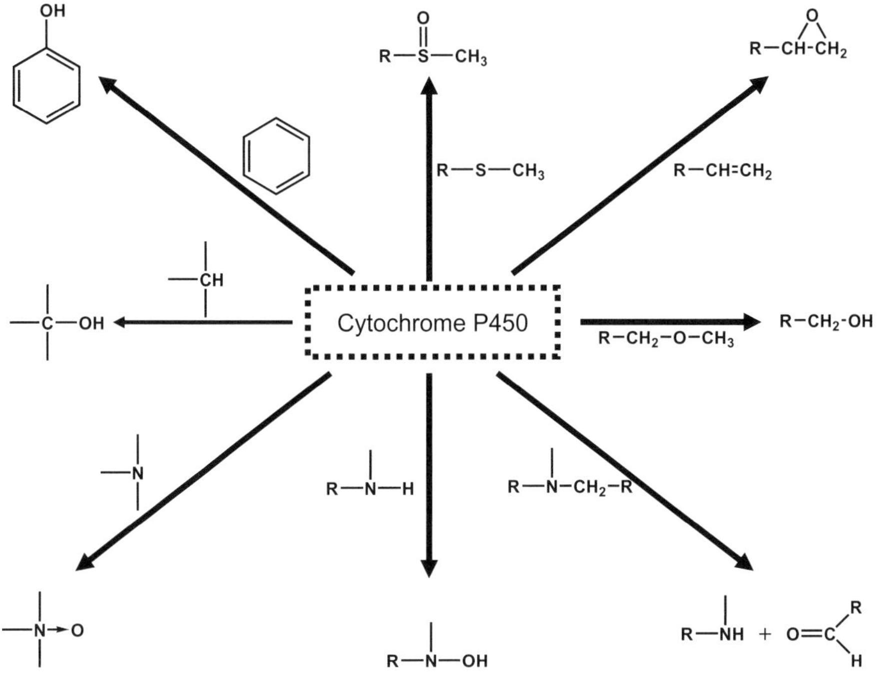

Figure 18.6   A brief listing of common P450 oxidation reactions.

mics agent and also a stereoisomer of quinine. Both isomers bind to and inhibit P450 2D6, albeit with markedly different affinities (quinidine has submicromolar K_i values while quinine has a K_i of about 50 μM). Besides possessing radically different affinities for the same enzyme, the isomers also bind completely different within the P450 2D6 active site; quinidine binding orientation yields a type II binding spectra while quinine generates a type I (Hutzler, Walker and Wienkers, 2003).

In simple terms, an enzyme inhibitor is a molecule that is able to prevent a substrate from entering the enzyme's active site and/or preclude the enzyme from catalyzing its reaction. As mentioned earlier, inhibitor binding is either reversible or irreversible. Reversible inhibitors bind with the enzyme through a variety of non-covalent interactions and based upon these binding motifs, elicit different types of inhibition, depending on whether the inhibitor binds the enzyme, the enzyme–substrate complex, or both. In contrast, an irreversible (mechanism-based) inhibitor is an enzyme inhibitor that contains a latent functional group that is chemically unreactive in starting material, but can be activated to a highly reactive intermediate by the enzyme through some part of the normal catalytic mechanism that is effected by the enzyme. The activated species then binds irreversibly to the enzyme and thereby inactivates it.

Given that the reactive groups present within the enzyme's active site are nucleophiles (such as hydroxyl or sulfhydryl groups), irreversible inhibition via covalent addition of reactive species to the enzymes requires that the reactive species needs to be some form of electrophilic moiety. For many enzymes, the formation of a reactive species (using the catalytic mechanism of the enzyme) falls into a few distinct categories, such as the generation of Michael acceptors, haloalkyl derivatives, and rearrangements leading to an acyl-enzyme intermediate (e.g. proteases) (Silverman, 1988).

Initially, characterization of MBIs followed an intuitive kinetic and chemical criteria (Wang and Walsh, 1978). Briefly an MBI was required to be a substrate, where physical binding to the enzyme active site precedes catalysis. The interaction between substrate and enzyme should reflect a binding equilibrium followed by first-order chemical process leading to inactivation. To this end, as inhibitor concentration is increased, a greater fraction of total enzyme will be occupied with inhibitor and that saturation of enzyme with inhibitor will be consistent with the kinetic specificity of the initial binding step. In addition, if covalent modification of the enzyme is a mechanism-based process which proceeds via adduct formation to a key amino acid residue within the enzyme active site, the chemical stoichiometry of modification should be unity (1:1 adduct/enzyme ratio) (Wang and Walsh, 1978).

As the understanding and laboratory experience evolved in this area, researchers expanded and refined the criteria applied to assess whether a substrate is a mechanism-based inactivator (Wang and Walsh, 1978; Walsh, 1984; Kent, Juschyshyn and Hollenberg, 2001). Briefly, the criteria include the following:

1. The loss of enzyme activity must exhibit time dependence.
2. Enzyme inactivation should exhibit saturation kinetics with respect to the concentration of the inhibitor.
3. The inactivation occurs in a catalytically competent system (e.g. necessary cofactors are present and that metabolism is occurring).
4. The enzyme should be protected from inactivation upon co-incubation with a competitive substrate/inhibitor.
5. Lack of suppression of inactivation by reactive intermediate scavengers.
6. The inactivation should be irreversible and activity should not return upon dialysis or gel filtration.
7. Following inactivation, it should be possible to demonstrate a 1:1 stoichiometry of inactivator to enzyme molecule inactivated.

A generalized kinetic scheme for MBI can be simply described (Fig. 18.7) as a competition or partition between the catalytic turnover of the inhibitor and enzyme inactivation (Walsh, 1984). Through the integration of the MBI's partition ratio, a mechanistic insight into the catalytic branch points associated with the enzyme reaction

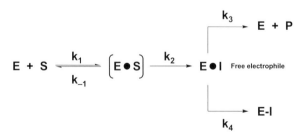

Figure 18.7 Reaction scheme of a mechanism-based enzyme inactivator. Rate constants k_1 and k_{-1} represent a reversible binding event between the enzyme (E) and substrate (S) to form the enzyme–substrate complex (ES), and k_2 represents catalysis of substrate to a reactive intermediate (I). k_3 represents the rate constant defining the release of the reactive intermediate represented as product (P), whereas k_4 represents inactivation of E by I forming the EI complex. k_3/k_4 = partition ratio = amount of formed P/inactivation, EI.

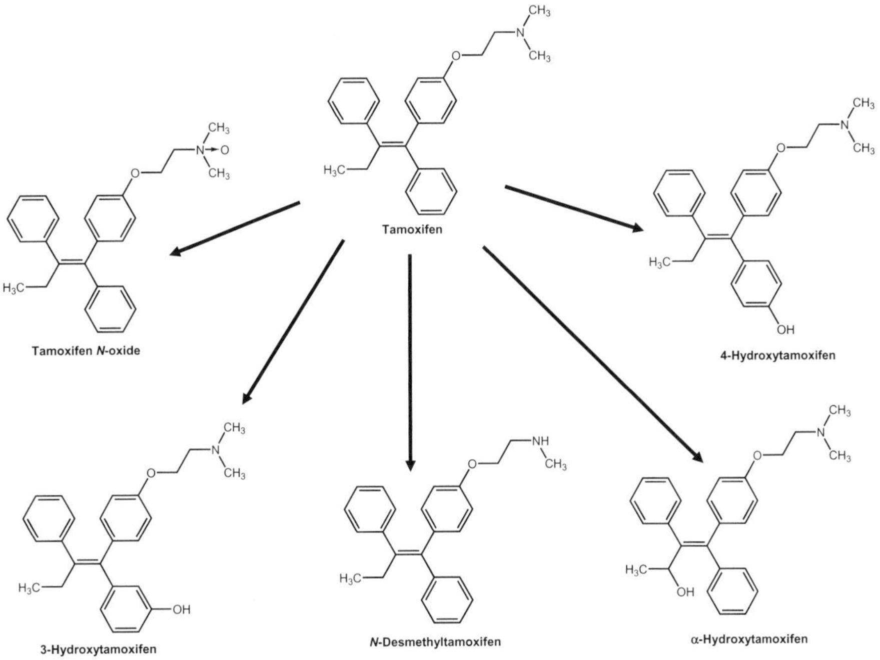

Figure 18.8   Multiple P450 3A4 oxidation reactions toward a single substrate, tamoxifen.

as well as the chemical nature of the reactive species in the enzyme active site can be gleaned. In this case, the lower the partition ratio (e.g. the closer to 1), the more efficient the inhibitor is at destroying the enzyme.

For cytochrome P450 MBIs, while metabolic intermediates are quite reactive, low partition ratios associated with MBIs are uncommon. This anomaly between the chemical reactivity of the activated species and partition ratio is explained by the fact that there are normally multiple metabolic pathways associated with many P450 substrates. For many traditional MBIs, the mechanism of enzyme inactivation takes advantage of a single catalytic step between substrate and enzyme. As stated earlier, cytochrome P450 enzymes have large active sites which allow for multiple substrate orientations and thus are susceptible to multiple oxidative reaction. An example of the P450 enzymes' ability to accommodate multiple substrate orientations and carry out a variety of energetically diverse oxidative reactions is seen with the P450 3A4 metabolism of tamoxifen (Fig. 18.8). In this example, P450 3A4 is able to metabolize tamoxifen to yield five products through four different oxidative reactions (Desta *et al.*, 2004). Adding to the complexity surrounding P450 mechanism-based inhibition is the fact that in some cases, the underlying mechanism of enzyme inactivation from one P450 enzyme does not extrapolate to a different P450 enzyme. For example, subtle sequence differences between P450 3A4 and P4503A5 (e.g. these enzymes share greater than 95% amino acid homology) result in profound differences in susceptibility to mechanism-based inhibition by several inhibitors including verapamil (Wang, Jones and Hall, 2005), tamoxifen (Zhao *et al.*, 2002), and erythromycin (McConn *et al.*, 2004).

As previously described, there are three general classes of mechanism-based inactivators of the cytochrome P450s. They are (1) inhibitors that bind covalently to the apoprotein; (2) inhibitors that interact in a pseudo-irreversible manner with heme iron;

and (3) inhibitors that cause destruction of the prosthetic heme group, oftentimes leading to heme-derived products that covalently modify the apoprotein (Fig. 18.3). The classes of cytochrome P450 mechanism-based inhibition can be further refined into six common features which serve as latent chemical groups associated with enzyme activation. These include substituted imidazoles, furan rings, thiophenes, alkylamines, methylenedioxyphenyl groups, and acetylenes. The following is a list of the types of P450 mechanism-based inhibition and the chemistry associated with these latent chemical moieties which leads to enzyme inactivation.

18.2 INHIBITORS THAT UPON ACTIVATION, BIND COVALENTLY TO THE P450 APOPROTEIN

18.2.1 Substituted Imidazoles

Putative bioactivation of this latent group requires formation of an imidazomethide, as the electrophilic species (Fig. 18.9) (Kunze and Trager, 1993). The mechanism(s) for the generation of this reactive species involves initial P450-mediated electron abstraction leading to the formation of a radical intermediate that can undergo oxygen rebound to form the hydroxyl metabolite. Alternatively, the radical may proceed through a second electron or hydrogen atom abstraction to form a reactive imidazomethide species, which can adduct to nucleophilic sites with the P450 enzyme active site or react with water (Hutzler et al., 2004).

18.2.2 Furans

Furans are electron-rich aromatic groups that are readily oxidized by P450 enzymes to form electrophilic species which have been linked to drug-related toxicities (Kouzi, McMurtry and Nelson, 1994) as well as serve as MBIs (Sahali-Sahly et al., 1996). The current hypothesis regarding P450-mediated bioactivation of furan rings (Fig. 18.10) involves epoxide formation that could either be deactivated via hydrolysis to yield the diol (Khojasteh-Bakht et al., 1999) or react irreversibly with the P450 enzyme, either through direct adduction with the epoxide or through an intramolecular rearrangement of the epoxide to form an α,β-unsaturated carbonyl which serves as a Michael acceptor for protein adduction (Chen, Hecht and Peterson, 1997).

18.2.3 Thiophenes

Currently, the cytochrome P450-mediated bioactivation of thiophene-containing drugs is thought to proceed via two possible mechanisms (Fig. 18.11). The first mechanism proceeds via P450 oxidation of the thiophene sulfur atom leading to formation of a reactive thiophene-S-oxide intermediate. The thiophene-S-oxide is susceptible to a Michael-type addition with nucleophilic amino acids within the P450 active site (Lopez-Garcia, Dansette and Mansuy, 1994). The second mechanism stems from a P450-mediated oxidation of the thiophene ring to form a reactive thiophene epoxide metabolite which may react directly with the P450 enzyme (Dansette, Bertho and Mansuy, 2005). Alternatively, the epoxide in a mechanism analogous to furan bioactivation generates a cis-2-butene-1, 4-dialdehyde reactive intermediate. In this instance, the thioketo-α,β-unsaturated aldehyde serves as the electrophilic intermediate that covalently binds to P450 enzyme (O'Donnell et al., 2003).

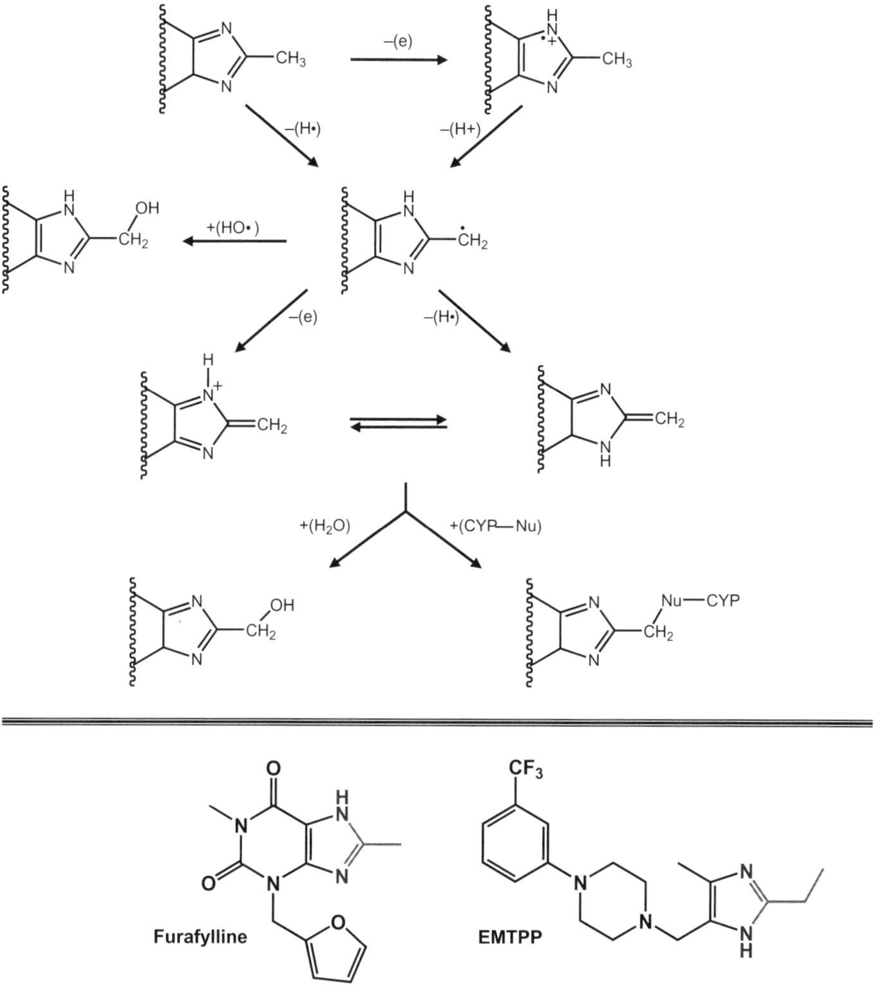

Figure 18.9 Mechanistic scheme for P450 enzyme-mediated bioactivation and subsequent enzyme inactivation of a substituted imidazole group. Structure of known P450 MBIs which possess a substituted imidazole as a latent chemical moiety: furafylline (Kunze and Trager, 1993) and 1-[(2-ethyl-4-methyl-1H-imidazol-5-yl)methyl]-4-[4-(trifluoromethyl)-2-pyridinyl]piperazine (EMTPP) (Hutzler *et al.*, 2004).

18.3 INHIBITORS THAT INTERACT IN A PSEUDO-IRREVERSIBLE MANNER WITH HEME IRON

18.3.1 Alkylamines

There are a relatively large number of drugs that contain primary, secondary, or tertiary alkylamines which are also known to be MBIs of cytochrome P450 enzymes (Yamazaki and Shimada, 1998; Jones *et al.*, 1999). The mechanism of inactivation for these molecules proceeds through a pseudo-irreversible interaction with the heme iron described

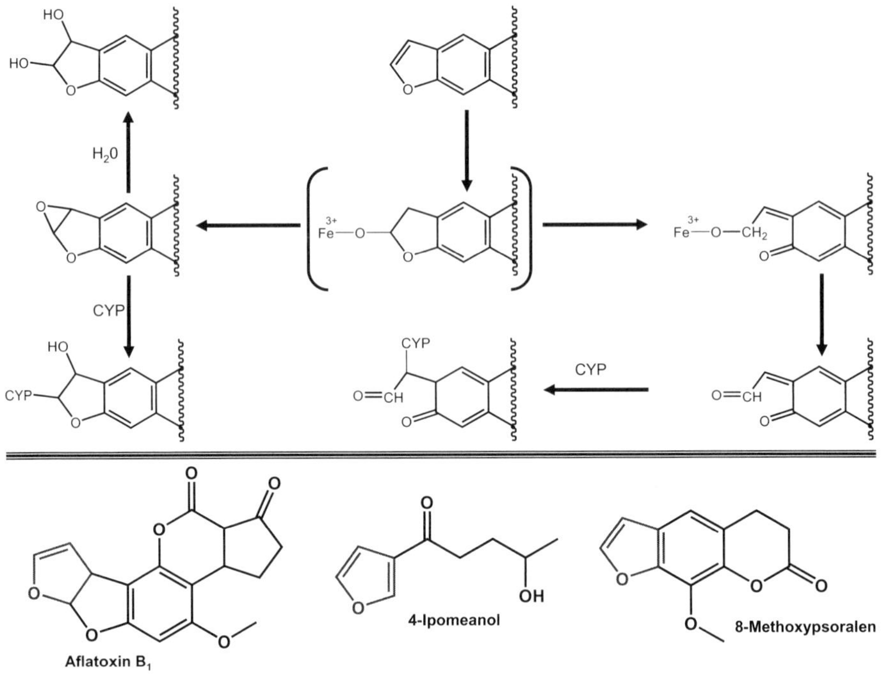

Figure 18.10 Mechanistic scheme for P450 enzyme-mediated bioactivation and subsequent enzyme inactivation of a furan group. Structures of known P450 MBIs which possess a furan as a latent chemical moiety: L-754,394 (Sahali-Sahly *et al.*, 1996), 8-methoxypsoralen (Koenigs and Trager, 1998a), and 4-ipomeanol (Alvarez-Diez and Zheng, 2004).

as a metabolite intermediate complex (MIC) (Pershing and Franklin, 1982; Ernest, Hall and Jones, 2005). The salient reactive species in this reaction is generally regarded to be a transient nitrosoalkane (Bensoussan, Delaforge and Mansuy, 1995). To generate the proposed reactive species requires a primary or secondary amine to serve as a precursor to a nitroso species. This involves an initial P450-catalyzed dealkylation step, followed by hydroxylation of the nitrogen to form a secondary hydroxylamine. The subsequent hydroxylamine is then further oxidized to yield a nitrone, which can undergo hydrolysis or be converted to the reactive nitroso species through two-electron oxidation process (Bensoussan, Delaforge and Mansuy, 1995). The schematic for MIC formation and examples of MBIs, which react through this pathway, are provided in Fig. 18.12.

18.3.2 Methylenedioxyphenyls

The P450 mechanism for bioactivation of this latent chemical moiety is postulated to proceed via proton abstraction of the dioxymethylene bridge which encounters a meta-bolic branch point of undergoing either typical oxidative demethylation to generate the catechol and formic acid, or intramolecular rearrangement to yield a carbine that is able to coordinate with the heme iron (Fig. 18.13). Formation of a carbene can arise by abstraction of a hydrogen atom from the methylenic carbon, or by elimination of water from a hydroxymethylene intermediate (Lin *et al.*, 1996; Hutzler *et al.*, 2006). The

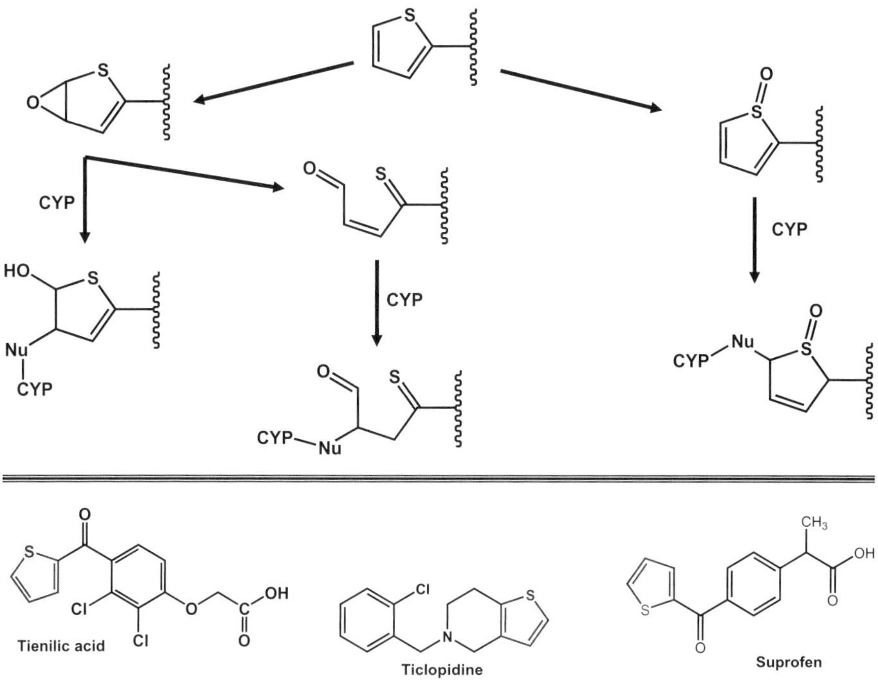

Figure 18.11 Mechanistic scheme for P450 enzyme-mediated bioactivation and subsequent enzyme inactivation of a thiophene group. Structure of known P450 MBIs which possess a thiophene as a latent chemical moiety: tienilic acid (Beaune *et al.*, 1987), ticlopidine (Ha-Duong *et al.*, 2001), and suprofen (O'Donnell *et al.*, 2003).

resulting dioxymethylene MI complex with P450 enzymes appears to be stable when the heme is in both the ferric or ferrous state. However, MI complexes with ferrous iron are thought to be more stable than those with ferric iron, due to the potential for back-bonding of electrons from the ferrous iron to the carbine carbon (Dickins *et al.*, 1979).

18.4 INHIBITORS THAT CAUSE DESTRUCTION OF THE PROSTHETIC HEME GROUP, OFTENTIMES LEADING TO HEME-DERIVED PRODUCTS THAT COVALENTLY MODIFY THE APOPROTEIN

18.4.1 Acetylenes

Many compounds containing an acetylenic group have been shown to be mechanism-based inactivators of P450 enzymes (Foroozesh *et al.*, 1997). For this functional moiety, two mechanisms for the inactivation of P450 have been described in detail (Chan, Sui and Ortiz de Montellano, 1993). The first involves the transfer of oxygen from the P450-activated oxygen intermediate to the internal carbon of the acetylene which generates a reactive intermediate that leads to heme alkylation and destruction of the heme chromophore. Alternatively, the transfer of oxygen from the activated oxygen intermediate of the P450 to the terminal carbon of the acetylene would result in an

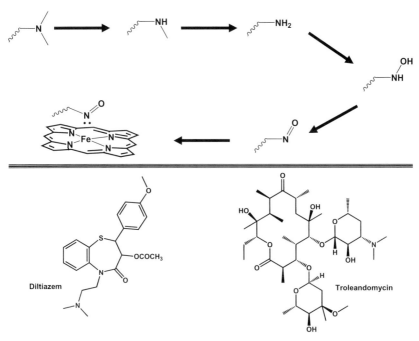

Figure 18.12 Mechanistic scheme for P450 enzyme-mediated bioactivation and subsequent enzyme inactivation of alkylamines. Structure of known P450 MBIs which possess an alkylamine as a latent chemical moiety: diltiazem (Jones *et al.*, 1999) and troleandomycin (Yamazaki and Shimada, 1998).

intermediate that would rearrange via a 1,2-shift of the terminal hydrogen to the vicinal carbon to generate a ketene. The reactive ketene species produced by this rearrangement can then be hydrolyzed to produce the carboxylic acid product, or it can acylate nucleophilic residues within the P450 active site to inactivate the protein (Fig. 18.14).

The objectives of the first portion of this chapter were to serve as a primer to highlight the uniqueness of cytochrome P450 reactions compared with many other enzyme systems, and to develop the relationship between structure of the inhibitor and catalytic function of the enzyme as it relates to P450 mechanism-based inhibition. The remainder of this chapter will focus on expanding these concepts as it relates to conducting studies to characterize the nature and inhibitory potential of mechanism-based inactivators which lead to irreversible destruction of the P450 enzymes.

Characterizing the mechanisms by which a new chemical entity (NCE) inhibits P450s is critical for the development of safe therapeutics as the magnitude of a drug interaction will typically be increased for an MBI (Fig. 18.15). In all cases, the precursor to P450 MBI results from bioactivation to generate a reactive chemical intermediate and therefore most MBIs can be linked to the generation of reactive metabolites (Evans *et al.*, 2004; Masubuchi, Makino and Murayama, 2007). As a result, MBIs increase the risk to patient safety through three basic pathways: (i) producing unexpectedly high drug levels of the MBI itself or an alternate drug cleared through the P450 affected by the MBI (Fig. 18.2); (ii) eliciting immune-induced hepatitis potentially as a result of the direct alkylation event; and (iii) oxidative stress through redox cycling

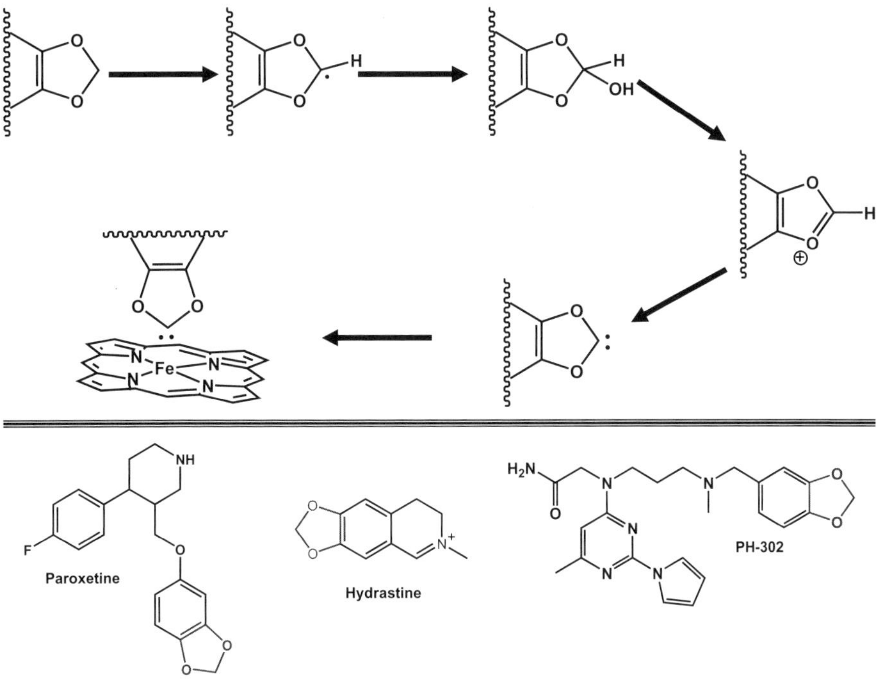

Figure 18.13 Mechanistic scheme for P450 enzyme-mediated bioactivation and subsequent enzyme inactivation of methylenedioxyphenyl groups. Structure of known P450 MBIs which possess a methylenedioxyphenyl as a latent chemical moiety: paroxetine (Hara *et al.*, 2005) and PH-302 (Hutzler *et al.*, 2006).

and depletion of endogenous glutathione. Regardless of the mode of action, the presence of MBI increases the potential safety risk associated with an NCE.

The most favorable scenario is to eliminate the structural feature(s) responsible for irreversible P450 inhibition, thereby reducing the potential for P450 inhibition and minimizing the risk of additional adverse drug reactions. The selective modification of a chemical series to eliminate MBI requires high-resolution structural information in order to pinpoint the mechanism by which the reactive intermediate forms. However, the inability to remove bioactivation may not preclude an NCE from development but will partially depend upon the severity and progression of the disease the NCE is aimed to treat (Kalgutkar and Soglia, 2005). In addition, a major focus in progressing with an MBI is to provide a risk assessment of the potential effects of the MBI by predicting the expected magnitude of drug interactions. This requires the functional enzymology to be defined including the inactivation parameters of MBI and how they are expected to translate *in vivo* based upon the expected clearance mechanisms, dose, and frequency of administration.

The balance between structural and functional characterization tools for interrogating MBI depends upon when MBI for a particular compound was discovered. For example, if the liability is revealed during lead optimization in discovery, a predominantly structural approach could guide chemistry efforts toward design strategies aimed at the removal or reduction in the extent of MBI. However, if the MBI cannot

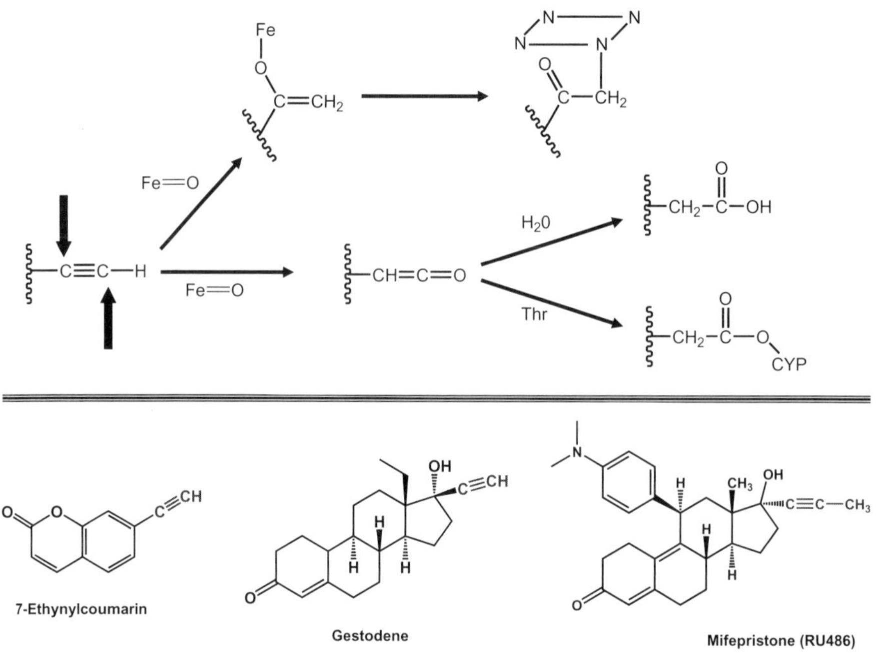

Figure 18.14 Mechanistic scheme for P450 enzyme-mediated bioactivation and subsequent enzyme inactivation of an alkyne group. Structure of known P450 MBIs which possess an alkyne group as a latent chemical moiety: gestodene (Guengerich, 1990), mifepristone (RU486) (He, Woolf and Hollenberg, 1999), and 7-ethynylcoumarin (Regal *et al.*, 2000).

be removed or in the unfortunate circumstance that MBI was discovered in preclinical development, a functional approach will dominate research efforts to define the potential impact of the MBI on patient population required to support continuation of the clinical candidate. Both structural and functional characterization is required not only for proper assessment of safety, but also in order to avoid incorporation of similar structural motifs prone to MBI into future chemical templates.

Given the complexity of P450 MBI characterization, a plethora of experimental protocols have evolved with the specific aim of identifying irreversible inhibition. When possible, the primary focus should be on removing MBI from a chemical series which requires the experimental procedures to focus on several key structural and functional end points (Fig. 18.16). This schematic provides an experimental guidance for the assessment of MBI from which the initial results provide the directive for additional experiments and end points for the continued development of an NCE with MBI liabilities. Therefore, Fig. 18.16 will serve as an outline for the remainder of the chapter and will cover in detail structural tools used to identify the mechanistic details of MBI and subsequently the functional tools used to determine the kinetics of MBI. A specific focus is given on the different assays used through the progression of a developing NCE, from early removal of the liability in discovery to risk assessment in later stages of development, through the generation of *in vitro* functional data which are used to extrapolate the expected *in vivo* pharmacokinetic impact of MBI on a developing clinical candidate.

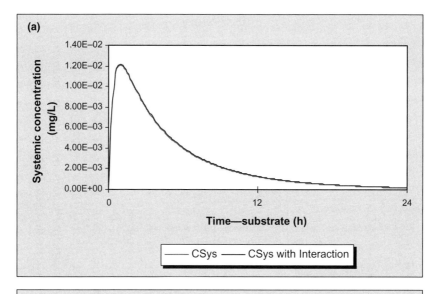

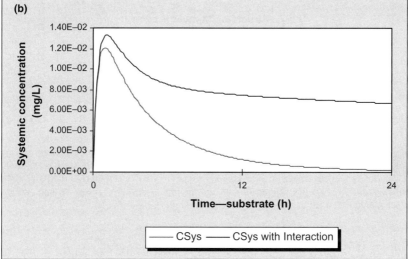

Figure 18.15 Simulation of single oral dose pharmacokinetic model of midazolam (5 mg) coadministered with erythromycin (500 mg) using Simcyp (Sheffield, UK). **(a)** Inactivation kinetics were omitted in this simulation resulting in no observed drug interaction. **(b)** K_I and k_{inact} were included in the prediction, highlighting the potential for drug interaction on the basis of erythromycin inactivation of CYP3A4.

Prior to discussing experimental procedures used to study MBI, it is imperative to consider the interactions leading to irreversible P450 inhibition. This provides a conceptual understanding of the MBI process and helps visualize the experimental design requirements necessary to elucidate and remove MBI from a chemical scaffold. The foremost precursor to MBI is that a substrate contains a functional group capable of

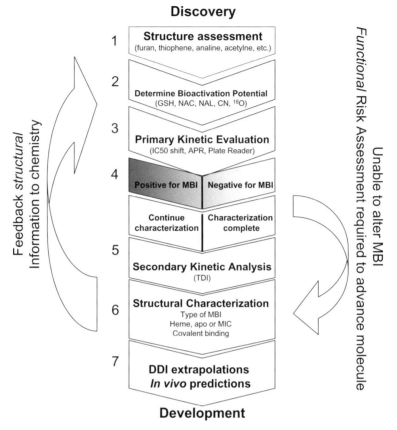

Figure 18.16   A schematic of the MBI testing funnel describing the use of structural screening-based assays in discovery which are highlighted by steps 1–6. Structural learnings are then fed back into the chemical design of new compounds until MBI is eliminated. If MBI cannot be attenuated, the program will attempt to manage the risk in development based upon steps 7–8. This schematic will serve as the outline for the remainder of the chapter.

bioactivation as discussed with the seven structural motifs highlighted in Figs 18.9–18.14. For P450 MBI is conceptually interesting because the catalytic step is activation of molecular oxygen. This implicitly means substrate oxidation is a by-product of P450 catalysis and consequently substrates do not typically possess a single enzyme–substrate conformation. The repercussions of this trait are that oxidation reactions can occur at multiple positions within a single molecule, favoring the most facile oxidative process unless the substrate is stabilized in a particular binding conformation (Fig. 18.17). Flurbiprofen metabolism by P450 2C9 can be used to illustrate this point (Fig. 18.18). The benzyl alcohol of flurbiprofen is the most facile oxidation site; however, the electrostatic interactions of carboxylic acid in flurbiprofen with Arg108 of P450 2C9 position flurbiprofen such that the only 4′-hydroxylation is observed (Fig. 18.18) (Wester *et al.*, 2004). This particular example shows how P450–substrate interactions may impact the site of metabolism.

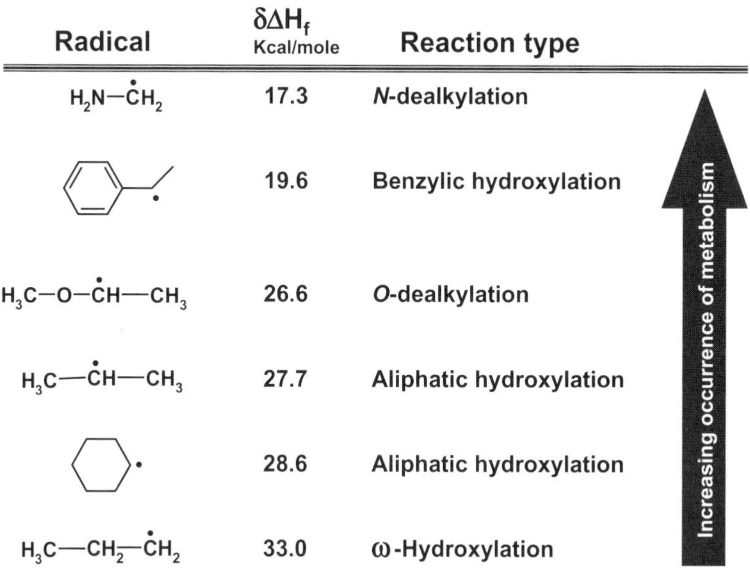

Radical	$\delta\Delta H_f$ Kcal/mole	Reaction type
$H_2N-\overset{\bullet}{C}H_2$	17.3	*N*-dealkylation
(phenyl)$\overset{\bullet}{}$	19.6	Benzylic hydroxylation
$H_3C-O-\overset{\bullet}{C}H-CH_3$	26.6	*O*-dealkylation
$H_3C-\overset{\bullet}{C}H-CH_3$	27.7	Aliphatic hydroxylation
(cyclohexyl)$\bullet$	28.6	Aliphatic hydroxylation
$H_3C-CH_2-\overset{\bullet}{C}H_2$	33.0	ω-Hydroxylation

Increasing occurrence of metabolism ↑

Figure 18.17   The energy differences between different P450-catalyzed reactions based upon hydrogen atom abstraction. The differences in energy have the potential to influence metabolite regioselectivity in P450-mediated metabolism (Jones, Shou and Korzekwa, 1996; Higgins *et al.*, 2001).

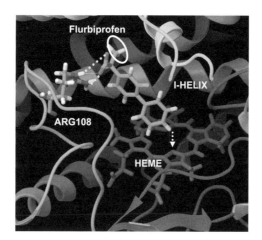

Figure 18.18   Crystal structure of P450 2C9 with flurbiprofen bound (1R90.pdb) highlighting the electrostatic interaction of the carboxylic acid of flurbiprofen with the quanidine moiety of side chain from Arg108. The electrostatic interaction is depicted in diagonal dashed line. The site of metabolism is indicated with the white arrow and is positioned directly above the heme, whereas the benzylic site is circled (white) (Wester *et al.*, 2004).

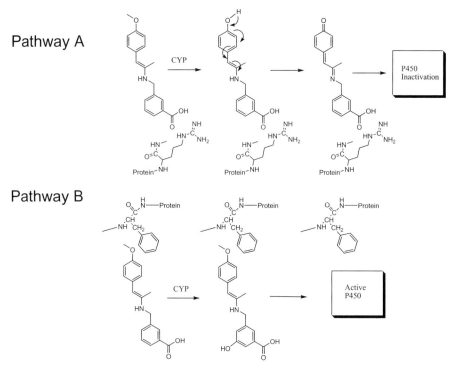

Figure 18.19 Hypothetical drug molecule metabolized by two different P450 enzymes. In pathway A, the molecule is positioned such that *O*-dealkylation occurs and leads to the generation of reactive quinone methide metabolite, whereas the metabolism in pathway B leads to formation of benign metabolite because of different metabolite regioselectivity.

As such, specific P450–substrate interactions may dictate whether or not bioactivation occurs. Take the compound in Fig. 18.19 for example, which has the ability to form a reactive quinone methide upon *O*-dealkylation in pathway A. The compound may be prone to such a reaction if the P450 responsible for metabolism interacts with compound's carboxylic acid group and positions the methoxy group for metabolism. Alternatively, if the compound is metabolized by a different P450 where pi stacking and hydrogen bonding interactions predominate, which favorably orient the same compound for metabolism away from *O*-dealkylation, then the bioactivation step may not occur resulting in the formation of alternate metabolites void of MBI (Fig. 18.19, pathway B). Therefore, P450–substrate interaction may direct the site of oxidation and influence to the potential propensity for bioactivation significantly contributing to whether or not MBI occurs.

Finally, the active site architecture dictates the ability for MBI. Each P450 has a unique composition of active site amino acids and thus has different nucleophiles capable of trapping reactive metabolites. Structural information offered from P450 crystal structures can provide crucial information regarding the propensity for MBI. To date, there are now structures for most of the key drug-metabolizing enzymes including P450 1A2 (2HI4.pdb) (Sansen *et al.*, 2007), P450 2D6 (2F9Q.pdb) (Rowland

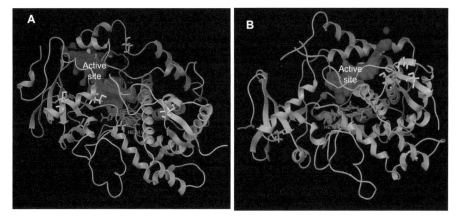

Figure 18.20   Comparison of P450 active sites **(a)** P450 3A4 (1TQN.pdb) and **(b)** P450 2C9 (1R9O.pdb). Representative active site volumes are depicted by the central shaded area. Heme is beneath the shaded area and cysteins are highlighted in white.

et al., 2006), P450 2C9 (1R9O.pdb) (Wester *et al.*, 2004), P4502C8 (1PQ2.pdb) (Schoch *et al.*, 2004), P4502A6 (1Z10.pdb) (Yano *et al.*, 2005), and P450 3A4 (1TQN.pdb) (Yano *et al.*, 2004). These three-dimensional representations provide visual coordinates to corroborate experimental findings, including the site of alkylation providing valuable insight to the relevance of adducts with respect to MBI. For example, the selective alkylation of Cys239 was observed with raloxifene (Baer, Wienkers and Rock, 2007). These findings are supported by the crystal structure which shows Cys239 distal to the heme in the enzyme active site (Fig. 18.20a). Experimental evidence of covalent adducts with P450 2C9 have illustrated the potential role of Ser365 in trapping the glutathione adducting compound, tienilic acid (Melet *et al.*, 2003) suggesting the absence of solvent-accessible cysteines in P450 2C9 which is confirmed upon analysis of its three-dimensional structure in Fig. 18.20b. Combining structural information of adducts with crystallography data will undoubtedly improve the speed at which future structural information regarding P450 MBI is obtained.

Furanocoumarins commonly found in natural foods such as celery, parsnips, grapefruit juice, and figs have been shown to undergo bioactivation (Fig. 18.21). Metabolism of the furan moiety leads to bioactivation and is the penultimate step prior to irreversible inhibition of several P450s (Koenigs and Trager, 1998a, b; von Weymarn *et al.*, 2005). The table in Fig. 18.21 lists three different P450 enzymes affected by the furanocoumarin, 8-methoxypsoralen (8-MOP). Of particular interest from these data is the K_I, or the rate of inactivation, at half maximal inhibitor concentration which appears similar for two separate P450 enzymes, P450 2B1 and P450 2A6. This suggests that the affinity and the forward commitment to irreversible inhibition combine in such a way that both enzymes are affected similarly with respect to the bioactivation and inactivation of 8-MOP. However, the k_{inact}, or rate at which maximal inactivation occurs, is six times faster with P450 2B1, suggesting either a more reactive residue toward the activated furan or a more proximal orientation of the reactive metabolite with a key active site nucleophile that promotes rapid inactivation versus release of product in the form of a

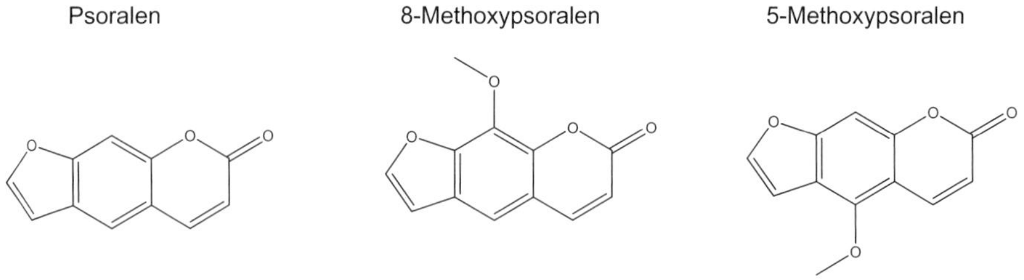

Psoralen   8-Methoxypsoralen   5-Methoxypsoralen

P450	K_I (µM)	K_{inact} (/min)	Partition ratio
1A2	1	NA	NA
2A6	1.9	2.0	11.2
2A6*	12	0.3	25.5
2A13	0.1	NA	NA
2B1	2.9	0.34	1.3
2B1* 5mop	33	0.13	14.9

Figure 18.21   The structures of psoralen, 8-methoxypsoralen, and 5-methoxypsoralen are illustrated with the inactivation rates for 8-methoxypsoralen listed for several different P450 enzymes. K_I values were measured to determine the inactivation at half the concentration required to achieve maximal inactivation and k_{inact}, or the maximal inactivation, was also determined. The partition ratio was included when measured. The inactivation parameters were not detected for P450 1A2 (Tantcheva-Poor et al., 2001) and the inactivation rate constants for P4502A13 could not be determined due to the high-affinity complex between the P4502A13 and 8-methoxypsoralen (von Weymarn et al., 2005). The close analog 5-methoxypsoralen is shown for comparison.

reactive metabolite. This is consistent with the partition ratios measured for the two enzymes, in which P450 2B1 reacts nearly one to one compared with P450 2A6 which undergoes nearly 10 turnover events to every one inactivation. Furthermore, based upon the proposed mechanism of irreversible inhibition, it may be plausible that one enzyme favors one reactive metabolite pathway versus the other. In combination with different regioselectivity between the two systems, the active site nucleophile may also be unique and could contribute to the different inactivation efficiencies.

5-Methoxypsoralen was also studied with P450 2A6 and P450 2B1, and in both cases showed a decrease in the efficiency of inactivation across all kinetic parameters, K_I, k_{inact}, and partition ratio (Fig. 18.21). Interestingly, the 5- and 8-psoralen derivatives were not substrates for P450 2B1 and therefore showed no irreversible inhibition. This provides a nice example of modulation of MBI without drastically changing the chemical scaffold (furan still present). Furthermore, the chemical properties were reasonably conserved with a modest 15-Da mass shift and 0.2 difference in CLogP from the methoxy derivatives. Alternatively in P450 2A6, the introduction of the hydroxyl derivatives led to a reduction in irreversible inhibition the loss of MBI in P450 2B1, indicating the likeliness of a specific enzyme–substrate interaction with 8-MOP which does not exist with the hydroxyl derivative. This type of interaction may prevent bioactivation of the furan and eliminates the potential for MBI in P450 2B1 (Koenigs and Trager, 1998b).

As highlighted with 8-MOP, the chemical structure provides the first indication for the potential of P450 bioactivation and thus the possibility for P450 MBI to occur (Fig. 18.10 and 18.21). In many instances, these structures are easily spotted (Figs 18.9–18.14). However, because of the robust oxidative chemistry associated with P450 reactions, many novel reaction schemes continue to be discovered (Isin and Guengerich, 2007). For example, the small-molecule ligand for gamma subtype peroxisome proliferator acceptor receptor for the regulation of glucose levels in type II diabetes, troglitazone, was withdrawn from the market after reports of liver failure (Watkins and Whitcomb, 1998; Graham *et al.*, 2003). While the mechanism of hepatoxicity has yet to be clearly defined, significant efforts have focused on the metabolism of troglitazone (Masubuchi, 2006). The structure contains a thiazolidinedione ring which undergoes oxidation at the sulfur, resulting in a ring opening of the thiazolidinedione ring which is then susceptible to nucleophilic attack (Fig. 18.22) (Kassahun *et al.*, 2001). In addition, the chromane ring structure also forms a GSH adduct after a P450-mediated hydrogen atom abstraction of the phenolic hydroxyl group to produce the ortho-quinone methide which is reactive toward GSH (Fig. 18.22) (Yamazaki *et al.*, 1999; Kassahun *et al.*, 2001). While it is not believed that the troglitazone adducts function independently to cause the observed hepatoxicity, they may present a contributing factor.

Furafylline serves as an interesting example where the furan structure readily stands out as a potential source of bioactivation and causative ingredient for MBI. In fact it has been highlighted as the site responsible for MBI in a review of P450 bioactivation (Fontana, Dansette and Poli, 2005). However, investigation into the mechanism of P450 1A2 inhibition by furafylline strongly supports the fact that inactivation occurs as a result of oxidative attack at the 8-methyl position of furafylline. A twofold reduction in activation results from the selective incorporation of deuterium replacing the C8 protons supporting the formation of a methide as the precursor to enzyme inactivation (Fig. 18.9) (Kunze and Trager, 1993). In addition, the use of a furafylline analog where the furan is replaced by a cyclohexyl derivative still yields P450 1A2 inactivation (Racha, Rettie and Kunze, 1998). Similar intermediates have been implicated as the precursor to P450 inactivation from 3-methylindole and 1-[(2-ethyl-4-methyl-1H-imidazol-5-yl)methyl]-4-[4-(trifluoromethyl)-2-pyridinyl]piperazine (Skiles and Yost, 1996; Hutzler *et al.*, 2004).

Aristolochic acid is a known human nephrotoxicant in humans (Bieler *et al.*, 1997). Similar to the activation of furafylline by P450 1A2, aristolochic acid bioactivation also occurs with P450 1A2 but does so in a reductive manner to produce the reactive nitreunium species. The chemical structure possesses a methylene dioxy derivative, known to produce metabolic intermediate complexes. The reduction of the 6-nitro group produces a cyclic nitreunium ion as determined from DNA adducts and appears to be the predominant bioactivation pathway (Fig. 18.23) (Chae *et al.*, 1999).

The ability to misjudge the safety of a chemical series has driven the development of multiple strategies to facilitate early indicators of MBI. In discovery, the focus is on providing filter points for the high volume of compounds that are screened in early-stage projects; thus, throughput is an important factor and will be a consistent theme within the early screening steps 1, 2, and 3 found in Fig. 18.16.

The most widely accepted surrogate marker for MBI is the presence of glutathionyl adducts which are generated from hepatic tissue fractions supplemented with the NADPH cofactor plus the addition of glutathione (GSH) (exogenous nucleophile), in order to mimic *in vivo* metabolism with the ability to trap any reactive metabolites during the incubation. In addition to GSH, several additional surrogate nucleophiles can be used and are depicted in Fig. 18.24. Most commonly, GSH is co-incubated with

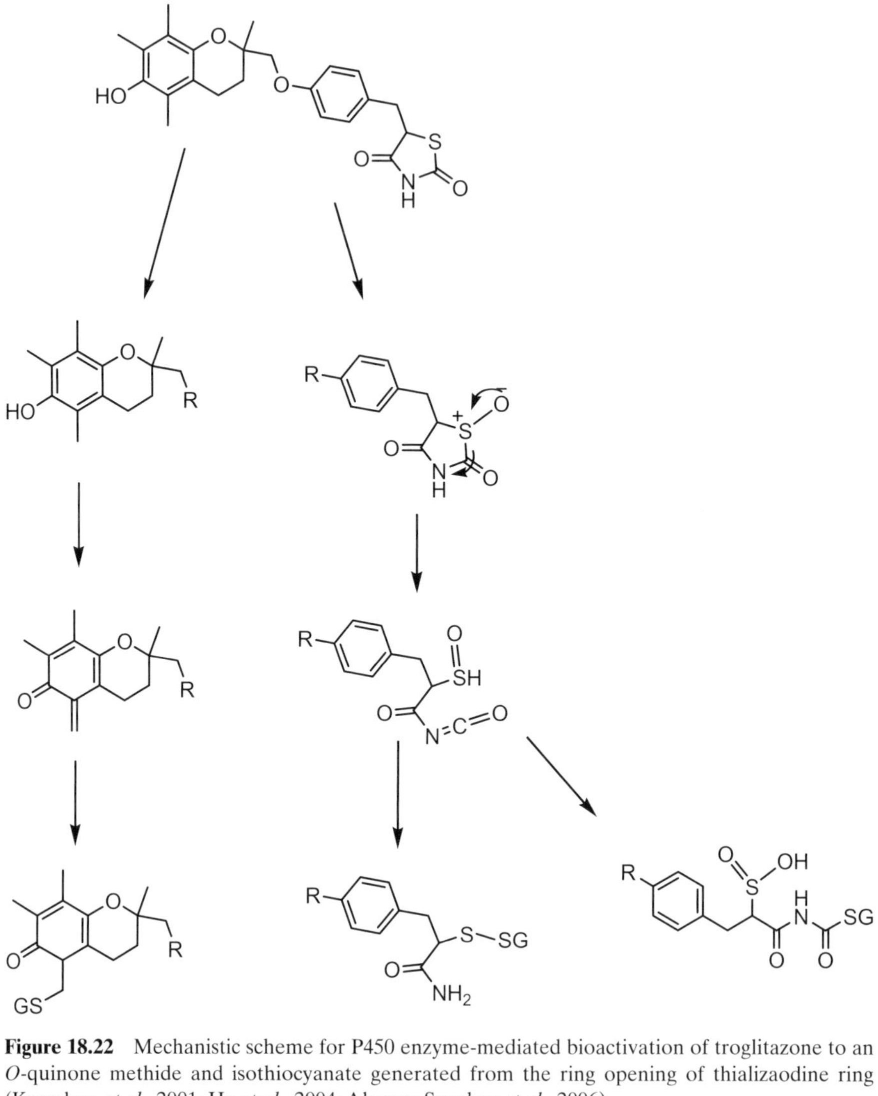

Figure 18.22 Mechanistic scheme for P450 enzyme-mediated bioactivation of troglitazone to an *O*-quinone methide and isothiocyanate generated from the ring opening of thializaodine ring (Kassahun *et al.*, 2001; He *et al.*, 2004; Alvarez-Sanchez *et al.*, 2006).

drug and hepatic tissue upon which, drug that undergoes bioactivation to a reactive intermediate can react with the nucleophilic sulfur of cysteine in glutathione. The assay is amenable to liquid chromatography–mass spectrometry/mass spectrometry (LC-MS/MS) and therefore can be designed for high throughput. For example, one of the first high-throughput versions of GSH screening took advantage of the neutral loss of pyroglutamate (129 Da) upon collision-induced dissociation (CID) fragmentation of the GSH conjugate (Baillie and Davis, 1993). This facilitates the identification of GSH-related material without prior knowledge of the parent molecular weight. Unfortunately, not all GSH conjugates produce this fragment ion upon CID which readily confounds attempts to identify GSH conjugates based upon a single LC-MS

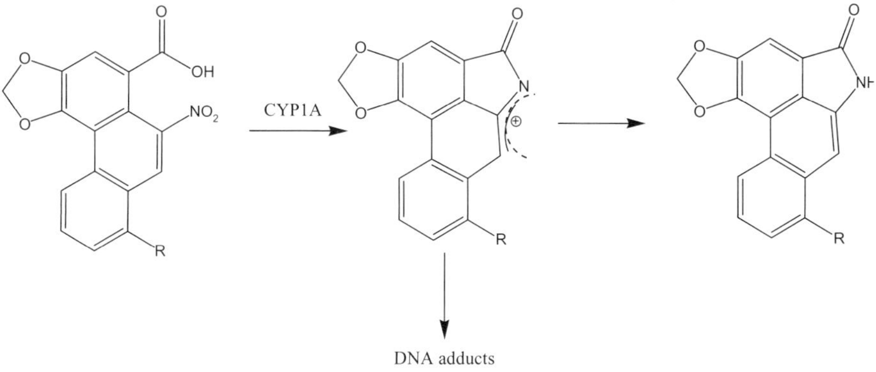

Figure 18.23 Mechanistic scheme for P450 enzyme-mediated bioactivation of aristolochic acid (Stiborova *et al.*, 2001). The reductive mechanism is consistent with the previous reports on the P450-mediated reduction of nitropyrenes (Chae *et al.*, 1999).

Trapping agent	Physical properties	Scan type	Diagnostic fragment	Limitation (s)
Glutathione (GSH) 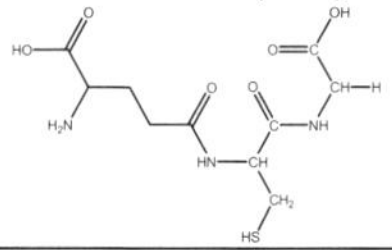	307 Da pKa ClogP	Neutral loss (NL) Precursor (PC)	129 Da m/z -272 Da	Low sensitivity High background interference Irreproducible fragmentation
N-acetyl lysine (NAL) 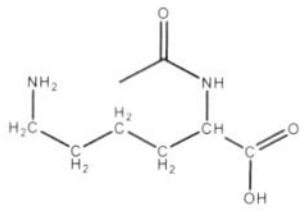	307 Da pKa ClogP			Reacts with hard electrophiles Reactivity may not represent that of endogenous proteins
N-acetyl cysteine (NAC) 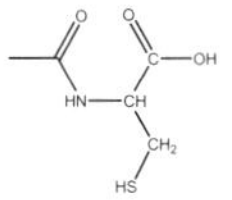	307 Da pKa ClogP	NL	129 Da	Reacts with soft electrophiles Reactivity may not represent that of endogenous proteins
Cyanide (CN) 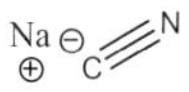	307 Da pKa ClogP	NL	27 Da	Reacts with hard electrophiles Reverisible binding

Figure 18.24 GSH and other nucleophiles used in trapping electrophilic intermediates from *in vitro* incubations.

scan (Fig. 18.24). Instead, different chemically generated conjugates such as aliphatic and benzyl thioether conjugates are known to undergo elimination of 307 Da and produce m/z 308 product ions consistent with the GSH_2^+ (Grillo *et al.*, 2003). Alternatively, thioesters can readily lose glutamic acid producing a fragment less than 147 Da from the parent protonated species (Baillie and Davis, 1993; Grillo *et al.*, 2003). The use of negative ion precursor has proven to be less sensitive to the chemical structure of the conjugate itself and serves as a robust method for producing GSH-related fragments (Dieckhaus *et al.*, 2005). As a result of the GSH dominant fragment patterns generated from the negative ion scan, the method serves as a good initial survey scan for the presence of GSH conjugated drug material with a low incidence of false negatives. Under these conditions, the CID fragmentation generally produces fragments related only to the GSH portion of the adduct and thus, additional MS/MS experiments are typically needed to explore the chemical nature of adduct to support rational chemical design strategies based upon structural information. The recent advent of the linear ion trap mass spectrometer continues to provide additional scanning functionality including enhanced product ion (EPI) scans which have been employed to maximize the information obtained during LC-MS runs used to screen for glutathione conjugates (Zheng *et al.*, 2007).

An important trait to bear in mind with respect to GSH conjugates is not all reactive intermediates are susceptible to attack from the soft electrophile as in the case of GSH. For example, compounds metabolized to hard electrophilic intermediates like iminium ions more readily react with hard nucleophiles like cyanide (Fig. 18.10) (Nguyen, Gruenke and Castagnoli, 1976). These electrophiles are capable of forming DNA adducts which are a known pathway for carcinogenicity (Stark, Harris and Juchau, 1989). Therefore, the use of additional nucleophiles with alternate reactive properties may be required. Cyanide (CN) or *N*-acetyl lysine represent two nucleophilic surrogates whose reactivity should be considered in order to rule out reactive metabolite formation (Fig. 18.24) (Argoti *et al.*, 2005; Druckova and Marnett, 2006). Furthermore, screening for cyanide adducts can screen relatively high throughput similar to those efforts described for GSH by monitoring the neutral loss of 27 Da (Argoti *et al.*, 2005).

A major drawback to screening with surrogate electrophiles lies in the assumption that the isolated conjugates relate to P450 inactivation (Baer, Wienkers and Rock, 2007). For example, the formation of multiple glutathione adducts confounds clear interpretation of the adduct responsible for P450 inactivation. Furthermore, many times glutathione offers no protection from MBI, thus increasing the uncertainty that the observed glutathione adducts represent the structural species leading to P450 inactivation. For example, the presence of glutathione in an incubation with P450 2C9 and tienilic acid reduces the partition ratio and stoichiometry of P450 2C9 inactivation, but the adduct itself may not represent the reactive species ultimately responsible for P450 2C9 inactivation (Koenigs *et al.*, 1999). Despite this discrepancy, it has become apparent with the quantification of GSH adducts that the amount of conjugated drug appears highly correlated to covalent binding results in many cases (Masubuchi, Makino and Murayama, 2007). Overall, screening with surrogate electrophiles provides an early complement to covalent binding studies where radiolabeled compound is required (Evans *et al.*, 2004). In a subsequent section of this chapter, the use of mass spectrometry of intact protein is described to elucidate drug-based adducts as a complement to GSH screening procedures.

The third step in the screening progression for MBI (Fig. 18.16) is to investigate the potential impact of the bioactivation pathway on P450 kinetics. In many cases, a negative finding from this assay provides a decision point at which MBI is no longer considered a safety concern for a particular compound. In some cases, GSH screening is omitted in favor of rapid kinetic screens focused on elucidating the potential for MBI. This choice of tactic may be especially poignant for P450 inhibitors that have multiple modes or mixed modes of inhibition such as 17α-ethynylestradiol (Lin and Hollenberg, 2007). This is feasible partly due to advances in laboratory automation (Lim *et al.*, 2005;

DeWitte and Robins, 2006), which makes the decision to move to a more informative kinetic assay (and in some cases, may replace the GSH LC-MS/MS approach outright) a cost-effective decision. The choice may reflect the stage of a project and if a team is still focused on improving absorption, distribution, metabolism, and excretion (ADME) properties with structure–activity relationships (SARs), then the use of GSH screen should provide mechanistic details necessary to eliminate drug bioactivation. However, in most cases, the inclusion of both screens is recommended as it heightens the awareness of the project team to the potential for MBI, showcases the susceptibility of P450s to generate reactive chemistries leading to complex mechanisms of P450 inhibition, and highlights any potential for other adverse drug reactions.

A straightforward method to assess the impact of bioactivation is performed with a pre-incubation step of the inhibitor in the presence of NADPH. In theory, the pre-incubation step provides a straightforward method to identify if pre-incubating alters the inhibitor through metabolism to increase the potency of inhibition of the compound of interest (Obach, Walsky and Venkatakrishnan, 2007). Two assay formats utilize a pre-incubation to assess the time and concentration effect of the inhibitor on P450 inhibition: (i) the IC_{50} shift assay and (ii) the apparent partition ratio screen (APR) (Fig. 18.25). Both formats rely upon a pre-incubation step which provides the necessary metabolism to assess whether or not inhibitor potencies increase upon metabolism (Obach, Walsky and Venkatakrishnan, 2007). A schematic of the experimental setup is illustrated in Fig. 18.26. Based upon the subtle difference in setup, it is clear that the left shift from a pre-incubation in the IC_{50} shift experiment does *not* distinguish the potential contribution of a reversible inhibitor that can occur from a metabolite formed during the initial incubation step.

To minimize the potential for inhibitory metabolites generated during the pre-incubation from contributing to the change in observed inhibition, large dilutions from the primary incubation are utilized to reduce reversible inhibition from inhibitory metabolites (minimum of one-tenth). Moreover, success of this experiment is predicated on the ability of the incubation system (inhibitor concentration, time, and enzyme source) to metabolize the compound. Therefore, an important control is to measure

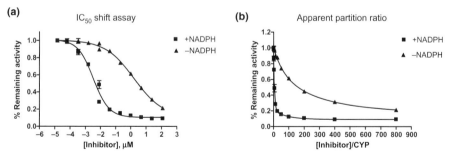

Figure 18.25 Data transformed from the experimental design in Figure 18.26 can be plotted one of two ways depending on the choice of control. For the IC_{50} shift experiment **(a)**, the control consists of the inhibitor without NADPH. This allows the formation of IC_{50} that can be compared with the pre-incubation with NADPH. The diagnostic sign of TDI is related to a left shift in the IC_{50} or increase in potency when the inhibitor is pre-incubated with NADPH. Alternatively, the apparent partition ratio **(b)** control utilizes NADPH but no inhibitor. The only other difference is the data on the abscissa is inhibitor concentration in the case of the IC_{50} shift compared with inhibitor concentration versus enzyme concentration in the apparent partition ratio experiment.

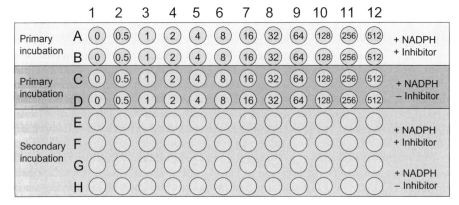

Figure 18.26 96-well plate pre-incubation format for determination of IC$_{50}$ shift or apparent partition ratio. Primary incubations with varying concentrations of suspected MBI are conducted for a finite time period with excess human liver microsomes (0.5 or 1 mg/mL) depending upon probe substrate reaction velocities and dilution scheme of the secondary incubation. The secondary incubations contain reaction buffer plus NADPH and the appropriate probes substrate for the P450 under investigation at four to five times the K$_m$ and are initiated with one-tenth to one-hundredth of the primary incubation volume.

the percent turnover of the inhibitor during the pre-incubation step. This provides confidence that the pre-incubation conditions are sufficient to generate the turnover required to properly investigate the potential for MBI. In the absence of sufficient turnover, it may be necessary to increase the pre-incubation time, change protein concentration, tissue fraction (S-9, microsomes, or hepatocytes), or any combination thereof.

One difference exists in the graphical representation of the data between the IC$_{50}$ shift and the APR formats which provides a possible advantage to representing data as an APR. The x-axis of the IC$_{50}$ shift is represented as the log concentration of inhibitor tested in the experiment whereas for the APR, the x-axis is represented in linear fashion by concentration of the inhibitor relative to P450 levels used in the incubation (Fig. 18.25). The net effect of this subtle experimental difference is that it provides a clear distinction from the control incubation especially as the concentrations of inhibitor increase, since the altered trajectory of NADPH supported pre-incubation in both the IC$_{50}$ shift and the APR reactions are highly dependent upon inhibitor concentration (k$_{inact}$) for the manifestation of MBI. Furthermore, representing the data as APR provides a direct quantitative measure between different inhibitors by providing an efficiency value for inactivation which is useful for SAR purposes.

Up to this point in Figure 18.16, the focus has been on Tier 1 or screening-based assays which would often be implemented in the early discovery stage of a project. If the results from these assays point to MBI, then the next step is to define the MBI potency. That is, if the inactivation rate is 1% of the natural degradation rate of the P450, then inactivation will not likely increase the drug interaction beyond that described by reversible inhibition. Therefore, investing in the technical and laborious structural investigations required to define the mechanism(s) behind MBI is likely not warranted.

In order to contextualize the significance of MBI, one must define properties of inactivation including the maximal possible rate of enzyme inactivation at a saturating

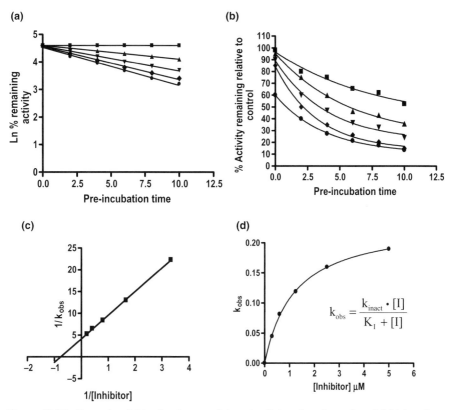

Figure 18.27 Example of kinetic plots used for visualizing time-dependent inhibition data. **(a)** Log-linear plot representing the remaining enzyme activity after various pre-incubation times with an MBI, **(b)** linear-linear plot representing the remaining enzyme activity after various pre-incubation times with an MBI, **(c)** linear plot (Kitz Wilson) represented by $1/k_{obs}$ and $1/[inhibitor]$, and **(d)** non-linear regression of k_{obs} versus inhibitor concentration.

amount of MBI, k_{inact}, and the concentration at which half of the maximal inactivation is achieved which is defined as the K_I. In doing so, the rate of inactivation can be directly compared with what is anticipated as the natural rate of P450 degradation. This assay commonly referred to as the time-dependent inhibition (TDI) assay includes multiple pre-incubation times and various concentrations of the inhibitor to yield individual inactivation rates for each concentration, which are combined and fit using nonlinear regression to determine the K_I and k_{inact} (Fig. 18.27) (Mayhew, Jones and Hall, 2000).

The assay requires two steps, starting with a primary incubation to activate the inhibitor (suspected MBI). From the primary incubation, aliquots are transferred into a secondary incubation (10- to 100-fold) containing a marker substrate selective for the particular P450 isoform under investigation (Fig. 18.28). Several variables must be considered to ensure accurate rates of inactivation are measured: (i) that the competitive inhibition at the secondary stage of the incubation is minimized by sufficiently diluting the primary to secondary incubations; (ii) an appropriate number of inhibitor concentrations are used; (iii) protein concentration is optimized to minimize nonspecific binding yet enough to produce sufficient turnover in the primary and secondary

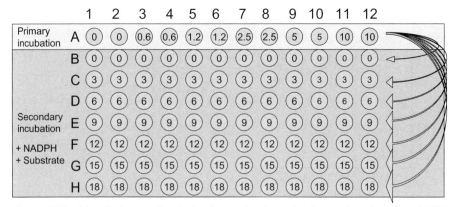

Figure 18.28 96-well plate pre-incubation format for determination of time-dependent inhibition. Primary incubations containing varying concentrations of suspected MBI are conducted in 200 μL final volume with excess human liver microsomes (0.5 or 1 mg/mL) depending upon probe substrate reaction velocities and dilution scheme of the secondary incubation. The secondary incubations contain reaction buffer plus NADPH and the appropriate probes substrate for the P450 under investigation at ~5× the K_m and are initiated with one-tenth to one-hundredth of the primary incubation volume. Transfers are made in time intervals consistent with the expected potency of the TDI and are made easily with 12-channel pipette or automation. Each subsequent row corresponds to unique time interval for each concentration under investigation.

incubations; and (iv) the secondary incubations are carried out under linear conditions with saturating amounts (V_{max} conditions) of the appropriate probe substrates. Careful attention to these design features helps reduce bias in the assay format and provides the necessary end points to accurately quantify the effect of time and metabolism on the inhibitor in question (Yang et al., 2005).

To ascertain the appropriateness of the experimental data, the kinetic results should be examined with several different plots to determine if further kinetic analysis is required. First, each inhibitor concentration and its time course should be plotted as the percent remaining relative to control on the ordinate axis versus the pre-incubation time on the abscissa. These data can be presented in a semilog plot or if the changes are small with a linear plot to facilitate visualization of any aberrant data points (Fig. 18.27a,b). Furthermore, this plot provides a semiquantitative view to the extent of competitive inhibition observed in the secondary incubations. For example, the highest concentration inhibitor sample in Fig. 18.27b indicates substantial competitive inhibition occurred at the highest inhibitor concentration. This is diagnosed by the first time point (no pre-incubation) at which the remaining activity should be close to 100%, but in these examples, is only 60% of control. Fitting inactivation data in the presence of significant competitive inhibition will reduce slope of the fit and lead to underestimation of the inactivation rate, confounding the accurate determination of MBI rates. Next, the data should be transformed to a linear plot (Kitz–Wilson) by plotting $1/k_{obs}$ on the ordinate versus 1/inhibitor on the abscissa (Kitz and Wilson, 1962) (Fig. 18.27c). The intercept of the ordinate axis defines $1/k_{inact}$ which can be used to approximate k_{inact}. The slope of the line estimates the k_{inact}/K_I. The range in data should be such that a single point does not dominate the linear regression. Secondly, any deviation from linearity may indicate limitations of inhibitor solubility or a mixed mechanism of inhibition. The K_I and k_{inact} values should be derived from nonlinear regression (Fig. 18.27d) (Maurer and Fung, 2000). At the end of a TDI experiment, there are at least four graphs that can be used to interrogate the integrity of the data from the TDI experiment. Once this has been completed, the significance of MBI in addition to reversible inhibition can be determined.

Once MBI has been estimated to lead to a significant interaction, it is paramount to structurally characterize the mechanism(s) responsible for irreversible inhibition to define and potentially eliminate the responsible chemical substructure. Three distinct mechanisms of MBI exist for P450 inactivation: (i) heme iron coordination through an MI complex, (ii) heme alkylation, and (iii) apoprotein alkylation. Prior to initiating studies aimed at defining the mechanism of inactivation, it is useful to try and define which P450s are affected by MBI through phenotyping efforts. Phenotyping is presented in detail in Chapter 16. Defining the metabolic pathways helps define what P450s need to be studied and the experimental tools available to study the structural aspects of MBI. Specifically, it enables the use of recombinant P450 enzymes and while these systems have been shown to produce unique kinetics properties in certain situations, their simplified matrix enables the use of a wider range of applications aiding in the structural characterization and mechanistic details regarding the nature of P450 enzyme inactivation.

Heme iron coordination is a result of P450-mediated metabolite which subsequently coordinates as the sixth axial ligand to the heme iron. This is commonly referred to as MIC formation or MI complex and leads to P450 inactivation. The formation and stability of the MIC is directly related to the chemical substructure of the metabolite and is commonly referred to as a quasi-irreversible inhibitor. Two substructures are prone to MIC formation: the tertiary amine and the methylene dioxy species. The most useful diagnostic for MIC formation is performed by measuring the increase in absorbance at 455 nm over time with a UV-Vis spectrophotometer (Franklin, 1995). In these studies, the enzyme, substrate, and NADPH are combined into a cuvette and pre-incubated at 37 °C. The absorbance at 455 nm is measured kinetically to determine the presence of MIC formation (Fig. 18.29) (Hutzler *et al.*, 2006). The confirmation of MIC formation as the sole contributor to P450 inactivation can be assessed by the ability to reverse the inhibition upon the addition of potassium ferricyanide (Sharma, Roberts and Hollenberg, 1996).

Direct elucidation of the chemical species responsible for MIC formation represents the most challenging out of the three forms of inactivation because the complex is not physically isolable. Instead, chemistry support is required to develop structural analogs to indirectly show the lack of MIC in the absence of the proposed culprit substructure. For example, for a methylene dioxy to form an MIC, initial hydrogen atom abstraction must occur which can be blocked by introducing a CF_2 at the bridge of the two oxygen atoms, or instead the methyl group can be replaced with an ethyl derivative eliminating the ability of the carbene to form after hydrogen atom abstraction (Fig. 18.13). Alternatively, if the tertiary amine is the precursor to MIC formation, a fluorinated derivative or alpha substitution can be used to deter *N*-dealkylation at the site of MIC formation. If the substructure causing the MIC is also required for therapeutic activity, a more subtle change in structure may need to be explored. For example, with the aid of P450 crystal structures, docking tools may be used to infer enzyme substrate interactions that contribute to the particular substrate regioselectivity. Specifically targeting these interactions may provide a different P450 regioselectivity that reduces MIC formation while maintaining the biological potency.

Heme alkylation is the second mechanism responsible for P450 inactivation and occurs via the bioactivation of a chemical species, which in turn alkylates the heme substructure itself and irreversibly inactivates the affected P450. Substructures prone

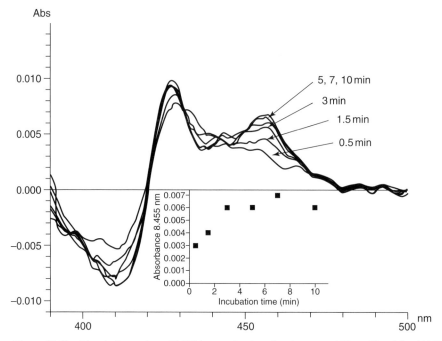

Figure 18.29   Kinetic formation of MIC by monitoring absorbance at 455 nm (Franklin, 1995).

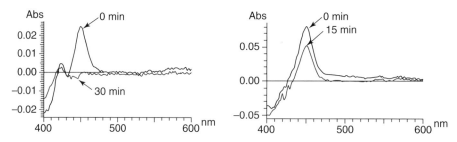

Figure 18.30   Measurement of carbon monoxide binding pre- and post-incubation with MBI to qualitatively determine the extent of heme-related adduct formation (Hutzler *et al.*, 2004).

to heme adduct formation include, but are not limited to, furans, aldehydes, and alkynes. Again, the UV-Vis spectrophotometer has been classically used to differentiate a heme adduct from an apoprotein adduct. The saturation of CO in a reduced solution of P450 affords a characteristic absorbance at 450 nm which is used to determine P450 enzyme levels from a complex solution (Omura and Sato, 1964). Heme alkylation prevents the binding of CO to the heme iron and therefore has been used to measure the P450 concentration pre- and post-incubation, whereby the loss in CO binding at the end of incubations represents the amount of alkylated heme. However, experimental data have shown the disappearance of CO binding may not always be linked to heme alkylation (Fig. 18.30). For example, CO spectral binding revealed ~60% loss of CO binding upon incubation with an MBI tienilic acid, which has been shown to form a P450 apoprotein adduct, indicating the decreased CO binding can occur through alternate mechanism(s) (Koenigs *et al.*, 1999). In addition, raloxifene has been shown to exclusively alkylate the apoprotein of P450 3A4 and yet after incubation, significant CO binding is lost (Baer, Wienkers and Rock, 2007). Finally, the photoaffinity probe, lapachenole, of which alkylates proteins via Michael addition to selectively label cysteine residues also produced a significant loss of CO binding (Wen *et al.*, 2005). Therefore, CO binding should not be used as the sole indicator of heme alkylation.

Heme alkylated products can also be measured directly using LC-UV-MS. For example, 1-ABT forms a covalent adduct with the heme of P450. The 1-ABT heme adduct was measured directly using mass spectrometry (Ortiz de Montellano and Mathews, 1981). This, combined with NMR data, was used to give the exact details of the adduct and its structure, which provided a basis for the mechanism for 1-ABT heme alkylation (Fig. 18.31). The advantage of this technique is that information regarding inactivation based on heme adduct formation is determined directly and therefore, the structure can be characterized affording mechanistic hypothesis for its formation leading to the ability to modify the chemistry responsible for inactivation. An alternative method that does not provide mechanistic detail but is not confounded by

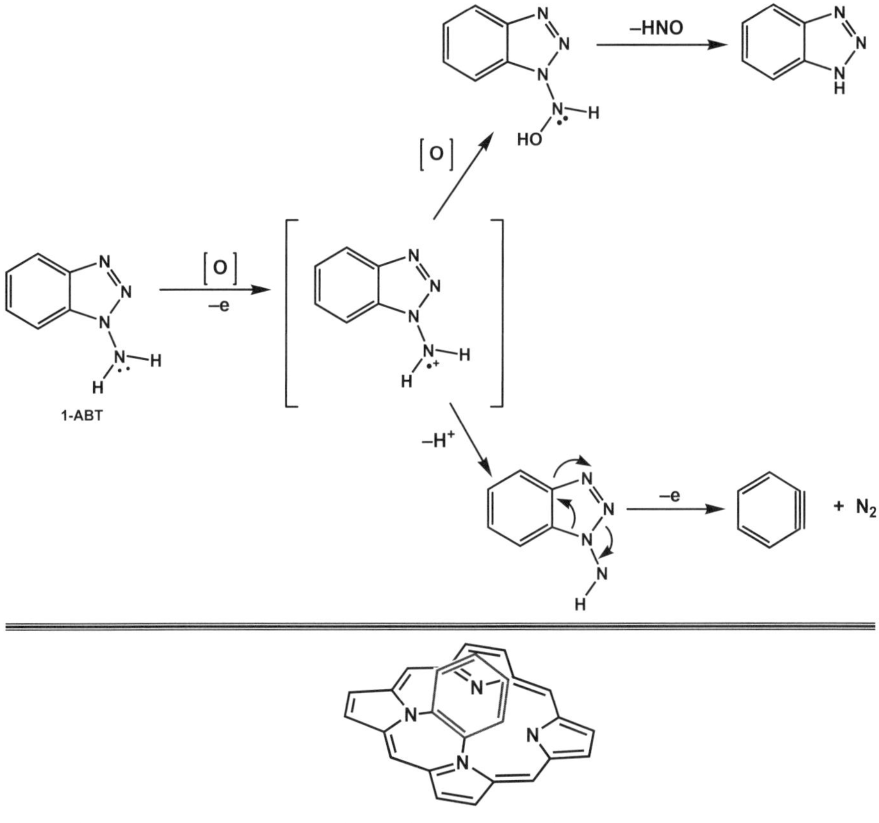

Figure 18.31 Mechanistic scheme for P450 enzyme-mediated bioactivation and subsequent enzyme inactivation of 1-aminobenzotriazole (ABT). Structure of ABT rearranges to benzyne upon metabolism and alkylated the heme. Other alkynes have shown similar reactivity toward the heme.

apoprotein adduct formation as the CO measurement is the direct quantification of heme itself pre- and post-incubation. This requires the use of recombinant expressed P450 enzymes without cytochrome b_5 so as to measure the disappearance of heme directly in order to implicate heme alkylation or destruction as a mechanism for inactivation.

Apoprotein adducts represent the third mechanism of inactivation and have classically been identified based upon the use of radiolabeled ligand in combination with gel electrophoresis, where P450 binding was implicated through the migration of radioactivity near 55 kDa. More recently, mass spectrometry has been used to characterize adducts of intact proteins which favorably combines the level of detail accustomed to mass spectral data, in addition to providing information regarding covalent binding stoichiometry and confirming the adduct is irreversible in nature. Figure 18.32 illustrates the use of intact protein mass spectrometry to identify the raloxifene P450 3A4 adduct producing mass shift of 472 Da compared with control P450 3A4 protein without

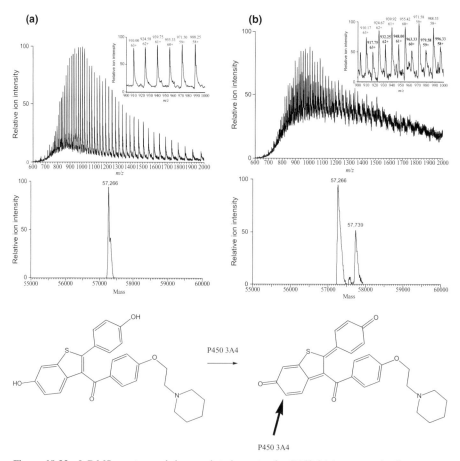

Figure 18.32 LC-MS spectra and deconvoluted spectra for P450 3A4 apoprotein. Spectra are shown for unmodified P450 3A4 **(a)**; P450 3A4 following the reconstitution and incubation with raloxifene **(b)** Raloxifene biotransformation to the reactive diquinone methide is shown highlighting one of the nucleophilic sites generated (Baer, Wienkers and Rock, 2007).

raloxifene present. These findings confer the diquinone methide is responsible for P450 3A4 inactivation as opposed to the ortho-quinone reactive metabolite which would have had a mass increase of 488 Da (Baer, Wienkers and Rock, 2007). Presumably, this information could be used to generate raloxifene analogs incapable of forming the diquinone methide preempting MBI formation in a similar fashion to what has been achieved with the close analog, arzoxifene (Liu *et al.*, 2005).

Another interesting example is provided from 1-[(2-ethyl-4-methyl-1H-imidazol-5-yl)methyl]-4-[4-(trifluoromethyl)-2-pyridinyl]piperazine (EMTPP), which was shown to form an apoprotein adduct to P450 2D6 (Fig. 18.33). The mass difference of the P450 adduct from the unlabeled P450 was 353 Da which was consistent with the parent compound. Interestingly, metabolism data including NMR and CID mass spectrometry were used to support the mechanism of adduct formation that likely resulted from multiple oxidations to form dehydrated methide of EMTPP similar to what has been proposed for 3-methylindole (Nocerini *et al.*, 1985). This example also nicely illustrates the potential limitation of intact protein mass spectrometry and how it must still be combined with additional experimental data in many cases to support the nature of the inactivating species, based upon the mass spectral resolution limitations and the difficulty in differentiating 2 Da with ~60,000 Da proteins. However, with the continued advances in mass spectrometry instrumentation such as the LTQ-Orbitrap and other hybrid mass spectrometry equipment, the use of this technique is rapidly spreading and may, in the near future, have the resolution to differentiate adducts by as little as 2 Da.

The structural limitations of intact protein mass spectrometry can be greatly improved with subsequent proteolytic digest of the adducted protein. This additional step also has the benefit of potentially providing greater detail with respect to the exact nature of the protein adduct, including the ability to provide the site of protein modification and the potential for CID spectral data to be obtained, which increases the potential structural knowledge about the chemistry behind the irreversible binding. The ability to elucidate covalently bound drug adducts to P450s is laden with technical challenges largely associated with difficulty in isolating the membranous P450s and obtaining sufficient sequence coverage from digests necessary to detail the location and nature of MBI adduct (Lightning *et al.*, 2000). To assist in this process, several techniques have been employed to locate nucleophilic residues including photoaffinity labeling (Gartner *et al.*, 2005; Wen *et al.*, 2006), alkylation with xenobiotic linked to biotin (Liu *et al.*, 2005), and direct alkylation of purified P450 with reactive electrophiles such as iodoacetamide (Baer, Wienkers and Rock, 2007). In combination with the nucleophilicity of active site amino acids, these tools provide a focal point for future efforts when trying to identify drug adducted P450s.

Photoaffinity labels have been used to probe enzyme structure function of cytosolic and transmembrane proteins (Wang, Bauman and Colman, 1998; Tessier *et al.*, 2005). The use of photoaffinity probes in P450 labeling has also been explored. Initial efforts were performed with 1-(4-azidophenyl)imidazole and P450 cam (Swanson and Dus, 1979). More recently, lapachenole was selected to investigate the alkylation of P450 3A4 (Fig. 18.34a) (Gartner *et al.*, 2005). The photo-activation of lapachenole produced two distinct peptide adducts. Two cysteine residues were identified, Cys98 and Cys468. Cys98 is in close proximity to the B-B'loop region located near the substrate recognition site 1 (SRS-1) (Roussel, Khan and Halpert, 2000). Moreover, the region near Cys98 and the B-C loop is postulated to be the access point to the heme iron and small changes in this region impact active site volumes (Williams *et al.*, 2004). Based

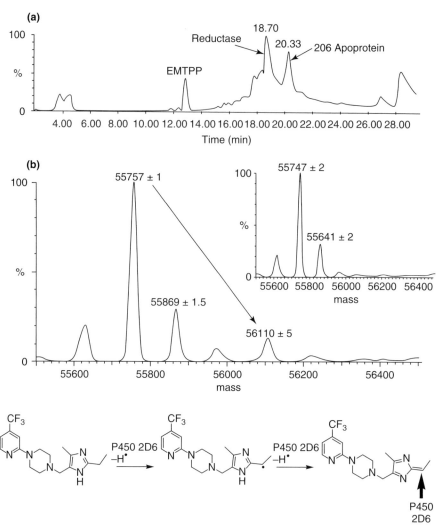

Figure 18.33 **(a)** Chromatogram illustrating separation of reductase and apoprotein from an incubation with recombinant P450 2D6. **(b)** Deconvoluted mass spectrum of EMTPP-inactivated P450 2D6 illustrates an EMTPP adduct ($56,110 \pm 5$ Da) consistent with a mass 353 Da. Inset shows the mass of the—NADPH control incubations. Proposed inactivation species of EMTPP by P450 2D6 involving the formation of a radical intermediate that undergoes additional hydrogen atom abstraction from the imidazole nitrogen to form a reactive imidazo-methide-like species (Hutzler *et al.*, 2004).

upon the crystal structure, the second adducted species, Cys468, appears to be a solvent-accessible cysteine on the exterior of the protein.

Alternatively, the use of biotin-linked electrophiles holds promise for use in identifying key enzyme electrophiles. This technique was extended by linking biotin with raloxifene (Fig. 18.34b) and subsequently used to illustrate binding of the activated raloxifene moiety to Cys47 of GST P1-1 (Liu *et al.*, 2005). When employed against a

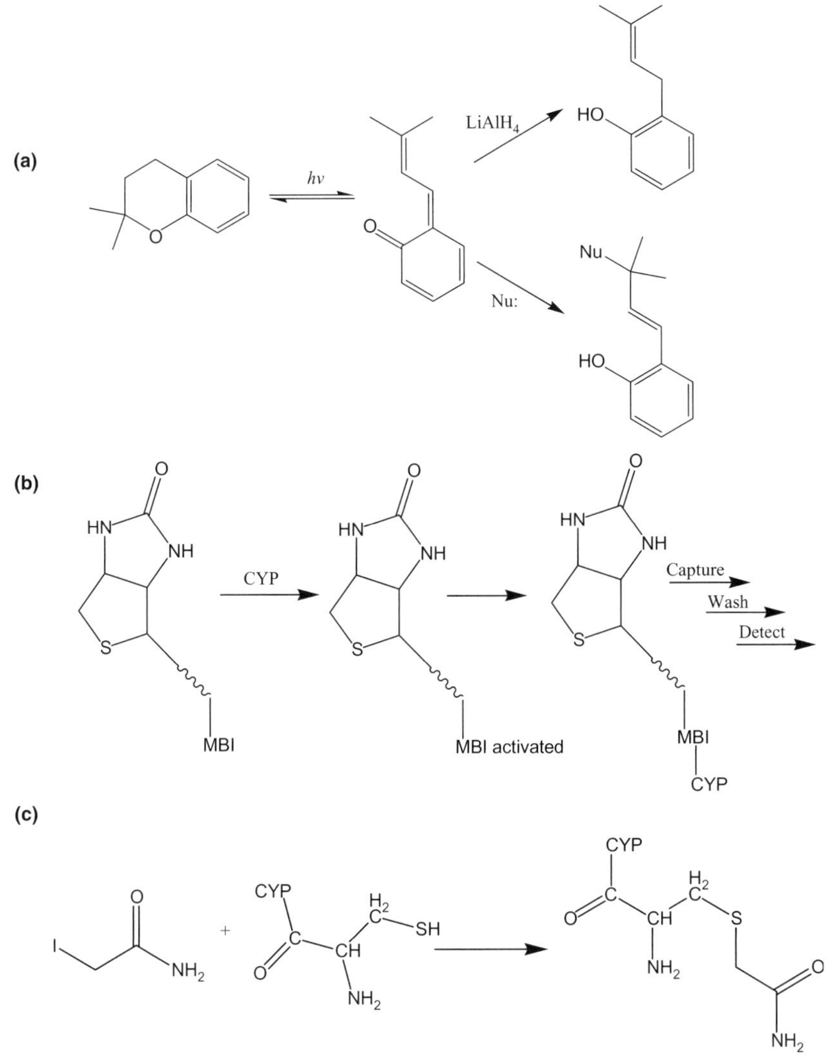

Figure 18.34 Different mechanistic approaches to locating P450 nucleophiles including **(a)** photoaffinity labels (Gartner *et al.*, 2005), **(b)** biotin-linked electrophiles (Liu *et al.*, 2005), and **(c)** direct alkylation with reactive electrophile (Baer, Wienkers and Rock, 2007).

dexamethasone-treated preparation of rat liver microsomes, five adducts were identified, none of which corresponded to P450s. The capture and isolation examples provide proof of concept that the technology is capable of isolating nucleophilic amino acids from a complex mixture, and if employed against a simplified *in vitro* system akin to the GST example, key active site residues may be more readily identified from P450s.

Direct alkylation with reactive electrophile iodoacetic acid (IA) and pyrenyliodo-acetic acid (PIA) offers a simple approach to locate solvent-accessible and reactive cysteine residues (Fig. 18.34c). Interestingly, in P450 3A4, the use of IA or PIA produced single stoichiometric reactions when run for a finite time (Baer, Wienkers and

Rock, 2007). Upon digestion with proteinase K, the alkylated residue was identified by LC-MS/MS as Cys239. Upon analysis of the crystal structure of P450 3A4 (1TQN.pdb), Cys239 resides above the B-C loop and may serve as a perfect hook for electrophilic products as they egress the active site of P450 3A4. Interestingly, the label of P450 3A4 by photoaffinity label lapachenole and that of IA produced distinct adducts. It is possible that IA and lapachenole exert structural changes to the P450 exposing unique nucleophiles. Alternatively, the reactivity of Cys239 and Cys98 may chemically favor the two different electrophilic intermediates. Either way, future efforts to identify MBI of P450 3A4 should focus on locating peptides containing cysteine residues 98 and 239.

The use of different digesting reagents has been explored to provide insight to MBI in P450. Cyanogen bromide chemically digests proteins to produce large peptide fragments and has been used to locate the regional location of adduct formation (Lightning et al., 2000). Lysyl endopeptidase also produces rather large fragments of the resultant protein as protein is cleaved at the *C*-terminus of lysine residues. Both of these digestion procedures are very amenable to time of flight (TOF) technology given their large molecular weights and have the advantage of producing fewer fragments of the protein to search. Trypsin digests cleave proteins at the *C*-terminus of lysine and arginine residues. The fragments produced from trypsin are very suitable for LC-MS/MS using a variety of mass spectrometry platforms given their tendency to carry multiple charges ($[M+2H]^{+2}$). Recently, proteinase K has been used to take advantage of the indiscriminant protein cleavage sites to produce small peptide fragments. When separating a proteinase K digested sample on reverse-phase column, the bulk of the small peptides elute quickly, whereas drug adducted peptide creates increased hydrophobic character creating separation from un-adducted peptides (Fig. 18.35). In addition, the use of proteinase K should improve ionization efficiencies by reduced hydrophobic aggregation and less ion suppression from the reduced number of co-eluting peptides. The tools described for identifying active site nucleophiles provide a focal point for future studies aimed at locating drug-based peptide adducts and over time should lead to improved ability to generate high-quality structural data in a more timely fashion.

Radiolabel is used as a quantitative and qualitative tool to rapidly assess irreversible binding of protein, locate peptides that have radiolabel drug incorporated, and define binding stoichiometry (Fig. 18.36). An additional advantage of using radiolabel is that *in vivo* and *in vitro* experiments can be used to better understand the reactivity of the bioactivated mechanism (Evans et al., 2004; Baillie, 2006; Williams et al., 2007). Covalent binding results are typically reported as pmol drug bound per milligram of protein. Based upon a meta-analysis, Evans et al. issued a perspective highlighting 50 pmol of drug/mg of protein as a reasonable cutoff for continued development of a molecule that possesses covalent binding (Evans et al., 2004). For MBI, the measured quantity of covalent binding may be compared with the relative quantities of P450 enzymes expected from a preparation of human liver microsomes. For example, the amount of P450 in a particular enzyme preparation can be measured (Alterman et al., 2005). The combination of enzyme abundance with covalent binding and MBI may provide additional knowledge regarding the predicted exposure to reactive metabolites. For example, Fig. 18.37 shows the average P450 abundance data compiled from the Caucasian physiological database from within Simcyp V7.01 (www.simcyp.com, Sheffield, UK). Provided that the bioactivation mechanism produces an MBI with a partition ratio close to 1 and the P450 pathway is known, the quantities of P450 shown in Fig. 18.37 serve

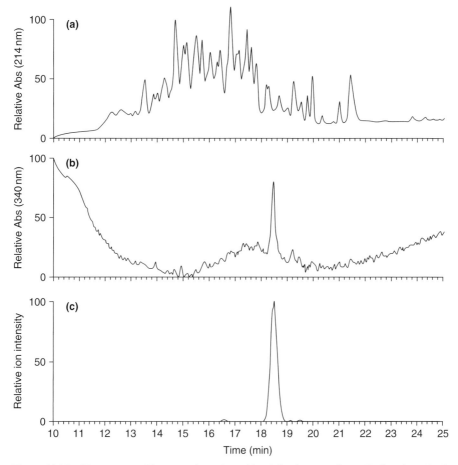

Figure 18.35 Chromatographic separation of peptides following proteinase K digestion of ral-oxifene-adducted P450 3A4: **(a)** UV 214 nm trace; **(b)** UV 340 nm trace; and **(c)** mass spectral trace of adduct (Baer, Wienkers and Rock, 2007).

as a benchmark for the expected amount of radiolabel incorporation into human liver microsome. For example, L-754,394 has a low partition ratio between 1 and 4, is metabolized by P450 3A4, and labeled approximately 120 pmol equivalents per mg of protein when incubated with human liver microsomes (Sahali-Sahly *et al.*, 1996; Lightning *et al.*, 2000). Based upon P450 levels, this is well within the expected range of P450 3A levels from a human liver microsome preparation and is consistent with a low partition ratio where less of the bioactivated drug would be expected to react with extracellular protein.

Alternatively, a series of dihydrobenzoxathiin selective estrogen receptors have been shown to undergo bioactivation via P450 3A4 (Zhang *et al.*, 2005). Two primary analogs were studied to reveal that compound I showed 1106 pmol equivalents per milligram versus compound II which yielded 461 pmol equivalents per milligram of protein (Fig. 18.38). Based upon the structures in Fig. 18.38, it is clear that a subtle

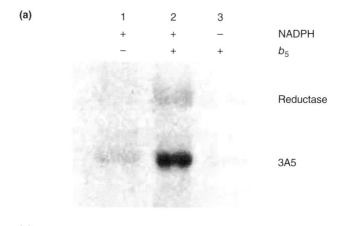

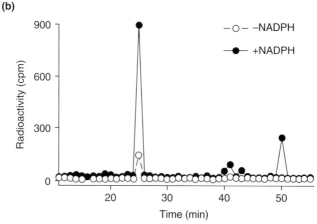

Figure 18.36 Covalent binding of EE to P450 3A5. **(a)** SDS-PAGE separation of 3A5 apoprotein and reductase in the reconstituted system incubated with [³H]EE with NADPH but no b₅ (Wienkers and Heath, 2005), with NADPH and b₅ (Guengerich, 1997), and with b₅ but no NADPH (Friedman *et al.*, 1999). **(b)** HPLC separation and analysis by liquid scintillation counting of the fractions after the [³H]EE-inactivated P450 3A5 was digested by Lys C (Lin and Hollenberg, 2007).

difference can have a dramatic effect on the level of covalent binding. More importantly, this example shows despite being P450 3A4-mediated, significant higher levels of protein adduction are observed from covalent binding studies. Unfortunately, these investigations did not include a partition ratio but based upon the covalent binding and P450 3A4 levels, the ratio may be expected to be estimated >10. This example illustrates that the disproportionate covalent binding levels relative to P450 levels should indicate an increase in extracellular microsomal protein and should be reflected by large partition ratios. Of course, heme-based adduct or small peptide fragments (<1000 Da) may not be effectively captured by the typical covalent binding filters which have a pore size of 0.7–1.0 µM. Therefore, MBI which occurs by MIC or heme alkylation may produce significantly lower binding values compared with apoprotein adducts.

CYP	Enzyme abundance (pmol/mg protein)	%CV
1A2	52	67
2A6	20.1	173
2B6	11	143
2C8	24	81
2C9	73	54
2C18	1	106
2C9	14	106
2D6	8	61
2E1	61	61
3A	210	163

Figure 18.37   Average enzyme abundances taken from SimP450 V7.01 illustrating the various levels of P450 from the human liver. Values are represented as picomoles per milligram of protein with coefficient of variance (%CV) reported (Sheffield, UK).

I

II

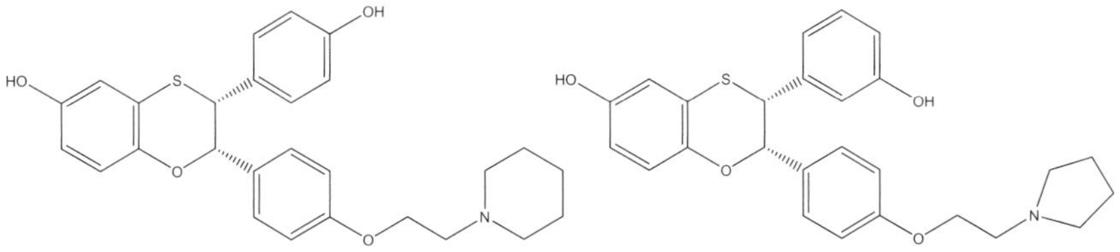

(2S,3R)-(+)-3-(4-hydroxyphenyl)-2-
[4-(2-piperidin-1-ylethoxy)-phenyl]-2,3-dihydro-1,4-benzoxathiin-6-ol

(2S,3R)-(+)-3-(3-hydroxyphenyl)-
2-[4-(2-pyrrolidin-1-ylethoxy)phenyl]-2,3-dihydro-1,4-
benzoxathiin-6-ol

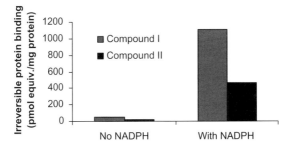

Figure 18.38   Structural comparison of two compounds aimed as selective estrogen modulators were studied. Based upon the data from compound I, compound II was developed partially due to the reduction observed in covalent binding (Zhang *et al.*, 2005).

The integration of the covalent binding data with the described experiments (Fig. 18.16) provides the potential for an integrated picture of the structural and functional information as it relates to P450 MBI for a particular compound. However, despite the low levels of binding that may be observed, even low levels of protein adducts may be sufficient to induce an immune response detrimental to patients. Therefore, results from covalent binding experiments are critical to context against additional risk factors related to the intended therapy including frequency, dose, and disease state the intended therapeutic is aimed to treat. A thorough review and discussion of the risk/benefit factors related to covalent binding are provided elsewhere (Baillie, 2006; Doss and Baillie, 2006).

Implementing the structural and functional tools discussed should help to develop the most appropriate clinical candidate. If MBI still persists, it is imperative that the potential impact on the metabolism of other medications is quantitatively assessed in order to define the clinical limitations of the therapeutic agent including the ability to coadminister other medications, define at-risk patient population (e.g. age or pharmacogenetics), expected exposure of the MBI, and duration of treatment. This requires the combination of much of the data already accumulated from Fig. 18.16 including the K_I and k_{inact} with the endogenous enzyme degradation rate in order to describe the extent of drug interaction that could result based upon irreversible inhibition kinetics (Fig. 18.27). Additional information regarding the expected metabolism and disposition of the substrate responsible for producing irreversible inhibitor will typically serve to increase the predictive power of the anticipated changes in AUC, including an understanding of the overall metabolic pathway (existence of multiple inactivation pathways) and potential for gut metabolism may be incorporated in an attempt to refine the *in vivo* extrapolation. Furthermore, the fraction of total hepatic metabolism expected for a victim drug through the affected P450 may be considered for improved predictability of the magnitude of the drug interaction (Wang, Jones and Hall, 2004). These additional properties will be individually discussed.

For most purposes, an AUC change of greater than twofold can define a drug interaction (Kalgutkar, Obach and Maurer, 2007). The extrapolation of *in vitro* MBI data has been described and makes use of the general equation described in Fig. 18.39 (Eq. 1). This equation is the same for reversible inhibition with the exception of the K_I/k_{inact} addition in the denominator of the equation, which for reversible inhibition reduces to 1 since there is no additional enzyme inactivation. The impact of this difference is illustrated in Fig. 18.15a, where the difference in midazolam AUC based upon coadministration of erythromycin is calculated without inactivation parameters compared with Fig. 18.15b, where the inactivation parameters are included in prediction.

Another important factor to consider is the $f_{m(P450)}$ which takes into account the extent to which the MBI-affected P450 contributes to the overall metabolism of the victim drug (AUC_{po}). Without accounting for the fraction metabolized, there may be a tendency to overpredict the clinically observed change in AUC. For P450 3A substrates, considering the extent of gut metabolism is an important factor in predicting the magnitude of an anticipated drug interaction. Based upon the inactivation kinetics, the percent of inhibition of gut metabolism results in an exponential relationship; therefore, when gut metabolism is inhibited substantially (>60%), the effect on AUC becomes more pronounced. When this is expected, Equation 1 from Fig. 18.39 takes on a new form represented by Equation 2.

Understanding the metabolite species responsible for P450 MBI may improve the accuracy of drug interaction predictions. This is especially poignant for MI complex

Expected *in vivo* interaction

1
$$\frac{AUCi}{AUC} = \frac{1}{1 + \dfrac{k_{inact} \cdot [I]}{K_I + [I]}}$$

2
$$\frac{AUCi}{AUC} = \frac{1}{1 + \left(\dfrac{f_{m\,(CYP)}}{\left(\dfrac{[I]}{K_I} \cdot \dfrac{k_{inact}}{k_{deg}} \right)} \right) + \left(1 - f_{m\,(CYP)} \right)}$$

3
$$\frac{AUCi}{AUC} = \frac{1}{F_g + \left(1 - F_g \cdot \dfrac{1}{\left(\dfrac{[I]}{([I] + KI)_I} \cdot \dfrac{k_{inact}}{k_{deg}} \right)} \right) + \left(1 - f_{m\,(CYP)} \right)} \cdot \frac{1}{\left(\dfrac{f_{m\,(CYP)}}{\left(\dfrac{[I]}{K_I} \cdot \dfrac{k_{inact}}{k_{deg}} \right)} \right) + \left(1 - f_{m\,(CYP)} \right)}$$

Figure 18.39 Kinetic equations used to extrapolate *in vitro* MBI into clinical predicted effect on AUC (1) general equation taking into account K_I, k_{inact}, $f_{m(P450)}$ and (2) same equation taking into consideration the gut metabolism by P4503A (Kalgutkar, Obach and Maurer, 2007).

formation from tertiary alkyl amine drugs, where multiple precursors to MBI may form. For example, verapamil exerted an unexpected drug interaction based upon the low plasma concentrations in combination with the initial assessment of P450 inhibition which was estimated to be low (Ma, Prueksaritanont and Lin, 2000). However, after further investigation, the *N*-dealkylation of verapamil occurs at two different sites producing two metabolites capable of MIC formation with P4503A (Wang, Jones and Hall, 2005). Furthermore, verapamil is administered as a racemic mixture and yields a mixture of metabolites from which the *S*-enantiomer of norverapamil is predicted to have the most pronounced effect with an K_{inact}/K_I of 164 min/nmol and offers an explanation for the observed interactions with various drugs (Wang, Jones and Hall, 2005).

Choosing the inhibitor concentration is another variable that can impact the final estimate of a drug interaction. The most accurate inhibitor concentration would be that available to the liver hepatocyte but is unmeasurable (Obach *et al.*, 2006). Alternatively, inhibitor concentrations that could be used in addition to total systemic inhibitor concentrations are free systemic or total hepatic inlet concentrations represented again either as free or total (Fig. 18.40). The hepatic inlet concentrations can be estimated based upon the fraction of the inhibitor which is absorbed and unchanged in the intestine in combination with the absorption rate constant, projected dose, and human hepatic blood flow (Kanamitsu, Ito and Sugiyama, 2000). However, the absorption rate constant and fraction absorbed unchanged are challenging values to obtain and potentially represent a source of error. Bearing patient safety in mind, the most conservative approach is to use total inhibitor concentration. In many cases, using all four inhibitor concentrations for the extrapolation may provide an upper and lower bounds of the potential clinical manifestation of the drug interaction (Fig. 18.40).

Inhibiting drug	IC$_{50}$ (µM)				Affected drug	Prediction based on most sensitive IC$_{50}$				
	Testosterone	Midazolam	Felodipine	fu		[I]= Csys	[I]= fu*Csys	[I]= Chep	[I]= fu*Chep	Actual drug interaction
Buspirone	70	59	>300	0.05	Alprazolam	1	1	1.04	1	1.08
Cyclosporin	4.5	1.4	300	0.07	Simvastatin	2.52	1.54	4.83	1.74	2.55
Diltiazem	60	83	>300	0.22	Buspirone	1.4	1.39	1.76	1.48	5.33
Disulfiram	6.9	3.6	7.9	0.04	Midazolam	1.74	1.74	13.8	2.7	0.94
Erythromycin	220	18	16	0.16	Buspirone	4.64	3.42	12.2	4.95	5.91
Fluconazole	17	6.8	23	0.89	Midazolam	7.64	7.15	10.5	9.86	3.6
Fluoxetine	16	36	88	0.05	Alprazolam	1.04	1.01	1.16	1.01	1.32
Fluvoxamine	28	43	62	0.23	Buspirone	2.41	2.38	3.4	2.61	2.4
Itraconazole	0.01	0.045	0.021	0.002	Buspirone	59.5	5.56	67.3	14.3	18.2
Ketoconazole	0.012	0.011	0.019	0.01	Midazolam	24.5	8.5	25	18.2	17
Mibefradil	0.33	0.19	0.68	0.005	Midazolam	10.3	1.82	20	2.16	8.86
Nefazodone	1.4	0.9	8.6	0.009	Midazolam	10.3	1.87	19.6	2.42	5.44
Nelfinavir	0.65	0.17	0.18	0.015	Simvastatin	19.3	3.54	21.4	12.7	6.07
Paroxetine	23	23	17	0.05	Alprazolam	1.02	1.01	1.12	1.01	0.99
Ritonavir	0.0088	0.0042	0.0037	0.015	Triazolam	22.2	17	22.3	20.3	20.3
Saquinavir	2	0.82	0.51	0.02	Midazolam	16.9	2.58	24.4	10.9	5.17
Sertraline	8.6	16	18	0.02	Alprazolam	1.05	1.01	1.52	1.02	1
Simvastatin	2	0.77	2.6	0.06	Midazolam	1.65	1.49	2.32	1.53	1.13
Troleandomycin	9.3	2.6	12	0.15	Midazolam	3.72	2.05	14.8	5.54	4.1
Verapamil	23	23	50	0.1	Midazolam	1.45	1.4	2.62	1.52	2.91

Figure 18.40 A table of *in vitro* IC$_{50}$ generated against three different P4503A probe substrates with numerous chemical inhibitors. The subsequent extrapolated DDI expected based upon different inhibitor concentrations which can then be compared with the actual clinically observed interaction in the last column.

The successful extrapolation of drug interaction with an irreversible inhibitor is also reliant upon an accurate rate of *de novo* P450 biosynthesis and degradation. Experimentally measured values for such a parameter in humans are not obtainable. Furthermore, the degradation rates are known to fluctuate based upon environmental factors and are also unique for individual P450s. To date, the best expected outcome is an estimate of the degradation value based upon existing experimental data. Degradation values for different P450 enzyme have been estimated using numerous techniques including (i) modeling the time course of P450 to return to baseline post-induction, (ii) turnover in primary human hepatocytes using pulse-chase methodology, (iii) Caco-2 cell line to estimate k_{deg} of P4503A, and (iv) the degradation of rat P450 3A. Clinical pharmacokinetic data for P450 1A2 (Faber and Fuhr, 2004), P450 2D6 (Venkatakrishnan and Obach, 2005), and intestinal P450 3A (Greenblatt *et al.*, 2003) based on the time course of de-induction following smoking cessation, recovery following paroxetine inactivation, and recovery following inactivation by grapefruit juice, respectively, have been utilized to determine degradation rate constants. The resulting k_{deg} estimates (per minute) were 0.000296, 0.000226, and 0.000481 for P450 1A2, P450 2D6, and intestinal P450 3A, respectively. For hepatic P450 3A, an initial estimate of 0.000321/ min was used based on the kinetics of de-induction of the oral clearance of verapamil (Fromm *et al.*, 1996), general clinical pharmacologic understanding of the kinetics of induction and de-induction of P450 3A (Lin *et al.*, 2001), and additionally supported by *in vitro* estimates of P450 3A turnover in primary human hepatocytes using pulse-chase methodology (Pichard *et al.*, 1992).

The integration of most of these kinetic factors has been successful within the simulation software provided by Simcyp. The program provides a platform to integrate the aforementioned concepts in order to interrogate a compound's ability to impact the AUC of other drugs based upon different dosing regimens, inactivation rates, fraction metabolized, and gut metabolism to successfully extrapolate the DDI potential of a new clinical candidate. In addition, Simcyp has developed the complexity necessary to investigate induction combined with reversible or irreversible P450 inhibition.

The potential for pharmacogenetics to impact MBI has only recently been investigated (McConn *et al.*, 2004; Pinto *et al.*, 2005; Op den Buijsch *et al.*, 2007). Hall *et al.* have shown numerous examples where P450 3A4 and P4503A5 display different kinetics with respect to inactivation of each P450, predominantly with MIC-forming compounds including diltiazem and verapamil. Differences in metabolism and generation of reactive metabolites could also contribute to different levels of exposure to reactive metabolites. Furthermore, with the advances in genotyping, it may become more apparent that some patients are prone to SNP (single nucleotide polymorphism) clusters. A cluster of SNPs in drug-metabolizing enzymes could exacerbate MBI and the potential for other adverse drug reactions. A recent retrospective study examined individuals linked to the ingestion of cooking oil contaminated with polychlorinated biphenyls (PCBs) and polychlorinated dibenzofurans (PCDFs) in 1979. Subjects genotyped to have the combination of the P4501A1-MspI mutant and GSTM1-null genotype were associated with increased risk of chloracne from the exposure of PCBs and PCDFs (odds ratio 2.8, 95% confidence interval 1.1–7.6) (Tsai *et al.*, 2006). A similar study found higher levels of BPDE-DNA adducts in individuals with the combined P4501A1(1/*2 or *2A/*2A)-GSTM1*0/*0 genotype suggest that these genotype combinations are at increased risk for contracting lung cancer when exposed to PAH (Goth-Goldstein *et al.*, 2000). As illustrated in the latter example, benzo[*a*]pyrene metabolism and moreover the generation of its DNA reactive metabolite were correlated to SNPs, and increased risk of cancer illustrates how the merging of experimental technologies and bioinformatics can provide exciting new opportunities to explore SNPs and disease states as they relate to xenobiotic metabolism.

Given the poor predictability of preclinical *in vivo* models for toxicity, *in vitro* systems continue to provide important experimental support for safe therapeutics. The integration of new *in vitro* test systems for toxicity (sandwich-cultured hepatocytes) may be able to provide a more global assessment of a compound's safety profile by combining transporters, phase I and phase II metabolism into a single functioning assay. This may expose new mechanism of complex metabolism-based inhibition, for example, the possibility of phase II conjugates capable of eliciting P450 inhibition. Only recently have other glucuronides been examined for their potential to elicit an adverse effect on drug-metabolizing enzymes as in the case of the gemfibrazil glucuronide which functions as an MBI against P450 2C8 (Ogilvie *et al.*, 2006). Interestingly, the unique bioactivation of troglitazone was postulated to have the potential to elicit a multiple drug binding event if disulfide formation occurred from the isocyanate intermediate followed by subsequent bioactivation of the chromane ring to the quinone; this could inactivate other systems such as transporters (Kassahun *et al.*, 2001). The combination of technology and biology should provide methods and insight to the structural and functional aspects of MBI including a better understanding of idiosyncratic drug reactions.

REFERENCES

Alterman, M.A., Kornilayev, B., Duzhak, T., Yakovlev, D. (2005). Quantitative analysis of cytochrome p450 isozymes by means of unique isozyme-specific tryptic peptides: a proteomic approach. *Drug Metabolism and Disposition, 33,* 1399–1407.

Alvarez-Diez, T.M., Zheng, J. (2004). Mechanism-based inactivation of cytochrome P450 3A4 by 4-ipomeanol. *Chemical Research in Toxicology, 17,* 150–157.

Alvarez-Sanchez, R., Montavon, F., Hartung, T., Pahler, A. (2006). Thiazolidinedione bioactivation: a comparison of the bioactivation potentials of troglitazone, rosiglitazone, and pioglitazone using stable isotope-labeled analogues and liquid chromatography tandem mass spectrometry. *Chemical Research in Toxicology, 19,* 1106–1116.

Argoti, D., Liang, L., Conteh, A., Chen, L., Bershas, D., Yu, C.P., Vouros, P., Yang, E. (2005). Cyanide trapping of iminium ion reactive intermediates followed by detection and structure identification using liquid chromatography-tandem mass spectrometry (LC-MS/MS). *Chemical Research in Toxicology, 18,* 1537–1544.

Axelrod, J. (1955). The enzymatic conversion of codeine to morphine. *The Journal of Pharmacology and Experimental Therapeutics, 115,* 259–267.

Baer, B.R., Wienkers, L.C., Rock, D.A. (2007). Time-dependent inactivation of P450 3A4 by raloxifene: identification of Cys239 as the site of apoprotein alkylation. *Chemical Research in Toxicology, 20,* 954–964.

Baillie, T.A. (2006). Future of toxicology-metabolic activation and drug design: challenges and opportunities in chemical toxicology. *Chemical Research in Toxicology, 19,* 889–893.

Baillie, T.A., Davis, M.R. (1993). Mass spectrometry in the analysis of glutathione conjugates. *Biological Mass Spectrometry, 22,* 319–325.

Bartkowski, R.R., Goldberg, M.E., Larijani, G.E., Boerner, T. (1989). Inhibition of alfentanil metabolism by erythromycin. *Clinical Pharmacology and Therapeutics, 46,* 99–102.

Beaune, P., Dansette, P.M., Mansuy, D., Kiffel, L., Finck, M., Amar, C., Leroux, J.P., Homberg, J.C. (1987). Human anti-endoplasmic reticulum autoantibodies appearing in a drug-induced hepatitis are directed against a human liver cytochrome P-450 that hydroxylates the drug. *Proceedings of the National Academy of Sciences of the United States of America*, *84*, 551–555.

Bensoussan, C., Delaforge, M., Mansuy, D. (1995). Particular ability of cytochromes P450 3A to form inhibitory P450-iron-metabolite complexes upon metabolic oxidation of aminodrugs. *Biochemical Pharmacology*, *49*, 591–602.

Bertz, R.J., Granneman, G.R. (1997). Use of in vitro and in vivo data to estimate the likelihood of metabolic pharmacokinetic interactions. *Clinical Pharmacokinetics*, *32*, 210–258.

Bieler, C.A., Stiborova, M., Wiessler, M., Cosyns, J.P., van Ypersele de Strihou, C., Schmeiser, H.H. (1997). 32P-post-labelling analysis of DNA adducts formed by aristolochic acid in tissues from patients with Chinese herbs nephropathy. *Carcinogenesis*, *18*, 1063–1067.

Bjornsson, T.D., Callaghan, J.T., Einolf, H.J., Fischer, V., Gan, L., Grimm, S., Kao, J., King, S.P., Miwa, G., Ni, L., Kumar, G., McLeod, J., Obach, R.S., Roberts, S., Roe, A., Shah, A., Snikeris, F., Sullivan, J.T., Tweedie, D., Vega, J.M., Walsh, J., Wrighton, S.A. (2003). The conduct of in vitro and in vivo drug-drug interaction studies: a Pharmaceutical Research and Manufacturers of America (PhRMA) perspective. *Drug Metabolism and Disposition: The Biological Fate of Chemicals*, *31*, 815–832.

Black, D.J., Kunze, K.L., Wienkers, L.C., Gidal, B.E., Seaton, T.L., McDonnell, N.D., Evans, J.S., Bauwens, J.E., Trager, W.F. (1996). Warfarin-fluconazole. II. A metabolically based drug interaction: in vivo studies. *Drug Metabolism and Disposition: The Biological Fate of Chemicals*, *24*, 422–428.

Brodie, B.B., Gillette, J.R., La Du, B.N. (1958). Enzymatic metabolism of drugs and other foreign compounds. *Annual Review of Biochemistry*, *27*, 427–454.

Chae, Y.H., Thomas, T., Guengerich, F.P., Fu, P.P., El-Bayoumy, K. (1999). Comparative metabolism of 1-, 2-, and 4-nitropyrene by human hepatic and pulmonary microsomes. *Cancer Research*, *59*, 1473–1480.

Chan, W.K., Delucchi, A.B. (2000). Resveratrol, a red wine constituent, is a mechanism-based inactivator of cytochrome P450 3A4. *Life Sciences*, *67*, 3103–3112.

Chan, W.K., Sui, Z. and Ortiz de Montellano, P.R. (1993). Determinants of protein modification versus heme alkylation: inactivation of cytochrome P450 1A1 by 1-ethynylpyrene and phenylacetylene. *Chemical Research in Toxicology*, *6*, 38–45.

Chen, L.J., Hecht, S.S., Peterson, L.A. (1997). Characterization of amino acid and glutathione adducts of cis-2-butene-1,4-dial, a reactive metabolite of furan. *Chemical Research in Toxicology*, *10*, 866–874.

Chen, Q., Ngui, J.S., Doss, G.A., Wang, R.W., Cai, X., DiNinno, F.P., Blizzard, T.A., Hammond, M.L., Stearns, R.A., Evans, D.C., Baillie, T.A., Tang, W. (2002). Cytochrome P450 3A4-mediated bioactivation of raloxifene: irreversible enzyme inhibition and thiol adduct formation. *Chemical Research in Toxicology*, *15*, 907–914.

Dansette, P.M., Bertho, G., Mansuy, D. (2005). First evidence that cytochrome P450 may catalyze both S-oxidation and epoxidation of thiophene derivatives. *Biochemical and biophysical Research Communications*, *338*, 450–455.

Davit, B., Reynolds, K., Yuan, R., Ajayi, F., Conner, D., Fadiran, E., Gillespie, B., Sahajwalla, C., Huang, S.M., Lesko, L.J. (1999). FDA evaluations using in vitro metabolism to predict and interpret in vivo metabolic drug-drug interactions: impact on labeling. *Journal of Clinical Pharmacology*, *39*, 899–910.

Desta, Z., Ward, B.A., Soukhova, N.V., Flockhart, D.A. (2004). Comprehensive evaluation of tamoxifen sequential biotransformation by the human cytochrome P450 system in vitro: prominent roles for CYP3A and CYP2D6. *The Journal of Pharmacology and Experimental Therapeutics*, *310*, 1062–1075.

DeWitte, R.S., Robins, R.H. (2006). A hierarchical screening methodology for physicochemical/ADME/Tox profiling. *Expert Opinion on Drug Metabolism & Toxicology*, *2*, 805–817.

Dickins, M., Elcombe, C.R., Moloney, S.J., Netter, K.J., Bridges, J.W. (1979). Further studies on the dissociation of the isosafrole metabolite-cytochrome P-450 complex. *Biochemical Pharmacology*, *28*, 231–238.

Dieckhaus, C.M., Fernandez-Metzler, C.L., King, R., Krolikowski, P.H., Baillie, T.A. (2005). Negative ion tandem mass spectrometry for the detection of glutathione conjugates. *Chemical Research in Toxicology*, *18*, 630–638.

Doss, G.A., Baillie, T.A. (2006). Addressing metabolic activation as an integral component of drug design. *Drug Metabolism Reviews*, *38*, 641–649.

Druckova, A., Marnett, L.J. (2006). Characterization of the amino acid adducts of the enedial derivative of teucrin A. *Chemical Research in Toxicology*, *19*, 1330–1340.

Ernest, C.S., II, Hall, S.D., Jones, D.R. (2005). Mechanism-based inactivation of CYP3A by HIV protease inhibitors. *The Journal of Pharmacology and Experimental Therapeutics*, *312*, 583–591.

Estabrook, R.W., Hildebrandt, A.G., Baron, J., Netter, K.J., Leibman, K. (1971). A new spectral intermediate associated with cytochrome P-450 function in liver microsomes. *Biochemical and Biophysical Research Communications*, *42*, 132–139.

Evans, D.C., Watt, A.P., Nicoll-Griffith, D.A., Baillie, T.A. (2004). Drug-protein adducts: an industry perspective on minimizing the potential for drug bioactivation in drug discovery and development. *Chemical Research in Toxicology*, *17*, 3–16.

Faber, M.S., Fuhr, U. (2004). Time response of cytochrome P450 1A2 activity on cessation of heavy smoking. *Clinical Pharmacology and Therapeutics*, *76*, 178–184.

Fabre, I., Fabre, G., Maurel, P., Bertault-Peres, P., Cano, J.P. (1988). Metabolism of cyclosporin A. III. Interaction of the macrolide antibiotic, erythromycin, using rabbit hepatocytes and microsomal fractions. *Drug Metabolism and Disposition: The Biological Fate of Chemicals*, *16*, 296–301.

Fersht, A.R. (1974). Catalysis, binding and enzyme-substrate complementarity. *Proceedings of the Royal Society of London. Series B, Containing Papers of a Biological Character*, *187*, 397–407.

Fontana, E., Dansette, P.M., Poli, S.M. (2005). Cytochrome p450 enzymes mechanism based inhibitors: common sub-structures and reactivity. *Current Drug Metabolism*, *6*, 413–454.

Foroozesh, M., Primrose, G., Guo, Z., Bell, L.C., Alworth, W.L., Guengerich, F.P. (1997). Aryl acetylenes as mechanism-based inhibitors of cytochrome P450-dependent monooxygenase enzymes. *Chemical Research in Toxicology*, *10*, 91–102.

Franklin, M.R. (1995). Enhanced rates of cytochrome P450 metabolic-intermediate complex formation from nonmacrolide amines in rifampicin-treated rabbit liver microsomes. *Drug Metabolism and Disposition*, *23*, 1379–1382.

Friedman, M.A., Woodcock, J., Lumpkin, M.M., Shuren, J.E., Hass, A.E., Thompson, L.J. (1999). The safety of newly approved medicines: do recent market removals mean there is a problem? *JAMA*, *281*, 1728–1734.

Fromm, M.F., Busse, D., Kroemer, H.K., Eichelbaum, M. (1996). Differential induction of prehepatic and hepatic metabolism of verapamil by rifampin. *Hepatology*, *24*, 796–801.

Gartner, C.A., Wen, B., Wan, J., Becker, R.S., Jones, G., 2nd, Gygi, S.P., Nelson, S.D. (2005). Photochromic agents as tools for protein structure study: lapachenole is a photoaffinity ligand of cytochrome P450 3A4. *Biochemistry*, *44*, 1846–1855.

Ghanbari, F., Rowland-Yeo, K., Bloomer, J.C., Clarke, S.E., Lennard, M.S., Tucker, G.T., Rostami-Hodjegan, A. (2006). A critical evaluation of the experimental design of studies of mechanism based enzyme inhibition, with implications for in vitro-in vivo extrapolation. *Current Drug Metabolism*, *7*, 315–334.

Goth-Goldstein, R., Stampfer, M.R., Erdmann, C.A., Russell, M. (2000). Interindividual variation in CYP1A1 expression in breast tissue and the role of genetic polymorphism. *Carcinogenesis*, *21*, 2119–2122.

Graham, D.J., Green, L., Senior, J.R., Nourjah, P. (2003). Troglitazone-induced liver failure: a case study. *American Journal of Medicine*, *114*, 299–306.

Greenblatt, D.J., von Moltke, L.L., Harmatz, J.S., Chen, G., Weemhoff, J.L., Jen, C., Kelley, C.J., LeDuc, B.W., Zinny, M.A. (2003). Time course of recovery of cytochrome p450 3A function after single doses of grapefruit juice. *Clinical Pharmacology and Therapeutics*, *74*, 121–129.

Grillo, M.P., Hua, F., Knutson, C.G., Ware, J.A., Li, C. (2003). Mechanistic studies on the bioactivation of diclofenac: identification of diclofenac-S-acyl-glutathione in vitro in incubations with rat and human hepatocytes. *Chemical Research in Toxicology*, *16*, 1410–1417.

Groves, J.T., McClusky, G.A. (1978). Aliphatic hydroxylation by highly purified liver microsomal cytochrome P-450. Evidence for a carbon radical intermediate. *Biochemical and Biophysical Research Communications*, *81*, 154–160.

Guengerich, F.P. (1990). Mechanism-based inactivation of human liver microsomal cytochrome P-450 IIIA4 by gestodene. *Chemical Research in Toxicology*, *3*, 363–371.

Guengerich, F.P. (1997). Role of cytochrome P450 enzymes in drug-drug interactions. *Advances in Pharmacology (San Diego, California)*, *43*, 7–35.

Ha-Duong, N.T., Dijols, S., Macherey, A.C., Goldstein, J.A., Dansette, P.M., Mansuy, D. (2001). Ticlopidine as a selective mechanism-based inhibitor of human cytochrome P450 2C19. *Biochemistry*, *40*, 12112–12122.

Hanioka, N., Ozawa, S., Jinno, H., Tanaka-Kagawa, T., Nishimura, T., Ando, M., Sawada, J.-I. (2002). Interaction of irinotecan (CPT-11) and its active metabolite 7-ethyl-10-hydroxycamptothecin (SN-38) with human cytochrome P450 enzymes. *Drug Metabolism and Disposition: The Biological Fate of Chemicals*, *30*, 391–396.

Hara, Y., Nakajima, M., Miyamoto, K.I., Yokoi, T. (2005). Inhibitory effects of psychotropic drugs on mexiletine metabolism in human liver microsomes: prediction of in vivo drug interactions. *Xenobiotica; the Fate of Foreign Compounds in Biological Systems*, *35*, 549–560.

Hasemann, C.A., Kurumbail, R.G., Boddupalli, S.S., Peterson, J.A., Deisenhofer, J. (1995). Structure and function of cytochromes P450: a comparative analysis of three crystal structures. *Structure*, *3*, 41–62.

He, K., Iyer, K.R., Hayes, R.N., Sinz, M.W., Woolf, T.F., Hollenberg, P.F. (1998). Inactivation of cytochrome P450 3A4 by bergamottin, a component of grapefruit juice. *Chemical Research in Toxicology*, *11*, 252–259.

He, K., Woolf, T.F., Hollenberg, P.F. (1999). Mechanism-based inactivation of cytochrome P-450-3A4 by mifepristone (RU486). *The Journal of Pharmacology and Experimental Therapeutics*, *288*, 791–797.

He, K., Talaat, R.E., Pool, W.F., Reily, M.D., Reed, J.E., Bridges, A.J., Woolf, T.F. (2004). Metabolic activation of troglitazone: identification of a reactive metabolite and mechanisms involved. *Drug Metabolism and Disposition*, *32*, 639–646.

Higgins, L., Korzekwa, K.R., Rao, S., Shou, M., Jones, J.P. (2001). An assessment of the reaction energetics for cytochrome P450-mediated reactions. *Archives of Biochemistry and Biophysics*, *385*, 220–230.

Hildebrandt, A., Estabrook, R.W. (1971). Evidence for the participation of cytochrome b 5 in hepatic microsomal mixed-function oxidation reactions. *Archives of Biochemistry and Biophysics*, *143*, 66–79.

Honig, P.K., Woosley, R.L., Zamani, K., Conner, D.P., Cantilena, L.R., Jr (1992). Changes in the pharmacokinetics and electrocardiographic pharmacodynamics of terfenadine with concomitant administration of erythromycin. *Clinical Pharmacology and Therapeutics*, *52*, 231–238.

Hutzler, J.M., Melton, R.J., Rumsey, J.M., Schnute, M.E., Locuson, C.W., Wienkers, L.C. (2006). Inhibition of cytochrome P450 3A4 by a pyrimidineimidazole: evidence for complex heme interactions. *Chemical Research in Toxicology*, *19*, 1650–1659.

Hutzler, J.M., Steenwyk, R.C., Smith, E.B., Walker, G.S., Wienkers, L.C. (2004). Mechanism-based inactivation of cytochrome P450 2D6 by 1-[(2-ethyl-4-methyl-1H-imidazol-5-yl)methyl]- 4-[4-

(trifluoromethyl)-2-pyridinyl]piperazine: kinetic characterization and evidence for apoprotein adduction. *Chemical Research in Toxicology*, *17*, 174–184.

Hutzler, J.M., Walker, G.S., Wienkers, L.C. (2003). Inhibition of cytochrome P450 2D6: structure-activity studies using a series of quinidine and quinine analogues. *Chemical Research in Toxicology*, *16*, 450–459.

Isin, E.M., Guengerich, F.P. (2007). Complex reactions catalyzed by cytochrome P450 enzymes. *Biochimica et Biophysica Acta*, *1770*, 314–329.

Ito, K., Ogihara, K., Kanamitsu, S., Itoh, T. (2003). Prediction of the in vivo interaction between midazolam and macrolides based on in vitro studies using human liver microsomes. *Drug Metabolism and Disposition: The Biological Fate of Chemicals*, *31*, 945–954.

Jones, D.R., Gorski, J.C., Hamman, M.A., Mayhew, B.S., Rider, S., Hall, S.D. (1999). Diltiazem inhibition of cytochrome P-450 3A activity is due to metabolite intermediate complex formation. *The Journal of Pharmacology and Experimental Therapeutics*, *290*, 1116–1125.

Jones, J.P., Shou, M., Korzekwa, K.R. (1996). Predicting the regioselectivity and stereoselectivity of cytochrome P450-mediated reactions: structural models for bioactivation reactions. *Advances in Experimental Medicine and Biology*, *387*, 355–360.

Jushchyshyn, M.I., Hutzler, J.M., Schrag, M.L., Wienkers, L.C. (2005). Catalytic turnover of pyrene by CYP3A4: evidence that cytochrome b5 directly induces positive cooperativity. *Archives of Biochemistry and Biophysics*, *438*, 21–28.

Kalgutkar, A.S., Obach, R.S., Maurer, T.S. (2007). Mechanism-based inactivation of cytochrome P450 enzymes: chemical mechanisms, structure-activity relationships and relationship to clinical drug-drug interactions and idiosyncratic adverse drug reactions. *Current Drug Metabolism*, *8*, 407–447.

Kalgutkar, A.S., Soglia, J.R. (2005). Minimising the potential for metabolic activation in drug discovery. *Expert Opinion on Drug Metabolism & Toxicology*, *1*, 91–142.

Kanamitsu, S., Ito, K., Green, C.E., Tyson, C.A., Shimada, N., Sugiyama, Y. (2000). Prediction of in vivo interaction between triazolam and erythromycin based on in vitro studies using human liver microsomes and recombinant human CYP3A4. *Pharmaceutical Research*, *17*, 419–426.

Kanamitsu, S., Ito, K., Sugiyama, Y. (2000). Quantitative prediction of in vivo drug-drug interactions from in vitro data based on physiological pharmacokinetics: use of maximum unbound concentration of inhibitor at the inlet to the liver. *Pharmaceutical Research*, *17*, 336–343.

Karuzina, I.I., Archakov, A.I. (1994). The oxidative inactivation of cytochrome P450 in monooxygenase reactions. *Free Radical Biology & Medicine*, *16*, 73–97.

Kassahun, K., Pearson, P.G., Tang, W., McIntosh, I., Leung, K., Elmore, C., Dean, D., Wang, R., Doss, G., Baillie, T.A. (2001). Studies on the metabolism of troglitazone to reactive intermediates in vitro and in vivo. Evidence for novel biotransformation pathways involving quinone methide formation and thiazolidinedione ring scission. *Chemical Research in Toxicology*, *14*, 62–70.

Kent, U.M., Juschyshyn, M.I., Hollenberg, P.F. (2001). Mechanism-based inactivators as probes of cytochrome P450 structure and function. *Current Drug Metabolism*, *2*, 215–243.

Khojasteh-Bakht, S.C., Chen, W., Koenigs, L.L., Peter, R.M., Nelson, S.D. (1999). Metabolism of (R)-(+)-pulegone and (R)-(+)-menthofuran by human liver cytochrome P-450s: evidence for formation of a furan epoxide. *Drug Metabolism and Disposition: The Biological Fate of Chemicals*, *27*, 574–580.

Kitz, R., Wilson, I.B. (1962). Esters of methanesulfonic acid as irreversible inhibitors of acetylcholinesterase. *Journal of Biological Chemistry*, *237*, 3245–3249.

Klingenberg, M. (1958). Pigments of rat liver microsomes. *Archives Biochemistry and Biophysics*, *75*, 376–386.

Koenigs, L.L., Trager, W.F. (1998a). Mechanism-based inactivation of P450 2A6 by furanocoumarins. *Biochemistry*, *37*, 10047–10061.

Koenigs, L.L., Trager, W.F. (1998b). Mechanism-based inactivation of cytochrome P450 2B1 by 8-methoxypsoralen and several other furanocoumarins. *Biochemistry*, *37*, 13184–13193.

Koenigs, L.L., Peter, R.M., Hunter, A.P., Haining, R.L., Rettie, A.E., Friedberg, T., Pritchard, M.P., Shou, M., Rushmore, T.H., Trager, W.F. (1999). Electrospray ionization mass spectrometric analysis of intact cytochrome P450: identification of tienilic acid adducts to P450 2C9. *Biochemistry*, *38*, 2312–2319.

Koshland, D.E., Jr, Neet, K.E. (1968). The catalytic and regulatory properties of enzymes. *Annual Review of Biochemistry*, *37*, 359–410.

Kouzi, S.A., McMurtry, R.J., Nelson, S.D. (1994). Hepatotoxicity of germander (Teucrium chamaedrys L.) and one of its constituent neoclerodane diterpenes teucrin A in the mouse. *Chemical Research in Toxicology*, *7*, 850–856.

Kunze, K.L., Eddy, A.C., Gibaldi, M., Trager, W.F. (1991). Metabolic enantiomeric interactions: the inhibition of human (S)-warfarin-7-hydroxylase by (R)-warfarin. *Chirality*, *3*, 24–29.

Kunze, K.L., Trager, W.F. (1993). Isoform-selective mechanism-based inhibition of human cytochrome P450 1A2 by furafylline. *Chemical Research in Toxicology*, *6*, 649–656.

Lightning, L.K., Jones, J.P., Friedberg, T., Pritchard, M.P., Shou, M., Rushmore, T.H., Trager, W.F. (2000). Mechanism-based inactivation of cytochrome P450 3A4 by L-754,394. *Biochemistry*, *39*, 4276–4287.

Lim, H.K., Duczak, N., Jr, Brougham, L., Elliot, M., Patel, K., Chan, K. (2005). Automated screening with confirmation of mechanism-based inactivation of CYP3A4, CYP2C9, CYP2C19, CYP2D6, and CYP1A2 in pooled human liver microsomes. *Drug Metabolism and Disposition*, *33*, 1211–1219.

Lin, H.L., Hollenberg, P.F. (2007). The inactivation of cytochrome P450 3A5 by 17alpha-ethynylestradiol is cytochrome b5-dependent: metabolic activation of the ethynyl moiety leads to the formation of glutathione conjugates, a heme adduct, and covalent binding to the apoprotein. *Journal of Pharmacology and Experimental Therapeutics*, *321*, 276–287.

Lin, L.Y., Fujimoto, M., Distefano, E.W., Schmitz, D.A., Jayasinghe, A., Cho, A.K. (1996). Selective mechanism-based inactivation of rat CYP2D by 4-allyloxymethamphetamine. *The Journal of Pharmacology and Experimental Therapeutics*, *277*, 595–603.

Lin, Y.S., Lockwood, G.F., Graham, M.A., Brian, W.R., Loi, C.M., Dobrinska, M.R., Shen, D.D., Watkins, P.B., Wilkinson, G.R., Kharasch, E.D., Thummel, K.E. (2001). In-vivo phenotyping for CYP3A by a single-point determination of midazolam plasma concentration. *Pharmacogenetics*, *11*, 781–791.

Liu, H., Liu, J., van Breemen, R.B., Thatcher, G.R., Bolton, J.L. (2005). Bioactivation of the selective estrogen receptor modulator desmethylated arzoxifene to quinoids: 4'-fluoro substitution prevents quinoid formation. *Chemical Research in Toxicology*, *18*, 162–173.

Liu, J., Li, Q., Yang, X., van Breemen, R.B., Bolton, J.L., Thatcher, G.R. (2005). Analysis of protein covalent modification by xenobiotics using a covert oxidatively activated tag: raloxifene proof-of-principle study. *Chemical Research in Toxicology*, *18*, 1485–1496.

Lopez-Garcia, M.P., Dansette, P.M., Mansuy, D. (1994). Thiophene derivatives as new mechanism-based inhibitors of cytochromes P-450: inactivation of yeast-expressed human liver cytochrome P-450 2C9 by tienilic acid. *Biochemistry*, *33*, 166–175.

Lu, A.Y., Coon, M.J. (1968). Role of hemoprotein P-450 in fatty acid omega-hydroxylation in a soluble enzyme system from liver microsomes. *The Journal of Biological Chemistry*, *243*, 1331–1332.

Lu, P., Schrag, M.L., Slaughter, D.E., Raab, C.E., Shou, M., Rodrigues, A.D. (2003). Mechanism-based inhibition of human liver microsomal cytochrome P450 1A2 by zileuton, a 5-lipoxygenase inhibitor. *Drug Metabolism and Disposition: The Biological Fate of Chemicals*, *31*, 1352–1360.

Ma, B., Prueksaritanont, T., Lin, J.H. (2000). Drug interactions with calcium channel blockers: possible involvement of metabolite-intermediate complexation with CYP3A. *Drug Metabolism and Disposition*, 28, 125–130.

Masubuchi, N., Makino, C., Murayama, N. (2007). Prediction of in vivo potential for metabolic activation of drugs into chemically reactive intermediate: correlation of in vitro and in vivo generation of reactive intermediates and in vitro glutathione conjugate formation in rats and humans. *Chemical Research in Toxicology*, 20, 455–464.

Masubuchi, Y. (2006). Metabolic and non-metabolic factors determining troglitazone hepatotoxicity: a review. *Drug Metabolism and Pharmacokinetics*, 21, 347–356.

Maurer, T., Fung, H.L. (2000). Comparison of methods for analyzing kinetic data from mechanism-based enzyme inactivation: application to nitric oxide synthase. *AAPS PharmSci*, 2, E8.

Mayhew, B.S., Jones, D.R., Hall, S.D. (2000). An in vitro model for predicting in vivo inhibition of cytochrome P450 3A4 by metabolic intermediate complex formation. *Drug Metabolism and Disposition*, 28, 1031–1037.

McConn, D.J., II, Lin, Y.S., Allen, K., Kunze, K.L., Thummel, K.E. (2004). Differences in the inhibition of cytochromes P450 3A4 and 3A5 by metabolite-inhibitor complex-forming drugs. *Drug Metabolism and Disposition: The Biological Fate of Chemicals*, 32, 1083–1091.

Melet, A., Assrir, N., Jean, P., Pilar Lopez-Garcia, M., Marques-Soares, C., Jaouen, M., Dansette, P.M., Sari, M.A., Mansuy, D. (2003). Substrate selectivity of human cytochrome P450 2C9: importance of residues 476, 365, and 114 in recognition of diclofenac and sulfaphenazole and in mechanism-based inactivation by tienilic acid. *Archives of Biochemistry and Biophysics*, 409, 80–91.

Michalets, E.L. (1998). Update: clinically significant cytochrome P-450 drug interactions. *Pharmacotherapy*, 18, 84–112.

von Moltke, L.L., Durol, A.L., Duan, S.X., Greenblatt, D.J. (2000). Potent mechanism-based inhibition of human CYP3A in vitro by amprenavir and ritonavir: comparison with ketoconazole. *European Journal Clinical Pharmacology*, 56, 259–261.

Monahan, B.P., Ferguson, C.L., Killeavy, E.S., Lloyd, B.K., Troy, J., Cantilena, L.R., Jr (1990). Torsades de pointes occurring in association with terfenadine use. *JAMA*, 264, 2788–2790.

Murray, M., Murray, K. (2003). Mechanism-based inhibition of CYP activities in rat liver by fluoxetine and structurally similar alkylamines. *Xenobiotica; the Fate of Foreign compounds in Biological Systems*, 33, 973–987.

Nebert, D.W., Adesnik, M., Coon, M.J., Estabrook, R.W., Gonzalez, F.J., Guengerich, F.P., Gunsalus, I.C., Johnson, E.F., Kemper, B., Levin, W. *et al.* (1987). The P450 gene superfamily: recommended nomenclature. *DNA (Mary Ann Liebert. Inc.)*, 6, 1–11.

Nelson, D.R. (1999). Cytochrome P450 and the individuality of species. *Archives of Biochemistry and Biophysics*, 369, 1–10.

Nelson, D.R., Koymans, L., Kamataki, T., Stegeman, J.J., Feyereisen, R., Waxman, D.J., Waterman, M.R., Gotoh, O., Coon, M.J., Estabrook, R.W., Gunsalus, I.C., Nebert, D.W. (1996). P450 superfamily: update on new sequences, gene mapping, accession numbers and nomenclature. *Pharmacogenetics*, 6, 1–42.

Nguyen, T.L., Gruenke, L.D., Castagnoli, N., Jr (1976). Metabolic N-demethylation of nicotine. Trapping of a reactive iminium species with cyanide ion. *Journal of Medicinal Chemistry*, 19, 1168–1169.

Nocerini, M.R., Yost, G.S., Carlson, J.R., Liberato, D.J., Breeze, R.G. (1985). Structure of the glutathione adduct of activated 3-methylindole indicates that an imine methide is the electrophilic intermediate. *Drug Metabolism and Disposition*, 13, 690–694.

O'Donnell, J.P., Dalvie, D.K., Kalgutkar, A.S., Obach, R.S. (2003). Mechanism-based inactivation of human recombinant P450 2C9 by the nonsteroidal anti-inflammatory drug suprofen. *Drug Metabolism and Disposition: The Biological Fate of Chemicals*, 31, 1369–1377.

Obach, R.S., Walsky, R.L., Venkatakrishnan, K. (2007). Mechanism-based inactivation of human cytochrome p450 enzymes and the prediction of drug-drug interactions. *Drug Metabolism and Disposition*, *35*, 246–255.

Obach, R.S., Walsky, R.L., Venkatakrishnan, K., Gaman, E.A., Houston, J.B., Tremaine, L.M. (2006). The utility of in vitro cytochrome P450 inhibition data in the prediction of drug-drug interactions. *Journal of Pharmacology and Experimental Therapeutics*, *316*, 336–348.

Ogilvie, B.W., Zhang, D., Li, W., Rodrigues, A.D., Gipson, A.E., Holsapple, J., Toren, P., Parkinson, A. (2006). Glucuronidation converts gemfibrozil to a potent, metabolism-dependent inhibitor of CYP2C8: implications for drug-drug interactions. *Drug Metabolism and Disposition*, *34*, 191–197.

Omura, T., Sato, R. (1962). A new cytochrome in liver microsomes. *The Journal of Biological Chemistry*, *237*, 1375–1376.

Omura, T., Sato, R. (1964). The carbon monoxide-binding pigment of liver microsomes. I. Evidence for its hemoprotein nature. *Journal of Biological Chemistry*, *239*, 2370–2378.

Op den Buijsch, R.A., Christiaans, M.H., Stolk, L.M., de Vries, J.E., Cheung, C.Y., Undre, N.A., van Hooff, J.P., van Dieijen-Visser, M.P., Bekers, O. (2007). Tacrolimus pharmacokinetics and pharmacogenetics: influence of adenosine triphosphate-binding cassette B1 (ABCB1) and cytochrome (CYP) 3A polymorphisms. *Fundamental & Clinical Pharmacology*, *21*, 427–435.

Ortiz de Montellano, P.R., Mathews, J.M. (1981). Autocatalytic alkylation of the cytochrome P-450 prosthetic haem group by 1-aminobenzotriazole. Isolation of an NN-bridged benzyne-protoporphyrin IX adduct. *The Biochemical Journal*, *195*, 761–764.

Pershing, L.K., Franklin, M.R. (1982). Cytochrome P-450 metabolic-intermediate complex formation and induction by macrolide antibiotics; a new class of agents. *Xenobiotica; the Fate of Foreign Compounds in Biological Systems*, *12*, 687–699.

Pichard, L., Fabre, I., Daujat, M., Domergue, J., Joyeux, H., Maurel, P. (1992). Effect of corticosteroids on the expression of cytochromes P450 and on cyclosporin A oxidase activity in primary cultures of human hepatocytes. *Molecular Pharmacology*, *41*, 1047–1055.

Pinto, A.G., Wang, Y.H., Chalasani, N., Skaar, T., Kolwankar, D., Gorski, J.C., Liangpunsakul, S., Hamman, M.A., Arefayene, M., Hall, S.D. (2005). Inhibition of human intestinal wall metabolism by macrolide antibiotics: effect of clarithromycin on cytochrome P450 3A4/5 activity and expression. *Clinical Pharmacology and Therapeutics*, *77*, 178–188.

Polasek, T.M., Elliot, D.J., Lewis, B.C., Miners, J.O. (2004). Mechanism-based inactivation of human cytochrome P4502C8 by drugs in vitro. *The Journal of Pharmacology and Experimental Therapeutics*, *311*, 996–1007.

Poulos, T.L. (1996). Ligands and electrons and haem proteins. *Nature Structural Biology*, *3*, 401–403.

Racha, J.K., Rettie, A.E., Kunze, K.L. (1998). Mechanism-based inactivation of human cytochrome P450 1A2 by furafylline: detection of a 1:1 adduct to protein and evidence for the formation of a novel imidazomethide intermediate. *Biochemistry*, *37*, 7407–7419.

Regal, K.A., Schrag, M.L., Kent, U.M., Wienkers, L.C., Hollenberg, P.F. (2000). Mechanism-based inactivation of cytochrome P450 2B1 by 7-ethynylcoumarin: verification of apo-P450 adduction by electrospray ion trap mass spectrometry. *Chemical Research in Toxicology*, *13*, 262–270.

Rettie, A.E., Korzekwa, K.R., Kunze, K.L., Lawrence, R.F., Eddy, A.C., Aoyama, T., Gelboin, H.V., Gonzalez, F.J., Trager, W.F. (1992). Hydroxylation of warfarin by human cDNA-expressed cytochrome P-450: a role for P-4502C9 in the etiology of (S)-warfarin-drug interactions. *Chemical Research in Toxicology*, *5*, 54–59.

Riley, R.J., Grime, K., Weaver, R. (2007). Time-dependent CYP inhibition. *Expert Opinion on Drug Metabolism & Toxicology*, *3*, 51–66.

Rodrigues, A.D., Mulford, D.J., Lee, R.D., Surber, B.W., Kukulka, M.J., Ferrero, J.L., Thomas, S.B., Shet, M.S., Estabrook, R.W. (1995). In vitro metabolism of terfenadine by a purified recombinant fusion protein containing cytochrome P4503A4 and NADPH-P450 reductase. Comparison to human liver microsomes and precision-cut liver tissue slices. *Drug Metabolism and Disposition: The Biological Fate of Chemicals*, *23*, 765–775.,

Roussel, F., Khan, K.K., Halpert, J.R. (2000). The importance of SRS-1 residues in catalytic specificity of human cytochrome P450 3A4. *Archives of Biochemistry and Biophysics*, *374*, 269–278.

Rowland, P., Blaney, F.E., Smyth, M.G., Jones, J.J., Leydon, V.R., Oxbrow, A.K., Lewis, C.J., Tennant, M.G., Modi, S., Eggleston, D.S., Chenery, R.J., Bridges, A.M. (2006). Crystal structure of human cytochrome P450 2D6. *Journal of Biological Chemistry*, *281*, 7614–7622.

Sahali-Sahly, Y., Balani, S.K., Lin, J.H., Baillie, T.A. (1996). In vitro studies on the metabolic activation of the furanopyridine L-754,394, a highly potent and selective mechanism-based inhibitor of cytochrome P450 3A4. *Chemical Research in Toxicology*, *9*, 1007–1012.

Sansen, S., Yano, J.K., Reynald, R.L., Schoch, G.A., Griffin, K.J., Stout, C.D., Johnson, E.F. (2007). Adaptations for the oxidation of polycyclic aromatic hydrocarbons exhibited by the structure of human P450 1A2. *Journal of Biological Chemistry*, *282*, 14348–14355.

Schoch, G.A., Yano, J.K., Wester, M.R., Griffin, K.J., Stout, C.D., Johnson, E.F. (2004). Structure of human microsomal cytochrome P450 2C8. Evidence for a peripheral fatty acid binding site. *Journal of Biological Chemistry*, *279*, 9497–9503.

Sharma, U., Roberts, E.S., Hollenberg, P.F. (1996). Formation of a metabolic intermediate complex of cytochrome P4502B1 by clorgyline. *Drug Metabolism and Disposition*, *24*, 1247–1253.

Silverman, R.B. (1988). The potential use of mechanism-based enzyme inactivators in medicine. *Journal of Enzyme Inhibition*, *2*, 73–90.

Skiles, G.L., Yost, G.S. (1996). Mechanistic studies on the cytochrome P450-catalyzed dehydrogenation of 3-methylindole. *Chemical Research in Toxicology*, *9*, 291–297.

Sligar, S.G. (1976). Coupling of spin, substrate, and redox equilibria in cytochrome P450. *Biochemistry*, *15*, 5399–5406.

Stark, K.L., Harris, C., Juchau, M.R. (1989). Influence of electrophilic character and glutathione depletion on chemical dysmorphogenesis in cultured rat embryos. *Biochemical Pharmacology*, *38*, 2685–2692.

Stiborova, M., Frei, E., Wiessler, M., Schmeiser, H.H. (2001). Human enzymes involved in the metabolic activation of carcinogenic aristolochic acids: evidence for reductive activation by cytochromes P450 1A1 and 1A2. *Chemical Research in Toxicology*, *14*, 1128–1137.

Swanson, R.A., Dus, K.M. (1979). Specific covalent labeling of cytochrome P-450CAM with 1-(4-azidophenyl)imidazole, an inhibitor-derived photoaffinity probe for P-450 heme proteins. *Journal of Biological Chemistry*, *254*, 7238–7246.

Tantcheva-Poor, I., Servera-Llaneras, M., Scharffetter-Kochanek, K., Fuhr, U. (2001). Liver cytochrome P450 CYP1A2 is markedly inhibited by systemic but not by bath PUVA in dermatological patients. *Br. J. Dermatol.*, *144*, 1127–1132.

Tessier, S., Boivin, S., Aubin, J., Lampron, P., Detheux, M., Fournier, A. (2005). Transmembrane domain V of the endothelin-A receptor is a binding domain of ETA-selective TTA-386-derived photoprobes. *Biochemistry*, *44*, 7844–7854.

Thummel, K.E., Wilkinson, G.R. (1998). In vitro and in vivo drug interactions involving human CYP3A. *Annual Review Pharmacology and Toxicology*, *38*, 389–430.

Trager, W.F. (1982). The postenzymatic chemistry of activated oxygen. *Drug Metabolism Reviews*, *13*, 51–69.

Tsai, P.C., Huang, W., Lee, Y.C., Chan, S.H., Guo, Y.L. (2006). Genetic polymorphisms in CYP1A1 and GSTM1 predispose humans to PCBs/PCDFs-induced skin lesions. *Chemosphere*, *63*, 1410–1418.

Van, L.M., Swales, J., Hammond, C., Wilson, C., Hargreaves, J.A., Rostami-Hodjegan, A. (2007). Kinetics of the time-dependent inactivation of CYP2D6 in cryopreserved human hepatocytes by methylenedioxymethamphetamine (MDMA). *European Journal of Pharmaceutical Sciences*, *31*, 53–61.

Venkatakrishnan, K., Obach, R.S. (2005). In vitro-in vivo extrapolation of CYP2D6 inactivation by paroxetine: prediction of nonstationary pharmacokinetics and drug interaction magnitude. *Drug Metabolism and Disposition*, *33*, 845–852.

Vermilion, J.L., Coon, M.J. (1978). Identification of the high and low potential flavins of liver microsomal NADPH-cytochrome P450 reductase. *The Journal of Biological Chemistry*, *253*, 8812–8819.

Voorman, R.L., Maio, S.M., Payne, N.A., Zhao, Z., Koeplinger, K.A., Wang, X. (1998). Microsomal metabolism of delavirdine: evidence for mechanism-based inactivation of human cytochrome P450 3A. *The Journal of Pharmacology and Experimental Therapeutics*, *287*, 381–388.

Walsh, C.T. (1984). Suicide substrates, mechanism-based enzyme inactivators: recent developments. *Annual Review of Biochemistry*, *53*, 493–535.

Wang, E., Walsh, C. (1978). Suicide substrates for the alanine racemase of Escherichia coli B. *Biochemistry*, *17*, 1313–1321.

Wang, J., Bauman, S., Colman, R.F. (1998). Photoaffinity labeling of rat liver glutathione S-transferase, 4-4, by glutathionyl S-[4-(succinimidyl)-benzophenone]. *Biochemistry*, *37*, 15671–15679.

Wang, Y.H., Jones, D.R., Hall, S.D. (2004). Prediction of cytochrome P450 3A inhibition by verapamil enantiomers and their metabolites. *Drug Metabolism and Disposition*, *32*, 259–266.

Wang, Y.H., Jones, D.R., Hall, S.D. (2005). Differential mechanism-based inhibition of CYP3A4 and CYP3A5 by verapamil. *Drug Metabolism and Disposition: The Biological Fate of Chemicals*, *33*, 664–671.

Watkins, P.B., Whitcomb, R.W. (1998). Hepatic dysfunction associated with troglitazone. *N. Engl. J. Med.*, *338*, 916–917.

Wen, B., Doneanu, C.E., Gartner, C.A., Roberts, A.G., Atkins, W.M., Nelson, S.D. (2005). Fluorescent photoaffinity labeling of cytochrome P450 3A4 by lapachenole: identification of modification sites by mass spectrometry. *Biochemistry*, *44*, 1833–1845.

Wen, B., Lampe, J.N., Roberts, A.G., Atkins, W.M., David Rodrigues, A., Nelson, S.D. (2006). Cysteine 98 in CYP3A4 contributes to conformational integrity required for P450 interaction with CYP reductase. *Archives of Biochemistry and Biophysics*, *454*, 42–54.

Wester, M.R., Yano, J.K., Schoch, G.A., Yang, C., Griffin, K.J., Stout, C.D., Johnson, E.F. (2004). The structure of human cytochrome P450 2C9 complexed with flurbiprofen at 2.0-A resolution. *Journal of Biological Chemistry*, *279*, 35630–35637.

von Weymarn, L.B., Zhang, Q.Y., Ding, X., Hollenberg, P.F. (2005). Effects of 8-methoxypsoralen on cytochrome P450 2A13. *Carcinogenesis*, *26*, 621–629.

White, R.E., Coon, M.J. (1982). Heme ligand replacement reactions of cytochrome P-450. Characterization of the bonding atom of the axial ligand trans to thiolate as oxygen. *The Journal of Biological Chemistry*, *257*, 3073–3083.

Wienkers, L.C., Heath, T.G. (2005). Predicting in vivo drug interactions from in vitro drug discovery data. *Nature Reviews*, *4*, 825–833.

Williams, D.P., Antoine, D.J., Butler, P.J., Jones, R., Randle, L., Payne, A., Howard, M., Gardner, I., Blagg, J., Park, B.K. (2007). The metabolism and toxicity of furosemide in the Wistar Rat and CD-1 mouse: a chemical and biochemical definition of the toxicophore. *Journal of Pharmacology and Experimental Therapeutics*, *322*, 1208–1220.

Williams, P.A., Cosme, J., Vinkovic, D.M., Ward, A., Angove, H.C., Day, P.J., Vonrhein, C., Tickle, I.J., Jhoti, H. (2004). Crystal structures of human cytochrome P450 3A4 bound to metyrapone and progesterone. *Science*, *305*, 683–686.

Wolfenden, R., Snider, M.J. (2001). The depth of chemical time and the power of enzymes as catalysts. *Accounts of Chemical Research*, *34*, 938–945.

Yamazaki, H., Shibata, A., Suzuki, M., Nakajima, M., Shimada, N., Guengerich, F.P., Yokoi, T. (1999). Oxidation of troglitazone to a quinone-type metabolite catalyzed by cytochrome P-450 2C8 and P-450 3A4 in human liver microsomes. *Drug Metabolism and Disposition*, *27*, 1260–1266.

Yamazaki, H., Shimada, T. (1998). Comparative studies of in vitro inhibition of cytochrome P450 3A4-dependent testosterone 6beta-hydroxylation by roxithromycin and its metabolites, troleandomycin, and erythromycin. *Drug Metabolism and Disposition: The Biological Fate of Chemicals*, *26*, 1053–1057.

Yang, J., Jamei, M., Yeo, K.R., Tucker, G.T., Rostami-Hodjegan, A. (2005). Kinetic values for mechanism-based enzyme inhibition: assessing the bias introduced by the conventional experimental protocol. *European Journal of Pharmaceutical Sciences*, *26*, 334–340.

Yano, J.K., Hsu, M.H., Griffin, K.J., Stout, C.D., Johnson, E.F. (2005). Structures of human microsomal cytochrome P450 2A6 complexed with coumarin and methoxsalen. *Nature Structural & Molecular Biology*, *12*, 822–823.

Yano, J.K., Wester, M.R., Schoch, G.A., Griffin, K.J., Stout, C.D., Johnson, E.F. (2004). The structure of human microsomal cytochrome P450 3A4 determined by X-ray crystallography to 2.05-A resolution. *Journal of Biological Chemistry*, *279*, 38091–38094.

Zhang, Z., Chen, Q., Li, Y., Doss, G.A., Dean, B.J., Ngui, J.S., Silva Elipe, M., Kim, S., Wu, J.Y., Dininno, F., Hammond, M.L., Stearns, R.A., Evans, D.C., Baillie, T.A., Tang, W. (2005). In vitro bioactivation of dihydrobenzoxathiin selective estrogen receptor modulators by cytochrome P450 3A4 in human liver microsomes: formation of reactive iminium and quinone type metabolites. *Chemical Research in Toxicology*, *18*, 675–685.

Zhao, X.J., Jones, D.R., Wang, Y.H., Grimm, S.W., Hall, S.D. (2002). Reversible and irreversible inhibition of CYP3A enzymes by tamoxifen and metabolites. *Xenobiotica; the Fate of Foreign Compounds in Biological Systems*, *32*, 863–878.

Zheng, J., Ma, L., Xin, B., Olah, T., Humphreys, W.G., Zhu, M. (2007). Screening and identification of GSH-trapped reactive metabolites using hybrid triple quadruple linear ion trap mass spectrometry. *Chemical Research in Toxicology*, *20*, 757–766.

Zhou, S., Yung Chan, S., Cher Goh, B., Chan, E., Duan, W., Huang, M., McLeod, H.L. (2005). Mechanism-based inhibition of cytochrome P450 3A4 by therapeutic drugs. *Clinical Pharmacokinetics*, *44*, 279–304.

CHAPTER 19

Clinical Drug–Drug Interactions

SUJAL V. DESHMUKH

19.1 INTRODUCTION

In the current health-care system, patients are typically on multiple medications for the treatment of one or more medical conditions. It is thus plausible that drugs may interact with each other in the human body. When these *in vivo* interactions result in a significant change in the pharmacokinetic or pharmacodynamic properties of at least one of the drugs, it is considered a clinical drug–drug interaction (DDI). Pharmacokinetic interactions are often multifactoral; these involve alteration of drug metabolism, drug transporter functions, percent of drug bound to plasma proteins, and drug excretion. Pharmacokinetic interactions that lead to adverse reaction or lack of efficacy are extremely detrimental to the patients and constitute as a major concern to the health-care providers. Pharmacodynamic interactions are those where the drug levels at the site of action are unaffected but its effects are altered; these may be synergistic, additive, or antagonistic in nature. Given the scope of this book, the following chapter will focus on the pharmacokinetic-based DDIs.

Recognizing and remembering pharmacokinetic-based DDIs are a daily challenge for many clinicians. It is thus important that drugmakers identify potential interactions that may occur when a drug is coadministered to humans with other medications. Several *in vitro* studies conducted at the drug discovery/development stages can alert against the possible risk for drug interaction and can help designing appropriate clinical studies to fully address the potentials of pharmacokinetic-based DDI. Early clinical studies directed toward studying potential DDIs are an integral and critical component of the drug discovery/development processes. Although the occurrence of a DDI does not preclude the use of a drug, it alerts the medical practitioner against the potential for DDI when used in combination with other drugs.

The factors that influence DDIs can be categorized as patient-related and drug-related. The patient-related factors are age, sex, disease condition, diet/nutrition, genetics, and environment. The drug-related factors include dose, dosing frequency, pharmacokinetic parameters like half-life, volume of distribution, oral availability, route of elimination, and interaction mechanism (Herman, 1999) (Fig. 19.1).

Drug Metabolism Handbook: Concepts and Applications, Edited by Ala F. Nassar, Paul F. Hollenberg, and JoAnn Scatina
Copyright © 2009 by John Wiley & Sons, Inc.

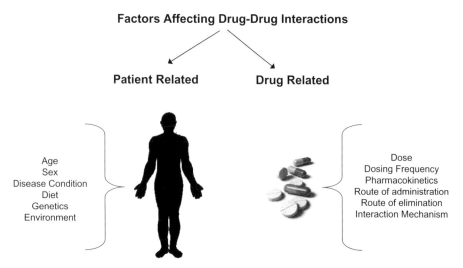

Figure 19.1   Patient and drug-related factors affecting DDIs.

When a DDI occurs, the drug causing the interaction is commonly referred to as the perpetrator and the drug that undergoes a change is the victim drug. Typically, a perpetrator is a potent inhibitor or inducer of drug-metabolizing enzymes and/or drug transporters, whereas a victim is an enzyme or transporter substrate.

Most commonly observed clinical DDIs are due to changes in the metabolism of the drugs. At the center of these DDIs are the phase I enzymes, predominantly the cytochrome P450s (CYPs). Among the various CYPs, CYP3A4 is responsible for metabolism of approximately 50% of the marketed drugs (Williams *et al.*, 2004). Drugs that cause inhibition or induction of CYP3A4 are most likely to cause DDIs. Additionally, drugs that are solely dependent on CYP3A4 for their elimination have a likelihood of being significantly affected by CYP3A4 inducers or inhibitors. Table 19.1 summarizes the marketed drugs as substrates, inhibitors, or inducers of CYPs. The second most common mechanism of DDIs is inhibition of drug transporters such as multidrug resistance transporter (MDR1)/P-glycoprotein (P-gp), organic anion-transporting polypeptide (OATP), breast cancer-resistant protein (BCRP), and organic anion transporter (OAT), which are found in the gut, liver, kidney, and brain (Ayrton and Morgan, 2001; Ho and Kim, 2005; Marchetti *et al.*, 2007). Table 19.2 provides a summary of various transporter-mediated DDIs.

The liver is a major organ for drug metabolism and thus DDIs predominantly occur in this organ. Furthermore, the hepatobiliary system has several transporters responsible for excreting drugs into the bile (Shitara, Sato and Sugiyama, 2005). Several drugs can compete for these transporters for biliary excretion, leading to DDIs. For orally administered drugs, the intestinal drug-metabolizing enzymes and transporters are sites for DDIs. Another major organ where DDIs commonly occur is the kidney (Fig. 19.2).

Although DDIs are generally associated with undesirable outcomes and in some cases lead to detrimental adverse events, there may be some benefits of DDIs as well. The modulation of drug transporters or enzymes is a strategy that has been explored to increase drug exposure. Inhibition of P-gp can be used to increase oral bioavailability and tumor levels of some anticancer drugs; for example, valspodar in combination with paclitaxel is being tested in clinical trails (Chico *et al.*, 2001).

TABLE 19.1   Substrates, inhibitors and inducers of CYPs.

CYP isoform	Substrates	Inhibitors	Inducers
CYP1A2	Caffeine, clozapine, imipramine, phenacetin, propranolol, R-warfarin, theophylline	Cimetidine, ciprofloxacin, fluvoxamine, furafylline, grepafloxacin	Barbiturates, omeprazole, phenytoin
CYP2C9	Diclofenac, dofetilide, fluvastatin, ibuprofen, mefanamic acid, naproxen, phenytoin, piroxicam, S-warfarin, tobutamide	Fluconazole, fluoxetine, fluvoxamine, ritonavir	Barbiturates, rifampicin
CYP2C19	Clomipramine, diazepam, hexobarbital, imipramine, mephobarbital, omeprazole, phenytoin, propranolol, S-mephenytoin	Fluoxetine, fluvoxamine, omeprazole	
CYP2D6	Amitriptyline, amphetamine, clomipramine, codeine, desipramine, dextromethorphran, dihydrocodeine, diphenylhydramine, fluoxetine, haloperidol, hydrocodone, imipramine, metoprolol, nortriptyline, ondensatron, oxycodone, paroxetine, penbutolol, perphenazine, propranolol, thioridazine, timolol, trimiramine	Fluoxetine, haloperidol, paroxetine, quinidine, ritonavir, sertraline, thioridazine	
CYP3A4	Amiodarone, amitriptyline, alprazolam, astemizole, carbamazepine, cyclosporine, cisapride, clindamycin, clomipramine, clonazepam, dapsone, dexamethasone, dextromethorphan, diazepam, diltiazem, erythromycin, felodipine, hydrocortisone, imipramine, indinavir, lidocaine, lovastatin, midazolam, nefazoldone, nelifinavir, nevirapine, nifedapine, nimodipine, propafenone, quinidine, R-warfarin, ritonavir, saquinavir, sertraline, simvastatin, tamoxifen, terfenadine, triazolam, verapamil, zolpidem	Cimetidine, clarithromycin, erythromycin, fluvoxamine, itraconazole, ketoconazole, miconazole, nefazodone, nelfinavir, remacemide, ritonavir	Barbiturates, carbamazepine, dexamethasone, phenytoin, rifabutin, rifampicin

From FDA (2006) and Indiana University School of Medicine (2008).

19.2   MECHANISMS OF DDIs

In order to find ways to mitigate against DDIs, it is important to understand the mechanism(s) by which these interactions occur. Drugs may interact with each other by one or more of the following mechanisms.

19.2.1   Interactions Affecting Oral Absorption

For orally administered drugs, if one of the drugs has an effect on the gut absorption, it may result in alteration of drug levels. The interaction may cause changes in either the rate of absorption or the amount of drug absorbed. The former affect the maximum drug concentration (C_{max}) and time to reach maximum drug concentration (t_{max}) but may not have an effect on the total drug absorbed, while the latter may affect both the

TABLE 19.2 Transporter-mediated DDIs.

Transporter	Victim	Perpetrator	Mechanism	Clinical outcome	Reference
MDR1	Digoxin	Quinidine	Inhibition of MDR1	Decreased renal clearance, increase in plasma levels	Doering (1979)
MDR1	Digoxin	Ritonavir	Inhibition of MDR1	Increased plasma levels and terminal half-life	Phillips, Rachlis and Ito (2003)
MDR1	Paclitaxel	Cyclosporin A	Inhibition of MDR1	Increased bioavailability	Meerum Terwogt et al. (1998)
MDR1	Talinolol	Rifampin	Induction of MDR1	Lowered plasma AUC	Westphal et al. (2000)
MDR1	Fexofenadine	Rifampin	Induction of MDR1	Decreased plasma levels	Hamman et al. (2001)
BCRP	Methotrexate	Omeprazole	Inhibition of BCRP	Increased plasma AUC and low clearance	Reid et al. (1993); Breedveld et al. (2004)
OATP C	Rosuvastatin	Gemfibrozil	Inhibition of OATP	Increased C_{max} and AUC	Schneck et al. (2004)
OAT	Cephalosporin (cefadroxil)	Probenecid	Inhibition of OAT	Decreased renal clearance, increase in C_{max} and half-life	Somogyi (1996)
OAT	ACE inhibitors (enalapril, lisinopril)	Probenecid	Inhibition of OAT	Decreased renal clearance, prolonged half-life	Somogyi (1996)
OAT, OCT, OATP	Procainamide, levofloxacin, dofetilide	Cimetidine	Inhibition of OAT, OCT, OATP	Decreased renal clearance, increased AUC	Christian, Meredith and Speeg (1984)
OAT3	Methotrexate	NSAIDs	Inhibition of OAT3	Decreased renal clearance	Takeda et al. (2002)
BSEP	Digoxin	Cyclosporine	Inhibition of BSEP	Causes cholestasis, digoxin toxicity	Dorian et al. (1988), Stieger et al. (2000)

C_{max} and the exposure (area under curve, AUC). Most oral DDIs that alter absorption result in decreased drug levels. Examples of DDI due to effects on oral absorption are presented in Table 19.3.

19.2.1.1 Changes in Gastrointestinal pH   Absorption of drug from the gastrointestinal tract is dependent on the pH of the gut, pKa of the drug, and the drugs' physiochemical properties. Changes in pH due to drugs such as antacids, or proton pump inhibitors can affect systemic levels of drugs whose absorption is pH-dependent. Omeprazole, a proton pump inhibitor, which causes a reduction in gastric acid secretion, increases the exposure of saquinavir (Winston *et al.*, 2006).

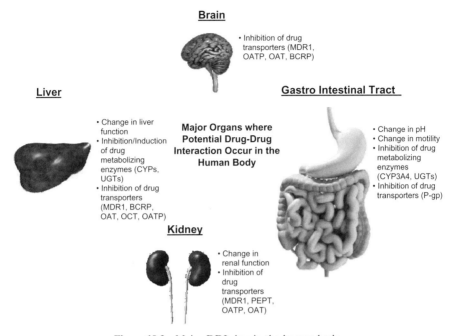

Figure 19.2   Major DDI sites in the human body.

19.2.1.2 Change in Gastric Motility

Gastric emptying time is an important factor that affects drug absorption; thus, drugs that alter gastric emptying will also affect drugs that are predominantly absorbed from the upper small intestine. Metoclopramide is a classical example in this category since it is a gastrointestinal prokinetic drug that increases the rate of gastric emptying, causing a reduction in the t_{max} and an increase in the C_{max} of several drugs (Johnson *et al.*, 1984; Greiff and Rowbotham, 1994).

19.2.1.3 Complex Formation/Chelation

Drugs that can potentially form a complex with other drugs can affect absorption. Activated charcoal is used to treat drug overdose. Charcoal can adsorb drugs and prevent them from being absorbed. Tetracycline antibiotics form complexes with metals such as calcium, aluminum, and bismuth, thus resulting in a reduction in their oral absorption (Tatro, 1972).

19.2.1.4 Malabsorption

Certain drugs may cause a malabsorption syndrome that results in decreased absorption from the gastrointestinal tract. Neomycin can cause this syndrome and is known to decrease the absorption of penicillin V and digoxin (Rothfeld and Osborne, 1963).

19.2.1.5 Effect on Gut Metabolizing Enzymes

Metabolism of drugs by gut enzymes is a significant component of first-pass effect for some orally absorbed drugs. CYP3A4 has been identified as one of the major enzymes in human enterocytes (Zhang *et al.*, 1999). Coadministration of potent CYP3A4 inhibitors can increase bioavailability of drugs which undergo substantial CYP3A4-mediated first-pass metabolism in the gut.

TABLE 19.3   Clinical DDIs affecting oral absorption.

Mechanism	Victim	Perpetrator	Effect	Clinical outcome	Reference
Change in gastric pH	Indinavir	Didanosine	Impaired indinavir absorption	Low plasma indinavir levels leading to failure of treatment	Bristol-Myers Squibb Company (2001)
	Saquinavir	Omeprazole	Increased oral exposure	No toxicity observed	Winston *et al.* (2006)
	Atazanavir	Omeprazole/famotidine	Impaired atazanavir absorption	Decrease in plasma levels not significantly affecting treatment	Bristol-Myers Squibb Company (2008)
Complex formation/chelation	Quinolone antibiotic (fluoroquinolones)	Antacids	Reduction in quinolone oral exposure	Reduced antimicrobial activity	Polk (1989)
	Tetracyclines	Antacids (containing Al3+, Ca2+, Mg2+, Bi2+, Zn2+, Fe2+)	Reduced antibiotic exposure	Reduction in antibiotic activity	Tatro (1972)
	Penicillamine	Antacids	Reduced absorption of penicillamine	Decreased therapeutic effects of penicillamine	Osman *et al.* (1983)
Change in gut motility	Digoxin	Metoclopramide	Reduced absorption	No effect of digoxin	Johnson *et al.* (1984)

It is, however, difficult to demonstrate this mechanism in the clinic since CYP-mediated gut metabolism cannot be isolated from metabolism that occurs in the liver.

19.2.1.6   Effect on Drug Transporters in the Gut   Enterocytes are known to harbor drug transporters such as MDR1, multidrug resistance-associated protein (MRP)1/2/3, BCRP, and OATP2, which actively uptake or efflux drugs (Ho and Kim, 2005). Potent inhibitors of these transporters can thus affect oral bioavailability of drugs which are substrates for these transporters.

19.2.2   Interaction Affecting Hepatic Metabolism

The liver is a major drug-metabolizing organ responsible for the elimination of drugs. Several drug-metabolizing enzymes of the liver transform drugs into more polar entities that can be readily excreted from the body. Alteration in the levels of these enzymes or their function will change the metabolic profile of drugs and thus will affect their blood levels. In addition to the liver's role as a drug-metabolizing organ, it also excretes drugs via the biliary system.

19.2.2.1 *Enzyme Inhibition*   Inhibition of drug-metabolizing enzymes is one of the most common mechanisms observed in clinical DDIs. Drugs that are potent inhibitor of drug-metabolizing enzymes cause severe DDIs. CYPs are responsible for the metabolism of approximately 70% of the drugs on the market, of this, CYP3A4 is involved in the metabolism of about 50% of the drugs followed by CYP2C9 and CYP2D6, each metabolizing about 20% drugs (Williams *et al.*, 2004). Several *in vitro* approaches can be used to demonstrate the inhibitory effect of a drug and based on this information, predictions can be made regarding the potential for clinical DDIs. There are two distinct mechanisms by which drugs inhibit CYP enzymes: (i) reversible and (ii) irreversible (time-dependent/mechanism-based).

Reversible inhibitors compete with the victim drug at the site of the enzyme and prevent it from being metabolized. There are several drugs in the clinic that are reversible inhibitors of CYPs. Drugs such as ketoconazole, quinidine, omeprazole, and methadone are some of the examples of reversible CYP inhibitors (Table 19.1). Interactions due to reversible inhibitors are clinically manageable.

On the other hand, irreversible/time-dependent inhibition has emerged as a greater concern in the clinic. These inhibitors cause inactivation of the enzyme due to formation of reactive species that irreversibly bind to the enzyme(s). The effect of these inhibitors persists long after the inhibitor is completely eliminated from the body. The rate of enzyme synthesis determines when the CYP activity is restored. Examples of irreversible inhibitors known to cause clinical DDIs include clarithromycin, fluoxetine, and ritonavir (Zhou *et al.*, 2004).

19.2.2.2 *Enzyme Induction*   Certain drugs can increase levels of drug-metabolizing enzymes by causing an induction in the expression of the gene associated with the respective enzyme. Several drugs have been shown to bind to response elements in the promoter region of genes and increase the expression of one or more enzymes depending on the transcription factor involved. Some of the nuclear receptors involved in expression of CYP enzymes are aryl hydrocarbon receptor (AHR), constitutive androstane receptor (CAR), and pregnane X receptor (PXR) (Lin, 2006). The extent to which enzyme induction occurs is generally dependent on the dose of the drug. DDIs arising from enzyme induction take a few days to a few weeks to develop fully. Since the effects originate at the molecular level, they also persist for a similar time after the drug is discontinued. Interaction resulting from enzyme induction can be managed by increasing the dose of the victim drug, but this has to be closely monitored so as to avoid toxicity due to overdose. Rifampin is a classical example of an enzyme inducer of hepatic and intestinal CYPs, and is often used as an *in vitro* and *in vivo* probe to investigate DDIs (Finch *et al.*, 2002).

19.2.2.3 *Alteration of Hepatic Blood Flow*   Hepatic clearance is dependent on hepatic blood flow and is an important factor for drugs with a high hepatic extraction ratio. Changes in hepatic blood flow will have an effect on the hepatic clearance. Cimetidine is known to decreases hepatic blood flow, it has been observed that it increases the bioavailability of propranolol via its effect on blood flow (Baciewicz and Baciewicz, 1989). Another example is that of verapamil and metoprolol which have been shown to affect hepatic blood flow and thus cause DDI (Bauer *et al.*, 2000).

19.2.2.4 Interactions Affecting Hepatobiliary Excretion Drugs are excreted in the bile either unchanged or as conjugated metabolites. Transporters located in the hepatocytes around the bile canaliculus can actively excrete drugs into the bile. Transporters identified so far include BCRP, MDR1/3, and MRP (Shitara, Sato and Sugiyama, 2005). Drugs can inhibit such transporters and affect elimination of drugs that are excreted by these transporters, causing an increase in the levels of the drug or its metabolite(s).

19.2.3 Interactions Due to Changes in Renal Excretion

Several drugs are excreted unchanged through the kidney and eliminated from the body along with urine. Some drugs are eliminated from the glomerular membrane via passive diffusion and filtered into the lumen of the tubules. In the tubules, active energy transport is responsible for excretion of some drugs. Active and passive transport in tubule cells can also play a role in the reabsorption of drugs back into the blood. The interplay between drug excretion and reabsorption determines the fate of a drug in the body. If any one of these processes is affected, it will directly affect drug exposure.

19.2.3.1 Changes in Urine pH Similar to oral drug absorption, reabsorption of drug from the urine depends on the ionization of the drugs. The non-ionized form of the drug undergoes passive diffusion back into the blood. This diffusion process is dependent on the pKa of the drug and the pH of the urine. In a highly alkaline pH, weakly acidic drugs exist predominantly in the ionized form and will thus be poorly absorbed from the urine. Conversely, weakly basic drugs are readily eliminated in the urine at acidic pH. Altering the pH of urine is a commonly used practice in treatment of overdose of drugs mainly eliminated via kidney, as it can facilitate the elimination of drugs from the body.

19.2.3.2 Changes in Active Tubule Excretion Active transport/secretion across the tubules is an important mechanism for drug elimination (Kosoglou and Vlasses, 1989; Shitara, Sato and Sugiyama, 2005). Transporters identified in the renal tubule include OAT1/3/4, MDR1, and MRP2 (Ho and Kim, 2005). Drugs that use the same transporters for their excretion will compete with each other. These interactions can affect their elimination from the body. A classical example is the interaction between penicillin and probenecid. The elimination of penicillin is affected by probenecid as they both compete for the same transporter, resulting in decreased elimination of penicillin (Barza and Weinstein, 1976). Thus, probenecid is used in conjunction to reduce the dose of penicillin.

19.2.3.3 Changes in Renal Blood Flow Factors that regulate renal blood flow such as prostaglandins can be affected by drugs. These changes will cause altered renal blood flow, resulting in changes in the rate at which drugs are eliminated from the body. Indomethacin inhibits the synthesis of prostaglandins and reduces renal blood flow. As a result, elimination of drugs such as lithium is reduced, resulting in increased blood levels of lithium (Weinblatt, 1989).

19.2.4 Interactions Affecting Plasma Protein Binding

In blood, drugs bind to plasma proteins and exist in a state of equilibrium between the bound and unbound form. The unbound drug is responsible for the pharmacological activity; hence, if levels of unbound drug increase, there may be enhanced therapeutic or toxicological effects. Drugs can bind to several plasma proteins such as alpha-1-glycoprotein, albumin, lipoprotein, and globulins. If two drugs compete to bind to the same protein, one drug could displace the other drug, increasing its free fraction. For tightly bound drugs, small changes in the binding cause a large change in the unbound fraction. For example, a drug that is 99.9% bound (i.e. 0.1% unbound) is displaced by another drug such that it is now 99% bound. The unbound fraction increases to 1%, that is, a change of 10-fold. However, this change in unbound fraction affects drugs with low distribution volume where the drugs remain in the plasma compartment rather than distributing into the tissue. In 1960s, it was reported that plasma protein binding may be responsible for sulfaphenazole-induced hypoglycemia in tolbutamide-treated patients (Christensen, Hansen and Kristensen, 1963) and potentiation of anti-coagulants by pyrazole compounds (Fox, 1964). However, the scientific rationale for the involvement of plasma protein binding has been challenged. Several reports have presented theoretical arguments about the clinical relevance of plasma protein binding on the exposure of patients to a drug (Sellers, 1979; Rolan, 1994; Benet and Hoener, 2002). Clinical effects attributed to plasma protein binding are not common and this phenomenon is not as profound as it was previously thought.

19.3 AVOIDING DDI

There are several different approaches that the medical practitioners can use to avoid DDIs. They involve use of resources that list DDIs, basic knowledge of pharmacology and pharmacokinetics, clinical judgment, and problem-solving skills. Information on clinical DDIs can be obtained from well-established drug interaction studies or new clinical data as they emerge. The Food and Drug Administration (FDA) published its first guidance on DDIs in 1997 (FDA, 1997), followed by a second guidance in 1999 (FDA, 1999). These included best practices for *in vitro* and *in vivo* drug interaction studies, including study design, dosing, statistical analysis, and guidance for appropriate labeling. *In vivo* and *in vitro* studies are mainly performed for metabolism and trans-porter-based DDIs. Investigational drugs are studied for their potential to inhibit or induce CYP enzymes or P-gp, or to know if drugs are substrates for CYPs or P-gp or any other transporter. Research in pharmaceutical industry has focused on screening compounds for possible DDI liability based on the limited *in vitro–in vivo* correlation models currently available. Models that can accurately predict clinical outcomes are currently lacking. Additional research targeted toward refining these models to improve clinical predictability is an impending need. In summary, understanding the mechanism of DDIs and the interplay of various factors affecting these interactions can assist in managing the risk associated with these DDIs so as to minimize undesirable clinical outcomes.

REFERENCES

Ayrton, A., Morgan, P. (2001). Role of transport proteins in drug absorption, distribution and excretion. *Xenobiotica, 31*, 469–497.

Baciewicz, A.M., Baciewicz, F.A., Jr (1989). Effect of cimetidine and ranitidine on cardiovascular drugs. *Am. Heart J., 118*, 144–154.

Barza, M., Weinstein, L. (1976). Pharmacokinetics of the penicillins in man. *Clin. Pharmacokinet.*, *1*, 297–308.

Bauer, L.A., Horn, J.R., Maxon, M.S., Easterling, T.R., Shen, D.D., Strandness, D.E., Jr (2000). Effect of metoprolol and verapamil administered separately and concurrently after single doses on liver blood flow and drug disposition. *J. Clin. Pharmacol.*, *40*, 533–543.

Benet, L.Z., Hoener, B.A. (2002). Changes in plasma protein binding have little clinical relevance. *Clin. Pharmacol. Ther.*, *71*, 115–121.

Breedveld, P., Zelcer, N., Pluim, D., Sonmezer, O., Tibben, M.M., Beijnen, J.H., Schinkel, A.H., van Tellingen, O., Borst, P., Schellens, J.H. (2004). Mechanism of the pharmacokinetic interaction between methotrexate and benzimidazoles: potential role for breast cancer resistance protein in clinical drug-drug interactions. *Cancer Res.*, *64*, 5804–5811.

Bristol-Myers Squibb Company (2001). *Didanosine Information*, http://www.fda.gov/cder/foi/label/2001/20155s25lbl.pdf.

Bristol-Myers Squibb Company (2008). *Highlights of Prescribing Information for Atazanavir Sulfate*, http://packageinserts.bms.com/pi/pi_reyataz.pdf.

Chico, I., Kang, M.H., Bergan, R., Abraham, J., Bakke, S., Meadows, B., Rutt, A., Robey, R., Choyke, P., Merino, M., Goldspiel, B., Smith, T., Steinberg, S., Figg, W.D., Fojo, T., Bates, S. (2001). Phase I study of infusional paclitaxel in combination with the P-glycoprotein antagonist PSC 833. *J. Clin. Oncol.*, 19, 832–842.

Christensen, L.K., Hansen, J.M., Kristensen, M. (1963). Sulphaphenazole-induced hypoglycaemic attacks in tolbutamide-treated diabetics. *Lancet*, 2, 1298–1301.

Christian, C.D., Jr, Meredith, C.G., Speeg, K.V., Jr (1984). Cimetidine inhibits renal procainamide clearance. *Clin. Pharmacol. Ther.*, *36*, 221–227.

Doering, W. (1979). Quinidine-digoxin interaction: pharmacokinetics, underlying mechanism and clinical implications. *N. Engl. J. Med.*, *301*, 400–404.

Dorian, P., Strauss, M., Cardella, C., David, T., East, S., Ogilvie, R. (1988). Digoxincyclosporine interaction: severe digitalis toxicity after cyclosporine treatment. *Clin. Invest. Med.*, *11*, 108–112.

FDA (1997). *Guidance for Industry: Drug Metabolism/Drug Interactions in the Drug Development Process: Studies In Vitro*, http://www.fda.gov/Cder/guidance/clin3.pdf (accessed April 1997).

FDA (1999). *Guidance for Industry: In Vivo Metabolism/Drug Interactions: Study Design, Data Analysis and Recommendation for Dosing and Labeling*, http://www.fda.gov/Cder/guidance/2635fnl.pdf (accessed December 1999).

FDA (2006). *Table of Substrate, Inhibitors and Inducers*, http://www.fda.gov/cder/drug/drugInteractions/tableSubstrates.htm.

Finch, C.K., Chrisman, C.R., Baciewicz, A.M., Self, T.H. (2002). Rifampin and rifabutin drug interactions: an update. *Arch. Intern. Med.*, *162*, 985–992.

Fox, S.L. (1964). Potentiation of anticoagulants caused by pyrazole compounds. *JAMA*, *188*, 320–321.

Greiff, J.M., Rowbotham, D. (1994). Pharmacokinetic drug interactions with gastrointestinal motility modifying agents. *Clin. Pharmacokinet.*, *27*, 447–461.

Hamman, M.A., Bruce, M.A., Haehner-Daniels, B.D., Hall, S.D. (2001). The effect of rifampin administration on the disposition of fexofenadine. *Clin. Pharmacol. Ther.*, *69*, 114–121.

Herman, R.J. (1999). Drug interactions and the statins. *CMAJ*, *161*, 1281–1286.

Ho, R.H., Kim, R.B. (2005). Transporters and drug therapy: implications for drug disposition and disease. *Clin. Pharmacol. Ther.*, *78*, 260–277.

Indiana University School of Medicine (2008). *Cytochrome P450 Drug Interaction Table*, http://medicine.iupui.edu/flockhart/clinlist.htm.

Johnson, B.F., Bustrack, J.A., Urbach, D.R., Hull, J.H., Marwaha, R. (1984). Effect of metoclopramide on digoxin absorption from tablets and capsules. *Clin. Pharmacol. Ther.*, 36, 724–730.

Kosoglou, T., Vlasses, P.H. (1989). Drug interactions involving renal transport mechanisms: an overview. *DICP*, 23, 116–122.

Lin, J.H. (2006). CYP induction-mediated drug interactions: in vitro assessment and clinical implications. *Pharm. Res.*, 23, 1089–1116.

Marchetti, S., Mazzanti, R., Beijnen, J.H., Schellens, J.H. (2007). Concise review: clinical relevance of drug–drug and herb–drug interactions mediated by the ABC transporter ABCB1 (MDR1, P-glycoprotein). *Oncologist*, 12, 927–941.

Meerum Terwogt, J.M., Beijnen, J.H., ten Bokkel Huinink, W.W., Rosing, H., Schellens, J.H. (1998). Co-administration of cyclosporin enables oral therapy with paclitaxel. *Lancet*, 352, 285.

Osman, M.A., Patel, R.B., Schuna, A., Sundstrom, W.R., Welling, P.G. (1983). Reduction in oral penicillamine absorption by food, antacid, and ferrous sulfate. *Clin. Pharmacol. Ther.*, 33, 465–470.

Phillips, E.J., Rachlis, A.R., Ito, S. (2003). Digoxin toxicity and ritonavir: a drug interaction mediated through p-glycoprotein? *AIDS*, 17, 1577–1578.

Polk, R.E. (1989). Drug-drug interactions with ciprofloxacin and other fluoroquinolones. *Am. J. Med.*, 87, 76S–81S.

Reid, T., Yuen, A., Catolico, M., Carlson, R.W. (1993). Impact of omeprazole on the plasma clearance of methotrexate. *Cancer Chemother. Pharmacol.*, 33, 82–84.

Rolan, P.E. (1994). Plasma protein binding displacement interactions—why are they still regarded as clinically important? *Br. J. Clin. Pharmacol.*, 37, 125–128.

Rothfeld, B., Osborne, D. (1963). Malabsorption syndrome produced by neomycin. *Am. J. Dig. Dis.*, 8, 763–768.

Schneck, D.W., Birmingham, B.K., Zalikowski, J.A., Mitchell, P.D., Wang, Y., Martin, P.D., Lasseter, K.C., Brown, C.D., Windass, A.S., Raza, A. (2004). The effect of gemfibrozil on the pharmacokinetics of rosuvastatin. *Clin. Pharmacol. Ther.*, 75, 455–463.

Sellers, E.M. (1979). Plasma protein displacement interactions are rarely of clinical significance. *Pharmacology*, 18, 225–227.

Shitara, Y., Sato, H., Sugiyama, Y. (2005). Evaluation of drug-drug interaction in the hepatobiliary and renal transport of drugs. *Annu. Rev. Pharmacol. Toxicol.*, 45, 689–723.

Somogyi, A. (1996). Renal transport of drugs: specificity and molecular mechanisms. *Clin. Exp. Pharmacol. Physiol.*, 23, 986–989.

Stieger, B., Fattinger, K., Madon, J., Kullak-Ublick, G.A., Meier, P.J. (2000). Drug- and estrogen-induced cholestasis through inhibition of the hepatocellular bile salt export pump (Bsep) of rat liver. *Gastroenterology*, 118, 422–430.

Takeda, M., Khamdang, S., Narikawa, S., Kimura, H., Hosoyamada, M., Cha, S.H., Sekine, T., Endou, H. (2002). Characterization of methotrexate transport and its drug interactions with human organic anion transporters. *J. Pharmacol. Exp. Ther.*, 302, 666–671.

Tatro, D.S. (1972). Tetracycline-antacid interactions. *JAMA*, 220, 586.

Weinblatt, M.E. (1989). Drug interactions with non steroidal anti-inflammatory drugs (NSAIDs). *Scand. J. Rheumatol. Suppl.*, 83, 7–10.

Westphal, K., Weinbrenner, A., Zschiesche, M., Franke, G., Knoke, M., Oertel, R., Fritz, P., von Richter, O., Warzok, R., Hachenberg, T., Kauffmann, H.M., Schrenk, D., Terhaag, B., Kroemer, H.K., Siegmund, W. (2000). Induction of P-glycoprotein by rifampin increases intestinal secretion of talinolol in human beings: a new type of drug/drug interaction. *Clin. Pharmacol. Ther.*, 68, 345–355.

Williams, J.A., Hyland, R., Jones, B.C., Smith, D.A., Hurst, S., Goosen, T.C., Peterkin, V., Koup, J. R., Ball, S.E. (2004). Drug-drug interactions for UDP-glucuronosyltransferase substrates: a pharmacokinetic explanation for typically observed low exposure (AUCi/AUC) ratios. *Drug Metab. Dispos.*, *32*, 1201–1208.

Winston, A., Back, D., Fletcher, C., Robinson, L., Unsworth, J., Tolowinska, I., Schutz, M., Pozniak, A.L., Gazzard, B., Boffito, M. (2006). Effect of omeprazole on the pharmacokinetics of saquinavir-500 mg formulation with ritonavir in healthy male and female volunteers. *AIDS*, *20*, 1401–1406.

Zhang, Q.Y., Dunbar, D., Ostrowska, A., Zeisloft, S., Yang, J., Kaminsky, L.S. (1999). Characterization of human small intestinal cytochromes P-450. *Drug Metab. Dispos.*, *27*, 804–809.

Zhou, S., Chan, E., Lim, L.Y., Boelsterli, U.A., Li, S.C., Wang, J., Zhang, Q., Huang, M., Xu, A. (2004). Therapeutic drugs that behave as mechanism-based inhibitors of cytochrome P450 3A4. *Curr. Drug Metab.*, *5*, 415–442.

FURTHER READING

Stockley, I.H. (1999). *Drug Interactions: A Source Book of Adverse Interactions, Their Mechanisms, Clinical Importance and Management*, 5th edn, Pharmaceutical Press, London.

Goodman, L.S. (2005). *Goodman & Gilman's the Pharmacological Basis of Therapeutics*, 11th edn, McGraw-Hill, New York.

CHAPTER 20

Pharmaceutical Excipients in Drug–Drug Interaction

CHUAN CHEN and ABU J.M. SADEQUE

20.1 INTRODUCTION

Drug formulations such as tablets, capsules, or injectable solutions consist of active pharmaceutical ingredients and other "inactive" ingredients that are added for many purposes. For example, surfactants are often used in formulations to solubilize water-insoluble drugs, and methylcellulose may be used to prepare drug suspensions or added to tablets as a disintegrating agent or binder. During the development of a given drug formulation, it is well understood that the interactions between the drug and excipients need to be carefully studied to ensure that the drug is stable in the formulation and that the solubility, as well as the dissolution properties in the case of orally administered drugs, is satisfactory (Kalasz and Antal, 2006).

Not all of the excipients in approved drug formulations are "inactive;" some are known to alter drug metabolism and/or transporter activities. Therefore, in the drug formulation development process, one also needs to consider whether the chosen excipients in drug formulations can alter systemic exposure of the drug itself or other potentially coadministered therapeutic agents due to interactions with drug-metabolizing enzymes and/or transporters, especially under situations where the excipients are not approved by the regulatory agencies, or if the amounts of the excipients used are higher than that used for previously approved drug formulations.

Interestingly, among the many formulation components found in marketed drug products (see FDA web site: http://www.fda.gov/cder/drug/iig/default.htm), most of the excipients that interact with drug biotransformation processes and/or transporters belong to a group known as solubilizing excipients, that is, additives used to solubilize drugs in oral and injectable dosage forms. Solubilizing excipients include *water-soluble organic solvents* such as polyethylene glycol 300 (PEG300), polyethylene glycol 400 (PEG400), ethanol, propylene glycol, glycerin, *N*-methyl-2-pyrrolidone, dimethylacetamide, and dimethylsulfoxide (DMSO); *nonionic surfactants* such as Cremophor EL, Cremophor RH40, Cremophor EH60, D-α-tocopherol polyethylene glycol

Drug Metabolism Handbook: Concepts and Applications, Edited by Ala F. Nassar,
Paul F. Hollenberg, and JoAnn Scatina

1000 succinate (TPGS1000), polysorbate 20 (Tween 20), polysorbate 80 (Tween 80), Solutol HS15, sorbitan monooleate, Poloxamer 407, Labrafil M-1944CS, Labrafil M-2125CS, Labrasol, Gellucire 44/14, Softigen 767, and mono- and di-fatty acid esters of PEG300, 400, or 1750; *water-insoluble lipids* such as castor oil, corn oil, cottonseed oil, olive oil, peanut oil, peppermint oil, safflower oil, sesame oil, soybean oil, hydrogenated vegetable oils, hydrogenated soybean oil, and medium-chain triglycerides of coconut oil and palm seed oil; *organic liquids/semisolids* such as beeswax, d-α-tocopherol, oleic acid, medium-chain mono- and diglycerides; various *complexing agents* such as α-cyclodextrin, β-cyclodextrin, hydroxypropyl-β-cyclodextrin, and sulfobutylether-β-cyclodextrin; and *phospholipids* such as hydrogenated soy phosphatidylcholine, distearoylphosphatidylglycerol, L-α-dimyristoylphosphatidylcholine, and L-α-dimyristoylphosphatidylglycerol (Strickley, 2004).

In this chapter, the literature on the drug–drug interaction potential of pharmaceutical excipients is reviewed, with a focus on solubilizing excipients.

20.2 INTERACTION OF PHARMACEUTICAL EXCIPIENTS WITH CYTOCHROME P450s (CYPs)

The majority of drugs are metabolized by one or more of the five major cytochrome P450 (CYP) isoforms, namely, CYP1A2, CYP2A6, CYP2C9, CYP2C19, CYP2D6, and CYP3A4, of which CYP3A4 is predominant. A drug can interact with CYP-mediated biotransformation of itself or other drugs via either inhibition of CYP activities or induction of CYP expression. *In vitro* studies aiming to assess the potential of CYP-mediated drug–drug interactions are routinely conducted during the drug discovery and development phases. Sometimes, clinical drug–drug interaction studies are necessary to fully address the possibility of CYP-mediated drug–drug interactions. Compared with drugs, less attention has been paid to the potential of CYP-mediated interactions on drug metabolism due to the excipients present in drug formulations. CYP isoforms are involved in the metabolism of some excipients. For example, vitamin E is metabolized via side-chain degradation initiated by CYP-catalyzed ω-hydroxylation followed by β-oxidation (Brigelius-Flohe, 2003). Therefore, some excipients are likely to interact with CYP enzymes. There are a few examples in which excipients interact with CYPs in *in vitro* studies and/or animal studies. However, conclusive findings from clinical studies that link excipients to CYP-mediated drug–drug interaction are still lacking. The interactions of different excipients with the CYP isoforms responsible for drug metabolism are summarized in Table 20.1.

20.2.1 CYP Inhibition

Mountfield *et al.* (2000) investigated the CYP inhibition potentials of 16 commonly used excipients in recombinant human CYP3A4, human liver microsomes, and intestinal microsomes from dogs and monkeys. Their data indicated that CYP3A4 activity was inhibited by *N*-methylpyrrolidone, dimethylacetamide, propylene glycol, PEG400, soybean oil, miglyol, glycocholic acid/lecithin, Tween 20, oleic acid, and a mixture of taurocholic acid/lecithin. However, the IC_{50}/K_i values were in the high μM to mM range, indicating the inhibition is not likely of significance from a practical drug–drug interaction point of view.

Solutol HS15 is used in intravenous and oral formulations in various countries. It was found to inhibit recombinant human CYP3A activity (IC_{50} value of $5\,\mu M$) (Bittner *et al.*, 2003a). Moreover, Solutol HS15 significantly reduced the metabolism of the CYP3A4 substrates colchicine and midazolam in rat hepatocytes at concentrations below its critical micellar concentration of 0.021% (Gonzalez *et al.*, 2004). Intravenous coadministration of Solutol HS15 and colchicine to rats resulted in a significant reduction in colchicine clearance as well as an increase in C_{max} values (Bittner *et al.*, 2003b). However, intravenous injection of Solutol HS15 had little effect on the pharmacokinetics of coadministered midazolam in rats (Bittner *et al.*, 2003a). In addition to Solutol HS15, 0.03% (w/v) of Cremophor EL and Tween 80 decreased the intrinsic clearance of midazolam in rat hepatocytes by 30% and 25%, respectively, without causing overt cytotoxicity (Gonzalez *et al.*, 2004). Cremophor RH40 inhibited CYP3A-mediated nifedipine oxidation in human liver microsomes ($IC_{50} = 0.3\%$, v/v) (Wandel, Kim and Stein, 2003).

Cyclodextrins are cyclic oligosaccharides with lipophilic inner cavities that are capable of forming non-covalent inclusion complexes with a large variety of molecules (Strickley, 2004). Although only a few approved drug formulations contain cyclodextrins, there is a growing interest in using cyclodextrins as a solubilizer in drug formulations (Challa *et al.*, 2005). Ishikawa, Yoshii and Furuta (2005) reported that hydroxypropyl-β-cyclodextrin inhibited recombinant CYP2C19 and CYP3A4 activities, while methyl-β-cyclodextrin inhibited recombinant CYP1A2, 2D6, 2C9, and 3A4. However, the IC_{50} values were in the mM range, indicating that the inhibition may not be potent enough to cause any clinically meaningful impact on CYP-mediated drug metabolism (Table 20.1).

Propylene glycol and DMSO are classified as water-soluble organic solvents (Strickley, 2004). Propylene glycol is used in some approved oral and injectable formulations and is also commonly used as a dosing vehicle in animal studies. DMSO is used in only some oral formulations. In a mouse model of acetaminophen-induced liver injury, the dosing vehicle propylene glycol was found to attenuate hepatotoxicity (Snawder *et al.*, 1993). The underlying mechanism was thought to be due to inhibition of CYP2e1 (Snawder *et al.*, 1993; Thomsen *et al.*, 1995), which is the major pathway for acetaminophen bioactivation in rodents. DMSO also diminished acetaminophen-induced liver damage in mice. This protective effect of DMSO was correlated with the inhibition of CYP-mediated aminopyrine-*N*-demethylation (Park *et al.*, 1988).

In summary, although some excipients inhibit CYP activities, the inhibition potency appears to be not high enough to be clinically relevant. However, in preclinical PK and toxicity studies, such excipients can potentially change the clearance of drug candidates by inhibiting CYP activities, because the injection volume in these studies can be up to 1% of body weight. So far, the underlying mechanisms for CYP inhibition by the aforementioned excipients have not been fully investigated.

20.2.2   CYP Induction

CYP3A4 and CYP4F2 are involved in vitamin E metabolism (Kikuta *et al.*, 1999; Parker, Sontag and Swanson, 2000). CYP3A induction is mainly mediated by the nuclear pregnane X receptor (PXR). PXR is activated by a large number of structurally unrelated xenobiotics. Interestingly, different forms of vitamin E were reported to activate a PXR-driven reporter gene in HepG2 cells (Landes *et al.*, 2003). The rank order of the potency of PXR activation was α- and γ-tocotrienol > δ-, α-, γ-tocopherol.

TABLE 20.1 Summary of literature findings: effects of excipients on function of CYPs and transporters.

Excipient category	Excipient	CYP-related effects	Transporter-related effects
Water-soluble organic solvents	Dimethylacetamide	Inhibits recombinant CYP3A4 (IC$_{50}$: 29.1 mM) (Mountfield et al., 2000)	
	DMSO	Inhibits Cyp2e1 (Park et al., 1988)	
	Propylene glycol	Inhibits Cyp2e1 (Snawder et al., 1993; Thomsen et al., 1995)	
		Inhibits recombinant CYP3A4 (IC$_{50}$: >50 mM) (Mountfield et al., 2000)	
	PEG400	Inhibits recombinant CYP3A4 (IC$_{50}$: 16.5 mM) (Mountfield et al., 2000)	
Nonionic surfactants	Cremophor RH40	Inhibits CYP3A4 in liver microsomes (IC$_{50}$: 0.3%, v/v) (Wandel, Kim and Stein, 2003)	Causes 20% increase in systemic exposure of the P-gp substrate digoxin in human (Tayrouz et al., 2003); Inhibits P-gp in Caco-2 cell (Wandel, Kim and Stein, 2003)
	Cremophor EL	Reduces metabolism of midazolam in rat hepatocytes and liver microsomes (Gonzalez et al., 2004)	Inhibits monocarboxylic acid transporter in Caco-2 cells (Rege, Kao and Polli, 2002); Inhibits P-gp in rat everted gut sac study (Cornaire et al., 2004)
	Labrasol		Inhibits BCRP in cell-based assay (Yamagata et al., 2007b); No effect on MRP2 function (Bogman et al., 2003); Inhibits P-gp in rat everted gut sac study (Cornaire et al., 2004)
	Pluronic P85		Inhibits Bcrp function in animal study and in cell-based assay (Yamagata et al., 2007a, b)
	Solutol HS15	Inhibits recombinant CYP3A4 (IC$_{50}$: 5 μM) (Bittner et al., 2003a); Reduces metabolism of midazolam in rat hepatocytes and liver microsomes (Gonzalez et al., 2004)	Inhibits P-gp in in vitro assays (Coon et al., 1991; Cornaire et al., 2004)

Category	Compound	CYP effect	Transporter effect
Organic liquids/semisolids	TPGS1000		Potent P-gp inhibitor, causing clinically relevant interaction with drugs (Chang, Benet and Hebert, 1996; Yu et al., 1999; Bogman et al., 2005) No effect on MRP2 function (Bogman et al., 2003)
	Tween 20	Inhibits recombinant CYP3A4 (IC$_{50}$: 3.8 µM) (Mountfield et al., 2000)	Inhibits digoxin transport in rat everted gut sac study (Cornaire et al., 2004) Inhibits Bcrp function in animal study and in cell-based assay (Yamagata et al., 2007a, b)
	Tween 80		Inhibits peptide transporter in Caco-2 assay (Rege, Kao and Polli, 2002) Inhibits digoxin transport in rat everted gut sac study (Cornaire et al., 2004) No effect on MRP2 function (Bogman et al., 2003)
	Oleic acid	Inhibits recombinant CYP3A4 (IC$_{50}$: 2.0 mM) (Mountfield et al., 2000)	
	Vitamin E	Induces CYP3A via PXR activation (Landes et al., 2003; Kluth et al., 2005)	
Complexing agents	β-Cyclodextrin		No effect on absorption of rhodamine 123 in rat jejunal loops (Oda et al., 2004)
	HP-β-CD	Inhibits recombinant CYP2C19 (IC$_{50}$: 6.81 mM) and CYP3A4 (IC$_{50}$: 2.68 mM) (Ishikawa, Yoshii and Furuta, 2005)	
	Methyl-β-cyclodextrin	Inhibits recombinant CYP2C19 (IC$_{50}$: 6.45 mM) and CYP3A4 (IC$_{50}$: 0.69 mM) (Ishikawa, Yoshii and Furuta, 2005)	Inhibits P-gp in Caco-2 cells (Arima et al., 2001)
Water-insoluble lipids	Soybean oil	Inhibits recombinant CYP3A4 (IC$_{50}$: >50 mM) (Mountfield et al., 2000)	

Furthermore, CYP3a11 mRNA was induced by α-tocopherol but not by γ-tocotrienol in mice (Kluth *et al.*, 2005). These findings indicate that at least some forms of vitamin E may increase their own metabolism and may also impact the metabolism of other CYP substrates.

20.3 INTERACTION OF PHARMACEUTICAL EXCIPIENTS WITH TRANSPORTERS

Transporter-based drug–drug interactions have been increasingly documented, leading to the inclusion of transporter studies in the recent FDA guidance on drug interaction studies (FDA Draft Guidance for Industry on Drug Interaction Studies, 2006). Some solubilizing excipients such as TPGS1000 have been shown to cause clinically relevant transporter-based drug interactions (Bogman *et al.*, 2005). Therefore, both formulation and drug metabolism/PK scientists involved in drug development need to be aware of the potential interactions between excipients and membrane-associated transporters when developing a drug formulation. Literature findings on the effects of excipients on transporter functions are summarized in Table 20.1.

20.3.1 P-glycoprotein (P-gp)

P-gp is a 170-kDa plasma membrane-associated transporter belonging to the superfamily of ATP-binding cassette transporters. It was originally discovered in multidrug-resistant tumor cells as a drug efflux transporter. This transporter was later found to be highly expressed in normal tissues. P-gp is localized on the columnar epithelial cells of the intestine, the canalicular membrane of hepatocytes in the liver, the brush-border domain of the proximal tubular cells in the kidneys, and the luminal surface of the blood–brain barrier. The anatomical localization of P-gp suggests that this transporter functions to protect the body against toxic xenobiotics by excreting these compounds into the intestinal lumen, bile, and urine, and by limiting their access to the brain (Ambudkar *et al.*, 2003). P-gp transports many structurally diverse drugs such as digoxin, vinca alkaloids, and HIV protease inhibitors. A number of drugs such as verapamil (Pedersen *et al.*, 1981; Ito *et al.*, 1993), clarithromycin (Wakasugi *et al.*, 1998), and quinidine (Sadeque *et al.*, 2000) are potent P-gp inhibitors. Coon *et al.* (1991) reported that Solutol HS15 renders multidrug-resistant tumor cells more susceptible to cytotoxic agents by inhibiting their efflux transport. Thereafter, a plethora of *in vitro* and *in vivo* studies demonstrated that many other solubilizing excipients inhibit P-gp function to different extents. For example, in a rat everted gut sac study, various excipients such as TPGS1000, Cremophor EL, Labrasol, Solutol HS15, Tween 20, Tween 80, Softigen 767, Miglyol, sucrose monolaurate, Imwitor742, and Acconon E inhibited the transport of digoxin across the intestinal mucosa (Cornaire *et al.*, 2004). Similarly, TPGS, Cremophor EL, Cremophor RH40, Tween 80, PEG300, Pluronic PE8100 and PE6100 inhibited P-gp function in several *in vitro* studies using cell-based transport assays (Hugger *et al.*, 2002; Rege, Kao and Polli, 2002; Bogman *et al.*, 2003). 2,6-D-*O*-methyl-β-cyclodextrin, a hydrophilic complexing agent, was found to decrease the efflux of two P-gp substrates, tacrolimus and rhodamine 123, across Caco-2 monolayers (Arima *et al.*, 2001), whereas β-cyclodextrin (1%) has little effect on the absorption of rhodamine 123 from *in situ* jejunal loops from rats (Oda *et al.*, 2004).

TPGS1000 is a water-soluble derivative of vitamin E comprised of a hydrophilic polar (water-soluble) head and a lipophilic (lipid-soluble) alkyl tail. Sokol *et al.* (1991)

reported that treatment with TPGS1000 (25 mg/kg, daily) enhanced absorption of the highly lipophilic drug cyclosporine in liver transplant patients. While the authors originally suggested that the improved oral bioavailability was due to the solubility enhancing effect of TPGS1000, Chang, Benet and Hebert (1996) later demonstrated that enhanced permeability due to P-gp inhibition also played a role. TPGS1000 (12%) is one of the components in the Agenerase oral formulation of the poorly water-soluble HIV protease inhibitor amprenavir from Glaxo Wellcome (Yu *et al.*, 1999). Amprenavir in capsule or tablet formulations is barely orally available, but formulations containing 20% and 50% TPGS1000 increase its oral bioavailability in beagle dogs at 25 mg/kg to 70% and 80%, respectively (Yu *et al.*, 1999). TPGS1000 was identified as a potent inhibitor of an active efflux system in Caco-2 cell monolayers at concentrations much lower than its critical micelle concentration (0.2 mg/mL) (Yu *et al.*, 1999). Paclitaxel, an anticancer drug, is typically administered intravenously due to its poor solubility and poor permeability. A recent study showed that TPGS1000 enhanced oral paclitaxel absorption by improving both parameters (Varma and Panchagnula, 2005). In these three examples, TPGS1000 enhances the oral bioavailability of poorly water-soluble drugs that are P-gp substrates not only by solubility-enhancing micelle formation, but also by increasing the permeability via inhibiting efflux transporter function.

Bogman *et al.* (2005) reported that the $AUC_{0-\infty}$ and C_{max} of the P-gp substrate talinolol were increased by 39% and 100%, respectively, in healthy human volunteers perfused intraduodenally with talinolol and 0.04% TPGS1000, whereas another solubilizing excipient, Poloxamer 188 (0.8%), had no effect on talinolol pharmacokinetics. Additionally, TPGS1000, but not Poloxamer 188, inhibited the P-gp-mediated talinolol transport in Caco-2 cells (Bogman *et al.*, 2005). These findings further support that TPGS1000 can cause clinically relevant P-gp-mediated drug interactions.

TPGS analogs with different PEG chain length, and/or a modified hydrophobic core were found to exhibit differential effects on rhodamine123 dye transport in Caco-2 monolayers (Collnot *et al.*, 2006). Moreover, the inhibitory potency of these TPGS analogs on P-gp-mediated efflux in Caco-2 cells highly correlated with that of P-gp ATPase activity but not with changes in membrane fluidity determined by electron spin resonance (Collnot *et al.*, 2007). These *in vitro* studies shed important light on the mechanism(s) through which TPGS modulates P-gp function, supporting the notion that efflux pump energy source depletion is a major factor in the inhibitory mechanism of TPGS.

In a clinical study, a high dose of oral Cremophor RH40 (600 mg three times daily) in combination with a single oral dose of 0.5 mg digoxin in a hard gelatin capsule resulted in an increase in the systemic exposure of digoxin by approximately 22% (Tayrouz *et al.*, 2003). This finding is in agreement with *in vitro* observation that Cremophor RH40 inhibited P-gp function in Caco-2 cells ($IC_{50} = 0.03\%$, v/v) (Wandel, Kim and Stein, 2003). Although this *in vivo* and *in vitro* correlation suggests that P-gp inhibition led to the pharmacokinetic interaction between Cremophor RH40 and digoxin, the influence of Cremophor on the pharmacokinetics of P-gp substrates is not straightforward. For example, although Cremophor EL in the intravenous formulation can result in greater than dose-proportional increases in paclitaxel exposure in cancer patients (Sparreboom *et al.*, 1996), a study by the same authors in mice suggests that P-gp inhibition may not be the mechanism for the observed decrease in clearance. At comparable blood levels achieved in the human study, tissue levels of Cremophor EL were below the limit of detection, and presumably too low to inhibit excretion from organs such as the liver and/or kidney (Sparreboom *et al.*, 1996). Further investigation by the same research group showed that addition of Cremophor EL at a clinically achievable concentration (0.5%, v/v) to human blood decreased the whole blood to plasma concentration ratio of paclitaxel from 1.07 to 0.69 (Sparreboom et al., 1999). These authors concluded that Cremophor EL formed micelles in blood that entrap compounds such as paclitaxel, resulting in an altered cellular distribution and a concomitant increase in plasma concentration. Furthermore, Cremophor EL in an oral paclitaxel formulation was found to reduce absorption of paclitaxel from the gut, presumably by micellar entrapment of the drug (Bardelmeijer et al., 2002).

20.3.2 Breast Cancer Resistance Protein (BCRP)

BCRP also belongs to the ATP-binding cassette family of drug efflux transporters. The tissue distribution of BCRP is similar to that of P-gp. BCRP transports a wide variety of compounds including anticancer drugs, food carcinogens, antibiotics, and conjugated metabolites. Therefore, BCRP, like P-gp, plays an important role in the absorption, distribution, and elimination of many drugs (Krishnamurthy and Schuetz, 2006). Yamagata et al. (2007a) reported that Pluronic P85 and Tween 20 given orally along with topotecan increased the AUC of topotecan by approximately twofold in wild-type mice but not in Bcrp knockout mice. Pluronic P85 and Tween 20 did not affect the AUC of topotecan after intravenous administration. These findings suggest that Pluronic P85 and Tween 20 can improve the oral bioavailability of BCRP substrates by inhibiting BCRP function in the gut. The same research group also reported that Cremophor EL, Tween 20, Span 20, Pluronic P85, and Brij30 enhanced the uptake of the BCRP substrate mitoxantrone into BCRP-expressing cells (Yamagata et al., 2007b). A recent study showed that in BCRP-expressing MDCKII cells, cholesterol depletion with methyl-β-cyclodextrin decreased BCRP activity by 40%, whereas cholesterol repletion with methyl-β-cyclodextrin/cholesterol-inclusion complexes restored BCRP function (Storch et al., 2007). These findings suggest that cyclodextrin affects BCRP function by altering the cholesterol composition of cellular membranes. Whether the BCRP inhibitory effects of these excipients can alter the pharmacokinetics of drugs in a clinical situation is still unknown.

20.3.3 Other Transporters

In contrast to P-gp and BCRP, there are only a few studies examining the effects of excipients on the function of other transporters. For example, Rege, Kao and Polli (2002) reported that Cremophor EL inhibited the monocarboxylic acid transporter function as measured by benzoic acid permeability, and Tween 80 inhibited the permeability of glycyl sarcosin, a substrate of peptide transporter. Bogman et al. (2003) showed that TPGS1000, Pluronic PE8100, Cremophor EL, Pluronic PE6100, and Tween 80 had little effects on multidrug resistance-associated protein 2 (MRP2)-mediated efflux in MRP2-overexpressing MDCK cells.

20.4 SUMMARY

The reasons to investigate the potential of drug candidates to cause enzyme- and transporter-mediated drug–drug interaction are well understood. Similarly, during formulation development, formulation and PK scientists should be equipped with information regarding the potential of selected excipients to cause transporter- and/or enzyme-based drug interaction. Such information is important, because (i) the amount of excipients in dosage formulations often far exceeds that of the active ingredients and (ii) some solubilizing excipients (e.g. TPGS1000) have been convincingly shown to affect the clinical PK of drugs by inhibiting P-gp function. The P-gp inhibition properties of excipients have been applied to increase the oral bioavailability of drugs that are P-gp substrates as exemplified by the development of the TPGS1000-containing Agenerase oral formulation of the HIV protease inhibitor amprenavir (Yu *et al.*, 1999). For many clinical indications and therapeutic areas, however, a dose formulation containing a large amount of a P-gp-inhibiting excipient such as TPGS1000 may result in undesirable product labeling and should be carefully considered. Other transporters, such as BCRP have been increasingly recognized as important factors that contribute to the absorption, distribution, and elimination of drugs. More basic and clinical research is needed in order to understand whether the functions of these transporters can be affected by excipients and whether the effects are clinically relevant. Despite many *in vitro* and *in vivo* studies showing that a number of excipients affect CYP activity and expression, clinically relevant data are still lacking to the best of our knowledge. Therefore, enzyme-based drug interaction involving excipients appears to occur predominantly in preclinical studies where the amount of excipients used in the dose formulations is much higher than that in the formulations intended for clinical usage.

REFERENCES

Ambudkar, S.V., Kimchi-Sarfaty, C., Sauna, Z.E., Gottesman, M.M. (2003). P-glycoprotein: from genomics to mechanism. *Oncogene, 22*, 7468–7485.

Arima, H., Yunomae, K., Hirayama, F., Uekama, K. (2001). Contribution of P-glycoprotein to the enhancing effects of dimethyl-beta-cyclodextrin on oral bioavailability of tacrolimus. *J. Pharmacol. Exp. Ther., 297*, 547–555.

Bardelmeijer, H.A., Ouwehand, M., Malingre, M.M., Schellens, J.H., Beijnen, J.H., van Tellingen, O. (2002). Entrapment by Cremophor EL decreases the absorption of paclitaxel from the gut. *Cancer Chemother. Pharmacol., 49*, 119–125.

Bittner, B., Gonzalez, R.C., Isel, H., Flament, C. (2003a). Impact of Solutol HS 15 on the pharmacokinetic behavior of midazolam upon intravenous administration to male Wistar rats. *Eur. J. Pharm. Biopharm., 56*, 143–146.

Bittner, B., Gonzalez, R.C., Walter, I., Kapps, M., Huwyler, J. (2003b). Impact of Solutol HS 15 on the pharmacokinetic behavior of colchicine upon intravenous administration to male Wistar rats. *Biopharm. Drug Dispos., 24*, 173–181.

Bogman, K., Erne-Brand, F., Alsenz, J., Drewe, J. (2003). The role of surfactants in the reversal of active transport mediated by multidrug resistance proteins. *J. Pharm. Sci., 92*, 1250–1261.

Bogman, K., Zysset, Y., Degen, L., Hopfgartner, G., Gutmann, H., Alsenz, J., Drewe, J. (2005). P-glycoprotein and surfactants: effect on intestinal talinolol absorption. *Clin. Pharmacol. Ther., 77*, 24–32.

Brigelius-Flohe, R. (2003). Vitamin E and drug Metabolism. *Biochem. Biophys. Res. Comm., 305*, 737–740.

Challa, R., Ahuja, A., Ali, J., Khar, R.K. (2005). Cyclodextrins in drug delivery: an updated review. *AAPS Pharm. Sci. Tech., 6*, E329–E357.

Chang, T., Benet, L.Z., Hebert, M.F. (1996). The effect of water-soluble vitamin E on cyclosporine pharmacokinetics in healthy volunteers. *Clin. Pharmacol. Ther., 59*, 297–303.

Collnot, E.M., Baldes, C., Wempe, M.F., Hyatt, J., Navarro, L., Edgar, K.J., Schaefer, U.F., Lehr, C.M. (2006). Influence of vitamin E TPGS poly(ethylene glycol) chain length on apical efflux transporters in Caco-2 cell monolayers. *J. Control. Release*, *111*, 35–40.

Collnot, E.M., Baldes, C.B., Wempe, M.F., Kappl, R., Hutermann, J., Hyatt, J.A., Edgar, K.J., Schaefer, U.F., Lehr, C.M. (2007). Mechanism of inhibition of P-glycoprotein mediated efflux by vitamin E TPGS: influence on ATPase activity and membrane fluidity. *Mol. Pharm.*, *4*, 465–474.

Coon, J.S., Knudson, W., Clodfelter, K., Lu, B., Weinstein, R.S. (1991). Solutol HS15, nontoxic polyoxyethylene esters of 12-hydroxystearic acid, reverses multidrug resistance. *Cancer Res.*, *51*, 897–902.

Cornaire, G., Woodley, J., Hermann, P., Cloarec, A., Arellano, C., Houin, G. (2004). Impact of excipients on the absorption of P-glycoprotein substrates *in vitro* and *in vivo*. *Inter. J. Pharm.*, *278*, 119–131.

FDA Draft Guidance for Industry on Drug Interaction Studies. (2006). www.fda.gov/cder/guidance/6695dft.htm.

Gonzalez, R.C., Huwyler, J., Boess, F., Walter, I., Bittner, B. (2004). In vitro investigation on the impact of the surface-active excipients Cremophor EL, Tween 80 and Solutol HS 15 on the metabolism of midazolam. *Biopharm. Drug Dispos.*, *25*, 37–49.

Hugger, E.D., Novak, B.L., Burton, P.S., Audus, K.L., Borchardt, R.T. (2002). A comparison of commonly used polyethoxylated pharmaceutical excipients on their ability to inhibit P-glycoprotein activity in vitro. *J. Pharm. Sci.*, *91*, 1991–2002.

Ishikawa, M., Yoshii, H., Furuta, T. (2005). Interaction of modified cyclodextrins with cytochrome P-450. *Biosci. Biotechnol. Biochem.*, *69*, 246–248.

Ito, S., Woodland, C., Harper, P.A., Koren, G. (1993). The mechanism of the verapamil-digoxin interaction in renal tubular cells (LLC-PK1). *Life Sci.*, *53*, 399–403.

Kalasz, H., Antal, I. (2006). Drug excipients. *Curr. Med. Chem.*, *13*, 2535–2563.

Kikuta, Y., Miyauchi, Y., Kusunose, E., Kusunose, M. (1999). Expression and molecular cloning of human liver leukotriene B4 omega-hydroxylase (CYP4F2) gene. *DNA Cell Biol.*, *18*, 723–730.

Kluth, D., Landes, N., Pfluger, P., Muller-Schmehl, K., Weiss, K., Bumke-Vogt, C., Ristow, M., Brigelius-Flohe, R. (2005). Modulation of CYP3a11 mRNA expression by a-tocopherol but not α-tocotrienol in mice. *Free Radic. Biol. Med.*, *38*, 507–514.

Krishnamurthy, P., Schuetz, J.D. (2006). Role of ABCG2/BCRP in biology and medicine. *Annu. Rev. Pharmacol. Toxicol.*, *46*, 381–410.

Landes, N., Pfluger, P., Kluth, D., Birringer, M., Ruhl, R., Bol, G.F., Glatt, H.R., Brigelius-Flohe, R. (2003). Vitamin E activates gene expression *via* the pregnane X receptor. *Biochem. Pharmacol.*, *65*, 269–273.

Mountfield, R.J., Senepin, S., Schleimer, M., Walter, I., Bittner, B. (2000). Potential inhibitory effects of formulation ingredients on intestinal Cytochrome P450. *Int. J. Pharm.*, *211*, 89–92.

Oda, M., Saitoh, H., Kobayashi, M., Aungst, B.J. (2004). Beta-cyclodextrin as a suitable solubilizing agent for *in situ* absorption study of poorly water-soluble drugs. *Int. J. Pharm.*, *280*, 95–102.

Park, Y., Smith, R.D., Combs, A.B., Kehrer, J.P. (1988). Prevention of acetaminophen-induced hepatotoxicity by dimethyl sulfoxide. *Toxicology*, *52*, 165–175.

Parker, R.S., Sontag, T.J., Swanson, J.E. (2000). Cytochrome P4503A-dependent metabolism of tocopherols and inhibition by sesamin. *Biochem. Biophys. Res. Commun.*, *277*, 531–534.

Pedersen, K.E., Dorph-Pedersen, A., Hvidt, S., Klitgaard, N.A., Nielsen-Kudsk, F. (1981). Digoxin-verapamil interaction. *Clin. Pharmacol. Ther.*, *30*, 311–316.

Rege, B.D., Kao, J.P., Polli, J.E. (2002). Effects of nonionic surfactants on membrane transporters in Caco-2 cell monolayers. *Eur. J. Pharm. Sci.*, *16*, 237–246.

Sadeque, A.J., Wandel, C., He, H., Shah, S., Wood, A.J. (2000). Increased drug delivery to the brain by P-glycoprotein inhibition. *Clin. Pharmacol. Ther.*, *68*, 231–237.

Snawder, J.E., Benson, R.W., Leaky, J.E., Roberts, D.W. (1993). The effects of propylene glycol on the P450-dependent metabolism of acetaminophen and other chemicals in subcellular fractions of mouse liver. *Life Sci.*, *52*, 183–189.

Sokol, R.J., Johnson, K.E., Karrer, F.M., Narkewicz, M.R., Smith, D., Kam, I. (1991). Improvement of cyclosporine absorption in children after liver transplantation by means of water-soluble vitamin E. *Lancet*, *338*, 212–214.

Sparreboom, A., van Tellingen, O., Nooijen, W.J., Beijnen, J.H. (1996). Nonlinear pharmacokinetics of paclitaxel in mice results from the pharmaceutical vehicle Cremophor EL. *Cancer Res.*, *56*, 2112–2115.

Sparreboom, A., van Zuylen, L., Brouwer, E., Loos, W.J., de Bruijn, P., Gelderblom, H., Pillay, M., Nooter, K., Stoter, G., Verweij, J. (1999). Cremophor EL-mediated alteration of paclitaxel distribution in human blood: clinical pharmacokinetics implications. *Cancer Res.*, *59*, 1454–1457.

Storch, C.H., Ehehalt, R., Haefeli, W.E., Weiss, J. (2007). Localization of the human breast cancer resistance protein (BCRP/ABCG2) in lipid rafts/caveolae and modulation of its activity by cholesterol *in vitro*. *J. Pharmacol. Exp. Ther.*, *323*, 257–264.

Strickley, R.G. (2004). Solubilizing excipients in oral and injectable formulations. *Pharm. Res.*, *21*, 201–230.

Tayrouz, Y., Ding, R., Burhenne, J., Riedel, K., Weiss, J., Hoppe-Tichy, T., Haefeli, W.E., Mikus, G. (2003). Pharmacokinetics and pharmaceutic interaction between digoxin and Cremophor RH40. *Clin. Pharmacol. Ther.*, *73*, 397–405.

Thomsen, M.S., Loft, S., Roberts, D.W., Poulsen, H.E. (1995). Cytochrome P450 2E1 inhibition by propylene glycol prevents acetaminophen hepatotoxicity in mice without Cytochrome P450 1A2 inhibition. *Pharmacol. Toxicol.*, *76*, 395–399.

Varma, M.V., Panchagnula, R. (2005). Enhanced oral paclitaxel absorption with vitamin E-TPGS: effect on solubility and permeability *in vitro, in situ* and *in vivo*. *Eur. J. Pharm. Sci.*, *25*, 445–453.

Wakasugi, H., Yano, I., Ito, T., Hashida, T., Futami, T., Nohara, R., Sasayama, S., Inui, K. (1998). Effect of clarithromycin on renal excretion of digoxin: interaction with P-glycoprotein. *Clin. Pharmacol. Ther.*, *64*, 123–128.

Wandel, C., Kim, R.B., Stein, C.M. (2003). "Inactive" excipients such as Cremophor can affect in vivo drug disposition. *Clin. Pharm. Ther.*, *73*, 394–396.

Yamagata, T., Kusuhara, H., Morishita, M., Takayama, K., Benameur, H., Sugiyama, Y. (2007a). Improvement of the oral drug absorption of topotecan through the inhibition of intestinal xenobiotics efflux transporter, breast cancer resistance protein, by excipients. *Drug Metab. Dispos.*, *35*, 1142–1148.

Yamagata, T., Kusuhara, H., Morishita, M., Takayama, K., Benameur, H., Sugiyama, Y. (2007b). Effect of excipients on breast cancer resistance protein substrate uptake activity. *J. Control. Release*, *124*, 1–5.

Yu, L., Bridgers, A., Polli, J., Vickers, A., Long, S., Roy, A., Winnike, R., Coffin, M. (1999). Vitamin E-TPGS increases absorption flux of an HIV protease inhibitor by enhancing its solubility and permeability. *Pharm. Res.*, *16*, 1812–1817.

PART IV

TOXICITY

CHAPTER 21

The Role of Drug Metabolism in Toxicity

CARL D. DAVIS and UMESH M. HANUMEGOWDA

21.1 INTRODUCTION

The term "toxicity" in general denotes any adverse event that results from exposure to an agent, which includes a wide variety of substances: natural or man-made, physical, chemical, biological or radiological, and endogenous or exogenous. Toxicity can result from exaggerated on-target pharmacology, unrelated off-target activity, or biotransformation to a metabolite that is reactive and/or more potent in its on- or off-target activity. Evolutionary pressures have equipped species, from bacteria to humans, with a complement of enzymes that help to regulate and limit the exposure and/or accumulation of xenobiotic agents, such as those encountered in the diet, the environment, accidental or deliberate exposure to poisons, or from administration of therapeutic agents. Usually these drug-metabolizing enzymes (DMEs) facilitate the elimination of xenobiotics from the body and serve principally as a detoxification process. However, in some cases, the same enzymes can be involved in the bioactivation of a xenobiotic and/or its metabolite(s) to facilitate adverse reactions and toxicity. In addition to the toxic agent itself, several important factors also contribute to or even determine the interspecies and interindividual variability and susceptibility to toxicity. These include differences in the expression of metabolizing enzymes, population differences or polymorphisms in the enzymes expressed, physiological state, underlying disease, age, and interactions associated with concurrently administered drugs.

There are many textbooks and reviews on both toxicity and the role of biotransformation; however, very few articles have reviewed the specific role of DMEs and susceptibility to toxicity. This chapter is intended for the practicing pharmaceutical scientist and provides real-world examples potentially encountered when evaluating and developing drug candidates. We describe the drug toxicity that is mediated predominantly through the formation of reactive metabolites. To extend the scope beyond the specific case to the general, we also summarize some key characteristics of the common enzymes encountered in metabolic clearance of drugs and highlight elements to consider, such as differences in tissue expression between species and/or gender. Examples are offered

Drug Metabolism Handbook: Concepts and Applications, Edited by Ala F. Nassar, Paul F. Hollenberg, and JoAnn Scatina
Copyright © 2009 by John Wiley & Sons, Inc.

to illustrate how DMEs are involved in the bioactivation of drugs to form reactive metabolites. Included are many of the well-studied drugs that provide molecular mechanisms of toxicity and examples that describe toxicity by target organ. This is followed by a brief overview of experimental methods that can be used to evaluate drug metabolism and toxicity. Lastly, a brief discussion is offered to describe the pharmaceutical industry's progress to date, with some case examples to illustrate how drug metabolism and toxicity of one drug can influence the development and success of subsequent drugs.

21.2   DRUG-METABOLIZING ENZYMES

Drug toxicity and the factors that govern it are quite complex. In drug discovery, toxicity can depend on factors that may be species-specific, thus forcing toxicologists to determine the relevance of toxicity in animals to that in humans. The balance of risk versus benefit is always weighed prior to exposing humans to a new chemical entity (NCE), with the pharmaceutical industry and regulatory agencies favoring safety over uncertainty. Fortunately, by combining an understanding of drug metabolism and toxicology, some common adverse effects in humans—with the possible exception of idiosyncratic drug reactions—can be predicted and to some extent prevented. This allows the safety and efficacy of NCEs to be assessed in clinical trials with minimal risk to the humans tested.

The efficacy and toxicity of a drug are both dependent on its overall exposure (including at the target site) and its intrinsic potency at the target. Adverse effects of an otherwise well-tolerated drug can be seen after its accidental or deliberate overdose (e.g. acetaminophen-induced liver toxicity). There are several mechanisms that help to regulate exposure to drugs: biliary, renal, and metabolic clearances are three such processes by which well-absorbed drugs can be eliminated. The net result is that drug accumulation is avoided and toxicity limited. These same mechanisms, when operating effectively, also make it difficult to achieve oral bioavailability or sustain an exposure that is adequate for a therapeutic effect. Therefore, the objective in discovery and optimization of drug candidates is to achieve adequate exposure for efficacy while staying below concentrations that potentially result in toxicity. Pragmatically, the emphasis therefore is not to eliminate the potential for toxicity per se, but rather to find drug candidates that offer the widest separation in exposure between therapeutic and adverse effects.

Biliary and renal clearance can be indirectly associated to the etiology of certain types of toxicity. The bile duct canaliculi (liver), nephron/proximal tubules (kidney), or the bladder, in particular, can all be exposed to relatively high local concentrations of drug and/or metabolites associated with their elimination. If biliary or renal toxicity are a direct result of unchanged drug levels, then, generally speaking, these adverse effects would simply be predicted to occur in a human population and thus provide evidence against the subsequent development of that compound as a drug. In contrast, toxicity associated with metabolic clearance is more complex (the net result of potential contributions from numerous and diverse enzymes and transporters that show tremendous differences between tissues and species) and usually much harder to extrapolate from animals to humans. With an NCE, one needs a good understanding of the enzymes involved in its metabolism and the functional activity and reactivity of its metabolite(s) to predict any consequences in humans.

DMEs are characterized by broad substrate specificity and tend not to be named by a specific reaction. Instead they have been identified and grouped into families based on functional activity and their shared homology in the amino acid sequence of the functional protein. Table 21.1 lists many of the major families of human enzymes involved in metabolic clearance and often implicated in adverse drug reactions. Most DMEs are expressed ubiquitously in tissues, usually with the liver expressing the highest activities (there are notable exceptions, a few of which are indicated in Table 21.1).With metabolism-related toxicity most often observed in the liver, hepatocytes and liver subfractions have become central to drug lead identification and optimization. Adverse effects and toxicity occurring specifically in other cells, tissues, and/or organ systems may require a tailored screening paradigm.

21.2.1   Cytochrome P450 Enzymes

The cytochrome P450s (CYPs) (EC 1.14.14.1) arguably are the single most important family of enzymes involved in regulating exposure to pharmacologically active, toxic, or potentially toxic xenobiotics. These enzymes exhibit highly conserved regions in their amino acid sequence across different species including humans. The CYPs are membrane-bound hemoproteins found in the smooth endoplasmic reticulum and can be prepared as microsomal fractions for use *in vitro*. Individual CYPs show some specialization with almost distinctive substrate selectivity that partly reflects roles in endogenous metabolism. As a family, CYPs catalyze oxidation and reduction reactions of a highly diverse collection of substrates providing metabolic clearance for compounds not readily eliminated in urine or bile. In general, CYP-mediated oxidation is the more relevant reaction for generating reactive or toxic metabolites.

CYP-mediated oxidation proceeds via redox cycling between the ferric to ferrous states of the heme protoporphyrin group in the active site of the enzyme, with the NAD(P)H oxidoreductase and cytochrome b5 accessory flavoproteins involved in the electron transfer from nicotinamide adenine dinucleotide phosphate (NADPH) to the heme and the hemo-substrate complex. Ultimately, this process catalyzes the

TABLE 21.1   Families and major distribution of common human drug-metabolizing enzymes.

Enzyme family	Significant family members	Expression
Cytochrome P450s (membrane-bound; microsomal and mitochondrial)	CYP1A1, 1A2, 2A6, 2A13, 2B6, 2C8, 2C9, 2C18, 2C19, 2D6, 2E1, 2F1, 2J2, 3A4, 3A5, 3A7; 4A11, 4F2, 4F3a, 4F3b, and 4F12	Skin, brain, kidney, heart, lung, nasal mucosa, intestine, with major expression of most commonly encountered forms in the liver
FMO (microsomal)	FMO1, FMO2, FMO3, FMO4, FMO5	Diverse tissues. Major expression FMO1—kidney; FMO2—lung (inactive Caucasians, Asians), FMO3, 4 and 5—all liver
Uridine glucuronosyl-S-Transferases (microsomal)	UGT1A1, 1A3, 1A4, 1A6, 1A7, 1A8, 1A9, 1A10, 2B4, 2B7, 2B10, 2B11, 2B15, 2B17	Kidney, intestine, lung, with major expression in the liver
N-acetyl transferases (cytosolic)	NAT1, NAT2	Liver, intestine, and others

TABLE 21.1 *Continued*

Enzyme family	Significant family members	Expression
Monoamine/diamine oxidase (mitochondrial)	MAO-A, MAO-B; DAO (histaminase)	Outer mitochondrial membrane with ubiquitous expression: MAO-A > MAO-B in lung, duodenal mucosa, placenta, adult brain; MAO-B > MAO-A in myocardium, platelets, lymphocytes, fetus brain; diamine oxidase highest in the intestinal mucosa
Alcohol dehydrogenases (cytosolic)	ADH classes 1–5. ADH1A, ADH1B, ADH1C;	ADH1A, -1B, -1C (liver); ADH4—stomach
Aldehyde dehydrogenases (cytosolic. microsomal, mitochondrial)	ALDH 10 families, 13 subfamilies: ALDH1A1, 1A2, 1A3, 3A1, 3A2, 4A1, 5A1, 8A1, 9A1, 1L1 (many others)	Most tissues
Aldehyde oxidase (cytosolic)	AO	Liver, kidney, lung, brain
Sulfotransferases (cytosolic)	SULT1A1, 1A2, 1A3, 1A4, 1B1, 1C2, 1C4, 1E1, 2A1, 2B1a, 2B1b, 4A1	Ubiquitous, with highest expression usually in liver, intestine, and placenta
Esterases (cytosolic and microsomal)	CE1, CE2, AChE, PON1, PON2, PON3	Ubiquitous expression as a family with tissue-specific expression characteristic of some enzymes (e.g. AChE in brain, muscle, red blood cells; PON1, PON3 in liver and blood; PON2 brain, liver, kidney, testes but not blood)
Epoxide hydrolases (cytosolic and microsomal)	mEH (HYL1), cEH (HYL2)	Mainly liver with a relatively low expression in other tissues, for example, lung, brain, blood, and ovary
Glutathione-*S*-transferases (cytosolic)	GSTA1-A4, GSTM1-M5, GSTP,Zeta-,Omega-,Sigma-	GSTA1-1, A2-2, A4-4, M1-1, M3-3, M4-4, K1-1,O1-1, T1-1, and Z1-1 are all expressed in the liver; GSTA1-1, A2-2, A4-4, M1-1, M3-3, K1-1,O101, P1-1, T1-1, and Z1-1 are expressed in the small intestine; and GSTP1-1 and T1-1 are found in erythrocytes
Peroxidases (cytosolic)	MPO, EPO, catalase; COX-1 (PHS-1), COX-2 (PHS-2)	Neutrophils, eosinophils, platelets, intestine, uterus, mammary gland, thyroid; COX-1 is expressed ubiquitously (housekeeping gene); COX-2 is an inducible form

CYPs, cytochrome P450s; FMO, flavin-containing monooxygenase; UGT, uridine diphosphate glucuronosyltransferases; NAT, *N*-acetyl transferase; MAO, monoamine oxidase; DAO, diamine oxidase; ADH, alcohol dehydrogenase; ALDH, aldehyde dehydrogenase; AO, aldehyde oxidase; SULT, sulfotransferase; CE, carboxylesterase; AChE, acetylcholinesterase; PON, paraoxonase; mEH, microsomal epoxide hydrolase; cEH, cytosolic epoxide hydrolase; GST, glutathione *S*-transferase; MPO, myeloperoxidase; EPO, eosinophil peroxidase; COX, cylcooxygenase; PHS, prostaglandin H synthase.

incorporation of oxygen from molecular oxygen producing water and a metabolite. This reaction usually results in a more polar molecule that can be eliminated directly in bile or urine, and/or subjected to further metabolism such as conjugation with sulfate or glucuronide as mediated by the sulfotransferases (SULTs) or uridine diphosphate glucuronosyltransferases (UGTs), respectively. Occasionally a less polar metabolite can be formed, for example, the CYP-mediated formation of an *N*-oxide metabolite of clozapine in human liver microsomes (Pirmohamed *et al.*, 1995). Peroxidase-like reactions can also occur in the CYP catalytic cycle, with the ferrous-O_2 complex decomposing to form a superoxide anion or hydrogen peroxide that potentially can react with drug and/or its metabolites.

In some instances, the transition intermediates formed in the CYP catalytic cycle can prove to be highly reactive and bind with the heme (diltiazem, tamoxifen or troleandomycin and CYP3A4) or apoprotein (dihydralazine and CYP1A2; tamoxifen and CYP2B6; tienilic acid and CYP2C9). Mechanism-based inactivation then occurs that results in partial or total irreversible inactivation of the CYP enzyme. Mechanism-based inactivation can exacerbate the risk of drug–drug interactions and toxicity from coadministered drugs by increasing their exposure for a prolonged period (i.e. until new enzyme has been synthesized to replace that which is inactivated). If the reactive species formed in the CYP is able to escape the active site and bind covalently with nucleophilic macromolecules such as proteins, RNA, or DNA, this can culminate in adverse drug reactions and toxicity (Kalgutkar, Obach, and Maurer, 2007; Kalgutkar and Soglia, 2005). The chemistry of these reactions has been well studied and structural alerts identified (Table 21.2). Based on such anecdotal experiences, medicinal chemistry often is guided by blocking the potential site of metabolism that could result in a problematic reactive intermediate or avoiding certain functional groups entirely. Table 21.2 lists the most common structural alerts encountered in medicinal chemistry.

CYPs are divided into subfamilies based on their shared sequence in protein homology, with >40% sequence similarity designating the family and >55% shared sequence homology designating the subfamily. Over 50 different CYPs have been identified in humans, with the CYP1A1, 1A2, 2A6, 2A13, 2B6, 2C8, 2C9, 2C18, 2C19, 2D6, 2E1, 2F1, 2J2, 3A4, 3A5, 3A7; 4A11, 4F2, 4F3a, 4F3b, and 4F12 forms responsible for the metabolic clearance of most drugs (CYP3A4, 2C9, 2D6, 2C19, 1A2, and 2C8 being most often encountered). Although CYPs are quite well conserved across species, it is important to note that they are not identical. Even a small difference in amino acid sequence can affect substrate affinity and the capacity to metabolize a substrate. This is exemplified by allelic variants of CYP2C9 in humans where, for instance, individuals possessing the CYP2C9*3 allele (an A1075→C transition in exon 7 changes the amino acid sequence from Ile^{359}→Leu^{359}) result in a form of CYP2C9 that demonstrates a lower clearance of several substrates, including (*S*)-warfarin (Takanashi *et al.*, 2000; Scordo *et al.*, 2002; Xie *et al.*, 2002). With even greater differences existing between orthologues in humans and other species, caution must be exercised in extrapolating from preclinical animals models and when interpreting the relevance of animal toxicity or lack thereof to humans.

21.2.2 Flavin-Containing Monooxygenases

Flavin-containing monooxygenases (EC 1.14.13.8) (FMOs) are another class of membrane-bound enzymes found in the endoplasmic reticulum. In common with CYPs, FMOs also mediate NADPH-dependent reactions that lead to oxidation of xenobiotics,

TABLE 21.2   Fuctional groups and drugs associated with reactive metabolite formation.

Functional Group	Example
Aniline (e.g. procainamide)	
Thiophene (e.g. tienilic acid)	
Hydrazines (e.g. isoniazid)	
Nitroarene (e.g. choramphenicol)	
Quinones (e.g. diethylstilbestrol)	

with FMOs favoring more specific *N*- and *S*-oxidation reactions. FMOs are otherwise quite distinct from CYPs: they are flavoproteins that contain a single flavin adenine dinucleotide (FAD) have a redox cycle that does not depend on a transition state metal for binding of molecular oxygen and generally a narrower substrate selectivity that is limited to soft nucleophiles, and, preferably, uncharged or single positive charged compounds with an electron-rich center (usually nitrogen or sulfur, but phosphorous and selenium can also provide the electrons) (Ziegler, 1993). The so-called "external monooxygenases" are the most common form of FMO and depend on NADPH or nicotinamide adenine dinucleotide (NADH) as external factors to reduce the flavin (flavin mononucleotide, FMN, or FAD) to the reactive form that binds molecular oxygen. In contrast, the rarely encountered "internal monooxygenases" can use the substrate itself

Functional Group	Example
Thiols (e.g. methimazole)	
Benzylamine (e.g. terbinafine)	
Thiazole (e.g. sudoxicam)	
Carboxylic acid (e.g. diclofenac)	
Epoxide (e.g. phenytoin)	

to reduce the flavin (lactate monooxygenase is one example of this type of FMO) (van Berkel *et al.*, 2006).

In humans, five functional FMOs are known and six pseudogenes, which are considerably fewer than the ~57 CYPs, are reported to date. The FMO family is expressed at the highest levels in the liver. The individual FMOs, however, show distinctive differences in their relative expression across tissues. FMO1 is predominantly expressed in fetal liver and adult human kidney and intestine. In kidney microsomes, FMO1 is expressed usually at the highest levels (FMO1 > FMO5 > FMO3) and FMO1 is seemingly greater in kidney microsomes prepared from African-Americans than from Caucasians (Krause, Lash and Elfarra, 2003). FMO2 is inactive in Caucasians and Asians, while an active full-length form of FMO2 is expressed polymorphically in the lung of

26% of African-Americans and 4.5% of Hispanics (Cashman, 2000; Whetstine *et al.*, 2000). In contrast, FMO3 and FMO5 are expressed at the highest levels in the liver and are the forms most likely to be involved in FMO-mediated drug metabolism (Cashman, 2000; Whetstine *et al.*, 2000; Krueger and Williams, 2005). The relevance of these findings is that if tissue-specific toxicity is indicated with a known FMO substrate, then tissue-specific differences in the expression of FMOs could underlie the mechanism in the detoxication or bioactivation of the drug. The ability to predict these findings in humans depends on correlating animal *in vitro* models with *in vivo* toxicity in those species, and finding similarities between those mechanisms and the biochemical toxicology indicated in human *in vitro* models.

Far fewer drugs are metabolized by FMOs than by CYPs and very few drugs are cleared primarily by FMOs—the exceptions include compounds like xanomeline, itopride, benzydamine, and olopatiadine, with each drug cleared by formation of its *N*-oxide metabolite catalyzed by the human FMO1 and FMO3 forms (Ring *et al.*, 1999; Lang and Rettie, 2000; Störmer *et al.*, 2000; Mushiroda *et al.*, 2001; Kajita *et al.* 2002). Consequently, the role of FMOs in toxicity is limited when compared with CYPs or other DMEs; however, there are notable examples that exemplify their importance on a case-by-case basis. For instance, monoamine oxidases (MAOs) catalyze the oxidation of the neurotoxin MPTP (1-methyl-4-phenyl-1,2,3,6-tetrahydropyridine) to form the highly toxic MPDP+ (1-methyl-4-phenyl-2,3-dihydropyridine) and MPP+ (1-methyl-4-phenylpyridine) metabolites that block mitochondrial transport, destroy the nigrastriatal dopaminergic pathways, and elicit a rapid onset of Parkinsonism (Davis *et al.*, 1979; Langston *et al.*, 1983; Williams and Ramsden, 2005). CYP2D6, CYP1A2 and, to some extent, CYP3A4 have been shown to form the *N*-desmethyl metabolite (4-phenyl-1,2,3,6-tetrahydropyridine, PTP) and the *N*-oxide metabolite (1-methyl-4-phenyl-1,2,3,6-tetrahydropyridine *N*-oxide) that serve to inactivate MPTP (Weissman *et al.*, 1985; Coleman *et al.*, 1996). FMO in the brain has also been shown to inactivate MPTP via formation of the *N*-oxide metabolite (Weissman *et al.*, 1985; Mushiroda *et al.*, 2001). It is possible that FMO-mediated clearance, in addition to supporting CYPs in protecting against neurotoxicity and hepatotoxicity from MPTP exposure, could potentially play an even more significant role in the inactivation of MPTP in CYP2D6 poor metabolizers. FMOs are also involved in the hepatotoxicity associated with *N*-deacetylketoconazole and the sulfoxidation of thioureas to form sulfenic acids (the latter subsequent to the formation of the reactive sulfinic acid, ethionamide) (Rodriguez and Miranda, 2000; Rodriguez and Buckholz, 2003; Henderson *et al.*, 2004). These reactions lead to redox cycling and depletion of glutathione and NADPH cofactors that affect the metabolism of the culpable agents and potentially that of other xenobiotics (Henderson *et al.*, 2004). FMOs also oxidize secondary and primary amines to form hydroxylated metabolites that can form a complex with hemoproteins in rat tissues, with inactivation of CYPs and ensuing drug–drug interactions and concomitant issues (Jönsson and Lindeke, 1992).

There are few drug–drug interactions associated with FMOs since no inducers have been found for FMOs and few drugs are metabolized exclusively by FMOs. Variability in FMO activity tends to be associated with polymorphic expression, with the best known example being FMO3. This form exclusively catalyzes the *N*-oxidation and clearance of trimethylamine (TMA), a product of choline (Cashman *et al.*, 1997; Lang *et al.*, 1998). In Caucasians, about 0.1% to 1% of the population is deficient in FMO3 and exhibits a "slow" phenotype for clearance of TMA. In these individuals, TMA is secreted in urine, sweat, and breath giving rise to the "fish odor" syndrome. The next most significant polymorphism is linked to FMO2 and the polymorphic expression of an active form in human lung. If a polymorphic form of FMO is significantly involved in the elimination and/or bioactivation of a drug, then it is plausible that certain individuals will have an inherently greater predisposition toward toxicity, either via the accumulation of toxic drug or formation of higher levels of reactive metabolites.

21.2.3 Uridine Diphosphate Glucuronosyltransferases

The UGTs are another class of membrane-bound enzymes that are expressed in the liver and intestine and often play an important role in the first-pass effect. UGTs catalyze the transfer of glucuronic acid from a nucletoide cofactor, uridine diphosphate glucuronic acid (UDPGA), to specific functionalities (-NH$_2$, -OH, and -COOH) on substrates. This typically generates a more polar metabolite that is amenable to biliary and/or renal elimination. Since UGTs are located on the luminal side of the endoplasmic reticulum, microsomal preparations require "activation" by pore-forming agents such as the antibacterial alamethacin, the detergent Brij58, or by mechanical disruption (e.g. sonication). The use of alamethacin is preferred since it is nonspecific and activates all UGTs.

UGTs, like CYPs, are classified by primary amino acid sequence into families and subfamilies. In humans, two gene families have been identified with a growing number of members that are currently grouped into three subfamilies (UGT1A, UGT2A, and UGT2B). Unlike CYPs, selective inhibitors (chemical or antibody) or substrates for a specific UGT are lacking, which makes it very difficult to identify the exact UGTs involved and their contribution to the total metabolic clearance. This apparent lack of selectivity of UGT substrates does, however, prove advantageous in clinical practice, with a relatively low incidence of UGT-related drug–drug interactions reported in humans (Williams et al., 2004).

Both UGTs and CYPs are co-localized in the endoplasmic reticulum and evidence exists for protein interactions between CYPs and UGTs that can affect glucuronidation—for instance, the ratio of morphine-3-glucuronide to morphine-6-glucuronide is affected by CYP3A4 content (Takeda et al., 2005). Complexities like this, coupled with the atypical kinetics observed with UGT reactions, issues with optimizing protein and phospholipid content, inhibition by fatty acids and the need for activation, and other conditions for metabolic turnover, make it difficult to accurately extrapolate from in vitro systems to in vivo models (Miners et al., 2006). This is extremely unfortunate because, in addition to substrates cleared directly via UGT-mediated conjugation (e.g. the 1,4-benzodiazepine anxiolytic, (S)-oxazepam and the HIV integrase anti-AIDS drug, raltegravir), many metabolites that are formed by CYPs (e.g. phenols and carboxylic acids) are themselves substrates for UGT-mediated glucuronidation (Abernethy et al., 1983; Kassahun et al., 2007). What this means is that even though UGTs may not be directly responsible for the clearance and/or therapeutic dose of a drug, they are often involved in the overall metabolism. This becomes especially important should the metabolites be implicated in adverse reactions, for example, diclofenac acyl-glucuronide and its potential to react with proteins.

Although UGT2B7 is more often implicated in drug clearance than any other UGT enzyme, the most frequent issue concerning UGTs involves the potential for inhibition of UGT1A1 and effects on the clearance of endogenous substrates such as estradiol, estriol, and the blood breakdown product, bilirubin (Ritter, Crawford and Owens, 1991; Court, 2005). UGT1A1 inhibition can slow the clearance of bilirubin causing an accumulation of unconjugated bilirubin, which leads to hyperbilirubinemia and toxicity. People with genetic polymorphisms of UGT1A1 expression exhibit increased levels of unconjugated bilirubin. A profound (90%) reduction in bilirubin clearance is seen in patients with the rare Crigler–Najjar syndrome, whereas a moderate (40%) decrease is reported on average for patients with the more commonplace Gilbert's syndrome, which is seen in 10% of the US population (associated with the UGT1A1*28 homozygous form) (Tukey and Strassburg, 2000). Drug-induced inhibition of UGT1A1 in humans can present symptomatically like Crigler–Najjar or Gilbert's syndrome and profound hyperbilirubinemia is observed in UGT1A1-deficient individuals. An example of where UGT1A1 genotype or phenotype has implications on human safety and drug tolerability is irinotecan, an ester camptothecin derivative used to treat metastatic colon cancer. Irinotecan is first metabolized by carboxyesterases to form SN-38, a carboxylic acid and the active chemotherapeutic agent that is subsequently cleared by glucuronidation mediated by UGT1A1. UGT1A1-deficient patients show increased incidence and severity of neutropenia and diarrhea likely stemming from prolonged exposure to SN-38 (Toffoli and Cecchin, 2004; Toffoli *et al.*, 2006).

Acyl glucuronides are an interesting class of metabolite, formed by glucuronidation of carboxylic acids, because of their (i) propensity to form covalent adducts with proteins and (ii) potential to hydrolyze and regenerate the carboxylic acid. Intramolecular chemical rearrangement of the biosynthetic 1-*O*-β-acyl glucuronide can form 2-, 3-, and 4-*O*-β-isomers via migration of the acyl group. These acyl glucuronide β-isomers can subsequently anomerize via ring opening and formation of planar aldehyde intermediates followed by recyclization to the 2-*O*-, 3-*O*-, and 4-*O*-α acyl glucuronide isomers. Reactivity can occur via transacylation reactions with the initial acyl glucuronide or via glycation of target proteins such as albumin. The proposed reaction scheme is shown in Fig. 21.1 (Garlick and Mazer, 1983; Smith, McDonagh and Benet, 1986; Smith, Benet and McDonagh, 1990). Adverse drug reactions associated with acyl glucuronides are seen with many nonsteroidal anti-inflammatory drugs (NSAIDs), including ibuprofen, diclofenac, naproxen, and aspirin (Sanford-Driscoll and Knodel, 1986). In fact, Bailey and Dickinson reported that carboxylic acids account for 25% of drug withdrawals associated with severe toxicity, including zomepirac, ibufenac, and benoxaprofen (Bailey and Dickinson, 2003).

The propensity for acyl glucuronides to cause toxicity and/or immune responses necessitates diligent analysis of reactivity *in vitro*. Recently, lysine-phenylalanine dipeptide was shown to act as a possible surrogate for proteins and offers a useful *in vitro* tool for screening acyl glucuronides for reactivity (Wang *et al.*, 2004). In terms of structure–activity relationships and optimization of medicinal chemistry, the reactivity can be circumvented by substituting a potentially more stable group for the carboxylic acid moiety (e.g. an amide) and/or developing a drug that can be administered at low doses and/or cleared by multiple mechanisms other than UGT.

Another complexity associated with acyl glucuronides is the tendency of 1-*O*-β-acyl glucuronides to be amenable to hydrolysis, more so than the transacylated isomers. Serum albumin, hydroxyl anion, and esterases can hydrolyze acyl glucuronides back to the drug and this requires additional precautions to be taken in bioanalysis and increases the cost of sample analysis and potential inaccuracies. Moreover, β-glucuronidases in the intestine can hydrolyze acyl glucuronides to regenerate drug or metabolite, which if reabsorbed results in the phenomenon of enterohepatic recirculation (EHR), such as seen with valproic acid in the rat (Pollack and Brouwer, 1991; Slattum, Cato and Brouwer, 1995). EHR can significantly prolong the residence time of a drug and add to difficulties in predicting exposure in humans from preclinical animal data. Lastly, consideration also must be made toward pH-dependent hydrolysis

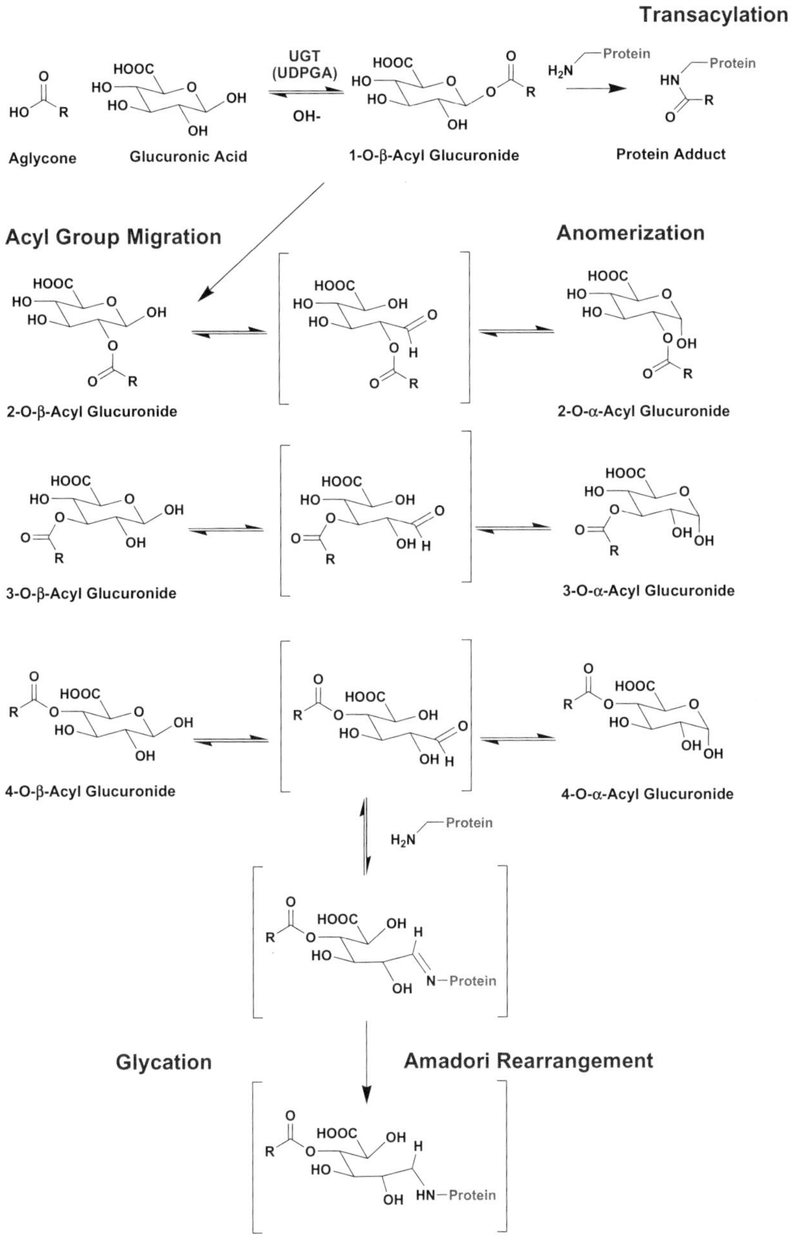

Figure 21.1 Reactivity scheme for acyl glucuronides. Acyl glucuronide can react with target nucleophiles directly via transacylation or by glycation via acyl group migration, with formation of an intermediate aldehyde and Amadori rearrangement.

and its effect on anomers. Blood and excreta samples may need to be acidified to allow an accurate quantitation of drug, acyl glucuronide metabolite, and its isomers, which adds bioanalytical cost and increases variability and uncertainty of the results.

21.2.4 Arylamine *N*-Acetyl Transferases

Arylamine *N*-acetyl transferases (EC 2.3.1.5) (NATs) are expressed in humans in two forms, NAT1 and NAT2, each the product of a distinct gene locus. Unlike the DMEs discussed thus far, NATs are soluble, cytosolic enzymes (*in vitro* assays need to be optimized accordingly). NAT1 and NAT2 share 87% identity in their nucleotide sequence and 81% homology in their amino acid sequence. At least 26 allelic variants of NAT1 and 36 variants of NAT2 have been identified to date (Butcher *et al.*, 2002). NAT1 is expressed ubiquitously in humans, with its highest levels of mRNA found in the urothelium and colon epithelial cells, whereas NAT2 is found mainly in the liver and intestine (Debiec-Rychter, Land and King, 1999). With the exception of *N*-acetylation of cysteine conjugates to form mercapturic acids, NAT1 and NAT2 metabolize predominantly aromatic amines. Both forms use acetyl CoA as the cofactor and have different but overlapping substrate specificities. For instance, *p*-aminobenzoic acid and sulfamethazine are relatively selective probe substrates of NAT1 and NAT2, respectively.

Understanding polymorphisms and interindividual differences in expression of both human NAT forms is an important part of the safe evaluation of drugs that serve as their substrates. Although NAT1 was initially considered to be a monomorphic enzyme—albeit one with a wide range in expression and activity between patients—it is now known to have a bimodal polymorphic distribution in humans, with approximately 8% of individuals found to be slow acetylators (Vatsis and Weber, 1993). The slow phenotype is attributed to NAT1*14 and NAT1*17 alleles and lower expression/reduced active form, with NAT1*14 the more frequently occurring (up to 3.8% in Caucasians). A higher occurrence of the NAT1*14 allele was seen in a Lebanese population, where up to 50% of the population is a slow acetylator (Dhaini and Levy, 2000).

When compared with NAT1, the NAT2 form is more distinctive in its polymorphism. The observed bimodal distribution in the plasma exposure of isoniazid—now associated with NAT2—is one of the earliest reported drug phenotypes (Peters, Miller and Brown, 1965). This bimodal distribution is very well illustrated in a relatively recent study of isoniazid plasma exposure in genetically unrelated German subjects (Weinshilboum and Wang, 2004). NAT2 displays an interesting distribution across different ethnicities with a slow acetylator phenotype of 5% in Canadian Eskimos, 10–20% in Asians, 40–70% in Europeans/Caucasians, 80% in Egyptians, and 90% in Moroccans (Evans, 1989). Drugs such as isoniazid, sulfamethazine, sulfadiazine, diaminodiphenyl sulfone, hydralazine, phenelzine, sulfapyridine, sulfamethoxypyridazine, and dapsone are metabolically cleared by NAT2 and should be administered with some regard to the target population to avoid overexposure and minimize the incidence of adverse effects.

NAT1 and NAT2 both catalyze the acetylation of arylamines or hydralazine, or catalyze the *O*-acetylation of a hydroxylamine. This usually serves as a detoxification mechanism in limiting exposure of xenobiotics, illustrated by the greater incidence of bladder cancer, isoniazid-induced neuropathy, and sulfamethoxazole- or dapsone-induced Stevens–Johnson syndrome and quicker onset of positive antinuclear antibody (ANA) syndrome following procainamide therapy in NAT2 slow acetylators. However,

NATs are also implicated directly in the bioactivation of procarcinogens, for example, biotransformation of aromatic and heterocyclic amines (e.g. 4-aminobiphenyl, 2-naphthylamine, benzidine, and cooked meat-derived heterocyclic amines) and also an increased risk of amonafide-induced myelotoxicity in fast acetylators (Innocenti, Iyer and Ratain, 2001).

Bioactivation of aromatic amines is an example where the metabolite of one enzyme reaction can often serve as the substrate for another as part of the mechanism of drug toxicity. Two mechanisms are proposed by which aryl amines can be bioactivated by NATs to an ultimate electrophilic aryl nitrenium ion, each requiring a sequential role for CYP and NAT. The first mechanism involves CYP-mediated *N*-hydroxylation to form a hydroxylamine with subsequent NAT-mediated *O*-acetylation to form *N*-acetoxyarylamine. The second mechanism proposed is initiated by direct *N*-acetylation to form an *N*-acetyl arylamine that undergoes CYP-mediated *N*-hydroxylation to form an *N*-acetohydroxamic acid, followed by *N,O*-transacetylation to form the *N*-acetoxyarylamine. In each case, heterolytic cleavage of the N–O bond of the common *N*-acetoxyarylamine product leads to aryl nitrenium formation and reactivity with target nucleophiles such as DNA and proteins—such adducts thus have potential to cause toxicity in humans and animals (Fig. 21.2). Today, there is a plenitude of evidence marking this functionality as a clear structural alert. Therefore, even though steric and electronic factors can block the reactive sites of metabolism and low-dose and acute administration regimens may obviate the risk of bioactivation, unembedded aryl amines are usually avoided in drug discovery and development.

NATs provide an interesting example of a distinct species difference in drug metabolism. The liver cytosol from domestic dogs and other canids (e.g. wolf, jackal, and fox) does not show any expression of NATs and this species appears to be NAT-deficient (Trepanier *et al.*, 1997). Consequently, the same idiosyncratic sulfonamide toxicities seen in human can be observed in dogs (Cribb *et al.*, 1996). In the event that an NAT substrate is developed as a drug, the dog is worth consideration as a preclinical model for the evaluation of *N*-acetylation and its role in pharmacokinetics (PK), pharmacodynamics (PD), and/or toxicity (Lavergne *et al.*, 2006).

21.2.5 Sulfotransferases

SULTs (EC 2.8.2) are another class of cytosolic enzymes that catalyze metabolic clearance via conjugation of xenobiotic and endogenous substrates. There is an overlap between SULTs and UGTs in the substrates they metabolize, with SULTs mediating the sulfation of hydroxy or amine functional group present in compounds such as catecholamines, neurotransmitters, steroids, and thyroid hormones. There are also

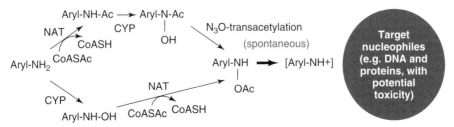

Figure 21.2 Reactivity scheme for *N*-acetyl transferases.

membrane-bound SULTs that are found in the Golgi apparatus: these tend not to be significant with respect to the metabolism of typical small-molecular-weight drugs; however, they are involved in the sulfonation and regulation of activity associated with proteins, peptides, lipids, and glycosaminoglycans (Baeuerle and Huttner, 1987; Huttner, 1988).

SULTs are found in many species including plants, birds, and mammals. To date, there are four families of human SULTs: SULT1 has four subfamilies and eight members (SULT1A1, 1A2, 1A3, 1A4, 1B1, 1C2, 1C4, 1E1); SULT2 has two subfamilies and three members (2A1 and 2B1a, 2B1b, which result from alternate slicing of the SULT2B1 gene); SULT4 has one subfamily and one member (SULT4A1); and SULT6 has one subfamily and one known member (SULT6B1). (Families share ≥45% amino acid sequence identity; subfamilies share ≥60% amino acid sequence identity.) Recently, 217 polymorphisms were found in the sequences of 118 DNA samples, including 64 polymorphisms in the 8785-bp coding regions of all 12 SULT genes. Of these, 39 were non-synonymous polymorphisms that alter the functional activity of the expressed enzyme, with evidence for lowered expression of immunoreactive protein rather than expression of mutant forms with altered enzyme kinetics (e.g. SULT1E1*2 showed 13% expression of the wild-type protein and 7% of its activity, SULT1C2*5 showed no expression, whereas SULT1A3/4*5 showed 120% of wild-type protein expression with 110% of its activity) (Hildebrandt et al., 2007).

SULTs are expressed in many tissues with some differentiation seen between isozymes: SULT1A1 is expressed in adult liver and platelets; SULT1A3 is expressed in placenta, fetal liver, and adult platelets; and SULT 4A1 has been found only in the brain (Coughtrie, 2002). An interesting difference between SULTs and enzymes such as CYPs or UGTs is a relatively high expression of SULTs in the fetus, which points to a role in embryogenesis and/or development—a potential consideration should embryotoxicity be indicated in safety studies.

SULTs are not commonly involved as the principal clearance mechanism yet they do play a significant role in the overall biotransformation and excretion balance of a drug. With regard to direct clearance, 17α-ethinyl estradiol (EE), troglitazone, and acetaminophen are three drugs that involve SULTs in their O-sulfation (they also show a possible role for SULTs in their toxicity); cisapride and DPC-423 are examples of drugs that undergo N-sulfonation reactions (Back et al., 1982; Slattery et al., 1987; Meuldermans et al., 1988; Kawai et al., 1997; Mutlib et al., 2002). It should be noted that sulfonation does not necessarily inactivate a drug—morphine-6-sulfate is an example of an active sulfonate conjugate (Zuckerman et al., 1999).

PAPS (3'-phosphoadenosine-5'-phosphosulfate) serves as the typical donor of the sulfuryl or sulfonate group (SO_3^-). In most tissues, only 4–80 μM of PAPS is found, which is low compared with the 200 μM levels reported for UDPGA and even lower in comparison with the 2–10 mM typical of GSH. Drugs such as acetaminophen that are given at high doses can quickly deplete the sulfonation pathway and dose-dependent kinetics are seen in its biotransformation. This phenomenon was shown in rats dosed with acetaminophen. In humans, however, increased plasma levels of acetaminophen were not seen and induction of the glucuronidation pathway seemingly compensates for the depletion of PAPS (Galinsky and Levy, 1981; Gelotte et al., 2007).

SULTs are able to bioactivate procarcinogens and genotoxic agents including N-hydroxyacetyl aminofluorene and 1-(α-hydroxyethyl) pyrene. The latter provides an example of enantioselective metabolism and bioactivation, with 160-fold ratio between the S- and R-enantiomers for mutagenicity in Salmonella typhimurium TA1538 expressing human SULT1E1 (Glatt, 2000). Sulfuric acid esters of benzylic and allylic alcohols, in particular, can be susceptible to bioactivation as a result of heterolytic cleavage of the electron-withdrawing sulfate group and formation of an electrophilic cation that is stabilized by mesomerism. This same mechanism is also implicated in the activation of hydroxylamines and hydroxamic acids, which suggests a possible interplay between SULTs and NATs in the metabolism-mediated toxicity associated with aryl amines (Glatt, 1997; Glatt *et al.*, 1998a, b).

SULTs can play a key role in the PK/PD of a new drug and should not be overlooked when assessing drug metabolism and predicting its effects in humans. For instance, hydroxytamoxifen is a potent active metabolite of tamoxifen inactivated by SULT1A1-mediated sulfonation. The variability in efficacy and apparent resistance to tamoxifen in breast cancer patients is attributed, in part, to interindividual differences in functional activity associated with sulfonate conjugation (Hayes and Pulford, 1995). In contrast, the hair-growth stimulant minoxidil is an *N*-oxide prodrug activated by SULTs specifically, to form a sulfonate conjugate that is more hydrophobic and functionally active than minoxidil as a result of an intramolecular H-bond (Falany and Kerl, 1990).

21.2.6 Peroxidases

Peroxidases (EC 1.11.1) involved in drug metabolism are typically heme-containing DMEs that have a characteristic ability to catalyze one-electron oxidations of substrates (e.g. aromatic amines) that produce free radicals with potential to propagate and terminate reactions. Peroxidases are classified into two types based on the mechanism of inactivation elicited by the common substrate phenylhydrazine. Phenylhydrazine forms phenyl radicals that can characteristically bind to the heme-iron of peroxidases such as CYPs, prostaglandin H synthase, catalase, and hemoglobin *inter alia*, or characteristically bind instead to the δ-meso heme edge and protein in the case of peroxidase such as myeloperoxidase (MPO), eosinophil peroxidase (EPO), and lactoperoxidase (LPO) (Tafazoli and O'Brien, 2005).

MPO is a lysosomal peroxidase found at high levels in the primary granules of neutrophils (2–5% of the dry weight) and monocytes, and is also found at significant levels in tissue macrophages (e.g. Kupffer cells in the liver). MPO is distinguished from the other peroxidases in its unique ability to oxidize chlorine to hypochlorous acid, a potent oxidizing agent with a key role in phagocytosis and function as an antimicrobial agent. Known substrates of MPO/HOCl include procainamide, clozapine, chlorpromazine, lamotrigine, dapsone, vesnarinone, sulfonamides, 3-hydroxy carbamazepine, 4-hydroxy phenytoin, and 7-hydroxy fluperlapine (Lai *et al.*, 2000; O'Brien, 2000; Lu and Uetrecht, 2007). MPO is released extracellularly into phagocytic vacuoles upon activation of neutrophils in an immune response (*in vitro* phorbol myristate acetate is often used to activate neutrophils). Hydrogen peroxide, formed by NADPH oxidase and superoxide dismutase (SOD), is also released and acts with chloride ion as cofactors in the MPO-mediated formation of hypochlorous acid. Hypochlorous acid itself is a powerful two-electron oxidant that can nonenzymatically oxidize many drugs and be directly responsible for tissue damage (Klebanoff, 1999). Hypothetically, this can elicit a "danger signal" and trigger idiosyncratic drug toxicity in susceptible individuals.

Infiltration of tissues by neutrophils or eosinophils can lead to increased local MPO activity and toxicity linked with hypochlorous acid and/or reactive oxygen species (ROS) formation, including oxygen radicals (e.g. indomethacin-induced kidney toxicity) (Basivireddy *et al.*, 2004). For instance, a 50- to 100-fold increase in MPO activity was reported in the liver following infiltration of neutrophils triggered by the injury sustained by ischemia/reperfusion (Kato *et al.*, 2000). Pulmonary insults, such as tobacco smoke, cause neutrophil infiltration into the lung (Schmekel *et al.*, 1990). Benzo[*a*]pyrene is one of several toxins found in cigarette smoke; its 7,8-diol metabolite can be bioactivated by MPO to form the very reactive and carcinogenic diol-epoxide metabolite (Mallet, Mosebrook and Trush, 1991). MPO release may play a direct role in the bioactivation and toxicity of procarcinogens.

A ⁻463G→A polymorphism in the promoter region of the MPO gene is associated with decreased transcriptional activity and could result in lower expression and improved risk in lung cancer (Feyler *et al.*, 2002). Unfortunately, identifying a significant role in cancer for low-penetrance genes has proven to be notoriously difficult and counterevidence is available with no association found in the ⁻463G→A polymorphism based on data from 988 cases of lung cancer in Caucasians and 1128 control subjects (Xu *et al.*, 2002). The evidence does, however, suggest that measuring MPO activity in target tissues of toxicity could provide a useful marker of oxidative stress and a possible biomarker of bioactivation.

Activated neutrophils, MPO-H_2O_2-Cl and HOCl are useful tools for generating and identifying potential metabolites and often HOCl alone can be used as a facile method to form metabolites. DMP 406 (a novel antipsychotic) and mianserin (a tetracyclic antidepressant) are two examples for which HOCl alone does not replicate the metabolism seen with activated neutrophils and *in vivo* their oxidation may proceed instead by sequential MPO-mediated single-electron mechanisms (Iverson, Zahid and Uetrecht, 2002).

There are peroxidases other than MPO to consider in drug metabolism and mechanistic studies of biochemical toxicology. For instance, EPO is involved in the formation of hypobromous acid and associated with an increase in hepatotoxicity seen with acetaminophen and high bromide plasma levels. EPO activity is not limited to the blood and the presence of eosinophils in the rat intestine accounts for most of the peroxidase activity seen in that tissue. Just as an inflammatory response can increase MPO activity in the liver, infiltration by eosinophils in the liver can also occur (e.g. triggered by parasite infections) and effect a profound increase in EPO activity (De, De and Banerjee, 1986).

Catalases, another family of peroxidases distinctive in being able to disproportionate hydrogen peroxide to oxygen and water, play a regulatory role in control of hydrogen peroxide levels formed by MPO and EPO and so on. Catalases also have a direct role in drug metabolism; they are able to oxidize low-molecular-weight compounds such as alcohols (e.g. methanol and ethanol) and hydrazines (e.g. phenylhydrazine, phenelzine, and isoniazid) when concentrations of hydrogen peroxide are low (Artemchik, Kurchenko and Metelitsa, 1985).

Peroxidases are not selective in their substrate affinity: ellipticine, for example, is a topoisomerase inhibitor that forms covalent DNA adducts implicated in the cytotoxicity and anticancer effects. These adducts are formed by oxidation reactions that involve multiple peroxidases including human CYP1A1, 1A2 and 3A4, MPO, human COX-2, human renal microsomes incubated in the presence of arachidonic acid (COX activity), bovine lactate peroxidase, ovine COX-1, and horseradish peroxidase (Stiborová *et al.*, 2004, 2007). With regard to drug therapy, tumor and tissue differences in the expression of peroxidases could offer an opportunity to exploit local bioactivation (e.g. tumor-specific cytotoxicity) and possibly predict the potential susceptibility of humans to cancer or therapy.

21.2.7   Prostaglandin H Synthases/Cyclooxygenases

Prostaglandin H synthase (PHS), also known as cyclooxygenase (COX), is named for its role in the biosynthesis of prostanoids from arachidonic acid. PHS catalyzes the oxidation of arachidonic acid to form PGG_2 and then a peroxidase-like reduction to PGH_2. Subsequent synthases lead to the formation of various prostaglandins and thromboxanes that feature in cell regulation and differentiation. In humans, there are two forms of PHS, each encoded by a single distinct gene: PHS-1 (COX-1) is found on the luminal surface of the endoplasmic reticulum and PHS-2 (COX-2) is found in the endoplasmic reticulum as well as in the inner and outer membranes of the nuclear envelope. They share approximately 60% of their amino acid sequence identity and have overlapping substrate affinities.

PHS-1 is considered to be a housekeeping gene with well-controlled basal levels expressed in most cells. In contrast, PHS-2 has significant basal levels in select cells only (e.g. brain hippocampus), yet upon inflammation, it is highly inducible in several cell types, a mechanism believed involved in the development of inflammatory responses such as pain and fever. Although NSAIDs work by inhibiting PHS-2, the NSAIDs (ibuprofen and sulindac) and peroxisome proliferators also cause its induction, which could influence the metabolism of coadministered drugs. The regulation of PHS is complex and reviewed in depth by Vogel (2000).

PHS is implicated in the neurodegeneration associated with the amphetamine derivatives, MDMA ("ecstasy") and methamphetamine ("speed"). One hypothesis involves their bioactivation by CYP2D6 in the brain to a catechol intermediate that subsequently forms a quinone that engages in redox cycling (Tucker *et al.*, 1994; Lin *et al.*, 1997). A second hypothesis proposes that PHS in the brain functions as a hydroperoxidase, metabolizing methylenedioxy amphetamine derivatives such as MDMA and forming free radicals and ROS that elicit tissue damage and irreversible loss of monoaminergic nerve terminals and long-term neurodegeneration (Jeng *et al.*, 2006). In addition to these drugs, various other substrates can serve as electron-donating redox partners for PHS and in the process are metabolized to radical cations, for example, arylamines, indoles, estrogens, and heterocyclic amines (e.g. methylenedioxy amphetamine derivatives). Phenytoin, Phenobarbital, and *N*-desmethylated metabolites of mephenytoin and trimethadione are all teratogenic via embryonic PHS bioactivation (Parman, Chen, and Wells, 1998).

21.2.8   Glutathione Transferases

Glutathione transferases, also known as glutathione *S*-transferases (EC 2.5.1.18) (GSTs), are dimeric proteins that are primarily cytosolic but also have a mitochondrial form and a membrane-bound form known as "membrane-associated proteins in eicosanoid and glutathione metabolism" (Jakobsson *et al.*, 1999). These enzymes catalyze the conjugation of glutathione (γ-glutamyl-cysteinyl-glycine; GSH) with reactive soft electrophiles such as epoxides, haloalkanes, alkenes, and aromatic nitro groups. GSTs with $\geq 60\%$ shared amino acid sequence identity are grouped into the same class and those with $<30\%$ shared identity are assigned their own class. Subfamilies of GSTs exist and share up to 90% sequence identity. The sequence at the *N*-terminus includes an important component of the active site and is also taken into consideration when classifying the GSTs. The Alpha-, Mu-, Pi-, and Theta-class of cytosolic GSTs are encoded by distinctive genes. (Zeta-, Omega-, Sigma-, and Kappa-GST forms also exist and are found in mitochondria and peroxisomes.) The Alpha-, Mu-, and Pi- forms are most similar in terms of structure, substrate affinity, and shared effects of inhibitors and are the classes most frequently involved in drug metabolism reactions (Sheehan *et al.*, 2001). The human forms identified to date include GSTA1-A4, GSTM1-M5, and GSTP. Various allelic forms exist for the GSTs including null allele polymorphisms linked with GSTM1-1 and GSTT1-1. In the case of GSTM1*0, no GSTM1-1 protein is expressed in approximately 50% of the Caucasian and Asian populations and for GSTT1*0, no GSTT1-1 protein is expressed in 10% of Mexican-Americans and up to 60% of East Asians. These individuals are potentially more susceptible to toxicity from substrates otherwise detoxified by these GST enzymes (Nakajima *et al.*, 1995; To-Figueras *et al.*, 1996; Bolt and Thier 2006).

There are some differences known in the tissue expression of UGTs, which could prove useful in understanding the mechanism of toxicity. GSTA1-1, A2-2, A4-4, M1-1, M3-3, M4-4, K1-1, O1-1, T1-1, and Z1-1 are all expressed in the liver; GSTA1-1, A2-2, A4-4, M1-1, M3-3, K1-1, O101, P1-1, T1-1, and Z1-1 are expressed in the small intestine and GSTP1-1 and T1-1 are found in erythrocytes (Mannervik *et al.*, 2005). Based on its known selectivity, the GSH conjugation of methyl halides in erythrocytes is a useful peripheral marker of GSTT-1 phenotype and genotype (Pemble *et al.*, 1994).

Most substrates of GSTs are also able to undergo a nonenzymatic reaction with GSH, which needs to be accounted for when predicting the role and contribution of the GST-mediated reaction to the overall metabolic clearance of a new drug. In drug discovery, screens are often used to identify early on potential issues with a new drug or chemotype. One such assay uses liver microsomal incubations performed in the presence of GSH as a trapping agent for the formation of reactive electrophiles. Although it is rather difficult to predict the consequences of forming GSH conjugates *in vitro* (vis-à-vis, many drugs form GSH adducts yet are generally considered to be safe), since this is principally a detoxification mechanism the presence of adducts is taken as evidence of formation of a reactive species and the key assumption made in medicinal chemistry is that when it comes to GSH adducts, less is better and none is ideal. GSH conjugates usually are excreted in the bile and/or urine either intact or as cysteine-glycine, cysteine, *N*-acetyl cysteine (NAC), and mercapturic acid metabolites. Therefore, when evaluating the *in vivo* biotransformation of a drug in human or animal models, the presence of any one of these metabolites can serve as a marker of GSH conjugation and potential formation of reactive drug-related species. In drug discovery, usually these studies are supported by *in vitro* liver microsomal covalent-binding experiments performed in the absence and presence of GSH. Since GST expression and activity can be induced (e.g. by phenobarbital, 3-methylcholanthrene, and β-naphthoflavone), pretreated *in vitro* or *in vivo* systems can provide additional insight into the mechanism of toxicity associated with the formation and detoxification of reactive electrophiles.

With GSH measured at up to 10 mM in cells, including hepatocytes, GSTs serve as a high-capacity detoxification mechanism that works well except in the event of significant drug overdose, such as acetaminophen toxicity and mitochondrial GSH depletion. The risk of toxicity is potentiated when acetaminophen is combined with ethanol use and alcoholics are more susceptible to acetaminophen toxicity than nonalcoholics (Seeff *et al.*, 1986; Zimmerman and Maddrey, 1995). The toxicity of acetaminophen is very well studied and known to result from the interplay of several factors. Ethanol is especially effective at potentiating adverse reactions since it has been shown to affect both the import of GSH from the cytosol into the mitochondria and also induces

CYP2E1 activity, with the latter responsible for an increase in the formation of the reactive electrophile *N*-acetyl-*p*-benzoquinoneimine (NAPQI) metabolite of acetaminophen that, when unchecked by conjugation with GSH, reacts with mitochondrial proteins leading to toxicity (Zhao and Slattery, 2002).

Depletion of GSH is associated with toxicity from pyrrolizidine alkaloids, clozapine and amodiaquine (Williams *et al.*, 1997; Naisbitt *et al.*, 1998; Fu *et al.*, 2004). Although not itself a drug, acrolein is an end product of lipid peroxidation and a potentially toxic fragment of several drugs, including cyclophosphamide. Acrolein is potently toxic to the mitochondria, affecting respiratory function and cell function; however, it is also a substrate of GSTP1-1 (Hayes, Flanagan and Jowsey, 2005). As an added complexity, when GSH levels are low, acrolein was shown to inactivate GSTP1 which could have consequences for other coadministered substrates of GSTs (van Iersel *et al.*, 1997).

The glutathione redox cycle represents a protective mechanism helping to regulate the levels of hydrogen peroxide formed in inflammatory response. Glutathione peroxidase catalyzes the oxidation of GSH by H_2O_2 to form a glutathione disulfide (GSSG). Glutathione reductase can then reduce GSSG back to GSH to begin a new cycle (Fig. 21.3). This redox cycle can be disrupted in several ways: for example, GSH depletion by reactive electrophiles, as described previously; inhibition of the synthesis of GSH (e.g. buthionine sulfoxime); and/or inhibition of the reduction of GSSG (e.g. 1,3-*bis*-(chloroethyl)-1-nitrosourea) (Harlan *et al.*, 1984). Interindividual variability in these processes, bioactivation mechanisms, and intrinsic susceptibility to drugs and/or toxins could factor into the expression of adverse effects including idiosyncratic drug reactions.

Induction of GSTs, GSTP1-1 in particular (characterized by significant basal expression in most tissues except hepatocytes), is linked to an increased resistance to chemotherapeutic agents (Mannervik *et al.*, 1987; Hamada *et al.*, 1994). This provides an opportunity in drug discovery to exploit GSTs in the bioactivation of chemotherapeutic prodrugs, such as described for the thiopurine prodrugs, cis-6-(2-acetylvinylthio)purine (cAVTP) and trans-6-(2-acetylvinylthio)guanine (tAVTG), which become conjugated with GSH prior to formation of the cytotoxic metabolites, 6-mercaptopurine (6-MP) or 6-thioguanine (6-TG), respectively, and release of *cis*-, *trans*-, and *bis*-glutathione-

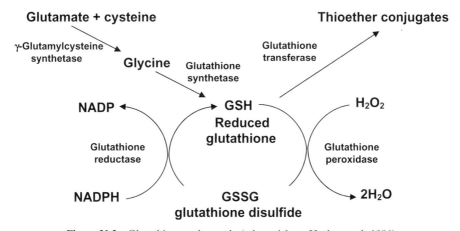

Figure 21.3 Glutathione redox cycle (adapted from Harlan *et al.*, 1984).

butenone (Eklund *et al.*, 2007). A similar reaction was reported with azathioprine, with 6-MP released and associated with cytotoxic effects and adverse reactions in individuals with high GST activity (Eklund *et al.*, 2006). Because GSTs are involved in the biosynthesis of endogenous molecules (e.g. steroids, leukotrienes, and prostaglandins) and help regulate cell signaling and expression of genes trans-activated by peroxisome proliferators, control of GST activity is a potential target for pharmacology. Perturbation of these processes by GSH depletion, GST inhibition, or GST induction could lead to toxicity following exposure to a GST substrate.

21.2.9 Other Drug-Metabolizing Enzymes

CYPs, FMOs, UGTs, NATs, SULTs, and peroxidases such as MPO are all part of a detoxification defense system that has evolved to protect humans and other species from accumulation of xenobiotics and endobiotics. Most of the time these enzymes effectively limit exposure of organs to xenobiotics, including drugs. Yet examples exist for all these enzymes being involved directly in the development of drug toxicity. GSTs generally represent essentially a detoxification pathway catalyzing the conjugation of GSH with many of the reactive electrophilic species formed in reactions mediated by the aforementioned enzymes. In contrast, toxicity associated with GSTs mostly concerns their deficiency (e.g. null polymorphism of GSTM1-1) or depletion of the GSH cofactor, rather than any specific metabolite that is formed via GST-mediated catalysis.

The aforementioned families of DMEs are those most commonly encountered in adverse reactions and can offer a significant challenge in understanding toxicity and the risk associated with a drug and, although not discussed in any detail, the relevance of animal models to human toxicity. Many other enzymes (e.g. epoxide hydrolases and monoamine oxidases) are known to be involved in drug metabolism; this adds to the difficulty faced by toxicologists and medicinal chemists. Fortunately, many of these enzymes are quite selective in their substrates. Therefore, the structure of the drug and/or its metabolite(s) provides an initial, simple indication of whether these enzymes participate in biotransformation, elimination, and/or toxicity.

For instance, epoxides represent an electrophilic species commonly formed from drugs and environmental agents, and often are subject to nucleophilic attack from GSH and, in parallel, hydrolysis catalyzed by the epoxide hydrolases (microsomal, mEH or HYL1; and cytosolic, cEH or HYL2) (Arand *et al.*, 2005). Epoxide hydrolases can react with epoxides to form polar dihydrodiol metabolites that tend to be unreactive and readily excreted. Epoxide hydrolases therefore usually represent an effective detoxification pathway, with the exception of epoxides of polycyclic aromatic compounds (e.g. benzo[*a*]pyrene) that become highly reactive and carcinogenic diol epoxides. Other examples are the mitochondrial monoamine oxidases (MAO-A, MAO-B) that can metabolize primary, secondary, and tertiary amines (e.g. sumatriptan). The MAOs are also able to form reactive aldehydes and metabolize the inert tertiary amine MPTP to the reactive MPP$^+$ metabolite that destroys the brain's dopaminergic pathways and induces Parkinson's disease in humans (Davis *et al.*, 1979; Langston *et al.*, 1983).

Other classes of DMEs include alcohol dehydrogenase (ADH) and aldehyde dehydrogenase (ALDH) (cytosolic enzymes that metabolize ethanol in a coupled reaction to form acetic acid), aldehyde oxidase (AO), xanthine oxidase (XO), carboxyesterases, superoxide dismutase, DT-diaphorase, NAD(P)H oxidase, carbonyl reductase, aldose reductase, and aldehyde reductase. The metabolite of one enzyme can become the substrate for another and their interplay can ultimately determine the individual risk associated with reactive metabolites from drugs. For example, carbonyl reductase can reduce a carbonyl group to an alcohol that can theoretically undergo an oxidation reaction catalyzed by ADH to form an aldehyde. This aldehyde can then either be further oxidized by ALDH to a polar carboxylic acid metabolite (which is excreted directly and/or conjugated with glucuronic acid by UGTs), or the aldehyde is oxidized by AO to a species that could be more or less reactive than its precursors, including the parent drug. Any of these steps could determine the individual risk of drug-related toxicity.

21.3 CLASSIFICATION OF TOXICITY

Drug-related toxicity or adverse drug reactions can be categorized in general terms such as based on target organ (e.g. hepatotoxicity) or based on mechanism of action (e.g. oxidative stress). Adverse drug reactions can be categorized specifically as related to (i) the exaggerated pharmacology of the compound (on-target effects)—for example, hypotension from beta-blockers; (ii) a consequence of off-target pharmacology—for example, QT prolongation by terfenadine; (iii) immunological and hypersensitivity reactions—for example, halothane hepatitis; (iv) formation of reactive metabolites—for example, agranulocytosis with clozapine; and (v) idiosyncratic reactions—for example, hepatotoxicity with carbamazepine. The general scheme of toxicity is depicted in Fig. 21.4.

Adverse effects can be anticipated in the first two categories and are generally quick to onset. Such effects if reversible are reasonably well tolerated. In such cases, subjects

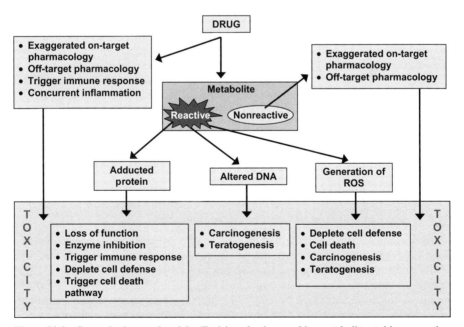

Figure 21.4 General scheme of toxicity. Toxicity of a drug and its metabolite, stable or reactive, is complicated and in parallel could precipitate various responses that result in toxicity.

can be monitored closely during clinical trials (e.g. ECG recording and blood-clotting times), with the exposure and/or doses titrated to maximize safety and tolerability. In contrast, those cases that fall into the other three categories of toxicity are more complicated and harder to evaluate and often require a good understanding of the biotransformation of the drug and the enzymes involved in its clearance. It is important to note that drugs can exhibit toxicities that fall into more than one category—their fate ultimately depends on the risk–benefit analysis in the target patient populations.

21.4 MOLECULAR MECHANISMS OF TOXICITY

In this section, we provide a general overview of the molecular mechanisms of toxicity that range from loss of function of cellular macromolecules, oxidative stress, balance between survival and cell death pathways, to antibody-driven and inflammation-influenced toxicities. It should be noted that at any given time with any given compound, there can be more than one mechanism active in precipitating toxicity.

Under normal physiological state, cellular functions such as energy metabolism, survival, and cell death signals are in homeostasis. Dysregulation is usually due to the deterioration in function of critical cellular macromolecules resulting in cell death. Such events can be triggered by the drugs themselves and/or by formation of reactive metabolites that bind with essential cellular macromolecules, effectively compromising cell survival and initiating oxidative stress and/or events that culminate in cell death (apoptosis, necrosis) or an immune response.

Some drugs (e.g. acetaminophen, adriamycin, nilutamide, and bleomycin) can cause toxicity by formation of oxyradicals that deplete cellular defenses. With drugs and reactive metabolites that cause toxicity directly, the number of protein adducts formed and the related incidence of toxicity increases proportionally with time and/or increasing dose. A few drugs are known to generate reactive intermediates that are not directly toxic when they bind with macromolecules but in doing so, form haptens that activate the immune system and cause potentially serious or fatal immune reactions. Lastly, there are a few drugs whose toxicity is termed "idiosyncratic," which means they do not display an obvious dose–response in animals or humans nor can their toxicity be predicted based on known pharmacology of the drug. Unfortunately this kind of toxicity is observed usually in late-stage development or post-marketing of a drug, only after a much bigger patient population has been exposed to the drug. When idiosyncratic drug reactions (IDRs) are observed drug withdrawals are likely to follow, unless an appropriate biomarker or predisposition to the adverse reactions is identified and can be monitored safely and cost-effectively in patients.

21.4.1 Loss of Function of Cellular Macromolecules

Many drugs (e.g. carbamazepine, diclofenac, and valproic acid) form reactive metabolites; however, not all reactive metabolites are quantitatively or qualitatively the same in risk. In fact, only a few drugs cause reactions that are significant enough to manifest as clinical symptoms. The working hypothesis for reactive intermediate-driven toxicity is that the reactive entity binds (usually by covalent bond formation) to critical cellular macromolecules, compromising the physiological function and eventually resulting in cell death and tissue damage. Given the broad range of macromolecules to which the reactive entity can theoretically bind, toxicity is seldom triggered by a single cascade of events and instead is usually a consequence of multiple parallel events (reviewed in Park, Pirmohamed, and Kitteringham, 1995).

Generally, the enzyme that mediates formation of the reactive metabolite is the target of covalent modification and subsequent inactivation. Such a phenomenon is seen as a time-dependent loss of enzyme function *in vitro* and is an indicator of potential clinical drug–drug interactions. A few examples include inactivation of CYP1A2 by dihydralazine (Masubuchi and Horie, 1999), CYP2B6 by tamoxifen (Sridar *et al.*, 2002) and CYP2C9 by tienilic acid (Melet *et al.*, 2003), CYP2C11 by diclofenac (Masubuchi *et al.*, 2001), and CYP3A4 by diltiazem (Jones *et al.*, 1999).

Other proteins, usually in the vicinity of the reactive intermediate generating system, can become a target for covalent modification. A good example is that of cisplatin, a commonly used chemotherapeutic agent, which produces a selective destruction of renal proximal tubule cells after bioactivation (cysteine conjugate followed by generation of a reactive platinum-thiol compound by cysteine *S*-conjugate beta-lyase). This reactive platinum-thiol binds to proteins in the mitochondria (e.g. alpha-ketoglutarate dehydrogenase and aconitase) that are in close proximity to cysteine *S*-conjugate beta-lyase (Zhang *et al.*, 2006). Similarly, methapyrilene, an antihistamine associated with hepatotoxicity and liver tumors in rats, is metabolized by CYPs to a reactive intermediate that covalently modifies mitochondrial proteins (Anderson *et al.*, 1992; Ratra *et al.*, 1998).

Sometimes a reactive metabolite can react with proteins in several cellular compartments. The classic example is the bioactivation of acetaminophen. Acetaminophen hepatotoxicity is precipitated by hepatic glutathione depletion by the reactive intermediate NAPQI which binds to and modifies many cellular proteins in most intracellular compartments of hepatocytes. Examples of enzymes inhibited by a toxic dose of acetaminophen include glyceraldehyde-3-phosphate dehydrogenase and Ca2+/Mg2+ ATPase, thus affecting ATP production and Ca homeostasis and initiating a cascade of events that ends in hepatocellular necrosis (Dietze *et al.*, 1997; Park *et al.*, 2005) Similarly, zomiperac, an NSAID, is metabolized to a chemically reactive acyl glucuronide that covalently modifies essential proteins like dipeptidyl peptidase IV and tubulin in liver in rats (Bailey *et al.*, 1998; Wang *et al.*, 2001).

The chemical stability of the drug-related reactive species is an important determinant in how far removed from the site of formation its protein adducts can be seen. There are many drugs whose reactive intermediates target not only the macromolecules in the cells that formed them, but also proteins in distant tissues. For example, NSAID acyl glucuronide metabolites are formed by UGTs in the liver, yet are stable enough to be secreted in the bile to form adducts with intestinal proteins (Atchison *et al.*, 2000; Aithal *et al.*, 2004). The reactive acyl glucuronide of diclofenac is implicated in intestinal injury in rats and through formation of protein adducts (Seitz and Boelsterli, 1998).

21.4.2 Oxidative Stress

Cellular stress and damage induced by ROS is a well-studied phenomenon and numerous compounds (e.g. acetaminophen, adriamycin, nilutamide, and bleomycin) induce cell death by this mechanism, either directly or via their reactive metabolites. The ROS (e.g. superoxide radical, hydrogen peroxide, and the highly reactive hydroxyl radical) initiate toxicity through oxidative damage of DNA, proteins, and lipids. Phospholipids, for instance, are highly unsaturated fatty acids and an integral part of the biological membranes; peroxidation of these lipids thus results in cell damage.

Because ROS are by-products of many enzyme mechanisms (e.g. CYPs and MPO), cells are equipped with defense mechanisms to counter the toxic effects of oxyradicals. These defenses include enzymes such as superoxide dismutase, catalase, glutathione peroxidase and stores of alpha-tocopherol, and reduced GSH and ascorbate. Toxicity usually results when these defenses are overwhelmed. In alcoholic liver disease, for example, the increased expression of CYP2E1 in hepatocytes leads to the generation of ROS and the subsequent depletion of GSH, which thus increases the vulnerability of cells (Jones *et al.*, 2002). Many anticancer drugs are known to cause redox cycling and thereby generate oxygen radicals that elicit cellular damage. For example, anthracycline-induced cardiotoxicity is attributed to the free radicals formed during redox cycling (Doroshow, Locker and Myers, 1980).

Hepatocytes are equipped with the ability to sense and respond to oxidative stress. Nuclear factor-erythroid 2-related factor 2 (Nrf2) is a redox-sensitive transcription factor that is normally kept inactive by Kelch-like (ECH)-associated protein 1 (Keap1), an adaptor of the proteasome (Jaiswal, 2004). Under oxidative stress, Nrf-2 ubiquitination is perturbed leading to saturation of Keap1 and newly synthesized Nrf2 translocates to the nucleus where it initiates the antioxidant response element-mediated gene transcription (Okawa *et al.*, 2006; Tong *et al.*, 2006). Nrf-2-responsive genes are enzymes involved in the metabolism of reactive intermediates (e.g. epoxide hydrolase) and phase II conjugation (e.g. GSTs) (Copple *et al.*, 2008). In the case of acetaminophen, for example, hepatic nuclear translocation of Nrf-2 was seen at nontoxic doses and before overt hepatotoxicity in mice (Goldring *et al.*, 2004). In fact, Nrf2-null mice are more susceptible to acetaminophen toxicity and hepatocyte-specific Keap1-null mice are highly resistant to doses of acetaminophen that are lethal to wild-type mice (Chan, Han and Kan, 2001; Okawa *et al.*, 2006). These suggest the critical roles of Nrf2 and Keap1 in protecting against oxidative stress. Similarly, in the case of anticonvulsant drugs felbamate, carbamazepine, phenobarbital, and phenytoin that are associated with varying degrees of effects on liver, pronounced effects on oxidative stress-reactive metabolite-responsive genes (aflatoxin B1 aldehyde reductase, epoxide hydrolase 1 and 2, DT-diaphorase, GST Ya subunit, glucose-6-phosphatase catalytic subunit) were seen in the liver of rats (Leone *et al.*, 2007).

Markers of oxidative stress that occur during the course of toxicity are demonstrated at the organ level. For example, valproic acid is an antiepileptic agent that is associated with cases of hepatotoxicity and hepatic failure in humans. When given to rats, valproic acid increased plasma and liver levels of 15-F(2t)-isoprostane, a marker of oxidative stress that precedes the onset of liver necrosis (Tong *et al.*, 2005). Similarly, a high level of plasma total peroxide concentration, indicative of oxidative stress, is seen in HIV-infected patients on protease inhibitor therapy. Oxidative stress also is implicated in the premature development of atherosclerosis and coronary heart disease in patients on antiretroviral therapy (Masiá *et al.*, 2007).

Oxidative damage to proteins and DNA is also associated with carcinogenesis and teratogenesis. For example, use of diethylstilbestrol during pregnancy was associated with adenocarcinoma of the vagina in daughters at puberty; ROS generation in embryonic tissue is one of the many hypotheses of carcinogenesis associated with diethylstilbestrol (Kodama *et al.*, 1993). Teratogenicity of thalidomide is also linked to ROS generation in embryonic tissue. Substantial DNA oxidation is seen in rabbits following thalidomide; however, pretreatment with a free radical-trapping agent can prevent the teratogenicity, supporting the hypothesis that free radicals are involved (Wells *et al.*, 1997; Parman, Wiley and Wells, 1999). Inhibition of gene expression related to DNA methylation during oxidative stress has been linked to altered function of protein kinases and transcription factors. Impaired signal transduction is one of the many factors implicated in teratogenesis and carcinogenesis (Fratelli *et al.*, 2005; López-Neblina and Toledo-Pereyra, 2006). In addition to initiating cellular damage, ROS are also implicated in the activation of alternative pathways that can contribute to adverse effects. Tamoxifen, for example, is an agent used in the treatment of breast cancer with an increased risk of thrombosis in humans. This has been attributed to reactive metabolites of tamoxifen, with *in vitro* evidence that these metabolites can increase superoxide release by platelets, thereby increasing platelet aggregation (Vitseva *et al.*, 2005).

21.4.3 Altered Balance of Cell Survival and Death Pathways

Pertubation of cellular homeostasis is a general sequel to changes initiated by functional compromise of critical cellular proteins or cellular defense systems. This ultimately results in cell death, modes of which are commonly described as apoptosis or necrosis. Numerous examples exist of toxic chemicals that can alter cellular homeostasis by affecting the balance between the cell survival and cell death pathways. A few drugs will be discussed as examples.

Apoptosis—or programmed cell death—involves activation of "death pathways," one component of which is p53, a critical death effector protein. Disulfiram, used to treat alcoholism, can induce apoptosis *in vitro*. The reactive metabolites of disulfiram that carbamoylate and inhibit aldehyde dehydrogenase also induce p53-dependent apoptosis in HepG2 cells (Liu *et al.*, 1998). Similarly, cisplatin-induced apoptosis in a dorsal root ganglion hybrid cell line is by accumulation of p53, with protection supplied by pre-incubation with antioxidants (Park *et al.*, 2000).

A portion of acetaminophen toxicity occurs via the initiation of cell death pathways, as demonstrated by some protection against the early apoptotic processes (and the later necrotic processes) conferred by neutralization of Fas ligand and tumor necrosis factor (TNF) (Blazka *et al.*, 1996; Zhang *et al.*, 2000). Neuronal loss in HIV dementia after treatment with the nucleoside reverse transcriptase inhibitor, 2′, 3′-dideoxycytidine, was linked to induction of apoptosis in neuronal cells. In gerbils, the drug reduces expression of the anti-apoptotic protein Bcl-2 and increases expression of pro-apoptotic protein caspase-3 in brain mitochondria (Opii *et al.*, 2007).

Troglitazone, a peroxisome proliferator-activated receptor gamma agonist in the treatment of type 2 diabetes, was removed from the market due to concerns over hepatotoxicity. In rat hepatocytes and HepG2 cells, troglitazone induced apoptosis (Toyoda *et al.*, 2001; Yamamoto *et al.*, 2001). Pretreatment of HepG2 cells with a c-Jun *N*-terminal protein kinase inhibitor blocked the troglitazone-induced change in pro-apoptotic protein levels with a reduction in apoptotic cell death (which suggests a role exists for signaling pathways in troglitazone-induced apoptosis) (Bae and Song, 2003).

Toxic epidermal necrolysis is a type of reactive metabolite-triggered immune-mediated cutaneous adverse reaction characterized by epidermal detachment. The incidence of such a reaction is rare but has been described for many drugs (e.g. sulfonamides, lamotrigine, carbamazepine, and nevirapine). Mechanistic studies have revealed apoptosis of keratinocytes as the critical event leading to necrosis, a classic histological feature of this condition. Unlike keratinocytes in normal, healthy skin, cells in this condition express high levels of Fas ligand at the cell surface (Viard-Leveugle *et al.*, 2003; French, 2006). Fas-mediated keratinocyte apoptosis can be inhibited *in vitro* by monoclonal antibodies to Fas and *in vivo* by immunoglobulins containing anti-Fas antibodies suggesting a prominent role for apoptosis in this condition (Tan *et al.*, 2005; French, Trent, and Kerdel, 2006).

21.4.4 Immune-Mediated

In contrast to the drugs that cause direct toxicity, including many of those described earlier, there are compounds that are able to form protein conjugates that can initiate toxicity via activation of the immune system. Well-known examples are cases of hepatitis following administration of halothane, tienilic acid, dihydralazine, or diclofenac. The incidences of such adverse effects are rare and often unpredictable, hence their designation as idiosyncratic adverse drug reactions.

The reasons for this type of reaction are far from understood and several hypotheses are proposed. The first is the "hapten hypothesis," which suggests that antibodies raised against drug/metabolite–protein (typically to the enzyme that produced the metabolite or proteins in the vicinity)—adducts seen by the immune system as "neoantigens"— subsequently recognize the modified or even the native proteins. These autoantibodies lead to cytolytic toxicity with further or repeated exposure to the same or similar drug that generated the neoantigens (Padovan *et al.*, 1997). A second hypothesis is the "danger hypothesis," which argues that the reactive metabolite causes damage to the tissue and this leads to activation of antigen-presenting cells in an immune response (Matzinger, 1994). A third hypothesis is "pharmacological interaction" which involves a complex formed between the reactive metabolite and the major histocompatibility complex class 1 (MHC-1) leading to a cell-mediated immune response (Schnyder *et al.*, 1997). Given an absence of overwhelming evidence in favor of just one hypothesis, it seems reasonable to assume that any combination of one or all and possibly other mechanisms, as yet unknown, could well be involved in the pathogenesis of immune-mediated drug toxicity.

Many drugs that initiate an immune response are demonstrated to follow the hapten or neoantigen hypothesis. Target proteins are generally localized within the cell. For example, halothane, a volatile anesthetic, is metabolized primarily by CYP2E1, generating a reactive trifluoroacetyl halide metabolite that acetylates various proteins including CYP2E1 (Spracklin *et al.*, 1997). Consequently, many patients suffering with halothane hepatitis have autoantibodies to CYP2E1 (Bourdi *et al.*, 1996). Halothane produces adducts that are almost exclusively found in the liver microsomal subfraction. These modified proteins are accessible to the immune system by exposure on the plasma membrane. Similarly dihydralazine, an antihypertensive drug associated with hepatitis, can trigger an immune response producing antibodies against CYP1A2 (Beaune *et al.*, 1993). Tienilic acid, a diuretic associated with hepatitis is metabolized by CYP2C9 to a reactive intermediate that in turn covalently modifies the same enzyme and eventually leads to the generation of antibodies to CYP2C9 (Lecoeur *et al.*, 1994). These mechanisms are not just limited to reactive metabolites formed via CYP-mediated reactions; theoretically, any reactive metabolite that can form a protein adduct is potentially able to initiate an immune response. For instance, diclofenac-induced hepatitis is also in part immune-mediated. In the rat, both CYPs and UGTs are involved in forming reactive intermediates of diclofenac, with a majority of protein adducts detected on the plasma membrane and localized within the bile canalicular membrane (Hargus *et al.*, 1994). A similar pattern of covalent binding is seen with other NSAIDs such as diflunisal, zomepirac, sulindac, and ibuprofen (Bailey and Dickinson, 1996; Wade, Kenna and Caldwell, 1997). Metabolic activation of abacavir to an aldehyde intermediate is thought to be involved in covalent binding to proteins leading to hypersensitive reactions in a small percent of patients (Walsh, Reese and Thurmond, 2002).

Haptenation in itself is not always sufficient to provoke an immune response and requires co-stimulatory signals to initiate activation of antigen-presenting cells and a subsequent immune response (which is evidence in support of the danger hypothesis). For example, the cutaneous adverse reactions seen with dapsone and sulfamethoxazole are attributed to an up-regulation of danger signals by their reactive arylhydroxylamine metabolites. In human epidermal keratinocytes, arylhydroxylamine metabolites up-regulate stress signals such as pro-inflammatory cytokines and heat shock protein 70—this can subsequently activate dendritic cells and initiate an immune reaction with the skin (Khan *et al.*, 2007).

The hapten hypothesis, and its basis in reactive metabolite formation as a trigger for immune response, was challenged when an immune response was observed for drugs such as lidocaine, lamotrigine, carbamazepine, and sulfamethoxazole but not their metabolites. The pharmacological interaction hypothesis explains this phenomenon by stating that, in this case, the drugs bind directly and reversibly to an immune receptor (e.g. the major histocompatibility complex or the T cell receptor), which then stimulates the cells in a manner similar to a pharmacological activation of the receptors (Pichler *et al.*, 2006; Roujeau, 2006).

21.4.5 Influence of Inflammatory Mediators and Other Cells on Toxicity

The pathophysiological role of inflammatory mediators in the manifestation of toxicity is well known and generally considered to be an integral part of the pathology. In the context of idiosyncratic reactions, however, substantial evidence exists that shows that animals with a low-grade inflammation develop a predisposition to drug toxicity from low dosages that otherwise are not toxic. Best known examples are acetaminophen-, ranitidine-, and chlorpromazine-associated hepatotoxicity. Experimentally, toxicities of these compounds have been shown to be precipitated by an ongoing low-grade inflammation initiated by treatment with lipopolysaccharide (LPS) (Ganey and Roth, 2001). It should be noted that these are examples of idiosyncratic toxicity and not reactive metabolite-driven adverse reactions.

Non-parenchymal cells of the liver, such as sinusoidal endothelial cells, stellate cells, and Kupffer cells, along with infiltrating neutrophils, can influence the toxicities of some drugs. Kupffer cells are the resident macrophages in the liver and during hepatic inflammation are activated to release pro-inflammatory cytokines (e.g. TNFalpha), in addition to a barrage of destructive mediators (e.g. ROS, reactive nitrogen species, and proteases) that could influence the toxicity of other compounds exposed at the same time. In the case of acetaminophen toxicity, for example, evidence that Kupffer cells play a direct role in acetaminophen toxicity is the protection seen in mice depleted of their Kupffer cells by dichloromethylene diphosphonate (Goldin *et al.*, 1996). Inflammatory mediators can also activate other cells, such as sinusoidal cells, with the subsequent expression of adhesion molecules and stellate cells causing a contraction of the sinusoids, which leads to the recruitment of neutrophils and facilitates their transmigration (Ganey *et al.*, 1994; Kharbanda *et al.*, 2004).

In addition to their hypothesized role in idiosyncratic toxicity, inflammatory mediators play a role in the precipitation of toxicity, as seen with compounds such as halothane and acetaminophen. In Balb/c mice, halothane increased the hepatic messenger levels of the inflammatory mediators TNF-alpha, IL-1beta, IL-6, and IL-8; this is in addition to recruiting neutrophils to the liver. These events suggest a role for pro-inflammatory mediators in the halothane-initiated hepatotoxicity observed in this strain of mice (You *et al.*, 2006). Similarly, an increase in macrophage migration inhibitory factor, a pro-inflammatory signal, was seen in guinea pigs with hepatotoxicity from halothane or acetaminophen. Migration inhibitory factor null mice were significantly less susceptible to acetaminophen toxicity, suggesting a role for this inflammatory mediator in the predisposition toward toxicity (Bourdi *et al.*, 2002). In addition to their role in inflammation, cytokines such as interleukins, TNF-alpha, and interferons can down-regulate CYP enzymes and thus affect the extent of bioactivation and/or clearance of drugs (Bertini *et al.*, 1988; Renton and Knickle, 1990; Abdel-Razzak *et al.*, 1993).

21.5 ORGAN SYSTEMS TOXICOLOGY

In this section, we describe the toxicity and mechanisms of several well-known drugs whose adverse reactions are mediated primarily by the formation of reactive intermediates. These drugs demonstrate distinctive specificity toward target organs and the examples are grouped accordingly.

21.5.1 Hepatotoxicity

Liver is the primary organ of metabolism of most compounds and the primary target for drug toxicity. The majority of compounds that form reactive metabolites are hepatotoxic. Apart from the innate reactivity of the reactive intermediate, its local concentration and the proximity of proteins, such as the enzyme that helped form it, are other factors that predispose toward injury of the hepatocyte. Since biliary clearance of endogenous and exogenous substrates is a key function of the liver, bile may therefore contain a relatively high concentration of reactive metabolites that can damage the cholangiocyte and, in severe cases, be seen as vanishing bile duct syndrome. The hepatotoxicities of selected drugs (e.g. acetaminophen, diclofenac, carbamazepine, valproic acid, and troglitazone), pathological features of their toxicities (e.g. hepatic necrosis, microvascular steatosis, and cholestasis), and the importance of risk–benefit analysis in the fate of drugs (e.g. bromfenac and troglitazone) are discussed in the following section. A comprehensive but certainly not exhaustive list of drugs associated with organ-specific toxicity is provided in Table 21.3. This serves to illustrate some of the issues and complexities that face scientists working in drug discovery and development.

Acetaminophen is a classic example of a drug that causes hepatotoxicity, which can present as acute liver failure and accounts for as much as 39% of the liver failure cases reported in tertiary care centers in the United States (Ostapowicz *et al.*, 2002). Acetaminophen hepatotoxicity is dose-dependent and is precipitated by loss of hepatic defenses. The pathology is characterized by hepatic necrosis, predominantly in the centrilobular region and is mediated by its reactive intermediate NAPQI. Within the therapeutic dose range, acetaminophen is detoxified by glucuronidation and sulfation; however, a portion of the drug is also metabolized primarily by CYP3A4 to the reactive NAPQI, which is immediately quenched by mitochondrial glutathione (Davis *et al.*, 1974). At toxic doses of the drug, CYP2E1, which is also ethanol inducible, plays a major role in the generation of NAPQI (CYP1A2 and 2A6 are also involved in the generation of NAPQI *in vitro*) (Patten *et al.*, 1993; Chen *et al.*, 1998). In the event of an overdose of acetaminophen, and/or situations in which glutathione is depleted, NAPQI is not quenched and instead binds covalently to critical cellular proteins. Pos-

TABLE 21.3 Drugs, putative reactive intermediates, implicated enzymes, and associated toxicity.

Drug	Putative reactive intermediate	Enzyme	Adverse reaction	References
Abacavir	Aldehyde	Alcohol dehydrogenase	Hypersensitivity	Walsh, Reese and Thurmond, 2002
Acetaminophen	Quinoneimine	CYP3A4, 2E1, 1A2, 2A6	Hepatotoxicity, nephrotoxicity	Chen et al., 1998; Patten et al., 1993; Davis et al., 1974; Hart et al., 1994
Adriamycin	Quinones, ROS	Redox cycling	Cardiotoxicity, neurotoxicity	Berthiaume and Wallace, 2007; Joshi et al., 2005
Amodiaquine	Quinoneimine	Peroxidases	Hepatotoxicity, agranulocytosis	Tingle et al., 1995; Maggs et al., 1988
Alclofenac	Epoxide	CYP	Hepatotoxicity, skin rash	Hort, 1975; Slack et al., 1981
Aminoglutethimide	Hydroxylamine	MPO	Agranulocytosis	Coleman, Khalaf and Nicholls, 2003; Siraki et al., 2007
Benoxaprofen	Acyl glucuronide?	UGT	Hepatotoxicity, photosensitivity	Halsey and Cardoe, 1982
Bromfenac	Acyl glucuronide?	UGT	Hepatotoxicity	Fontana et al., 1999
Carbamazepine	Arene oxide, quinoneimine	CYP2D, 1A2	Hepatotoxicity	Masubuchi et al., 2001
Chloramphenicol	Nitroso		Aplastic anemia	Yunis et al., 1980
Cisplatin	ROS, platinum-thiol	Redox cycling; GGT	Neurotoxicity, nephrotoxicity, ototoxicity	Townsend et al., 2003; Smyth et al., 1997
Clozapine	Nitrenium	MPO, CYP	Agranulocytosi, hepatotoxicity, myocarditis	Uetrecht, Zahid and Whitfield, 1994; Pirmohamed et al., 1995; Killian et al., 1999
Cyclophosphamide	Acrolein	CYP2B1, 2C6, 2C11	Bladder toxicity, ovarian toxicity	Stillwell and Benson, 1988; Chang and Waxman, 1993; Clarke and Waxman, 1989; Plowchalk and Mattison, 1991
Diethylstilbestrol	Quinones	CYP	Carcinogenesis	Cavalieri, Rogan and Chakravarti, 2002; Chae et al., 1998
Dapsone	Nitrenium, hydroxylamine	CYP3A4, MPO	Methemoglobinemia, agranulocytosis, cutaneous reactions	Grossman and Jollow, 1988; Uetrecht et al., 1988; Vyas et al., 2005
Daunorubicin	Quinones, ROS	Redox cycling; xanthine oxidase?	Cardiotoxicity	Gustafson, Swanson and Pritsos, 1991
Dihydralazine	Diazene or diazonium	CYP1A2	Hepatitis	Bourdi et al., 1992; Masubuchi and Horie, 1999

TABLE 21.3 *Continued*

Drug	Putative reactive intermediate	Enzyme	Adverse reaction	References
Diclofenac	Quinoneimine, arene oxide, acyl glucuronide	CYP3A4, 2C9, 2C11, MPO, UGT	Hepatotoxicity	Tang et al., 1999; Masubuchi et al., 2001; Poon et al., 2001
Erythromycin	Nitrosoalkane	CYP3A4	DDI	Periti et al., 1992
Fluperlapine	Quinoneimine	MPO	Agranulocytosis	Lai et al., 2000
Flutamide	Electrophilic intermediate?	CYP3A4, 1A2	Hepatotoxicity	Berson et al., 1993
Felbamate	2-Phenylpropenal	CYP	Aplastic anemia, hepatotoxicity	O'Neil et al., 1996; Kaufman et al., 1997; Popovic et al., 2004
Haloperidol	Pyridinium	CYP3A4	Parkinsonism, tardive dyskinesia	Usuki, Van der Schyf and Castagnoli, 1998
Halothane	Trifluoroacetyl chloride	CYP2E1	Hepatotoxicity	Spracklin et al., 1997
Hydralazine	Diazene or diazonium	CYP?, MPO	Autoimmunological syndrome	Hofstra, Matassa and Uetrecht, 1991
Ibufenac	Acyl glucuronide	UGT	Hepatotoxicity	Castillo and Smith, 1995
Irinotecan	SN-38	Esterases	Intestinal toxicity	Takasuna et al., 1998; Wadkins et al., 2004
Indomethacin	Quinoneimine	Amidase, MPO, CYP	Aplastic anemia, agranulocytosis	Ju and Uetrecht, 1998
Isoniazid	N-Acetylhydrazine	Amidase, CYP1A1, 1A2, 2E1	Hepatotoxicity	Sarich et al., 1999
Lamotrigine	Arene oxide	CYP	Skin rashes; toxic epidermal necrolysis	Maggs et al., 2000
Metronidazole	5-Aminoimidazole, hydroxylamine	CYP	Genotoxicity	Perez-Reyes, Kalyanaraman and Mason, 1980; Kedderis, Argenbright and Miwa, 1989
Methimazole	Sulfenic acid, sulfinic acid	CYP, FMO	Agranulocytosis, hepatotoxicity	Decker and Doerge, 1992; Woeber, 2002; Oh et al., 2007
Mianserin	Iminium	CYP2D	Skin rashes, hepatotoxicity, blood dyscrasias	Masubuchi, Konishi and Horie, 1999; Iverson, Zahid and Uetrecht, 2002
Nilutamide	Nitroanion, ROS	CYP450 reductase, redox cycling	Pulmonary toxicity, hepatotoxicity	Berger et al., 1992; Fau et al., 1992
Nitrofurantoin	Nitroanion, ROS	CYP450 reductase, redox cycling	Pulmonary toxicity, hepatotoxicity	Wang et al., 2008

Drug	Reactive metabolite	Enzyme	Toxicity	Reference
Procainamide	Nitrenium	CYP2D6	Lupus erythematosus; agranulocytosis	Lessard et al., 1999
Paroxetine	Catechol	CYP2D6	DDI	Bertelsen et al., 2003
Phenytoin	Unstable nitrogen-centered radical	COX, peroxidases	Teratogenesis	Wells et al., 1997; Parman, Wiley and Wells, 1999
Ritonavir	Ring scission	CYP3A4	Enzyme inactivation	Kempf et al., 1997
Raloxifene	Arene epoxide; quinone	CYP3A4	Enzyme inactivation	Yukinaga et al., 2007
Sudoxicam	Ring scission		Hepatotoxicity	Hobbs and Twomey, 1977
Suprofen	Epoxide, acyl glucuronide	CYP2C9	Nephrotoxicity	O'Donnell et al., 2003; Smith and Liu, 1993
Sulfamethoxazole	Hydroxylamine	CYP2C9	Hypersensitivity	Miller and Trepanier, 2002
Terbinafine	Aldehyde metabolite?	CYP	Hepatotoxicity	Mallat et al., 1997; Iverson and Uetrecht, 2001
Tacrine	Quinone methide	CYP1A2	Hepatotoxicity	Madden et al., 1993; Spaldin et al., 1995
Thalidomide	Unstable nitrogen-centered radical	COX	Teratogenesis	Wells et al., 1997; Parman, Wiley and Wells, 1999
Tolbutamide	n-Butylisocyanate	Degradation?	Teratogenesis	Guan et al., 1999
Tolcapone	Quinoneimine	CYP2E1, CYP1A2	Hepatotoxicity	Smith et al., 2003
Troglitazone	Quinone, quinone methide	CYP3A4	Hepatotoxicity	Kassahun et al., 2001; Yamamoto et al., 2002
Tamoxifen	Quinone methide	CYP	Carcinogenesis	Fan and Bolton, 2001
Trimethoprim	Quinone methide	CYP	Hepatotoxicity	Lai, Zahid and Uetrecht, 1999
Ticlopidine		CYP2C19, 2B6	Agranulocytosis, aplastic anemia, thrombocytopenia	Liu and Uetrecht, 2000
Tienilic acid	Tienilic acid-S-oxide	CYP2C9	Hepatitis	Melet et al., 2003
Thiabendazole	Thioformamide	CYP	Nephrotoxicity	Mizutani, Yoshida and Kawazoe, 1993
Valproic acid	4-ene	CYP2C9	Hepatotoxicity	Stephens and Levy, 1992; Kiang et al., 2006
Vesnarinone	Imidoiminium ion	CYP3A4, 2E1	Agranulocytosis	Uetrecht, Zahid and Whitfield, 1994; Frye et al., 1999
Zomepirac	Acyl glucuronide?	UGT?	Allergic reactions	Bailey and Dickinson, 1999; Olsen et al., 2005

sible mechanisms of hepatocellular necrosis include modification of sulfhydryl groups, oxidative stress, lipid peroxidation, and arylation of essential proteins. In rodents, more than 30 hepatic enzymes are shown to be covalently modified by NAPQI with many of them functionally compromised as a result (Park *et al.*, 2005).

NSAIDs as a class are associated with some degree of hepatotoxicity; however, due to the structural diversity of NSAIDs, the toxicity presented is not uniform. For example, a higher incidence of liver toxicity is reported for nimesulide compared with diclofenac or ibuprofen (Traversa *et al.*, 2003). Diclofenac is associated with hepato-toxicity with a low incidence rate of 6 to 18 cases/100,000 person-years (Walker, 1997). The pattern of toxicity is predominantly hepatocellular and to a lesser extent choles-tatic (Banks *et al.*, 1995). Histologically, acute lobular hepatitis and in severe cases bridging necrosis and fibrous expansion of portal tracts are observed (Chitturi and George, 2002). Both immunological and nonimmunological mechanisms are proposed for diclofenac hepatotoxicity. UGT-mediated glucuronic acid conjugation of diclofenac results in an acyl glucuronide that forms protein adducts either directly or after acyl migration, forming a reactive aldehyde. The mechanisms of acyl glucuronidation and reactivity are shown in Fig. 21.1. The reactive acyl glucuronide of diclofenac is shown to form covalent adducts with hepatic and circulating proteins. A similar pattern of covalent binding was reported for zomepirac, diflunisal, sulindac, and ibuprofen (Bailey and Dickinson, 1996; Wade, Kenna and Caldwell, 1997). Adverse reactions with diclof-enac are not limited to formation of a reactive acyl glucuronide since diclofenac also undergoes CYP3A4 and CYP2C9 oxidation to metabolites such as 5-hydroxydiclofe-nac that are processed further by oxidation to form a reactive iminoquinone. CYPs and MPO are implicated in the formation of microsomal protein adducts and postu-lated to play a role in hepatotoxicity (Hargus *et al.*, 1994; Miyamoto, Zahid and Uetrecht, 1997; Tang *et al.*, 1999; Poon *et al.*, 2001). *In vitro*, an arene oxide intermedi-ate generated during the process of diclofenac hydroxylation is shown to inactivate CYP2C11 (Masubuchi, Ose, and Horie, 2001). Furthermore, covalent modification of proteins on the cell surface is possibly involved in an immune-mediated hepatitis associated with diclofenac therapy.

Carbamazepine is an anticonvulsant linked to a variety of understood and idiosyn-cratic toxicities, including skin rash, hepatotoxicity, and hematotoxicity, with adverse reactions reported in 33–50% of patients (reviewed in Kalgutkar *et al.*, 2005). Carbamazepine is metabolized to 2-hydroxycarbamazepine and subsequently to 2-hydroxyiminostilbene and finally to a putative reactive iminoquinone, primarily via CYP3A4 metabolism (Pearce, Uetrecht and Leeder, 2005). Time-dependent inactivation of CYP2D in rat liver microsomes and CYP1A2 in human liver microsomes was observed which, although not conclusive, does implicate these enzymes as directly involved in the formation of reactive metabolites of carbamazepine (Masubuchi *et al.*, 2001).

Valproic acid is another anticonvulsant associated with a large incidence of hepa-totoxicity in patients. An asymptomatic dose-dependent increase in serum markers of hepatotoxicity was seen in about 44% of patients on valproic acid therapy and non-dose-dependent liver failure in rare instances (Parra, Iriarte and Pierre-Louis, 1996). Reactive intermediates of valproic acid, such as the 4-ene and 2,4-diene, inhibit fatty acid metabolism by alkylation of enzymes such as 3-ketoacyl-coenzyme A thiolase. Hepatotoxicity is therefore linked to inhibition of mitochondrial beta-oxidation by its metabolites, as demonstrated *in vitro* and *in vivo*. Consequently, liver injury is charac-terized by microvesicular steatosis and hepatocellular necrosis (Kassahun, Farrell and Abbott, 1991; Stephens and Levy, 1992; Kossak *et al.*, 1993).

Isoniazid (antituberculosis agent) causes hepatotoxicity, usually seen as massive necrosis or chronic hepatitis and is related to dose and duration of exposure. Mild elevations of plasma liver enzyme activities have been reported in 20% of the patients (Garibaldi *et al.*, 1972). The mechanism of isoniazid hepatotoxicity is attributed to its metabolism to an intermediate acetylhydrazine, whose bioactivation in turn leads to the formation of reactive acetylating species that can covalently modify hepatic proteins (Timbrell *et al.*, 1980).

Troglitazone (Rezulin), a peroxisome proliferator-activated receptor gamma agonist, effective in the treatment of type 2 diabetes, was associated with cases of severe hepatotoxicity and liver failure that led to its withdrawal from the US market. Troglitazone-associated liver injury, in some instances continued despite discontinuation of troglitazone treatment (Menon, Angulo and Lindor, 2001). Histologically massive necrosis with post-collapse scarring and bile duct proliferation were seen (Gitlin *et al.*, 1998; Vella, de Groen and Dinneen, 1998). When introduced, it was considered a novel drug with a unique mechanism of action that benefited certain patients. The withdrawal was therefore not immediate but was subsequent to the introduction to the market of rosiglitazone and pioglitazone, considered safer drugs in the same class (Lee, 2003). Troglitazone is metabolized by CYP3A4 to several metabolites, one of which is a reactive *O*-quinone methide derivative, which suggests a role for bioactivation (Kassahun *et al.*, 2001; Yamamoto *et al.*, 2002). The actual mechanism of hepatotoxicity of troglitazone is, however, unclear, with several factors apparently involved—including bioactivation, mitochondrial dysfunction by troglitazone itself, and inhibition of bile salt export pump by troglitazone sulfate (Masubuchi, 2006).

Lastly, bromfenac, an NSAID approved as a short-term analgesic for orthopedic pain for use for periods of 10 days or less, is another example that illustrates the importance of risk–benefit analysis. Bromfenac was withdrawn from the market within a year of release following the many reports of severe liver injury (all in patients taking the drug for more than 30 days) (Fontana *et al.*, 1999; Lee, 2003). Although the NSAID class has toxicity issues (to a varying degree) possibly due to reactive intermediates such as acyl glucuronides, the precise mechanism by which bromfenac causes hepatotoxicity is not clear.

21.5.2 Hematotoxicity

Blood cells and hematopoietic tissue is, besides the liver, the organ system most exposed to chemicals. Neutropenia and agranulocytosis (a common manifestation of toxicity to neutrophils/its precursors), methemoglobinemia and anemia (effects reflecting toxicity to erythrocytes/its precursors), and thrombocytopenia (toxicity to platelets/its precursors) are adverse reactions often reported for drugs that are known to form reactive intermediates. Reactive intermediates that are formed in the liver are to some extent responsible for toxicity to the blood cells; however, blood cells such as neutrophils can also metabolize drugs that escape first-pass metabolism and it is this mechanism that contributes most to the majority of examples of agranulocytosis described in the literature (Uetrecht, 1992).

As described earlier, MPO is the major oxidizing enzyme in neutrophils, with hypochlorous acid (HOCl) the major oxidant produced by activated neutrophils. The MPO–HOCl system is involved in generation of reactive intermediates for the vast number of compounds that cause agranulocytosis (Uetrecht, 1992). For example, clozapine, an atypical antipsychotic drug is associated with neutropenia and agranulocytosis in human. The toxicity of clozapine is attributed to a reactive nitrenium ion metabolite formed by MPO in neutrophils (Williams *et al.*, 1997). Similarly, fluperlapine, another antipsychotic structurally related to clozapine, is also associated with agranulocytosis. The major metabolite of fluperlapine, 7-hydroxyfluperlapine, is converted to a reactive iminoquinone by neutrophil MPO and is the likely proximate cause of toxicity (Lai *et al.*, 2000). Carbamazepine-induced agranulocytosis is also linked to reactive metabolites, in this case the metabolite 9-acridine carboxaldehyde, which is generated by MPO in neutrophils and monocytes (Furst and Uetrecht, 1995). Bioactivation of amodiaquine, tebuquine, cycloquine, and pyronaridine to respective chemically reactive quinoneimine metabolites occurs in the presence of horseradish peroxidase and hydrogen peroxide, suggesting a mechanism for the agranulocytosis seen with some of these drugs (Naisbitt *et al.*, 1998). Diclofenac is also associated with bone marrow toxicity, with a reactive iminoquinone formed by HOCl or by activated neutrophil oxidation of an intermediate 5-hydroxydiclofenac (Miyamoto, Zahid and Uetrecht, 1997). Similarly, indomethacin also demonstrates agranulocytosis associated with a reactive iminoquinone formed by activated neutrophils, in this case via oxidation of its major metabolite desmethyldeschlorobenzoylindomethacin (Ju and Uetrecht, 1998). Trimethoprim, an antibacterial drug used synergistically with sulfonamides, is associated with agranulocytosis—a reactive pyrimidine iminoquinone formed in activated neutrophils may be responsible (Lai, Zahid and Uetrecht, 1999). Similarly, agranulocytosis induced by vesnarinone, an agent used in the treatment of congestive heart failure, is linked to a reactive quinoneimine formed by neutrophils or neutrophil precursors in bone marrow (Uetrecht, Zahid and Whitfield, 1994). Activated neutrophils are also shown to metabolize phenytoin to reactive intermediates that bind irreversibly to macromolecules in neutrophils (Mays *et al.*, 1995).

There are some examples that link reactive metabolite formation to toxicity to erythrocytes or its precursors leading to methemoglobinemia and anemia. Ticlopidine, an ADP-receptor antagonist used in the prevention of atherothrombosis, is associated with incidences of agranulocytosis and aplastic anemia, with a putative thiophene-*S*-chloride formed by HOCl or activated neutrophils implicated (Liu and Uetrecht, 2000). Dapsone, a drug used to treat leprosy, is associated with methemoglobinemia and agranulocytosis, with CYP3A4 and/or MPO forming a hydroxylamine metabolite that is toxic to erythrocytes and leukocytes (Zuidema, Hilbers-Modderman and Merkus, 1986; Grossman and Jollow, 1988; Uetrecht *et al.*, 1988). Nomifensine is an antidepressant drug that was withdrawn from the market due to allergic reactions including hemolytic anemia. The arylhydroxylamine metabolite and nitroso metabolites of nomifensine are thought to be responsible for its toxicity (Salama and Mueller-Eckhardt, 1985; Lindberg and Syvälahti, 1986). The arylhydroxylamine metabolite of phenacetin (*N*-hydroxyphenetidine) through direct actions in the erythrocytes is responsible for phenacetin-induced hemolytic anemia and methemoglobinemia (Jensen and Jollow, 1991). Similarly, the arylhydroxylamine metabolite, 6-methoxy-8-hydroxylaminoquinoline of the antimalarial drug primaquine, is toxic to erythrocytes (Bolchoz *et al.*, 2001). Aplastic anemia associated with the use of the antibiotic chloramphenicol is believed to be due to its nitroso metabolite that is formed in the bone marrow (Yunis *et al.*, 1980). Lastly, aplastic anemia reported with felbamate toxicity is thought to be immune-mediated and associated with its reactive metabolite, 2-phenylpropenal, a potent immunogen (Popovic *et al.*, 2004).

Drug-induced thrombocytopenia can result from either a decrease in platelet production (e.g. from bone marrow toxicity) or destruction of platelets (e.g. from an immune response). Antibodies that are drug-dependent, hapten-dependent, and platelet-reactive are involved in drug-induced thrombocytopenia (van den Bemt, Meyboom and Egberts, 2004). Quinine and sulfonamides are examples of drugs that induce immune thrombocytopenia. Aminoglutethimide is an aromatase inhibitor used in the treatment of breast cancer. Thrombocytopenia is one of the adverse reactions seen with this drug and possibly due to its hydroxylamine metabolite (Stuart-Harris and Smith, 1984; Dalrymple and Nicholls, 1988).

21.5.3 Gastrointestinal Toxicity

Gastrointestinal toxicity, primarily of the intestines and as a consequence of reactive intermediates, is described for many compounds. Intestinal toxicity occurs more frequently from reactive metabolites formed in the liver and then secreted in bile than from direct bioactivation within enterocytes (Treinen-Moslen and Kanz, 2006). This is linked directly to the significant difference that exists between enterocytes and hepatocytes in their metabolic capacity at the organ level. It should be recognized that orally ingested drugs can have very high local concentrations in the intestine and toxicity can occur directly as a result of toxicodynamics related to the drug itself.

Carboxylic acid-containing NSAIDs may be extensively metabolized by UGTs in the liver to produce reactive acyl glucuronides that are secreted in the bile and are implicated in NSAID enteropathy. One example is diclofenac, which was shown to cause ulceration significantly in the small intestine of rats given bile from diclofenac-treated rats compared with rats that received diclofenac alone. Further, hepatocanalicular conjugate export pump deficient (TR-) rats were refractory to intestinal injury by diclofenac and therefore intestinal injury of diclofenac therapy was linked to reactive acyl glucuronide metabolites of diclofenac secreted in the bile (Seitz and Boelsterli, 1998). Similarly, the reactive acyl glucuronide metabolite of the immunosuppressant drug mycophenolic acid (MPA) is thought to contribute to intestinal injury. Although the reactive acyl glucuronide is formed to a lesser extent than the more stable ether glucuronide metabolites, significant levels are still reported in MPA-treated patients (Shipkova *et al.*, 2002). Furthermore, the reactive metabolite of MPA is also formed in human intestinal microsomes, suggesting a possible role for *in situ* formation of this metabolite in intestinal injury (Shipkova *et al.*, 2001).

Irinotecan is a semisynthetic derivative of camptothecin used in cancer chemotherapy and another example of a drug associated with intestinal injury. As described earlier, irinotecan and its more potent metabolite, SN-38, cause irreversible DNA damage and cell death and concomitant anticancer activity. Probably the same mechanisms are also involved in the intestinal toxicity. Irinotecan administered intravenously is bioactivated by carboxylesterase in the liver to SN-38, which in turn undergoes glucuronidation to an acyl glucuronide. Glucuronidated SN-38, unconjugated SN-38, and unchanged irinotecan are secreted in the bile into the intestine, where more SN-38 is formed by two processes: (i) intestinal carboxylesterase metabolism of irinotecan and (ii) deconjugation of glucuronidated SN-38 by bacterial beta-glucuronidase (Takasuna *et al.*, 1998; Wadkins *et al.*, 2004). SN-38 is also delivered to intestine by direct secretion across intestinal exporters (Itagaki *et al.*, 2005). Collectively, it seems that exposure of the intestines to SN-38 is most likely the cause of intestinal toxicity associated with irinotecan therapy, with interindividual variation in the expression of the enzymes involved potentially linked to the relative susceptibility of patients to adverse drug reactions.

21.5.4 Kidney and Bladder Toxicity

The kidneys are similar to the intestines in that they are not only exposed to a wide array of metabolites generated in the liver and excreted (in this case in the urine), but are also capable of metabolizing drugs *in situ*. Many DMEs including CYPs (e.g. CYP1A1, 2D6, and 3A4), prostaglandin synthetase, GST, and epoxide hydrolase are expressed in kidneys, although their distribution varies markedly between different cell types (Korashy, Elbekai and El-Kadi, 2004). In addition to effects related directly to the unchanged drug itself, renal toxicity can therefore occur either as a consequence of metabolism in the kidneys (especially in cells of the proximal renal tubules), or as a secondary event from exposure to drug metabolites formed in the liver.

An example that shows a role for liver metabolism in renal toxicity is methoxyflurane, a general anesthetic that undergoes extensive CYP-mediated metabolism in the liver, producing fluoride ions that cause renal toxicity by inhibiting chloride transport in the ascending loop of Henle (Mazze, Trudell and Cousins, 1971; Roman *et al.*, 1977). Another example is cyclophosphamide, which causes hemorrhagic cystitis and increases the risk of bladder cancer. Cyclophosphamide is metabolized in the liver by CYPs to a metabolite that undergoes spontaneous decomposition to yield phosphoramide mustard and acrolein (Sladek, 1988). Acrolein is a toxic fragment excreted in the urine and responsible for bladder injury following prolonged exposure to high concentrations in urine (Stillwell and Benson, 1988).

There are a few drugs that illustrate how *in situ* metabolism in the renal cells can cause toxicity. Efavirenz, a widely prescribed non-nucleoside reverse transcriptase inhibitor for the treatment of HIV infection, produces renal tubular epithelial cell necrosis in rats but not in humans. Detailed studies revealed that rats produce a unique glutathione conjugate that is further processed by the renal gamma-glutamyltranspeptidase (GGT) to yield reactive metabolites and renal toxicity. Pretreatment with acivicin, a potent and selective inhibitor of GGT, prevented or reduced the nephrotoxicity of efavirenz in rats, which supports the hypothesis that renal processing of glutathione conjugate is responsible for the adverse drug reactions (Mutlib *et al.*, 2000). Cisplatin is another example of a drug that undergoes bioactivation in kidneys and treatment selectively destroys renal proximal tubule cells. The glutathione conjugate of cisplatin is further processed in kidney proximal tubular cells by enzymes, GGT, aminopeptidase, and cysteine *S*-conjugate beta-lyase, sequentially to a cysteinyl-glycine-conjugate, a cysteine *S*-conjugate and finally to a reactive platinum-thiol compound that is likely responsible for the kidney-specific toxicity (Townsend *et al.*, 2003). Pretreatment with acivicin or aminooxyacetic acid (an inhibitor of cysteine *S*-conjugate beta-lyase) protected mice from cisplatin nephrotoxicity (Townsend and Hanigan, 2002). We should dispel the notion that toxicity is always as simplistic as a reactive metabolite functionally inactivating a single essential target molecule—clearly other parallel processes can be involved. In the case of cisplatin, for instance, the selective deposition of platinum in mitochondria in kidneys, as seen in mice, along with a selective inactivation of alpha-ketoglutarate dehydrogenase enzyme complex, as demonstrated in LLC-PK1 cells, could also contribute to renal injury from cisplatin therapy (Zhang *et al.*, 2006). Acetaminophen is also associated with nephrotoxicity. Bioactivation of acetaminophen to the reactive intermediate NAPQI by renal CYP 2E1 and/or processing of glutathione conjugates of acetaminophen by renal GGT is implicated in its toxicity (Hart *et al.*, 1994).

21.5.5 Pulmonary Toxicity

The lung consists of various cell types, of which type II alveolar epithelial cells, pulmonary alveolar macrophages, capillary endothelial cells, and non-ciliated bronchiolar epithelial cells (Clara cells) are especially important as they express many enzymes of drug metabolism. Esterases, acid phosphatases, CYPs, and conjugating enzymes are all found in type II alveolar epithelial cells. Levels of CYPs are substantially higher ($\leq 10\times$) in Clara cells than in other pulmonary cell types (reviewed in Gram, 1997).

Nitrofurantoin, an anti-infective used to treat urinary tract infections, with chronic use is associated with pneumotoxicity, as evidenced by cough, dyspnea, infiltrates in lung, and pulmonary fibrosis (Witten, 1989). It displays covalent binding *in vitro*, to a greater extent under anaerobic conditions than under the aerobic conditions necessary for CYP reactions. When nitrofurantoin is incubated with lung microsomes and NADPH under aerobic conditions, large amounts of superoxide are formed. Recombinant cytochrome P450 reductase was found to support redox cycling of nitrofurantoin and to generate ROS. The evidence collectively points to the formation of ROS (e.g. superoxide, H_2O_2, and hydroxyl radical) as the cause of pneumotoxicity from nitrofurantoin, rather than a result of protein covalent binding from putative reactive intermediates (Martin, 1983; Wang *et al.*, 2008). Nilutamide, another nitroaromatic drug (used in the treatment of prostatic carcinoma), also causes pulmonary interstitial fibrosis by ROS formation (Seigneur *et al.*, 1988; Berger *et al.*, 1992). *In vitro*, nilutamide is reduced by rat lung microsomes NADPH-cytochrome reductase into a nitroanion free radical which under aerobic conditions further undergoes redox cycling with the generation of ROS (Berger *et al.*, 1992).

Bleomycin, an antineoplastic agent used in the treatment of lymphomas and squamous cell tumors in humans, is limited by a dose-dependent pneumonitis often progressing to interstitial pulmonary fibrosis. The cytotoxicity of bleomycin is thought to be due to single- and double-strand cleavage of DNA by the bleomycin-Fe(II)-oxygen complex (Scheulen *et al.*, 1981). *In vivo*, bleomycin is inactivated by bleomycin hydrolase, whose level of activity is inversely correlated with the susceptibility of tissue or species to bleomycin toxicity. For example, mice lack bleomycin hydrolase activity in lung and hence are susceptible to toxicity, in contrast to rabbits that have high levels of bleomycin hydrolase activity and are relatively resistant to lung damage. The lung and skin, target sites for bleomycin toxicity, are low in bleomycin hydrolase activity compared with resistant tissues such as the liver and kidney (Lazo and Humphreys, 1983; Sebti, DeLeon and Lazo, 1987).

21.5.6 Neurotoxicity

Neurons are capable of metabolism and hence susceptible to toxicity from drugs that are bioactivated. It should be noted that such an occurrence is relatively rare given the predominance of hepatic metabolism, as well as the limited access to the central nervous system imposed by the blood–brain barrier when functioning under normal physiological conditions. Haloperidol, a neuroleptic agent, is associated with adverse reactions that include Parkinsonism and tardive dyskinesia. The intermediate pyridinium metabolite, 4-(4-chlorophenyl)-1-(4-fluorophenyl)-4-oxobutylpyridinium [HPP(+)], is formed by CYP3A4 from haloperidol and can be detected in the urine and brain of humans on haloperidol therapy. HPP+ is a potent inhibitor of complex I of the mitochondrial respiratory chain and causes ATP depletion with subsequent neurotoxicity (Subramanyam *et al.*, 1991; Eyles *et al.*, 1997; Avent, DeVoss and Gillam, 2006).

Additionally, ROS-mediated toxicities are also described for neurons. For example, peripheral neuropathy is seen in some patients on isoniazid therapy that is attributed primarily to the perturbation of vitamin B6 metabolism; however, some of the intermediates of isoniazid metabolism are also thought to contribute to neuropathy via formation of ROS (Sanfeliu, Wright and Kim, 1999). Similarly, induction of ROS via redox cycling of quinone-containing anthracycline adriamycin is implicated in somnolence or chemobrain (loss of memory or executive functions) in patients on chemotherapy—since ROS-mediated protein oxidation and lipid peroxidation is seen in the brain of treated mice (Joshi *et al.*, 2005). Cisplatin is associated with sensory neuropathy in patients on chemotherapy. The neuropathy induced by cisplatin can be explained by apoptosis of neurons in the dorsal root ganglion induced by oxygen radicals, as seen by accumulation of p53 and Fas/Fas-L proteins; since antioxidants can effectively protect the neurons from cisplatin-induced neuropathy, ROS are implicated in the pathogenesis (Smyth *et al.*, 1997; Gill and Windebank, 1998).

21.5.7 Reproductive Organ Toxicity

Spironolactone is a diuretic associated with inhibition of steroidogenesis in testes. Its metabolite, 7-alpha-thiospironolactone, is converted to a reactive intermediate that results in destruction of testicular CYPs. SU-10603, an inhibitor of 17-alpha-hydroxylase, prevented degradation of testicular CYP by 7-alpha-thiospironolactone, which indicates a prominent role for 17-alpha hydroxylase in the formation of a reactive intermediate (Menard *et al.*, 1979; Kossor *et al.*, 1992).

Cyclophosphamide is an ovarian toxicant that is shown to target dormant primordial ovarian follicles in animals and induce apoptosis of human granulosa cell line (Plowchalk and Mattison, 1991; Tsai-Turton *et al.*, 2007). Phosphoramide mustard, a metabolite of cyclophosphamide, is likely involved in this toxicity and is toxic *in vitro* to mouse ovary cultures as well as *in vivo* in mice. Acrolein, another reactive fragment from cyclophosphamide metabolism, was not found to cause ovarian toxicity in mice (Plowchalk and Mattison, 1991; Desmeules and Devine, 2006).

21.5.8 Adverse Drug Reactions in Skin

Cutaneous adverse drug reactions are seen with many compounds and present symptomatically as a broad spectrum, ranging from minor rash to epidermal necrosis. Stevens–Johnson syndrome and toxic epidermal necrolysis are two extreme conditions of adverse drug reactions, both characterized morphologically by apoptosis and the rapid onset of keratinocyte cell death. Use of therapeutic drugs was reported in over 95% of cases with toxic epidermal necrolysis and in about 50% of cases with Stevens–Johnson syndrome (French, 2006). More than 100 drugs, including sulfonamides, quinolones, cephalosporins, and tetracyclines are associated with these conditions (French, 2006). Generally, these reactions are considered to be immune-mediated and initiated by the modification of proteins by reactive intermediates of drugs. Keratinocyte apoptosis mediated by Fas is the pathological feature of this condition. Intravenous immunoglobins containing anti-Fas antibodies has been used with some success in treating toxic epidermal necrolysis (Tan *et al.*, 2005; French, Trent, and Kerdel, 2006). Reactions with sulfonamides are attributed to its reactive metabolites and the resultant oxidative stress in cells. Intracellular protein adducts were seen in human dermal fibroblasts incubated with sulfamethoxazole, for instance, which suggests a mechanistic role for reactive intermediates in the observed cutaneous toxicity (Bhaiya *et al.*, 2006). Similarly, dapsone is associated with cutaneous drug reactions via its bioactivation to a reactive arylhydroxylamine metabolite in human epidermal keratinocytes (Vyas *et al.*, 2005).

The relative expression of CYPs in the skin is notably distinct from that in the liver, with inducible forms such as CYP1A1, 1B1, and 2E1 much more dominant even at basal levels (Bergström *et al.*, 2007). The metabolic reaction phenotype of a drug, especially when it involves the generally regarded "lesser" hepatic enzyme forms, should always be considered whenever adverse reactions are encountered and marked by a distinctive tissue specificity.

21.5.9 Other Organ Systems

Any organ or cell system can be a target for toxicity (usually involving the pharmacology of the drug), although the majority of cases in drug development will involve the aforementioned tissues and organs. A few examples that describe metabolism-related adverse drug reactions and toxicity in other organs now follow.

Ototoxicity is one of the limiting side effects of cisplatin and is characterized by bilateral high-frequency hearing loss in patients (Anniko and Sobin, 1986). Oxidative stress is implicated in cisplatin ototoxicity. In the rat, chochlear damage induced by cisplatin is associated with the generation of ROS and probably nitric oxide, as inferred from up-regulation of an inducible form of nitric oxide synthase in damaged cochlear tissue (Kopke *et al.*, 1997; Li, Liu and Frenz, 2006). Protection from ototoxicity is seen in rats given cytoprotective antioxidants, like D- or L-methionine, which further supports the hypothesis that oxidative stress is responsible for cisplatin ototoxicity (Li *et al.*, 2001).

In addition to its hematotoxicity, clozapine is also associated with myocarditis. This toxicity is likely a result of a reactive nitrenium ion metabolite, such as formed by microsomes prepared from cardiac tissue (Iverson, Zahid and Uetrecht, 2002). Anthracycline quinine-containing anticancer agents adriamycin and daunorubicin (anthracyclines) exert their therapeutic effect by generating ROS via quinone-hydroxyquinone redox cycling. Toxicity is therefore dependent on the tissue's ability to defend against ROS. For example, cardiotoxicity seen with anthracycline therapy is attributed to an inability of the myocytes to defend against the redox cycling (Doroshow, Locker and Myers, 1980; Berthiaume and Wallace, 2007).

Spironolactone can inhibit steroidogenesis in adrenal cortex leading to toxicity (Menard *et al.*, 1979). The mechanism involves a series of enzymes with *in situ* metabolism of spironolactone in the adrenal gland. Enzymes in adrenal microsomes deacetylate spironolactone to form 7-alpha-thiospironolactone, which is subsequently metabolized by 17-alpha-hydroxylase to form a reactive entity that covalently modifies and inhibits CYP and 17-alpha hydroxylase and their functional role in steroidogenesis (Colby *et al.*, 1991; Kossor *et al.*, 1991).

21.6 CARCINOGENESIS

Carcinogenesis is a widely studied and specialized toxicity. Significant time and resources are expended during the drug development process to assess the potential of drug candidates to cause cancer. The mechanisms of bioactivation and DNA adduct formation are well studied for carcinogens such as aflatoxin. Fortunately, relatively few examples exist of drugs used in therapy that demonstrate carcinogenesis mediated through metabolic activation. Screening for this toxicity is a requisite part of drug development and the Ames assay with and without liver S9 for bioactivation helps to eliminate many drugs that would otherwise pose a risk to humans.

Tamoxifen is a nonsteroidal antiestrogen used in the treatment of breast cancer. Tamoxifen is clastogenic in lymphoblastoid cell line expressing CYPs and hepatocarcinogenic in animal models, including rats (Dragan *et al.*, 1994; Styles *et al.*, 1994). However, even though tamoxifen is usually given as long-term therapy, there is no conclusive evidence of hepatocarcinogenesis in women. The apparent species difference seen in the toxicity of tamoxifen appears to be explained by differences in how animals and humans biotransform and eliminate the drug and its metabolites. Tamoxifen is metabolized to an alpha-hydroxylated intermediate that is subsequently either conjugated by glucuronidation, or converted to a genotoxic sulfate metabolite that collapses to a reactive carbocation that forms DNA adducts (Martin *et al.*, 1995; Park *et al.*, 2005). *In vitro*, rat liver enzymes favor the sulfonation reaction, thereby bioactivating tamoxifen, whereas human liver enzymes favor the glucuronidation pathway and elimination (Boocock *et al.*, 1999, 2000). The difference in relative rates of hepatic bioactivation and bioinactivation of tamoxifen in rats and humans seems to explain the difference in susceptibility of rodents to a genotoxic insult. This provides a good example of how species can differ in their susceptibility to toxicity and, if evidence can be provided that the mechanisms involved are specific to an animal species and not present in humans, it is possible that a drug candidate can progress in clinical trials and gain marketing approval. Although there is not enough evidence for hepatocarcinogenesis, there is an association with long-term tamoxifen therapy and an increased incidence of endometrial cancer in patients (Colacurci *et al.*, 2000). The precise mechanism of carcinogenesis in the endometrium and the relative susceptibility is unknown, but with metabolic activation seen in explant cultures of human endometrium it is speculated that bioactivation is involved, similar to that involving liver CYPs (Sharma, Shubert, Sharma *et al.*, 2003).

Estrogens are known to cause cancer (e.g. breast cancer). Natural estrogens (estrone and estradiol) have been shown to form quinone metabolites that can react with DNA and initiate carcinogenesis. It should, however, be noted that estrogen receptor-mediated tumor promotion is a better understood mechanism of carcinogenesis. Diethylstilbestrol, a synthetic nonsteroidal estrogen, was used in the United States until the 1970s to prevent threatened miscarriage. It worked by stimulating placental synthesis of progesterone and estrogen. Maternal use of diethylstilbestrol prior to gestational age of 18 weeks was associated later with incidences (<0.1%) of vaginal clear cell adenocarcinoma of female offspring at puberty and non-adenocarcinoma reproductive tract problems in an estimated 95% of female offspring (Herbst, Ulfelder and Poskanzer, 1971; Melnick *et al.* 1987). A genotoxic epoxide intermediate of diethylstilbestrol formed by fetal CYPs during gestation is thought to be responsible for the subsequent carcinogenesis at puberty (Balling *et al.*, 1985).

The anticonvulsant phenytoin (diphenylhydantoin) is genotoxic *in vitro*. Its major nontoxic metabolite, 5-(p-hydroxyphenyl)-5-phenylhydantoin (HPPH), is conjugated by UGTs and eliminated; however, it is also bioactivated by peroxidases such as prostaglandin H synthase to form a reactive intermediate that initiates DNA oxidation and genotoxicity (e.g. micronuclei formation in rat skin fibroblasts) (Kim *et al.*, 1997). An individual deficiency in UGT activity (e.g. via depletion of UDPGA and/or polymorphism) therefore plays an important role in a patient's susceptibility to phenytoin genotoxicity (Kim *et al.*, 1997).

Nitrofurazone is an antimicrobial drug used in veterinary medicine and a carcinogen in experimental animals and also a suspected human carcinogen (IARC, 1990). The carcinogenesis of nitrofurazone possibly results from the initiation of redox cycling during metabolic activation, with generation of ROS and resultant oxidative damage to the DNA (Hiraku *et al.*, 2004). Hydroxylamine derivatives can also form DNA adducts that may play a role in carcinogenesis (Fan, Schut and Cytotoxicity, 1995).

21.7 TERATOGENESIS

Teratogenesis is another widely studied and specialized toxicity, with a major effort and investment made in the drug development process to assess the potential of drug candidates to cause birth defects. The expressions of DMEs in the fetus and embryo are notably different to that seen in adult humans and a developmental role is suggested (e.g. NAT1 expression in the fetus is implicated in neurulation, with deficiencies associated with cleft palate formation). There is a marked ontogeny that can factor significantly in bioactivation and the mechanism involved in adverse drug reactions. CYP3A7, for instance, is expressed as the major CYP form in the human fetus, but within days of birth it is virtually absent in the liver of neonates. The liver of human fetus or embryo has the capacity to metabolize compounds, and other fetal extrahepatic tissues such as adrenal gland also express CYP enzymes. The human placenta also expresses CYPs and contributes to the metabolism of xenobiotics, albeit to a minor extent (Hakkola *et al.*, 1998). Collectively, the feto-placental unit contributes to the bioactivation of several known teratogens. Although there are many compounds—including drugs—that are known to be teratogenic, relatively few undergo bioactivation to embryotoxic reactive intermediates. Carbamazepine, phenytoin, and tolbutamide are presented as examples of such teratogens.

Carbamazepine when used during pregnancy is associated with fetal malformations such as craniofacial defects, spina bifida, and fingernail hypoplasia (Rosa, 1991). The teratogenicity of carbamazepine is associated with the metabolite carbamazepine-10,11-epoxide and also the oxidation of the epoxide or carbamazepine at positions on the aromatic ring that can form putative reactive arene oxide or quinone intermediates (Bennett *et al.*, 1996). Coadministration of stiripentol, a broad spectrum CYP inhibitor, protects mice fetus from carbamazepine effects and provides evidence that CYP bioactivation is required for teratogenesis (Finnell *et al.*, 1995).

Phenytoin is teratogenic in animals and human. The teratogenicity of phenytoin is associated with its bioactivation to embryotoxic reactive intermediates. As discussed previously, the major nontoxic metabolite of phenytoin, HPPH, is conjugated by UGTs and eliminated, thus offering protection from alternative pathways including bioactivation. HPPH is also bioactivated by peroxidases (e.g. PHS) to form a reactive intermediate that is genotoxic and implicated in teratological initiation (Kim *et al.*, 1997). UGTs play a role in protecting against the teratogenic effects of phenytoin.

Thalidomide is now used as an immunomodulatory agent but when originally introduced, it was as a sedative/hypnotic (used as a sleeping aid and to ameliorate morning sickness in pregnant women). It was tragic that thalidomide use was linked to numerous cases of birth defects—estimated to be as high as 7000 worldwide (Lenz, 1966). Thalidomide is metabolized by PHS to a teratogenic intermediate as demonstrated by significant reduction in birth defects when rabbits were pretreated with acetylsalicylic acid, an inhibitor of PHS (Arlen and Wells, 1996). One of the proposed molecular mechanisms of thalidomide teratogenicity is the oxidative damage to cellular macromolecules in the embryo. Thalidomide causes a substantial increase in embryonic DNA oxidation in rabbits but not in mice, which do not exhibit thalidomide-induced teratogenicity. When rabbits are pretreated with a free radical spin-trapping agent, alpha-phenyl-N-tert-butylnitrone, prior to thalidomide administration, the embryonic DNA oxidation is blocked and teratogenicity is not observed (Wells *et al.*, 2005).

As discussed earlier, glutathione is present in cells usually at relatively high concentrations and serves to protect tissues from oxidative stress. This nucleophile must be continuously replenished by reduction of oxidized glutathione by glutathione reductase. If glutathione homeostasis is compromised, it can lead to an increased risk of teratogenicity (Hales and Brown, 1991). Tolbutamide, for instance, is a hypoglycemic agent used in the treatment of non-insulin-dependent diabetes mellitus and another drug known to be teratogenic in animals and humans. Tolbutamide teratogenicity is attributed to the reactive metabolites n-butyl isocyanate and S-(n-butylcarbamoyl) glutathione, which can carbamoylate and inhibit glutathione reductase (Guan *et al.*, 1999).

21.8   ABROGATION/MITIGATION OF BIOACTIVATION—CASE EXAMPLES

Drug evaluation is conducted on a case-by-case basis, which can prove costly and time-consuming. Fortunately, historical evidence has led to the compilation of "structural alerts" that are used to focus and optimize chemistry efforts in drug discovery and maximize the opportunity for success in development (Table 21.2). One strategy adopted in the drug discovery stage is to avoid these features entirely, while another is to evaluate their safety specifically and with full awareness of anecdotal issues related to such structural motifs. The latter strategy is supported by the many examples that exist of drugs that required only relatively minor structural changes to improve their safety profile and yet retain the desired pharmacological properties. This is not necessarily an easy task.

In this section, clozapine and olanzapine, captopril and enalapril, tolcapone and entacapone, and ritonavir and indinavir are discussed as examples (Table 21.4).

Functional groups such as thiophenes can be oxidized to form reactive species that have a propensity to react with nucleophilic macromolecules. In the case of tienilic acid, its metabolite is sufficiently reactive to inactivate the CYP2C9 enzyme involved in its formation (Melet *et al.*, 2003). Cases involving nitroaromatics, hydrazines, and anilines have been implicated in toxicity, including mutagenicity, carcinogenicity, and hepatotoxicity and issues with drugs like isoniazid and procainamide (Freeman *et al.*, 1981; Blanco *et al.*, 1998; Bomhard and Herbold, 2005; Maeda *et al.*, 2007). Molecules with quinone or quinoid functional groups can conjugate in a manner that leads to covalent binding with nucleophiles, as exemplified by acetaminophen and troglitazone, both drugs possessing marked hepatotoxicity at toxic dosages (He *et al.*, 2004; Shin *et al.*, 2007). The hydrazine-based drugs, hydralazine and isoniazid, are associated with autoimmune disease and hepatotoxicity, respectively (Hari, Raza and Clayton, 1998; Sarich *et al.*, 1999). Drugs containing benzylamine (such as terbinafine), a metabolic soft spot, can undergo a variety of biotransformations, forming aldehydes, carboxylic acids, hydroxylamines, and GSH conjugates (Mallat *et al.*, 1997). Thiol-containing drugs such as methamizole, penicillamine, and captopril cause toxicity by covalently modifying proteins by reacting with cysteinyl-disulfide linkages or by oxidation to reactive sulfenic acid (Jaffe, 1986; Decker and Doerge, 1992; Piepho, 2000). The thiazole ring can undergo oxidative scission to form reactive intermediates, as seen with sudoxicam and associated with its hepatotoxicity (Hobbs and Twomey, 1977). Carboxylic acid-containing drugs are predisposed to form acyl glucuronides, which are potentially reactive and toxic metabolites and known products of carboxylic acid NSAIDs (e.g. diclofenac and zomepirac) (Bailey and Dickinson, 1996; Seitz and Boelsterli, 1998). Unsaturated carbon–carbon double and triple bonds are electron-rich targets and subject to oxidation and formation of epoxides or oxirene metabolites, respectively—these are associated with subsequent reactions including toxicity and mechanism-dependent CYP inhibition. For instance, 17alpha-ethynylestradiol is metabolized by CYP3A4, a process that results in the inactivation of this enzyme by reactive acetylenic intermediate (Guengerich, 1990).

The clinical utility of clozapine is limited by its known propensity to cause agranulocytosis, hepatotoxicity, and cardiotoxicity. The reactive nitrenium ion of clozapine is thought to be responsible for its toxicity. Olanzapine is a close structural analog of clozapine, but associated with much lower incidences of toxicities than clozapine. Although olanzapine can be metabolized to a reactive nitrenium ion, the significantly lower clinical dose of olanzapine compared with clozapine provides a lower overall systemic exposure to the nitrenium ion, which presumably explains its low incidence of agranulocytosis (Gardner *et al.*, 1998; Uetrecht, 2001). In addition, olanzapine is also metabolized to an *N*-glucuronide, which likely circumvents the bioactivation pathway and contributes to its overall improved safety relative to clozapine (Kalgutkar *et al.*, 2005). Similarly, captopril, an angiotensin-converting enzyme (ACE) inhibitor, is associated with bone marrow toxicity and skin rashes, linked to the thiol moiety. Enalapril, a subsequently developed ACE inhibitor, lacks the thiol goup and is not associated with as many of the toxicities seen with captopril (Piepho, 2000; Kalgutkar *et al.*, 2005).

Tolcapone, a catechol-*O*-methyltransferase inhibitor used in the treatment of Parkinson's disease, is associated with a relatively high incidence of liver toxicity. Reduction of the nitro group in tolcapone and subsequent generation of the reactive quinoneimine is implicated in its hepatotoxicity (Smith *et al.*, 2003). In contrast, entacapone is structurally related to tolcapone but is not associated with hepatotoxicity. The metabolism of entacapone is governed primarily by glucuronidation and not reduction of the nitro group, which results in an overall low exposure to its reactive quinoneimine (Lautala *et al.*, 2000; Smith *et al.*, 2003). In addition to metabolic activation, direct mitochondrial toxicity of tolcapone but not entacapone is also implicated in tolcapone hepatotoxicity, based on studies with isolated mitochondria (Boelsterli *et al.*, 2006). Ritonavir, an HIV protease inhibitor, is a known mechanism-based inactivator of CYP3A4 through oxidative thiazole ring opening to form reactive intermediatcs. The HIV protease inhibitors indinavir and saquinavir lack the thiazole ring and do not inactivate CYP3A4 (Kempf *et al.*, 1997; Koudriakova *et al.*, 1998).

These examples illustrate the benefit of understanding the mechanism of toxicity, the biotransformation pathways, and the reaction phenotype involved in the adverse drug reactions. Relatively small structural changes that minimize or block the formation of reactive intermediates can profoundly affect the safety of a drug, especially when those changes are coupled with increased potency at the target, reduced metabolic clearance, and overall lower dosages.

TABLE 21.4 Structural modification and mitigation of toxicity.

Drug with relatively greater liability due to reactive intermediate generation	Drug with mitigated effect
Clozapine	Olanzapine
Captopril	Enalapril
Tolcapone	Entacapone
Ritonavir	Indinavir

21.9 EXPERIMENTAL METHODS FOR SCREENING

The early detection of reactive metabolites and prediction of an associated potential liability in drug development is useful in deciding the fate of a novel drug candidate and/or optimizing the structure-activity relationship (SAR) of the chemotype. Part of the repertoire available to medicinal chemistry includes reactivity screens that can be performed *in silico, in vitro*, or *in vivo*.

A number of software are available commercially that can help predict metabolites, potential for mutagenicity and carcinogenicity (Caldwell and Yan, 2006). Systems such as METEOR and MetabolExpert exist for prediction of metabolites and DEREK and HazardExpert exist to help identify potential genotoxicants. These programs rely on a set of rules derived from the available metabolism and toxicological knowledge (e.g. structural alerts). There are also systems like TOPKAT and MCASE that use toxicity data from known drugs to help determine the potential genotoxicity (CompuDrug International, Inc.; Lhasa Limited; Accelrys; Multicase Inc.). *In silico* predictions are very attractive and easy to perform; however, such predictions are not always accurate and should be used to provide alerts for a chemotype and subsequent testing in relevant assays (e.g. miniAmes).

There are several methods available to identify and quantify reactive metabolites *in vitro*. Time-dependent CYP inhibition is one commonly used *in vitro* method. Another is the qualitative or quantitative determination of adduct formation in liver microsomes. This is usually performed using either a co-incubated soft nucleophile (e.g. glutathione, *N*-acetyl-cysteine, or *N*-acetyl-lysine) to trap soft electrophiles (e.g. epoxides, arene oxides, nitrenium, and/or quinoneimines), or a hard nucleophile (e.g. potassium cyanide and sodium cyanide) to trap hard electrophiles. If a radiolabeled drug candidate is available, then its covalent binding to protein can be measured *in vitro* using liver microsomes from human or any other species of interest (usually done with and without NADPH and with and without a co-incubated nucleophile). Based on liver microsomal covalent binding, Evans *et al.* (2004) concluded that a binding activity of >50 pmol compound per mg of total liver protein poses an increased risk of reactive metabolite formation *in vivo* and associated adverse drug reactions. Non-liver microsomal *in vitro* assays also exist, including MPO for reactive metabolite generation; fluorometric probes such as 2′,7′-dichlorodihydrofluorescein for ROS detection. There are also cell-based assays to evaluate the functional consequence of reactive metabolite formation, and primary hepatocytes or immortalized cell lines expressing CYPs are routinely used in drug discovery to evaluate the hepatotoxic potential. Lastly, the Ames test in *S. typhimurium* and *E. coli* using Arochlor-induced rat liver S9 enzyme fraction, is a well-established *in vitro* assay to evaluate the potential for mutagenicity.

The *in vivo* determination of glutathione adducts, as well as qualitative and quantitative determination of covalently bound proteins, either by a radiolabeled compound or immunological methods, can be useful—albeit relatively labor- and resource-intensive. There are also *in vivo* clastogenicity assays, such as the comet assay, that are an important part of the drug development process. These *in vivo* assays are probably best used to show the mitigation of an *in vitro* liability (e.g. by drug clearance) and/or to identify extrahepatic and non-CYP-mediated toxicity that may not be indicated by routine *in vitro* screens and assays.

When toxicity indicates a role for DMEs in the pathogenesis, reaction phenotyping and drug–drug interaction studies can prove informative in evaluating and potentially even predicting the susceptibility of specific patient populations to adverse drug reactions (e.g. CYP3A4 interactions and terfenadine cardiac arrhythmias; CYP2E1 induction and acetaminophen hepatotoxicity; UGT1A1 polymorphism and irnitoecan gastrointestinal toxicity).

21.10   SUMMARY

A reasonable prediction of a drug candidate's safety, safety or exposure multiples, and/or safe starting dose for clinical trials can be made if the exposure limits and mechanism of action are known for its pharmacodynamic and toxicodynamic effects. The fate and effects of a drug and its metabolites can be measured *in vivo* using animal models as surrogates for human anatomy and physiology, and also *in vitro* using tissue available from animals and human donors; interspecies scaling and extrapolation factors are then applied to optimize the prediction of the drug's behavior in humans. Normally, preclinical animal models provide the no observed effect level (NOEL) and/or the no observed adverse effect level (NOAEL), which are combined with the target efficacious exposure to predict safety/exposure multiples in humans and drive a safe and viable evaluation of drug candidates in human volunteers and patients.

None of these *a priori* predictions is guaranteed and the estimates serve instead to provide a conservative and ethically acceptable safe starting dose. Human clinical trials are typically conducted with the drug candidate formulated at a starting dose that is a significantly low fraction of the preclinical NOEL or NOAEL. Evaluating the potential toxicity and predicting the margins of safety in humans helps to mitigate the risk in human clinical trials and has become *sine qua non* in the modern drug discovery and development process. Achieving that objective is not always easy and is often iterative. In many cases, one or more backup drug candidates will be required to address observed issues encountered with lead compounds.

While potential adverse drug reactions can be predicted for the majority of drug candidates that advance through the optimization process, such predictions cannot be made as easily when compounds form reactive metabolites. Drugs such as acetaminophen, diclofenac, and ibuprofen are known to produce reactive intermediates and are associated with occasional toxicities (overdose, idiosyncratic), and yet these and many others like them occupy a prominent position in the market place being invaluable in their therapeutic potential and use. If the benefits outweigh the risks, then a properly characterized and monitored drug candidate with potential for human toxicity can still become a successful therapeutic. A risk–benefit analysis, therefore, is particularly critical in deciding the fate of a drug or a drug candidate that forms a reactive intermediate. This analysis should consider a wide variety of factors that include the dose, the therapeutic area, target population, urgency of medical need, duration of treatment, availability of safer alternatives, and post-marketing surveillance. This is often a dynamic process within and between the industry and regulatory agencies.

A low predicted human dose (e.g. 10 mg) is a major factor in mitigating the risk associated with a reactive metabolite (Uetrecht, 2001). For example, 17alpha-ethynylestradiol is metabolized to reactive intermediates that can inactivate CYP3A4 but it is not associated with idiosyncratic toxicity in humans, probably because of its very low therapeutic dose (<0.035 mg). The hepatotoxicity associated with troglitazone (200–400 mg) but not pioglitazone (<10 mg) is very likely because of the difference in therapeutic dose.

Pragmatically, a certain level of risk must be accepted when the benefits are known to outweigh the risk, such as in the treatment of unmet medical conditions or serious medical conditions. For example, felbamate is an antiepileptic drug that shortly after its introduction was reported to be associated with multiple cases of aplastic anemia, hepatotoxicity, and deaths. Felbamate is, however, an effective therapy in refractory patients and instead of withdrawing the drug entirely, its use is limited to only that population of patients. Amodiaquine is associated with cases of life-threatening agranulocytosis and hepatotoxicity, yet as a malarial therapy it is effective against chloroquine-resistant isolates of *Plasmodium falciparum*. Its clinical use is thus limited to the treatment of acute malaria and not for prophylaxis.

A careful monitoring program can also help manage the risk of useful drugs, as in the case of bosentan, a drug used in treatment of pulmonary hypertension. Bosentan's approval as a drug included a risk management program, limited distribution, and monthly monitoring for hepatotoxicity (Temple, 2006). Measures such as these are clearly expensive but do allow potentially dangerous drugs to be used optimally in those patients who will benefit the most. Thalidomide, because of teratogenicity, is now restricted in its use and prescribed for immunomodulation only when the prescribing physicians and patients are enrolled in the System for Thalidomide Education and Prescribing Safety (STEPS). This is an oversight program that helps to ensure that pregnancy does not occur during therapy.

The availability of safer alternatives is an important factor that decides the fate of drugs. For example, bromfenac was withdrawn from the market because of its higher incidence of severe hepatotoxicity relative to the already existing alternatives (e.g. diclofenac) and because it offered no significant advantage over the others. When troglitazone was introduced to the market, it was unique in its mechanism of action and appropriately labeled for its risk; however, troglitazone was soon withdrawn after the introduction of rosiglitazone, which was without any reports of hepatotoxicity and deemed to be a safer drug in the same class.

There are numerous factors to consider in finding and developing a novel drug that are not limited to drug metabolism, although often this is key to understanding the potential fate and risks predicted for that drug in the patient population. There are also many things that are unknown about toxicity and drug metabolism, especially as it relates to differences between individuals in their susceptibility to adverse effects. And in many cases, the exact role of a given enzyme in the mechanism of toxicity is still poorly defined, if defined at all.

Some considerations of safety have become standardized, such as the evaluation of carcinogenesis; however, other factors encountered on a more singular basis often can become an issue and a potential showstopper to the success of a drug. Given that very few, if any, drugs have identical properties and issues, to properly understand the risk–benefit ratio of an experimental drug candidate it is prudent to approach its characterization on an individual level. This can be a daunting goal in the fast-paced, ever-changing and always competitive landscape of modern pharmaceutical research and development. Fortunately, the structure of a molecule can offer many clues to its fate. When this is combined with knowledge of the properties the experimental drug shares with its established ilk, a comprehensive understanding of the therapeutic target and the various mechanisms of toxicities, and also taking into consideration the characteristics of the enzymes commonly involved in metabolism and adverse drug reactions, a modern safety assessment can be tailored somewhat specifically to each new drug to help accelerate its development.

ACKNOWLEDGMENT

The authors acknowledge the many thoughtful comments and suggestions from Dr. Stephen P. Adams.

REFERENCES

Abdel-Razzak, Z., Loyer, P., Fautrel, A., Gautier, J.C., Corcos, L., Turlin, B., Beaune, P., Guillouzo, A. (1993). Cytokines down-regulate expression of major cytochrome p-450 enzymes in adult human hepatocytes in primary culture. *Mol. Pharmacol.*, *44* (4), 707–715.

Abernethy, D.R., Greenblatt, D.J., Divoll, M., Shader, R.I. (1983). Enhanced glucuronide conjugation of drugs in obesity: studies of lorazepam, oxazepam and acetaminophen. *J. Lab. Clin. Med.*, *101* (6), 873–880.

Aithal, G.P., Ramsay, L., Daly, A.K., Sonchit, N., Leathart, J.B., Alexander, G., Kenna, J.G., Caldwell, J., Day, C.P. (2004). Hepatic adducts, circulating antibodies and cytokine polymorphisms in patients with diclofenac hepatotoxicity. *Hepatology*, *39* (5), 1430–1440.

Anderson, N.L., Copple, D.C., Bendele, R.A., Probst, G.S., Richardson, F.C. (1992). Covalent protein modifications and gene expression changes in rodent liver following administration of methapyrilene: a study using two-dimensional electrophoresis. *Fundam. Appl. Toxicol.*, *18* (4), 570–580.

Anniko, M., Sobin, A. (1986). Cisplatin: evaluation of its ototoxic potential. *Am. J. Otolaryngol.*, *7* (4), 276–293.

Arand, M., Cronin, A., Adamska, M., Oesch, F. (2005). Epoxide hydrolases: structure, function, mechanism and assay. *Methods Enzymol.*, *400*, 569–588. Review.

Arlen, R.R., Wells, P.G. (1996). Inhibition of thalidomide teratogenicity by acetylsalicylic acid: evidence for prostaglandin H synthase-catalyzed bioactivation of thalidomide to a teratogenic reactive intermediate. *J. Pharmacol. Exp. Ther.*, *277* (3), 1649–1658.

Artemchik, V.D., Kurchenko, V.P., Metelitsa, D.I. (1985). Peroxidase activity of catalase with respect to aromatic amines. *Biokhimiia*, *50* (5), 826–832.

Atchison, C.R., West, A.B., Balakumaran, A., Hargus, S.J., Pohl, L.R., Daiker, D.H., Aronson, J.F., Hoffmann, W.E., Shipp, B.K., Treinen-Moslen, M. (2000). Drug enterocyte adducts: possible causal factor for diclofenac enteropathy in rats. *Gastroenterology*, *119* (6), 1537–1547.

Avent, K.M., DeVoss, J.J., Gillam, E.M. (2006). Cytochrome P450-mediated metabolism of haloperidol and reduced haloperidol to pyridinium metabolites. *Chem. Res. Toxicol.*, *19* (7), 914–920.

Back, D.J., Breckenridge, A.M., MacIver, M., Orme, M., Purba, H.S., Rowe, P.H., Taylor, I. (1982). The gut wall metabolism of ethinyloestradiol and its contribution to the pre-systemic metabolism of ethinyloestradiol in humans. *Br. J. Clin. Pharmacol.*, *13* (3), 325–330.

Bae, M.A., Song, B.J. (2003). Critical role of c-Jun N-terminal protein kinase activation in troglitazone-induced apoptosis of human HepG2 hepatoma cells. *Mol. Pharmacol.*, *63* (2), 401–408.

Baeuerle, P.A., Huttner, W.B. (1987). Tyrosine sulfation is a trans-Golgi-specific protein modification. *J. Cell Biol.*, *105* (6 Pt 1), 2655–2664.

Bailey, M.J., Dickinson, R.G. (1996). Chemical and immunochemical comparison of protein adduct formation of four carboxylate drugs in rat liver and plasma. *Chem. Res. Toxicol.*, *9* (3), 659–666.

Bailey, M.J., Dickinson, R.G. (1999). Limitations of hepatocytes and liver homogenates in modelling *in vivo* formation of acyl glucuronide-derived drug-protein adducts. *J. Pharmacol. Toxicol. Methods*, *41* (1), 27–32.

Bailey, M.J., Dickinson, R.G. (2003). Acyl glucuronide reactivity in perspective: biological consequences. *Chem. Biol. Interact.*, *145* (2), 117–137. Review.

Bailey, M.J., Worrall, S., de Jersey, J., Dickinson, R.G. (1998). Zomepirac acyl glucuronide covalently modifies tubulin *in vitro* and *in vivo* and inhibits its assembly in an *in vitro* system. *Chem. Biol. Interact.*, *115* (2), 153–166.

Balling, R., Haaf, H., Maydl, R., Metzler, M., Beier, H.M. (1985). Oxidative and conjugative metabolism of diethylstilbestrol by rabbit preimplantation embryos. *Dev. Biol.*, *109* (2), 370–374.

Banks, A.T., Zimmerman, H.J., Ishak, K.G., Harter, J.G. (1995). Diclofenac-associated hepatotoxicity: analysis of 180 cases reported to the Food and Drug Administration as adverse reactions. *Hepatology*, *22* (3), 820–827.

Basivireddy, J., Jacob, M., Pulimood, A.B., Balasubramanian, K.A. (2004). Indomethacin-induced renal damage: role of oxygen free radicals. *Biochem. Pharmacol.*, *67* (3), 587–599.

Beaune, P., Bourdi, M., Belloc, C., Gautier, J.C., Guengerich, F.P., Valadon, P. (1993). Immunotoxicology and expression of human cytochrome P450 in microorganisms. *Toxicology*, *82* (1–3), 53–60.

van den Bemt, P.M., Meyboom, R.H., Egberts, A.C. (2004). Drug-induced immune thrombocytopenia. *Drug Saf.*, *27* (15), 1243–1252.

Bennett, G.D., Amore, B.M., Finnell, R.H., Wlodarczyk, B., Kalhorn, T.F., Skiles, G.L., Nelson, S.D., Slattery, J.T. (1996). Teratogenicity of carbamazepine-10, 11-epoxide and oxcarbazepine in the SWV mouse. *J. Pharmacol. Exp. Ther.*, *279* (3), 1237–1242.

Berger, V., Berson, A., Wolf, C., Chachaty, C., Fau, D., Fromenty, B., Pessayre, D. (1992). Generation of free radicals during the reductive metabolism of nilutamide by lung microsomes: possible role in the development of lung lesions in patients treated with this anti-androgen. *Biochem. Pharmacol.*, *43* (3), 654–657.

Bergström, M.A., Ott, H., Carlsson, A., Neis, M., Zwadlo-Klarwasser, G., Jonsson, C.A., Merk, H.F., Karlberg, A.T., Baron, J.M. (2007). A skin-like cytochrome P450 cocktail activates prohaptens to contact allergenic metabolites. *J. Invest. Dermatol.*, *127* (5), 1145–1153.

Berson, A., Wolf, C., Chachaty, C., Fisch, C., Fau, D., Eugene, D., Loeper, J., Gauthier, J.C., Beaune, P., Pompon, D. *et al.* (1993). Metabolic activation of the nitroaromatic antiandrogen flutamide by rat and human cytochromes P-450, including forms belonging to the 3A and 1A subfamilies. *J. Pharmacol. Exp. Ther.*, *265* (1), 366–372.

Bertelsen, K.M., Venkatakrishnan, K., Von Moltke, L.L., Obach, R.S., Greenblatt, D.J. (2003). Apparent mechanism-based inhibition of human CYP2D6 in vitro by paroxetine: comparison with fluoxetine and quinidine. *Drug Metab. Dispos.*, *31* (3), 289–293.

Berthiaume, J.M., Wallace, K.B. (2007). Adriamycin-induced oxidative mitochondrial cardiotoxicity. *Cell Biol. Toxicol.*, *23* (1), 15–25. Review.

Bertini, R., Bianchi, M., Villa, P., Ghezzi, P. (1988). Depression of liver drug metabolism and increase in plasma fibrinogen by interleukin 1 and tumor necrosis factor: a comparison with lymphotoxin and interferon. *Int. J. Immunopharmacol.*, *10* (5), 525–530.

Bhaiya, P., Roychowdhury, S., Vyas, P.M., Doll, M.A., Hein, D.W., Svensson, C.K. (2006). Bioactivation, protein haptenation and toxicity of sulfamethoxazole and dapsone in normal human dermal fibroblasts. *Toxicol. Appl. Pharmacol.*, *215* (2), 158–167.

Blanco, M., Martínez, A., Urios, A., Herrera, G., O'Connor, J.E. (1998). Detection of oxidative mutagenesis by isoniazid and other hydrazine derivatives in Escherichia coli WP2 tester strain IC203, deficient in OxyR: strong protective effects of rat liver S9. *Mutat. Res.*, *417* (1), 39–46.

Blazka, M.E., Elwell, M.R., Holladay, S.D., Wilson, R.E., Luster, M. (1996). Histopathology of acetaminophen-induced liver changes: role of interleukin 1 alpha and tumor necrosis factor alpha. *Toxicol. Pathol.*, *24* (2), 181–189.

Boelsterli, U.A., Ho, H.K., Zhou, S., Leow, K.Y. (2006). Bioactivation and hepatotoxicity of nitroaromatic drugs. *Curr. Drug Metab.*, *7* (7), 715–727.

Bolchoz, L.J., Budinsky, R.A., McMillan, D.C., Jollow, D.J. (2001). Primaquine-induced hemolytic anemia: formation and hemotoxicity of the arylhydroxylamine metabolite 6-methoxy-8-hydroxylaminoquinoline. *J. Pharmacol. Exp. Ther.*, *297* (2), 509–515.

Bolt, H.M., Thier, R. (2006). Relevance of the deletion polymorphisms of the glutathione S-transferases GSTT1 and GSTM1 in pharmacology and toxicology. *Curr. Drug Metab.*, *7* (6), 613–628. Review.

Bomhard, E.M., Herbold, B.A. (2005). Genotoxic activities of aniline and its metabolites and their relationship to the carcinogenicity of aniline in the spleen of rats. *Crit. Rev. Toxicol.*, *35* (10), 783–835.

Boocock, D.J., Maggs, J.L., Brown, K., White, I.N., Park, B.K. (2000). Major inter-species differences in the rates of O-sulphonation and O-glucuronylation of alpha-hydroxytamoxifen *in vitro*: a metabolic disparity protecting human liver from the formation of tamoxifen-DNA adducts. *Carcinogenesis*, *21* (10), 1851–1858.

Boocock, D.J., Maggs, J.L., White, I.N., Park, B.K. (1999). Alpha-hydroxytamoxifen, a genotoxic metabolite of tamoxifen in the rat: identification and quantification *in vivo* and *in vitro*. *Carcinogenesis*, *20* (1), 153–160.

Bourdi, M., Gautier, J.C., Mircheva, J., Larrey, D., Guillouzo, A., Andre, C., Belloc, C., Beaune, P.H. (1992). Anti-liver microsomes autoantibodies and dihydralazine-induced hepatitis: specificity of autoantibodies and inductive capacity of the drug. *Mol. Pharmacol.*, *42* (2), 280–285.

Bourdi, M., Chen, W., Peter, R.M., Martin, J.L., Buters, J.T., Nelson, S.D., Pohl, L.R. (1996). Human cytochrome P450 2E1 is a major autoantigen associated with halothane hepatitis. *Chem. Res. Toxicol.*, *9* (7), 1159–1166.

Bourdi, M., Reilly, T.P., Elkahloun, A.G., George, J.W., Pohl, L.R. (2002). Macrophage migration inhibitory factor in drug-induced liver injury: a role in susceptibility and stress responsiveness. *Biochem. Biophys. Res. Commun.*, *294* (2), 225–230.

Butcher, N.J., Boukouvala, S., Sim, E., Minchin, R.F. (2002). Pharmacogenetics of the arylamine N-acetyltransferases. *Pharmacogenomics J.*, *2* (1), 30–42. Review.

Caldwell, G.W., Yan, Z. (2006). Screening for reactive intermediates and toxicity assessment in drug discovery. *Curr. Opin. Drug Discov. Devel.*, *9* (1), 47–60. Review.

Cashman, J.R. (2000). Human flavin-containing monooxygenase: substrate specificity and role in drug metabolism. *Curr. Drug Metab.*, *1*, 181–191.

Cashman, J.R., Bi, Y.A., Lin, J., Youil, R., Knight, M., Forrest, S., Treacy, E. (1997). Human flavin-containing monooxygenase form 3: cDNA expression of the enzymes containing amino acid substitutions observed in individuals with trimethylaminuria. *Chem. Res. Toxicol.*, *10* (8), 837–841.

Castillo, M., Smith, P.C. (1995). Disposition and reactivity of ibuprofen and ibufenac acyl glucuronides *in vivo* in the rhesus monkey and *in vitro* with human serum albumin. *Drug Metab. Dispos.*, *23* (5), 566–572.

Cavalieri, E.L., Rogan, E.G., Chakravarti, D. (2002). Initiation of cancer and other diseases by catechol ortho-quinones: a unifying mechanism. *Cell Mol. Life Sci.*, *59* (4), 665–681. Review.

Chae, K., Lindzey, J., McLachlan, J.A., Korach, K.S. (1998). Estrogen-dependent gene regulation by an oxidative metabolite of diethylstilbestrol, diethylstilbestrol-4′,4″-quinone. *Steroids*, *63* (3), 149–157.

Chan, K., Han, X.D., Kan, Y.W. (2001). An important function of Nrf2 in combating oxidative stress: detoxification of acetaminophen. *Proc. Natl. Acad. Sci. U.S.A.*, *98* (8), 4611–4616.

Chang, T.K., Waxman, D.J. (1993). Cyclophosphamide modulates rat hepatic cytochrome P450 2C11 and steroid 5 alpha-reductase activity and messenger RNA levels through the combined action of acrolein and phosphoramide mustard. *Cancer Res.*, *53* (11), 2490–2497. Erratum in: *Cancer Res* (1993) 53 (16):3846.

Chen, W., Koenigs, L.L., Thompson, S.J., Peter, R.M., Rettie, A.E., Trager, W.F., Nelson, S.D. (1998). Oxidation of acetaminophen to its toxic quinoneimine and nontoxic catechol metabolites by baculovirus-expressed and purified human cytochromes P450 2E1 and 2A6. *Chem. Res. Toxicol.*, *11* (4), 295–301.

Chitturi, S., George, J. (2002). Hepatotoxicity of commonly used drugs: nonsteroidal anti-inflammatory drugs, antihypertensives, antidiabetic agents, anticonvulsants, lipid-lowering agents, psychotropic drugs. *Semin. Liver Dis.*, *22* (2), 169–183. Review.

Clarke, L., Waxman, D.J. (1989). Oxidative metabolism of cyclophosphamide: identification of the hepatic monooxygenase catalysts of drug activation. *Cancer Res.*, *49* (9), 2344–2350.

Colacurci, N., De Seta, L., De Franciscis, P., Mele, D., Fortunato, N., Cassese, S. (2000). Tamoxifen effects on endometrium. *Panminerva Med.*, *42*, 45–47.

Colby, H.D., O'Donnell, J.P., Flowers, N.L., Kossor, D.C., Johnson, P.B., Levitt, M. (1991). Relationship between covalent binding to microsomal protein and the destruction of adrenal cytochrome P-450 by spironolactone. *Toxicology*, *67* (2), 143–154.

Coleman, T., Ellis, S.W., Martin, I.J., Lennard, M.S., Tucker, G.T. (1996). 1-Methyl-4-phenyl-1,2,3,6-tetrahydropyridine (MPTP) is N-demethylated by cytochromes P450 2D6, 1A2 and 3A4—implications for susceptibility to Parkinson's disease. *J. Pharmacol. Exp. Ther.*, *277* (2), 685–690.

Coleman, M.D., Khalaf, L.F., Nicholls, P.J. (2003). Aminoglutethimide-induced leucopenia in a mouse model: effects of metabolic and structural determinates. *Environ. Toxicol. Pharmacol.*, *15*, 27–32.

Copple, I.M., Goldring, C.E., Kitteringham, N.R., Park, B.K. (2008). The Nrf2-Keap1 defence pathway: role in protection against drug-induced toxicity. *Toxicology*, *246* (1), 24–33.

Coughtrie, M.W. (2002). Sulfation through the looking glass—recent advances in sulfotransferase research for the curious. *Pharmacogenomics J.*, *2* (5), 297–308. Review.

Court, M.H. (2005). Isoform-selective probe substrates for *in vitro* studies of human UDP-glucuronosyltransferases. *Methods Enzymol.*, *400*, 104–116.

Cribb, A.E., Lee, B.L., Trepanier, L.A., Spielberg, S.P. (1996). Adverse reactions to sulphonamide and sulphonamide-trimethoprim antimicrobials: clinical syndromes and pathogenesis. *Adverse Drug React. Toxicol. Rev.*, *15* (1), 9–50.

Dalrymple, P.D., Nicholls, P.J. (1988). Metabolism profiles and excretion of 14C-aminoglutethimide in several animal species and man. *Xenobiotica*, *18* (1), 75–81.

Davis, D.C., Potter, W.Z., Jollow, D.J., Mitchell, J.R. (1974). Species differences in hepatic glutathione depletion, covalent binding and hepatic necrosis after acetaminophen. *Life Sci.*, *14* (11), 2099–2109.

Davis, G.C., Williams, A.C., Markey, S.P., Ebert, M.H., Caine, E.D., Reichert, C.M., Kopin, I.J. (1979). Chronic Parkinsonism secondary to intravenous injection of meperidine analogues. *Psychiatry Res.*, *1* (3), 249–254.

De, S.K., De, M., Banerjee, R.K. (1986). Localization and origin of the intestinal peroxidase—effect of adrenal glucocorticoids. *J. Steroid Biochem.*, *24* (2), 629–635.

Debiec-Rychter, M., Land, S.J., King, C.M. (1999). Histological localization of acetyltransferases in human tissue. *Cancer Lett.*, *143* (2), 99–102.

Decker, C.J., Doerge, D.R. (1992). Covalent binding of 14C- and 35S-labeled thiocarbamides in rat hepatic microsomes. *Biochem. Pharmacol.*, *43* (4), 881–888.

Desmeules, P., Devine, P.J. (2006). Characterizing the ovotoxicity of cyclophosphamide metabolites on cultured mouse ovaries. *Toxicol. Sci.*, *90* (2), 500–509.

Dhaini, H.R., Levy, G.N. (2000). Arylamine N-acetyltransferase 1 (NAT1) genotypes in a Lebanese population. *Pharmacogenetics*, *10* (1), 79–83.

Dietze, E.C., Schäfer, A., Omichinski, J.G., Nelson, S.D. (1997). Inactivation of glyceraldehyde-3-phosphate dehydrogenase by a reactive metabolite of acetaminophen and mass spectral characterization of an arylated active site peptide. *Chem. Res. Toxicol.*, *10* (10), 1097–1103.

Doroshow, J.H., Locker, G.Y., Myers, C.E. (1980). Enzymatic defenses of the mouse heart against reactive oxygen metabolites: alterations produced by doxorubicin. *J. Clin. Invest.*, *65* (1), 128–135.

Dragan, Y.P., Fahey, S., Street, K., Vaughan, J., Jordan, V.C., Pitot, H.C. (1994). Studies of tamoxifen as a promoter of hepatocarcinogenesis in female Fischer F344 rats. *Breast. Cancer Res. Treat*, *31* (1), 11–25.

Eklund, B.I., Moberg, M., Bergquist, J., Mannervik, B. (2006). Divergent activities of human glutathione transferases in the bioactivation of azathioprine. *Mol. Pharmacol.*, *70* (2), 747–754.

Eklund, B.I., Gunnarsdottir, S., Elfarra, A.A., Mannervik, B. (2007). Human glutathione transferases catalyzing the bioactivation of anticancer thiopurine prodrugs. *Biochem. Pharmacol.*, *73* (11), 1829–1841.

Evans, D.A. (1989). N-acetyltransferase. *Pharmacol. Ther.*, *42* (2), 157–234.

Evans, D.C., Watt, A.P., Nicoll-Griffith, D.A., Baillie, T.A. (2004). Drug-protein adducts: an industry perspective on minimizing the potential for drug bioactivation in drug discovery and development. *Chem. Res. Toxicol.*, *17* (1), 3–16. Erratum in: *Chem Res Toxicol.* (2005) *18* (11):1777.

Eyles, D.W., Avent, K.M., Stedman, T.J., Pond, S.M. (1997). Two pyridinium metabolites of haloperidol are present in the brain of patients at post-mortem. *Life Sci.*, *60* (8), 529–534.

Falany, C.N., Kerl, E.A. (1990). Sulfation of minoxidil by human liver phenol sulfotransferase. *Biochem. Pharmacol.*, *40* (5), 1027–1032.

Fan, L., Schut, H.A., Snyderwine, E.G. (1995). Cytotoxicity DNA adduct formation and DNA repair induced by 2-hydroxyamino-3-methylimidazo[4,5-f]quinoline and 2-hydroxyamino-1-methyl-6-phenylimidazo[4,5-b]pyridine in cultured human mammary epithelial cells. *Carcinogenesis*, *16* (4), 775–779.

Fan, P.W., Bolton, J.L. (2001). Bioactivation of tamoxifen to metabolite E quinone methide: reaction with glutathione and DNA. *Drug Metab. Dispos.*, *29* (6), 891–896.

Fau, D., Berson, A., Eugene, D., Fromenty, B., Fisch, C., Pessayre, D. (1992). Mechanism for the hepatotoxicity of the antiandrogen, nilutamide. Evidence suggesting that redox cycling of this nitroaromatic drug leads to oxidative stress in isolated hepatocytes. *J. Pharmacol. Exp. Ther.*, *263* (1), 69–77.

Feyler, A., Voho, A., Bouchardy, C., Kuokkanen, K., Dayer, P., Hirvonen, A., Benhamou, S. (2002). Point myeloperoxidase $^{-463}$G → a polymorphism and lung cancer risk. *Cancer Epidemiol. Biomarkers Prev.*, *11* (12), 1550–1554.

Finnell, R.H., Bennett, G.D., Slattery, J.T., Amore, B.M., Bajpai, M., Levy, R.H. (1995). Effect of treatment with phenobarbital and stiripentol on carbamazepine-induced teratogenicity and reactive metabolite formation. *Teratology*, *52* (6), 324–332.

Fontana, R.J., McCashland, T.M., Benner, K.G., Appelman, H.D., Gunartanam, N.T., Wisecarver, J.L., Rabkin, J.M., Lee, W.M. (1999). Acute liver failure associated with prolonged use of bromfenac leading to liver transplantation. The Acute Liver Failure Study Group. *Liver Transpl. Surg.*, *5* (6), 480–484.

Fratelli, M., Goodwin, L.O., Ørom, U.A., Lombardi, S., Tonelli, R., Mengozzi, M., Ghezzi, P. (2005). Gene expression profiling reveals a signaling role of glutathione in redox regulation. *Proc. Natl. Acad. Sci. U.S.A.*, *102* (39), 13998–14003.

Freeman, R.W., Uetrecht, J.P., Woosley, R.L., Oates, J.A., Harbison, R.D. (1981). Covalent binding of procainamide *in vitro* and *in vivo* to hepatic protein in mice. *Drug Metab. Dispos.*, *9* (3), 188–192.

French, L.E. (2006). Toxic epidermal necrolysis and Stevens-Johnson syndrome: our current understanding. *Allergol. Int.*, *55* (1), 9–16.

French, L.E., Trent, J.T., Kerdel, F.A. (2006). Use of intravenous immunoglobulin in toxic epidermal necrolysis and Stevens-Johnson syndrome: our current understanding. *Int. Immunopharmacol.*, *6* (4), 543–549. Review.

Frye, R.F., Tammara, B., Cowart, T.D., Bramer, S.L. (1999). Effect of disulfiram-mediated CYP2E1 inhibition on the disposition of vesnarinone. *J. Clin. Pharmacol.*, *39* (11), 1177–1183.

Fu, P.P., Xia, Q., Lin, G., Chou, M.W. (2004). Pyrrolizidine alkaloids—genotoxicity, metabolism enzymes, metabolic activation and mechanisms. *Drug Metab. Rev.*, *36* (1), 1–55. Review.

Furst, S.M., Uetrecht, J.P. (1995). The effect of carbamazepine and its reactive metabolite, 9-acridine carboxaldehyde, on immune cell function *in vitro*. *Int. J. Immunopharmacol.*, *17* (5), 445–452.

Galinsky, R.E., Levy, G. (1981). Dose- and time-dependent elimination of acetaminophen in rats: pharmacokinetic implications of cosubstrate depletion. *J. Pharmacol. Exp. Ther.*, *219* (1), 14–20.

Ganey, P.E., Bailie, M.B., VanCise, S., Colligan, M.E., Madhukar, B.V., Robinson, J.P., Roth, R.A. (1994). Activated neutrophils from rat injured isolated hepatocytes. *Lab. Invest.*, *70* (1), 53–60.

Ganey, P.E., Roth, R.A. (2001). Concurrent inflammation as a determinant of susceptibility to toxicity from xenobiotic agents. *Toxicology*, *169* (3), 195–208. Review.

Gardner, I., Leeder, J.S., Chin, T., Zahid, N., Uetrecht, J.P. (1998). A comparison of the covalent binding of clozapine and olanzapine to human neutrophils *in vitro* and *in vivo*. *Mol. Pharmacol.*, *53* (6), 999–1008.

Garibaldi, R.A., Drusin, R.E., Ferebee, S.H., Gregg, MB. (1972). Isoniazid-associated hepatitis. Report of an outbreak. *Am. Rev. Respir. Dis.*, *106* (3), 357–365.

Garlick, R.L., Mazer, J.S. (1983). The principal site of nonenzymatic glycosylation of human serum albumin *in vivo*. *J. Biol. Chem.*, *258* (10), 6142–6146.

Gelotte, C.K., Auiler, J.F., Lynch, J.M., Temple, A.R., Slattery, J.T. (2007). Disposition of acetaminophen at 4, 6 and 8 g/day for 3 days in healthy young adults. *Clin. Pharmacol. Ther.*, *81* (6), 840–848.

Gill, J.S., Windebank, A.J. (1998). Cisplatin-induced apoptosis in rat dorsal root ganglion neurons is associated with attempted entry into the cell cycle. *J. Clin. Invest.*, *101* (12), 2842–2850.

Gitlin, N., Julie, N.L., Spurr, C.L., Lim, K.N., Juarbe, H.M. (1998). Two cases of severe clinical and histologic hepatotoxicity associated with troglitazone. *Ann. Intern. Med.*, *129* (1), 36–38.

Glatt, H. (1997). Sulfation and sulfotransferases 4: bioactivation of mutagens via sulfation. *FASEB. J.*, *11* (5), 314–321. Review.

Glatt, H. (2000). Sulfotransferases in the bioactivation of xenobiotics. *Chem. Biol. Interact.*, *129* (1–2), 141–170. Review.

Glatt, H., Bartsch, I., Christoph, S., Coughtrie, M.W., Falany, C.N., Hagen, M., Landsiedel, R., Pabel, U., Phillips, D.H., Seidel, A., Yamazoe, Y. (1998a). Sulfotransferase-mediated activation of mutagens studied using heterologous expression systems. *Chem. Biol. Interact.*, *109* (1–3), 195–219.

Glatt, H., Davis, W., Meinl, W., Hermersdörfer, H., Venitt, S., Phillips, D.H. (1998b). Rat, but not human, sulfotransferase activates a tamoxifen metabolite to produce DNA adducts and gene mutations in bacteria and mammalian cells in culture. *Carcinogenesis*, *19* (10), 1709–1713.

Goldin, R.D., Ratnayaka, I.D., Breach, C.S., Brown, I.N., Wickramasinghe, S.N. (1996). Role of macrophages in acetaminophen (paracetamol)-induced hepatotoxicity. *J. Pathol.*, *179* (4), 432–435.

Goldring, C.E., Kitteringham, N.R., Elsby, R., Randle, L.E., Clement, Y.N., Williams, D.P., McMahon, M., Hayes, J.D., Itoh, K., Yamamoto, M., Park, B.K. (2004). Activation of hepatic Nrf2 *in vivo* by acetaminophen in CD-1 mice. *Hepatology*, *39* (5), 1267–1276.

Gram, T.E. (1997). Chemically reactive intermediates and pulmonary xenobiotic toxicity. *Pharmacol. Rev.*, *49* (4), 297–341.

Grossman, S.J., Jollow, D.J. (1988). Role of dapsone hydroxylamine in dapsone-induced hemolytic anemia. *J. Pharmacol. Exp. Ther.*, *244* (1), 118–125.

Guan, X., Davis, M.R., Tang, C., Jochheim, C.M., Jin, L., Baillie, T.A. (1999). Identification of S-(n-butylcarbamoyl)glutathione, a reactive carbamoylating metabolite of tolbutamide in the rat and evaluation of its inhibitory effects on glutathione reductase *in vitro*. *Chem. Res. Toxicol.*, *12* (12), 1138–1143.

Guengerich, F.P. (1990). Mechanism-based inactivation of human liver microsomal cytochrome P-450 IIIA4 by gestodene. *Chem. Res. Toxicol.*, *3* (4), 363–371.

Gustafson, D.L., Swanson, J.D., Pritsos, C.A. (1991). Role of xanthine oxidase in the potentiation of doxorubicin-induced cardiotoxicity by mitomycin C. *Cancer Commun.*, *3* (9), 299–304.

Hakkola, J., Pelkonen, O., Pasanen, M., Raunio, H. (1998). Xenobiotic-metabolizing cytochrome P450 enzymes in the human feto-placental unit: role in intrauterine toxicity. *Crit. Rev. Toxicol.*, *28* (1), 35–72. Review.

Hales, B.F., Brown, H. (1991). The effect of *in vivo* glutathione depletion with buthionine sulfoximine on rat embryo development. *Teratology*, *44* (3), 251–257.

Halsey, J.P., Cardoe, N. (1982). Benoxaprofen side-effect profile in 300 patients. *Br. Med. J. (Clin. Res. Ed.)*, *284* (6326), 1365–1368.

Hamada, S., Kamada, M., Furumoto, H., Hirao, T., Aono, T. (1994). Expression of glutathione S-transferase-pi in human ovarian cancer as an indicator of resistance to chemotherapy. *Gynecol. Oncol.*, *52* (3), 313–319.

Hargus, S.J., Amouzedeh, H.R., Pumford, N.R., Myers, T.G., McCoy, S.C., Pohl, L.R. (1994). Metabolic activation and immunochemical localization of liver protein adducts of the nonsteroidal anti-inflammatory drug diclofenac. *Chem. Res. Toxicol.*, *7* (4), 575–582.

Hari, C.K., Raza, S.A., Clayton, M.I. (1998). Hydralazine-induced lupus and vocal fold paralysis. *J. Laryngol. Otol.*, *112* (9), 875–877.

Harlan, J.M., Levine, J.D., Callahan, K.S., Schwartz, B.R., Harker, L.A. (1984). Glutathione redox cycle protects cultured endothelial cells against lysis by extracellularly generated hydrogen peroxide. *J. Clin. Invest.*, *73* (3), 706–713.

Hart, S.G., Beierschmitt, W.P., Wyand, D.S., Khairallah, E.A., Cohen, S.D. (1994). Acetaminophen nephrotoxicity in CD-1 mice. I. Evidence of a role for in situ activation in selective covalent binding and toxicity. *Toxicol. Appl. Pharmacol.*, *126* (2), 267–275.

Hayes, J.D., Flanagan, J.U., Jowsey, I.R. (2005). Glutathione transferases. *Annu. Rev. Pharmacol. Toxicol.*, *45*, 51–88. Review.

Hayes, J.D., Pulford, D.J. (1995). The glutathione S-transferase supergene family: regulation of GST and the contribution of the isoenzymes to cancer chemoprotection and drug resistance. *Crit. Rev. Biochem. Mol. Biol.*, *30* (6), 445–600. Review.

He, K., Talaat, R.E., Pool, W.F., Reily, M.D., Reed, J.E., Bridges, A.J., Woolf, T.F. (2004). Metabolic activation of troglitazone: identification of a reactive metabolite and mechanisms involved. *Drug Metab. Dispos.*, *32* (6), 639–646.

Hearse, D.J., Weber, W.W. (1973). Multiple N-acetyltransferases and drug metabolism. Tissue distribution, characterization and significance of mammalian N-acetyltransferase. *Biochem. J.*, *132* (3), 519–526.

Henderson, M.C., Krueger, S.K., Stevens, J.F., Williams, D.E. (2004). Human flavin-containing monooxygenase form 2 S-oxygenation: sulfenic acid formation from thioureas and oxidation of glutathione. *Chem. Res. Toxicol.*, *17*, 633–640.

Herbst, A.L., Ulfelder, H., Poskanzer, D.C. (1971). Adenocarcinoma of the vagina. Association of maternal stilbestrol therapy with tumor appearance in young women. *N. Engl. J. Med.*, *284* (15), 878–881.

Hildebrandt, M.A., Carrington, D.P., Thomae, B.A., Eckloff, B.W., Schaid, D.J., Yee, V.C., Weinshilboum, R.M., Wieben, E.D. (2007). Genetic diversity and function in the human cytosolic sulfotransferases. *Pharmacogenomics J.*, *7* (2), 133–143.

Hiraku, Y., Sekine, A., Nabeshi, H., Midorikawa, K., Murata, M., Kumagai, Y., Kawanishi, S. (2004). Mechanism of carcinogenesis induced by a veterinary antimicrobial drug, nitrofurazone, via oxidative DNA damage and cell proliferation. *Cancer Lett.*, *215* (2), 141–150.

Hobbs, D.C., Twomey, T.M. (1977). Metabolism of sudoxicam by the rat, dog and monkey. *Drug Metab. Dispos.*, *5* (1), 75–81.

Hofstra, A.H., Matassa, L.C., Uetrecht, J.P. (1991). Metabolism of hydralazine by activated leukocytes: implications for hydralazine induced lupus. *J. Rheumatol.*, *18* (11), 1673–1680.

Hort, J.F. (1975). Adverse reactions to alclofenac. *Curr. Med. Res. Opin.*, *3* (5), 333–337.

Huttner, W.B. (1988). Tyrosine sulfation and the secretory pathway. *Annu. Rev. Physiol.*, *50*, 363–376. Review.

IARC Working Group (1990). Nitrofural (nitrofurazone), in *IARC Monographs on the Evaluation of the Carcinogenic Risks of Chemicals to Humans*, *50*, IARC, Lyon, pp. 195–209.

van Iersel, M.L., Ploemen, J.P., Lo Bello, M., Federici, G., van Bladeren, P.J. (1997). Interactions of alpha, beta-unsaturated aldehydes and ketones with human glutathione S-transferase P1-1. *Chem. Biol. Interact.*, *108* (1-2), 67–78.

Innocenti, F., Iyer, L., Ratain, M.J. (2001). Pharmacogenetics of anticancer agents: lessons from amonafide and irinotecan. *Drug Metab. Dispos.*, *29* (4 Pt 2), 596–600.

Itagaki, S., Sumi, Y., Shimamoto, S., Itoh, T., Hirano, T., Takemoto, I., Iseki, K. (2005). Secretory transport of irinotecan metabolite SN-38 across isolated intestinal tissue. *Cancer Chemother. Pharmacol.*, *55* (5), 502–506.

Iverson, S.L., Uetrecht, J.P. (2001). Identification of a reactive metabolite of terbinafine: insights into terbinafine-induced hepatotoxicity. *Chem. Res. Toxicol.*, *14* (2), 175–181.

Iverson, S., Zahid, N., Uetrecht, J.P. (2002). Predicting drug-induced agranulocytosis: characterizing neutrophil-generated metabolites of a model compound, DMP 406 and assessing the relevance of an *in vitro* apoptosis assay for identifying drugs that may cause agranulocytosis. *Chem. Biol. Interact.*, *142* (1–2), 175–199.

Jaffe, I.A. (1986). Adverse effects profile of sulfhydryl compounds in man. *Am. J. Med.*, *80* (3), 471–476.

Jaiswal, A.K. (2004). Nrf2 signaling in coordinated activation of antioxidant gene expression. *Free Radic. Biol. Med.*, *36* (10), 1199–1207. Review.

Jakobsson, P.J., Morgenstern, R., Mancini, J., Ford-Hutchinson, A., Persson, B. (1999). Common structural features of MAPEG—a widespread superfamily of membrane associated proteins with highly divergent functions in eicosanoid and glutathione metabolism. *Protein Sci.*, *8* (3), 689–692.

Jeng, W., Ramkissoon, A., Parman, T., Wells, P.G. (2006). Prostaglandin H synthase-catalyzed bioactivation of amphetamines to free radical intermediates that cause CNS regional DNA oxidation and nerve terminal degeneration. *FASEB J.*, *20* (6), 638–650.

Jensen, C.B., Jollow, D.J. (1991). The role of N-hydroxyphenetidine in phenacetin-induced hemolytic anemia. *Toxicol. Appl. Pharmacol.*, *111* (1), 1–12.

Jones, D.R., Gorski, J.C., Hamman, M.A., Mayhew, B.S., Rider, S., Hall, S.D. (1999). Diltiazem inhibition of cytochrome P-450 3A activity is due to metabolite intermediate complex formation. *J. Pharmacol. Exp. Ther.*, *290* (3), 1116–1125.

Jones, B.E., Liu, H., Lo, C.R., Koop, D.R., Czaja, M.J. (2002). Cytochrome P450 2E1 expression induces hepatocyte resistance to cell death from oxidative stress. *Antioxid. Redox. Signal.*, *4* (5), 701–709.

Jönsson, K.H., Lindckc, B. (1992). Cytochrome P-455 nm complex formation in the metabolism of phenylalkylamines. XII. Enantioselectivity and temperature dependence in microsomes and reconstituted cytochrome P-450 systems from rat liver. *Chirality*, *4* (8), 469–477.

Joshi, G., Sultana, R., Tangpong, J., Cole, M.P., St Clair, D.K., Vore, M., Estus, S., Butterfield, D.A. (2005). Free radical mediated oxidative stress and toxic side effects in brain induced by the anti cancer drug adriamycin: insight into chemobrain. *Free Radic. Res.*, *39* (11), 1147–1154.

Ju, C., Uetrecht, J.P. (1998). Oxidation of a metabolite of indomethacin (Desmethyldeschlorobenzoylindomethacin) to reactive intermediates by activated neutrophils, hypochlorous acid and the myeloperoxidase system. *Drug Metab. Dispos.*, *26* (7), 676–680.

Kajita, J., Inano, K., Fuse, E., Kuwabara, T., Kobayashi, H. (2002). Effects of olopatadine, a new antiallergic agent, on human liver microsomal cytochrome P450 activities. *Drug Metab. Dispos.*, *30* (12), 1504–1511.

Kalgutkar, A.S., Gardner, I., Obach, R.S., Shaffer, C.L., Callegari, E., Henne, K.R., Mutlib, A.E., Dalvie, D.K., Lee, J.S., Nakai, Y., O'Donnell, J.P., Boer, J., Harriman, S.P. (2005). A comprehensive listing of bioactivation pathways of organic functional groups. *Curr. Drug Metab.*, *6* (3), 161–225.

Kalgutkar, A.S., Obach, R.S., Maurer, T.S. (2007). Mechanism-based inactivation of cytochrome P450 enzymes: chemical mechanisms, structure-activity relationships and relationship to clinical drug-drug interactions and idiosyncratic adverse drug reactions. *Curr. Drug Metab.*, *8* (5), 407–447.

Kalgutkar, A.S., Soglia, J.R. (2005). Minimising the potential for metabolic activation in drug discovery. *Expert Opin. Drug Metab. Toxicol.*, *1* (1), 91–142.

Kassahun, K., Farrell, K., Abbott, F. (1991). Identification and characterization of the glutathione and N-acetylcysteine conjugates of (E)-2-propyl-2,4-pentadienoic acid, a toxic metabolite of valproic acid, in rats and humans. *Drug Metab. Dispos.*, *19* (2), 525–535.

Kassahun, K., McIntosh, I., Cui, D., Hreniuk, D., Merschman, S., Lasseter, K., Azrolan, N., Iwamoto, M., Wagner, J.A., Wenning, L.A. (2007). Metabolism and disposition in humans of raltegravir (MK-0518), an anti-AIDS drug targeting the human immunodeficiency virus 1 integrase enzyme. *Drug Metab. Dispos.*, *35* (9), 1657–1663.

Kassahun, K., Pearson, P.G., Tang, W., McIntosh, I., Leung, K., Elmore, C., Dean, D., Wang, R., Doss, G., Baillie, T.A. (2001). Studies on the metabolism of troglitazone to reactive intermediates *in vitro* and *in vivo*. Evidence for novel biotransformation pathways involving quinone methide formation and thiazolidinedione ring scission. *Chem. Res. Toxicol.*, *14* (1), 62–70.

Kato, A., Yoshidome, H., Edwards, M.J., Lentsch, A.B. (2000). Regulation of liver inflammatory injury by signal transducer and activator of transcription-6. *Am. J. Pathol.*, *157* (1), 297–302.

Kaufman, D.W., Kelly, J.P., Anderson, T., Harmon, D.C., Shapiro, S. (1997). Evaluation of case reports of aplastic anemia among patients treated with felbamate. *Epilepsia*, *38* (12), 1265–1269. Review.

Kawai, K., Kawasaki-Tokui, Y., Odaka, T., Tsuruta, F., Kazui, M., Iwabuchi, H., Nakamura, T., Kinoshita, T., Ikeda, T., Yoshioka, T., Komai, T., Nakamura, K. (1997). Disposition and metabolism of the new oral antidiabetic drug troglitazone in rats, mice and dogs. *Arzneimittelforschung*, *47* (4), 356–368.

Kedderis, G.L., Argenbright, L.S., Miwa, G.T. (1989). Covalent interaction of 5-nitroimidazoles with DNA and protein *in vitro*: mechanism of reductive activation. *Chem. Res. Toxicol.*, *2* (3), 146–149.

Kempf, D.J., Marsh, K.C., Kumar, G., Rodrigues, A.D., Denissen, J.F., McDonald, E., Kukulka, M.J., Hsu, A., Granneman, G.R., Baroldi, P.A., Sun, E., Pizzuti, D., Plattner, J.J., Norbeck, D.W., Leonard, J.M. (1997). Pharmacokinetic enhancement of inhibitors of the human immunodeficiency virus protease by coadministration with ritonavir. *Antimicrob. Agents Chemother.*, *41* (3), 654–660.

Khan, F.D., Vyas, P.M., Gaspari, A.A., Svensson, C.K. (2007). Effect of arylhydroxylamine metabolites of sulfamethoxazole and dapsone on stress signal expression in human keratinocytes. *J. Pharmacol. Exp. Ther.*, *323* (3), 771–777.

Kharbanda, K.K., Rogers, D.D., II, Wyatt, T.A., Sorrell, M.F., Tuma, D.J. (2004). Transforming growth factor-beta induces contraction of activated hepatic stellate cells. *J. Hepatol.*, *41* (1), 60–66.

Kiang, T.K., Ho, P.C., Anari, M.R., Tong, V., Abbott, F.S., Chang, T.K. (2006). Contribution of CYP2C9, CYP2A6 and CYP2B6 to valproic acid metabolism in hepatic microsomes from individuals with the CYP2C9*1/*1 genotype. *Toxicol. Sci.*, *94* (2), 261–271.

Killian, J.G., Kerr, K., Lawrence, C., Celermajer, D.S. (1999). Myocarditis and cardiomyopathy associated with clozapine. *Lancet*, *354* (9193), 1841–1845.

Kim, P.M., Winn, L.M., Parman, T., Wells, P.G. (1997). UDP-glucuronosyltransferase-mediated protection against *in vitro* DNA oxidation and micronucleus formation initiated by phenytoin and its embryotoxic metabolite 5-(p-hydroxyphenyl)-5-phenylhydantoin. *J. Pharmacol. Exp. Ther.*, *280* (1), 200–209.

Klebanoff, S.J. (1999). Myeloperoxidase. *Proc. Assoc. Am. Physicians*, *111* (5), 383–389. Review.

Kodama, M., Inoue, F., Kaneko, M., Oda, T., Sato, Y. (1993). Formation of free radicals and active oxygen species from diethylstilbestrol and its derivatives. *Anticancer Res.*, *13* (4), 1209–1213.

Kopke, R.D., Liu, W., Gabaizadeh, R., Jacono, A., Feghali, J., Spray, D., Garcia, P., Steinman, H., Malgrange, B., Ruben, R.J., Rybak, L., Van de Water, T.R. (1997). Use of organotypic cultures of Corti's organ to study the protective effects of antioxidant molecules on cisplatin-induced damage of auditory hair cells. *Am. J. Otol.*, *18* (5), 559–571.

Korashy, H.M., Elbekai, R.H., El-Kadi, A.O. (2004). Effects of renal diseases on the regulation and expression of renal and hepatic drug-metabolizing enzymes: a review. *Xenobiotica*, *34* (1), 1–29.

Kossak, B.D., Schmidt-Sommerfeld, E., Schoeller, D.A., Rinaldo, P., Penn, D., Tonsgard, J.H. (1993). Impaired fatty acid oxidation in children on valproic acid and the effect of L-carnitine. *Neurology*, *43* (11), 2362–2368.

Kossor, D.C., Kominami, S., Takemori, S., Colby, H.D. (1991). Role of the steroid 17 alpha-hydroxylase in spironolactone-mediated destruction of adrenal cytochrome P-450. *Mol. Pharmacol.*, *40* (2), 321–325.

Kossor, D.C., Kominami, S., Takemori, S., Colby, H.D. (1992). Destruction of testicular cytochrome P-450 by 7 alpha-thiospironolactone is catalyzed by the 17 alpha-hydroxylase. *J. Steroid Biochem. Mol. Biol.*, *42* (3–4), 421–424.

Koudriakova, T., Iatsimirskaia, E., Utkin, I., Gangl, E., Vouros, P., Storozhuk, E., Orza, D., Marinina, J., Gerber, N. (1998). Metabolism of the human immunodeficiency virus protease inhibitors indinavir and ritonavir by human intestinal microsomes and expressed cytochrome P4503A4/3A5: mechanism-based inactivation of cytochrome P4503A by ritonavir. *Drug Metab. Dispos.*, *26* (6), 552–561.

Krause, R.J., Lash, L.H., Elfarra, A.A. (2003). Human kidney flavin-containing monooxygenases and their potential roles in cysteine s-conjugate metabolism and nephrotoxicity. *J. Pharmacol. Exp. Ther.*, *304*, 185–191.

Krueger, S.K., Williams, D.E. (2005). Mammalian flavin-containing monooxygenases: structure/function, genetic polymorphisms and role in drug metabolism. *Pharmacol. Ther.*, *106*, 357–387.

Lai, W.G., Gardner, I., Zahid, N., Uetrecht, J.P. (2000). Bioactivation and covalent binding of hydroxyfluperlapine in human neutrophils: implications for fluperlapine-induced agranulocytosis. *Drug Metab. Dispos.*, *28* (3), 255–263.

Lai, W.G., Zahid, N., Uetrecht, J.P. (1999). Metabolism of trimethoprim to a reactive iminoquinone methide by activated human neutrophils and hepatic microsomes. *J. Pharmacol. Exp. Ther.*, *291* (1), 292–299.

Lang, D.H., Rettie, A.E. (2000). *In vitro* evaluation of potential *in vivo* probes for human flavin-containing monooxygenase (FMO): metabolism of benzydamine and caffeine by FMO and P450 isoforms. *Br. J. Clin. Pharmacol.*, *50* (4), 311–314.

Lang, D.H., Yeung, C.K., Peter, R.M., Ibarra, C., Gasser, R., Itagaki, K., Philpot, R.M., Rettie, A.E. (1998). Isoform specificity of trimethylamine N-oxygenation by human flavin-containing monooxygenase (FMO) and P450 enzymes: selective catalysis by FMO3. *Biochem. Pharmacol.*, *56*, 1005–1012.

Langston, J.W., Ballard, P., Tetrud, J.W., Irwin, I. (1983). Chronic Parkinsonism in humans due to a product of meperidine-analog synthesis. *Science, 219* (4587), 979–980.

Lautala, P., Ethell, B.T., Taskinen, J., Burchell, B. (2000). The specificity of glucuronidation of entacapone and tolcapone by recombinant human UDP-glucuronosyltransferases. *Drug Metab. Dispos., 28* (11), 1385–1389.

Lavergne, S.N., Danhof, R.S., Volkman, E.M., Trepanier, L.A. (2006). Association of drug-serum protein adducts and anti-drug antibodies in dogs with sulphonamide hypersensitivity: a naturally occurring model of idiosyncratic drug toxicity. *Clin. Exp. Allergy, 36* (7), 907–915.

Lazo, J.S., Humphreys, C.J. (1983). Lack of metabolism as the biochemical basis of bleomycin-induced pulmonary toxicity. *Proc. Natl. Acad. Sci. U.S.A., 80* (10), 3064–3068.

Lecoeur, S., Bonierbale, E., Challine, D., Gautier, J.C., Valadon, P., Dansette, P.M., Catinot, R., Ballet, F., Mansuy, D., Beaune, P.H. (1994). Specificity of *in vitro* covalent binding of tienilic acid metabolites to human liver microsomes in relationship to the type of hepatotoxicity: comparison with two directly hepatotoxic drugs. *Chem. Res. Toxicol., 7* (3), 434–442.

Lee, W.M. (2003). Drug-induced hepatotoxicity. *N. Engl. J. Med., 349* (5), 474–485.

Lenz, W. (1966). Malformations caused by drugs in pregnancy. *Am. J. Dis. Child., 112* (2), 99–106.

Leone, A.M., Kao, L.M., McMillian, M.K., Nie, A.Y., Parker, J.B., Kelley, M.F., Usuki, E., Parkinson, A., Lord, P.G., Johnson, M.D. (2007). Evaluation of felbamate and other antiepileptic drug toxicity potential based on hepatic protein covalent binding and gene expression. *Chem. Res. Toxicol., 20* (4), 600–608.

Lessard, E., Hamelin, B.A., Labbé, L., O'Hara, G., Bélanger, P.M., Turgeon, J. (1999). Involvement of CYP2D6 activity in the N-oxidation of procainamide in man. *Pharmacogenetics, 9* (6), 683–696.

Li, G., Frenz, D.A., Brahmblatt, S., Feghali, J.G., Ruben, R.J., Berggren, D., Arezzo, J., Van De Water, T.R. (2001). Round window membrane delivery of L-methionine provides protection from cisplatin ototoxicity without compromising chemotherapeutic efficacy. *Neurotoxicology, 22* (2), 163–176.

Li, G., Liu, W., Frenz, D. (2006). Cisplatin ototoxicity to the rat inner ear: a role for HMG1 and iNOS. *Neurotoxicology, 27* (1), 22–30.

Lin, L.Y., Di Stefano, E.W., Schmitz, D.A., Hsu, L., Ellis, S.W., Lennard, M.S., Tucker, G.T., Cho, A.K. (1997). Oxidation of methamphetamine and methylenedioxymethamphetamine by CYP2D6. *Drug Metab. Dispos., 25* (9), 1059–1064.

Lindberg, R.L., Syvälahti, E.K. (1986). Metabolism of nomifensine after oral and intravenous administration. *Clin. Pharmacol. Ther., 39* (4), 378–383.

Liu, G.Y., Frank, N., Bartsch, H., Lin, J.K. (1998). Induction of apoptosis by thiuramdisulfides, the reactive metabolites of dithiocarbamates, through coordinative modulation of NFkappaB, c-fos/c-jun and p53 proteins. *Mol. Carcinog., 22* (4), 235–246.

Liu, Z.C., Uetrecht, J.P. (2000). Metabolism of ticlopidine by activated neutrophils: implications for ticlopidine-induced agranulocytosis. *Drug Metab. Dispos., 28* (7), 726–730.

López-Neblina, F., Toledo-Pereyra, L.H. (2006). Phosphoregulation of signal transduction pathways in ischemia and reperfusion. *J. Surg. Res., 134* (2), 292–299.

Lu, W., Uetrecht, J.P. (2007). Possible bioactivation pathways of lamotrigine. *Drug Metab. Dispos., 35* (7), 1050–1056.

Madden, S., Woolf, T.F., Pool, W.F., Park, B.K. (1993). An investigation into the formation of stable, protein-reactive and cytotoxic metabolites from tacrine in vitro. Studies with human and rat liver microsomes. *Biochem. Pharmacol., 46* (1), 13–20.

Maeda, T., Nakamura, R., Kadokami, K., Ogawa, H.I. (2007). Relationship between mutagenicity and reactivity or biodegradability for nitroaromatic compounds. *Environ. Toxicol. Chem., 26* (2), 237–241.

Maggs, J.L., Naisbitt, D.J., Tettey, J.N., Pirmohamed, M., Park, B.K. (2000). Metabolism of lamotrigine to a reactive arene oxide intermediate. *Chem. Res. Toxicol.*, *13* (11), 1075–1081.

Maggs, J.L., Tingle, M.D., Kitteringham, N.R., Park, B.K. (1988). Drug-protein conjugates–XIV. Mechanisms of formation of protein-arylating intermediates from amodiaquine, a myelotoxin and hepatotoxin in man. *Biochem. Pharmacol.*, *37* (2), 303–311.

Mallat, A., Zafrani, E.S., Metreau, J.M., Dhumeaux, D. (1997). Terbinafine-induced prolonged cholestasis with reduction of interlobular bile ducts. *Dig. Dis. Sci.*, *42* (7), 1486–1488.

Mallet, W.G., Mosebrook, D.R., Trush, M.A. (1991). Activation of (+-)-trans-7,8-dihydroxy-7,8-dihydrobenzo[a]pyrene to diolepoxides by human polymorphonuclear leukocytes or myeloperoxidase. *Carcinogenesis*, *12* (3), 521–524.

Mannervik, B., Board, P.G., Hayes, J.D., Listowsky, I., Pearson, W.R. (2005). Nomenclature for mammalian soluble glutathione transferases. *Methods Enzymol.*, *401*, 1–8.

Mannervik, B., Castro, V.M., Danielson, U.H., Tahir, M.K., Hansson, J., Ringborg, U. (1987). Expression of class Pi glutathione transferase in human malignant melanoma cells. *Carcinogenesis*, 8 (12), 1929–1932.

Martin, E.A., Rich, K.J., White, I.N., Woods, K.L., Powles, T.J., Smith, L.L. (1995). 32P-postlabelled DNA adducts in liver obtained from women treated with tamoxifen. *Carcinogenesis*, *16* (7), 1651–1654.

Martin, W.J., II (1983). Nitrofurantoin: evidence for the oxidant injury of lung parenchymal cells. *Am. Rev. Respir. Dis.*, *127* (4), 482–486.

Masiá, M., Padilla, S., Bernal, E., Almenar, M.V., Molina, J., Hernández, I., Graells, M.L., Gutiérrez, F. (2007). Influence of antiretroviral therapy on oxidative stress and cardiovascular risk: a prospective cross-sectional study in HIV-infected patients. *Clin. Ther.*, *29* (7), 1448–1455.

Masubuchi, Y. (2006). Metabolic and non-metabolic factors determining troglitazone hepatotoxicity: a review. *Drug Metab. Pharmacokinet.*, *21* (5), 347–356. Review.

Masubuchi, Y., Horie, T. (1999). Mechanism-based inactivation of cytochrome P450s 1A2 and 3A4 by dihydralazine in human liver microsomes. *Chem. Res. Toxicol.*, *12* (10), 1028–1032.

Masubuchi, Y., Konishi, M., Horie, T. (1999). Imipramine- and mianserin-induced acute cell injury in primary cultured rat hepatocytes: implication of different cytochrome P450 enzymes. *Arch. Toxicol.*, *73* (3), 147–151.

Masubuchi, Y., Nakano, T., Ose, A., Horie, T. (2001). Differential selectivity in carbamazepine-induced inactivation of cytochrome P450 enzymes in rat and human liver. *Arch. Toxicol.*, *75* (9), 538–543.

Masubuchi, Y., Ose, A., Horie, T. (2001). Mechanism-based inactivation of CYP2C11 by diclofenac. *Drug Metab. Dispos.*, *29* (9), 1190–1195.

Matzinger, P. (1994). Tolerance, danger and the extended family. *Annu. Rev. Immunol.*, *12*, 991–1045.

Mays, D.C., Pawluk, L.J., Apseloff, G., Davis, W.B., She, Z.W., Sagone, A.L., Gerber, N. (1995). Metabolism of phenytoin and covalent binding of reactive intermediates in activated human neutrophils. *Biochem. Pharmacol.*, *50* (3), 367–380.

Mazze, R.I., Trudell, J.R., Cousins, M.J. (1971). Methoxyflurane metabolism and renal dysfunction: clinical correlation in man. *Anesthesiology*, *35* (3), 247–252.

Melet, A., Assrir, N., Jean, P., Pilar Lopez-Garcia, M., Marques-Soares, C., Jaouen, M., Dansette, P.M., Sari, M.A., Mansuy, D. (2003). Substrate selectivity of human cytochrome P450 2C9: importance of residues 476, 365 and 114 in recognition of diclofenac and sulfaphenazole and in mechanism-based inactivation by tienilic acid. *Arch. Biochem. Biophys.*, *409* (1), 80–91.

Melnick, S., Cole, P., Anderson, D., Herbst, A. (1987). Rates and risks of diethylstilbestrol-related clear-cell adenocarcinoma of the vagina and cervix. An update. *N. Engl. J. Med.*, *316* (9), 514–516.

Menard, R.H., Guenthner, T.M., Kon, H., Gillette, J.R. (1979). Studies on the destruction of adrenal and testicular cytochrome P-450 by spironolactone. Requirement for the 7alpha-thio group and evidence for the loss of the heme and apoproteins of cytochrome P-450. *J. Biol. Chem.*, *254* (5), 1726–1733.

Menon, K.V.N., Angulo, P., Lindor, K.D. (2001). Severe cholestatic hepatitis from troglitazone in a patient with nonalcoholic steatohepatitis and diabetes mellitus. *Am. J. Gastroenterol.*, *96* (5), 1631–1634.

Meuldermans, W., Van Peer, A., Hendrickx, J., Lauwers, W., Swysen, E., Bockx, M., Woestenborghs, R., Heykants, J. (1988). Excretion and biotransformation of cisapride in dogs and humans after oral administration. *Drug Metab. Dispos.*, *16* (3), 403–409.

Miller, J.L., Trepanier, L.A. (2002). Inhibition by atovaquone of CYP2C9-mediated sulphamethoxazole hydroxylamine formation. *Eur. J. Clin. Pharmacol.*, *58* (1), 69–72.

Miners, J.O., Knights, K.M., Houston, J.B., Mackenzie, P.I. (2006). *In vitro-in vivo* correlation for drugs and other compounds eliminated by glucuronidation in humans: pitfalls and promises. *Biochem. Pharmacol.*, *71*, 1531–1539.

Miyamoto, G., Zahid, N., Uetrecht, J.P. (1997). Oxidation of diclofenac to reactive intermediates by neutrophils, myeloperoxidase and hypochlorous acid. *Chem. Res. Toxicol.*, *10* (4), 414–419.

Mizutani, T., Yoshida, K., Kawazoe, S. (1993). Possible role of thioformamide as a proximate toxicant in the nephrotoxicity of thiabendazole and related thiazoles in glutathione-depleted mice: structure-toxicity and metabolic studies. *Chem. Res. Toxicol.*, *6* (2), 174–179.

Mushiroda, T., Ariyoshi, N., Yokoi, T., Takahara, E., Nagata, O., Kato, H., Kamataki, T. (2001). Accumulation of the 1-methyl-4-phenylpyridinium ion in suncus (Suncus murinus) brain: implication for flavin-containing monooxygenase activity in brain microvessels. *Chem. Res. Toxicol.*, *14* (2), 228–232.

Mutlib, A.E., Gerson, R.J., Meunier, P.C., Haley, P.J., Chen, H., Gan, L.S., Davies, M.H., Gemzik, B., Christ, D.D., Krahn, D.F., Markwalder, J.A., Seitz, S.P., Robertson, R.T., Miwa, G.T. (2000). The species-dependent metabolism of efavirenz produces a nephrotoxic glutathione conjugate in rats. *Toxicol. Appl. Pharmacol.*, *169* (1), 102–113.

Mutlib, A.E., Shockcor, J., Chen, S.Y., Espina, R.J., Pinto, D.J., Orwat, M.J., Prakash, S.R., Gan, L.S. (2002). Disposition of 1-[3-(aminomethyl)phenyl]-N-[3-fluoro-2′- (methylsulfonyl)-[1,1′-biphenyl]-4-yl]-3-(trifluoromethyl)- 1H-pyrazole-5-carboxamide (DPC 423) by novel metabolic pathways. Characterization of unusual metabolites by liquid chromatography/mass spectrometry and NMR. *Chem. Res. Toxicol.*, *15* (1), 48–62.

Naisbitt, D.J., Williams, D.P., O'Neill, P.M., Maggs, J.L., Willock, D.J., Pirmohamed, M., Park, B.K. (1998). Metabolism-dependent neutrophil cytotoxicity of amodiaquine: a comparison with pyronaridine and related antimalarial drugs. *Chem. Res. Toxicol.*, *11* (12), 1586–1595.

Nakajima, T., Elovaara, E., Anttila, S., Hirvonen, A., Camus, A.M., Hayes, J.D., Ketterer, B., Vainio, H. (1995). Expression and polymorphism of glutathione S-transferase in human lungs: risk factors in smoking-related lung cancer. *Carcinogenesis*, *16* (4), 707–711.

O'Brien, P.J. (2000). Peroxidases. *Chem. Biol. Interact.*, *129* (1–2), 113–139. Review.

O'Donnell, J.P., Dalvie, D.K., Kalgutkar, A.S., Obach, R.S. (2003). Mechanism-based inactivation of human recombinant P450 2C9 by the nonsteroidal anti-inflammatory drug suprofen. *Drug Metab. Dispos.*, *31* (11), 1369–1377.

O'Neil, M.G., Perdun, C.S., Wilson, M.B., McGown, S.T., Patel, S. (1996). Felbamate-associated fatal acute hepatic necrosis. *Neurology*, *46* (5), 1457–1459.

Oh, E.J., Chae, H.J., Park, Y.J., Park, J.W., Han, K. (2007). Agranulocytosis, plasmacytosis and thrombocytosis due to methimazole-induced bone marrow toxicity. *Am. J. Hematol.*, *82* (6), 500.

Okawa, H., Motohashi, H., Kobayashi, A., Aburatani, H., Kensler, T.W., Yamamoto, M. (2006). Hepatocyte-specific deletion of the keap1 gene activates Nrf2 and confers potent resistance against acute drug toxicity. *Biochem. Biophys. Res. Commun.*, *339* (1), 79–88.

Olsen, J., Li, C., Bjørnsdottir, I., Sidenius, U., Hansen, S.H., Benet, L.Z. (2005). *In vitro* and *in vivo* studies on acyl-coenzyme A-dependent bioactivation of zomepirac in rats. *Chem. Res. Toxicol.*, *18* (11), 1729–1736.

Opii, W.O., Sultana, R., Abdul, H.M., Ansari, M.A., Nath, A., Butterfield, D.A. (2007). Oxidative stress and toxicity induced by the nucleoside reverse transcriptase inhibitor (NRTI)—2′,3′-dideoxycytidine (ddC): relevance to HIV-dementia. *Exp. Neurol.*, *204* (1), 29–38.

Ostapowicz, G., Fontana, R.J., Schiødt, F.V., Larson, A., Davern, T.J., Han, S.H., McCashland, T.M., Shakil, A.O., Hay, J.E., Hynan, L., Crippin, J.S., Blei, A.T., Samuel, G., Reisch, J., Lee, WM, U.S. Acute Liver Failure Study Group (2002). Results of a prospective study of acute liver failure at 17 tertiary care centers in the United States. *Ann. Intern. Med.*, *137* (12), 947–954. Summary for patients in: Ann Intern Med. 2002 Dec 17;137(12):I24.

Padovan, E., Bauer, T., Tongio, M.M., Kalbacher, H., Weltzien, H.U. (1997). Penicilloyl peptides are recognized as T cell antigenic determinants in penicillin allergy. *Eur. J. Immunol.*, *27* (6), 1303–1307.

Park, B.K., Pirmohamed, M., Kitteringham, N.R. (1995). The role of cytochrome P450 enzymes in hepatic and extrahepatic human drug toxicity. *Pharmacol. Ther.*, *68* (3), 385–424.

Park, S.A., Choi, K.S., Bang, J.H., Huh, K., Kim, S.U. (2000). Cisplatin-induced apoptotic cell death in mouse hybrid neurons is blocked by antioxidants through suppression of cisplatin-mediated accumulation of p53 but not of Fas/Fas ligand. *J. Neurochem.*, *75* (3), 946–953.

Park, B.K., Kitteringham, N.R., Maggs, J.L., Pirmohamed, M., Williams, D.P. (2005). The role of metabolic activation in drug-induced hepatotoxicity. *Annu. Rev. Pharmacol. Toxicol.*, *45*, 177–202. Review.

Parman, T., Chen, G., Wells, P.G. (1998). Free radical intermediates of phenytoin and related teratogens. Prostaglandin H synthase-catalyzed bioactivation, electron paramagnetic resonance spectrometry, and photochemical product analysis. *J. Biol. Chem.*, *273* (39), 25079–25088.

Parman, T., Wiley, M.J., Wells, P.G. (1999). Free radical-mediated oxidative DNA damage in the mechanism of thalidomide teratogenicity. *Nat. Med.*, *5* (5), 582–585.

Parra, J., Iriarte, J., Pierre-Louis, S.J. (1996). Valproate toxicity. *Neurology*, *47* (6), 1608.

Patten, C.J., Thomas, P.E., Guy, R.L., Lee, M., Gonzalez, F.J., Guengerich, F.P., Yang, C.S. (1993). Cytochrome P450 enzymes involved in acetaminophen activation by rat and human liver microsomes and their kinetics. *Chem. Res. Toxicol.*, *6* (4), 511–518.

Pearce, R.E., Uetrecht, J.P., Leeder, J.S. (2005). Pathways of carbamazepine bioactivation *in vitro*: II. The role of human cytochrome P450 enzymes in the formation of 2-hydroxyiminostilbene. *Drug Metab. Dispos.*, *33* (12), 1819–1826.

Pemble, S., Schroeder, K.R., Spencer, S.R., Meyer, D.J., Hallier, E., Bolt, H.M., Ketterer, B., Taylor, J.B. (1994). Human glutathione S-transferase theta (GSTT1): cDNA cloning and the characterization of a genetic polymorphism. *Biochem. J.*, *300* (Pt 1), 271–276.

Pemble, S.E., Wardle, A.F., Taylor, J.B. (1996). Glutathione S-transferase class Kappa: characterization by the cloning of rat mitochondrial GST and identification of a human homologue. *Biochem. J.*, *319* (Pt 3), 749–754.

Perez-Reyes, E., Kalyanaraman, B., Mason, R.P. (1980). The reductive metabolism of metronidazole and ronidazole by aerobic liver microsomes. *Mol. Pharmacol.*, *17* (2), 239–244.

Periti, P., Mazzei, T., Mini, E., Novelli, A. (1992). Pharmacokinetic drug interactions of macrolides. *Clin. Pharmacokinet.*, *23* (2), 106–131. Review. Erratum in: *Clin Pharmacokinet*, (1993) *24* (1):70.

Peters, J.H., Miller, K.S., Brown, P. (1965). Studies on the metabolic basis for the genetically determined capacities for isoniazid inactivation in man. *J. Pharmacol. Exp. Ther.*, *150* (2), 298–304.

Pichler, W.J., Beeler, A., Keller, M., Lerch, M., Posadas, S., Schmid, D., Spanou, Z., Zawodniak, A., Gerber, B. (2006). Pharmacological interaction of drugs with immune receptors: the p-i concept. *Allergol. Int.*, *55* (1), 17–25. Review.

Piepho, R.W. (2000). Overview of the angiotensin-converting-enzyme inhibitors. *Am. J. Health Syst. Pharm.*, *57*, S3–7.Suppl 1.

Pirmohamed, M., Williams, D., Madden, S., Templeton, E., Park, B.K. (1995). Metabolism and bioactivation of clozapine by human liver *in vitro*. *J. Pharmacol. Exp. Ther.*, *272* (3), 984–990.

Plowchalk, D.R., Mattison, D.R. (1991). Phosphoramide mustard is responsible for the ovarian toxicity of cyclophosphamide. *Toxicol. Appl. Pharmacol.*, *107* (3), 472–481.

Pollack, G.M., Brouwer, K.L. (1991). Physiologic and metabolic influences on enterohepatic recirculation: simulations based upon the disposition of valproic acid in the rat. *J. Pharmacokinet. Biopharm.*, *19* (2), 189–225.

Poon, G.K., Chen, Q., Teffera, Y., Ngui, J.S., Griffin, P.R., Braun, M.P., Doss, G.A., Freeden, C., Stearns, R.A., Evans, D.C., Baillie, T.A., Tang, W. (2001). Bioactivation of diclofenac via benzoquinone imine intermediates-identification of urinary mercapturic acid derivatives in rats and humans. *Drug Metab. Dispos.*, *29* (12), 1608–1613.

Popović, M., Nierkens, S., Pieters, R., Uetrecht, J. (2004). Investigating the role of 2-phenylpropenal in felbamate-induced idiosyncratic drug reactions. *Chem. Res. Toxicol.*, *17* (12), 1568–1576.

Ratra, G.S., Morgan, W.A., Mullervy, J., Powell, C.J., Wright, M.C. (1998). Methapyrilene hepatotoxicity is associated with oxidative stress, mitochondrial disfunction and is prevented by the Ca2+ channel blocker verapamil. *Toxicology*, *130* (2–3), 79–93.

Renton, K.W., Knickle, L.C. (1990). Regulation of hepatic cytochrome P-450 during infectious disease. *Can. J. Physiol. Pharmacol.*, *68* (6), 777–781. Review.

Ring, B.J., Wrighton, S.A., Aldridge, S.L., Hansen, K., Haehner, B., Shipley, L.A. (1999). Flavin-containing monooxygenase-mediated N-oxidation of the M(1)-muscarinic agonist xanomeline. *Drug Metab. Dispos.*, *27* (10), 1099–1103.

Ritter, J.K., Crawford, J.M., Owens, I.S. (1991). Cloning of two human liver bilirubin UDP-glucuronosyltransferase cDNAs with expression in COS-1 cells. *J. Biol. Chem.*, *266* (2), 1043–1047.

Rodriguez, R.J., Buckholz, C.J. (2003). Hepatotoxicity of ketoconazole in Sprague-Dawley rats: glutathione depletion, flavin-containing monooxygenases-mediated bioactivation and hepatic covalent binding. *Xenobiotica*, *33* (4), 429–441.

Rodriguez, R.J., Miranda, C.L. (2000). Isoform specificity of N-deacetyl ketoconazole by human and rabbit flavin-containing monooxygenases. *Drug Metab. Dispos.*, *28* (9), 1083–1086.

Roman, R.J., Carter, J.R., North, W.C., Kauker, M.L. (1977). Renal tubular site of action of fluoride in Fischer 344 rats. *Anesthesiology*, *46* (4), 260–264.

Rosa, F.W. (1991). Spina bifida in infants of women treated with carbamazepine during pregnancy. *N. Engl. J. Med.*, *324* (10), 674–677. Review.

Roujeau, J.C. (2006). Immune mechanisms in drug allergy. *Allergol. Int.*, *55* (1), 27–33. Review.

Salama, A., Mueller-Eckhardt, C. (1985). The role of metabolite-specific antibodies in nomifensine-dependent immune hemolytic anemia. *N. Engl. J. Med.*, *313* (8), 469–474.

Sanfeliu, C., Wright, J.M., Kim, S.U. (1999). Neurotoxicity of isoniazid and its metabolites in cultures of mouse dorsal root ganglion neurons and hybrid neuronal cell line. *Neurotoxicology*, *20* (6), 935–944.

Sanford-Driscoll, M., Knodel, L.C. (1986). Induction of hemolytic anemia by nonsteroidal antiinflammatory drugs. *Drug Intell. Clin. Pharm.*, *20* (12), 925–934. Review.

Sarich, T.C., Adams, S.P., Petricca, G., Wright, J.M. (1999). Inhibition of isoniazid-induced hepato-toxicity in rabbits by pretreatment with an amidase inhibitor. *J. Pharmacol. Exp. Ther.*, *289* (2), 695–702.

Scheulen, M.E., Kappus, H., Thyssen, D., Schmidt, C.G. (1981). Redox cycling of Fe(III)-bleomycin by NADPH-cytochrome P450 reductase. *Biochem. Pharmacol.*, *30* (24), 3385–3388.

Schmekel, B., Karlsson, S.E., Linden, M., Sundström, C., Tegner, H., Venge, P. (1990). Myeloperoxidase in human lung lavage. I. A marker of local neutrophil activity. *Inflammation*, *14* (4), 447–454.

Schnyder, B., Mauri-Hellweg, D., Zanni, M., Bettens, F., Pichler, W.J. (1997). Direct MHC-dependent presentation of the drug sulfamethoxazole to human alphabeta T cell clones. *J. Clin. Invest.*, *100* (1), 136–141.

Scordo, M.G., Pengo, V., Spina, E., Dahl, M.L., Gusella, M., Padrini, R. (2002). Influence of CYP2C9 and CYP2C19 genetic polymorphisms on warfarin maintenance: dose and metabolic clearance. *Clin. Pharmacol. Ther.*, *72* (6), 702–710.

Sebti, S.M., DeLeon, J.C., Lazo, J.S. (1987). Purification, characterization and amino acid composi-tion of rabbit pulmonary bleomycin hydrolase. *Biochemistry*, *26* (14), 4213–4219.

Seeff, L.B., Cuccherini, B.A., Zimmerman, H.J., Adler, E., Benjamin, S.B. (1986). Acetaminophen hepatotoxicity in alcoholics. A therapeutic misadventure. *Ann. Intern. Med.*, *104* (3), 399–404. Review.

Seigneur, J., Trechot, P.F., Hubert, J., Lamy, P. (1988). Pulmonary complications of hormone treat-ment in prostate carcinoma. *Chest*, *93* (5), 1106.

Seitz, S., Boelsterli, U.A. (1998). Diclofenac acyl glucuronide, a major biliary metabolite, is directly involved in small intestinal injury in rats. *Gastroenterology*, *115* (6), 1476–1482.

Sharma, M., Shubert, D.E., Sharma, M., Lewis, J., McGarrigle, B.P., Bofinger, D.P., Olson, J.R. (2003). Biotransformation of tamoxifen in a human endometrial explant culture model. *Chem. Biol. Interact.*, *146* (3), 237–249.

Sheehan, D., Meade, G., Foley, V.M., Dowd, C.A. (2001). Structure, function and evolution of glu-tathione transferases: implications for classification of non-mammalian members of an ancient enzyme superfamily. *Biochem. J.*, *360* (Pt 1), 1–16. Review.

Shin, N.Y., Liu, Q., Stamer, S.L., Liebler, D.C. (2007). Protein targets of reactive electrophiles in human liver microsomes. *Chem. Res. Toxicol.*, *20* (6), 859–867.

Shipkova, M., Armstrong, V.W., Weber, L., Niedmann, P.D., Wieland, E., Haley, J., Tönshoff, B., Oellerich, M. (2002). German Study Group on Mycophenolate Mofetil Therapy in Pediatric Renal Transplant Recipients. Pharmacokinetics and protein adduct formation of the pharma-cologically active acyl glucuronide metabolite of mycophenolic acid in pediatric renal trans-plant recipients. *Ther. Drug Monit.*, *24* (3), 390–399.

Shipkova, M., Strassburg, C.P., Braun, F., Streit, F., Gröne, H.J., Armstrong, V.W., Tukey, R.H., Oellerich, M., Wieland, E. (2001). Glucuronide and glucoside conjugation of mycophenolic acid by human liver, kidney and intestinal microsomes. *Br. J. Pharmacol.*, *132* (5), 1027–1034.

Siraki, A.G., Bonini, M.G., Jiang, J., Ehrenshaft, M., Mason, R.P. (2007). Aminoglutethimide-induced protein free radical formation on myeloperoxidase: a potential mechanism of agranu-locytosis. *Chem. Res. Toxicol.*, *20* (7), 1038–1045.

Slack, J.A., Ford-Hutchinson, A.W., Richold, M., Choi, B.C. (1981). Some biochemical and phar-macological properties of an epoxide metabolite of alclofenac. *Chem. Biol. Interact.*, *34* (1), 95–107.

Sladek, N.E. (1988). Metabolism of oxazaphosphorines. *Pharmacol. Ther.*, *37* (3), 301–355. Review.

Slattery, J.T., Wilson, J.M., Kalhorn, T.F., Nelson, S.D. (1987). Dose-dependent pharmacokinetics of acetaminophen: evidence of glutathione depletion in humans. *Clin. Pharmacol. Ther.*, *41* (4), 413–418.

Slattum, P.W., Cato, A.E., III, Pollack, G.M., Brouwer, K.L. (1995). Age-dependent intestinal hydrolysis of valproate glucuronide in rat. *Xenobiotica*, *25* (3), 229–237.

Smith, P.C., Liu, J.H. (1993). Covalent binding of suprofen acyl glucuronide to albumin in vitro. *Xenobiotica*, *23* (4), 337–348.

Smith, P.C., Benet, L.Z., McDonagh, A.F. (1990). Covalent binding of zomepirac glucuronide to proteins: evidence for a Schiff base mechanism. *Drug Metab. Dispos.*, *18* (5), 639–644.

Smith, P.C., McDonagh, A.F., Benet, L.Z. (1986). Irreversible binding of zomepirac to plasma protein *in vitro* and *in vivo*. *J. Clin. Invest.*, *77* (3), 934–939.

Smith, K.S., Smith, P.L., Heady, T.N., Trugman, J.M., Harman, W.D., Macdonald, T.L. (2003). *In vitro* metabolism of tolcapone to reactive intermediates: relevance to tolcapone liver toxicity. *Chem. Res. Toxicol.*, *16* (2), 123–128.

Smyth, J.F., Bowman, A., Perren, T., Wilkinson, P., Prescott, R.J., Quinn, K.J., Tedeschi, M. (1997). Glutathione reduces the toxicity and improves quality of life of women diagnosed with ovarian cancer treated with cisplatin: results of a double-blind, randomized trial. *Ann. Oncol.*, *8* (6), 569–573.

Spaldin, V., Madden, S., Adams, D.A., Edwards, R.J., Davies, D.S., Park, B.K. (1995). Determination of human hepatic cytochrome P4501A2 activity in vitro use of tacrine as an isoenzyme-specific probe. *Drug Metab. Dispos.*, *23* (9), 929–934.

Spracklin, D.K., Hankins, D.C., Fisher, J.M., Thummel, K.E., Kharasch, E.D. (1997). Cytochrome P450 2E1 is the principal catalyst of human oxidative halothane metabolism *in vitro*. *J. Pharmacol. Exp. Ther.*, *281* (1), 400–411.

Sridar, C., Kent, U.M., Notley, L.M., Gillam, E.M., Hollenberg, P.F. (2002). Effect of tamoxifen on the enzymatic activity of human cytochrome CYP2B6. *J. Pharmacol. Exp. Ther.*, *301* (3), 945–952.

Stephens, J.R., Levy, R.H. (1992). Valproate hepatotoxicity syndrome: hypotheses of pathogenesis. *Pharm. Weekbl Sci.*, *14* (3A), 118–121. Review.

Stiborová, M., Poljaková, J., Ryslavá, H., Dracínský, M., Eckschlager, T., Frei, E. (2007). Mammalian peroxidases activate anticancer drug ellipticine to intermediates forming deoxyguanosine adducts in DNA identical to those found *in vivo* and generated from 12-hydroxyellipticine and 13-hydroxyellipticine. *Int. J. Cancer*, *120* (2), 243–251.

Stiborová, M., Sejbal, J., Borek-Dohalská, L., Aimová, D., Poljaková, J., Forsterová, K., Rupertová, M., Wiesner, J., Hudecek, J., Wiessler, M., Frei, E. (2004). The anticancer drug ellipticine forms covalent DNA adducts, mediated by human cytochromes P450, through metabolism to 13-hydroxyellipticine and ellipticine N2-oxide. *Cancer Res.*, *64* (22), 8374–8380.

Stillwell, T.J., Benson R.C., Jr (1988). Cyclophosphamide-induced hemorrhagic cystitis. A review of 100 patients. *Cancer*, *61* (3), 451–457.

Störmer, E., Roots, I., Brockmöller, J. (2000). Benzydamine N-oxidation as an index reaction reflecting FMO activity in human liver microsomes and impact of FMO3 polymorphisms on enzyme activity. *Br. J. Clin. Pharmacol.*, *50* (6), 553–561.

Stuart-Harris, R.C., Smith, I.E. (1984). Aminoglutethimide in the treatment of advanced breast cancer. *Cancer Treat. Rev.*, *11* (3), 189–204.

Styles, J.A., Davies, A., Lim, C.K., De Matteis, F., Stanley, L.A., White, I.N., Yuan, Z.X., Smith, L.L. (1994). Genotoxicity of tamoxifen, tamoxifen epoxide and toremifene in human lymphoblastoid cells containing human cytochrome P450s. *Carcinogenesis*, *15* (1), 5–9.

Subramanyam, B., Pond, S.M., Eyles, D.W., Whiteford, H.A., Fouda, H.G., Castagnoli N., Jr (1991). Identification of potentially neurotoxic pyridinium metabolite in the urine of schizophrenic patients treated with haloperidol. *Biochem. Biophys. Res. Commun.*, *181* (2), 573–578.

Tafazoli, S., O'Brien, P.J. (2005). Peroxidases: a role in the metabolism and side effects of drugs. *Drug Discov. Today*, *10* (9), 617–625. Review.

Takanashi, K., Tainaka, H., Kobayashi, K., Yasumori, T., Hosakawa, M., Chiba, K. (2000). CYP2C9 Ile359 and Leu359 variants: enzyme kinetic study with seven substrates. *Pharmacogenetics*, *10* (2), 95–104.

Takasuna, K., Hagiwara, T., Hirohashi, M., Kato, M., Nomura, M., Nagai, E., Yokoi, T., Kamataki, T. (1998). Inhibition of intestinal microflora beta-glucuronidase modifies the distribution of the active metabolite of the antitumor agent, irinotecan hydrochloride (CPT-11) in rats. *Cancer Chemother. Pharmacol.*, *42* (4), 280–286.

Takeda, S., Ishii, Y., Iwanaga, M., Mackenzie, P.I., Nagata, K., Yamazoe, Y., Oguri, K., Yamada, H. (2005). Modulation of UDP-glucuronosyltransferase function by cytochrome P450: evidence for the alteration of UGT2B7-catalyzed glucuronidation of morphine by CYP3A4. *Mol. Pharmacol.*, *67* (3), 665–672.

Tan, A.W., Thong, B.Y., Yip, L.W., Chng, H.H., Ng, S.K. (2005). High-dose intravenous immunoglobulins in the treatment of toxic epidermal necrolysis: an Asian series. *J. Dermatol.*, *32* (1), 1–6.

Tang, W., Stearns, R.A., Wang, R.W., Chiu, S.H., Baillie, T.A. (1999). Roles of human hepatic cytochrome P450s 2C9 and 3A4 in the metabolic activation of diclofenac. *Chem. Res. Toxicol.*, *12* (2), 192–199.

Temple, R. (2006). Hy's law: predicting serious hepatotoxicity. *Pharmacoepidemiol. Drug Saf.*, *15* (4), 241–243.

Timbrell, J.A., Mitchell, J.R., Snodgrass, W.R., Nelson, S.D. (1980). Isoniazid hepatoxicity: the relationship between covalent binding and metabolism *in vivo*. *J. Pharmacol. Exp. Ther.*, *213* (2), 364–369.

Tingle, M.D., Jewell, H., Maggs, J.L., O'Neill, P.M., Park, B.K. (1995). The bioactivation of amodiaquine by human polymorphonuclear leucocytes in vitro: chemical mechanisms and the effects of fluorine substitution. *Biochem. Pharmacol.*, *50* (7), 1113–1119.

Toffoli, G., Cecchin, E. (2004). Uridine diphosphoglucuronosyl transferase and methylenetetrahydrofolate reductase polymorphisms as genomic predictors of toxicity and response to irinotecan-, antifolate- and fluoropyrimidine-based chemotherapy. *J. Chemother.*, *4*, 31–35.

Toffoli, G., Cecchin, E., Corona, G., Russo, A., Buonadonna, A., Pasetto, L.M., D'Andrea, M., Pessa, S., Errante, D., De Pangher, V., Giusto, M., Medici, M., Gaion, F., Sandri, P., Galligioni, E., Bonura, S., Boccalon, M., Biason, P., Frustaci, S. (2006). The role of UGT1A1*28 polymorphism in the pharmacodynamics and pharmacokinetics of irinotecan in patients with metastatic colorectal cancer. *J. Clin. Oncol.*, *24*, 3061–3068.

To-Figueras, J., Gene, M., Gomez-Catalan, J., Galan, C., Firvida, J., Fuentes, M., Rodamilans, M., Huguet, E., Estape, J., Corbella, J. (1996). Glutathione-S-Transferase M1 and codon 72 p53 polymorphisms in a northwestern Mediterranean population and their relation to lung cancer susceptibility. *Cancer Epidemiol Biomarkers Prev.*, *5* (5), 337–42.

Tong, K.I., Kobayashi, A., Katsuoka, F., Yamamoto, M. (2006). Two-site substrate recognition model for the Keap1-Nrf2 system: a hinge and latch mechanism. *Biol. Chem.*, *387* (10–11), 1311–1320. Review.

Tong, V., Teng, X.W., Karagiozov, S., Chang, T.K., Abbott, F.S. (2005). Valproic acid glucuronidation is associated with increases in 15-F2t-isoprostane in rats. *Free Radic. Biol. Med.*, *1*; 38 (11), 1471–1483.

Townsend, D.M., Deng, M., Zhang, L., Lapus, M.G., Hanigan, M.H. (2003). Metabolism of cisplatin to a nephrotoxin in proximal tubule cells. *J. Am. Soc. Nephrol.*, *14* (1), 1–10.

Townsend, D.M., Hanigan, M.H. (2002). Inhibition of gamma-glutamyl transpeptidase or cysteine S-conjugate beta-lyase activity blocks the nephrotoxicity of cisplatin in mice. *J. Pharmacol. Exp. Ther.*, *300* (1), 142–148.

Toyoda, Y., Tsuchida, A., Iwami, E., Miwa, I. (2001). Toxic effect of troglitazone on cultured rat hepatocytes. *Life Sci.*, *68* (16), 1867–1876.

Traversa, G., Bianchi, C., Da Cas, R., Abraha, I., Menniti-Ippolito, F., Venegoni, M. (2003). Cohort study of hepatotoxicity associated with nimesulide and other non-steroidal anti-inflammatory drugs. *BMJ*, *327* (7405), 18–22.

Treinen-Moslen, M., Kanz, M.F. (2006). Intestinal tract injury by drugs: importance of metabolite delivery by yellowbile road. *Pharmacol. Ther.*, *112* (3), 649–667. Review.

Trepanier, L.A., Ray, K., Winand, N.J., Spielberg, S.P., Cribb, A.E. (1997). Cytosolic arylamine N-acetyltransferase (NAT) deficiency in the dog and other canids due to an absence of NAT genes. *Biochem. Pharmacol.*, *54* (1), 73–80.

Tsai-Turton, M., Luong, B.T., Tan, Y., Luderer, U. (2007). Cyclophosphamide-induced apoptosis in COV434 human granulosa cells involves oxidative stress and glutathione depletion. *Toxicol. Sci.*, *98* (1), 216–230.

Tucker, G.T., Lennard, M.S., Ellis, S.W., Woods, H.F., Cho, A.K., Lin, L.Y., Hiratsuka, A., Schmitz, D.A., Chu, T.Y. (1994). The demethylenation of methylenedioxymethamphetamine ("ecstasy") by debrisoquine hydroxylase (CYP2D6). *Biochem. Pharmacol.*, *47* (7), 1151–1156.

Tukey, R.H., Strassburg, C.P. (2000). Human UDP-glucuronosyltransferases: metabolism, expression and disease. *Annu. Rev. Pharmacol. Toxicol.*, *40*, 581–616.

Uetrecht, J. (2001). Prediction of a new drug's potential to cause idiosyncratic reactions. *Curr. Opin. Drug Discov. Devel.*, *4* (1), 55–59. Review.

Uetrecht, J., Zahid, N., Shear, N.H., Biggar, W.D. (1988). Metabolism of dapsone to a hydroxylamine by human neutrophils and mononuclear cells. *J. Pharmacol. Exp. Ther.*, *245* (1), 274–279.

Uetrecht, J.P. (1992). The role of leukocyte-generated reactive metabolites in the pathogenesis of idiosyncratic drug reactions. *Drug Metab. Rev.*, *24* (3), 299–366. Review.

Uetrecht, J.P. (1999). New concepts in immunology relevant to idiosyncratic drug reactions: the "danger hypothesis" and innate immune system. *Chem. Res. Toxicol.*, *12* (5), 387–395. Review.

Uetrecht, J.P., Zahid, N., Whitfield, D. (1994). Metabolism of vesnarinone by activated neutrophils: implications for vesnarinone-induced agranulocytosis. *J. Pharmacol. Exp. Ther.*, *270* (3), 865–872.

Usuki, E., Van der Schyf, C.J., Castagnoli N., Jr (1998). Metabolism of haloperidol and its tetrahydropyridine dehydration product HPTP. *Drug Metab. Rev.*, *30* (4), 809–826.

van Berkel, W.J., Kamerbeek, N.M., Fraaije, M.W. (2006). Flavoprotein monooxygenases, a diverse class of oxidative biocatalysts. *J. Biotechnol.*, *124* (4), 670–689.

Vatsis, K.P., Weber, W.W. (1993). Structural heterogeneity of Caucasian N-acetyltransferase at the NAT1 gene locus. *Arch. Biochem. Biophys.*, *301* (1), 71–76.

Vella, A., de Groen, P.C., Dinneen, S.F. (1998). Fatal hepatotoxicity associated with troglitazone. *Ann. Intern. Med.*, *129* (12), 1080.

Viard-Leveugle, I., Bullani, R.R., Meda, P., Micheau, O., Limat, A., Saurat, J.H., Tschopp, J., French, L.E. (2003). Intracellular localization of keratinocyte Fas ligand explains lack of cytolytic activity under physiological conditions. *J. Biol. Chem.*, *278* (18), 16183–16188.

Vitseva, O., Flockhart, D.A., Jin, Y., Varghese, S., Freedman, J.E. (2005). The effects of tamoxifen and its metabolites on platelet function and release of reactive oxygen intermediates. *J. Pharmacol. Exp. Ther.*, *312* (3), 1144–1150.

Vogel, C. (2000). Prostaglandin H synthases and their importance in chemical toxicity. *Curr. Drug Metab.*, *1* (4), 391–404. Review.

Vyas, P.M., Roychowdhury, S., Woster, P.M., Svensson, C.K. (2005). Reactive oxygen species generation and its role in the differential cytotoxicity of the arylhydroxylamine metabolites of sulfamethoxazole and dapsone in normal human epidermal keratinocytes. *Biochem. Pharmacol.*, *70* (2), 275–286.

Wade, L.T., Kenna, J.G., Caldwell, J. (1997). Immunochemical identification of mouse hepatic protein adducts derived from the nonsteroidal anti-inflammatory drugs diclofenac, sulindac and ibuprofen. *Chem. Res. Toxicol.*, *10* (5), 546–555.

Wadkins, R.M., Hyatt, J.L., Yoon, K.J., Morton, C.L., Lee, R.E., Damodaran, K., Beroza, P., Danks, M.K., Potter, P.M. (2004). Discovery of novel selective inhibitors of human intestinal carboxylesterase for the amelioration of irinotecan-induced diarrhea: synthesis, quantitative structure-activity relationship analysis and biological activity. *Mol. Pharmacol.*, *65* (6), 1336–1343.

Walker, A.M. (1997). Quantitative studies of the risk of serious hepatic injury in persons using nonsteroidal antiinflammatory drugs. *Arthritis. Rheum.*, *40* (2), 201–208.

Walsh, J.S., Reese, M.J., Thurmond, L.M. (2002). The metabolic activation of abacavir by human liver cytosol and expressed human alcohol dehydrogenase isozymes. *Chem. Biol. Interact.*, *142* (1–2), 135–154.

Wang, J., Davis, M., Li, F., Azam, F., Scatina, J., Talaat, R. (2004). A novel approach for predicting acyl glucuronide reactivity via Schiff base formation: development of rapidly formed peptide adducts for LC/MS/MS measurements. *Chem. Res. Toxicol.*, *17* (9), 1206–1216.

Wang, M., Gorrell, M.D., McGaughan, G.W., Dickinson, R.G. (2001). Dipeptidyl peptidase IV is a target for covalent adduct formation with the acyl glucuronide metabolite of the anti-inflammatory drug zomepirac. *Life Sci.*, *68* (7), 785–797.

Wang, Y., Gray, J.P., Mishin, V., Heck, D.E., Laskin, D.L., Laskin, J.D. (2008). Role of cytochrome P450 reductase in nitrofurantoin-induced redox cycling and cytotoxicity. *Free Radic. Biol. Med.*, *44* (6), 1169–1179.

Weinshilboum, R., Wang, L. (2004). Pharmacogenomics: bench to bedside. *Nat. Rev. Drug Discov.*, *3* (9), 739–748. Review.

Weissman, J., Trevor, A., Chiba, K., Peterson, L.A., Caldera, P., Castagnoli, N.Jr, , Baillie, T. (1985). Metabolism of the nigrostriatal toxin 1-methyl-4-phenyl-1,2,3,6-tetrahydropyridine by liver homogenate fractions. *J. Med. Chem.*, *28* (8), 997–1001.

Wells, P.G., Bhuller, Y., Chen, C.S., Jeng, W., Kasapinovic, S., Kennedy, J.C., Kim, P.M., Laposa, R.R., McCallum, G.P., Nicol, C.J., Parman, T., Wiley, M.J., Wong, A.W. (2005). Molecular and biochemical mechanisms in teratogenesis involving reactive oxygen species. *Toxicol. Appl. Pharmacol.*, *207* (Suppl. 2), 354–366.

Wells, P.G., Kim, P.M., Laposa, R.R., Nicol, C.J., Parman, T., Winn, L.M. (1997). Oxidative damage in chemical teratogenesis. *Mutat. Res.*, *396* (1–2), 65–78. Review.

Whetstine, J.R., Yueh, M.F., McCarver, D.G., Williams, D.E., Park, C.S., Kang, J.H., Cha, Y.N., Dolphin, C.T., Shephard, E.A., Phillips, I.R., Hines, R.N. (2000). Ethnic differences in human flavin-containing monooxygenase 2 (FMO2) polymorphisms: detection of expressed protein in African-Americans. *Toxicol. Appl. Pharmacol.*, *168*, 216–224.

Williams, A.C., Ramsden, D.B. (2005). Autotoxicity, methylation and a road to the prevention of Parkinson's disease. *J. Clin. Neurosci.*, *12* (1), 6–11. Review.

Williams, D.P., Pirmohamed, M., Naisbitt, D.J., Maggs, J.L., Park, B.K. (1997). Neutrophil cytotoxicity of the chemically reactive metabolite(s) of clozapine: possible role in agranulocytosis. *J. Pharmacol. Exp. Ther.*, *283* (3), 1375–1382.

Williams, J.A., Hyland, R., Jones, B.C., Smith, D.A., Hurst, S., Goosen, T.C., Peterkin, V., Koup, J.R., Ball, S.E. (2004). Drug-drug interactions for UDP-glucuronosyltransferase substrates: a pharmacokinetic explanation for typically observed low exposure (AUCi/AUC) ratios. *Drug Metab. Dispos.*, *32* (11), 1201–1208.

Witten, C.M. (1989). Pulmonary toxicity of nitrofurantoin. *Arch. Phys. Med. Rehabil.*, *70* (1), 55–57.

Woeber, KA. (2002). Methimazole-induced hepatotoxicity. *Endocr. Pract.*, *8* (3), 222–224. Review.

Xie, H.G., Prasad, H.C., Kim, R.B., Stein, C.M. (2002). CYP2C9 allelic variants: ethnic distribution and functional significance. *Adv. Drug Deliv. Rev.*, *54* (10), 1257–1270.

Xu, L.L., Liu, G., Miller, D.P., Zhou, W., Lynch, T.J., Wain, J.C., Su, L., Christiani, D.C. (2002). Counterpoint: the myeloperoxidase $^{-463}$G → a polymorphism does not decrease lung cancer susceptibility in Caucasians. *Cancer Epidemiol. Biomarkers Prev.*, *11* (12), 1555–1559.

Yamamoto, Y., Nakajima, M., Yamazaki, H., Yokoi, T. (2001). Cytotoxicity and apoptosis produced by troglitazone in human hepatoma cells. *Life Sci.*, *70* (4), 471–482.

Yamamoto, Y., Yamazaki, H., Ikeda, T., Watanabe, T., Iwabuchi, H., Nakajima, M., Yokoi, T. (2002). Formation of a novel quinone epoxide metabolite of troglitazone with cytotoxicity to HepG2 cells. *Drug Metab. Dispos.*, *30* (2), 155–160.

You, Q., Cheng, L., Reilly, T.P., Wegmann, D., Ju, C. (2006). Role of neutrophils in a mouse model of halothane-induced liver injury. *Hepatology*, *44* (6), 1421–1431.

Yukinaga, H., Takami, T., Shioyama, S.H., Tozuka, Z., Masumoto, H., Okazaki, O., Sudo, K. (2007). Identification of cytochrome P450 3A4 modification site with reactive metabolite using linear ion trap-Fourier transform mass spectrometry. *Chem. Res. Toxicol.*, *20* (10), 1373–1378.

Yunis, A.A., Miller, A.M., Salem, Z., Corbett, M.D., Arimura, GK. (1980). Nitroso-chloramphenicol possible mediator in chloramphenicol-induced aplastic anemia. *J. Lab. Clin. Med.*, *96* (1), 36–46.

Zhang, H., Cook, J., Nickel, J., Yu, R., Stecker, K., Myers, K., Dean, N.M. (2000). Reduction of liver Fas expression by an antisense oligonucleotide protects mice from fulminant hepatitis. *Nat. Biotechnol.*, *18* (8), 862–867.

Zhang, L., Cooper, A.J., Krasnikov, B.F., Xu, H., Bubber, P., Pinto, J.T., Gibson, G.E., Hanigan, M.H. (2006). Cisplatin-induced toxicity is associated with platinum deposition in mouse kidney mitochondria *in vivo* and with selective inactivation of the alpha-ketoglutarate dehydrogenase complex in LLC-PK1 cells. *Biochemistry*, *45* (29), 8959–8971.

Zhao, P., Slattery, J.T. (2002). Effects of ethanol dose and ethanol withdrawal on rat liver mitochondrial glutathione: implication of potentiated acetaminophen toxicity in alcoholics. *Drug Metab. Dispos.*, *30* (12), 1413–1417.

Ziegler, D.M. (1993). Recent studies on the structure and function of multisubstrate flavin-containing monooxygenases. *Annu. Rev. Pharmacol. Toxicol.*, *33*, 179–199.

Zimmerman, H.J., Maddrey, W.C. (1995). Acetaminophen (paracetamol) hepatotoxicity with regular intake of alcohol: analysis of instances of therapeutic misadventure. *Hepatology*, *22* (3), 767–773. Erratum in Hepatology 1995 Dec, 22 (6), 1898.

Zuckerman, A., Bolan, E., de Paulis, T., Schmidt, D., Spector, S., Pasternak, G.W. (1999). Pharmacological characterization of morphine-6-sulfate and codeine-6-sulfate. *Brain Res.*, *842* (1), 1–5.

Zuidema, J., Hilbers-Modderman, E.S., Merkus, F.W. (1986). Clinical pharmacokinetics of dapsone. *Clin. Pharmacokinet.*, *11* (4), 299–315.

CHAPTER 22

Allergic Reactions to Drugs

MARK P. GRILLO

> Now this is not the end. It is not even the beginning of the end. But it is, perhaps, the end of the beginning.[1]

22.1 INTRODUCTION

There are many pharmaceutical scientists today who say that one of the most difficult obstacles to overcome in drug discovery is the prediction, and therefore the knowledge-able prevention, of idiosyncratic allergic reactions to drugs. The immune system is a highly complex and developed system and is our natural defense in fighting varied pathogenic organisms. The more we have learned about the immune system over the last two decades, due in large part to molecular advancements in immunology, the more it has revealed for researchers to come closer to understanding the mechanism(s) of allergic reactions to drugs (Uetrecht, 2008). Unfortunately, the immune system is susceptible to devastating diseases. Such diseases can occur when the immune system does not differentiate between host cells/macromolecules and foreign substances which leads to autoimmune-type disorders like multiple sclerosis, systemic lupus, rheumatoid arthritis, type I diabetes, and inflammatory bowel disease. In this chapter, some of the proposed chemical mechanisms by which example drugs cause allergic reactions will be discussed. Allergic reactions to drugs can lead to potentially life-threatening reactions such as asthma, anaphylaxis, dermatitis, hepatitis, hemolytic anemia, lupus, nephritis, Stevens–Johnson syndrome, toxic epidermal necrolysis, urticaria, and vasculitis (Dansette *et al.*, 1998; Ewan, 1998; Zimmerman, 1999; Fung *et al.*, 2001). These drug-induced immune-based diseases can create an enormous burden to affected patients and to the respective pharmaceutical companies involved in their manufacturing. There

---

[1] A quotation from Winston Churchill—*Lord Mayor's Luncheon, Mansion House following the victory at El Alameinin North Africa, London, 10 November 1942* (http://www.winstonchurchill.org/).

Drug Metabolism Handbook: Concepts and Applications, Edited by Ala F. Nassar,
Paul F. Hollenberg, and JoAnn Scatina
Copyright © 2009 by John Wiley & Sons, Inc.

were 121 drugs withdrawn from the worldwide market between 1960 and 1999, where hepatic (26.2%), hematologic (10.5%), dermatologic (6.3%), and anaphylaxis (3.3%) were among the top six toxicological reasons for their withdrawal (Fung et al., 2001). Such toxicities can deprive patients of organ function that sometimes results in fatalities. A primary reason why drugs fail today in drug development is due to animal toxicity and human adverse events (Guengerich and MacDonald, 2007). A functional understanding of the association between the use of some drugs and the unpredictable formation of drug-induced hypersensitivity reactions is of major importance in the pharmaceutical industry today (Baillie, 2006; Guengerich and MacDonald, 2007; Kumar et al., 2008). The medical and clinical implications are clear. Immune-based drug-induced allergic diseases may also present scientific opportunities. Studying them could answer critical questions about the nature of the immune system and yield increasingly valuable information as it relates to the discovery of safer drugs. In this chapter, an examination of a set of drug-induced allergic reactions will be reviewed, each mediated by different mechanisms that are driven by the inherently different chemistry and biochemistry of the drug. Examples of drug-induced allergic reactions reviewed in the present chapter include penicillin, a β-lactam antibiotic; felbamate, an anticonvulsant drug; halothane, an inhalation anesthetic; tienilic acid, a diuretic drug; sulfamethoxazole, a sulfonamide-containing antibiotic drug; and benoxaprofen, a carboxylic acid-containing nonsteroidal anti-inflammatory drug (NSAID).

22.2   IMMUNE SYSTEM: A BRIEF OVERVIEW

Both innate (non-memory) and adaptive (specificity and memory) immunity lines of defense against foreign antigens are involved in the formation of allergic reactions (Goldsby et al., 2003). An allergic reaction can be divided into two primary stages, namely sensitization and elicitation. The sensitization stage occurs when a foreign substance, for example a drug–protein adduct or a pathogenic organism, is dealt with by a mechanism known as antigen processing, sometimes referred to as the "gateway" to the immune response. Here the antigen becomes engulfed and then degraded by antigen-presenting cells (macrophages), resulting in peptide fragments (approximately 8–15 amino acids) of the foreign substance (or microorganism) being presented on the surface of antigen-presenting cells, and specifically on the major histocompatibility complex (MHC) for T lymphocyte recognition (Kalish, 1995; Park, Pirmohamed and Kitteringham, 1998). Antigen-specific helper and killer T cells (T lymphocytes from the thymus) recognize, and then become activated by, these antigen-presenting cells. The activated helper and killer T cells then replicate, including the production specialized memory T cells, which are cells that respond to subsequent exposure to a foreign substance or infection. Importantly, killer T cells (which destroy cells containing the foreign antigen, i.e. cellular-mediated immunity) require helper T cells for activation. Helper T cells also function in the activation of B cells which are important in the elicitation stage of an allergic reaction. The elicitation (inflammation) stage begins with immature B cells (lymphocytes that produce antibodies) found in bone marrow. Once B cells become mature, which occurs in secondary lymphoid organs, they interact with, and become further activated by, free specific antigens in the blood. B cells are important because they function in the development of antibody-mediated immunity (i.e. humoral immunity). Just as in the case of killer T cell activation, helper T cells also activate B cells, leading to B cell replication and subsequent production of antibodies that bind to specific antigens. Then, either a complement system destroys the specific antigen, or phagocytic cells engulf the antigen prior to their degradation. Memory B cells, also formed upon B cell activation by helper T cells, are very important because they can respond to subsequent exposures of an antigen, resulting in an immune response sometimes associated with immune-mediated drug toxicity. The present chapter is confined primarily to the metabolism of drugs as it applies to the elicitation of allergic reactions; therefore, the reader is directed to a more in-depth review of the cellular basis of hypersensitivity reactions (Park, Pirmohamed and Kitteringham, 1998; Janeway *et al.*, 2005) for further reading.

22.3 DRUG METABOLISM AND THE HAPTEN HYPOTHESIS

A principal working hypothesis still used today since its inception in 1935 is the hapten hypothesis, where a drug or reactive drug metabolite (see Chapter 23) becomes covalently bound to protein, is recognized by the immune system as foreign, and leads to an allergic-type reaction (Landsteiner and Jacobs, 1935; Park, Coleman and Kitteringham, 1987; Uetrecht, 2008) (Fig. 22.1). For example, there are four stages of allergic hepatitis mediated by reactive metabolites formed in the liver. The first stage is metabolic activation of a xenobiotic to a reactive metabolite leading to the formation of a hepatocellular protein adduct. The second stage is subcellular trafficking and then to

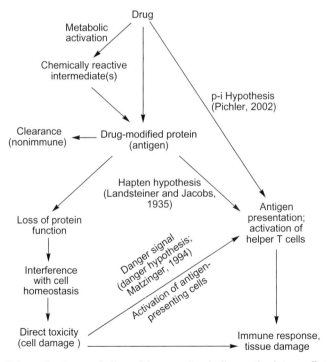

Figure 22.1 Scheme for the metabolism of drugs to chemically reactive intermediates leading to direct toxicity or immune-mediated toxicity (adapted from Pirmohamed, Madden and Park, 1996 and Dansette *et al.*, 1998).

cell surface expression of protein adduct fragments. Then, in the third stage, antigen presentation and immune recognition promote an antibody response and cellular sensitization. Finally, in the fourth stage, an immune attack on adducted hepatocytes leads to immune-mediated liver injury.

Other more recent hypotheses for the cause of drug-induced allergic reactions include the danger hypothesis and the p-i (pharmacological interaction) hypothesis. The danger hypothesis follows a scheme where a reactive drug metabolite causing cellular damage induces an endogenous signal, referred to as the "danger" signal, which leads to the activation of antigen-presenting cells as a major factor in an immune response (Matzinger, 1994; Uetrecht, 1999). An example of a drug proposed to mediate allergic reactions in this manner is the NSAID diclofenac (see Section 22.4.6). The p-i hypothesis proposes that immune system T cells recognize the parent drug as foreign, rather than drug–protein adducts, leading to an immune reaction (Pichler, 2002). The anticancer drug oxaliplatin provides an example where the parent molecule is believed to have a direct effect on the immune system leading to an allergic reaction (Maindrault-Goebel *et al.*, 2005). Another example is for the anticonvulsant drug lamotrigine, where it was shown that T cells from lamotrigine-hypersensitive patients recognize the drug in the absence of drug metabolism, covalent binding to protein, or antigen processing (Naisbitt *et al.*, 2003).

Allergic-based toxicities that might result from the haptenization of proteins can lead to a triggering of the immune system in hypersensitive individuals (Ju and Uetrecht, 2002; Uetrecht, 2003; Kaplowitz, 2005). These toxicities, which are sometimes fatal, can show an extremely low incidence, ranging from 1 in 1000 to as low as 1 in 100,000 patients (Zimmerman, 1999; Uetrecht, 2008). This is a very serious concern in the pharmaceutical industry, since there have been several products withdrawn from the market because of immune-based toxicities (Fung *et al.*, 2001; Walgren, Mitchell and Thompson, 2005; Guengerich and MacDonald, 2007). It is of especially strong concern because there are almost no experimental models for the human immune system (Shenton, Chen and Uetrecht, 2004; van Wijk and Nierkens, 2006).

22.4   ALLERGIC REACTIONS TO DRUGS (EXAMPLES)

22.4.1   Penicillin

Penicillin (benzylpenicillin; Fig. 22.2) is a β-lactam antibiotic that was discovered serendipitously in 1928 by the Scottish biologist Sir Alexander Fleming (Fleming, 1929). The mechanism of action of penicillin occurs through inhibition of the growth of gram-positive bacteria by interrupting the biosynthesis of cell wall peptidoglycan proteins that are important for membrane shape and integrity (Zubay, 1988). Of course, the selective toxicity of penicillin for bacteria is due to the fact that animal cells do not possess a comparable structure. Penicillin is a structural analog to that of the D-alanyl-D-alanine *C*-terminal end of peptidoglycan proteins that are normally cross-linked by transpeptidase enzymes. Because of the affinity of penicillin for this transpeptidase enzyme, and due to the high chemical reactivity of the strained 4-membered β-lactam ring, penicillin covalently binds to an active site serine, via transacylation, thereby irreversibly inactivating the enzyme (Wise and Park, 1965; Yocum *et al.*, 1979).

The first treatment and cure of streptococcal infection by penicillin occurred in 1942 in New Haven, Connecticut (Saxon, 1999; Markel, 2004). Since then, penicillin has

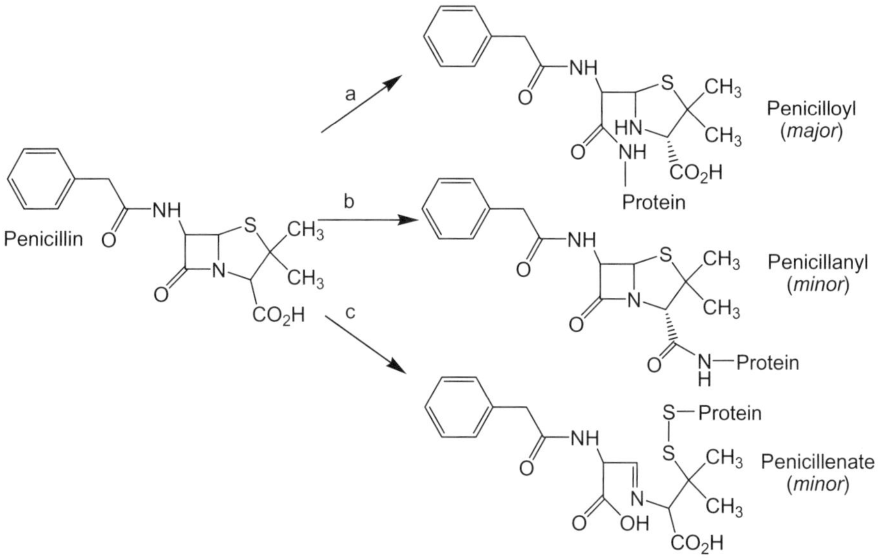

Figure 22.2 Chemical mechanisms for the covalent binding of penicillin to protein by **(a)** reaction of a protein-lysinyl amine with the carbonyl carbon of the β-lactam moiety; **(b)** reaction of a protein-lysinyl amine with a reactive acylating metabolite of penicillin; and **(c)** reaction of a protein-cysteinyl sulfhydryl with the thiazolidine ring sulfhydryl group (adapted from Gruchalla, 2003).

become the most widely used antibiotic worldwide for gram-positive bacterial infections (Wright, 1999).

Allergic reactions to penicillin are the most common adverse side effect of the drug and also represent the most common cause of drug hypersensitivity reactions in humans (Mandell and Petri, 1996; Neugut, Ghatak and Miller, 2001; Solensky, 2003). An allergic reaction to penicillin, where an IgE-mediated response occurs, has long been considered a prototype for drug-mediated allergic reactions. Allergic reactions to penicillin can lead to mild side effects such as skin rashes, but also to a potentially life-threatening, and sometimes fatal, anaphylactic shock.

The mechanism of penicillin allergic reaction in patients is believed to be due to the inherent chemical reactivity of the drug with nucleophiles such as lysine groups on proteins that react with the carbonyl carbon of the β-lactam ring, leading to the formation of protein adducts. Such adducts then function as antigens and trigger the immune system response (Fig. 22.2) (Levine, 1966; Josephson, 2004). The ε-lysyl–amide-linked penicillin adduct is referred to as the penicilloyl–protein conjugate and is designated the major antigenic determinant with respect to its quantitative formation, since approximately 95% of penicillin molecules are bound to protein in this manner (Levine, 1966; Gruchalla, 2003). Other penicillin–protein adducts, and therefore putative penicillin antigens, have been identified as penicillenate– and penicillanyl–protein adducts (Levine and Redmond, 1969; Baldo, 1999). Penicillenate–protein adducts are proposed to be formed by a mechanism where penicillin first undergoes isomerization to penicillenic acid, a thiazolidine ring opened structure bearing a free sulfhydryl group, and secondly reacts with protein sulfhydryls to form stable drug–protein mixed disulfide linkages (Parker *et al.*, 1962). The penicillenate–protein adduct is known to be inherently more immunogenic than penicilloyl–protein adduct (Park, Pirmohamed and Kitteringham, 1998). The third known antigenic determinant is the penicillanyl–protein adduct, which is formed when penicillin becomes adducted to ε-lysyl amine residues on proteins via the carbonyl carbon of the carboxylic acid moiety of penicillin. The mechanism of formation of the penicillanyl–protein adduct, to my knowledge, has not been shown, but presumably occurs by reaction of an acyl-linked metabolite of penicillin with protein lysine groups forming the penicillanyl–protein amide linkage. Both the penicillenate– and penicillanyl–protein adducts are known as minor antigenic determinants, together comprising approximately 5% of total penicillin-related protein adducts. In summary, penicillin-mediated hypersensitivity reactions go together with the hapten hypothesis, in which the inherent chemical reactivity and instability of the β-lactam moiety of penicillin allows it to function as the hapten (Uetrecht, 2007). The mechanism of the idiosyncratic nature of the IgE response that some patients have to penicillin is not yet known.

22.4.2 Felbamate

Felbamate (2-phenyl-1,3-propanediol dicarbamate; Fig. 22.3) is a broad-spectrum antiepileptic drug that was approved by the Food and Drug Administration and introduced into the market in 1993, where it was the first new anticonvulsant in 20 years (Wagner, 1994). This was an exciting medication for the treatment of epilepsy because it was effective in patients refractory to other types of anticonvulsant drugs, and also because

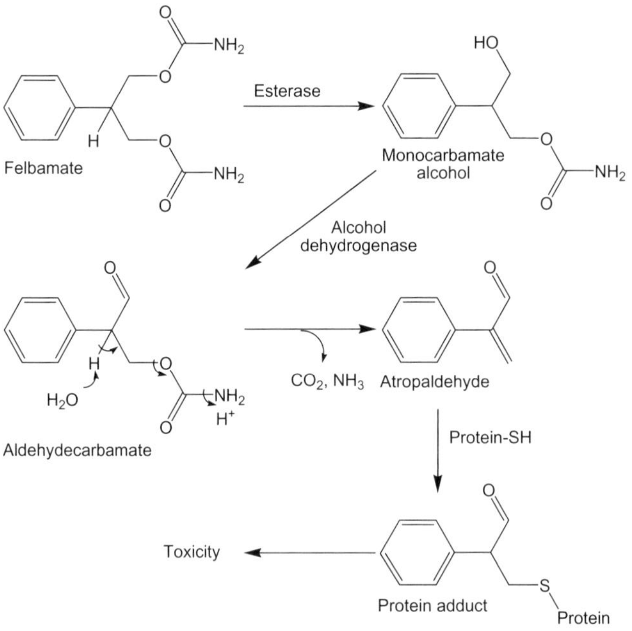

Figure 22.3 Proposed chemical mechanism for the metabolic activation of felbamate to the reactive intermediate, atropaldehyde, which reacts covalently with protein leading to toxicity (Dieckhaus *et al.*, 2002).

it was non-sedating. Unfortunately, the use of felbamate was severely curtailed when black box warnings were given after only one year on the market. It was found that out of the 120,000 patients that had received the drug, 33 cases of aplastic anemia and 18 cases of hepatic failure occurred (Pellock, 1999). Since 1994, the Food and Drug Administration has restricted felbamate for compassionate use only. The mechanism of hepatotoxicity is proposed to be mediated by reactive metabolites of the drug (Dieckhaus et al., 2002; Kapetanovic et al., 2002). Bioactivation of felbamate has been shown to occur through a mechanism where it is first metabolized by an esterase to provide a monocarbamate alcohol intermediate. This alcohol metabolite is further oxidized by alcohol dehydrogenase to the corresponding aldehyde carbamate intermediate, which decomposes by the loss of CO_2 and ammonia to the reactive metabolite, atropaldehyde. Atropaldehyde is then proposed to react with protein cysteinyl sulfhydryls, in a Michael acceptor-type fashion, resulting in the formation of covalent protein adducts that are immunogenic (Dieckhaus et al., 2000) (Fig. 22.3).

In a recent toxicogenomic study, felbamate was evaluated along with other anticonvulsant drugs for potential toxicity based on in vitro and in vivo covalent binding to protein and corresponding gene expression. In that study, felbamate was shown to produce robust effects on oxidative stress/reactive metabolite reporter genes in rat liver. However, the drug was relatively weak in terms of covalent binding to protein in model in vivo and in vitro systems. For instance, in incubations with rat liver microsomes (100 µM for 2 hours), felbamate-derived covalent binding to protein was measured to be 80 pmol equivalents/mg protein versus 523 pmol equivalents/mg protein detected for acetaminophen in the same study (Leone et al., 2007). These toxicogenomic studies showed that gene expression results from rat liver (as indicators of oxidative stress/reactive metabolite formation) can be very useful for screening candidate drugs for potential idiosyncratic drug toxicity (McMillian et al., 2004).

Fluorofelbamate, a fluorinated analog of felbamate (where the hydrogen atom attached at the benzylic carbon is replaced by a fluorine atom), was designed to block the breakdown to the atropaldehyde reactive metabolite (Hovinga, 2002; Roecklein et al., 2007). The carbon–fluorine bond is chemically inert, such that the corresponding aldehyde carbamate metabolite of fluorofelbamate does not lose hydrofluoric acid leading to the formation of atropaldehyde. Fluorofelbamate, which retains the broad-spectrum activity of felbamate, is a much more potent anticonvulsant than felbamate, shows no evidence of glutathione conjugation in vivo, and is currently a phase I drug candidate (Bialer, 2006; Roecklein et al., 2007).

22.4.3 Halothane

Halothane (2-bromo-2-chloro-1,1,1-trifluoroethane) is a volatile inhalation anesthetic used for general anesthesia that has been shown to lead to cases of hepatotoxicity resembling an allergic reaction (Fig. 22.4) (Davies, 1973; Satoh et al., 1985; Kenna et al., 1988). Halothane is metabolized to reactive intermediates (Fig. 22.4) that bind covalently to hepatic proteins (van Dyke and Gandolfi, 1974; Gandolfi et al., 1980). In some cases, halothane causes a fulminant hepatitis that is associated with an immune response to halothane-derived protein adducts (Pohl et al., 1988). Increases in plasma transaminases are detected in less than 20% of patients, indicating a mild hepatotoxicity which is proposed to be due to reductive metabolism of halothane by cytochrome P450 (P450 2D6 and P450 3A4), leading to free radical formation causing lipid peroxidation and covalent binding to protein (Fig. 22.4) (Spracklin, Thummel and Kharasch,

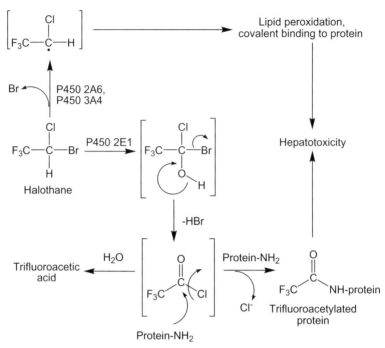

Figure 22.4   Chemical mechanism for the P450-mediated bioactivation of halothane, leading to the formation of a reactive free radical intermediate and a trifluoroacetyl chloride acylating species that covalently binds to protein nucleophiles (Spracklin, Thummel and Kharasch, 1996).

1996). Fulminant hepatic necrosis, also referred to as "halothane hepatitis," is rare (1:3500 patients after multiple exposures to halothane) but can be fatal (National Halothane Study, 1966). Halothane hepatitis is proposed to be immunologically mediated by the reactive trifluoroacetylchloride metabolite hapten which is formed from P450 2E1-mediated oxidative metabolism. Chemically reactive trifluoroacetyl chloride reacts, in a transacylation fashion, with lysine residues on hepatic proteins leading to the formation of antigenic trifluoroacetylated-protein adducts. These antigens are proposed to initiate humoral and cellular immune responses that lead to hepatitis. Possible immunogens causing halothane hepatitis include protein disulfide isomerase, carboxylesterase, Erp72 heat shock protein, immunoglobulin heavy chain-binding protein/glucose-regulated protein 78 (BiP/GRP78), Erp99/GRP94 chaperone, and P450 2E1 (Martin, Reed, and Pohl, 1993; Pumford *et al.*, 1993; Bourdi *et al.*, 1996). Sevoflurane (2,2,2-trifluoro-1-[trifluoromethyl]ethyl fluoromethyl ether), which does not form reactive metabolites that lead to trifluoroacetylated-protein adducts and has a much less risk of hepatotoxicity, is used instead of halothane for inhalation anesthesia today (Stoelting, 1999).

22.4.4   Tienilic Acid

Tienilic acid (Fig. 22.5) is a potent diuretic that was introduced into the market in 1979. Several reports of immune-mediated hepatotoxicity, which was sometimes fatal,

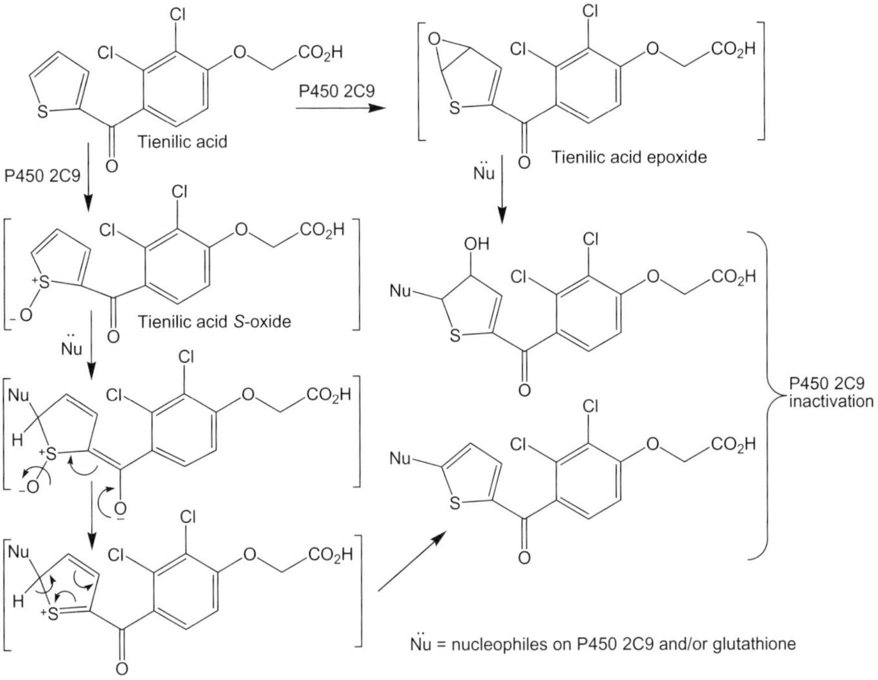

Figure 22.5 Proposed chemical mechanism for the metabolic activation of tienilic acid to epoxide- and S-oxide-type reactive intermediates that undergo covalently binding reactions with P450 2C9 apoprotein (Koenigs *et al.*, 1999).

occurred and led to its rapid removal from the market in 1980 (Barclay, 1980; Zimmerman *et al.*, 1984; Dansette *et al.*, 1998; Fung *et al.*, 2001). Tienilic acid undergoes P450 2C9-mediated metabolic activation to epoxide- and S-oxide-type reactive intermediates (Fig. 22.5) that inactivate the P450 2C9 enzyme (López-Garcia, *et al.*, 1994), covalently bind to protein (Lecoeur *et al.*, 1994; Koenigs *et al.*, 1999), and react with glutathione (Fig. 22.5) (Valadon *et al.*, 1996). Anti-P450 2C9 antibodies have been detected in patients with tienilic-induced hepatotoxicity, which suggests that metabolic activation of the drug is related to the mechanism of hepatotoxicity (Beaune *et al.*, 1987; Neuberger and Williams, 1989; Mitchell *et al.*, 1990; Beaune *et al.*, 1994; Robin *et al.*, 1996).

Tienilic acid also contains a carboxylic acid structural alert (Chapter 23). As discussed later, carboxylic acid-containing drugs can be metabolized to reactive acyl-linked metabolites that are capable of undergoing covalent binding to protein. Tienilic acid is a phenoxyacetic acid derivative, and these types of carboxylic acids have been shown to lead to highly reactive acyl-coenzyme A (acyl-CoA) thioester-linked metabolites (Li, Grillo and Benet, 2003). However, the corresponding routes of metabolism for tienilic acid have not yet been investigated.

22.4.5 Sulfamethoxazole

Sulfonamide-containing antibiotic drugs are derivatives of *p*-aminobenzene-sulfonamide (sulfanilamide; Mandell and Petri, 1996). These drugs inhibit the

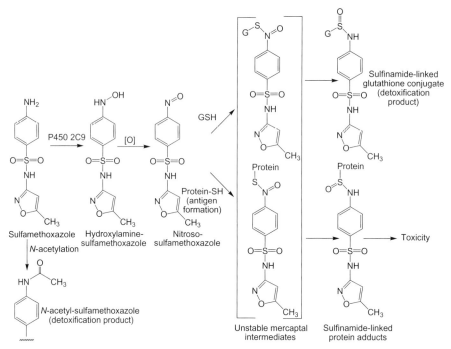

Figure 22.6 Chemical mechanism for the P450-mediated bioactivation of sulfamethoxazole to nitroso-sulfamethoxazole followed by subsequent reactions with glutathione and protein nucleophiles (Pirmohamed, Madden and Park, 1996).

biosynthesis of folic acid in bacteria by competitively inhibiting dihydropteroate synthase, the enzyme that converts p-aminobenzoic acid into folic acid, a necessary cofactor (Mandell and Petri, 1996). The chemical structure of sulfonamides contains a primary aromatic amine that can be metabolized by oxidation to N-hydroxylation metabolites. These oxidized metabolites then can be oxidized further to chemically reactive nitrosoamine products, which are known to react with protein nucleophiles forming drug–protein adducts that serve as antigens for hypersensitivity reactions (Fig. 22.6). Many sulfonamide-type drugs cause untoward allergic reactions such as Stevens–Johnson syndrome and toxic epidermal necrolysis (Meekins, Sullivan and Gruchalla, 1994; Mandell and Petri, 1996; French, 2006).

The antibiotic drug sulfamethoxazole (4-amino-N-[5-methylisoxazol-3-yl]-benzenesulfonamide) is known to cause idiosyncratic hypersensitivity reactions which resemble immunopathologic reactions. These allergic reactions occur in 2% to 4% of normal individuals, but in as much as 60% in AIDS patients (Gruchalla and Sullivan, 1991). The allergic reaction is a classic idiosyncratic drug hypersensitivity that is rarely experimentally reproduced in animals. The toxicity has a delayed onset of 7–14 days after starting therapy. The clinical symptoms are fever, severe cutaneous skin reactions (Stevens–Johnson syndrome and toxic epidermal necrolysis), and multi-organ toxicity. The toxicity is rare, unpredictable, and potentially fatal. Sulfamethoxazole undergoes oxidative metabolism in dendritic cells by cyclooxygenase enzymes, in lymphocytes by myeloperoxidase enzymes, and in the liver (Cribb, Spielberg and Griffin, 1995) and skin (Reilly *et al.*, 2000) by cytochrome P450 2C9 yielding nitroso metabolites that react with glutathione, or with proteins forming adducts leading to immune-mediated hypersensitivity reactions (Fig. 22.6). Direct evidence of an immune-mediated response to sulfamethoxazole comes from studies showing that IgE antibodies against cellular proteins or drug are present in the serum of hypersensitive patients having immediate-type allergic reaction to sulfamethoxazole (Carrington *et al.*, 1987; Meekins, Sullivan and Gruchalla, 1994). In addition, sulfamethoxazole has been shown to activate T cells in patients with previous sulfamethoxazole-induced hypersensitivity reactions (Pichler, 2002; Sanderson, Naisbitt and Park, 2006).

22.4.6 Carboxylic Acid-Containing Drugs

Increasing evidence suggests that many carboxylic acid-containing drugs are metabolized to chemically reactive intermediates that form covalent adducts with proteins (Faed, 1984; Boelsterli, 2002; Li, Benet and Grillo 2002a, b). Such adducts have been proposed to act as immune complexes leading to immunotoxic reactions (Spahn-Langguth and Benet, 1992; Zimmerman, 1994). Of the 121 drugs withdrawn from the worldwide market for hepatotoxic, anaphylactic, dermatologic, gastrointestinal, and nephrotoxic reactions that were sometimes fatal, 16 were carboxylic acid-containing drugs (Fung *et al.*, 2001), some of which are shown in Table 22.1. Many of these carboxylic acid-containing drugs are metabolized to two prospective reactive metabolites (Fig. 22.7), namely acyl glucuronides (Stogniew and Fenselau, 1982) and acyl-CoA thioesters (Caldwell, 1984; Boelsterli, 2002). Acyl glucuronides, which can be major metabolites of carboxylic acid-containing drugs, have been shown to covalently bind to protein by two alternative mechanisms. The first mechanism occurs when, because of the electrophilic nature of the carbonyl carbon of the 1-*O*-acyl linkage, the acyl glucuronide reacts with protein nucleophiles (e.g. lysinyl-NH_2, serine-OH, and cysteinyl-SH) in a nucleophilic displacement-type reaction leading to the transacylation of cellular proteins (Fig. 22.8). The second mechanism occurs when the 1-*O*-acyl glucuronide undergoes acyl migration, where the aglycone migrates from hydroxyl group to hydroxyl group around the glucuronic acid ring (Hasegawa, Smith and Benet, 1982). These acyl migration isomers then undergo ring-chain tautomerism leading to an open-chain aldehyde that is able to form Schiff-base adducts with protein-lysinyl residues. These Schiff-base-linked adducts are stabilized by undergoing an Amadori rearrangement that yields stable 1-amino-2-keto-linked acyl glucuronide protein adducts (Smith, Benet and McDonagh, 1990; Ding *et al.*, 1993) (Fig. 22.9). It is important to mention here that some of the carboxylic acid-containing drugs shown in Table 22.1 may also be metabolized by cytochrome P450 to chemically reactive metabolites. For example, zomepirac, in addition to being metabolized to acyl glucuronide and acyl-CoA reactive metabolites, has been shown recently to be bioactivated on the pyrrole structural moiety leading to a reactive epoxide metabolite (Chapter 23, Section 23.5.9, Fig. 23.16) (Chen *et al.*, 2006). Another example is for the drug alclofenac, which contains an alkene group that has been shown to form a reactive epoxide metabolite as well (Brown and Ford-Hutchinson, 1982). A final example is the NSAID bromfenac which was on the market in the United States for about nine months when there were many cases of liver damage and sudden death, before being pulled from the market in 1998 (Fontana *et al.*, 1999). The chemical structure of bromfenac contains at least two other structural alerts, namely the parachlorophenyl group and the aniline moiety, in terms of the potential for the formation of chemically reactive metabolites that have been associated with toxicological problems and liabilities (Nelson, 1995).

TABLE 22.1 Partial list of carboxylic-acid containing drugs withdrawn from clinical use and/ or also associated with allergic reactions.

Withdrawn	Known to cause allergic reactions
Alclofenac	Ibuprofen
Bromfenac	Diclofenac
Ibufenac	Fenoprofen
Zomepirac	Tolmetin
Tienilic acid	Clofibric acid
Benoxaprofen	Probenecid
Flunoxaprofen	Diflunisal
Indoprofen	Valproic acid

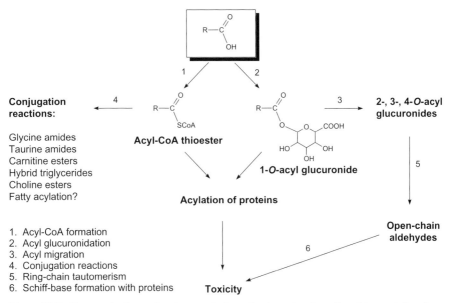

Figure 22.7 Proposed scheme for the metabolic activation of carboxylic acid-containing drugs to reactive, and hence potentially toxic, intermediates, and to nontoxic metabolites.

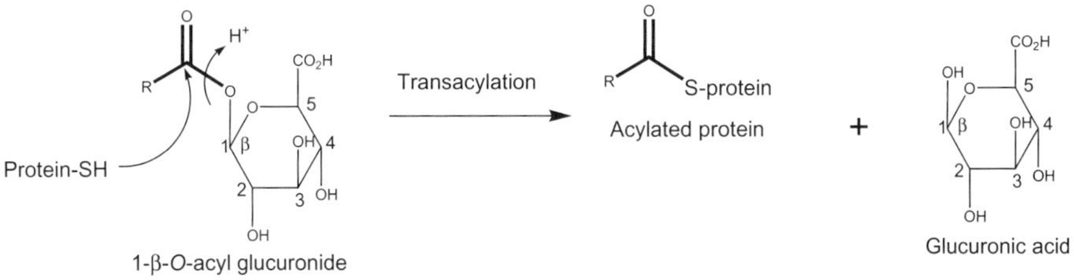

Figure 22.8 Mechanism for transacylation of protein nucleophiles by 1-β-O-acyl-linked glucuronides.

Thioester-linked acyl-CoA intermediary metabolites of carboxylic acid-containing drugs possess an electrophilic carbonyl carbon at the thioester linkage (Huxtable, 1986). Because of their large size and amphipathic nature, xenobiotic acyl-CoA thioesters do not diffuse out of the cell, but are substrates for further metabolism such as amino acid conjugation, carnitine ester formation, and hybrid triglyceride formation, all of which occur by transacylation-type reactions, albeit enzyme-mediated, which require a reactive thioester carbonyl carbon. Xenobiotic acyl-CoA thioesters are reactive species that have been shown to transacylate endogenous nucleophiles. For example, salicylic acid acyl-CoA thioester is able to react with glycine nonenzymatically *in vitro*, leading to the formation of a salicyluric glycine amide conjugate (Tishler and Goldman, 1970). Other studies have shown that endogenous long-chain acyl-CoA derivatives such as palmitoyl-CoA, arachidonyl-CoA, and retinoyl-CoA react in an "autoacylation" fashion with the cysteinyl-sulfhydryls of proteins and peptides (Yamashita *et al.*, 1995; Duncan and Gilman, 1996; Kubo *et al.*, 2005). Recent examples

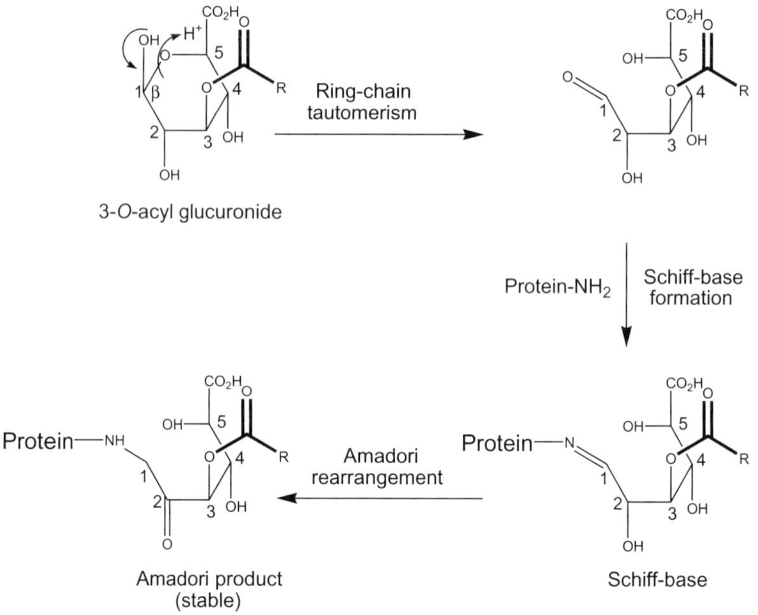

Figure 22.9 Mechanism for stable adduct formation with lysine residues on proteins by acyl glucuronide migration isomers.

of reactive acyl-CoA thioesters include those formed from the endogenous bile acid cholic acid (Mano *et al.*, 2001), the lipid-lowering drug clofibric acid (Grillo and Benet, 2002), the NSAIDs tolmetin (Olsen *et al.*, 2005) and zomepirac (Olsen *et al.*, 2007), and nafenopin (Sallustio *et al.*, 2000). The prediction of acyl glucuronides (Benet *et al.*, 1993; Bolze *et al.*, 2002) and acyl-CoA thioester (Sidenius *et al.*, 2004) metabolites for their ability to covalently bind to protein based on their chemical structures has been shown. In general, the chemical reactivity of acyl glucuronides and acyl-CoA thioester derivatives with biological nucleophiles, such as glutathione and human serum albumin, decreases with increasing substitution at the carbon adjacent to the carbonyl carbon of the carboxylic acid moiety.

An extensively studied example of a drug known to cause idiosyncratic toxicities is diclofenac (Boelsterli, 2003; Tang, 2003). Diclofenac (2-[2-(2,6-dichlorophenyl)-aminophenyl]ethanoic acid; see Chapter 23, Fig. 23.21) is an NSAID used for the treatment of osteoarthritis, rheumatoid arthritis, ankylosing spondylitis, and acute muscle pain (Small, 1989). Adverse reactions to diclofenac include enteropathy, which is common during prolonged use of the drug, acute and chronic hepatitis (15% of patients have increased plasma transaminase levels), hemolytic anemia (autoantibodies have been detected), and fatal anaphylaxis which is related to an IgE- or IgG-mediated immune response. One molecular mechanism that is proposed to contribute to diclofenac-induced hepatitis includes the recognition of plasma membrane drug–protein adducts by the immune system, leading to an immunotoxic response. Another mechanism is when the cytotoxic reaction leads to the release of reactive metabolite–covalently bound protein adducts and or immune "danger" signals (Uetrecht, 1999) (Fig. 22.1). The cytotoxicity is proposed to be caused by mitochondrial damage by the drug, reactive diclofenac metabolites, or by oxidative stress mediated by free radical-type metabolites (Masubuchi, Yamada and Horie, 2000; Galati *et al.*, 2002). Reactive quinone imine metabolites formed by P450, for example, can lead to the formation of a P450 2C11 adduct with diclofenac (Tang *et al.*, 1999; Masubuchi, Ose and Horie, 2001). In addition, diclofenac acyl glucuronide reacts with protein, forming adducts with dipeptidyl peptidase IV (110 kDa) as well as with unidentified plasma membrane proteins that weigh 140 and 200 kDa (Kretz-Rommel and Boelsterli, 1994; Hargus *et al.*, 1995). Results from one study showed the formation of an IgM autoantibody to the acyl glucuronide conjugate of the 4′-hydroxydiclofenac occurring in a patient dosed with diclofenac (Bougie *et al.*, 1997). The glucuronide metabolite was shown also to promote agglutination of normal red blood cells which was proposed to potentially lead to immune hemolytic anemia.

Diclofenac-induced small intestinal injury has been proposed to occur by enterocyte cytotoxicity mediated by the reactive acyl glucuronide of diclofenac, or the acyl glucuronide of one of its oxidative metabolites (Seitz and Boelsterli, 1998). In another study, diclofenac was administered to rats by gastric gavage, and immunoblot analysis of small intestine homogenates and isolated enterocyte subcellular fractions with drug-specific antiserum revealed 142-, 130-, 110-, and 55-kDa protein adducts mediated by diclofenac (Ware *et al.*, 1998). The 142- and 130-kDa protein adducts of diclofenac were identified as aminopeptidase N (CD13) and sucrase-isomaltase, respectively. These authors proposed that oral tolerance to intestinal drug–protein adducts can be formed in gut-associated lymphoid tissue, leading to the down-regulation of drug-induced allergic reactions in many individuals that do not have gut toxicity to diclofenac.

An example of an NSAID carboxylic acid-containing drug removed from the market due to toxicity is benoxaprofen (Table 22.1). Benoxaprofen (2-[2-(4-chlorophenyl)-1,3-benzoxazol-5-yl]propanoic acid; Fig. 22.10) had been shown to cause a rare hepatotoxicity which led to its withdrawal from the market in 1982. The proposal was that benoxaprofen-induced toxicity was related to the formation of covalent adducts to proteins via a reactive acyl glucuronide metabolite (Fig. 22.10). In covalent binding studies in rat, benoxaprofen was shown to become irreversibly bound to plasma and liver proteins. An immunochemical approach was used to detect hepatic protein targets and showed adduct masses of 110 and 70 kDa as the major liver protein targets modified by benoxaprofen (Dong, Liu and Smith, 2005). In other studies, tandem mass spectrometry was used to identify the structures and specific covalent binding sites for protein adducts formed upon the reaction of benoxaprofen acyl glucuronide and human serum albumin *in vitro* (Qiu, Burlingame and Benet, 1998). Results showed that benoxaprofen glucuronide forms covalent adducts with protein nucleophiles on human serum albumin by both nucleophilic displacement (transacylation) and Schiff-base formation with acyl migration isomers with ε-amino-lysyl residues, where Lys-159 was determined to be the major covalent binding site. Similar investigations with the acyl glucuronide metabolite of the NSAID tolmetin (Table 22.1) showed covalent binding to human serum albumin by both transacylation- and Schiff-base-type mechanisms; however, in this case Lys-199 was the major site of protein adduct formation (Ding *et al.*, 1995). Although the acyl-CoA thioester intermediary metabolite of benoxaprofen has not been identified, its formation has been shown indirectly by the detection of the corresponding taurine conjugate (Mohri, Okada and Benet, 2005). Therefore, benoxaprofen-*S*-acyl-CoA may also contribute to covalent binding to hepatic proteins *in vivo* that leads to an immune response and subsequent liver injury (Boelsterli, 2002). For benoxaprofen, the benzoxazole substructure might be metabolized to a reactive intermediate that could explain some of the covalent binding to hepatic protein; however,

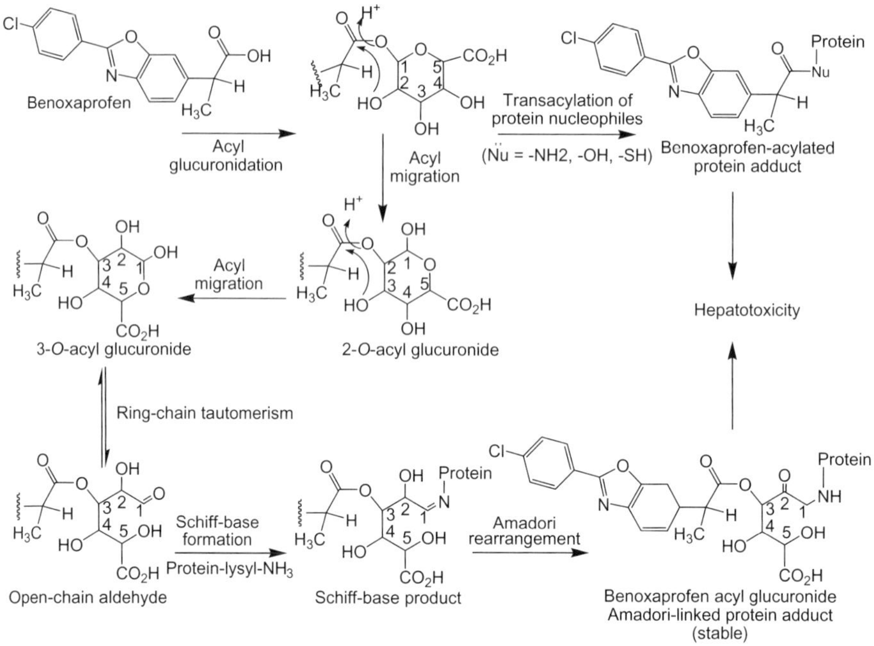

Figure 22.10 Proposed chemical mechanism for the metabolic activation of benoxaprofen to reactive acyl glucuronide metabolites that undergo covalent binding to protein (Qiu, Burlingame and Benet, 1998; Dong, Liu and Smith, 2005).

no work has been done with this agent in this regard because the drug was not made readily available for study.

In summary, many carboxylic acid-containing drugs (Table 22.1), including the discontinued drugs ibufenac (Castillo and Smith, 1995), zomepirac (Smith, McDonagh and Benet, 1986), and suprofen (Smith and Liu, 1995), and currently used drugs such as diclofenac (Boelsterli, 2003), diflunisal (McKinnon and Dickinson, 1989), valproic acid (Williams *et al.*, 1992), tolmetin (Zia-Amirhosseini *et al.*, 1995), clofibric acid (Sallustio *et al.*, 1991), and salicylic acid (Dickinson, Baker and King, 1994) form reactive acyl-linked metabolites that lead to covalent binding to protein and may be responsible for mediating hypersensitivity reactions related to their use (Fung *et al.*, 2001).

22.5 CONCLUSIONS

Pharmaceutical and academic scientists have been trying to put together a picture of the phenomenon of drug-induced allergic reactions in terms of what has happened in the pharmaceutical industry over the past 40 or so years. Things got started with allergic reactions to drugs like sulfonamides and penicillins, but which are now used to understand allergic reactions related to the hapten hypothesis. Pharmaceutical scientists are continuously searching for ways to design drugs that do not cause allergic reactions and to be able to identify those that do at an early stage in drug discovery in order to have significant impact on candidate drug selection. As pharmaceutical scientists from

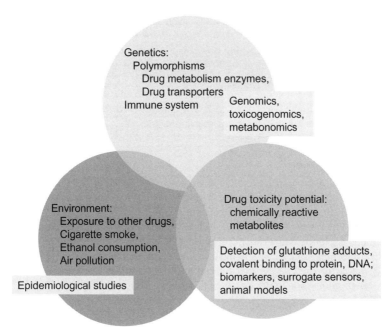

Figure 22.11 Risk of allergic reactions to drugs (circles) and corresponding areas of ongoing research (squares).

drug metabolism, biology, toxicology, and immunology disciplines dig deeper and deeper, they have realized that there is a lot more work to do. Such research includes the generation of predictive animal models of drug hypersensitivity reactions, the application of genomics and toxicogenomics related to drug allergic reactions, the increased study of the influence of environmental factors (such as other drugs, toxins, or disease) on allergic diseases, and on the role of reactive drug metabolites and covalent binding to protein on subsequent immunotoxicity (Fig. 22.11).

In the future, animal models should become very important for toxicological research into drug allergic reactions, because these models are essentially the only way to study the currently available hypotheses (Dansette *et al.*, 1998; Uetrecht, 2006). However, useful animal models for allergic reactions are very few because allergic reactions to drugs are not predictable in animals either (Shenton, Chen and Uetrecht, 2004). Some examples of drugs that have been studied in animal models for drug-induced immune reactions are penicillamine, halothane, isoniazid, and sulfamethoxazole. Penicillamine, a drug sometimes used for the treatment of rheumatoid arthritis, can cause autoimmune diseases such as lupus-like syndrome and skin rash (Stein *et al.*, 1980) and can also induce autoimmunity in Brown Norway rats (Mason and Uetrecht, 2004). Halothane has been shown to demonstrate a cellular immune response in halothane-exposed guinea pigs (Furst *et al.*, 1997). Isoniazid is a drug known to cause a high incidence of hypersensitivity reactions leading to liver injury (Zimmerman, 1993). An animal model for isoniazid has been reported to be the rabbit, where treatment with isoniazid led to elevation in transaminase enzyme levels and the formation of liver necrosis in some test animals (Sarich *et al.*, 1995). Sulfonamide-type antimicrobials, such as sulfamethoxazole, have been shown to lead to drug hypersensitivity reactions in dogs, especially in larger dogs such as Doberman pinschers (Lavergne *et al.*, 2006). Increased numbers of animal models for drug hypersensitivity reactions are necessary in order to speed up the progress toward understanding the mechanisms of drug-induced allergies (Uetrecht, 2006). An increased understanding of how and why allergic reactions to drugs occur should assist in improving decision-making processes for the selection of safer drug candidates. Hopefully, in the future, this greater and greater understanding will lead to fewer and fewer drugs being developed that might someday cause allergic reactions in patients.

ACKNOWLEDGMENTS

I would like to thank Dr. Lixia Jin and Judy Huang, B.A. (Department of Pharmacokinetics and Drug Metabolism, Amgen Inc.), Dr. Kathila Rajapaksa (Toxicology, Amgen Inc.), Dr. Joseph A. Ware (Late Stage Pharmacokinetics and Pharmacodynamics, Genentech, South San Francisco), and Dr. Christian Skonberg (Department of Pharmaceutics and Analytical Chemistry, University of Copenhagen, Denmark) for critically reviewing this chapter. I also thank the constructive criticism and helpful suggestions of an anonymous reviewer, which has been included in the references.

REFERENCES

Baillie, T.A. (2006). Future of toxicology-metabolic activation and drug design: challenges and opportunities in chemical toxicology. *Chem. Res. Toxicol.*, *19*, 889–893.

Baldo, B. (1999). Penicillins and cephalosporins as allergens-structural aspects of recognition and cross-reactions. *Clin. Exp. Allergy.*, *29*, 744–749.

Barclay, W.R. (1980). Ticrynafen's withdrawal from the market. *JAMA*, *243*, 771.

Beaune, P., Dansette, P.M., Mansuy, D., Kiffel, L., Finck, M., Amar, C., Leroux, J.P., Homber, J.C. (1987). Human anti-endoplasmic reticulum autoantibodies appearing in a drug-induced hepatitis are directed against a human liver cytochrome P-450 that hydroxylates the drug. *Proc. Natl. Acad. Sci. U.S.A.*, *84*, 551–555.

Beaune, P., Pessayre, D., Dansette, P.M., Mansuy, D., Manns, M. (1994). Autoantibodies against cytochromes P450: role in human diseases. *Adv. Pharmacol.*, *30*, 199–245.

Benet, L.Z., Spahn-Langguth, H., Iwakawa, S., Volland, C., Mizuma, T., Mayer, S., Mutschler, E., Lin, E.T. (1993). Predictability of the covalent binding of acidic drugs in man. *Life Sci.*, *53*, 141–146.

Bialer, M. (2006). New antiepileptic drugs that are second generation to existing antiepileptic drugs. *Expert Opin. Investig. Drugs*, *15*, 637–647.

Boelsterli, U.A. (2002). Xenobiotic acyl glucuronides and acyl CoA thioesters as protein-reactive metabolites with the potential to cause idiosyncratic drug reactions. *Curr. Drug Metab.*, *3*, 439–450.

Boelsterli, U.A. (2003). Diclofenac-induced liver injury: a paradigm of idiosyncratic drug toxicity. *Toxicol. Appl. Pharmacol.*, *192*, 307–322.

Bolze, S., Bromet, N., Gay-Feutry, C., Massiere, F., Boulieu, R., Hulot, T. (2002). Development of an in vitro screening model for the biosynthesis of acyl glucuronide metabolites and the assessment of their reactivity toward human serum albumin. *Drug Metab. Dispos.*, *30*, 404–413.

Bougie, D., Johnson, S.T., Weitekamp, L.A., Aster, R.H. (1997). Sensitivity to a metabolite of diclofenac as a cause of acute immune hemolytic anemia. *Blood*, *90*, 407–413.

Bourdi, M., Chen, W., Peter, R.M., Martin, J.L., Buters, J.T., Nelson, S.D., Pohl, L.R. (1996). Human cytochrome P450 2E1 is a major autoantigen associated with halothane hepatitis. *Chem. Res. Toxicol.*, *9*, 1159–1166.

Brown, L.M., Ford-Hutchinson, A.W. (1982). The destruction of cytochrome P-450 by alclofenac: possible involvement of an epoxide metabolite. *Biochem. Pharmacol.*, *31*, 195–199.

Caldwell, J. (1984). Xenobiotic acyl-coenzyme A: critical intermediates in the biochemical pharmacology and toxicology of carboxylic acids. *Biochem. Soc. Trans.*, *12*, 9–11.

Carrington, D., Earl, H., Sullivan, T. (1987). Studies of human IgE to a sulfonamide determinant. *J. Allergy. Clin. Immunol.*, *79*, 442–447.

Castillo, M., Smith, P.C. (1995). Disposition and reactivity of ibuprofen and ibufenac acyl glucuronides in vivo in rhesus monkey and in vitro with human serum albumin. *Drug Metab. Dispos.*, *23*, 566–572.

Chen, Q., Doss, G.A., Tung, E.C., Liu, W., Tang, Y.S., Braun, M.P., Didolka, V., Strauss, J.R., Wang, R.W., Stearns, R.A., Evans, D.C., Baillie, T.A., and Tang, W. (2006). Evidence for the bioactivation of zomepirac and tolmetin by an oxidative pathway: identification of glutathione adducts in vitro in human liver microsomes and in vivo in rats. *Drug Metab. Dispos.*, *34*, 145–151.

Cribb, A.E., Spielberg, S.P., Griffin, G.P. (1995). N4-hydroxylation of sulfamethoxazole by cytochrome P450 of the cytochrome P4502C subfamily and reduction of sulfamethoxazole hydroxylamine in human and rat hepatic microsomes. *Drug Metab. Dispos.*, *23*, 406–414.

Dansette, P.M., Bonierbale, E., Minoletti, C., Beaune, P.H., Pessayre, D., Mansuy, D. (1998). Drug-induced immunotoxicity. *Eur. J. Drug Metab. Pharmacokinet.*, *23*, 443–451.

Davies, G.E. (1973). Halothane hepatotoxicity: is halothane hepatotoxicity an allergic reaction to halothane. *Proc. Roy. Soc. Med.*, *66*, 1–2.

Dickinson, R.G., Baker, P.V., King, A.R. (1994). Studies on the reactivity of acyl glucuronides—VII. Salicyl acyl glucuronide reactivity in vitro and covalent binding of salicylic acid to plasma protein of humans taking aspirin. *Biochem. Pharmacol.*, *47*, 469–476.

Dieckhaus, C.M., Miller, T.A., Sofia, R.D., Macdonald, T.L. (2000). A mechanistic approach to understanding species differences in felbamate bioactivation: relevance to drug-induced idiosyncratic reactions. *Drug Metab. Dispos.*, *28*, 814–822.

Dieckhaus, C.M., Thompson, C.D., Roller, S.G., Macdonald, T.L. (2002). Mechanisms of idiosyncratic drug reactions: the case of felbamate. *Chem. Biol. Interact.*, *142*, 99–117.

Ding, A., Ojingwa, J.C., McDonagh, A.F., Burlingame, A.L., Benet, L.Z. (1993). Evidence for covalent binding of acyl glucuronides to serum albumin via an imine mechanism as revealed by tandem mass spectrometry. *Proc. Natl. Acad. Sci. U.S.A.*, *90*, 3797–3801.

Ding, A., Zia-Amirhosseini, P., McDonagh, A.F., Burlingame, A.L., Benet, L.Z. (1995). Reactivity of tolmetin glucuronide with human serum albumin. Identification of binding sites and mechanisms of reaction by tandem mass spectrometry. *Drug Metab. Dispos.*, *23*, 369–376.

Dong, J.Q., Liu, J., Smith, P.C. (2005). Role of benoxaprofen and flunoxaprofen acyl glucuronides in covalent binding to rat plasma and liver proteins in vivo. *Biochem. Pharmacol.*, *70*, 937–948.

Duncan, J.A., Gilman, A.G. (1996). Autoacylation of G protein alpha subunits. *J. Biol. Chem.*, *271*, 23594–23600.

van Dyke, R.A., Gandolfi, A.J. (1974). Studies on irreversible binding of radioactivity from (14C)halothane to rat hepatic microsomal lipids and protein. *Drug Metab. Dispos.*, *2*, 469–476.

Ewan, P.W. (1998). ABC of allergies: anaphylaxis. *BMJ*, *316*, 1442–1445.

Faed, E.M. (1984). Properties of acyl glucuronides: implications for studies of the pharmacokinetics and metabolism of acidic drugs. *Drug Metab. Rev.*, *15*, 1213–1249.

Fleming, A. (1929). On the antibacterial action of cultures of a penicillium, with special reference to their use in the isolation of *B. influenzae*. *Br. J. Pathol.*, *10*, 226–236.

Fontana, R.J., McCashland, T.M., Benner, K.G., Appelman, H.D., Gunartanam, N.T., Wisecarver, J.L., Rabkin, J.M., Lee, W.M. (1999). Acute liver failure associated with prolonged use of bromfenac leading to liver transplantation. The Acute Liver Failure Study Group. *Liver Transpl. Surg.*, *5*, 480–484.

French, L.E. (2006). Toxic epidermal necrolysis and Stevens Johnson syndrome: our current understanding. *Allergol. Int.*, *55*, 9–16.

Fung, M., Thornton, A., Mybeck, K., Wu, J.H., Hornbuckle, K., Muniz, E. (2001). Evaluation of the characteristics of safety withdrawal of prescription drugs from world wide pharmaceutical markets—1960 to 1999. *Drug Inform. J.*, *35*, 293–317.

Furst, S.M., Luedke, D., Gaw, H.H., Reich, R., Gandolfi, A.J. (1997). Demonstration of a cellular immune response in halothane-exposed guinea pigs. *Toxicol. Appl. Pharmacol.*, *143*, 245–255.

Galati, G., Tafazoli, S., Sabzevari, O., Chan, T.S., O'Brien, P.J. (2002). Idiosyncratic NSAID drug induced oxidative stress. *Chem. Biol. Interact.*, *142*, 25–41.

Gandolfi, A.J., White, R.D., Sipes, I.G., Pohl, L.R. (1980). Bioactivation and covalent binding of halothane in vitro: studies with [3H]- and [14C]halothane. *J. Pharmacol. Exp. Ther.*, *214*, 721–725.

Goldsby, R.A., Kindt, T.K., Osborne, B.A., Kubym, J. (2003). *Immunology*, 5th edn, W.H. Freeman and Company, New York.

Grillo, M.P., Benet, L.Z. (2002). Studies on the reactivity of clofibryl-S-acyl-CoA thioester with glutathione in vitro. *Drug Metab. Dispos.*, *30*, 55–62.

Gruchalla, R.S. (2003). 10. Drug allergy. *J. Allergy Clin. Immunol.*, *111*, S548–559.

Gruchalla, R.S., Sullivan, T.J. (1991). Detection of human IgE to sulfamethoxazole by skin testing with sulfamethoxazoyl-poly-L-tyrosine. *J. Allergy Clin. Immunol.*, *88*, 784–792.

Guengerich, F.P., MacDonald, J.S. (2007). Applying mechanisms of chemical toxicity to predict drug safety. *Chem. Res. Toxicol.*, *20*, 344–369.

Hargus, S.J., Martin, B.M., George, J.W., Pohl, L.R. (1995). Covalent modification of rat liver dipeptidyl peptidase IV (CD26) by the nonsteroidal anti-inflammatory drug diclofenac. *Chem. Res. Toxicol.*, *8*, 993–996.

Hasegawa, J., Smith, P.C., Benet, L.Z. (1982). Apparent intramolecular acyl migration of zomepirac glucuronide. *Drug Metab. Dispos.*, *10*, 469–473.

Hovinga, C.A. (2002). Novel anticonvulsant medications in development. *Expert Opin. Investig. Drugs*, *11*, 1387–1406.

Huxtable, R. (1986). Thiols, disulfides and thioesters, in *Biochemistry of Sulfur* (ed. E. Frieden), Plenum Press, New York, 230–245.

Janeway, C., Travers, P., Walport, M., Shlomchik, M. (2005). *Immunobiology*, 6th edn, Garland Science. New York.

Josephson, A.S. (2004). Penicillin allergy: a public health perspective. *J. Allergy Clin. Immunol.*, *113*, 605–606.

Ju, C., Uetrecht, J.P. (2002). Mechanism of idiosyncratic drug reactions: reactive metabolite formation, protein binding and the regulation of the immune system. *Curr. Drug Metab.*, *3*, 367–377.

Kalish, R.S. (1995). Antigen processing: the gateway to the immune response. *J. Am. Acad. Dermatol.*, *31*, 640–652.

Kapetanovic, I.M., Torchin, C.D., Strong, J.M., Yonekawa, W.D., Lu, C., Li, A.P., Dieckhaus, C.M., Santos, W.L., Macdonald, T.L., Sofia, R.D., Kupferberg, H.J. (2002). Reactivity of atropaldehyde, a felbamate metabolite in human liver tissue in vitro. *Chem. Biol. Interact*, *142*, 119–134.

Kaplowitz, N. (2005). Idiosyncratic drug hepatotoxicity. *Nat. Rev. Drug Discov.*, *4*, 489–499.

Kenna, J.G., Satoh, H., Christ, D.D., Pohl, L.R. (1988). Metabolic basis for a drug hypersensitivity: antibodies in sera from patients with halothane hepatitis recognize liver neoantigens that contain the trifluoroacetyl group derived from halothane. *J. Pharmacol. Exp. Ther.*, *245*, 1103–1109.

Koenigs, L.L., Peter, R.M., Hunter, A.P., Haining, R.L., Rettie, A.E., Friedberg, T., Pritchard, M.P., Shou, M., Rushmore, T.H., Trager, W.F. (1999). Electrospray ionization mass spectrometric analysis of intact cytochrome P450: identification of tienilic acid adducts to P450 2C9. *Biochemistry*, *38*, 2312–2319.

Kretz-Rommel, A., Boelsterli, U.A. (1994). Mechanism of covalent adduct formation of diclofenac to rat hepatic microsomal proteins. Retention of the glucuronic acid moiety in the adduct. *Drug Metab. Dispos.*, *22*, 956–961.

Kubo, Y., Wada, M., Ohba, T., Takahashi, N. (2005). Formation of retinoylated proteins from retinoyl-CoA in rat tissues. *J. Biochem. (Tokyo)*, *138*, 493–500.

Kumar, S., Kassahun, K., Tschirret-Guth, R.A., Mitra, K., Baillie, T.A. (2008). Minimizing metabolic activation during pharmaceutical lead optimization: progress, knowledge gaps and future directions. *Curr. Opin. Drug Discov. Devel.*, *11*, 43–52.

Landsteiner, K., Jacobs, J. (1935). Studies on the sensitization of animals with simple chemical compounds. *J. Exp. Med.*, *61*, 643–656.

Lavergne, S.N., Danhof, R.S., Volkman, E.M., Trepanier, L.A. (2006). Association of drug-serum protein adducts and anti-drug antibodies in dogs with sulfonamide hypersensitivity: a naturally occurring model of idiosyncratic drug toxicity. *Clin. Exp. Allergy*, *36*, 907–915.

Lecoeur, S., Bonierbale, E., Challine, D., Gautier, J.C., Valadon, P., Dansette, P.M., Catinot, R., Ballet, F., Mansuy, D., Beaune, P.H. (1994). Specificity of in vitro covalent binding of tienilic acid metabolites to human liver microsomes in relationship to the type of hepatotoxicity: comparison with two directly hepatotoxic drugs. *Chem. Res. Toxicol.*, *7*, 434–442.

Leone, A.M., Kao, L.M., McMillian, M.K., Nie, A.Y., Parker, J.B., Kelley, M.F., Usuki, E., Parkinson, A., Lord, P.G., Johnson, M.D. (2007). Evaluation of felbamate and other antiepileptic drug toxicity potential based on hepatic protein covalent binding and gene expression. *Chem. Res. Toxicol.*, *20*, 600–608.

Levine, B. (1966). Immunologic mechanisms of penicillin allergy. A haptenic model system for the study of allergic disease in man. *N. Engl. J. Med.*, *275*, 1115–1125.

Levine, B., Redmond, A. (1969). Minor haptenic determinant-specific regains of penicillin hypersensitivity in man. *Int. Arch. Allergy Appl. Immunol.*, *35*, 445–455.

Li, C., Benet, L.Z., Grillo, M.P. (2002a). Studies on the chemical reactivity of 2-phenylpropionic acid 1-O-acyl glucuronide and S-acyl-CoA thioester metabolites. *Chem. Res. Toxicol.*, *15*, 1309–1317.

Li, C., Benet, L.Z., Grillo, M.P. (2002b). Enantioselective covalent binding of 2-phenylpropionic Acid to protein in vitro in rat hepatocytes. *Chem. Res. Toxicol.*, *15*, 1480–1487.

Li, C., Grillo, M.P., Benet, L.Z. (2003). In vitro studies on the chemical reactivity of 2,4-dichlorophenoxyacetyl-S-acyl-CoA thioester. *Toxicol. Appl. Pharmacol.*, *187*, 101–109.

López-Garcia, M.P., Dansette, P.M., Mansuy, D. (1994). Thiophene derivatives as new mechanism-based inhibitors of cytochromes P-450: inactivation of yeast-expressed human liver cytochrome P-450 2C9 by tienilic acid. *Biochemistry*, *33*, 166–175.

McKinnon, G.E., Dickinson, R.G. (1989). Covalent binding of diflunisal and probenecid to plasma protein in humans: persistence of the adducts in the circulation. *Res. Commun. Chem. Pathol. Pharmacol.*, *66*, 339–354.

McMillian, M., Nie, A.Y., Parker, J.B., Leone, A., Bryant, S., Kemmerer, M., Herlich, J., Liu, Y., Yieh, L., Bittner, A., Liu, X., Wan, J., Johnson, M.D. (2004). A gene expression signature for oxidant stress/reactive metabolites in rat liver. *Biochem. Pharmacol.*, *68*, 2249–2261.

Maindrault-Goebel, F., André, T., Tournigand, C., Louvet, C., Perez-Staub, N., Zeghib, N., de Gramont, A. (2005). Allergic-type reactions to oxaliplatin: retrospective analysis of 42 patients. *Eur. J. Cancer*, *41*, 2262–2267.

Mandell, G.L., Petri, W.A. (1996). Antimicrobial agents, in *Goodman and Gilman's the Pharmacological Basis of Therapeutics*, 9th edn (eds J.G. Hardman, A.G. Gilman and L.E. Limbard), McGraw-Hill, New York, pp. 1073–1101.

Mano, N., Uchida, M., Okuyama, H., Sasaki, I., Ikegawa, S., Goto, J. (2001). Simultaneous detection of cholyl adenylate and coenzyme A thioester utilizing liquid chromatography/electrospray ionization mass spectrometry. *Anal. Sci.*, *17*, 1037–1042.

Markel, H. (2004). Shaping the mold, from lab glitch to lifesaver. The New York Times (Science Times), April 20, 2004, p. D6.

Martin, J.L., Reed, G.F., Pohl, L.R. (1993). Association of anti-58 kDa endoplasmic reticulum antibodies with halothane hepatitis. *Biochem. Pharmacol.*, *46*, 1247–1250.

Mason, M.J., Uetrecht, J.P. (2004). Tolerance induced by low dose D-penicillamine in the Brown Norway rat model of drug-induced autoimmunity is immune-mediated. *Chem. Res. Toxicol.*, *17*, 82–94.

Masubuchi, Y., Ose, A., Horie, T. (2001). Mechanism-based inactivation of CYP2C11 by diclofenac. *Drug Metab. Dispos.*, *29*, 1190–1195.

Masubuchi, Y., Yamada, S., Horie, T. (2000). Possible mechanism of hepatocyte injury induced by diphenylamine and its structurally related nonsteroidal anti-inflammatory drugs. *J. Pharmacol. Exp. Ther.*, *292*, 982–987.

Matzinger, P. (1994). Tolerance, danger and the extended family. *Annu. Rev. Immunol.*, *12*, 991–1045.

Meekins, C.V., Sullivan, T.J., Gruchalla, R.S. (1994). Immunochemical analysis of sulfonamide drug allergy: identification of sulfamethoxazole-substituted human serum proteins. *J. Allergy Clin. Immunol.*, *94*, 1017–1024.

Mitchell, J.A., Gillam, E.M., Stanley, L.A., Sim, E. (1990). Immunotoxic side-effects of drug therapy. *Drug Saf.*, *5*, 168–178.

Mohri, K., Okada, K., Benet, L.Z. (2005). Stereoselective taurine conjugation of (R)-benoxaprofen enantiomer in rats: in vivo and in vitro studies using rat hepatic mitochondria and microsomes. *Pharm. Res.*, *22*, 79–85.

Naisbitt, D.J., Farrell, J., Wong, G., Depta, J.P., Dodd, C.C., Hopkins, J.E., Gibney, C.A., Chadwick, D.W., Pichler, W.J., Pirmohamed, M., Park, B.K. (2003). Characterization of drug-specific T cells in lamotrigine hypersensitivity. *J. Allergy Clin. Immunol.*, *111*, 1393–1403.

National Halothane Study (1966). Summary of the National Halothane Study. Possible association between halothane anesthesia and post-operative hepatic necrosis. *JAMA*, *197*, 775–788.

Nelson, S.D. (1995). Mechanisms of the formation and disposition of reactive metabolites that can cause acute liver injury. *Drug Metab. Rev.*, *27*, 147–177.

Neuberger, J., Williams, R. (1989). Immune mechanisms in tienilic acid associated hepatotoxicity. *Gut*, *30*, 515–519.

Neugut, A., Ghatak, A., Miller, R. (2001). Anaphylaxis in the United States: an investigation into its epidemiology. *Arch. Intern. Med.*, *161*, 15–21.

Olsen, J., Li, C.C., Bjørnsdottir, I., Sidenius, U., Benet, L.Z., Hansen, S.H., Benet, L.Z. (2005). In vitro and in vivo studies on acyl-coenzyme A-dependent bioactivation of zomepirac in rats. *Chem. Res. Toxicol.*, *18*, 1729–1736.

Olsen, J., Li, C., Skonberg, C., Bjørnsdottir, I., Sidenius, U., Benet, L.Z., Hansen, S.H. (2007). Studies on the metabolism of tolmetin to the chemically reactive acyl-coenzyme A thioester intermediate in rats. *Drug Metab. Dispos.*, *35*, 758–764.

Park, B.K., Coleman, J.W., Kitteringham, N.R. (1987). Drug disposition and drug hypersensitivity. *Biochem. Pharmacol.*, *36*, 581–590.

Park, B.K., Pirmohamed, M., Kitteringham, N.R. (1998). Role of drug disposition in drug hypersensitivity: a chemical, molecular, and clinical perspective. *Chem. Res. Toxicol.*, *11*, 969–988.

Parker, C.W., Deweck, A.L., Kern, M., Eisen, H.N. (1962). The preparation and some properties of penicillenic acid derivatives relevant to penicillin hypersensitivity. *J. Exp. Med.*, *115*, 803–819.

Pellock, J.M. (1999). Felbamate. *Epilepsia*, *40*, 57–62.

Pichler, W.J. (2002). Pharmacological interaction of drugs with antigen-specific immune receptors: the p-i concept. *Curr. Opin. Allergy Clin. Immunol.*, *2*, 301–305.

Pirmohamed, M., Madden, S., Park, K. (1996). Idiosyncratic drug reactions: metabolic bioactivation as a pathogenic mechanism. *Clin. Pharmacokinet.*, *31*, 215–230.

Pohl, L.R., Satoh, H., Christ, D.D., Kenna, J.G. (1988). The immunologic and metabolic basis of drug hypersensitivities. *Annu. Rev. Pharmacol. Toxicol.*, *28*, 367–387.

Pumford, N.R., Martin, B.M., Thomassen, D., Burris, J.A., Kenna, J.G., Martin, J.L., Pohl, L.R. (1993). Serum antibodies from halothane hepatitis patients react with the rat endoplasmic reticulum protein ERp72. *Chem. Res. Toxicol.*, *6*, 609–615.

Qiu, Y., Burlingame, A.L., Benet, L.Z. (1998). Mechanisms for covalent binding of benoxaprofen glucuronide to human serum albumin: studies by tandem mass spectrometry. *Drug Metab. Dispos.*, *26*, 246–256.

Reilly, T.P., Lash, L.H., Doll, M.A., Hein, D.W., Woster, P.M., Svensson, C.K. (2000). A role for bioactivation and covalent binding within epidermal keratinocytes in sulfonamide-induced cutaneous drug reactions. *J. Invest. Dermatol.*, *114*, 1164–1173.

Robin, M.A., Maratrat, M., Le Roy, M., Le Breton, F.P., Bonierbale, E., Dansette, P., Ballet, F., Mansuy, D., Pessayre, D. (1996). Antigenic targets in tienilic acid hepatitis. Both cytochrome P450 2C11 and 2C11-tienilic acid adducts are transported to the plasma membrane of rat hepatocytes and recognized by human sera. *J. Clin. Invest.*, *98*, 1471–1480.

Roecklein, B.A., Sacks, H.J., Mortko, H., Stables, J. (2007). Fluorofelbamate. *Neurotherapeutics*, *4*, 97–101.

Sallustio, B.C., Knights, K.M., Roberts, B.J., Zacest, R. (1991). In vivo covalent binding of clofibric acid to human plasma proteins and rat liver proteins. *Biochem. Pharmacol.*, *42*, 1421–1425.

Sallustio, B.C., Nunthasomboon, S., Drogemuller, C.J., Knights, K.M. (2000). In vitro covalent binding of nafenopin-CoA to human liver proteins. *Toxicol. Appl. Pharmacol.*, *163*, 176–182.

Sanderson, J.P., Naisbitt, D.J., Park, B.K. (2006). Role of bioactivation in drug-induced hypersensitivity reactions. *AAPS J.*, *8*, 55–64.

Sarich, T.C., Zhou, T., Adams, S.P., Bain, A.I., Wall, R.A., Wright, J.M. (1995). A model of isoniazid-induced hepatotoxicity in rabbits. *J. Pharmacol. Toxicol. Methods*, *34*, 109–116.

Satoh, H., Fukuda, Y., Anderson, D.K., Ferrans, V.J., Gillette, J.R., Pohl, L.R. (1985). Immunological studies on the mechanisms of halothane induced hepatotoxicity: immunohistochemical evidence of trifluoroacetylated hepatocytes. *J. Pharmacol. Exp. Ther.*, *223*, 857–862.

Saxon, W. (1999). Anne Miller, 90, first patient who was saved by penicillin. The New York Times, June 9.

Seitz, S., Boelsterli, U.A. (1998). Diclofenac acyl glucuronide, a major biliary metabolite, is directly involved in small intestinal injury in rats. *Gastroenterology*, *115*, 1476–1482.

Shenton, J.M., Chen, J., Uetrecht, J.P. (2004). Animal models of idiosyncratic drug reactions. *Chem. Biol. Interact.*, *150*, 53–70.

Sidenius, U., Skonberg, C., Olsen, J., Hansen, S.H. (2004). In vitro reactivity of carboxylic acid-CoA thioesters with glutathione. *Chem. Res. Toxicol.*, *17*, 75–81.

Small, R.E. (1989). Diclofenac sodium. *Clin. Pharm.*, *8*, 545–558.

Smith, P.C., Benet, L.Z., McDonagh, A.F. (1990). Covalent binding of zomepirac glucuronide to proteins: evidence for a Schiff base mechanism. *Drug Metab. Dispos.*, *18*, 639–644.

Smith, P.C., Liu, J.H. (1995). Covalent binding of suprofen to renal tissue of rat correlates with excretion of its acyl glucuronide. *Xenobiotica*, *25*, 531–540.

Smith, P.C., McDonagh, A.F., Benet, L.Z. (1986). Irreversible binding of zomepirac to plasma protein in vitro and in vivo. *J. Clin. Invest.*, 77, 934–939.

Solensky, R. (2003). Hypersensitivity reactions to beta-lactam antibiotics. *Clin. Rev. Allergy Immunol.*, 24, 201–220.

Spahn-Langguth, H., Benet, L.Z. (1992). Acyl glucuronides revisited: is the glucuronidation process a toxification as well as a detoxification mechanism? *Drug Metab. Rev.*, 24, 5–47.

Spracklin, D.K., Thummel, K.E., Kharasch, E.D. (1996). Human reductive halothane metabolism in vitro is catalyzed by cytochrome P450 2A6 and 3A4. *Drug Metab. Dispos.*, 24, 976–983.

Stein, H.B., Patterson, A.C., Offer, R.C., Atkins, C.J., Teufel, A., Robinson, H.S. (1980). Adverse effects of D-penicillamine in rheumatoid arthritis. *Ann. Intern. Med.*, 92, 24–29.

Stoelting, R.K. (1999). Inhaled anesthetics, in *Pharmacology and Physiology in Anesthetic Practice*, Lippincott-Raven Publishers, pp. 36–76.

Stogniew, M., Fenselau, C. (1982). Electrophilic reactions of acyl-linked glucuronides. *Drug Metab. Dispos.*, 10, 609–613.

Tang, W. (2003). The metabolism of diclofenac—enzymology and toxicology perspectives. *Curr. Drug Metab.*, 4, 319–329.

Tang, W., Stearns, R.A., Bandiera, S.M., Zhang, Y., Raab, C., Braun, M.P., Dean, D.C., Pang, J., Leung, K.H., Doss, G.A., Strauss, J.R., Kwei, G.Y., Rushmore, T.H., Chiu, S.H., Baillie, T.A. (1999). Studies on cytochrome P-450-mediated bioactivation of diclofenac in rats and in human hepatocytes: identification of glutathione conjugated metabolites. *Drug Metab. Dispos.*, 27, 365–372.

Tishler, S.L., Goldman, P. (1970). Properties and reactions of salicyl-coenzyme A. *Biochem. Pharmacol.*, 19, 143–150.

Uetrecht, J. (2006). Role of animal models in the study of drug-induced hypersensitivity reactions. *AAPS J.*, 7, 914–921.

Uetrecht, J.P. (1999). New concepts in immunology relevant to idiosyncratic drug reactions: the "danger hypothesis" and innate immune system. *Chem. Res. Toxicol.*, 12, 387–395.

Uetrecht, J.P. (2003). Screening for the potential of a drug candidate to cause idiosyncratic drug reactions. *Drug Discov. Today*, 8, 832–837.

Uetrecht, J.P. (2008). Idiosyncratic drug reactions: past, present, and future. *Chem. Res. Toxicol*, 21, 84–92.

Valadon, P., Dansette, P.M., Girault, J.P., Amar, C., Mansuy, D. (1996). Thiophene sulfoxides as reactive metabolites: formation upon microsomal oxidation of a 3-aroylthiophene and fate in the presence of nucleophiles in vitro and in vivo. *Chem. Res. Toxicol.*, 9, 1403–1413.

Wagner, M.L. (1994). Felbamate: a new antiepileptic drug. *Am. J. Hosp. Pharm.*, 51, 1657–1666.

Walgren, J.L., Mitchell, M.D., Thompson, D.C. (2005). Role of metabolism in drug-induced idiosyncratic hepatotoxicity. *Crit. Rev. Toxicol.*, 35, 325–361.

Ware, J.A., Graf, M.L., Martin, B.M., Lustberg, L.R., Pohl, L.R. (1998). Immunochemical detection and identification of protein adducts of diclofenac in the small intestine of rats: possible role in allergic reactions. *Chem. Res. Toxicol.*, 11, 164–171.

van Wijk, F., Nierkens, S. (2006). Assessment of drug-induced immunotoxicity in animal models. *Drug Discov. Today*, 3, 103–109.

Williams, A.M., Worrall, S., de Jersey, J., and Dickinson, R.G. (1992). Studies on the reactivity of acyl glucuronides—III. Glucuronide-derived adducts of valproic acid and plasma protein and anti-adduct antibodies in humans. *Biochem. Pharmacol.* 43, 745–755.

Wise, E.M., Jr, Park, J.T. (1965). Penicillin: its basic site of action as an inhibitor of a peptide cross-linking reaction in cell wall mucopeptide synthesis. *Proc. Natl. Acad. Sci. U.S.A.*, 54, 5–81.

Wright, A.J. (1999). Penicillins. *Mayo Clin. Proc.*, 74, 290–307.

Yamashita, A., Watanabe, M., Tonegawa, T., Sugiura, T., Waku, K. (1995). Acyl-CoA binding and acylation of UDP-glucuronosyltransferase isoforms of rat liver: their effect on enzyme activity. *Biochem J.*, *312*, 301–308.

Yocum, R.R., Waxman, D.J., Rasmussen, J.R., Stromincser, J.L. (1979). Mechanism of penicillin action: penicillin and substrate bind covalently to the same active site serine in two bacterial D-alanine carboxypeptidases. *Biochemistry*, *76*, 2730–2734.

Zia-Amirhosseini, P., Ding, A., Burlingame, A.L., McDonagh, A.F., Benet, L.Z. (1995). Synthesis and mass-spectrometric characterization of human serum albumins modified by covalent binding of two non-steroidal anti-inflammatory drugs: tolmetin and zomepirac. *Biochem. J.*, *311*, 431–435.

Zimmerman, H. (1999). *Hepatotoxicity: The Adverse Effects of Drugs and Other Chemicals on The Liver*, Lippincott Williams & Wilkins, Philadelphia, PA.

Zimmerman, H.J. (1993). Hepatotoxicity. *Disease-A-Month*, *39*, 673–788.

Zimmerman, H.J. (1994). Hepatic in jury associated with nonsteroidal anti-inflammatory drugs, in *Nonsteroidal Antiinflammatory Drugs. Mechanisms and Clinical Use* (eds A.J. Lewis and D.E. Furst), Marcel Dekker Inc, New York, pp. 171–194.

Zimmerman, H.J., Lewis, J.H., Ishak, K.G., Maddrey, W.C. (1984). Ticrynafen-associated hepatic injury: analysis of 340 cases. *Hepatology*, *4*, 315–323.

Zubay, G.L. (1988). Carbohydrates, in *Biochemistry*, Macmillan Publishing Co., New York, pp. 663–667.

CHAPTER 23

Chemical Mechanisms in Toxicology

MARK P. GRILLO

We pray that every field of science may contribute in bringing happiness—not disaster—to human beings.[1]

23.1 INTRODUCTION

This chapter deals with the mechanisms involved in the biotransformation of drugs to chemically reactive metabolites, current experimental techniques used to identify reactive metabolites, and the potential consequences of reactive metabolite formation as it relates to the discovery of nontoxic drugs. Many pharmaceutical researchers in the fields of drug metabolism and toxicology currently perceive chemically reactive metabolites as an unwanted feature of any drug or drug candidate (Baillie, 2008; Kumar et al., 2008). Therefore, an imperative goal in drug discovery is to eliminate, or to substantially decrease, the metabolic activation liability of drug candidates leading to the increased probability of safer drugs being successfully developed (Baillie and Kassahun, 2001; Kalgutkar and Soglia, 2005; Park et al., 2005b; Baillie, 2006; Doss and Baillie, 2006; Williams, 2006). Drug metabolism scientists today are trained in taking closer and closer looks at the metabolism of new chemical entities sometimes having greater and greater complexity. One of the most important objectives in the pharmaceutical industry is to be able to make knowledgeable judgments about hidden risks so that new drugs that could be potentially toxic to patients are never released (Evans et al., 2004; Nassar and Lopez-Anaya, 2004; Evans and Baillie, 2005).

The history of the subject discussed in this chapter may have started more than 200 years ago with a report published by the English surgeon Percivall Pott (1714–1788; Pott, 1775). Pott is credited with recognizing one of the first known industrial-linked diseases when he described the association between working as a chimney sweep and

---

[1] A quotation from Kenichi Fukui. He was corecipient of the Nobel Prize in Chemistry in 1981 for investigations into the mechanisms of chemical reactions.

Drug Metabolism Handbook: Concepts and Applications, Edited by Ala F. Nassar, Paul F. Hollenberg, and JoAnn Scatina
Copyright © 2009 by John Wiley & Sons, Inc.

the development of scrotal carcinoma (Waldron, 1983; Greene, 1990; Ottoboni, 1991). Pott was impressed by the idea that coal soot contained some substance that was causing the cancer, thereby linking this occupational-linked chemical exposure with formation of the unusual tumor (Melicow, 1975). This was the dawn of a whole new era of the way that we now think about substance-induced carcinogenesis. During that time in the rest of Europe, scrotal cancer was not occurring at the same rate, presumably because of the reduced carcinogenicity of soot produced from wood-burning furnaces (IARC, 1985), and also because of the practice of wearing protective clothing (Davis, 2007).

Progress in understanding carcinogenesis sort of sat there for a very long period of time until the early 1900s, when it began to be critically analyzed again by a research group in Japan studying the cancer-causing property of coal tar. A scientific breakthrough in understanding carcinogenesis occurred when, for the first time, scientists were able to experimentally induce cancer in laboratory animals (Yamagiwa and Ichikawa, 1918). In their experiments, Yamagiwa and Ichikawa showed that after applying coal tar to the ear skin of rabbits, tumor formation would occur at the site of exposure.

The next major development was when the isolation and identification of the carcinogen benzo[a]pyrene from coal tar occurred (Cook, Hewett and Hieger, 1933; Berenblum and Schoental, 1947; Kennaway, 1955) (Fig. 23.1). It was shown that when radiolabeled benzo[a]pyrene was applied on mouse skin (Pereira, Burns and Albert, 1979), added to mouse embryo cell cultures (Duncan, Brookes and Dipple, 1969), or incubated in the presence of rat liver microsomes, NADPH, and calf thymus nucleic acids (Gelboin, 1969), electrophilic metabolites were formed that became covalently bound to DNA. From these observations, the initiation of carcinogenesis by polycyclic aromatic hydrocarbons was proposed to involve covalent binding of benzo[a]pyrene to DNA macromolecules (Brookes and Lawley, 1964). This covalent binding phenomenon was considered very peculiar at the time, because benzo[a]pyrene was known to

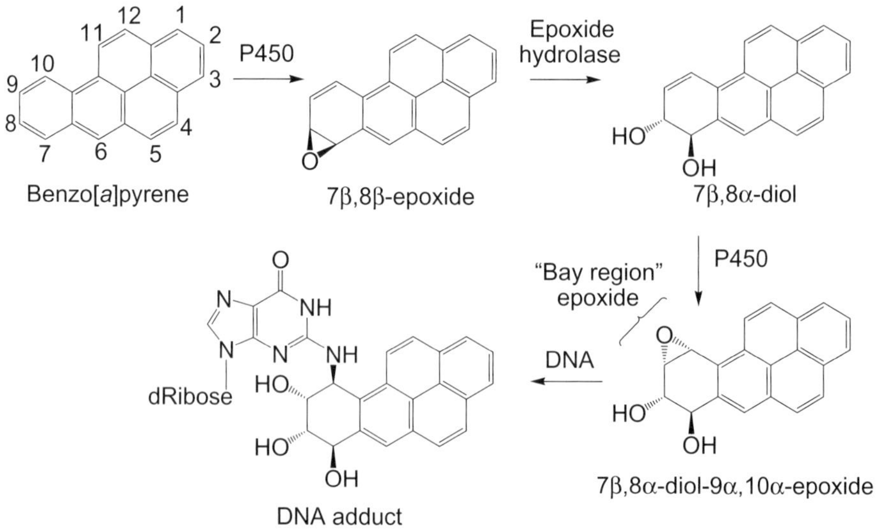

Figure 23.1 Metabolic activation of benzo[a]pyrene leading to covalent binding to DNA.

TABLE 23.1   **Hard versus soft nucleophiles and electrophiles.**

Hard nucleophiles	Intermediate	Soft nucleophiles
Oxygen in purines and pyrimidines of nucleic acids; phosphate oxygen in DNA; OH⁻; H_2O; ROH; NH_3	Nitrogen in primary and secondary amino groups of proteins; nitrogen in amino groups in purine bases in DNA; CN⁻	Sulfur in thiols (e.g. cysteine residues in proteins and peptides like glutathione)

Hard electrophiles	Intermediate	Soft electrophiles
Alkyl carbonium ions	Aryl carbonium ions; aryl halides; benzylic carbonium ions; nitrenium ions; carbon of epoxides	Carbon in polarized double bonds (e.g. α,β-unsaturated ketones, quinones, and quinone imines)

Adapted from Carlson (1990) and Gregus and Klaassen (2001).

be inherently unreactive, and the mechanism whereby it became irreversibly bound to DNA was not understood. But we now know that the bioactivation of benzo[*a*]pyrene follows a mechanism where it undergoes metabolism by cytochrome P450 to a chemically reactive, electrophilic, bay region diol epoxide that binds covalently to DNA (Sims *et al.*, 1974) (Fig. 23.1).

Observations from collected mechanistic studies were combined into a working hypothesis during the 1960s and 1970s from the innovative research and major insights of Miller and Miller (Miller and Miller, 1974, 1979, 1981; Heidelberger, 1975). The hypothesis that was developed stated that for some compounds to become carcinogenic, they first must be metabolically activated by membrane-bound cytochrome P450-dependent monooxygenases to chemically reactive species (Miller and Miller, 1974). Predominantly, it was viewed that chemically reactive hard electrophiles (those species that have a concentrated charge such as a carbocation, a nitrenium ion, or an iminium ion) react covalently with hard nucleophiles such as those located in nucleic acids (Carlson, 1990; Gregus and Klaassen, 2001) (Table 23.1). This mechanistic proposal, which became the framework by which we now study reactive drug metabolites, is considered the beginning of the field of xenobiotic bioactivation and led to a transformation of thinking about chemical carcinogenesis (Mitchell *et al.*, 1976; Nelson and Pearson, 1990).

These approaches continued to be advanced at the National Institute of Health by research from Brodie, Gillette, Mitchell, and Nelson, and their colleagues which concentrated on the mechanistic toxicology of drugs (Jollow *et al.*, 1973; Gillette and Pohl, 1977; Nelson *et al.*, 1978; Gillette, 1981; Baillie, 2006). They showed that drug-induced hepatotoxicity could be associated with the metabolic activation of therapeutic agents to chemically reactive soft electrophiles. Soft electrophiles are often uncharged and less electronegative than hard, highly polarized electrophiles. Soft electrophiles react with soft nucleophiles, such as the amino acid side chains of proteins (e.g. the sulfur atom of thiols found in cysteinyl residues of proteins and glutathione) (Table 23.1). In general, soft electrophiles react with soft nucleophiles, whereas hard electrophiles react more readily with hard nucleophiles (Carlson, 1990).

A focus of this chapter is on the implication of metabolic activation in the discovery of drugs. This is a topic of substantial interest in the pharmaceutical industry due to the major concern for the potential of some drugs to form reactive metabolites that

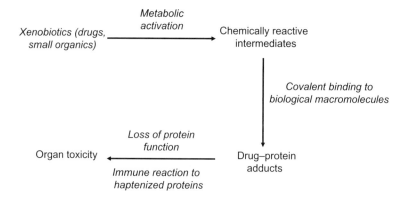

Figure 23.2   Metabolic activation of xenobiotics to chemically reactive and potentially toxic products.

TABLE 23.2   Drugs or chemicals known to cause idiosyncratic hepatotoxicity.

Acetaminophen[a]	Diclofenac	Isoxanine[b]	Tacrine[a]
Alpidem[b]	Dihydralazine[b]	Ketoconazole	Tamoxifen[a]
Amineptine[b]	Disulfiram[a]	Leflunomide[a]	Terbinafine[a]
Amodiaquine	Ebrotidine[b]	Methyldopa[a]	Tienilic acid[b]
Benoxaprofen[b]	Ethanol	Nefazadone[b]	Tilbroquinol[b]
Bromefenac[b]	Felbamate[a]	Niperotidien[b]	Troglitazone[b]
Carbamazepine[a]	Halothane[a]	Phenytoin	Trovofloxacin[b]
Cinchophen[b]	Ibufenac[b]	Pirprofen[b]	Valproic acid[a]
Clozapine[a]	Iproniazid[b]	Rifampin[a]	Zileuton[a]

[a]Drugs with black box warnings.
[b]Withdrawn drugs.
Adapted from Guengerich and MacDonald (2007) and Walgren, Mitchell and Thompson (2005).

can react with macromolecules in the body causing immune-based idiosyncratic toxicities (Uetrecht, 2003; Kalgutkar and Soglia, 2005; Baillie, 2006; Williams, 2006) (Fig. 23.2). These toxicities sometimes are caused by the haptenization of protein that leads to an untoward "triggering" event in the immune system of hypersensitive individuals (Ju and Uetrecht, 2002; Uetrecht, 2003; Kaplowitz, 2005). These severe, and occasionally fatal, immunologic-based toxic reactions resulting from this abnormality show a very low incidence of 1 in 1000 or even 1 in 100,000 patients (Zimmerman, 1999; Calvey, 2005; Pirmohamed, 2006). This is a tremendous concern in the pharmaceutical industry, since there have been several drugs that have been withdrawn from the market for this reason (Walgren, Mitchell and Thompson, 2005; Guengerich and MacDonald, 2007) (Table 23.2). It is especially of frustrating concern because there are almost no animal models available today to be able to predict toxic interactions of drugs with the human immune system (Shenton, Chen and Uetrecht, 2004; van Wijk and Nierkens, 2006).

After almost four decades of research showing strong indirect evidence that many types of reactive intermediates may be cytotoxic, pharmaceutical scientists are still trying to understand the significance of chemically reactive metabolites in drug toxicity (Kalgutkar and Soglia, 2005; Baillie, 2006). Over the last 10 to 20 years, there have been substantial improvements in understanding how reactive metabolites are formed and how they behave chemically. These improvements occurred because the disciplines of drug metabolism and analytical chemistry became increasingly capable at characterizing the chemical structures of reactive intermediates (Baillie, 2006; Liebler, 2006). These advancements are primarily due to considerable discoveries in liquid chromatography–tandem mass spectrometry (LC-MS/MS) techniques, such as the advent of the electrospray ionization interface, accurate mass measurement to enhance metabolite structural identification, and recent improvements in data acquisition and processing for accelerating metabolite identification. However, "Advances in the field of biochemical toxicology have not kept pace with those in drug metabolism and bioanalytical chemistry, such that we are unable to predict, *a priori*, which reactive intermediates may be toxic to the host cell -*vs* those that will be relatively benign" (Baillie, 2006). One accomplishable path forward used by drug metabolism scientists today has been to try to "minimize" the formation of reactive metabolites before taking drug discovery candidates into development, but unfortunately, today, still not being able to predict which drug candidate(s) may potentially cause idiosyncratic toxicity in patients (Evans and Baillie, 2005; Kumar *et al.*, 2008). This pathway forward currently involves two approaches that require a team of experts from drug metabolism, medicinal chemistry, and toxicology working closely together within a drug discovery program (Evans *et al.*, 2004; Mutlib *et al.*, 2005). The first approach currently used by many pharmaceutical companies involves screening techniques to find glutathione adducts formed when candidate drugs are incubated with liver microsomal preparations from a range of species usually including rat, dog, monkey, and human (Dieckhaus *et al.*, 2005; Kalgutkar and Soglia, 2005; Kalgutkar *et al.*, 2005).

23.2   GLUTATHIONE ADDUCTS

In vitro experiments in liver microsomes fortified with NADPH (the cofactor necessary for P450-mediated oxidation) are conducted in the presence of nucleophilic trapping agents such as glutathione (a soft nucleophile that reacts with soft electrophiles) or cyanide anion (a hard nucleophile that can react with hard electrophiles), in order to trap chemically reactive species that may have been formed in these incubations. Glutathione reacts with, and thereby captures, electrophilic metabolites by both nucleophilic substitution and nucleophilic addition reactions as depicted in Fig. 23.3.

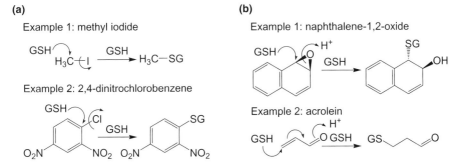

Figure 23.3   Chemically reactive compounds react with GSH by one of two general mechanisms: **(a)** nucleophilic substitution-type and **(b)** nucleophilic addition-type reactions.

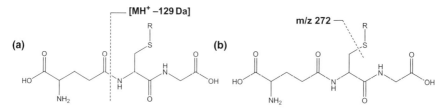

Figure 23.4 LC-MS/MS screening for glutathione adducts present in biological extracts by **(a)** constant neutral loss scanning (−129 Da in positive ion scan mode) and **(b)** precursor ion scanning (m/z 272 in negative ion scan mode).

If a conjugate is detected, then the focus is to elucidate the adduct structure by LC-MS/MS characterization, and, when sufficient amount of adduct has been isolated, by nuclear magnetic resonance (NMR) analysis to confirm the atomic site of adduct linkage. Once the site(s) of metabolic activation of a compound is (are) determined, then medicinal chemists can modify the drug candidate (e.g. by fluorine atom substitution) to block, or at least hinder, bioactivation. LC-MS/MS methods are now routinely used to examine biological extracts for the presence of glutathione adducts produced *in vitro* or *in vivo* using specialized scanning techniques. The LC-MS/MS detection techniques employed most often for glutathione conjugate detection are traditional constant neutral loss scanning for 129 Da, which is based on the loss of the elements of pyroglutamic acid from the glutathione adduct protonated molecular ion upon collision-induced dissociation (Fig. 23.4a) (Baillie and Davis, 1993), and by negative ion tandem LC-MS/MS for precursors of the product ion at m/z 272 (Fig. 23.4b) (Dieckhaus *et al.*, 2005). Therefore, new chemical entities that form glutathione adducts can, in most cases, be detected by these two LC-MS/MS screening methodologies which work independent of compound structure, can be detected without exact knowledge of the adduct structure, and are amenable to high-throughput screening (Baillie and Davis, 1993; Kalgutkar and Soglia, 2005).

The capturing of chemically reactive intermediates with glutathione, and subsequent identification and structural characterization by LC-MS/MS and NMR, is an advantageous methodology because it does not require often expensive and time-consuming synthesis of radioactive drug candidates (Zheng *et al.*, 2007). Other useful strategies include the use of an equal mixture of stable isotope-labeled (e.g. ^{13}C, ^{15}N) glutathione and unlabeled glutathione to trap reactive metabolites which assists in adduct detection by LC/MS analysis as an "isotope doublet" (Yan and Caldwell, 2004; Mutlib *et al.*, 2005), as well as the use of derivatized glutathione for semiquantitative determination of glutathione adduct levels (Gan *et al.*, 2005; Soglia *et al.*, 2006). However, these methods employing glutathione stable isotope labeled or derivatized analogs are not suitable for the trapping and identification of hard electrophilic intermediates, where a mixture of KCN/K^{13}C^{15}N should be used to capture, for example, reactive carbocation or iminium ion intermediates (Evans *et al.*, 2004; Argoti *et al.*, 2005).

For successful NMR analysis of purified conjugates, up to milligram amounts of adduct may be necessary for structural characterization. Therefore, *in vivo* studies in preclinical species can be performed in bile duct cannulated animals, where glutathione or mercapturic acid adducts can be isolated by HPLC purification from bile or urine, respectively (Fig. 23.5).

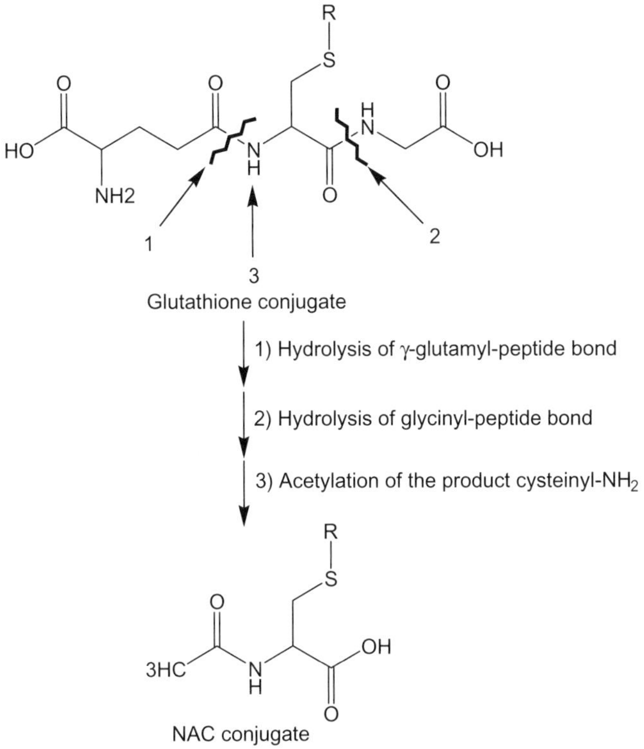

Figure 23.5 Enzyme-mediated degradation of glutathione conjugates in the kidney leading to the corresponding mercapturic acid conjugate (*N*-acetylcysteine [NAC] conjugate).

This first approach analysis is performed to provide an initial mechanistic understanding of the route of metabolic activation and is of express interest early in the discovery program when changes can be made to a drug structure due to enhanced program resources. These issues regarding chemically reactive metabolite are inherently important in a drug discovery program and require rapid answers. With the knowledge of reactive metabolite structure, medicinal chemists can, in most cases, modify lead structures to block, or at least substantially decrease, the potential for metabolic activation. Glutathione adduct detection is performed as an iterative process in the early stages of drug discovery and lead candidate optimization. In summary, reactive metabolite-trapping experiments with glutathione, *N*-acetylcysteine (NAC), or cyanide anion can provide stable adducts for structural elucidation to assist in the mechanistic understanding of metabolic activation of candidate drugs (Evans *et al.*, 2004).

Mercapturic acid adducts excreted in urine can also be used as biomarkers for exposure to chemically reactive metabolites formed *in vivo* in preclinical species and in humans (van Welieet *et al.*, 1992; Scholz *et al.*, 2005). Rapid and sensitive detection of mercapturic acid conjugates of unknown structure can be analyzed by negative ion mode LC-MS/MS constant neutral loss scanning for 129 Da from urine extracts, or from extracts obtained from liver microsomal incubations fortified with NADPH and NAC (Scholz *et al.*, 2005) (Fig. 23.6).

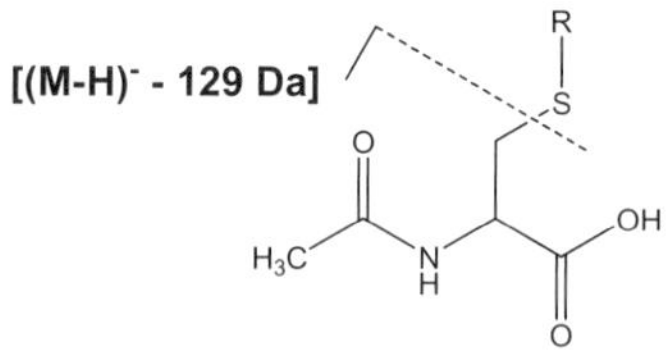

Figure 23.6   LC-MS/MS screening for mercapturic acid adducts present in biological extracts by constant neutral loss scanning (–129 Da in negative ion scan mode).

23.3   COVALENT BINDING

Drug candidate optimization during late-stage discovery often leads to only a few qualified compounds remaining that retain desirable drug properties such as high target potency, excellent receptor selectivity, and appropriate animal pharmacokinetics. A key decision to be made at this stage is in the selection of the best compound to take forward into development. Making this decision can be assisted by evaluating drug candidate covalent binding potential, especially if glutathione conjugates were detected during preliminary *in vitro* and *in vivo* metabolic activation studies. Once a compound is introduced into drug development, there is major economic expenditure made in making bulk drug to conduct preclinical toxicity studies. Therefore, the obvious need is to perform experiments on metabolic activation rapidly so that medicinal chemists can make the necessary structural changes such that bioactivation properties of the compound are eliminated, of course without dramatically affecting potency, receptor selectivity, and preclinical species pharmacokinetics (Doss and Baillie, 2006; Guengerich and MacDonald, 2007; Kumar *et al.*, 2008). This second experimental approach, which has been embraced by many pharmaceutical companies, is to measure covalent binding of radioactivity from radiolabeled [^{14}C-, ^{3}H-, or ^{35}S-labeled (if applicable)] drug candidates formed *in vitro* in microsomal liver incubations (Evans *et al.*, 2004; Kalgutkar and Soglia, 2005). The utmost impediment for these covalent binding studies is the availability of radiolabeled drug candidates. However, once radiolabeled derivatives are made available, liver microsomal incubations are performed in the presence and absence of NADPH (to provide a negative control) and are compared directly with results from a positive control compound studied in the same experiment. Covalent binding of radiolabeled compounds to liver microsomal proteins is measured under standardized conditions (Day *et al.*, 2005). Such studies are performed in microsomal preparations cross-species, including human. If covalent binding to protein is detected *in vitro*, then *in vivo* covalent binding to liver protein can also be measured to evaluate *in vivo* exposure to reactive metabolites, although usually only studied in rats. The highest levels of covalent binding to protein are usually detected in the liver due to the high concentration of P450 enzymes in hepatic tissue (Waring and Anderson, 2005). The focal point of this second approach is to assess the exposure to reactive metabolites, but it does not foretell toxicity or potential toxicity of the drug candidate. To date, it is not known what extent of covalent binding to protein correlates with the risk of a drug candidate being able to cause an adverse idiosyncratic drug reaction (Evans *et al.*, 2004; Baillie, 2006; Uetrecht, 2006).

If covalent binding to protein is detected, then the goal is to block exposure to these reactive intermediates by making strategic changes to the chemical structure of the candidate molecule to hinder bioactivation pathway(s) (Evans *et al.*, 2004; Doss and

Baillie, 2006). What covalent binding amount is considered acceptable? This important question has been prudently and carefully addressed by the Preclinical Drug Metabolism department at Merck Research Laboratories (Evans *et al.*, 2004; Evans and Baillie, 2005; Doss and Baillie, 2006). They came up with a target covalent binding number of 50 pmol equivalents of drug residue/mg protein/h of incubation. It was proposed that this value be used as a target covalent binding level that discovery programs would want their candidate drugs to be below. At Merck Research Laboratories, they stress that this target value should not be considered a threshold value (Evans *et al.*, 2004). The number of 50 pmol equivalents of the drug residue/mg protein/h was not arbitrarily derived, but came from a thorough literature search for the levels of covalent binding to liver proteins in animals dosed with known hepatotoxins, for example bromobenzene (Monks, Hinson and Gillette, 1982), isoniazid (Nelson *et al.*, 1978), and acetaminophen (Matthews *et al.*, 1997), under conditions where these drugs induced hepatotoxicity (Evans *et al.*, 2004). When the values of covalent binding to protein for these compounds were measured, the levels were as high as 1000 to 2000 pmol equivalents/mg liver protein. Therefore, the covalent binding target adopted by Merck Research Laboratories is about 5% as much as that caused by many of these model hepatotoxic compounds. In addition, an experimental reason for this value is that the number of 50 pmol equivalents of the drug residue/mg protein/h is about 10-fold higher than the limits of detection of standard liquid scintillation counters used today (Evans *et al.*, 2004). Covalent binding studies can be very helpful because they aid drug discovery scientists in screening out drugs candidates that form reactive intermediates from going forward into late drug discovery or early drug development.

Having obtained results on covalent binding for a drug candidate, a drug discovery team can put this data into perspective using a set of qualifying considerations as proposed by Merck Research Laboratories (Evans *et al.*, 2004; Doss and Baillie, 2006). The qualifying considerations reached by Evans *et al.* (2004) state that if a drug is designed for the treatment of a disabling or life-threatening disease, if there is an unmet medical need, or if the duration of the use of the drug is acute rather than chronic, then the drug discovery program might decide to take the drug forward into development. In addition, if the drug is designed for a novel therapeutic target and to be used only for proof-of-concept purposes, then the compound might still be considered appropriate for progression into target validation-type studies in humans. When pursuing to develop drugs for pediatric indications, the compounds would probably have to be as "issue-free" as possible (i.e. no detectable covalent binding or metabolic activation potential). Another important aspect is related to the clinical dose. There have been very few drugs taken off the market for toxicity reasons when the daily dose was less than 10 mg (Uetrecht, 2001; Kalgutkar *et al.*, 2002). Therefore, if candidate drugs have high potency and excellent pharmacokinetic properties, such that these drugs could be given to patients at low daily doses, then a project team might decide to advance such compounds into development, even when faced with covalent binding issues.

23.4 STRUCTURAL ALERTS

Hypotheses dealing with direct (intrinsic) or immunologic toxicities mediated by chemically reactive species were increasingly being developed and tested during the 1990s (Tirmenstein and Nelson, 1989; Nelson and Pearson, 1990; Rashed, Myers and Nelson, 1990; Roberts, Price and Jollow, 1990; Myers *et al.*, 1995; Nelson, 1995). Since then, there has been a merging of the chemical, biochemical, and toxicologic information concerning metabolic activation which has afforded hypotheses to address the roles of chemically reactive intermediates as mediators in drug-induced toxicity. This work has resulted in the generation of a list of "structural alerts" which are structural moieties within larger drug structures that seem to be problematic for drugs based on a historical perspective (Nelson, 1995; Kalgutkar *et al.*, 2005; Guengerich and MacDonald, 2007). These structural alerts include, for example, aromatics, 3-methylindoles, hydrazines and hydrazides, carboxylic acids (especially arylacetic and arylpropionic acids), thiophenes, furans, pyrroles, anilines and anilides, moieties that form quinones and quinone imines, halogenated hydrocarbons and some halogenated aromatics, nitroaromatics, and moieties that form reactive α,β-unsaturated ketone-like structures, thiazolidinediones, alkenes, alkynes, and finally thioureas (Table 23.3).

As shown in Table 23.3, xenobiotics can be biotransformed to chemically reactive intermediates by a wide variety of mechanisms leading to products that can covalently bind to macromolecules such as nucleic acids and proteins. Covalent binding in some cases corresponds to toxicity (Nelson and Pearson, 1990; Cohen *et al.*, 1997). Toxic xenobiotics can be categorized into two basic groups: intrinsic toxins and idiosyncratic toxins (Pumford and Halmes, 1997). Intrinsic toxins are substances that have a very high potential to cause organ damage at increased doses, but will not be toxic if used appropriately. These compounds can produce a predictable, dose-dependent, and acute toxicity. In other words, intrinsic toxicity is demonstrated by the concentration of a substance that produces a toxic effect in an organism. One example of an intensely studied intrinsic toxin is acetaminophen, a commonly used over-the-counter medication.

Idiosyncratic toxins are xenobiotics that lead to unpredictable and unexpected toxic drug reactions. Idiosyncratic reactions to drugs can be defined as reactions that do not occur in most patients at any dose, they do not relate to the pharmacological activity of the drug, and the onset of reaction usually takes more than one week after initial exposure to the xenobiotic (Ju and Utrecht, 2002). These types of toxic reactions are commonly referred to as hypersensitivity, or type-B, reactions (Uetrecht, 1999). A well-studied example of a drug known to cause idiosyncratic toxicity is the nonsteroidal anti-inflammatory drug (NSAID) diclofenac (Boelsterli, 2003; Tang, 2003).

23.5   EXAMPLES OF METABOLIC ACTIVATION

23.5.1   Aromatics

Naphthalene (Fig. 23.7) is an aromatic hydrocarbon that is made up of two fused benzene rings and can be found in cigarette smoke (Schmeltz *et al.*, 1978), petroleum-based fuels (Clark *et al.*, 1982), and in mothballs. Naphthalene is a lung toxicant that undergoes P450 2F-, 1A1- and 1A2-mediated bioactivation forming a chemically reactive epoxide, namely naphthalene-1,2-oxide. This epoxide reacts with glutathione, leading to the formation of three separate thioether-linked glutathione adducts (Buona-rati, Jones and Buckpitt, 1990; Shultz *et al.*, 1999). Radiolabeled [14C]naphthalene is commercially available and is often used as a positive control in standardized covalent binding studies (Day *et al.*, 2005). *In vitro* covalent binding studies performed under standardized conditions show that [14C]naphthalene (10 μM) covalently binds to microsomal protein (1 mg/mL) in the presence of NADPH (1 mM) at ~900 and ~1200 pmol equivalents/mg protein/h for rat liver microsomal and human liver microsomal incubations, respectively (Day *et al.*, 2005).

23.5.2 Thiophenes

Suprofen (Fig. 23.8), an NSAID, was used in the United States for about three years before several cases of nephrotoxicity occurred prior to the drug being removed from the market (Hart, Ward and Lifschitz, 1987; Fung *et al.*, 2001). Suprofen contains the thiophene structural alert and is metabolically activated to thiophene *S*-oxide, thiophene epoxide, and γ-thioketo-α,β-unsaturated aldehyde reactive intermediates. The drug also causes a mechanism-based inactivation of P4502C9, where the electrophilic γ-thioketo-α,β-unsaturated aldehyde intermediate was considered the most likely species to covalently bind to, and inactivate, the enzyme (O'Donnell, *et al.*, 2003) (Fig. 23.8). Another thiophene-containing drug that forms chemically reactive metabolites is diuretic drug tienilic acid, where it undergoes P4502C9-mediated metabolic activation to epoxide- and *S*-oxide-type reactive intermediates (see Chapter 22; Fig. 22.5).

23.5.3 Furans

Furosemide (Fig. 23.9) is an example of a drug that contains the furan-type structural alert (Kalgutkar *et al.*, 2005; Williams, 2006) (Fig. 23.9). A major use of the drug is for pulmonary edema, and is widely and safely used for this purpose (Jackson, 1996). Furosemide is metabolically activated by P450 leading to the formation of chemically reactive epoxide and α,β-unsaturated dicarbonyl intermediates. Both of these reactive species lead to covalent binding to protein (Mitchell, Snodgrass, Gillette, 1976; Wirth, Bettis and Nelson, 1976; Kalgutkar *et al.*, 2005), and to the production of glutathione adducts formed *in vitro* in microsomal incubations and *in vivo* in rats (Williams *et al.*, 2007). Other published examples of furan metabolic activation include the novel 5-lipoxygenase inhibitor, L-739,010 (Zhang *et al.*, 1996) and the potent HIV-1 encoded furano-pyridine-type protease inhibitor, L-754,394 (Sahali-Sahly *et al.*, 1996). Although not used as drugs, other furan-containing molecules known to undergo bioactivation are the furanocoumarin, bergamottin (Kent *et al.*, 2006), the dihydrofuran, aflatoxin (Degen and Neumann, 1978), and the lung-toxic furanoterpenoid, 4-ipomeanol (Boyd, 1976).

23.5.4 Thiazolidinediones

Troglitazone (Fig. 23.10) is an oral antidiabetic drug that was used effectively to treat type-2 diabetes and which contains the thiazolidinedione structural alert. However, the drug was removed from the market in March 2000, when it was shown to cause a severe hepatotoxicity that resulted in some fatalities (Isley, 2003). Cytochrome P450 mediates the opening of the thiazolidinedione ring system with the production of isothiocyanate- and sulfenic acid-type electrophilic moieties formed in the same product (Kassahun *et al.*, 2001). Troglitazone is also metabolized by P450 on the chromane moiety, giving rise to a quinone methide-type reactive metabolite (Table 23.3) (Kassahun *et al.*, 2001). Therefore, more than one pathway is involved in the metabolic activation of troglitazone, but whether reactive intermediates are involved in troglitazone-induced liver failure is not yet understood. There are other thiazolidinedione-containing antidiabetic drugs on the market, including pioglitazone (therapeutic dose 15–30 mg/day) and

TABLE 23.3 **Examples of chemicals containing structural alerts and their corresponding reactive metabolites.**

Structural alert	Example chemical	Proposed reactive metabolite
Hydrazines	Procarbazine	Methyl diazonium
Hydrazides	Iproniazid	Radical and diazonium ion
Arylacetic acids	Zomepirac	Acyl glucuronide and acyl-CoA
Arylpropionic acids	Benoxaprofen	Acyl glucuronide and acyl-CoA
Thiophene	Tienilic acid	Thioohene-*S*-oxide and a,β-unsaturated dicarbonyl
Furans	Furosemide	Epoxide and a,β-unsaturated dicarbonyl

Structural alert(s)	Example chemical	Proposed reactive metabolite(s)
Halogenated hydrocarbons	Halothane	Acyl halide
Halogenated aromatics	Bromobenzene	Epoxides, quinones
Nitro-aromatics	Dantrolene	Hydroxylamine, nitroso
Thioureas	1-Methylthiourea	Sulfenic, sulfinic, and sulfonic acids
Moieties that form reactive a,β-unsaturated ketone-like structures	Felbamate	Michael acceptor
Thiazolidinediones	Troglitazone	Isocyanate, *S*-oxide

TABLE 23.3 *Continued*

Structural alert	Example chemical	Proposed reactive metabolite
Anilines	Aniline	Hydroxylamines, nitroso and quinone imine
Anilides	Acetaminophen	Quinone imine
Phenols	Phenol	*p*-Quinone, *o*-quinone
Moieties that form a reactive quinone imine	Diclofenac	Quinone imines
Alkenes	4-Vinylcyclohexene	Epoxide

Adapted from Kalgutkar *et al.* (2005) and Guengerich and MacDonald (2007).

rosiglitazone (therapeutic dose 4–8 mg/day), both of which have been shown to cause hepatotoxicity. However, the incidence rate of hepatotoxicity is very low compared with troglitazone (therapeutic dose of 600 mg/day), potentially due to the comparatively low doses used for these new-generation thiazolidinediones (Scheen, 2001a, b).

23.5.5 Compounds That Form Reactive Quinones

23.5.5.1 *Halogenated Aromatics* The organic solvent bromobenzene (Fig. 23.11) is a halogenated aromatic molecule that undergoes P450-mediated bioactivation to the reactive intermediates, namely bromobenzene epoxide and bromobenzene *p*-quinone,

Structural alert(s)	Example chemical	Proposed reactive metabolite(s)
Pyrroles	1,3,4-Trimethylpyrrole	Imine methide
Moieties that form a reactive quinone methide	Troglitazone	Quinone methide
Aromatics	Napthalene	Naphthalene oxide
Indoles	3-Methylindole	Imine methide, indole epoxide
Alkynes	17a-Ethynylestradiol	Oxirene and ketene

both of which are able to arylate tissue macromolecules *in vivo* (Bambal and Hanzlik, 1995), arylate proteins *in vitro* (Narasimhan *et al.*, 1988), and react with glutathione forming the corresponding adducts as depicted in Fig. 23.11 (Fakjian and Buckpitt, 1984). Bromobenzene has been shown to cause hepatotoxicity and nephrotoxicity which is proposed to be mediated by the formation of these chemically reactive metabolites (Reid *et al.*, 1971; Gregus and Klaassen, 2001).

23.5.5.2   Phenols   Raloxifene (Fig. 23.12) is a benzothiophene-containing selective estrogen receptor modulator used for the treatment of osteoporosis in postmenopausal women (Lufkin, Wong and Deal, 2001). Raloxifene contains both aromatic- and

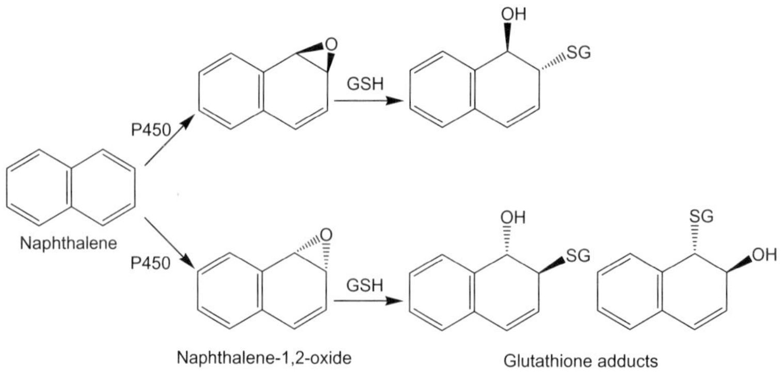

Figure 23.7   Scheme for the metabolic activation of naphthalene and formation of corresponding glutathione adducts.

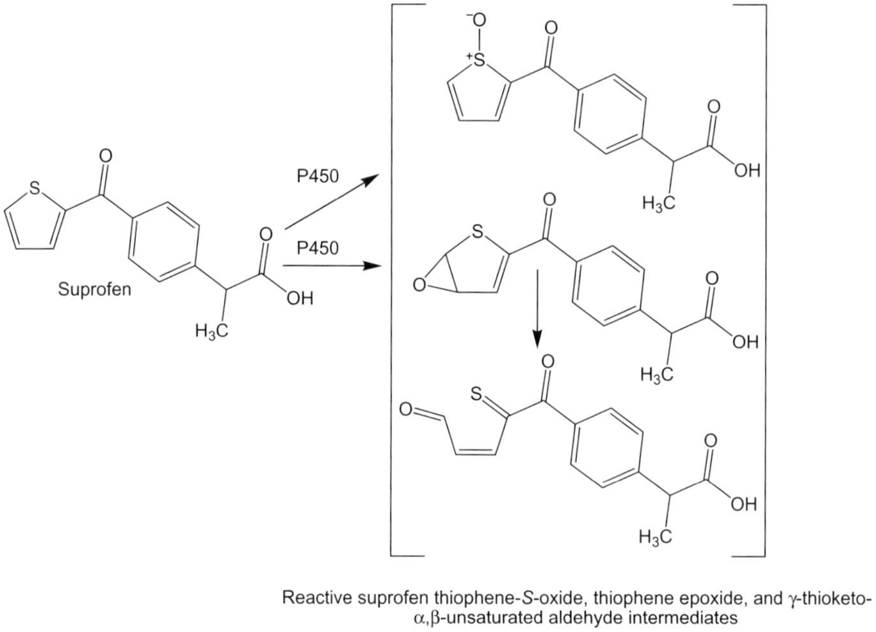

Reactive suprofen thiophene-*S*-oxide, thiophene epoxide, and γ-thioketo-α,β-unsaturated aldehyde intermediates

Figure 23.8   Scheme for the metabolic activation of suprofen to electrophilic thiophene *S*-oxide-, a thiophene epoxide-, and an γ-thioketo-α,β-unsaturated aldehyde-type intermediates.

phenol-type structural alerts. The drug was recently shown to undergo metabolic activation in incubations with human liver microsomes, leading to the proposed formation of arene oxide(s) and a quinone-type reactive metabolites that react with glutathione, but also lead to the time-dependent inactivation of P450 3A4 (Chen *et al.*, 2002) (Fig. 23.12). Mass spectrometric analysis was used to show that raloxifene becomes covalently bound to intact P450 3A4 apoprotein (Baer, Wienkers and Rock, 2007). Further mass spectrometric studies on the adducted apoprotein provided strong evidence (mass increase by 471 Da) indicating that raloxifene-mediated P450 3A4 inactivation occurred

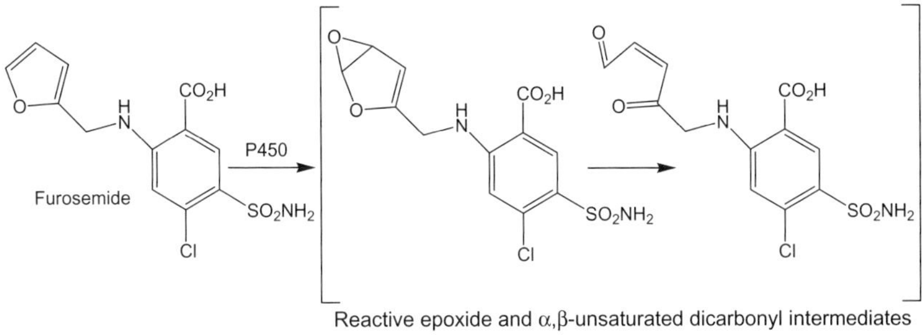

Reactive epoxide and α,β-unsaturated dicarbonyl intermediates

Figure 23.9   Scheme for the metabolic activation of furosemide to reactive furan epoxide and α,β-unsaturated dicarbonyl intermediates.

Troglitazone glutathione adducts formed from the isocyanate intermediate

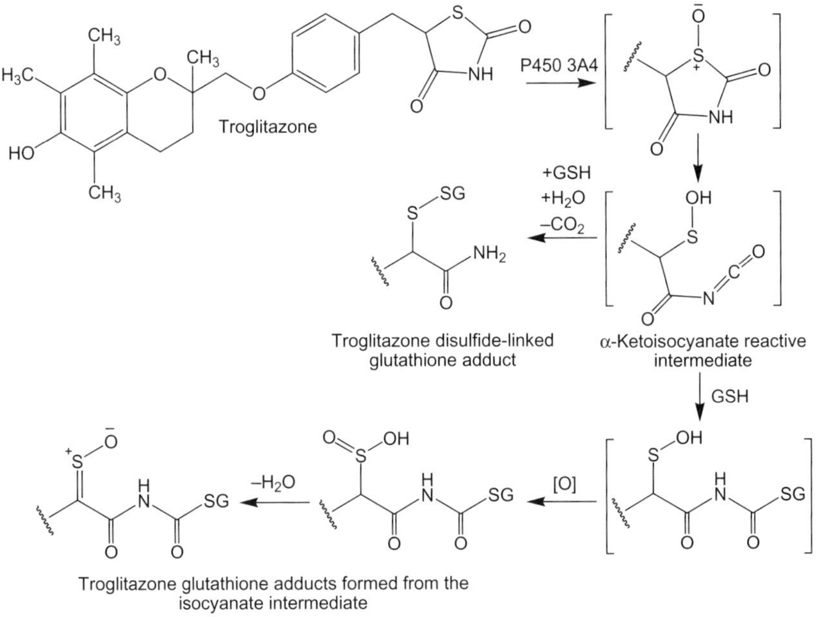

Figure 23.10   Scheme for the metabolic activation of the thiazolidinedione moiety of troglitazone leading to the formation of glutathione adducts (adapted from Kassahun *et al.*, 2001).

via bioactivation to the extended quinone reactive metabolite (Fig. 23.12) with subsequent arylation of the Cys239 active-site residue.

23.5.6   3-Methylindoles

Zafirlukast, a leukotriene receptor antagonist used for the treatment of asthma, contains a 3-methylindole-like structural alert (Fig. 23.13). The use of zafirlukast has been associated with an idiosyncratic hepatotoxicity that is consistent with a hypersensitivity reaction to the drug (Reinus *et al.*, 2000). Recently, it was shown that zafirlukast is metabolically activated by P450 3A4 in human liver microsomes to a highly reactive electrophilic α,β-unsaturated iminium methide intermediate that forms adducts with

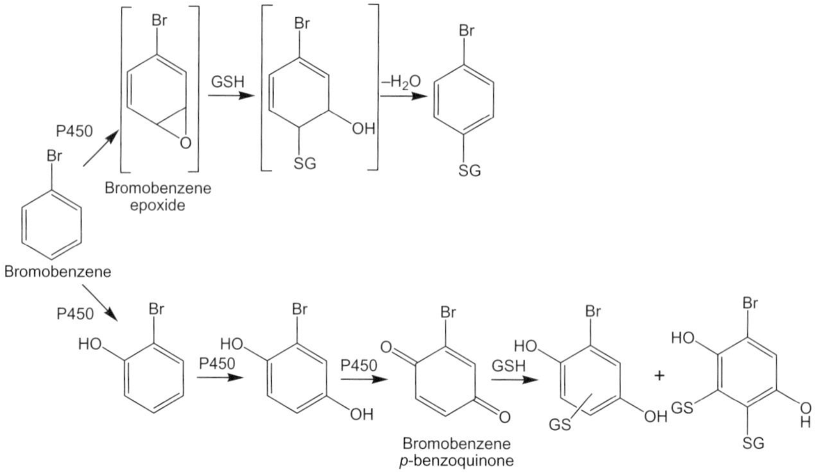

Figure 23.11   Scheme for the metabolic activation of bromobenzene and formation of the corresponding glutathione adducts (adapted from Kalgutkar *et al.*, 2005).

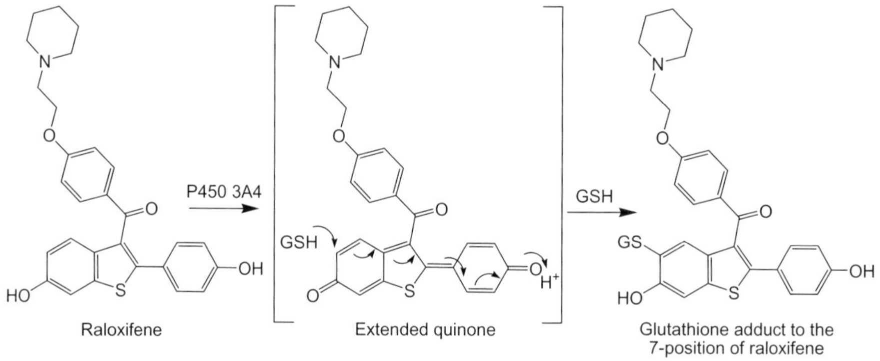

Figure 23.12   Proposed scheme for the metabolic activation of the raloxifene to the chemically reactive extended quinone intermediate followed by reaction with glutathione to form the most abundant 7-glutathionyl-raloxifene adduct.

glutathione, is proposed to mediate detected time-dependent inactivation of P450 3A4, and is proposed to be responsible for zafirlukast-induced liver injury (Fig. 23.13) (Kassahun *et al.*, 2005). The bioactivation of zafirlukast follows a scheme very consistent with the mechanism of P450-mediated metabolic activation of the lung toxicant 3-methylindole (Yan *et al.*, 2007) (Table 23.3).

23.5.7   Drugs Metabolized to α,β-Unsaturated Ketones

Valproic acid (VPA, Fig. 23.14) is an anticonvulsant drug that has been shown to induce a rare, but sometimes fatal, hepatotoxicity characterized by microvesicular steatosis (Zimmerman and Ishak, 1982). VPA is metabolized by P450 to an unsaturated metabolite, namely 4-ene-VPA (Rettie *et al.*, 1987), which undergoes further metabolism by the enzymes of fatty acid β-oxidation, leading to the formation of the reactive

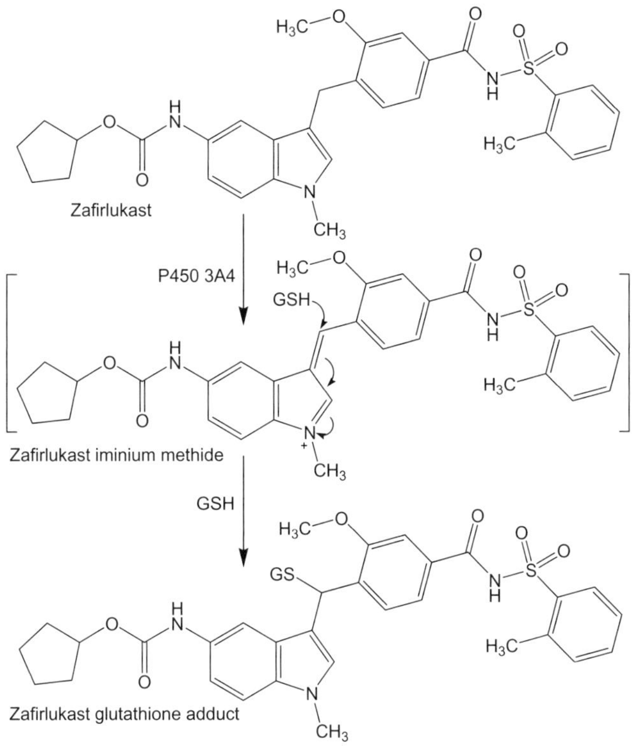

Figure 23.13 Scheme for the metabolic activation of zafirlukast to an electrophilic α,β-unsaturated iminium methide intermediate and formation of the corresponding glutathione adduct.

intermediates, (E)-2,4-diene-VPA-CoA and 3-keto-4-ene-VPA-CoA. These intermediates can react with glutathione (Kassahun and Abbott, 1993), but also are proposed to inhibit fatty acid metabolism by covalent binding to, and thereby irreversibly inhibiting, fatty acid β-oxidation enzymes (Baillie, 1988; Porubek *et al.*, 1991) (Fig. 23.14). Studies in rat have shown that α-fluorination of 4-ene-VPA resulted in a non-hepatotoxic derivative (Tang *et al.*, 1995). Further mechanistic studies comparing VPA with its α-fluorinated derivative (F-VPA) for their abilities to form acyl-CoA thioester derivatives *in vivo* in rat liver showed that F-VPA did not form the corresponding acyl-CoA metabolite (Grillo *et al.*, 2001). From these observations, and from related studies showing the lack of toxicity due to α-fluoro-substitution, it was proposed that metabolism of VPA by acyl-CoA (see Section 23.5.8 and Section 22.4.6 in Chapter 22) formation may mediate, at least in part, the idiosyncratic hepatotoxicity of the drug.

23.5.8 Carboxylic Acids

Zomepirac (5-[*p*-chlorobenzoyl]-1,4-dimethylpyrrole-2-acetic acid, Fig. 23.15) is a carboxylic acid-containing NSAID that was withdrawn from clinical use in March 1983 due to severe, and sometimes fatal, hepatotoxic reactions (Kiani and Kushner, 1983; Levy and Vasilomanolakis, 1984). Zomepirac is metabolized primarily to an unstable

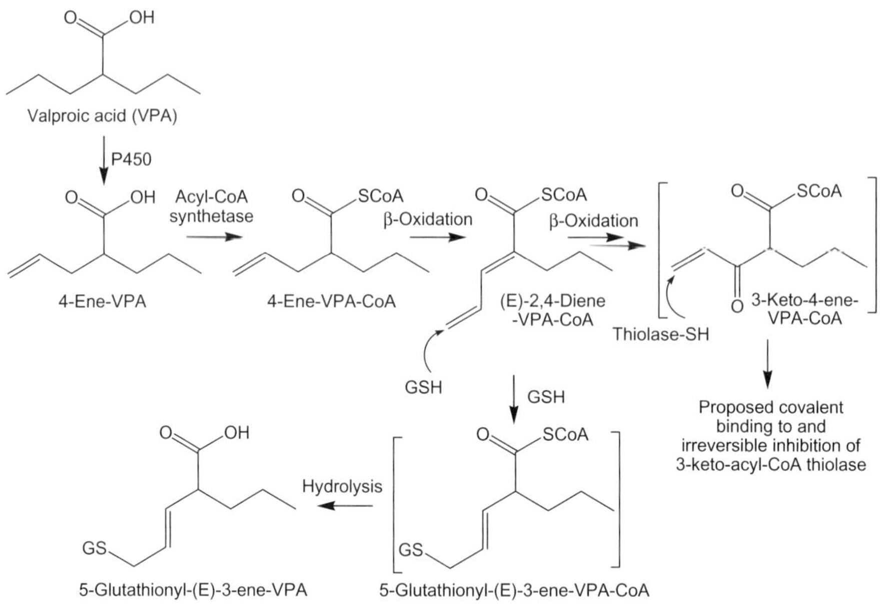

Figure 23.14 Proposed scheme for the concerted metabolic activation of valproic acid by P450 and fatty acyl-CoA β-oxidation enzymes.

and chemically reactive acyl glucuronide, namely zomepirac-1-O-acyl glucuronide (Smith, Benet and McDonagh, 1990), which has been implicated in the toxicity of the drug. Zomepirac-1-O-acyl glucuronide is proposed to covalently bind to protein by two different mechanisms, including transacylation-type reactions with protein nucleophiles by the 1-O-acyl glucuronide isomer, and through a glycation mechanism involving the reaction of protein amino-groups with open-chain aldehyde forms of the acyl migration isomers of the acyl glucuronide (Fig. 23.15) (Spahn-Langguth and Benet, 1992). Drug–protein adducts resulting from these reactions then are proposed to be recognized by the immune system as foreign, thereby eliciting an untoward immune-mediated hepatotoxicity (Zia-Amirhosseini, Spahn-Langguth and Benet, 1994). Experiments with zomepirac have shown that the drug becomes covalently bound to protein *in vivo* in human plasma (Smith, Benet and McDonagh, 1990) and *in vivo* in rat liver (Wang and Dickinson, 2000). Recently, zomepirac was shown to form another reactive acylating species, namely zomepirac-S-acyl-CoA (Olsen *et al.*, 2005), and that the acyl-CoA metabolite may be more important than the corresponding acyl glucuronide in contributing to the acylation of GSH and protein nucleophiles in liver tissue (Grillo and Hua, 2003).

23.5.9 Pyrroles

Zomepirac (Fig. 23.16) also contains a pyrrole-type structural alert that has been shown recently to undergo metabolic activation by P450 (Chen *et al.*, 2006). In incubations with human liver microsomes, and *in vivo* in rat, zomepirac is metabolized by P450 forming a proposed reactive epoxide intermediate that reacts with glutathione forming 5-[4′-chlorobenzoyl]-1,4-dimethyl-3-glutathionyl-pyrroleacetic acid. In these studies,

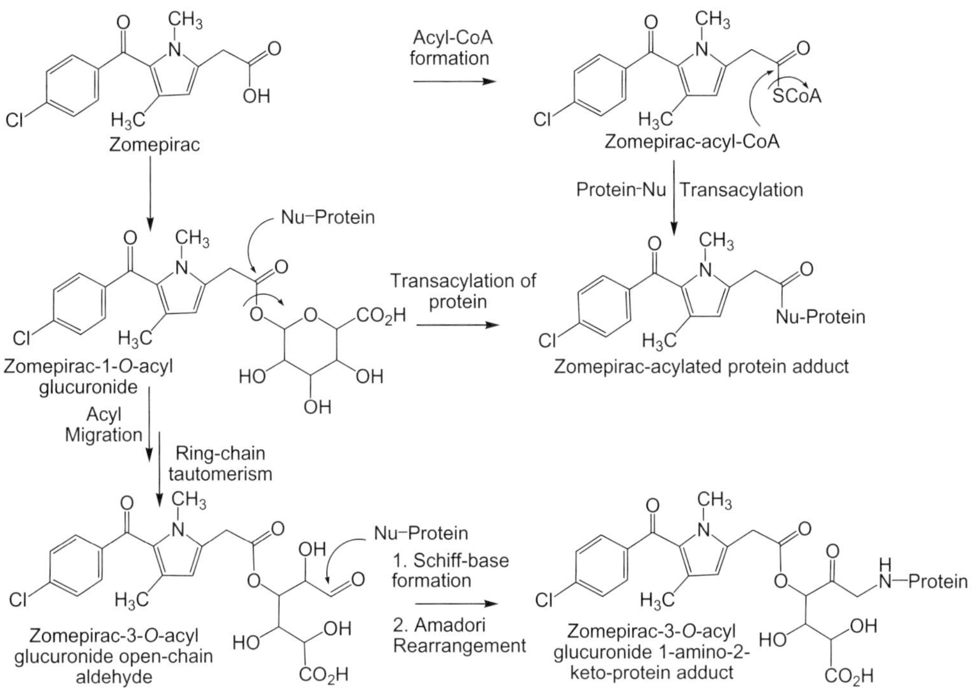

Figure 23.15   Proposed scheme for the metabolic activation of the zomepirac by acyl glucuronidation and acyl-CoA formation.

the corresponding glutathione adduct was also shown to be formed for the drug tolmetin, a structurally similar pyrrole-containing NSAID. The authors from these studies proposed that, in addition to the formation of reactive acyl glucuronides, P450-dependent metabolism of zomepirac and tolmetin leads to the production of a reactive epoxide-type intermediate that may form covalent adducts with protein, and thereby contribute to potential immune-mediated hepatotoxic reactions.

23.5.10   Hydrazines

Procarbazine, *N*-isopropyl-α-(2-methylhydrazine)-*p*-toluamide hydrochloride (Fig. 23.17), is a hydrazine-containing drug that is an effective anticancer agent used for the treatment of several types of cancer including Hodgkin's disease (Calabresi and Chabner, 1996). Procarbazine undergoes oxidative dealkylation, via the formation of an azo-intermediate (spontaneously or catalyzed by P450), which ultimately leads to the production of a methyl diazonium reactive species which methylates, and subsequently inactivates, DNA (Horstman, Meadows and Yost, 1987; Swaffar *et al.*, 1992) (Fig. 23.17).

23.5.11   Hydrazides

Isoniazid (Fig. 23.18) is an antituberculosis drug that is well known to be able to cause drug-induced liver toxicity and that contains a hydrazide structural alert (Zimmerman,

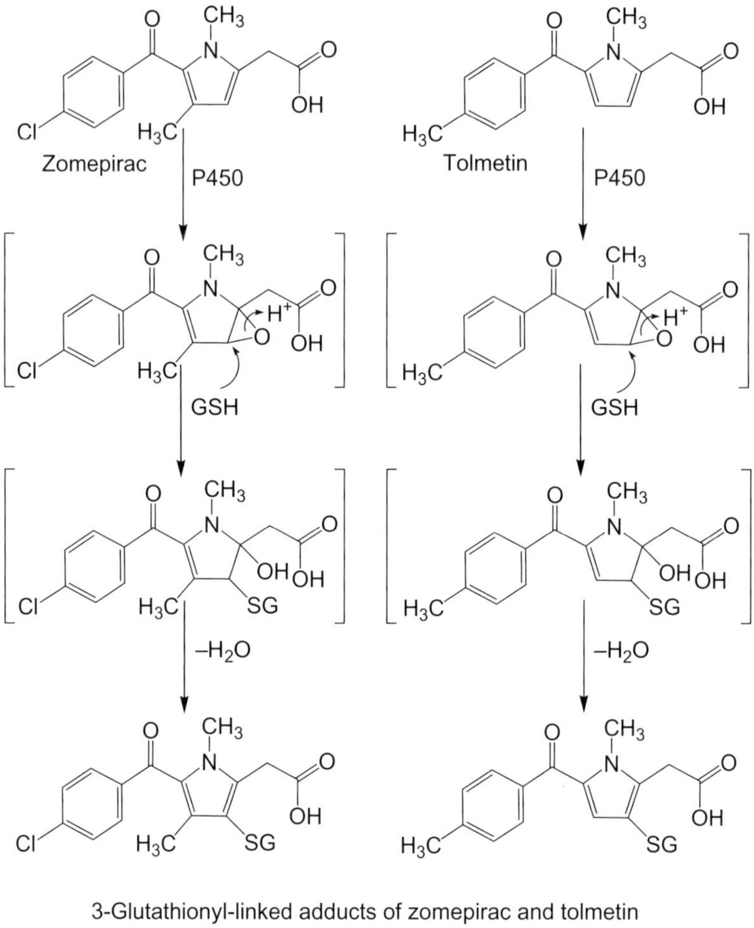

3-Glutathionyl-linked adducts of zomepirac and tolmetin

Figure 23.16 Proposed scheme for the metabolic activation of the zomepirac and tolmetin by P450-mediated epoxidation of the pyrrole moiety (adapted from Chen *et al.*, 2006).

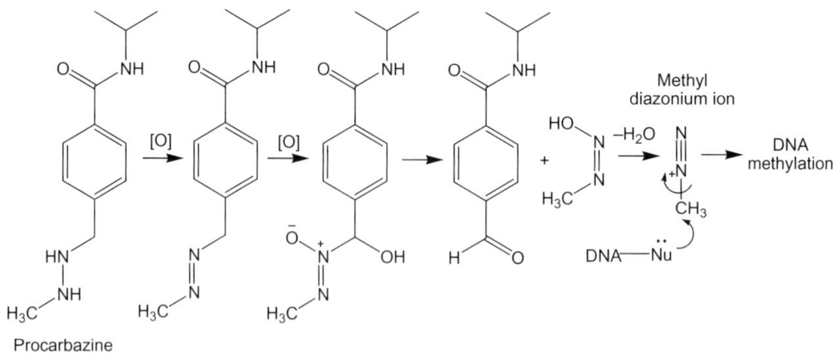

Figure 23.17 Proposed scheme for the metabolic activation of procarbazine leading to the generation of a reactive methyl diazonium ion intermediate that methylates DNA nucleophiles.

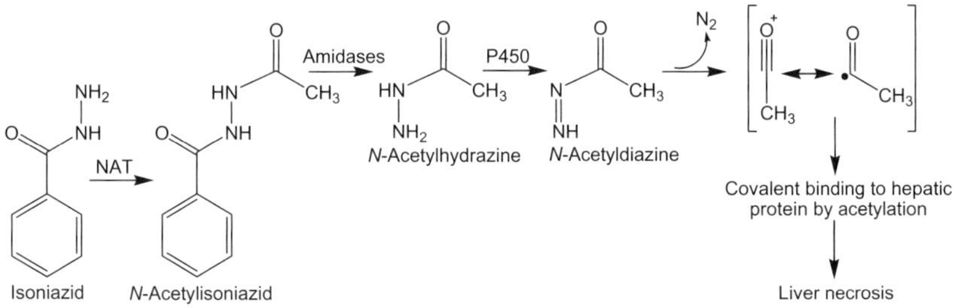

Figure 23.18 Proposed mechanism in the metabolic activation of isoniazid (adapted from Timbrell *et al.*, 1980 and Kalgutkar *et al.*, 2005).

1993). The proposed mechanism of bioactivation of isoniazid involves the initial acetylation of the primary amine moiety by *N*-acetyltransferase enzymes leading to the formation of *N*-acetylisoniazid. This product then is hydrolyzed by amidase-mediated cleavage to give *N*-acetylhydrazine, which serves as a substrate for P450. The P450-oxidized product, namely *N*-acetyldiazine, is unstable and decomposes releasing nitrogen gas and forming both electrophilic and free radical-type intermediates that acetylate hepatic protein and which are proposed to induce liver necrosis (Timbrell *et al.*, 1980; Kalgutkar *et al.*, 2005).

23.5.12 Drugs That Form a Reactive Quinone Imine

23.5.12.1 Acetaminophen Acetaminophen (*N*-acetyl-*p*-aminophenol, APAP; Fig. 23.19) is a commonly used over-the-counter analgesic and antipyretic medication that, when consumed in an overdose fashion, has been shown to produce serious and sometimes fatal hepatic necrosis (Nelson, 1990; Pumford and Halmes, 1997). APAP is a very safe drug when taken at therapeutic doses (~15 mg/kg), but can cause elevation in aminotransferases in healthy adults receiving 4 g (~57 mg/kg) daily (Watkins *et al.*, 2006). Centrilobular hepatic necrosis occurs at doses of 150 mg/kg and higher (>10 g). At a dose of 350 mg/kg, APAP can induce hepatic failure. There are ~9000 reported cases/year in the United States alone, and approximately 60 deaths annually (Holubek *et al.*, 2006). Many of these overdoses are due to suicide attempts, unintentional overdosing in infants, and from adverse drug interactions in alcoholics. Early on in the testing of APAP toxicity, Jollow *et al.* (1973) detected covalent binding of radioactivity to protein in hepatocytes in the centrilobular location of the liver, where liver cells became necrotic post-APAP overdose. The proposed mechanism of bioactivation of APAP is shown in Fig. 23.19.

The metabolic activation of APAP, which contains an anilide-type structural alert, occurs by cytochrome P450 2E1-, 2D6-, 3A4-, and 1A2-mediated biotransformation, leading to the formation of a reactive *N*-acetyl-*p*-benzoquinone imine (NAPQI) intermediate that can spontaneously, or catalytically (via glutathione *S*-transferase π), lead to the formation of a glutathione-thioether-linked conjugate detoxification product, namely 3-(glutathione-*S*-yl)APAP (Dahlin *et al.*, 1984; Nelson, 1990; Dong *et al.*, 2000) (Fig. 23.19). This same adduct can then be degraded to the corresponding mercapturic acid conjugate in the kidney, followed by excretion into urine (Wilson *et al.*, 1982). In an overdose situation, the detoxification of APAP by glucuronidation and/or sulfation becomes saturated such that increased amounts of drug are converted to the reactive

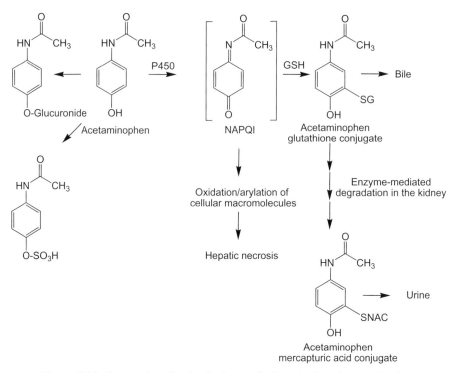

Figure 23.19 Proposed mechanism in the metabolic activation of acetaminophen.

NAPQI electrophile. In addition, at overdose levels of the drug, stores of GSH can be depleted such that the amount of NAPQI formed can exhaust hepatocyte defenses through covalent binding to, and destruction of, critical proteins (Park *et al.*, 2005a). An important critical protein and part of this cellular defense is the Keap1:Nrf2 signaling pathway (Goldring *et al.*, 2004; Okawa *et al.*, 2006). Keap1 and Nrf2 form a heterodimer in cytosol of cells. Keap1 is a small protein that is rich in cysteinyl sulfhydryl groups. These free sulfhydryls are targets for alkylation, arylation, or oxidation by many electrophiles such as NAPQI. The mechanism follows on such that when reactive electrophiles bind covalently to, or oxidize, cysteine thiols of Keap1, the heterodimer dissociates, modified Keap1 is ubiquinated and degraded, and the Nrf2 moiety translocates to the nucleus where it binds to the antioxidant response element (ARE) which triggers the transcription of those genes that are cytoprotective such as glutathione *S*-transferase and glutathione peroxidase. (Fig. 23.20) (Lee *et al.*, 2005). Therefore, Keap1:Nrf2 is proposed to be a protective signaling pathway for reactive electrophiles and oxidizing reagents, and it is believed to be activated by APAP through metabolic activation of the drug to NAPQI (Goldring *et al.*, 2004).

Comparative mechanistic studies have been performed with 3-hydroxyacetanalide (AMAP), which is a simple positional isomer of APAP (Tirmenstein and Nelson, 1989; Nelson and Pearson, 1990; Rashed, Myers and Nelson, 1990; Roberts, Price and Jollow, 1990; Myers *et al.*, 1995). Under conditions where APAP causes hepatotoxicity, AMAP was shown to undergo metabolic activation and subsequent covalently binding to liver tissue, without eliciting hepatotoxicity in test animals. This was, and still is, a confusing

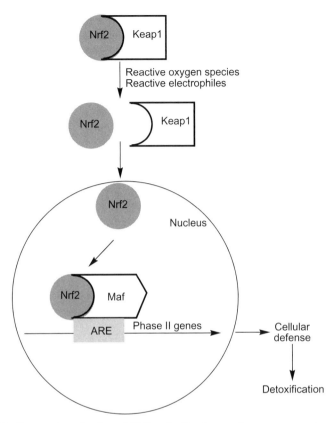

Figure 23.20 Proposed mechanism of Nrf2 activation by reactive oxygen species and reactive electrophilic species (adapted from Lee *et al.*, 2005).

result since both APAP and AMAP are converted to very similar reactive electrophiles. APAP is converted to NAPQI, and AMAP is converted to a *p*-benzoquinone intermediate. Both of these reactive species have been shown to arylate cysteine residues on liver proteins. The reactive metabolites of AMAP formed in the liver covalently bind to protein at approximately the same level as acetaminophen. However, it was found that the relative distribution of covalent binding to protein in subcellular locations was different (Tirmenstein and Nelson, 1989). The AMAP regioisomer, which again elicits no toxicity, was shown to covalently bind to cytoplasmic proteins and a few endoplasmic reticulum proteins, many of which were found to be the same proteins are as determined for APAP. The major difference was in the covalent binding to mitochondria, where it was shown that APAP binds extensively to many critical proteins (Tirmenstein and Nelson, 1989). By contrast, the reactive metabolites of AMAP were not found to arylate these same critical proteins of this important organelle. It was proposed that the electrophilic metabolites of AMAP are more reactive than NAPQI formed from acetaminophen, because they bind closer to where they are formed by P450s located in the endoplasmic reticulum. These *p*-benzoquinone AMAP metabolites may be too reactive to escape reactions with protein and other nucleophiles in the endoplasmic reticulum or cytosol and therefore cannot reach, covalently bind to, and subsequently injure critical mitochondrial proteins. Using two-dimensional gel electrophoresis and mass spectrometric methods, more than 20 APAP-labeled hepatic protein adducts were identified from [^{14}C]APAP-treated mice (Qiu, Benet and Burlingame, 1998). The challenge continues to be able to identify the protein targets that are critical for the onset of hepatotoxicity caused by APAP (Nelson, 1995; Qiu, Benet and Burlingame, 2001; Welch *et al.*, 2005).

23.5.12.2 Diclofenac Diclofenac (2-[2-(2,6-dichlorophenyl)aminophenyl]ethanoic acid; Fig. 23.21), an NSAID, is metabolically activated by cytochrome P450 to chemically reactive species that react with endogenous nucleophiles such as glutathione and protein, and that have been proposed to play a role in the rare, but sometimes severe, hepatotoxicity associated with use of the drug (Banks *et al.*, 1995; Boelsterli, 2003; Tang, 2003) (Fig. 23.21). Diclofenac is used for the treatment of osteoarthritis, rheumatoid arthritis, ankylosing spondylitis, and for acute muscle pain (Small, 1989). Adverse reactions to diclofenac include enteropathy (common with long-term use), acute and chronic hepatitis (15% of patients have increased plasma transaminase levels), hemolytic anemia, and fatal anaphylaxis (i.e. IgE- or IgG-mediated immune response); however, no animal models are available to predict these toxicities (Boelsterli, 2003; Shenton, Chen and Uetrecht, 2004).

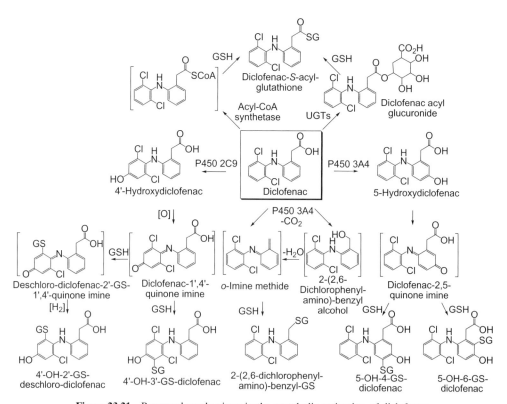

Figure 23.21 Proposed mechanisms in the metabolic activation of diclofenac.

Diclofenac is metabolized by cytochrome P450s 2C9 and 3A4 by oxidation of the aromatic ring to 4'- and 5-hydroxylated products, respectively. Both of these hydroxylated metabolites can undergo further P450-mediated oxidation to electrophilic quinone imine metabolites that react, in a Michael acceptor fashion, with glutathione forming derivatives easily detected by LC-MS/MS analysis (Tang *et al.*, 1999). An additional glutathione conjugate of diclofenac was identified as 4'-hydroxy-2'-glutathione-deschloro-diclofenac, which is formed from the reaction of diclofenac-1',4'-quinone imine, the product of 4'-hydroxydiclofenac oxidation, with glutathione via an *ipso* substitution at the carbon attached to the chlorine atom (Yu, Chen and Deninno, 2005). The formation of this glutathione adduct was shown to be mediated by P450 2C9 bioactivation of diclofenac (Yan *et al.*, 2007) (Fig. 23.21). Yet another glutathione was recently detected *in vitro* in rat and human hepatocyte incubations, and rat and human liver microsomes, as well as *in vivo* as detected in bile from diclofenac-treated rats (Grillo *et al.*, 2008; Teffera *et al.*, 2008). This adduct, namely 2-(2,6-dichloro-phenylamino)-benzyl-*S*-thioether glutathione is proposed to be formed from P450-mediated oxidative decarboxylation of diclofenac leading to an intermediate 2-(2,6-dichlorophenylamino)-benzyl-carbon centered radical, which could either recombine with hydroxyl radical forming 2-(2,6-dichlorophenylamino)benzyl alcohol and undergo dehydration leading to the *o*-imine methide, or undergo further oxidation leading directly to the *o*-imine methide reactive intermediate. Reaction of the *o*-imine methide species at the benzylic position with glutathione then forms the stable conjugate. Therefore, diclofenac undergoes varied types of P450-mediated bioactivation mechanisms leading to reactive metabolite formation and hence potential toxicity. Diclofenac, in standardized covalent binding assays in human liver microsomes, becomes irreversibly bound to protein at a level of ~300 pmol equivalents/mg protein/h (Evans *et al.*, 2004). The reactive metabolite responsible for the bulk of the covalent binding to protein *in vitro* in incubations with human liver microsome remains to be determined.

In addition to metabolism by P450 enzymes, diclofenac is metabolized by acyl glucuronidation to diclofenac-1-*O*-acyl glucuronide (Fig. 23.21) (Boelsterli, 2003), an unstable and chemically reactive metabolite (Grillo *et al.*, 2003a) that has also been implicated as playing a role in diclofenac-induced hepatotoxicity (Kretz-Rommel and Boelsterli, 1993; Boelsterli, 2003). Acyl glucuronide metabolites of acidic drugs are proposed to bind covalently to protein by two different mechanisms (Spahn-Langguth and Benet, 1992) (see Section 23.5.8). These include transacylation-type reactions with protein nucleophiles by the 1-*O*-acyl glucuronide isomer, and by a glycation mechanism that involves the reaction of open-chain aldehyde forms of the acyl migration glucuronide isomers with protein amino groups (Ding *et al.*, 1993; Zia-Amirhosseini, Spahn-Langguth and Benet, 1994). Then it is proposed that the drug–protein adducts are recognized by the immune system as foreign, resulting in an immune response leading to the associated idiosyncratic hepatotoxicity (Spahn-Langguth and Benet, 1992). Diclofenac acyl glucuronide was shown to react irreversibly with the canalicular protein dipeptidyl peptidase IV, and which corresponded to a decrease in functional activity of this protein in diclofenac-treated rats (Hargus *et al.*, 1995).

The *S*-acyl-glutathione thioester-linked metabolite, namely diclofenac-*S*-acyl-glutathione, has been detected *in vivo* in rats and *in vitro* in rat and human hepatocyte incubations (Fig. 23.21) (Grillo *et al.*, 2003a, b). It was shown that inhibition of diclofenac acyl glucuronide production in incubations with rat hepatocytes had no effect on diclofenac-*S*-acyl-glutathione formation. Therefore, another potential reactive metabolite, namely diclofenac-*S*-acyl-CoA thioester, was proposed. Acyl-CoA metabo-lites of carboxylic acid-containing drugs are being increasingly investigated as poten-tially toxicologically important reactive metabolites of carboxylic acid-containing drugs (Li, Benet and Grillo, 2002; Sidenius *et al.*, 2004; Olsen *et al.*, 2007).

In summary, chemically reactive metabolites of diclofenac have been proposed to mediate the idiosyncratic hepatotoxicity associated with the use of the drug (Boelsterli, 2003; Tang, 2003), but which one if any is important in this respect remains to be answered. One recent study showed that glucuronidation-inhibited rat hepatocytes were more susceptible to diclofenac-induced cytotoxicity than vehicle-treated cells (Siraki, Chevaldina and O'Brien, 2005). Metabolic activation studies in rat hepatocytes showed that diclofenac acyl glucuronide was the most important mediator of covalent binding to protein. However, cytotoxicity was attributed primarily to P450-mediated oxidative products (Kretz-Rommel and Boelsterli, 1993).

A proteomic profiling study was conducted for diclofenac to examine the effects of the drug on proteins excreted in rat bile (Jones *et al.*, 2003). Six proteins were detected in bile from diclofenac-treated rats that were not present in untreated animals. These proteins included dipeptidyl peptidase IV, dihydropyrimidinase, apoptosis-inducing factor, α-1-antitrypsin precursor, vitamin D-binding protein prepeptide, and 60S ribo-somal protein L4. The most important change observed was the increased presence of dipeptidyl peptidase IV, which is an integral membrane protein having a homodimeric mass of 110 kDa (Hargus *et al.*, 1995). However, no data were collected that confirmed the presence of diclofenac-derived protein adducts excreted in bile in the same study. To date, convincing evidence for the causative hepatotoxic reactive metabolite(s) of diclofenac remains elusive (Tang, 2003).

23.5.13 Alkenes

4-Vinylcyclohexene (Fig. 23.22) is a chemical that contains the alkene structural alert and is used in the manufacture of flame-retardants, insecticides, and plasticizers. Expo-sure of mice to 4-vinylcyclohexene results in a loss of small preantral (primordial and primary) follicles in the mouse ovary (Rajapaksa *et al.*, 2007). 4-Vinylcyclohexene-

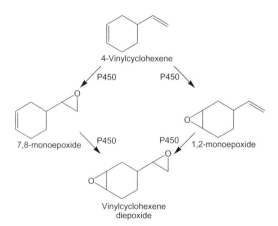

Figure 23.22 Proposed mechanisms in the metabolic activation of the ovotoxin 4-vinylcyclohexene (adapted from Rajapaksa *et al.*, 2007).

induced ovarian toxicity has been attributed to bioactivation of the compound to a chemically reactive diepoxide metabolite (Fig. 23.22). Therefore, this occupational chemical is metabolically activated to either a 1,2- or 7,8-monoepoxide, and subsequently to the ovotoxic metabolite vinylcyclohexene diepoxide via cytochrome P450 enzymes. Results from mechanistic structure–activity studies suggest that bioactivation to the diepoxide is necessary for vinylcyclohexene-induced ovarian toxicity (Rajapaksa *et al.*, 2007).

23.5.14 Alkynes

17α-Ethynylestradiol is an estrogenic oral contraceptive that contains an alkyne structural alert moiety (Fig. 23.23). 17α-Ethynylestradiol is known to inactivate cytochrome P450 in a mechanism-based fashion. Bioactivation is proposed to occur via P450-mediated oxidation to 17α-oxirene-related reactive intermediates (oxirene and ketene species; Fig. 23.23) that react with and form heme and apoprotein adducts, resulting in the inactivation of P4503A5. In addition, the compound was shown to form two glutathione adducts in incubations with the reconstituted cytochrome P4503A5 system (Lin and Hollenberg, 2007).

23.5.15 Anilines

Dapsone (diamino-diphenyl sulphone; Fig. 23.24) is an antimicrobial drug used for the treatment of leprosy (Mandell and Petri, 1996). Use of dapsone has been associated

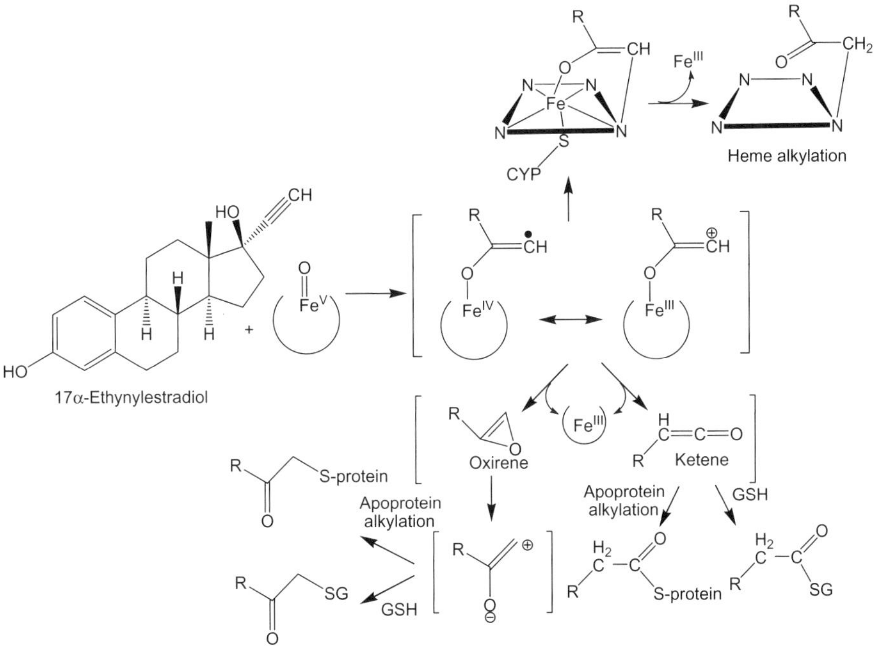

Figure 23.23 Proposed mechanisms in the metabolic activation of the estrogenic oral contraceptive 17α-ethynylestradiol to reactive oxirene- and ketene-type intermediates (adapted from Kalgutkar *et al.*, 2005).

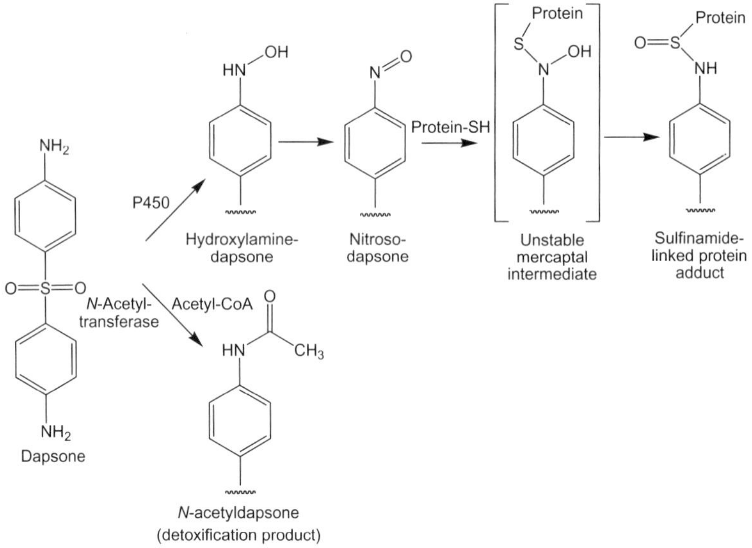

Figure 23.24 Proposed mechanism in the metabolic activation of dapsone to a reactive aromatic nitroso-type intermediate (adapted from Vyas *et al.*, 2006).

with the formation of allergic reactions sometimes leading to hepatitis (Joseph, 1985). The chemical structure of dapsone contains the aniline structural alert which is metabolized by P450-mediated oxidation to an *N*-hydroxylated product. This metabolite then can be further oxidized to a chemically reactive nitroso product, which is known to react with protein nucleophiles forming drug–protein adducts that can serve as antigens that trigger hypersensitivity reactions (Vyas *et al.*, 2006) (Fig. 23.24). The *N*-hydroxylated product might also serve as a substrate for sulfation leading to the formation of a reactive hydroxylamine-sulfate conjugate (see Section 23.5.16). Another route of dapsone metabolism occurs by acetylation of the aromatic amine moiety which is mediated by *N*-acetyltransferase (utilizing the cofactor acetyl-CoA) leading to the formation of a nontoxic metabolite, namely *N*-acetyldapsone. It has been proposed that the relative extents of these two routes of metabolism occurring in a patient undergoing dapsone therapy could determine the onset of an adverse allergic reaction (Bluhm *et al.*, 1999).

23.5.16 Nitroaromatics

The nitroaromatic moiety is another structural alert that has been associated with the induction of liver toxicity by some drugs. One proposed pathway of bioactivation of the aromatic nitro group is through reductive metabolism generating nitroanion and hydronitroxide radicals which can lead to DNA damage (Hewick, 1982) (Fig. 23.25). Another bioactivation mechanism occurs by additional one-electron reduction of the nitroanion radical to a nitroso-benzene intermediate which reacts with protein nucleophiles, leading to the formation of sulfinamide-linked protein adducts (Fig. 23.25). Additional one-electron reduction of the nitroso-benzene intermediate leads to a hydroxyl-amine intermediate which can be further metabolized to a reactive hydroxylamine-sulfate conjugate that is able to react with and damage DNA

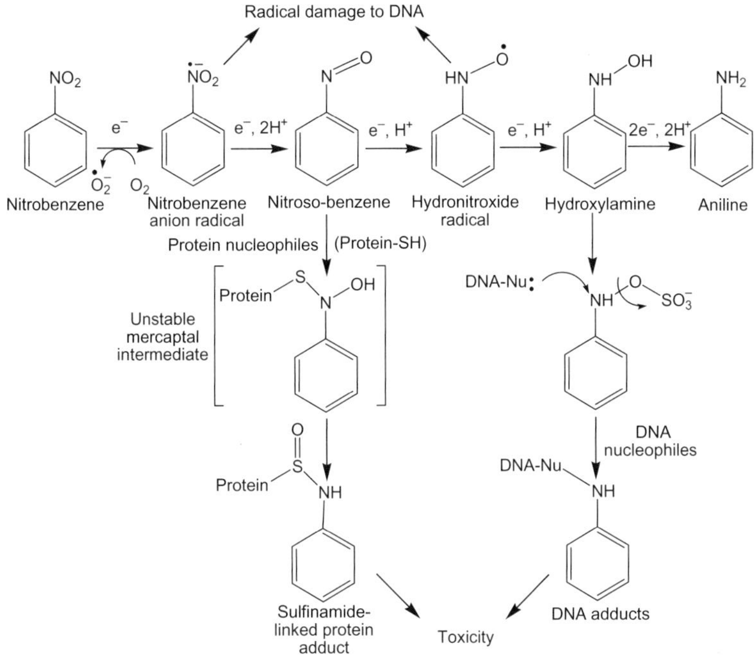

Figure 23.25 Proposed mechanism in the metabolic activation of nitrobenzene to reactive aromatic hydroxylamine- and nitroso-type intermediates (adapted from Hewick, 1982).

(Kalgutkar *et al.*, 2005). Nitroaromatic-containing drugs known to be associated with toxicity, including rare cases of idiosyncratic liver injury, are tolcapone, nimesulide, nilutamide, flutamide, and nitrofurantoin (Boelsterli *et al.*, 2006). The reduction, and hence bioactivation, of the nitro group can be catalyzed P450 enzymes, xanthine and aldehyde oxidases, and quinone reductases (Kedderis, Argenbright and Miwa, 1989; Boelsterli *et al.*, 2006).

23.6   CONCLUSIONS

Pharmaceutical scientists working in drug metabolism and toxicology today consider reactive metabolites of drugs or drug candidates an important potential toxicological issue (Kalgutkar and Soglia, 2005; Park *et al.*, 2005b; Baillie, 2006; Doss and Baillie, 2006). Many pharmaceutical companies are dealing with this concern by evaluating data harvested from varied types of metabolic activation studies. Experimental approaches (from glutathione trapping of reactive intermediates to covalent binding of drug-related radioactivity to protein) to assess drug candidates for their ability to form reactive intermediates, both *in vitro* and *in vivo*, are performed routinely by many pharmaceutical companies today (Kalgutkar and Soglia, 2005; Mutlib *et al.*, 2005; Baillie, 2006). Skilled in ferreting out glutathione adducts, and probable structures of reactive intermediate, as well as in modifying drug candidates accordingly, drug metabolism scientists know full well that no matter how capable we have become at describing and finding reactive metabolites, we still struggle to understand why an individual comes down with a toxic reaction to a certain drug. The causes of toxicity are still difficult to decipher, even when using up-to-date technologies. Nonetheless, it is hoped that by having closer interactions between the fields of drug metabolism, medicinal chemistry, biology, and toxicology that it will play a key role in eliminating potentially toxic compounds from entering development. Importantly, mechanistic metabolic activation studies should be performed prior to the selection of drug candidates due to the tremendous economic investment in further stages of development (Guengerich and MacDonald, 2007; Kumar *et al.*, 2008). It is an ongoing goal in the pharmaceutical industry to decrease the metabolic activation properties of drug candidates in the hope that this may increase the probability of compounds being successfully developed as nontoxic drugs (Baillie, 2006). With the recognition of metabolic activation occurring for a candidate drug, pharmaceutical companies will have to make knowledgeable judgments about potential risks to patients so that toxic drugs are never released onto the market.

In terms of emerging trends, there will be a substantial application of mass spectrometry-based proteomic techniques to identify critical targets of reactive metabolites (Qiu, Benet and Burlingame, 1998; Liebler, 2002; Liebler, 2006). The important challenge then is to be able to use these new techniques to assist in providing convincing evidence that modification of pivotal proteins may have toxic consequences. Coincident advancement of toxicogenomic methodologies, in terms of observing both covalent binding data in juxtaposition with gene expression data, should be very promising in determining the toxic significance of covalent binding (Guengerich and MacDonald, 2007; Leone *et al.*, 2007).

Representative indicators of cellular electrophilic exposure related to the Keap1/Nrf2 signaling system, which is important for triggering cellular defense mechanisms (Enomoto *et al.*, 2001; Dinkova-Kostova, Holtzclaw and Kensler, 2005; Hong *et al.*, 2005), are increasingly being studied. Keap1/Nrf2 could be used as a surrogate marker of exposure to electrophilic intermediates or to oxidative stress that occurs within the cell (Hong *et al.*, 2005). It is envisioned that the accumulation of Nrf2 as a marker for the evaluation of potential toxicity could be developed in order to assess compounds for their potential to trigger a warning signal (Baillie, 2006).

The development of computer simulation techniques to anticipate chemical structures susceptible to bioactivation is becoming more effective (Roberts, 2001; Gombar, Silver and Zhao, 2003). Structural alerts are functional groups of molecules that pharmaceutical chemists attempt not to use because these moieties have a background of forming chemically reactive, potentially toxic intermediates. Hopefully, in the near future, drug metabolism scientists will be able to predict with *in silico* tools, together with knowledge of the tertiary structures of the various enzymes and the electronic characteristics of substrates, the tendency of a molecule to be metabolized by certain enzymes (Korzekwa and Jones, 1993) to chemically reactive species.

Finally, there is a critical demand for increased research on the mechanisms of immune-mediated toxicities at the molecular level (Baillie, 2006; Liebler, 2006; Guengerich and MacDonald, 2007; Kumar *et al.*, 2008; Uetrecht, 2008). The maturation of consilient approaches toward the use of new technologies which are finding immediate applications includes metabonomics (Nicholson *et al.*, 2002; Mortishire-Smith *et al.*, 2004), proteomics (Jones *et al.*, 2003; Bradshaw and Burlingame, 2005; Welch *et al.*, 2005), and toxicogenomics (Martin *et al.*, 2006; Leone *et al.*, 2007), each being brought together to aid in someday understanding the implications of metabolic activation in drug-mediated toxicities (Baillie, 2006, 2008; Liebler, 2006; Tang, 2007).

ACKNOWLEDGMENTS

I would like to thank Dr. Christian Skonberg (Department of Pharmaceutics and Analytical Chemistry, University of Copenhagen, Denmark), Dr. Kathila Rajapaksa (Toxicology, Amgen Inc.), and Judy Huang, B.A. (Pharmacokinetics and Drug Metabolism, Amgen Inc.) for critically reviewing this chapter. I also thank the constructive criticism and helpful suggestions of two anonymous reviewers, which have been included in the references.

REFERENCES

Argoti, D., Liang, L., Conteh, A., Chen, L., Bershas, D., Yu, C.P., Vouros, P., Yang, E. (2005). Cyanide trapping of iminium ion reactive intermediates followed by detection and structure identification using liquid chromatography-tandem mass spectrometry (LC-MS/MS). *Chem. Res. Toxicol.*, *18*, 1537–1544.

Baer, B.R., Wienkers, L.C., Rock, D.A. (2007). Time-dependent inactivation of P450 3A4 by raloxifene: identification of Cys239 as the site of apoprotein alkylation. *Chem. Res. Toxicol.*, *20*, 954–964.

Baillie, T.A. (1988). Metabolic activation of valproic acid and drug-mediated hepatotoxicity. Role of the terminal olefin, 2-n-propyl-4-pentenoic acid. *Chem. Res. Toxicol.*, *1*, 195–199.

Baillie, T.A. (2006). Future of toxicology-metabolic activation and drug design: challenges and opportunities in chemical toxicology. *Chem. Res. Toxicol.*, *19*, 889–893.

Baillie, T.A. (2008). Metabolism and toxicity of drugs. Two decades of progress in industrial drug metabolism. *Chem. Res. Toxicol.*, *21*, 129–137.

Baillie, T.A., Davis, M.R. (1993). Mass spectrometry in the analysis of glutathione conjugates. *Biol. Mass Spectrom.*, *22*, 319–325.

Baillie, T.A., Kassahun, K. (2001). Biological reactive intermediates in drug discovery and development: a perspective from the pharmaceutical industry. *Adv. Exp. Med. Biol.*, *500*, 45–51.

Bambal, R.B., Hanzlik, R.P. (1995). Bromobenzene 3,4-oxide alkylates histidine and lysine side chains of rat liver proteins in vivo. *Chem. Res. Toxicol.*, *8*, 729–735.

Banks, A.T., Zimmerman, H.J., Ishak, K.G., Harter, J.G. (1995). Diclofenac-associated hepatotoxicity: analysis of 180 cases reported to the Food and Drug Administration as adverse reactions. *Hepatology*, *22*, 820–827.

Berenblum, I., Schoental, R. (1947). Carcinogenic constituents of coal tar. *Br. J. Cancer*, *1*, 157–165.

Bluhm, R.E., Adedoyin, A., McCarver, D.G., Branch, R.A. (1999). Development of dapsone toxicity in patients with inflammatory dermatoses: activity of acetylation and hydroxylation of dapsone as risk factors. *Clin. Pharmacol. Ther.*, *65*, 598–605.

Boelsterli, U.A. (2003). Diclofenac-induced liver injury: a paradigm of idiosyncratic drug toxicity. *Toxicol. Appl. Pharmacol.*, *192*, 307–322.

Boelsterli, U.A., Ho, H.K., Zhou, S., Leow, K.Y. (2006). Bioactivation and hepatotoxicity of nitroaromatic drugs. *Curr. Drug Metab.*, *7*, 715–727.

Boyd, M.R. (1976). Role of metabolic activation in the pathogenesis of chemically induced pulmonary disease: mechanism of action of the lung-toxic furan, 4-ipomeanol. *Environ. Health Perspect.*, *16*, 127–138.

Bradshaw, R.A., Burlingame, A.L. (2005). From proteins to proteomics. *IUBMB Life*, *57*, 267–272.

Brookes, P., Lawley, P.D. (1964). Evidence for the binding of polynuclear aromatic hydrocarbons to the nucleic acids of mouse skin: relation between carcinogenic power of hydrocarbons and their binding to DNA. *Nature (Lond.)*, *202*, 781–784.

Buonarati, M., Jones, A.D., Buckpitt, A. (1990). In vivo metabolism of isomeric naphthalene oxide glutathione conjugates. *Drug Metab. Dispos.*, *18*, 183–189.

Calabresi, P., Chabner, B.A. (1996). Chemotherapy of nepoplastic diseases, in *Goodman and Gilman's The Pharmacological Basis of Therapeutics*, 9th edn (eds M.J. Wonsiewicz and P. McCurdy), McGraw-Hill, New York, pp. 1225–1287.

Calvey, N. (2005). Adverse drug reactions. *Anaesth. Int. Care Med.*, *6*, pp. 245–249.

Carlson, R.M. (1990). Assessment of the propensity for covalent binding of electrophiles to biological substrates. *Environ. Health Perspect.*, *87*, 227–232.

Chen, Q., Doss, G.A., Tung, E.C., Liu, W., Tang, Y.S., Braun, M.P., Didolka, V., Strauss, J.R., Wang, R.W., Stearns, R.A., Evans, D.C., Baillie, T.A., Tang, W. (2006). Evidence for the bioactivation of zomepirac and tolmetin by an oxidative pathway: identification of glutathione adducts in vitro in human liver microsomes and in vivo in rats. *Drug Metab. Dispos.*, *34*, 145–151.

Chen, Q., Ngui, J.S., Doss, G.A., Wang, R.W., Cai, X., DiNinno, F.P., Blizzard, T.A., Hammond, M.L., Stearns, R.A., Evans, D.C., Baillie, T.A., Tang, W. (2002). Cytochrome P450 3A4-mediated bioactivation of raloxifene: irreversible enzyme inhibition and thiol adduct formation. *Chem. Res. Toxicol.*, *15*, 907–914.

Clark, C.R., Henderson, T.R., Royer, R.E., Brooks, A.L., McClellan, R.O., Marshall, W.F., Naman, T.M. (1982). Mutagenicity of diesel exhaust particle extracts: influence of fuel composition in two diesel engines. *Fundam. Appl. Toxicol.*, *2*, 38–43.

Cohen, S.D., Pumford, N.R., Khairallah, E.A., Boekelheide, K., Pohl, L.R., Amouzadeh, H.R., Hinson, J.A. (1997). Selective protein covalent binding and target organ toxicity. *Toxicol. Appl. Pharmacol.*, *143*, 1–12.

Cook, J.W., Hewett, C.R., Hieger, I. (1933). The isolation of cancer-producing hydrocarbon from coal tar. *J. Chem. Soc. (Lond.)*, *1*, 395–405.

Dahlin, D.C., Miwa, G.T., Lu, A.Y., Nelson, S.D. (1984). N-acetyl-p-benzoquinone imine: a cytochrome P-450-mediated oxidation product of acetaminophen. *Proc. Natl. Acad. Sci. U.S.A.*, *81*, 1327–1331.

Davis, D. (2007). *The Secret History of The War on Cancer*, Basic Books, New York, p. 27.

Day, S.H., Mao, A., White, R., Schulz-Utermoehl, T., Miller, R., Beconi, M.G. (2005). A semi-automated method for measuring the potential for protein covalent binding in drug discovery. *J. Pharmacol. Toxicol. Methods*, *52*, 278–285.

Degen, G.H., Neumann, H.G. (1978). The major metabolite of aflatoxin B1 in the rat is a glutathione conjugate. *Chem. Biol. Interact.*, *22*, 239–255.

Dieckhaus, C.M., Fernández-Metzler, C.L., King, R., Krolikowski, P.H., Baillie, T.A. (2005). Negative ion tandem mass spectrometry for the detection of glutathione conjugates. *Chem. Res. Toxicol.*, *18*, 630–638.

Ding, A., Ojingwa, J.C., McDonagh, A.F., Burlingame, A.L., Benet, L.Z. (1993), Evidence for covalent binding of acyl glucuronides to serum albumin via an imine mechanism as revealed by tandem mass spectrometry. *Proc. Natl. Acad. Sci. U.S.A.*, *90*, 3797–3801.

Dinkova-Kostova, A.T., Holtzclaw, W.D., Kensler, T.W. (2005). The role of Keap1 in cellular protective responses. *Chem. Res. Toxicol.*, *18*, 1779–1791.

Dong, H., Haining, R.L., Thummel, K.E., Rettie, A.E., Nelson, S.D. (2000). Involvement of human cytochrome P450 2D6 in the bioactivation of acetaminophen. *Drug Metab. Dispos.*, *28*, 1397–1400.

Doss, G.A., Baillie, T.A. (2006). Addressing metabolic activation as an integral component of drug design. *Drug Metab. Rev.*, *38*, 641–649.

Duncan, M., Brookes, P., Dipple, A. (1969). Metabolism and binding to cellular macromolecules of a series of hydrocarbons by mouse embryo cells in culture. *Int. J. Cancer*, *4*, 818–819.

Enomoto, A., Itoh, K., Nagayoshi, E., Haruta, J., Kimura, T., O'Connor, T., Harada, T., Yamamoto, M. (2001). High sensitivity of Nrf2 knockout mice to acetaminophen hepatotoxicity associated with decreased expression of ARE-regulated drug metabolizing enzymes and antioxidant genes. *Toxicol. Sci.*, *59*, 169–177.

Evans, D.C., Baillie, T.A. (2005). Minimizing the potential for metabolic activation as an integral part of drug design. *Curr. Opin. Drug Discov. Devel.*, *8*, 44–50.

Evans, D.C., Watt, A.P., Nicoll-Griffith, D.A., Baillie, T.A. (2004). Drug-protein adducts: an industry perspective on minimizing the potential for drug bioactivation in drug discovery and development. *Chem. Res. Toxicol.*, *17*, 3–16.

Fakjian, N., Buckpitt, A.R. (1984). Metabolism of bromobenzene to glutathione adducts in lung slices from mice treated with pneumotoxicants. *Biochem. Pharmacol.*, *33*, 1479–1486.

Fung, M., Thornton, A., Mybeck, K., Wu, J.H., Hornbuckle, K., Muniz, E. (2001). Evaluation of the characteristics of safety withdrawal of prescription drugs from world wide pharmaceutical markets-1960 to 1999. *Drug Inform. J.*, *35*, 293–317.

Gan, J., Harper, T.W., Hsueh, M.M., Qu, Q., Humphreys, W.G. (2005). Dansyl glutathione as a trapping agent for the quantitative estimation and identification of reactive metabolites. *Chem. Res. Toxicol.*, *18*, 896–903.

Gelboin, H.V.A. (1969). Microsomal-dependent binding of benzo(a)pyrene to DNA. *Cancer Res.*, *29*, 1272–1276.

Gillette, J.R. (1981). An integrated approach to the study of chemically reactive metabolites of acetaminophen. *Arch. Intern. Med.*, *23*, 375–379.

Gillette, J.R., Pohl, L.R. (1977). A prospective on covalent binding and toxicity. *J. Toxicol. Environ. Health*, *2*, 849–871.

Goldring, C.E., Kitteringham, N.R., Elsby, R., Randle, L.E., Clement, Y.N., Williams, D.P., McMahon, M., Hayes, J.D., Itoh, K., Yamamoto, M., Park, B.K. (2004). Activation of hepatic Nrf2 in vivo by acetaminophen in CD-1 mice. *Hepatology*, *39*, 1267–1276.

Gombar, V.K., Silver, I.S., Zhao, Z. (2003). Role of ADME characteristics in drug discovery and their in silico evaluation: in silico screening of chemicals for their metabolic stability. *Curr. Top. Med. Chem.*, *3*, 1205–1225.

Greene, F.L. (1990). Environmental carcinogens–Percivall Pott revisited. *J. S. C. Med. Assoc.*, *86*, 328.

Gregus, Z., Klaassen, C.D. (2001). Mechanisms of toxicity, in *Casarett and Doull's Toxicology (2001) The Basic Science of Poisons*, 6th edn (ed. C.D. Klaassen), The McGraw-Hill Company, Inc., pp. 35–82.

Grillo, M.P., Chiellini, G., Tonelli, M., Benet, L.Z. (2001). Effect of alpha-fluorination of valproic acid on valproyl-S-acyl-CoA formation in vivo in rats. *Drug Metab. Dispos.*, *29*, 1210–1215.

Grillo, M.P., Hua, F. (2003). Identification of zomepirac-S-acyl-glutathione in vitro in incubations with rat hepatocytes and in vivo in rat bile. *Drug Metab. Dispos.*, *31*, 1429–1436.

Grillo, M.P., Knutson, C.G., Sanders, P.E., Waldon, D.J., Hua, F., Ware, J.A. (2003a). Studies on the chemical reactivity of diclofenac acyl glucuronide with glutathione: identification of diclofenac-S-acyl-glutathione in rat bile. *Drug Metab. Dispos.*, *31*, 1327–1336.

Grillo, M.P., Hua, F., Knutson, C.G., Ware, J.A., Li, C. (2003b). Mechanistic studies on the bioactivation of diclofenac: identification of diclofenac-S-acyl-glutathione in vitro in incubations with rat and human hepatocytes. *Chem. Res. Toxicol.*, *16*, 1410–1417.

Grillo, M.P., Ma, J., Teffera, Y., Waldon, D.J. (2008). A novel bioactivation pathway for diclofenac initiated by P450-mediated oxidative decarboxylation. *Drug Metab. Dispos.*, *36*, 1740–1744.

Guengerich, F.P., MacDonald, J.S. (2007). Applying mechanisms of chemical toxicity to predict drug safety. *Chem. Res. Toxicol.*, *20*, 344–369.

Hargus, S.J., Martin, B.M., George, J.W., Pohl, L.R. (1995). Covalent modification of rat liver dipeptidyl peptidase IV (CD26) by the nonsteroidal anti-inflammatory drug diclofenac. *Chem. Res. Toxicol.*, *8*, 993–996.

Hart, D., Ward, M., Lifschitz, M.D. (1987). Suprofen-related nephrotoxicity. A distinct clinical syndrome. *Ann. Intern. Med.*, *106*, 235–238.

Heidelberger, C. (1975). Chemical carcinogenesis. *Ann. Rev. Biochem.*, *44*, 79–121.

Hewick, D.S. (1982). Reductive metabolism of nitrogen-containing functional groups, in *Metabolic Basis of Detoxication: Metabolism of Functional Groups* (eds W.B. Jakoby, J.R. Bend and J. Caldwell), Academic Press, Inc., New York, pp. 151–170.

Holubek, W.J, Kalman, S., and Hoffman, R.S. (2006). Acetaminophen-induced acute liver failure: results of a United States multicenter, prospective study. *Hepatology*, *43*, 880–882.

Hong, F., Sekhar, K.R., Freeman, M.L., Liebler, D.C. (2005). Specific patterns of electrophile adduction trigger Keap1 ubiquitination and Nrf2 activation. *J. Biol. Chem.*, *280*, 31768–31775.

Horstman, M.G., Meadows, G.G., Yost, G.S. (1987). Separate mechanisms for procarbazine spermatotoxicity and anticancer activity. *Cancer Res.*, *47*, 1547–1550.

IARC (1985). Polynuclear aromatic 4. bitumens, coal tars and derived products, shale oils and soots. IARC Monographs on the Evaluation of Carcinogenic Risk of Chemicals to Humans, International Agency for Research on Cancer, pp. 271.

Isley, W.L. (2003). Hepatotoxicity of thiazolidinediones. *Expert Opin. Drug Saf.*, *2*, 581–586.

Jackson, E.K. (1996). Diuretics, in *Goodman and Gilman's. The Pharmacological Basis of Therapeutics*, 9th edn (eds M.J. Wonsiewicz and P. McCurdy), McGraw-Hill, New York, pp. 685–713.

Jollow, D.J., Mitchell, J.R., Potter, W.Z., Davis, D.C., Gillette, J.R., Brodie, B.B. (1973). Acetaminophen-induced hepatic necrosis. II. Role of covalent binding in vivo. *J. Pharmacol. Exp. Ther.*, *187*, 195–202.

Jones, J.A., Kaphalia, L., Treinen-Moslen, M., Liebler, D.C. (2003). Proteomic characterization of metabolites, protein adducts, and biliary proteins in rats exposed to 1,1-dichloroethylene or diclofenac. *Chem. Res. Toxicol.*, *16*, 1306–1317.

Joseph, M.S. (1985). Hypersensitivity reaction to dapsone. *Lepr. Rev.*, *56*, 315–320.

Ju, C., Uetrecht, J.P. (2002). Mechanism of idiosyncratic drug reactions: reactive metabolite formation, protein binding and the regulation of the immune system. *Curr. Drug Metab.*, *3*, 367–377.

Kalgutkar, A.S., Dalvie, D.K., O'Donnell, J.P., Taylor, T.J., Sahakian, D.C. (2002). On the diversity of oxidative bioactivation reactions on nitrogen-containing xenobiotics. *Curr. Drug Metab.*, *3*, 379–424.

Kalgutkar, A.S., Gardner, I., Obach, R.S., Shaffer, C.L., Callegari, E., Henne, K.R., Mutlib, A.E., Dalvie, D.K., Lee, J.S., Nakai, Y., O'Donnell, J.P., Boer, J., Harriman, S.P. (2005). A comprehensive listing of bioactivation pathways of organic functional groups. *Curr. Drug Metab.*, *6*, 161–225.

Kalgutkar, A.S., Soglia, J.R. (2005). Minimising the potential for metabolic activation in drug discovery. *Expert Opin. Drug Metab. Toxicol.*, *1*, 91–142.

Kaplowitz, N. (2005). Idiosyncratic drug hepatotoxicity. *Nat. Rev. Drug Discov.*, *4*, 489–499.

Kassahun, K., Abbott, F. (1993). In vivo formation of the thiol conjugates of reactive metabolites of 4-ene VPA and its analog 4-pentenoic acid. *Drug Metab. Dispos.*, *21*, 1098–1106.

Kassahun, K., Pearson, P.G., Tang, W., McIntosh, I., Leung, K., Elmore, C., Dean, D., Wang, R., Doss, G., Baillie, T.A. (2001). Studies on the metabolism of troglitazone to reactive intermediates in vitro and in vivo. Evidence for novel biotransformation pathways involving quinone methide formation and thiazolidinedione ring scission. *Chem. Res. Toxicol.*, *14*, 62–70.

Kassahun, K., Skordos, K., McIntosh, I., Slaughter, D., Doss, G.A., Baillie, T.A., Yost, G.S. (2005). Zafirlukast metabolism by cytochrome P450 3A4 produces an electrophilic alpha,beta-unsaturated iminium species that results in the selective mechanism-based inactivation of the enzyme. *Chem. Res. Toxicol.*, *9*, 1427–1437.

Kedderis, G.L., Argenbright, L.S., Miwa, G.T. (1989). Covalent interaction of 5-nitroimidazoles with DNA and protein in vitro: mechanism of reductive activation. *Chem. Res. Toxicol.*, *2*, 146–149.

Kennaway, E. (1955). The identification of a carcinogenic compound in coal tar. *Br. Med. J.*, *2*, 749–752.

Kent, U.M., Lin, H.L., Noon, K.R., Harris, D.L., Hollenberg, P.F. (2006). Metabolism of bergamottin by cytochromes P450 2B6 and 3A5. *J. Pharmacol. Exp. Ther.*, *318*, 992–1005.

Kiani, R., Kushner, M. (1983). Zomepirac-induced serum sickness. A report of two cases. *JAMA*, *249*, 2812–2813.

Korzekwa, K.R., Jones, J.P. (1993). Predicting the cytochrome P450 mediated metabolism of xenobiotics. *Pharmacogenetics*, *3*, 1–18.

Kretz-Rommel, A., Boelsterli, U.A. (1993). Diclofenac covalent protein binding is dependent on acyl glucuronide formation and is inversely related to P450-mediated acute cell injury in cultured rat hepatocytes. *Toxicol. Appl. Pharmacol.*, *120*, 155–161.

Kumar, S., Kassahun, K., Tschirret-Guth, R.A., Mitra, K., Baillie, T.A. (2008). Minimizing metabolic activation during pharmaceutical lead optimization: progress, knowledge gaps and future directions. *Curr. Opin. Drug Discov. Devel.*, *11*, 43–52.

Lee, J.M., Li, J., Johnson, D.A., Stein, T.D., Kraft, A.D., Calkins, M.J., Jakel, R.J., Johnson, J.A. (2005). Nrf2, a multi-organ protector? *FASEB J.*, *19*, 1061–1066.

Leone, A.M., Kao, L.M., McMillian, M.K., Nie, A.Y., Parker, J.B., Kelley, M.F., Usuki, E., Parkinson, A., Lord, P.G., Johnson, M.D. (2007). Evaluation of felbamate and other antiepileptic drug toxicity potential based on hepatic protein covalent binding and gene expression. *Chem. Res. Toxicol.*, *20*, 600–608.

Levy, D.B., Vasilomanolakis, E.C. (1984). Anaphylactic reactions due to zomepirac. *Drug Intell. Clin. Pharm.*, *18*, 983–984.

Li, C., Benet, L.Z., Grillo, M.P. (2002). Enantioselective covalent binding of 2-phenylpropionic acid to protein in vitro in rat hepatocytes. *Chem. Res. Toxicol.*, *15*, 1480–1487.

Liebler, D.C. (2002). Proteomic approaches to characterize protein modifications: new tools to study the effects of environmental exposures. *Environ. Health Perspect.*, *110* (Suppl. 1), 3–9.

Liebler, D.C. (2006). The poisons within: application of toxicity mechanisms to fundamental disease processes. *Chem. Res. Toxicol.*, *19*, 610–613.

Lin, H.L., Hollenberg, P.F. (2007). The inactivation of cytochrome P450 3A5 by 17alpha-ethynylestradiol is cytochrome b5-dependent: metabolic activation of the ethynyl moiety leads to the formation of glutathione conjugates, a heme adduct, and covalent binding to the apoprotein. *J. Pharmacol. Exp. Ther.*, *321*, 276–287.

Lufkin, E.G., Wong, M., Deal, C. (2001). The role of selective estrogen receptor modulators in the prevention and treatment of osteoporosis. *Rheum. Dis. Clin. North. Am.*, *27*, 163–185.

Mandell, G.L., Petri, W.A. (1996). Antimicrobial agents, in *Goodman and Gilman's The Pharmacological Basis of Therapeutics*, 9th edn (eds J.G. Hardman, A.G. Gilman and L.E. Limbard), McGraw-Hill, New York, pp. 1155–1174.

Martin, R., Rose, D., Yu, K., Barros, S. (2006). Toxicogenomics strategies for predicting drug toxicity. *Pharmacogenomics*, *7*, 1003–1016.

Matthews, A.M., Hinson, J.A., Roberts, D.W., Pumford, N.R. (1997). Comparison of covalent binding of acetaminophen and the regioisomer 3'-hydroxyacetanilide to mouse liver protein. *Toxicol. Lett.*, *90*, 77–82.

Melicow, M.M. (1975). Percivall Pott (1713–1788): 200th anniversary of first report of occupation-induced cancer scrotum in chimney sweepers (1775). *Urology*, *6*, 745–749.

Miller, E.C., Miller, J.A. (1974). Biochemical mechanisms of chemical carcinogenesis, in *The Molecular Biology of Cancer* Academic Press, New York, pp. 377–402.

Miller, E.C., Miller, J.A. (1979). Milestones in chemical carcinogenesis. *Oncology*, *6*, 445–456.

Miller, E.C., Miller, J.A. (1981). Reactive metabolites as key intermediates in pharmacologic and toxicologic responses: examples from chemical carcinogenesis. *Adv. Exp. Med. Biol.*, *136*, 1–21.

Mitchell, J.R., Nelson, W.L., Potter, W.Z., Sasame, H.A., Jollow, D.J. (1976). Metabolic activation of furosemide to a chemically reactive, hepatotoxic metabolite. *J. Pharmacol. Exp. Ther.*, *199*, 41–52.

Mitchell, J.R., Snodgrass, W.R., Gillette, J.R. (1976). The role of biotransformation in chemical-induced liver injury. *Environ. Health Perspect.*, *15*, 27–38.

Monks, T.J., Hinson, J.A., Gillette, J.R. (1982). Bromobenzene and p-bromophenol toxicity and covalent binding in vivo. *Life Sci.*, *30*, 841–848.

Mortishire-Smith, R.J., Skiles, G.L., Lawrence, J.W., Spence, S., Nicholls, A.W., Johnson, B.A., Nicholson, J.K. (2004). Use of metabonomics to identify impaired fatty acid metabolism as the mechanism of a drug-induced toxicity. *Chem. Res. Toxicol.*, *17*, 165–173.

Mutlib, A., Lam, W., Atherton, J., Chen, H., Galatsis, P., Stolle, W. (2005). Application of stable isotope labeled glutathione and rapid scanning mass spectrometers in detecting and characterizing reactive metabolites. *Rapid Commun. Mass Spectrom.*, *19*, 3482–3492.

Myers, T.G., Dietz, E.C., Anderson, N.L., Khairallah, E.A., Cohen, S.D., Nelson, S.D. (1995). A comparative study of mouse liver proteins arylated by reactive metabolites of acetaminophen and its nonhepatotoxic regioisomer, 3'-hydroxyacetanilide. *Chem. Res. Toxicol*, *8*, 403–413.

Narasimhan, N., Weller, P.E., Buben, J.A., Wiley, R.A., Hanzlik, R.P. (1988). Microsomal metabolism and covalent binding of [3H/14C]-bromobenzene. Evidence for quinones as reactive metabolites. *Xenobiotica*, *18*, 491–499.

Nassar, A.E., Lopez-Anaya, A. (2004). Strategies for dealing with reactive intermediates in drug discovery and development. *Curr. Opin. Drug Discov. Devel.*, *7*, 126–136.

Nelson, S.D. (1990). Molecular mechanisms of the hepatotoxicity caused by acetaminophen. *Semin. Liver Dis.*, *10*, 267–278.

Nelson, S.D. (1995). Mechanisms of the formation and disposition of reactive metabolites that can cause acute liver injury. *Drug Metab. Rev.*, *27*, 147–177.

Nelson, S.D., Pearson, P.G. (1990). Covalent and noncovalent interactions in acute lethal cell injury caused by chemicals. *Ann. Rev. Pharmacol. Toxicol.*, *30*, 169–195.

Nelson, S.D., Mitchell, J.R., Snodgrass, W.R., Timbrell, J.A. (1978). Hepatotoxicity and metabolism of iproniazid and isopropylhydrazine. *J. Pharmacol. Exp. Ther.*, *206*, 574–585.

Nicholson, J.K., Connelly, J., Lindon, J.C., Holmes, E. (2002). Metabonomics: a platform for studying drug toxicity and gene function. *Nat. Rev. Drug Discov.*, *1*, 153–161.

O'Donnell, J.P., Dalvie, D.K., Kalgutkar, A.S., Obach, R.S. (2003). Mechanism-based inactivation of human recombinant P450 2C9 by the nonsteroidal anti-inflammatory drug suprofen. *Drug Metab. Dispos.*, *31*, 1369–1377.

O'Neill, P.J., Yorgey, K.A., Renzi, N.L., Williams, R.L., Benet, L.Z. (1982). Disposition of zomepirac sodium in man. *J. Clin. Pharmacol.*, *22*, 470–476.

Okawa, H., Motohashi, H., Kobayashi, A., Aburatani, H., Kensler, T.W., Yamamoto, M. (2006). Hepatocyte-specific deletion of the keap1 gene activates Nrf2 and confers potent resistance against acute drug toxicity. *Biochem. Biophys. Res. Commun.*, *339*, 79–88.

Olsen, J., Li, C.C., Bjørnsdottir, I., Sidenius, U., Benet, L.Z., Hansen, S.H., Benet, L.Z. (2005). In vitro and in vivo studies on acyl-coenzyme A-dependent bioactivation of zomepirac in rats. *Chem. Res. Toxicol.*, *18*, 1729–1736.

Olsen, J., Li, C., Skonberg, C., Bjørnsdottir, I., Sidenius, U., Benet, L.Z., Hansen, S.H. (2007). Studies on the metabolism of tolmetin to the chemically reactive acyl-coenzyme A thioester intermediate in rats. *Drug Metab. Dispos.*, *35*, 758–764.

Ottoboni, M.A. (1991). *The Dose Makes The Poison: A Plain-Language Guide to Toxicology*, 2nd edn, Van Nostrand Reinhold, New York, p. 96.

Park, B.K., Kitteringham, N.R., Maggs, J.L., Pirmohamed, M., Williams, D.P. (2005a). The role of metabolic activation in drug-induced hepatotoxicity. *Ann. Rev. Pharmacol. Toxicol.*, *45*, 177–202.

Park, K., Williams, D.P., Naisbit, D.J., Kitteringham, N.R., Pirmohamed, M. (2005b). Investigation of toxic metabolites during drug development. *Toxicol. Appl. Pharmacol.*, *207*, 425–434.

Pereira, M.A., Burns, F.J., Albert, R.E. (1979). Dose response for benzo(a)pyrene adducts in mouse epidermal DNA. *Cancer Res.*, *39*, 2556–2559.

Pirmohamed, M. (2006). Genetic factors in the predisposition to drug-induced hypersensitivity reactions. *AAPS J*, *8*, 20–26.

Porubek, D.J., Grillo, M.P., Olsen, R.K., Baillie, T.A. (1991). Toxic metabolites of valproic acid: inhibition of rat liver acetoacetyl-CoA thiolase by 2-*n*-propyl-⁴-pentenoic acid (Δ^4-VPA) and related branched-chain carboxylic acids, in *Idiosyncratic Reactions to Valproate Clinical Risk Patterns and Mechanisms of Toxicity* (ed. R.H. Levy, J.K. Penry), Raven Press, Ltd., New York, pp. 53–58.

Pott, P. (1775). *Cancer Scroti, Chiurgical Observations. The Chiurgical Works of Percival Pott*, Hawes, Clark, and Collings, London.

Pumford, N.R., Halmes, N.C. (1997). Protein targets of xenobiotic reactive intermediates. *Annu. Rev. Pharmacol. Toxicol.*, *37*, 91–117.

Qiu, Y., Benet, L.Z., Burlingame, A.L. (1998). Identification of the hepatic protein targets of reactive metabolites of acetaminophen in vivo in mice using two-dimensional gel electrophoresis and mass spectrometry. *J. Biol. Chem.*, *273*, 17940–17953.

Qiu, Y., Benet, L.Z., Burlingame, A.L. (2001). Identification of hepatic protein targets of the reactive metabolites of the non-hepatotoxic regioisomer of acetaminophen, 3′-hydroxyacetanilide, in the mouse in vivo using two-dimensional gel electrophoresis and mass spectrometry. *Adv. Exp. Med. Biol.*, *500*, 663–673.

Rajapaksa, K.S., Cannady, E.A., Sipes, I.G., Hoyer, P.B. (2007). Involvement of CYP 2E1 enzyme in ovotoxicity caused by 4-vinylcyclohexene and its metabolites. *Toxicol. Appl. Pharmacol.*, *221*, 215–221.

Rashed, M.S., Myers, T.G., Nelson, S.D. (1990). Hepatic protein arylation, glutathione depletion, and metabolite profiles of acetaminophen and a non-hepatotoxic regioisomer, 3′-hydroxyacetanilide, in the mouse. *Drug Metab. Dispos.*, *18*, 765–770.

Reid, W.D., Christie, B., Krishna, G., Mitchell, J.R., Moskowitz, J., Brodie, B.B. (1971). Bromobenzene metabolism and hepatic necrosis. *Pharmacology*, *6*, 41–55.

Reinus, J.F., Persky, S., Burkiewicz, J.S., Quan, D., Bass, N.M., Davern, T.J. (2000). Severe liver injury after treatment with the leukotriene receptor antagonist zafirlukast. *Ann. Intern. Med.*, *133*, 964–968.

Rettie, A.E., Rettenmeier, A.W., Howald, W.N., Baillie, T.A. (1987). Cytochrome P-450-catalyzed formation of delta 4-VPA, a toxic metabolite of valproic acid. *Science*, *235*, 890–893.

Roberts, S.A. (2001). High-throughput screening approaches for investigating drug metabolism and pharmacokinetics. *Xenobiotica*, *31*, 557–589.

Roberts, S.A., Price, V.F., Jollow, D.J. (1990). Acetaminophen structure-toxicity studies: in vivo covalent binding of a nonhepatotoxic analog, 3-hydroxyacetanilide. *Toxicol. Appl. Pharmacol.*, *105*, 195–208.

Sahali-Sahly, Y., Balani, S.K., Lin, J.H., Baillie, T.A. (1996). *In vitro* studies on the metabolic activation of the furanopyridine L-754,394, a highly potent and selective mechanism-based Inhibitor of cytochrome P450 3A4. *Chem. Res. Toxicol.*, *9*, 1007–1012.

Scheen, A.J. (2001a). Hepatotoxicity with thiazolidinediones: is it a class effect? *Drug Saf.*, *24*, 873–888.

Scheen, A.J. (2001b). Thiazolidinediones and liver toxicity. *Diabetes Metab*, *27*, 305–333.

Schmeltz, I., Tosk, J., Hilfrich, J., Jirota, N., Hoffman, D., Wynder, E. (1978). Bioassays of naphthalene for co-carcinogenic activity. Relation to tobacco carcinogenesis, in *Polynuclear Aromatic Hydrocarbons* (eds P. Jones, R. Freudenthal), Raven Press, New York, Vol. 3, pp. 47–60.

Scholz, K., Dekant, W., Völkel, W., Pähler, A. (2005). Rapid detection and identification of N-acetyl-L-cysteine thioethers using constant neutral loss and theoretical multiple reaction monitoring combined with enhanced product-ion scans on a linear ion trap mass spectrometer. *J. Am. Soc. Mass Spectrom.*, *16*, 1976–1984.

Shenton, J.M., Chen, J., Uetrecht, J.P. (2004). Animal models of idiosyncratic drug reactions. *Chem. Biol. Interact.*, *150*, 53–70.

Shultz, M.A, Choudary, P.V, and Buckpitt A.R. (1999). Role of murine cytochrome P-450 2F2 in metabolic activation of naphthalene and metabolism of other xenobiotics. *J. Pharmacol. Exp. Ther.*, *290*, 281–288.

Sidenius, U., Skonberg, C., Olsen, J., Hansen, S.H. (2004). In vitro reactivity of carboxylic acid-CoA thioesters with glutathione. *Chem. Res. Toxicol.*, *17*, 75–81.

Sims, P., Grover, P.L., Swaisland, A., Pal, K., Hewer, A. (1974). Metabolic activation of benzo(a)pyrene proceeds by a diol-epoxide. *Nature*, *22*, 326–328.

Siraki, A.G., Chevaldina, T., O'Brien, P.J. (2005). Application of quantitative structure-toxicity relationships for acute NSAID cytotoxicity in rat hepatocytes. *Chem. Biol. Interact.*, *151*, 177–191.

Small, R.E. (1989). Diclofenac sodium. *Clin. Pharm.*, *8*, 545–558.

Smith, P.C., Benet, L.Z., McDonagh, A.F. (1990). Covalent binding of zomepirac glucuronide to proteins: evidence for a Schiff base mechanism. *Drug Metab. Dispos.*, *18*, 639–644.

Soglia, J.R., Contillo, L.G., Kalgutkar, A.S., Zhao, S., Hop, C.E., Boyd, J.G., Cole, M.J. (2006). A semiquantitative method for the determination of reactive metabolite conjugate levels in vitro utilizing liquid chromatography-tandem mass spectrometry and novel quaternary ammonium glutathione analogues. *Chem. Res. Toxicol.*, *19*, 480–490.

Spahn-Langguth, H., Benet, L.Z. (1992). Acyl glucuronides revisited: is the glucuronidation process a toxification as well as a detoxification mechanism? *Drug Metab. Rev.*, *24*, 5–47.

Swaffar, D.S., Pomerantz, S.C., Harker, W.G., Yost, G.S. (1992). Non-enzymatic activation of procarbazine to active cytotoxic species. *Oncol. Res.*, *4*, 49–58.

Tang, W. (2003). The metabolism of diclofenac—enzymology and toxicology perspectives. *Curr. Drug Metab.*, *4*, 319–329.

Tang, W. (2007). Drug metabolite profiling and elucidation of drug-induced hepatotoxicity. *Expert Opin. Drug Metab. Toxicol.*, *3*, 407–420.

Tang, W., Borel, A.G., Fujimiya, T., Abbott, F.S. (1995). Fluorinated analogues as mechanistic probes in valproic acid hepatotoxicity: hepatic microvesicular steatosis and glutathione status. *Chem. Res. Toxicol*, *8*, 671–782.

Tang, W., Stearns, R.A., Bandiera, S.M., Zhang, Y., Raab, C., Braun, M.P., Dean, D.C., Pang, J., Leung, K.H., Doss, G.A., Strauss, J.R., Kwei, G.Y., Rushmore, T.H., Chiu, S.H., Baillie, T.A. (1999). Studies on cytochrome P-450-mediated bioactivation of diclofenac in rats and in human hepatocytes: identification of glutathione conjugated metabolites. *Drug Metab. Dispos.*, *27*, 365–372.

Teffera, Y., Waldon, D.J., Colletti, A.E., Albrecht, B.K., Zhao, Z. (2008). Identification of a novel glutathione conjugate of diclofenac by LTQ-orbitrap. *Drug Metab. Lett.*, *2*, 35–40.

Timbrell, J.A., Mitchell, J.R., Snodgrass, W.R., Nelson, S.D. (1980). Isoniazid hepatotoxicity: the relationship between covalent binding and metabolism in vivo. *J. Pharmacol Exp Ther.*, *213*, 364–369.

Tirmenstein, M.A., Nelson, S.D. (1989). Subcellular binding and effects on calcium homeostasis produced by acetaminophen and a nonhepatotoxic regioisomer, 3'-hydroxyacetanilide, in mouse liver. *J. Biol. Chem.*, *264*, 9814–9819.

Uetrecht, J. (2001). Prediction of a new drug's potential to cause idiosyncratic reactions. *Curr. Opin. Drug Discov. Devel.*, *4*, 55–59.

Uetrecht, J. (2006). Evaluation of which reactive metabolites, if any, is responsible for a specific idiosyncratic reaction. *Drug Metab. Rev.*, *38*, 745–753.

Uetrecht, J.P. (1999). New concepts in immunology relevant to idiosyncratic drug reactions: the "danger hypothesis" and innate immune system. *Chem. Res. Toxicol.*, *12*, 387–395.

Uetrecht, J.P. (2003). Screening for the potential of a drug candidate to cause idiosyncratic drug reactions. *Drug Discov. Today.*, *8*, 832–837.

Uetrecht, J.P. (2008). Idiosyncratic drug reactions: past, present and future. *Chem. Res. Toxicol.*, *21*, 84–92.

Vyas, P.M., Roychowdhury, S., Khan, F.D., Prisinzano, T.E., Lamba, J., Schuetz, E.G., Blaisdell, J., Goldstein, J.A., Munson, K.L., Hines, R.N., Svensson, C.K. (2006). Enzyme-mediated protein haptenation of dapsone and sulfamethoxazole in human keratinocytes: I. expression and role of cytochromes P450. *J. Pharmacol. Exp. Ther.*, *319*, 488–496.

Waldron, H.A. (1983). A brief history of scrotal cancer. *Br. J. Ind. Med.*, *40*, 390–401.

Walgren, J.L., Mitchell, M.D., Thompson, D.C. (2005). Role of metabolism in drug-induced idiosyncratic hepatotoxicity. *Crit. Rev. Toxicol.*, *35*, 325–361.

Wang, M., Dickinson, R.G. (2000). Bile duct ligation promotes covalent drug-protein adduct formation in plasma but not in liver of rats given zomepirac. *Life Sci.*, *68*, 525–537.

Waring, J.F., Anderson, M.G. (2005). Idiosyncratic toxicity: mechanistic insights gained from analysis of prior compounds. *Curr. Opin. Drug Discov. Devel.*, *8*, 59–65.

Watkins, P.B., Kaplowitz, N., Slattery, J.T., Colonese, C.R., Colucci, S.V., Stewart, P.W., Harris, S.C. (2006). Aminotransferase elevations in healthy adults receiving 4 grams of acetaminophen daily: a randomized controlled trial. *JAMA.*, *296*, 87–93.

Welch, K.D., Wen, B., Goodlett, D.R., Yi, E.C., Lee, H., Reilly, T.P., Nelson, S.D., Pohl, L.R. (2005). Proteomic identification of potential susceptibility factors in drug-induced liver disease. *Chem. Res. Toxicol.*, *18*, 924–933.

van Welie, R.T., van Dijck, R.G., Vermeulen, N.P., van Sitter, N.J. (1992). Mercapturic acids, protein adducts, and DNA adducts as biomarkers of electrophilic chemicals. *Crit. Rev. Toxicol.*, *22*, 271–306.

van Wijk, F., Nierkens, S. (2006). Assessment of drug-induced immunotoxicity in animal models. *Drug Discov. Today*, *3*, 103–109.

Williams, D.P. (2006). Toxicophores: investigations in drug safety. *Toxicology*, *226*, 1–11.

Williams, D.P., Antoine, D.J., Butler, P.J., Jones, R., Randle, L., Payne, A., Howard, M., Gardner, I., Blagg, J., Park, B.K. (2007). The metabolism and toxicity of furosemide in the Wistar rat and CD-1 mouse: a chemical and biochemical definition of the toxicophore. *J. Pharmacol. Exp. Ther.*, *322*, 1208–1220.

Wilson, J.M., Slattery, J.T., Forte, A.J., Nelson, S.D. (1982). Analysis of acetaminophen metabolites in urine by high-performance liquid chromatography with UV and amperometric detection. *J. Chromatogr.*, *227*, 453–462.

Wirth, P.J., Bettis, C.J., Nelson, W.L. (1976). Microsomal metabolism of furosemide evidence for the nature of the reactive intermediate involved in covalent binding. *Mol. Pharmacol.*, *12*, 759–768.

Yamagiwa, K., Ichikawa, K. (1918). Experimental study of the pathogenesis of carcinoma. *J. Cancer Res*, *3*, 1–13.

Yan, Z., Caldwell, G.W. (2004). Stable-isotope trapping and high-throughput screenings of reactive metabolites using the isotope MS signature. *Anal. Chem.*, *76*, 6835–6847.

Yan, Z., Easterwood, L.M., Maher, N., Torres, R., Huebert, N., Yost, G.S. (2007). Metabolism and bioactivation of 3-methylindole by human liver microsomes. *Chem. Res. Toxicol.*, *20*, 140–148.

Yu, L.J., Chen, Y., Deninno, M.P., O'Connell, T.N., Hop, C.E. (2005). Identification of a novel glutathione adduct of diclofenac, 4'-hydroxy-2'-glutathion-deschloro-diclofenac, upon incubation with human liver microsomes. *Drug Metab. Dispos.*, *33*, 484–488.

Zhang, K.E., Naue, J.E., Arison, B., Vyas, K.P. (1996). Microsomal metabolism of the 5-lipoxygenase inhibitor L-739,010: evidence for furan bioactivation. *Chem. Res. Toxicol.*, *9*, 547–554.

Zheng, J., Ma, L., Xin, B., Olah, T., Humphreys, W.G., Zhu, M. (2007). Screening and identification of GSH-trapped reactive metabolites using hybrid triple quadruple linear ion trap mass spectrometry. *Chem. Res. Toxicol.*, *20*, 757–766.

Zia-Amirhosseini, P., Spahn-Langguth, H., Benet, L.Z. (1994). Bioactivation by glucuronide-conjugate formation. *Adv. Pharmacol.*, *27*, 385–397.

Zimmerman, H. (1999). *Hepatotoxicity: The Adverse Effects of Drugs and Other Chemicals on the Liver*, Lippincott Williams & Wilkins, Philadelphia, PA.

Zimmerman, H., Ishak, K.G. (1982). Valproate-induced hepatic injury: analysis of 23 fatal cases. *Hepatology*, *2*, 591–597.

Zimmerman, H.J. (1993). Hepatotoxicity. *Disease-a-Month*, *39*, 673–788.

CHAPTER 24

Mechanisms of Reproductive Toxicity

AILEEN F. KEATING and PATRICIA B. HOYER

24.1 INTRODUCTION

Today, women are entering the workplace in record numbers. Many are postponing childbearing to establish a career. For this reason and because people are exposed to ever-increasing concentrations of environmental chemicals, there is an increased awareness of the impact of environmental chemicals on reproductive health. The female must perform two distinct reproductive functions: development and support of female germ cells and maintenance of the fetus until it can survive in the outside world. The reproductive functions of the male include production of the male germ cell (spermatogenesis) and male steroid hormones (steroidogenesis). Therefore, reproductive toxicology involves detecting and understanding potentially detrimental environmental influences on reproductive success in females and males.

This field has been developing in response to observations linking clustered effects in humans to specific types of exposures. There have been several examples of how xenobiotics in the form of pharmaceuticals can impact reproductive function in humans. One is the effect of the sedative thalidomide, prescribed to women in the 1950s for morning sickness during early pregnancy (Seegmiller, 1997). A greatly increased incidence of children born with developmental organ and limb malformations was traced to *in utero* exposure to thalidomide. This drug was reported to have been responsible for 8000 malformed children over a two-year period. Whereas thalidomide is an example involving developmental limb defects, reproductive effects in humans have been seen with the synthetic estrogen, diethylstilbestrol (DES). From the 1940s to 1960s, DES was widely prescribed for women with high-risk pregnancies. In 1971, Herbst, Ulfelder and Poskanzer (1971) observed an increased incidence of rare vaginal clear cell adenocarcinoma in their daughters who had been exposed *in utero* to DES. Subsequent research has demonstrated that prenatal exposure to DES can also cause fertility defects, teratogenesis, and neoplasia throughout the male and female reproductive tracts (Hendry *et al.*, 1999).

The following chapter will outline what is currently known about how xenobiotics can impact reproductive function in females and males. The information is based

Drug Metabolism Handbook: Concepts and Applications, Edited by Ala F. Nassar, Paul F. Hollenberg, and JoAnn Scatina
Copyright © 2009 by John Wiley & Sons, Inc.

largely on results from animal studies. Furthermore, how this information in some cases has been translated into effects in humans will also be discussed. To facilitate an understanding of how these chemicals can compromise reproductive function, an overview of reproductive physiology in females and males will first be presented.

24.2   OVERVIEW—FEMALE REPRODUCTION

The major functions of the female gonad, the ovary, are production of the female germ cell, the oocyte, and production of female sex steroid hormones. The ovary of the mature mammalian female contains a heterogeneous mixture of structures that undergo dynamic changes during the estrous/menstrual cycle (Fig. 24.1). Specifically, the ovary contains two endocrine glands, the follicle and the corpus luteum. The follicle is responsible for gametogenesis (oogenesis) and production of the hormone 17β-estradiol (steroidogenesis). The other endocrine system, the corpus luteum, is derived from follicular tissue following ovulation and is present in the ovary of the nonpregnant female only during the latter part of the estrous/menstrual cycle, but is maintained if pregnancy occurs. The corpus luteum produces the steroid hormone progesterone.

17β-Estradiol and progesterone comprise the major ovarian steroids. 17β-Estradiol is responsible for follicular maturation and hyperplasia of the endometrium and uterine

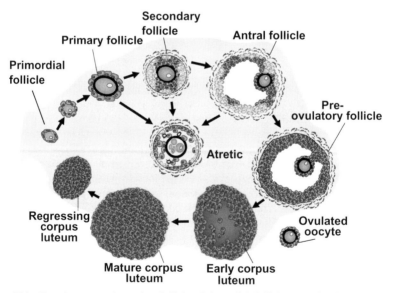

Figure 24.1   Development of ovarian follicles. Primordial follicles are signaled to grow and develop toward ovulation. Following ovulation, the remaining tissue differentiates to become a corpus luteum, which undergoes regression if a pregnancy has not been established. The majority (>99%) of follicles die by the natural process of atresia (adapted from Hoyer and Devine, 2002).

vasculature during the follicular phase of the cycle. Additionally, 17β-estradiol provides both negative and positive feedbacks on the hypothalamus and pituitary for regulation of secretion of the gonadotropins, luteinizing hormone (LH) and follicle-stimulating hormone (FSH). Progesterone facilitates implantation by preparing the uterus to accommodate a blastocyst, and provides maintenance of pregnancy by inhibiting uterine contractions and endometrial sloughing. Additionally, progesterone provides negative feedback on the hypothalamus and pituitary for inhibition of the release of gonadotropins.

LH regulates androgen production in the follicle and progesterone production in the corpus luteum, whereas, FSH regulates follicle maturation and estrogen production. Tonic secretion of LH and FSH from the anterior pituitary is affected by the inhibitory actions of 17β-estradiol and progesterone in a classical loop of negative feedback. Because both steroids participate in the inhibition of gonadotropin release, the absence of 17β-estradiol, progesterone, or both, causes basal LH and FSH to increase. Additionally, a protein hormone, inhibin, produced by granulosa cells in developing follicles contributes to the selective inhibition of FSH release. In contrast to tonic LH secretion, which is controlled by a negative feedback mechanism, an LH surge that triggers ovulation is produced by a neuroendocrine reflex arc of positive feedback, which is stimulated by increasing, marked elevations in circulating 17β-estradiol levels produced by the largest developing preovulatory (Graafian) follicle. As a result, when levels become sufficiently elevated, 17β-estradiol shifts from its inhibitory role in regulating LH and FSH release and initiates an LH surge which triggers ovulation. In contrast to 17β-estradiol, progesterone inhibits the LH surge. This ensures that estrogen output during early pregnancy does not stimulate ovulation.

As mentioned earlier, one of the primary functions of the ovary is to provide mature oocytes for successful reproduction. Development and maturation of oocytes occurs within ovarian follicles. Immediately following formation during fetal development, follicles enter the most immature stage of development, termed "primordial." Successful ovulation requires appropriate follicular development, during which the follicle has passed through a number of distinct developmental stages (Hirshfield, 1991). Primordial follicles provide the pool for recruitment of developing follicles; therefore, they are the fundamental reproductive unit within the ovary.

24.2.1   Prenatal Follicle Formation

During fetal development of the ovary, primordial germ cells (oogonia) are formed and become oocytes when they stop dividing, and become arrested at the diplotene stage (prophase) of the first meiotic division. The oocyte does not commence meiosis again unless triggered to ovulate, should that occur. As a result, the lifetime supply of oocytes is set at the time of birth. Around the time of birth, individual oocytes within the ovary become surrounded by a single layer of flattened somatic cells (pre-granulosa cells) and a basement membrane to form primordial follicles (Hirshfield, 1991). Association of the granulosa cells with the oocyte is critical at all subsequent times for maintenance of viability and follicle development (Buccione, Schroeder and Eppig, 1990).

24.2.2   Follicular Development in Adults

Puberty, the time at which sexual reproduction becomes possible, generally occurs between the ages of 9 and 16 in humans. Puberty is identified in women as the first menstruation, which usually occurs prior to the first ovulation. From birth and throughout the prepubertal period, waves of follicular development in the ovary occur; however, all follicles become atretic. Thus, puberty marks the time at which oocytes can be recruited and developed to ovulation (reviewed in Devine and Hoyer, 2005; Hoyer, 2005).

In humans, 1 to 2 preovulatory follicles develop approximately every 28 days, whereas in rats, 6 to 12 follicles develop every 4 to 5 days (Richards, 1980). Primordial follicles form the pool from which these preovulatory follicles develop. Throughout the reproductive life span, the total number of primordial follicles that ovulate is small compared with the total population. Instead, the vast majority of follicles are lost to attrition in various early stages of development by a process called atresia. The exact determinant for selection of a follicle for ovulation is not understood, but is believed to be under intra-ovarian control (Hirshfield, 1991; Devine and Hoyer, 2005).

The first sign of oocyte growth in primordial follicles is alteration of the surrounding squamous (flattened) granulosa cells into cuboidal shaped cells, followed by initiation of proliferation of these cells (Hirshfield, 1991). Once a follicle makes the transition from primordial to primary, other structural changes occur such as development of the zona pellucida, a protective glycoprotein matrix (Richards, 1980; Hirshfield, 1991). At this stage, another layer of specialized somatic cells, designated theca interna cells, begins to proliferate outside the basement membrane enclosing the oocyte and granulosa cell layer. Theca cells express receptors for LH which stimulates production of androstenedione, the substrate for aromatization to 17β-estradiol by neighboring granulosa cells (Richards, 1980).

As follicles continue to develop, the layers of granulosa cells surrounding the oocyte increase rapidly to become large pre-antral, growing follicles, with diameters reaching 250 μm. The somatic cells acquire receptors for FSH to enhance follicle growth, and develop steroidogenic capacity for synthesis of 17β-estradiol.

The number of follicles that reach the final stage of development is quite small compared with those that began development from the primordial pool. As the follicle develops beyond the pre-antral stage, it acquires a fluid-filled cavity, called the antrum, formed by separations within the granulosa cell layer. The antral follicle continues to grow, and at its most mature stage prior to ovulation is known as a preovulatory (Graafian) follicle. During the final phase of development, a preovulatory follicle becomes more sensitive to the gonadotropins FSH and LH than smaller antral and pre-antral follicles (Richards, 1980). Prior to ovulation, granulosa cells begin to express receptors for LH in readiness for receiving a signal for ovulation, and progression of the oocyte through to the second meiotic division. Following ovulation, the cells remaining, which formed the structure of the follicle, infiltrate and differentiate (luteinize) to form a solid gland, the corpus luteum (Fig. 24.1). The second meiotic division is only completed if fertilization of the oocyte occurs (reviewed in Devine and Hoyer, 2005 and Hoyer, 2005).

24.2.3 Follicular Atresia

The number of oocytes present in ovaries is dynamic and varies with age, with a peak in the total number of oocytes occurring during embryonic development. In humans, that number (about 7 million) occurs at five months' gestation; at birth the number has dropped to 2 million; 250,000 to 400,000 at puberty; and no viable follicles remain at menopause (Mattison and Schulman, 1980; Hirshfield, 1991). During the lifetime of a woman, ovulation accounts for only 400 to 600 oocytes; the others have been lost at various stages of development by the process of atresia which occurs via programmed cell death, apoptosis (Tilly *et al.*, 1991). Therefore, atresia is the natural fate of the vast majority of ovarian follicles (>99%), since only a select few will ever be ovulated (Hirshfield, 1991).

24.2.4 Menopause

Depletion of functional primordial follicles from the ovary in women is the underlying cause of ovarian failure (menopause), because this dormant follicle pool represents the cohort for recruitment of all developing follicles. Thus, absence of primordial follicles ultimately leads to the complete loss of follicles of all sizes. The average age of menopause in the United States is 51, and this is a direct consequence of depletion of the follicular reserve. As a result, estrogen-producing (granulosa) cells in preovulatory follicles also become depleted. Menopause is preceded by a period of increasingly irregular cycles (the perimenopausal period). This progressive failure in ovarian function is accompanied by a gradual increase in circulating levels of FSH (and eventually LH), and a decline in circulating 17β-estradiol concentrations. Besides the loss of fertility, menopause has been associated with a variety of health problems in women, such as osteoporosis and increased risk of cardiovascular disease (Mattison *et al.*, 1989). It is thought that the absence of 17β-estradiol is the underlying cause of most of the clinical symptoms associated with menopause. Recently, this loss of 17β-estradiol has been implicated in other health risks such as depression (Sowers and LaPietra, 1995), colon cancer (Fernandez *et al.*, 1998), Alzheimer's disease (Waring *et al.*, 1999), and macular degeneration (Smith, Mitchell and Wang, 1997). Thus, identifying and protecting against environmental or occupational exposures that may accelerate the onset of menopause in women becomes even more critical (Devine and Hoyer, 2005; Hoyer, 2005).

24.3 OVERVIEW—MALE REPRODUCTION

The purpose of the male reproductive system is to produce sperm and deliver them to the oocyte for fertilization. The male gonad is the testis. The two major functions of the testis are production of the male germ cell, sperm (spermatogenesis), and production of male sex steroid hormones (steroidogenesis). These two functions are compartmentalized within the testis. The testes in man are paired organs located in the scrotum, outside the abdominal cavity. This location is essential for spermatogenesis because this process is temperature-sensitive and cannot occur at normal body temperature. Each testis is divided into a number of pyramidal septa which are separated by a connective tissue capsule. Each septum contains a large number of tubules, the seminiferous tubules, which are convoluted such that they occupy a minimum amount of space within the septum. The seminiferous tubules are avascular and are the site of spermatogenesis and sperm transport. The interstitial compartment contains the cells of Leydig, lymphatics, blood vessels, and various connective tissue elements, and occupies the space between the seminiferous tubules. The interstitium is the site of steroidogenesis (in Leydig cells) and is highly vascularized. Therefore, testicular compartmentalization affords the ability of mature sperm to be selectively transported outside the body, and the steroid hormones to have localized effects on spermatogenesis, but also to communicate with distant sites within the body via the circulation (Herbert, Suprakar and Roy, 1995).

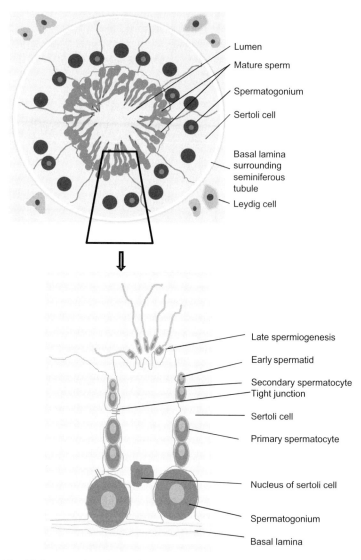

Figure 24.2 **(a)** Cross section of a seminiferous tubule (adapted from Martini and Bartholomew, 1997). **(b)** Spermatogenesis occurring in the Sertoli cell (adapted from Sherwood, 1997).

Approximately 80% of testicular mass consists of seminiferous tubules (Fig. 24.2a). Examination of these tubules in cross section reveals anatomical progression that parallels the temporal sequence of spermatogenesis. Sertoli cells which line the basement membrane of the tubules provide crucial support for spermatogenesis. Also lining the basement membrane are undifferentiated spermatogonia with each clone of differentiating daughter cells migrating toward the tubular lumen as it moves progressively through the stages of development. A major function of Sertoli cells is to maintain tight junctions between themselves and thereby form the blood–testis barrier. This barrier separates the basal and adluminal compartments within the seminiferous tubules, and provides avascular compartmentalization for the tubular lumen. The barrier prevents movement of proteins, charged organic molecules, and ions from interstitial fluid into seminiferous tubules. Consequently, the composition of intratubular fluid differs markedly from interstitial fluid, and this composition may be critical for later stages of spermatogenesis. The blood–testis barrier also provides important immunological protection, preventing the formation of antibodies against the highly differentiated spermatozoa (Constanzo, 1998).

24.3.1 Spermatogenesis

Spermatogenesis involves the complex process by which relatively undifferentiated germ cells containing a diploid complement of 46 chromosomes are turned into extremely specialized sperm in which the chromosome number is reduced by half (haploid) through meiotic division. This halving of the diploid chromosome number occurs prior to the complete differentiation of spermatozoa. Meiosis also occurs in the female ovum. Therefore, subsequent pairing of male and female chromosomes at the time of fertilization has the effect of giving each individual a unique genotype. Spermatogenesis (formation of spermatozoa from spermatogonia) in humans requires 64 days; however, the time between initiations of two successive spermatogenic generations is approximately 16 days. The rate of spermatogenesis is hormone-independent and quite consistent within a species; however, the yield is hormone-dependent and varies from one individual to another (Sherwood, 1997).

Spermatogenesis can be divided into three major stages: (i) mitotic proliferation, (ii) meiosis, and (iii) packaging (also called spermiogenesis). It begins with mitosis of a spermatogonium to form two daughter cells. One of these cells then progresses to form mature sperm, while the other is retained at the basement membrane as a ready source for maintaining the germ cell line. The mitotic capabilities of spermatogonia are demonstrated by the number of spermatogonia in the testes at birth (0.6×10^6), compared with the number present in the testes at puberty (1.2×10^6). In fact, a normal adult male produces $100–200 \times 10^6$ sperm per day (Fig. 24.2b). After undergoing a variable number of mitotic divisions, the daughter cell becomes a primary spermatocyte and enters a resting phase during which chromosomes are duplicated in preparation for meiosis. In man, the daughter cell divides once to become a secondary spermatocyte, and again to become a spermatid. These two meiotic divisions are without DNA replication; therefore, four haploid (23 chromosomes) spermatids are produced from one primary spermatocyte. In addition to the reduction in chromosome number, meiosis allows some chromosomes to exchange segments of DNA, thus mixing their genetic information. After meiosis, spermatids resemble normal epithelial cells. Production of mature spermatozoa requires extensive remodeling of both nucleus and cytoplasm. This packaging (spermiogenesis) involves (i) condensation of chromatin into a tight inert packet, (ii) formation of the acrosome by aggregation of Golgi-produced enzymatic vesicles, (iii) growth of the tail out of one centriole, and (iv) extrusion of most of the cytoplasm (Sherwood, 1997).

24.3.2 Somatic Cells

The hypothalamic hormone gonadotropin-releasing hormone (GnRH) stimulates secretion of FSH and LH from gonadotrophes in the anterior pituitary. The primary testicular target for FSH is Sertoli cells, located within the seminiferous tubules, which perform several tasks essential for normal spermatogenesis. Sertoli cells are located in an avascular compartment; however, since they provide formation of the blood–testis barrier, the basal face communicates with the vascularized interstitium. The major functions of the Sertoli cells are to (i) maintain the blood–testis barrier, (ii) clean up the damaged cells by phagocytosis, (iii) provide nourishment for developing sperm, (iv) maintain appropriate ionic composition of the seminal fluid, (v) serve as a target cell for hormonal regulation of spermatogenesis, (vi) produce estrogens from androgens, and (vii) synthesize and secrete specific proteins such as inhibin. Inhibin is a protein that feeds back to selectively inhibit secretion of FSH, but not LH, from the anterior pituitary. This provides a specific form of regulation that can preferentially alter secretion of gonadotropins on an individual basis (Sherwood, 1997).

The somatic cells of the testes provide a crucial function to male reproduction. Leydig cells located in the vascularized interstitial compartment are the primary target of LH and the site of synthesis and secretion of the male sex steroid hormones, the androgens (testosterone and androstenedione). LH is also required for maintenance of testicular tissue. In males deprived of adequate LH levels (such as those on anabolic steroids which provide negative feedback to the pituitary), reversible testicular atrophy can occur.

Testosterone is the major male sex steroid produced in the testes, and is lipophilic. Therefore, once released into the blood, it circulates largely (97%) bound to serum-binding proteins. Once secreted, testosterone can arrive at distant sites and be further metabolized. Testosterone can be converted to two active metabolites, as well as an inactive metabolite, androsterone 17-keto-steroid. In sex accessory tissues, testosterone is converted to dihydrotestosterone (DHT) by the enzyme 5α-reductase. DHT is known to be more effective than testosterone in producing fetal development of the prostrate, penis, and scrotum, and in promoting sperm maturation and transport in the adult. Conversion of testosterone by the enzyme aromatase to another active metabolite, 17β-estradiol occurs in Sertoli cells and peripheral tissues; therefore, in addition to its important physiological role, testosterone can serve as a precursor for two other active steroid hormones (Constanzo, 1998). Testosterone also provides negative feedback on LH and FSH release at the level of the hypothalamus and pituitary (Rhoads and Pflanzer, 1996).

In addition to its absolute requirement for spermatogenesis, testosterone provides a variety of reproductive and nonreproductive effects. Testosterone promotes differentiation of the reproductive tract in the developing male fetus, maturation of the reproductive tract at puberty, and maintenance of the reproductive tract in adults. At puberty, testosterone is responsible for development of secondary sex characteristics in the male and development of the libido. Nonreproductive effects of testosterone include anabolic effects on muscle mass. Although the physiological significance of androstenedione is not fully known, it may be required for growth of the secondary sex structures (Sherwood, 1997).

24.4 SITES OF TARGETING AND OUTCOMES

In evaluating reproductive effects in toxicological studies, it is important to consider whether gender differences can be seen. In some cases, females may be susceptible and males not, or vice versa. That is, chemicals that produce a selective reduction in gonadotropin secretion from the pituitary may affect 17β-estradiol versus testosterone production in a selective manner. As a result, ovary-dependent end points could be impaired, while testes-dependent end points may remain unaffected. Additionally, the impact may be reversible in one gender, and irreversible in the other. For example, in the case of chemicals that destroy germ cells in the early stages of meiotic division, extensive damage may cause irreversible ovarian failure in females, and reversible reductions in spermatogenesis in males (4-vinylcyclohexene diepoxide, VCD). This is because within the ovary the oocyte is arrested in prophase of the first meiotic division, and once destroyed no more can be formed, whereas within the testis, spermatogonia (the most immature form of sperm) can divide by mitosis to maintain a continuously renewable source of germ cells. Because of these possibilities, it is important to investigate the effects of reproductive toxicants in both genders and determine the targeted site(s) in order to evaluate and predict different levels of risk between women and men (Hoyer, 2001). There is currently no general understanding as to what underlies a specific chemical causing reproductive toxicity. In some cases, a chemical may be known to have widespread effects that are not selective for only reproductive tissues, such as polycyclic aromatic hydrocarbons (PAHs). Whereas in other cases, a chemical may selectively target the gonads as with VCD (Richards, 1980; Hoyer and Devine, 2002). Certain aspects of reproductive physiology may render reproductive systems more susceptible to toxicity, such as germ cell-specific growth factors or signaling pathways. Additionally, target tissue-specific expression of xenobiotic-metabolizing enzymes may contribute to localized bioactivation.

24.4.1 Female Effects

Exposure to environmental or occupational chemicals can disrupt female reproductive function (Mattison, 1985). These chemicals can have their effects at a variety of sites in the reproductive axis, including the hypothalamus, pituitary, ovary, or reproductive tract (Mattison and Schulman, 1980; Thomas, 1993; Hoyer and Sipes, 1996; Hoyer, 1997). Disruption of any of these sites can ultimately manifest as a disruption of ovarian function, resulting in infertility; however, certain classes of compounds have been identified that directly affect the ovary. A number of studies have shown that exposure to direct ovarian toxicants often leads to destruction of oocytes, premature ovarian failure, and development of neoplasms (Krarup, 1967, 1969, 1970; Beamer and Tennent, 1986; Tennent and Beamer, 1986; Maronpot, 1987; Melnick *et al.*, 1990; Hoyer and Sipes, 1996). How these effects are produced is generally not well understood, but may be due to one of several possible mechanisms. Oocyte destruction can result from a direct effect of the toxic chemical. Conversely, toxicity can occur via one of several routes. The ovary contains enzymes responsible for biotransformation and detoxification of many xenobiotics. Both the rat and mouse ovary express epoxide hydrolase, glutathione-*S*-transferases, and cytochrome P450s, which metabolize known ovarian toxicants (Mukhtar, Philpot and Bend, 1978; Mattison and Thorgeirsson, 1979; Heinrichs and Juchau, 1980; Bengtsson and Rydstrom, 1983; Bengtsson, Reinholt and Rydstrom, 1992; Flaws *et al.*, 1994; Cannady *et al.*, 2002; Rajapaksa, Sipes and Hoyer, 2007). Therefore, biotransformation of a chemical within the ovary might occur within selected compartments and thereby provide specific exposure of certain classes of oocytes to toxicity. Additionally, since oocytes at all stages of follicular development are surrounded by granulosa cells, these mechanisms might be indirect and involve effects within the granulosa cell, resulting in loss of its ability to maintain viability of the oocyte (Buccione, Schroeder and Eppig, 1990).

For chemicals that destroy oocytes, the stage of development at which the follicle is lost determines the impact on reproduction. Chemicals that selectively damage large growing or antral follicles only temporarily interrupt reproductive function because these follicles can be replaced by recruitment from the pool of primordial follicles (Hoyer and Sipes, 1996). Thus, these chemicals produce a readily reversible form of infertility that manifests itself relatively soon after exposure (Generoso, Stout and Huff, 1971; Mattison and Schulman, 1980; Jarrell *et al.*, 1991; Davis, Maronpot and Heindel, 1994). Conversely, chemicals that destroy oocytes contained in primordial and primary follicles can lead to permanent infertility and premature ovarian failure (early menopause in humans) because once a primordial follicle is destroyed, it cannot be replaced. Selective destruction of oocytes contained in primordial follicles may have a delayed effect on reproduction until such a time that recruitment for the number of growing and antral follicles can no longer be supported (Generoso, Stout and Huff, 1971; Mattison and Schulman, 1980; Hooser *et al.*, 1994; Mayer *et al.*, 2002). Ultimately, as a result of loss of negative feedback on the hypothalamus and pituitary normally imposed by the ovary (steroids and inhibin), circulating levels of FSH increase (Jarrell *et al.*, 1991; Hooser *et al.*, 1994; Mayer *et al.*, 2002), followed eventually by increases in circulating levels of LH. The increase in release of these hormones from the pituitary will not rise until a sufficient loss of ovarian cyclicity has resulted (several months after disrupted cyclicity is observed).

The level and duration of exposure to an environmental toxicant can determine its effects on reproduction. Individuals are only rarely exposed acutely to high levels of xenobiotic chemicals. These exposures can usually be readily identified. However, the possible effects of chronic, low-dose exposures are of particular concern, because they are more likely to be environmental and are more difficult to identify. These types of exposures may cause reproductive or fertility problems that go unrecognized for years due to the potential for cumulative damage. For example, cigarette smoking in women appears to accelerate the onset of menopause. This is probably due to exposure to PAHs contained in cigarette smoke (Mattison and Schulman, 1980; Mattison *et al.*, 1989; Devine and Hoyer, 2005). Therefore, ongoing selective PAH-induced damage of small pre-antral follicles may not initially cause detectable effects that signal the onset of premature ovarian failure (early menopause). In addition to reproductive effects caused by environmental exposures in fertile women, the age at which exposure occurs can impact the outcome. For example, exposure during childhood might cause sterility by chemical-induced destruction of germ cells. Furthermore, exposures *in utero* may cause improper development of ovarian follicles or permanent alterations in the reproductive tract (Fig. 24.3) (Hoyer and Sipes, 1996; Hoyer and Devine, 2002). This might result from greater sensitivity of developing fetal organs to xenobiotics or increased metabolic contributions from the placenta. Finally, the increased risk of ovarian cancer in postmenopausal women might be potentiated by a lifetime of exposure to environmental chemicals and/or by increased sensitivity of the senescent ovary (Hoyer, 1997).

24.4.2 Male Effects

Paternal exposure to toxic agents is of concern due to the potential for congenital abnormalities, carcinogenesis, and developmental delays in offspring. Effects of xeno-

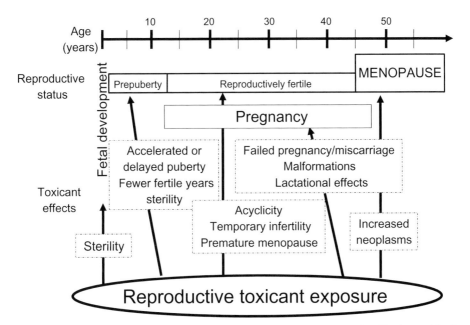

Figure 24.3   Potential age-related effects of reproductive toxicants in females. The impact(s) of reproductive toxicants is partially dependent upon the reproductive status of the exposed individual. In most cases, direct ovarian toxicity can lead to premature ovarian failure and infertility (menopause).

biotic exposure on the male reproductive system are also dependent on the site of exposure. Some of the potential direct effects include reduced sperm concentrations and production of compromised spermatozoa. Indirect effects include impaired hormone production of testosterone and the gonadotropins. Xenobiotic exposure to developing germ cells in the male reproductive tract can result in temporary or permanent infertility. Metabolism of chemicals can cause production of highly reactive free radicals which can damage the germinal epithelium due to the lack of catalase in these cells (Buhrley and Ellis, 1973; Docampo, 1987). Examples of xenobiotics whose metabolism results in free radical production include nitrofurans, adriamycin, and bleiomycins (Docampo, 1987). Another anti-spermatic agent is gossypol, isolated from cotton plants, which produces free radicals and also interacts with membrane proteins or the cellular lipid bilayer (Sundaram and Witorsch, 1995). Additionally, xenobiotic exposures can damage the germinal epithelium by initiating apoptosis.

Toxic agents which target developing germ cells (PAHs, cisplatin) can inhibit mitosis (spermatogonia) or meiosis (spermatocytes) and lead to azoospermia or oligospermia. A common side effect due to chemotherapeutic treatment for cancer in males is fertility impairment. Mechanisms by which this occurs include inhibition of DNA synthesis, direct DNA damage, or changes to the kinetics of DNA division. Exposure of germ cells to toxicants can result in irreversible infertility if spermatogonia in the germinal epithelium are targeted (Sundaram and Witorsch, 1995).

Direct targeting of Sertoli cells can damage germ cells by an indirect mechanism (Vinclozolin, VCD, di-(2-ethylhexyl)phthalate; DEHP). Exposure of male rodents to Sertoli cell toxicants results in a rapid reduction in fertility which is prolonged and sometimes permanent. This toxicity is thought to be due to the dependency of spermatogenesis on the Sertoli cell (Chapin *et al.*, 1988; Foster, 1988).

The major function of the Leydig cell is the production of testosterone, the main circulating androgen in males. This production is regulated by pituitary-derived LH, which is regulated by GnRH from the hypothalamus. Testosterone is important for the reproductive health of the adult male, and plays roles in both sexual behavior and maintenance of spermatogenesis. Thus, chemicals that impair Leydig cell function (2,3,7,8-tetrachlorodibenzo-*p*-dioxin; TCDD, cadmium, lead, mercury, methoxychlor) can have widespread consequences in the male (Sundaram and Witorsch, 1995). Such impairment could result from exposure to chemicals that interfere with GnRH or LH production or by direct targeting of Leydig cells.

The blood–testes barrier is another potential target for xenobiotic-induced effects. Breakdown of the blood–testes barrier by toxic agents can be caused by disruption of tight junctions that form between Sertoli cells (Huang *et al.*, 1988; Weber *et al.*, 1988; Kopf-Maier, 1992).

24.5   EFFECTS OF SPECIFIC CHEMICALS

A number of studies have demonstrated reproductive effects of a wide variety of classes of chemicals in females and males. For many, the results have suggested a mechanism of action (Table 24.1).

24.5.1   Chemotherapeutics

Roughly 25 females per 1000 will develop some type of cancer before the age of 35 (Statistics Canada), and approximately 1 in 1000 adults are survivors of childhood cancers (Revel and Laufer, 2002). Concerns over side effects of cancer therapy have increased as survival rates of cancer patients improve (currently ~56% overall survival rate; Byrne, 1990). Thus, the toxic effects of chemotherapeutic drugs in women cancer survivors have become important issues. Since the beginning of the use of antineoplastic agents to treat a variety of diseases and malignancies, the ability of these agents to produce reproductive failure has been reported. These effects have been described in patients treated with cyclophosphamide (CPA), nitrogen mustard, chlorambucil, cisplatin, or vinblastine (Miller, Williams and Leissring, 1971; Sobrinho, Levine and DeConti, 1971; Warne *et al.*, 1973; Koyama *et al.*, 1977; Chapman, 1983; Dnistrian *et al.*, 1983; Damewood and Grochow, 1986). In some cases, combined treatment with CPA and methotrexate and/or 5-fluorouracil produced the same effect; however, the individual contribution to ovarian failure by the latter compounds alone was not clearly identified (Koyama *et al.*, 1977; Dnistrian *et al.*, 1983). Effects of chemotherapeutic exposures on reproductive function are primarily a concern for those under the age of 40 who may wish to have children. Children exposed to chemotherapy prior to puberty are less likely to become permanently infertile than adults (Blumenfeld *et al.*, 2002); however, the Childhood Cancer Survivor Study reported that 6.3% of girls will undergo acute ovarian failure, and 8% will experience premature menopause (Garcia, 2007). Retention of ovarian function following chemotherapy does not mean ovarian damage did not occur, and premature menopause is still likely. Furthermore, undergoing early reproductive failure also increases the risks for developing other diseases. Studies on male fertility after cancer treatment have shown that chemotherapeutic agents are

TABLE 24.1 Summary of targeted sites and predicted effects of specific chemicals on reproductive function in females and males.

Chemical	Small pre-antral follicles	Growing follicles	Spermatogenesis	Hormone levels Male	Hormone levels Female
Cisplatin	−	−	−	+	−
CPA	+	+	−	−	−
BaP	+	+/−	+	−	+
DMBA	+	+	+	−	+
3-MC	+	−	−	−	−
1,3-BD	+	+	+	+	−
VCH/VCD	+	−	+	−	−
1-BP	−	+	+	−	−
2-BP	+	+	+	−	−
DBP	−	−	+	−	−
MEHP	−	+	+	−	+
DEPH	−	+	+	−	+/−
Cadmium	−	+	+	+	+
Lead	−	−	+	+	+
Mercury	+	+	+	+	+
TCDD	−	−	+	+	+
PCBs	+	+	+	−	−
BPA	+	−	+	−	−
DDT/DDE	−	−	+	−	−
Methoxychlor	−	+	−	+	+
Vinclozolin	−	−	+	−	−
Psoralens	−	−	+	−	+
Atrazine	−	−	+	−	+

+, effect observed; −, no effect observed; +/−, possible effect.

CPA, cyclophosphamide; BaP, benzo[a]pyrene; DMBA, 9,10-dimethylbenz[a]anthracene; 3-MC, 3-methylcholanthrene; 1,3-BD, 1,3-butadiene; VCH, 4-vinylcyclohexene; VCD, 4-vinylcyclohexene diepoxide; 1-BP, 1-bromopropane; 2, BP-2-bromopropane; DBP, di-butylphthalate; MEHP, mono-(2-ethylhexyl)phthalate; DEHP, di-(2-ethylhexyl)phthalate; TCDD, 2,3,7,8-tetrachlorodibenzo-p-dioxin; PCB, polychlorinated biphenyl; BPA, bisphenol A; DDT, dichlorodiphenyltrichloroethane; DDE, dichlorodiphenyldichloroethene.

particularly harmful to male germ cells. A study of testicular cancer survivors found that there was a 30% reduction in fertility after chemotherapeutic treatment (Huyghe *et al.*, 2004). Susceptibility to treatment is dependent on both drug and dose of treatment (Howell and Shalet, 2002a, b). In addition, mutagenic changes in spermatozoa have also been observed; however, they may not persist after treatment (Thomas *et al.*, 2004).

These observations in humans have motivated a variety of studies with CPA in rodents to better elucidate the mechanism(s) by which chemotherapeutic agents cause reproductive toxicity (Miller and Cole, 1970; Botta, Hawkins and Weikel, 1974; Jarrell *et al.*, 1987; Ataya *et al.*, 1995). Plowchalk and Mattison (1992) observed a time- and dose-dependent relationship between CPA and ovarian toxicity by looking at changes in ovarian structure and function in C57BL/6N mice. Loss of primordial follicles was essentially complete in three days and the estimated ED50 (concentration that produced 50% follicle loss) was 122 mg/kg body weight. From these results, it appears that premature ovarian failure in women following chemotherapy is likely via destruction of primordial follicles. Phosphoramide mustard has been determined to be the antineoplastic and ovotoxic form of this chemical (Plowchalk and Mattison, 1992). Mice were dosed with chemicals capable of forming specific metabolites of CPA. Only chemicals that released phosphoramide mustard induced ovarian toxicity. The greater potency of phosphoramide mustard-producing chemicals compared with CPA in mice was attributed to a bypassing of detoxification steps, allowing more toxic metabolite to reach the ovary (Plowchalk and Mattison, 1992).

In addition to the widespread destruction of primordial follicles by CPA, growing follicles have also been reported to be affected. Ataya et al. (1995) described specific losses of follicles >30 μm in rhesus monkeys given multiple doses of CPA. Also, in rats, CPA caused destruction of antral follicles at doses that did not affect primordial follicles (Jarrell et al., 1987). In contrast, in mice under conditions that completely destroyed primordial follicles, only partial destruction of antral follicles was observed (Plowchalk and Mattison, 1992). This demonstrated the greater sensitivity of primordial germ cells to this compound in mice as compared with rats. However, in both studies in rats and mice, lower ovarian weight, reduced follicular and luteal volume, and lower circulating 17β-estradiol levels were most highly associated with the loss of antral follicles (Jarrell et al., 1987; Plowchalk and Mattison, 1992). The effect on antral follicle numbers in both species was reversible. This provided a demonstration that unlike irreversible effects resulting from ovotoxicity that impacts the primordial follicle pool, damage to larger follicles can have a temporary impact on cyclicity.

Cisplatin is a chemotherapeutic agent used in the treatment of testicular cancers. This therapy can have long-lasting effects on spermatogenesis (Petersen et al., 1994). In mice, cisplatin at low doses has been shown to be selectively toxic for spermatogonia, while at high doses all cell types were affected (Meistrich et al., 1982; Boekelheide, 2005). Thus, even low-dose exposure might cause sterility in males.

24.5.2 Polycyclic Aromatic Hydrocarbons

PAHs are widespread in the environment from various combustion processes (including automobile exhaust). Additionally, cigarette smoke contains high levels of PAHs. In women, a positive connection between smoking and early menopause has been established (Jick and Porter, 1977). In animal studies, three PAHs have been shown to be carcinogenic and ovotoxic; benzo[a]pyrene (BaP), 3-methylcholanthrene (3-MC), and 7,12-dimethylbenz[a]anthracene (DMBA) (Mattison and Thorgeirsson, 1978; Vahakangas, Rajaniemi and Pelkonen, 1985). Initially, it was reported that DMBA depletes oocytes and produces ovarian tumors in mice (Krarup, 1967, 1969, 1970). Subsequently, these effects have also been reported in a number of studies for BaP and 3-MC. These three compounds have been shown to destroy oocytes in small follicles in Sprague-Dawley (SD) rats and in B6 and D2 mice within 14 days following a single dose (Mattison and Thorgeirsson, 1979). The order of potency for oocyte destruction by PAH was determined to be DMBA > 3-MC > BaP. This order of potency was similar to the relative toxicities determined for these compounds in experiments exposing mice and rats to multiple doses at lower levels of these compounds (Borman et al., 2000).

A direct relationship between the dose of PAHs and destruction of primordial follicles has been shown in the mouse ovary (Mattison and Thorgeirsson, 1979). Interestingly, significant oocyte destruction was demonstrated following a single high dose of BaP (100 mg/kg), whereas the same level of oocyte loss was observed with a low dose (10 mg/kg) given daily for 10 days (Mattison and Nightingale, 1980). This observation was supported in another study in mice in which, relative to a single high dose of DMBA, there was approximately 250 times greater ovotoxicity from repeated low-dose exposure (Borman *et al.*, 2000). Taken together, these observations provide support for a cumulative effect of chronic exposures to low doses. In addition to pre-antral follicles, an apparent effect of DMBA and BaP on antral follicles has also been shown in mice (Mattison, 1980). Beside effects in adults, *in utero* exposure to PAHs during fetal development can also affect reproductive potential of offspring. Daily oral exposure (10 mg/kg) of pregnant mice between 7 and 16 days of gestation with BaP caused severely compromised fertility of female offspring (MacKenzie and Angevine, 1981).

In male rats, treatment of seminiferous tubule segments with BaP or DMBA resulted in inhibition of DNA replication during spermatogenesis, with DMBA being 10-fold more potent than BaP (Georgellis, Parvinen and Rydström, 1989). Additionally, high concentrations of BaP and DMBA inhibited progression of spermatocytes through meiotic division (Georgellis *et al.*, 1990; Schrader and Kanitz, 1994).

PAHs must become bioactivated to arene oxides to become ovotoxic. Inhibition of PAH metabolism with α-naphthoflavone prevented oocyte destruction observed in mice (Mattison, 1983; Shiromizu and Mattison, 1984). Expression of enzymes involved in the bioactivation of PAHs (i.e. aryl hydrocarbon hydroxylase and epoxide hydrolase) has been identified in ovaries of mice and rats (Mattison and Thorgeirsson, 1977, 1979; Mattison and Nightingale, 1980; Bengtsson *et al.*, 1983; Bengtsson, Hamberger and Rydstrom, 1988; Bengtsson and Mattison, 1989; Rajapaksa, Sipes and Hoyer, 2007), monkeys (Bengtsson and Mattison, 1989), and humans (Bengtsson, Hamberger and Rydstrom, 1988). Therefore, localized ovarian metabolism of these compounds may be involved in induction of their ovotoxic effects. In males, BaP metabolites have also been detected in the testes (Smith *et al.*, 2007). Overall, it is the combined processes of bioactivation, distribution, and detoxification both throughout the body and locally in the reproductive tract that determine susceptibility to these compounds.

24.5.3 Occupational Chemicals

In general, compounds known to contain epoxide moieties (or which are capable of bioactivation by epoxidation) have been shown to affect reproductive function in laboratory animals. These compounds include butadiene and its derivatives and 4-vinylcyclohexene (VCH) and its respective derivatives (reviewed in National Toxicology Program, 1986, 1989 and Hoyer and Sipes, 1996). Many epoxidated compounds have been associated with increased mutagenicity in *in vitro* bacterial assays. The ability of epoxides to produce DNA adducts and induce sister chromatin exchanges has also demonstrated effects at the molecular level (James, Fouts and Bend, 1976; DeRaat, 1978; Citti *et al.*, 1984). The role of enzymatic epoxidation in bioactivation of parent compounds to these toxic metabolites has been implicated in a variety of studies (reviewed in Hoyer and Sipes, 1996). Because of the ability of these compounds to become epoxidated, they have the potential to be reproductive toxicants and carcinogenic agents.

24.5.3.1 Butadiene 1,3-Butadiene (BD) and the related olefins, isoprene and styrene, are released during the manufacture of synthetic rubber and thermoplastic resins, and the occupational exposure of US employees has been estimated to be 3700 to 1,000,000 people annually (IARC 1994). These chemicals have also been reported in cigarette smoke and automobile exhaust. Chronic animal inhalation studies have shown that carcinogenesis caused by BD is higher in mice than in rats. At low doses, female mice exposed daily to BD by inhalation for as long as two years exhibited ovarian atrophy, granulosa cell hyperplasia, and benign and malignant cell tumors (Melnick *et al.*, 1990). Furthermore, ovarian effects of BD appear at lower concentrations than are required to produce effects in other tissues.

In one study, the metabolite of BD, 1,3-butadiene monoepoxide (1.43 mmole/kg), depleted small follicles by 98% and growing follicles by 87% in female B6C3F$_1$ mice dosed daily for 30 days (Doerr *et al.*, 1995). Whereas, the diepoxide of 1,3-butadiene at a much lower dose (0.14 mmole/kg) depleted small follicles by 85% and growing follicles by 63%. The results of this study support that the diepoxide formed in the metabolism of BD is more potent than the monoepoxide at inducing follicle loss (Doerr *et al.*, 1995).

Male rats exposed to 0, 200, 1000, or 8000 ppm of 1,3-BD by inhalation for five days showed concentration-related increases in the frequency of abnormal sperm morphology at five weeks after exposure (Morrissey *et al.*, 1990). Another study measured the effect of diepoxybutane, a metabolite of butadiene on reproductive function. Mice were given a single i.p. dose at varying concentrations and tissues were collected 7, 14, 21, 28, and 35 days after treatment. At 7, 21, and 28 days after dosing, there were dose-dependent reductions in round and elongated spermatids; however, this was not detected at 35 days after dosing, indicating transient damage to differentiating spermatogonia (Spano *et al.*, 1996).

24.5.3.2 4-Vinylcyclohexene

The dimerization of BD forms VCH. The VCH family of compounds are occupational chemicals released at low concentrations during the manufacture of rubber tires, plasticizers, and pesticides (IARC 1994). VCH and its diepoxide metabolite VCD have been shown in mice and rats to (i) produce extensive destruction of small pre-antral follicles (Smith, Mattison and Sipes, 1990), (ii) cause premature ovarian failure (Hooser *et al.*, 1994; Mayer *et al.*, 2002), (iii) increase the risk for development of ovarian tumors (Hoyer and Sipes, 1996), and (iv) affect normal ovarian development of female offspring exposed *in utero* (Hoyer and Sipes, 1996; Devine and Hoyer, 2005). Because no significant effects on other tissues have been reported in studies with this class of chemicals, the damage they produce appears to be highly selective, and does not involve widespread toxicity. Ovarian damage caused by VCH and its related epoxide metabolites has been demonstrated by a variety of exposure routes, including dermal (National Toxicology Program, 1989), oral (Grizzle *et al.*, 1994), inhalation (Bevan *et al.*, 1996), and i.p. injection (Smith, Mattison and Sipes, 1990). It is therefore important to consider the potential risks for human exposure.

The role of biotransformation in VCH-induced ovarian toxicity has been well established. Following 30 days of daily dosing of female B6C3F$_1$ mice and Fischer 344 rats with VCH, primordial and primary follicles were reduced in ovaries of mice, but not of rats (Smith, Mattison and Sipes, 1990). The greater sensitivity of mice compared with rats was partly due to different capabilities for bioactivation of VCH. Studies have shown that the diepoxidation of VCH to VCD represents bioactivation, and it has been concluded that the species variability in susceptibility to VCH is, in part, due to differences in the capacity of mice and rats to form VCD. However, the increased sensitivity of rats versus mice to the ovotoxicity of VCD has also suggested that detoxification of VCD may play a role in species specificity to VCH-induced ovotoxicity. It was shown that the rat had greater capacity for conversion of VCD to its inactive tetrol, as compared with the mouse, and that only rats possessed detectable ovarian VCD-hydrolytic enzymatic activity (Keller *et al.*, 1997). Therefore, the greater susceptibility of mice versus rats relates to both enhanced bioactivation to and reduced detoxification of the ovotoxic epoxides.

Ovarian follicle destruction by VCD is selective for primordial and primary follicles. Mechanistic studies in rats have determined that VCD causes ovotoxicity by accelerating the natural process of atresia (apoptosis) and this requires repeated exposures (Springer *et al.*, 1996; Borman *et al.*, 1999). Pro-apoptotic signaling events in the Bcl-2 and mitogen-activated protein kinase families have been shown to be selectively activated in fractions of small pre-antral follicles (targets for VCD; Hu *et al.*, 2002a, b). Thus, because VCD causes a physiological form of ovotoxicity, follicle loss is "silent" and mimics normal cellular atresia. Therefore, damage caused by VCD would go unnoticed in exposed individuals that are affected. As a result, chronic exposure in women to low levels of this chemical may represent a risk for early menopause (Hoyer and Sipes, 2007).

In males, VCD targets testicular germ cells; however, unlike in females this effect is reversible. Male mice were dosed daily with 320 mg/kg/day i.p. and tissues collected after 5, 10, 15, 20, 25, or 30 days of dosing (Hooser *et al.*, 1995). Two groups were also dosed daily for 30 days and allowed to recover for an additional 30 or 60 days. Decreases in testicular weight were observed by day 5 and this continued to day 30. The decrease in weight corresponded to necrosis of germ cells. After day 5 of VCD treatment, there was a loss of spermatogonia, and after 30 days of dosing, seminiferous tubules were devoid of germ cells. In mice that were allowed to recover for 30 days, however, there was complete repopulation of seminiferous tubules, and epididymal spermatogonia were present 60 days after dosing (Hooser *et al.*, 1995). This finding supported VCD targeting of cells undergoing meiotic division since the stem cell population of spermatogonia remained viable and could support renewal of spermatogenesis.

24.5.3.3 2-Bromopropane   Due to their detrimental effects on the ozone layer, chlorofluorocarbons were banned and 1- and 2-bromopropane (BP) were proposed as substitute industrial propellants and cleaning solvents. These chemicals were found to have adverse reproductive effects on both male and female workers in a Korean factory (Park *et al.*, 1997). Subsequent animal studies have determined in female rats that 2-BP causes destruction of ovarian follicles in all stages of development, substantiating the cause of amenorrhea and increased hormone levels in the female workers (Yu *et al.*, 1999). For 1-BP, reproductive studies of long-term inhalation exposures (400 to 800 ppm) suggest that growing and antral follicles, and not primordial follicles are the target of this chemical (Yamada *et al.*, 2003). 2-BP (500, 1000 mg/kg, i.p., once every 2 to 3 days for 15 to 17 days) was found to prolong estrous cycles in rats, decrease ovulations, and alter preovulatory follicle morphology (Sekiguchi *et al.*, 2001). Direct effects of 2-BP on the oocyte might also occur, as suggested by increased micronuclei and decreased cell numbers observed in preimplantation mouse embryos following exposures to 2-BP (Ishikawa, Tian and Yamauchi, 2001).

In male rats, spermatogonia have been identified as targets of 1- and 2-BP, thus providing an explanation for reduced sperm production detected in the male workers (Omura *et al.*, 1999). Exposure of male rats to 1-BP reduced epididymal sperm count and motility and increased percentages of abnormal sperm (Ichihara *et al.*, 2000). Further treatment of male rats with 2-BP over a time course resulted in reductions in number and increases in degenerating spermatogonia, oligospermia, and testicular atrophy (Son *et al.*, 1999). A transplacental effect of 2-BP exposure has also been shown (Kang *et al.*, 2002). There was a decrease in the pup survival due to 2-BP exposure to the dam. In addition, atrophy of the uterus and testes were present in the offspring (Kang *et al.*, 2002).

24.5.4 Phthalates

One class of chemicals receiving much recent attention in reproductive toxicology is the phthalates. Phthalates are the most abundant class of synthetic chemical in the environment. Detrimental reproductive effects have been reported in both human epidemiological and rodent studies (Kavlock *et al.*, 2002). These chemicals, which are diesters of *o*-phthalic acid, are utilized in the plastics industry to enhance the flexibility of polyvinyl chloride products. They are included in such products as cosmetics, lubricants, plastic tubing, medical devices, vinyl upholstery, surgical gloves, toys, solvents, and pesticides. Phthalates are of particular concern because they can leach out of plastics into air, water, or food (Kavlock *et al.*, 2002; Calafat and McKee, 2006).

DEHP is widely used in the production of many polyvinyl chloride-based plastics, including medical and food packages. Annual production of DEHP is 1–4 million tons, making it the most commercially important phthalate ester plasticizer (Mylchreest *et al.*, 2000). In a reference human population, multiple phthalate metabolites were detected in the urine of approximately 75% of those tested (Blount *et al.*, 2000), demonstrating that humans in the general population are exposed to significant levels. Daily exposure can vary greatly among individuals (3–30 μg/kg body weight/day), but is estimated to be 2 mg/day for DEPH alone (Fay, Donohue and DeRosa, 1999; Kavlock *et al.*, 2002). Additionally, workers in both the phthalate and PVC industries are estimated to have exposures of 143 and 286 μg/kg body weight/workday, respectively (Kavlock *et al.*, 2002). Correlations were identified between urinary concentrations of phthalates in humans and pregnancy complications (Tabacova, Little and Balabaeva, 1999) or rates of unsuccessful pregnancies (Aldyreva *et al.*, 1975). Additionally, decreased rates of pregnancy, increased rates of miscarriage, and anovulation have been associated with occupational exposure of Russian women to phthalates.

DEHP and its monoester metabolite (MEHP) are likely the most widely studied of the phthalate esters. Repeated oral exposure of female SD rats to DEHP caused delayed ovulation, reduced granulosa cell size in antral follicles, decreased circulating 17β-estradiol, progesterone and decreased LH levels, and increased FSH (Davis, Maronpot and Heindel, 1994). Mechanistic analyses have linked effects of phthalates to peroxisome proliferator-activated receptors (PPARs; reviewed in Lovekamp-Swan and Davis, 2003). It is known that multiple phthalates can induce hepatic 17β-hydroxysteroid dehydrogenase (17β-HSD IV), which metabolizes estradiol to estrone (Fan, Cattley and Corton, 1998). Further evidence demonstrates that this up-regulation of 17β HSD occurs through the PPAR-α receptor. In addition to PPAR-α, MEHP can also activate PPAR-γ, which suppresses expression of aromatase in granulosa cells of antral follicles (Lovekamp and Davis, 2001; Lovekamp-Swan, Jetten and Davis, 2003). From the responses to DEHP in rats, it can be concluded that these reproductive effects resulted from a specific targeting of large antral follicles, because of suppressed granulosa cell 17β-estradiol production. An *in vitro* study supported this conclusion because granulosa cells collected from diethylstilbestrol (DES)-primed female rats produced lower 17β-estradiol as increasing MEHP was added to the medium (Davis, Maronpot and Heindel, 1994). By evaluating the effect of FSH and dibutyryl-cAMP, it was concluded that inhibition of steroidogenesis was via inhibition of aromatase activity.

In males, DEHP caused a decrease in Sertoli cell proliferation *in vivo* (Heindel and Chapin, 1989). This chemical has been shown to inhibit FSH signal transduction in cultured rat Sertoli cells, and this may provide a mechanism for the *in vivo* observation. Furthermore, *in vitro* experiments have demonstrated that the effect results from inhibition of FSH-stimulated cAMP production. Based on these observations, the phthalate target is likely to be located in the Sertoli cell membrane. Therefore, DEHP and MEHP appear to have analogous targets within the testis (Sertoli cells) and ovary (granulosa cells). Male rat pups were dosed with increasing concentrations of the diester, di-*n*-pentylphthalate (DPP) (0, 0.25, 1, or 2 g/kg), and tissues collected weekly over a 10-week period. In the high-dose group, there was reduced fertility, and epididymal sperm count from three weeks of treatment, persisting until seven weeks after treatment. Testicular lesions were also reported in the high-dose (2 g/kg) animals (Lindström *et al.*, 1988; Sundaram and Witorsch, 1995).

Another phthalate, di-butylphthalate (DBP), is a component of adhesives, paper coatings, printing inks, aerosols, nail polishes, and hairsprays. In Western industrialized countries, annual production of DBP is 10,000–50,000 tons (Mylchreest *et al.*, 2000). Animal studies investigating transplacental exposure to DBP have demonstrated effects on sexual differentiation in male offspring. The effects include missing epididymides and vasa deferentia, effects on seminal vesicles, induction of hypospadias, decreased anogenital distance, retained thoracic nipples, and cryptorchidism (Mylchreest *et al.*, 2000). DBP exposure to male rats *in utero* has been shown to cause increases in the transcription factor Oct4 from embryonic days 15.5 to 17.5. Additionally, decreases in testicular germ cell number and testes descent were seen (Ferrara *et al.*, 2006). Although transplacental effects of DBP on reproductive tract development have been widely reported in male offspring, analogous effects in female offspring have not been observed. During fetal development, androgenic influences are critical in sexual differentiation in males, but not in females (Hoyer, 2001). Thus, this is likely due to the fact that male effects are thought to result from the anti-androgenic activity of DBP.

In summary, comparing male and female reproductive responses to phthalates provides interesting insight into gender differences and similarities. In rats, DEHP and MEHP are both male and female reproductive toxicants with opposing effects on FSH. Conversely, DBP and its active metabolite, monobutyl phthalate, produce developmental effects mainly in males. Thus, in males and females, within the class of phthalates, different compounds display similar as well as different effects.

24.5.5 Metals

Animal studies have provided evidence that several metals found in the environment have the potential to affect ovarian function either directly or indirectly via hypothalamic-pituitary targeting (Hoyer, 2005). There is evidence for testicular and ovarian accumulation of these metals and in some cases, they have been detected in seminal and follicular fluid. While there is little concrete evidence for adverse reproductive effects in humans caused by metals, future studies are required to determine whether environmental exposures to metals actually pose a risk to reproductive function.

24.5.5.1 Cadmium Cadmium is one of the most extensively reported metals with respect to effects on reproductive function (Hoyer, 2005). Cadmium exposure in the environment is widespread and occurs via food, modern industrial processes, waste disposal, and terrestrial and aquatic ecosystems (Nath *et al.*, 1984; Piasek and Laskey, 1994). One particular source is cigarette smoke (Zenzes *et al.*, 1995; Piasek *et al.*, 2001; Younglai *et al.*, 2002) and cadmium concentrations have been demonstrated to be higher in follicular fluid of smokers versus nonsmokers (Zenzes *et al.*, 1995). It was suggested that this accumulation might compromise the quality of oocytes, becoming a risk factor for fertility. However, the findings from two studies in which fertility was evaluated in women with high versus low levels of cadmium in follicular fluid did not support this conclusion (Drbohlav *et al.*, 1998; Younglai *et al.*, 2002). Cadmium has also been identified in seminal fluid of smokers and positive associations were observed between increased cadmium levels and reduction in testes size, along with increases in 17β-estradiol, FSH, and testosterone (Jurasovic *et al.*, 2004).

Ovarian accumulation of cadmium in animals can cause direct adverse effects. In mice, atretic follicles and degenerating corpora lutea were detected following exposure to cadmium (Godowicz and Pawlus, 1985). Cadmium exposure has also been reported to impact ovarian steroidogenesis. Acute exposure of rats to cadmium (up to 5 mg/kg) resulted in decreases in circulating progesterone and 17β-estradiol, and this was cycle stage-dependent (Piasek *et al.*, 1996). Support that the effects of cadmium on steroidogenesis result from direct ovarian targeting is provided by *in vitro* studies. Whole ovarian cell cultures from rats demonstrated particular susceptibility to *in vitro* cadmium exposure when tissues were collected during proestrous or early pregnancy. During those stages, progesterone and testosterone (Piasek and Laskey, 1999; Piasek *et al.*, 2002) as well as 17β-estradiol (Piasek and Laskey, 1994) decreased when cadmium was added to the medium.

Cadmium in male rats at two levels of exposure showed irreversible impairment of the germinal epithelium with toxic effects on spermatogenesis at high doses (Xu *et al.*, 2001). Additionally, daily oral dosing of male sheep with cadmium chloride (3 mg/kg) resulted in reductions in semen density and number of spermatozoa (Lymberopoulos *et al.*, 2000).

24.5.5.2 Lead

Exposure to lead is common as it is also widespread in the environment. The greatest source of exposure is from commercial products such as paint, printing material, and acid batteries (Dearth *et al.*, 2002). High levels of exposure can also occur in industrial settings, such as mining and refining plants (Taupeau *et al.*, 2001). A major concern related to lead exposure is its impact on fertility in women, fetal development, and prepubertal growth. Thus, lead exposure and toxicity are a leading environmental health concern for children, as well as women of childbearing age (Dearth *et al.*, 2002; Taupeau *et al.*, 2003; Hoyer, 2005).

One human study provided evidence for direct ovarian impairment of steroid production. Expression of aromatase and the estrogen receptor-beta mRNA and protein were significantly reduced by *in vitro* lead acetate exposure of granulosa cells collected from women undergoing *in vitro* fertilization (IVF; Taupeau *et al.*, 2003). Exposure to lead in animals has been shown to have a transplacental effect on the female F1 offspring. *In utero* exposure to lead chloride in SD rats resulted in decreased ovarian conversion of progesterone to androstenedione in prepubertal and pubertal females (Wiebe, Barr and Buckingham, 1988). It was concluded that prenatal exposure to lead significantly altered subsequent ovarian steroid production and gonadotropin binding. *In utero* exposure to lead acetate also resulted in delayed timing of puberty in female rat offspring, which was associated with suppressed serum levels of LH and 17β-estradiol (Dearth *et al.*, 2002). Another study showed that prepubertal exposure of female rats to lead acetate resulted in delayed vaginal opening and disrupted estrous cyclicity (Ronis *et al.*, 1996).

In males, rat Leydig cells cultured with lead acetate showed reduced levels of testosterone and progesterone, along with reductions in steroidogenic enzymes (Thoreux-Manlay *et al.*, 1995). Epidemiological studies of men with high body blood lead levels have shown effects on sperm morphology (Telisman *et al.*, 2007). Another study of males exposed to high levels of occupational lead found structural damage to sperm (Naha and Chowdhury, 2006). Moreover, high seminal lead concentrations in sub-fertile males were found to be associated with decreased sperm motility and concentration (Pant *et al.*, 2003).

24.5.5.3 *Mercury* Agriculture, consumption of fossil fuels, and industry are the main sources of mercury pollution in the environment. These are largely in the form of metallic mercury as inorganic or organic compounds. Additionally, ingestion of organomercurials from eating fish can be absorbed via the gastrointestinal tract, and eating fish from contaminated waters is a current concern (Hoyer, 2005). All chemical forms of mercury given to animals have been shown to cause female reproductive problems including infertility, disturbances in the estrous cycle, and inhibition of ovulation—all potentially arising from ovarian targeting (Schuurs, 1999). Mercuric chloride has been shown to accumulate in the female hamster pituitary (Lamperti and Printz, 1974); thus, exposure to mercury may cause hormonal effects by targeting a hypothalamic-pituitary site.

Low-dose mercuric chloride exposure in male mice exhibited reduced epididymal sperm number, dissociation of spermatozoa from the basement membrane, and necrosis. There were additional effects on Leydig cells and epididymis (Orisakwe *et al.*, 2001). A study of mercuric chloride treatment on bull sperm samples showed dose-dependent increases in DNA damage and decreases in cellular viability (Arabi, 2005). Furthermore, methylmercuric chloride exposure to male rats showed reductions in testes size and testosterone concentrations, and increases in testicular cell apoptosis (Homma-Takeda *et al.*, 2001).

24.5.6 Endocrine Disruptors

Endocrine disruptors that display estrogenic/antiestrogenic effects have been widely studied in their ability to target developmental and uterine sites of action, although intracellular mechanisms involved in these actions remain poorly understood. The effects of TCDD on sexual development and fertility have been studied in depth (Safe and Krishnan, 1995). There is much evidence for a hypothalamic-pituitary site of action, although this compound also has been shown to cause direct ovotoxicity (Li, Johnson and Rozman, 1995). The effect of a single oral dose of TCDD on numbers of oocytes ovulated and estrous cyclicity was observed in female rats (Li, Johnson and Rozman, 1995). Exposure to TCDD prolonged the diestrous stage and reduced the time in pro-estrus and estrus. Additionally, there was a reduction in the number of oocytes ovulated in treated rats. Thus, these findings provide strong evidence that TCDD impairs ovulation (Li, Johnson and Rozman, 1995). The data from this study were consistent with reduced ovulation via a hypothalamic-pituitary effect. However, ovulation was also reduced in hypophysectomized animals, suggesting a direct ovarian effect as well. The mechanism by which TCDD directly targets the ovary might be due to inhibition of 17β-estradiol production in granulosa cells, as suggested by *in vitro* studies (Moran *et al.*, 1997). Further investigation demonstrated that TCDD produced these effects via interactions with the EGF linked-mitotic signaling pathway involving mitogen-activated protein kinase and protein kinase A. TCDD also affects male reproduction, and has been shown to cause reduced levels of plasma testosterone, reduced testis and epididymis weights, reduced sperm production, and alteration of the number of GnRH receptors in the pituitary gland (Bookstaff, Moore and Peterson, 1990; Bookstaff *et al.*, 1990; Mably, Moore and Peterson, 1992; Mably *et al.*, 1992). Additionally, a recent study showed that male rat pups exposed *in utero* to TCDD had increased testicular CYP1A1 and reduced CYP17 mRNA levels (Taketoh *et al.*, 2007).

The polychlorinated biphenyl (PCB) compound 3,4,5,5-tetrachlorobiphenyl (TCB) has been shown to be teratogenic in the mouse and embryolethal in the rat, as well as to cause transplacental ovarian toxicity in the mouse (Ronnback, 1991). Follicles at all stages of development were reduced 40% to 50% in female offspring at 28 days of age when mice were exposed *in utero* on day 13 of gestation. Interestingly, during a five-month period of testing, this extent of follicular damage did not adversely affect reproductive function in these offspring. Exposure of male rats to another PCB, 2,2′,3,4′,5′,6-hexachlorobiphenyl (PCB 132) has been shown to cause reductions in sperm count and motility (Hsu *et al.*, 2007).

Another widely studied endocrine disruptor has been bisphenol A (BPA; Papaconstantinou *et al.*, 2000). BPA is frequently used in the manufacture of polycarbonate plastics, epoxy resins, dental sealants, and as a stabilizing agent in plastics such as polyvinyl chloride. Possible exposure of humans to BPA has been reported due to its leaching from laboratory flasks, baby-feeding bottles, epoxy resins used for lining food cans, several types of dental sealants, and plastic waste samples. BPA has been detected in human follicular fluid in 1–2 ng/mL concentrations (Ikezuki *et al.*, 2002). *In vitro* animal studies have determined that BPA can bind the estrogen receptor (α and β), and induce estrogen-dependent gene expression/responses. Estrogenicity associated with BPA has been demonstrated in a number of *in vitro* and *in vivo* assays. *In vitro* assay end points include proliferation of MCF-7 human breast cancer cells, and the activation of estrogen response element (ERE)-driven reporter gene constructs. *In vivo* effects include the induction of vaginal cornification, growth and differentiation of the mammary gland, decrease in serum cholesterol levels, and increases in prolactin levels, uterine vascular permeability, and c-fos mRNA levels in the uterus and vagina (Papaconstantinou *et al.*, 2000). In mice, BPA exposure caused disruptions in puberty onset, regularity of estrous cyclicity (Markey *et al.*, 2002), and development of polycystic ovaries (Kato *et al.*, 2003). It has also been hypothesized that BPA may interfere with FSH-regulated progesterone synthesis in granulosa cells (Mlynarcikova, Fickova and Scsukova, 2005). *In utero* exposure of mice to BPA resulted in disturbances to recombination events during oogenesis (Susiarjo *et al.*, 2007).

Male mice exposed to BPA *in utero* or postnatally were shown to have reduced testes weights and lower testicular catalase activity (Kabuto, Amakawa and Shishibori, 2004); however, another study which exposed rats to BPA neonatally showed no effect on testicular development (Kato *et al.*, 2006). In spite of compelling animal data, to date, there have been no convincing demonstrations of human effects caused by BPA exposure.

24.5.7 Agricultural Compounds

24.5.7.1 Pesticides   A particular area of concern has focused on the effects of endocrine-disrupting chemical exposure on fish and wild life (Peterson *et al.*,

1997; Thomas, 1999). Feminization of male birds, alligators, and fish and the production of the estrogen-induced yolk precursor, vitellogenin, in male freshwater fish have been reported after exposure to xenobiotic estrogens, such as o,p'-dichlorodiphenyltrichloroethane (DDT), kepone, pulp kraft mill effluent, and sewage containing nonylphenols. DDT was used in the 1940s to control wartime typhus and agricultural pests (De Jager *et al.*, 2006). Juvenile male alligators exposed to dicofol and DDT in Florida had depressed plasma testosterone concentrations, poorly organized testes, and micropenises (Guillette *et al.*, 1995). Injection of fertilized gull eggs with DDT caused male embryos to develop ovarian tissue and oviducts (De Jager *et al.*, 2006). DDT pollution in California has been associated with feminization of male gulls. Dall's purpoises have elevated blubber concentrations of dichlorodiphenyldichloroethene (DDE) that are correlated with depressed plasma testosterone concentrations. An epidemiological study of men in Mexico showed that DDT exposure was associated with increased sperm tail defects and decreased sperm motility (De Jager *et al.*, 2006).

Methoxychlor (MXC) is an organochlorine pesticide that is presently used on agricultural crops as a replacement for DDT (Murono and Derk, 2005). MXC is believed to be metabolized by the liver to 2,2-bis(p-hydroxyphenyl)-1,1,1-trichloroethane (HPTE), which in some tissues is a more active metabolite than MXC (Murono, Derk and Akgul, 2006). MXC has been shown to cause reproductive effects in female as well as in male rodents. Early studies reported that MXC induces ovarian atrophy in mice (Eroschenko, Abuel-Atta and Grober, 1995) and decreases steroidogenesis in rat ovarian cells (Bal, 1984). More recently, *in vitro* studies have shown that HPTE in cultured rat granulosa cells inhibits progesterone and 17β-estradiol production by interfering with FSH signaling (Zachow and Uzumcu, 2006). Mechanistic investigations into ovarian effects of MXC on isolated mouse antral follicles have indicated that MXC accelerates atresia via increased apoptosis involving the Bcl-2 proto-oncogene family (Borgeest *et al.*, 2002; Miller *et al.*, 2005). In determining pathways with which MXC might interact, incubation of antral follicles with 17β-estradiol protected against MXC or HPTE-induced atresia (Miller, Gupta and Flaws, 2006). This finding was further supported in a study in which atresia was enhanced in antral follicles from MXC-exposed ERβ-overexpressing mice compared with their wild-type counterparts (Tomic *et al.*, 2006). Thus, it appears that ER-mediated pathways may mediate the ovarian toxicity of MXC. MXC has also been shown to exert its effects via increases in oxidative stress-induced mitochondrial damage (Gupta *et al.*, 2006a, b). Whether the mechanisms involving Bcl-2, ERβ and oxidative stress pathways interact in MXC-induced atresia is of interest.

In males, an early study reported that exposure to MXC caused testicular atrophy in rats (Tullner and Edgcomb, 1962). More information related to this effect has been provided by the demonstration that repeated exposure of rats to MXC (p.o.) decreased seminal vesicle weights, and circulating testosterone and dehydroepiandrosterone levels (Eroschenko, Abuel-Atta and Grober, 1995). Furthermore, in incubations of Leydig cells cultured from those animals, testosterone production and P450 cholesterol side-chain cleavage activity were reduced in cells from MXC-treated rats. In one *in vitro* study, HPTE inhibited basal and hCG-stimulated testosterone production in cultured fetal rat Leydig cells (Murono and Derk, 2005). In another *in vitro* study, HPTE inhibited LH-stimulated testosterone production in cultured Leydig cells from immature rats and the estrogen receptor antagonist ICI 182,780 reversed this inhibition (Akingbemi *et al.*, 2000). Long-term effects in male offspring were seen following transplacental exposure to MXC (Amstislavsky *et al.*, 2006). At six months of age, testosterone levels were reduced, seminal vesicle weight was increased, and sexual arousal was compromised in males exposed *in utero* to MXC. Thus, MXC and its metabolite HPTE cause reproductive effects in females as well as in males, and these effects relate to compromised gonadal steroidogenesis. Interestingly, these effects appear to be mediated by the estrogen receptor in both genders; thus, MXC can be considered an endocrine disruptor.

24.5.7.2 Fungicides   Vinclozolin is a dicarboximide fungicide used to treat fruits, vegetables, ornamental plants, and turfgrass. Administration of vinclozolin to pregnant rats at 100 or 200 mg/kg/day demasculinizes and feminizes male offspring (Gray *et al.*, 1999). This effect is thought to be the result of vinclozolin interacting with the androgen receptor, and in that manner acting as an endocrine disruptor. Earlier studies reported the lowest observed effect level (LOEL) for vinclozolin ranged from 50 to 111 g/kg/day (Van Ravenzwaay, 1992). However, in a study in which even lower doses of vinclozolin were given to pregnant rats (3.125 and 6.25 mg/kg/day), subtle alterations in the differentiation of the external genitalia, ventral prostate, and nipple tissue in male offspring were observed. This is an order of magnitude below previously reported developmental no observed effect levels (NOELs, 22–77 mg/kg/day; Amstislavsky *et al.*, 2006). Thus, it has been proposed that the anti-androgenic effects of vinclozolin can be produced at very low levels, and studies that do not include androgen-sensitive end points might yield NOELs that are at least an order of magnitude too high. Vinclozolin also has an *in utero* effect on male offspring, with reductions in numbers of spermatids and epididymal sperm, sperm motility, and increased spermatogenic cell apoptosis (Anway *et al.*, 2006).

24.5.7.3 Psoralens   The psoralens (or linear furanocoumarins) are photoactivated (Cole, 1970; Murray, Mendez and Brown, 1982; Trumble *et al.*, 1991) naturally occurring plant biosynthetic metabolites that are found in several common fruits and vegetables (Beier, 1990; Berenbaum, 1991; Diawara and Trumble, 1997). Synthetic 5-methoxypsoralen (bergapten) and 5- and 8-methoxypsoralen (5-MOP; 8-MOP; xanthotoxin or methoxsalen) are widely used in the treatment of several skin disorders including psoriasis, conditions of skin depigmentation (such as leprosy, vitiligo, and leucoderma), mycosis fungoides, polymorphous dermatitis, and eczema (Musajo and Rodighiero, 1962; Scott, Pathak and Mohm, 1976; Van Scott, 1976; Pathak and Fitzpatrick, 1992; Chadwick *et al.*, 1994; Ehrsson *et al.*, 1994). However, the psoralens have been linked to a higher incidence of skin cancer (Musajo and Rodighiero, 1962; Scott, Pathak and Mohm, 1976; Van Scott, 1976; Stern *et al.*, 1979; Grekin and Epstein, 1981) and other disorders such as sister chromatid exchange, gene mutation, and chromosomal aberrations in humans (Calzavara-Pinton *et al.*, 1994). Studies on the effect of psoralens on reproductive function have reported a dose-dependent effect of 5-MOP and 8-MOP in the presence of UVA light on birth rate in female rats. The number of offspring was reduced along with a reduction in uterine weight and serum estradiol levels (Diawara *et al.*, 1999). This was determined to be due to reduced ovulations resulting from targeting of granulosa cells. Additionally, male rats treated with 5-MOP or 8-MOP in the absence of UVA light showed a decrease in pituitary gland weight and number of sperm, with an increase in relative testicular weights and serum testosterone. Subsequent matings in this study resulted in a reduced number of pregnancies in females (Diawara *et al.*, 2001).

24.5.7.4 Atrazine   Atrazine is a chlorotriazine and was introduced in the 1950s as a broad-spectrum herbicide and is widely used worldwide (Eldridge, Wetzel and Tyrey, 1999; Cooper *et al.*, 2007). Total annual use of atrazine is approximately 167 kg of active ingredient. In female SD rats, atrazine caused mammary tumors which were attributed to a hormonal influence due to reproductive aging (Wetzel *et al.*, 1994). Hysterectomy prevented mammary tumor development due to atrazine exposure. It is thought that atrazine prevents the LH surge associated with ovulation, thus supporting a hypothalamic/pituitary site of atrazine action. In addition, atrazine exposure leads to significant increases in serum levels of estrone and 17β-estradiol in male and ovariectomized female rats (Cooper *et al.*, 2007). Male rats exposed to atrazine *in utero* had lower body weights, experienced delayed preputial separation, and increased prostate inflammation compared with controls (Rayner *et al.*, 2007). Additionally, atrazine exposure resulted in delayed puberty onset in both male and female rats (Stoker *et al.*, 2002; Laws *et al.*, 2003).

24.6 TRANSLATIONAL APPLICATIONS

A major goal of toxicological research is to enhance the ability to predict and evaluate potential risk of the environment to human health. However, controlled toxicological studies are usually carried out in laboratory rodents because of the economy of using them in large numbers, along with the fact that controlled, manipulative studies in humans are rarely feasible. A reasonable criticism of this approach is the validity of extrapolating responses in rodents to humans. This is especially true of reproductive physiology, in which species variation is the hallmark. Thus, the ideal goal is to coordinate human information related to exposure outcomes with that obtained from animal studies.

Assessment of reproductive effects in humans usually begins with a recognizable outcome in women (acyclicity, delayed time to conception, early menopause), pregnancy outcome (spontaneous abortion) or in men (reduced sperm production, impotence, infertility). Generally, identification of reproductive effects in humans progresses slowly because it relies on population-based and clinical observations (MacKenzie and Angevine, 1981; Whelan, 1997). There are, however, several situations in which reproductive effects in humans have been linked to exposures before animal studies have predicted the outcomes. For example, a number of infertility cases were discovered in men working in a California pesticide factory (Whorton *et al.*, 1977). The suspected cause was exposure to 1,2-dibromo-3-chloropropane (DBCP). Upon evaluation of these men, major effects including azoospermia or oligospermia were accompanied by increased circulating levels of FSH and LH. Follow-up testing in rats has demonstrated that DBCP is directly toxic to sperm (Kluwe *et al.*, 1983). Another example is the case of 2-BP which was discovered to have toxic effects on the reproductive systems of both males and females after occupational exposure of workers in South Korea. Evaluation of workers exposed in factories to this industrial propellant and cleaning solvent uncovered a high incidence of bone marrow effects as well as secondary amenorrhea accompanied by increased circulating FSH and LH levels and hot flashes in 16 of 25 exposed women. Additionally, six of eight exposed men displayed azoospermia or oligospermia, and reduced sperm motility. Based on these observations, animal studies were designed and have since shown that 2-BP causes destruction of ovarian follicles and spermatogonia (Omura *et al.*, 1999; Yu *et al.*, 1999).

Epidemiological studies conducted over the last four decades have demonstrated a relationship between smoking and impaired fertility. Cigarette smoke is a well-known reproductive toxicant. One study reported that rates of pregnancy were reduced to 75% in light smokers and 57% in heavy smokers when compared with nonsmokers. Furthermore, smokers required one year longer to conceive than did nonsmokers (Baird and Wilcox, 1985). Along with the impact on fertility, there were also effects of cigarette smoking on pregnancy and the fetus. Prenatal exposure to cigarette smoke has been associated with retarded intrauterine growth and premature deliveries (Mattison *et al.*, 1989). Additionally, conception in women whose mothers smoked while pregnant was significantly reduced when compared with women whose mothers had not (Weinberg, Wilcox and Baird, 1989). Decreased follicular levels of 17β-estradiol were observed in smoking women when compared with nonsmokers, and cultured human granulosa cells were reported to secrete decreased amounts of 17β-estradiol in the presence of an extract of cigarette smoke (Van Voorhis *et al.*, 1992). In a study of females undergoing IVF, increased levels of the PAH, BaP, were found in serum and follicular fluid in those who were smokers, and this was associated with lower rates of implantation and successful pregnancies compared with control nonsmoking women (Neal *et al.*, 2007). Thus, these effects may relate to the infertility associated with cigarette smoking. When the ovaries of women are depleted of primordial follicles, the result is menopause. Because the ovary contains a finite number of oocytes at birth, human exposure to environmental chemicals that destroy primordial follicles can cause early menopause. Female smokers have also been reported to experience a one- to four-year-earlier age for the onset of menopause (Fig. 24.4) (Jick and Porter, 1977). Therefore, in women, cigarette smoking may impair fertility as well as accelerate the onset of menopause.

Cigarette smoke is a complex mixture of alkaloids (nicotine), PAHs, nitroso compounds, aromatic amines, and protein pyrosylates, many of which are carcinogenic (Stedman, 1968). Several ovotoxic PAHs are found in cigarette smoke, including DMBA, 3-MC, and BaP. Because these PAHs are known to damage primordial and primary follicles in mice and rats, direct oocyte destruction caused by one or more of these chemicals is the most logical mechanism by which cigarette smoke induces early menopause (Mattison *et al.*, 1989). Of additional concern is the finding in animal studies

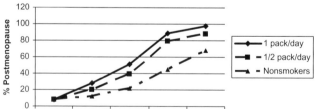

Figure 24.4   Effect of smoking on age of menopause. Toxicants in cigarette smoke appear to destroy small oocytes. No major effects on fertility are noted during the reproductive life span, but shortening of reproductive life span occurs (adapted from Jick and Porter, 1977).

that exposure of mice *in utero* to cigarette smoke resulted in a smaller number of ovarian primordial follicles in female offspring (Vahakangas, Rajaniemi and Pelkonen, 1985). Secondary follicles in laboratory animals are also susceptible to PAHs (Mattison, 1980; Borman *et al.*, 2000). Thus, in women smokers, prior to early menopause, damage to larger follicles (secondary through antral) caused by exposure to PAHs could also cause problems with fertility. Another component of cigarette smoke, nicotine, acting on the central nervous system, may affect the secretion of hormones involved in regulation of ovarian function and in that way may be responsible for impaired fertility in female smokers (Jick and Porter, 1977). In addition to ovarian effects, cigarette smoke may also induce liver-metabolizing enzymes which may accelerate metabolism of steroid hormones thereby compromising gonadal function (Jick and Porter, 1977; Yeh and Barbieri, 1989). Information obtained from animal studies may be useful for planning epidemiological investigations about potential reproductive effects resulting from exposure of women to ovotoxic chemicals in cigarette smoke.

24.7   CONCLUSIONS AND FUTURE DIRECTIONS

Once the reproductive effects of a chemical have been identified and characterized, it becomes of interest to elucidate mechanisms by which they occur. It is important, therefore, to determine cellular targets and intracellular pathways that are involved. In the case of some chemicals, mechanistic insight has been gained; however, this is generally limited in humans and more research is required to further develop the field of reproductive toxicology. The application of widely available molecular technologies that have been emerging over the past decade can help achieve this aim. Three of these hold particular promise for the field: DNA microarray, laser capture microdissection (LCM), and proteomic analyses tools. DNA microarray involves examining changes that occur in gene expression during xenobiotic exposure. Profiling DNA microarrays can be used to evaluate changes in expression of genes that occur due to a specific chemical exposure. A second use of this method is in single nucleotide polymorphism (SNP) profiling. This area will be useful for determining individual susceptibility to reproductive toxicants. This application also allows genetic screening for mutations, which may impair reproductive capacity as well as enhance susceptibility to cancer. A third use of DNA microarrays involves identifying effects of chemical exposures on regulatory regions of genes that are important in cell signaling using chromatin immunoprecipitation assay (ChIP-on-chip). LCM is a tool that allows specific regions of heterogeneous tissues to be isolated for evaluation of changes in a specific cell population. Proteomics consists of an array of techniques that allows proteins to be separated from complex mixtures, and analyzed for post-transcriptional/translational modifications caused by chemical exposures. In addition, transgenic animals that either over- or underexpress genes of importance in xenobiotic metabolism are proving to play vital roles in determining pathways that are involved in effects on reproductive toxicology.

In conclusion, environmental exposure to xenobiotics that have the potential to impact human reproduction has increased. Research in this field in the next decade is likely to see an increase in mechanistic studies that will identify cellular and molecular events involved in producing effects caused by a wide variety of environmental chemicals using the emerging technologies that have been discussed. Gaining improved mechanistic insight into the effects of reproductive toxicants in the laboratory will provide a greater understanding for translating these effects into more accurate assessments of human risk.

REFERENCES

Akingbemi, B.T., Ge, R.S., Klinefelter, G.R., Gunsalus, G.L., Hardy, M.P. (2000). A metabolite of methoxychlor, 2,2-bis(p-hydroxyphenyl)-1,1,1-trichloroethane, reduces testosterone biosynthesis in rat leydig cells through suppression of steady-state messenger ribonucleic acid levels of the cholesterol side-chain cleavage enzyme. *Biol. Reprod.*, *62*, 571–578.

Aldyreva, M.V., Klimova, T.S., Iziumova, A.S., Timofeevskaia, L.A. (1975). The effect of phthalate plasticizers on the generative function. *Gig. Tr. Prof. Zabol.*, *12*, 25–29.

Amstislavsky, S.Y., Amstislavskaya, T.G., Amstislavsky, V.S., Tibeikina, M.A., Osipov, K.V., Eroschenko, V.P. (2006). Reproductive abnormalities in adult male mice following preimplantation exposures to estradiol or pesticide methoxychlor. *Reprod. Toxicol.*, *21*, 154–150.

Anway, M.D., Memon, M.A., Uzumcu, M., Skinner, M.K. (2006). Transgenerational effect of the endocrine disruptor vinclozolin on male spermatogenesis. *J. Androl.*, *27*, 868–879.

Arabi, M. (2005). Bull spermatozoa under mercury stress. *Reprod. Domest. Anim.*, *40*, 454–459.

Ataya, K., Rao, L.V., Lawrence, E., Kimmel, R. (1995). Luteinizing hormone-releasing hormone agonist inhibits cyclophosphamide-induced ovarian follicular depletion in rhesus monkeys. *Biol. Reprod.*, *52*, 365–372.

Baird, D.D., Wilcox, A.J. (1985). Cigarette smoking associated with delayed conception. *JAMA*, *253*, 2979–2983.

Bal, H.S. (1984). Effect of methoxychlor on reproductive systems of the rat. *Proc. Soc. Exp. Biol. Med.*, *176*, 187–196.

Beamer, W.G., Tennent, B.J. (1986). Gonadotropin uptake in genetic and irradiation models of ovarian tumorigenesis. *Biol. Reprod.*, *34*, 761–770.

Beier, R.C. (1990). Natural pesticides and bioactive components in food. *Rev. Environ. Contam. Toxicol.*, *113*, 47–137.

Bengtsson, M., Hamberger, L., Rydstrom, J. (1988). Metabolism of 7,12-dimethylbenz[a]anthracene by different types of cells in the human ovary. *Xenobiotica*, *18*, 1255–1270.

Bengtsson, M., Mattison, D.R. (1989). Gonadotropin-dependent metabolism of 7,12-dimethylbenz(a)anthracene in the ovary of rhesus monkey. *Biochem. Pharmacol.*, *38*, 1869–1872.

Bengtsson, M., Montelius, J., Mankovitz, L., Rydstrom, J. (1983). Metabolism of polycyclic aromatic hydrocarbons in the rat ovary: comparison with metabolism in adrenal and liver tissues. *Biochem. Pharmacol.*, *32*, 129–136.

Bengtsson, M., Reinholt, F.P., Rydstrom, J. (1992). Cellular localization and hormonal regulation of 7,12-dimethylbenzanthracene mono-oxygenase activity in the rat ovary. *Toxicology*, *71*, 203–222.

Bengtsson, M., Rydstrom, J. (1983). Regulation of carcinogen metabolism in the rat ovary by the estrous cycle and gonadotropin. *Science*, *219*, 1437–1438.

Berenbaum, M.R. (1991). Coumarins, in *Herbivores: Their Interactions with Secondary Plant Metabolites, the Chemical Participants*, Vol. 1 (eds G.A. Rosenthal and M.R. Berenbaum), Academic Press, New York, pp. 221–249.

Bevan, C., Stadler, J.C., Elliot, G.S., Frame, S.R., Baldwin, J.K., Leung, H.W., Moran, E., Panepinto, A.S. (1996). Subchronic toxicity of 4-vinylcyclohexene in rats and mice by inhalation exposure. *Fundam. Appl. Toxicol.*, *32*, 1–10.

Blount, B.C., Silva, M.J., Caudill, S.P., Needham, L.L., Pirkle, J.L., Sampson, E.J., Lucier, G.W., Jackson, R.J., Brock, J.W. (2000). Levels of seven urinary phthalate metabolites in a human reference population. *Environ. Health Perspect.*, *108*, 979–982.

Blumenfeld, Z., Dann, E., Avivi, I., Epelbaum, R., Rowe, J.M. (2002). Fertility after treatment for Hodgkins disease. *Ann. Oncol.*, *13*, 138–147.

Boekelheide, K. (2005). Mechanisms of toxic damage to spermatogenesis. *J. Natl. Cancer Inst. Monogr.*, *34*, 6–8.

Bookstaff, R.C., Kamel, F., Moore, R.W., Bjerke, D.L., Peterson, R.E. (1990). Altered regulation of pituitary gonadotropin-releasing hormone (GnRH) receptor number and pituitary responsiveness to GnRH in 2,3,7,8-tetrachlorodibenzo-p-dioxin-treated male rats. *Toxicol. Appl. Pharmacol.*, *105* (1), 78–92.

Bookstaff, R.C., Moore, R.W., Peterson, R.E. (1990). 2,3,7,8-tetrachlorodibenzo-p-dioxin increases the potency of androgens and estrogens as feedback inhibitors of luteinizing hormone secretion in male rats. *Toxicol. Appl. Pharmacol.*, *104* (2), 212–224.

Borgeest, C., Symonds, D., Mayer, L.P., Hoyer, P.B., Flaws, J.A. (2002). Methoxychlor may cause ovarian follicular atresia and proliferation of the ovarian epithelium in the mouse. *Toxicol. Sci.*, *68*, 473–478.

Borman, S.M., Christian, P.J., Sipes, I.G., Hoyer, P.B. (2000). Ovotoxicity in female Fischer rats and B6 mice induced by low-dose exposure to three polycyclic aromatic hydrocarbons: comparison through calculation of an ovotoxic index. *Toxicol. Appl. Pharmacol.*, *167*, 191–198.

Borman, S.M., VanDePol, B.J., Kao, S., Thompson, K.E., Sipes, I.G., Hoyer, P.B. (1999). A single dose of the ovotoxicant 4-vinylcyclohexene diepoxide is protective in rat primary ovarian follicles. *Toxicol. Appl. Pharmacol.*, *158*, 244–252.

Botta, J.A., Jr, Hawkins, H.C., Weikel, J.H., Jr (1974). Effects of cyclophosphamide on fertility and general reproductive performance in rats. *Toxicol. Appl. Pharmacol.*, *27*, 602–611.

Buccione, R., Schroeder, A.C., Eppig, J.J. (1990). Interactions between somatic cells and germ cells throughout mammalian oogenesis. *Biol. Reprod.*, *43*, 543–547.

Buhrley, L.E., Ellis, L.C. (1973). Catalase activity in rat testicular preparations: distribution and changes in activity induced by hypophysectomy and x-irradiation. *Fertil. Steril.*, *24*, 956–961.

Byrne, J. (1990). Fertility and pregnancy after malignancy. *Semin. Perinatol.*, *14*, 423–429.

Calafat, A.M., McKee, R.H. (2006). Integrating biomonitoring exposure data into the risk assessment process: phthalates [diethyl phthalate and di(2-ethylhexyl) phthalate] as a case study. *Environ. Health Perspect.*, *114*, 1783–1789.

Calzavara-Pinton, P., Carlino, A., Manfedi, E., Semeraro, F., Zane, C., Panfilis, G. (1994). Ocular side-effects of PUVA-treated patients refusing eye sun protection. *Acta Derm. Venereol. Suppl. (Stockh)*, *186*, 164–165.

Cannady, E.A., Dyer, C.A., Christian, P.J., Sipes, I.G., Hoyer, P.B. (2002). Expression and activity of microsomal epoxide hydrolase in follicles isolated from mouse ovaries. *Toxicol. Sci.*, *68*, 24–31.

Chadwick, C.A., Potten, C.S., Cohen, A.J., Young, A.R. (1994). The time of onset and duration of 5-methoxypsoralen photochemoprotection from UVR-induced DNA damage in human skin. *Br. J. Dermatol.*, *131*, 483–494.

Chapin, R.E., Gray, T.J., Phelps, J.L., Dutton, S.L. (1988). The effects of mono-(2-ethylhexyl)-phthalate on rat Sertoli cell-enriched primary cultures. *Toxicol. Appl. Pharmacol.*, *92*, 467–479.

Chapman, R.M. (1983). Gonadal injury resulting from chemotherapy. *Am. J. Ind. Med.*, *4*, 149–161.

Citti, L., Gervasi, P.G., Turchi, G., Bellucci, G., Bianchini, R. (1984). The reaction of 3,4-epoxy-1-butene with deoxyguanosine and DNA in vitro: synthesis and characterization of the main adducts. *Carcinogenesis*, *5*, 47–52.

Cole, R.S. (1970). Light-induced cross-linking of DNA in the presence of a furocoumarin (psoralen). Studies with phase lambda, Escherichia coli and mouse leukemia cells. *Biochem. Biophys. Acta*, *217*, 30–39.

Constanzo, L.S. (1998). Reproductive physiology, in *Physiology* (eds W.R. Schmitt and A.M. Shaw), W.B. Saunders Co., Philadelphia, PA, pp. 393–411.

Cooper, R.L., Laws, S.C., Das, P.C., Narotsky, M.G., Goldman, J.M., Lee Tyrey, E., Stoker, T.E. (2007). Atrazine and reproductive function: mode and mechanism of action studies. *Birth Defects Res. B Dev. Reprod. Toxicol.*, *80*, 98–112.

Damewood, M.D., Grochow, L.B. (1986). Prospects for fertility after chemotherapy or radiation for neoplastic disease. *Fertil. Steril.*, *45*, 443–459.

Davis, B.J., Maronpot, R.R., Heindel, J.J. (1994). Di-(2-ethylhexyl)phthalate suppresses estradiol and ovulation in cycling rats. *Toxicol. Appl. Pharmacol.*, *128*, 216–223.

De Jager, C., Farias, P., Barraza-Villarreal, A., Avila, M.H., Ayotte, P., Dewailly, E., Dombrowski, C., Rousseau, F., Sanchez, V.D., Bailey, J.L. (2006). Reduced seminal parameters breakthroughs in andrology associated with environmental DDT exposure and p,p′-DDE concentrations in men in Chiapas, Mexico: a cross-sectional study. *J. Androl.*, *27* (1), 16–27.

Dearth, R.K., Hiney, J.K., Srivastave, V., Burdick, S.B., Bratton, G.R., Dees, W.L. (2002). Effects of lead (Pb) exposure during gestation and lactation on female prepubertal development in the rat. *Reprod. Toxicol.*, *16*, 343–352.

DeRaat, W.K. (1978). Induction of sister chromatid exchanges by styrene and its presumed metabolite styrene oxide in the presence of rat liver homogenate. *Chem. Biol. Interact.*, *20*, 163–170.

Devine, P.J., Hoyer, P.B. (2005). Ovotoxic environmental chemicals: indirect endocrine disruptors, in *Endocrine Disruptors: Effects on Male and Female Reproductive System* (ed. R. Nog), CRC Press, Boca Raton, FL, pp. 67–100.

Diawara, M.M., Chavez, K.J., Hoyer, P.B., Williams, D.E., Dorsch, J., Kulkosky, P., Franklin, M.R. (1999). A novel group of ovarian toxicants: the psoralens. *J. Biochem. Mol. Toxicol.*, *13* (3/4), 195–203.

Diawara, M.M., Chavez, K.J., Simpleman, D., Williams, D.E., Franklin, M.R., Hoyer, P.B. (2001). The psoralens adversely affect reproductive function in male wistar rats. *Reprod. Toxicol.*, *15*, 137–144.

Diawara, M.M., Trumble, J.T. (1997). Linear furanocoumarins, in *Handbook on Plant and Fungal Toxicants* (ed. J.P.F. D'Mello), CRC Press, Boca Raton, FL, pp. 175–189.

Dnistrian, A.M., Schwartz, M.K., Fracchia, A.A., Kaufman, R.J., Hakes, T.B., Currie, V.E. (1983). Endocrine consequences of CMF adjuvant therapy in premenopausal and postmenopausal breast cancer patients. *Cancer*, *51*, 803–807.

Docampo, R. (1987). Male contraception and chemotherapy for trypanosomiasis and leishmaniasis: an analysis of biochemical similarities and considerations for future research, in *Cellular and Molecular Events in Spermiogenesis*, Cambridge University Press, Oaxtepec, Mexico.

Doerr, J.K., Hooser, S.B., Smith, B.J., Sipes, I.G. (1995). Ovarian toxicity of 4-vinylcyclohexene and related olefins in B6C3F1 mice: role of diepoxides. *Chem. Res. Toxicol.*, *8*, 963–969.

Drbohlav, P., Bencko, V., Masata, J., Bendl, J., Rezacova, J., Zouhar, T., Cerny, V., Halkova, E. (1998). Detection of cadmium and zinc in the blood and follicular fluid of women in the IVF and ET program. *Ceska Gynekol.*, *63*, 292–300.

Ehrsson, H., Wallin, I., Ros, A.M., Eksborg, S., Berg, M. (1994). Food-induced increase in bioavailability of 5-methoxypsoralen. *Eur. J. Clin. Pharmacol.*, *46*, 375–377.

Eldridge, J.C., Wetzel, L.T., Tyrey, L. (1999). Estrous cycle patterns of Sprague-Dawley rats during acute and chronic atrazine administration. *Reprod. Toxicol.*, *13*, 491–199.

Eroschenko, V.P., Abuel-Atta, A.A., Grober, M.S. (1995). Neonatal exposures to technical methoxychlor alters ovaries in adult mice. *Reprod. Toxicol.*, *9*, 379–387.

Fan, L.Q., Cattley, R.C., Corton, J.C. (1998). Tissue-specific induction of 17 beta-hydroxysteroid dehydrogenase type IV by peroxisome proliferator-activated receptor alpha. *J. Endocrinol.*, *158*, 237–246.

Fay, M., Donohue, J.M., DeRosa, C. (1999). ATSDR evaluation of health effects of chemicals. VI. Di(2-ethylhexyl)phthalate. Agency for toxic substances and disease registry. *Toxicol. Ind. Health*, *15*, 651–746.

Fernandez, E., La Vecchia, C., Braga, C., Talamini, R., Negri, E., Parazzini, F., Franceschi, S. (1998). Hormone replacement therapy and risk of colon and rectal cancer. *Cancer Epidemiol. Biomarkers Prev.*, *7*, 329–333.

Ferrara, D., Hallmark, N., Scott, H., Brown, R., McKinnell, C., Mahood, I.K., Sharpe, R.M. (2006). Acute and long-term effects of in utero exposure of rats to di(n-butyl) phthalate on testicular germ cell development and proliferation. *Endocrinology*, *147* (11), 5352–5362.

Flaws, J.A., Salyers, K.L., Sipes, I.G., Hoyer, P.B. (1994). Reduced ability of rat preantral ovarian follicles to metabolize 4-vinyl-1-cyclohexene diepoxide in vitro. *Toxicol. Appl. Pharmacol.*, *126*, 286–294.

Foster, P.M. (1988). M-dinitrobenzene: studies on its toxicity to the testicular sertoli cell. *Arch. Toxicol.* Suppl., *13*, 3–17.

Garcia, C.R. (2007). Fertility risk in pediatric and adolescent cancers. XVIth Ovarian Workshop. Ovarian Differentiation, development, function and persistence. San Antonio, TX, Biosymposia, Inc.

Generoso, W.M., Stout, S.K., Huff, S.W. (1971). Effects of alkylating chemicals on reproductive capacity of adult female mice. *Mutat. Res.*, *13*, 172–184.

Georgellis, A., Parvinen, M., Rydström, J. (1989). Inhibition of stage-specific DNA synthesis in rat spermatogenic cells by polycyclic aromatic hydrocarbons. *Chem. Biol. Interact.*, *72*, 79–92.

Georgellis, A., Toppari, J., Veromaa, T., Rydstrom, J., Parvinen, M. (1990). Inhibition of meiotic divisions of rat spermatocytes in vitro by polycyclic aromatic hydrocarbons. *Mutat. Res.*, *231*, 125–135.

Godowicz, P., Pawlus, M. (1985). Effect of cadmium chloride on the ovulation and structure of ovary in the inbred KP and CBA mice strains. *Folia Histochem.*, *23*, 209–215.

Gray, L.E., Jr, Ostby, J., Monosson, E., Kelce, W.R. (1999). Environmental antiandrogens: low doses of the fungicide vinclozolin alter sexual differentiation of the male rat. *Toxicol. Ind. Health*, *15*, 48–64.

Grekin, D.A., Epstein, J.H. (1981). Psoralens, UVA (PUVA) and photocarcinogenesis. *Photochem. Photobiol.*, *33*, 957–960.

Grizzle, T.B., George, J.D., Fail, P.A., Seely, J.C., Heindel, J.J. (1994). Reproductive effects of 4-vinylcyclohexene in Swiss mice assessed by a continuous breeding protocol. *Fundam. Appl. Toxicol.*, *22*, 122–129.

Guillette, L.J., Jr, Pickford, D.B., Crain, D.A., Rooney, A.A., Percival, H.F. (1995). Reduction in penis size and plasma testosterone concentrations in juvenile alligators living in a contaminated environment. *Gen. Comp. Endocrinol.*, *101*, 32–42.

Gupta, R.K., Miller, K.P., Babus, J.K., Flaws, J.A. (2006a). Methoxychlor inhibits growth and induces atresia of antral follicles through an oxidative stress pathway. *Toxicol. Sci.*, *93*, 382–389.

Gupta, R.K., Schuh, R.A., Fiskum, G., Flaws, J.A. (2006b). Methoxychlor causes mitochondrial dysfunction and oxidative damage in the mouse ovary. *Toxicol. Appl. Pharmcol.*, *216*, 436–445.

Heindel, J.J., Chapin, R.E. (1989). Inhibition of FSH-stimulated cAMP accumulation by mono(2-ethylhexyl) phthalate in primary rat sertoli cell cultures. *Toxicol. Appl. Pharmacol.*, *97*, 377–385.

Heinrichs, W.L., Juchau, M.R. (1980). *Extrahepatic Metabolism of Drugs and Other Foreign Compounds*, Vol. 1, SP Medical and Scientific Books, New York, p. 319.

Hendry, W.J., DeBrot, B.L., Zheng, X.L., Branham, W.S., Sheehan, D.M. (1999). Differential activity of diethylstilbestrol versus estradiol as neonatal endocrine disrupters in the female hamster (Mesocricetus auratus) reproductive tract. *Biol. Reprod.*, *61*, 91–100.

Herbert, D.C., Suprakar, P.C., Roy, A.K. (1995). Male reproduction, in *Reproductive Toxicology* (ed. R.J. Witorsch), Raven Press, New York, pp. 3–22.

Herbst, A.L., Ulfelder, H., Poskanzer, D.C. (1971). Adenocarcinoma of the vagina. Association of maternal stilbestrol therapy with tumor appearance in young women. *N. Engl. J. Med.*, *284*, 878–881.

Hirshfield, A.N. (1991). Development of follicles in the mammalian ovary. *Int. Rev. Cytol.*, *124*, 43–101.

Homma-Takeda, S., Kugenuma, Y., Iwamuro, T., Kumagai, Y., Shimojo, N. (2001). Impairment of spermatogenesis in rats by methylmercury: involvement of stage- and cell-specific germ cell apoptosis. *Toxicology*, *169*, 25–35.

Hooser, S.B., DeMerell, D.G., Douds, D.A., Hoyer, P.B., Sipes, I.G. (1995). Testicular germ cell toxicity caused by vinylcyclohexene diepoxide in mice. *Reprod. Toxicol.*, *9* (4), 359–367.

Hooser, S.B., Douds, D.A., Hoyer, P.B., Sipes, I.G. (1994). Long-term ovarian and hormonal alterations due to the ovotoxin, 4-vinylcyclohexene. *Reprod. Toxicol.*, *8*, 315–323.

Howell, S.J., Shalet, S.M. (2002a). Testicular function following chemotherapy. *Hum. Reprod. Update*, *7*, 363–369.

Howell, S.J., Shalet, S.M. (2002b). Effect of cancer therapy on pituitary-testicular axis. *Int. J. Androl.*, *25*, 269–276.

Hoyer, P.B. (1997). Female reproductive toxicology: introduction and overview, in *Comprehensive Toxicology: Reproductive and Endocrine Toxicology* (eds K. Boekelheide, R. Chapin, P.B. Hoyer, C. Harris, I.G. Sipes, C.A. McQueen and A.J. Gandolfi), Elsevier, p. 10.

Hoyer, P.B. (2001). Reproductive toxicology: current and future directions. *Biochem. Pharmacol.*, *62*, 1557–1564.

Hoyer, P.B. (2005). Impact of metals on ovarian function, in *Metals, Fertility and Reproductive Toxicity*, CRC Press, Boca Raton, FL, pp. 155–173.

Hoyer, P.B., Devine, P.J. (2002). Endocrine toxicology: the female reproductive system, in *Handbook of Toxicology* (eds M.J. Derelanko and M.A. Hollinger), CRC Press, pp. 573–595.

Hoyer, P.B., Sipes, I.G. (1996). Assessment of follicle destruction in chemical-induced ovarian toxicity. *Ann. Rev. Pharmacol. Toxicol.*, *36*, 307–331.

Hoyer, P.B., Sipes, I.G. (2007). Development of an animal model for ovotoxicity using 4-vinylcyclohexene diepoxide: a case study. *Birth Defects Res. Part B Dev. Reprod. Toxicol.*, *80*, 113–125.

Hsu, P.C., Pan, M.H., Li, L.A., Chen, C.J., Tsai, S.S., Guo, Y.L. (2007). Exposure in utero to 2,2',3,3',4,6'-hexachlorobiphenyl (PCB 132) impairs sperm function and alters testicular apoptosis-related gene expression in rat offspring. *Toxicol. Appl. Pharmacol.*, *221*, 68–75.

Hu, X., Christian, P.J., Sipes, I.G., Hoyer, P.B. (2002a). Expression and redistribution of cellular Bad, Bax, and Bcl-X(L) protein is associated with VCD-induced ovotoxicity in rats. *Biol. Reprod.*, *65*, 1489–1495.

Hu, X., Flaws, J.A., Sipes, I.G., Hoyer, P.B. (2002b). Activation of mitogen-activated protein kinases and AP-1 transcription factor in ovotoxicity induced by 4-vinyl cyclohexene diepoxide in rats. *Biol. Reprod.*, *67*, 718–724.

Huang, H.F., Yang, C.S., Meyenhofer, M., Gould, S., Boccabella, A.V. (1988). Disruption of sustentacular (Sertoli) cell tight junctions and regression of spermatogenesis in vitamin-A-deficient rats. *Acta Anat. (Basel)*, *133*, 10–15.

Huyghe, E., Matsuda, T., Daudin, M., Chevreau, C., Bachaud, J.M., Plante, P., Bujan, L., Thonneau, P. (2004). Fertility after testicular cancer treatments: results of a large multicenter study. *Cancer*, *100*, 732–737.

Ichihara, G., Yu, X., Kitoh, J., Asaeda, N., Kumazawa, T., Iwai, H., Shibata, E., Yamada, T., Wang, H., Xie, Z., Maeda, K., Tsukmura, H., Takeuchi, Y. (2000). Reproductive toxicity of 1-bromopropane, a newly introduced alternative to ozone layer depleting solvents, in male rats. *Toxicol. Sci.*, *54*, 416–423.

Ikezuki, Y., Tsutsumi, O., Takai, Y., Kamei, Y., Taketani, Y. (2002). Determination of bisphenol A concentrations in human biological fluids reveals significant early prenatal exposure. *Hum. Reprod.*, *17*, 2839–2841.

International Agency for Research on Cancer (IARC) (1994). 4-Vinylcyclohexene. IARC monographs on the evaluation of carcinogenic risks to humans: some industrial chemicals. Lyon, France. *60*, 347.

Ishikawa, H., Tian, Y., Yamauchi, T. (2001). Induction of micronuclei formation in preimplantation mouse embryos after maternal treatment with 2-bromopropane. *Reprod. Toxicol.*, *15*, 81–85.

James, M.O., Fouts, J.R., Bend, J.R. (1976). Hepatic and extrahepatic metabolism, in vitro, of an epoxide (8-14C-styrene oxide) in the rabbit. *Biochem. Pharmacol.*, *25*, 187–193.

Jarrell, J., Lai, E.V., Barr, R., McMahon, A., Belbeck, L., O'Connell, G. (1987). Ovarian toxicity of cyclophosphamide alone and in combination with ovarian irradiation in the rat. *Cancer Res.*, *47*, 2340–2343.

Jarrell, J.F., Bodo, L., Younglai, E.V., Barr, R.D., O'Connell, G.J. (1991). The short-term reproductive toxicity of cyclophosphamide in the female rat. *Reprod. Toxicol.*, *5*, 481–485.

Jick, H., Porter, J. (1977). Relation between smoking and age of natural menopause. *Lancet*, *1*, 1354–1355.

Jurasovic, J., Cvitkovic, P., Pizent, A., Colak, B., Telisman, S. (2004). Semen quality and reproductive endocrine function with regard to blood cadmium in Croatian male subjects. *Biometals*, *17*, 735–743.

Kabuto, H., Amakawa, M., Shishibori, T. (2004). Exposure to bisphenol A during embryonic/fetal life and infancy increases oxidative injury and causes underdevelopment of the brain and testis in mice. *Life Sci.*, *74*, 2931–2940.

Kang, K.S., Li, G.X., Che, J.H., Lee, Y.S. (2002). Impairment of male rat reproductive function in F1 offspring from dams exposed to 2-bromopropane during gestation and lactation. *Reprod. Toxicol.*, *16*, 151–159.

Kato, H., Furuhashi, T., Tanaka, M., Katsu, Y., Watanabe, H., Ohta, Y., Iguchi, T. (2006). Effects of bisphenol A given neonatally on reproductive function of male rats. *Reprod. Toxicol.*, *22*, 20–29.

Kato, H., Ota, T., Furuhashi, T., Ohta, Y., Iguchi, T. (2003). Changes in reproductive organs of female rats treated with bisphenol A during the neonatal period. *Reprod. Toxicol.*, *17*, 283–288.

Kavlock, R., Boekelheide, K., Chapin, R., Cunningham, M., Faustman, E., Foster, P., Golub, M., Henderson, R., Hinberg, I., Little, R., Seed, J., Shea, K., Tabacova, S., Tyl, R., Williams, P., Zacharewski, T. (2002). NTP center for the evaluation of risks to human reproduction: phthalates expert panel report on the reproductive and developmental toxicity of di(2-ethylhexyl) phthalate. *Reprod. Toxicol.*, *16*, 529–653.

Keller, D.A., Carpenter, S.C., Cagen, S.Z., Reitman, F.A. (1997). In vitro metabolism of 4-vinylcyclohexene in rat and mouse liver, lung, and ovary. *Toxicol. Appl. Pharmacol.*, *144*, 36–44.

Kluwe, W.M., Lamb, J.C., IV, Greenwell, A.E., Harrington, F.W. (1983). 1,2-dibromo-3-chloropropane (DBCP)-induced infertility in male rats mediated by a post-testicular effect. *Toxicol. Appl. Pharmacol.*, *71*, 294–298.

Kopf-Maier, P. (1992). Effects of carboplatin on the testis. A histological study. *Cancer Chemother. Pharmacol.*, *29*, 227–235.

Koyama, H., Wada, T., Nichizawa, Y., Iwanaga, T., Aoki, Y. (1977). Cyclophosphamide-induced ovarian failure and its therapeutic significance in patients with breast cancer. *Cancer, 39,* 1403–1409.

Krarup, T. (1967). 9,10-Dimethyl-1,2-benzanthracene induced ovarian tumors in mice. *Acta Pathol. Microbiol. Scand., 70,* 241–248.

Krarup, T. (1969). Oocyte destruction and ovarian tumorigenesis after direct application of a chemical carcinogen (9,10-dimethyl-benzanthracene) to the mouse ovary. *Int. J. Cancer, 4,* 61–75.

Krarup, T. (1970). Oocyte survival in the mouse ovary after treatment with 9,10-dimethyl-1,2-benzanthracene. *J. Endocrinol., 46,* 483–495.

Lamperti, A.A., Printz, R.H. (1974). Localization, accumulation and toxic effects of mercuric chloride on the reproductive axis of the female hamster. *Biol. Reprod., 11,* 180–186.

Laws, S.C., Ferrell, J.M., Stoker, T.E., Cooper, R.L. (2003). Prepubertal development in female Wistar rats following exposure to propazine and atrazine biotransformation by-products, diamino-S-chlorotriazine and hydroxyatrazine. *Toxicol. Sci., 76,* 190–200.

Li, X., Johnson, D.C., Rozman, K.K. (1995). Reproductive effects of 2,3,7,8-tetrachlorodibenzo-p-dioxin (TCDD) in female rats: ovulation, hormonal regulation, and possible mechanism(s). *Toxicol. Appl. Pharmacol., 133,* 321–327.

Lindström, P., Harris, M., Ross, M., Lamb, J.C., IV, Chapin, R.E. (1988). Comparison of changes in serum androgen binding protein with germinal epithelial damage and infertility induced by di-n-pentyl phthalate. *Fundam. Appl. Toxicol., 11,* 528–539.

Lovekamp, T.N., Davis, B.J. (2001). Mono-(2-ethylhexyl) phthalate suppresses aromatase transcript levels and estradiol production in cultured rat granulosa cells. *Toxicol. Appl. Pharmacol., 172,* 217–224.

Lovekamp-Swan, T., Davis, B.J. (2003). Mechanisms of phthalate ester toxicity in the female reproductive system. *Environ. Health Perspect., 111,* 139–145.

Lovekamp-Swan, T., Jetten, A.M., Davis, B.J. (2003). Dual activation of PPARalpha and PPAR-gamma by mono-(2-ethylhexyl) phthalate in rat ovarian granulosa cells. *Mol. Cell. Endocrinol., 201,* 133.

Lymberopoulos, A.G., Kotsaki-Kovatsi, V.P., Taylor, A., Papaioannou, N., Brikas, P. (2000). Effects of cadmium chloride administration on the macroscopic and microscopic characteristics of ejaculates from Chios ram-lambs. *Theriogenology, 54,* 1145–1157.

Mably, T.A., Bjerke, D.L., Moore, R.W., Gendron-Fitzpatrick, A., Peterson, R.E. (1992). In utero and lactational exposure of male rats to 2,3,7,8-tetrachlorodibenzo-p-dioxin. 3. Effects on spermatogenesis and reproductive capacity. *Toxicol. Appl. Pharmacol., 114* (1), 118–126.

Mably, T.A., Moore, R.W., Peterson, R.E. (1992). In utero and lactational exposure of male rats to 2,3,7,8-tetrachlorodibenzo-p-dioxin. 1. Effects on androgenic status. *Toxicol. Appl. Pharmacol., 114* (1), 97–107.

MacKenzie, K.M., Angevine, D.M. (1981). Infertility in mice exposed in utero to benzo[a]pyrene. *Biol. Reprod., 24,* 183–191.

Markey, C.M., Rubin, B.S., Soto, A.M., Sonnenschein, C. (2002). Endocrine disruptors: from wingspread to environmental developmental biology. *J. Steroid. Biochem. Mol. Biol., 83,* 235–244.

Maronpot, R.R. (1987). Ovarian toxicity and carcinogenicity in eight recent National Toxicological Program studies. *Environ. Health Perspect., 73,* 125–130.

Martini, F.H., Bartholomew, E.F. (1997). The reproductive system of the male, in *Essentials of Anatomy and Physiology* (ed. D.K. Brake), Prentice Hall, New Jersey, pp. 516–523.

Mattison, D.R. (1980). Morphology of oocyte and follicle destruction by polycyclic aromatic hydrocarbons in mice. *Toxicol. Appl. Pharmacol., 53,* 249–259.

Mattison, D.R. (1983). The mechanisms of action of reproductive toxins. *Am. J. Ind. Med., 4,* 65–79.

Mattison, D.R. (1985). Clinical manifestation of ovarian toxicity, in *Reproductive Toxicology* (ed. R.L. Dixon), Raven, New York, pp. 109–130.

Mattison, D.R., Nightingale, M.R. (1980). The biochemical and genetic characteristics of murine aryl ovarian hydrocarbon (benzo[a]pyrene) hydroxylase activity and its relationship to primordial oocyte destruction by polycyclic aromatic hydrocarbons. *Toxicol. Appl. Pharmacol.*, 56, 399–408.

Mattison, D.R., Schulman, J.D. (1980). How xenobiotic chemicals can destroy oocytes. *Contemp. Obstet. Gynecol.*, 15, 157.

Mattison, D.R., Thorgeirsson, S.S. (1977). Genetic differences in mouse ovarian metabolism of benzo[a]pyrene and oocyte toxicity. *Biochem. Pharmacol.*, 26, 909–912.

Mattison, D.R., Thorgeirsson, S.S. (1978). Gonadal aryl hydrocarbon hydroxylase in rats and mice. *Cancer Res.*, 38, 1368–1373.

Mattison, D.R., Thorgeirsson, S.S. (1979). Ovarian aryl hydrocarbon hydroxylase activity and primordial oocyte toxicity of polycyclic aromatic hydrocarbons in mice. *Cancer Res.*, 39, 3471–3475.

Mattison, D.R., Plowchalk, B.S., Meadows, M.J., Miller, M.M., Malek, A., London, S. (1989). The effect of smoking on oogenesis, fertilization, and implantation. *Semin. Reprod. Endocrinol.*, 7, 291.

Mayer, L.P., Pearsall, N.A., Christian, P.J., Devine, P.J., Payne, C.M., McCuskey, M.K., Marion, S.L., Sipes, I.G., Hoyer, P.B. (2002). Long-term effects of ovarian follicular depletion in rats by 4-vinylcyclohexene diepoxide. *Reprod. Toxicol.*, 16, 775–781.

Meistrich, M.L., Finch, M., da Cunha, M.F., Hacker, U., Au, W.W. (1982). Damaging effects of fourteen chemotherapeutic drugs on mouse testis cells. *Cancer Res.*, 42, 122–131.

Melnick, R.L., Huff, J., Chou, B.J., Miller, R.A. (1990). Carcinogenicity of 1,3-butadiene in C57BL/6 X C3HF1 mice at low exposure concentrations. *Cancer Res.*, 50, 6592–6599.

Miller, J.J., Cole, L.J. (1970). Changes in mouse ovaries after prolonged treatment with cyclophosphamide. *Proc. Soc. Exp. Biol. Med.*, 133, 190–193.

Miller, J.J., Williams, G.F., Leissring, J.C. (1971). Multiple late complications of therapy with cyclophosphamide, including ovarian destruction. *Am. J. Med.*, 50, 530–535.

Miller, K.P., Gupta, R.K., Flaws, J.A. (2006). Methoxychlor metabolites may cause ovarian toxicity through estrogen-regulated pathways. *Toxicol. Sci.*, 93, 180–188.

Miller, K.P., Gupta, R.K., Greenfeld, C.R., Babus, J.K., Flaws, J.A. (2005). Methoxychlor directly affects ovarian antral follicle growth and atresia through Bcl-2- and Bax-mediated pathways. *Toxicol. Sci.*, 88, 213–221.

Mlynarcikova, A., Fickova, M., Scsukova, S. (2005). Ovarian intrafollicular processes as a target for cigarette smoke components and selected environmental reproductive disruptors. *Endocr. Regul.*, 39, 20–31.

Moran, F.M., Enan, E., Vandevoort, C.A., Stewart, D.R., Conley, A.J., Overstreet, J.W., Lasley, B.L. (1997). 2,3,7,8-Tetrachlorodibenzo-p-dioxin (TCDD) effects on steroidogenesis of human luteinized granulosa cells in vitro. *Biol. Reprod. Suppl.*, 56, 65.

Morrissey, R.E., Schwetz, B.A., Hackett, P.L., Sikov, M.R., Hardin, B.D., McClanahan, B.J., Decker, J.R., Mast, T.J. (1990). Overview of reproductive and developmental toxicity studies of 1,3-butadiene in rodents. *Environ. Health Perspect.*, 86, 79–84.

Mukhtar, H., Philpot, R.M., Bend, J.R. (1978). The postnatal development of microsomal epoxide hydrase, cytosolic glutathione S-transferase, and mitochondrial and microsomal cytochrome P-450 in adrenals and ovaries of female rats. *Drug Metab. Dispos.*, 6, 577–583.

Murono, E.P., Derk, R.C. (2005). The reported active metabolite of methoxychlor, 2,2-bis(p-hydroxyphenyl)-1,1,1-trichloroethane, inhibits testosterone formation by cultured Leydig cells from neonatal rats. *Reprod. Toxicol.*, 20, 503–513.

Murono, E.P., Derk, R.C., Akgul, Y. (2006). In vivo exposure of young adult male rats to methoxychlor reduces serum testosterone levels and ex vivo Leydig cell testosterone formation and cholesterol side-chain cleavage activity. *Reprod. Toxicol.*, *21*, 148–153.

Murray, R.D.H., Mendez, J., Brown, S.A. (1982). *The Natural Coumarins*, John Wiley & Sons, Ltd, Chichester.

Musajo, L., Rodighiero, G. (1962). The skin-photosensitizing furanocoumarins. *Experientia*, *18*, 153–200.

Mylchreest, E., Wallace, D.G., Cattley, R.C., Foster, P.M. (2000). Dose-dependent alterations in androgen-regulated male reproductive development in rats exposed to di(n-butyl) phthalate during late gestation. *Toxicol. Sci.*, *55*, 143–151.

Naha, N., Chowdhury, A.R. (2006). Inorganic lead exposure in battery and paint factory: effect on human sperm structure and functional activity. *J. UOEH*, *28* (2), 157–171.

Nath, R., Prasad, R., Palinal, V.K., Chopra, R.K. (1984). Molecular basis of cadmium toxicity. *Prog. Food Nutr. Sci.*, *8*, 109–163.

National Toxicology Program (1986). Toxicology and carcinogenesis studies of 4-vinylcyclohexene in F344/N rats and B6C3F1 mice. NTP Tech. Rep. No. 303. Research Triangle Park, NC., US Dep. Health Hum. Serv. Public Health Serv. Natl. Inst. Health. Public Inf. Natl. Toxicol. Program.

National Toxicology Program (1989). Toxicology and carcinogenesis studies of 4-vinylcyclohexene diepoxide in F344/N rats and B6C3F1 mice. NTP technical report: 362. Research Triangle Park, NC., US Dep. Health Hum. Serv. Public Health Serv. Natl. Inst. Health. Public Inf. Natl. Toxicol. Program.

Neal, M.S., Zhu, J., Holloway, A.C., Foster, W.G. (2007). Follicle growth is inhibited by benzo-[a]-pyrene at concentrations representative of human exposure, in an isolated rat follicle culture assay. *Hum. Reprod.*, *22*, 961–967.

Omura, M., Romero, Y., Zhao, M., Inoue, N. (1999). Histopathological evidence that spermatogonia are the target cells of 2-bromopropane. *Toxicol. Lett.*, *104*, 19–26.

Orisakwe, O.E., Afonne, O.J., Nwobodo, E., Asomugha, L., Dioka, C.E. (2001). Low-dose mercury induces testicular damage protected by zinc in mice. *Eur. J. Obstet. Gynecol. Reprod. Biol.*, *95*, 92–96.

Pant, N., Upadhyay, G., Pandey, S., Mathur, N., Saxena, D.K., Srivastava, S.P. (2003). Lead and cadmium concentration in the seminal plasma of men in the general population: correlation with semen quality. *Reprod. Toxicol.*, *17*, 447–450.

Papaconstantinou, A.D., Umbreit, T.H., Fisher, B.R., Goering, P.L., Lappas, N.T., Brown, K.M. (2000). Bisphenol A-induced increase in uterine weight and alterations in uterine morphology in ovariectomized B6C3F1 mice: role of the estrogen receptor. *Toxicol. Sci.*, *56*, 332–339.

Park, J., Kim, Y., Park, D., Choi, K., Park, S., Moon, Y. (1997). An outbreak of hemapoietic and reproductive disorders due to solvents containing 2-bromopropane in an electronic factory, South Korea: epidemiological study. *J. Occup. Health*, *39*, 138–143.

Pathak, M.A., Fitzpatrick, T.B. (1992). The evolution of photochemotherapy with psoralens and UVA (PUVA): 200 BC to 1992 AD. *J. Photochem. Photobiol. B*, *14*, 3–33.

Petersen, P.M., Hansen, S.W., Giwercman, A., Rørth, M., Skakkebaek, N.E. (1994). Dose-dependent impairment of testicular function in patients treated with cisplatin-based chemotherapy for germ cell cancer. *Ann. Oncol.*, *5*, 355–358.

Peterson, R.E., Cooke, P.S., Klece, W.R., Gray, L.E. (1997). Environmental endocrine disruptors, in *Comprehensive Toxicology*, Vol. 10 (eds I.G. Sipes, C.A. McQueen and A.G. Gandolfi), Elsevier Science, New York, pp. 181–191.

Piasek, M., Blanusa, M., Kostial, K., Laskey, J.W. (2001). Placental cadmium and progesterone concentrations in cigarette smokers. *Reprod. Toxicol.*, *15*, 673–681.

Piasek, M., Laskey, J.W. (1994). Acute cadmium exposure and ovarian steroidogenesis in cycling and pregnant rats. *Reprod. Toxicol.*, *8*, 495–507.

Piasek, M., Laskey, J.W. (1999). Effects of in vitro cadmium exposure on ovarian steroidogenesis in rats. *J. Appl. Toxicol.*, *19*, 211–217.

Piasek, M., Laskey, J.W., Kostial, K., Blanusa, M. (2002). Assessment of steroid disruption using cultures of whole ovary and/or placenta in rat and in human placental tissue. *Int. Arch. Occup. Environ. Health*, *75*, S36–44.

Piasek, M., Schonwald, N., Blanusa, M., Kostial, K., Laskey, J.W. (1996). Biomarkers of heavy metal reproductive effects and interaction with essential elements in experimental studies on female rats. *Arh. Hig. Rada Toksikol.*, *47*, 245–259.

Plowchalk, D.R., Mattison, D.R. (1992). Reproductive toxicity of cyclophosphamide in the C57GBL/6N mouse. 1. Effects on ovarian structure and function. *Reprod. Toxicol.*, *6*, 411–421.

Rajapaksa, K.S., Sipes, I.G., Hoyer, P.B. (2007). Involvement of microsomal epoxide hydrolase in ovotoxicity caused by 7,12-dimethylbenz[a]anthracene. *Toxicol. Sci.*, *96*, 327–334.

Rayner, J.L., Enoch, R.R., Wolf, D.C., Fenton, S.E. (2007). Atrazine-induced reproductive tract alterations after transplacental and/or lactational exposure in male Long-Evans rats. *Toxicol. Appl. Pharmacol.*, *218*, 238–248.

Revel, A., Laufer, N. (2002). Protecting female fertility from cancer therapy. *Mol. Cell. Endocrinol.*, *187*, 83–91.

Rhoads, R., Pflanzer, R. (1996). Reproductive physiology, in *Human Physiology* (ed. J.L. Alexander), Saunders College Publishing, Orlando, FL, pp. 900–950.

Richards, J.S. (1980). Maturation of ovarian follicles: actions and interactions of pituitary and ovarian hormones on follicular cell differentiation. *Physiol. Rev.*, *60*, 51–89.

Ronis, M.J., Badger, T.M., Shema, S.J., Roberson, P.K., Shaikh, F. (1996). Reproductive toxicity and growth effects in rats exposed to lead at different periods during development. *Toxicol. Appl. Pharmacol.*, *136*, 361–371.

Ronnback, C. (1991). Effect of 3,3',4,4'-tetrachlorobiphenyl (TCB) on ovaries of foetal mice. *Pharm. Toxicol.*, *69*, 340.

Safe, S., Krishnan, V. (1995). Chlorinated hydrocarbons: estrogens and antiestrogens. *Toxicol. Lett.*, *82*, 731–736.

Schrader, S.M., Kanitz, M.H. (1994). Occupational hazards to male reproduction, in *Occupational Medicine: State of the Art Reviews: Reproductive Hazards*, Vol. 9 (eds E. Gold, M. Schenker and B. Lasley), Hanley & Bilfus, Philadelphia, PA, pp. 405–414.

Schuurs, A.H. (1999). Reproductive toxicity of occupational mercury. A review of the literature. *J. Dent.*, *27*, 249–256.

Scott, B.R., Pathak, M.A., Mohm, G.R. (1976). Molecular and genetic basis of furanocoumarin reactions. *Mutat. Res.*, *39*, 29–74.

Seegmiller, R.E. (1997). Selected examples of developmental toxicants, in *Reproductive and Endocrine Toxicology*, Vol. 10 (eds K. Boekelheide, R.E. Chapin, P.B. Hoyer and C. Harris), Elsevier Science, New York, Chapter 45.

Sekiguchi, S., Asano, G., Suda, M., Honma, T. (2001). Influence of 2-bromopropane on reproductive system-short-term administration of 2-bromopropane inhibits ovulation in F344 rats. *Toxicol. Ind. Health*, *16*, 277–283.

Sherwood, L. (1997). The reproductive system, in *Human Physiology: From Cells to Systems* (ed. P. Lewis), Wadsworth Publishing Co., pp. 700–750.

Shiromizu, K., Mattison, D.R. (1984). The effect of intraovarian injection of benzo(a)pyrene on primordial oocyte number and ovarian aryl hydrocarbon [benzo(a)pyrene] hydroxylase activity. *Toxicol. Appl. Pharmacol.*, *76*, 18–25.

Smith, B.J., Mattison, D.R., Sipes, I.G. (1990). The role of epoxidation in 4-vinylcyclohexene-induced ovarian toxicity. *Toxicol. Appl. Pharmacol.*, *105*, 372–381.

Smith, T.L., Merry, S.T., Harris, D.L., Joe Ford, J., Ike, J., Archibong, A.E., Ramesh, A. (2007). Species-specific testicular and hepatic microsomal metabolism of benzo(a)pyrene, an ubiquitous toxicant and endocrine disruptor. *Toxicol. In Vitro*, *21*, 753–738.

Smith, W., Mitchell, P., Wang, J.J. (1997). Gender, oestrogen, hormone-replacement and age-related macular degeneration: results from the blue mountains eye study. *Aust. N. Z. J. Ophthalmol.*, *25*, S13–S15.

Sobrinho, L.G., Levine, R.A., DeConti, R.C. (1971). Amenhorrhea in patients with Hodgkin's disease treated with antineoplastic agents. *Am. J. Obstet. Gynecol.*, *109*, 135–139.

Son, H.Y., Kim, Y.B., Kang, B.H., Cho, S.W., Ha, C.S., Roh, J.K. (1999). Effects of 2-bromopropane on spermatogenesis in the Sprague-Dawley rat. *Reprod. Toxicol.*, *13* (3), 179–187.

Sowers, M.R., LaPietra, M.T. (1995). Menopause: its epidemiology and potential association with chronic diseases. *Epidemiol. Rev.*, *17*, 287–302.

Spano, M., Bartoleschi, C., Cordelli, E., Leter, G., Segre, L., Mantovani, A., Fazzi, P., Pacchierotti, F. (1996). Flow cytometric and histological assessment of 1,2:3,4-diepoxybutane toxicity on mouse spermatogenesis. *J. Toxicol. Environ. Health*, *47*, 423–441.

Springer, L.N., McAsey, M.E., Flaws, J.A., Tilly, J.L., Sipes, I.G., Hoyer, P.B. (1996). Involvement of apoptosis in 4-vinylcyclohexene diepoxide-induced ovotoxicity in rats. *Toxicol. Appl. Pharmacol.*, *139*, 394–401.

Stedman, R.L. (1968). The chemical composition of tobacco and tobacco smoke. *Chem. Rev.*, *68*, 153–207.

Stern, R.S., Thibodeau, L.A., Kleinerman, R.A., Parrish, J.A., Fitzpatrick, T.B. (1979). Risk of cutaneous carcinoma in patients treated with oral methoxsalen photochemotherapy for psoriasis. *N. Engl. J. Med.*, *300*, 809–813.

Stoker, T.E., Guidici, D.L., Laws, S.C., Cooper, R.L. (2002). The effects of atrazine metabolites on puberty and thyroid function in the male Wistar rat. *Toxicol. Sci.*, *67*, 198–206.

Sundaram, K., Witorsch, R.J. (1995). Toxic effects on the testes, in *Reproductive Toxicology* (ed. R.J. Witorsch), Raven Press, New York, pp. 99–122.

Susiarjo, M., Hassold, T.J., Freeman, E., Hunt, P.A. (2007). Bisphenol A exposure in utero disrupts early oogenesis in the mouse. *PLoS Genet.*, *3* (1), 63–70.

Tabacova, S., Little, R., Balabaeva, L. (1999). Maternal exposure to phthalates and complications of pregnancy. *Epidemiology*, *10*, S127.

Taketoh, J., Mutoh, J., Takeda, T., Ogishima, T., Takeda, S., Ishii, Y., Ishida, T., Yamada, H. (2007). Suppression of fetal testicular cytochrome P450 17 by maternal exposure to 2,3,7,8-tetrachlorodibenzo-p-dioxin: a mechanism involving an initial effect on gonadotropin synthesis in the pituitary. *Life Sci.*, *80*, 1259–1267.

Taupeau, C., Poupon, J., Nome, F., Lefevre, B. (2001). Lead accumulation in the mouse ovary after treatment-induced follicular atresia. *Reprod. Toxicol.*, *15*, 385–391.

Taupeau, C., Poupon, J., Treton, D., Brosse, A., Richard, Y., Machelon, V. (2003). Lead reduces messenger RNA and protein levels of cytochrome P450 aromatase and estrogen receptor beta in human ovarian granulosa cells. *Biol. Reprod.*, *68*, 1982–1988.

Telisman, S., Colak, B., Pizent, A., Jurasovic, J., Cvitkovic, P. (2007). Reproductive toxicity of low-level exposure in men. *Environ. Res.*, *105*, 256–266.

Tennent, B.J., Beamer, W.G. (1986). Ovarian tumors not induced by irradiation and gonadotropins in hypogonadal (hpg) mice. *Biol. Reprod.*, *34*, 751–769.

Thomas, C., Cans, C., Pelletier, R., De Robertis, C., Hazzouri, M., Sele, B., Rousseaux, S., Hennebicq, S. (2004). No long-term increase in sperm aneuploidy rates after anticancer therapy: sperm fluorescence in situ hybridization analysis in 26 patients treated for testicular cancer or lymphoma. *Clin. Cancer Res.*, *10*, 6535–6543.

Thomas, J.A. (1993). Toxic responses of the reproductive system, in *Toxicology* (eds M.O. Andue, J. Doull and C.D. Klassen), McGraw Hill, New York, pp. 484–520.

Thomas, P. (1999). Nontraditional sites of endocrine disruption by chemicals on the hypothalamus-pituitary-gonadal axis: interactions with steroid membrane receptors, monoaminergic pathways, and signal transduction systems, in *Endocrine Disruptors: Effects on Male and Female Reproductive Systems* (ed. R.K. Naz), CRC Press, Boca Raton, FL, pp. 3–38.

Thoreux-Manlay, A., Le Goascogne, C., Segretain, D., Jegou, B., Pinon-Lataillade, G. (1995). Lead affects steroidogenesis in rat Leydig cells in vivo and in vitro. *Toxicology*, *103*, 53–62.

Tilly, J.L., Kowalski, K.I., Johnson, A.L., Hsueh, A.J. (1991). Involvement of apoptosis in ovarian follicular atresia and postovulatory regression. *Endocrinology*, *129*, 2799–2801.

Tomic, D., Frech, M.S., Babus, J.K., Gupta, R.K., Furth, P.A., Koos, R.D., Flaws, J.A. (2006). Methoxychlor induces atresia of antral follicles in ER alpha-overexpressing mice. *Toxicol. Sci.*, *93*, 196–204.

Trumble, J.T., Moar, W.J., Brewer, M.J., Carson, W.G. (1991). Impact of UV radiation on activity of linear furanocoumarins and Bacillus thuringiensis var kurstaki against Spodoptera exigua: implications for tritrophic interactions. *J. Chem. Ecol.*, *17*, 973–987.

Tullner, W.W., Edgcomb, J.H. (1962). Cystic tubular nephropathy and decrease in testicular weight in rats following oral methoxychlor treatment. *J. Pharmacol. Exp. Ther.*, *138*, 126–130.

Vahakangas, K., Rajaniemi, H., Pelkonen, O. (1985). Ovarian toxicity of cigarette smoke during pregnancy in mice. *Toxicol. Lett.*, *25*, 75–80.

van Ravenzwaay, B. (1992). Discussion of prenatal and reproductive toxicity to Reg. No. 83-258 (Vinclozolin). Data submission to US Environmental Protection Agency from BASF, MRIA 425813-02.

van Scott, E.J. (1976). Therapy of psoriasis 1975. *JAMA*, *235*, 197–198.

VanVoorhis, B.J., Syrop, C.H., Hammitt, D.G., Dunn, M.S., Snyder, G.D. (1992). Effects of smoking on ovulation induction for assisted reproduction techniques. *Fertil. Steril.*, *58*, 981–985.

Waring, S.C., Rocca, W.A., Peterson, R.C., O'Brien, P.C., Tangalos, E.G., Kokmen, E. (1999). Postmenopausal estrogen replacement therapy and risk of AD: a population-based study. *Neurology*, *52*, 965–970.

Warne, G.L., Fairley, K.F., Hobbs, J.B., Martin, F.I. (1973). Cyclophosphamide-induced ovarian failure. *N. Engl. J. Med.*, *289*, 1159–1162.

Weber, J.E., Turner, T.T., Tung, K.S., Russell, L.D. (1988). Effects of cytochalasin D on the integrity of the Sertoli cell (blood-testis) barrier. *Am. J. Anat.*, *182*, 130–147.

Weinberg, C.R., Wilcox, A.J., Baird, D.D. (1989). Reduced fecundibility in women with prenatal exposure to cigarette smoking. *Am. J. Epidemiol.*, *129*, 1072–1078.

Wetzel, L.T., Luempert, L.G., III, Breckenridge, C.B., Tisdel, M.O., Stevens, J.T., Thakur, A.K., Extrom, P.J., Eldridge, J.C. (1994). Chronic effects of atrazine on estrus and mammary tumor formation in female Sprague-Dawley and Fischer 344 rats. *J. Toxicol. Environ. Health*, *43*, 169–182.

Whelan, E.A. (1997). Risk assessment studies: epidemiology, in *Comprehensive Toxicology*, Vol. 10 (I.G. Sipes, C.A. McQueen and A.J. Gandolfi), Elsevier Science, New York, pp. 359–366.

Whorton, D., Krauss, R.M., Marshall, S., Milby, T.H. (1977). Infertility in male pesticide workers. *Lancet*, *2*, 1259–1261.

Wiebe, J.P., Barr, K.J., Buckingham, K.D. (1988). Effect of prenatal and neonatal exposure to lead on gonadotropin receptors and steroidogenesis in rat ovaries. *J. Toxicol. Environ. Health*, *24*, 461–476.

Xu, L.C., Wang, S.Y., Yang, X.F., Wang, X.R. (2001). Effects of cadmium on rat sperm motility evaluated with computer assisted sperm analysis. *Biomed. Environ. Sci.*, *14*, 312–317.

Yamada, T., Ichihara, G., Wang, H., Yu, X., Maeda, K., Tsukamura, H., Kamaijima, M., Nakajima, T., Takeuchi, Y. (2003). Exposure to 1-bromopropane causes ovarian dysfunction in rats. *Toxicol. Sci.*, *71*, 96–103.

Yeh, J., Barbieri, R.L. (1989). Twenty-four-hour urinary-free cortisol in premenopausal cigarette smokers and nonsmokers. *Fertil. Steril.*, *52*, 1067–1069.

Younglai, E.V., Foster, W.G., Hughes, E.G., Trim, K., Jarrell, J.F. (2002). Levels of environmental contaminants in human follicular fluid, serum and seminal plasma of couples undergoing in vitro fertilization. *Arch. Environ. Contam. Toxicol.*, *43*, 121–126.

Yu, X., Kamijima, M., Ichihara, G., Li, W., Kitoh, J., Xie, Z., Shibata, E., Hisanaga, N., Takeuchi, Y. (1999). 2-Bromopropane causes ovarian dysfunction by damaging primordial follicles and their oocytes in female rats. *Toxicol. Appl. Pharmacol.*, *159*, 185–193.

Zachow, R., Uzumcu, M. (2006). The methoxychlor metabolite, 2,2-bis-(p-hydroxyphenyl)-1,1,1-trichloroethane, inhibits steroidogenesis in rat ovarian granulosa cells in vitro. *Reprod. Toxicol.*, *22*, 659–665.

Zenzes, M.T., Krishnan, S., Krishnan, B., Zhang, H., Casper, R.F. (1995). Cadmium accumulation in follicular fluid of women in in vitro fertilization-embryo transfer is higher in smokers. *Fertil. Steril.*, *64*, 599–603.

CHAPTER 25

An Introduction to Toxicogenomics

AARON L. VOLLRATH and CHRISTOPHER A. BRADFIELD

25.1 INTRODUCTION

Toxicogenomics can be viewed as the assessment of chemically induced biological change through a merger of classical toxicology, high-throughput genomics, and bioinformatics. The growth of toxicogenomics is largely due to the complete sequencing of a growing number of genomes, coupled with the development of novel technologies that allow us to detect chemically induced changes in global gene transcription. The product of toxicogenomics is often a high-density data set that expands our understanding of chemical–biological interactions. Through toxicogenomics approaches, the ability to predict the toxicities of a chemical compound and classify the hazard of a large number of chemicals is now possible. This ability is derived from the idea that chemicals with similar toxicological mechanisms will be revealed by their common influence on subsets of the global genomic response. Moreover, the high-throughput nature of toxicogenomics results in a shorter time for chemical testing which in turn allows for a more timely understanding of toxicities that arise from exposure to chemicals in commerce and those from our natural environment. This emerging scientific discipline is poised to have a significant impact on a number of important toxicological questions.

25.2 MICROARRAYS: GENOMICS-BASED METHODS FOR TRANSCRIPTIONAL PROFILING

The methodological domain of toxicogenomics is quite broad and growing rapidly. This basic introduction to the field will focus primarily on the use of microarrays to assess transcriptional events associated with exposure to a toxicant of interest. This is a method commonly known as transcriptional profiling or toxicant profiling. Microarray-based methods to detect whole-genome transcriptional changes or transcriptional profiling provide a means to identify how a chemical affects cells at the level of gene transcription. In this review, we define a transcriptional profile as the set of genes represented on the microarray platform and their associated measured relative

Drug Metabolism Handbook: Concepts and Applications, Edited by Ala F. Nassar, Paul F. Hollenberg, and JoAnn Scatina

transcription values as compared with a control state. Thus, a transcriptional profile can be thought of as a vector of values in which the number of values is dependent on the number of individual probes on the array. In the case where a transcriptional profile is generated on an array of all known open reading frames, this transcriptional profile would represent a global transcriptional profile. While microarray-based methods are not the only means to assess changes in transcriptional events, they are currently one of the more widely used methods.

25.3 CHEMICAL CLASSIFICATION—IDENTIFYING CHEMICALLY INDUCED SIGNATURES AND THEIR POTENTIAL USE IN RISK ASSESSMENT

One of the applications of toxicogenomics is to classify hazardous chemicals based upon their transcriptional profiles. Once profiling is performed on a large number of potentially toxic chemicals, a chemical class can be defined as a set of chemicals that induce transcriptional responses that are highly similar. It follows that once classes of known chemicals are established by the identification of unique subsets of genes, or chemical signatures, derived from transcriptional profiles, chemicals novel to commerce or understudied natural products may then be classified based on their comparison with the set of established classes (Hayes and Bradfield, 2005). If an unknown chemical is then associated with a known class of chemicals based on the utilization of chemical signatures, predictions can be made regarding the effects of the exposure and, in the case of well-characterized classes of chemicals, mechanisms of action can be predicted.

Recently, *The Committee on Applications of Toxicogenomic Technologies to Predict Toxicology and Risk Assessment* from the National Research Council released a report suggesting toxicogenomic methods be integrated by regulatory agencies into the risk assessment process (Committee on Applications of Toxicogenomic Technologies to Predictive Toxicology and Risk Assessment, 2007). Prior to the release of this report, a number of studies in the field of toxicogenomics have been conducted based on this premise with promising results providing proof of principle (Thomas *et al.*, 2001b; Bushel *et al.*, 2002; Hamadeh *et al.*, 2002a, b, 2004; Steiner *et al.*, 2004) In fact, some studies report that their use of chemically induced signatures has a greater degree of predictive accuracy relative to current short-term pathology observations or genomic biomarkers. Additionally, the identified chemically induced signatures are superior to estimates based on longer-term sub-chronic studies (Elcombe *et al.*, 2002; Allen *et al.*, 2004; Heinloth *et al.*, 2004; Fielden *et al.*, 2007).

In theory, the identification of chemical signatures and the feasibility of their subsequent use in predicting or classifying the effects of a chemical exposure and risk assessment analyses could be put into practice in the coming years. However, the Committee asserts that "Fully integrating toxicogenomic technologies into predictive toxicology will require a coordinated effort approaching the scale of the Human Genome Project." Given that assessment, there are some limiting factors that prevent the use of transcriptional profiles and associated chemical signatures in these capacities. One limiting factor is the lack of a database containing a large number of transcriptional profiles of chemicals and associated chemical classes from which the chemical signatures are derived. This is largely due to a lack of a well-organized and executed plan to generate the necessary data. The lack of a large-scale approach is due to a number of unresolved issues. Examples of these issues include a lack of consensus on the utilization of a model system (e.g. organisms such as the mouse and rat), tissues/organs to analyze, and types of experiments to perform. The question of what array platform to utilize is also a prominent issue.

We have organized the rest of this review into the following sections. First, we provide a basic description of microarray technologies consisting of the two most popular approaches to estimate relative transcriptional changes: one-color and two-color arrays. Second, statistical issues associated with microarrays will be presented, specifically identifying and establishing the level of variation within a microarray experiment. Third, we present an established standard protocol used to describe microarray experiments and address the important decisions that influence the choice of microarray-based approaches. Specifically, regarding the latter, we present some basic experimental design methods utilized to obtain the relative changes in transcription. Fourth, we address the types of questions that can be answered utilizing microarrays and the issues related to interpretation and representation of large data sets. In other words, what bioinformatics methods can be utilized to answer questions regarding microarray data? Finally, we will address how transcriptional events elucidated by microarray technologies can be utilized to gain insight into establishing a mechanism of action.

25.3.1 Basics of Microarray Gene Expression Analyses

Currently, it is now feasible to simultaneously monitor gene expression levels on a scale encompassing all the known gene products of toxicological importance of organisms including *Saccharomyces cerevisiae*, *Mus musculus*, *Rattus norvegicus*, *Canis familiari*, and *Homo sapiens*. There are two popular microarray-based approaches, one-color and two-color, utilized to measure gene expression. In both methods, array platforms have been developed and utilized to assay gene expression in this wide range of organisms. Conceptually, both approaches are similar as each requires a step to incorporate a measurable fluorescent label and these methods can be utilized to quantify changes in gene expression in a relative manner.

In one popular one-color-based approach used by Affymetrix (Fig. 25.1a), total RNA is first isolated and then reverse-transcribed into double-stranded cDNA. A subsequent *in vitro* transcription step is performed and a biotin tag is attached to the antisense cRNAs. After incorporation of the biotin tag, the labeled cRNAs are fragmented into 25- to 200-bp pieces and dispensed on an array so that they can hybridize to their complementary probes on the gene chip. After hybridization, a staining process (i.e. fluorescent labeling) occurs whereby streptavidin-phycoerythrin biotinylated anti-streptavidin antibodies bind to the biotin-tagged cRNAs hybridized to the chip. A laser excites the phycoerythrin bound to the cRNA at each spot producing a signal. The data are derived from a scanner that measures the corresponding fluorescent signal at each spot or DNA probe location on the array. The relative transcriptional differences between the treated and control samples can be determined by taking the ratio of the signals at corresponding spots on each array.

In the two-color approach, two distinct fluorescent dyes, commonly Cy3 (fluorescein/Cyanine 3) and Cy5 (rhodamine/Cyanine 5) are used (Fig. 25.1b), such that the treated and control samples are labeled with different dyes (e.g. treated sample labeled with Cy3 and a reference sample/control labeled with Cy5). The labeled cDNA or cRNA samples are mixed, and both are allowed to competitively hybridize to DNA probes on a single array. To capture data, this approach requires that every spot on the array

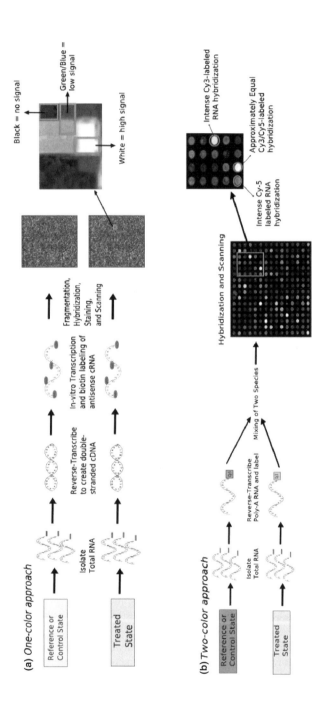

Figure 25.1 **(a)** A flowchart depicting the one-color approach used by Affymetrix (*Array images courtesy of Affymetrix*). In this approach, when comparing gene expression between two samples, control and treated, a separate array is required for each sample. For both, total RNA is isolated and reverse-transcribed to create double-stranded cDNA. The double-stranded cDNA serves as a template for the *in vitro* transcription of antisense cRNA. A biotin label is added at this step to facilitate the detection of those cRNAs that bind to complementary probes on the array. After the hybridization step, the array is washed to remove unbound cRNA, stained with streptavidin phycoerythrin conjugate and subsequently scanned with a laser emitting light at 570nm. The strength of signal is measured by the amount of light emitted. When looking at the array, white, red, and yellow indicate a strong signal at a probe location while green, light blue, and dark blue indicate a weak signal. Black represents no signal at that probe location. These signals are quantified using specialized software to determine the level of expression. Different samples can be compared by taking the ratio of the signals at corresponding probes on each array. **(b)** A flowchart depicting the two-color approach. To obtain the relative expression ratios between an RNA sample from a control/reference state and an RNA sample from a treated state, RNA is first isolated and then the poly-A RNA is reverse-transcribed into cRNA or cDNA while being labeled with a fluor. In this case, the control sample is labeled with Cy5 and the treated sample is labeled with Cy3. After labeling, the labeled species are mixed together and subsequently hybridized to a microarray. During hybridization, the two labeled species bind to the probes and relative differences can be detected by scanning the array with a laser at two different wavelengths. One wavelength excites the Cy5 fluor and the other excites the Cy3 fluor, resulting in the emission of light which is quantified for each by the use of specialized software.

be excited with a laser at two distinct frequencies to generate a signal based on the amount of Cy3- and Cy5-labeled targets that hybridized. Relative differences in transcription can then be calculated based on the raw fluorescence values obtained for each laser-induced signal. This value is usually represented as a log ratio or as a fold change. As an example, assuming no dye bias, if the Cy3 channel was measured at 500 fluorescent units and the Cy5 channel was measure at 1000 fluorescent units, the ratio (Cy5/Cy3) would be 1000/500, representing a twofold change.

25.3.2 Statistical Issues

Variation can affect the results of a microarray experiment. There are two types of variation: biological and technical. Biological variation is variation that mainly occurs prior to RNA extraction. It can be thought of as variation arising because two samples are biologically different. Technical variation occurs as the result of differences that arise during amplification, labeling, and hybridization of the source RNA. Additionally, technical variation is introduced during the measurement of the signals produced from the scanning of the array and in data analysis. Each step in the microarray experiment is a potential source of variation in the final value used to represent the relative difference in transcription (Churchill, 2002).

Biological sources of variation include environment and genetic factors that influence the levels of a transcript and its response to chemical treatment (Oleksiak *et al.*, 2002; Spruill *et al.*, 2002). Even animals on the same genetic background raised in the same environment will exhibit some variation in transcriptional response to treatment with a chemical. Any attempt to assess relative transcriptional differences must take into consideration the underlying biological variation within the experiment being conducted. These estimations are accomplished through the utilization of biological replicates. Examples of such replicates include multiple independent samples for control and treated sources of RNA obtained under the same experimental condition (Fig. 25.2). A large number of biological replicate comparisons are the ideal, but in cases when the RNA source or cost of microarray analysis is the limiting experimental factor, it is common practice to pool RNA from replicates. However, pooling comes at the cost of reducing the number of independent biological replicates. When large numbers of individual, unique samples obtained under the same experimental conditions are analyzed, it is easier to detect anomalous samples and obtain a better understanding of the inherent biological variation. An example of an anomalous sample would be when one of five samples came from an animal model that has experienced an inflammatory response to a pathogen or wound specific to the individual in question and not the experiment. A discussion of the determination of the required number of biological replicates that should be utilized to obtain statistically significant results is beyond the scope of this work, but has been discussed elsewhere (Churchill, 2002).

The second source of variation, technical variation, is introduced when the RNA is isolated, processed (i.e. reverse-transcribed and fluorescently labeled), and hybridized to the microarray. Utilizing technical replicates allows for the estimation of the amount of technical variation present within the experiment (i.e. running the same sample multiple times). Technical replicates can be RNA from the same source (and control) hybridized multiple times on unique microarrays from the same array platform (i.e. two or more different arrays that have the exact probe set). Ideally, the measurements from the same samples should be able to be replicated with high fidelity from slide to slide if the same protocols and practices are utilized (Fig. 25.3). In our own experience

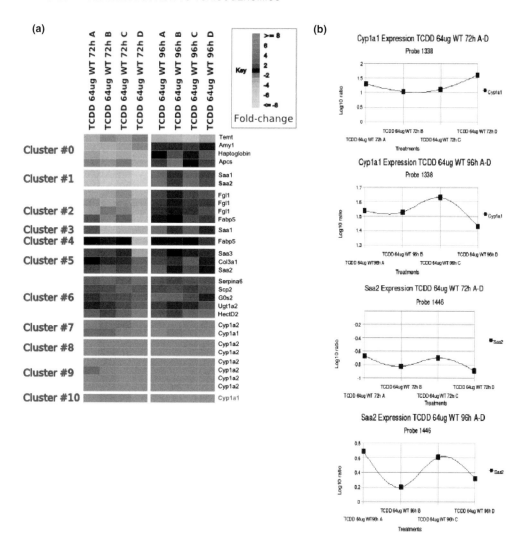

and that of others, technical variance is typically much smaller than biological variance when utilizing commercial arrays (Yauk *et al.*, 2004; Ayroles and Gibson, 2006). Thus, we are advocates of performing biological replicates in the hope that these samples also capture the technical variance.

In the case of the two-color approach, a dye swap can be performed to estimate technical error and account for any dye bias. Dye bias is defined as an intensity difference between samples labeled with different dyes attributable to the dyes instead of a difference in the actual gene expression of the samples (Dobbin *et al.*, 2005). Dye swap is a simple process where the originally Cy5-labeled sample is labeled with Cy3 and the originally Cy3 sample is labeled with Cy5; the samples are then rehybridized to another array. In addition to a single dye-swap approach, if the necessary amount of

Figure 25.2   Identifying the biological variation of two sets of treatments. **(a)** This figure represents a k-means (where k = 11, with 11 arbitrarily chosen) clustering analysis of two sets of biological replicates at two different time points with a fold-change threshold of ±3 for genes in at least one of the biological replicate treatments in either set. The first set, TCDD 64 µg WT 72 h A–D (listed at top), composes the left-hand side of the heat map and is separated by a vertical white-space margin from the second set, TCDD 64 µg WT 96 h A–D (listed at top), composing the right-hand side of the heat map. The genes that met the query criteria are listed vertically on the right of the heat map. The heat map is meant to give a qualitative assessment of the biological variation of the biological replicates. **(b)** To obtain a quantitative understanding of biological variation, it is important to look at biological variation at each probe across the replicates. This can be done by utilizing statistical methods. The four charts show the Log_{10} expression values of two genes across the biological replicate sets. These charts represent a qualitative and quantitative assessment of the biological variation of the biological replicates at the two different time points of 72 and 96 hours. The top two charts are for the gene Cyp1a1, labeled in the gene list of the heat map in bold, green text. For the top chart, the average Log_{10} ratio value for Cyp1a1 expression of probe 1338 across the TCDD 64 µg WT 72 h A–D biological replicates is 1.26 and the calculated standard deviation is 0.22. The average Log_{10} ratio value for Cyp1a1 expression of probe 1338 across the TCDD 64 µg WT 96 h A–D biological replicates is 1.53 and the calculated standard deviation is 0.07. The simple statistics suggest that the range of expression for the TCDD 64 µg WT 72 h biological replicates has more biological variation than the TCDD 64 µg WT 96 h biological replicates based on the standard deviation values. The bottom two charts are for the gene Saa2 in the gene list of the heat map in bold, blue text. For the top chart, the average Log_{10} ratio value for Saa2 expression of probe 1446 across the TCDD 64 µg WT 72 h A–D biological replicates is −0.77 and the calculated standard deviation is 0.09. The average Log_{10} ratio value for Saa2 expression of probe 1446 across the TCDD 64 µg WT 96 h A–D biological replicates is 0.46 and the calculated standard deviation is 0.2. The simple statistics suggest that the range of expression for the TCDD 64 µg WT 72 h biological replicates has less biological variation than the TCDD 64 µg WT 96 h biological replicates based on the standard deviation values.

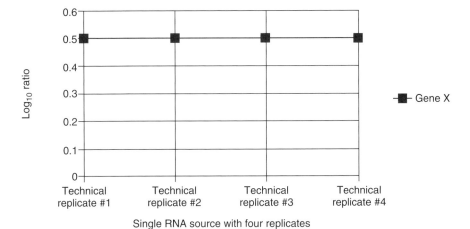

Figure 25.3   Ideal set of technical replicates. This figure shows the expression of gene X from a hypothetical set of technical replicates from a single, unique RNA source. Ideally, technical variation would be zero and all of the expression values for gene X across the technical replicates would be the same for this probe and all other probes on the array.

sample is available, multiple dye-swap hybridizations of the same sample/control can be performed.

Finally, the arrangement and number of spots of each probe on the array can also allow for the estimation of technical variation. Placing replicate probe spots sharing the same sequence on the array and randomizing the placement of each probe on the array can address the variance introduced by spot-to-spot printing variation. In practice, assessing variation of probe placement is not always feasible due to the complications of printing probes randomly on an array, but replicate spotting of some or all unique probe sequences is common.

In summary, the final fluorescence value calculated for each probe on the array ideally represents the level of gene transcription. The calculated expression values can be affected by two different sources of variation: biological and technical. It is important to understand both forms of variation occur within the experiment. Biological variation can be estimated by utilizing multiple individual sources of RNA obtained under the same experimental conditions. Technical variation can be assessed by repeating experiments using the same sample sources and can be minimized by experience and use of standardized protocols. In fact, technical variance can be minimized to the extent that cross-platform comparisons can begin to be made due to increased sensitivity allowing for measurement of meaningful biological differences (Bammler *et al.*, 2005; Dobbin *et al.*, 2005; Irizarry *et al.*, 2005). Understanding the biological variation while limiting the technical variation can increase the fidelity of the results obtained. Put simply, technical variation can be minimized by technique, but biological variation is often inherent in the experiment.

25.4 EXPERIMENTAL DESIGNS OF MICROARRAY-BASED TECHNOLOGIES

We have presented a brief introduction to the basic idea of microarrays as a means to identify gene transcription changes on a genomic scale. Though microarrays represent a powerful tool, when toxicogenomics data are published, it is imperative to provide a detailed description of the entire experimental process to ensure the results can be confirmed and reanalyzed by scientific peers. Additionally, proper experimental design and analysis methods must be employed to ensure that those data obtained are the most informative. To that end, the need for standardizing the description of a microarray experiment has been carefully examined (Brazma *et al.*, 2001). In the following sections, we will give a brief overview of some of the basic experimental designs utilized with microarray-based gene experiments, as well as methods for their reporting to the scientific community.

25.4.1 MIAME—A Way to Standardize the Description of Microarray-Based Experiments

After the publication of early microarray studies, it was apparent that there was a need to develop a set of standards to ensure that the necessary technical information was available to interpret or reproduce the reported results. A set of guidelines, called the "Minimum Information About Microarray Experiment" (MIAME) was established as a standard (Brazma *et al.*, 2001). The agreed-upon minimum information consists of six descriptions (Fig. 25.4). This standard has been accepted by numerous editorial boards and consequently adherence to these guidelines is necessary in order for microarray

MIAME guidelines
• Experimental Design—A formal description of the design utilized
• Array Design—A formal description of the array and the probes
• Experimental Samples—A formal description of the samples utilized and the methods used for RNA extraction and labeling
• Hybridizations—A description of the protocol utilized
• Measurements—A formal description of the quantifications utilized as well as images of the actual arrays
• Normalization—A formal description of the types of normalization utilized

Figure 25.4   MIAME guidelines.

data to be published by most peer-reviewed journals. For a more detailed analysis, the reader is highly encouraged to refer to Brazma *et al.* (2001) and http://www.mged.org/Workgroups/MIAME/miame.html. An extension of the MIAME guidelines applicable to toxicogenomics, MIAME/Tox, has been designed to address toxicogenomic-specific requirements (http://www.mged.org/Workgroups/rsbi/rsbi.html).

In addition to adhering to the MIAME standards, some peer-reviewed journals in biology require that microarray data must be uploaded into a publicly available database. The purpose of this is to provide transparency and allow other scientists to analyze those data using their own methods. Two of the main databases that accept submission of MIAME-compliant microarray data are Gene Expression Omnibus (GEO) (http://www.ncbi.nlm.nih.gov/geo/) and Array Express (http://www.ebi.ac.uk/microarray-as/aer/). Knowing the requirements for publication of studies utilizing microarray platforms ahead of time will make things easier when submitting those microarray data.

25.4.2   Design of Microarray Gene Expression Experiments

In this section, we will address the design of microarray-based experiments to assess relative changes in gene transcription. One important design choice is identifying the proper control/reference state(s) with which to relate the treated states(s). To that end, the basis of the design is predicated on the question posed and the resources available. To address specific types of experimental questions, "reference," "direct," and "loop" designs can be employed (Fig. 25.5).

The *reference* design is applicable to the two-color approach and utilizes a reference sample against which the control and experimental samples are compared. A reference sample is a common control RNA source that might come from a historical control sample or a representation of all possible RNAs that might be analyzed in the future. For large-scale experiments, where adequate amounts of reference sample are available, the reference design provides a means of common comparison and historical

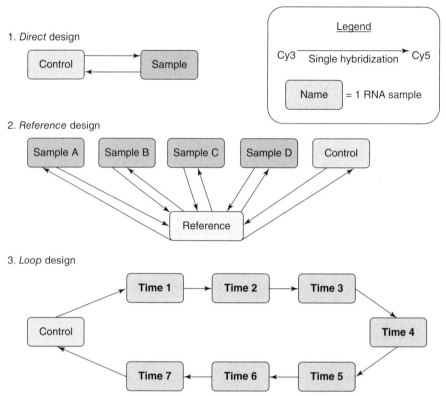

Figure 25.5 Three representations of the basic types of microarray design. **1.** The direct design is the most straightforward and utilizes a dye swap to compare the expression of a treated sample directly with a control sample for a total of two microarrays used. **2.** The reference example depicted here is comprised of 10 arrays where four separate treated samples and the control state for this experiment are hybridized against a reference sample. Dye swap is utilized. **3.** The loop design depicts a time series experiment where seven samples and a control state (time = 0) are hybridized in a clockwise loop. This design does not utilize dye swap. Additionally, if one of the arrays were to fail, the loop would be broken affecting the results. None of the representations utilize biological replicates, limiting the ability to identify the level of biological variance.

reflection. From this design, differences in expression can be calculated and compared by subtracting log ratios. As an example, if your reference sample is labeled with the same fluor for all comparisons, different samples can be compared utilizing the following equation:

$$\log_x(\mathbf{Y}/\mathbf{Z}) = \log_x(\mathbf{Y}/\mathbf{R}) - \log_x(\mathbf{Z}/\mathbf{R}) = \log_x(\mathbf{Y}/\mathbf{R})/\log_x(\mathbf{Z}/\mathbf{R}),$$

where $\mathbf{x}$ is the base of the log, $\mathbf{R}$ is the fluorescence measurement value of the reference in one channel, and $\mathbf{Y}$ and $\mathbf{Z}$ are the fluorescent measurement values in the other channel of the samples whose comparison is being computed. The use of a common reference facilitates the identification of chemical signatures, because the chemically treated samples are all compared with the same reference over the duration of the study (which could be years). An element of inefficiency inherent to the reference design is that half of the hybridization resources are dedicated to a reference sample.

The *direct* design entails the direct comparison of a treated sample with a control sample, allowing for the identification of relative changes in transcription. When coupled with biological replicates, this is often the most sensitive approach to identify changes in transcription. This is due to the fact that one can compare samples of most interest directly without having a comparison with intermediate reference sample that has the potential to add technical variation. Additionally, subsequent mathematical calculations are not required to obtain the relative differences between two different samples hybridized against a reference on separate arrays.

When the number of conditions being tested is equal to or greater than three, more complex designs can be helpful. The *loop* design involves the sequential hybridization of one sample against another until all the samples have been connected. The loop design is commonly utilized in time-series analyses. In some ways the loop design represents an extension of the basic direct design, because under most circumstances the most interesting samples are compared directly and in sequence. Thus, chronologically proximate samples are usually hybridized against each other to obtain the most temporally sensitive changes in gene transcription. The loop design should incorporate redundancy (i.e. replicates) to ameliorate the effects of a microarray failure that would otherwise break the chain of hybridizations.

In summary, efficiency in sample comparisons is dictated by the length of the path connecting the samples of interest and the number of paths connecting them (Kerr and Churchill, 2001; Yang and Speed, 2002). The direct design offers the shortest path between two samples of interest. Although it is the most simple of the experimental designs, it is limited to the comparison of two conditions. The loop and reference designs require additional data-processing steps to obtain a relative comparison between the different conditions; however, they are amenable to more than two conditions such as in time-series and dose–response studies. The loop design can be more cost-efficient than the reference design (Vinciotti *et al.*, 2005) because in the reference design, 50% of the hybridization resources are directed toward the reference sample. It is important to choose the most efficient design that has the potential to provide the highest degree of statistical significance. However, other factors such as cost and sample resources can affect the choice of experimental design and the level of biological and technical replication. For a more in-depth analysis of these experimental designs and suggestions for when to use them, the reader is directed to the following references: Kerr and Churchill (2001), Yang and Speed (2002), and Hayes and Bradfield (2005).

25.5 THE INTERPRETATION AND ANALYSIS OF DATA GENERATED BY MICROARRAY METHODOLOGIES

One of the major goals of toxicogenomics is to identify and establish global expression profiles resulting from the treatment of a particular chemical or group of chemicals. The common premise of such an experiment is that this profile can be utilized as a means to classify a chemical based on knowledge about related or similar chemicals of known toxicity. We have defined the concept of transcriptional profiles and chemical signatures. We have also highlighted the importance of choosing the correct experimental design to address the particular question being asked. However, simply generating high-density data is not enough. In order to gain biological insight, one must utilize informatics approaches to analyze and extract meaningful information. Next, we will describe four microarray-based approaches used to analyze microarray data.

25.6 FOUR BASIC MICROARRAY-BASED APPROACHES

Four basic microarray-based approaches deserve special attention: class discovery, class comparison, class prediction, and pathway/network analysis (Quackenbush, 2006). In using these approaches, a basic framework of data analysis takes shape. In general, class discovery approaches are utilized first to look for interesting patterns in the data that may be biologically relevant. Used in conjunction with additional data such as pathology, similarly grouped transcriptional profiles can begin to be assigned to distinct classes. After classes of toxicants are established based on objective criteria, differentially expressed genes (i.e. chemical signatures) can then be identified using class comparison methods to determine how to best distinguish classes in the data. In addition, established classes allow for the methods of class prediction to be utilized to identify the most probable class of unassigned transcriptional profiles. Finally, the incorporation of pathway/network analysis can be utilized for mechanistic analyses.

25.6.1 Class Discovery

Unsupervised methods are utilized during class discovery experiments when one is at the early stages of classification. Unsupervised methods are informatics methods that do not take into consideration *a priori* knowledge about the samples and provide the potential to organize data in an unbiased manner. Common unsupervised approaches used in transcriptional profiling are the clustering approaches. Clustering approaches commonly utilize metrics such as Euclidean distance or correlation coefficients to determine which genes and treatments are most similar to each other. The underlying mathematics of these algorithms is beyond the scope of this work, but the reader is encouraged to consult the following for a more rigorous mathematical description: Eisen *et al.* (1998), Sherlock (2000), and Quackenbush (2001).

There are a large number of different methods to cluster data generated by microarray analyses and this review will concentrate on two of the more common clustering methods: Hierarchical and k-means. Hierarchical clustering is probably the most common means of depicting gene expression profiles (Fig. 25.6). In this method, genes, as well as the treatments themselves, can be grouped based solely on the underlying changes in gene transcription. In hierarchical clustering, clustered data are represented by a heat map and dendrograms, or hierarchical trees, to represent gene expression changes and similarity relations between the genes and treatments. The most similar treatments and genes can be visually identified as being the most proximate relative to other treatments and genes. The branches within the dendrogram can be assigned a value whose magnitude is utilized to determine relative differences between genes and/or treatments by simply summing the distances along the branches.

k-means clustering allows the researcher to specify the number of distinct groups he or she would like the data to be partitioned into prior to the execution of the clustering algorithm (Fig. 25.7). Normally, before the portioning process begins, genes are randomly assigned to each cluster to "seed" the algorithm. At that point, the algorithm reassigns genes to different cluster groups such that within each group, the distance between the genes is minimized and between each group the distance between the

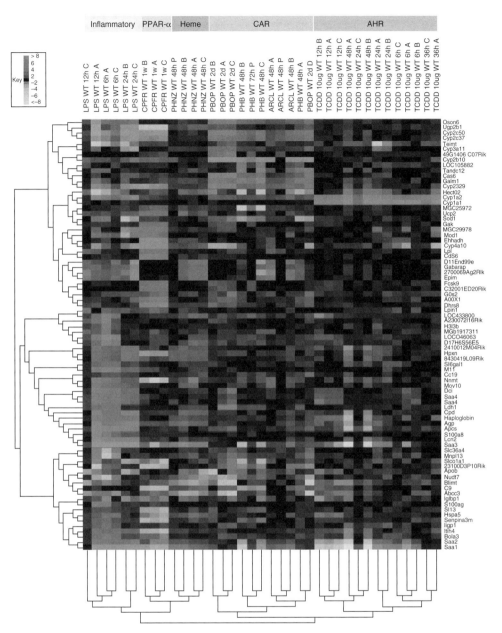

Figure 25.6 Hierarchical clustering. This figure represents an example of the class discovery step. Transcription profiles of 37 different C57BL/6 mice treated with either lipopolysaccharide (LPS), a known inflammatory agent; ciprofibrate (CPFR), a known PPAR-α agonist; phenylhydrazine (PHNZ), a known inducer of hemolytic anemia; 1,4-bis(2-(3,5-dichloropyridyloxy))benzene (PBOP), a known CAR agonist; phenobarbital (PHB), a known CAR agonist; Aroclor (ARCL), a known CAR agonist; or 2,3,7,8-tetrachlorodibenzo-p-dioxin (TCDD), were clustered using hierarchical clustering. A total of 80 genes for each profile were included based on a magnitude fold change greater than or equal to 3.5 (±3.5). The classes of the transcriptional profiles are listed across the top of the cluster. Though the classes of the transcriptional profiles were known beforehand, by looking at the dendrogram for the treatments at the bottom of the cluster, it is evident that the algorithm was able to accurately cluster the transcriptional profiles based on their classes.

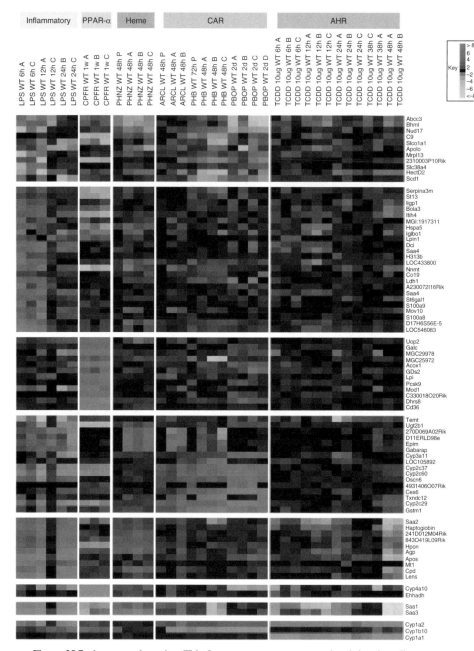

Figure 25.7 k-means clustering. This figure represents an example of the class discovery step. Transcription profiles of the same transcriptional profiles from Fig. 25.5 were clustered using k-means clustering. Again, a total of 80 genes for each profile were included based on a magnitude fold change greater than or equal to 3.5 (±3.5). The classes of the transcriptional profiles are listed across the top of the cluster and were ordered based upon their assigned class. An arbitrary number for k, 8, was chosen to segregate the genes based on their expression levels. This allows for the qualitative identification of distinct clusters of gene expression patterns allowing for discrimination between the classes. Clustering also allows for the qualitative examination of disparities in biological replicates.

gene groups is maximized. The algorithm runs a predetermined amount of time until a particular threshold (e.g. algorithm running time) is reached. The end result is a heat map that consists of a predefined number of groups of genes that have been deemed to be most similar within each group and most disparate between groups.

The class discovery or clustering methods presented are useful in identifying potential classes of chemicals and biologically interesting patterns in gene expression. Additionally, they provide a means of generating hypotheses. However, they are limited in their ability to identify chemically induced signatures due to their unsupervised nature. On the other hand, class comparison methods can be utilized to identify chemically induced signatures.

25.6.2   Class Comparison

Class comparison approaches can be utilized to analyze transcriptional profiles and identify chemically induced signatures based on differences in expression between predefined classes of transcriptional profiles. A goal of toxicogenomics is to utilize transcriptional profiles to identify chemically induced signatures. To that end, statistical methods have been developed to discover chemically induced signatures. Some examples of these methods include statistical analysis of microarrays (SAM) (Tusher *et al.*, 2001), *t*-tests (Baggerly *et al.*, 2001), analysis of variance (Shannon *et al.*, 2003), regression modeling (Thomas *et al.*, 2001a), the Empirical Bayes method (Efron and Tibshirani, 2001), and the mixture model (Pan *et al.*, 2003). When looking to identify chemically induced signatures that can discriminate between different classes, class comparison approaches start with a set of transcriptional profiles that have been assigned to predefined classes based on *a priori* knowledge. Then, statistical methods are able to compare the gene expression profiles of different groups and establish sets of genes (i.e. chemically induced signatures) that can be utilized to discriminate between classes of toxins. These chemically induced signatures provide a basis for the classification of novel chemicals through the utilization of class prediction approaches. These signatures have been referred to as predictive gene sets, because they can subsequently be utilized in class prediction approaches (Thomas *et al.*, 2001b).

Unlike class discovery methods, class comparison requires the intervention of expert knowledge to assign transcriptional profiles to distinct classes. This aspect represents a supervised approach to the analysis of microarray data. In assigning the classes, an inherent bias is entered into the class comparison approach. If the classes of the chemicals are well known and characterized, theoretically the identified chemically induced signatures can be considered to be accurate representations of the chemicals in the assigned class. However, improper assignment of classes to transcriptional profiles can adversely affect the results and interpretation of methods utilizing supervised approaches.

25.6.3   Class Prediction

Class prediction is a supervised approach allowing for the incorporation of *a priori* information into algorithms so that members in a class can be predicted based on transcriptional profiles. In general, class prediction approaches require the utilization of a training set of data and a test set of data. Prior to the actual classification of an unknown transcriptional profile, some class prediction methods incorporate the methods of the class comparison approach to determine chemically induced signatures

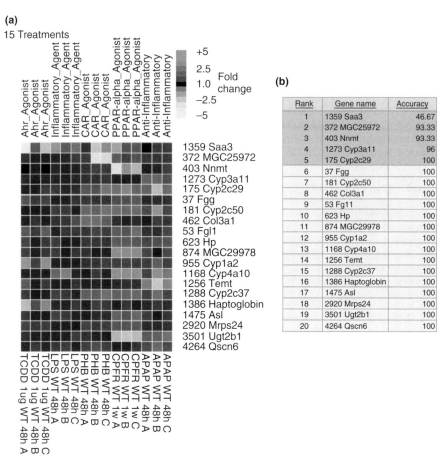

(a)
15 Treatments

Rank	Gene name	Accuracy
1	1359 Saa3	46.67
2	372 MGC25972	93.33
3	403 Nnmt	93.33
4	1273 Cyp3a11	96
5	175 Cyp2c29	100
6	37 Fgg	100
7	181 Cyp2c50	100
8	462 Col3a1	100
9	53 Fgl1	100
10	623 Hp	100
11	874 MGC29978	100
12	955 Cyp1a2	100
13	1168 Cyp4a10	100
14	1256 Temt	100
15	1288 Cyp2c37	100
16	1386 Haptoglobin	100
17	1475 Asl	100
18	2920 Mrps24	100
19	3501 Ugt2b1	100
20	4264 Qscn6	100

Figure 25.8 An example of Naive Bayes classification. This figure represents the output of a Naive Bayes-based classification algorithm that first identifies a predefined number of genes constituting a subset of the genes in a transcriptional profile produced by the array platform. In this case, the transcriptional profiles represent a set of gene probes printed on a custom cDNA array. RNA was obtained from murine livers and then competitively hybridized with control RNA. Three different treatments from five different class designations defined by the mechanism of action were selected for the generation of a set of genes that could be used to discriminate between the different classes. The classes and respective treatments are (i) Ahr agonist class consisting of three biological replicate treatments of 1-μg TCDD (dioxin) at 48 hours after dosing; (ii) inflammatory agent class consisting of three biological replicate treatments of 1000-μg lipopolysaccharide (LPS) at 48 hours after exposure; (iii) CAR agonist class consisting of three biological replicates of phenobarbital (PHB) at 48 hours after dosing; (iv) PPAR-alpha agonist class consisting of three biological replicate treatments of ciprofibrate (CPFR) at seven days after dosing; and (v) anti-inflammatory class consisting of three biological replicate treatments of 500μg/kg acetaminophen (APAP) at 48 hours after dosing. **(a)** A heat map representation of the 20 genes selected by information gain. The 20 selected genes are placed along the right-hand side of the heat map in the order of their selection. The numbers preceding the gene names represent the probe number on the array. The classes of treatments are labeled across the top of the heat map and the corresponding treatments are labeled across the bottom. **(b)** A table depicting the order in which the genes were selected by the algorithm based on their ability to classify the

that can be used to discriminate between the unknown transcriptional profiles. The training set and test set of data should represent sets of chemically induced signatures or transcriptional profiles with well-known characteristics and an appropriate class assignment based on expert knowledge and objective criteria. The supervised algorithm then develops a model based on the training set of data and utilizes the generated model to determine the classification of profiles in the test set. In general, the accuracy of the algorithm is based on the number of correctly assigned transcriptional profiles in the test set of data. The applicability of a generated model to future classification of novel transcriptional profiles is dependent on the robustness (i.e. number of transcription profiles, number of unique chemical classes, etc.) of the original transcriptional profiles it was built from. As an example, the broader the application, say for instance in classifying a random chemical drawn from the currently 70,000–80,000 chemicals in commerce for which toxicity is not known, the greater the requirement for a more robust training set to effectively assign a class.

Some examples of class prediction methods utilized to classify chemical profiles include k-nearest neighbors (Theilhaber *et al.*, 2002), weighted voting (Golub *et al.*, 1999), artificial neural networks (Bloom *et al.*, 2004), discriminant analysis (Nguyen and Rocke, 2002; Orr and Scherf, 2002; Antoniadis *et al.*, 2003; Le *et al.*, 2003), classification and regression trees (Boulesteix *et al.*, 2003), Naive Bayes (Thomas *et al.*, 2001b), and support vector machines (Brown *et al.*, 2000; Ramaswamy *et al.*, 2001, 2003).

Proof of principle for utilizing classification methods in toxicogenomics has support from the field of medical oncology (Ross *et al.*, 2000; Scherf *et al.*, 2000; Nielsen *et al.*, 2002; Ramaswamy *et al.*, 2003). Not only have transcriptional profiles been utilized to differentiate tumor types, but studies have correlated transcriptional profiles with clinical outcomes (Dhanasekaran *et al.*, 2001; Sorlie *et al.*, 2001; West *et al.*, 2001; Pomeroy *et al.*, 2002; Singh *et al.*, 2002; van 't Veer *et al.*, 2002; van de Vijver *et al.*, 2002; Williams *et al.*, 2004; Dyrskjot *et al.*, 2005).

To provide a brief example of a class prediction-based algorithm, the output of a Naive Bayes-based classification algorithm used in our laboratory is presented in Fig. 25.8. The purpose of this class prediction approach is to identify a set of genes that can predict the chemical class based on a model generated from a set of transcriptional profiles and their assigned classes. For an in-depth explanation of this approach and the mathematical basis for this algorithm, the reader is encouraged to refer to Thomas *et al.* (2001b).

Figure 25.8 *Continued*
treatments correctly. After the initial selection of genes utilizing information gain, each gene is tested for its ability to classify the treatments with the greatest degree of accuracy using a "leave-one-out cross validation" approach. The first gene is removed from the initial set of 20 and then subsequently tested with all the other genes to see which two provide the greatest predictive accuracy. The second is removed and the three genes that give the best predictive accuracy are selected and the iterative process of removing one gene and adding it to the most predictive list continues until all 20 genes have been placed in order. As can be seen here, five genes (highlighted in red) are enough to give 100% predictive accuracy with the five classes of treatments. Any additional genes (highlighted in gray) added to the list after the first five do not improve or degrade the predictive accuracy. The reader is encouraged to consult the following for further insight: Thomas *et al.* (2001b).

25.6.4   Mechanistic Analysis—Methods Used for Pathway/Network Analysis

In addition to the identification of toxic signatures, toxicogenomics can provide evidence for the identification of the underlying mechanisms of a chemical's toxicity. Utilizing "pathway mapping," toxicogenomic methods can assess the effects of a poison on known toxic signaling pathways and elucidate potential novel mechanisms when a signature cannot be ascribed to any previously identified mechanism. In the context of transcriptional profiling, pathway mapping can be defined as the process of overlaying data derived from transcription studies onto their respective established networks. This process can allow for the generation of hypotheses that are testable by utilizing complementary functional genomics methods to selectively target members of an identified signaling pathway for perturbation.

Class comparison and class prediction approaches can, respectively, identify groups of genes that are differentially expressed and provide a means for classifying unknown transcriptional profiles. However, alone they do not provide utility in terms of identifying the underlying biological mechanisms of response. Pathway/network analysis methods used in combination with transcriptional profiles can help to establish a biological context. Pathway/network analysis methods are in the early stages of development at the moment. The major limiting factor in these analyses is the lack of knowledge of the underlying biology in higher organisms such as the mouse, rat, and human. This factor is being addressed as interactions between biomolecules of all types—proteins, genes, noncoding DNA, and metabolites—are now being elucidated. Much like the genomic projects that have been utilized to catalog every known gene and its regulatory elements, there are a number of projects with the goal of cataloging the functional role of the product(s) of every gene and the role of every metabolite within cells of higher organisms. Some of these projects include those of the Gene Ontology Consortium (GO) (Ashburner, Ball, *et al.*, 2000) and the Kyoto Encyclopedia of Genes and Genomes (KEGG) (Kanehisa, Goto, *et al.*, 2006). With these annotations and interaction data, it is now possible to probe transcriptional profiles for potential biological pathways affected by a toxic exposure. There are tools available for the analysis, visualization, and modeling of these annotation and molecular interaction data. These tools include Cytoscape (Shannon, Markiel, *et al.*, 2003), Osprey (Breitkreutz, Stark, *et al.*, 2003), PathwayAssist (Nikitin, Egorov, *et al.*, 2003), Pathways Database System (Krishnamurthy, Nadeau, *et al.*, 2003), GeneGO (www.genego.com), VisANT (Hu, Mellor, *et al.*, 2005), and BiologicalNetworks (Baitaluk, Sedova, *et al.*, 2006).

At the simplest level, interrogation of transcriptional profiles via integration with these pathway/network data involves using a list of genes of interest and selecting from the interaction data all interactions associated with those genes, and overlaying the gene expression values onto the members of the network. A gene network is a graphical representation of the molecular relationships between genes or gene products. Genes or gene products are represented as nodes, and the biological relationship between two nodes is represented as a line or edge. The nodes can be color-coded based upon the relative expression values assigned to them by microarray analyses.

Another method utilizes already defined signaling pathways and networks and overlays the data from transcriptional profiles (Dahlquist *et al.*, 2002; Doniger *et al.*, 2003). When used in conjunction with metabolic pathways, these methods can elucidate chemically induced perturbations. This method may be more appropriate for identifying dose–response relationships when dealing with well-characterized toxins.

More sophisticated approaches use probabilistic/statistical methods to determine or predict the metabolic pathways, transcriptional networks, or protein interaction networks affected based on the transcriptional profile(s) used as input. Examples of these type of methods include simple Boolean Networks (Akutsu *et al.*, 2000; Savoie *et al.*, 2003; Soinov, 2003), Probabilistic Boolean Networks (Shmulevich *et al.*, 2002a, b; Datta *et al.*, 2004; Hashimoto *et al.*, 2004), and Bayesian Networks (Friedman *et al.*, 2000; Imoto *et al.*, 2003; Savoie *et al.*, 2003; Tamada *et al.*, 2003; Zou and Conzen, 2005). In-depth analysis of these types of models is beyond the scope of this work.

In summary, pathway/network analysis holds great promise. However, the major limitation of these methods is the current overall paucity of interaction data. Currently, these methods are utilized more as a descriptive tool. However, in "simpler" organisms such as *Escherichia coli*, predictive models are being generated (Famili, Mahadevan, *et al.*, 2005; Wiback, Mahadevan, *et al.*, 2004; Palsson, 2006).

25.7 INFORMATICS TOOLS UTILIZED BY THESE INFORMATICS APPROACHES

Most software required to perform these algorithms is commercially available and new algorithms appear in the literature with great frequency. Commercial options can be expensive for a smaller lab, but can offer a good customer support system to answer questions as well as a "turn-key" approach helpful to a smaller lab. An alternative solution is to use an open-source software solution that can be customized to specific user objectives. There are a number of robust statistical packages available in the public domain. The "R" statistical package in conjunction with Bioconductor is a freely available software package project to provide tools for the analysis and comprehension of genomic data. Bioconductor is a community-supported effort and provides fairly detailed documentation of its features. Both commercial and free software options will require some effort on the part of the end user to implement.

25.8 CONCLUSION

Within this review we have focused on how microarray gene expression analyses can be utilized in the context of toxicogenomics. We began by introducing fundamental aspects of the microarray assay for detecting gene transcriptional changes and conceptualized transcriptional profiles and chemically induced signatures. Because there are a number of different microarray platforms available and the scope of their utilization is applicable to a large number of different fields of study, we presented the importance of standardized description of microarray experiments as a way to provide transparency and a means for others to confirm the results. From there we stressed the importance of a well-thought-out experimental design to generate data that can be used to answer the question posed while taking into consideration limiting factors such as cost and resources. Finally, a basic bioinformatic framework was presented to facilitate the process of using transcriptional profiles to identify potential classes of toxic signatures, to identify genes that are differentially expressed between classes of transcriptional profiles, to identify the class of an unknown transcriptional profile based on a model generated by supervised methods, and to utilize transcriptional profiles as a means of probing known biological pathways or networks. We provided examples of some of the basic algorithms associated with the different methods of this framework.

At the moment, toxicogenomic methods provide an attractive complement to classical toxicology experiments. However, it is thought that at some point, enough data can be generated by these methods to begin to utilize known toxic signatures to streamline and expedite the testing process of the current large number of outstanding uncharacterized chemicals. This could entail the utilization of *in vitro* cell or tissue culture methods to predict the toxic effects of a chemical without the costly utilization of animal models. In general, the dose of a particular toxicant determines whether or not it will have adverse effects. As the sensitivity and fidelity of toxicogenomic methods increase, the ability of these methods to identify subtle changes will enhance our ability to identify thresholds of toxic exposure, as well as make more informed regulatory decisions. Finally, as the functional annotation of gene products progresses and the interactions between biomolecules are elucidated, the ability to use pathway/network analysis methods to identify mechanisms of action for toxins will be enhanced.

REFERENCES

Akutsu, T., Miyano, S. *et al.* (2000). Algorithms for identifying Boolean networks and related biological networks based on matrix multiplication and fingerprint function. *J. Comput. Biol.*, *7* (3–4), 331–343.

Allen, D.G., Pearse, G. *et al.* (2004). Prediction of rodent carcinogenesis: an evaluation of prechronic liver lesions as forecasters of liver tumors in NTP carcinogenicity studies. *Toxicol. Pathol.*, *32* (4), 393–401.

Antoniadis, A., Lambert-Lacroix, S. *et al.* (2003). Effective dimension reduction methods for tumor classification using gene expression data. *Bioinformatics*, *19* (5), 563–570.

Ashburner, M., Ball, C.A. *et al.* (2000). Gene ontology: tool for the unification of biology. The Gene Ontology Consortium. *Nat. Genet.*, *25* (1): 25–29.

Ayroles, J.F., Gibson, G. (2006). Analysis of variance of microarray data. *Methods Enzymol.*, *411*, 214–233.

Baggerly, K.A., Coombes, K.R. *et al.* (2001). Identifying differentially expressed genes in cDNA microarray experiments. *J. Comput. Biol.*, *8* (6), 639–659.

Baitaluk, M., M. Sedova, *et al.* (2006). BiologicalNetworks: visualization and analysis tool for systems biology. *Nucleic Acids Res.*, *34* (Web Server issue): W466–471.

Bammler, T., Beyer, R.P. *et al.* (2005). Standardizing global gene expression analysis between laboratories and across platforms. *Nat. Methods*, *2* (5), 351–356.

Bloom, G., Yang, I.V. *et al.* (2004). Multi-platform, multi-site, microarray-based human tumor classification. *Am. J. Pathol.*, *164* (1), 9–16.

Boulesteix, A.L., Tutz, G. *et al.* (2003). A CART-based approach to discover emerging patterns in microarray data. *Bioinformatics*, *19* (18), 2465–2472.

Brazma, A., Hingamp, P. *et al.* (2001). Minimum information about a microarray experiment (MIAME)-toward standards for microarray data. *Nat. Genet.*, *29* (4), 365–371.

Breitkreutz, B.J., Stark, C. *et al.* (2003). Osprey: a network visualization system. *Genome Biol.*, *4* (3): R22.

Brown, M.P., Grundy, W.N. *et al.* (2000). Knowledge-based analysis of microarray gene expression data by using support vector machines. *Proc. Natl. Acad. Sci. U.S.A.*, *97* (1), 262–267.

Bushel, P.R., Hamadeh, H.K. *et al.* (2002). Computational selection of distinct class- and subclass-specific gene expression signatures. *J. Biomed. Inform.*, *35* (3), 160–170.

Churchill, G.A. (2002). Fundamentals of experimental design for cDNA microarrays. *Nat. Genet.*, *32* (Suppl.), 490–495.

Committee on Applications of Toxicogenomic Technologies to Predictive Toxicology and Risk Assessment, N. R. C. (2007). *Applications of Toxicogenomic Technologies to Predictive Toxicology and Risk Assessment*, The National Academies Press.

Dahlquist, K.D., Salomonis, N. *et al.* (2002). GenMAPP, a new tool for viewing and analyzing microarray data on biological pathways. *Nat. Genet.*, *31* (1), 19–20.

Datta, A., Choudhary, A. *et al.* (2004). External control in Markovian genetic regulatory networks: the imperfect information case. *Bioinformatics*, *20* (6), 924–930.

Dhanasekaran, S.M., Barrette, T.R. *et al.* (2001). Delineation of prognostic biomarkers in prostate cancer. *Nature*, *412* (6849), 822–826.

Dobbin, K.K., Kawasaki, E.S. *et al.* (2005). Characterizing dye bias in microarray experiments. *Bioinformatics*, *21* (10), 2430–2437.

Doniger, S.W., Salomonis, N. *et al.* (2003). MAPPFinder: using Gene Ontology and GenMAPP to create a global gene-expression profile from microarray data. *Genome Biol.*, *4* (1), R7.

Dyrskjot, L., Zieger, K. *et al.* (2005). A molecular signature in superficial bladder carcinoma predicts clinical outcome. *Clin. Cancer Res.*, *11* (11), 4029–4036.

Efron, B., Tibshirani, R. (2001). Microarrays, empirical Bayes methods, and false discovery rates. *Genet. Epidemiol.*, *23* (1), 70–86.

Eisen, M.B., Spellman, P.T. *et al.* (1998). Cluster analysis and display of genome-wide expression patterns. *Proc. Natl. Acad. Sci. U.S.A.*, *95* (25), 14863–14868.

Elcombe, C.R., Odum, J. *et al.* (2002). Prediction of rodent nongenotoxic carcinogenesis: evaluation of biochemical and tissue changes in rodents following exposure to nine nongenotoxic NTP carcinogens. *Environ. Health Perspect.*, *110* (4), 363–375.

Famili, I., Mahadevan, R. *et al.* (2005). k-Cone analysis: determining all candidate values for kinetic parameters on a network scale. *Biophys. J.*, *88* (3): 1616–1625.

Fielden, M.R., Brennan, R. *et al.* (2007). A gene expression biomarker provides early prediction and mechanistic assessment of hepatic tumor induction by nongenotoxic chemicals. *Toxicol. Sci.*, *99* (1), 90–100.

Friedman, N., Linial, M. *et al.* (2000). Using Bayesian networks to analyze expression data. *J. Comput. Biol.*, *7* (3–4), 601–620.

Golub, T.R., Slonim, D.K. *et al.* (1999). Molecular classification of cancer: class discovery and class prediction by gene expression monitoring. *Science*, *286* (5439), 531–537.

Hamadeh, H.K., Bushel, P.R. *et al.* (2002a). Prediction of compound signature using high density gene expression profiling. *Toxicol. Sci.*, *67* (2), 232–240.

Hamadeh, H.K., Bushel, P.R. *et al.* (2002b). Gene expression analysis reveals chemical-specific profiles. *Toxicol. Sci.*, *67* (2), 219–231.

Hamadeh, H.K., Jayadev, S. *et al.* (2004). Integration of clinical and gene expression endpoints to explore furan-mediated hepatotoxicity. *Mutat. Res.*, *549* (1–2), 169–183.

Hashimoto, R.F., Kim, S. *et al.* (2004). Growing genetic regulatory networks from seed genes. *Bioinformatics*, *20* (8), 1241–1247.

Hayes, K.R., Bradfield, C.A. (2005). Advances in toxicogenomics. *Chem. Res. Toxicol.*, *18* (3), 403–414.

Heinloth, A.N., Irwin, R.D. *et al.* (2004). Gene expression profiling of rat livers reveals indicators of potential adverse effects. *Toxicol. Sci.*, *80* (1), 193–202.

Hu, Z., Mellor, J. *et al.* (2005). VisANT: data-integrating visual framework for biological networks and modules. *Nucleic Acids Res.*, *33* (Web Server issue): W352–357.

Imoto, S., Kim, S. *et al.* (2003). Bayesian network and nonparametric heteroscedastic regression for nonlinear modeling of genetic network. *J. Bioinform. Comput. Biol.*, *1* (2), 231–252.

Irizarry, R.A., Warren, D. *et al.* (2005). Multiple-laboratory comparison of microarray platforms. *Nat. Methods*, *2* (5), 345–350.

Kanehisa, M., Goto, S. *et al.* (2006). From genomics to chemical genomics: new developments in KEGG. *Nucleic Acids Res.*, *34* (Database issue): D354–357.

Kerr, M.K., Churchill, G.A. (2001). Experimental design for gene expression microarrays. *Biostatistics*, *2* (2), 183–201.

Krishnamurthy, L., Nadeau, J. *et al.* (2003). Pathways database system: an integrated system for biological pathways. *Bioinformatics*, *19* (8): 930–937.

Le, Q.T., Sutphin, P.D. *et al.* (2003). Identification of osteopontin as a prognostic plasma marker for head and neck squamous cell carcinomas. *Clin. Cancer Res.*, *9* (1), 59–67.

Nguyen, D.V., Rocke, D.M. (2002). Multi-class cancer classification via partial least squares with gene expression profiles. *Bioinformatics*, *18* (9), 1216–1226.

Nielsen, T.O., West, R.B. *et al.* (2002). Molecular characterisation of soft tissue tumours: a gene expression study. *Lancet*, *359* (9314), 1301–1307.

Nikitin, A., Egorov, S., Daraselia, N., Mazo, I. (2003). Pathway studio—the analysis and navigation of molecular networks. *Bioinformatics*, *19* (16): 2155–2157.

Oleksiak, M.F., Churchill, G.A. *et al.* (2002). Variation in gene expression within and among natural populations. *Nat. Genet.*, *32* (2), 261–266.

Orr, M.S., Scherf, U. (2002). Large-scale gene expression analysis in molecular target discovery. *Leukemia*, *16* (4), 473–477.

Palsson, B. (2006). *Systems Biology: Properties of Reconstructed Networks*. Cambridge University Press, New York.

Pan, W., Lin, J. *et al.* (2003). A mixture model approach to detecting differentially expressed genes with microarray data. *Funct. Integr. Genomics*, *3* (3), 117–124.

Pomeroy, S.L., Tamayo, P. *et al.* (2002). Prediction of central nervous system embryonal tumour outcome based on gene expression. *Nature*, *415* (6870), 436–442.

Quackenbush, J. (2001). Computational analysis of microarray data. *Nat. Rev. Genet.*, *2* (6), 418–427.

Quackenbush, J. (2006). Computational approaches to analysis of DNA microarray data. *Methods Inf. Med.*, *45* (Suppl. 1), 91–103.

Ramaswamy, S., Ross, K.N. *et al.* (2003). A molecular signature of metastasis in primary solid tumors. *Nat. Genet.*, *33* (1), 49–54.

Ramaswamy, S., Tamayo, P. *et al.* (2001). Multiclass cancer diagnosis using tumor gene expression signatures. *Proc. Natl. Acad. Sci. U.S.A.*, *98* (26), 15149–15154.

Ross, D.T., Scherf, U. *et al.* (2000). Systematic variation in gene expression patterns in human cancer cell lines. *Nat. Genet.*, *24* (3), 227–235.

Savoie, C.J., Aburatani, S. *et al.* (2003). Use of gene networks from full genome microarray libraries to identify functionally relevant drug-affected genes and gene regulation cascades. *DNA Res.*, *10* (1), 19–25.

Scherf, U., Ross, D.T. *et al.* (2000). A gene expression database for the molecular pharmacology of cancer. *Nat. Genet.*, *24* (3), 236–244.

Shannon, P., Markiel, A. *et al.* (2003). Cytoscape: a software environment for integrated models of biomolecular interaction networks. *Genome Res.*, *13* (11), 2498–2504.

Sherlock, G. (2000). Analysis of large-scale gene expression data. *Curr. Opin. Immunol.*, *12* (2), 201–205.

Shmulevich, I., Dougherty, E.R. *et al.* (2002a). Probabilistic Boolean Networks: a rule-based uncertainty model for gene regulatory networks. *Bioinformatics*, *18* (2), 261–274.

Shmulevich, I., Dougherty, E.R. *et al.* (2002b). Gene perturbation and intervention in probabilistic Boolean networks. *Bioinformatics*, *18* (10), 1319–1331.

Singh, D., Febbo, P.G. *et al.* (2002). Gene expression correlates of clinical prostate cancer behavior. *Cancer Cell*, *1* (2), 203–209.

Soinov, L.A. (2003). Supervised classification for gene network reconstruction. *Biochem. Soc. Trans.*, *31* (Pt 6), 1497–1502.

Sorlie, T., Perou, C.M. *et al.* (2001). Gene expression patterns of breast carcinomas distinguish tumor subclasses with clinical implications. *Proc. Natl. Acad. Sci. U.S.A.*, *98* (19), 10869–10874.

Spruill, S.E., Lu, J. *et al.* (2002). Assessing sources of variability in microarray gene expression data. *Biotechniques*, *33* (4), 916–920, 922–923.

Steiner, G., Suter, L. *et al.* (2004). Discriminating different classes of toxicants by transcript profiling. *Environ. Health Perspect.*, *112* (12), 1236–1248.

Tamada, Y., Kim, S. *et al.* (2003). Estimating gene networks from gene expression data by combining Bayesian network model with promoter element detection. *Bioinformatics*, *19* (Suppl. 2), ii227–ii236.

Theilhaber, J., Connolly, T. *et al.* (2002). Finding genes in the C2C12 osteogenic pathway by k-nearest-neighbor classification of expression data. *Genome Res.*, *12* (1), 165–176.

Thomas, J.G., Olson, J.M. *et al.* (2001a). An efficient and robust statistical modeling approach to discover differentially expressed genes using genomic expression profiles. *Genome Res.*, *11* (7), 1227–1236.

Thomas, R.S., Rank, D.R. *et al.* (2001b). Identification of toxicologically predictive gene sets using cDNA microarrays. *Mol. Pharmacol.*, *60* (6), 1189–1194.

Tusher, V.G., Tibshirani, R. *et al.* (2001). Significance analysis of microarrays applied to the ionizing radiation response. *Proc. Natl. Acad. Sci. U.S.A.*, *98* (9), 5116–5121.

van 't Veer, L.J., Dai, H. *et al.* (2002). Gene expression profiling predicts clinical outcome of breast cancer. *Nature*, *415* (6871), 530–536.

van de Vijver, M.J., He, Y.D. *et al.* (2002). A gene-expression signature as a predictor of survival in breast cancer. *N. Engl. J. Med.*, *347* (25), 1999–2009.

Vinciotti, V., Khanin, R. *et al.* (2005). An experimental evaluation of a loop versus a reference design for two-channel microarrays. *Bioinformatics*, *21* (4), 492–501.

West, M., Blanchette, C. *et al.* (2001). Predicting the clinical status of human breast cancer by using gene expression profiles. *Proc. Natl. Acad. Sci. U.S.A.*, *98* (20), 11462–11467.

Wiback, S.J., Mahadevan, R. *et al.* (2004). Using metabolic flux data to further constrain the metabolic solution space and predict internal flux patterns: the Escherichia coli spectrum. *Biotechnol. Bioeng.*, *86* (3): 317–331.

Williams, R.D., Hing, S.N. *et al.* (2004). Prognostic classification of relapsing favorable histology Wilms tumor using cDNA microarray expression profiling and support vector machines. *Genes Chromosomes Cancer*, *41* (1), 65–79.

Yang, Y.H., Speed, T. (2002). Design issues for cDNA microarray experiments. *Nat. Rev. Genet.*, *3* (8), 579–588.

Yauk, C.L., Berndt, M.L. *et al.* (2004). Comprehensive comparison of six microarray technologies. *Nucleic Acids Res.*, *32* (15), e124.

Zou, M., Conzen, S.D. (2005). A new dynamic Bayesian network (DBN) approach for identifying gene regulatory networks from time course microarray data. *Bioinformatics*, *21* (1), 71–79.

CHAPTER 26

Role of Bioactivation Reactions in Chemically Induced Nephrotoxicity

LAWRENCE H. LASH

26.1 OVERVIEW OF RENAL STRUCTURE AND FUNCTION: TOXICOLOGICAL IMPLICATIONS

The kidneys are uniquely sensitive to toxicants because they receive and filter a large quantity of blood relative to their weight. On average, the two kidneys receive 25% of the cardiac output, while comprising only about 1% of total body weight. This fact, along with the presence of a myriad of plasma membrane transport proteins, which can result in a high degree of accumulation and concentration of chemicals within the tubular epithelial cells, and drug metabolism enzymes that can result in bioactivation of inert or nontoxic chemicals, contribute to the susceptibility of the kidneys. The major function of the kidneys is to excrete waste products while maintaining total body electrolytes, water, and acid–base balance. Thus, the kidneys are the primary organs responsible for maintaining the constancy of the internal environment. They accomplish this task by three general mechanisms: (i) glomerular filtration, (ii) tubular reabsorption, and (iii) tubular secretion. Although the focus of this chapter is on the role of metabolism in chemically induced nephrotoxicity, it is important to appreciate the role that normal renal physiology plays in the intoxication process.

The mammalian kidney is a complex organ that possesses structural divisions at both the whole organ level and the level of the nephron, which is the basic functional unit of the kidney. A detailed consideration of renal physiology and morphology is beyond the scope of this chapter. However, some key points will be made as they critically impact renal handling of drugs and patterns of chemically induced injury that are observed. Interested readers are referred to a variety of reviews and other publications that discuss renal physiology and nephron structure in greater depth (e.g. Horster, 1978; Jacobson, 1981; Walker and Valtin, 1982; Guder and Ross, 1984; Brenner, 1996; Guder, Wagner and Wirthensohn, 1986; Sands and Verlander, 2005).

At the whole organ level, the kidney is subdivided into cortex and medulla, with the latter being further divided into outer and inner stripes of the outer medulla and inner medulla or papilla. Each of these zones is comprised of distinctive cell types and

Drug Metabolism Handbook: Concepts and Applications, Edited by Ala F. Nassar, Paul F. Hollenberg, and JoAnn Scatina
Copyright © 2009 by John Wiley & Sons, Inc.

receives distinctive patterns of blood flow. For example, the renal medulla receives much less blood supply than the cortex, generally contains lower as well as different profiles of activities of drug-metabolizing enzymes and membrane transporters than the cortex, and exhibits a higher degree of anaerobic metabolism than the cortex. Some of the consequences of these differences are that the renal medulla is particularly sensitive to hypoxia, ischemia, or drugs that result in inhibition of energy metabolism (Shanley *et al.*, 1988; Epstein *et al.*, 1989), and that most drugs exhibit regional selectivity in how they are accumulated in and metabolized by the kidneys.

The nephron, which is the major functional unit of the renal epithelium, is comprised of multiple and diverse cell types. Depending on the detail by which one classifies nephron cell types, at least 12 distinct cell populations are found: early and late proximal convoluted tubule, proximal straight tubule, descending and ascending thin limbs of the loop of Henle, medullary and cortical thick ascending limb of the loop of Henle, distal convoluted tubule, connecting tubule, cortical collecting tubule, and medullary and papillary collecting ducts. Each of these cell types possesses characteristic morphology, cellular energetics, transporters, and drug metabolism enzymes. The primary nephron cell type that is of interest for drug metabolism is the proximal tubule. Reasons for this include that this is the first cell population that becomes exposed to either filtered or blood-borne chemicals and that, in contrast to many other cell types in the kidney, such as distal tubules or medullary thick ascending limb cells, proximal tubules possess relatively high activities of plasma membrane transporters and intracellular drug metabolism enzymes.

The following four factors are, therefore, critical determinants of susceptibility to chemically induced nephrotoxicity: (i) high renal blood flow relative to organ weight delivers a disproportionately high amount of blood-borne chemicals to the kidneys; (ii) glomerular filtration, which concentrates chemicals in the tubular fluid; (iii) the presence of a large array of transporters on the basolateral and brush-border plasma membranes, particularly in proximal tubular epithelial cells, for organic anions and cations, amino acids, glucose, and other chemicals; this enables accumulation of many drugs and xenobiotics inside the cell; and (iv) once inside the cell (primarily proximal tubular epithelial cells), accumulated chemicals can be substrates for a wide array of both phase I and phase II drug metabolism enzymes, some of which can lead to bioactivation by formation of reactive intermediates. The last factor, that of bioactivation of accumulated drugs and xenobiotics, is the subject of this chapter.

26.2   INTERORGAN AND INTRA-RENAL BIOACTIVATION PATHWAYS

Taking into account renal blood flow, glomerular filtration, and the presence of several active transport proteins on basolateral and brush-border plasma membranes of proximal tubular epithelial cells, patterns of interorgan and intra-renal pathways for metabolism and transport of drugs and other xenobiotics that highlight all the key functions of the kidneys in general, and the proximal tubular epithelial cells in particular, are shown in Figs 26.1 and 26.2. For both drugs that are metabolized by the cytochrome P450 (CYP) system and the glutathione (GSH) conjugation pathway, most of the parent compound is metabolized in the liver. In many cases, the liver efficiently catalyzes efflux of the metabolites into bile or plasma and, through enterohepatic and renal-hepatic circulations, delivers a variety of metabolites to the kidneys. The ability of the liver to translocate chemicals to the circulatory system and that of the kidneys

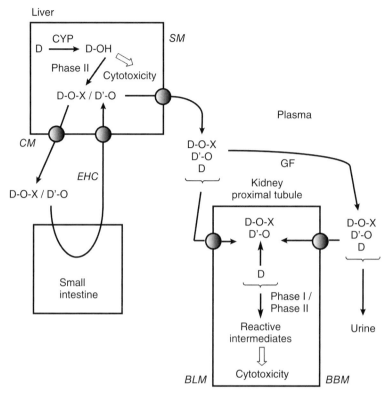

Figure 26.1 Scheme of interorgan pathways involving cytochrome P450 (CYP). Metabolism and transport of drugs (D) in the liver, kidneys, and small-intestinal epithelium. Processes depicted in the liver include metabolism of a drug by CYP enzymes to yield the hydroxylated metabolite (D-OH), phase II metabolism to yield either a conjugate (D-O-X) or another species (D′-O). These metabolites can be exported into bile by transport across the canalicular plasma membrane (CM), where it may undergo additional metabolism in the intestines and then be returned to the liver via enterohepatic circulation (EHC). The various metabolites are then exported across the sinusoidal plasma membrane (SM) to be delivered to the kidneys, where the drugs may either undergo glomerular filtration (GF) or be absorbed from the renal periplasmic space by transport across the basolateral plasma membrane (BLM). Filtered drugs and their metabolites may either be excreted in the urine or reabsorbed by transport into the cell across the brush-border plasma membrane (BBM). Inside the proximal tubular cell, the drugs and their various metabolites may undergo various bioactivation reactions to yield reactive intermediates and cause cytotoxicity.

to concentrate both blood-borne and filtered chemicals, play a major role in delivery of chemicals to bioactivation enzymes in the proximal tubular epithelium.

For drugs that are metabolized by CYP reactions (Fig. 26.1), the initial reaction occurs predominantly in liver, where it may either exert hepatotoxicity or be metabolized by a phase II conjugation reaction. The conjugate thus formed (glucuronide, sulfate, or GSH conjugate) is readily excreted into bile and eventually gets to the kidneys by enterohepatic and renal-hepatic circulations. The parent compound that is not absorbed and metabolized initially in the liver may be metabolized by renal CYPs,

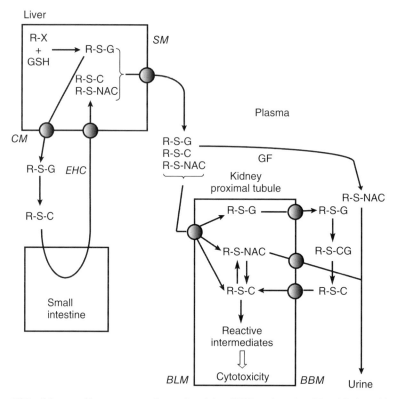

Figure 26.2   Scheme of interorgan pathways involving GSH conjugation. Xenobiotics with good leaving groups (R-X) may undergo conjugation with GSH to form the GSH conjugate (R-S-G), which occurs primarily in the liver. The liver readily exports conjugates into bile by transport across the canalicular plasma membrane (CM), where they are largely metabolized to the corresponding cysteine conjugate (R-S-C) and mercapturate (R-S-NAC). Enterohepatic circulation (EHC) delivers these conjugates back to the liver, which then exports them into plasma by transport across the sinusoidal plasma membrane (SM) for delivery to the kidneys. The renal proximal tubular cell is exposed to the various conjugates via either glomerular filtration (GF), which delivers the conjugates to the tubular lumen, or via the renal periplasmic space, from which they can be transported into the cell by carriers on the basolateral plasma membrane (BLM). GSH conjugate in the lumen is subject to degradation, forming the cysteinylglycine conjugate (R-S-CG) and then R-S-C. The latter is transported into the cell by transport across the brush-border plasma membrane (BBM), where they can either be *N*-acetylated to form the mercapturate, or can be a substrate for bioactivation enzymes that generate reactive intermediates that cause cytotoxicity. R-S-NAC can either be secreted into the lumen and excreted in the urine or can be deacetylated in the cell to regenerate R-S-C.

although this occurs to a much smaller extent than in the liver. Various CYP-derived metabolites may elicit toxicity within the kidney or be excreted into the urine.

For drugs that are metabolized by the GSH conjugation pathway (Fig. 26.2), all pathways of metabolism, transport, and interorgan translocation eventually lead to the kidneys, regardless of whether the initial reaction occurs in the kidneys or in the liver. If the GSH conjugate is formed in the liver, it is primarily excreted into bile, where it can be degraded by biliary or subsequently intestinal γ-glutamyltransferase (GGT) and dipeptidase (DP) to the corresponding cysteine conjugate. This cysteine conjugate is ultimately delivered to the kidneys by enterohepatic and renal-hepatic circulations. However, most of the degradation of GSH conjugate to the corresponding cysteine conjugate occurs in the kidneys as this is the predominant site of GGT in the body (Hinchman and Ballatori, 1990). Once the cysteine conjugate is formed in the kidneys, it has several potential fates, which will be discussed in detail in Section 26.4.

Regardless of whether the key metabolite is derived from CYP activity or GSH conjugation, these interorgan pathways illustrate the close interdependence of metabolism and transport processes in determining target organ specificity. As will be discussed later, similar activities that lead to bioactivation of cysteine conjugates and formation of reactive species in the kidneys are also present in the liver. No hepatotoxicity, however, is observed because of the efficiency of transport processes that mediate excretion of these metabolites from the liver and uptake by the kidneys.

26.3 CYTOCHROME P450 (CYP)-DEPENDENT BIOACTIVATION IN RENAL PROXIMAL TUBULAR CELLS

The basic biochemistry of the CYP monooxygenase system in the kidneys is essentially the same as that in the more studied liver, although important differences exist (Lock and Reed, 1997; Lohr, Willsky and Acara, 1998). First, amounts of CYP enzymes are generally only 5% to 20% of those in the liver. Second, CYP enzymes are not uniformly distributed throughout the cell types of the nephron. Rather, most of the CYP enzymes are found in the proximal tubular epithelium, although some are also found in other cell types, such as the glomerular epithelium or distal tubular cells (Guder and Ross, 1984; Lock and Reed, 1997; Lohr, Willsky and Acara, 1998). Third, the pattern of expression of CYP enzymes differs between the liver and kidney, with the liver expressing a more extensive array of individual enzymes. Finally, another tissue-specific difference is that for many CYP enzymes that are expressed in both liver and kidney, substrate specificity and inducibility differ (Anders, 1980; Jones, Orrenius and Jakobson, 1980; Lock and Reed, 1997; Lohr, Willsky and Acara, 1998; Ronis *et al.*, 1998). This has important implications when considering metabolism of a specific chemical in tissue samples from animals exposed either prior to or simultaneously with a CYP inducer.

When considering CYP-dependent metabolism in the kidneys, species-dependent differences are very prominent (Table 26.1). This has to be taken into account when extrapolating metabolism data from rodents to humans. For example, the environmental contaminant and "probable human carcinogen" trichloroethylene (TRI) (NTP, 2005) is metabolized by several CYP enzymes, although CYP2E1 is the primary enzyme that catalyzes TRI oxidation (Lash *et al.*, 2000). CYP-dependent metabolism of TRI in rat kidney occurs primarily by CYP2E1 and secondarily by CYP2C11 (Cummings, Parker and Lash, 2001), but is undetectable in human kidney (Cummings and Lash, 2000). Another factor that is important when considering renal CYP metabolism in humans is the existence of interindividual variability due to genetic polymorphisms (Lash, Putt and Cai, 2008).

As in the liver, the kidney contains four major families of CYPs, CYP1–CYP4. Although endogenous substrates exist for several enzymes, particularly for enzymes of the CYP2, CYP3, and CYP4 families, enzymes of the CYP1, CYP2, and CYP3 families are those that are involved in xenobiotic metabolism. A brief summary of the

TABLE 26.1 Major cytochrome P450 (CYP) enzymes in rat and human kidney.

CYP enzyme	Rats	Humans
CYP1A1/2	Present	Absent or low levels
CYP1B1/2	Present	Present
CYP2B1/2	Present	Absent
CYP2C11 (CYP2C19)	Present	Absent
CYP2D6	Present	Present at low levels
CYP2E1	Present	Absent
CYP3A1/2 (CYP3A4/5)	Present	Present
CYP4A2/3 (CYP4A11)	Present	Present

CYP enzymes present in rats are listed with the human orthologues, when appropriate, listed in parentheses.

properties of enzymes of each CYP family, with an emphasis on species and tissue differences and their role in metabolism of endogenous as opposed to exogenous, or xenobiotic, chemicals will be presented later. In Sections 26.5 and 26.6, some examples of drugs and xenobiotics that are activated by these and other pathways will be given.

The CYP1 family consists of CYP1A1/2 and CYP1B1/2, with the former being that which is inducible by 3-methylcholanthrene, 2,3,7,8-tetrachlorodibenzo-p-dioxin (TCDD), and β-naphthoflavone. In the rat, CYP1A1 is only detected after exposure to inducers.

The CYP2 family is a highly diverse group of enzymes that metabolize a broad group of substrates, including endogenous compounds such as androgens and estrogens, and xenobiotics such as ethanol, halogenated solvents, and debrisoquine. Enzymes of the CYP2A family are expressed in mouse, but not in rat or human, kidney (Lock and Reed, 1997). CYP2B1/2 are inducible by phenobarbital in rat liver but not in rat kidney. In contrast, clofibrate induces expression of CYP2B1/2 in rat kidney (Cummings *et al.*, 1999) but cannot be detected in human kidney (Lash, Putt and Cai, 2008). Enzymes of the CYP2C family are generally considered to be constitutively expressed and exhibit sex specificity and developmental differences. For drug and xenobiotic metabolism, the major CYP2C enzyme of interest in rats is CYP2C11 (Cummings *et al.*, 1999; Cummings, Parker and Lash, 2001). There is no evidence for expression of the human orthologue CYP2C19 in kidney (Lash, Putt and Cai, 2008). Enzymes of the CYP2D family have not been readily detected in the kidney, although there is evidence that some family members are present in rat (Gonzalez, 1990) and CYP2D6 is present at low levels in human kidney microsomes (Lash, Putt and Cai, 2008). The CYP2E family, the most prominent of which is CYP2E1, metabolizes a large array of small chemicals, including ethanol, acetaminophen, halogenated solvents such as TRI and chloroform, and is inducible by ethanol, acetone, pyridine, pyrazole, and 4-methylpyrazole. Although CYP2E1 is expressed in rat kidney (Ronis *et al.*, 1998; Cummings *et al.*, 1999), its expression has not been detected in human kidney (Amet *et al.*, 1997; Cummings and Lash, 2000; Cummings, Lasker and Lash, 2000; Lash, Putt and Cai, 2008). In the rat and mouse kidney, CYP2E1 expression is under testosterone regulation and males have significantly higher levels than females. Accordingly, male mice are markedly more sensitive to the nephrotoxicity of certain CYP2E1 substrates than are females (Hu, Rhoten and Yang, 1990). This is a clear example of where extrapolation of metabolism and toxicity data from rodents to humans is not possible because of qualitative differences in metabolism.

The CYP3 gene family, of which CYP3A1 and CYP3A2 are the major enzymes in rats and CYP3A4 and CYP3A5 are the major orthologous enzymes in humans, is prominent in the liver, is inducible by glucocorticoids, and is involved in steroid hydroxylation and metabolism of macrolide antibiotics such as cyclosporine. CYP3A1/2 was detected in microsomes from rat kidney cortex homogenates (Schuetz *et al.*, 1992; Bebri *et al.*, 1995; Cummings *et al.*, 1999) and CYP3A4/5 was detected in microsomes from human kidney cortex homogenates (Schuetz *et al.*, 1992; Lash, Putt and Cai, 2008). CYP3A1/2 protein expression, however, was not detected in microsomes from renal proximal tubular cells (Cummings *et al.*, 1999), suggesting that these CYPs may be localized primarily in other regions of the nephron besides the proximal tubule, such as the glomerulus. In human kidney microsomes, CYP3A4/5 were readily detected in some, but not all samples, consistent with the presence of genetic polymorphisms, which is well known for the liver (Lash, Putt and Cai, 2008).

Enzymes of the CYP4A family are primarily involved in metabolism of fatty acids and arachidonic acid, are prominently expressed in the kidneys (Okita *et al.*, 1997; Ito *et al.*, 1998; Stec *et al.*, 2003), and are important in renal physiology, because of the generation of metabolites such as hydroxyeicosatetraenoic acid that influence hemodynamics. CYP4A enzymes are induced by hypolipidemic drugs, such as the fibrates, and other chemicals that cause peroxisome proliferation. These inducing effects, however, are more prominent in the liver than in the kidney and in rodents than in humans (Goldsworthy and Popp, 1987). The major CYP4A enzymes in rat kidney are CYP4A2 and CYP4A3. They hydroxylate fatty acids such as lauric acid and catalyze ω- and ω-1 hydroxylation of arachidonic acid, generating metabolites that, as noted earlier, are important in the regulation of blood pressure and renal hemodynamics. The major CYP4A enzyme in human kidney is CYP4A11, which also metabolizes lauric and arachidonic acids. CYP4A11 is localized exclusively in the S2 and S3 segments of the proximal tubule (Lasker *et al.*, 2000) and its expression in primary cultures of human proximal tubular cells is induced by ethanol and dexamethasone (Cummings, Lasker and Lash, 2000). CYP4A11 is considered the human orthologue of rat CYP4A2.

26.4 FLAVIN-CONTAINING MONOOXYGENASE (FMO)-DEPENDENT BIOACTIVATION IN RENAL PROXIMAL TUBULAR CELLS

FMOs are a multigene family of microsomal enzymes that catalyze oxidation of sulfur-, selenium-, and nitrogen-containing drugs and xenobiotics (Ziegler, 1993; Cashman and Zhang, 2006). Five active isoforms (FMO1–5) have been identified and are present in most mammalian tissues, although species-, sex-, tissue-, and developmental-dependent differences exist in expression of these isoforms (Hines *et al.*, 1994; Dolphin *et al.*, 1998; Ripp *et al.*, 1999; Koukouritaki *et al.*, 2002). Like the CYPs, FMOs are localized in the endoplasmic reticulum (microsomes), catalyze oxidation of a broad range of xenobiotics as well as endogenous compounds, and are most prominently expressed in the liver, although extrahepatic expression occurs.

Primary cultures of human proximal tubular cells express only three isoforms of FMO: FMO1, FMO3, and FMO5 (Krause, Lash and Elfarra, 2003). In contrast to human liver, which expresses primarily FMO3 as well as several other FMO enzymes at significant levels, human kidney expresses primarily FMO1, somewhat lower levels of protein for FMO5, and very low levels of protein for FMO3 (Krause, Lash and Elfarra, 2003). Nishimura and Naito (2006) determined mRNA expression profiles of

FMOs in human kidney, and found that by far the major mRNA expressed was that of FMO1; mRNAs for FMO2, FMO3, FMO4, and FMO5 were detected at 4.0%, 0.09%, 24.7%, and 12.6%, respectively, of the levels found for FMO1. Additionally, FMO1 protein expression levels varied significantly among human kidney samples, suggesting the presence of polymorphisms (Krause, Lash and Elfarra, 2003).

Although cysteine conjugate β-lyase (CCBL) was the first discovered enzymatic activity of the GSH-dependent bioactivation pathway that acts on cysteine conjugates (see Section 26.5), is the primary catalyst of cysteine conjugate bioactivation in many systems, and is the best and most frequently studied reaction, it became increasingly evident around 1990 that other metabolites could be formed and that additional enzymes besides the CCBL were involved in their formation. Sulfoxides have long been known as metabolites of many cysteine conjugates and are often recovered in the urine of animals treated with cysteine conjugates or their precursors, as stable end products. Elfarra and colleagues identified and subsequently purified an enzymatic activity from rat liver and kidney microsomes that metabolizes several cysteine conjugates, including S-(1,2-dichlorovinyl)-L-cysteine (DCVC), to sulfoxides (Sausen and Elfarra, 1990; Sausen, Duescher and Elfarra, 1993). Early evidence from these studies implicating FMOs included inhibition of S-oxidation of cysteine conjugates by other FMO substrates (e.g. N,N-dimethylaniline, methimazole, and n-octylamine), NADPH dependence, restoration of activity of flavin-depleted microsomes by exogenous flavin adenine dinucleotide (FAD) or flavin mononucleotide (FMN), and immunoreactivity of purified S-oxidase with antibodies to FMOs.

The role of FMOs in xenobiotic bioactivation in the kidneys is not well characterized, with the exception of its role in the metabolism of certain nephrotoxic cysteine conjugates to reactive sulfoxides (Lash et al., 1994, 2003; Cummings and Lash, 2000; Elfarra and Krause, 2007). As described in Section 26.5, the cysteine conjugate of TRI, DCVC, is a substrate for FMOs, which produce the reactive DCVC sulfoxide that can then react with GSH and other cellular nucleophiles to produce cytotoxicity (Sausen and Elfarra, 1991). Incubations of human cDNA-expressed FMO1, FMO3, FMO4, and FMO5 with DCVC showed that DCVC sulfoxide was only detected with FMO3 (Krause, Lash and Elfarra, 2003). The observation that FMO1, rather than FMO3, is the major isoform present in rat and human proximal tubular cells raises the question of the importance of this pathway in DCVC bioactivation. In support of a role for FMO in DCVC-induced nephrotoxicity, however, is the observation of significant protection by pretreatment with the FMO substrate methimazole (Lash et al., 1994, 2003; Cummings and Lash, 2000). Moreover, DCVC sulfoxide is a more potent nephrotoxicant in vivo and cytotoxicant in vitro than DCVC (Lash et al., 1994, 2003). Further studies are needed to resolve this uncertainty about the relative importance of FMO-dependent bioactivation.

26.5 RENAL BIOACTIVATION OF CYSTEINE CONJUGATES: GSH CONJUGATION AS A BIOACTIVATION RATHER THAN A DETOXICATION PATHWAY

26.5.1 GSH Conjugation

GSH conjugation of reactive electrophiles is classically considered the first step in the mercapturic acid pathway, which is a prominent detoxification mechanism for many xenobiotics. With studies beginning in the 1960s, however, it became apparent that several classes of chemicals, chief among them halogenated alkanes and alkenes, are bioactivated by GSH conjugation. Although there are some reports of toxicity in other target organs, by and large toxicity from xenobiotics that are bioactivated by the GSH conjugation pathway is restricted to the kidneys. Factors responsible for this target-tissue specificity are those that are discussed in Section 26.2, and include the discrete tissue localization of some of the enzymes of the GSH conjugation pathway and of membrane transporters for efflux and uptake of various metabolic intermediates in the pathway.

Multiple families of glutathione S-transferases (GSTs) exist in mammals. GSTs can be found primarily in three subcellular compartments: cytoplasm, mitochondria, and endoplasmic reticulum (or microsomes). The isoforms that are important for xenobiotic metabolism in the kidneys are those found in the cytoplasm and microsomes. All cytoplasmic GSTs are dimeric with subunits of 199 to 244 amino acids in length. Based on amino acid sequence, seven classes of cytoplasmic GSTs exist in mammals, and are designated Alpha, Mu, Pi, Sigma, Theta, Omega, and Zeta. In rodents and humans, GST classes are defined as those enzymes that share >40% amino acid sequence identity. The convention used is to refer to rodent GSTs by Greek letters (i.e. GSTα, μ, π, σ, τ, φ, and ζ) and human GSTs by capital Arabic letters (i.e. GSTA, M, P, S, T, O, and Z). At least 16 cytoplasmic GST subunits exist in humans, many of which exist as heterodimers. Consequently, a large variety of isoenzymes can be generated from the various subunits.

In the rat kidney, three families of cytoplasmic GSTs have been studied in some detail and have been found to be expressed: GSTα, GSTπ, and GSTμ. Immunolocalization studies (Rozzel *et al.*, 1993) and Western blot analyses in renal cortical microsomes and isolated renal proximal tubular and distal tubular cells (Cummings, Parker and Lash, 2000) showed that only GSTα is expressed in the proximal tubules whereas GSTπ and GSTμ are expressed in the distal nephron. In contrast to this distribution of GST isoforms, human renal proximal tubular cells express GSTA, GSTP, and GSTM (Cummings, Lasker and Lash, 2000; Lash, Putt and Cai, 2008). Inasmuch as the various GST isoforms have broad and often overlapping substrate specificities, it is difficult to assess the impact of this difference in GST isoform expression pattern on extrapolation of renal metabolism data from rats to humans. Suffice it to say, however, the markedly distinct profiles of GST isoform expression will have an impact on substrate specificities and relative activities.

Besides the more commonly studied families of cytoplasmic GSTs, the Zeta class was recently discovered by a bioinformatics approach using human expressed sequence tag databases (Board *et al.*, 1997, 2001; Board and Anders, 2005). GSTZs are widely distributed in eukaryotes and were shown to be identical to maleylacetoacetate isomerase, which catalyzes the penultimate step in tyrosine catabolism. The substrate specificity pattern of GSTZ differs from the other GST families in lacking significant activity toward the prototypical substrate 1-chloro-2,4-dinitrobenzene. Although GSTZ exhibits some activity with organic peroxides, it plays a critical role in the metabolism of various α-haloacids, including dichloroacetic acid (DCA), which is metabolized to glyoxylic acid. Biotransformation of DCA is important because DCA is used in the clinical management of congenital lactic acidosis, is a common drinking water contaminant, and is a metabolite of the environmental contaminant and probable human carcinogen TRI (Lash *et al.*, 2000). At least four polymorphic variants of GSTZ are known in humans, each exhibiting markedly distinct catalytic activities.

The MAPEG (membrane-associated proteins in eicosanoid and glutathione metabolism) represent a unique family of GSTs that share no sequence identity with either the cytoplasmic or mitochondrial GSTs. These proteins are found in the endoplasmic reticulum or microsomes and have the distinct property of being involved in the production of eicosanoids. Four subgroups of MAPEG are known, with proteins within each subgroup sharing at least 20% sequence identity. In humans, six MAPEGs have been identified, which fall into subgroups I, II, and IV (Jakobsson, Morgenstern and Mancini, 1999). Two of the proteins are critical for leukotriene biosynthesis, and three are cytoprotective, exhibiting GSH S-transferase and peroxidase activities. Evidence suggests that MGST1 (microsomal GSH S-transferase 1) functions solely as a detoxication enzyme, whereas human MGST2 and MGST3 can both detoxify foreign compounds and synthesize leukotriene C_4. MGSTs have been immunolocalized to several rat tissues, including the kidneys (Otieno et al., 1997). The role of MAPEG in bioactivation of chemicals that ultimately become cysteine conjugate substrates for other bioactivation enzymes, such as the CCBL or FMO, is less clear, although TRI is metabolized to DCVG in the presence of GSH and liver or kidney microsomal fractions from rats or humans (Lash et al., 1995, 1998, 1999).

26.5.2 γ-Glutamyltransferase and Cysteinylglycine Dipeptidase Activities

Whether the GSH conjugate is formed in the liver or in the kidneys, the kidneys are the primary sites for its further processing because of the interorgan translocation pathways described in Section 26.2. The next two steps in the metabolism of GSH S-conjugates involve successive cleavage of the γ-glutamyl isopeptide and cysteinylglycine peptide bonds, to release the corresponding cysteine S-conjugate as the product. The first step, catalyzed by γ-glutamyltransferase (GGT), is found on the luminal membrane of several epithelial tissues, although most prominently in the renal proximal tubule (Hinchman and Ballatori, 1990). Hence, although GGT activity is found on extrarenal sites such as the hepatic canalicular plasma membrane and the jejunal brush-border plasma membrane, its extremely high activity on the renal proximal tubular brush-border plasma membrane coupled with the high activity of plasma membrane transporters and the glomerular filtration process, serves to deliver GSH conjugates primarily to the kidneys (cf. Fig. 26.2). By these mechanisms, most of the whole-body turnover of GSH and GSH conjugates can be accounted for by renal metabolism. The importance of GGT and subsequent DP activity in the turnover of GSH (and by analogy, GSH S-conjugates) is highlighted by the profound glutathionuria that occurs when GGT activity is inhibited (Griffith and Meister, 1979).

26.5.3 Cysteine Conjugate β-Lyase Activities

Metabolism to the cysteine conjugate represents a branch point because alternative reactions can lead to either detoxication and excretion or bioactivation and generation of reactive intermediates. Multiple enzymes can bioactivate those cysteine conjugates whose properties make them substrates for these enzymes in preference to the microsomal N-acetyltransferase that forms mercapturates. This concept is illustrated in Fig. 26.3. After processing of the GSH conjugate by the actions of GGT and DP activities, the cysteine conjugate thus formed can be N-acetylated by the cysteine conjugate N-acetyltransferase (NAT) to yield the mercapturate. Mercapturates are highly polar and are generally readily excreted in urine. However, they may also be acted on by a deacetylase (or aminoacylase) to regenerate the cysteine conjugate. As a consequence of this activity, many N-acetyl-L-cysteine-S-conjugates of nephrotoxic haloalkenes or

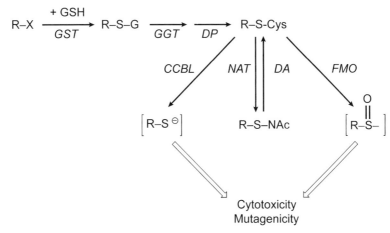

Figure 26.3   Bioactivation of GSH conjugates. Drug/xenobiotic (R-X) forms a GSH conjugate (R-S-G) that undergoes degradation by γ-glutamyltransferase (GGT) and dipeptidase (DP) to yield the cysteine conjugate (R-S-Cys). R-S-Cys functions as a branch point in that it is subject to either detoxication by action of the *N*-acetyltransferase (NAT) to form the mercapturate (R-S-NAc) or bioactivation by the action of either the cysteine conjugate β-lyase (CCBL) or flavin-containing monooxygenase (FMO), which generate reactive thiolates and sulfoxides, respectively. Mercapturates may also regenerate the cysteine conjugate by deacetylation, catalyzed by a deacetylase (DA) or aminoacylase.

haloalkanes may be deacetylated and exhibit toxicity (Wolfgang *et al.*, 1989; Zhang and Stevens, 1989; Birner *et al.*, 1998). Alternatively, certain cysteine conjugates can be substrates of either the CCBL or FMO, which generate reactive and chemically unstable metabolites.

One of the most studied activities for bioactivation of cysteine conjugates is the CCBL. The most recently published review by Cooper and Pinto (2006) highlights the multiplicity of enzymes that can catalyze this overall reaction. The various mammalian enzymes that have been identified as possessing CCBL activity in both the cytoplasm and mitochondria are listed in Table 26.2. The various CCBL activities are all pyridoxal 5′-phosphate (PLP)-containing enzymes and can catalyze either a direct β-elimination reaction to yield a reactive thiolate, or, in some cases, a transamination reaction. The transamination reaction generally yields an unstable α-keto acid that spontaneously rearranges to yield the reactive thiolate. In the kidneys, there are two known CCBL enzymes in the cytoplasm and two in the mitochondria, each capable of catalyzing both β-elimination and transamination reactions. The high-molecular-weight forms in renal cytoplasm and mitochondria have not been extensively characterized, but it has been suggested that at least in the mitochondria, this form is the primary enzyme catalyzing CCBL activity (Abraham, Patel and Cooper, 1995; Abraham, Thomas and Cooper 1995).

One consequence of the ability of the various renal CCBL enzymes to catalyze transamination reactions is that this reaction can compete with the β-elimination reaction, so that maximal CCBL activity requires the presence of an α-keto acid in the reaction mixture (Stevens, Robbins and Byrd, 1986; Elfarra, Lash and Anders, 1987). In the absence of an α-keto acid, therefore, the PLP prosthetic group of the CCBL

TABLE 26.2 **Mammalian enzymes that catalyze cysteine conjugate β-lyase activity.**

Enzyme	Tissue	Reaction	Subunit MW
Cytoplasmic enzymes			
GTK (EC 2.6.1.64)/KAT (EC 2.6.1.7)	Kidney, choroid plexus	BE, TA	45 kDa; homodimer
Kynureninase (EC 3.7.1.3)	Liver	BE	55 kDa; homodimer
cytASAT (EC 2.6.1.1)	Heart	BE	45 kDa; homodimer
ALAT (EC 2.6.1.2)	Heart	BE	45 kDa; homodimer
BCAT$_c$ (EC 2.6.1.42)	Brain	BE	44 kDa; homodimer
High-MW β-lyase	Liver, kidney	BE, TA	330 kDa
Mitochondrial enzymes			
mtASAT (EC 2.6.1.1)	Liver	BE, TA	45 kDa; homodimer
BCAT$_m$ (EC 2.6.1.42)	Brain	BE	44 kDa; homodimer
GTK (EC 2.6.1.64)/KAT (EC 2.6.1.7)	Brain, kidney	BE, TA	45 kDa; homodimer
High-MW β-lyase	Liver, kidney	BE, TA	330 kDa

ALAT, alanine aminotransferase; BCAT$_{c/m}$, branched-chain aminotransferase (cytoplasm/mitochondria); BE, beta-elimination; cyt/mitASAT, cytosolic and mitochondrial aspartate aminotransferase; GTK, glutamine transaminase K; KAT, kynurenine aminotransferase; TA, transamination.

becomes trapped in the pyridoxamine 5′-phosphate (PMP) form and activity is very low or non-detectable.

The most commonly studied nephrotoxic cysteine conjugates are those of halogenated alkanes and alkenes (Fig. 26.4). Each of the cysteine conjugates has in common the presence of a good leaving group for the CCBL reaction. The haloalkenyl conjugates are all derived from environmental contaminants that are of interest for human health risk assessment because of their prevalence in the environment, their widespread use in industry, and their known carcinogenicity in animals and "probable" carcinogenicity in humans (e.g. Lash, Parker and Scott, 2000; Lash and Parker, 2001; NTP, 2005). The examples shown here are the cysteine *S*-conjugates of trichloroethylene (DCVC), perchloroethylene (TCVC), 1,1,2,2-tetrafluoroethylene (TFEC), and 1-chloro-1,2,2-trifluoroethylene (CTFC). The haloalkenyl cysteine conjugates (DCVC, TCVC), when metabolized by the CCBL, generate reactive thiol-containing species that can cause both cytotoxicity and mutagenicity. In contrast, the haloalkyl cysteine conjugates (TFEC, CTFC) that contain only fluorines or chlorines are cytotoxic but not mutagenic, whereas those containing a bromine atom are also mutagenic (Finkelstein *et al.*, 1994). These other conjugates include several environmentally and clinically important compounds, such as chlorofluorocarbons that have been used as refrigerants (Yin, Jones and Anders, 1995) and degradation products of anesthetic agents such as sevoflurane (Iyer *et al.*, 1998; Altuntas and Kharasch, 2002; Anders, 2005).

Besides these types of halogenated solvent substrates, studies by Hanigan and colleagues (Hanigan *et al.*, 1994, 2001; Townsend and Hanigan, 2002; Townsend *et al.*, 2003a, b; Zhang and Hanigan, 2003) demonstrated that the mono-platinum-monoglutathione conjugate is a substrate for GGT, that the cysteine conjugate of cisplatin is a substrate for CCBL activities, and that nephrotoxicity due to these conjugates in mice and cytotoxicity in various *in vitro* proximal tubular cell models is variously inhibited by inhibitors of GGT and/or the CCBL (see further discussion).

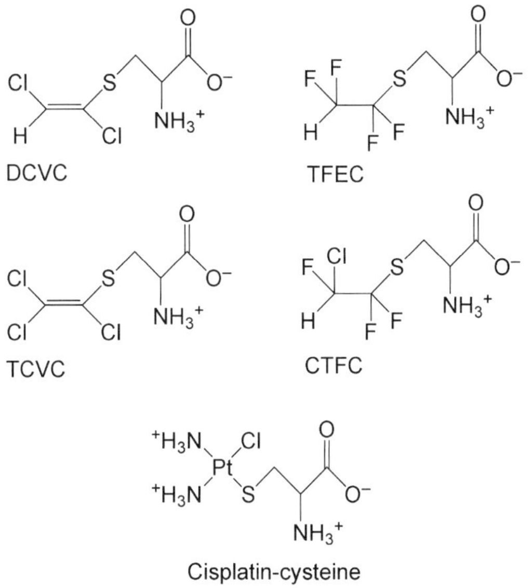

Figure 26.4 Structures of representative cysteine conjugates that undergo renal bioactivation. CTFC, *S*-(1-chloro-1,2,2-trifluoroethyl)-L-cysteine; DCVC, *S*-(1,2-dichlorovinyl)-L-cysteine; TFEC, *S*-(1,1,2,2-tetrafluoroethyl)-L-cysteine; TCVC, *S*-(1,2,2-trichlorovinyl)-L-cysteine.

26.5.4 Validation of GSH-Dependent Bioactivation Pathways

The critical importance of the CCBL activity in determining the metabolism and toxicity of various haloalkyl and haloalkenyl cysteine conjugates has been demonstrated in whole animals and in various *in vitro* preparations of renal proximal tubular cells or renal subcellular fractions (Fig. 26.5). Thus, inhibition of the CCBL activity with aminooxyacetic acid (AOAA) prevents some or all of the toxicity associated with the cysteine conjugate. Addition of appropriate α-keto acids to maximize CCBL activity also enhances cysteine conjugate-induced cytotoxicity. For those cysteine conjugates that are also substrates for FMO, alternative FMO substrates such as methimazole are protective both *in vivo* and *in vitro*, as discussed earlier. The scheme also illustrates how the toxicity of conjugates at earlier steps in the pathway can be modulated by the use of inhibitors or co-substrates. Thus, acivicin, which is an irreversible inhibitor of GGT, prevents the toxicity of GSH conjugates and addition of glycylglycine (GlyGly) or another suitable γ-glutamyl group acceptor, can increase GGT activity and enhance the toxicity of GSH conjugates. Addition of either 1,10-phenanthroline or phenylalanylglycine (PheGly), which is an irreversible or competitive inhibitor of the dipeptidase activity, respectively, protects from the toxicity of both GSH and cysteinylglycine conjugates. Finally, several studies have demonstrated, in both *in vivo* studies and in isolated proximal tubular cell preparations, that probenecid can protect against the cytotoxicity of either GSH or cysteine conjugates, implicating function of the basolateral organic anion transporters in the uptake and accumulation of these conjugates in proximal tubular cells. Hence, modulation of both transport and metabolism can result in alteration of conjugate nephrotoxicity *in vivo* and cytotoxicity *in vitro*.

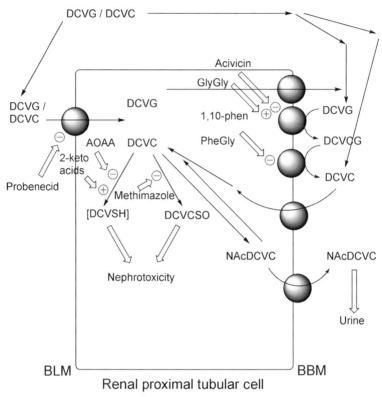

Figure 26.5   Validation of role of enzymatic and transport processes in the renal proximal tubule producing nephrotoxicity. Using the GSH-derived conjugates of the environmental contaminant trichloroethylene as an example, transport and metabolism of the various conjugates are illustrated. The GSH conjugate S-(1,2-dichlorovinyl)glutathione (DCVG) is either transported across the basolateral plasma membrane (BLM) by probenecid-sensitive organic anion transporters or undergoes glomerular filtration, which delivers it to the lumen for metabolism by γ-glutamyltransferase (GGT) to the cysteinylglycine conjugate S-(1,2-dichlorovinyl)-L-cysteinylglycine (DCVCG) and the dipeptidase (DP) to the cysteine conjugate S-(1,2-dichlorovinyl)-L-cysteine (DCVC). DCVC enters the renal proximal tubular cell by either transport across the BLM by probenecid-sensitive organic anion or amino acid transporters or transport across the brush-border plasma membrane (BBM). GGT and DP activities can be modulated by acivicin or glycylglycine (GlyGly) and 1,10-phenanthroline (1,10-phen) or phenylalanylglycine (PheGly), respectively. Once inside the proximal tubular cell, DCVC can be N-acetylated to form the mercapturate S-(1,2-dichlorovinyl)-N-acetyl-L-cysteine (NAcDCVC), which can either be excreted in the urine or deacetylated to regenerate DCVC. DCVC can also be a substrate for cysteine conjugate β-lyase (CCBL) or flavin-containing monooxygenase (FMO) activities to generate reactive 1,2-dichlorovinylthiol (DCVSH) or DCVC sulfoxide (DCVCSO), respectively. CCBL activity can be modulated by aminooxyacetic acid (AOAA) or 2-keto acids, and FMO activity can be modulated by co-substrates such as methimazole. Correspondence between inhibition and stimulation of each transport or metabolic step and nephrotoxicity *in vivo* or cytotoxicity *in vitro* validated these pathways. –, inhibition; +, stimulation.

26.6 RENAL BIOACTIVATION OF ACETAMINOPHEN

Although acetaminophen (APAP) is a commonly used analgesic and is considered safe to use under appropriate conditions, it represents a significant drug of concern for public health as it is a frequent cause of poisonings due to overdose. Whereas the liver is typically the most prominent target of toxicity in APAP overdose, the kidneys can also significantly metabolize APAP to both detoxication and bioactivation products (Newton *et al.*, 1982), which can result in analgesic nephropathy (Duggin, 1980). The overall metabolic pathway for APAP, which occurs in both liver and kidneys (Fig. 26.6), illustrates the presence of several competing pathways for both detoxication and bioactivation. Quantitatively, the predominant pathways are either glucuronidation or sulfation, which lead to highly polar metabolites that are readily excreted. A lower activity pathway is mediated by CYP (predominantly CYP2E1 but also CYP1A2). CYP-dependent metabolism produces a reactive intermediate, *N*-acetyl-*p*-benzoquinoneimine (NAPQI), which readily reacts with nucleophiles such as protein sulfhydryl groups and GSH. GSH conjugation of APAP or its deacetylated congener *p*-aminophenol (PAP) has usually been viewed as a detoxication route, whereby the GSH conjugates are processed to mercapturates, which are readily excreted in the

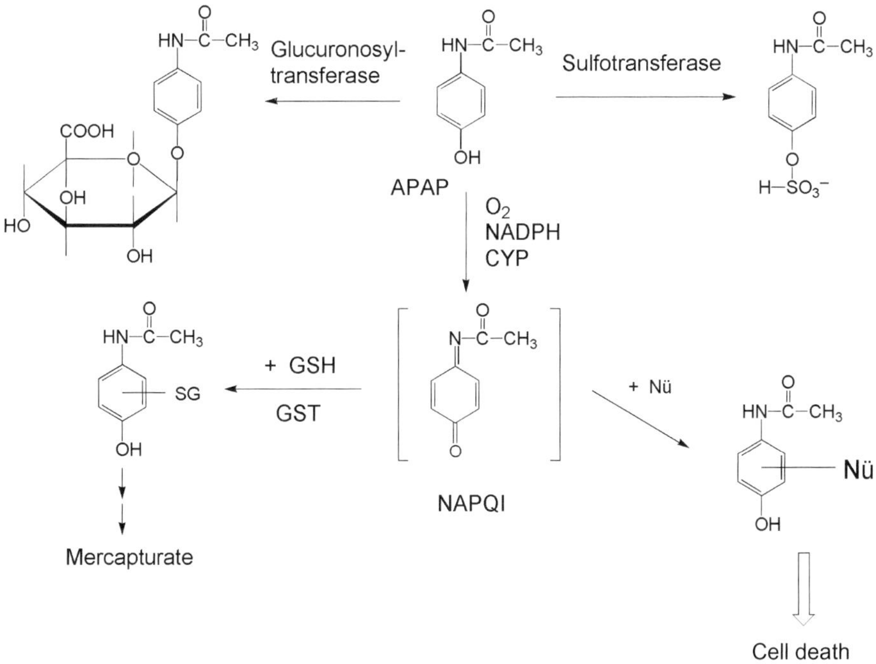

Figure 26.6 Renal metabolism of acetaminophen (APAP). Acetaminophen may undergo either glucuronidation, sulfation, or oxidation by cytochrome P450 (CYP). The latter reaction yields a reactive intermediate, *N*-acetyl-*p*-benzoquinoneimine (NAPQI). NAPQI reacts with either GSH, as catalyzed by GSH *S*-transferases (GSTs) to ultimately yield a mercapturate, or other nucleophiles (Nü). The latter reaction may lead to cytotoxicity.

urine. Several studies, however, have found that GSH conjugation of APAP or PAP leads to nephrotoxicity in rats or mice that could be prevented by pre-incubation with acivicin or probenecid (Klos *et al.*, 1992; Emeigh Hart *et al.*, 1996; Trumper, Monasterolo and Elias, 1996). While these findings are similar to those with conjugates of chemicals such as TRI that are ultimately bioactivated by CCBL or FMO, there is no evidence that these latter two bioactivation enzymes are involved. Nonetheless, the findings suggest that GGT-dependent metabolism is required to elicit nephrotoxicity for APAP or PAP.

26.7 SUMMARY AND CONCLUSIONS

The kidneys express a diverse array of enzymes that can catalyze the bioactivation of many drugs and xenobiotics. Unlike the liver, which is comprised predominantly of hepatocytes that express high levels of many phase I and phase II enzymes, the kidneys are comprised of numerous cell types that differ with respect to morphology, physiological functions, and expression of drug metabolism enzymes and membrane transporters, among other factors. Thus, most of the drug metabolism enzymes that are critical for bioactivation are found in the proximal tubular cells. One implication of this selective localization along the nephron is that when one looks to quantify many drug metabolism enzymes in renal homogenates, the apparent activity may be low because of dilution. In other words, the enzyme of interest may only be present in a few discrete cell types and the homogenate will likely include those cell types as well as several others that either do not express the enzyme or express it at low levels. Nonetheless, drug metabolism capability for the kidneys is typically much lower than that for the liver, even when cell type distribution is taken into account.

The kidneys are particularly susceptible to chemically induced toxicity because of their basic physiology and biochemistry. In terms of physiology, the high rate of renal blood flow delivers chemicals to the kidneys where these chemicals can undergo glomerular filtration or absorption from the periplasmic space by uptake across the basolateral membrane. The high flow rate of renal blood flow and glomerular filtration serve to deliver blood-borne chemicals to plasma membrane transporters that mediate their uptake and intracellular accumulation. Once inside the cell (primarily the proximal tubular cell), bioactivation enzymes similar to those found in hepatocytes are present to convert drugs to reactive and toxic species.

This chapter reviewed three major bioactivation pathways in the kidneys: the CYP, FMO, and GSH conjugation pathways. Many of the same CYP enzymes that are found in the liver are also found in the kidneys, although predominantly in the proximal tubules. Even with the same enzymes, some differences exist between the liver and kidney with respect to substrate specificity and inducibility. Significant species differences also exist for the CYPs, making conclusions made in studies with rodents not always readily applicable to humans. The GSH conjugation pathway initially leads to generation of cysteine conjugates by processing of the GSH conjugates by GGT and DP activities on the brush-border membrane of the renal proximal tubule. The cysteine conjugates thus formed serve as a branch point for several types of GST substrates, including halogenated solvents and the chemotherapeutic agent cisplatin, as they can either be metabolized to mercapturates and excreted or bioactivated by CCBL or FMO. The final example, that of APAP, illustrates the interaction and competition between alternate pathways for the given chemical.

REFERENCES

Abraham, D.G., Patel, P.P., Cooper, A.J.L. (1995). Isolation from rat kidney of a cytosolic high molecular weight cysteine-S-conjugate β-lyase with activity toward leukotriene E₄. *J. Biol. Chem.*, *270*, 180–188.

Abraham, D.G., Thomas, R.J., Cooper, A.J.L. (1995). Glutamine transaminase K is not a major cysteine S-conjugate β-lyase of rat kidney mitochondria: evidence that a high-molecular weight enzyme fulfills this role. *Mol. Pharmacol.*, *48*, 855–860.

Altuntas, T.G., Kharasch, E.D. (2002). Biotransformation of L-cysteine S-conjugates and N-acetyl-L-cysteine S-conjugates of the sevoflurane degradation product fluoromethyl-2,2-difluoro-1-(trifluoromethyl)vinyl ether (Compound A) in human kidney in vitro: interindividual variability in N-acetylation, N-deacetylation, and β-lyase-catalyzed metabolism. *Drug Metab. Dispos.*, *30*, 148–154.

Amet, Y., Berthou, F., Fournier, G., Dreano, Y., Bardou, L., Cledes, J., Menez, J.-F. (1997). Cytochrome P-450 4A and 2E1 expression in human kidney microsomes. *Biochem. Pharmacol.*, *53*, 765–771.

Anders, M.W. (1980). Metabolism of drugs by the kidney. *Kidney Int.*, *18*, 636–647.

Anders, M.W. (2005). Formation and toxicity of anesthetic degradation products. *Annu. Rev. Pharmacol. Toxicol.*, *45*, 147–176.

Bebri, K., Boobis, A.R., Davies, D.S., Edward, R.J. (1995). Distribution and induction of CYP3A1 and CY3A2 in rat liver and extrahepatic tissue. *Biochem. Pharmacol.*, *50*, 2047–2056.

Birner, G., Werner, M., Rosner, E., Mehler, C., Dekant, W. (1998). Biotransformation, excretion, and nephrotoxicity of the hexachlorobutadiene metabolite (E)-N-acetyl-S-(1,2,3,4,4-pentachlorobutadienyl)-L-cysteine sulfoxide. *Chem. Res. Toxicol.*, *11*, 750–757.

Board, P.G., Anders, M.W. (2005). Human glutathione transferase-Zeta. *Methods Enzymol.*, *401*, 61–77.

Board, P.G., Baker, R.T., Chelvanayagam, G., Jermiin, L.S. (1997). Zeta, a novel class of glutathione transferases in a range of species from plants to humans. *Biochem. J.*, *328*, 929–935.

Board, P.G., Chelvanayagam, G., Jermiin, L.S., Tetlow, N., Tzeng, H.F., Anders, M.W., Blackburn, A.C. (2001). Identification of novel glutathione transferases and polymorphic variants by expressed sequence tag database analysis. *Drug Metab. Dispos.*, *29*, 544–547.

Brenner, B.M. (1996). *The Kidney*, 5th edn, W.B. Saunders Company, Philadelphia.

Cashman, J.R., Zhang, J. (2006). Human flavin-containing monooxygenases. *Annu. Rev. Pharmacol. Toxicol.*, *46*, 65–100.

Cooper, A.J.L., Pinto, J.T. (2006). Cysteine S-conjugate β-lyases. *Amino Acids*, *30*, 1–15.

Cummings, B.S., Lash, L.H. (2000). Metabolism and toxicity of trichloroethylene and S-(1,2-dichlorovinyl)-L-cysteine in freshly isolated human proximal tubular cells. *Toxicol. Sci.*, *53*, 458–466.

Cummings, B.S., Lasker, J.M., Lash, L.H. (2000). Expression of glutathione-dependent enzymes and cytochrome P450s in freshly isolated and primary cultures of proximal tubular cells from human kidney. *J. Pharmacol. Exp. Ther.*, *293*, 677–685.

Cummings, B.S., Parker, J.C., Lash, L.H. (2000). Role of cytochrome P450 and glutathione S-transferase α in metabolism and cytotoxicity of trichloroethylene in rat kidney. *Biochem. Pharmacol.*, *59*, 531–543.

Cummings, B.S., Parker, J.C., Lash, L.H. (2001). Cytochrome P450-dependent metabolism of trichloroethylene in rat kidney. *Toxicol. Sci.*, *60*, 11–19.

Cummings, B.S., Zangar, R.C., Novak, R.F., Lash, L.H. (1999). Cellular distribution of cytochromes P-450 in rat kidney. *Drug Metab. Dispos.*, *27*, 542–548.

Dolphin, C.T., Beckett, D.T., Janmohamed, A., Cullingford, T.E., Smith, R.L., Shephard, E.A., Phillips, I.R. (1998). The flavin-containing monooxygenase 2 gene (FMO2) of humans, but not of other primates, encodes a truncated, nonfunctional protein. *J. Biol. Chem.*, *273*, 30599–30607.

Duggin, G.G. (1980). Mechanisms in the development of analgesic nephropathy. *Kidney Int.*, *18*, 553–561.

Elfarra, A.A., Krause, R.J. (2007). *S*-(1,2,2-Trichlorovinyl)-L-cysteine sulfoxide, a reactive metabolite of S-(1,2,2-trichlorovinyl)-L-cysteine formed in rat liver and kidney microsomes, is a potent nephrotoxicant. *J. Pharmacol. Exp. Ther.*, *321*, 1095–1101.

Elfarra, A.A., Lash, L.H., Anders, M.W. (1987). Alpha-keto acids stimulate rat renal cysteine conjugate β-lyase activity and potentiate the cytotoxicity of *S*-(1,2-dichlorovinyl)-L-cysteine. *Mol. Pharmacol.*, *31*, 208–212.

Emeigh Hart, S.G., Wyand, D.S., Khairallah, E.A., Cohen, S.D. (1996). Acetaminophen nephrotoxicity in the CD-1 mouse. II. Protection by probenecid and AT-125 without diminution of renal covalent binding. *Toxicol. Appl. Pharmacol.*, *136*, 161–169.

Epstein, F.H., Silva, P., Spokes, K., Brezis, M., Rosen, S. (1989). Renal medullary Na-K-ATPase and hypoxic injury in perfused rat kidneys. *Kidney Int.*, *36*, 768–772.

Finkelstein, M.B., Vamvakas, S., Bittner, D., Anders, M.W. (1994). Structure-mutagenicity and structure-cytotoxicity studies on bromine-containing cysteine S-conjugates and related compounds. *Chem. Res. Toxicol.*, *7*, 157–163.

Goldsworthy, T.L., Popp, J.A. (1987). Chlorinated hydrocarbon-induced peroxisomal enzyme activity in relation to species and organ carcinogenicity. *Toxicol. Appl. Pharmacol.*, *88*, 225–233.

Gonzalez, F.J. (1990). Molecular genetics of the P-450 superfamily. *Pharmacol. Ther.*, *45*, 1–38.

Griffith, O.W., Meister, A. (1979). Translocation of intracellular glutathione to membrane-bound γ-glutamyl transpeptidase as a discrete step in the γ-glutamyl cycle: glutathionuria after inhibition of transpeptidase. *Proc. Natl. Acad. Sci. U.S.A.*, *76*, 268–272.

Guder, W.G., Ross, B.D. (1984). Enzyme distribution along the nephron. *Kidney Int.*, *26*, 101–111.

Guder, W.G., Wagner, S., Wirthensohn, G. (1986). Metabolic fuels along the nephron: pathways and intracellular mechanisms of interaction. *Kidney Int.*, *29*, 41–45.

Hanigan, M.L., Gallagher, B.C., Taylor, P.T., Jr, Large, M.K. (1994). Inhibition of gamma-glutamyl transpeptidase activity by acivicin in vivo protects the kidney from cisplatin-induced toxicity. *Cancer Res.*, *54*, 5925–5929.

Hanigan, M.L., Lykissa, E.D., Townsend, D.M., Ou, C., Barrios, R., Lieberman, M.W. (2001). Gamma-glutamyl transpeptidase-deficient mice are resistant to the nephrotoxicity of cisplatin. *Am. J. Pathol.*, *159*, 1889–1894.

Hinchman, C.A., Ballatori, N. (1990). Glutathione-degrading capacities of liver and kidney in different species. *Biochem. Pharmacol.*, *40*, 1131–1135.

Hines, R.N., Cashman, J.R., Philpot, R.M., Williams, D.E., Ziegler, D.M. (1994). The mammalian flavin-containing monooxygenases: molecular characterization and regulation of expression. *Toxicol. Appl. Pharmacol.*, *125*, 1–6.

Horster, M. (1978). Principles of nephron differentiation. *Am. J. Physiol.*, *235*, F387–F393.

Hu, J.J., Rhoten, W.B., Yang, C.S. (1990). Mouse renal cytochrome P450IIE1: immunocytochemical localization, sex-related differences and regulation by testosterone. *Biochem. Pharmacol.*, *40*, 2597–2602.

Ito, O., Alonso-Galicia, M., Hopp, K.A., Roman, R.J. (1998). Localization of cytochrome P-450 4A isoforms along the rat nephron. *Am. J. Physiol.*, *274*, F395–F404.

Iyer, R.A., Frink, E.J., Jr, Ebert, T.J., Anders, M.W. (1998). Cysteine conjugate β-lyase-dependent metabolism of Compound A (2-[fluoromethoxy]-1,1,3,3,3-pentafluoro-1-propene) in human subjects anesthetized with sevoflurane and in rats given Compound A. *Anesthesiology*, *88*, 611–618.

Jacobson, H.R. (1981). Functional segmentation of the mammalian nephron. *Am. J. Physiol.*, *241*, F203–F218.

Jakobsson, P.-J., Morgenstern, R., Mancini, J., Ford-Hutchinson, A., Persson, B. (1999). Common structural features of MAPEG—a widespread superfamily of membrane associated proteins with highly divergent functions in eicosanoid and glutathione metabolism. *Protein Sci.*, *8*, 689–692.

Jones, D.P., Orrenius, S., Jakobson, S.W. (1980). Cytochrome P-450-linked monooxygenase systems in the kidney, in *Extrahepatic Metabolism of Drugs and Other Foreign Compounds* (ed. T.E. Gram), Spectrum Publications, New York, pp. 123–158.

Klos, C., Koob, M., Kramer, C., Dekant, W. (1992). p-Aminophenol nephrotoxicity: biosynthesis of toxic glutathione conjugates. *Toxicol. Appl. Pharmacol.*, *115*, 98–106.

Koukouritaki, S.B., Simpson, P., Yeung, C.K., Rettie, A.E., Hines, R.N. (2002). Human hepatic flavin-containing monooxygenase 1 (FMO1) and 3 (FMO3) developmental expression. *Pediatr. Res.*, *51*, 236–243.

Krause, R.J., Lash, L.H., Elfarra, A.A. (2003). Human kidney flavin-containing monooxygenases and their potential roles in cysteine S-conjugate metabolism and nephrotoxicity. *J. Pharmacol. Exp. Ther.*, *304*, 185–191.

Lash, L.H., Fisher, J.W., Lipscomb, J.C., Parker, J.C. (2000). Metabolism of trichloroethylene. *Environ. Health Perspec.*, *108* (Suppl. 2), 177–200.

Lash, L.H., Lipscomb, J.C., Putt, D.A., Parker, J.C. (1999). Glutathione conjugation of trichloroethylene in human liver and kidney: kinetics and individual variation. *Drug Metab. Dispos.*, *27*, 351–359.

Lash, L.H., Parker, J.C. (2001). Hepatic and renal toxicities associated with perchloroethylene. *Pharmacol. Rev.*, *53*, 177–208.

Lash, L.H., Parker, J.C., Scott, C.S. (2000). Modes of action of trichloroethylene for kidney tumorigenesis. *Environ. Health Perspec.*, *108* (Suppl. 2), 225–240.

Lash, L.H., Putt, D.A., Cai, H. (2008). Drug metabolism enzyme expression and activity in primary cultures of human proximal tubular cells. *Toxicology*, *244*, 56–65.

Lash, L.H., Putt, D.A., Hueni, S.E., Krause, R.J., Elfarra, A.A. (2003). Roles of necrosis, apoptosis, and mitochondrial dysfunction in *S*-(1,2-dichlorovinyl)-L-cysteine sulfoxide-induced cytotoxicity in primary cultures of human renal proximal tubular cells. *J. Pharmacol. Exp. Ther.*, *305*, 1163–1172.

Lash, L.H., Qian, W., Putt, D.A., Jacobs, K., Elfarra, A.A., Krause, R.J., Parker, J.C. (1998). Glutathione conjugation of trichloroethylene in rats and mice: sex-, species-, and tissue-dependent differences. *Drug Metab. Dispos.*, *26*, 12–19.

Lash, L.H., Sausen, P.J., Duescher, R.J., Cooley, A.J., Elfarra, A.A. (1994). Roles of cysteine conjugate β-lyase and *S*-oxidase in nephrotoxicity: studies with *S*-(1,2-dichlorovinyl)-L-cysteine and *S*-(1,2-dichlorovinyl)-L-cysteine sulfoxide. *J. Pharmacol. Exp. Ther.*, *269*, 374–383.

Lash, L.H., Xu, Y., Elfarra, A.A., Duescher, R.J., Parker, J.C. (1995). Glutathione-dependent metabolism of trichloroethylene in isolated liver and kidney cells of rats and its role in mitochondrial and cellular toxicity. *Drug Metab. Dispos.*, *23*, 846–853.

Lasker, J.M., Chen, W.B., Wolf, I., Bloswick, B.P., Wilson, P.D., Powell, P.K. (2000). Formation of 20-hydroxyeicosatetraenoic acid, a vasoactive and natriuretic eicosanoid, in human kidney: role of Cyp4F2 and Cyp4A11. *J. Biol. Chem.*, *275*, 4118–4126.

Lock, E.A., Reed, C.J. (1997). Renal xenobiotic metabolism, in *Comprehensive Series in Toxicology, Vol. 7: Kidney Toxicology* (ed. R.S. Goldstein), Elsevier, Oxford, pp. 77–97.

Lohr, J.W., Willsky, G.R., Acara, A. (1998). Renal drug metabolism. *Pharmacol. Rev.*, *50*, 107–141.

National Toxicology Program (NTP) (2005). *Report on Carcinogens*, 11th edn, U.S. Department of Health and Human Services, Public Health Service, National Toxicology Program, Research Triangle Park, NC.

Newton, J.F., Braselton, W.E., Kuo, C.-H., Kluwe, W.M., Gemborys, M.W., Mudge, G.H., Hook, J.B. (1982). Metabolism of acetaminophen by the isolated perfused kidney. *J. Pharmacol. Exp. Ther.*, *221*, 76–79.

Nishimura, M., Naito, S. (2006). Tissue-specific mRNA expression profiles of human phase I metabolizing enzymes except cytochrome P450 and phase II metabolizing enzymes. *Drug Metab. Pharmacokinet.*, *21*, 357–374.

Okita, J.R., Johnson, S.B., Castle, P.J., Dezellem, S.C., Okita, R.T. (1997). Improved separation and immunodetection of rat cytochrome P450 4A forms in liver and kidney. *Drug Metab. Dispos.*, *25*, 1008–1012.

Otieno, M.A., Baggs, R.B., Hayes, J.D., Anders, M.W. (1997). Immunolocalization of microsomal glutathione *S*-transferases in rat tissues. *Drug Metab. Dispos.*, *25*, 12–20.

Ripp, S.L., Itagak, I.K., Philpot, R.M., Elfarra, A.A. (1999). Species and sex differences in expression of flavin-containing monooxygenase form 3 in liver and kidney microsomes. *Drug Metab. Dispos.*, *27*, 46–52.

Ronis, M.J.J., Huang, J., Longo, V., Tindberg, B., Ingelman-Sundberg, M., Badger, T.M. (1998). Expression and distribution of cytochrome P450 enzymes in male rat kidney: effects of ethanol, acetone and dietary conditions. *Biochem. Pharmacol.*, *55*, 123–129.

Rozzel, B., Hansson, H.-A., Guthenberg, C., Tahir, M.K., Mannervik, B. (1993). Glutathione transferases of classes αμ and π show selective expression in different regions of rat kidney. *Xenobiotica*, *23*, 835–849.

Sands, J.M., Verlander, J.W. (2005). Anatomy and physiology of the kidneys, in *Toxicology of the Kidney*, 3rd edn (eds J.B. Tarloff and L.H. Lash), CRC Press, Boca Raton, Chapter 1.

Sausen, P.J., Duescher, R.J., Elfarra, A.A. (1993). Further characterization and purification of the flavin-dependent *S*-benzyl-L-cysteine *S*-oxidase activities of rat liver and kidney microsomes. *Mol. Pharmacol.*, *43*, 388–396.

Sausen, P.J., Elfarra, A.A. (1990). Cysteine conjugate S-oxidase: characterization of a novel enzymatic activity in rat hepatic and renal microsomes. *J. Biol. Chem.*, *265*, 6139–6145.

Sausen, P.J., Elfarra, A.A. (1991). Reactivity of cysteine S-conjugates sulfoxides: formation of S-[1-chloro-2-(S-glutathionyl)vinyl]-L-cysteine sulfoxide by the reaction of S-(1,2-dichlorovinyl)-L-cysteine sulfoxide with glutathione. *Chem. Res. Toxicol.*, *4*, 655–660.

Schuetz, E.G., Schuetz, J.D., Grogan, W.M., Naray-Fejes-Toth, A., Fejes-Toth, G., Raucy, J., Guzelian, P., Gionela, K., Watlington, C.O. (1992). Expression of cytochrome P450 3A in amphibian, rat, and human kidney. *Arch. Biochem. Biophys.*, *294*, 206–217.

Shanley, P.F., Shapiro, J.I., Chan, L., Burke, T.J., Johnson, G.C. (1988). Acidosis and hypoxic medullary injury in the isolated perfused kidney. *Kidney Int.*, *34*, 791–796.

Stec, D.E., Flasch, A., Roman, R.J., White, J.A. (2003). Distribution of cytochrome P-450 4A and 4F isoforms along the nephron in mice. *Am. J. Physiol.*, *284*, F95–F102.

Stevens, J.L., Robbins, J.D., Byrd, R.A. (1986). A purified cysteine conjugate β-lyase from rat kidney cytosol: requirement for an α-keto acid or an amino acid oxidase for activity and identity with soluble glutamine transaminase K. *J. Biol. Chem.*, *261*, 15529–15537.

Townsend, D.M., Deng, M., Zhang, L., Lapus, M.G., Hanigan, M.H. (2003a). Metabolism of cisplatin to a nephrotoxin in proximal tubule cells. *J. Am. Soc. Nephrol.*, *14*, 1–10.

Townsend, D.M., Hanigan, M.H. (2002). Inhibition of gamma-glutamyl transpeptidase or cysteine conjugate β-lyase activity blocks the nephrotoxicity of cisplatin in mice. *J. Pharmacol. Exp. Ther.*, *300*, 142–148.

Townsend, D.M., Marto, J.A., Deng, M., Macdonald, T.J., Hanigan, M.H. (2003b). High pressure liquid chromatography and mass spectrometry characterization of the nephrotoxic biotransformation products of cisplatin. *Drug Metab. Dispos.*, *31*, 705–713.

Trumper, L., Monasterolo, L.A., Elias, M.M. (1996). Nephrotoxicity of acetaminophen in male Wistar rats: role of hepatically derived metabolites. *J. Pharmacol. Exp. Ther.*, *279*, 548–554.

Walker, L.A., Valtin, H. (1982). Biological importance of nephron heterogeneity. *Annu. Rev. Physiol.*, *44*, 203–219.

Wolfgang, G.H.I., Gandolfi, A.J., Stevens, J.L., Brendel, K. (1989). N-Acetyl S-(1,2-dichlorovinyl)-L-cysteine produces a similar toxicity to S-(1,2-dichlorovinyl)-L-cysteine in rabbit renal slices: differential transport and metabolism. *Toxicol. Appl. Pharmacol.*, *101*, 205–219.

Yin, H., Jones, J.P., Anders, M.W. (1995). Metabolism of 1-fluoro-1,1,2-trichloroethane, 1,2-dichloro-1,1-difluoroethane, and 1,1,1-trifluoro-2-chloroethane. *Chem. Res. Toxicol.*, *8*, 262–268.

Zhang, G., Stevens, J.L. (1989). Transport and activation of S-(1,2-dichlorovinyl)-L-cysteine and N-acetyl-S-(1,2-dichlorovinyl)-L-cysteine in rat kidney proximal tubules. *Toxicol. Appl. Pharmacol.*, *100*, 51–61.

Zhang, L., Hanigan, M.H. (2003). Role of cysteine S-conjugate β-lyase in the metabolism of cisplatin. *J. Pharmacol. Exp. Ther.*, *306*, 988–994.

Ziegler, D.M. (1993). Recent studies on the structure and function of multisubstrate flavin-containing monooxygenases. *Annu. Rev. Pharmacol. Toxicol.*, *33*, 179–199.

PART V

REGULATORY PERSPECTIVES

CHAPTER 27

Drug Metabolism in Regulatory Guidances, Clinical Trials, and Product Labeling

MARK N. MILTON

27.1 INTRODUCTION

The role of the drug metabolism scientist has changed markedly over the past 20 years with our enhanced understanding of the enzymology of drug metabolism, the importance of drug metabolism in the efficacy and safety of drugs, and the increasingly interdisciplinary nature of the drug development process. Nowadays, the drug metabolism scientist will need to interact with many different disciplines within a pharmaceutical company. These disciplines include toxicology, pharmacology, formulations, chemistry, clinical, clinical pharmacology, regulatory affairs, marketing, medical affairs, competitive intelligence, and business development. Additionally, they may be responsible for representing the multiple subdisciplines (e.g. bioanalytical, *in vitro* metabolism, *in vivo* metabolism, transporters, and pharmacokinetics) that contribute to the understanding of the absorption, distribution, metabolism, and elimination (ADME) properties of a drug. Therefore, the drug metabolism scientist need not only be an expert in drug metabolism and its related disciplines, but also be a generalist who understands the need of his or her customers, whether they be medicinal chemists, toxicologists, regulatory agencies, or clinicians. They will need to possess a basic understanding of how knowledge of the ADME properties of a drug can be used to efficiently and effectively develop a drug and the regulatory environment that pharmaceutical companies operate within.

The objective of this chapter is to give the reader a high-level overview of how drug metabolism contributes to the risk–benefit analysis of a novel pharmaceutical; the role of drug metabolism in *clinical trials* and the contribution of drug metabolism to the *product label*. Readers should not rely on this chapter as a definitive source of information. They should read the relevant source information for themselves and come to their own conclusions regarding the interpretation of a given regulatory guidance. A

Drug Metabolism Handbook: Concepts and Applications, Edited by Ala F. Nassar, Paul F. Hollenberg, and JoAnn Scatina
Copyright © 2009 by John Wiley & Sons, Inc.

more detailed review of the drug metabolism and pharmacokinetics in drug development has been performed by Bonate and Howard (2004) and Rogge and Taft (2005). The role of drug metabolism in the development of agricultural chemicals, veterinary products, and food additives will not be addressed in this chapter. It is difficult, at times, to separate drug metabolism from pharmacokinetics and their contributions to safety and efficacy. Therefore, this chapter will view drug metabolism, pharmacokinetics, and drug transporters as being inexorably linked and will attempt to give drug metabolism scientists an overview of how their discipline relates to, and interacts with, the regulatory environment, clinical trials, and the product label.

Certain parts of this chapter will contain materials that have been lifted verbatim from regulatory guidances. Since the exact wording of regulatory guidances can be of paramount importance, it should be noted that such extracts of the guidances will no longer be in the context of the full guidance and therefore the reader is encouraged to use this chapter as a starting point for understanding the role of drug metabolism in regulatory guidances and to review the source documentation for themselves. As all guidances are open to interpretation, the views expressed here are those of the author and may differ from the views of the reader, colleagues, or the regulatory agencies. It should also be noted that the interpretation of the guidance documents may change with time based on the experience of the regulatory agencies, may vary between countries, and even within a given country's regulatory agency.

Subsections have been included in this chapter for ease of navigation, although it should be noted that it can be difficult to place a given guidance in a given category due to the multidisciplinary nature of many of these guidances and the overlap between nonclinical and clinical development.

27.2   THE REGULATORY ENVIRONMENT

27.2.1   Why Do We Have Drug Regulations?

The regulations for the pharmaceutical industry have tended to be created in reaction to tragedy rather than being created proactively in order to avoid a tragedy such as what occurred with sulfanilamide, TGN1412, thalidomide, terfenadine, the floxacins, FIAU, Phen-fen, Ephedra, encainide/flecainide, propulsid, Redux, and Duract. The subsequently created guidances are an effort in order to avoid further and greater tragedies. Drug metabolism has been implicated to varying extents in these tragedies that have shaped the regulatory environment for pharmaceuticals. Many of the regulatory safeguards in the United States came about as part of the Kefauver–Harris drug amendments to the Food, Drug and Cosmetic Act in 1962. Interestingly, the original purpose of these amendments was to introduce price-cutting but in the wake of the Thalidomide tragedy, the price-cutting elements of the amendments were downplayed and a greater importance on safety was introduced.

27.2.2   The Global Nature of Drug Development

The regulation of the development of pharmaceuticals will be broadly similar throughout the world, although each country or region will have its own specific guidances or areas of interest. This chapter will focus primarily on regulatory submissions to the FDA (United States) and EMEA (European Union), since these regulatory agencies

TABLE 27.1   Web sites containing regulatory guidances.

Country or region	Web site
United States	http://www.fda.gov
Canada	http://www.hc-sc.gc.ca
EU	http://www.emea.europa.eu
Japan	http://www.mhlw.go.jp/english/index.html
Australia	http://www.tga.gov.au
Global (ICH)	http://www.ich.org
Global (OECD)	http://www.oecd.org

have created the majority of the existing regulatory guidances or guidelines. However, additional guidances have been created in other regions (e.g. Canada and Japan). Efforts have been made to globalize certain guidances under the auspices of the International Conference on Harmonization of Technical Requirements for Registration of Pharmaceuticals for Human Use (ICH). This is a unique project that brings together the regulatory authorities of Europe, Japan, and the United States, and experts from the pharmaceutical industry in the three regions to discuss scientific and technical aspects of product registration. The purpose of the ICH is to make recommendations on ways to achieve greater harmonization in the interpretation and application of technical guidelines and requirements for product registration in order to reduce or obviate the need to duplicate the testing carried out during the research and development of new medicines. The objective of such harmonization is a more economical use of human, animal, and material resources, and the elimination of unnecessary delay in the global development and availability of new medicines while maintaining safeguards on quality, safety, and efficacy, and regulatory obligations to protect public health. It should be noted that the different countries that form the European Medicines Agency (EMEA) may have different local practices with respect to the implementation of the EMEA guidelines or EU directives. Sources of information on regulatory guidances are summarized in Table 27.1.

The regulatory process and requirements in Japan can be quite different to the requirements in other parts of the world. An overview of the regulatory process and guidelines in Japan was published by the Japan Pharmaceutical Manufacturers Association in 2007 (Japan Pharmaceutical Manufacturers Association, 2007) and serves as a very good reference source for drug development in Japan. Additionally, a very detailed document was issued in 1995 (Japanese Guidelines for Nonclinical Studies of Drugs Manual 1995) under the editorial supervision of the Pharmaceuticals and Cosmetics Division, Pharmaceutical Affairs Bureau, Japanese Ministry of Health and Welfare (Ministry of Health and Welfare, Pharmaceutical Affairs Bureau, 1995). This document contains guidelines for toxicity studies of drugs, guidelines for general pharmacology studies, and guidelines for nonclinical pharmacokinetic studies. This document is a very good resource for a general description of the design and conduct of nonclinical studies.

27.2.3   The Process of Drug Development

The development of new drugs and the expense of clinical trials to demonstrate the safety and efficacy of innovative drugs is rewarded through granting of a period of marketing exclusivity that shields the product from competition. The purpose of the regulatory agencies, such as the FDA, is to protect the safety of the patient from people that would want to sell them snake oil (drugs that are non-efficacious and harmful) while providing an avenue for drug developers to demonstrate efficacy, assess toxicity, and secure marketing approval.

Since the development of a new drug is a costly and timely effort and since it is of paramount importance to protect the safety of the patient or volunteer, it is important to build a strong working relationship with the regulatory agencies. For the FDA, this relationship can be initiated prior to the initiation of the first clinical trial by holding a Pre-IND meeting (US FDA, 2000a). The focus of the regulatory agency during its review of an IND application (Investigational New Drug) for a first-in-human (FIH) clinical trial is to ensure that the drug is safe to administer to the proposed population (patients or normal human volunteers) and not to determine whether or not there is a market need for the drug (US FDA, 2007o). At this stage of development, the FDA's focus is not on efficacy. This statement applies to regulatory agencies worldwide, and not just to the FDA. Prior to the submission of the IND, most sponsors will have taken advantage of holding a Pre-IND meeting with the FDA (US FDA, 2005a). The pre-IND meeting can be very valuable in planning a drug development program, especially if sponsors' questions are not fully answered by guidances and other information provided by the FDA. Early interactions with FDA staff can help to prevent clinical hold issues from arising. A pre-IND meeting can also provide sponsors information that will assist them in preparing to submit complete IND applications. For a Pre-IND meeting, the sponsors will submit specific questions that they wish to have answered by the FDA along with a briefing book that will provide sufficient background information for the FDA to provide its response. It is recommended that the sponsor does not ask open-ended questions or ask whether a certain study should be performed or not. The most successful meetings are those in which there is a distinct objective and which proposals are made that the regulatory agency can respond to. This approach will apply to all regulatory agencies (not just the FDA) and to all types of meetings.

In the EU, the process for interacting with the EMEA is, in general, similar to the process for interacting with the FDA. The EMEA has created a guideline for obtaining scientific advice (EMEA, 2007c). However, few sponsors request scientific advice from the EMEA prior to the FIH study due to the length of time and cost of obtaining scientific advice.

At the time of the FDA's review of the sponsor's marketing application (NDA; New Drug Application), the focus will be on safety and efficacy (i.e. the drug must be safe and efficacious). However, the FDA may also choose to take into account the current standard of care for the disease, the benefit that the new drug offers over the current therapies, and projected cost of the therapy when approving the drug. This latter approach has recently been criticized by Bob Essner, the outgoing chief executive of Wyeth. In an article that was published in the *Financial Times*, he expressed concerns that the FDA assessments are exceeding their legal mandate and such reviews now include whether a drug can work better than an existing product, instead of allowing the market to determine what it wants (Bowe, 2007). Concern has been expressed that the FDA has become ultracautious in the past few years which may be in response to high-visibility safety concerns with drugs such as Vioxx. Similarly, regulatory agencies in the EU have become cautious in the wake of the "TeGenero incident" (Wikipedia, 2008a). Recently, the osteoarthritis drug Prexige has been withdrawn from the market in several countries including Canada and Australia amid concerns about the drug increasing the risk for serious liver-related damage including hepatitis (Health Canada, 2007b; Novartis Australia Pty Limited, 2007).

In anticipation of the expiration of marketing exclusivity and applicable patent protections, potential generic manufacturers begin the scientific and technical process of generic drug development. To receive approval, generic drug applicants must demonstrate that their products are pharmaceutically equivalent and bioequivalent to the reference product. Pharmaceutically equivalent products have the same active ingredient(s) in the same strength in the same dosage form. Bioequivalent products show no significant difference in the rate and extent of absorption of the active ingredient at the site of action. For many drug products, demonstrating pharmaceutical equivalence and bioequivalence is straightforward. Analytical chemistry can identify and quantitate the active ingredient. Comparison of pharmacokinetic parameters is used to evaluate bioequivalence. Bioequivalence based on plasma drug concentration has been identified as the most commonly used and successful biomarker of safety and efficacy. When common approaches to the assessment of bioequivalence and pharmaceutical equivalence are not applicable, as is the case for complex drug products and locally acting drugs, scientific challenges have presented barriers to the development and approval of generic drugs. Examples of where these challenges have limited development include topical and inhaled drug products.

27.2.4 The Nature of Regulatory Guidances

The requirements for the development of a pharmaceutical are primarily managed by guidance (FDA terminology) or guideline (EMEA terminology) documents and to a far lesser extent by legislation. Guidances or guidelines are not legally enforceable documents but instead provide the regulated community with a better understanding of agency positions on specific issues. This is illustrated by a statement that is found on most, if not all, of the FDA guidance documents: "This guidance represents the Food and Drug Administration's (FDA's) current thinking on this topic. It does not create or confer any rights for or on any person and does not operate to bind FDA or the public. You can use an alternative approach if the approach satisfies the requirements of the applicable statutes and regulations. If you want to discuss an alternative approach, contact the FDA staff responsible for implementing this guidance." Alternative statements include "FDA's guidance documents, including this guidance, do not establish legally enforceable responsibilities. Instead, guidances describe the Agency's current thinking on a topic and should be viewed only as recommendations, unless specific regulatory or statutory requirements are cited." The lack of binding nature of the guidances is again emphasized in the FDA's Good Guidance Practice rules (U.S. Environmental Protection Agency, 2008), guidance documents "do not legally bind the public or FDA" and entities "may choose to use an approach other than the one set forth in a guidance document." In addition to guidance documents, the FDA will occasionally issue Q&A documents that provide clarification to a given regulatory guidance document. It should be noted that the Q&A documents do not appear on the FDA's CDER guidances web page. This lack of coordination is unfortunate and will be described further in Section 27.9, "Procedural Guidances." Guidelines issued by the EMEA (excepting those generated by the ICH process) will contain a statement regarding the legal status of the guideline and indicate whether the guideline should be read in conjunction with other documentation such as directives (laws) or guidelines.

It is important to note that guidance documents are essentially recommendations and not mandates. This point is illustrated by the following statement that is commonly found in FDA guidance documents: "The use of the word should in Agency guidances means that something is suggested or recommended, but not required." However, once agencies commit to a position in a guidance document, they may become inflexible in applying it, and may apply it in such a way as to create binding regulatory obligations. The flexibility in the regulatory guidances can be a double-edged sword. Not only do the guidances allow the sponsors the latitude to propose alternative strategies to the regulatory agency, they also allow for great variability in interpretation between review divisions at the FDA and between reviewers within a given organization. The guidances are not intended to provide a "one size fits all" approach but are intended to be implemented on a "case-by-case basis" that takes into account the nature of the therapeutic and the disease.

Guidances have been used by regulatory agencies in an attempt to nudge sponsors in the "correct" direction and to provide some level of standardization of the materials that are provided to the regulatory agencies for review. The regulatory reviewers have a high workload and have to make very impactful decisions in a very short period of time. These time frames for review of an application for a FIH clinical trial are very short (30 days in the United States and 60 days in the United Kingdom, although the goal for the review in the United Kingdom is 14 days). The reviews are based on the quality and the content of the written documentation that is submitted to the regulatory agencies. The regulatory agencies will not take into account the scientific expertise of the individual(s) involved in the conduct and interpretation of these studies, even if they did know the identities of these individuals. It should be noted that a sponsor could have performed an excellent job of understanding the properties of the drug, only for it not to be approved (either for entry in to clinical trials or for marketing) based on the level and quality of the documentation. The topic of documentation will be addressed in Section 27.9, "Procedural Guidances."

27.2.5 Other Important Sources of Information

In addition to regulatory guidances, there are other sources of information that will be of great use to the drug metabolism scientists in their role of designing a robust plan for the characterization of the ADME properties of a drug. These sources include the regulatory agency's review of the sponsor's marketing application. These reviews are known as the Summary Basis of Approval (SBA) for drugs approved by the US FDA and the European Public Assessment Report (EPAR) for drugs approved by the Committee for Medicinal Products for Human Use (CHMP) which is part of the EMEA (EMEA, 2008a; US FDA, 2008c). The European Medicines Agency (EMEA) is a decentralized body of the European Union with headquarters in London. Both the SBAs and EPARs will contain a summary of the information used for the regulatory agency's decision to approve the marketing application. There is a greater level of detail in the SBA than in the EPAR, although both documents will not contain commercially confidential information (usually related to chemistry, manufacturing, and controls or CMC). The Division of Oncology Drug Products of the FDA has recently begun to publish an "approval summary" for a newly approved oncologic agent in the peer-reviewed literature. For example, the FDA has published a paper on the approval of Vorinostat for the treatment of advanced primary cutaneous T-cell lymphoma (Mann *et al.*, 2007). Other sources of pertinent information include "withdrawal reports" that are published by the EMEA which are similar in detail to the EPAR but also containing the reason why the drug's marketing application was not approved (EMEA, 2008f), scientific publications or presentations by regulatory agency personnel, meeting presentations and transcripts from FDA Advisory Committee meetings, approval letters published by the FDA and product labels, and web sites for regional pharmaceutical manufacturer's associations. As can be seen, there is a large amount of information that is available to the drug metabolism scientist.

27.2.6 The Process of Creating a New Regulatory Guidance

When FDA plans to issue a new regulation or revise an existing one, it places an announcement in the Federal Register on the day the public comment period begins. Published every weekday, the Federal Register is available at many public libraries and colleges, and on the FDA web site (www.fda.gov). In the Federal Register, the "notice of proposed rulemaking" describes the planned regulation and provides background on the issue. It also gives the address for submitting written comments and the name of the person to contact for more information. The draft guidance is also made available on the FDA's guidance web page. Also noted is the "comment period," which specifies how long the agency will accept public comments. Usually, the file (or docket) stays open for comments at least 60 days, though some comment periods have been as short as 10 days or as long as nine months. Weekends and holidays are included in the comment period. This comment period offers the drug metabolism scientists in the regulated industry the opportunity to influence the final guidance. All drug metabolism scientists are highly encouraged to keep themselves abreast of the appropriate regulatory guidances (final or draft) as well as participate in providing comments on the draft guidances. A similar review and comment approach is taken by the EMEA. However, the EMEA will often first issue a "concept paper" that outlines what the draft guideline will contain. In addition to publishing draft guidances, the FDA and EMEA will issue lists of guidances that they are planning on issuing in the current year. This list can provide a useful insight into the issues that these regulatory agencies view as being unaddressed. It should be noted that the FDA and EMEA handle draft guidances/guidelines in a very different manner once the guidance/guideline is finalized. The FDA updates its web site with the finalized version of the guidance and removes the draft guidance. Occasionally, the draft guidance can be located by searching the FDA web site, although in most cases it is no longer available to the general public. The EMEA takes a different approach. When the draft guideline is finalized, it will be added to the relevant web page. However, the web page will also contain a link to the draft guidance as well as a summary of the comments that were received on the draft guideline. Therefore, it is possible to readily determine what changed between the draft and final versions of the guideline.

In recent years, both the FDA and EMEA have made available to the public the comments on guidances made by interested parties. These documents can help the reader understand the extent and depth of comments from individuals, pharmaceutical companies, industry groups, and others. The comments on EMEA guidelines are colocated with the final guideline. The EMEA has organized the guidelines for a given topic (e.g. nonclinical) in a comprehensive and easily understandable tabular format (EMEA, 2008h). The comments are tabulated and include the EMEA's responses to the individual comments, noting whether they will be accepted or not. Comments on draft FDA guidances can be found in the docket (e.g. Amgen's comments on the draft guidance on metabolites in safety testing) (Amgen, 2005) but cannot be easily navigated to and from the main guidance document page (US FDA, 2008g).

From time to time, the pharmaceutical industry may publish a position paper on a topic that describes an approach to a certain element of the drug development process. Such position papers are often created with the hope that the existence of a standardized industry approach may negate the need for the creation of a regulatory guidance or to proactively influence the FDA's thinking on a particular topic. While this approach may be successful in certain circumstances, it may not work in others. For example, in 2002, a proposal was published on "metabolites in safety testing" (MIST) by Baillie *et al.* (2002), which suggested some guidelines regarding when it is necessary to provide greater assessment of the safety of metabolites. This position paper addressed the circumstances under which the toxicity of a metabolite should be evaluated. The cutoff below (in terms of exposure of humans to the metabolite relative to the exposure to total compound-related material) which no testing would be required was far greater than the FDA was comfortable with. This resulted in a response to the original article in terms of a letter to the editor by Hastings *et al.* (2003) and a response by Baillie *et al.* (2003) and a subsequent draft guidance that was issued for review in June 2005 (US FDA, 2005c). It has been speculated that the draft guidance was created in response to the PhRMA white paper and may not have been created if the white paper had not been published. The final guidance was issued in February 2008 and the draft guidance is no longer available on the FDA's Guidance Documents web page. As a note of caution, the web page on the FDA web site that is intended to redirect the reader from the draft guidance to the final guidance does not (as of 12 March 2008) direct the reader to the finalized version of the guidance "Safety Testing of Drug Metabolites" but instead redirects the reader to a different guidance, namely "Clinical Trial Endpoints for the Approval of Cancer Drugs and Biologics" (US FDA, 2007i).

There is an increasingly common phenomenon, namely the "permanently draft" guidance. A draft guidance may be issued for review but never finalized. There are draft guidances issued by the FDA that are over a decade old. This can lead to draft guidances becoming *de facto* finalized guidances. It would be preferable if draft guidances were finalized in a timely manner or moved to a separate section where they are not intermixed with finalized guidances. A table of finalized and draft guidances can be found on the FDA's web site (US FDA, 2008a) and are summarized by type in Table 27.2. A review of this list on May 1, 2008 revealed that there were 460 documents listed (FDA guidances, ICH guidances, and industry letters). The vast majority (approximately 33%) of these documents were still in draft form.

Twenty-one (~13%) of the draft documents were issued in 2007. The oldest of the draft documents dates from January 1985. The FDA is currently developing additional draft guidances and as the projects under the Critical Path Initiative (US FDA, 2006a) progress, even more guidances are likely to be created. Based on data from 1998 to May 2008, it is predicted that in 2022 there will be approximately 900 FDA guidances of which approximately 50% will be draft. However, given funding and resource constraints, it is possible that the rate of growth of the number of guidances may slow and the percentage of draft guidances may increase.

The FDA routinely publishes a list of guidances under development (US FDA, 2008f). Guidances that CDER (Center for Drug Evaluation and Research) is planning to develop during 2008 that are of interest to the drug metabolism scientist include "Immunogenicity Assessment for Therapeutic Proteins," "PET CGMPs," "Determining Whether Human Research Studies Can Be Conducted Without an IND,"

TABLE 27.2   Draft and final FDA guidances and letters by type.

Type	Number of guidances		Percent of guidances	
	Final	Draft	Final	Draft
Advertising	3	5	38	63
Biopharmaceutics	11	3	79	21
Chemistry	31	9	78	23
Clinical antimicrobial	5	26	16	84
Clinical medical	51	37	58	42
Clinical pharmacology	8	3	73	27
Combination products	1	3	25	75
Compliance	17	7	71	29
Current good manufacturing practices	7	5	58	42
Drug safety	1	1	50	50
Electronic submissions	6	7	46	54
Generic drugs	19	4	83	17
Good review practices	3	0	100	0
ICH—efficacy	19	4	83	17
ICH—multidisciplinary	9	2	82	18
ICH—quality	23	4	85	15
ICH—safety	15	1	94	6
INDs	1	0	100	0
Industry letters	14	0	100	0
Labelling	7	10	41	59
OTC	5	6	45	55
Pharm/Tox	14	4	78	22
Procedural	31	9	78	23
Small entity compliance guide	1	0	100	0
User fee	7	1	88	13
Total	309	151	67	33

"Content and Formatting of the Clinical Pharmacology Section of a Product Label," "Biotechnology-Derived Pharmaceuticals: Nonclinical Safety Evaluation," "Genotoxic and Carcinogenic Impurities in Drug Substances and Products: Recommended Approaches," "Assessment of Abuse Potential of Drugs," and "Determining Whether Human Research With a Radioactive Drug Can Be Conducted Under a Radioactive Drug Research Committee (RDRC)." Two guidances on the list, "Nonclinical Safety Evaluation of Reformulated Drug Products Intended for Administration by an Alternate Route" and "Diabetes Mellitus: Developing Drugs and Therapeutic Biologics for Treatment and Prevention" were issued in early March 2008 (US FDA, 2008h, 2008i). The ICH web site (ICH, 2008) provides insight into proposed topics (concept papers). Similarly, the EMEA (CHMP) publishes work plans indicating the status of ongoing projects. However, these work plans are somewhat buried on its web site as it is presently structured (EMEA, 2008i).

Recently, there has been an interesting turn of events that has led to the "release" of a draft guidance that did not appear to be on the list of guidances that the FDA is developing. This draft guidance, "Nonclinical Safety Evaluation of Biotechnology-Derived Pharmaceuticals," is dated August 2006 and was not officially released for review but nevertheless had been widely distributed within the pharmaceutical industry. However, this topic is now on the list of guidances that the FDA intends to develop in 2008 (US FDA, 2008f).

Over the past few years, there seems to have been a level of unofficial collaboration between the FDA and EMEA in terms of which guidances are being developed. Logistically, it makes sense for only one agency to develop guidance on a given topic in order to conserve resources. For example, the EMEA has issued a guideline on the analysis of control samples from toxicology studies. Since sponsors may potentially submit a marketing application in both the European Union and United States, they will follow this guidance even if they are presently only developing the drug within the United States. Similarly, the FDA has issued a draft guidance on the safety testing of drug metabolites. At present, the EMEA does not appear to be planning to create an analogous guideline.

In addition to regulatory guidances, there are other important sources of information that the drug metabolism scientist should pay attention to. For example, the Oncology Division of the FDA has published articles in peer-reviewed journals or as book chapters. These published articles by DeGeorge et al. (1997) and Farrell et al. (2003) have become de facto guidances. These articles are often used by the FDA reviewers to supplement the existing regulatory guidances. Recently, the FDA published an article that provided an update on CYP enzymes, transporters, and the guidance process (Huang et al., 2008). This article provides a good overview of the topic and how the FDA arrived at its current position.

In addition to regulatory guidances, certain aspects of the drug development process have been codified into law. A list of pertinent regulations for the United States can be found on the FDA web site (US FDA, 2008k). The legislation tends to be high level compared with the regulatory guidances and the information contained in the legislation is objective rather than subjective as is the case with the guidances. Examples of US legislation that pertains to drug metabolism in the development of a novel pharmaceutical include Good Laboratory Practice (21 CFR Part 58), Product Labeling (21 CFR Part 201), Bioavailability and Bioequivalence (21 CFR Part 320), and Radioactive Drugs for Certain Research Purposes (21 CFR Part 361.1) (US FDA, 2007a, b, c, d). In the EU, the legislative documents are known as directives (European Commission, 2008). One example of a directive is the Clinical Trials Directive on the approximation of the laws, regulations, and administrative provisions of the Member States relating to the implementation of good clinical practice in the conduct of clinical trials on medicinal products for human use (European Commission, 2001).

27.2.7 The Challenge of Finding the Information That You Require

The drug metabolism scientist can face several challenges when attempting to locate guidances or guidelines that mention drug metabolism. This challenge is not unique to drug metabolism and pharmaceutical scientists because many disciplines may find similar challenges. However, the challenges for the drug metabolism scientist may be greater than for other disciplines since, although drug metabolism is a very important factor in the evaluation of the safety and efficacy of drugs, it has very rarely been the main topic of regulatory guidances or guidelines. The lack of visibility of the contribution of drug metabolism to the risk–benefit analysis for a new pharmaceutical is reflected in the fact that the nonclinical section of the US IND is known as "Pharm/Tox" and the reviewers at the FDA are called Pharm/Tox reviewers.

The guidances on the FDA web site are arranged according to organizational areas such as advertising, chemistry, clinical pharmacology, pharmacology/toxicology, and procedural. There are 22 such categories. The guidelines on the EMEA web site are organized into five categories: Quality (Chemical and Herbal), Biologics, Non-Clinical, Clinical Efficacy and Safety, and Multidisciplinary. The ICH guidances are organized into four categories: Safety (S), Efficacy (E), Multidisciplinary (M), and Quality (Q). In all cases, there presently is no simple way to find all guidances that relate to drug metabolism by searching for keywords or the content of each guidance.

27.2.8   Using Regulatory Guidances

As mentioned previously, it is not mandatory to follow regulatory guidances, although it is most often advisable to do so. Utilizing information in finalized regulatory guidances is relatively straightforward. However, the existence of a large number of draft guidances creates a conundrum for the sponsor. Since the guidance has not been finalized, it may be difficult to use the draft guidance as precedent as it may no longer represent the regulatory agency's current thoughts on a given topic. Interestingly, since there appears to be no obligation for the draft guidances to be finalized within a defined time frame, or to be withdrawn if they have not been finalized, it leaves open the possibility that regulatory guidances could be released as draft guidances without the need to take the extra step to finalize them. Since these draft guidances can be found on the FDA's guidance web page, many sponsors will attempt to comply with them even though they are not final guidances. This may be risky, since the sponsor will not know *a priori* why the guidance has not been finalized. The reasons could range from workload to difficulty in coming to consensus. The draft guidance could contain outdated or inaccurate information, although it should be noted that finalized guidances may also contain outdated information. Another confusing point is that some guidances state that they are "guidance for industry" and others state that they are "guidance for industry and review staff." It is unclear whether the FDA is trying to make a subtle distinction or not.

27.3   BACKGROUND INFORMATION IN REGULATORY GUIDANCES

Several regulatory guidances provide some basic drug metabolism information in order to establish the context for the guidance. Such information is useful to the nonexpert and helps the expert better understand the rationale behind the guidance. Several examples of such background information that pertains to drug metabolism are presented further in the discussion. One such example of the background information that may be found in a guidance document comes from the 1997 FDA guidance, "Drug Metabolism/Drug Interaction Studies in the Drug Development Process: Studies In Vitro" (US FDA, 1997a). This guidance states:

> After entering the body, a drug is eliminated either by excretion or by metabolism to one or more active or inactive metabolites. When elimination occurs primarily by metabolism, the routes of metabolism can significantly affect the drug's safety and efficacy and the directions for use. When elimination occurs via a single metabolic pathway, individual differences in metabolic rates can lead to large differences in drug and metabolite concentrations in the blood and tissue. In some instances, differences exhibit a bimodal distribution

indicative of a genetic polymorphism (e.g. CYP450 2D6, CYP450 2C19, N-acetyl trans-ferase). When a genetic polymorphism affects an important metabolic route of elimination, large dosing adjustments may be necessary to achieve the safe and effective use of the drug. Pharmacogenetics already has influenced therapeutics. For a drug that is primarily metabolized by CYP450 2D6, approximately 7% of Caucasians will not be able to metabo-lize the drug, but the percentage for other racial populations is generally far lower. Similar information is known for other pathways, prominently, CYP450 2C19 and N-acetyl-transferase. Equally important, if not more so, many enzymatic metabolic routes of elimina-tion, including most of those occurring via the CYP450 enzymes, can be inhibited or induced by concomitant drug treatment. As a result, abrupt changes can occur with a co-administered agent in a single individual. Such interactions can lead to a substantial decrease or increase in the blood and tissue concentrations of a drug or metabolite or cause the accumulation of a toxic substance (e.g. certain antihistamine-antifungal interactions). These types of changes can alter a new drug's safety and efficacy profile in important ways, particularly for a drug with a narrow therapeutic range.

An understanding of metabolic pathways and potential interactions sometimes allows the use of a drug that would produce an unacceptable level of toxicity if blood levels were not predictable. For these reasons, it is important to learn at an early stage of development whether a drug is eliminated primarily by excretion of unchanged drug or by one or more routes of metabolism. If elimination is primarily by metabolism, the principal metabolizing route(s) should be understood. This information will help identify the implications of metabolic differences between and within individuals and the importance of certain drug-drug and other interactions. Having such information also will aid in determining whether the pharmacologic properties of certain metabolites should be explored further.

This background information clearly communicates the FDA's rationale for the development of this guidance.

Similarly, the draft guidance on "Drug Interaction Studies—Study Design, Data Analysis, and Implications for Dosing and Labeling" that was issued in 2006 (US FDA, 2006g) states:

The desirable and undesirable effects of a drug arising from its concentrations at the sites of action are usually related either to the amount administered (dose) or to the resulting blood concentrations, which are affected by its absorption, distribution, metabolism, and/or excretion. Elimination of a drug or its metabolites occurs either by metabolism, usually by the liver or gut mucosa, or by excretion, usually by the kidneys and liver. In addition, protein therapeutics may be eliminated through a specific interaction with cell surface receptors, followed by internalization and lysosomal degradation within the target cell. Hepatic elimination occurs primarily by the cytochrome P450 family (CYP) of enzymes located in the hepatic endoplasmic reticulum, but may also occur by non-P450 enzyme systems, such as N-acetyl and glucuronosyl transferases. Many factors can alter hepatic and intestinal drug metabolism, including the presence or absence of disease and/or concomi-tant medications, or even some foods, such as grapefruit juice. While most of these factors are usually relatively stable over time, concomitant medications can alter metabolism abruptly and are of particular concern. The influence of concomitant medications on hepatic and intestinal metabolism becomes more complicated when a drug, including a prodrug, is metabolized to one or more active metabolites. In this case, the safety and effi-cacy of the drug/prodrug are determined not only by exposure to the parent drug but by exposure to the active metabolites, which in turn is related to their formation, distribution, and elimination. Therefore, adequate assessment of the safety and effectiveness of a drug includes a description of its metabolism and the contribution of metabolism to overall elimination. For this reason, the development of sensitive and specific assays for a drug and its important metabolites is critical to the study of metabolism and drug-drug interactions.

A third guidance, the 1999 FDA guidance on "In Vivo Drug Metabolism/Drug Interaction Studies—Study Design, Data Analysis, and Recommendations for Dosing and Labeling" also makes some general statements and additionally provides some context for the guidance (US FDA, 1999a). The guidance states:

> No distinction is made in this document between the effects of concomitant drugs and/or alterations in metabolism on gastrointestinal absorption and hepatic elimination, although the pharmacokinetic effects of the two may be different. The desirable and undesirable effects of a drug arising from its concentrations at the sites of action are usually related either to the amount administered (dose) or to the resulting blood concentrations, which are affected by its absorption, distribution, metabolism and/or excretion.
>
> Elimination of a drug or its metabolites occurs either by metabolism, usually by the liver, or by excretion, usually by the kidneys and liver. In addition, protein therapeutics may be eliminated via a specific interaction with cell surface receptors, followed by internalization and lysosomal degradation within the target cell. Hepatic elimination occurs primarily by the cytochrome P450 family of enzymes located in the hepatic endoplasmic reticulum but may also occur by non-P450 enzyme systems, such as N-acetyl and glucuronosyl transferases. P450 enzyme systems located in gut mucosa can also significantly affect the amount of drug absorbed into the systemic circulation. Many factors can alter hepatic and intestinal drug metabolism, including the presence or absence of disease and/or concomitant medications. While most of these factors are usually relatively stable over time, concomitant medications can alter metabolic routes of absorption and elimination abruptly and are of particular concern. The influence of concomitant medications on hepatic and intestinal metabolism becomes more complicated when a drug, including a prodrug, is metabolized to one or more active metabolites. In this case, the safety and efficacy of the drug/prodrug are determined not only by exposure to the parent drug but by exposure to the active metabolites, which in turn is related to their formation, distribution, and elimination.

These latter two guidances not only give a background rationale for the guidance, but they also make the reader recognize the fact that not all metabolism occurs in the liver and not all metabolisms are CYP-mediated. These are important points to recognize since the details of this (and other guidances) do give the strong impression that only CYP-mediated hepatic metabolism needs to be taken into account when evaluating the metabolism of a drug. This background information also raises the topic of prodrugs which is a topic that is mentioned in passing in several guidances but for which there is not a specific guidance.

A final example comes from a guidance document in which metabolism is mentioned in passing and is not the main thrust of the guidance. ICHS3b was issued in 1995 and is titled "Pharmacokinetics: Guidance for Repeated Dose Tissue Distribution Studies" (ICH, 1994a). In the introduction to this guidance, it states:

> A comprehensive knowledge of the absorption, distribution, metabolism and elimination of a compound is important for the interpretation of pharmacology and toxicology studies. Tissue distribution studies are essential in providing information on distribution and accumulation of the compound and/or metabolites, especially in relation to potential sites of action; this information may be useful for designing toxicology and pharmacology studies and for interpreting the results of these experiments.

Interestingly, there is no such guidance for single-dose tissue distribution studies nor is there a clear suggestion that such studies are a crucial part of the drug development process in any regulatory guidance. This topic will be discussed further later on in this chapter.

27.4 NONCLINICAL GUIDANCES AND EARLY-STAGE DRUG DEVELOPMENT

27.4.1 Introduction

Drug metabolism scientists work closely with their nonclinical (toxicology and pharmacology) colleagues to design studies and interpret the data from these studies. The vast majority of nonclinical guidances primarily relate to toxicology, several to drug metabolism and few, if any, to pharmacology. It should be noted that the line between toxicology and pharmacology can become blurred. Safety pharmacology has been defined as "those studies that investigate the potential undesirable effects of a substance on physiological functions in relation to exposure in the therapeutic range and above." These studies are most often performed by toxicologists and are part of the safety assessment of a drug, but will be found in the pharmacology section of a regulatory submission. It is not surprising that there are a large number of guidances related to toxicology since protecting the safety of the patient is of paramount importance and the fact that all drugs have side effects. This latter fact was noted as long ago as the 1500s by Paracelsus, who stated that "All things are poison and nothing is without poison, only the dose permits something not to be poisonous" (Wikipedia, 2008b).

Toxicology guidances cover a range of topics including species selection, duration of studies and types of studies that may be conducted in order to evaluate the potential for a drug to cause a particular toxicity. Drug metabolism and pharmacokinetics are frequently mentioned in such guidances, although often only in passing. The FDA guidance on "Safety Testing of Drug Metabolites" is an exception where the whole premise of the guidance relates to drug metabolism (US FDA, 2008j). The broader topic of preclinical considerations for regulatory submissions has been reviewed by Baldrick (2001) and Larson (2005).

Similarly, drug metabolism scientists provide very similar support to both nonclinical and clinical functions. This subsection of the chapter will focus on those items that are clearly nonclinical in nature (i.e. supporting animal pharmacology and toxicology). Items that clearly relate to support to clinical studies will be addressed later in the chapter. However, there will be some element of redundancy between the two subsections. It is recommended that the nonclinical and clinical subsections be read in concert in order to obtain the fullest understanding of the role of drug metabolism in nonclinical and clinical development.

27.4.2 Good Laboratory Practice

Good Laboratory Practice (GLP) is a concept that drug metabolism scientists should ensure that they are familiar with. Good Laboratory Practice embodies a set of principles that provides a framework within which laboratory studies are planned, performed, monitored, recorded, reported, and archived. These studies are undertaken to generate data by which the hazards and risks to users, consumers and third parties, including the environment, can be assessed for pharmaceuticals, agrochemicals, cosmetics, food and feed additives and contaminants, novel foods, and biocides. The GLPs help assure regulatory authorities that the data submitted are a true reflection of the results obtained during the study and can therefore be relied upon when making risk/ safety assessments. GLP regulations have been implemented globally and are very consistent from country to country, although there may exist some local requirements. Details regarding GLPs can be found in country-specific legislation and on regulatory agencies' web sites (European Parliament, 2004; US FDA, 2007j; OECD, 2007, 2008; U.S. Environmental Protection Agency, 2008).

The GLP regulations became effective in 1979 under 21 CFR Part 58.1 which applies to all nonclinical safety studies intended to support research permits or marketing authorizations for products regulated by the FDA (Dannan, 2003; US FDA, 2007a). The regulations were established after FDA inspected several research laboratories during the mid-1970s, which revealed serious problems with the conduct of safety studies submitted to the agency. The violations included poor record keeping and storage of raw data, lack of proper personnel training and handling of test facilities, and fraud. To apply the fundamental concepts of data quality and integrity, FDA requires that the regulations cover all operations in facilities conducting nonclinical studies and, most importantly, it requires that all facility operations and procedures be strictly documented.

The interpretation of the GLP regulations, particularly related to the subject of which studies should be compliant with GLPs, has engendered much debate. The regulations (in 21 CFR Part 58) state that the GLPs apply to "all non-clinical laboratory studies that support or intend to support applications for research or marketing permits for products regulated by FDA." The FDA defines nonclinical laboratory studies as "in vivo or in vitro experiments in which a test article is studied under laboratory conditions to determine its safety." As can be seen, this definition leaves room for interpretation. GLP compliance will be required for all toxicology studies (except mechanistic or investigative studies) and for the toxicokinetics support (including the bioanalysis) for these studies. Pharmacology (excepting safety pharmacology), drug metabolism, and pharmacokinetic studies do not need to be GLP-compliant since such studies do not directly provide information related to the safety of the test article. However, there will be some circumstances under which some ADME studies should be conducted in a GLP-compliant manner. The first is for the tissue distribution studies that are conducted for a pharmaceutical for the purpose of dosimetry (i.e. the confirmation that the dose of radioactivity that is proposed to be given to humans falls below the maximum allowable limits). Although this is an ADME study, it is clear that it is also a safety study and therefore should be conducted in accordance with GLP regulations. The second circumstance is for veterinary products or agrochemicals. For these products, studies to determine tissue accumulation and depletion should be GLP-compliant since the intent of these studies is to predict the exposure of humans to compound (and/or metabolites) due to ingestion of food that may contain residues of veterinary products or agrochemicals. The FDA has stated that "For drugs and feed additives used in food producing animals, metabolism studies come under the GLPs. In these cases, the studies are intended to define the tissue residues of toxicological concern as well as to estimate tissue depletion. Such studies on other regulated products are usually conducted as part of the pharmacological evaluation and would not be covered. However, metabolism studies on food additives are covered" (US FDA, 2007j). Despite the fact that most drug metabolism studies are not required to be GLP-compliant, they should still be conducted with high standards of quality ensuring reproducibility of data (i.e. "in the spirit of GLP"). Examples of other studies that are outside the scope of

GLP regulations include those utilizing human subjects, which are covered by good manufacturing practice (GMP) and good clinical practice (GCP) guidelines and efficacy studies (covered by GCP guidelines). Although clinical studies fall outside the scope of the GLP regulations, the bioanalytical support to these studies is conducted in compliance with the GLP regulations.

27.4.3 Bioanalytical

The measurement of parent drug and relevant metabolites (however they are defined) is an integral part of understanding the risk–benefit for a drug. This topic is relevant to both nonclinical and clinical studies and is mentioned in a large number of guidances. For brevity, not all guidances that mention the measurement of parent drug and/or metabolites will be covered here. Instead, unique or interesting aspects will be highlighted.

27.4.3.1 Bioanalytical Method Validation   In May 2001, the FDA issued a guidance document, "Bioanalytical Method Validation" that describes some of the criteria that should be applied to the validation of "bioanalytical procedures such as gas chromatography (GC), high-pressure liquid chromatography (LC), combined GC and LC mass spectrometric (MS) procedures such as LC-MS, LC-MS-MS, GC-MS, and GC-MS-MS performed for the quantitative determination of drugs and/or metabolites in biological matrices such as blood, serum, plasma, or urine" (US FDA, 2001b). This guidance also applies to other bioanalytical methods, such as immunological and microbiological procedures, and to other biological matrices, such as tissue and skin samples. The guidance was created after much discussion between the FDA, pharmaceutical companies, and contract research organizations (CROs). This dialog was initiated in 1990 at the first American Association of Pharmaceutical Scientists (AAPS)/Food and Drug Administration (FDA) Bioanalytical Workshop which focused on key issues relevant to bioanalytical methodology and provided a platform for scientific discussions and deliberations. The workshop, and the report, raised awareness of the need for validated bioanalytical methods for the regulatory acceptance of bioequivalence and pharmacokinetic data (Shah *et al.*, 1992). Just over eight years later (in January 1999), the FDA issued a draft guidance on bioanalytical validation based on the workshop report and recent FDA experience. A second workshop was held in January 2000 at which the draft guidance was discussed (Shah *et al.*, 2000). The guidance was finalized in May 2001.

The guidance addresses when a full or partial validation should be performed. The guidance notes that the circumstances that would dictate that a full validation would be required include "when developing and implementing a bioanalytical method for the first time" and "A full validation of the revised assay is important if metabolites are added to an existing assay for quantification." Partial validations are "modifications of already validated bioanalytical methods. Typical bioanalytical method changes that fall into this category include, selectivity demonstration of an analyte in the presence of specific metabolites."

The guidance advises that "the stability of the analyte (drug and/or metabolite) in the matrix during the collection process and the sample storage period should be assessed, preferably prior to sample analysis." This is of particular importance for potentially labile metabolites. The guidance notes that "the stability of analyte in matrix from dosed subjects (or species) should be confirmed." It should be noted that such potentially labile metabolites could degrade during the sample processing or in the mass spectrometer itself. The conversion of a labile metabolite to the parent drug could potentially lead to the overestimation of the concentrations of the parent drug. The topic of potentially labile metabolites is one that will plague the determination of the metabolite profile in *in vitro* and *in vivo* studies. It should be noted that if metabolites are labile, they may not be detected and identified in the metabolism studies.

The guidance notes that "The accuracy, precision, reproducibility, response function, and selectivity of the method for endogenous substances, metabolites, and known degradation products should be established for the biological matrix. For selectivity, there should be evidence that the substance being quantified is the intended analyte." As mentioned previously, the selectivity of the assays is extremely important. The guidance states that "As with chromatographic methods, microbiological and ligand-binding assays should be shown to be selective for the analyte. Cross-reactivity of metabolites, concomitant medications, or endogenous compounds should be evaluated individually and in combination with the analyte of interest." The guidance includes a couple of interesting and important definitions:

"Analyte: A specific chemical moiety being measured, which can be intact drug, biomolecule or its derivative, metabolite, and/or degradation product in a biologic matrix."

"Selectivity: The ability of the bioanalytical method to measure and differentiate the analytes in the presence of components that may be expected to be present. These could include metabolites, impurities, degradants, or matrix components."

It is interesting to note that the guidance mentions "degradation product" and "impurities" since such substances are very rarely quantitated routinely, unless there was the specific need to qualify (in a toxicology study) a degradation product or impurity that was present in a given batch of clinical trial material.

It can be speculated that the guidance was written to include the possibility that it may be necessary to quantitate degradation products or impurities on a case-by-case basis or on a more routine basis in the future.

The development of bioanalytical assays for metabolites can be important for several reasons. These reasons include:

- Confirmation that the selected toxicology species have been exposed to "relevant" human metabolites to a sufficient extent (either after the administration of the parent drug or the metabolite itself)
- Confirmation (or refutation) of misdosing of control group animals in toxicology studies
- Develop sensitive assays for the evaluation of drug–drug interactions (DDIs) (change in metabolite concentrations as well as, or instead of, parent concentrations).

One element that is not discussed in great, if any, detail in the regulatory guidances is the selection of the matrix for the bioanalytical assay. The default matrix is plasma which is prepared from whole blood samples. This matrix is selected due to its ease of collection and the assumption that the plasma concentrations of drug and/or metabolites are in equilibrium with the tissue concentrations and therefore the plasma concentrations of the drug and/or metabolite can act as a surrogate for the tissue concentrations. Despite the caveats that can be made to this assumption, the exposure to the drug in plasma can be correlated with efficacy or toxicity in the majority of cases. Serum is sometimes selected as the matrix instead of plasma, particularly for protein-based therapeutics. A word of caution should be made regarding the use of plasma as the matrix for the quantitation of drug. If the drug partitions into the red blood cells to a marked extent, an erroneous estimate of the plasma concentration of the drug and/or metabolites may be obtained if the blood sample undergoes hemolysis (i.e. the drug and/or metabolites are released from the RBCs after the collection of the blood sample). For drugs that partition heavily into the RBCs, consideration should be given to the use of whole blood as the matrix for the quantitation of drug and/or metabolites.

In May 2006, the third AAPS/FDA Bioanalytical Workshop was held. The report from this workshop, along with several related articles, was published in a themed issue of *The AAPS Journal* (Rocci *et al.*, 2006; Viswanathan *et al.*, 2007). The report (Scharberg, 2007) addressed "Determination of Metabolites During Drug Development." The report from the workshop states: "A draft FDA Guidance for Industry, entitled 'Safety Testing of Drug Metabolites' was issued in June 2005 by the Center for Drug Evaluation and Research (CDER)" (US FDA, 2005c). "There is general support from the pharmaceutical community for the idea that a more extensive characterization of the pharmacokinetics of unique and/or major human metabolites (UMMs) would provide greater insight into the connection between metabolites and toxicological observations. This information would be best generated by the use of rugged, bioanalytical methods applied at appropriate times in drug development. Characterization of UMMs should proceed using a flexible, 'tiered' approach to bioanalytical methods validation. This tiered approach would allow metabolite screening studies to be performed in early drug development using bioanalytical methods with limited validation, with validation criteria increasing as a product moves into clinical trials. A tiered validation approach to metabolite determination would defer bioanalytical resource allocation to later in the drug development timeline when there is a greater likelihood of drug success. As a minimum, the specifics of this tiered validation process should be driven by scientifically appropriate criteria, established a priori." The finalized guidance that was issued in February 2008 does not support the concept of deferring the validation of bioanalytical methods for metabolites until later in the development process (US FDA, 2008j). The topic of "Safety Testing of Drug Metabolites" is discussed in great depth in Section 27.4.5.

Despite the fact that there were many areas of consensus from the third workshop, consensus could not be achieved on all topics. One such area was "Cross-Validation Required When Using Different Strains or Sexes of a Species." The report states that "No decisive arguments came forward in support of this activity although it was agreed that there could be some differences in biological matrix originating from the different strains or sexes. The general trend of the debate was that validation experiments to address such differences should not be considered the norm and should be performed when there are method-related concerns that can be attributed to a specific strain or sex-related difference." Although not mentioned here, it is possible for a drug to be metabolized differently in males and females which may mean that the original evaluation of the selectivity of the assay may not have been sufficiently comprehensive if the validation was performed for one gender only. An obvious extension of this topic is the effect of age or provider of the animals. However, this topic did not appear to be addressed in this report.

27.4.3.2 *Analysis of Samples* The potential for misdosing of control group animals was addressed in an EMEA guideline (Guideline on the Evaluation of Control Samples in Nonclinical Safety Studies: Checking for Contamination with the Test Substance) that was issued in 2005 (EMEA, 2005c). The FDA has not issued guidance on this topic and does not appear to be developing such guidance. There most probably is no need to issue such guidance since sponsors will most likely follow the EMEA guidance in order to be able to file for marketing application in a major market. Hence, this EMEA guideline is *de facto* also an FDA guidance. The guideline provides some clear rationale for the creation of this guideline. "Since 1999 when the first situation was identified, several applications have revealed that samples from control animals contained levels of test substance, or antibodies against the test substance, suggesting that contamination of control animals or samples from control animals with the test substance had occurred. The causes for contamination of controls samples may be multiple and in the majority of the situations the source of contamination has not been identified. The levels of contamination of controls samples were considered relevant/ significant in some studies (e.g. carcinogenicity and reproductive toxicity) and led to their invalidation or impacted on a negative opinion for the Marketing Authorisation Application. A survey conducted by the European Federation of Pharmaceutical Industries Association (EFPIA) shows that contamination of controls with different levels of test compound often occurs during toxicology studies, regardless of the route of administration used, the dose levels and duration of treatment. The CPMP/ICH/384/95 Note for Guidance, Toxicokinetics: A Guidance Assessing Systemic Exposure in Toxicology Studies reads in its Note 8 that 'it is often considered unnecessary to assay samples from control groups. Samples from controls may be collected and then assayed if it is deemed that this may help in the interpretation of the toxicity findings, or in the validation of the assay methods.' Significant contamination of controls may lead to the invalidation of studies due to their poor overall quality or to inappropriateness of toxicokinetic data to allow adequate interpretation of non-clinical safety studies and human risk assessment."

The EMEA guideline states that "When significant contamination of control samples with the test compound at a level susceptible to impact on the validity of the study is observed, the sources of contamination should be investigated and identified. The applicant should make appropriate efforts to clarify whether contamination occurred in vivo or ex-vivo. Approaches may include, but are not exclusive to, identification of test substance in control tissues, identification of plasma metabolites in controls, or detection of antibodies against the test substance in case of a biotechnology product. Corrective measures should be taken accordingly." As can be seen, the existence of an assay for the quantitation of metabolites in plasma samples may be extremely useful to help prove that the presence of test article in plasma samples obtained from control group animals was due to either endogenous interference or *ex vivo* contamination. Despite the clear intent of this guideline, there can be great differences in opinion regarding how it should be implemented. The guideline states: "For all pivotal studies that include a toxicokinetic evaluation, control samples should be collected and analysed" but does not define what pivotal is. The term "pivotal" is not used in the ICHM3(R1) guidance. This term could be interpreted as being the longest duration toxicology study that is submitted with the marketing application. However, it is best defined as the GLP toxicology studies that are used to support a given clinical trial, which would imply that samples should be collected and analyzed from control animals in all GLP-compliant toxicology studies.

The guideline further states that "in non-rodent studies, control samples should be collected and analysed in the same way as treated samples. In rodent studies, control samples should be collected and analysed in at least the general proximity of T_{max} of the test substance." Unfortunately, this will add significant cost (potentially 25% for a traditional study design that includes one control group and three test article groups) to the bioanalytical portion of a non-rodent toxicology study, although the impact on the bioanalytical costs of a rodent toxicology study will be less due to the fewer number of samples that will require analysis. However, there will be a significant increase in the costs of the toxicology portion of a rodent toxicology studies in which samples are collected from control animals. For rodent studies, the toxicokinetics of the drug is most often conducted in "satellite" groups of animals. By using satellite animals, the sponsor is able to avoid the repeated blood draws required to collect the toxicokinetics samples having an impact on certain toxicology tests such as hematology and clinical chemistry assessments. In the past, the satellite groups did not include a control group. However, a control group is now frequently included in the satellite groups which leads to increased costs and increased use of animals. It should be noted that the number of animals in the control satellite group of a rodent study will be less than in a drug-dosed satellite group due to the fact that samples need only be collected in the region of the T_{max}. Although the guidance implies that one sample needs to be collected from a control group rodent, the best practice is most probably to collect two samples from the same rat so that the sponsor can distinguish between misdosed animals and samples that have been contaminated post-sampling. It should be noted that for mice, only one blood sample can be collected per mouse in a toxicology study. The ability to be able to distinguish between misdosing and post-sampling contamination or bioanalytical carryover is important. If a control sample is analyzed immediately after a sample containing high levels of drug, it may appear that the control sample contains drug due to sample "carryover" which may be due to small amounts of sample remaining on the injection needle prior to injection of the sample from the control animal. Care should therefore be taken when analyzing samples from control animals to minimize or eliminate sample carryover.

The report from the third AAPS/FDA Bioanalytical Workshop identified several areas of consensus but also illustrated areas in which consensus could not be achieved (Scharberg, 2007). Some of the items of interest to the drug metabolism scientist/pharmacokineticist in this report include the dynamic range of the standard curve and the analysis of incurred samples. The dynamic range for the standard curve was an area of consensus. The report stated that "If a narrow range of analysis values is unanticipated, but observed after start of the sample analysis, it is recommended that the analysis be stopped and either the standard curve narrowed, existing QC concentrations revised, or QC samples at additional concentrations added to the original curve prior to continuing with sample analysis. It is not necessary to reanalyze samples analyzed prior to optimizing the standard curve or QC concentrations." This statement reflects what had been the FDA's current practice over the years prior to the third workshop. The FDA had issued 483s (the form used by the FDA to record observations during an FDA inspection of a facility or study) when bioanalytical methods used to support bioequivalence studies had too great an analytical range. While laboratories like to make calibration ranges as wide as linearity permits, it is clear that the FDA expects calibration ranges to be equal to the actual range of concentrations found in the study samples and that the low, medium, and high QC samples represent the actual concentration range of samples analyzed within the batch. This means that labs will have to take one of two approaches: validate a series of ranges, each with three levels of QC concentrations; or validate a very wide range but with multiple QC concentrations spread over the entire range. Another implication is that the pharmacokineticist will need to work very closely with the bioanalytical scientist in order to accurately predict the required dynamic range for the calibration curve and/or provide input regarding the extent of dilution that the samples may require.

During inspections by FDA investigators, large inconsistencies between original and repeated results from acceptable runs were found. For example, during validation, inter-day precision and accuracy was less than 6% but original and repeated C_{max} values differed by 30–80%. Such discrepancies are often noted upon repeat analysis of samples that have been identified as requiring repeat analysis by the pharmacokineticist. The analysis of incurred samples is thus addressed:

> There are several situations where the performance of standards and QCs may not adequately mimic that of study samples from dosed subjects (incurred samples). Examples include metabolites converting to the parent species, protein binding differences in patient samples, recovery issues, sample inhomogeneity, and mass spectrometric ionization matrix effects. These factors can affect both the reproducibility and accuracy of the concentration determined in incurred samples. While these effects are often characterized and minimized during method development using QC samples, it is important to ensure that they are under control when the method is applied to the analysis of incurred samples.

> A proper evaluation of incurred sample reproducibility and accuracy needs to be performed on each species used for Good Laboratory Practice (GLP) toxicology experiments. It is not necessary for additional incurred sample investigations to be performed in toxicology species once the initial assessment has been performed. Incurred sample evaluations performed using samples from one study would be sufficient for all other studies using that same species. It is generally accepted that the chance of incurred sample variability is greater in humans than in animals, so the following discussion pertains primarily to clinical studies. The final decision as to the extent and nature of the incurred sample testing is left to the analytical investigator, and should be based on an in-depth understanding of the method, the behavior of the drug, metabolites, and any concomitant medications in the matrices of interest. There should be some assessment of both reproducibility and accuracy of the reported concentration. Sufficient data should be generated to demonstrate that the current matrix produces results similar to those previously validated. It is recognized that accuracy of the result generated from incurred samples can be more difficult to assess. It requires evaluation of any additional factors besides reproducibility upon storage, which could perturb the reported concentration. These could include metabolites converted to parent during sample preparation or LC-MS/MS analysis, matrix effects from high concentrations of metabolites, or variable recovery between analyte and internal standard. If a lack of accuracy is not a result of assay performance (ie, analyte instability or interconversion) then the reason for the lack of accuracy should be investigated and its impact on the study assessed. The extent and nature of these experiments is dependent on the specific sample being addressed and should provide sufficient confidence that the concentration being reported is accurate. The results of incurred sample reanalysis studies may be documented in the final bioanalytical or clinical report for the study, and/or as an addendum to the method validation report. In selecting samples to be reassayed, it is encouraged that issues such as concentration, patient population, and special populations (eg, renally impaired) be considered, depending on what is known about the drug, its metabolism, and its clearance. First-in-human, proof-of-concept in patients, special population, and bioequivalence studies are examples of studies that should be considered for incurred-sample

concentration verification. The study sample results obtained for establishing incurred sample reproducibility may be used for comparison purposes, and do not necessarily have to be used in calculating reported sample concentrations.

It will be interesting to see how the analysis of incurred samples will be implemented. Before the analysis of incurred samples becomes a routine part of the bioanalytical support for a program, it may be beneficial for the bioanalytical community to hold a dialog with the pharmacokinetic community, who are the customers for the data generated by the bioanalytical groups.

Although the drug metabolism scientists or pharmacokineticists may not be directly responsible for the bioanalytical support to a project, they should be aware of the FDA's close scrutiny of bioanalytical data since the bioanalytical data are a key element in their understanding of the ADME properties of a drug. As was mentioned earlier, much can be learned from warning letters issued by the FDA regarding the FDA's position on a topic and other company's approaches to drug development. This can be illustrated by the series of warning letters issued by the FDA to MDS Pharma Services in Montreal, Canada. There were several different issues that were identified in the initial warning letter that was issued in April 2004 as a result of the FDA's audit of a bioequivalence study of loratadine in July 2003 (US FDA, 2004e). In December 2004, a second warning letter was issued as a result of the inspection of a larger number of BE studies (US FDA, 2004e). The FDA asked MDS Pharma Services to conduct a five-year retrospective review of all BE studies to ensure that the issues that had been identified in the original warning letter were isolated incidents and not part of a larger pattern. A third warning letter was issued in May 2006 and clearly indicates that the FDA was not pleased with MDS Pharma Services' response to the FDA's original warning letter and the quality of its follow-through on this topic (US FDA, 2006k). In January 2007, the FDA sent a letter to pharmaceutical companies that had used MDS between January 2000 and December 2004 to perform bioanalysis in support of bio-equivalence studies (Buehler, 2007). The letter noted that serious questions remain about the validity of bioequivalence data generated by MDS in studies during this time period. The letter stated: "In view of these findings, FDA is informing holders of approved ANDA(s) of these issues and would like to know what steps are being taken by you to assure the accuracy of data submitted in these applications and confirm the validity of MDS's analytical studies that were conducted from January 2000 through December 2004 and subsequently submitted to the FDA. Accordingly, with respect to these studies submitted in your applications, we recommend that within 6 months of the date of this letter you do one of the following, in order of FDA preference:

1. Repeat the bioequivalence studies.
2. Re-assay the samples at a different bioanalytical facility. For this option, the integrity of the original samples must be demonstrated for the frozen storage period.
3. Commission a scientific audit by a qualified independent expert, who is knowledgeable in the area of bioequivalence studies and bioanalytical data, and selected by your company rather than by MDS, to verify the results obtained by MDS.

In addition, because one of the agency's significant findings for the inspected MDS studies was the presence of anomalous results, we are recommending for all of the above options that the blood/plasma level results obtained in the studies be compared to any published literature or other relevant information that is publicly available."

This clearly places a great burden on the sponsor. At the present time, it is unknown as to whether further deficiencies have been found and whether any approved generics have had to be removed from the market as a consequence of these deficiencies. It is clear that the sponsor should play a more active role in the oversight of the bioanalytical CRO (Scharberg, 2008).

27.4.4 Pharmacokinetics and Toxicokinetics

An early guidance from Europe addresses the topic of "Pharmacokinetics and Metabolic Studies in the Safety Evaluation of New Medicinal Products in Animals" (EMEA, 1994b). This guidance states: "Data on the levels of substance and metabolites in blood, body fluids, organs and in the excreta can be obtained by physical, chemical or biological methods. The investigator should justify the details of the methods used, their validity and reproducibility, including the specificity, precision and accuracy. (The study of the time course of its pharmacodynamic effects may provide useful additional information)." The guidance further states: "When using labeled substances attention must be given to the fact that the measured label in body fluids may not correspond to that of the unmodified substance, but may include labeled metabolites and conjugates." This point is a very important one that is often not appreciated by many pharmaceutical scientists outside of the ADME community and is a key limitation in the interpretation of data from tissue distribution studies. The final point of note is that "Attention should be given to the possibility of isotope exchange with endogenous compounds." In particular, care should be taken when conducting studies with compounds labeled with tritium, since the tritium can exchange with hydrogen atoms from water.

An assessment of the systemic exposure to parent drug and metabolites is a key part of a toxicology study and is important in putting any observed toxicities in the context of the exposure of humans to the parent drug and/or metabolites. This topic is covered in another ICH guidance, "ICHS3A—Toxicokinetics: The assessment of Systemic Exposure in Toxicity Studies" that was issued in 1995 (ICH, 1994b). This guidance addresses the possibility of establishing a maximum dose to be administered in the toxicology study if "toxicokinetic data indicate that absorption of a compound limits exposure to parent compound and/or metabolite(s)." However, it should be recognized that the use of a dose "cap" may not address local (GI) toxicity or the effect of metabolites generated by the GI mucosa or microflora for orally administered drugs. The guidance also advocates the use of metabolism, tissue distribution data, and plasma protein-binding data in the interpretation of the exposure data. For highly protein-bound drugs, the guidance suggests that the exposure to drug be expressed in terms of free (unbound) drug, although it does not address whether concentrations of free drug should be directly measured or can be calculated in other ways as is suggested in other guidances. Exposure to metabolites is addressed thus: "Measurement of metabolite concentrations may be especially important when documentation of exposure to human metabolite(s) is needed in the non-clinical toxicity studies in order to demonstrate adequate toxicity testing of these metabolites. It is recognized that measurement of metabolite(s) as a part of toxicokinetic evaluation serves only to assess exposure and cannot account for possible reactive intermediate metabolites." This latter statement emphasizes the importance of the availability of bioanalytical assays to quantitate metabolites (as appropriate). This guidance encourages the sponsor to perform "concomitant toxicokinetics" (i.e. describe exposure at the same time that the toxicology study is conducted rather than at a later time). The guidance notes that the most commonly described PK parameters are T_{max}, C_{max}, and AUC_{0-t}. The guidance further states that half-life or $AUC_{0-infinity}$ is rarely calculated. This latter point is important since the standard sample collection scheme (five or six samples) is not usually sufficient for the calculation of these PK parameters, especially for drugs that are administered extravascularly (i.e. oral or subcutaneous). The inappropriate calculation of TK parameters (e.g. AUC, half-life, and clearance) is a common pitfall for the inexperienced pharmaceutical scientist.

The guidance also notes that "Increases in exposure may arise unexpectedly as a result of nonlinear kinetics due to saturation of a clearance process. Increasing exposure may also occur during the course of a study for those compounds which have a particularly long plasma half-life." It should be noted that the pharmacokinetics of most drugs are determined at a low dose level (e.g. 10 mg/kg) whereas the toxicokinetics of the drug may be described at doses that are often one to two orders of magnitude greater than the dose used in the PK study. This discrepancy may lead to a species as being viewed as inappropriate for use in toxicology studies for a given compound due to its low bioavailability and hence potentially high compound requirements. However, if the clearance is saturable (e.g. due to saturation of metabolism), it may be feasible to achieve adequate exposure to parent drug in that species. Further points made by the guidance include "Careful attention should also be paid to compounds which achieve high C_{max} values over comparatively short time periods within the dosing interval" and "Conversely, unexpectedly low exposure may occur during a study as a result of auto-induction of metabolizing enzymes." The former point speaks to the appropriateness of selection of the time points at which blood is collected. The latter point pertains primarily to toxicology studies conducted in the rat. Auto-induction of metabolism will, as pointed out in the guidance, lead to low exposure to parent compound but will lead to high exposure to metabolites. This point will be further elucidated later on in this chapter.

27.4.5 Metabolism

One of the most controversial draft guidances that has been issued in the past decade has been the FDA draft guidance "Safety Testing of Drug Metabolites" that was issued for comment in June 2005 (US FDA, 2005c). This guidance was finalized in February 2008 (US FDA, 2008j). The history of the genesis of this guidance has been described previously elsewhere. Since approximately 30 months elapsed between the issuance of the draft and final guidances, the draft guidance was essentially being used as a *de facto* finalized guidance. This section of the chapter will compare the draft and final guidance. Both documents have the same title, are 11 pages in length, and contain a decision tree for determining when to evaluate the safety of a drug metabolite. Despite these similarities, the content of the two versions of the guidance including the content of the decision tree is quite different.

The draft guidance provided some key terms, some of which were retained in the finalized guidance and some of which were revised. The terms used in the draft guidance were:

"Major metabolite—A metabolite in humans that accounts for plasma levels greater than 10% of the administered dose or systemic exposure, whichever is less.

Metabolite—A compound derived from the parent compound through Phase I and/or Phase II metabolic pathways.

Pharmacologically active metabolite—A metabolite that has pharmacological activity at the target receptor that is greater than, equal to, or less than the parent compound.

Unique human metabolite—A metabolite produced only in humans."

The finalized guidance retained the terminology for "metabolite" and "pharmacologically active metabolite" but dropped the terminology for "major metabolite" and "unique human metabolite." These dropped terms were replaced with the term "disproportionate drug metabolite."

The terminology used in the final guidance is as follows:

"Disproportionate drug metabolite—A metabolite present only in humans or present at higher plasma concentrations in humans than in the animals used in nonclinical studies. In general, these metabolites are of interest if they account for plasma levels greater than 10% of parent systemic exposure, measured as area under the curve (AUC) at steady state.

Metabolite—A compound derived from the parent drug through Phase I and/or Phase II metabolic pathways.

Pharmacologically active metabolite—A metabolite that has pharmacological activity at the target receptor. The activity may be greater than, equal to, or less than that of the parent drug."

The move away from the terms "major metabolite" and "unique human metabolite" is important since it emphasizes that metabolites are unlikely to be unique and also downplays the importance of whether a metabolite was major or not and softens the % exposure limit for safety testing by using the term "generally." The finalized guidance states that:

An active metabolite may bind to the therapeutic target receptors or other receptors, interact with other targets (e.g. enzymes, proteins), and cause unintended effects. This is a particularly important problem when such a metabolite is formed in humans and not in animals, but the occurrence of a metabolite only in humans and not in any animal test species is rare. A more common situation is the formation of a metabolite at disproportionately higher levels in humans than in the animal species used in safety testing of the parent drug. This disproportionality stems from the typical qualitative and/or quantitative differences in metabolic profiles between humans and animals. If at least one animal test species forms this drug metabolite at adequate exposure levels (approximately equal to or greater than human exposure), as determined during toxicology testing of the parent drug, it can be assumed that the metabolite's contribution to the overall toxicity assessment has been established.

The last statement acknowledges the fact that the doses used in animal toxicity studies are far higher than those used in human clinical trials and consequently the safety of a given metabolite may have been assessed in the animal toxicity studies by virtue of the large doses used (and hence large exposures obtained), even if such a metabolite is present in humans on a higher percentage basis than in animals.

The terminology (or similar terminology) described earlier has been used in other guidances, although it is unknown whether the same terminology has the same or different meanings.

The intent and key conclusions of the guidance can be found in the introduction of both the draft and the finalized guidance: "This guidance makes recommendations on when and how to identify, characterize, and evaluate the safety of unique human metabolites and major metabolites of small molecule (nonbiologic) drug products. These metabolites may not be adequately assessed during standard nonclinical studies because they occur only in humans (unique metabolite), or at much higher levels (major metabolite) in humans than in the species used during standard nonclinical toxicology testing. If such metabolites are identified, they should be evaluated as early as possible during the clinical development program."

The draft guidance also states that "This guidance defines major metabolites primarily as those identified in human plasma that account for greater than 10% of drug related material (administered dose or systemic exposure whichever is less) and that were not present at sufficient levels to permit adequate evaluation during standard nonclinical animal studies." This statement does not appear in the finalized guidance. However, an additional statement has been added in the finalized guidance: "This guidance applies to small molecule nonbiologic drug products. This guidance does not apply to some cancer therapies where a risk-benefit assessment is considered. The Food and Drug Administration (FDA) is developing more specific guidance for the safety testing of drug metabolites in cancer therapies." This last statement is interesting, since such a draft guidance is not included in the current (March 2008) list of guidances that are under development for 2008 (US FDA, 2008f). This is not the only guidance that states that a separate guidance will be developed for oncology drugs (US FDA, 2006h). However, no such oncology-specific guidances have been created to date.

The draft guidance also gives a fairly succinct summary of the history of the use of metabolism data in safety evaluation:

> Traditionally, drug metabolites in general have not been routinely evaluated in cross-species safety assessments because their specific contribution to the overall toxicological potential of the parent drug has been unknown. With the availability during the past decade of technologies that can identify, measure, and characterize metabolites, we have gained a better understanding of the role metabolites play in drug safety assessment. Generally, we have used measurements of circulating concentrations of a parent drug in animals as an index of systemic exposure in humans. Quantitative and qualitative differences in metabolite profiles are important when comparing exposure and safety of a drug in a nonclinical species relative to humans during risk assessment. Based on data obtained from in vitro and in vivo metabolism studies, when the metabolic profile of a parent drug is similar qualitatively and quantitatively across species, we can generally assume that potential clinical risks of the parent drug and its metabolites have been adequately characterized during standard nonclinical safety evaluations. However, metabolic profiles and metabolite concentrations can vary across species, and there are cases when clinically relevant metabolites have not been identified or adequately evaluated during nonclinical safety studies. This may be because the metabolite being formed in humans was absent in the animal test species (unique human metabolite) or because the metabolite was present at much higher levels in humans (major metabolite) than in the species used during standard toxicity testing.

The concepts of these statements have been retained in the finalized guidance, although the wording has been changed slightly to read:

> Generally, drug plasma concentration and systemic exposure in the nonclinical studies are compared with systemic exposure in humans to assess the potential risks suggested by nonclinical findings and guide monitoring in clinical trials. This testing paradigm usually is

sufficient when the metabolic profile in humans is similar to that in at least one of the animal species used in nonclinical studies, but metabolic profiles can vary across species both quantitatively and qualitatively, and there are cases when clinically relevant metabolites have not been identified or adequately evaluated during nonclinical safety studies. This situation can occur if the metabolite is formed only in humans and is absent in the animal test species or if the metabolite is present at disproportionately higher levels in humans than in the animal species used in the standard toxicity testing with the parent drug.

It is not standard practice for drug metabolites to be evaluated separately in a cross-species safety assessment. As a result, their specific contribution to the overall toxicity of the parent drug has often remained unknown. This lack of appreciation of the role of metabolites in drug toxicity may be partly because of the inadequate sensitivity of the analytical methods used to detect and characterize metabolites derived from the parent drug. Technological advances during the past decade have greatly improved the analytical capabilities to detect, identify, and characterize metabolites and allow for better understanding of the role metabolites play in drug safety assessment.

The statement "This lack of appreciation of the role of metabolites in drug toxicity may be partly because of the inadequate sensitivity of the analytical methods used to detect and characterize metabolites derived from the parent drug. Technological advances during the past decade have greatly improved the analytical capabilities to detect, identify, and characterize metabolites and allow for better understanding of the role metabolites play in drug safety assessment" is magnanimous since the approach used to the detection of metabolites has not changed significantly for many decades, although advances in LC-MS/MS have improved our ability to determine the structure of metabolites and to detect metabolites at very low levels. It would be fair to state that the relative lack of evaluation of the toxicity of drug metabolites in the past has been due to a scientific blind spot that has caused many people to assume that the parent drug is responsible for both toxicity and efficacy. Similarly, the pharmaceutical industry has not paid great attention to the contribution of active metabolites to the overall therapeutic effect until recently.

The draft guidance placed a special emphasis on certain types of metabolites: "It is especially important to identify metabolites that may be unique to humans." The guidance also states that "compounds with the following characteristics are of particular concern and may warrant additional investigation:

- Narrow therapeutic indices
- Significant toxicity
- Significantly diverse metabolic profiles between human and nonclinical species
- Irreversible toxicity, or adverse effects not readily monitored in the clinic"

This statement helps the reader to understand some of the sources of the FDA's concerns. However, this section was not included in the finalized guidance. The finalized guidance does provide some statements regarding general considerations for nonclinical study designs. The finalized guidance states: "When designing a nonclinical study for a disproportionate drug metabolite, it is important to consider the following factors:

- Similarity of the metabolite to the parent molecule
- Pharmacological or chemical class

- Solubility
- Stability in stomach pH
- Phase I versus Phase II metabolite
- Relative amounts detected in humans versus the amounts detected in animals"

These are all factors worth considering, but the guidance does not give insight as to how such information would be used in designing the nonclinical study.

The FDA proposed a cutoff for exposure to a metabolite in humans above which additional safety testing may be required. The draft guidance states that:

> Generally, we recommend that metabolites identified in human plasma that account for greater than 10% of drug related material (administered dose or systemic exposure whichever is less) be considered for safety assessment. The rationale for setting the level at greater than 10% for characterization of metabolites reflects consistency with other FDA and EPA regulatory guidances (U.S. Food and Drug Administration 2002; U.S. Environmental Protection Agency 1998) and is supported by actual cases, described below, in which it has been determined that the toxicity of a drug could be attributed to one or more metabolites present at greater than 10% of the administered dose. Of the cases that follow, the last two are examples of a situation when a metabolite present at less than 10% caused toxicity. As a result, depending on the situation, some metabolites present at less than 10% should also be tested.

The draft guidance provides examples (halothane, felbamate, cyclophosphamide, and acetaminophen) to support the selection of 10% as a cutoff. The cutoff is one of the greatest sources of contention between the FDA and the pharmaceutical industry on this draft guidance. The PhRMA white paper proposed a cutoff of 25% for the definition of a major metabolite (Baillie *et al.*, 2002). In response to this proposed cutoff, Hastings *et al.* wrote a letter to the editor (Hastings *et al.*, 2003) that noted that "we want to emphasize that this threshold carries no regulatory authority or implications with respect to the need for safety assessment." They further pointed out that the EPA has established a limit of 5% of administered compound as the threshold for metabolite testing in the toxicological evaluation of pesticides, that ICHQ3A (R2) (Impurities in New Drug Substances) (ICH, 2006a) states that a contaminant/degradant should be "qualified" if present in drug product at >0.1% of the drug substance, and that the FDA Center for Veterinary Medicine has used 10% of total drug residue present in edible tissue as the definition of "major metabolite" in food-producing animals. These points clearly indicate the direction that the FDA was heading and therefore it was not surprising when the draft guidance was issued that the cutoff for the definition of a major metabolite was 10% of the exposure to total compound-related material. As an aside to the letter to the editor created by Hastings *et al.* (2003), it should be noted that ICHQ3A(R2) states that "Excluded from this document are extraneous contaminants that should not occur in new drug substances and are more appropriately addressed as Good Manufacturing Practice (GMP) issues." An "Extraneous Contaminant" is defined as "An impurity arising from any source extraneous to the manufacturing process" (ICH, 2006a).

An alternative approach to the use of a relative abundance was proposed by Smith and Obach (2005). Their proposal was that the absolute abundance of a metabolite in plasma or excreta should be combined with other information regarding the chemical structure of the metabolite. The approach of Smith and Obach carries some merit since the use of relative abundance could potentially lead to some bizarre situations. For example, duloxetine is extensively metabolized *in vivo* in humans (Lantz *et al.*, 2003). The extent of metabolism is so great that only 2.7% of the plasma exposure to total compound-related material can be attributed to duloxetine. Interestingly, duloxetine itself does not meet either the FDA's or PhRMA's criteria for the definition of a major metabolite. There appear to be four potential major metabolites (using the FDA's criteria) in the plasma of humans dosed with duloxetine, although the exposure to these metabolites in plasma was not reported in this publication.

The finalized guidance adhered to the 10% limit, although the rationale for selecting this limit was changed slightly. The examples that were provided in the draft guideline are not present in the final guidance. The rationale for the selection of the 10% cutoff is now given as being harmonization with EPA guidances. The finalized guidance states that:

> Human metabolites that can raise a safety concern are those formed at greater than 10% of parent drug systemic exposure at steady state (Exposure should be at steady state unless there is some justification for a different measure of exposure. Comparison between human and animal exposure generally is based on area under the curve, but sometimes it may be more appropriate to use C_{max}). The choice of a level of greater than 10% for characterization of drug metabolites reflects consistency with FDA and Environmental Protection Agency guidances.

The final guidance provides some caveats regarding the 10% cutoff. It states that:

> Although the drug metabolite of toxicological concern usually is one circulating in plasma at greater than 10% of parent systemic exposure, other metabolites also can elicit safety concern. For example, a drug metabolite representing greater than 10% of urinary excretion relative to the bioavailable dose, or a human fecal metabolite in cases where biliary elimination is the predominant route of excretion in humans, may reflect potential localized renal or bile duct toxicity, respectively. Further characterization in these instances should be addressed on a case-by-case basis with the review division.

The draft guidance also provides some background information on drug metabolism and its role in the inactivation of drugs:

> Drugs entering the body undergo biotransformation via Phase I and Phase II metabolic pathways. Based on the nature of the chemical reactions involved, metabolites formed from Phase I reactions (e.g. oxidation, reduction) are more likely to be pharmacologically active, and require safety evaluation, than Phase II products (e.g. glucuronidation, sulfation). Although conjugated metabolites from Phase II reactions are generally pharmacologically inactive, more water soluble, and readily eliminated from the body, some are toxic. Sulfate and some glucuronide metabolites (e.g. acyl glucuronides of carboxylic acids) may retain pharmacological activity as well as toxicity of the parent drug and may require toxicological evaluation. Demonstration that a metabolite is pharmacologically inactive at the target receptor does not guarantee that it is not toxic, however. If the unique or major metabolites are suspected to contain a reactive functional group, it is important to assess the toxicity potential of these reactive metabolites. Chemically reactive intermediates are rarely detectable due to their short half-life, although stable products (i.e. glutathione conjugates) resulting from such intermediates can provide some indication of exposure to these potentially toxic species.

The finalized guidance does not use the same text, but embodies the same concepts, stating that:

Metabolites that form chemically reactive intermediates can be difficult to detect and measure because of their short half-lives. However, they can form stable products (e.g. glutathione conjugates) that can be measured and, therefore, may eliminate the need for further evaluation. Phase II conjugation reactions generally render a compound more water soluble and pharmacologically inactive, thereby eliminating the need for further evaluation. However, if the conjugate forms a toxic compound such as acylglucuronide, additional safety assessment may be needed.

The points that are made here serve to illustrate that not all metabolites are inert. It is fairly well appreciated in the pharmaceutical industry that metabolites may be toxic, but the concept that metabolites may retain (or have enhanced) pharmacological activity is not as well understood. This latter point can be illustrated by the cholesterol absorption inhibitor, ezetimibe. After oral administration, ezetimibe is absorbed and extensively conjugated to a pharmacologically active phenolic glucuronide (ezetimibe-glucuronide) (Merck/Schering-Plough Pharmaceuticals, 2008). Ezetimibe is primarily metabolized in the small intestine and liver via glucuronide conjugation (a phase II reaction) with subsequent biliary and renal excretion. Minimal oxidative metabolism (a phase I reaction) has been observed in all species evaluated. In humans, ezetimibe is rapidly metabolized to ezetimibe-glucuronide. Ezetimibe and ezetimibe-glucuronide are the major drug-derived compounds detected in plasma, constituting approximately 10–20% and 80–90% of the total drug in plasma, respectively. Both ezetimibe and ezetimibe-glucuronide are slowly eliminated from plasma with a half-life of approximately 22 hours for both ezetimibe and ezetimibe-glucuronide. Plasma concentration–time profiles exhibit multiple peaks, suggesting enterohepatic recycling. Ezetimibe not only emphasizes the point that metabolites can be pharmacologically active and may be the major pharmacologically active component in the circulation, it also illustrates that not all meaningful metabolisms occur in the liver.

The timing of the generation of an understanding of the metabolism of a drug candidate is also addressed in the draft guidance: "The Agency recommends that— and this guidance encourages—attempts be made to identify as early as possible during the drug development process differences in drug metabolism in animals used in nonclinical safety assessments compared to humans." The guidance further states that "The discovery of unique or major human metabolites late in drug development can cause development delays and could have possible implications for marketing approval. Early identification of unique or major metabolites will allow for timely assessment of potential safety issues." These points are further elaborated as follows:

The in vitro studies are generally conducted prior to the in vivo studies and provide an initial comparative metabolic profile. Results from these studies can assist in the selection of the appropriate animal species for toxicological assessments, should qualitative interspecies differences in metabolism be detected.

Identifying a major metabolite in animals that does not exist in humans can mean that a toxicity observed in that animal species may not be relevant to humans. Conversely, identifying a human metabolite during clinical development that did not form at appreciable levels in animals would raise safety concerns because it probably was not evaluated in the nonclinical studies due to inadequate exposure. Additionally, when a potentially clinically relevant toxicity is observed during standard nonclinical studies, it is prudent to determine if metabolites contribute to that finding. In such cases, we recommend that the metabolites be synthesized and directly administered to the appropriate animal species for further pharmacological/toxicological evaluation. When qualitative and/or quantitative species differences in metabolite profiles are discovered, we also recommend investigation of different routes of administration or use of alternative animal species for safety assessments.

The guidance goes as far as stating that "Discovery of such a metabolite could delay development until the relationship between metabolite exposure and toxicity is understood."

These statements serve to emphasize the potential impact of metabolites on the safety assessment of a drug and the timelines for the development of a drug. The finalized guidance clearly states that the clinical program may be delayed if adequate evaluation of the nonclinical safety of disproportionate drug metabolites is not performed in a timely manner. Implicit in these statements is the fact that it is essential to determine the metabolism of an NCE early on in clinical development rather than at a later stage. The addition of the statement regarding the timing of such safety evaluation studies for drugs that are intended to treat life-threatening diseases is a great addition to the guidance. This enables the sponsor and the FDA to perform a true risk–benefit analysis rather than apply a "one size fits all approach" to the safety assessment of disproportionate drug metabolites.

The finalized guidance provides an overview of the general approaches for assessing metabolite safety. This section was not included in the draft guidance and represents a useful addition. The finalized guidance states:

A metabolite identified in animals that is not present in humans can mean that a toxicity observed in that animal species, attributed to the metabolite, may not be relevant to humans. Conversely, a drug metabolite identified during clinical development that is not present in animal test species or is present at much lower levels in animals than in humans can suggest the need for further studies in animals to determine the potential toxicity of the metabolite. In such cases, two approaches can be considered to assess the drug metabolite. The first approach is to identify an animal species routinely used in toxicity studies that forms the metabolite at adequate exposure levels (equivalent to or greater than the human exposure), and then investigate the drug's toxicity in that species. The second approach, if a relevant animal species that forms the metabolite cannot be identified, is to synthesize the drug metabolite and directly administer it to the animal for further safety evaluation. In this approach, analytical methods that are capable of identifying and measuring the metabolite in nonclinical toxicity studies should be developed.

The difficulties associated with synthesizing a specific metabolite as well as the inherent complexities that accompany its direct administration are acknowledged. Direct dosing of a metabolite to animals may lead to subsequent metabolism that may not reflect the clinical situation and thus may complicate the toxicity evaluation. Moreover, new and different toxicities may arise from administration of the metabolite that were not observed with the parent drug. However, notwithstanding these possible complications, identification and evaluation of the potential toxicity of the drug metabolite is considered important to ensure clinical safety, and the decision to conduct direct safety testing of a metabolite should be based on a comprehensive evaluation of the data on the parent drug and any information available for the metabolite. Appendix B provides three case examples when drug metabolites were formed at disproportionately higher levels in humans than in test animals used in the nonclinical studies and how the safety evaluation was approached. In Case 1, testing of the drug metabolite was not needed because the metabolite was adequately characterized in nonclinical toxicity studies with the parent drug. However, in Cases 2 and 3, the drug metabolites had to be tested in toxicity studies by direct administration to the animal. In Case 3, the drug metabolite was pharmacologically inactive at the therapeutic target receptor but showed a unique toxicity not observed with the parent molecule.

The acknowledgment of the difficulty in synthesizing drug metabolites as well as the inherent complexities of the administration of these metabolites is welcomed. Additionally, the creation of examples greatly helps the reader understand the situations under which the FDA may reasonably expect that safety testing of drug metabolites be warranted.

The case studies in the finalized guidance are as follows.

"**Case 1.** From an initial mass balance study, a metabolite represented 1 to 2 percent of total radioactive dose in rat plasma, 5 percent in dogs, and 20 percent in humans (radioactivity of this metabolite in urine and/or feces was minimal). Based on the up-to-20 fold greater exposure in humans than in animals, nonclinical safety testing was recommended. However, the data generated in the general toxicology studies with the parent drug in the rat and dog suggested that the maximum doses tested produced metabolite exposures that represented at least the therapeutic exposure at the maximum recommended human dose. Also, the plasma concentrations of this metabolite measured in the in vivo genetic toxicity study, embryo-fetal development toxicity study, and carcinogenicity studies conducted with the parent drug provided adequate exposure and characterization of the metabolite. Therefore, no additional testing with the metabolite was needed."

This case implies the *in vivo* metabolism data were available relatively early in development since the exposure to the metabolite was addressed in the *in vivo* genetic toxicity study and the embryo-fetal development study. However, it is possible that the sponsor re-conducted these studies in order to show the exposure to the metabolite in these studies. This case also serves to emphasize the estimation, however crude, of the maximum recommended human dose early in the development plan. The case implies, but does not specifically state, that mean exposure data were used.

"**Case 2.** Two primary hydroxylated metabolites, M1 and M2, were shown to undergo further oxidation to form secondary metabolites M3 and M4 using hepatic microsomes and hepatocytes from human, monkey, rat, dog, rabbit, and mouse. This metabolic profile was later confirmed by in vivo data. The results showed the following:

- M1 and M4 were the predominant metabolites in human, monkey, and dog microsomes, whereas rat, mouse, and rabbit formed M2 and M3.
- M4 was formed in humans at 4 fold higher levels than parent drug, but M4 was formed at very low levels in rodents and only represented one-third of the parent exposure in monkey.

AUC_{0-24hr} at the Maximum Dose

	Human (MRHD)	Monkey	Rat
Parent	1,800	15,000	12,500
M4	7,700	5,000	135

MRHD = Maximum Recommended Human Dose

Severe drug-related and novel target organ toxicities were observed with the parent drug in monkeys but not in rats.

- M4 was pharmacologically inactive at the drug target receptors.

The following additional studies were done with M4:

- Subchronic toxicity study: 3 months in the rat
- Embryo-fetal development study in the rat
- In vitro genotoxicity testing: M4 was positive for point mutation and chromosomal aberration; the parent drug was negative
- Because of the positive genotoxicity, a carcinogenicity study that included M4 was recommended."

This case is interesting, and should have warranted further discussion in order to help the readers. The statement that M4 was not pharmacologically active against drug target receptors seems somewhat inconsistent with the statement that "Severe drug-related and novel target organ toxicities were observed with the parent drug in monkeys but not in rats." It could be concluded from these data that M4 is not toxic based on the observed toxicities in the monkey and apparent lack of toxicity in the rat. However, there may be additional data that are not included in this case study that would support the course of action taken. A more fully fleshed out description of the case study would have been more useful to the reader. It would have also been enlightening to mention whether M4 was generated by Aroclor-induced rat liver S9 or not.

"**Case 3.** M2 is a Phase I oxidative metabolite that formed up to 50 percent of parent drug exposure in humans, 10 percent of parent drug exposure in mice, 15 percent of parent drug exposure in dogs, and only trace amounts in rats. In vitro metabolism studies in these species supported the in vivo findings. Based on structure activity relationship (SAR) analyses, there was no reason to anticipate any difference or exaggeration in toxicity of the metabolite compared to parent molecule. The parent drug showed no significant toxicity or identifiable target organ of toxicity in any of the animal species tested in safety assessment studies. Because disproportionate human exposure was identified, further safety testing was needed. When M2 was tested in a short-term tolerance study in the dog, it produced unexpected and significant cardiotoxicity at all doses and in all of the dogs. M2 was pharmacologically inactive at the therapeutic target receptor."

This case also raises some interesting points. Firstly, the existence of SAR data, as well as the acceptance of such data by the FDA, is uncommon. It implies the existence of a validated hypothesis for the mechanism of toxicity, the existence of a series of analogs of the compound under development as well as the existence of an *in vitro* model that could be used to expeditiously and efficiently generate the SAR data. The observation that there was no significant toxicity observed in the dog with the parent compound could have implications for the validity of the safety assessment program for this compound. A successful program would have identified the potential human toxicities. A study that shows a lack of toxicity could be regarded as a failed study. The observation of an unanticipated toxicity with M2 raises questions regarding the relevance of the study based on the lack of toxicity of the parent drug and the ability of the dog to produce this metabolite. Additionally, the case does not mention whether such toxicities were observed in clinical trials. This case could benefit from a discussion of how the FDA viewed the data and whether the program was able to progress or not.

Despite these reservations, the inclusion of case studies is a great step forward and case studies should be used, as appropriate, in future guidances.

The identification of metabolites is addressed in relatively great detail compared with other guidances. The draft guidance states:

> In vitro studies using liver slices, microsomes, or hepatocytes from animals and humans to identify the drug metabolic profile are generally conducted before initiation of clinical trials. It is important to also try to determine whether the concomitant use of drugs results in the inhibition or the induction of common metabolic pathways. In vivo metabolic profiles in nonclinical test species are generally available early in drug development, and their results may reveal significant quantitative and/or qualitative differences in metabolism across species. However, a unique metabolite may only be recognized after completion of in vivo metabolic profiling in humans. Therefore, we recommend the in vivo metabolic evaluation in humans be performed as early as feasible.

This statement emphasizes the limitations of *in vitro* studies and strongly advises that *in vivo* human metabolism data be generated as soon as is reasonable.

Once again, the finalized guidance uses different wording but retains the same general content. The finalized guidance states:

> Metabolite concentrations cannot be inferred by measurement of parent drug concentrations. The metabolic profile of the drug should be identified during the drug development process. This identification can be accomplished at different stages of development using in vitro and in vivo methods. In vitro studies can use liver microsomes, liver slices, or hepatocytes from animals and humans and generally should be conducted before initiation of clinical trials. In vivo metabolism study results in nonclinical test species generally should be available early in drug development, and their results will either confirm the results obtained from the in vitro studies or reveal quantitative and/or qualitative differences in metabolism across species. It is the latter situation that may pose a safety concern. Human in vivo metabolism studies usually have been performed relatively later in drug development, but we strongly recommend in vivo metabolic evaluation in humans be performed as early as feasible. Adequacy of exposure to drug metabolites that are present at disproportionately lower levels in animals used in nonclinical studies should be considered on a case-by-case basis. Generally, systemic exposure is assessed by measuring the concentration of the parent drug at steady state, in serum or plasma. However, when measurements cannot be made in plasma of the test species for any reason, verification of adequate exposure can be made in other biological matrices such as urine, feces, or bile. We encourage contacting the FDA early in drug development to discuss these issues.

This statement is not as clear as it could be. It accidentally implies that exposure to a metabolite can be assessed by measuring the concentrations of the parent drug at steady state. It also implies that plasma exposure to a metabolite in humans can be compared with the exposure to that metabolite in the excreta in animals, although it does not provide information as to how such a comparison can be made. Care must be taken when determining that steady state has been achieved in animals (and to a lesser extent in humans) since the achievement of steady state may not readily be determined based on the single-dose pharmacokinetics of the drug. Changes in clearance mechanisms with time due to aging or induction of the clearance mechanisms can confound the prediction of when steady state will have been achieved. It is common practice to determine the exposure to drug and/or metabolites at the end of the safety assessment study, regardless of whether steady state may have been achieved before that point in time.

The design and conduct of human ADME studies are described in Section 27.5.7.10. Although this guidance does give some insight into how the metabolism of a drug could be determined, it does not give much guidance on the logistics of determining the metabolism *in vivo*. The use of radiotracers in drug development has been reviewed by Lappin and Tracey (2006). There are several ways in which the *in vivo* metabolism of a drug can be determined. The metabolism and mass balance are best determined using a radioactive tracer (usually ^{14}C but occasionally ^{3}H). In most cases, the metabolism is determined after the administration of a single dose of drug. However, the exposure to metabolites after a single dose may not reflect the exposure after repeat since the exposure to parent drug and metabolites can change with duration of dosing (e.g. auto-induction or auto-inactivation of metabolism) and the age of the animals. Consideration should therefore be given to determining the *in vivo* metabolism after repeat-dose administration. One such design would be to administer unlabeled drug for a period of 7–14 days followed by the administration of a dose containing the radiotracer. It should also be noted that most *in vivo* metabolism studies are conducted at low doses (e.g. 10 mg/kg) since this is the dose range that the drug metabolism scientist or pharmacokineticist usually works at. However, this may not be appropriate since the purpose of identifying the metabolites in animals is to provide information for the safety assessment of the drug. Such studies are conducted at doses that are one or two orders of magnitude higher than the dose at which the PK studies are conducted. The metabolism and the pharmacokinetics of the drug may be markedly different at these higher dose levels and consequently, the *in vivo* metabolic profile may be markedly different. The drug metabolism scientist is strongly encouraged to consider conducting the radiolabeled *in vivo* metabolism studies in animals after repeat-dose administration and at toxicologically relevant doses. The use of the plural in the previous sentence is deliberate since the metabolism should be described at multiple dose levels since the metabolism could be dose-dependent. Once the relevant metabolites are identified, bioanalytical assays should be developed and the exposure to these metabolites determined in the subsequent toxicology studies (as necessary). The characterization of the *in vitro* metabolism of a drug is a task that is increasingly performed during the discovery phase of drug development. These studies tend to be conducted using non-radiolabeled material which can lead to limitations in the interpretation of the data from these studies. It is often not feasible to determine the relative percentages of metabolites and parent drug due to differences in ionization potential in the mass spectrometer. Additionally, the *in vitro* system is a static system and cannot take into account differences in rates of clearance of metabolites and parent drug *in vivo*. Therefore, data generated *in vitro* can be valuable in showing qualitative similarities or differences but should not be used to show quantitative differences. One product of the early investigation of the *in vitro* metabolism of a drug is a graphic that shows the metabolites of the drug and potential pathways and enzymes involved in the metabolism of the drug. This scheme may be very complicated, depending upon the amount of effort that the drug metabolism scientist has expended in identifying metabolites that are present in low amounts. Therefore, the metabolism *in vitro* may appear to be very complex and raise some potential concerns that are not realized once data on the metabolism of the drug *in vivo* are obtained. Another area of caution is the determination of the enzymes involved in the metabolism of the drug. Such a study may report that 80% of the metabolism is mediated by CYP3A4/5, which may cause some concerns regarding the potential for DDIs or interindividual variability in exposure to the parent drug. However, such data are out of context since it is important to understand whether metabolism is the major route of clearance of the drug and the contribution of hepatic microsomal metabolism to total metabolism of the drug. To date, none of the guidances or publications give much insight into how to determine the extent of metabolism of a drug. Strategies for determining the metabolism of a drug *in vivo* are described later in Section 27.5.7.10, "Human ADME Studies."

The draft guidance also makes some general statements regarding species differences in metabolism and how exposure to metabolites is assessed:

> In general, systemic exposure to metabolites varies among species, and it is uncommon for humans to form unique metabolites. Therefore, identification of major human metabolites at levels higher than those measured in the test species used for toxicological assessment is of serious concern. For metabolites detected in humans as well as in nonclinical species (although at lower levels in the latter), adequacy of exposure should be considered on a case-by-case basis. Generally, systemic exposure is assessed by measuring the concentration of the compound in serum or plasma. However, when measurements cannot be made in plasma for any one or a number of reasons, measurements can be made in other biological matrices such as urine, feces, or bile. Noncirculating metabolites (i.e. excreted in bile, urine) are sometimes identified before clinical trials, but are not usually monitored. It is quite likely that excreted metabolite levels may be a more appropriate metric in many instances. For example, if Phase II conjugation products of a metabolite are present in the excreta, it can be assumed that systemic exposure to the metabolite has occurred. We recommend consulting the ICH Q3A guidance with regard to the development of analytical methods for measuring metabolites in selected matrices. If the systemic exposure in nonclinical species is equivalent to human exposure when measured in plasma and/or excreta, levels may be considered sufficient and alleviate the need for additional toxicity testing. We encourage contacting the Agency early in drug development to discuss these issues.

The latter part of the quoted statement can be used to illustrate the care that needs to be taken in the review of draft guidances. The guidance states: "We recommend consulting the ICH Q3A guidance with regard to the development of analytical methods for measuring metabolites in selected matrices." It should be noted that the ICHQ3A guidance is titled "Impurities in New Drug Substances" (ICH, 2006a). The title of the guidance is noted as a footnote in the "Safety Testing of Drug Metabolites" draft guidance (US FDA, 2005c). This guidance does not really address the measurement of metabolites in selected matrices. The only statement related to metabolites is the following statement: "Impurities that are also significant metabolites present in animal and/or human studies are generally considered qualified. A level of a qualified impurity higher than that present in a new drug substance can also be justified based on an analysis of the actual amount of impurity administered in previous relevant safety studies." It is unclear whether this was the intended reference or whether the FDA intended to cite a different guidance. It is unfortunate that this "potential error" could actually become an ingrained fact if the draft guidance does not become a final guidance and the draft guidance remains on the FDA web site. The existence of this "potential error" could raise the question as to how many other "potential errors" exist in the large number of draft guidances that exist. This error was corrected in the final guidance.

The guidance is one of the few guidances (draft or final) that encourage the use of *in silico* predictive tools for the prediction of potential toxicity. The guidance makes that point thus: "It is now possible to identify the molecular structures of metabolites early in drug development. With the availability of computational software designed to predict activity relative to a known structure, the mutagenic, carcinogenic, or teratogenic potential of a drug or a metabolite can be evaluated as soon as a structure is identified. Although structure activity relationship analyses are not considered a substitute for actual testing, we encourage submission of the results from these analyses."

The guidance gives insight into the types of studies that should be conducted in order to evaluate the potential toxicity of a unique or major human metabolite. The guidance provides the following general considerations:

When designing a nonclinical study for a unique or major metabolite, it is important to consider physicochemical characteristics of the metabolite, including solubility, permeability, extent of absorption, route of administration, and exposure. The indicated patient population, duration of use, and exposures at the therapeutic dose are also important considerations for the risk assessment. Another important consideration is the potential for biotransformation of directly administered human metabolites in animals as well as the presence of impurities in the synthesized metabolites.

It is important to consider combined exposure to parent and pharmacologically active metabolites in safety assessments. A pharmacologically active metabolite can be more, equal, or less active than the parent drug at the target receptor. Similarly, a metabolite may cause toxicity by (1) eliciting exaggerated pharmacological effects via the target receptor, (2) activating receptors different from the parent drug target receptors, or (3) through nonreceptor mediated mechanisms (e.g. physico-chemical)."

The relevance of conducting studies with metabolites has been questioned. For example, if a metabolite is administered orally, there is the possibility that the metabolite will be poorly absorbed and therefore limit the validity of the study. The metabolite could be administered intravenously to avoid such issues, but this approach will still not address all potential concerns. Other concerns include that the metabolite may not readily be taken up into cells and may distribute differently throughout the body and that the metabolite undergoes sequential metabolism to a downstream product that is toxic, leading to differences in tissue-specific toxicity. Additionally, the testing of a reactive metabolite may be an exercise in futility, even if it could be synthesized, due to its likely short half-life and the fact that the safety of the metabolite will be tested out of context (i.e. not at the site of its generation *in vivo* from the parent drug. The challenges of detecting reactive metabolites will be addressed in the section of this chapter that covers drug metabolism in clinical trials and clinical guidances. The complicating factors in safety testing of drug metabolites have recently been addressed by Prueksaritanont, Lin and Baillie (2006). It should also be noted that by administering only the metabolite of interest, the sponsor will not be able to address the potential for additive or synergistic toxicity between the parent drug (or other metabolites) and the administered metabolite.

The statement that the metabolite may cause toxicity through activating receptors different from the parent drug has some large implications. This statement can be taken as implying that the metabolite should be evaluated for potential off-target effects in the same *in vitro* panel that the parent drug was assessed. It may not be sufficient to determine that the metabolite does not have the same intended pharmacology as the parent compound.

Both the draft and final guidances provided a list of studies that may be required for the assessment of the safety of metabolites. The information provided in both the draft and final guidances is essentially the same, although there were subtle differences. The finalized guidance states that "Good laboratory practice guidelines apply to the nonclinical studies with the drug metabolite designed to evaluate safety" (21 CFR Part 58).

The guidance describes the types of studies (general toxicity, genotoxicity, embryo-fetal development, and carcinogenicity studies) that should be conducted in order to assess the safety of the disproportionate drug metabolite. In essence, the studies are similar to those that would be conducted for the parent drug. However, the guidance does make some important statements that pertain particularly to the safety testing of drug metabolites. For example, in the section on General Toxicity Studies, the guidance states: "The toxicity of the drug metabolite should be investigated at multiples of the human exposure or at least at levels comparable to those measured in humans. We also recommend using the parent drug's intended clinical route of administration. However, with justification, other routes can be used to achieve sufficient exposure to the dispro-portionate metabolite. If the clinical route is oral, it is important to verify the stability of the metabolite in the stomach environment. It is crucial to gather toxicokinetic data from this type of study to ensure adequate exposure." The concern that the adminis-tered metabolite may not be stable in the stomach environment is a valid one. If the metabolite is not stable in the GI tract, then the animal may not be exposed systemi-cally to the metabolite. It is unclear whether the concern is related to chemical stability, metabolism by GI microflora, or metabolism by the GI mucosa. It is unclear how the FDA would wish the sponsor to address the stability of the metabolite in the stomach environment. The guidance notes that "Sometimes the conduct of an embryo-fetal development toxicity study in only one species that forms the drug metabolite can be justified." However, the guidance does not address in which species the embryo-fetal development toxicity study should be conducted, but it would seem prudent to conduct the study in rat if at all possible due to the lower amount of chemical required com-pared with the rabbit and the fact that general toxicity studies would also be conducted in the rat. Finally, the guidance addresses carcinogenicity studies by stating that "Car-cinogenicity studies should be conducted on metabolites of drugs that are administered continuously for at least 6 months, or that are used intermittently in the treatment of chronic or recurrent conditions when the carcinogenic potential of the metabolite cannot be adequately evaluated from carcinogenicity studies conducted with the parent drug. A single carcinogenicity study or an alternative bioassay should be conducted." Again, the guidance does not address which species (rat or mouse) should be used for the carcinogenicity assessment of a metabolite. Although a study in the mouse would use less chemical than a study in the rat, additional range-finding studies would need to be conducted in order to establish the doses for the carcinogenicity assessment whereas the general rat toxicity studies could be utilized to establish the doses for a rat carcinogenicity study.

The timing of the safety assessments is an important topic and is addressed in both the draft and final guidances. Both guidances state: "Early identification of unique human or major metabolites can provide clear justification for nonclinical testing in animals, assist in planning and interpreting clinical studies, and prevent delays in drug development." The draft guidance then proceeds to go into more detail, stating:

Sponsors are encouraged to conduct in vitro studies to identify and characterize unique human or major metabolites early in drug development. If toxicity studies of a human metabolite are warranted, we recommend studies be completed and the study reports be submitted to the Agency before beginning large-scale phase 3 trials. In some cases, it may be appropriate for these nonclinical safety studies with unique human metabolites to be conducted before phase 3 studies; for example, (1) if the metabolite belongs to a chemical class with known toxicity; (2) if the metabolite has positive structural alerts for genotoxicity, carcinogenicity, or reproductive toxicity; or (3) if clinical findings suggest the metabolite or related compounds have indicated special clinical safety concerns, such as QT prolongation.

To optimize and expedite development of drugs for serious or life-threatening diseases that lack an approved effective therapy, the number of nonclinical studies for the unique or major human metabolites may be limited on a case-by-case basis. We recommend sponsors contact the relevant review division to discuss such situations.

The duration of the toxicology studies that are required to support clinical trials of a particular duration or stage of development are covered by ICH guidances, ICHM3(R) and ICHS6. The final guidance is more succinct without the caveats of the draft guidance. It states that:

If toxicity studies of a drug metabolite are warranted, studies should be completed and study reports provided to the FDA before beginning large-scale clinical trials. To optimize and expedite drug development for serious or life-threatening diseases other than cancer (e.g. ALS, stroke, human immunodeficiency virus), the number and type of nonclinical studies for the drug metabolites can be modified on a case-by-case basis for those drugs with major beneficial therapeutic advances, and for drugs for illnesses that lack an approved effective therapy. Sponsors should contact the appropriate review division to discuss such situations.

The FDA issued, in October 2007, a draft guidance related to the testing of drugs for the treatment of HIV. This guidance "Role of HIV Resistance Testing in Antiretroviral Drug Development" notes the importance of active metabolites by stating that "Any metabolite that exerts inhibitory activity should be delineated and its specific target shown." It is interesting that the draft guidance states "any metabolite" and does not mention "major metabolite" (US FDA, 2007k). Therefore, this draft guidance does not provide any clear guidance on the cutoff level for the characterization of the activity of human metabolites.

The hot nature of this topic can be seen by the relatively large number of publications in the scientific literature over the past two years. These publications have come from industry (Davis-Bruno and Atrakchi, 2006; Fura, 2006; Humphreys and Unger, 2006; Prueksaritanont, Lin and Baillie, 2006; Smith and Obach, 2006; Naito et al., 2007). The publication by Davis-Bruno and Atrakchi (2006) includes two examples where safety testing of metabolites was performed. The SBA and EPAR for Nexavar (sorafenib) provide an illustration of the studies that may be performed to evaluate the safety of a metabolite. M-2 (an N-oxide) appears to be the major metabolite of sorafenib in humans. Since the metabolic profile of sorafenib in rats and dogs differed from that in humans, the sponsor conducted a four-week toxicology study with the metabolite M-2 in rats. An Ames test (for genotoxicity) was also performed with M-2 (US FDA, 2005h; EMEA, 2006i).

As can be gathered, the costs of evaluating the safety of metabolites can be very high, regardless of whether the assessment of the toxicity is performed as part of the assessment of the toxicity of the parent drug or as separate studies of the assessment of the toxicity of metabolites. In both situations, there will be the need to synthesize the metabolites for use as standards in the bioanalytical assay, the need to synthesize internal standards for use in the bioanalytical assay, the need to develop and validate methods for the metabolites, the need to re-evaluate the specificity of the existing bioanalytical methodology for parent drug, and the need to analyze the samples from toxicology or other studies. In the situation that the major human metabolite is generated in the animal species for the toxicology evaluation of the parent drug, it will be important to show exposure to these metabolites in the toxicology studies (particularly the chronic or carcinogenicity assessment studies). If identification of major human metabolites (i.e. the conduct of the human ADME study) is conducted late in the drug development process, the situation may arise where the chronic toxicology studies have been conducted but the exposure to the major human metabolites has not been assessed. In this situation, it may be possible to analyze banked samples from the chronic study for the metabolite(s) of interest, although this would be dependent upon the metabolites being stable in the matrix during their long-term storage. More likely, a bridging study would need to be conducted to describe the exposure in those species. Since the exposure to parent drug and metabolites can change with duration of dosing (e.g. auto-induction or auto-inactivation of metabolism), age of the animals, and source of the animals, it is possible that the sponsor may be requested to reconduct the chronic toxicology studies. Additionally, the sponsor may need to consider conducting studies to assess the exposure to metabolites in other studies such as the reproductive toxicology and safety pharmacology studies. It should be noted that the FDA's draft guidance states that the safety testing of metabolites should be conducted prior to large (phase III) clinical trials which would imply that the *in vivo* metabolism of the drug in humans should have been assessed long before that point in time.

In the situation where toxicology studies need to be conducted in animals dosed with the metabolite(s) of interest, there will obviously be additional costs for the synthesis of these metabolites and the conduct of the toxicology studies as well as the bioanalytical costs outlined herein. It should be noted that there may be additional bioanalytical costs if the administered metabolite is not stable by the route that it is administered and can only be quantitated indirectly (e.g. by quantitating a secondary metabolite). Synthesis of metabolites, particularly glucuronide metabolites, is a far from trivial task and may not be achievable in a reasonable time frame. It is speculated that the development of at least one drug has been discontinued due to difficulty in generating the metabolites required for the additional safety evaluation studies that were requested by a regulatory agency.

The scenarios noted here are complicated but it is possible that even more complicated scenarios exist. For example, it may be possible that there are two major metabolites identified in humans, only one of which is present in the toxicology species. This raises the question of what is the appropriate toxicity testing schema. Is it a study in which the unrepresented major human metabolite is administered to animals or is it a study that involves the administration of both the parent drug and the underrepresented metabolite? The worst-case scenario may be that both types of study are required.

Given the large impact to budgets and timelines, it can be seen that it may be prudent to obtain an understanding of the *in vivo* metabolism of the drug in humans at the earliest opportunity. Although such a study may incur short-term costs that could be viewed by some people as being unnecessary, such knowledge could be very cost-effective in the long term. It is also interesting to speculate whether extensively metabolized compounds should be bought into development since they may have the potential to generate "unique" human metabolites.

It should be noted that genotypic differences in drug-metabolizing enzymes (DMEs) can be found between strains of rats and within colonies of dogs (Morita *et al.*, 1998; Paulson *et al.*, 1999; Kamimura, 2006). Such differences may have a profound effect on the metabolism of certain drugs and hence complicate the design and interpretation of animal studies with these drugs. Less understood is the potential for phenotypic "drift" or variation that is not related to any genetic change. Robosky *et al.* (2005, 2006) recently published on this phenomenon. They noted that "stock Sprague-Dawley (Crl: CD(SD)) rats obtained from the Charles River Raleigh facility demonstrated a distinct endogenous urinary metabonomic profile that differed from historical control SD urine spectral profiles obtained over the past several years in our laboratory. In follow-up studies, the origin of the variant phenotype was narrowed down to animals of both sexes that were housed in one specific room (Room 9) in the Raleigh facility. It is likely that the two phenotypes are related to distinct populations of gut flora that particularly impact the metabolism of aromatic molecules." It is possible that differences in GI microflora may cause variability in the metabolism and pharmacokinetics of certain drugs, although this is an under-researched topic (Holmes and Nicholson, 2005; Nicholson, Holmes and Wilson, 2005; Li *et al.*, 2008; Qin, 2008).

27.4.6   Distribution

27.4.6.1   Plasma Protein Binding   The topic of plasma protein binding is an interesting one. There are no guidances for which plasma protein biding is the main topic. However, it is addressed in some guidances (US FDA, 2002a). Plasma protein binding is usually only addressed *in vitro* using plasma obtained from the blood of supposedly healthy individuals. These individuals are usually employees of the laboratory conducting the work and are deemed as being healthy because they are not being actively treated for a disease. Additionally, these volunteers are often asked to fast overnight prior to the blood draw. Plasma is prepared from the blood and the binding of the drug to this plasma is determined. As can be seen, such an evaluation will not give insight into the role of lipoproteins on the binding of the drug and hence the effect of variations in levels of lipoproteins on the pharmacokinetics of a drug. The role of lipoproteins in modifying the biological activity of hydrophobic drugs such as aminoglycosides and cyclosporine A has been reviewed by several authors (Wasan and Cassidy, 1998; Chung and Wasan, 2004; Wasan *et al.*, 2008). Lipoprotein-based formulations of anti-cancer drugs are being evaluated as potential ways by which to minimize some of the side effects of these drugs (Lacko *et al.*, 2007).

27.4.6.2   Tissue Distribution   The final topic to be discussed in this section is that of tissue distribution. The ICHS3b guidance, "Pharmacokinetics: Guidance for Repeated Dose Tissue Distribution Studies" was issued in 1995 (ICH, 1994a). This guidance states that "A comprehensive knowledge of the absorption, distribution, metabolism and elimination of a compound is important for the interpretation of pharmacology and toxicology studies. Tissue distribution studies are essential in providing information on distribution and accumulation of the compound and/or metabolites, especially in relation to potential sites of action; this information may be useful for designing toxicology and pharmacology studies and for interpreting the results of these experiments." Interestingly, there is no such guidance for single-dose tissue distribution studies nor is there a clear suggestion that such studies are a crucial part of the drug development process in any regulatory guidance. This guidance describes circumstances under which repeated-dose tissue distribution studies should be considered. These include "When single dose tissue distribution studies suggest that the apparent half-life of the test compound (and/or metabolites) in organs or tissues significantly exceeds the apparent half-life of the elimination phase in plasma and is also more than twice the dosing interval in the toxicity studies, repeated dose tissue distribution studies may be appropriate" and "When steady-state levels of a compound/metabolite in the circulation, determined in repeated dose pharmacokinetic or toxicokinetic studies, are markedly higher than those predicted from single dose kinetic studies, then repeated dose tissue distribution studies should be considered." Despite the fact that this guidance has been in place for over a decade, very few repeat-dose tissue distribution studies are performed, even though the criteria mentioned may have been met for a given drug. The guidance gives input on the study design, "Information from previous pharmacokinetic and toxicokinetic studies should be used in selecting the duration of dosing in repeated dose tissue distribution studies. One week of dosing is normally considered to be a minimum period. A longer duration should be selected when the blood/plasma concentration of the compound and/or its metabolites does not reach steady state. It is normally considered unnecessary to dose for longer than three weeks." The guidance further states that "Consideration should be given to measuring unchanged compound and/or metabolites in organs and tissues in which extensive accumulation occurs or if it is believed that such data may clarify mechanisms of organ toxicity."

Despite the fact that there is no guidance that specifically relates to single-dose tissue distribution studies, such studies are commonly conducted as part of the nonclinical plan. The primary reason for conducting such studies is to determine the exposure of specific organs and organ systems to radioactivity in order to calculate the maximum amount of radioactivity that can be administered in a human ADME study. The human ADME study will be described in greater detail in the next section of this chapter. Occasionally, as mentioned earlier, the toxicologist may request that the drug metabolism scientist perform a tissue distribution study to aid in this prediction. When discussing accumulation or tissue distribution with a toxicologist, the drug metabolism scientist (or pharmacokineticist) should seek to understand the concern of the toxicologist (i.e. if they are concerned about high exposure of a drug in a particular organ) and explain to them the limitations of the techniques used to assess the tissue distribution of a drug.

Tissue distribution studies are relatively straightforward in design, reasonably easy to conduct, but can be very difficult to interpret. These studies are usually conducted in pigmented animals and involve an evaluation of the distribution at a range of time points in a limited number (often N = 1 / time point) of animals and usually only in male animals. The study may be a "tissue excision" study in which the tissues are removed from the body or a "quantitative whole body autoradiography (QWBA)" study. In the former type of study, either unlabeled drug can be administered with quantitation of drug concentrations by techniques such as LC-MS/MS or radiolabeled drug can be administered with the concentration of drug being determined by liquid scintillation counting (LSC) after combustion of the tissues. Tissue distribution studies utilizing unlabeled drug are uncommon, but this study design is appropriate when the question being posed is "What is the concentration of the drug (or a particular metabolite) in a given tissue?" For the QWBA study, a radiolabeled drug is administered, the animals are frozen, embedded in a matrix, and the radioactivity determined in the tissue section by an imaging technique. A limitation of the QWBA studies is that the equipment used for imaging can accommodate rodents and small monkeys but is inadequate to accommodate large monkeys or dogs, unless the animals are further sectioned and placed in multiple detection cassettes. In the studies that utilize the administration of radioactivity, the concentrations of radioactivity represent the sum of the concentration of parent drug plus the concentration of all metabolites. Therefore, a tissue distribution study may be able to provide information as to which tissues the drug did not distribute to (at least, to a minimum level), but the study will not be able to provide information as to whether the radioactivity was present as parent drugs and/or metabolites. Therefore, it can be easy to assume that the organs that contain high levels of radioactivity will contain high levels of parent drug when, in reality, the radioactivity may represent metabolites and not parent drug. Another pitfall of tissue distribution studies is the assumption that organs that show high levels of radioactivity will show toxicity in animal studies. The resolution of autoradiography studies is usually limited to the tissue level, although it is possible to also use autoradiography to determine the location of drugs within cells (Stumpf, 2005). Recently, noninvasive imaging techniques, such as positron emission tomography (PET) or magnetic resonance spectroscopy (MRS), have become available for *in vivo* drug distribution assessment. These techniques will be able to provide distribution of compound-related material over time within a given animal or person which will provide an advantage over QWBA. However, these techniques have several disadvantages (cost, availability of equipment, limited number of drugs that can be used, and duration of the study) compared with QWBA and suffer from some of the same limitations as QWBA (low spatial resolution and inability to discriminate between bound and unbound drug or parent compound and metabolite) (Brunner and Langer, 2006).

In addition to the caveats regarding the lack of ability to discriminate between parent drug and metabolites, it should be recognized that the presence of parent drug and/or metabolites in a given tissue does not mean that there will be a toxic insult in that organ. A common conclusion drawn from the observation of the retention of radioactivity in a tissue is that the retained radioactivity is due to the covalent binding of the drug or metabolite which would raise a red flag regarding the potential for toxicity. However, it should be noted that the retained radioactivity may represent the incorporation of the drug and/or metabolite into lipids. The incorporation of compounds containing carboxylic acids into lipids has long been known but is often overlooked (Dodds, 1991, 1995). A common finding in tissue distribution studies conducted in pigmented animals is the retention of radioactivity by melanin in the eye. The correlation of binding of drugs or their metabolites to melanin and subsequent toxicity is equivocal, and it is possible that the retained radioactivity represents an animal-specific metabolite and therefore be a false-positive result (Salazar-Bookaman, Wainer and Patil, 1994; Leblanc *et al.*, 1998). Another consequence of the detection of binding of parent drug and/or metabolites to melanin is the need to conduct a photosensitivity study as described previously. Therefore, unless a tissue distribution study is required for dosimetry purposes, tissue distribution studies should be only conducted if there is a very specific question to be answered and then the appropriate study design should be selected in order to avoid the generation of potentially misleading information. For example, if the intent of the study is to determine the concentrations of parent drug and/or a specific metabolite in a given tissue, an appropriate approach may be to dose the animals with unlabeled drug, remove the tissue (or tissues) of interest, and quantitate the concentrations of the analyte of interest using LC-MS/MS. Variations on this approach include the analysis of tissue sections by mass spectrometric techniques and the use of microdialysis (Joukhadar and Müller, 2005; Rohner, Staab and Stoeckli, 2005; Brunner and Langer, 2006; Chaurand, Corentt and Caprioli, 2006). The analysis of tissue sections will allow for spatial resolution of the distribution of the analyte within a tissue which may be important for correlating exposure to the analyte with a histopathological observation. Microdialysis offers the ability to obtain an estimate of the tissue distribution of a drug across time within a single animal. This latter technique can also be used in humans. It should be noted that both of these techniques have their limitations and are not widely used.

27.4.7 Toxicology

27.4.7.1 General Toxicology The primary guidance that pertains to toxicology studies is the ICH guidance, ICHM3(R1), on "Non-Clinical Safety Studies for the Conduct of Human Clinical Trials for Pharmaceuticals" (ICH, 2000a). The purpose of this document is to recommend international standards for, and promote harmonization, of the nonclinical safety studies needed to support human clinical trials of a given scope and duration. Although it does not address what metabolism-related studies should be performed, it does make some statements that are pertinent to the drug metabolism statement: "Exposure data in animals should be evaluated prior to human clinical trials. Further information on absorption, distribution, metabolism and excretion in animals should be made available to compare human and animal metabolic pathways. Appropriate information should usually be available by the time the Phase I (Human Pharmacology) studies have been completed." The subject of the timing of the availability of ADME data will be addressed later in this chapter.

A concept paper for the revision of ICH M3 was issued by the ICH in September 2006 (ICH, 2006c). The rationale for the revision is that the approved text of the existing ICH-M3 guideline includes significant regional differences needing to be revisited in the light of accumulated information. In February 2008, the EMEA issued a draft guideline, "Guideline on Repeated Dose Toxicity" (EMEA, 2008c). The executive summary of this draft guideline states that "The purpose of testing toxicity after repeated dosing is to contribute to the development of safe medicinal products that need repeated administration to patients. General principles are provided on substance quality and excipients. The criteria discussed takes into account the choice of animal species, the size of groups and animal husbandry. Dose regimen, duration and route of administration should be selected based on the intended clinical use. Guidance is given on the parameters to be monitored during the in-life phase, and which special studies may be needed in case of a special activity of a certain medicinal product. A list of recommended tissues to be studied histopathologically is attached." This draft guideline complements the ICHM3 guidance by providing more detail regarding the conduct of the studies. It addresses the selection of the species and the safety testing of drug metabolites stating:

> Within the usual spectrum of laboratory animals used for toxicity testing, the species should be chosen based on their similarity to humans with regard to pharmacokinetic profile including biotransformation.

Exposure to the main human metabolite(s) should be ensured. If this can not be achieved in toxicity studies with the parent compound, specific studies with the metabolite(s) should be considered. When the product administered is a prodrug, its conversion to the active substance should be demonstrated in the species under study. Whenever possible, the selected species should be responsive to the primary pharmacodynamic effect of the substance. In certain cases for example when the pharmacodynamic effect by itself will cause toxicity, studies in disease models may be warranted.

27.4.7.2 *Safety Pharmacology*   Safety pharmacology is an area of toxicology that has been addressed in two separate ICH guidances, "ICHS7A—Safety Pharmacology Studies for Human Pharmaceuticals" and "ICHS7B—The Nonclinical Evaluation of the Potential for Delayed Ventricular Repolarization (QT Interval Prolongation) by Human Pharmaceuticals" that were issued in 2001 and 2005, respectively (ICH, 2000b, 2005a). ICHS7A states that "Generally, any parent compound and its major metabolites that achieve, or are expected to achieve, systemic exposure in humans should be evaluated in safety pharmacology studies. Evaluation of major metabolites is often accomplished through studies of the parent compound in animals. If the major human metabolites are found to be absent or present only at relatively low concentrations in animals, assessment of the effects of such metabolites on safety pharmacology endpoints should be considered. Additionally, if metabolites from humans are known to substantially contribute to the pharmacological actions of the therapeutic agent, it could be important to test such active metabolites. When the in vivo studies on the parent compound have not adequately assessed metabolites, as discussed above, the tests of metabolites can use in vitro systems based on practical considerations." The role of drug metabolism in the selection of the species used in the safety pharmacology testing is addressed thus: "Data from humans (e.g., in vitro metabolism), when available, should also be considered in the test system selection. The time points for the measurements should be based on pharmacodynamic and pharmacokinetic considerations. Justification should be provided for the selection of the particular animal model or test system." The timing of the availability of metabolism data is also addressed: "During early development, sufficient information (e.g., comparative metabolism) may not always be available to rationally select or design the studies in accordance with the points stated above; in such circumstances, a more general approach in safety pharmacology investigations can be applied." The emphasis on the relative lack of data on species differences in metabolism and the emphasis on the use of *in vitro* is a reflection of the fact that safety pharmacology studies should be conducted prior to the FIH study. The guidance also addresses the exposure to metabolites in the following manner: "Regardless of the route of administration, exposure to the parent substance and its major metabolites should be similar to or greater than that achieved in humans when such information is available." These statements appear to set up a circular argument. Since the objective is to conduct the safety pharmacology studies before the phase I study, there will be no information available on the *in vivo* exposure of humans to metabolites of the drug that is being administered. Such information will usually only be available after the human ADME study has been conducted. This then raises the possibility that the sponsor will need to conduct a battery of safety pharmacology studies prior to phase I and potentially need to reconduct selected studies based on the outcome of the human ADME study. At least, the sponsor may be required to conduct a PK study in the species used for the safety pharmacology studies in order to describe the exposure to selected human metabolites and therefore retrospectively confirm the relevance of the original safety pharmacology studies.

ICHS7B addresses the topic of cardiovascular safety pharmacology studies and addresses the conduct of an integrated risk assessment. The guidance states that the objectives of the studies covered by this guidance are to "(1) identify the potential of a test substance and its metabolites to delay ventricular repolarization, and (2) relate the extent of delayed ventricular repolarization to the concentrations of a test substance and its metabolites." This guidance notes that "Cardiac cells and tissues have limited capacity for drug metabolism; therefore, in vitro studies using the parent substance do not provide information on the effects of metabolites. When in vivo nonclinical or clinical studies reveal QT interval prolongation that is not consistent with data from in vitro studies using the parent substance, testing metabolites in the in vitro test systems should be considered."

The integrated risk assessment can include information from "Pharmacokinetic studies, including plasma levels of parent substance and metabolites (including human data if available); drug interaction studies; and tissue distribution and accumulation studies." Since some of the studies are conducted *in vitro* in systems that have little, if any, drug metabolism capacity, the guidance implies that synthesis and testing of metabolites will need to be performed. The guidance encourages the sponsor to perform S7B nonclinical studies prior to the FIH study which would imply that additional studies may be required later on in development for certain human metabolites. The guidance also suggests that the use of metabolic inducers or inhibitors be considered for studies to determine the relative role of the parent drug and the metabolites in any observed positive results.

27.4.7.3 *Reproductive Toxicology*

Drug metabolism is mentioned to a limited extent in guidances that pertain to the assessment of reproductive toxicity. The area of reproductive toxicology in the large part evolved from the thalidomide tragedy of the late 1950s and early 1960s (Stephens and Brynner, 2001). It took many decades for the role of drug metabolism in the teratogenicity of thalidomide to be recognized (Arlen and Wells, 1996; Ando, Fuse and Figg, 2002).

The ICHS5(R2) guidance, "Detection of Toxicity to Reproduction for Medicinal Products & Toxicity to Male Fertility" contains some interesting statements that should pertain to each and every guidance: "To employ this concept successfully, flexibility is needed. No guideline can provide sufficient information to cover all possible cases, all persons involved should be willing to discuss and consider variations in test strategy according to the state of the art and ethical standards in human and animal experimentation. Areas where more basic research would be useful for optimization of test designs are male fertility assessment, and kinetic and metabolism in pregnant/lactating animals" (ICH, 2005b). The potential for the need to evaluate the toxicity of certain metabolites is described thus: "For secondary testing or further substance characterization other test systems offer the possibility to study some of the observable developmental processes in detail, for example to reveal specific mechanisms of toxicity, to establish concentration-response relationships, to select 'sensitive periods', or to detect effects of defined metabolites." The concept of body burden is used in this guidance and is defined as being "The total internal dosage of an individual arising from the administration of a substance, comprising parent compound and metabolites, taking distribution and accumulation into account." This definition implies that a tissue distribution study is performed in the pregnant animal.

In 2001, the FDA issued a draft guidance for reviewers on the "Integration of Study Results to Assess Concerns about Human Reproductive and Developmental Toxicities" (US FDA, 2001e). There are several statements of note in this draft guidance. These include "The evaluation should also compare animal and human pharmacodynamic effects, animal and human metabolism and disposition, animal and human pharmacologic and toxic effects, and drug exposures in animal studies in relation to the highest proposed dose in humans." The guidance further states that:

> Drug distribution, elimination, and biotransformation (pathways and metabolites) in the test species and in humans should be compared. Quantitative differences in metabolic/drug distribution profiles between the test species and humans are often seen, and may not have important implications and should not be overemphasized. Reproductive and developmental toxicities induced by compounds whose metabolic and distribution profiles are very similar in animals and humans increases concern for reproductive or developmental toxicity in humans. For compounds with highly dissimilar metabolic or tissue distribution profiles in animals and humans, there is less concern if the toxic effect seen in the test species can be attributed to a metabolite or tissue distribution profile not seen in humans. For any other scenario, concern is unchanged. When there are significant differences in drug distribution or metabolic profiles between several species, yet each test species demonstrates a positive signal for a reproductive or developmental toxicity, the toxicity is assumed to be attributable to the parent drug or a common bio-transformed product and concern is increased.

Finally, the guidance states that:

> Comparison of systemic drug exposure at the NOEL for the reproductive or toxicity class in the test species to that in humans at the maximum recommended dose is a critical determination. This comparison should be based on the most relevant metric (e.g., AUC, C_{max}, C_{min}, BSA [body surface area] adjusted dose). In general, there is increased concern for reproductive or developmental toxicity in humans for relative exposure ratios (animal: human) that are <10, decreased concern for exposure ratios >25, and no change in concern for ratios between 10 and 25. When applicable, the relative exposure ratio should consider both the parent compound and its metabolites. For example, it is appropriate to combine parent and metabolite when both are pharmacologically active and the activity relates to the reproductive or developmental toxicity.

The EMEA issued a draft guideline (Guideline on Risk Assessment of Medicinal Products on Human Reproduction and Lactation: From Data to Labelling) in 2006 that covers similar topics to the two FDA guidances described earlier (EMEA, 2006c). This draft guideline states that "Information about the excretion into milk of the active substance and/or its metabolites should be available" and "The exposure in pregnant animals measured by plasma concentrations of the compound and/or metabolites should be assessed."

The points in the FDA draft guidance and the EMEA guideline imply that the secretion of drug and/or metabolites should be evaluated in animals and/or humans, and that the metabolism of the drug in animals be determined prior to the determination of the exposure of the pregnant animal to the drug and/or metabolites. In most cases, the toxicology study is performed prior to the determination of the metabolism in the pregnant animal, if the metabolism is described in the pregnant animal. If the metabolism of the drug is determined in the pregnant animal, it is often only done in one species (usually rabbit). It is assumed that the metabolisms in the nonpregnant and pregnant rat are identical, although that is not always the case. Additionally, distribution of drug and/or metabolites to the fetus is only occasionally performed and is often only performed in one species (it is assumed that the placental transfer is similar in both the rat and rabbit). The timing of when (in the course of the pregnancy) to address the metabolism of the drug and fetal exposure to the drug and/or metabolites is a subject that can engender great debate. It is well known that the metabolic capacity and capabilities of the fetus will change with the stage of pregnancy; that pregnancy can affect the clearance of the drug by the mother and that the placental transfer of the drug and/or metabolites can change with the stage of the pregnancy. Therefore, it can be seen that the description of the metabolism in pregnant animals is a complicated matter that can easily become resource- and time-intensive.

The utilization of animal studies to determine the risk of birth defects and other reproductive effects has been reviewed in depth by Brent (2004).

It should also be noted that none of the aforementioned guidances emphasize the possibility of a drug causing toxicity to the female when the drug has been administered to a male due to excretion of the drug in the semen. This route of excretion has been observed for several drugs such as antiviral drugs, chloroquine, and caffeine but the potential for a drug to be excreted in the semen is very rarely evaluated proactively (Pichini, Zuccaro and Pacifici, 1994; Cao *et al.*, 2007).

In summary, it appears fair to state that, at present, there appear to be gaps in our knowledge regarding how to determine the metabolism of drugs during pregnancy and that this may be an area that may warrant further discussion.

27.4.7.4 Organ-Specific Toxicity
Both the FDA and EMEA have issued draft guidances on hepatoxicity. The EMEA guideline (Non-Clinical Guideline on Drug-Induced Hepatotoxicity) was issued in 2006, with the comment period ending in January 2007. This version of the draft guideline is no longer available. The guideline was re-released for comments in January 2008 and it is anticipated that the guideline will be finalized in 2008–2009 (EMEA, 2008g).

The 2008 draft guideline states that "A drug can cause liver toxicity *via* several mechanisms. For instance, it can be directly acting or indirectly through reactive metabolites. The drug or its metabolites may cause liver toxicity after specific receptor binding, or reactive metabolites can react with hepatic macromolecules leading to direct cytotoxicity." It also notes that important signals related to hepatoxicity may be obtained from many areas, including bioaccumulation, expression of xenobiotic-metabolizing enzymes, and generation of reactive metabolites. The guidance further addresses these topics by stating that "In the absence of other signals of hepatotoxicity, accumulation of the drug in the liver is not a cause of concern. Investigation of metabolic pathways provides clues to the production of (possible reactive) metabolites. At therapeutic dose levels, drugs may affect expression of xenobiotic metabolizing enzymes (induction or inhibition) and potentially increase the concentration of metabolites. In the presence of hepatotoxicity signals, possible reactive metabolites should be identified."

Accumulation is a mass balance effect where input exceeds output. Factors involved in accumulation include selective binding of the drug to tissue molecules, concentration of lipophilic drugs in body fat, absent or slow metabolism of the drug, and slow excretion of the drug. Drug accumulation is a topic that can engender great debate between toxicologists and pharmacokineticists. The toxicologist may inquire about accumulation of drug in order to aid them in determining potential organs of toxicity. The toxicologist may request that the drug metabolism scientist perform a tissue distribution study to aid in this prediction. When discussing accumulation or tissue distribution with a toxicologist, the drug metabolism scientist (or pharmacokineticist) should seek to understand the concern of the toxicologist (i.e. if they are concerned about high exposure of a drug in a particular organ) and explain to them any limitations of the techniques used to assess the tissue distribution of a drug. Tissue distribution studies and their limitations have been previously discussed in Section 27.4.6.2.

The EMEA guideline addresses the topic of the cytotoxicity of metabolites by stating that "If hepatotoxic signals occur *in vivo* (Step I) or signals are described in literature, parent compound or (reactive) metabolites if feasible, should be screened in *in vitro* mechanistic tests for potential hepatotoxicity." This may be easier said than done due to the difficulty in generating sufficient amounts of a reactive metabolite for testing in cytotoxicity assays. It should also be noted that the reactive metabolite may be administered to the cell culture medium and be detoxified before it enters the cell. Therefore, testing the cytotoxicity of a reactive metabolite will have many caveats. Several of these caveats will also apply to nonreactive metabolites and have been addressed earlier in the discussion of the safety testing of drug metabolites.

The first version of the draft guideline addressed the role of metabolism in immune-mediated mechanisms of hepatotoxicity. It stated that:

> The mechanism underlying immune-based liver toxicity is often triggered by the formation of reactive metabolite(s) that covalently bind(s) to proteins. However, as this event is considered insufficient to cause pathogenic immune reaction, a second co-stimulatory trigger is considered to be required (e.g. mild hepatic injury, inflammatory conditions, viral or bacterial infection). Immune-based liver damage could also result from an autoimmune response against hepatic antigens, due to loss of tolerance towards those antigens (secondary to cell damage or cross-reaction with the drug/drug-protein adduct), or from bystander damages to a non-antigen bearing hepatocytes. Moreover, a general inflammatory reaction in response to previous liver damage or unspecific activation of resident or accumulated leukocytes could occur. Inflammatory processes produce cytokines that can affect liver damage or repair. For example an activation Kupffer cells would in turn secrete pro inflammatory cytokines like IL-1, TNF alpha or nitric oxide (NO) which may induce apoptosis, stimulate cell growth or initiate inflammatory processes.

However, this mechanism of immune-mediated hepatoxicity is not discussed in the most recent version of the draft guideline. The reason for this change is unclear. The generation of autoantibodies against CYP isoforms has been observed for several drugs including cyclosporine, tacrolimius, hydralazine, and tienilic acid (Lytton *et al.*, 2002; Villeneuve and Pichette, 2004). Anti-CYP2D6 antibodies have been observed to be present in the serum of patients with type II autoimmune hepatitis (Miyakawa *et al.*, 2000). The role of bioactivation in drug-induced hypersensitivity has recently been reviewed by Sanderson, Naisbitt and Park (2006).

In October 2007, the FDA released its own draft guidance on hepatotoxicity, "Drug-Induced Liver Injury: Premarketing Clinical Evaluation" (US FDA, 2007l). This draft guidance, as can be seen from its title, is focused on the clinical aspects of hepatotoxicity. The guidance notes the limitations of animal models and states:

> The drugs that have caused severe DILI in humans have not shown clear hepatotoxicity in animals, generally have not shown dose-related toxicity, and, as noted, generally have caused low rates of severe injury in humans (1 in 5,000 to 10,000 or less). These reactions

thus appear to reflect host factors and individual susceptibility. Consequently, they have been termed *idiosyncratic*, meaning dependent upon the individual person's particular constitution. Whether they are the result of genetic or acquired differences has not yet been established, and to date no genetic, metabolic, or other characteristic has been found to predict severe DILI in an individual.

The timing of the evaluation of hepatoxicity is addressed thus: "Based on our experience, we recommend that the following analyses related to liver injury potential be carried out and included in an NDA or BLA, or included in an investigational new drug application when DILI is suspected and being evaluated."

The role of drug metabolism in hepatotoxicity is recognized by this draft guidance and is described thus: "The metabolism of a drug can have serious consequences for the safety profile of the drug. A drug may be metabolized to a hepatotoxic metabolite (e.g., acetaminophen, halothane, and isoniazid). Most hepatotoxic drugs have been oxidatively metabolized by the CYP450 system. Several in vitro methods are available to detect and quantify binding for a drug or its metabolites to liver proteins, including radiochemical and immunological methods." It should be noted that the guidance does not address exactly how the evaluation of the binding of the drug and/or its metabolites should be performed.

In addition to the aforementioned guidances, there have been several recent publications on the topic of drug-induced hepatoxicity by both industry and the FDA (Senior, 2007; Tang, 2007). The FDA has a web page dedicated to drug-induced liver toxicity (US FDA, 2008b).

27.4.7.5 Genotoxicity Testing

Genotoxicity has been addressed in several regulatory guidances, including the ICHS2A guidance "Specific Aspects of Regulatory Genotoxicity Tests for Pharmaceuticals" that was issued in 1996 (ICH, 1995). The guidance states that, in general, only one gender (male) of rodent (rat or mouse) needs to be evaluated in the *in vivo* bone marrow micronucleus test or other established *in vivo* tests. The guidance further states that "If there is a clear qualitative difference in metabolites between male and female rodents, then both sexes should be used." The clear implication is that the sponsor should have an understanding of the whether there is a gender-related difference in the metabolism of their drug before conducting the *in vivo* study. This study is not required before the phase I clinical studies unless there is a positive result in an *in vitro* genetic toxicity test. In any circumstance, the *in vivo* study should be conducted prior to phase II studies. The determination that there are gender-related differences in metabolism is usually made based on the observation of gender-related differences in exposure to parent drug in general toxicity studies rather than on *in vivo* metabolism studies conducted in both genders of rodent. It should also be noted that, even if there is clear evidence of gender-related difference in metabolism, the sponsor does not always perform the *in vivo* study in both genders.

Metabolism plays a key role in the *in vitro* assessment of the genotoxicity potential of a drug. The standard bacterial reverse mutation assay (Ames assay) is conducted in the presence or absence of a metabolic activation system. The most commonly used metabolic activation system is Aroclor-induced male rat liver S9. Interestingly, the sponsors will very rarely, if ever, conduct studies to describe the metabolism of their drug in this test system. Instead, it is assumed that by using an induced system, a large number of metabolites will have been generated. Very few, if any, Ames assays are performed using human liver S9, although it can be argued that this would be the more relevant metabolic activation system. The reason for the use of rat liver S9 rather than human liver S9 is that the rat system will have less lot-to-lot variability than the human system and hence give more reproducible results.

The guidance addresses the role of metabolism in understanding negative results in the *in vitro* assays. It states: "For in vitro negative results, special attention should be paid to the following considerations (the examples given are not exhaustive, but are given as an aid to decision-making): Does the structure or known metabolism of the compound indicate that standard techniques for in vitro metabolic activation (e.g., rodent liver S9) may be inadequate? Does the structure or known reactivity of the compound indicate that the use of other test methods/systems may be appropriate?" However, if a sponsor obtains a negative result in the *in vitro* test, they tend to move the program forward rather than spend time in determining whether the result was a false negative or not.

At times, the *in vitro* and *in vivo* tests can produce discordant results. This guidance gives some possible explanations for the discordant results: "They include: (i) An active metabolite produced in vitro may not be produced in vivo, (ii) an active metabolite may be rapidly detoxified in vivo but not in vitro, and (iii) rapid and efficient excretion of a compound may occur in vivo." The guidance appears to clearly emphasize the role of drug metabolism in genotoxicity testing and that the drug metabolism scientist and toxicologist should work closely together in this area.

There are a few additional words of caution to be made regarding drug metabolism in genotoxicity testing. One element of the *in vitro* battery of tests includes the evaluation of the potential for the induction of chromosomal aberrations in cellular systems. It should be noted that these systems may not have a large capacity for drug metabolism. In fact, the metabolism of drugs by these cellular systems is very rarely, if ever, evaluated. Additionally, the stability of the drug in the buffer or media used for these studies is very rarely, if ever, determined. This last caveat would also apply to other *in vitro* studies.

In 2006, the ICH issued a final concept paper for a revision to ICHS2 (ICH, 2006d). This concept paper does not contain any mention of the role of metabolism in the activation of genotoxic chemicals.

27.4.7.6 *Carcinogenicity Assessment* Drug metabolism is discussed to a small extent in several guidances issued by the FDA and ICH on the subject of the assessment of carcinogenic potential for a drug (list guidances). The FDA's 2002 guidance on "Carcinogenicity Study Protocol Submissions" states that "Exposure (steady state $AUC_{(0-24)}$) data should be provided for the parent drug and for the major metabolites from clinical trials conducted at the maximum recommended human dose (MRHD) or other appropriate human reference dose if the MRHD exposure data are unavailable" (US FDA, 2002a). It further states that "Where pharmacokinetics differ significantly between genders, data from males and females should be reported separately, as a gender difference can modify the study approach and conclusions. It is our experience that gender differences occur rarely." The guidance also states that "Plasma protein binding data should be provided for the parent drug and the major human metabolites (to the extent feasible) in the rodent test species over the range of concentrations encountered in the dose-rangefinding experiment and in humans at concentrations encountered in clinical trials conducted at the reference dose."

The availability of metabolism data is addressed as follows: "Metabolic profiles should be provided for the drug in humans and in the species employed for assessment of carcinogenic potential. In cases where in vivo data are unavailable, in vitro data can be used (see the recommendations in ICH S1C)."

The statements in this guidance have some interesting implications. Since carcinogenicity assessment studies take approximately three years from first dose to final report, they are often critical path to the submission of the marketing application. Therefore, the protocol for the carcinogenicity study may need to be submitted to the FDA before the maximum recommended human dose has been determined. In this case, the sponsors would base their calculations on the highest dose that will be evaluated in the phase III clinical trials. Another implication of these statements is that the exposure to the major human metabolites has been determined prior to the submission of the carcinogenicity protocol to the FDA. Therefore, the implicit assumption is that the human ADME has been conducted, any major metabolites identified, bioanalytical methods to quantitate such metabolites have been developed and validated, and the exposure to these metabolites determined at the MRHD prior to the submission of the carcinogenicity protocol. If this assumption were true, then this would mean that the human ADME study should have been conducted relatively early on in the course of the development program. However, the guidance does allow for the provision of in vitro data in lieu of in vivo data. Caution should be employed in using this approach since the use of such data relies on the assumption that the in vitro data are qualitatively and quantitatively similar to the in vivo data. If in vitro data are used in lieu of in vivo data, the sponsor runs the risk of discovering, at a later date, that the in vivo data do not support the doses selected for the carcinogenicity assessment study, with the possible need to repeat this very expensive and time-consuming study. The timing of the human ADME study will be covered in the sections on drug metabolism in clinical studies (Sections 27.4.5 and 27.5.7.10).

This guidance uses the term "major" metabolite in a manner that is subtly different from other guidances. The guidance states: "For determining the appropriateness of using AUC/limit dose approaches, major metabolites are defined as metabolites that, if excluded from the analysis, significantly change the comparison ratios between species. Data (point estimates as well as individual animal values) should be reported separately for males and females from the same strain as proposed for the bioassay." This definition is different from that used in the guidance "Drug Safety Testing of Metabolites" (US FDA, 2008j). This guidance is one of the few guidances that addresses the potential for strain-related differences in the metabolism of a drug and notes that the metabolism should be determined in the appropriate strain. Although not specifically mentioned in this guidance document, the metabolism of the drug should be assessed in the transgenic mouse if they are being used for carcinogenicity assessment. Simply evaluating the metabolism in the background strain of mouse may be insufficient due to potential differences in the metabolism of the drug in the parental and transgenic strain of mouse (Ariyoshi et al., 2001; Sanders et al., 2001).

The ICH guidance, "Dose Selection for Carcinogenicity Studies of Pharmaceuticals & Limit Dose S1C(R1)," is one of the few guidances to acknowledge that metabolism after a single dose can be different to the metabolism after repeat-dose administration (ICH, 2005c). The guidance states: "Changes in metabolite profile or alterations in metabolising enzyme activities (induction or inhibition) over time, should be understood to allow for appropriate interpretation of studies." Auto-induction of metabolism may markedly change the exposure to metabolites upon repeat-dose phenomenon. Auto-induction of metabolism of a drug is not an uncommon phenomenon and is often observed in rats for compounds that are CYP3A substrates. Occasionally, this phenomenon is observed in dog, monkey, and human. Therefore, the assessment of the exposure to metabolites in the rat after a single dose may underestimate the true exposure to metabolites and may lead the sponsor to make an incorrect assessment that a given human metabolite is under-represented in the rat, leading to the potential for the need for safety testing of that metabolite or incorrect dose setting for the carcinogenicity assessment studies. The concept of linearity in the metabolism of a drug is also discussed in this guidance. The guidance states that "The mid and low doses selected for the carcinogenicity study should take into account saturation of metabolic and elimination pathways." The implication of this statement is that the metabolism of the drug may need to be assessed at multiple dose levels rather than the one dose level that is traditionally used.

The guidance also provides an appropriate caveat regarding the correlation of plasma and tissue exposure to drug or metabolites by stating: "It is recognised that the doses administered to different species may not correspond to tissue concentrations because of different metabolic and excretory patterns. Comparability of systemic exposure is better assessed by blood concentrations of parent drug and metabolites than by administered dose. The unbound drug in plasma is thought to be the most relevant indirect measure of tissue concentrations of unbound drug. The AUC is considered the most comprehensive pharmacokinetic endpoint since it takes into account the plasma concentration of the compound and residence time in vivo." The guidance does not address how unbound drug concentrations are determined (i.e. directly or indirectly). The standard practice is to determine total drug concentrations and then correct the concentrations for the extent of plasma protein binding. This is a relatively simple task if the drug has concentration-independent plasma protein binding. However, if the plasma protein binding is concentration-dependent, the task can be a very laborious one. Additionally, the relationship between plasma protein binding and concentration should be well established in order to provide an accurate assessment of the exposure to unbound drug. This guidance appears to use the terms "blood" and "plasma" interchangeably and the reader should not necessarily assume that this guidance advocates the quantitation of drug in whole blood samples rather than in plasma samples.

27.4.7.7 *Combination Drugs*

The subject of drug combinations was addressed in an FDA guidance that was issued in 2006 (US FDA, 2006h). This guidance, "Nonclinical Safety Evaluation of Drug or Biologic Combinations," is not intended to cover fixed dose combination and co-packaged drug products for the treatment of HIV, since these drugs are covered in a separate guidance. Additionally, the guidance states that the Agency is developing a draft guidance specifically addressing oncologic drug combinations. However, this guidance has not yet been issued and does not appear on the list of guidances that the FDA plans to work on in 2007. Drug metabolism is mentioned in this guidance in the context of the possibility of a PK interaction. The guidance states: "A PK interaction can manifest in several ways, some of which can be monitored in vivo and some of which cannot. One drug/biologic product may alter the absorption or excretion of another product, change its distribution into one or more tissues, or change its pattern or rate of metabolism. Drugs may compete for serum protein binding, resulting in an increase in circulating free levels and tissue uptake of one drug. Metabolic interactions have been seen for combinations of two biologics." The guidance further states: "The FDA recommends that sponsors conduct combination PK/ADME studies to assess the potential for a PK interaction between the drugs/ biologics. These data are valuable for supporting the safety profile and guiding the drug/biologic development process. The FDA further recommends that PK/ADME combination studies (e.g., in vitro drug metabolism studies) be conducted early in drug development. The FDA encourages sponsors to evaluate serum protein binding and to monitor plasma concentrations of each drug in the toxicology studies. It may be possible to collect PK data as part of the toxicology studies instead of in a separate study." One of the many challenges in the development of drug combinations is the fact that many of the drugs that are used in the combinations may have been approved for marketing several decades ago and much of the basic ADME data that are contained in present day submissions will not exist. Therefore, a sponsor that is developing a drug combination may be required to determine the metabolism of these older drugs. The development of combination therapies is becoming more common and therefore this guidance will begin to assume an ever-increasing importance. A recent example of an approved combination therapy is Zetia (ezetimibe and simvastatin). Although this drug was approved prior to the FDA's guidance being issued, the SBA for Zetia is informative regarding the approach taken to the development of a combination therapy (US FDA, 2002e). In 2007, the FDA approved Janumet which is a combination of the recently approved DPP-4 inhibitor sitagliptin and the generic drug metformin (US FDA, 2007p).

In approximately the same time frame, the EMEA issued a draft guideline "Guideline on the Non-Clinical Development of Fixed Combinations of Medicinal Products" (EMEA, 2005f). The guideline was finalized in January 2008 with an effective date of August 1, 2008 (EMEA, 2008e). This guideline covers very similar ground to the FDA's guidance.

This guideline does not make mention of metabolism, other than related to the selection of the species to be used in the toxicology studies, saying that "For general toxicity studies usually one species only may be sufficient. The selection of the species used should be scientifically justified based on pharmacodynamic responsiveness to both components, pharmacokinetics, metabolism, target organs and sensitivity to the toxicological aspects." The rationale for combination therapy is described thus: "The rationale for a combination therapy and a fixed combination is often that pharmacological or pharmacokinetic interactions, leading to improved efficacy or safety profiles, compared with the single components. The main aim with the non-clinical studies to support the clinical development of a fixed combination is to characterise potential additive, synergistic, potentiation or antagonistic effects of the compounds when used together and to characterise the pharmacology, pharmacokinetics and toxicology of the combination under development. Additionally, these studies may identify toxicity unique to the combination and previously not seen for either compound administered alone."

The finalized guideline addresses fixed combination of compounds already approved as free combination therapy by stating that "When the fixed combination under development includes compounds for which there is sufficiently documented human experience of their individual and combined use, safety studies in animals are in general not required (CPMP/EWP/240/95)."

The EMEA document cited here is the "Note for guidance on fixed dose combination medicinal products" that was issued in 1996. This document does not appear to be available on the EMEA web site. However, the EMEA did issue a draft guideline in February 2008 (Guideline on Fixed Combination Medicinal Products) that has the same document reference number (CPMP/EWP/240/95) as the 1996 note for guidance

(EMEA, 2008b). However, this draft guideline does not refer to the 1996 note for guidance.

The guideline further states that "If the combination contains compounds from the same classes, as other compounds in well established combinations, for which there is considerable clinical experience and there are no pharmacokinetic interactions identified, further non-clinical studies may not be needed. A justification should in such case be provided. However in some situations certain non-clinical testing may still be warranted for example when the non-clinical data package does not fulfil the recommendations outlined on the draft Note for Guidance on the Non-Clinical Documentation of Medicinal Products for Mixed Marketing Applications (CPMP/SWP/799/95)."

This statement seems to imply that there may be class effects for pharmacokinetic interactions. This would appear to be a novel concept that is not supported by specific guidances or guidelines on drug–drug interactions. Additionally, the guideline does not define what is meant by "class." This terminology could be used to describe compounds with the same mechanism of action (pharmacological class), similar chemical structure (chemical class), or even BCS classification. It should be noted that the document "Draft Note for Guidance on the Non-Clinical Documentation of Medicinal Products for Mixed Marketing Applications (CPMP/SWP/799/95)" is no longer available. However, the finalized guidance does list (in the Reference section) the document "Guidance on the Non-Clinical Documentation of Medicinal Products for Mixed Marketing Applications (CPMP/SWP/799/95)" which is available on the EMEA web site (EMEA, 2005g).

The guideline also addresses fixed combination of approved compounds not approved as free combination therapy by stating that:

> For combination of approved compounds not approved as combination therapy, even though the safety and efficacy of the individual substances have been considered adequately documented for approval, some aspects regarding expected as well as potential unexpected interactions may still need to be addressed (CPMP/EWP/240/95). If pharmacodynamic interactions are the rationale for the combination, such effects should be documented. Relevant non-clinical studies may be part of the proof of principle. Pharmacodynamic data with the combination may also be justified to allow assessment of unexpected/undesirable interactions. Non-clinical (*in vitro* or *in vivo*) or clinical pharmacokinetic interaction studies with the combination should be considered, when appropriate. Provided that the pharmacokinetics of the single components are adequately characterised in animals, including the profile for enzyme induction and inhibition and drug–drug interactions, additional non-clinical documentation on pharmacokinetic interactions is generally not needed (CPMP/EWP/240/95).

> If a pharmacokinetic interaction constitutes the rationale for the combination, such interaction should be documented and appropriate non-clinical data may be needed.

> For the safety evaluation of the fixed combination, safety pharmacology and toxicity studies should be considered. The need for (combination) studies will depend on the type of the anticipated interactions between the components and on the range of concentrations and exposures covered in the available studies with the single components. If the systemic exposure expected from the use of the fixed combination is not covered adequately by the existing data, additional studies may be warranted.

> Additional testing may also be justified if the compounds to be combined target the same organ system(s), or belong to a class of compounds associated with a specific type of toxicity. The toxicological profile of the combination may be studied in non-clinical bridging studies, to support the safe human use and identify potential interactions. In these bridging

studies, specific endpoints addressing safety concerns based on the characteristics (pharmacology, toxicology) of the individual components may be included in addition to the conventional toxicological evaluation. If sufficient systemic exposure can be obtained, identified relevant safety pharmacology endpoints to be studied may also be included into the (bridging) toxicity studies. In addition, special and/or mechanistic studies may be warranted, for example studies addressing immunotoxicity or dependence, depending on the properties of the individual components in the combination.

The statement "Provided that the pharmacokinetics of the single components are adequately characterised in animals, including the profile for enzyme induction and inhibition and drug-drug interactions, additional non-clinical documentation on pharmacokinetic interactions is generally not needed (CPMP/EWP/240/95)" appears to be confusing for several different reasons. Firstly, it implies that enzyme induction and inhibition and drug interaction studies should be performed in animals and that such studies are predictive of the human situation. Secondly, it cites the document CPMP/EWP/240/95 as the source for this statement. However, this document (Draft Guideline on Fixed Combination Medicinal Products) does not seem to address these topics (EMEA, 2008b).

This draft guideline does address the topic of PD and PK interactions, stating that "The possibility of interactions between the substances should always be considered. The applicant should submit data either to establish that such interactions do not occur or that they are clearly recognized and defined." The guideline further addresses PK studies, stating:

> In general, the applicant must demonstrate that the various substances do not affect each others respective pharmacokinetic patterns. In some cases, however, a pharmacokinetic interaction (i.e. combination with a metabolism inhibitor) constitutes the rationale of the fixed combination.
>
> These interactions should be studied in healthy volunteers but also in patients if the disease modifies the pharmacokinetics of one substance and in high risk subgroups (elderly, patients with renal failure or hepatic impairment).

27.4.7.8 *Special Populations*

27.4.7.8.1 Anticancer Drugs As mentioned previously, the FDA does not have a guidance document on the preclinical development of anticancer drugs, although the EMEA does have one on this topic. Instead, the FDA utilizes an article published by DeGeorge *et al.* (1997) as well as a book chapter written by Farrell *et al.* (2003). The use of combinations of drugs for the treatment of cancer is extremely common. The DeGeorge article notes that "When synergistic effects may be anticipated such as when one agent interferes with the metabolism or elimination of the other agent or both cytotoxic agents target the same metabolic pathway or cellular function, pre-clinical testing of the combination is desirable."

Changes in the route of administration or in the formulation of anticancer drugs are often pursued with a goal of improving drug utility (DeGeorge *et al.*, 1997). If a clinical trial is proposed by the oral route for a drug that has already been investigated by intravenous administration, then additional preclinical studies should address whether there is enhanced liver toxicity, direct gastrointestinal toxicity, or altered metabolism (due to microflora in the gastrointestinal tract, the intestinal wall, or a first-pass effect through the liver). This is one of the few documents or guidances that acknowledges the potential role of the gut microflora in the metabolism of drugs.

The book chapter written by Farrell *et al.* (2003) provides the clearest indication of what preclinical information is recommended for inclusion in the initial IND filing for an oncologic drug. These studies are single-dose toxicity, repeat-dose toxicity and genotoxicity, although the latter may not be required at the IND filing for studies in patients with advanced cancers. What is enlightening is not so much what is included in the list but what is not included. For example, it would appear that the FDA is recommending that pharmacology or ADME data are not required for the initial IND filing for an oncologic drug. However, many such IND submissions contain a great deal of pharmacology and ADME information and these IND submissions look very similar to an IND submission for a non-life-threatening disease. This chapter further states that "Preclinical combination testing may not be necessary provided that each agent's toxicities have been fully characterized, and data do not indicate that the combination use would be unsafe. Concern may increase if one agent interferes with the metabolism or elimination of another, or if both cytotoxic agents target the same metabolic or cellular pathway, or cellular function."

An additional useful source of information is the material (presentations and transcripts) from the FDA Oncologic Drugs Advisory Committee meeting that was held on March 13, 2006 (US FDA CDER, 2006). This meeting discussed preclinical requirements and phase I trial design issues for the development of oncologic products. At this meeting, John Leighton, a Pharm/Tox reviewer from the FDA stated that "Pharmacokinetic studies provide information on absorption, distribution, metabolism and excretion. These studies are strongly encouraged, particularly for drugs with extended expected duration of exposure, for example drugs administered by depot formulation." It is interesting that the word "encouraged" was used rather than "should" or "expected," indicating the somewhat secondary nature of pharmacokinetics and metabolism in the development of novel oncologic agents.

It should be noted that the EU issued "Note for Guidance on the Pre-Clinical Evaluation of Anticancer Medicinal Products" in 1999 (EMEA, 1998). The purpose of this guideline is to "define the preclinical data which are considered obtainable from preclinical studies with respect to pharmacodynamic, pharmacokinetic and toxicological properties of new anticancer drugs and which are considered relevant with respect to Phase I (Human Pharmacology), Phase II (Therapeutic Exploratory) and Phase III (Therapeutic Confirmatory) Clinical Trials and Marketing Applications. Furthermore, the guideline serves the purpose of avoiding unnecessary tests, thus enabling the promptest possible introduction of newly developed anticancer medicinal products into clinical trials without compromising safety." This note for guidance does not contain any specific mention of metabolism or metabolites. However, it does note that the mechanism of resistance could be investigated, "In parallel with the characterization of the mechanism(s) of action, the corresponding profile with respect to possible mechanism(s) of resistance (e.g. overexpression of P-glycoprotein/multidrug resistance protein/glutathione, changes in topoisomerase I and II) can be obtained. Observed resistance could be investigated for its circumvention by resistance modulating agents. Investigation of the possible induction of resistance by long-term exposure of cell lines to the new drug and further characterization of mechanism(s) of resistance are encouraged." Pharmacokinetic/toxicokinetics studies are mentioned, although the extent of such studies are limited. "The evaluation of limited kinetic parameters, for example peak plasma levels and AUC, at doses around the MTD in the animal species used for preclinical studies may facilitate dose escalation during Phase I studies. Further information on ADME in animals should normally be made available prior to Phase II/III studies."

The topic of nonclinical safety evaluation of oncology drugs has also been taken up by the ICH based on a proposal by PhRMA in October 2006. In May 2007, the ICH issued a concept paper entitled "S9: Preclinical Guideline on Oncology Therapeutic Development" (ICH, 2007). The rationale for the need for an ICH guidance on this topic is described thus:

> The approaches to the preclinical development of oncology drug products have been and continue to be independently discussed and developed in Europe, the USA and Japan. The preclinical approaches are not agreed on across product classes, between small molecules of different molecular mechanisms and across product classes such as biologics and drugs. The available disharmonized guidance has resulted in inefficient use of animal resources, and ineffective drug development in a critical area of human health. The European Medicines Evaluation Agency (EMEA) has official preclinical guidance for the development of cytotoxic cancer treatments. While cytotoxic agents constitute a decreasing class of agents in development, guidance is needed to address the newer, non-cytotoxic (signaling pathways) and context dependent (tumor mutation specific) mechanisms for treatment of cancer now in development. Nor are biologic approaches that include both cytotoxic and non-cytotoxic mechanism addressed in these guidances. The FDA is developing guidances for biologic therapeutics and drugs that will broadly address preclinical development of the various mechanisms of cancer therapy. Further, in the context of FDA's Exploratory IND Guidance, some preclinical approaches to early investigative oncology drug development are briefly outlined, although not limited to this therapeutic application. The MHLW is currently developing preclinical guidance that will address various mechanisms of anti-cancer therapy, but will not include biologics in its scope. Thus, there is substantial concern that, when these guidances are completed, they will not offer a harmonized approach for preclinical development of drugs needed for the treatment of cancer.
>
> This independent proliferation of guidances focused on preclinical issues in oncology drug development highlights the critical need for guidance on this topic and indicates the inadequacy of the current state. While these current independent approaches have their merits, the dissociated efforts will likely result in disparate recommendations that will adversely impact global development of needed cancer therapies. The result will be both delays in the availability of needed agents and waste of valuable resources, including test animals.
>
> Amplifying this problem, is the fact that most ICH S guidances and M3 either explicitly or implicitly exclude cancer therapies from their recommendations, yet there is no alternative offered for a harmonized product development strategy. As a result, M3 or S6 guidance is often inappropriately relied on (or expected) particularly for non-cytotoxic and biologic agents in development of cancer therapeutics.

These are very strong statements and indicate that this topic is one that will likely advance quickly. The issues to be resolved include:

1. Scope of the guidance: Therapeutic classes to be included; preclinical development approach in relation to product class and how to cover the breadth of the product classes in the oncology guidance; classes to be excluded are therapeutic vaccines, cell and gene therapy.
2. Necessity for long-term preclinical testing to support early clinical trials.
3. Duration, species, type of nonclinical safety studies to support the early and late development phases.

4. Harmonization of approaches to the extent justified for biological and small molecules.

It is interesting to note that the ICH will aggressively develop this guideline. However, the aggressive timeline calls for final guidance to be issued in early 2010.

In 2004, the FDA issued a guidance on "IND Exemptions for Studies of Lawfully Marketed Drug or Biological Products for the Treatment of Cancer" (US FDA, 2004b). This guidance identifies the circumstances under which a sponsor may be able to claim an exemption from the requirements for the submission of an IND. It includes a statement that is very similar to that made in the DeGeorge article: "Because of the danger of synergistic toxicity (i.e., enhanced effects from the combination) occurring with a new drug combination, if there are no data from the literature on its safety, the initial study of a new drug combination should ordinarily be performed under an IND. Synergistic toxicity may be anticipated when one agent interferes with the metabolism or elimination of the other agent; when both agents target the same metabolic pathway or cellular function; or when one agent targets signaling pathways that are reasonably expected to modulate sensitivity to the other agent. If it is determined that synergistic toxicity is likely, animal studies should be considered for determining a safe starting dose for the drug combination in humans" (DeGeorge *et al.*, 1997). This statement is important since it emphasizes the understanding of the enzymology of the metabolism of the drug. However, it should be noted that the enzyme systems involved in the metabolism of many of the drugs that were approved before the early to mid-1990s may not be well characterized.

27.4.7.8.2 Diabetes In February 2008, the FDA issued a draft guidance "Diabetes Mellitus: Developing Drugs and Therapeutic Biologics for Treatment and Prevention" (US FDA, 2008i). Appendix A of this draft guidance addresses preclinical considerations for peroxisome proliferator-activated receptor antagonists. Of interest are the statements "The ICH guidance regarding the duration of chronic toxicity studies in rodents and nonrodents has been adopted and for the nonrodent chronic toxicity study, a 9-month duration generally is appropriate for supporting chronic human use. However, since the no observed adverse effect levels for some of the toxicities associated with PPAR agonists can be adequately defined only after chronic administration, a 1-year study in nonrodents is recommended for drugs in the PPAR class." The peroxisome proliferators have long been known to cause liver enlargement and be non-genotoxic carcinogens. Therefore, it is not surprising to see the hepatic effects discussed in this Appendix: "The cause of any liver enlargement observed should be determined (peroxisome proliferation, mitochondrial proliferation/swelling). Liver tissues should be stained to detect the presence of fatty changes. The sections can be stained with Sudan IV or Oil Red-O. Liver enzyme levels and biochemical markers of peroxisome proliferation (Acyl CoA and CYP 4A) should be analyzed in rodents and nonrodents." The last statement should be clarified in the final guidance since the terminology used is not as accurate as it should be.

27.4.7.8.3 Pediatrics The safety testing of drug products for use in a pediatric population has been addressed in a guidance (Nonclinical Safety Evaluation of Pediatric Drug Products) issued by the FDA in 2006 (US FDA, 2006i).

This guidance notes that there may be differences in drug safety profiles between mature and immature systems, and that metabolism may be involved in these differences. The guidance states:

> Some therapeutics have shown different safety profiles in pediatric and adult patients. Inherent differences between mature and immature systems introduce the possibility of drug toxicity, or resistance to toxicity in immature systems that are not observed in mature systems. Several factors contribute to these potential differences. Postnatal growth and development can affect drug disposition and action. Examples include developmental changes in metabolism (including the maturation rate of Phase I and II enzyme activities), body composition (i.e., water and lipid partitions), receptor expression and function, growth rate, and organ functional capacity. These developmental processes are susceptible to modification or disruption by drugs. Although some age-dependent effects can be largely predicted by knowledge of the changes in drug metabolic pathways during development, others cannot.

The guidance then proceeds to provide some examples of drugs that exhibit differences in toxicity between adult and pediatric patients. These include acetaminophen. "Acute acetaminophen toxicity is a classic example of how maturation can affect the toxicity profile of a drug. Young children are far less susceptible to acute acetaminophen toxicity than adults because children possess a higher rate of glutathione turnover and more active sulfation. Thus, they have a greater capacity to metabolize and detoxify an overdose of acetaminophen when compared to adults."

The selection of the species is also addressed by the guidance:

> The species of the juvenile animal tested should be appropriate for evaluating toxicity endpoints important for the intended pediatric population. Traditionally, rats and dogs have been the rodent and nonrodent species of choice. In some circumstances, however, other species may be more appropriate. For example, when drug metabolism in a particular species differs significantly from humans, an alternative species (e.g., minipigs, pigs, monkeys) may be more appropriate for testing.

The implication of this statement is that the sponsor may need to perform a wider "species screen" for the metabolically relevant species. The guidance does not address whether both the rodent and non-rodent species need to be metabolically relevant or whether only one species needs to be metabolically relevant. The availability of suitable materials (e.g. hepatic microsomes obtained from human infants) may make an *in vitro* comparison difficult. Interestingly, the guidance intimates that the non-rodent species will be the species in which the metabolism will most likely be different from that in humans. The guidance does not address whether such a species screen is performed *in vivo* or *in vitro*. Additionally, the search for alternative species for testing of the potential for toxicity in juveniles may have strategic implications if it is found that the minipig would be a more relevant species for toxicology studies than the dog. If this is the case, an argument could be made that the general non-rodent toxicology studies should be performed in minipigs. The guidance further states that "Assessment of developmental differences in parent drug disposition and profiles of significant metabolites in juvenile animals should be made according to established guidelines (see the ICH guideline for industry S3A Toxicokinetics: Assessment of Systemic Exposure in Toxicity Studies)." Significant metabolite is not defined in this guidance. The guidance makes the point that the dose frequency should "be relevant to the intended clinical use of the drug. In some cases, however, the use of dosing frequencies similar to those anticipated for clinical administration are not feasible because of technical considerations for the animal models used. Changes in frequency can be made when variables such as metabolic and kinetic differences are considered." This implies that there the sponsor may need to conduct several dose-ranging studies that involve an evaluation of the exposure top both parent and relevant metabolites. Finally, the guidance provides, in Table 7, an overview of the developmental modulation of phase I/II metabolism. This table lists a range of enzyme activities and their expression in human, rat, and rabbit. No information is provided regarding other species such as dogs, minipigs, pigs, or monkeys.

In 2005, the CHMP issued a draft guidance entitled "Guideline on the Need for Non-Clinical Testing in Juvenile Animals on Human Pharmaceuticals for Paediatric Indications" (EMEA, 2005e). This guidance was finalized in January 2008 and came into effect in August 2008 (EMEA, 2008d). The guideline notes that the development of certain functions related to the pharmacokinetic handling of an active substance may take up to several years stating that "The development of the major systems is age dependent, for example:

- Nervous system: Development up to adulthood
- Reproductive system: Development up to adulthood
- Pulmonary system: Development up to two years old
- Immune system: Development up to 12 years old
- Renal system: Development up to one year of age
- Skeletal system: Development up to adulthood
- Organs and/or systems involved in absorption and metabolism of drugs. Development of biotransformation enzymes up to adolescence."

The guideline also addresses pre- and postnatal reproduction studies as follows: "Before performing a juvenile animal toxicity study, it should be considered whether a developmental toxicity issue could be addressed in a modified pre- and postnatal development study in rats. Key factors that need to be examined include, but are not restricted to, the amount of the active substance and/or relevant metabolites excreted via the milk and the resulting plasma exposure of the pups, which organs under development that will be exposed during the pre-weaning period, physical development and histopathological investigations."

The importance of pharmacokinetic data is described:

Pharmacokinetic data (including in vitro data, if relevant) contributes to the evaluation of the relevance of the animal model for human risk assessment. It is recognised that the collection of blood samples to obtain a full kinetic profile of a test compound under study in juvenile animals might sometimes be difficult. However, sampling at a few time points, using pooled samples if necessary, should be performed to obtain an estimate of basic kinetic characteristics, for example Cmax and AUC. Inclusion of satellite groups of dosed animals, for example excess offspring from a pre-postnatal study, for blood sampling may be considered. The use of methodology allowing population pharmacokinetic/toxicokinetic determinations may also be considered. Toxicokinetics should also be considered to confirm appropriate exposure levels in different treatment groups.

Finally, the guideline states that "under special circumstances, data on absorption, distribution, metabolism and/or excretion in juvenile animals may be valuable to further investigate and for example understand a specific safety concern."

27.4.7.9   Other Topics

27.4.7.9.1   Dependence Potential   The topic of dependence potential was addressed by the EMEA in its "Guideline on the Non-Clinical Investigation of the Dependence Potential of Medicinal Products" that was issued in March 2006 (EMEA, 2006g). This guidance states that the impetus for conducting such studies will include early indicators that may come from pharmacokinetic studies (brain penetration, metabolites entering or formed in the brain). However, it should be noted that the brain penetration of parent drug or metabolites, or the metabolism of a drug by the brain is rarely, if ever, determined.

Consistent with the FDA's guidance on Safety Testing of Drug Metabolites (US FDA, 2008j), the guideline states: "Separate studies evaluating the dependence potential of metabolites would be needed when there is reasonable evidence that such metabolites are formed in humans, are suspected to penetrate the human CNS in significant concentrations and are not adequately evaluated in non-clinical studies with the parent."

The guideline addresses the timing of these studies in the following statements:

> Receptor binding studies and safety pharmacology studies regarding the CNS as outlined under ICH S7A should be completed before phase 1 clinical trials. ... In general, metabolites do not need to be evaluated for dependence potential prior to human trials. Separate receptor binding studies on metabolites should be initiated as soon as human pharmacokinetic data are available and such studies are considered appropriate. When studies on metabolite receptor affinity and other studies described under the first tier are deemed relevant, data from these studies should become available before large populations are exposed to the investigational medicinal product, generally before Phase III. All relevant data should be available at the time of a Marketing Authorisation Application.

The same topic was addressed by Health Canada in a draft guidance "Clinical Assessment of Abuse Liability for Drugs with Central Nervous System Activity" that was also issued in March 2006 (Health Canada, 2007a).

27.4.7.9.2   Photosafety Testing   The 2003 FDA guidance on "photosafety testing" is one of several guidances that points out that toxicity may be due to a metabolite rather than (or as well as) due to the parent drug itself (US FDA, 2003a). In addition, this guidance makes the statement that "Tissue distribution studies of systemically administered drug products, usually included in IND submissions, can be used to assess the extent of partitioning into the skin or eyes. In the absence of partitioning into light-exposed compartments, photoirritation testing is unlikely to be informative and need not be conducted." However, there is no mandate for tissue distribution studies to be included in the IND submission. It would be interesting to conduct a survey of pharmaceutical and biotechnology companies to determine the frequency of inclusion of such studies in their IND submissions and the reason for the conduct of these studies.

27.4.7.9.3 Excipients Excipients were the subject of a guidance entitled "Nonclinical Studies for the Safety Evaluation of Pharmaceutical Excipients" that was issued by the FDA in 2005 (US FDA, 2005d). The intent of this guidance is to outline the process for the safety evaluation of excipients that are not fully qualified by existing safety data with respect to the currently proposed level of exposure, duration of exposure, or route of administration. The basis for this guidance is the observation that not all excipients are inert substances and that some have been shown to be potential toxicants. The Federal Food, Drug, and Cosmetic Act of 1938 (the Act) was enacted after the tragedy of the elixir of sulfanilamide in 1937 in which an untested excipient was responsible for the death of many children who consumed the pharmaceutical. Essentially, the guidance states that you would characterize the ADME properties of the excipient in a similar manner to a drug. The topic of the toxicity of excipients has been extensively reviewed by Weiner and Kotkoskte (2007).

27.4.7.9.4 Immunotoxicology The assessment of the distribution of drug and/or metabolites to a particular tissue is mentioned in the ICHS8 guidance "Immunotoxicity Studies for Human Pharmaceuticals" that was issued in April 2006 (ICH, 2005d). The guidance states that "If the compound and/or its metabolites are retained at high concentrations in cells of the immune system, additional immunotoxicity testing should be considered." However, the assessment of the retention of compound-related material in the cells of the immune system is rarely, if ever, performed unless there is a toxicity signal that such information would be useful in interpreting.

27.4.7.9.5 Alternate Routes of Administration In March 2008, the FDA released a draft guidance entitled "Nonclinical Safety Evaluation of Reformulated Drug Products Intended for Administration by an Alternate Route" (US FDA, 2008h). Metabolism is addressed in a couple of places in the draft guidance. The guidance states that:

> All routes of administration can result in systemic exposure. Therefore, the adequacy of the available systemic toxicity information should be evaluated based on the systemic exposure obtained after administration of a proposed new formulation or of a previous formulation by a new route. Additional toxicity studies might be recommended if the available toxicity information is not sufficient to support the exposure measured with the new formulation or if a significantly different pattern of exposure results from the new formulation. An adequate evaluation of the pharmacokinetics and absorption, distribution, metabolism, and elimination (PK/ADME) of the drug substance is recommended for new formulations. These data and any available human data can be helpful in determining what additional nonclinical toxicity data, if any, are recommended. When comparing the PK/ADME of a new formulation with a previously approved formulation, it is important to examine the shape of the concentration/time curve and not just the total area under curve. For example, alterations in absorption or the dosing frequency can produce significantly different concentration/time profiles that might lead to different toxicological effects. Changes in the vehicle composition or form also can alter the PK of active ingredients. In some cases, PK/ADME for the new formulation might not be available. In these cases, an assumption of 100% bioavailability from the proposed clinical dose might be used to judge the adequacy of available systemic toxicity information."

The guidance describes many different routes of administration. It is only for the ocular route that pharmacokinetics is expressly mentioned "If the active ingredient has not been used by the ocular route, then toxicity studies in two species with complete eye and systemic evaluation for the appropriate duration should be carried out with the new formulation. In certain cases, studies in one most appropriate species may be adequate. Optimal design of these studies would include the evaluation of ocular and systemic PK."

27.4.7.10 Support to First-in-Human Studies   In March 2006, the "TeGenero incident" occurred in which very serious adverse reactions occurred in a first-in-human clinical trial of TGN412 which is a monoclonal antibody that was being developed as a medicine to treat leukemia and autoimmune diseases such as rheumatoid arthritis (Wikipedia, 2008a). In the clinical trial, six healthy male volunteers experienced severe systemic adverse reactions soon after receiving TGN1412 intravenously. All six volunteers developed a cytokine release syndrome with multi-organ failure and required intensive treatment and supportive measures that were provided by the Intensive Therapy Unit at Northwick Park Hospital (the site of the clinical trial). As a result of this incident, an Expert Scientific Group (ESG), led by Prof. Gordon W. Duff, was set up following this incident. The group received input from a variety of groups and individuals and issued their final report in December 2006 (Expert Group on Phase One Clinical Trials, 2006). The TeGenero incident also led to the Association of the British Pharmaceutical Industry updating its "Guidelines for Phase I Clinical Trials" in 2007 (Association of the British Pharmaceutical Industry, 2007).

As a result of the TeGenero incident, the EMEA issued a controversial guideline in September 2007, entitled "Guideline on Strategies to Identify and Mitigate Risks for First-In Human Clinical Trials with Investigational Medicinal Products" (EMEA, 2007d). Originally, the intent of the guideline was to address potential high-risk biological products. However, the final guidance covers both biologicals and small molecules.

Metabolism is only mentioned in the guideline in terms of the selection of the relevant species for toxicology testing. However, this guidance will be discussed in some depth due to the potential profound impact of the guidance on the nonclinical groups and the potential need for information that may not routinely be generated.

This guideline acknowledges that:

> For many new investigational medicinal products, the non-clinical safety pharmacology and toxicology program provides sufficient safety data for estimating risk prior to first administration in humans. However, for some novel medicinal products this non-clinical safety program might not be sufficiently predictive of serious adverse reactions in man and the non-clinical testing and the design of the first-in-human study requires special consideration. When planning a first-in-human clinical trial, sponsors and investigators should identify the factors of risk and apply risk mitigation strategies accordingly as laid down in this guideline. In addition to the principles expressed in this guideline, some special populations such as paediatrics may deserve specific considerations.

The focus of risk identification and risk mitigation before performing an FIH study is a somewhat new approach. Previously, the approach had been primarily focused on hazard identification with a lesser emphasis on risk identification and risk mitigation.

The risk factors that should be taken into account are described in the following manner: "Predicting the potential severe adverse reactions for the first-in-human use of an investigational medicinal product, involves the identification of the factors of risk. Concerns may be derived from particular knowledge or lack thereof regarding (1) the mode of action, (2) the nature of the target, and/or (3) the relevance of animal models."

This statement makes the point regarding lack of knowledge and starts to lay the groundwork for the conclusion that new drugs should be regarded as having high risk unless data exist to the contrary. The guidance further states, "The Sponsor should discuss the following criteria for all first-in-human trials in their clinical trial authorisation application. These criteria should be taken into account on a case-by-case basis."

Mode of action

While a novel mechanism of action might not necessarily add to the risk per se, consideration should be given to the novelty and extent of knowledge of the supposed mode of action. This includes the nature and intensity (extent, amplification, duration, reversibility) of the effect of the medicinal product on the specific target and non-targets and subsequent mechanisms, if applicable. The type and steepness of the dose response as measured in experimental systems, which may be linear within the dose range of interest, or non-linear (e.g. plateau with a maximum effect, over-proportional increase, U-shaped, bell-shaped), is of importance.

For example, the following modes of action might require special attention:

— A mode of action that involves a target which is connected to multiple signaling pathways (target with pleiotropic effects), e.g. leading to various physiological effects, or targets that are ubiquitously expressed, as often seen in the immune system.

— A biological cascade or cytokine release including those leading to an amplification of an effect that might not be sufficiently controlled by a physiologic feedback mechanism (e.g., in the immune system or blood coagulation system). CD3 or CD28 (super-) agonists might serve as an example.

When analysing risk factors associated with the mode of action, aspects to be considered may include:

— Previous exposure of human to compounds that have related modes of action.

— Evidence from animal models (including transgenic, knock-in, or knock-out animals) for the potential risk of serious, pharmacologically mediated toxicity

— Novelty of the molecular structure of the active substance(s), for example, a new type of engineered structural format, such as those with enhanced receptor interaction as compared to the parent compound.

Nature of the target

The target in human should be discussed in detail. Beyond the mode of action, the nature of the target itself might impact on the risk inherent to a first administration to humans, and sponsors should discuss the following aspects, based on the available data:

— the extent of the available knowledge on the structure, tissue distribution (including expression in/on cells of the human immune system), cell specificity, disease specificity, regulation, level of expression, and biological function of the human target including "downstream" effects, and how it might vary between individuals in different populations of healthy subjects and patients.

— If possible, a description of polymorphisms of the target in relevant animal species and humans, and the impact of polymorphisms on the pharmacological effects of the medicinal product.

Relevance of animal species and models

The sponsor should compare the available animal species to humans taking into account the target, its structural homology, distribution, signal transduction pathways and the nature of pharmacological effects.

The guidance then refers to the section on "Demonstration of relevance of the animal model." This section of the guideline states that "Qualitative and quantitative differences may exist in biological responses in animals compared to humans. For example, there might be differences in affinity for molecular targets, tissue distribution of the molecular target, cellular consequences of target binding, cellular regulatory mechanisms, metabolic pathways, or compensatory responses to an initial physiological perturbation.

Where there is evidence of species-specificity of action from *in vitro* studies with human cells compared with cells from a test species, the value of the *in vivo* response of the test species may be significantly reduced in terms of predicting the *in vivo* human response. It should be noted that a similar response in human and animal cells *in vitro* is not necessarily a guarantee that the *in vivo* response will be similar.

In practice this means that animal studies with highly species-specific medicinal products may:

- not reproduce the intended pharmacological effect in humans;
- give rise to misinterpretation of pharmacokinetic and pharmacodynamic results;
- not identify relevant toxic effects.

A weight-of-evidence approach should involve integration of information from *in vivo*, *ex vivo* and *in vitro* studies into the decision-making process.

High species-specificity of a medicinal product makes the non-clinical evaluation of the risk to humans much more difficult, but does not imply that there is always an increased risk in first-in-human trials.

The demonstration of relevance of the animal model(s) may include comparison with humans of:

- Target expression, distribution and primary structure. However, a high degree of homology does not necessarily imply comparable effects;
- Pharmacodynamics
  - Binding and occupancy, functional consequences, including cell signalling if relevant.
  - Data on the functionality of additional functional domains in animals, if applicable, e.g. Fc receptor system for monoclonal antibodies.
- Metabolism and other pharmacokinetic aspects
- Cross-reactivity studies using human and animal tissues (e.g. monoclonal antibodies).

The search for a relevant animal model should be documented and justified in detail. Where no relevant species exists, the use of homologous proteins or the use of relevant transgenic animals expressing the human target may be the only choice. The data gained is more informative when the interaction of the product with the target receptor has similar physiological consequences to those expected in humans. The use of *in vitro* human cell systems could provide relevant additional information. The relevance and limitations of all models used should be carefully considered and discussed fully in the supporting documentation.

The good news in these lengthy statements is that the drugs will be evaluated on a "case-by-case" basis and that the "weight of evidence" approach will be used. The bad news is that it is clear that a far greater understanding of the biology of the target and the pharmacology of the drug. A far more detailed understanding of the dose or concentration–effect relationship will be required as will be an understanding of the biology of the pharmacological target in humans.

In summary on the topic of species relevance, the guideline states that "Where available animal species/models or surrogates are perceived to be of questionable relevance for thorough investigation of the pharmacological and toxicological effects of the medicinal product, this should be considered as adding to the risk."

The guideline provides further detail on two topics (pharmacodynamics and pharmacokinetics) that are of interest to the drug metabolism scientists. Regarding pharmacodynamics, the guideline states that:

Pharmacodynamic studies should address the mode of action, and provide knowledge on the biology of the target. These data will help to characterise the pharmacological effects and to identify the most relevant animal models. The primary and secondary pharmacodynamics, should be conducted in *in vitro* animal and human systems and *in vivo* in the animal models. These studies should include target interactions preferably linked to functional response, e.g. receptor binding and occupancy, duration of effect and dose-response. A dose/concentration-response curve of the pharmacological effect(s) should be established with sufficient titration steps in order to increase the likelihood to detect significant pharmacological effects with low doses and to identify active substances with U-shaped or bell-shaped dose response curves. Such significant or even reverse effects have been reported with biological compounds. Since a low dose is to be administered to humans in the first-in-human trial, this is of high importance. Although GLP compliance is not mandatory for pharmacodynamic and pharmacokinetic studies, they should be of high quality and consistent with the principles of GLP.

The guideline also addresses pharmacokinetics. "Standard pharmacokinetic and toxicokinetic data should be available in all species used for safety studies before going into human (ICH S3, S6, M3). Exposures at pharmacodynamic doses in the relevant animal models should be determined especially when pharmacodynamic effects are suspected to contribute to potential safety concerns." If this statement is followed, there will be the need for an increased level of evaluation of the pharmacokinetics of the drug in the efficacy species. Presently, the exposure is usually determined only at one dose level (e.g. the ED_{50} dose) rather than at the multiple dose levels that are implied by the guideline. Additionally, since a theme of the guideline is the understanding of the concentration–effect relationship, there is the implication that the sponsor would have determined whether metabolites contribute meaningfully to the efficacy in the animal model. It also implies that if metabolites do contribute meaningfully to the efficacy, the sponsor will have described the exposure to the parent drug and relevant metabolites in the animal efficacy model.

The estimation of the first dose in humans is addressed by this guideline:

The estimation of the first dose in human is an important element to safeguard the safety of subjects participating in first-in-human studies. All available information has to be taken in consideration for the dose selection and this has to be made on a case-by-case basis. Different methods can be used.

In general, the No Observed Adverse Effect Level (NOAEL) determined in non-clinical safety studies performed in the most sensitive and relevant animal species, adjusted with allometric factors (see references "Other guidelines") or on the basis of pharmacokinetics gives the most important information. The relevant dose is then reduced/adjusted by appropriate safety factors according to the particular aspects of the molecule and the design of the clinical trials.

For investigational medicinal products for which factors influencing risk according to section 4.1 have been identified, an additional approach to dose calculation should be taken. Information about pharmacodynamics can give further guidance for dose selection. The 'Minimal Anticipated Biological Effect Level' (MABEL) approach is recommended. The MABEL is the anticipated dose level leading to a minimal biological effect level in humans. When using this approach, potential differences of sensitivity for the mode of action of the investigational medicinal product between humans and animals, need to be taken into consideration e.g. derived from *in-vitro* studies. A safety factor may be applied for the calculation of the first dose in human from MABEL as discussed below.

The calculation of MABEL should utilise all *in vitro* and *in vivo* information available from pharmacokinetic/pharmacodynamic (PK/PD) data such as:

   i) target binding and receptor occupancy studies *in vitro* in target cells from human and the relevant animal species;

   ii) concentration-response curves *in vitro* in target cells from human and the relevant animal species and dose/exposure-response *in vivo* in the relevant animal species.

   iii) exposures at pharmacological doses in the relevant animal species.

Wherever possible, the above data should be integrated in a PK/PD modeling approach for the determination of the MABEL.

In order to further limit the potential for adverse reactions in humans, a safety factor may be applied in the calculation of the first dose in human from the MABEL. This should take into account criteria of risks such as the novelty of the active substance, its biological potency and its mode of action, the degree of species specificity, and the shape of the dose-response curve and the degree of uncertainty in the calculation of the MABEL. The safety factors used should be justified. When the methods of calculation (e.g. NOAEL, MABEL) give different estimations of the first dose in man, the lowest value should be used, unless justified. Other approaches may also be considered in specific situations, e.g. for studies with conventional cytotoxic IMPs in oncology patients.

It is heartening to note that it has been recognized that the approach outlined in this guideline will not be applied across the board and should not inhibit the development of new anticancer drugs for which the FIH study will be conducted in a patient population, that is, in a patient population that clearly has a different risk–benefit than do normal healthy volunteers.

It can be seen from the statements in this guideline that there has been a shift in basic philosophy regarding the safety of new drugs. Previously, the mind-set was that medicinal products *are not* high risk unless we have evidence to the contrary. Now the mind-set seems to be that medicinal products *are* high risk unless we have evidence to the contrary.

Clearly, there are many potential implications of this guideline and many open questions. These implications and open questions occur for many different disciplines include, but are not restricted to the following:

Pharmacology

- How many animal models of efficacy should be evaluated?
  - — More models in order to better define MABEL?
  - — Fewer models due to increased workload?
  - — Fewer models but repeated more than once?
- What dosing regimens should be used?
- May need to increase the number of dose levels in animal efficacy studies
- May need increased PK assessment in animal efficacy studies
- May need robust PD markers for both NCEs and biologics
- May need to enhance understanding of variability in, and polymorphisms of, the target
- Need to better understand the pharmacology of the target in humans, including potential sources of variability of the target
- May need to conduct animal efficacy studies "in the spirit of GLP"
- How relevant are our animal models?
  - — for example, utility of CIA-induced arthritis model in rat.

Toxicology

- Selection of species for toxicology testing
  - — How do we address differences in target based on supplier, age, etc.?
  - — Need to determine tissue distribution of target for a NCE?
- Need robust PD markers in toxicology studies for both NCEs and biologics
- May need to include limited toxicity assessments in animal efficacy studies
- Design of Tox studies
  - — Using repeat-dose Tox studies to support single-dose clinical trials is already conservative
    - • Should we use single-dose Tox studies to support single-dose clinical studies?
      - — This approach may work in the United States but not in the EU

PK/ADME

- Greater need for PK/PD modelers in nonclinical
- Need robust PD markers for both NCEs and biologics
- Will need more studies to define PK/PD than are presently conducted
- Need to ensure relevance of *in vitro* metabolism data in selection of Tox species

Clinical

- Selection of starting dose
  - — What does "minimal" mean?
    - • 10% effect?
    - • Need to have robust assays that can distinguish small changes in response
  - — All FIH clinical trials may be (in essence) a microdose study

- Each cohort will take longer to perform due to sequential dosing
- Any adverse event that was not observed in animal toxicology studies will likely halt dose escalation
- Slow conduct of clinical trial due to lag time between cohorts due to stopping rules and increased reliance on PK/PD
- Need for real-time (bedside) PD assays?
- Need to develop "rescue" procedures for patients that develop adverse events
- Slow IV infusions will be used instead of bolus administration

Regulatory

- Presently, the EU regulatory agencies do not require study reports for review but they may now start to ask for them
- Gaining scientific advice from the EMEA prior to FIH used to be rare but may well become common (equivalent of Pre-IND meeting)
  — Can the EMEA handle an increased level of meetings?
- Will the FDA default to the EU guideline?
- FIH clinical trials may become rare in the EU
- Since some high-risk products are only high risk upon repeat-dose administration, will multiple-dose clinical trials also be placed under similar scrutiny?
- Will need better competitive intelligence regarding other compounds with similar MOAs

CMC

- May need to provide a greater range of dose strengths for the FIH study
- May need to provide solution doses to aid flexibility in dose escalation

Strategic

- The pharmaceutical industry may choose to no longer develop drugs with:
  — Steep dose/response curves
  — Novel targets
  — Big Pharma may leave this risk to the small biotech companies.
- Fewer FIH studies may be conducted in the EU
- Need for greater sharing of information between companies working on the same MOA?
- Need for greater interaction between pharmacology, toxicology, PK, and clinical groups within a company.

It should be noted that there are some strong similarities between this guideline and the FDA guidance on Exploratory INDs that will be discussed in Section 27.5.3.

In July 2005, the FDA issued a guidance, "Estimating the Maximum Safe Starting Dose in Initial Clinical Trials for Therapeutics in Adult Healthy Volunteers" (US FDA, 2005e). This guidance outlines a process (algorithm) and vocabulary for deriving the maximum recommended starting dose (MRSD) for FIH clinical trials of new molecular entities in adult healthy volunteers, and recommends a standardized process by which the MRSD can be selected. The purpose of this process is to ensure the safety of the human volunteers. The guidance states that:

> Although the process outlined in this guidance uses administered doses, observed toxicities, and an algorithmic approach to calculate the MRSD, an alternative approach could be proposed that places primary emphasis on animal pharmacokinetics and modeling rather than dose. In a limited number of cases, animal pharmacokinetic data can be useful in determining initial clinical doses. However, in the majority of investigational new drug applications (INDs), animal data are not available in sufficient detail to construct a scientifically valid, pharmacokinetic model whose aim is to accurately project an MRSD.

The following statement is included as a footnote to the previous paragraph:

> If the parent drug is measured in the plasma at multiple times and is within the range of toxic exposures for two or more animal species, it may be possible to develop a pharmacokinetic model predicting human doses and concentrations and to draw inferences about safe human plasma levels in the absence of prior human data. Although quantitative modeling for this purpose may be straightforward, the following points suggest this approach can present a number of difficulties when estimating a safe starting dose. Generally, at the time of IND initiation, there are a number of unknowns regarding animal toxicity and comparability of human and animal pharmacokinetics and metabolism: (1) human bioavailability and metabolism may differ significantly from that of animals; (2) mechanisms of toxicity may not be known (e.g., toxic accumulation in a peripheral compartment); and/ or (3) toxicity may be due to an unidentified metabolite, not the parent drug. Therefore, relying on pharmacokinetic models (based on the parent drug in plasma) to gauge starting doses would require multiple untested assumptions. Modeling can be used with greatest validity to estimate human starting doses in special cases where few underlying assumptions would be necessary. Such cases are exemplified by large molecular weight proteins (e.g., humanized monoclonal antibodies) that are intravenously administered, are removed from circulation by endocytosis rather than metabolism, have immediate and detectable effects on blood cells, and have a volume of distribution limited to the plasma volume. In these cases, allometric, pharmacokinetic, and pharmacodynamic models have been useful in identifying the human mg/kg dose that would be predicted to correlate with safe drug plasma levels in nonhuman primates. Even in these cases, uncertainties (such as differences between human and animal receptor sensitivity or density) have been shown to affect human pharmacologic or toxicologic outcomes, and the use of safety factors as described in this guidance is still warranted.

The guidance goes on to state:

> Toxicity should be avoided at the initial clinical dose. However, doses should be chosen that allow reasonably rapid attainment of the phase 1 trial objectives (e.g., assessment of the therapeutic's tolerability, pharmacodynamic or pharmacokinetic profile). All of the relevant preclinical data, including information on the pharmacologically active dose, the full toxicologic profile of the compound, and the pharmacokinetics (absorption, distribution, metabolism, and excretion) of the therapeutic, should be considered when determining the MRSD. Starting with doses lower than the MRSD is always an option and can be particularly appropriate to meet some clinical trial objectives.

These statements are confusing since they appear to encourage the sponsor to utilize nonclinical metabolism and pharmacokinetic data to establish the starting dose for the clinical trial, yet also make some strong caveats as to why the use of such data may not be appropriate.

The guidance further states that:

> Initial IND submissions for first-in-human studies by definition lack *in vivo* human data or formal allometric comparison of pharmacokinetics. Measurements of systemic levels or exposure (i.e., AUC or C_{max}) cannot be employed for setting a safe starting dose in humans, and it is critical to rely on dose and observed toxic response data from adequate and well-conducted toxicology studies. However, there are cases where nonclinical data on bioavailability, metabolite profile, and plasma drug levels associated with toxicity may influence the choice of the NOAEL. One such case is when saturation of drug absorption occurs at a dose that produces no toxicity. In this instance, the lowest saturating dose, not the highest (nontoxic) dose, should be used for calculating the HED.

The selection of the most appropriate species for calculating the MRSD is addressed thus:

> After the HEDs have been determined from the NOAELs from all toxicology studies relevant to the proposed human trial, the next step is to pick one HED for subsequent derivation of the MRSD. This HED should be chosen from the most appropriate species. In the absence of data on species relevance, a default position is that the most appropriate species for deriving the MRSD for a trial in adult healthy volunteers is the most sensitive species (i.e., the species in which the lowest HED can be identified). Factors that could influence the choice of the most appropriate species rather than the default to the most sensitive species include: (1) differences in the absorption, distribution, metabolism, and excretion (ADME) of the therapeutic between the species, and (2) class experience that may indicate a particular animal model is more predictive of human toxicity. Selection of the most appropriate species for certain biological products (e.g., human proteins) involves consideration of various factors unique to these products. Factors such as whether an animal species expresses relevant receptors or epitopes may affect species selection (refer to ICH guidance for industry S6 Preclinical Safety Evaluation of Biotechnology-Derived Pharmaceuticals for more details). When determining the MRSD for the first dose of a new therapeutic in humans, absorption, distribution, and elimination parameters will not be known for humans. Comparative metabolism data, however, might be available based on in vitro studies. These data are particularly relevant when there are marked differences in both the in vivo metabolite profiles and HEDs in animals."

These statements could be interpreted as implying that data on the ADME of the drug in animals are known at the time that the MRSD is calculated.

The FDA allows for the possibility of decreasing the safety factor used for the selection of the safe starting dose. The guidance states:

> Safety factors of less than 10 may be appropriate under some conditions. The toxicologic testing in these cases should be of the highest caliber in both conduct and design. Most of the time, candidate therapeutics for this approach would be members of a well-characterized class. Within the class, the therapeutics should be administered by the same route, schedule, and duration of administration; should have a similar metabolic profile and bioavailability; and should have similar toxicity profiles across all the species tested including humans.

However, it would appear unlikely that compounds of the same class of drugs would have a similar metabolic profile and bioavailability, even if one could agree what similar meant in this circumstance.

The subject of allometric scaling was raised in the FDA guidance, "Estimating the Maximum Safe Starting Dose in Initial Clinical Trials for Therapeutics in Adult Healthy Volunteers" (US FDA, 2005e). This topic is one for which there are no regulatory guidances and which can generate vigorous debate. Allometric scaling relates a pharmacokinetic parameter, such as clearance or volume of distribution, to total body weight (or other physiological parameter) using a power equation. The use of allometry makes many assumptions and has a mixed record in its ability to predict the pharmacokinetics of a drug in humans. It should be noted that allometry cannot actually be used to predict the pharmacokinetics of a drug specifically in humans since it is simply a means to relate a PK parameter to a physiological parameter such as body weight. In other words, allometry can be used to determine the PK parameters of a drug in a 70-kg animal, but that animal may be a human, a pig, or a small wildebeest. The nonclinical pharmacokineticist is often asked to predict the human pharmacokinetics of a drug before it goes into the clinic, although how such information is used will vary from company to company.

Some companies place great weight on the predicted human pharmacokinetics and expend great effort in trying to refine the calculations, while others use it to a far lesser extent and are comfortable with a broad estimate of the predicted human pharmacokinetics. It should be noted that pharmacokineticists may feel that their prediction of the pharmacokinetics was successful if the observed and the predicted results were within twofold (Rowland, 2007). It has been reported that even with such a wide acceptance range (i.e. the observed clearance is within 50–200% of the predicted clearance), the success rate is <53% (Nagilla and Ward, 2004, 2005a, b; Ward and Smith, 2004a, b; Mahmood, 2005). However, other groups within the company may expect that the predictions should be more accurate. No matter how much effort was expended to predict the human pharmacokinetics, the predicted pharmacokinetics will be of limited, if any, value as soon as the PK parameters are calculated from the first dose cohort in the FIH clinical trial. Despite these caveats, the predicted human PK parameters can be useful in helping to estimate the approximate efficacious dose (based on matching exposures in the animal efficacy model and the predicted human exposure at a given dose).

The predicted dose can be used to provide guidance to the formulation scientists in order for them to determine the appropriate dose strengths for the clinical trial materials (if a capsule or tablet is to be used) and to determine whether the dose will be feasible from a logistical or commercial perspective (e.g. if a dose of 500 mg t.i.d. is predicted, such a dose may have logistical and commercial challenges).

In some cases, the pharmacokineticist may perform a series of simulations in order to predict what the exposure to the parent drug will be at the dose levels that may be used in the FIH clinical trial. This information can be used to help select the doses to be used in the clinical trial, although the starting dose will most likely be based upon the toxicology data as indicated here. Additionally, these data may be used to give an estimate of the highest dose that may be administered in the FIH study if the toxicology data indicate that a dose cap is appropriate. In this scenario, the maximum dose would be re-evaluated on an ongoing basis based on the actual human pharmacokinetic data.

A simulation of the predicted human plasma concentration-versus-time curve could be useful in a couple of situations. Firstly, this information may be useful in the selection of the times at which to collect blood for quantitation of the plasma concentration of parent drug, although the selection of these time points is most commonly driven by logistical issues rather than by the predicted plasma concentration-versus-time curve. Secondly, and of greater utility, is the use of the simulated plasma concentrations to aid the bioanalytical scientist in the selection of the dynamic range that may be required for the human plasma assay and the potential dilution scheme that may be required for the analysis of the samples. It should be noted that the results of the simulation of the plasma concentration-versus-time curve will be dependent upon several assumptions including the linearity of the pharmacokinetics and the bioavailability of the drug. Usually, dose proportionality (linearity) of the PK parameters is assumed and the simulations are performed for a range of bioavailabilities.

As can be seen, the line between nonclinical and clinical ADME can be very fine, if it indeed exists at all. In some pharmaceutical companies, different groups will be responsible for the nonclinical and clinical work. In other companies, the same group will be responsible for both the nonclinical and clinical work.

The studies required to support FIH clinical trials is being addressed by the ICH as part of the maintenance of ICHM3 as ICHM3(R2). The EMEA is planning to issue a draft guideline in 4Q 2008 based on the 1Q 2006 concept paper, "Guideline on the Development of a CHMP Guideline on the Non-Clinical Requirements to Support Early Phase I Clinical Trials with Pharmaceutical Compounds" (EMEA, 2006d).

27.5 CLINICAL GUIDANCES AND CLINICAL TRIALS

Pharmacokinetics is a key element of clinical studies and is pivotal to understanding the relationship between dose or exposure and effect (either therapeutic or toxic). Like drug metabolism, pharmacokinetics is inherent in many guidances or guidelines and occasionally is the main subject of a guidance document. Since metabolism is one of the factors that shapes the pharmacokinetics of a drug, the drug metabolism scientist should pay attention to guidances that relate to pharmacokinetics as well as drug metabolism. The guidances that contain information about pharmacokinetics are many and varied. These guidances emphasize the importance of pharmacokinetics, the role of protein binding in understanding the PK or PK/PD relationships, the need to develop sensitive and specific assays, and the need to determine the pharmacokinetics of metabolites and/or enantiomers (as appropriate).

27.5.1 General

Many elements of drug metabolism and drug metabolites in clinical trials are covered in the ICH guidance, "E6(R1) Good Clinical Practice: Consolidated Guidance" that was issued in 1996 (ICH, 1996a). Among many other topics, this guidance describes the content of the investigator's brochure (IB). The guidance states that the IB should contain a brief summary (preferably not exceeding two pages), highlighting the significant physical, chemical, pharmaceutical, pharmacological, toxicological, pharmacokinetic, metabolic, and clinical information available that is relevant to the stage of clinical development of the investigational product. It also states that "The results of all relevant nonclinical pharmacology, toxicology, pharmacokinetic, and investigational product metabolism studies should be provided in summary form. This summary should address the methodology used, the results, and a discussion of the relevance of the findings to the investigated therapeutic and the possible unfavorable and unintended effects in humans." Further detail is provided for the different nonclinical disciplines as follows.

Nonclinical Pharmacology

A summary of the pharmacological aspects of the investigational product and, where appropriate, its significant metabolites studied in animals should be included. Such a summary should incorporate studies that assess potential therapeutic activity (e.g., efficacy models, receptor binding, and specificity) as well as those that assess safety (e.g., special studies to assess pharmacological actions other than the intended therapeutic effect(s)).

Pharmacokinetics and Product Metabolism in Animals

A summary of the pharmacokinetics and biological transformation and disposition of the investigational product in all species studied should be given. The discussion of the findings should address the absorption and the local and systemic bioavailability of the investigational product and its metabolites, and their relationship to the pharmacological and toxicological findings in animal species.

Pharmacokinetics and Product Metabolism in Humans

A summary of information on the pharmacokinetics of the investigational product(s) should be presented, including the following, if available: Pharmacokinetics (including metabolism, as appropriate, and absorption, plasma protein binding, distribution, and elimination).

It should be noted that the IB should be a summary of the relevant information and not an extensive overview of the complete knowledge of the ADME properties of the drug. Summarizing the ADME information in a few pages can be a challenge.

Further general points can be found in another ICH guidance, ICHE8, "General Considerations for Clinical Trials" that was issued in 1997 (ICH, 1997a). This guidance states that:

Pharmacokinetic studies are particularly important to assess the clearance of the drug and to anticipate possible accumulation of parent drug or metabolites and potential drug-drug interactions. Some pharmacokinetic studies are commonly conducted in later phases to answer more specialised questions. For many orally administered drugs, especially modified release products, the study of food effects on bioavailability is important. Obtaining pharmacokinetic information in sub-populations such as patients with impaired elimination (renal or hepatic failure), the elderly, children, women and ethnic subgroups should be considered. Drug-drug interaction studies are important for many drugs; these are generally performed in phases beyond Phase I but studies in animals and in vitro studies of metabolism and potential interactions may lead to doing such studies earlier.

The guidance also mentions some special considerations:

A number of special circumstances and populations require consideration on their own when they are part of the development plan.

Studies of Drug Metabolites

Major active metabolite(s) should be identified and deserve detailed pharmacokinetic study. Timing of the metabolic assessment studies within the development plan depends on the characteristics of the individual drug.

Drug-Drug Interactions

If a potential for drug–drug interaction is suggested by metabolic profile, by the results of non-clinical studies or by information on similar drugs, studies on drug interaction during clinical development are highly recommended. For drugs that are frequently co-administered it is usually important that drug-drug interaction studies be performed in non-clinical and, if appropriate in human studies. This is particularly true for drugs that are known to alter the absorption or metabolism of other drugs (see ICH E7), or whose metabolism or excretion can be altered by effects by other drugs.

Special Populations

Some groups in the general population may require special study because they have unique risk/benefit considerations that need to be taken into account during drug development, or because they can be anticipated to need modification of use of the dose or schedule of a drug compared to general adult use. Pharmacokinetic studies in patients with renal and hepatic dysfunction are important to assess the impact of potentially altered drug metabolism or excretion. Other ICH documents address such issues for geriatric patients (ICH E7) and patients from different ethnic groups (ICH E5). The need for non-clinical safety studies to support human clinical trials in special populations is addressed in the ICH M3 document.

Excretion of the drug or its metabolites into human milk should be examined where applicable. When nursing mothers are enrolled in clinical studies their babies should be monitored for the effects of the drug.

DDIs and PK in special populations will be addressed in greater depth later in this section. The mention of ICHE7 in the DDI section of the ICHE8 guidance is interesting since the main thrust of the ICHE7 guidance is the clinical evaluation of drugs in the elderly (ICH, 1994c).

The EMEA guideline (EMEA, 2005b), "Guideline on Risk Management Systems for Medicinal Products for Human Use" that was issued in 2005 states that:

Sometimes, potential risks or unforeseen benefits in special populations might be identified from preapproval clinical trials, but cannot be fully quantified due to small sample sizes or the exclusion of subpopulations of patients from these clinical studies. These populations might include the elderly, children, or patients with renal or hepatic disorder. Children, the elderly, and patients with co-morbid conditions might metabolize drugs differently from patients typically enrolled in clinical trials. Further clinical trials might be used to determine and to quantify the magnitude of the risk (or benefit) in such populations.

27.5.2 Pharmacokinetics and Pharmacodynamics

27.5.2.1 General One of the earliest EMEA guidelines related to drug metabolism was "Pharmacokinetic Studies in Man Legislative basis Directive 75/318/EEC as amended" which came into force in 1988 (European Economic Community, 1987). This directive contains some basic concepts such as "The percentage and characteristics of binding to serum proteins should be studied using appropriate ex vivo or in vitro methods. Particularly in the case of substances or their active metabolites of which a high percentage is bound to plasma proteins, factors which might alter protein binding and so alter therapeutic response should be studied. Binding to red blood cells and other blood components should also be known." Red blood cells and lipoproteins may be important factors for the disposition of certain drugs (Wasan and Cassidy, 1998; Wasan *et al.*, 2008). However, the contribution of these factors is rarely evaluated. Metabolism is addressed in the following manner in this guidance:

With a few exceptions substances are to a greater or lesser extent subject to metabolic breakdown within the human body. Pharmacokinetic studies should indicate whether the rate of biotransformation may be substantially modified in case of genetic enzymatic deficiency and whether within the dosage levels normally used, saturation of metabolism may occur, thereby inducing non-linear kinetics. The possibility of enzymatic induction should also be studied if metabolic clearance as a fraction of the systemic clearance is relatively high. If there is an indication that pharmacologically active metabolites (the qualitative activity of which may also occasionally differ from that of the parent substance) are formed, this should be ascertained and, if there is reason to suspect that they contribute to a significant extent to the therapeutic activity and/or adverse reactions in man, they should be examined in suitable animal models or if necessary in appropriate human clinical pharmacological studies. The pharmacokinetic data on such metabolites, the rate of their formation and elimination and their distribution and clearance characteristics should be known.

Excretion is another topic of relevance, although the directive only addresses urinary excretion and not fecal excretion. The reason for this omission is not clear, but it may be as a consequence of the lack of availability of assays to measure drug concentration in fecal samples and the lack of utility of such assays for orally administered drug. In this latter case, the quantitation of drug in the feces cannot provide insight as to whether the drug had been absorbed and eliminated in the bile without metabolism or simply had never been absorbed. The directive does make an interesting statement: "If urinary data are obtained, the urine should be collected until there is no further detectable excretion of parent substance or metabolites within the limits of the method used." What makes this statement interesting is that *a priori* you will not be able to determine how long it will take for the drug to be cleared and for there to be no further detectable excretion of parent substance or metabolites. Additionally, no statement is made as to whether the intent is to measure all metabolites or only major metabolites.

27.5.2.2 *Exposure–Response*

The contribution of metabolites to the exposure–response relationship is addressed in the 2003 FDA guidance, "Exposure-Response Relationships—Study Design, Data Analysis, and Regulatory Applications" (US FDA, 2003b). This guidance suggests that it is not only important to establish an exposure–response relationship but that it is also important to understand when there is no such response. The guidance states that "For example, it might be reassuring to observe that even patients with increased plasma concentrations (e.g., metabolic outliers or patients on other drugs in a study) do not have increased toxicity in general or with respect to a particular concern (e.g., QT prolongation)."

The concept that metabolism may be different depending upon the patient population is also introduced "In some cases, if there is a change in the mix of parent and active metabolites from one population (e.g., pediatric vs. adult), dosage form (e.g., because of changes in drug input rate), or route of administration, additional exposure-response data with short-term endpoints can support use in the new population, the new product, or new route without further clinical trials." However, the guidance does not address how the change in the mix of parent and active metabolites is determined. The implication is that active metabolites are identified in adults and then the exposure to these metabolites determined in the pediatric population. The role of drug metabolism and drug metabolites in the pediatric population is discussed later in this section.

The guidance also addresses the "Chemical Moieties for Measurement." For active moieties it states:

To the extent possible, it is important that exposure-response studies include measurement of all active moieties (parent and active metabolites) that contribute significantly to the effects of the drug. This is especially important when the route of administration of a drug is changed, as different routes of administration can result in different proportions of parent compound and metabolites in plasma. Similarly, hepatic or renal impairment or concomitant drugs can alter the relative proportions of a drug and its active metabolites in plasma.

The role of drug metabolism and drug metabolites in renal and hepatic impairment is discussed later in this section. This guidance states that "Blood concentrations can also be helpful when both parent drug and metabolites are active." It should be noted that this guidance seems to use the terms "blood concentrations" and "plasma concentrations" interchangeably, and does not seem to be attempting to imply that drug concentrations should be measured in whole blood samples.

27.5.2.3 *Bioequivalence*
One of the major topics for regulatory guidances that relate to clinical pharmacology is bioequivalence (BE). Such studies are required for the approval of generic products or when changes are made to formulations during the course of development of the originator product. BE studies are used to "bridge" between data sets and should only be conducted when absolutely necessary since they are highly regulated studies with rigid acceptance criteria. If a BE study fails to show comparability, there may be delays in the development program. At other times, a relative bioavailability study (without rigid acceptance criteria) may be acceptable. For example, if a phase I study is performed with a capsule formulation and the phase IIa study will be performed with a tablet formulation, a BE study would most likely not be required. Instead, a simple comparison of the relative BA of the two formulations (in a study design similar to a BE study but with fewer subjects and less stringent acceptance criteria) would enable the sponsor to determine whether any dosage adjustment is required in the planned doses for the phase IIa study. One such guidance was issued by the FDA in 2003, "Bioavailability and Bioequivalence Studies for Orally Administered Drug Products—General Considerations" (US FDA, 2003c). In the United States, bioequivalence (BE) is defined as the absence of a significant difference in the rate and extent to which an active ingredient or active moiety in pharmaceutical equivalents or pharmaceutical alternatives becomes available at the same molar dose under similar conditions in an appropriately designed study. The FDA's 2003 guidance states that "BA can be generally documented by a systemic exposure profile obtained by measuring drug and/or metabolite concentration in the systemic circulation over time." The guidance further states that:

> BA for orally administered drug products can be documented by developing a systemic exposure profile. A profile can be obtained by measuring the concentration of active ingredients and/or active moieties and, when appropriate, its active metabolites over time in samples collected from the systemic circulation. Systemic exposure patterns reflect both release of the drug substance from the drug product and a series of possible presystemic/systemic actions on the drug substance after its release from the drug product. We recommend that additional comparative studies be performed to understand the relative contribution of these processes to the systemic exposure pattern.

The topic of the quantitation of parent drug or metabolites is also covered:

> The moieties to be measured in biological fluids collected in BA and BE studies are either the active drug ingredient or its active moiety in the administered dosage form (parent

drug) and, when appropriate, its active metabolites (21 CFR 320.24(b)(1)(i)). This guidance recommends the following approaches for BA and BE studies.

For BA studies, we recommend that determination of moieties to be measured in biological fluids take into account both concentration and activity. Concentration refers to the relative quantity of the parent drug or one or more metabolites in a given volume of an accessible biological fluid such as blood or plasma. Activity refers to the relative contribution of the parent drug and its metabolite(s) in the biological fluids to the clinical safety and/or efficacy of the drug. For BA studies, we also recommend that both the parent drug and its major active metabolites be measured, if analytically feasible.

For BE studies, measurement of only the parent drug released from the dosage form, rather than the metabolite, is generally recommended. The rationale for this recommendation is that concentration-time profile of the parent drug is more sensitive to changes in formulation performance than a metabolite, which is more reflective of metabolite formation, distribution, and elimination. The following are exceptions to this general approach.

- Measurement of a metabolite may be preferred when parent drug levels are too low to allow reliable analytical measurement in blood, plasma, or serum for an adequate length of time. We recommend that the metabolite data obtained from these studies be subject to a confidence interval approach for BE demonstration.

- A metabolite may be formed as a result of gut wall or other presystemic metabolism. If the metabolite contributes meaningfully to safety and/or efficacy, we also recommend that the metabolite and the parent drug be measured. When the relative activity of the metabolite is low and does not contribute meaningfully to safety and/or efficacy, it does not have to be measured. We recommend that the parent drug measured in these BE studies be analyzed using a confidence interval approach. The metabolite data can be used to provide supportive evidence of comparable therapeutic outcome.

The guidance also states that "We recommend that under normal circumstances, blood, rather than urine or tissue, be used. In most cases, drug, or metabolites are measured in serum or plasma."

Racemates and enantiomers are discussed in this guidance, "Many drugs are optically active and are usually administered as the racemate. Enantiomers sometimes differ in both their pharmacokinetic and pharmacodynamic properties. Early elucidation of the PK and PD properties of the individual enantiomers can help in designing a dosing regimen and in deciding whether it can be of value to develop one of the pure enantiomers as the final drug product."

The significance of unbound drug and/or active metabolite (protein binding) is also discussed:

Most standard assays of drug concentrations in plasma measure the total concentration, consisting of both bound and unbound drug. Renal or hepatic diseases can alter the binding of drugs to plasma proteins. These changes can influence the understanding of PK and PK-PD relationships. Where feasible, studies to determine the extent of protein binding and to understand whether this binding is or is not concentration-dependent are important, particularly when comparing responses in patient groups that can exhibit different plasma protein binding (e.g., in various stages of hepatic and renal disease). For highly protein bound drugs, PK and PK-PD modeling based on unbound drug concentrations may be more informative, particularly if there is significant variation in binding among patients or in special populations of patients.

This guidance does not give any insight into what would constitute a "significant difference in binding between individuals." Some of the issues and challenges with plasma protein binding have been addressed previously. In addition to this guidance, bioequivalence studies are captured in the CFRs (21 CFR Part 320) (US FDA, 2007c).

Bioequivalence is a topic that is addressed extensively in the FDA's Critical Path Initiative documents, indicating that the pharmaceutical industry will likely receive more guidance on this topic (US FDA, 2008d).

At the present time, the EMEA is in the process of revising its guideline "Note for Guidance on the Investigation of Bioavailability and Bioequivalence CPMP/EWP/QWP/1401/98" that came into force in 2002 and replaced a previous guideline that was issued in 1991 (EMEA, 2001). This guideline includes mention of metabolites in the section "Characteristics to be Investigated." A Q&A document that included a discussion of the use of metabolites in BE testing was issued for this guideline in 2006 (EMEA, 2006j). The comment period on the document "Recommendation on the Need for Revision of ⟨Note for Guidance on the Investigation of Bioavailability and Bioequivalence⟩ CPMP/EWP/QWP/1401/98" ended in August 2007 (EMEA, 2007h). The topics addressed in the Q&A document are addressed in this latter document.

Points of interest in the Q&A and recommendation document include the following.

When should metabolite data be used to establish bioequivalence?

"According to the guideline, the only situations where metabolite data can be used to establish bioequivalence are if the concentration of the active substance is too low to be accurately measured in the biological matrix, thus giving rise to significant variability."

Comments: Metabolite data can only be used if the Applicant presents convincing, state-of-the art arguments that measurements of the parent compound are unreliable. Even so, it is important to point out that C_{max} of the metabolite is less sensitive to differences in the rate of absorption than C_{max} of the parent drug. Therefore, when the rate of absorption is considered of clinical importance, bioequivalence should, if possible, be determined for C_{max} of the parent compound, if necessary at a higher dose. Furthermore, when using metabolite data as a substitute for parent drug concentrations, the applicant should present data supporting the view that the parent drug exposure will be reflected by metabolite exposure.

If metabolites significantly contribute to the net activity of an active substance and the phar-macokinetic system is non-linear.

Comments: To evaluate the significance of the contribution of metabolites, relative AUCs and non-clinical or clinical pharmacodynamic activities should be compared with those of the parent drug. PK/PD modeling may be useful. If criteria for significant contribution to activity and pharmacokinetic non-linearity are met, then "it is necessary to measure both parent drug and active metabolite plasma concentrations and evaluate them separately." Any discrepancy between the results obtained with the parent compound and the metabo-lites should be discussed based on relative activities and AUCs. If the discrepancy lies in C_{max}, the results of the parent compound should usually prevail. Pooling of the plasma concentrations or pharmacokinetic parameters of the parent drug and its metabolite for calculation of bioequivalence is not acceptable.

When using metabolite data to establish bioequivalence, may one use the same justification for widening the C_{max} acceptance criteria as in the case of the parent compound?

Comments: In principle, the same criteria apply as for the parent drug (see Question on widening the acceptance range for C_{max}). However, as stated above (see Question regarding when metabolite data can be used), C_{max} of the metabolite is less sensitive to differences in the rate of absorption than C_{max} of the parent drug. Therefore, widening the C_{max} acceptance range when using metabolites instead of the parent compound is generally not accepted. When the metabolite has a major contribution to, or is completely responsible for, the therapeutic effect, and if it can be demonstrated that a widened acceptance range would not lead to any safety or efficacy concerns, which will usually prove more difficult than for the parent compound (see Question on widening the acceptance range for C_{max}), then a widened acceptance range for C_{max} of metabolite may be accepted.

It is interesting to note that the original guideline (Note for Guidance on the Investigation of Bioavailability and Bioequivalence CPMP/EWP/QWP/1401/98) (EMEA, 2001) is not located in the Efficacy section of the EMEA guidelines are as the subsequent documents that pertain to bioequivalence, although a hyperlink to the original guideline can be found in the Efficacy guidelines section. Instead, it resides in the Quality section of the guidelines.

The Q&A document also mentions the subject of bioequivalence of highly variable drugs. This topic has also been addressed in a discussion document by Health Canada (2003), at the Clinical Pharmacology Subcommittee of the FDA's Pharmaceutical Science Advisory Committee (US FDA, 2006e)and in recent publications by the FDA (Haidar *et al.*, 2008; Davit *et al.*, 2008). Haidar *et al.* (2008) summarize the challenges for highly variable drugs by stating that "Over the past decade, concerns have been expressed increasingly regarding the difficulty for highly variable drugs and drug products (%CV greater than 30) to meet the standard bioequivalence (BE) criteria using a reasonable number of study subjects. The topic has been discussed on numerous occasions at national and international meetings. Despite the lack of a universally accepted solution for the issue, regulatory agencies generally agree that an adjustment of the traditional BE limits for these drugs or products may be warranted to alleviate the resource burden of studying relatively large numbers of subjects in bioequivalence trials." Davit *et al.* (2008) reviewed 1010 acceptable bioequivalence studies for 180 drugs that were submitted to the FDA between 2003 and 2005. Of these drugs, 57/180 (31%) were highly variable. They concluded that about 60% of the highly variable drugs were highly variable due to drug substance pharmacokinetic characteristics. For about 20% of the highly variable drugs, it appeared that formulation performance contributed to the high variability. For the remaining 20% of the highly variable drugs, it was not possible to identify the factors that contributed to the high variability.

The subject of the effect of food on the pharmacokinetics of a drug has been addressed in the FDA's 2002 guidance, "Food-effect Bioavailability and Fed Bioequivalence Studies" and in the EMEA's 2000 guidance "Note for Guidance on Modified Release Oral and Transdermal Dosage Forms: Section II (Pharmacokinetic and Clinical Evaluation)" (EMEA, 1999; US FDA, 2002b). The FDA guidance notes that "Food can change the BA of a drug and can influence the BE between test and reference products. Food effects on BA can have clinically significant consequences. Food can alter BA by various means, including:

- Delay gastric emptying
- Stimulate bile flow
- Change gastrointestinal (GI) pH

- Increase splanchnic blood flow
- Change luminal metabolism of a drug substance
- Physically or chemically interact with a dosage form or a drug substance"

The FDA's guidance also describes a standard test meal which is a high-fat (approximately 50% of total caloric content of the meal) and high-calorie (approximately 800 to 1000 calories) meal. This test meal should derive approximately 150, 250, and 500–600 calories from protein, carbohydrate, and fat, respectively. An example test meal would be two eggs fried in butter, two strips of bacon, two slices of toast with butter, four ounces of hash brown potatoes, and eight ounces of whole milk. Substitutions in this test meal can be made as long as the meal provides a similar amount of calories from protein, carbohydrate, and fat, and has comparable meal volume and viscosity. This meal represents an "extreme" in terms of diet and will not reflect the standard meal in most places around the world.

Bioequivalence studies can be costly and take a long time to conduct. Dependent upon the physicochemical properties of the drugs, there may be a scientific rationale that such studies may not be required for certain drugs. This concept is embodied in the FDA's guidance, "Waiver of In Vivo Bioavailability and Bioequivalence Studies for Immediate-Release Solid Oral Dosage Forms Based on a Biopharmaceutics Classification System" (US FDA, 2000b).

> The BCS is a scientific framework for classifying drug substances based on their aqueous solubility and intestinal permeability. When combined with the dissolution of the drug product, the BCS takes into account three major factors that govern the rate and extent of drug absorption from IR solid oral dosage forms: dissolution, solubility, and intestinal permeability.
>
> The BCS classifies drug substances as follows:
>
> > Class I—high permeability, high solubility
> > Class II—high permeability, low solubility
> > Class III—low permeability, high solubility
> > Class IV—low permeability, low solubility.
>
> In addition, IR solid oral dosage forms are categorized as having rapid or slow dissolution. Within this framework, when certain criteria are met, the BCS can be used as a drug development tool to help sponsors justify requests for biowaivers for highly soluble and highly permeable drug substances (i.e. Class I) in IR solid oral dosage forms that exhibit rapid *in vitro* dissolution using the recommended test methods.

The drug metabolism scientist/pharmacokineticist should play a role in creating the application for a biowaiver by providing information on the permeability of the drug. However, this is not always the case and often the formulation scientist will make the assessment independent of input from ADME specialists. This is an area in which there should be close collaboration between the formulation and ADME groups.

The FDA guidance states:

> The permeability class boundary is based indirectly on the extent of absorption (fraction of dose absorbed, not systemic BA) of a drug substance in humans and directly on measurements of the rate of mass transfer across human intestinal membrane. Alternatively, nonhuman systems capable of predicting the extent of drug absorption in humans can be

used (e.g., in vitro epithelial cell culture methods). In the absence of evidence suggesting instability in the gastrointestinal tract, a drug substance is considered to be *highly permeable* when the extent of absorption in humans is determined to be 90% or more of an administered dose based on a mass balance determination or in comparison to an intravenous reference dose.

The guidance provides extensive details regarding the determination of the drug substance permeability class:

The permeability class of a drug substance can be determined in human subjects using mass balance, absolute BA, or intestinal perfusion approaches. Recommended methods not involving human subjects include in vivo or in situ intestinal perfusion in a suitable animal model (e.g., rats), and/or in vitro permeability methods using excised intestinal tissues, or monolayers of suitable epithelial cells. In many cases, a single method may be sufficient (e.g., when the absolute BA is 90% or more, or when 90% or more of the administered drug is recovered in urine). When a single method fails to conclusively demonstrate a permeability classification, two different methods may be advisable. Chemical structure and/or certain physicochemical attributes of a drug substance (e.g., partition coefficient in suitable systems) can provide useful information about its permeability characteristics. Sponsors may wish to consider use of such information to further support a classification.

1. Pharmacokinetic Studies in Humans

  a. Mass Balance Studies
     Pharmacokinetic mass balance studies using unlabeled, stable isotopes or a radiolabeled drug substance can be used to document the extent of absorption of a drug. Depending on the variability of the studies, a sufficient number of subjects should be enrolled to provide a reliable estimate of extent of absorption. Because this method can provide highly variable estimates of drug absorption for many drugs, other methods described below may be preferable.

  b. Absolute Bioavailability Studies
     Oral BA determination using intravenous administration as a reference can be used. Depending on the variability of the studies, a sufficient number of subjects should be enrolled in a study to provide a reliable estimate of the extent of absorption. When the absolute BA of a drug is shown to be 90% or more, additional data to document drug stability in the gastrointestinal fluid is not necessary.

2. Intestinal Permeability Methods

The following methods can be used to determine the permeability of a drug substance from the gastrointestinal tract: (1) in vivo intestinal perfusion studies in humans; (2) in vivo or in situ intestinal perfusion studies using suitable animal models; (3) in vitro permeation studies using excised human or animal intestinal tissues; or (4) in vitro permeation studies across a monolayer of cultured epithelial cells.

In vivo or in situ animal models and in vitro methods, such as those using cultured monolayers of animal or human epithelial cells, are considered appropriate for passively transported drugs. The observed low permeability of some drug substances in humans could be caused by efflux of drugs via membrane transporters such as P-glycoprotein (P-gp). When the efflux transporters are absent in these models, or their degree of expression is low compared to that in humans, there may be a greater likelihood of misclassification of permeability class for a drug subject to efflux compared to a drug transported passively. Expression of known transporters in selected study systems should be characterized.

Functional expression of efflux systems (e.g., P-gp) can be demonstrated with techniques such as bidirectional transport studies, demonstrating a higher rate of transport in the basolateral-to-apical direction as compared to apical-to-basolateral direction using selected model drugs or chemicals at concentrations that do not saturate the efflux system (e.g., cyclosporin A, vinblastine, rhodamine 123). An acceptance criterion for intestinal efflux that should be present in a test system cannot be set at this time. Instead, this guidance recommends limiting the use of nonhuman permeability test methods for drug substances that are transported by passive mechanisms. Pharmacokinetic studies on dose linearity or proportionality may provide useful information for evaluating the relevance of observed in vitro efflux of a drug. For example, there may be fewer concerns associated with the use of in vitro methods for a drug that has a higher rate of transport in the basolateral-to-apical direction at low drug concentrations but exhibits linear pharmacokinetics in humans.

For application of the BCS, an apparent passive transport mechanism can be assumed when one of the following conditions is satisfied:

- A linear (pharmacokinetic) relationship between the dose (e.g., relevant clinical dose range) and measures of BA (area under the concentration-time curve) of a drug is demonstrated in humans

- Lack of dependence of the measured in vivo or in situ permeability is demonstrated in an animal model on initial drug concentration (e.g., 0.01, 0.1, and 1 times the highest dose strength dissolved in 250 ml) in the perfusion fluid

- Lack of dependence of the measured in vitro permeability on initial drug concentration (e.g., 0.01, 0.1, and 1 times the highest dose strength dissolved in 250 ml) is demonstrated in donor fluid and transport direction (e.g., no statistically significant difference in the rate of transport between the apical-to-basolateral and basolateral-to-apical direction for the drug concentrations selected) using a suitable in vitro cell culture method that has been shown to express known efflux transporters (e.g., P-gp).

To demonstrate suitability of a permeability method intended for application of the BCS, a rank-order relationship between test permeability values and the extent of drug absorption data in human subjects should be established using a sufficient number of model drugs. For in vivo intestinal perfusion studies in humans, six model drugs are recommended. For in vivo or in situ intestinal perfusion studies in animals and for in vitro cell culture methods, twenty model drugs are recommended. Depending on study variability, a sufficient number of subjects, animals, excised tissue samples, or cell monolayers should be used in a study to provide a reliable estimate of drug permeability. This relationship should allow precise differentiation between drug substances of low and high intestinal permeability attributes.

For demonstration of suitability of a method, model drugs should represent a range of low (e.g., <50%), moderate (e.g., 50–89%), and high (≥90%) absorption. Sponsors may select compounds from the list of drugs and/or chemicals provided in Attachment A or they may choose to select other drugs for which there is information available on mechanism of absorption and reliable estimates of the extent of drug absorption in humans.

After demonstrating suitability of a method and maintaining the same study protocol, it is not necessary to retest all selected model drugs for subsequent studies intended to classify a drug substance. Instead, a low and a high permeability model drug should be used as internal standards (i.e., included in the perfusion fluid or donor fluid along with the test drug substance). These two internal standards are in addition to the fluid volume marker (or a zero permeability compound such as PEG 4000) that is included in certain types of perfusion techniques (e.g., closed loop techniques). The choice of internal standards should be based on compatibility with the test drug substance (i.e., they should not exhibit any significant physical, chemical, or permeation interactions). When it is not feasible to follow this protocol, the permeability of internal standards should be determined in the same subjects, animals, tissues, or monolayers, following evaluation of the test drug substance. The permeability values of the two internal standards should not differ significantly between different tests, including those conducted to demonstrate suitability of the method. At the end of an in situ or in vitro test, the amount of drug in the membrane should be determined.

For a given test method with set conditions, selection of a high permeability internal standard with permeability in close proximity to the low/high permeability class boundary may facilitate classification of a test drug substance. For instance, a test drug substance may be determined to be highly permeable when its permeability value is equal to or greater than that of the selected internal standard with high permeability.

3. Instability in the Gastrointestinal Tract

Determining the extent of absorption in humans based on mass balance studies using total radioactivity in urine does not take into consideration the extent of degradation of a drug in the gastrointestinal fluid prior to intestinal membrane permeation. In addition, some methods for determining permeability could be based on loss or clearance of a drug from fluids perfused into the human and/or animal gastrointestinal tract either in vivo or in situ. Documenting the fact that drug loss from the gastrointestinal tract arises from intestinal membrane permeation, rather than a degradation process, will help establish permeability. Stability in the gastrointestinal tract may be documented using gastric and intestinal fluids obtained from human subjects. Drug solutions in these fluids should be incubated at 37°C for a period that is representative of in vivo drug contact with these fluids; for example, 1 hour in gastric fluid and 3 hours in intestinal fluid. Drug concentrations should then be determined using a validated stability-indicating assay method. Significant degradation (>5%) of a drug in this protocol could suggest potential instability. Obtaining gastrointestinal fluids from human subjects requires intubation and may be difficult in some cases. Use of gastrointestinal fluids from suitable animal models and/or simulated fluids such as Gastric and Intestinal Fluids USP can be substituted when properly justified.

In May 2007, the CHMP issued a concept paper (Concept Paper on BCS-Based Biowaiver) for review (EMEA, 2007b). This concept paper describes some concerns that have risen regarding the use of the BCS system and aims to provide clarification leading to a harmonized approach to the use of the BCS-based biowaiver approach.

27.5.3 Phase 0 studies, Microdosing, and the Exploratory IND

In the past decade, a new phase of clinical development has been described. This phase of development is known as phase 0 and phase 0 studies are conducted before the traditional phase I studies are conducted. There is no universal definition of a phase 0 study, but the definition can include:

- Designed to study the PK/PD of a drug
- Studies a limited number of doses
- Studies lower doses than a traditional phase I study
- Utilizes fewer patients compared to a phase I study
- An evaluation of disease biomarkers.

Phase 0 studies are uncommon but can play a key role in the development of a drug. Phase 0 studies may be conducted under an exploratory IND which is a concept that came out of the FDA's "Critical Path Report" that was issued in 2004 and in which the Agency explained that to reduce the time and resources expended on candidate products that are unlikely to succeed, new tools are needed to distinguish earlier in the process those candidates that hold promise from those that do not (US FDA, 2004a). Prior to the issuance of a draft guidance on exploratory IND studies in April 2005 (US FDA, 2005b), the concept was known as a "screening IND." A MAPP (Manual of Policies and Procedures) was issued for screening INDs in 2001 (US FDA, 2001d). Over 350 comments were submitted to the docket. In 2006, the FDA issued the final guidance, "Exploratory IND Studies" that "describes some early phase 1 exploratory approaches that are consistent with regulatory requirements while maintaining needed human subject protection, but that involve fewer resources than is customary, enabling sponsors to move ahead more efficiently with the development of promising candidates" (US FDA, 2006f). Although this guidance does contain some procedural elements, it is located in the Pharm/Tox section of the FDA guidance page. The topic of exploratory INDs has recently been reviewed by Sarapa (2007).

The FDA's guidance contains the following introductory statement:

This guidance is intended to clarify what preclinical and clinical approaches, as well as chemistry, manufacturing, and controls information, should be considered when planning exploratory studies in humans, including studies of closely related drugs or therapeutic biological products, under an investigational new drug (IND) application (21 CFR 312). Existing regulations allow a great deal of flexibility in the amount of data that needs to be submitted with an IND application, depending on the goals of the proposed investigation, the specific human testing proposed, and the expected risks. The Agency believes that sponsors have not taken full advantage of that flexibility and often provide more supporting information in INDs than is required by regulations. This guidance is intended to clarify what manufacturing controls, preclinical testing, and clinical approaches can be considered when planning limited, early exploratory IND studies in humans.

For the purposes of this guidance the phrase *exploratory IND study* is intended to describe a clinical trial that:

— is conducted early in phase 1
— involves very limited human exposure, and
— has no therapeutic or diagnostic intent (e.g., screening studies, microdose studies).

Such exploratory IND studies are conducted prior to the traditional dose escalation, safety, and tolerance studies that ordinarily initiate a clinical drug development program. The duration of dosing in an exploratory IND study is expected to be limited (e.g., 7 days). This guidance applies to early phase 1 clinical studies of investigational new drug and biological products that assess feasibility for further development of the drug or biological product.

The guidance describes a traditional IND approach in order to contrast it to an exploratory IND approach:

Typically, during pharmaceutical development, large numbers of molecules are generated with the goal of identifying the most promising candidates for further development. These molecules are generally structurally related, but can differ in important ways. Promising candidates are often selected using in vitro testing models that examine binding to receptors, effects on enzyme activities, toxic effects, or other in vitro pharmacologic parameters; these tests usually require only small amounts of the drug. Candidates that are not rejected during these early tests are prepared in greater quantities for in vivo animal testing for efficacy and safety. Commonly, a single candidate is selected for an IND application and introduction into human subjects, initially healthy volunteers in most cases.

Before the human studies can begin, an IND must be submitted to the Agency containing, among other things, information on any risks anticipated based on the results of pharmacologic and toxicological data collected during studies of the drug in animals (21 CFR 312.23(a)(8)). These basic safety tests are most often performed in rats and dogs. The studies are designed to permit the selection of a safe starting dose for humans, to gain an understanding of which organs may be the targets of toxicity, to estimate the margin of safety between a clinical and a toxic dose, and to predict pharmacokinetic and pharmacodynamic parameters. These early tests are usually resource intensive, requiring significant investment in product synthesis, animal use, laboratory analyses, and time. Many resources are invested in, and thus wasted on, candidate products that subsequently are found to have unacceptable profiles when evaluated in humans—less than 10 percent of INDs for new molecular entities (NME) progress beyond the investigational stage to submission of a marketing application (NDA). In addition, animal testing does not always predict performance in humans, and potentially effective candidates may not be developed because of resource constraints.

Existing regulations allow a great deal of flexibility in terms of the amount of data that need to be submitted with any IND application, depending on the goals of the proposed investigation, the specific human testing proposed, and the expected risks. The Agency believes that sponsors have not taken full advantage of that flexibility. As a result, limited, early phase 1 studies, such as those described in this guidance, are often supported by a more extensive preclinical database than is required by the regulations.

This guidance describes preclinical and clinical approaches, and the chemistry, manufacturing, and controls information that should be considered when planning exploratory IND studies in humans, including studies of closely related drugs or therapeutic biological products, under a single IND application (21 CFR 312).

The Exploratory IND approach is described thus:

Exploratory IND studies usually involve very limited human exposure and have no therapeutic or diagnostic intent. Such studies can serve a number of useful goals. For example, an exploratory IND study can help sponsors

- Determine whether a mechanism of action defined in experimental systems can also be observed in humans (e.g., a binding property or inhibition of an enzyme)
- Provide important information on pharmacokinetics (PK)
- Select the most promising lead product from a group of candidates designed to interact with a particular therapeutic target in humans, based on PK or pharmacodynamic (PD) properties
- Explore a product's biodistribution characteristics using various imaging technologies

One benefit that the FDA identifies for the use of an exploratory IND is the ability to select "the most promising lead product from a group of candidates" based on PK or PD. In principle, this would be advantageous. However, in practice it is rare that a sponsor would have several potential clinical candidates that could be evaluated simultaneously in the clinic and for which there were no distinguishing features based on nonclinical data. Under these circumstances, it would be beneficial to select the candidate that had the best PK or PD properties in humans. It should be noted that even if the compound with the best PK or PD properties were selected from such a "screening" IND, it would not necessarily mean that the selected compound would not fail in a traditional phase I study that would be conducted subsequently. Since the intent of studies conducted an exploratory IND is not to cause toxicity, the exploratory IND clinical trial could not "weed out" compounds that may show toxicity at doses higher than those studies in the exploratory IND study and could only select compounds that had suboptimal PK or PD properties. It should be noted that if a drug were to fail due to inappropriate PK properties, it would fail regardless of whether an exploratory or traditional IND approach was used. As mentioned earlier, it is unlikely that all of the compounds would be available at exactly the same time and that the filing of the exploratory IND or the clinical study that is to be conducted under the exploratory IND may be delayed until all of the compounds had been synthesized and the appropriate studies conducted. This may introduce a delay into the overall development plan. Even if the company intended to take, for example, three compounds into an exploratory IND, it may be likely that one or more of them drops out of contention based on data generated in the toxicology studies that are conducted to support the exploratory IND filing. If this were to happen, some of the benefits of conducting a clinical trial under and exploratory IND would be lost. Although the costs of taking one compound into a clinical trial to be conducted under an exploratory IND will be lower than the costs of taking one compound into a clinical trial to be conducted under a traditional IND, the cost savings will dwindle as more compounds are included in the exploratory IND. Therefore, the sponsor should only consider evaluating those compounds which may have properties that would suggest that they could become marketed drugs.

The guidance further states:

Whatever the goal of the study, exploratory IND studies can help identify, early in the process, promising candidates for continued development and eliminate those lacking promise. As a result, exploratory IND studies may help reduce the number of human subjects and resources, including the amount of candidate product, needed to identify promising drugs. The studies discussed in this guidance involve dosing a limited number of subjects with a limited range of doses for a limited period of time. Existing regulations provide more flexibility with regard to the preclinical testing requirements for exploratory IND studies than for traditional IND studies. However, sponsors submitting the kinds of studies described in this guidance have not always taken full advantage of that flexibility. Sponsors often provide more supporting information in their INDs than is required by the regulations. Because exploratory IND studies involve administering either sub-pharmacologic doses of a product, or doses expected to produce a pharmacologic, but not a toxic, effect, the potential risk to human subjects is less than for a traditional phase 1 study that, for example, seeks to establish a maximally tolerated dose. Because exploratory IND studies present fewer potential risks than do traditional phase 1 studies that look for dose-limiting toxicities, such limited exploratory IND investigations in humans can be initiated with less, or different, preclinical support than is required for traditional IND studies.

The Agency expects that this early phase 1 exploratory IND approach will apply to a number of different study paradigms. Although his guidance explores several potential applications, many others can be proposed. The Agency believes that, consistent with its Critical Path Initiative, clarifying Agency thinking about how much and what kind of testing is needed to support early studies in humans will facilitate the entry of new products into clinical testing and speed product development.

Although exploratory IND studies may be used during development of products intended for any indication, it is particularly important for manufacturers to consider this approach when developing products to treat serious diseases. Because the approach can help identify promising candidates more quickly and precisely, exploratory IND studies could become an important part of the armamentarium when developing drug and biological products to treat a serious or life-threatening illness. The Agency has previously articulated its commitment to ensuring that appropriate flexibility is applied when patients with a serious disease and no satisfactory alternative therapies are enrolled in a trial with therapeutic intent.

The guidance described types of studies that can be conducted under the exploratory IND. "Potentially useful study designs include both single- and multiple-dose studies. In single-dose studies, a sub-pharmacologic or pharmacologic dose is administered to a limited number of subjects (healthy volunteers or patients). For example, microdose studies usually involve the single administration of a small dose with the goal of collecting pharmacokinetic information or performing imaging studies, or both." A footnote gives further clarification of a sub-pharmacologic study, "A radiolabeled candidate compound can be administered at doses that are known to have no pharmacologic effect in humans without an IND application in basic research studies when the compound has previously been studied in humans and the results published in the literature. These basic research investigations are conducted under the oversight of an institutional review board (IRB) and a radioactive drug research committee (21 CFR 361.1)." This statement could be interpreted as implying that a microdose study could be conducted without an IND, although to do so would require that the sponsor know that the doses would have no pharmacological effect.

Repeat-dose clinical studies can be designed with pharmacologic or pharmacodynamic end points. In exploratory IND studies, the duration of dosing should be limited (e.g. seven days). For escalating dose studies done under an exploratory IND, dosing should be designed to investigate a pharmacodynamic end point, not to determine the limits of tolerability.

A large part of the guidance is devoted to a description of the design of safety evaluation programs. A general statement is made regarding the nonclinical safety assessment program followed by specific examples for different clinical study designs:

Pharmacology and toxicology information is derived from preclinical safety testing performed in animals and in vitro. Preclinical studies for small molecules are described in ICH M3 while those for biologics follow guidance described in ICH S6. Some of the toxicology tests described in this guidance may not be appropriate for biologics. The toxicology evaluation recommended for an exploratory IND application is more limited than for a traditional IND application. The basis for the reduced preclinical package is the reduced scope of an exploratory IND clinical study. Although exploratory IND studies in some cases are expected to induce pharmacologic effects, they are not designed to establish maximally tolerated doses. Furthermore, the duration of drug exposure in exploratory IND studies is limited. The level of preclinical testing performed to ensure safety will depend on the scope and intended goals of the clinical trials.

There are a number of study objectives for which the preclinical safety programs may be tailored to the study design. Examples include: confirming that an expected mechanism of action can be observed in humans; measuring binding affinity or localization of drug; assessing PK and metabolism; comparing the effect on a potential therapeutic target with other therapies.

Clinical studies of pharmacokinetics or imaging

Microdose studies are designed to evaluate pharmacokinetics or imaging of specific targets and are designed not to induce pharmacologic effects. Because of this, the risk to human subjects is very limited, and information adequate to support the initiation of such limited human studies can be derived from limited nonclinical safety studies. A *microdose* is defined as less than 1/100th of the dose of a test substance calculated (based on animal data) to yield a pharmacologic effect of the test substance with a maximum dose of <100 micrograms (for imaging agents, the latter criterion applies). Due to differences in molecular weights as compared to synthetic drugs, the maximum dose for protein products is ≤30 nanomoles.

FDA currently accepts the use of extended single-dose toxicity studies in animals to support single-dose studies in humans. For microdose studies, a single mammalian species (both sexes) can be used if justified by in vitro metabolism data and by comparative data on in vitro pharmacodynamic effects. The route of exposure in animals should be by the intended clinical route. In these studies, animals should be observed for 14 days post-dosing with an interim necropsy, typically on day 2, and endpoints evaluated should include body weights, clinical signs, clinical chemistries, hematology, and histopathology (high dose and control only if no pathology is seen at the high dose). The study should be designed to establish a dose inducing a minimal toxic effect, or alternatively, establishing a margin of safety. To establish a margin of safety, the sponsor should demonstrate that a large multiple (e.g., 100X) of the proposed human dose does not induce adverse effects in the experimental animals. Scaling from animals to humans based on body surface area can be used to select the dose for use in the clinical trial. Scaling based on pharmacokinetic/pharmacodynamic modeling would also be appropriate if such data are available.

Because microdose studies involve only single exposures to microgram quantities of test materials and because such exposures are comparable to routine environmental exposures, routine genetic toxicology testing is not needed. For similar reasons, safety pharmacology studies are also not recommended.

The mention that an "extended single-dose toxicology study" can be used to support a single-dose study in humans is worthy of note (US FDA, 1996). Since many of the studies conducted under an exploratory IND may be single-dose clinical trials, it may be important for the sponsor to use an "extended single-dose toxicology study" to support the clinical trial rather than the more extensive set of studies outlined in the exploratory IND guidance. The more extensive toxicology studies described in this guidance will support some limited duration of dosing. However, many of the questions that could be answered under an exploratory IND could be answered using a single-dose clinical trial rather than a multiple-dose clinical trial.

Clinical trials to study pharmacologically relevant doses

A second example involves clinical trials designed to study pharmacologic effects of candidate products. More extensive preclinical safety data would be needed to support the safety of such studies. However, since the goal would not include defining a maximally tolerated dose, the evaluation can still be less extensive than typically needed to support a traditional IND application. See the flow chart in the Attachment to this document. Repeat dose clinical trials lasting up to 7 days can be supported by a 2-week repeat dose toxicology study in a sensitive species accompanied by toxicokinetic evaluations. The goal of such a study would be to select safe starting and maximum doses for the clinical trial. The rat is the usual species chosen for this purpose, but other species might be selected. In addition to studies in a rodent species, additional studies in nonrodents, most often dogs, can be used to confirm that the rodent is an appropriately sensitive species. If it is known that a particular species is most appropriate for a class of compounds, studies can be limited to that species. This confirmation can be approached in a number of ways. A lack of gender difference in the rodent study can serve as a basis for testing only a single sex in the second species if only a single sex will be studied in the clinical trial.

The numbers of animals used in the confirmatory study can be fewer than normally used to attain statistically meaningful comparisons, but of sufficient number to rule out any toxicologically significant difference in sensitivity compared with rodent (e.g. four non-rodents per treatment group). The confirmatory study could be a dedicated study involving repeat administrations of a single dose level approximating the rat NOAEL calculated on the basis of body surface area. Alternatively, the test in the second species could be incorporated as part of an exploratory, dose escalating study culminating in repeated doses equivalent to the rat NOAEL. The number of repeat administrations at the rat NOAEL should, at a minimum, be equal to the number of administrations, given with the same schedule, intended clinically. The route of administration should be the same as the expected clinical route, and toxicokinetic measurements should be used to assess exposure. The same endpoints assessed in the rodent study should be evaluated in the second species. If the data from the confirmatory study suggest that the rodent is not the more sensitive species, a 2-week repeated dose toxicity study should be performed in the second species to select doses for human trials. This study should include measurements of body weight, clinical signs, clinical chemistries, hematology, and histopathology.

In contrast to microdose studies, for clinical trials designed to evaluate higher or repeated doses, each candidate product to be tested should be evaluated for safety pharmacology. Evaluation of the central nervous and respiratory systems can be performed as part the rodent toxicology studies while safety pharmacology for the cardiovascular system can be assessed in the nonrodent species, generally the dog, and can be conducted as part of the confirmatory or dose-escalation study.

In general, each product in this type of exploratory IND should be tested for potential genotoxicity unless such testing is not appropriate for the population (e.g. terminally ill patients) or product to be studied. The genetic toxicology tests should include a bacterial mutation assay using all five tester strains with and without metabolic activation as well as a test for chromosomal damage either in vitro (cytogenetics assay or mouse lymphoma thymidine kinase gene mutation assay) or in vivo. The in vivo test can be a micronucleus assay performed in conjunction with the repeated dose toxicity study in the rodent species. The high dose in this case should be a maximally tolerated or limit dose.

The results from the preclinical program can be used to select starting and maximum doses for the clinical trials. The starting dose is anticipated to be no greater than $1/50$ of the NOAEL from the 2-week toxicology study in the sensitive species on a mg/m^2 basis. The maximum clinical dose would be the lowest of the following:

- $\frac{1}{4}$ of the 2-week rodent NOAEL on a mg/m^2 basis
- Up to $\frac{1}{2}$ of the AUC at the NOAEL in the 2-week rodent study, or the AUC in the dog at the rat NOAEL, whichever is lower
- The dose that produces a pharmacologic and/or pharmacodynamic response or at which target modulation is observed in the clinical trial
- Observation of an adverse clinical response

Escalation from the proposed maximal clinical dose should only be performed after consultation with and concurrence of the FDA.

It is recognized that the studies described above are most appropriate for chemical drugs. Other animal models (e.g. nonhuman primates) may be more appropriate for biologics, and some tests may be inappropriate (e.g. genetic toxicology testing) for proteins.

These stopping rules have several implications. The use of an AUC as a stopping rule implies real-time PK analysis which may slow down trial conduct. The guidance does not give insight whether the stopping rule would be based on mean data or data from individual subjects. It should be noted that stopping rules (based on PK or PD) are not routinely used in clinical trials unless there was a toxicity observed in animals that could not be monitored in humans.

The use of pharmacologic and/or pharmacodynamic response as a stopping rule raises the implied availability of precise, real-time biomarker assay. Again, waiting for PD data before dose escalating could delay the conduct of the clinical trial. There may be challenges in using a PD assay for dose escalation/stopping rules. The FIH clinical trial may be the first time that the PD assay has been used in humans and therefore the ruggedness and robustness of the assay will not have been fully tested. Therefore, it may not be feasible to utilize the PD assay as part of the stopping rules.

Another stopping rule advocated by the exploratory IND guidance is the observation of an adverse clinical response. However, the guidance does not indicate whether the adverse clinical response would need to be a serious adverse event or whether any adverse event (including a headache) would halt the study. The guidance allows for escalation from the proposed maximal clinical dose. However, this should only be performed after consultation with and concurrence of the FDA which will require close, real-time communication with the FDA and slow the conduct of the clinical trial.

Clinical studies of MOAs related to efficacy

A third example involves clinical studies intended to evaluate mechanisms of action (MOAs). To support this approach, the FDA will accept alternative, or modified, pharmacologic and toxicological studies to select clinical starting doses and dose escalation schemes. For example, short-term, modified toxicity or safety studies in two animal species based on a dosing strategy to achieve a clinical pharmacodynamic endpoint can in some instances serve as the basis for selecting the safe clinical starting dose for a new candidate drug. These animal studies would incorporate endpoints that are mechanistically based on the pharmacology of the new chemical entity and thought to be important to clinical effectiveness. For example, if the degree of saturation of a receptor or the inhibition of an enzyme were considered possibly related to effectiveness, this parameter would be characterized and determined in the animal study and then used as an endpoint in a subsequent clinical investigation. The dose and dosing regimen determined in the animal study would be extrapolated for use in the clinical investigation. In some cases, a single species could be used if it were established as the most relevant species based on scientific evidence using the specific candidate intended for the clinical investigation.

Although the production of frank toxicity is not the primary intended goal of the nonclinical study, relevant informative endpoints (e.g., hematology and histopathology) selected as important for clinical safety evaluation should be investigated. For example, an antibody that binds with a high degree of selectivity to a tumor-associated antigen could be studied in accordance with this third category. The mechanism of action of antibody-based products is generally associated with their binding properties and the effect on functions associated with immunoglobulins. Pharmacology and toxicology studies provide information about the selection of doses used in clinical studies through evidence of both a safe upper and potentially efficacious lower limit of exposure. These doses might be consistent with target plasma levels of the drug based on animal models of disease. The upper safe levels could be established in animal studies that show a lack of toxicity at these levels.

The exploratory IND guidance describes the toxicology studies that are required and provides a flowchart for these studies. Although the sequence of studies described in the flowchart provides for a minimization of the amount of work that is conducted to support the exploratory IND submission, the sequential nature of the tasks and the small difference in cost of the studies if they are conducted in parallel rather than in sequence, means that the guidance, if followed as written, may delay the initiation of the clinical trial. It should be noted that many pharmaceutical companies may have more stringent criteria (vis-à-vis toxicology) for nominating a compound to enter into clinical trials than are required by the exploratory IND guidance.

The guidance does not make any concessions regarding compliance of the nonclinical studies to GLP practices. "It is expected that all preclinical safety studies supporting the safety of an exploratory IND application will be performed in a manner consistent with good laboratory practices (GLP) (21 CFR Part 58). The GLP provisions apply to a broad variety of studies, test articles, and test systems. Sponsors are encouraged to discuss any need for an exemption from GLP provisions with the FDA prior to conducting safety related studies, for example, during a pre-IND meeting. Sponsors must justify any nonconformance with GLP provisions (21 CFR 312.23(a)(8)(iii))."

The content of the IND submissions is also addressed. In essence, there is little, if any, difference between the format and content of a traditional and an exploratory IND, although the type of information required may vary between these two types of IND as outlined in the guidance. "To begin any kind of testing in humans, applicants must submit an IND application to the Agency with certain types of information (see 21 CFR 312.23 IND Content and Format). The primary purpose of the IND submission is to ensure that subjects will not face undue risk of harm. The major information that must be submitted includes:

- Information on a clinical development plan
- Chemistry, manufacturing, and controls information
- Pharmacology and toxicology information
- Previous human experience with the investigational candidate or related compounds, if there is any.

Because the exploratory IND studies addressed by this guidance will be first in human studies, previous human experience is not pertinent."

With respect to clinical information, the guidance states:

A traditional IND application describes the rationale for the proposed clinical trial program and discusses the potential outcome of the clinical investigation. The exploratory IND studies discussed here focus on a circumscribed study or group of studies, and plans for further development cannot be formulated without the results of these studies. Therefore, an exploratory IND application should articulate the rationale for selecting a compound (or compounds) and for studying them in a single trial or related trials, as this represents all that is known about the overall development plan at this stage. This section should also make it clear that the IND is intended to be withdrawn after completion of the outlined study or studies. The withdrawn, or inactive, IND can be referenced in any subsequent traditional IND.

This point is an important one, and one that may lead to subsequent delays in the development of any drug candidate that has shown promise in an exploratory IND study. More chemical will need to be conducted for subsequent toxicology and clinical trials and a two-week non-rodent toxicology study. The clinical study report from the study or studies conducted under the exploratory IND will need to be written, the investigator's brochure will need to be updated and submitted with updated IND summaries to a new IND. The exploratory IND will need to be closed out per the FDA guidance. This will introduce a 6- to 12-month delay which could be stressful time within the pharmaceutical company if the drug had performed well in the clinical study conducted under the exploratory IND.

The Chemistry, Manufacturing and Controls (CMC) Information that is required for an exploratory IND is less than that is required for a traditional IND. The guidance describes in detail the CMC requirements, describing two scenarios under which CMC information can be provided to an IND application. The guidance states that:

> In the first scenario, the ***same batch*** of candidate product is used in both the toxicology studies and clinical trials. This material will be qualified for human use based on the CMC information (see III.B.1, above) and results of the toxicology studies described elsewhere in this guidance. Although we recommend establishing the impurity profile to the extent possible for future reference and/or comparison, not all impurities of the candidate product may need characterization at this stage of product development. If an issue arises during the toxicology qualification of the product, the appropriate parameters can be studied further, on an as-needed basis. Impurities (e.g., chemical and microbiological) should be characterized in accordance with recommendations in Agency guidance, if, and when, the sponsor files a traditional IND for further clinical investigation.

> In the second scenario, the batch of candidate drug product to be used in the clinical studies may not be the same as that used in the nonclinical toxicology studies. In such a case, the sponsor should demonstrate by analytical testing that the batch to be used is ***representative*** of batches used in the nonclinical toxicology studies. To achieve this, relevant analytical quality test results should be sufficient to enable comparison of different batches of the product. Tests to accomplish this include:
>
> - Identity
> - Structure (e.g., optical rotation (for chiral compounds), reducing/non-reducing electrophoresis (for proteins))
> - Assay for purity
> - Impurity profile (e.g., product- and process-related impurities, residual solvents, heavy metals)
> - Assay for potency (biologic)
> - Physical characteristics (as appropriate)
> - Microbiological characteristics (as appropriate).

The cost and time savings for an exploratory IND are predominantly found in the CMC area, and to a lesser extent in the toxicology area.

Despite the encouragement of the FDA, very few exploratory INDs have been received by the FDA (Anon, 2006). Those that have been received were not for the purposes originally envisioned by the FDA (Jacobson-Kram, 2007). The majority of the INDs have been submitted for imaging studies (National Cancer Institute, 2007). It has been stated that the exploratory IND is a new paradigm that is available to an established industry which has been slow to adopt this new paradigm. The slow adoption may be due to the fact that microdose studies may not be predictive of pharmacological dose studies and that the exploratory IND protracts the development timelines. Additionally, the exploratory IND is designed to kill drugs early that are likely to fail. Since no development team thinks that their drug is a loser, they will therefore push for full-blown development of the drug and not take advantage of the exploratory IND. Small companies may find that the exploratory IND approach may offer some benefit if the company will derive additional value in the eyes of the investment community if they can show that their compound has promise based on acceptable PK in humans or some evidence of a pharmacodynamic response. Another factor that may favor the use of a traditional approach rather than an exploratory IND is the fact that most companies will only have on compound/project that can be taken into clinical trials and that part of the decision to proceed into clinical trials was that the compound was predicted to have acceptable PK properties in humans. In this scenario, the risk–benefit to the company (as measured by overall cost and net present value based on time to market) will most likely favor the traditional IND approach.

One of the driving factors for the exploratory IND was to try to ensure that drugs that would fail due to unacceptable PK or PD could "fail fast," thereby reducing the overall cost of developing a drug.

The causes of attrition of drugs in the drug development process have garnered much interest over the past 10 years. In 2004, Kola and Landis published an article that indicated that PK/bioavailability accounted for approximately 40% of all causes of attrition in 1991, but only 10% of all causes of attrition in 2000 (Kola and Landis, 2004). This was taken by the ADME community as a validation that the high-throughput ADME screens that had been put in place in the early to mid-1990s had paid great dividends. While these screens no doubt did serve to identify drugs that had poor ADME properties, it is highly unlikely that these screens were the cause of such a large drop in the attrition rate due to PK/bioavailability issues. In 2005, Schuster *et al.* called attention to a little known article by Kennedy that was published in 1997 (Kennedy, 1997; Schuster, Laggner and Langer, 2005). Kennedy noted that the original data set generated by the Center for Medicines Research contained a large number of anti-infective drugs. When these drugs were excluded from the data set, the attrition due to PK/bioavailability issues dropped from 39% to 7%. This could be taken to indicate that the attrition rate due to PK/bioavailability issues did not change over the period 1991 to 2000, although a clear understanding of the true situation is not possible due to the lack of availability of the data sets used for these analyses to the scientific community.

Therefore, it can be seen that the exploratory IND approach will address only a relatively minor cause of drug attrition (i.e. unacceptable PK properties). Ironically, the use of an exploratory IND to evaluate the relative PK properties of several compounds will inevitably increase the attrition due to unacceptable PK properties. For example, if all INDs submitted to the FDA were exploratory INDs for the purpose of the selection of the best of three compounds, the attrition rate due to unacceptable PK properties would be 66%. This attrition rate will be decreased by the other sources of failure at other stages of development but will still be higher than the present attrition rate due to unacceptable PK properties of approximately 11%.

In September 2007, the NCI held a workshop on "Phase 0 Trials in Oncologic Drug Development" (National Cancer Institute, 2007). This meeting was held in order to raise the awareness of phase 0 studies and how they can contribute to the development of a novel anticancer agent. Despite the push to utilize phase 0 studies in anticancer drug development, concern has been raised regarding the appropriateness of phase 0/ exploratory IND studies for drugs to treat cancer. One concern is that since reliable and validated assays are not currently available even for most approved targeted cancer drugs, many experimental drugs will fail phase 0 testing. There is a chance that potentially active drugs could be misidentified as being inactive (Marchetti and Schellens, 2007). Ethical concerns regarding conducting phase 0 studies in cancer patients have been that the doses that will be administered will be subtherapeutic and the patient will derive no benefit from that study. Participation in such a study may deny them the possibility of taking part in a clinical trial in which they may receive a treatment that could provide therapeutic benefit (Hill, 2007; Marchetti and Schellens, 2007). Despite these concerns, phase 0 studies for imaging purposes in cancer patients may be appropriate.

Recently, the results of a phase 0 study conducted by the NCI were reported (Kummar *et al.*, 2007). The objective of the study was to determine a dose range at which ABT-888 inhibits PARP in tumor tissue and in peripheral blood mononuclear cells (PBMCs); and the PK of ABT-888. Patients with advanced solid tumors refractory to at least one line of therapy were eligible; patients with CLL or follicular lymphomas were also eligible if standard therapy was not currently indicated. A single oral dose of ABT-888 was administered per patient; dose escalations were planned in cohorts of three patients each (10, 25, 50, 100, and 150 mg). PBMC and tumor sampling were performed before and after drug administration for real-time PK and PD analyses. All patients underwent PBMC sampling; tumor biopsies were planned once significant inhibition of PARP activity in PBMCs was seen in one of three patients in a cohort or plasma C_{max} of 210 nM was achieved in at least one patient. Tumor biopsies were performed at baseline in the week prior to drug administration and then three to six hours post drug administration. Significant inhibition of PARP activity was defined as at least 0.69 reduction on the log scale, which also satisfied statistical significance. A total of six patients have been studied so far, three each for the 10- and 25-mg cohorts. No treatment-related adverse events have been observed. Target C_{max} was exceeded in the first cohort; all patients in the next cohort underwent tumor biopsies in addition to PBMC sampling. A trend toward inhibition of PARP activity in PBMCs was observed in the first cohort. Significant inhibition of PAR levels was observed in tumor biopsies from all three patients in the second cohort (92%, 99%, 100% reductions respectively, as compared with baseline). Greater than 85% reduction of PAR levels was observed in PBMCs from two of the three patients in the second cohort (one patient was not evaluable). ABT-888 is orally bioavailable and inhibits PARP activity in PBMCs and tumor cells. Target assay feasibility was established in human samples.

It is unclear whether this study was conducted under an exploratory IND or a traditional IND. It may be assumed that the latter approach was taken since the researchers did not discuss the use of stopping rules and that a maximum biologic effect was observed in the tumor biopsy. This study could be viewed as being a phase 0 study in the sense that one of the objectives was to evaluate the utility of the PD assay and that a single dose was administered (i.e. the patient would receive limited, if any, therapeutic benefit).

Microdosing studies are one example of a study that can be conducted under an exploratory IND. This type of study has garnered much interest over the past decade since regulatory agencies and CROs have proposed that such studies would enable a rapid identification of compounds with suboptimal PK properties. In the United States, microdose studies could be conducted under an exploratory IND or under a traditional IND, although strong emphasis has been placed upon conducting such studies under an exploratory IND. In either situation, the format of the IND submission would be identical and would include information on a clinical development plan; chemistry, manufacturing, and controls (CMC) information; pharmacology, toxicology, and ADME information; and previous human experience with the investigational candidate or related compounds, if there is any.

The CHMP issued "Position Paper on Non-Clinical Safety Studies to Support Clinical Trials with a Single Microdose" which came into effect in January 2003, with Revision 1 coming into effect in June 2004 (EMEA, 2004). The Position Paper defined "common standards of the non-clinical safety studies needed to support human clinical trials of a single dose of a pharmacologically active compound using microdose techniques." The term "microdose" was defined as being "less than 1/100th of the dose calculated to yield a pharmacological effect of the test substance based on primary pharmacodynamic data obtained in vitro and in vivo (typically doses in, or below, the low microgram range) and at a maximum dose of ≤100 microgram." Examples of the types of clinical trial included "the early characterization of a substance's pharmacokinetic-/distribution properties or receptor selectivity profile using positron emission tomography (PET) imaging, accelerator mass spectrometry (AMS) or other very sensitive analytical techniques."

As can be anticipated, the calculation of the dose required to yield a pharmacological effect will be based on many assumptions including the prediction of the pharmacokinetics of the drug in humans. It is also unclear as to how a pharmacological effect in humans would be defined (i.e. a full effect or a 10% effect) and implies that a PD assay would be available to measure that effect in humans in the microdose study. It should also be noted that a microdose study can be supported using a sensitive LC-MS/MS assay and does not necessarily have to use AMS as the bioanalytical tool.

The clinical trials that were covered in this Position Paper will be exploratory in nature (pre-phase 1) and may be conducted with a single test substance or with a number of closely related pharmaceutical candidates to choose the preferred candidate or formulation for further development. The CPMP proposed that "certain deviations from the existing CPMP/ICH notes for guidance to support pre-phase 1 clinical trials may be scientifically justified." The Position Paper noted that "In the European Union, repeated dose toxicity studies in two species (one non-rodent) for a minimum of 2 weeks are required to support a single, first human dose. However, in the United States of America, single dose acute toxicity studies are in some cases considered sufficient to support a single dose human clinical trial." The Position Paper noted that under the ICH M3 guidance, safety pharmacology, single-dose toxicity studies, and repeated-dose toxicity studies would be required for a single-dose human clinical trial. The Position Paper stated that "This set of studies may be replaced by an extended single-dose toxicity study in only one mammalian species if the choice in species could be justified based on comparative in vitro metabolism data and by comparative data on in vitro primary pharmacodynamics/biological activity." The design of the single-dose toxicity is described in some detail. Pertinent points include "The extended single-dose toxicity study should include a control group, and a sufficient number of treatment groups to allow the establishment of the dose inducing a minimal toxic effect. For compounds with low toxicity a limited dose approach could be used. Allometric scaling from animal species to man and using a safety factor of 1000 should be used to set the limit dose.

If a toxic effect is observed at the limit dose, the non-toxic dose level should be established." The Position Paper refers the reader to CPMP/ICH/283/95 for allometric scaling factors. It is interesting to note that the reference is to ICH Topic Q3C(R3) Impurities: Residual Solvents (ICH, 1997b). It should be noted that the allometric scaling factors and the assumptions on body weights differ between the ICHQ3C(R3) guidance and the FDA's guidance "Estimating the Maximum Safe Starting Dose in Initial Clinical Trials for Therapeutics in Adult Healthy Volunteers" that was issued in 2005. The FDA's guidance focuses on the use of toxicology data to establish the maximum safe starting dose. The guidance does mention metabolism and pharmacokinetic data in the following manner:

> Although the process outlined in this guidance uses administered doses, observed toxicities, and an algorithmic approach to calculate the MRSD (Maximum Recommended Starting Dose), an alternative approach could be proposed that places primary emphasis on animal pharmacokinetics and modeling rather than dose. In a limited number of cases, animal pharmacokinetic data can be useful in determining initial clinical doses. However, in the majority of investigational new drug applications (INDs), animal data are not available in sufficient detail to construct a scientifically valid, pharmacokinetic model whose aim is to accurately project an MRSD. Toxicity should be avoided at the initial clinical dose. However, doses should be chosen that allow reasonably rapid attainment of the phase 1 trial objectives (e.g., assessment of the therapeutic's tolerability, pharmacodynamic or pharmacokinetic profile). All of the relevant preclinical data, including information on the pharmacologically active dose, the full toxicologic profile of the compound, and the pharmacokinetics (absorption, distribution, metabolism, and excretion) of the therapeutic, should be considered when determining the MRSD. Starting with doses lower than the MRSD is always an option and can be particularly appropriate to meet some clinical trial objectives.

Interestingly, one of the references cited for the use of PK data to establish the starting dose for a clinical trial was published by FDA employees (Mahmood, Green and Fisher, 2003).

In 1Q 2006, the CHMP adopted a concept paper, "Concept Paper on the Development of a CHMP Guideline on the Non-Clinical Requirements to Support Early Phase I Clinical Trials with Pharmaceutical Compounds" (EMEA, 2006b). The Concept Paper states that:

> The requirements for non-clinical safety studies to support the conduct of human clinical trials for pharmaceuticals were largely addressed by the International Conference on Harmonization of Technical Requirements for the Registration of Pharmaceuticals for Human Use (ICH) Topic M3: Note for Guidance on Non-clinical Safety Studies for the Conduct of Human Clinical Trials for Pharmaceuticals (CPMP/ICH/286/95). This guidance aimed to facilitate the timely conduct of clinical trials and reduce the unnecessary use of animals and other resources. This was designed to promote safe and ethical development and availability of new pharmaceuticals. The development of a pharmaceutical is a stepwise process involving an evaluation of both the animal and human safety information. The overall goals of the non-clinical safety evaluation include a characterization of toxic effects with respect to target organs, dose dependence, relationship to exposure, and potential reversibility. This information is important for the estimation of an initial safe starting dose for the human trials and for the identification of parameters for clinical monitoring for potential adverse effects. The non-clinical safety studies at any stage should be adequate to characterise potential adverse effects under the conditions of the proposed clinical trial.

It is recognised that significant advances in harmonization of the timing of non-clinical safety studies for the conduct of human clinical trials for pharmaceuticals have already been achieved. However, differences remain in a few areas, including toxicity studies to support first entry into man. In 2003, the CPMP released a position paper on non-clinical studies to support clinical trials with a single microdose (CPMP/SWP/2599/02). More recently, the FDA have released guidance on conducting clinical trials under an exploratory IND.

The Concept Paper further states that:

The CHMP Safety Working Party (SWP) have discussed a revised toxicology package designed to support early clinical investigations in man. The objectives of these clinical studies would differ from the more classical phase I approach, in that they would investigate low doses and would not be designed to investigate tolerability. Instead, the trial would generate pharmacokinetic (PK)-related or pharmacodynamic (PD)-related data, possibly based on a sensitive biomarker of efficacy. These data could then be used for early decision making on whether to progress a particular compound or work on a potential therapeutic target. The clinical studies would require the administration of one or two doses that would achieve an exposure in the predicted pharmacological range, but which would be below the tolerability range. The initial dose administered to man would be very low, and the dose would be slowly escalated (using real time pharmacokinetic monitoring) until a predefined maximum exposure was achieved. Using this approach, a good understanding of dose exposure relationship in man could be obtained. In order to maintain international harmonization, the revised toxicology package will follow the general principles laid down by the ICH M3 guideline. However, the dose levels used in the nonclinical studies could be lower than usual, as they are only required to support the low intended clinical doses. Human safety will not be compromised by the revised package.

The current schedule calls for the draft guideline detailing what nonclinical data are required to be included in a clinical trials application for an early phase I study in humans to be issued in 1Q 2008. This guideline should allow for a flexibility of approaches, including those outlined in the EU microdose guideline or the FDA exploratory IND guideline. How this guideline will relate to the recently released EMEA guideline or the planned ICH guidance is presently unknown.

It should be recognized that as of November 2007, despite what is reported by several journals and CROs, there are no guidances or guidelines in the United States or EU that pertain solely to microdose studies. In the EU, there is a Position Paper and a Concept Paper but no guideline. In the United States, there is a guidance on exploratory INDs in which microdose studies are mentioned but are not the main topic of this guidance.

The utility of microdosing to predict the pharmacokinetics of a drug when that drug is administered at a pharmacologic dose has been much debated. Lappin *et al.* conducted a study (the CREAM study) in which the PK of five drugs (warfarin, ZK253, diazepam, midazolam, and erythromycin) were evaluated at a microdose and pharmacological dose level (Lappin *et al.*, 2006). This study found that the PK after a microdose of warfarin was different to that obtained at a pharmacological dose, possibly due to high-affinity, low-capacity tissue binding of warfarin. The oral microdose of erythromycin failed to provide detectable plasma levels as a result of possible acid lability in the stomach. The absolute bioavailability for the other three compounds showed good concordance between microdose and pharmacologic dose levels. These researchers concluded that "Overall, when used appropriately, microdosing offers the potential to aid early drug candidate selection." In 2007, the American College of Clinical Pharmacology published a position statement on the use of microdosing in the drug development process (Bertino, Greenberg and Reed, 2007a, b). This publication stated that:

> At present, it would appear that studies using microdosing methodology should not be relied on as the primary or sole approach to screen new drug candidates, as the potential exists with current methodologies to possibly reject important new drugs while possibly accepting drugs that could result in significant safety issues. Until more information is available and has undergone appropriate scrutiny, it would appear that a microdosing strategy could complement standard animal-to-human allometric scaling, refining current phase I study designs. Microdosing methodology appears to be one of the many new viable "tools" in the drug development "toolbox." The exact role and impact that microdosing methodologies will have on new drug development is yet to be fully realized but will surely be continuously refined with continued experience and, above all, data sharing through the publication of studies employing this novel methodologic approach.

The conclusions of the ACCP position paper have been challenged by Rowland (2007). Other researchers have stated that "microdose studies will have limited impact on the overall efficiency of drug development" (Boyd and Lalonde, 2007).

It has been reported that new paradigms are being considered in the maintenance of ICHM3R "Timing of preclinical studies in relation to clinical trials" (Jacobson-Kram, 2007). "These new paradigms include:

— Microdose up to 5 doses each not to exceed 1/100th the NOAEL determined in the toxicology study or 1/100th the anticipated pharmacodynamically active dose or a total dose of 100 μg whichever is lower.

— Microdose up to 5 doses of 100 μg each not to exceed 1/100th the NOAEL determined in the toxicology study or 1/100th the anticipated pharmacodynamically active dose and with a washout of six half-lives between doses. Repeat-dose exploratory studies up to 14 days in the therapeutic range but not to an MTD. Supported by safety studies using large multiples of clinical exposure but not based on toxicity."

It will be interesting to see how the revision to ICHM3 progresses (ICH, 2006c).

27.5.4 Ethnic Factors

The ICHE5 (R1) guidance addresses "Ethnic Factors in the Acceptability of Foreign Clinical Data" (ICH, 1998). The basis for this guidance is described thus:

> To assess a medicine's sensitivity to ethnic factors it is important that there be knowledge of its pharmacokinetic and pharmacodynamic properties and the translation of those properties to clinical effectiveness and safety. Some properties of a medicine (chemical class, metabolic pathway, pharmacologic class) make it more or less likely to be affected by ethnic factors. Characterization of a medicine as "ethnically insensitive", that is unlikely to behave differently in different populations, would usually make it easier to extrapolate data from one region to another and need less bridging data.

The guidance provides details of the characteristics of an "ethnically sensitive" and an "ethnically insensitive" drug. It states that "the following properties of a compound make it less likely to be sensitive to ethnic factors:

- Linear pharmacokinetics (PK)
- A flat pharmacodynamic (PD) (effect-concentration) curve for both efficacy and safety in the range of the recommended dosage and dose regimen (this may mean that the medicine is well-tolerated)
- A wide therapeutic dose range (again, possibly an indicator of good tolerability)
- Minimal metabolism or metabolism distributed among multiple pathways
- High bioavailability, thus less susceptibility to dietary absorption effects
- Low potential for protein binding
- Little potential for drug-drug, drug-diet and drug-disease interactions
- Non-systemic mode of action
- Little potential for inappropriate use

The following properties of a compound make it more likely to be sensitive to ethnic factors:

- Non-linear pharmacokinetics
- A steep pharmacodynamic curve for both efficacy and safety (a small change in dose results in a large change in effect) in the range of the recommended dosage and dose regimen
- A narrow therapeutic dose range
- Highly metabolized, especially through a single pathway, thereby increasing the potential for drug-drug interaction
- Metabolism by enzymes known to show genetic polymorphism
- Administration as a prodrug, with the potential for ethnically variable enzymatic conversion
- High inter-subject variation in bioavailability
- Low bioavailability, thus more susceptible to dietary absorption effects
- High likelihood of use in a setting of multiple co-medications
- High likelihood for inappropriate use, e.g., analgesics and tranquilizers."

Based on the given definitions, it would appear that most drugs would be viewed as being ethnically sensitive. The concept that "metabolism distributed among multiple pathways" is an interesting concept since it would imply that such drugs would have a low potential for DDIs based on the existence of the "safety valve" of alternative routes of metabolism. This is a concept that was proposed by Vickers *et al.* in 1999 but, to date, this concept has not been extensively validated by *in vivo* data (Vickers *et al.*, 1999). A Q&A document for this guidance was issued in 2006 (ICH, 2006e).

27.5.5 Pharmacogenomics

In recent years, the genotyping of DMEs in clinical trials for the purpose of understanding interindividual variability in pharmacokinetics and effect (either efficacy or side effects) has become more common. This information has been included in product labels and has been the subject of several regulatory guidances. Additionally, several genotyping tests have been developed and commercialized. The Amplichip CYP450 test is the first FDA-cleared test for analysis of CYP2D6 and CYP2C19 (Roche, 2008). The test identifies a patient's genotype and, based on this analysis, provides their predicted phenotype—either poor, intermediate, extensive, or ultrarapid metabolizer. These two genes code for enzymes that metabolize many antidepressants, antipsychotics, and ADHD drugs, as well as other medications. Patients seeking therapy for depression and other mood disorders may not respond to or may have adverse reactions to many antidepressants, antipsychotics, and ADHD drugs due to their ability to metabolize these drugs. Physicians have traditionally used a "start low, go slow" dosing approach with their patients to mitigate efficacy and side-effect issues. This test can help physicians adjust dosing and select drugs by predicting a phenotype based on a genotype so that patient treatment can be individualized to get the best therapeutic results possible.

The Invader® UGT1A1 Molecular Assay is cleared for use to identify patients who may be at increased risk of adverse reaction to the chemotherapy irinotecan (Camptosar®, Pfizer) (Third Wave Technologies, 2008). Irinotecan is indicated for use in the treatment of metastatic colorectal cancer. While irinotecan is effective, irinotecan-associated toxicity can result in both neutropenia (reduced white blood cell count) and severe diarrhea. The UGT1A1*28 allele that the Invader UGT1A1 Molecular Assay detects, homozygous in approximately 10% of the North American population, has been shown to be an effective genetic marker for predicting irinotecan-induced toxicity. In July 2005, the FDA revised the safety labeling for irinotecan, recommending that treatment be altered for individuals who are homozygous for the UGT1A1*28 allele. One month later, the FDA approved the genotyping test for this allele.

In August, FDA approved updated labeling for Coumadin (warfarin), explaining that people with variations of the genes CYP2C9 and VKORC1 may respond differently to the drug (US FDA, 2007g). Manufacturers of generic warfarin are adding similar information to their products' labeling. The next month, the FDA approved the Nanosphere Verigene Metabolism Nucleic Acids Test detects particular variations in two genes, CYP2C9 and VKORC1, which are involved in the metabolism and mechanism of action of warfarin, respectively (Nanosphere Inc., 2008). The Nanosphere test is not intended to be a stand-alone tool to determine optimum drug dosage, but should be used along with clinical evaluation and other tools, including INR, to determine the best treatment for patients. The FDA cleared the test based on results of a study conducted by the manufacturer of hundreds of DNA samples as well as on a broad range of published literature. In a three-site study, the test was accurate in all cases where the test yielded a result; 8% of the tests could not identify which genetic variants were present.

Although these tests will be primarily used for therapeutic decisions, it can be seen that such tests may have a role in the selection of patients for inclusion in a particular clinical trial. Clinical Research Organizations (CROs) may offer populations of volunteers that have been prescreened for certain genetic polymorphisms.

Prior to the creation of regulatory guidances on pharmacogenomics, a report of an FDA-PWG-PhRMA-DruSafe Workshop on Pharmacogenetics and Pharmacogenomics in Drug Development and Regulatory Decision Making was published in 2003 (Lesko et al., 2003). This workshop laid the foundations for the FDA guidances that followed. The dialog between the FDA and industry has continued through subsequent workshops (Frueh et al., 2006).

Guidances that pertain to genotyping include:

- ICH Topic E15. Establish definitions for genomic biomarkers, pharmacogenomics, pharmacogenetics, genomic data and sample coding categories (Step 5; approved in EU) (EMEA, 2007g)

- Reflection Paper on the Use of Pharmacogenetics in the Pharmacokinetic Evaluation of Medicinal Products (EMEA/128517/2006) (EMEA, 2007i)
- Pharmacogenomic Data Submissions (FDA; 2005) (US FDA, 2005f)
- Pharmacogenomic Data Submissions—Companion Guidance (FDA; Draft 2007) (US FDA, 2007m).

These documents make several general points on drug metabolism and genotyping. ICHE15 states that "The definition for a genomic biomarker does not include the measurement and characterization of proteins or low molecular weight metabolites."

The EMEA reflection paper was adopted by the CHMP in May 2007. The purpose of this reflection paper is to "target the place of PG in the clinical PK evaluation of medicinal products during drug development and use." Like most guidelines or related documents, the reflection paper notes that it should be read in conjunction with other guidelines. Among the topics addressed by this reflection paper are:

- in which situations should the effect of PG on PK of new or existing medicinal products be studied;
- at what stage in the clinical development program should PG/PK studies be performed;
- study design and methodology;
- evaluation of the clinical consequences of genetic differences in drug substance exposure;
- special considerations related to drug–drug interactions and impaired or immature organ functions.

The reflection paper addresses the situations in which the effect of PG on PK should be studied. It states:

Studies of the effect of PG on PK are required for PK evaluation of a new chemical entity if the genetic variation is likely to translate into important differences in the systemic and/or local exposure to this substance or its active or toxic metabolites, thereby potentially affecting safety and efficacy of the treatment. PG variants may impact on Absorption, Distribution, Metabolism and Excretion of the compound.

Furthermore, combined PG/PK studies that may contribute to the identification of novel polymorphic loci are encouraged if the compound exhibits important inter-individual PK variability, likely to affect clinical efficacy and/or safety. Technology has advanced to the point that analysis of genetic factors affecting safety and efficacy of medicines is now fast and reliable. If there is clinically important variability in PK and genetic causes for this may not be excluded, it is advisable to carry out genetic analysis of loci likely to be responsible for this variation. In all cases where unexplained PK variation has been identified, samples for pharmacogenetic purpose should be collected. This would allow a critical evaluation of the clinical and, if population PK is used, PK consequences of this polymorphism at a later stage.

Studies of PG differences in the activity and expression of transport proteins involved in drug distribution (efflux or influx) to target organs such as the central nervous system, possibly explaining adverse events or lack of therapeutic effect, are encouraged if there are indications of clinically important differences in these respects and if a relevant polymorphism can be studied using cohorts with sufficient power to reach conclusions.

Although some polymorphisms are extremely rare, the applicant should always consider studying the PK and clinical consequences of all potentially clinically relevant polymorphisms. If this is not feasible, the applicant should justify the lack of data and discuss the possible safety (or efficacy) consequences based on prior knowledge on the effect of the polymorphisms on protein activity. Lack of data should be reflected in the Summary of Product Characteristics (SPC) if considered clinically relevant.

The timing of PG/PK studies is also addressed thus:

In general, the importance of PG for the PK of a drug substance may be indicated by *in vitro* data where the enzymes involved in drug metabolism as well as formation and metabolism of pharmacologically active metabolites have been identified. Preclinical animal studies should be interpreted with great caution, as there are usually marked species differences. If human *in vitro* data suggest major involvement of a protein known to be subject to functionally important genetic polymorphism, inclusion of genotyping directed to the candidate gene is warranted in early phase I studies. When the involvement of the polymorphic gene has been verified, *in vivo* studies of the effects of specific polymorphisms on the PK of the pharmacologically active compounds likely to contribute to clinical efficacy and/or safety are recommended.

It should be noted that the reflection paper does not give insight into exactly how the importance of a polymorphic DME is determined.
Polymorphisms in drug transporters are also addressed:

Involvement of transporters may also be indicated by *in vitro* data but presently the knowledge in this field is not mature enough for early genotyping to be required only on the basis of *in vitro* data. However, if the *in vitro* data together with other data such as ADME or renal excretion data indicate that a polymorphic transporter has a major role in the PK of a drug, genotyping in the subsequent PK studies and clinical studies is encouraged.

The collection and analysis of samples is also addressed: "It is recommended that samples from the early phase I studies are stored to allow retrospective analysis, when more experience has been gained or new proteins have been shown to play a major role in the PK of the active substances." It is not always possible to obtain informed consent from the patients/subjects and the cost of storage until the samples are analyzed can become burdensome. When such analyses are performed at a later date, consideration should be given to providing the results of the genotyping to the patient/subject.
The reflection paper further states that:

When the genotype is predicted or known to markedly affect the PK of pharmacologically active compounds contributing to *in vivo* efficacy and/or safety of a medicinal product, genotyping is encouraged in as many of the phase I, II and III clinical studies as possible to increase the amount of data that will support the recommendations for use in the genetic subpopulation(s). Dose-response studies or other clinical studies covering the exposure of active substances obtained can be used to support safety and efficacy in a specific genetic subpopulation.

The effect of genotype in certain special populations is also recognized:

The enzymes and transport proteins involved in the PK of a drug substance may be quantitatively and qualitatively different in paediatric patients than in adults as a consequence of developmental changes in gene expression. The most marked differences are expected in newborn infants, infants and toddlers (0–2 year-old children). In the very elderly patients PG related differences in metabolism may occur and have an impact due to impaired functionality of system/organs; limited knowledge is available at present.

The role of PG in interpreting DDIs is another topic addressed by this reflection paper. The paper states:

> Genotyping of the population included in an interaction study is recommended when PG is expected to affect the PK of any of the active substances. Depending on the question investigated, directed inclusion or exclusion of specific genotypes may be useful.

> If well known major elimination pathways are absent in a subpopulation, the consequences of an inhibition of parallel pathways should be considered. The magnitude of such an interaction can be difficult to determine in the absence of an in vivo interaction study. However, a "worst case" scenario can usually be estimated. If the safety consequences of the interaction are predicted to give an unacceptable risk-benefit, or if sufficient safety data is lacking at predicted exposures, an interaction study in the subpopulation may be needed to reach satisfactory treatment recommendations.

> The effect of active substances, that are enzyme or protein inhibitors or inducers, may also be different if the contribution of an enzyme is absent in a subpopulation. The consequences of inhibition of a protein in a population with a lower but not abolished activity of that protein should also be considered. In case of induction, the net result will depend on the degree of induction -if any- of the parallel pathway. This should be considered especially when a dose recommendation for a certain drug-combination is studied and evaluated.

> An increased systemic exposure in genetic subpopulations may lead to more pronounced effects of the investigated drug on other drugs. This should be considered and if necessary reflected in the SPC.

The contribution of genetic polymorphisms to inter-individual variability in drug response was again emphasized in a recent reflection paper on the use of genomics in cardiovascular clinical intervention trials (EMEA, 2007j). This reflection paper provided an example of the role of genetic polymorphisms for cardiovascular drugs, stating that "Genes coding for enzymes involved in the metabolism of warfarin and clopidogrel, and those coding for platelet components involved in the antiplatelet action of drugs have been extensively studied in relationship to pharmacokinetic issues, and anti-clotting effect."

The FDA has a deep interest in pharmacogenomics and toxicogenomics which is reflected on the amount of information available on these topics on its web site (US FDA, 2008e). Toxicogenomics, which is the study of drug-related safety/toxicity and pathology at the gene expression level in preclinical studies, is also of great interest to the FDA. At the present, there are no guidances specific to toxicogenomics, although the expectation is that the data from toxicogenomic studies would be submitted in a similar manner to the data from pharmacogenomic studies. Since DMEs or transporters are not a common feature of toxicogenomic studies, there will be no further discussion of toxicogenomics in this chapter.

The FDA issued a guidance on "Pharmacogenomic Data Submissions" in 2005 (US FDA, 2005f). This guidance notes that "some pharmacogenetic tests—primarily those related to drug metabolism—have well-accepted mechanistic and clinical significance and are currently being integrated into drug development decision making and clinical practice." For the purposes of this guidance, a pharmacogenomic test result may be considered a valid biomarker if "(1) it is measured in an analytical test system with well-established performance characteristics and (2) there is an established scientific framework or body of evidence that elucidates the physiologic, pharmacologic, toxicologic, or clinical significance of the test results."

For example, the consequences for drug metabolism of genetic variation in the human enzymes CYP2D6 and thiopurine methyltransferase are well understood in the scientific community and are reflected in certain approved drug labels. The results of genetic tests that distinguish allelic variants of these enzymes are considered to be well established and, therefore, valid biomarkers.

The fact that most PG tests currently involve DMEs is illustrated by the examples given in this guidance. The guidance states:

> If a pharmacogenomic test shows promise for enhancing the dose selection, safety, or effectiveness of a drug, a sponsor may wish to fully integrate pharmacogenomic data into the drug development program. ... The pharmacogenomic data may be intended to be included in the drug labeling in an informational manner. For example, such data might be used to describe the potential for dose adjustment by drug metabolism genotype (e.g., CYP2D6*5) or to mention the possibility of a side effect of greater severity or frequency in individuals of a certain genotype or gene expression profile. In such cases, the pharmacogenomic test result would be considered a known valid biomarker. However, an FDA-approved pharmacogenomic test may not be available or required to be available, or a commercial pharmacogenomic test may not be widely available. Given this level of complexity, at the current time, sponsors should consult the relevant FDA review division for advice on how to proceed in a specific case.

The guidance provides a definition of a pharmacogenetic test: "An assay intended to study interindividual variations in DNA sequence related to drug absorption and disposition (pharmacokinetics) or drug action (pharmacodynamics), including polymorphic variation in the genes that encode the functions of transporters, metabolizing enzymes, receptors, and other proteins." It is interesting to note that drug transporters are only mentioned in the definition of a pharmacogenetics test, but not in the main body of the guidance.

The recent draft guidance, "Pharmacogenomic Data Submissions—Companion Guidance" (US FDA, 2007m) states that:

> This guidance is intended to be used as a companion to the guidance *Pharmacogenomic Data Submissions* (March 2005). It reflects experience gained since the issuance of that guidance with voluntary genomic data submissions as well as with review by the FDA of numerous protocols and data submitted under investigational new drug (IND) applications, new drug applications (NDAs), and biologics license applications (BLAs). The recommendations are intended to facilitate scientific progress in the field of pharmacogenomics and to facilitate the use of pharmacogenomic data in drug development. The FDA believes that the recommendations made in this companion guidance, together with the recommendations in the March 2005 guidance, will benefit sponsors considering the submission of either voluntary genomic data submissions or marketing submissions containing genomics data. As technology changes and more experience is gained, these recommendations may be updated.

The draft guidance discusses genotyping methods:

> Genetic differences among individuals occur in a variety of forms, from alterations in chromosomal arrangement or copy number to single base-pair changes. Much of the

genetic variation currently used in pharmacogenetics occurs at the level of individual genes (e.g., drug metabolizing enzymes) on a scale ranging from single base-pair changes to entire gene duplications or deletions. Examining genomic DNA is often the most reliable and practical method for characterizing genetic variation, although methods based on protein or mRNA expression levels can be preferable in some situations, such as when determining treatment-sensitivity of cancer or viral infection. Many methods are currently available for characterizing DNA variations, and new methods are rapidly being developed.

The reporting of the data in a genotyping report is also addressed: "We recommend that the following information be included in the genotyping report, regardless of the genomics submission type (see the *Pharmacogenomic Data Submissions* guidance for regulatory requirements):

- Description of assay platform or methodology
- Samples studied, including demographics and sample size justification for genotype/clinical phenotype correlation and adequate coverage for ethnic/racial groups; include expected allele frequency in different populations
- Alleles measured and correlation with metabolic status designation
- For metabolizing enzymes, how EM (extensive metabolizer), PM (poor metabolizer), IM (intermediate metabolizer), or UM (ultra rapid metabolizer) are determined
- Sample test report
- For new genes, correlation between gene variant and encoded protein activity.
- Whether the assay was performed in a CLIA-certified lab or research lab."

The reporting of genomic data in clinical reports is described:

There are many possible sources of data for genomic data submissions. Genomic data from clinical studies may result from microarray expression profiling experiments, genotyping or single-nucleotide polymorphism (SNP) experiments, or from other evolving analytical methodologies pertaining to drug dosing or metabolism, safety assessments, or efficacy evaluations. Genomic data may also be reported from studies where other data are also reported, such as with efficacy or safety data from clinical or nonclinical studies. However, these data can be reviewed only if the content of the clinical data report included in the submission contains *sufficient detail* regarding the sample selection.

The draft companion guidance does not appear to provide guidance that is markedly different from that previously provided but simply serves to emphasize the FDA's interest and commitment to the use of genomic data in the evaluation of the safety and efficacy of drugs.

The FDA has authored several papers on the implementation of the guidance on pharmacogenomics data submission, in which its members have summarized their experiences to date (Frueh *et al.*, 2006; Goodsaid and Frueh, 2007).

Although pharmacogenetics is important in the disposition of certain drugs, it should be noted that a large degree of variability can arise due to other factors other than genotype, such as induction and inhibition of the enzyme systems or transporters. Additionally, there may be ethical concerns regarding genotyping for polymorphisms in CYP isoforms since some polymorphisms have been associated with increased risk of certain diseases, including cancer (Agundez, 2004; Danko and Chaschin, 2005; Rodriguez-Antona and Ingelman-Sundberg, 2006). PhRMA has created a white paper describing the current best practices with respect to genotyping (Williams *et al.*, 2008).

27.5.6   Drug–Drug Interactions

The topic of unexpected drug–drug interactions first came to the forefront in the 1970s and 1980s. The history of the withdrawal of drugs from the market due to drug–drug interactions and the development of regulatory guidances has been summarized by Huang *et al.* (2008). Their summary is presented in abbreviated form in Table 27.3.

The example of Posicor (mibefradil) provides some insight into the process of the withdrawal of a drug from the market to the potential (and real) safety consequences of DDIs. Posicor was withdrawn from the market in June 1998, when it was learned that more than 25 drugs were found to be dangerous when used with Posicor. When Posicor entered the market in August of 1997, its enzyme-inhibiting properties were described in the labeling. The labeling specifically listed three drugs (astemizole, cisapride, and terfenadine) that could be expected to accumulate to dangerous levels if Posicor was coadministered (US FDA, 1998a). In less than a year, it became apparent that the liabilities caused by the inhibition of the metabolism of concomitant medications by Posicor meant that it would be extremely difficult, if not impossible, to safely prescribe Posicor.

It is clear that the FDA has a keen interest in DDIs and they have created a web page that contains a great deal of useful information on this topic (US FDA, 2007f). The vast majority of the information on DDIs relates to the interaction between two chemical drugs (NCEs). Comparatively little information is available regarding the potential for interaction between an NCE and a biologic, between two biologics, and between herbs and NCEs or biologics. The identification of herbs that interact with drugs can be of great importance given the extensive use of herbal medication by patients and the safety concerns that have arisen regarding the use of certain herbal supplements (Seeff, 2007; Zhou *et al.*, 2007a). The amount of literature that is available regarding the interaction between herbs and drugs is growing greatly (Hu *et al.*, 2005; Skalli, Zaid and Soulaymani, 2007; Zhou *et al.*, 2007b).

TABLE 27.3   Timeline of events related to the development of FDA guidances for the evaluation of the potential for DDI studies.

Year	Event
1970–1980s	Reports of unanticipated DDIS
1990	Terfenadine case study
1997	First *in vitro* DDI guidance posted by the FDA
1998	Withdrawal of terfenadine, mibefradil and bromfenac from US market
1999	First *in vivo* DDI guidance posted by the FDA
1999	Withdrawal of astemizole and grepafloxacin from US market
2000	Withdrawal of alosteron, cisapride and troglitazone from US market
2001	Withdrawal of cerivastatin and rapacuronium from US market
2003	RhRMA White white Paper paper on the conduct of *in vitro* and *in vivo* DDI studies
2006	FDA internet site devoted to DDIs launched
2006	Draft guidance on DDI studies issued by the FDA

There are several guidances on the topic of drug–drug interactions, one of which was issued by the FDA in 1997, "Drug Metabolism/Drug Interaction Studies in the Drug Development Process: Studies In Vitro" (US FDA, 1997a). The initial elements of the guidance provide some framework for the later points by stating that:

> This FDA guidance to industry provides suggestions on current approaches to studies in vitro of drug metabolism and interactions. The guidance is intended to encourage routine, thorough evaluation of metabolism and interactions in vitro whenever feasible and appropriate. The FDA also recognizes that clinical observations can address some of the same issues identified in this document as being susceptible to in vitro study. The suggested approaches delineated in this document, however, are efficient and inexpensive considering the breadth of information they can provide, and often can reduce or eliminate the need for further clinical investigations. This particular guidance is directed toward a broad class of drugs: molecules with a molecular weight below 10 kiloDaltons.

It is unclear how the cutoff of 10 kDa was derived or what types of molecules that the FDA was imagining would be of the range of 2–10 kDa. It may be an attempt to indicate that such *in vitro* studies are not required for biotechnology-derived products such as monoclonal antibodies, oligonucleotide drugs, or aptamers.

The level of interest of the FDA in DDIs is clear in this guidance. The guidance states:

> Review scientists at the FDA have long been interested in the impact of drug metabolism and drug–drug interactions on drug safety and efficacy. As a result, discussion of this topic also is contained in other FDA guidance documents, including *General Considerations for the Clinical Evaluation of Drugs* (FDA 77-3040), *Guideline for Studying Drugs Likely to be Used in the Elderly* (11/89), and *Guideline for the Study and Evaluation of Gender Differences in the Clinical Evaluation of Drugs* (58 FR 39406, July 22, 1993).

The guidance notes that DDIs may be two-sided, that is, the drug could be the perpetrator or the victim. The guidance states that "Even for drugs that are not substantially metabolized, the potential effect of that drug on the metabolism of concomitant drugs could be important." The effect of genetic polymorphisms on exposure and the increase in drug levels as a consequence of a DDI is also discussed:

> Large differences in blood levels can occur because of individual differences in metabolism. Some drugs, such as tricyclic antidepressants, exhibit order of magnitude differences in blood concentrations depending on the enzyme status of patients. Drug-drug interactions can have similarly large effects when one drug inhibits the metabolism of another. For example, ketoconazole greatly increases concentrations of parent terfenadine, leading to QT prolongation and *torsades de pointes*.

The guidance notes that *in vitro* studies should form only one part of the evaluation of the potential for a DDI:

> The studies in vitro described in this document are one set of approaches to developing information about drug metabolism and drug-drug interactions. Mechanistic and empirical clinical study approaches are available as well to provide further information. As always, a carefully designed mix of approaches is likely to yield optimal results in the shortest time and at the least cost.

The timing of DDI studies is also addressed:

> Metabolic effects and drug-drug interactions should be considered as early as possible as well as later in the drug development process. Appropriately designed pharmacokinetic/ phase 1 studies could provide important information about drug metabolism, relevant metabolites, and actual or potential drug interactions. Blood level data obtained during phase 2 and 3 clinical trials, for example, via a pharmacokinetic screen, also could reveal interactions or marked inter-individual differences. Because clinical trial protocols some-times limit concomitant drug use, some later studies may not be optimally informative about possible drug interactions. Decreasing exclusions of concomitant drug treatment and measurement of blood levels before and after treatment with a test drug (interaction screen), as well as testing drug blood levels more frequently, could make later phase clinical studies more useful. All of these studies could be more informative if significant metabo-lites and prodrugs could be identified and their pharmacological properties described.

The term "significant metabolites" is now outdated and does not fit with the current terminology used in the final guidance, "Drug Safety Testing of Metabolites" (US FDA, 2008j). It is also unclear exactly what is meant by the phrase "prodrugs could be metabolized." It could be speculated that this refers to the observation that a sponsor may have initially believed that the parent drug was active (i.e. terfenadine) but later discovered that a metabolite was actually responsible for the majority of the pharma-cological activity (i.e. fexofenadine).

The guidance gives an overview of the techniques and approaches for studies *in vitro* of drug metabolism and drug interactions. The guidance states that "The goals in evaluating in vitro drug metabolism are:

(1) to identify all of the major metabolic pathways that affect the test drug and its metabolites, including the specific enzymes responsible for elimination and the intermediates formed; and

(2) to explore and anticipate the effects of the test drug on the metabolism of other drugs and the effects of other drugs on its metabolism."

The guidance makes the interesting point that:

> Knowledge that a particular drug is not a substrate for certain metabolic pathways is helpful. For example, if it is learned early in drug development that a molecule is not a substrate for CYP450 3A4 or that this pathway represents only a minor contribution to overall metabolism, then concern is lessened or eliminated for possible inhibition of 3A4 metabolism by drugs such as ketoconazole and erythromycin or possible induction of metabolism by drugs such as rifampin and anticonvulsants. Studies in vitro also could indicate whether a drug itself is or is not an inhibitor of common metabolic pathways. The potential for a drug inhibiting the metabolism of other drugs is almost always present for drugs metabolized by the same pathway, but can also be present for entirely separate pathways, including the principal metabolic route for a compound. This potential was first appreciated for quinidine, which is a substrate for metabolism by CYP450 3A4 and is also a very potent inhibitor of CYP450 2D6.

Study designs are also addressed in this guidance. The guidance gives some very basic information regarding the cytochrome P450 system. This information is not pre-sented here due to its basic nature and the fact that the information is almost 10 years old and may no longer represent state of the art information. The guidance addresses the identification of the CYP isoforms involved in the metabolism of the drug as well as the ability of the drug to inhibit the activity of select CYP isoforms.

Although the main focus of the guidance is CYP mediated metabolism, the guidance does note that other drug metabolizing enzymes may play a role in the metabolism of a particular drug as may cytosolic enzymes.

The guidance acknowledges that the principal site of drug metabolism but also recognizes that "For particular drugs, however, other tissues may predominate (e.g., the kidney or gastrointestinal mucosa)." Furthermore, the guidance states: "Because most drugs are given orally, interest has been increasing in the effect of gastrointestinal mucosal enzymes on drug entry to the systemic circulation. Drugs susceptible to metabolism via CYP450 3A4 may exhibit low and/or variable bioavailability. Thus, determining the susceptibility of a drug to metabolism by CYP450 3A4 may be important not only in identifying routes of elimination but also in predicting the likelihood of significant first-pass metabolism."

The guidance notes that:

When a difference arises between findings in vitro and in vivo, the results in vivo should always take precedence over studies in vitro. In many cases, however, studies in vitro, which are inexpensive and readily carried out, will serve as an adequate screening mechanism that can rule out the importance of a metabolic pathway and make in vivo testing unnecessary.

For example:

If investigations in vitro suggest that the answer to the question "Does CYP450 2D6 metabolize this drug?" is "no," clinical studies to identify the impact of the slow metabolizer phenotype or to study the effect of CYP450 2D6 inhibitors will not be needed. Because studies in vitro, however, cannot adequately define the importance of a metabolic pathway, if the in vitro study answer is "yes," additional clinical studies will be important to answer whether CYP450 2D6 is clinically important to the elimination of the drug.

The guidance makes the important point that "although a drug may be extensively metabolized in vitro, a mass balance study in vivo may demonstrate that metabolism is less important than urinary or biliary excretion."

In 1999, the FDA issued the guidance, "In Vivo Drug Metabolism/Drug Interaction Studies—Study Design, Data Analysis, and Recommendations for Dosing and Labeling" (US FDA, 1999a).

In a manner similar to other guidances, the rationale for the conduct of DDI studies is laid out:

Examples of substantially changed exposure associated with administration of another drug include (1) increased levels of terfenadine, cisapride, or astemizole with ketoconazole or erythromycin (inhibition of CYP3A4); (2) increased levels of simvastatin andits acid metabolite with mibefradil or itraconazole (inhibition of CYP3A4); (3) increased levels of desipramine with fluoxetine, paroxetine, or quinidine (inhibition of CYP2D6); and (4) decreased carbamazepine levels with rifampin (induction of CYP3A4). These large changes in exposure can alter the safety and efficacy profile of a drug and/or its active metabolites in important ways. This is most obvious and expected for a drug with a narrow therapeutic range (NTR), but is also possible for non-NTR drugs as well (e.g., HMG CoA reductase inhibitors). Depending on the extent and consequence of the interaction, the fact that a drug's metabolism can be significantly inhibited by other drugs and that the drug itself can inhibit the metabolism of other drugs can require important changes in either its dose or the doses of drugs with which it interacts, that is, on its labeled conditions of use. Rarely, metabolic drug-drug interactions may affect the ability of a drug to be safely marketed.

Additionally, the guidance describes the underlying concepts for the guidance:

- Adequate assessment of the safety and effectiveness of a drug includes a description of its metabolism and the contribution of metabolism to overall elimination.

- Metabolic drug-drug interaction studies should explore whether an investigational agent is likely to significantly affect the metabolic elimination of drugs already in the market-place and, conversely, whether drugs in the marketplace are likely to affect the metabolic elimination of the investigational drug.

- Even drugs that are not substantially metabolized can have important effects on the metabolism of concomitant drugs. For this reason, metabolic drug-drug interactions should be explored, even for an investigational compound that is not eliminated signifi-cantly by metabolism.

- In some cases, metabolic drug-drug interaction studies cannot be informative unless metabolites and prodrugs have been identified and their pharmacological properties described.

- Identifying metabolic differences in patient groups based on genetic polymorphism, or on other readily identifiable factors, such as age, race, and gender, can aid in interpreting results.

- The impact of an investigational or approved interacting drug can be either to inhibit or induce metabolism.

- A specific objective of metabolic drug-drug interaction studies is to determine whether the interaction is sufficiently large to necessitate a dosage adjustment of the drug itself or the drugs it might be used with, or whether the interaction would require additional therapeutic monitoring.

- In some instances, understanding how to adjust dosage in the presence of an interacting drug, or how to avoid interactions, may allow marketing of a drug that would otherwise have been associated with an unacceptable level of toxicity. Sometimes a drug interaction may be used intentionally to increase levels or reduce elimination of another drug. Rarely, the degree of interaction caused by a drug, or the degree to which other drugs alter its metabolism, may be such that it cannot be marketed safely.

The guidance discusses the importance of "the development of sensitive and specific assays for a drug and its key metabolites." However, the guidance does not define what a "key metabolite" is. This terminology is no longer in vogue in current guidances.

The guidance makes some general statements about the magnitude of DDIs and the development of DDIs:

Unlike relatively fixed influences on metabolism, such as hepatic function or genetic char-acteristics, metabolic drug-drug interactions can lead to abrupt changes in exposure. Depending on the nature of the drugs, these effects could potentially occur when a drug is initially administered, when it has been titrated to a stable dose, or when an interacting drug is discontinued. Interactions can occur after even a single concomitant dose of an inhibitor.

The timing of understanding the potential for drug to be involved in DDIs (either as a perpetrator or as a victim) is also addressed:

> The effects of an investigational drug on the metabolism of other drugs and the effects of other drugs on an investigational drug's metabolism should be assessed relatively early in drug development so that the clinical implications of interactions can be assessed as fully as possible in later clinical studies.

The design of *in vivo* DDI studies is addressed in detail. This guidance advises that "If in vitro studies and other information suggest a need for in vivo metabolic drug-drug interaction studies, the following general issues and approaches should be considered." However, at times it may be prudent for the sponsor to conduct DDI studies to confirm the prediction of a lack of DDI. Such studies will be important when drugs will be administered in combination for an intended therapeutic effect or for marketing purposes to show that two drugs will not interact with one another:

> In vivo metabolic drug-drug interaction studies generally are designed to compare substrate levels with and without the interacting drug. Because a specific study may consider a number of questions and clinical objectives, no one correct study design for studying drug–drug interactions can be defined. A study can use a randomized crossover (e.g., S followed by S + I, S + I followed by S), a one-sequence crossover (e.g., S always followed by S + I or the reverse), or a parallel design (S in one group of subjects and S + I in another).

The point that there is no "one size fits all" approach is an important one and the guidance is consistent with the regulatory agencies point of view that drugs should be developed on a "case by case" basis:

> The following possible dosing regimen combinations for a substrate and interacting drug may also be used: single dose/single dose, single dose/multiple dose, multiple dose/single dose, and multiple dose/multiple dose. The selection of one of these or another study design depends on a number of factors for both the substrate and interacting drug, including (1) acute or chronic use of the substrate and/or interacting drug; (2) safety considerations, including whether a drug is likely to be an NTR (narrow therapeutic range) or non-NTR drug; (3) pharmacokinetic and pharmacodynamic characteristics of the substrate and interacting drugs; and (4) the need to assess induction as well as inhibition. The inhibiting/inducing drugs and the substrates should be dosed so that the exposure of both drugs are relevant to their clinical use.

The guidance then further describes selected situations, including:

> When both substrate and interacting drug are likely to be given chronically over an extended period of time, administration of the substrate to steady state with collection of blood samples over one or more dosing intervals could be followed by multiple dose administration of the interacting drug, again with collection of blood for measurement of both the substrate and the interacting drug (as feasible) over the same intervals. This is an example of a one-sequence crossover design.

At times an extended dosing period may be required as noted in the guidance:

> The time at steady state before collection of endpoint observations depends on whether inhibition or induction is to be studied. Inducers can take several days or longer to exert

their effects, while inhibitors generally exert their effects more rapidly. For this reason, a more extended period of time after attainment of steady state for the substrate and interacting drug may be necessary if induction is to be assessed.

However, repeat-dose administration is not always required. The guidance states that "For a rapidly reversible inhibitor, administration of the interacting drug either just before or simultaneously with the substrate on the test day might be the appropriate design to increase sensitivity."

In other circumstances, a nontraditional approach may be required:

When attainment of steady state is important and either the substrate or interacting drugs and/or their metabolites exhibit long half-lives, special approaches may be useful. These include use of a loading dose to achieve steady state conditions more rapidly and selection of a one-sequence crossover or a parallel design, rather than a randomized crossover study design.

The achievement of steady state should not be assumed, but should be proved, as indicated in this guidance:

When a substrate and/or an interacting drug is to be studied at steady state, documentation that near steady state has been attained is important both for each drug and its metabolites of interest. This documentation can be accomplished by sampling over several days prior to the periods when samples are collected. This is important for both metabolites and the parent drug, particularly when the half-life of the metabolite is longer than the parent, and is especially important if both parent drug and metabolites are metabolic inhibitors or inducers.

The guidance does not define what a "metabolite of interest" is. The design of DDI studies can become complicated for drugs that will be administered in combination. The guidance states:

If the drug interaction effects are to be assessed for both agents in a combination regimen, the assessment can be done in two separate studies. If the pharmacokinetic and pharmacodynamic characteristics of the drugs make it feasible, the dual assessment can be done in a single study. Some design options are randomized three-period crossover, parallel group, and one-sequence crossover.

These statements introduce the concept of DDIs as a consequence of induction of drug metabolizing agents, as is the case with St. John's wort. The prediction of the ability of a drug to induce DMEs is not a trivial task. In nonclinical studies, the induction of DMEs is usually only evaluated if auto-induction of the drug's metabolism is observed when the drug is repeatedly administered to the animal. Even if auto-induction of metabolism is observed, sponsors do not frequently identify which enzyme systems have been induced. The reason for this approach will vary from company to company, but is most likely due to the differences in substrate specificities between animal and human DMEs. Auto-induction of a drug's metabolism is a common phenomenon in rats, but rarely occurs in monkeys or dogs. Alternative methods for the prediction of induction of DMEs or transporters include human hepatocyte systems and assays to determine the binding of the candidate drug to certain nuclear receptors. Both of these techniques have their limitations. Even if induction of DMEs or transporters is predicted, this knowledge may not lead to the discontinuation of the development of a drug. However, such knowledge may lead to an assessment of the induction of DMEs or transporters in early clinical development. Such studies may involve a simple evaluation of the pharmacokinetics of the drug after multiple-dose administration, an evaluation of urinary cortisol/6β-OH cortisol ratio (as a crude marker for CYP3A activity), or by the use of a cocktail of substrates of different CYP isoforms. The urinary cortisol/6β-OH cortisol ratio assay is not widely accepted as a robust assay for the evaluation of CYP3A activity, although it can provide some useful information. The limitations of this assay have been described in the literature (Ohno *et al.*, 2000; Galteau and Shamsa, 2003; Micuda *et al.*, 2007). The different composition of the CYP cocktails, and the pros and cons of the use of a CYP cocktail, to evaluate the potential for drug–drug interactions have been debated in the literature (Chainuvati *et al.*, 2003; Zhou, Tong and McLeod, 2004; Ryu *et al.*, 2007).

The subject of the patient population to be used is also addressed:

> Clinical drug-drug interaction studies may generally be performed using healthy volunteers or volunteers drawn from the general population, on the assumption that findings in this population should predict findings in the patient population for which the drug is intended. Safety considerations, however, may preclude the use of healthy subjects. In certain circumstances, subjects drawn from the general population and/or patients for whom the investigational drug is intended offer certain advantages, including the opportunity to study pharmacodynamic endpoints not present in healthy subjects and reduced reliance on extrapolation of findings from healthy subjects. In either patient or healthy/general population subject studies, performance of phenotype or genotype determinations to identify genetically determined metabolic polymorphisms is often important in evaluating effects on enzymes with polymorphisms, notably CYP2D6 and CYP2C19.

Examples of where safety considerations preclude the use of healthy subjects include drugs for the treatment of cancer. However, it should be noted that noncytotoxic drugs can, and have, been safely administered to humans. For example, the interaction of Sutent (sunitinib) with ketoconazole and rifampin was evaluated in normal volunteers (US FDA, 2006l). It should be noted that conducting DDI studies in patient populations may be fraught with difficulties and the level of rigor that may be achieved in studies in normal healthy volunteers may not be able to be achieved in patients (e.g. cancer patients). The design of such studies in cancer patients may involve an abbreviated study design and the collection of fewer samples. It may also not be ethical to conduct a true DDI study since this would involve the withdrawal of all medications other than the drugs to be evaluated in the DDI study. It should also be noted that the use of nonprescription, herbal medications by cancer patients is not uncommon. Such substances would have a pronounced effect on the outcome of the DDI studies, as well as therapeutic response (Meijerman, Beijnen and Schellens, 2006; Marchetti *et al.*, 2007). However, many cancer (or other) patients do not think of these substances as having a potential effect and may not indicate their use of these substances when being screened for the study. Recent estimates suggest an overall prevalence for herbal preparation use of 13% to 63% among cancer patients and that 72% of the cancer patients do not inform their physicians of their use of complementary and alternative medicines (Sparreboom *et al.*, 2004).

It should also be noted that smoking and the ingestion of certain foods such as charbroiled meats or cruciferous vegetables may also impact the outcome of DDI (or other) studies. Food-derived chemicals, including constituents of cruciferous vegetables and fruits, modulate the expression of certain CYP isoforms and phase II enzymes. For example, some dietary indoles and flavonoids activate CYP1A expression. Apart from altered CYP regulation, a number of dietary agents also inhibit CYP enzyme activity, leading to pharmacokinetic interactions with coadministered drugs. A well-described example is that of grapefruit juice, which contains psoralens and possibly other chemicals, which inactivate intestinal CYP3A4. Chemicals in teas and cruciferous vegetables may also inhibit human CYP enzymes that have been implicated in the bioactivation of chemical carcinogens. Thus, food constituents modulate CYP expression and function by a range of mechanisms, with the potential for both deleterious and beneficial outcomes (Murray, 2006).

It is interesting to note that certain clinical trials or therapies may take advantage of DDIs. Kaletra is a co-formulation of lopinavir and ritonavir. The ritonavir inhibits the CYP3A-mediated metabolism of lopinavir, thereby providing increased plasma levels of lopinavir (Abbott Laboratories, 2007). Presently, there is an ongoing clinical trial of PKC412 with or without itraconazole in treating patients with acute myeloid leukemia or myelodysplastic syndrome (U.S. National Institutes of Health, 2008). Although not overtly stated, the rationale for this combination appears to be the need to maintain high levels of PKC412 which has previously been shown to induce its own metabolism upon repeated administration to cancer patients (Propper *et al.*, 2001). In this treatment paradigm, PKC412 would induce its own metabolism but the metabolism of PKC412 would be inhibited by itraconazole, leading to elevated plasma concentrations of PKC412 compared with patients that had received PKC412 alone.

The choice of substrate and interacting drugs is also dealt with in great detail (US FDA, 1999a). For clarity, the majority of this section of the guidance is repeated here verbatim. The choice of the substrates for an investigational drug is addressed thus:

> In contrast to earlier approaches that focused mainly on a specific group of approved drugs (digoxin, hydrochlorothiazide) where coadministration was likely or the clinical consequences of an interaction were of concern, improved understanding of the metabolic basis of drug-drug interactions enables more general approaches to and conclusions from specific drug-drug interaction studies. In studying an investigational drug as the interacting drug, the choice of substrates (approved drugs) for initial in vivo studies depends on the P450 enzymes affected by the interacting drug. In testing inhibition, the substrate selected should generally be one whose pharmacokinetics is markedly altered by coadministration of known specific inhibitors of the enzyme systems (i.e., a very sensitive substrate should be chosen) to assess the impact of the interacting investigational drug. Examples of substrates include, but are not limited to, (1) midazolam, buspirone, felodipine, simvastatin, or lovastatin for CYP3A4; (2) theophylline for CYP1A2; (3) S-warfarin for CYP2C9; and (4) desipramine for CYP2D6. If the initial study is positive for inhibition, further studies of other substrates may be useful, representing a range of substrates based on the likelihood of coadministration. For example, possible substrates for further study of a CYP3A4 interacting investigational drug might include dihydropyridine calcium channel blockers and triazolobenzodiazepines, or for a CYP2D6 inhibiting investigational drug might include metoprolol. If the initial study is negative with the most sensitive substrates, it can be presumed that less sensitive substrates will also be unaffected.

The subject of the investigational drug as a substrate is addressed in the following manner:

> In testing an investigational drug for the possibility that its metabolism is inhibited or induced (i.e., as a substrate), selection of the interacting drugs should be based on in vitro or other metabolism studies identifying the enzyme systems that metabolize the drug. The

choice of interacting drug should then be based on known, important inhibitors of the pathway under investigation. For example, if the investigational drug is shown to be metabolized by CYP3A4 and the contribution of this enzyme to the overall elimination of this drug is substantial, the choice of inhibitor and inducer could be ketoconazole and rifampin, respectively, because of the substantial effects of these interacting drugs on CYP3A4 metabolism (i.e., they are the most sensitive in identifying an effect of interest). If the study results are negative, then absence of a clinically important drug-drug interaction for the metabolic pathway could be claimed. If the clinical study of the most potent specific inhibitor/inducer is positive and the sponsor wishes to claim lack of an interaction between the test drug and other less potent specific inhibitors, or give advice on dosage adjustment, further clinical studies would generally be recommended. Certain approved drugs are not optimal selections as the interacting drug. For example, cimetidine is not considered an optimal choice to represent drugs inhibiting a given pathway because its inhibition affects multiple metabolic pathways as well as certain drug transporters.

This statement focuses on "class effects" where a class would be all substrates of a given CYP isoform. Although studies of this type are often conducted, studies using substrate or inhibitor drugs that would potentially be commonly coadministered with the test agent may be appropriate. When designing the DDI component of the clinical pharmacology plan, the sponsor should pay attention to the use of agents that may not routinely be recognized as concomitant medications. For example, if the test agent is a CYP3A substrate, the effect of St. John's wort or grapefruit juice on the metabolism of the test agent should be considered. Conversely, if the test agent will be used in women of childbearing potential, the effect of the test agent on the metabolism of oral contraceptives could be considered.

The guidance notes that the "for an investigational agent used as either an interacting drug or substrate, the route of administration should generally be the one planned for in product labeling. When multiple routes are being developed, the necessity for doing metabolic drug-drug interaction studies by all routes should be based on the expected mechanism of interaction and the similarity of corresponding concentration-time profiles for parent and metabolites." It further states that the dose selected for the DDI study should "maximize the possibility of finding an interaction. For this reason, the maximum planned or approved dose and shortest dosing interval of the interacting drug (as inhibitors or inducers) should be used. Doses smaller than those to be used clinically may be needed for substrates on safety grounds and may be more sensitive to the effect of the interacting drug." This latter point is one the sponsor should play close attention to.

The end points of the study, in terms of pharmacokinetic parameters are primarily measurements of exposure, such as AUC, C_{max}, time to C_{max} (T_{max}), and others as appropriate. Additionally, pharmacokinetic parameters such as clearance, volumes of distribution, and half-lives may be calculated. Additional measures may help in steady-state studies (e.g. trough concentration, C_{min}) to demonstrate that dosing strategies were adequate to achieve near steady state before and during the interaction. The guidance gives additional details regarding the calculation of PK parameters and the collection of samples.

The focus of the guidance is clearly illustrated in the following statement: "Because this guidance focuses on metabolic drug-drug interactions, protein binding determinations are considered unnecessary except for data interpretation." Additionally, this guidance does not address transporter-mediated DDIs. Since many drugs are substrates or inhibitors of both DMEs and transporters, it may be difficult to determine whether an observed interaction is due to an effect on the metabolism or transport of the drug. Although it would be ideal to be able to deconvolute the individual contributions of metabolism and transport to a given DDI, it is not essential to do so. The magnitude of effect should be reported in the label, regardless of whether the interaction is due to a metabolic, transport or plasma protein binding DDI.

The guidance has not addressed some factors (e.g. gender, age, the use of multiple inhibitors, effect on GI microflora, effect of GI motility, or the effect of pH) that could play a role in the extent of a DDI.

DDIs may arise as a consequence of the physicochemical properties of the drug. For drugs that show marked pH-dependent solubility, an evaluation of the effect of pH on the exposure to the drug should be conducted. This type of DDI may occur when the drug is taken with antacids. The interaction can be "traditional" (i.e. inhibition of CYP-mediated metabolism) or "nontraditional" (change in pH resulting in change of solubility of the drug or physical adsorption of the drug to the antacid) (Neuvonen and Kivisto, 1994; Sadowski, 1994). Additionally, it should be noted that the pH of the stomach can change with age. For example, approximately 25% of elderly subjects suffer from achlorhydria, a condition in which the pH of the stomach is elevated (Mallet, 2008). Achlorhydria can interfere with the absorption of ketoconazole (Hurwitz et al., 2003). This effect can be extremely pronounced and it has been noted that a manufacturer of ketoconazole has recommended that "in cases of achlorhydria, the patients should be advised to dissolve each tablet in 4 ml aqueous solution of 0.2 N HCl" and to "use a drinking straw so as to avoid contact with the teeth." On a related note, acidic beverages (e.g. Coca-Cola) have been shown to enhance the exposure to carbamazepine and itraconazole (Lange et al., 1997; Malhotra, Dixit and Garg, 2002). It is possible that the change in stomach pH (from pH 1.4–2.1 to pH 3–7) when food is administered may be a contributing factor to observed effect of food upon the absorption of a given drug (Dressman et al., 1998).

The guidance (US FDA, 1999a) notes that the desired goal of the DDI study is to "determine the clinical significance of any increase or decrease in exposure to the substrate in the presence of the interacting drug" and that:

> Results of drug-drug interaction studies should be reported as 90% confidence intervals about the geometric mean ratio of the observed pharmacokinetic measures with (S + I) and without the interacting drug (S). Confidence intervals provide an estimate of the distribution of the observed systemic exposure measure ratio of S + I versus S alone and convey a probability of the magnitude of the interaction. In contrast, tests of significance are not appropriate because small, consistent systemic exposure differences can be statistically significant ($p < 0.05$) but not clinically relevant.

The guidance further addresses this topic in great detail:

> When a drug-drug interaction is clearly present (e.g., comparisons indicate twofold or greater increments in systemic exposure measures for S + I) the sponsor should be able to provide specific recommendations regarding the clinical significance of the interaction based on what is known about the dose-response and/or PK/PD relationship for either the investigational agent or the approved drugs used in the study. ... The sponsor may wish to make specific claims in the package insert that no drug-drug interaction is expected. In these instances, the sponsor should be able to recommend specific no effect boundaries, or clinical equivalence intervals, for a drug-drug interaction. No effect boundaries define the interval within which a change in a systemic exposure measure is considered not clinically meaningful.

There are three approaches to define no effect boundaries.

Approach 1: No effect boundaries can be based on population (group) average dose and/or concentration-response relationships, PK/PD models, and other available information for the substrate drug. If the 90% confidence interval for the systemic exposure measurement in the drug-drug interaction study falls completely within the no effect boundaries, the sponsor may conclude that no clinically significant drug-drug interaction was present.

Approach 2: No effect boundaries may also be based on the concept that a drug-drug interaction study addresses the question of switchability between the substrate given in combination with an interacting drug (test) versus the substrate given alone. Based on this concept, the sponsor may wish to use an individual equivalence criterion to allow scaling of the no effect boundary and to determine other useful information as well. Sponsors who wish to use this approach are encouraged to contact the Office of Clinical Pharmacology and Biopharmaceutics to discuss approaches to study design and data analysis.

Approach 3: In the absence of no effect boundaries defined in (1) or (2) above, a sponsor may use a default no effect boundary of 80–125% for both the investigational drug and the approved drugs used in the study. When the 90% confidence intervals for systemic exposure ratios fall entirely within the equivalence range of 80–125%, standard Agency practice is to conclude that no clinically significant differences are present.

The guidance notes that:

The selection of the number of subjects for a given drug-drug interaction study will depend on how small an effect is clinically important to detect, or rule out, the inter- and intrasubject variability in pharmacokinetic measurements, and possibly other factors or sources of variability not well recognized. In addition, the number of subjects will depend on how the results of the drug-drug interaction study will be used, as described above. This guidance should not be interpreted by sponsors as generally recommending the inclusion of some number of subjects in a drug-drug interaction study such that the 90% confidence interval for the ratio of pharmacokinetic measurements falls entirely within the no effect boundaries of 80–125%. This approach, however, could be deemed appropriate by a sponsor, after considering the expected outcome of a drug-drug interaction study, the anticipated magnitude of variability in pharmacokinetic measurements, and the desired label claim that no clinically significant drug-drug interaction was present.

Since this guidance was issued in 1999, our knowledge of the design, conduct, and interpretation of DDIs studies has evolved. This has led to the FDA issuing a draft guidance "Drug Interaction Studies—Study Design, Data Analysis, and Implications for Dosing and Labeling" in 2006 (US FDA, 2006g). This draft guidance mainly focuses on *in vivo* studies but does also address some aspects of *in vitro* studies. It is a substantial guidance, totaling 52 pages in its draft form and containing a great deal of information in terms of basic information regarding substrates and inhibitors as well as detailed description of potential study designs. A large amount of this information can be found in the Appendices which account for over 66% of the page count for this draft guidance.

The major advance in this guidance is the recognition of the importance of drug transporters in DDIs. The guidance makes this point thus, "Furthermore, not every drug-drug interaction is metabolism-based, but may arise from changes in pharmacokinetics caused by absorption, distribution, and excretion interactions. Drug-drug interactions related to transporters are being documented with increasing frequency and are important to consider in drug development." The guidance notes the interrelated nature of guidances by stating that "Discussion of metabolic and other types of drug-drug interactions is also provided in other guidances, including the International Conference on Harmonization (ICH) E7 Studies in Support of Special Populations: Geriatrics, and E3 Structure and Content of Clinical Study Reports, and FDA guidances for industry on Studying Drugs Likely to be Used in the Elderly and Study and Evaluation of Gender Differences in the Clinical Evaluation of Drugs."

The guidance addresses several different types of DDIs. Firstly, the draft guidance addresses metabolism-based drug–drug interactions. The draft guidance provides a general introduction to this topic:

> Many metabolic routes of elimination, including most of those occurring through the P450 family of enzymes, can be inhibited or induced by concomitant drug treatment. Observed changes arising from metabolic drug-drug interactions can be substantial – an order of magnitude or more decrease or increase in the blood and tissue concentrations of a drug or metabolite – and can include formation of toxic and/or active metabolites or increased exposure to a toxic parent compound. These large changes in exposure can alter the safety and efficacy profile of a drug and/or its active metabolites in important ways. This is most obvious and expected for a drug with a narrow therapeutic range (NTR), but is also possible for non-NTR drugs as well (e.g., HMG CoA reductase inhibitors).

The purpose of the evaluation of metabolism-based DDIs is addressed:

> It is important that metabolic drug-drug interaction studies explore whether an investigational agent is likely to significantly affect the metabolic elimination of drugs already in the marketplace and likely in medical practice to be taken concomitantly and, conversely, whether drugs in the marketplace are likely to affect the metabolic elimination of the investigational drug. Even drugs that are not substantially metabolized can have important effects on the metabolism of concomitant drugs. For this reason, metabolic drug-drug interactions should be explored, even for an investigational compound that is not eliminated significantly by metabolism.

In this draft guidance, the applicability of the evaluation of metabolism-based DDIs for therapeutic biologics is described:

> Classical biotransformation studies are not a general requirement for the evaluation of therapeutic biologics (ICH guidance *S6 Preclinical Safety Evaluation of Biotechnology-Derived Pharmaceuticals*), although certain protein therapeutics modify the metabolism of drugs that are metabolized by the P450 enzymes. Type I interferons, for example, inhibit CYP1A2 production at the transcriptional and post-translational levels, inhibiting clearance of theophylline. The increased clinical use of therapeutic proteins may raise concerns regarding the potential for their impacts on drug metabolism. Generally, these interactions cannot be detected by in vitro assessment. Consultation with FDA is appropriate before initiating metabolic drug-drug interaction studies involving biologics.

It is interesting to see the mention of the potential effect of biologics on CYP expression. This topic has not previously been addressed in guidances related to DDIs. Additionally, the potential for DDIs between therapeutic monoclonal antibodies is one that has not been discussed in great depth by regulatory agencies. However, this subject was recently extensively reviewed by Seitz and Zhou (2007).

The possibility of a DDI-based on the effect of drugs on the degradation of CYP isoforms has not been addressed. Although the mechanisms of degradation of the CYP isoforms have been described, the relevance of this mechanism as a potential source of DDIs is unknown (Correia, 2003; Correia, Sadeghi and Mundo-Paredes, 2005).

The role of genotype on metabolism-based DDIs is also addressed:

Identifying metabolic differences in patient groups based on genetic polymorphism, or on other readily identifiable factors, such as age, race, and gender, can aid in interpreting results. The extent of interactions may be defined by these variables (e.g., CYP2D6 genotypes). Further, in subjects who lack the major clearance pathway, remaining pathways become important and should be understood and examined. A specific objective of metabolic drug-drug interaction studies is to determine whether the interaction is sufficiently large to necessitate a dosage adjustment of the drug itself or the drugs with which it might be used, or whether the interaction would require additional therapeutic monitoring. In some instances, understanding how to adjust dose or dosage regimen in the presence of an interacting drug, or how to avoid interactions, may allow marketing of a drug that would otherwise have been associated with an unacceptable level of toxicity. Sometimes a drug interaction can be used intentionally to increase levels or reduce elimination of another drug (e.g., ritonavir and lopinavir). Rarely, the degree of interaction caused by a drug, or the degree to which other drugs alter its metabolism, can be such that it cannot be marketed safely.

As can be seen, some of the wording of this draft guidance is taken verbatim from the 1999 guidance. It will be interesting to see whether the 1999 guidance is withdrawn once this draft guidance is finalized.

Transporter-based drug–drug interactions are mentioned, albeit with a lesser extent of detail than was provided for metabolism-based DDIs:

Transporter-based interactions have been increasingly documented. Examples of these include the inhibition or induction of transport proteins, such as P glycoprotein (P-gp), organic anion transporter (OAT), organic anion transporting polypeptide (OATP), organic cation transporter (OCT), multidrug resistance-associated proteins (MRP), and breast cancer resistant protein (BCRP). Examples of transporter-based interactions include the interactions between digoxin and quinidine, fexofenadine and ketoconazole (or erythromycin), penicillin and probenecid, and dofetilide and cimetidine. Of the various transporters, P-gp is the most well understood and may be appropriate to evaluate during drug development.

The draft guidance provides a useful table that lists some of the major human transporters and known substrates, inhibitors, and inducers.

This draft guidance provides some input into the strategy of the evaluation of DDI studies. Guidances have traditionally not provided much input into strategy and therefore the fact that this draft guidance provides thoughts on this topic is novel and refreshing. The draft guidance states:

To the extent possible, drug development should follow a sequence in which early in vitro and in vivo investigations can either fully address a question of interest or provide information to guide further studies. Optimally, a sequence of studies could be planned, moving from in vitro studies to in vivo human studies, including those employing special study designs and methodologies where appropriate. In many cases, negative findings from early in vitro and early clinical studies can eliminate the need for later clinical investigations. Early investigations should explore whether a drug is eliminated primarily by excretion or metabolism, with identification of the principal metabolic routes in the latter case. Using suitable in vitro probes and careful selection of interacting drugs for early in vivo studies, the potential for drug-drug interactions can be studied early in the development process, with further study of observed interactions assessed later in the process, as needed. These early studies can also provide information about dose, concentration, and response relationships in the general population, specific populations, and individuals, which can be useful in interpreting the consequences of a drug-drug interaction. Once potential drug-drug interactions have been identified, based on in vitro and/or in vivo studies, sponsors are encouraged to design and examine the safety and efficacy databases of larger clinical studies, as feasible, to (1) permit confirmation/discovery of the interactions predicted from earlier studies and/or (2) verify that dosage adjustments or other prescribing modifications made in response to the potential interaction(s) have been adequate to avoid undesired consequences of the drug-drug interaction.

The guidance clearly indicates that an early understanding of the major route of clearance would be useful, but does not provide insight as to how the relative contributions of metabolism and excretion are determined for an orally administered drug. The strategy for *in vitro* studies is outlined in the following manner:

A complete understanding of the quantitative relationship between the in vitro findings and in vivo results of metabolism/drug-drug interaction studies is still emerging. Nonetheless, in vitro studies can frequently serve as a screening mechanism to rule out the importance of a metabolic pathway and the drug-drug interactions that occur through this pathway so that subsequent in vivo testing is unnecessary. This opportunity should be based on appropriately validated experimental methods and rational selection of substrate/interacting drug concentrations. For example, if suitable in vitro studies at therapeutic concentrations indicate that CYP1A2, CYP2C8, CYP2C9, CYP2C19, CYP2D6, or CYP3A enzyme systems do not metabolize an investigational drug, then clinical studies to evaluate the effect of CYP2D6 inhibitors or CYP1A2, CYP2C8, CYP2C9, CYP2C19, or CYP3A inhibitors/inducers on the elimination of the investigational drug will not be needed. Similarly, if in vitro studies indicate that an investigational drug does not inhibit CYP1A2, CYP2C8, CYP2C9, CYP2C19, CYP2D6, or CYP3A metabolism, then corresponding in vivo inhibition-based interaction studies of the investigational drug and concomitant medications eliminated by these pathways are not needed. Figure 1 in Appendix B shows a decision tree on when in vivo interaction studies are indicated based on in vitro metabolism, inhibition, and induction and in vivo metabolism data. The CYP2D6 enzyme has not been shown to be inducible. Recent data have shown co-induction of CYP2C, CYP2B and ABCB1 (P-gp) transporter with CYP3A. CYP3A appears to be sensitive to all known co-inducers. Therefore, to evaluate whether an investigational drug induces CYP1A2, CYP2C8, CYP2C9, CYP2C19, or CYP3A, the initial in vitro induction evaluation may include only CYP1A2 and CYP3A. If in vitro studies indicate that an investigational drug does not induce CYP3A metabolism, then in vivo induction-based interaction studies of the investigational drug and concomitant medications eliminated by CYP2C/CYP2B and CYP3A may not be needed. Drug interactions based on CYP2B6 are emerging as important interactions. When appropriate, in vitro evaluations based on this enzyme can be conducted. Other CYP enzymes, including CYP2A6 and CYP2E1, are less likely to be involved in clinically important drug interactions, but should be considered when appropriate.

This section identifies the inter-relatedness of the induction of selected CYP isoforms and drug transporters and a refinement in the understanding of the relative importance of the CYP isoforms (notably the CYP 2C family and CYP2B6). The induction of CYP isoforms is a topic that has not been well developed in previous guidances. The Appendices for this draft guidance describe general considerations in the *in vitro* evaluation of CYP-related metabolism and interactions; considerations in the experimental design, data analysis, and data interpretation in DME identification, including CYP enzymes (new drug as a substrate), CYP inhibition (new drug as an inhibitor), and CYP induction (new drug as an inducer), respectively. Additionally, the Appendices contain general considerations in the *in vitro* evaluation of P-gp substrates and inhibitors, including decision trees on when *in vivo* P-gp-based interaction studies are indicated based on *in vitro* evaluation. *In vivo* clinical investigations are also addressed in detail. The draft guidance notes that:

> In addition to in vitro metabolism and drug-drug interaction studies, appropriately designed pharmacokinetic studies, usually performed in the early phases of drug development, can provide important information about metabolic routes of elimination, their contribution to overall elimination, and metabolic drug-drug interactions. Together with information from in vitro studies, these in vivo investigations can be a primary basis of labeling statements and can often help avoid the need for further investigations.

The last statement is interesting, since it allows for the possibility that *in vitro* data can be used for labeling statements and that *in vivo* studies may not be required in all situations. The product label for Inspra (eplerenone) states that "Eplerenone is not an inhibitor of CYP1A2, CYP3A4, CYP2C19, CYP2C9, or CYP2D6. Eplerenone did not inhibit the metabolism of chlorzoxazone, diclofenac, methylphenidate, losartan, amiodarone, dexamethasone, mephobarbital, phenytoin, phenacetin, dextromethorphan, metoprolol, tolbutamide, amlodipine, astemizole, cisapride, 17α-ethinyl estradiol, fluoxetine, lovastatin, methylprednisolone, midazolam, nifedipine, simvastatin, triazolam, verapamil, and warfarin in vitro" (G.D. Searle LLC, a Division of Pfizer Inc. 2008). One point that is not addressed by this draft guidance is the assessment of the stability of the test article under the conditions that the *in vitro* study is performed. The Clinical Pharmacology Biopharmaceutics review of an *in vitro* induction study performed with Inspra in human hepatocytes notes that the data provided by the sponsor was not sufficient to determine whether the induction was caused by eplerenone or one of its metabolites (US FDA, 2002d). The implication was that it may be prudent to determine the metabolite profile in the *in vitro* incubations that are performed for the assessment of induction of CYP isoforms in human hepatocytes.

The draft guidance provides some general issues and approaches that should be considered if *in vitro* studies suggest that *in vivo* studies would be helpful. Consultation with the FDA regarding study protocols is recommended. In the following discussion, the term substrate (S) is used to indicate the drug studied to determine whether its exposure is changed by another drug, termed the interacting drug (I). Depending on the study objectives, the substrate and the interacting drug can be the investigational agents or approved products. Study designs are addressed as follows in the draft guidance:

> In vivo drug-drug interaction studies generally are designed to compare substrate concentrations with and without the interacting drug. Because a specific study can consider a number of questions and clinical objectives, many study designs for studying drug-drug interactions can be considered. A study can use a randomized crossover (e.g., S followed by S + I, S + I followed by S), a one-sequence crossover (e.g., S always followed by S + I or the reverse), or a parallel design (S in one group of subjects and S + I in another). The following possible dosing regimen combinations for a substrate and interacting drug can also be used: single dose/single dose, single dose/multiple dose, multiple dose/single dose, and multiple dose/multiple dose. The selection of one of these or another study design depends on a number of factors for both the substrate and interacting drug, including (1) acute or chronic use of the substrate and/or interacting drug; (2) safety considerations, including whether a drug is likely to be an NTR (narrow therapeutic range) or non-NTR

drug; (3) pharmacokinetic and pharmacodynamic characteristics of the substrate and interacting drugs; and (4) assessment of induction as well as inhibition. The inhibiting/inducing drugs and the substrates should be dosed so that the exposures of both drugs are relevant to their clinical use, including the highest doses likely to be used. Simulations can be helpful in selecting an appropriate study design. The following considerations may be useful:

- When attainment of steady state is important and either the substrate or interacting drugs and/or their metabolites have long half-lives and a loading dose to reach steady state promptly cannot be used, special approaches may be needed. These include the selection of a one-sequence crossover or a parallel design, rather than a randomized crossover study design.

- When it is important that a substrate and/or an interacting drug be studied at steady state because the effect of an interacting drug is delayed, as is the case for inducers and certain inhibitors, documentation that near steady state has been attained for the pertinent drug and metabolites of interest is critical. This documentation can be accomplished by sampling over several days prior to the periods when test samples are collected. This is important for both metabolites and the parent drug, particularly when the half-life of the metabolite is longer than the parent, and is especially important if both parent drug and metabolites are metabolic inhibitors or inducers.

- Studies can usually be open label (unblinded), unless pharmacodynamic endpoints (e.g., adverse events that are subject to bias) are critical to the assessment of the interaction.

- For a rapidly reversible inhibitor, administration of the interacting drug either just before or simultaneously with the substrate on the test day might increase sensitivity. For a mechanism-based inhibitor (a drug that requires metabolism prior to its inactivation of the enzyme; examples include erythromycin), administration of the inhibitor prior to the administration of the substrate drug can maximize the effect. If the absorption of an interacting drug (e.g., an inhibitor or an inducer) may be affected by other factors (e.g., the gastric pH), it may be appropriate to control the variables and confirm the absorption through plasma level measurements of the interacting drug.

- When the effects of two drugs on one another are of interest, the potential for interactions can be evaluated in a single study or two separate studies. Some design options are randomized three-period crossover, parallel group, and one-sequence crossover.

- To avoid variable study results because of uncontrolled use of dietary supplements, juices, or other foods that may affect various metabolizing enzymes and transporters during in vivo studies, it is important to exclude their use when appropriate.

Examples of statements in a study protocol could include "Participants will be excluded for the following reasons: Use of prescription or over-the-counter medications, including herbal products, or alcohol within two weeks prior to enrollment" and "For at least two weeks prior to the start of the study until its conclusion, volunteers will not be allowed to eat any food or drink any beverage containing alcohol, grapefruit or grapefruit juice, apple or orange juice, vegetables from the mustard green family (e.g., kale, broccoli, watercress, collard greens, kohlrabi, brussels sprouts, mustard) and charbroiled meats."

The last statement is a clear reminder that "natural" components in foods can have an effect on the metabolism and/or absorption of a drug.

The draft guidance notes that:

Clinical drug-drug interaction studies can generally be performed using healthy volunteers. Findings in this population should predict findings in the patient population for which the drug is intended. Safety considerations may preclude the use of healthy subjects, however, and in certain circumstances, subjects drawn from the population of patients for whom the investigational drug is intended offer advantages, including the opportunity to study pharmacodynamic endpoints not present in healthy subjects. Performance of phenotype or genotype determinations to identify genetically determined metabolic polymorphisms is important in evaluating effects on enzymes with polymorphisms, notably CYP2D6, CYP2C19, and CYP2C9. The extent of drug interactions (inhibition or induction) may be different depending on the subjects' genotype for the specific enzyme being evaluated. Subjects lacking the major clearance pathway, for example, cannot show metabolism and remaining pathways can become important and should be understood and examined.

The last statement could be taken as implying that the metabolism *in vivo* in humans may need to be determined for subjects of varying genotypes if the *in vitro* data indicate that the major route of metabolism of the drug is via a polymorphic DME.

The choice of substrate and interacting drugs for several different questions is also addressed. The section on "Investigational Drug as an Inhibitor or an Inducer of CYP Enzymes" updates the traditional thinking on the selection of drugs for use in DDI studies by noting that:

In contrast to earlier approaches that focused mainly on a specific group of approved drugs (digoxin, hydrochlorothiazide) where co-administration was likely or the clinical consequences of an interaction were of concern, improved understanding of the mechanistic basis of metabolic drug-drug interactions enables more general approaches to and conclusions from specific drug-drug interaction studies. In studying an investigational drug as the interacting drug, the choice of substrates (approved drugs) for initial in vivo studies depends on the P450 enzymes affected by the interacting drug. In testing inhibition, the substrate selected should generally be one whose pharmacokinetics are markedly altered by co-administration of known specific inhibitors of the enzyme systems to assess the impact of the interacting investigational drug. Examples of substrates include (1) midazolam for CYP3A; (2) theophylline for CYP1A2; (3) repaglinide for CYP2C8; (4) warfarin for CYP2C9 (with the evaluation of S-warfarin); (5) omeprazole for CYP2C19; and (6) desipramine for CYP2D6. Additional examples of substrates, along with inhibitors and inducers of specific CYP enzymes, are listed in Table 2 in Appendix A. If the initial study determines an investigation drug either inhibit or induce metabolism, further studies using other substrates, representing a range of substrates, based on the likelihood of co-administration, may be useful. If the initial study is negative with the most sensitive substrates (for sensitive substrates, see Tables 3 and 4 in Appendix A), it can be presumed that less sensitive substrates will also be unaffected.

CYP3A inhibitors can be classified based on their in vivo fold-change in the plasma AUC of oral midazolam or other CYP3A substrate, when given concomitantly. For example, if an investigational drug increases the AUC of oral midazolam or other CYP3A substrates by 5-fold or higher (>5-fold), it can be labeled as a strong CYP3A inhibitor. If an investigational drug, when given at the highest dose and shortest dosing interval, increases the AUC of oral midazolam or other sensitive CYP3A substrates by between 2- and 5-fold (>2- and <5-fold) when given together, it can be labeled as a moderate CYP3A inhibitor. Similarly, if an investigational drug, when given at the highest dose and shortest dosing interval, increases the AUC of oral midazolam or other sensitive CYP3A substrates by between 1.25- and 2-fold (>1.25- and <2-fold), it can be labeled as a weak CYP3A inhibitor. When an investigational drug is determined to be an inhibitor of CYP3A, its interaction with sensitive CYP3A substrates or CYP3A substrates with narrow therapeutic range (see Table 3 in Appendix A for a list) can be described in various sections of the labeling, as appropriate. Similar classifications of inhibitors of other CYP enzymes are discussed in section V.

When an in vitro evaluation cannot rule out the possibility that an investigational drug is an inducer of CYP3A (see Appendix C-3), an in vivo evaluation can be conducted using the most sensitive substrate (e.g., oral midazolam, see Table 3 in Appendix A). When midazolam has been co-administered orally following administration of multiple doses of the investigational drug, as may have been done as part of an in vivo inhibition evaluation, and the results are negative, it can be concluded that the investigational drug is not an inducer of CYP3A (in addition to the conclusion that it is not an inhibitor of CYP3A). In vivo induction evaluation has often been conducted with oral contraceptives. However, as they are not the most sensitive substrates, negative data may not exclude the possibility that the investigational drug may be an inducer of CYP3A.

The use of a CYP cocktail approach is also considered by this draft guidance:

Simultaneous administration of a mixture of substrates of CYP enzymes in one study (i.e., a "cocktail approach") in human volunteers is another way to evaluate a drug's inhibition or induction potential, provided that the study is designed properly and the following factors are present: (1) the substrates are specific for individual CYP enzymes; (2) there are no interactions among these substrates; and (3) the study is conducted in a sufficient number of subjects (see section IV.G). Negative results from a cocktail study can eliminate the need for further evaluation of particular CYP enzymes. However, positive results can indicate the need for further in vivo evaluation to provide quantitative exposure changes (such as AUC, C_{max}), if the initial evaluation only assessed the changes in the urinary parent to metabolite ratios. The data generated from a cocktail study can supplement data from other in vitro and in vivo studies in assessing a drug's potential to inhibit or induce CYP enzymes.

The use of CYP cocktails has been described previously in this chapter.

The evaluation of the investigational drug as a substrate of CYP enzymes is also a topic addressed by this comprehensive draft guidance. The draft guidance states:

In testing an investigational drug for the possibility that its metabolism is inhibited or induced (i.e., as a substrate), selection of the interacting drugs should be based on in vitro or in vivo studies identifying the enzyme systems that metabolize the drug. The choice of interacting drug can then be based on known, important inhibitors of the pathway under investigation. For example, if the investigational drug is shown to be metabolized by CYP3A and the contribution of this enzyme to the overall elimination of this drug is either substantial (>25% of the clearance pathway) or unknown, the choice of inhibitor and inducer could be ketoconazole and rifampin, respectively, because they are the most sensitive in identifying an effect of interest. If the study results are negative, then absence of a clinically important drug-drug interaction for the metabolic pathway would have been demonstrated. If the clinical study of the strong, specific inhibitor/inducer is positive and the sponsor wished to determine whether there is an interaction between the test drug and other less potent specific inhibitors or inducers, or to give advice on dosage adjustment, further clinical studies would generally be needed (see Table 2, Appendix A, for a list of CYP inhibitors and inducers; see Table 5, Appendix A, for additional 3A inhibitors). If a drug is metabolized by CYP3A and its plasma AUC is increased 5-fold or higher by CYP3A inhibitors, it is considered a sensitive substrate of CYP3A. The labeling can indicate that it is a "sensitive CYP3A substrate" and its use with strong or moderate inhibitors may call for caution, depending on the drug's exposure-response relationship. If a drug is metabolized by CYP3A and its exposure-response relationship indicates that increases in the exposure levels by the concomitant use of CYP3A inhibitors may lead to serious safety concerns (e.g., Torsades de Pointes), it is considered as a "CYP3A substrate with narrow therapeutic range" (see Table 3 of Appendix A for a list). Similar classifications of substrates of other CYP enzymes are discussed in section V and listed in Table 6, Appendix A.

The draft guidance then expands the concept of DDIs beyond interactions with other pharmaceuticals and addresses the potential for interaction with food constituents and herbal medications:

> If an orally administered drug is a substrate of CYP3A and has low oral bioavailability because of extensive presystemic extraction contributed by enteric CYP3A, grapefruit juice may have a significant effect on its systemic exposure. Use of the drug with grapefruit juice may call for caution, depending on the drug's exposure-response relationship (see section V for labeling implications). If a drug is a substrate of CYP3A or P-gp and co-administration with St. John's wort can decrease the systemic exposure and effectiveness, St John's wort may be listed in the labeling along with other known inducers, such as rifampin, rifabutin, rifapentin, dexamethasone, phenytoin, carbamazepine, or phenobarbital, as possibly decreasing plasma levels.

The role of polymorphic DMEs on DDIs is also covered:

> If a drug is metabolized by a polymorphic enzyme (such as CYP2D6, CYP2C9, or CYP2C19), the comparison of pharmacokinetic parameters of this drug in poor metabolizers versus extensive metabolizers may indicate the extent of interaction of this drug with strong inhibitors of these enzymes, and make interaction studies with such inhibitors unnecessary. When the above study shows significant interaction, further evaluation with weaker inhibitors may be necessary.

In recognition that polypharmacy is a common situation in many patients the guidance states:

> There may be situations when an evaluation of the effect of multiple CYP inhibitors on the drug can be informative. For example, it may be appropriate to conduct an interaction study with more that one inhibitor if all of the following conditions are met: (1) the drug exhibits blood concentration-dependent safety concerns; (2) multiple CYP enzymes are responsible for the metabolic clearance of the drug; (3) the residual or non-inhibitable drug clearance is low. Under these conditions, the effect of multiple, CYP-selective inhibitors on the blood AUC of a drug may be much greater than the product of the fold AUC changes observed when the inhibitors are given individually with the drug. The degree of uncertainty will depend on the residual fractional clearance (the smaller the fraction, the greater the concern) and the relative fractional clearances of the inhibited pathway. However, if results from a study with a single inhibitor trigger a safety concern (i.e., contraindication), no multiple inhibitor studies will be necessary. Additional considerations may include the likelihood of co-administration of the drug with multiple inhibitors. Before investigating the impact of multiple inhibitors on drug exposure, it is important to first characterize the individual effects of the CYP inhibitors and to estimate the combined effect of the inhibitors based on computer simulation. For safety concerns, lower doses of the investigational drug may be appropriate for evaluating the fold increase in systemic exposure when combined with multiple inhibitors.

Since many interacting agents may have an impact on both drug metabolism and transport of the drug, the guidance states:

The implications of simultaneous inhibition of a dominant CYP enzyme(s) and an uptake or efflux transporter that controls the availability of the drug to CYP enzymes can be just as profound as that of multiple CYP inhibitors. For example, the large effect of co-administration of itraconazole and gemfibrozil on the systemic exposure (AUC) of repaglinide may be attributed to collective effects on both enzyme and transporters. Unfortunately, current knowledge does not permit the presentation of specific guidance. The sponsor will need to use appropriate judgement when considering this situation.

The evaluation of transporter-based DDIs is also addressed. However, as can be seen by the relative brevity of the text, the importance of such interactions is not yet fully understood. Additionally, most focus (from both industry and regulatory agencies) has been on P-gp, although the importance of other transporters for certain classes of drugs is acknowledged. The draft guidance addresses investigational drug as an inhibitor or an inducer of P-gp transporter thus: "In testing an investigational drug for the possibility that it may be an inhibitor/inducer of P-gp, selection of digoxin or other known substrates of P-gp may be appropriate." In a similar brief manner, the evaluation of the investigational drug as a substrate of P-gp transporter is addressed as follows: "In testing an investigational drug for the possibility that its transport may be inhibited or induced (as a substrate of P-gp), an inhibitor of P-gp, such as ritonavir, cyclosporine, or verapamil, or an inducer, such as rifampin should be studied. In cases where the drug is also a CYP3A substrate, inhibition should be studied by using a strong inhibitor of both P-gp and CYP3A, such as ritonavir." The evaluation of interactions of investigational drugs as a substrate of other transporters is addressed: "In testing an investigational drug for the possibility that its disposition may be inhibited or induced (i.e., as a substrate of transporters other than or in addition to P-gp), it may be appropriate to use an inhibitor of many transporters (e.g., P-gp, OATP), such as cyclosporine. Recent interactions involving drugs that are substrates for transporters other than or in addition to P-gp include some HMG Co-A reductase inhibitors, rosuvastatin, and pravastatin." However, the draft guidance does not address the possibility that the investigational drug may be an inhibitor of drug transporters other than P-gp.

As was described in the previous (1999) guidance, "the route of administration chosen for a metabolic drug-drug interaction study is important. For an investigational agent, the route of administration should generally be the one planned for clinical use." The guidance makes additional statements that are similar, or identical to, those made in the 1999 guidance. The most pertinent of these statements are:

> Sometimes certain routes of administration can reduce the utility of information from a study. For example, intravenous administration of a substrate drug may not reveal an interaction for substrate drugs where intestinal CYP3A activity markedly alters bioavailability. For an approved agent used either as a substrate or interacting drug, the route of administration will depend on available marketed formulations.

Dose selection for DDI studies is extremely important in order to be able to position the findings of the DDI study:

> For both a substrate (investigational drug or approved drug) and interacting drug (investigational drug or approved drug), testing should maximize the possibility of finding an interaction. For this reason, we recommend that the maximum planned or approved dose and shortest dosing interval of the interacting drug (as inhibitors or inducers) be used. For example, when using ketoconazole as an inhibitor of CYP3A, dosing at 400 mg QD for

multiple days would be preferable to lower doses. When using rifampin as an inducer, dosing at 600 mg QD for multiple days would be preferable to lower doses. In some instances, doses smaller than those to be used clinically may be recommended for substrates on safety grounds. In such instances, any limitations of the sensitivity of the study to detect the drug-drug interaction due to the use of lower doses should be discussed by the sponsor in the protocol and study report.

The changes in pharmacokinetic parameters that may occur as a result of a DDI study can be used to assess the clinical importance of drug–drug interactions. The draft guidance notes that:

Interpretation of findings from these studies will be aided by a good understanding of dose/concentration and concentration/response relationships for both desirable and undesirable drug effects in the general population or in specific populations. In certain instances, reliance on endpoints in addition to pharmacokinetic measures/parameters may be useful. Examples include INR measurement (when studying warfarin interactions) or QT interval measurements.

The draft guidance reiterates the 1999 guidance in terms of the selection of PK end points for a DDI study:

The following measures and parameters of substrate PK should be obtained in every study: (1) exposure measures such as AUC, C_{max}, time to C_{max} (T_{max}), and others as appropriate; and (2) pharmacokinetic parameters such as clearance, volumes of distribution, and half-lives. In some cases, these measures may be of interest for the inhibitor or inducer as well, notably where the study is assessing possible effects on both study drugs. Additional measures may help in steady state studies (e.g., trough concentration) to demonstrate that dosing strategies were adequate to achieve near steady state before and during the interaction. In certain instances, an understanding of the relationship between dose, blood concentrations, and response may lead to a special interest in certain pharmacokinetic measures and/or parameters. For example, if a clinical outcome is most closely related to peak concentration (e.g., tachycardia with sympathomimetics), C_{max} or another early exposure measure might be most appropriate. Conversely, if the clinical outcome is related more to extent of absorption, AUC would be preferred. The frequency of sampling should be adequate to allow accurate determination of the relevant measures and/or parameters for the parent and metabolites. For the substrate, whether the investigational drug or the approved drug, determination of the pharmacokinetics of important active metabolites is important.

The guidance also notes that pharmacodynamic measures can sometimes provide additional useful information, although in most instances PK measurements will suffice:

Pharmacodynamic measures may be indicated when a pharmacokinetic/pharmacodynamic relationship for the substrate endpoints of interest is not established or when pharmacodynamic changes do not result solely from pharmacokinetic interactions (e.g., additive effect of quinidine and tricyclic antidepressants on QT interval). In most cases, when an approved drug is studied as a substrate, the pharmacodynamic impact of a given change in blood level (C_{max}, AUC) caused by an investigational interaction should be known from other data. If a PK/PD study is needed, it will generally need to be larger than the typical PK study (e.g., a study of QT interval effects).

Sample size and statistical considerations are also addressed in this draft guidance:

> The goal of the interaction study is to determine whether there is any increase or decrease in exposure to the substrate in the presence of the interacting drug. If there is, its implications must be assessed by an understanding of PK/PD relations both for C_{max} and AUC. Results of drug-drug interaction studies should be reported as 90% confidence intervals about the geometric mean ratio of the observed pharmacokinetic measures with (S + I) and without the interacting drug (S alone). Confidence intervals provide an estimate of the distribution of the observed systemic exposure measure ratio of (S + I) versus (S alone) and convey a probability of the magnitude of the interaction. In contrast, tests of significance are not appropriate because small, consistent systemic exposure differences can be statistically significant ($p < 0.05$) but not clinically relevant.

> When a drug-drug interaction of potential importance is clearly present (e.g., comparisons indicate twofold (or lower for certain NTR drugs) or greater increments in systemic exposure measures for (S + I)), the sponsor should provide specific recommendations regarding the clinical significance of the interaction based on what is known about the dose-response and/or PK/PD relationship for either the investigational agent or the approved drugs used in the study. For a new drug, the more difficult issue is the impact on the investigational drug as substrate. For inhibition or induction by the investigational drug, the main consequence of a finding will be to add the drug to the list of inhibitors or inducers likely already present in labeling of the older drug. This information can form the basis for reporting study results and for making recommendations in the package insert with respect to either the dose, dosing regimen adjustments, precautions, warnings, or contraindications of the investigational drug or the approved drug. FDA recognizes that dose-response and/or PK/PD information can sometimes be incomplete or unavailable, especially for an older approved drug used as S.

> The sponsor may wish to make specific claims in the package insert that no drug-drug interaction of clinical significance occurs. In these instances, it would be helpful for the sponsor to recommend specific no effect boundaries, or clinical equivalence intervals, for a drug-drug interaction. No effect boundaries represent the interval within which a change in a systemic exposure measure is considered not clinically meaningful.

Two approaches for defining no effect boundaries are presented.

In 2008, the FDA published an article that provided an overview of the drug interaction guidances and the steps that the FDA took to revise the original drug interaction guidance documents (Huang *et al.*, 2008). The article also summarizes and highlights updated sections in the current draft guidance (US FDA, 2006g). PhRMA has also made contributions on the subject of the prediction of drug–drug interactions by publishing an article in 2003 that described PhRMA's perspective on the conduct of *in vitro* and *in vivo* DDI studies (Bjornsson *et al.*, 2003). The content of this position paper did not engender the kind of response that was created by PhRMA's MIST document since the FDA and PhRMA seem to be pretty much aligned on the topic of DDIs. More recently, Hewitt *et al.* conducted a survey of the *in vitro* methodologies that are used by industry to predict the potential for the induction of DMEs (Hewitt, de Kanter and LeCluyse, 2007). The survey reviewed the methodologies, and the interpretation of the data generated, for compliance with the FDA's recommendations. The survey showed that, although the basic methods used were similar, no two companies or CROs performed and interpreted the data in the exactly the same manner. They also noted that no single method was superior to another, but that all methods used enzyme activities as the major end point for the assessment of induction. Currently, the use of human hepatocyte cultures appears to provide the best indication of the *in vivo* induction potential of an NCE. It was concluded that the use of a twofold induction threshold and a percent of positive control criteria may not be the best methods to accurately assess the *in vivo* induction potential of a drug.

27.5.7 Special Populations

The evaluation of the safety and ADME properties of a drug in special populations such as patients with impaired renal or hepatic function, the elderly, or pediatrics is of great importance in the clinical evaluation of a new drug. Therefore, special populations have been the topic of several regulatory guidances.

27.5.7.1 Cardiovascular Safety The corollary to the guidances on the nonclinical evaluation of cardiovascular safety pharmacology is the ICH guidance, "ICH E14 The Clinical Evaluation Of QT/QTc Interval Prolongation and Proarrhythmic Potential for Non-Antiarrhythmic Drugs" (ICH, 2005e). Among other things, this guidance describes study designs for the evaluation of the potential for QT prolongation. The guidance states "Parallel group studies might be preferred under certain circumstances:

- For drugs with long elimination half-lives for which lengthy time intervals would be required to achieve steady-state or complete washout
- If carryover effects are prominent for other reasons, such as irreversible receptor binding or long-lived active metabolites"

"The timing of the collection of ECGs and the study design (e.g., single or multiple dose, duration) of the 'thorough QT/QTc study' should be guided by the available information about the pharmacokinetic profile of the drug. For drugs with short half-lives and no metabolites, a single dose study might be sufficient. Studies should characterize the effect of a drug on the QT/QTc throughout the dosing interval. While the peak serum concentration does not always correspond to the peak effect on QT/QTc interval, care should be taken to perform ECG recordings at time points around the C_{max}." As noted in previous guidance, the exact words used in the guidance may not be critical. For example, this guidance mentions "serum concentrations" but does not necessarily advocate that measurements be made on serum samples rather than on plasma samples.

Although one would not suspect a prominent role of drug metabolism in this guidance based on the title of the guidance, drug metabolism does indeed play a significant role in this guidance. However, this is not surprising due to the role of drug–drug interactions in unexpected sudden cardiac death in a small number of overtly healthy patients who were taking the antihistamine drug, terfenadine (Seldane) (Monahan *et al.*, 1990). It was recognized that terfenadine alone could have a modest QT prolonging effect, but when taken together with certain antifungal agents that inhibit the hepatic metabolism of terfenadine, toxic levels of terfenadine could develop with marked QT prolongation and an increased probability for life-threatening ventricular tachyarrhythmias. Within a few years of these findings, terfenadine was removed from the marketplace and replaced with its metabolite, fexofenadine (Allegra), a drug with good antihistamine properties but no QT prolonging effects.

The guidance states:

An adequate drug development program should ensure that the dose-response and generally the concentration-response relationship for QT/QTc prolongation have been characterized, including exploration of concentrations that are higher than those achieved following the anticipated therapeutic doses. Data on the drug concentrations around the time of ECG assessment would aid this assessment. If not precluded by considerations of safety or tolerability due to adverse effects, the drug should be tested at substantial multiples of the anticipated maximum therapeutic exposure. Alternatively, if the concentrations of a drug can be increased by drug-drug or drug-food interactions involving metabolizing enzymes (e.g., CYP3A4, CYP2D6) or transporters (e.g., P-glycoprotein), these effects could be studied under conditions of maximum inhibition. This approach calls for an understanding of the pharmacokinetic and pharmacodynamic properties of the parent and significant human metabolites. In general, the duration of dosing or dosing regimen should be sufficient to characterize the effects of the drug and its active metabolites at relevant concentrations.

This approach to evaluating the safety of high levels of drug is an interesting one and may be applicable to drugs that are extensively metabolized. However, the extent of use of such an approach is not yet known. Although not addressed by the guidance, the use of inducers to increase the exposure to metabolites could be required to assess the potential of such metabolites to prolong the QT interval.

Metabolism is further mentioned in this guidance in the following statement: "If the 'thorough QT/QTc study' is positive, analyses of the ECG and adverse event data from certain patient sub-groups are of particular interest, such as:

• Patients with electrolyte abnormalities (e.g., hypokalemia);
• Patients with congestive heart failure
• Patients with impaired drug metabolizing capacity or clearance (e.g., renal or hepatic impairment, drug interactions)
• Female patients
• Patients aged <16 and over 65 years"

The potential for QT prolongation is routinely evaluated as part of the single and multiple ascending dose clinical trails that are conducted in normal healthy volunteers. This initial evaluation should be followed up by a "thorough QT/QTc study" which is intended to determine whether the drug has a threshold pharmacologic effect on cardiac repolarization, as detected by QT/QTc prolongation. The guidance states that:

The threshold level of regulatory concern, discussed further below, is around 5 ms as evidenced by an upper bound of the 95% confidence interval around the mean effect on QTc of 10 ms. The study is typically carried out in healthy volunteers (as opposed to individuals at increased risk of arrhythmias) and is used to determine whether or not the effect of a drug on the QT/QTc interval in target patient populations should be studied intensively during later stages of drug development. It is not intended to identify drugs as being pro-arrhythmic. Although data are limited, it is not expected that the results of the "thorough QT/QTc study" would be affected by ethnic factors.

The "thorough QT/QTc study" would typically be conducted early in clinical development to provide maximum guidance for later trials, although the precise timing will depend on the specifics of the drug under development. It would usually not be the first study, as it is important to have basic clinical data for its design and conduct, including tolerability and pharmacokinetics. Some drugs might not be suitable for study in healthy volunteers because of issues related to tolerability (e.g., neuroleptic agents, chemotherapeutics). The results of the "thorough QT/QTc study" will influence the amount of information collected in later stages of development:

- A negative "thorough QT/QTc study" will almost always allow the collection of on-therapy ECGs in accordance with the current practices in each therapeutic area to constitute sufficient evaluation during subsequent stages of drug development
- A positive "thorough QT/QTc study" will almost always call for an expanded ECG safety evaluation during later stages of drug development.

There could be very unusual cases in which the "thorough QT/QTc study" is negative but the available nonclinical data are strongly positive (e.g., hERG (human ether-a-go-go-related gene) positive at low concentrations and in vivo animal model results that are strongly positive). If this discrepancy cannot be explained by other data, and the drug is in a class of pharmacological concern, expanded ECG safety evaluation during later stages of drug development might be appropriate.

These statements have been included here since they raise the timing of the conduct of the "thorough QT/QTc study." The guidance clearly advocates that this study should be conducted early in clinical development and implies that extensive QT monitoring may be required in clinical trials if there are no data to support the assumption that the risk for QT prolongation is low (i.e. absence of evidence is not evidence of absence). Since extensive QT monitoring in a large clinical trial is expensive and logistically challenging, the sponsor should strongly consider conducting the "thorough QT/QTc study" in the early stages of the clinical program. Since one of the goals of this study is to determine a concentration–effect relationship, the sponsor should ensure that they have identified the correct analytes to be measured. Therefore, it may be necessary to identify the major metabolites in humans prior to conducting the "thorough QT/QTc study." This would imply that the human ADME study should also be conducted in early clinical development. However, arguments can be put forth for conducting the "thorough QT/QTc study" at a later stage of development in order to ensure that an appropriate concentration range has been identified and in order to avoid incurring the very high costs of the "thorough QT/QTc study" until the sponsor has some evidence of efficacy in humans. The E14 guidance has been reviewed by Darpo *et al.* (Darpo, Nebout and Sager, 2006).

Another challenge that the sponsor will face when conducting a "thorough QT/QTc study" is that the regulatory agencies will want to determine the concentration–effect relationship, regardless of the relationship of the concentration that causes an effect to the concentrations associated with efficacy. Such knowledge will be important for generating an understanding of high concentrations of drug (e.g. due to polymorphisms in metabolism; DDIs or overdose) on QT prolongation. However, the sponsors' goal may be to show that their drug does not cause QT prolongation at a certain concentration that is a certain fold greater than the C_{max} at the predicted efficacious dose. The sponsor may be reluctant to push to an effect since once such data have been generated, the stigma of QT prolongation will forever be associated with the drug, regardless of the therapeutic index of the drug. Therefore, it is clear that there may be a gap between the goals of the regulatory agencies and the sponsors.

The issue of drug–drug interactions has generated significant concern within the pharmaceutical industry and among regulatory authorities since the 1990s. This has arisen with respect to early termination of clinical development (e.g. furafylline), refusal of approval (e.g. mibefradil in Sweden), severe prescribing restrictions and withdrawal from the market (e.g. sorivudine, terfenadine, mibefradil, astemizole, and cisapride), and threatened litigation.

27.5.7.2 Oncology   The EMEA issued a guideline in 2006 entitled "Guideline on the Evaluation of Anticancer Medicinal Products in Man" (EMEA, 2006f). This guideline replaces "NfG on Evaluation of Anticancer Medicinal Products in Man (CPMP/EWP/205/95, Rev. 2)" (EMEA, 2003). This guideline states that "Consideration should be given to study high-risk patients (e.g. high risk with respect to target organ toxicity or compromised metabolic or excretory mechanisms for the experimental compound) separately." Although organ impairment is covered in other guidances or guidelines, this subject is dwelt upon in this guideline:

> Studies in patients with decompensated liver function are rarely indicated, but patients with liver metastases should normally be included in the development program. For compounds metabolized by the liver, PK studies are expected, exploring the relationship between, e.g. enzyme levels, or bilirubin increase and exposure. For compounds developed for use in late line therapies, the need for dose reductions in patients with impaired bone marrow reserve due to prior chemo/radio therapy may need special attention. Exploratory studies, including PK, in patients with malignant ascites or other third space conditions are encouraged. Effects of renal impairment should be studied as appropriate. If justified by the target indication, the studies referred to above may be conducted post licensing.

The latter point is a pertinent one that is discussed later in this chapter. It should also be noted that many cancer patients receive combinations of drugs. These combinations may be for the treatment of their disease, for the treatment of comorbidities, or as prophylactic treatment for the side effects of the therapeutic treatment. Riechelmann *et al.* recently reported that the average number of drugs prescribed to a cancer patient was 5, with a range of 0–23 (Riechelmann *et al.*, 2005, 2007; Riechelmann and Saad, 2006). This polypharmacy obviously will have an impact on the design and interpretation of drug–drug interaction studies. Haddad *et al.* have also recently reviewed the role of CYP3A4 in DDIs in cancer patients (Haddad, Davis and Langman, 2007). The integral role of CYP activity on toxicity and efficacy of anticancer drugs in patients has been shown by Li *et al.* (2006), who have used CYP3A phenotyping to predict the exposure to EGFR tyrosine kinase inhibitors and Alexandre *et al.* (2007), who have shown a link between CYP3A activity and docetaxel-induced febrile neutropenia due to cancer-related inflammation. Transcriptional repression of hepatic CYP3A4 expression by cancer has also been reported by Charles *et al.* (2006), who determined that the repression of CYP3A4 expression was due to cancer-associated inflammation. They noted that targeted therapy to reduce inflammation may provide clinical benefit for chemotherapy drugs metabolized by CYP3A4 by improving their PK profile. Since many chemotherapeutic drugs have a narrow therapeutic index, it may be necessary to "personalize" the dosing for each subject by using therapeutic drug monitoring (TDM) and dose adaptation. This topic has been reviewed by Dumez *et al.* (2004) and de Jonge *et al.* (2005).

An "Addendum on Paediatric Oncology" that was issued to CPMP/EWP/205/95, Rev. 2 (i.e. the version of the guideline "Evaluation of Anticancer Medicinal Products in Man" that was updated in the 2006 guideline) was issued in 2004 (EMEA, 2003). It is curious to note that this addendum is listed on the EMEA guidance page, although the guideline that it is an addendum to is no longer listed (EMEA, 2008h). It is unclear as to why the addendum was not included in the updated guideline on "Evaluation of Anticancer Medicinal Products in Man" that was issued in 2006 (EMEA, 2006f). This guideline illustrates the point that regulatory documents undergo periodic updating and that the drug development scientist needs to be continually aware of the changes of any updates or revisions to guidances as well as keeping abreast of the guidances that are under development or being planned.

27.5.7.3 Rheumatoid Arthritis The FDA's guidance on "Clinical Development Programs for Drugs, Devices, and Biological Products for the Treatment of Rheumatoid Arthritis (RA)" that was issued in 1999, introduces the timing of the conduct of drug–drug interaction studies (US FDA, 1999b). This guidance states:

> Before starting phase 3 trials, an evaluation of the test product's interaction with other agents likely to be used by the target population should be performed. Initial information can be established based on metabolic pathways, studies of in vitro systems, animal or human pharmacology studies, or drug interaction studies. This type of information helps in directing areas in need of clinical evaluation. When products are intended to be tested as combination therapy with the investigational agent, substantial information on interactions and safety of co-administration should be developed in phase 2.

The guidance clearly implies that the relevant DDI studies should be performed before the initiation of phase 3 clinical trials. The guidance also reminds the reader that DDIs may not only arise as a consequence of metabolic interactions but potentially as a consequence of displacement of a drug from plasma proteins. The guidance states: "Because polypharmacy is common during the treatment of rheumatic disorders, in vitro binding studies with blood from patients with active disease should be used as a preliminary screening tool for potential displacement reactions."

Many reports in the biomedical literature have documented that hepatic disease can alter the absorption and disposition of drugs (PK) as well as their efficacy and safety (PD). These reports have been based on studies in patients with common hepatic diseases, such as alcoholic liver disease and chronic infections with hepatitis viruses B and C, and less common diseases, such as acute hepatitis D or E, primary biliary cirrhosis, primary sclerosing cholangitis, and alpha1-antitrypsin deficiency. Liver disease may also alter kidney function, which can lead to accumulation of a drug and its metabolites even when the liver is not primarily responsible for elimination. Liver disease may also alter PD effects (e.g. increased encephalopathy with certain drugs in patients with hepatic failure). The specific impact of any disease on hepatic function is often poorly described and highly variable, particularly with regard to effects on the PK and PD of a drug.

27.5.7.4 Diabetes The draft FDA guidance on "Diabetes Mellitus: Developing Drugs and Therapeutic Biologics for Treatment and Prevention" was recently issued (February 2008) (US FDA, 2008i). This draft guidance has been in development since 2004. It addresses the pharmacokinetics of drugs for the treatment of diabetes mellitus by stating that:

> In general, pharmacokinetic parameters of noninsulin therapeutics should be evaluated in phase 1 studies. These studies can be performed in healthy volunteers to determine the basic pharmacokinetic parameters (e.g., absolute bioavailability, area under the curve (AUC), C_{max}, T_{max}, $T_{1/2}$). Additionally, pharmacokinetic studies also may be appropriate in

the intended patient population. We recommend that exposure-response data be obtained during the phase 2 dose-finding studies. (See the guidance for industry *Exposure-Response Relationships: Study Design, Data Analysis, and Regulatory Applications*.)

In patients with diabetes, the high prevalence of altered glomerular filtration rates, delayed or deficient gastrointestinal transit and absorption, and the potential for interactions with commonly used medications usually dictate the need for the evaluation of the pharmacokinetics of new agents in the target population, beyond investigations in healthy volunteers. It is important to evaluate the in vivo and in vitro mechanisms of drug absorption and disposition. This information will provide the basis for the design of the drug interaction studies addressing the class effects of oral antidiabetic drugs (e.g., addressing the induction potential of CYP enzymes by thiazolidinediones, CYP2C-based interactions with sulfonylureas, and interactions with renal tubular secretion of metformin). We also recommend interaction studies with drugs that have a narrow therapeutic index and with drugs likely to be co-administered in the diabetic population. (See the draft guidance for industry *Drug Interaction Studies—Study Design, Data Analysis, and Implications for Dosing and Labeling* for details.) Effects of food on pharmacokinetics should be evaluated in the development of therapeutic products that are intended to be administered orally in temporal proximity to meals (e.g., agents designed to exert effects on glycemia peri- or postprandially, such as meglitinides). Because patients with diabetes may be a particularly sensitive population in terms of polypharmacy and underlying, often subclinical, cardiac disease, we also encourage sponsors to address the effect of the drug on the QT interval by conducting a thorough QT study.

The statements in this draft guidance appear to be consistent with existing guidances and best practices.

27.5.7.5   Hepatic Impairment   In 2003, the FDA issued a guidance called "Pharmacokinetics in Patients with Impaired Hepatic Function: Study Design, Data Analysis, and Impact on Dosing and Labeling" (US FDA, 2003e). This guidance provides a general background on the subject of hepatic impairment, including how it is defined and measured. However, it notes that:

Despite extensive efforts, no single measure or group of measures has gained widespread clinical use to allow estimation in a given patient of how hepatic impairment will affect the PK and/or PD of a drug. Even though clinically useful measures of hepatic function to predict drug PK and PD are not generally available, clinical studies in patients with hepatic impairment, usually performed during drug development, can provide information that may help guide initial dosing in patients. This information can be appropriately used with the understanding that careful observation and dose titration are critical to achieve the optima.

This document provides guidance on deciding whether to conduct a study in patients with impaired hepatic function. This guidance recommends a PK study in patients with impaired hepatic function if:

hepatic metabolism and/or excretion accounts for a substantial portion (>20 percent of the absorbed drug) of the elimination of a parent drug or active metabolite. The guidance also recommends a hepatic impairment study even if the drug and/or active metabolite is eliminated to a lesser extent (<20 percent), if its labeling or literature sources suggest that it is a narrow therapeutic range drug. If the metabolism of the drug is unknown and other information is lacking to suggest that hepatic elimination routes are minor, the Agency recommends that the drug be considered extensively metabolized.

Conversely, for some drugs, hepatic functional impairment is not likely to alter PK sufficiently to require dosage adjustment. The guidance notes that, "In such cases, a study to confirm the prediction is generally not important. The following drug properties may support this conclusion:

- The drug is excreted entirely via renal routes of elimination with no involvement of the liver.
- The drug is metabolized in the liver to a small extent (<20 percent), and the therapeutic range of the drug is wide, so that modest impairment of hepatic clearance will not lead to toxicity of the drug directly or by increasing its interaction with other drugs.
- The drug is gaseous or volatile, and the drug and its active metabolites are primarily eliminated via the lungs."

In an additional caveat, the guidance states that "For drugs intended only for single-dose administration, a hepatic impairment study will generally not be useful, unless clinical concerns suggest otherwise."

The design of the study is addressed in the guidance which notes that the study can be a single-dose or multiple-dose study with PK assessment of the parent drug and any active metabolite(s). In a multiple-dose study, PK assessment is appropriately carried out at steady state. The guidance notes that:

> A single-dose study may be satisfactory for cases where prior evidence indicates that multiple-dose PK is accurately predicted by single-dose data for both parent drug and active metabolites. This would be the case when the drug and active metabolites exhibit linear and time-independent PK at the concentrations anticipated in the patients to be studied. A multiple-dose study is desirable when the drug or an active metabolite is known to exhibit nonlinear or time-dependent PK.

The mention of the measurement of metabolites clearly indicates that the sponsor should have identified the important metabolites prior to the conduct of a study in patients with hepatic impairment. This guidance appears to use the phrase "active metabolite" to mean metabolites that may have either pharmacological or toxicological activity.

The guidance provides some quite detailed information on sample collection and data analysis:

> The blood sampling duration should be adequate to determine the terminal half-life of the drug and its active metabolite(s), with the expectation that these times may be extended in the patient compared to the control population. For drugs that are highly extracted by the liver (extraction ratio > 0.7) and that are extensively bound to plasma proteins (fraction unbound < 10 percent), the Agency recommends that the unbound fraction be determined at least at trough and maximum plasma concentration. The clearance and volume parameters are appropriately expressed in terms of both unbound and total concentrations of drug in plasma/serum/blood.

The EMEA has also issued a guideline on this topic, "Guideline on the Evaluation of the Pharmacokinetics of Medicinal Products in Patients with Impaired Hepatic Function" (EMEA, 2005d). This guideline was issued in August 2005. Like the FDA guidance, this guideline provides some background information on hepatic function and hepatic disease, albeit with a different level of detail and emphasis. The guideline states:

> Hepatic function decreases with age, but due to the high capacity of the liver this is considered not to change the pharmacokinetics to a clinically relevant extent. Liver disease, however, is known to be a common cause of altered pharmacokinetics of drugs. Hepatic function can be decreased through different pathophysiological mechanisms. Worldwide, chronic infections with hepatitis B or C are the most common causes of chronic liver disease, whereas in the western world, chronic and excessive alcohol ingestion is one of the major causes of liver disease. Other causes are uncommon diseases such as primary biliary cirrhosis, primary sclerosing cholangitis and autoimmune chronic active hepatitis. Ongoing destruction of the liver parenchyma in chronic liver diseases ultimately leads to liver cirrhosis and the development of portal hypertension. However, even if liver cirrhosis is established, the residual metabolic function of the liver may be rather well preserved for many years because of regeneration of hepatocytes. Clinical symptoms related to hepatocellular failure and portal hypertensions are most importantly ascites, oesophageal varices and encephalopathy. Serum markers of liver failure are low serum albumin and a prothrombin deficiency. Serum bilirubin as well as other liver tests may or may not be affected to a varying degree, e.g. depending on the liver disease (cholestatic versus hepatocellular). Liver cirrhosis is irreversible in nature, but progression can be modified by e.g. abstinence of alcohol in alcohol liver cirrhosis. The pharmacokinetics and pharmacodynamics of drugs may be altered by liver disease through different mechanisms. The effect most often depends on the severity of hepatic impairment. The effects on pharmacokinetics can be difficult to predict due to consequences of shunting of blood past the liver (both portosystemic and intra-hepatic), impaired hepatocellular function, impaired biliary excretion and decreased protein binding. Factors that influence the need for pharmacokinetic data in patients with hepatic impairment, and interpretation of these data, are the intended use of the drug, pharmacokinetic characteristic features in otherwise healthy individuals and PK/PD relationships. Based on this, the major concern (side effects or lack of efficacy) should be identified.

The guideline provides some input into when to perform pharmacokinetic studies in patients with impaired hepatic function. It states: "A pharmacokinetic study in subjects with impaired hepatic function is recommended when:

- The drug is likely to be used in patients with impaired hepatic function and
- Hepatic impairment is likely to significantly alter the pharmacokinetics (especially metabolism and biliary excretion) of the drug and/or its active metabolites and a posology adjustment may be needed for such patients taking into account the PK/PD relationship.

Furthermore:

> If no study is performed in patients with hepatic impairment, a justification should be given. Lack of data may be justified if the drug is not intended to be used in patients with hepatic impairment. If the drug is likely to be used in these patients, the applicant should discuss the potential for hepatic impairment to influence the pharmacokinetics (of parent drug, active and "inactive" metabolites) and should include relevant information in the SPC. Lack of data may lead to restriction in the use of the drug (not only warnings but also contraindications).

The guidance provides input into methodologies to be used for the confirmation that the subjects to be studied actually have an impaired metabolic capacity. It may be useful to administer, for instance, a CYP3A4 probe drug (if the drug under investigation is a CYP3A4 substrate) to the subjects to be included to observe if the pharmacokinetics of the probe drug is altered (like a "positive control" known to be specially sensitive to liver impairment). This probe would have to be sensitive enough to identify a range of severity in hepatic dysfunction. The guideline also describes other approaches:

Exogenous markers that have been used to assess different hepatic drug elimination mechanisms are antipyrine, MEGX (lidocaine metabolite), ICG (indocyanine green) and galactose. Such markers may be used in parallel with the Child-Pugh classification and a justification for the choice of marker(s) should be given.

In conclusion, until optimal markers have been found, the Child-Pugh classification system can be used to categorise the degree of hepatic impairment of subjects included in a pharmacokinetic study and can, together with its individual components, be used when evaluating the pharmacokinetic results. The sponsor should submit all individual scores of the subjects included in the study, as well as other information on subjects characteristics, for example results of standard laboratory tests.

The collection and analysis of samples is also addressed:

"Plasma (or whole blood, as appropriate) samples should be analysed for parent drug and any metabolites with known or suspected activity (therapeutic or adverse). Metabolites, identified as toxic in preclinical studies, which could be affected by hepatic function should be evaluated. Also, metabolites that are considered relatively inactive in patients with normal hepatic function may reach active/toxic levels if the accumulation of the metabolites is substantial. Hence, evaluation of such metabolites should be considered. The frequency and duration of plasma sampling should be sufficient to accurately estimate relevant pharmacokinetic parameters for parent drug and metabolites.

If the drug or metabolites exhibit a high extent of plasma protein binding, the pharmacokinetics should be described and analysed with respect to the unbound concentrations of the drug and active metabolites in addition to total concentration.

For chiral drugs, the analysis of the enantiomers should be considered as the metabolic profile for each enantiomer may be different in subjects with hepatic impairment. … Consideration should also be given to possible consequences of altered importance of other elimination pathways and the interaction with concomitantly administered drugs. For pro-drugs (i.e., drugs with activity predominantly due to a hepatically generated metabolite), the plasma levels of the active substance may be decreased in patients with hepatic impairment and adjustments of the dose and/or dosing interval may be needed."

The importance of dose adaptation for liver disease in patients with antineoplastic drugs has been reviewed by Tchambaz *et al.* (2006). Dose adaptation may be required due to the high prevalence of impaired liver function in cancer patients and the narrow therapeutic index and the potential for serious adverse events with antineoplastic drugs. Recently, Nakai *et al.* (2008) have shown a correlation of decreased expression of certain CYP isoforms and drug transporters with the progression of liver fibrosis in chronic hepatitis C patients.

27.5.7.6 Renal Impairment   The effect of renal function on the pharmacokinetics of a drug has been addressed by the FDA in a guidance issued in 1998, entitled "Pharmacokinetics in Patients with Impaired Renal Function—Study Design, Data Analysis, and Impact on Dosing and Labeling" (US FDA, 1998b).

The format and content of this guidance is analogous to the FDA's guidance on the effect of hepatic impairment on the pharmacokinetics of a drug. The guidance provides some background information:

> Although the most obvious type of change arising from renal impairment is a decrease in renal excretion, or possibly renal metabolism, of a drug or its metabolites, renal impairment has also been associated with other changes, such as changes in absorption, hepatic metabolism, plasma protein binding, and drug distribution. These changes may be particularly prominent in patients with severely impaired renal function and have been observed even when the renal route is not the primary route of elimination of a drug. Thus, for most drugs that are likely to be administered to patients with renal impairment, PK characterization should be assessed in patients with renal impairment to provide rational dosing recommendations.

The guidance provides suggestions on when studies in patients with impaired renal function may be important:

> A PK study in patients with impaired renal function is recommended when the drug is likely to be used in these patients and (1) renal impairment is likely to significantly alter the PK of a drug and/or its active/toxic metabolites and (2) a dosage adjustment is likely to be necessary for safe and effective use in such patients. In particular, a study in patients with impaired renal function is recommended when the drug or its active metabolites exhibit a narrow therapeutic index and when excretion and/or metabolism occurs primarily via renal mechanisms (excretion or metabolism). A study also should be considered when a drug or an active metabolite exhibits a combination of high hepatic clearance (relative to hepatic blood flow) and significant plasma protein binding. In this setting, renal impairment could induce a significant increase in the unbound concentrations after parenteral administration due to a decreased plasma protein binding coupled with little or no change in the total clearance (decrease in unbound clearance).

Conversely, such studies may not be important when, "For some drugs, renal impairment is not likely to alter PK enough to justify dosage adjustment. In such cases, a study to confirm that prediction may be helpful but is not necessary. If a study is not conducted, the labeling should indicate that the impact of renal impairment was not studied, but that an effect requiring dosage adjustment is unlikely to be present. Current knowledge suggests that the following drug properties may justify this approach:

- Drug and active metabolites with a relatively wide therapeutic index and that are primarily eliminated via hepatic metabolism or biliary excretion;
- Gaseous or volatile drug and active metabolites that are primarily eliminated via the lungs;
- Drugs intended only for single-dose administration unless clinical concerns dictate otherwise.

Controversy exists regarding the impact of severe renal impairment on hepatic metabolism. For this reason, a renal impairment study is still considered desirable for a drug eliminated primarily via hepatic metabolism unless it also has a relatively wide therapeutic index."

The guidance provides input on the design of the study:

A single-dose study is satisfactory for cases where there is clear prior evidence that the multiple-dose PK is accurately predictable from single-dose data for all chemical species of interest (drug and potentially active metabolites). A multiple-dose PK is predictable from a single-dose PK when the drug and active metabolites exhibit linear and time-independent PK at the concentrations anticipated in the patients to be studied. A multiple-dose study is desirable when the drug or an active metabolite is known to exhibit nonlinear or time-dependent PK. ... For multiple-dose studies, lower or less frequent doses as renal function decreases may be important to prevent accumulation of drug and metabolites to unsafe levels. The dosage regimen may be adjusted based on the best available prestudy estimates of the PK of the drug and its active metabolites in patients with impaired renal function.

As was the case with the guidance on studies in hepatically impaired patients, this guidance provides input into sample collection, analysis, and data interpretation:

Plasma or whole blood, if appropriate, (and optionally urine) samples should be analyzed for parent drug and any metabolites with known or suspected activity (therapeutic or adverse). This is particularly important in patients with impaired renal function since renally excreted metabolites can accumulate to a much higher degree in such patients. The frequency and duration of plasma sampling and urine collection should be sufficient to accurately estimate the relevant pharmacokinetic parameters for the parent drug and its active metabolites.

Plasma protein binding is often altered in patients with impaired renal function. For systemically active drugs and metabolites, the unbound concentrations are generally believed to determine the rate and extent of delivery to the sites of action. This leads to the recommendation that the PK should be described and analyzed with respect to the unbound concentrations of the drug and active metabolites. Although unbound concentrations should be measured in each plasma sample, if the binding is concentration-independent and is unaffected by metabolites or other time-varying factors, the fraction unbound may be determined using a limited number of samples or even a single sample from each patient. The unbound concentration in each sample is then estimated by multiplying the total concentration by the fraction unbound for the individual patient. For drugs and metabolites with a relatively low extent of plasma protein binding (e.g., extent of binding less than 80%), alterations in binding due to impaired renal function are small in relative terms. In such cases, description and analysis of the PK in terms of total concentrations should be sufficient.

It should be noted that the direct measurement of unbound drug concentrations is rarely performed and can be extremely expensive. It should be noted that the guidance also mentions that it may be appropriate to measure "potentially active metabolites." This seems to imply that if one knew that a metabolite existed, but did not know whether it was active or not, that metabolite should be quantitated. This statement appears to be out of context of the exposure to this metabolite. The guidance does not appear to address the potential impact of metabolic genotype on the exposure to parent drug and/or metabolites in patients with renal impairment.

Patients with impaired renal function may undergo dialysis as a treatment for their impaired renal function. Dialysis may significantly affect the PK of a drug to an extent that dosage adjustment is appropriate. The guidance states: "The need for dosage adjustment results when a significant fraction of the drug or active metabolites in the body is removed by the dialysis process. In such cases, a change in the dosage regimen, such as a supplemental dose following the dialysis procedure, may be required." The guidance further states that:

> In general, a study of the effect of dialysis on PK may be omitted if the dialysis procedure is unlikely to result in significant elimination of drug or active metabolites. This is arguable for drugs and active metabolites that have a large unbound volume of distribution (V_u)or a large unbound nonrenal clearance ($CL_{NR,u}$).

> If the drug and metabolites have a large unbound volume of distribution, only a small fraction of the amounts in the body will be removed by dialysis. For example, if V_u were greater than 360 L, less than 10 percent of the amount initially in the body could be removed by 3 hours of high flux hemodialysis with an unbound dialysis clearance of 200 mL/min.

> If the drug and metabolites have a large unbound nonrenal clearance, dialysis contributes a relatively small amount to the overall unbound clearance. For example, if $CL_{NR,u}$ were greater than 125 mL/min, 3 hours of high-flux hemodialysis with an unbound dialysis clearance of 200 mL/min administered every 2 days would contribute less than 10 percent to the overall clearance.

Unlike the guidance on the evaluation in hepatically impaired patients, this guidance emphasizes the importance of pharmacodynamics. The guidance states:

> Whenever appropriate, pharmacodynamic assessment should be included in the studies of renal impairment. The selection of the pharmacodynamic endpoints should be discussed with the appropriate FDA review staff and should be based on the pharmacological characteristics of the drug and metabolites (e.g., extent of protein binding, therapeutic index, and the behavior of other drugs in the same class in patients with renal impairment).

The Data Analysis section of the guidance states that "the primary intent of the data analysis is to assess whether dosage adjustment is required for patients with impaired renal function, and, if so, to develop dosing recommendations for such patients based on measures of renal function. The data analysis typically consists of the following steps:

- Estimation of PK parameters;
- Mathematical modeling of the relationship between measures of renal function and the PK parameters;
- Development of dosing recommendations including an assessment of whether dosage adjustment is warranted in patients with impaired renal function."

Each of these points is further elaborated upon. For example, the guidance states that for Parameter Estimation:

> Plasma concentration data (and urinary excretion data if collected) should be analyzed to estimate various parameters describing the PK of the drug and its active metabolites. The PK parameters of a drug can include the area under the plasma concentration curve (AUC), peak concentration (C_{max}), apparent clearance (CL/F), renal clearance (CLR), apparent volume of distribution ($V_{z/F}$ or $V_{ss/F}$), terminal half-life ($t_{1/2}$). The PK parameters

of active metabolites can include the area under the plasma concentration curve (AUC), peak concentration (C_{max}), terminal half-life ($t_{1/2}$). If possible, parameters are preferably expressed in terms of unbound concentrations; for example, apparent clearance relative to the unbound drug concentrations ($CL_{u/F} = D/AUC_u$) where the subscript 'u' indicates unbound drug. Noncompartmental and/or compartmental modeling approaches to parameter estimation can be employed.

The modeling of the relationship between renal function and PK is described thus:

The objective of this step is to construct mathematical models for the relationships between the RF, the measures of renal function, particularly creatinine clearance (CL_{cr}) and relevant PK parameters. The PK parameters of greatest interest are usually the apparent unbound clearance (CL_u/F), or the dose-normalized area under the unbound concentration curve (AUC_u/D), and the dose-normalized peak unbound concentration ($C_{max,u}/D$) for the drug and active metabolites. The intended result is a model that can successfully predict K behavior given information about renal function.

Further details of this approach are given in the guidance but are not presented here.

Finally, the guidance addresses the development of dosing recommendations:

Specific dosing recommendations should be constructed based on the study results using the aforementioned model for the relationships between RF and relevant PK parameters. Typically the dose is adjusted to produce a comparable range of unbound plasma concentrations of drug or active metabolites in both normal patients and patients with impaired renal function. Simulations are encouraged as a means to identify doses and dosing intervals that achieve that goal for patients with different levels of renal function.

For some drugs, even severe renal impairment may not alter PK sufficiently to warrant dosage adjustment. A sponsor could make this claim by providing an analysis of the study data to show that the PK measurements most relevant to therapeutic outcome in patients with severe renal impairment are similar, or equivalent, to those in patients with normal renal function.

The guidance then proceeds to describe how prior to the conduct of the studies, specific "no effect" boundaries for the ratio of a PK measurement from patients with severe and normal renal functions respectively can be used.

27.5.7.7 Geriatrics   ICHE7, "Studies in Support of Special Populations: Geriatrics" was issued in 1994 (ICH, 1994c). The vast majority of this guidance was devoted to pharmacokinetics, pharmacodynamics, and DDIs. On the subject of pharmacokinetics, the guidance states that:

Most of the recognized important differences between younger and older patients have been pharmacokinetic differences, often related to impairment of excretory (renal or hepatic) function or to drug-drug interactions. It is important to determine whether or not the pharmacokinetic behavior of the drug in elderly subjects or patients is different from that in younger adults and to characterize the effects of influences, such as abnormal renal or hepatic function, that are more common in the elderly even though they can occur in any age group. Information regarding age-related differences in the pharmacokinetics of the drug can come, at the sponsor's option, either from a Pharmacokinetic Screen (as described subsequently) or from formal pharmacokinetic studies, in the elderly and in patients with excretory functional impairment. It is recognized that for certain drugs and applications (e.g., some topically applied agents, some proteins) technical limitations such as low systemic drug levels may preclude or limit exploration of age-related pharmacokinetic differences.

Formal Pharmacokinetic Studies are described as:

Formal PK studies can be done either in healthy geriatric subjects or in patient volunteers with the disease to be treated by the drug.

The initial PK study can be a pilot trial of limited size conducted under steady-state conditions to look for sizable differences between older and younger subjects or patients. A larger, single-dose PK study of sufficient size to permit statistical comparisons between geriatric and younger subjects' or patients' pharmacokinetic profiles is also acceptable. In either case, if large (i.e., potentially medically important) age-related differences are found, the initial PK study may need to be followed by a multiple-dose PK study of sufficient size to permit statistical comparisons (geriatric vs. younger) at steady-state.

The Pharmacokinetic Screening Approach is also addressed:

Sponsors may opt, instead of conducting a separate PK evaluation of the elderly, to utilize a pharmacokinetic screen in conjunction with the main Phase 3 (and Phase 2, if the sponsor wishes) clinical trials program. This screening procedure involves obtaining, under steady-state conditions, a small number (one or two) of drug blood level determinations at "trough" (i.e., just prior to the next dose) or other defined times from sufficient numbers of Phase 2/3 clinical trials patients, geriatric and younger, to detect age-associated differences in pharmacokinetic behavior, if they are present. It is important to record time of dosing prior to blood concentration measurements, and relation of dosing to meals, and to examine the influence of demographic and disease factors, such as gender renal function, presence of liver disease, gastrointestinal disease or heart disease, body size and composition, and concomitant illnesses.

Small differences are unlikely to be of medical importance. Where the screen detects large differences, formal pharmacokinetic studies may be indicated unless the screen's results are sufficiently informative.

The advantage of a pharmacokinetic screen is that it can assess the effects, not only of age itself, but also of other factors associated with age (altered body composition, other drugs, concomitant illness) and their interactions.

The topic of renally or hepatically impaired geriatric patients is also addressed. The guidance states:

Renal impairment is an aging-associated finding that can also occur in younger patients. Therefore, it is a general principle, not specific to these guidelines, that drugs excreted (parent drug or active metabolites) significantly through renal mechanisms should be studied to define the effects of altered renal function on their pharmacokinetics. Such information is needed for drugs that are the subject of this guideline but it can be obtained in younger subjects with renal impairment.

Similarly, drugs subject to significant hepatic metabolism and/or excretion, or that have active metabolites, may pose special problems in the elderly. Pharmacokinetic studies should be carried out in hepatically impaired young or elderly patient volunteers.

If a Pharmacokinetic Screen approach is chosen by the sponsor (Section VI, see above), and if patients with documented renal impairment or hepatic impairment (depending on the drug's elimination pattern) are included and the results indicate no medically important pharmacokinetic difference, that information may be sufficient to meet this Geriatric Guideline's purpose.

Pharmacodynamics is also addressed. The guidance notes that "The number of age-related pharmacodynamic differences (i.e., increased or decreased therapeutic response, or side effects, at a given plasma concentration of drug) discovered to date is too small to necessitate dose response or other pharmacodynamic studies in geriatric patients as a routine requirement. Separate studies are, however, recommended in the following situations:

- Sedative/hypnotic agents and other psychoactive drugs or drugs with important CNS effects, such as sedating antihistamines;
- Where subgroup comparisons (geriatric versus younger) in the Phase 2/3 clinical trials database indicate potentially medically significant age-associated differences in the drug's effectiveness or adverse reaction profile, not explainable by PK differences."

Drug–drug interactions are of particular importance to geriatric patients, who are more likely to be using concomitant medications than younger patients. The guidance states that "Therefore it is a general principle, not specific to these guidelines, that in cases where the therapeutic range (i.e., range of toxic to therapeutic doses) of the drug or likely concomitant drugs is narrow, and the likelihood of the concomitant therapy is great, that specific drug-drug interaction studies be considered. The studies needed must be determined case-by-case, but the following are ordinarily recommended:

- Digoxin and oral anticoagulant interaction studies, because so many drugs alter serum concentrations of these drugs, they are widely prescribed in the elderly, and they have narrow therapeutic ranges;
- For drugs that undergo extensive hepatic metabolism, determination of the effects of hepatic-enzyme inducers (e.g., phenobarbital) and inhibitors (e.g., cimetidine);
- For drugs metabolized by cytochrome P-450 enzymes, it is critical to examine the effects of known inhibitors, such as quinidine (for cytochrome P-450 2D6) or ketoconazole and macrolide antibiotics (for drugs metabolized by cytochrome P-450 3A4). There is a rapidly growing list of drugs that can interfere with other drugs that metabolize, and sponsors should remain aware of it;
- Interaction studies with other drugs that are likely to be used with the test drug (unless important interactions have been ruled out by a pharmacokinetic screen)."

The aforementioned changes in PK and metabolism with age may have a pronounced impact on the safety of narrow therapeutic index drugs (e.g. anticancer drugs). These effects have been reviewed by Wildiers *et al.* (2003). Other safety concerns in the elderly may be due to DDIs that arise due to the polypharmacy that is common in the elderly (Mallet, Spinewine and Huang, 2007).

In August 2006, the European Medicines Agency (EMEA) received a request from the European Commission (EC) for the Committee for Human Medicinal Products

(CHMP) to provide for an opinion on the adequacy of guidance on the elderly regarding medicinal products for human use, on the basis of Article 5(3) of Regulation (EC) No 724/2004. This led to the creation of a report, "Adequacy of Guidance on the Elderly Regarding Medicinal Products for Human Use" that made several recommendations (EMEA, 2006a). Although the exact recommendations contained in this report are not pertinent to the topic of drug metabolites in regulatory guidances, clinical trials, and product labeling, the report does discuss drug metabolism in several places.

> The "Note for Guidance on Studies in Support of Special Populations: Geriatrics" (CPMP/ICH/379/95) states that the information regarding age-related differences in the PK of a drug can come, at the sponsor's option, either from a PK screen in conjunction with the main phase 3 studies or from formal PK studies in the elderly (EMEA, 1994a). Where the screen detects large differences, then formal PK studies may be indicated. The formal PK studies can be done in healthy geriatric subjects or in patient volunteers with the disease to be treated by the drug. More restrictive in this sense, the guideline on "Pharmacokinetic Studies in Man" (EudraLex 3CC3a) also recommends that kinetics should be studied in the extreme ages but in patients and not in healthy subjects (European Economic Community, 1987). However, it acknowledges that this requires multiple, long, and expensive studies which cannot all be performed before authorization. The ICH E7 considers that specific drug-drug interaction studies should be planned in the cases where the therapeutic index of the drug or concomitant drugs is narrow (ICH, 1994c). Moreover, ICHE8 "General considerations for clinical trials" recommends that for drugs which are frequently co-administered, it is usually important that drug–drug interaction testing are performed in non-clinical and, if appropriate, in human studies (ICH, 1997a). The "Note for guidance on the investigations of drug interactions" (CPMP/EWP/560/95) outlines the requirements for interaction studies on new chemical entities (NCE) on the basis of their physico-chemical, pharmacokinetic and pharmacodynamic properties, but not on the basis of their clinical indication or the likelihood of the targeted population of having concomitant treatments (EMEA, 1997). Nevertheless, this is done intentionally, as mechanistic studies can be done to extrapolate to a variety of situations. Knowledge from adequate absorption/distribution/metabolism/excretion (ADME) studies and mechanistic interaction studies may help in predicting possible problems due to change in C_{max}, AUC or interaction. The fact that specific PK studies would probably be conducted in healthy elderly, and therefore may be less relevant for the targeted population, is also discussed. It is agreed that the place of population PK should be further explored.

27.5.7.8 Pediatrics The administration of drugs to the pediatric population is covered in several guidances or guidelines issued by the FDA, EMEA, Health Canada, and ICH. Many of these documents include information of relevance to the drug metabolism scientist, indicating the importance of drug metabolism in the pediatric population. In 1998, the FDA issued a draft guidance entitled "General Considerations for Pediatric Pharmacokinetic Studies for Drugs and Biological Products" (US FDA, 1998c). As of October 2007, this guidance has not been finalized and the reasons for it remaining in the draft state are unknown. The guidance provides extensive literature citations to support the points made in this draft guidance. However, the literature citations may now be slightly outdated. This guidance provides some background information, including:

> In the pediatric population, growth and developmental changes in factors influencing ADME also lead to changes in pharmacokinetic measures and/or parameters. To achieve AUC and C_{max} values in children similar to values associated with effectiveness and safety

in adults, it may be important to evaluate the pharmacokinetics of a drug over the entire pediatric age range in which the drug will be used. Where growth and development are rapid, adjustment in dose within a single patient over time may be important to maintain a stable systemic exposure.

The guidance then provides information on special areas of importance in planning pediatric pharmacokinetic studies:

Developmental changes in the pediatric population that can affect absorption include effects on gastric acidity, rates of gastric and intestinal emptying, surface area of the absorption site, gastrointestinal enzyme systems for drugs that are actively transported across the gastrointestinal mucosa, gastrointestinal permeability, and biliary function. Similarly, developmental changes in skin, muscle, and fat, including changes in water content and degree of vascularization, can affect absorption patterns of drugs delivered via intramuscular, subcutaneous, or percutaneous absorption. … Drug metabolism usually occurs in the liver, but may also occur in the blood, gastrointestinal wall, kidney, lung, and skin. Developmental changes in metabolizing capacity can affect both absorption and elimination, depending on the degree to which intestinal and hepatic metabolic processes are involved. Although developmental changes are recognized, information on drug metabolism of specific drugs in newborns, infants, and children is limited. In general, it can be assumed that children will form the same metabolites as adults via pathways such as oxidation, reduction, hydrolysis, and conjugation, but rates of metabolite formation can be different. In vitro studies performed early in drug development may thus be useful in focusing attention on metabolic pathways of elimination in both adults and children. … Protein binding may change with age and concomitant illness. In certain circumstances, an understanding of protein binding may be needed to interpret the data from a blood level measurement and to determine appropriate dose adjustments. In vitro plasma protein binding studies can determine the extent of binding of the parent and the major active metabolite(s) and identify specific binding proteins, such as albumin and alpha-1 acid glycoprotein. Optimal estimates of the degree to which protein binding is linear may be obtained by testing maximum and minimum observed concentrations.

Further information is provided upon factors to consider in designing a study in the pediatric population:

Because there may be limited information on the safety of the dose to be administered to a neonate or infant, doses in initial studies require careful consideration. Factors for consideration include (1) the relative bioavailability of the new formulation compared to the adult formulation; (2) the age of the pediatric population; (3) the therapeutic index of the drug; (4) pharmacokinetic data from the adult population; and (5) body size of the pediatric study population. Initial doses should be based on mg/kg of body weight or mg/m^2 of body surface area, extrapolated from adult doses. Knowledge of ADME in an adult population should be combined with an understanding of the physiologic development of the intended pediatric study population to modify the initial dose estimate. Consideration should initially be given to administering a fraction of the dose calculated from adult exposure, depending on the factors mentioned above and depending on whether there is any pediatric experience. Subsequent clinical observations and prompt assay of biological fluids for the drug and/or its metabolites should permit subsequent dose adjustment. The analytical method used to quantify the drug and metabolite(s) in the biological fluid of interest should be accurate, precise, sensitive, specific, and reproducible. Ideally, the method should be relatively rapid, readily adaptable, and use only minimum sample volumes. Protein binding studies may be performed if considered important.

The use of small sample volumes may require a revalidation of the bioanalytical assay.

In 2001, an ICH guidance (E11) called "Clinical Investigation of Medicinal Products in the Pediatric Population" came into operation (ICH, 2000c). This guidance makes some general statements regarding the role of metabolism in the clearance of drugs in the pediatric population. "Knowing the pathways of clearance (renal and metabolic) of the medicinal product and understanding the age-related changes of those processes will often be helpful in planning pediatric studies." The development of clearance mechanisms in different age groups is also described as follows.

Infants and toddlers (28 days to 23 months)

This is a period of rapid CNS maturation, immune system development and total body growth. Oral absorption becomes more reliable. Hepatic and renal clearance pathways continue to mature rapidly. By 1 to 2 years of age, clearance of many drugs on a mg/kg basis may exceed adult values. The developmental pattern of maturation is dependent on specific pathways of clearance. There is often considerable inter-individual variability in maturation.

Children (2 to 11 years)

Most pathways of drug clearance (hepatic and renal) are mature, with clearance often exceeding adult values. Changes in clearance of a drug may be dependent on maturation of specific metabolic pathways.

Puberty

The onset of puberty is highly variable and occurs earlier in girls, in whom normal onset of puberty may occur as early as 9 years of age. Puberty can affect the apparent activity of enzymes that metabolize drugs, and dose requirements for some medicinal products on a mg/kg basis may decrease dramatically (e.g., theophylline). In some cases, it may be appropriate to specifically assess the effect of puberty on a medicinal product by studying pre- and postpubertal pediatric patients.

The guidance also notes that "the volume of blood withdrawn should be minimized in pediatric studies." One suggestion is the use of sensitive assays for parent drugs and metabolites to decrease the volume of blood required per sample. However, it should be noted that the volume of plasma used for the bioanalytical assay should be small and that a sensitive assay does not necessarily mean that the volume of plasma required for the assay will be low.

In 2003, Health Canada issued an addendum to ICH E11 that clarifies that the Canadian research and regulatory environment is meant to more precisely define regulatory considerations for the timing and conduct of clinical studies in the pediatric population.

In 2007, the EMEA issued a guideline called "Guideline on the Role of Pharmacokinetics in the Development of Medicinal i the Paediatric Population" (EMEA, 2006h). In a similar manner to the draft FDA guidance, this guideline provides some general and specific information. The guideline states that:

> The design of pharmacokinetic and PK/PD studies in paediatric patients should be based on a number of factors including the known pharmacokinetic characteristics (dose- and time-dependency of pharmacokinetics, route of elimination, presence of active metabolites, protein binding, etc), route of administration, therapeutic index, paediatric group investi-

gated, possibility to collect blood samples, sensitivity of the analytical method, method for the pharmacokinetic data analysis, desired use of the pharmacokinetic data, etc.

It further states that:

If there are pharmacologically active metabolites significantly contributing to the efficacy or safety, the pharmacokinetics of such metabolites should be studied, unless it may be assumed that the exposure ratio of metabolite to parent drug is similar to the ratio in the "reference age group". Blood sampling may be difficult and the number of samples is usually limited, especially in younger age groups. Effort should be put into optimising study design and use of the available data as well as further developing the analytical methods to allow for small sample volumes to be used. If a medicinal product is not metabolised and elimination is predominantly renal, or if the drug is partly eliminated through renal excretion in a dose-linear fashion, data on urinary excretion may be used to describe elimination capacity.

The guideline provides some specific considerations for preterm and term newborn infants, infants and toddlers since these groups present the largest pharmacokinetic challenges. The guideline notes that "rapid developmental changes in absorption, distribution, metabolism and excretion, combined with all possible disease processes that might interfere with the developmental changes, necessitate a study design that is tailored to these populations." Two points of interest are the statements that "determining the protein binding of highly bound drugs and active metabolites should be considered when studying newborns" and "The effect of genetic polymorphisms in genes coding for drug metabolising enzymes, as well as hepatic or renal impairment on the pharmacokinetics within the paediatric age group, may be different from the effect in adults." How these points are addressed is not discussed in this guideline. It should be noted that in 2008, a corrigendum was issued to this guideline. However, it is not possible to determine what was corrected in this version since neither the guideline itself nor the EMEA web page gives any indication as to what was corrected and the uncorrected version of the document is no longer available on the EMEA web site.

One of the challenges in conducting pharmacokinetic or other studies in the pediatric population is the selection of the dose formulation. Small children may not be able to swallow the capsules or tablets that are administered to adults. Additionally, the dose strengths of such capsules or tablets may be inappropriate for pediatric studies. This topic was addressed in a reflection paper (Formulations of Choice for the Paediatric Population) issued by the EMEA in 2006 (EMEA, 2006k).

This reflection paper notes that "Gastric pH is relatively high in the neonatal period; gastric emptying increases; gut motility matures during early infancy and there are changes to splanchnic blood flow, intestinal drug-metabolising enzymes, microflora and transporters. There are few published bioavailability studies but in general, rate of absorption is slower in neonates and infants than in older children." The mention of differences in microflora between the pediatric and adult population acknowledges the role of the microflora as a drug-metabolizing system. The role of the GI microflora in the metabolism of certain drugs represents an understudied area. The guideline does not address some of the logistical challenges in the administration of drugs to the pediatric population. Often, drugs are dissolved in apple (or other juice) or mixed with food in order to ensure that the drug can be injested by the subject. Therefore, the stability of the drug in the dose formulation (e.g. apple juice or infant cereal) should be determined. Additionally, the effect of the juice components on the PK of the drug and the potential food effect when the drug is administered in food should be taken into account when interpreting data from pediatric studies that utilize these modes of drug administration. It should be noted that some of these modes of administration may also be required for administration to an elderly population.

The guideline further notes that "In general, clearance of substances metabolised in the liver is greater in children than in adults, requiring higher doses per kg body weight." The ontogeny of the development of DMEs is also described:

> The main pathway for phase 1 reactions is oxidation using the cytochrome P450 dependent (CYP) enzymes and these enzymes are generally immature at birth reaching maximum values at about 2 years of age. Hepatic clearance of some substances will be greater on a per kg body weight basis than for adults e.g. carbamazepine, theophylline. The different CYP families of enzymes mature at different rates and there may be significant inter-individual variation. Phase 2 reactions include glucuronidation and sulphation. Many different enzymes are involved and they develop at different rates such that metabolism of substances may vary considerably in infancy both qualitatively and quantitatively. Neonates are unable to conjugate benzoic acid efficiently and this is of great importance to the use of benzyl alcohol as an excipient in this age group since its metabolite benzoic acid can accumulate and is toxic. The ontogeny and pharmacogenomics of metabolising enzymes for active drugs is receiving increasing attention but there is little information about the effect on excipients. They may not be able to metabolise or eliminate an ingredient in a pharmaceutical product in the same manner as an adult. Several EU guidelines related to the use and declaration of excipients have been published and should be consulted. Additional information may be found in documents published by the US Food and Drug Administration (FDA).

In addition to making several important and interesting scientific points, this guideline is one of the few regulatory documents that explicitly advocates the use of guidances from other regulatory agencies.

The guideline provides examples of excipients whose use may be problematic in infants. Examples include benzyl alcohol/benzoic acid/sodium benzoate. "Benzyl alcohol is often used as a preservative in injectable medicines. It can be toxic in neonates and pre-term neonates due to their immature metabolism. In developing pharmaceutical preparations for use in pre-term infants, neonates and young children benzyl alcohol should be carefully evaluated and may best be avoided. It may cause pain on injection" and propylene glycol:

> Propylene glycol is used as a solvent in oral, topical and injectable medications, often for substances which are not highly soluble in water, e.g. phenobarbital, phenytoin and diazepam. It is also commonly used in injectable multivitamin concentrates. Children below 4 years have a limited metabolic pathway (alcohol dehydrogenase), therefore accumulation of propylene glycol can occur in the body. For example, it has been shown that neonates have a longer propylene glycol half-life (16.9 hours) compared to adults (5 hours). Products containing high levels of propylene glycol should not be administered to children below the age of 4 years. Main toxic action is depression of the central nervous system. High osmotic pressure may cause laxative effects. Topical administration has been reported to cause contact dermatitis.

The CHMP released a draft guideline on "The Investigation of Medicinal Products in the Term and Preterm Neonate" for review in September 2007, with final comments being due in May 2008 (EMEA, 2007k).

This draft guideline provides some useful information related to the physiology of organ systems such as the heart/lung, CNS, liver, kidney, and GI tract of the neonate, and how they relate to PK and metabolism in the neonate. The draft guideline states that:

> The post-natal cardiopulmonary system adaptation marks the most dramatic changes during and after birth. Some of these changes occur instantaneous with the first breath, whereas others occur within hours or days after birth. In general, the impact of lung and heart maturation on PK/PD relationship (e.g., closure of the ductus arteriosus) has to be considered. As adequate cardiopulmonary function is paramount to maintain organ function in general (e.g., renal blood flow, brain perfusion, liver function), any potential impact on either cardiac or pulmonary function needs to be carefully monitored in neonatal clinical trials. The influence of cardiopulmonary function as the basis to maintain hepatic drug metabolism and excretion as well as renal excretion has to be considered.

> Transport across the blood brain barrier by both passive diffusion and by active transporters is age-related and undergoes constant maturational changes in the neonate. This may contribute to a significantly altered distribution of active substances or metabolites into the CNS with a potential impact on both clinical efficacy and adverse effects. Medicinal products known or expected to be substrate for specific transporters (e.g., P glycoprotein, Pgp) require specific consideration.

> Hepatic blood flow, plasma protein binding and intrinsic clearance determining hepatic clearance undergo significant post-natal changes. Most enzymatic microsomal systems responsible for drug metabolism are present at birth and their activities increase with advancing post-natal and gestational age. Rapid maturational changes occur during the first weeks of life. Hepatic clearance may be influenced by premature birth, pathologic conditions of the neonate or administration of drugs to the mother or to the neonate.

> To predict the exact nature of these consequences requires an understanding of post-natal maturation and main involved enzymes. The development of specific enzymes is partly described in the scientific literature and may allow estimations of drug metabolism in the neonate. These data should be considered when planning neonatal studies.

> The main pathway responsible for metabolism may be different in neonates as compared to adults. The applicant should consider this when assessing exposure margins of metabolites to the animals used in preclinical studies and also when comparing human safety data obtained in adults and older children. The relevant hepatic phase I and II metabolic pathways should be identified.

> If pharmacologically active metabolites are known to be formed, potential differences in exposure of such metabolites should be considered. If feasible, the applicant is encouraged to perform studies investigating drug metabolism in vitro in neonatal hepatic material (microsomes, hepatocytes etc.).

> In utero exposure to enzyme inducing agents (e.g., antiepileptic drugs, barbiturates, glucocorticoids) and the potential to temporarily alter post-natal drug disposition need to be considered when planning a study in neonates and in the interpretation of data.

> If the drug investigated is likely to be eliminated mainly through hepatic metabolism, markers of hepatic function could be included as covariates in the pharmacokinetic data analysis (e.g., in population PK analysis) as well as included in the safety assessment. Monitoring could include standard laboratory and imaging procedures.

> Renal clearance mechanisms include glomerular filtration (GFR), tubular secretion and reabsorption. Glomerular filtration matures faster than the tubular function, and both depend not only on age and maturational status but also on adverse factors occurring in

the pre- and post-natal period, including for example intrauterine growth retardation or administration of nephrotoxic drugs to the mother and the neonate.

Data concerning maturational changes of the neonatal gastrointestinal tract that may influence drug bioavailability are still limited.

Gastrointestinal absorption is influenced by factors such as tissue perfusion, surface area, gastric and intestinal pH, intestinal mobility and transit time as well as maturation of transporters and receptors. In principle, all these factors are reduced or immature in the neonate. The post-natal developmental pattern of these factors may additionally be highly variable due to environmental factors (i.e., diet, drug administration), genetic factors and underlying pathophysiology. Changes in bioavailability during the early post-natal period have to be considered and need to be predicted as accurate as possible in clinical trials including drugs administered orally.

Gastric pH is neutral at birth with gastric acid secretory capacity appearing after the first 24 to 48 hours of life. Post-natal increases in gastric acid production generally correlate with post-natal age and adult levels are reached by approximately 2 years of age. High gastric pH in the neonate may lead to increased bioavailability of weakly basic compounds and reduced bioavailability of weakly acidic compounds. Additionally, in premature infants, gastric pH may remain elevated due to immature acid secretion. This may lead to higher serum concentrations of acid-labile drugs in the premature neonate.

As pancreatic and biliary functions are immature at birth, bioavailability of drugs requiring pancreatic exocrine and biliary function may have reduced bioavailability. Both functions develop rapidly in the neonatal period, requiring careful consideration of increased bioavailability of orally administered drugs in neonatal clinical trials.

Reduced gastrointestinal motility may have unpredictable effects on drug availability in neonates. It may reduce the rate of drug absorption or conversely improve drug bioavailability due to longer retention times in the small intestine. Additionally, maturation of intestinal metabolising enzymes and transport proteins remains largely unknown, further leading to the unpredictability of oral bioavailability and intestinal first-pass effect of orally administered drugs in the neonate. Drugs undergoing secondary metabolism and secretion into the gut, especially when glucuronidation with enterohepatic recirculation occurs in adults and older children, may have different bioavailability and exposure because of reduced glucuronidation and bacterial activity in the intestine of neonates. Reduced gastrointestinal mobility that is often present in sick neonates is therefore particularly important to consider.

Formulations and route of administration are also addressed as follows:

The choice of formulation and route of administration should depend on the condition to be treated and the clinical state of the neonate. Age-appropriate formulations using appropriate excipients must be developed to avoid extemporaneous preparations, even more so for neonates. Novel formulations should be evaluated through preclinical studies and in adults or older children as appropriate before consideration for administration to neonates.

Medication errors in neonatal practice are commonly due to use of inappropriate formulations requiring calculation and measurement of very small volumes or multiple dilutions. Prescribing software may not be appropriate for neonatal use. Excipients used for adults and older children may be toxic in neonates because of immature metabolism and elimination. The salt of the active ingredient and the chemical nature of the preparation must be carefully considered to avoid administration of excessive amounts of electrolytes.

In general, the IV route will normally be used in clinically unstable term and preterm neonates.

PK and PK/PD studies are described in detail. Reference is made to the "Guideline on the Role of Pharmacokinetics in the Development of Medicinal Products in the Paediatric Population," especially section 4.1. The current draft guideline states that:

Pharmacokinetic information is important to support adequate dosing in subpopulations of the clinically studied population and to assess the potential for clinical relevance of toxicity findings in the preclinical studies. In neonates, however, pharmacokinetics alone is of limited value in neonates for extrapolating efficacy and safety from other patient groups and extrapolation of efficacy will in general need PK/PD monitoring.

A population PK approach is preferable due to the importance of finding covariates related to dose-individualization between individuals and over time in the maturating individual. The analysis can be made on rich and/or sparse data depending on the number of patients available and the possibility of developing highly sensitive analytical methods where very small sample volumes could be used. The initial model could be based on rich data of a limited number of individuals and on other prior information, followed by a population PK approach.

It should be noted that population PK and modelling of oral administration require extra cautious consideration in the neonatal population as there may be marked absorption differences in neonates as compared to other age groups as well as very prolonged absorption in a subgroup of individuals.

In cases where C_{max} is clinically important for safety or efficacy reasons, efforts should be made to characterise this parameter satisfactorily due to the differences in volume of distribution between neonates and older children. If possible, the protein binding of highly protein bound active substances should be assessed to enable the measurement of free plasma concentrations. Immature expression of carrier proteins should also be considered. Special consideration should be given to drugs which are highly protein bound and fast metabolized in adults, since major differences can be assumed in newborns, as synthesis of binding proteins such as albumin could be lower in the neonate with consequences on drug binding and free bilirubin. The need for differentiation between a loading dose (large V_d) and smaller maintenance doses (low total body clearances) as important, e.g., for methylxanthines, aminoglycosides, and anticonvulsants, has to be identified.

Effort should be made to include the determination of potential covariates in the studies (PNA, PMA, GA, weight, body surface area [BSA], renal function, concomitant use of drugs, S-bilirubin, repeated feeding and feeding patterns etc.) for allowing covariates to be identified which may allow satisfactory dose individualization. Adjustments of the dose by covariates (e.g. bodyweight, BSA) should usually be based on the covariate with the highest correlation to the relevant PK parameters. However, the difficulties in determining the covariate should be taken into account. The determination of BSA is difficult in neonates and other covariates should be considered if their use gives an adequate dosing. Titration based on plasma concentration or a clinical safety or efficacy marker should also be considered. This is further described in the Guideline on the Role of Pharmacokinetics in the Development of Medicinal Products in the Paediatric Population.

Neonates treated in hospital and especially on NICUs often receive multiple drugs to treat different conditions. Therefore, any known or potential interactions of the medicinal product investigated should be carefully considered when planning a clinical study as well as during data analysis. Concomitantly used drugs should be included in the population pharmacokinetic analysis. In general, formal interaction studies should be performed in adults. However, if the main enzymes involved in the elimination of the drug are different in the neonate, results of adult interaction studies investigating effects of other drugs on the investigated medicinal product can not be directly extrapolated to neonates. In these cases, estimations based on *in vitro* metabolism data as well as other sources of information should if possible be performed. If a dosing recommendation is needed for a commonly used drug combination and if an interaction is expected, specific pharmacokinetic interaction studies should be considered.

Related guidances include the 2005 FDA draft guidance "Clinical Lactation Studies —Study Design, Data Analysis, and Recommendations for Labeling" (US FDA, 2005g). The draft guidance states that "Such a study usually enrolls mother-infant pairs who are planning to or are currently receiving study medication. Its hallmark is the frequent collection of corresponding maternal blood and milk samples as well as sampling of infant blood and/or urine. Infant sampling provides information regarding the fraction of drug that is systemically available to the breast-fed child. Total clearance of the drug or metabolite by the breast-fed child can be estimated as well. This design can be considered if information is already known about the extent of drug transfer into breast milk, but the amount absorbed by the breast-fed child is not known. Other drugs that can be considered for a mother-infant pair design include drugs already approved and known to be used by lactating women who continue to breast-feed and drugs used to treat chronic maternal conditions. Drug or metabolite characteristics that favor selection of this study design include:

- High lipophilicity (weak bases)
- Potential for accumulation in breast milk
- Likelihood of being well absorbed by the breast-fed child
- Wide distribution to multiple organs
- Long half-life"

An alternative study design is also described: "In a sequential or step-wise approach to lactation studies, the lactating women (plasma and milk) study design might be considered before the infant is exposed to drug via breast milk in a more complex study. Situations that might favor use of this design include newly approved drugs (especially for drugs with no pediatric data), short-term or acute maternal dosing, and unknown risk of exposure to the breast-fed child. Drug and metabolite characteristics that favor selection of this study design include:

- High lipophilicity (weak bases)
- Presence in milk
- Predictions that drug is present in milk
- Knowledge of a class effect"

The draft guidance states that "For drugs that are hepatically metabolized and known to exhibit genetic polymorphism (e.g., CYP2D6 or CYP2C19), the metabolic status of the enrolled subjects (maternal and infant) can be important factors when analyzing the results of the study." However, how such polymorphisms (which can be different in parent and child) are used in evaluating the data is not discussed.

The guidance makes the point that:

Because of varying lipophilicity among drugs, it is also important to assay milk samples for milk fat. Alternative, noninvasive pediatric sampling strategies (e.g., saliva, tears) might also be used to estimate drug levels in infants. However, drug concentrations obtained from alternative fluids (e.g., saliva, tears) might not be equivalent to those obtained from plasma. Sponsors are, therefore, encouraged to demonstrate the relationship of the drug concentration between plasma and alternative fluids in adults. Estimating infant drug exposure via breast milk solely from excretion of unchanged drug in infant urine can be of limited utility because of the difficulty with urine collection and the variability of renal clearance and urine production in infants.

As can be seen, the design, conduct and analysis of such studies will be complicated.

27.5.7.9 Reproduction and Pregnancy   A draft guideline (Guideline on Risk Assessment of Medicinal Products on Human Reproduction and Lactation: From Data to Labeling) was issued by the EMEA in 2006 which covers similar topics to the two FDA guidances described earlier (EMEA, 2006c). This draft guideline states that "Information about the excretion into milk of the active substance and/or its metabolites should be available" and "The exposure in pregnant animals measured by plasma concentrations of the compound and/or metabolites should be assessed." These points imply that the secretion of drug and/or metabolites should be evaluated in animals and/or humans and that the metabolism of the drug in animals be determined prior to the determination of the exposure of the pregnant animal to the drug and/or metabolites. In most cases, the toxicology study is performed prior to the determination of the metabolism in the pregnant animal, if the metabolism is described in the pregnant animal. If the metabolism of the drug is determined in the pregnant animal, it is often only done in one species (usually rabbit). It is assumed that the metabolism in the nonpregnant and pregnant rat are identical, although that is not always the case. Additionally, distribution of drug and/or metabolites to the fetus is only occasionally performed and is often only performed in one species (it is assumed that the placental transfer is similar in both the rat and rabbit). The timing of when (in the course of the pregnancy) to address the metabolism of the drug and fetal exposure to the drug and/or metabolites is a subject that can engender great debate. It is well known that the metabolic capacity and capabilities of the fetus will change with the stage of pregnancy; that pregnancy can affect the clearance of the drug by the mother and that the placental transfer of the drug and/or metabolites can change with the stage of the pregnancy. Therefore, it can be seen that the description of the metabolism in pregnant animals is a complicated matter that can easily become resource- and time-intensive.

In 2004, the FDA issued a draft guidance entitled "Pharmacokinetics in Pregnancy—Study Design, Data Analysis, and Impact on Dosing and Labeling" (US FDA, 2004c). The CHMP issued a draft guideline (Guideline on Risk Assessment of Medicinal Products on Human Reproduction and Lactation: From Data to Labelling) in 2006 that covers similar topics to the two FDA guidances described earlier (EMEA, 2006c). The guideline is expected to be finalized in 4Q 2007/1Q 2008. The FDA guidance provides far more details on impact of pregnancy on the pharmacokinetics of a drug. For example, the guidance states: "Extrapolation of PK data from studies performed in nonpregnant adults fails to take into account the impact of the many physiologic changes that occur during pregnancy. Most of the physiologic changes manifest during the first trimester and peak during the second trimester of pregnancy. Physiologic changes are not fixed throughout pregnancy but rather reflect a continuum of change as pregnancy progresses, with return to baseline at various rates in the postpartum period. The physiologic changes have the potential to alter the PK and/or PD of drugs. Some of these changes include:

- Changes in total body weight and body fat composition.
- Delayed gastric emptying and prolonged gastrointestinal transit time.
- Increase in extra cellular fluid and total body water.
- Increased cardiac output, increased stroke volume, and elevated maternal heart rate.
- Decreased albumin concentration with reduced protein binding.
- Increased blood flow to the various organs (e.g., kidneys, uterus).
- Increased glomerular filtration rate.
- Changed hepatic enzyme activity, including phase I CYP450 metabolic pathways (e.g., increased CYP2D6 activity), xanthine oxidase, and phase II metabolic pathways (e.g., N-acetyltransferase)."

It should be noted that the induction of CYP2D6 in pregnancy is somewhat unique and unexpected, since CYP2D6 does not appear to be inducible by xenobiotics. The guidance further states that:

A significant amount of pharmacologic research has been conducted to improve the quality and quantity of data available for other altered physiologic states (e.g., in patients with renal and hepatic disease) and for other patient subpopulations (e.g., pediatric patients). The need for PK/PD studies in pregnancy is no less than for these populations, nor is the need for the development of therapeutic treatments for pregnant women.

For drugs that are metabolized by enzymes known to exhibit genetic polymorphism (e.g., CYP2D6 or CYP2C19), the FDA recommends that the investigator consider the metabolic status of the enrolled subjects when analyzing the results of the study. Genotype has been shown to have an effect on pregnancy-related changes in metabolism. … Plasma protein binding, like renal function, is often altered in pregnancy. For example, albumin and alpha-1-acid glycoprotein levels are reduced in pregnancy, consequently the protein binding of drugs can be affected. With systemically active drugs and metabolites, the unbound concentrations are generally believed to determine the rate and extent of delivery to the sites of action. For drugs and metabolites with a relatively low extent of plasma protein binding (e.g., the extent of binding is less than 80 percent), alterations in binding due to pregnancy are small in relative terms. In such cases, description and analysis of the PK in terms of total concentrations would be sufficient. For drugs where the extent of protein binding is greater than 80 percent, primarily to albumin, it is recommended that the PK be described and analyzed with respect to the unbound concentrations of the drug and active metabolites. Although unbound concentrations should be measured in each plasma sample, if the binding is concentration-independent and unaffected by metabolites or other time-varying factors, the fraction unbound can be determined using a limited number of samples or even a single sample from each patient during each trimester. The unbound concentration in each sample should then be estimated by multiplying the total concentration by the fraction unbound for the individual patient.

The design of pharmacokinetic studies, as well as ethical concerns, is addressed in depth in this draft FDA guidance. Selected points of interest are as follows. The draft guidance notes that:

Ethical issues are important when considering studying drugs in pregnant women. Given the large number of pregnant women who need prescription medicines to maintain their health, some have argued that it is unethical not to obtain dosing information in this subpopulation. Others recommend that only pregnant women who need a drug for therapeutic reasons be included in clinical studies, citing that drug studies cannot be done in "normal pregnant volunteers."

It further notes that:

Although PK studies in pregnancy can be considered in Phase 3 development programs depending on anticipated use in pregnancy and the results of reproductive toxicity studies, the FDA anticipates that most PK studies in pregnant women will occur in the postmarketing period and will be conducted using pregnant women who have already been prescribed the drug as therapy by their own physician. An example of a minimal risk study would be one to determine PK/PD of an antihypertensive medication in pregnant women who are taking that medication to treat hypertension during pregnancy. The decision to use the antihypertensive medication is made by the patient and her physician independent of participation in the PK/PD study.

Generating a full understanding of the effect of pregnancy on a given drug is a far from trivial task as noted in the FDA's draft guidance:

Study design considerations are important when conducting a study in pregnant women to determine if the PK and/or PD are altered enough to require an adjustment from the established dosage. Ideally, PK studies in pregnancy would be done pre-pregnancy (for baseline comparison) and during all three trimesters, especially for chronically administered drugs. Given the constraints of a study design that enrolls women prior to pregnancy, an alternative can be to determine PK/PD in the second and third trimesters, with the baseline assessment for comparison to the pregnant state done in the postpartum period. The Agency recommends care be taken to select the most appropriate postpartum time for PK/PD determination, if known. Cardiovascular and renal changes do not return to the pre-pregnancy state until 3 months postpartum. Optimally, postpartum PK/PD assessments for comparative purposes to PK/PD in pregnancy would be done when the woman is neither pregnant nor lactating.

The guidance notes that:

It is possible to study drugs that have no intended direct therapeutic benefit to the pregnant woman provided that the risk to the fetus is minimal (45 CFR 46). For example, probe substrates can be used to investigate drug metabolism (e.g., cytochrome P-450 activity) or drug transporter status (e.g., p-glycoprotein). Data from these studies offer generalizable information to other pregnant women but do not offer direct therapeutic benefit to study participants. The Agency encourages sponsors or investigators to explore additional safeguards for human subject protection for this type of study. To minimize exposure to a nontherapeutic drug, each pregnant woman can be exposed to the drug once during pregnancy and in the postpartum period employing a nonlongitudinal design (e.g., one cohort of women sampled in second trimester and postpartum and another cohort of women sampled in third trimester and postpartum). Examples of additional safeguards include administering only products with a long or known record of safety in pregnancy, administering products using only a single dose of the drug, using lower doses of the drug, decreasing the number of drugs (probe substrates) used in any study subject, and limiting study participants to pregnant women only in second or third trimester.

Although some researchers (e.g. Hodge and Tracey, 2007) have summarized the literature related to the effect of pregnancy on the disposition on drugs, it appears fair to state that, at the present, there appear to be gaps in our knowledge regarding how to determine the metabolism of drugs during pregnancy and that this is an area that may warrant further research and understanding.

27.5.7.10 *Human ADME Studies*   As can be gathered from all of the guidances and guidelines discussed earlier, the determination of the metabolism of a drug in humans is of pivotal importance. However, there are no specific regulatory guidances or guidelines that pertain to the design and conduct of ADME studies in humans. However, legislation exists in the United States that is aimed at ensuring "safe and effective use of the radioactive drug" and in the EU that is aimed at "protection of individuals against the dangers of ionizing radiation."

Unlike the guidelines and guidances, the topic of radiation safety is written into law and therefore has a different level of enforcement than do the guidances or guidelines. The EU legislation enables individual member states to compose national regulations regarding exposure to radioactivity in medical research. There are two European Atomic Energy Community (EURATOM) directives that are pertinent to the conduct of human ADME studies. The first is 96/29/EURATOM which imposes the ALARA principle ("As Low as Reasonably Achievable"), the intent of which is to minimize the exposure of human subjects to radiation (The Council of the European Union, 1996). This directive is aimed at workers and the general public. The administration of radio-activity to subjects in clinical trials is addressed in 97/43/EURATOM (The Council of the European Union, 1997). This directive supplements the previous directive and addresses issues such as the role of the Ethics Committee and quality of the radioactive material, but does not address the level of radioactivity that can be administered.

The US legislation (21 CFR 361) addresses the use of drugs in humans "intended to obtain basic information regarding the metabolism (including kinetics, distribution and localization) of a radioactively labeled drug" (US FDA, 2007d). The legislation places the responsibility for the assessment of the safety of the radiolabeled material to be used in the clinical trial on the local Radioactive Drug Research Committee (RDRC). Echoing the EU directive, the US legislation requires that the administered dose is as low as possible without jeopardizing the feasibility of the study. Maximum allowable exposures (in terms of a single dose and annual exposure) are defined for the whole body, active blood-forming organs, the lens of the eye, and the gonads. For subjects that are less than 18 years of age, the limits are 10-fold lower.

Neither the EU nor US legislation addresses other elements of the design and conduct of human ADME studies. Additionally, there are no guidances or guidelines that address these topics. The design and conduct of human ADME (mass balance) studies have been reviewed in detail by Beumer, Beijnen and Schellens (2006) and Roffey *et al.* (2007). The majority of studies are conducted in the United States, some studies are conducted in the EU (e.g., United Kingdom, the Netherlands, and Sweden), and no studies are conducted in Japan. The studies tend to be conducted in normal healthy male volunteers, although some studies are conducted in normal healthy female volunteers and the *in vivo* metabolism/mass balance of cytotoxic oncology drugs are conducted in cancer patients (if conducted at all). Studies tend to be conducted in a "young" population 18–60 and are not conducted in the pediatric or geriatric populations. Additionally, the number of subjects per study tends to be low (6–8). The studies tend to be conducted at one dose level only, with the dose approximating to the maximum dose for which approval will be sought. Serial blood samples (for processing to plasma) are collected. Urine and feces are also collected over a period of 7–10 days. Subjects are usually not discharged until a certain mass balance has been achieved or the amount of radioactivity that is present in the collected samples has dropped below a threshold level. Additionally, the human ADME study tends to be conducted after the administration of a single dose, as is usually the case for these studies when they are performed in animals. However, it should be noted that alternative approaches should be considered for drugs that show nonlinear PK due to saturation of metabolism, for drugs that either induce or inhibit their own metabolism, for drugs that show polymorphic metabolism, and for drugs that show age-dependent differences in metabolism. A standard human ADME study will collect blood (which is subsequently processed to plasma), urine, and feces. In some circumstances, it may be useful to collect other matrices such as semen, hair, nails, and saliva. An understanding of the excretion of a drug in the semen may help in the risk–benefit evaluation with respect to reproductive toxicity. If drugs are excreted into the hair, nails, or saliva, it may be possible to utilize these matrices as means to noninvasively measuring exposure to drug and/or metabolites. For drugs that are administered orally, it will not be possible to determine whether parent drug that is found in the feces was never absorbed or absorbed and then eliminated in the bile without metabolism. Occasionally, it is possible to determine the extent of biliary elimination of a drug if the subject has an indwelling biliary T-tube or may be about to undergo surgery and will consent to the collection of bile samples. As can be anticipated, such opportunities rarely arise.

Roffey *et al.* (2007) reviewed 171 human mass balance studies and found that the average mass balance was $89 \pm 11\%$ (range = 39–113%). Eleven studies showed recoveries of >100%, which is scientifically impossible and most likely was due to errors in the determination of the amount of radioactivity in the dose or excreta or as a result of misdosing of the subject. Forty-two percent of the studies had recoveries that were <90%, and 15% of the studies had recoveries that were <80%. Usually, a recovery of $100 \pm 10\%$ is regarded as being acceptable. Low recovery of radioactivity can be due to methodological, analytical, and biological factors.

The biological factors include:

- Long elimination half-life
- Binding to specific proteins in tissues
- Binding to melanin
- Binding to phospholipids
- Other non-covalent binding to tissues
- Covalent binding
- Loss through expiration (^{14}C) or exchange with water (^{3}H).

Methodological factors include:

- Compliance
- Inadequate homogenization of feces.

Metabolite profiling tends to primarily be conducted on plasma samples, although profiling is also performed on urine and feces. Since the risk–benefit analysis for a drug will be based on the systemic (plasma) exposure to drug and/or metabolites, it can be seen that the primary focus should be on understanding the plasma metabolite profile. The number of samples that are profiled can vary from study to study and from sponsor to sponsor. Some sponsors will profile each plasma sample from each sample.

The concentration of parent and metabolite will be determined and the exposure (as measured by AUC) determined for parent drug and each metabolite and subject. The mean exposure to parent drug and each metabolite will then be calculated. Other sponsors may choose to pool samples across subjects at a given time point and then profile the pooled samples as indicated. Another approach is to pool samples across subjects and subjects. If the samples are pooled on a time-weighted basis, an estimate of the mean exposure (AUC) to the drug and metabolites can be determined based on the profiling of a single sample (Hop *et al.*, 1998; Cheung *et al.*, 2005; Han *et al.*, 2006). It may be appropriate to create two pooled samples: one that represents the samples collected over the proposed dosing interval and one based on the duration of time that the plasma samples contained measurable levels of radioactivity. Obviously, this approach means that it is not possible to make an estimate of the interindividual variability in exposure to a given metabolite. Therefore, another alternative is to create the "pseudo-AUC" pooled sample for each subject and profile that sample. This approach will allow the sponsor to begin to understand the interindividual variability in metabolic profile, albeit in a small population. A different approach is taken for the profiling of urine and fecal samples. For these matrices, a common practice is to determine the samples that cumulatively account for 90% or more of the radioactivity that is excreted in that matrix. The samples are then pooled (either per subject or across subjects) on a relative volume basis. The cutoff of 90% is somewhat arbitrary, but does ensure that the pooled sample is not diluted unnecessarily by samples that contain low amounts of radioactivity. The approaches described here are also used to varying extent for the analysis of samples from animal mass balance studies. As can be seen, there are many acceptable approaches to the profiling of samples. The manner in which the profiling is performed is more of a matter of a company's standard practices rather than an accepted required or best practice. Regardless of how the plasma profiling is performed, the outcome should be an estimate of the exposure to the parent drug, the contribution of the exposure to parent drug to exposure to total radioactivity, the contribution of a particular metabolite to the exposure to total radioactivity (for the purpose of identifying major or important metabolites), and an assessment of the exposure to selected metabolites. The FDA has not provided guidance, nor appears to have a position on, on whether a metabolite should be designated as a major metabolite based on mean or individual data (D. Jacobson-Kram, 2006, personal communication). Assuming that none of the subjects are outliers due to a genetic polymorphism in a DME involved in the metabolism of a drug, it is most probably prudent that the designation is made based upon mean data. The assessment of whether a subject is an outlier or not can be made based on exposure to parent drug or total radioactivity, or based upon the contribution of exposure to parent drug to the exposure to total radioactivity. Additionally, no insight has been provided regarding how to report the exposure to the parent or metabolite (i.e. number of significant figures and rounding of the data).

The amount of radioactivity that can be administered in a human ADME study can be contentious. Both the US and EU legislation state that the lowest amount of radioactivity that can be used to achieve the objective of the study should be used. This amount of radioactivity is determined based on the distribution of radioactivity in an animal study. These studies are most commonly conducted in pigmented rats, although other species have been used (Solon and Kraus, 2001). There is no guidance or legislation that addresses the conduct of these tissue distribution studies. From the data generated in the tissue distribution study, calculations are performed to determine the exposure of humans to a given amount of radioactivity (Solon and Lee, 2002). Interestingly, most human ADME studies use a dose of 100 μCi and very rarely, if ever, are calculations performed to determine whether it would be necessary to administer this amount of radioactivity to achieve the objectives of the study (Dain, Collins and Robinson, 1994; Beumer, Beijnen and Schellens, 2006; Roffey *et al.*, 2007). In many cases, if the calculations were performed to determine the minimum amount of radioactivity that would be required to achieve the objectives of the study, the amount of radioactivity required would be lower than 100 μCi.

Over the past decade, AMS (accelerator mass spectrometry) has increasingly been used to determine the amount of radioactivity in biological samples, including samples obtained from a human ADME study. Since this technology can detect far lower levels of radioactivity than traditional techniques such as LSC, the amount of radioactivity that can be used in a human ADME study can be reduced even further. It is a useful technique for measuring ^{14}C, but not for measuring ^{3}H. The use of AMS is described in detail in a previous chapter of this book.

To date, very few human ADME studies have been performed using AMS as the detection tool. Recently, ixabepilone was approved for the treatment of breast cancer. A human ADME study was performed using AMS. However, the rationale for the use of AMS was the instability of $[^{14}C]$ ixabepilone. Due to autoradiolysis, ixabepilone proved to be very unstable when labeled with conventional $[^{14}C]$-levels (100 μCi in a typical human radiotracer study). This necessitated the use of much lower levels of $[^{14}C]$-labeling and an ultrasensitive detection method, AMS (Beumer *et al.*, 2007). Regardless of the rationale for the use of AMS, it is clear that the use of AMS was acceptable to the FDA. This is consistent with statements that have been made that "As long as the approach used is scientifically valid, the FDA does not care what type of detection is used" (D. Jacobson-Kram, 2006, personal communication).

The adoption of the use of microradioactivity human ADME studies is slow. This is due partially due to institutional inertia (i.e. it takes a long time to change the traditional approaches of a pharmaceutical company), the cost of the AMS equipment, concerns about sample contamination, the cost of sample analysis ($/sample), the lack of general availability of the equipment, and a misunderstanding as to how AMS will be used for a microradioactivity study.

AMS has mainly been used primarily as a sensitive tool for the detection of the very low plasma concentrations of drug that occur during a course of a microdosing study, although it has also been used to measure femtomolar concentrations of doxorubicin in tumor cells and levels of zidovudine in peripheral blood monocytes (DeGregorio *et al.*, 2006; Vuong *et al.*, 2007). AMS may enable cancer patients to participate in a human ADME study while they are at home. Since microdose studies use μg amounts of drug, it has been mistakenly assumed that such low doses of drug would be used for a microradioactivity human ADME study. With such low levels of drug, concerns can be raised regarding the linearity in metabolism between these low doses and the pharmacologically relevant doses. However, a microradioactivity human ADME study will use pharmacologically relevant doses of drug but very low amounts of radioactivity (<200 nCi).

With respect to cost, an analysis of a traditional and microradioactivity study will most likely find that the costs for the two studies are similar in the long run. Although the analysis costs for a microradioactivity study will be higher than the costs for a traditional ADME study, there will be significant cost savings in other areas. For example, the microradioactivity study will require a smaller amount of radiolabeled drug.

Additionally, the low amount of radiolabeled drug that will be administered will be a trace amount and may not need to be subject to synthesis or release under GMP conditions. Another cost saving may be that it is likely that a tissue distribution study (for the purpose of dosimetry) will not be required before a human microradioactivity ADME study is conducted. In addition to the cost savings, there will also be time savings that would enable a microradioactivity dose human ADME study to start sooner than a traditional human ADME study could.

The US legislation states that the RDRC will determine the safety of the administered dose based on the data that they feel is appropriate. Using very conservative assumptions, it has been calculated that a dose of 200 nCi will represent 0.1% of the US whole body dose limit of 3000 mRem for a single dose. Therefore, it is likely that RDRCs will not require dosimetry studies for human ADME studies in which the dose of radioactivity is ≤200 nCi. This change in approach will potentially reduce the cost of conducting a human ADME study and give greater flexibility in the timing of such studies. It also opens the possibility that low doses of isotopically labeled drug can be administered as part of the initial human clinical trial and thereby enables a very rapid understanding of the metabolism of the NCE *in vivo* in humans. Having not to conduct the tissue distribution study also would enable sponsors to avoid some of the unintended consequences that often arise when attempting to interpret tissue distribution studies that utilize the determination of total radioactivity as the analyte. In a tissue distribution study, the concentrations of radioactivity represent the sum of the concentration of parent drug plus the concentration of all metabolites. Therefore, a tissue distribution study may be able to provide information as to which tissues the drug did not distribute to (at least, to a minimum level) but the study will not be able to provide information as to whether the radioactivity was present as parent drugs and/or metabolites. Therefore, it can be easy to assume that the organs that contain high levels of radioactivity will contain high levels of parent drug when, in reality, the radioactivity may represent metabolites and not parent drug. Another pitfall of tissue distribution studies is the assumption that organs that show high levels of radioactivity will show toxicity in animal studies. Not only do the aforementioned caveats regarding the lack of ability to discriminate between parent drug and metabolites apply, it should be recognized that the presence of parent drug and/or metabolites in a given tissue does not mean that there will be a toxic insult in that organ. Finally, a common finding in tissue distribution studies conducted in pigmented animals is the retention of radioactivity by melanin in the eye. The correlation of binding of drugs or their metabolites to melanin and subsequent toxicity is equivocal and it is possible that the retained radioactivity represents an animal-specific metabolite and therefore be a false-positive result (Roffey *et al.*, 2007). Another consequence of the detection of binding of parent drug and/or metabolites to melanin is the need to conduct a photosensitivity study (see earlier discussion). Since tissue distribution studies are not specifically recommended by guidances/guidelines or legislation for purposes other than dosimetry calculations in support of a radiolabeled study in humans and the technology exists to conduct such studies in humans at doses of radioactivity that do not require dosimetry studies, such studies may play a far lesser role in future development programs.

One element of the organizational inertia that may be responsible for the lack of adoption of the use of microradioactivity human ADME studies is not specific to this particular study design, but is a reflection of a reluctance to embark on a costly evaluation of the *in vivo* metabolism of a drug in humans until some signs of efficacy have been seen in clinical trials or until the latest possible time in the drug development plan. Another element that should not be underestimated is that of a desire not to be the department that was responsible for the discontinuation of the development of a drug.

Although it is recognized that these studies are costly and labor-intensive, this mindset no longer seems to be consistent with the message that is being delivered by regulatory agencies (particularly the FDA) through guidances and guidelines that an understanding of the role of metabolites in efficacy and toxicity is of paramount importance. A lack of knowledge regarding the safety of metabolites may result in potential delays to clinical trials and a lack of knowledge of regarding the contribution of metabolites to the efficacy of a drug may result in inaccurate PK/PD models being created and incorrect doses being selected for phase III clinical trials.

The following is a proposal that may enable a sponsor to understand the metabolism of a new drug at a very early stage of clinical development and then design the clinical pharmacology portion of the clinical plan based on this understanding. This proposal could be applicable to non-cytotoxic drugs.

The FIH study would be conducted, in large part, in the same manner as a traditional FIH study. However, each dose would include a small (<200 nCi) amount of radiolabeled drug. This is easily achievable if the dose is administered as "powder in a bottle" (i.e. the dose is dissolved at the clinical site and administered as an oral solution) and the compound has adequate solubility. Serial blood samples would be collected and plasma prepared. Urine and/or fecal samples could also be collected in order to evaluate the mass balance as necessary. Plasma concentrations of the parent drug will be determined using an appropriate validated method (e.g. LC-MS/MS). Once the pharmacokinetics of the drug and the safety of the drug have been reviewed and determined to be acceptable for further development, the plasma profiling can then occur. Based on the parent drug exposure, the dose that most closely approximates to the predicted efficacious dose is selected for further analysis. The plasma samples are pooled across subjects and across time points on a time-weighted basis for all samples collected through 24 hours post dose (for a drug that is intended to be administered qd). Total radioactivity is determined in these pooled samples. The AUC_{0-24h} of the parent drug is compared with that of total radioactivity determined in the pooled sample. If the AUC of parent drug represents >90% of the AUC of total radioactivity, then no profiling of the sample is warranted at this point in time since no single metabolite can represent >10% of the exposure to total radioactivity and hence be classified as a major metabolite as defined in the FDA's guidance on "Safety Testing of Drug Metabolites" (US FDA, 2008j). If the AUC of the parent drug is <90% of the exposure to total radioactivity, profiling of the plasma samples can be performed. The metabolites can be separated by chromatography and the fractions pooled for analysis by AMS. The exposure to the metabolites in humans can be then compared with the exposure in animals and decisions made regarding the need to perform any further testing based on the FDA's guidance on "Safety Testing of Drug Metabolites" (US FDA, 2008j). If nonlinear pharmacokinetics of the parent drug is observed, samples from additional dose levels may be analyzed. With this approach it will be feasible to very rapidly determine the extent of metabolism of a novel drug and therefore determine the optimal design for the clinical pharmacology program for this drug. It should be noted that the data generated in this early type of study will not necessarily generate all of the data that would be generated from a traditional human ADME study. The remaining data could be generated from the samples conducted in this early study or by a dedicated human ADME study (either microradioactivity or traditional study).

In order to describe the metabolism in special populations (e.g. renally impaired, hepatically impaired, geriatric, and pediatric patients), the metabolism may be first described in normal healthy volunteers and the exposure to the major/important metabolites determined subsequently in the special populations. An identical approach can be taken to estimating the effect of environmental factors (e.g. diet, alcohol, and smoking), the effect of food, and the effect of concomitant medications on the metabolism of a given drug.

27.6 BIOLOGICS

The development of a biological drug differs from the development of an NCE in several respects. The standard testing paradigms that are used for an NCE are not appropriate, and a customized approach should be taken. Several guidances and guidelines have been issued on the topic of safety testing of biological products. The primary guidance is the ICHS6 Guidance (Preclinical Safety Evaluation of Biotechnology-Derived Pharmaceuticals) that was issued in 1997 (ICH, 1996b). In 2006, a draft guidance entitled "Nonclinical Safety Evaluation of Biotechnology-Derived Pharmaceuticals" was made available to some pharmaceutical industry researchers. This guidance has not been officially released for review by the FDA but does appear on its list of ongoing or planned guidances for 2008 (US FDA, 2008f). However, this guidance has fostered much debate between the FDA and sponsors on the development of biologics. This "unofficial" draft guidance from the FDA states: "Unlike small molecule drugs, proteins are not metabolized by hepatic cytochrome P450 mechanisms. Instead, they are catabolized into individual amino acids which may then be excreted or reused for protein synthesis or energy production. This poses a problem for studies that rely on radiolabeling, such as mass balance assessments and tissue distribution studies typically performed for small molecules. The radiolabelled amino acid of the biopharmaceutical will be released through catabolism and possibly recycled into normal endogenous protein synthesis. Thus, measurements will likely reflect normal protein turnover. Studies assessing metabolism and excretion are therefore not generally warranted nor are they typically informative for biopharmaceuticals." This latter statement is even more strongly enforced in the ICHS6 guidance (Preclinical Safety Evaluation of Biotechnology-Derived Pharmaceuticals) that was issued in 1997 (ICH, 1996b). This guidance states: "The expected consequence of metabolism of biotechnology-derived pharmaceuticals is the degradation to small peptides and individual amino acids. Therefore, the metabolic pathways are generally understood. Classical biotransformation studies as performed for pharmaceuticals are not needed." The ICHS6 guidance also emphasizes some of the challenges of conducting radiolabeled ADME studies for biologics. The guidance states: "When using radiolabeled proteins, it is important to show that the radiolabeled test material maintains activity and biological properties equivalent to that of the unlabeled material." This is a concept that is not often discussed for small molecules that may be labeled with carbon-14 or tritium. However, it should be noted that the incorporation of a deuterium atom for a hydrogen atom in a small molecule could potentially change the metabolism of a drug and therefore caution should be employed in the use of certain stable isotopes in the assessment of the metabolism of a drug. The cautionary note in ICHS6 arises due to the fact that a common means of inserting a radiolabel into a biologic is by the incorporation of an iodine isotope such as Iodine-131. The effect of iodination on the PK of proteins has been shown by Bauer *et al.* (1996).

ICH S6 goes on to introduce a note of caution regarding the interpretation of radiolabeled studies of biologics by stating: "Tissue concentrations of radioactivity and/or autoradiography data using radiolabeled proteins may be difficult to interpret due to rapid *in vivo* metabolism or unstable radiolabeled linkage. Care should be taken in the interpretation of studies using radioactive tracers incorporated into specific amino acids because of recycling of amino acids into non-drug related proteins/ peptides."

The "unofficial" draft guidance from the FDA states that "tissue distribution studies that provide insight into the localization of the administered biopharmaceutical may be useful. For example, tissue distribution studies for imaging agents may be informative since the interest is in identifying organ systems at risk from the radiolabel moiety, not the protein component of the molecule. Thus, distribution of the radiolabel is important because it retains its pharmacological activity, regardless of whether it is maintained within the intact molecule or released as a consequence of protein degradation." The example given relates to a specific subclass of biologics. The vast majority of biologics that are in development at this point in time are monoclonal antibodies. For such agents, the identification of potential organs of toxicity is determined by an evaluation of the ability of the antibody to bind to a range of tissues rather than by an evaluation of the distribution of the monoclonal antibody. This is performed on histopathology slides and allows the relatively easy assessment of the binding of the antibody across a range of species. This is the preferred way in which to evaluate the potential tissue distribution of a monoclonal antibody, primarily due to its ease, low cost, and relatively high throughput. If a traditional tissue distribution study were to be performed for a monoclonal antibody, the study should be performed in the species selected for toxicity testing (i.e. monkey). Conducting a tissue distribution study in the monkey (particularly one that involves an evaluation of the tissue distribution over time) would require a large number of monkeys and would be extremely expensive. Additionally, the equipment that would allow the visualization of the tissue distribution (i.e. quantitative whole body autoradiography [QWBA]) is uncommon. Therefore, for ethical, economic, and practical reasons, classical tissue distribution studies for monoclonal antibodies are not conducted unless the monoclonal antibody is to be used as an imaging agent and an assessment of the potential radiation exposure for humans is required.

ICHS6 also addresses a parameter that is commonly determined for "small molecules" but is often underlooked for biologics. The guidance states: "Understanding the behaviour of the biopharmaceutical in the biologic matrix, (e.g., plasma, serum, cerebral spinal fluid) and the possible influence of binding proteins is important for understanding the pharmacodynamic effect." Exactly how one determines this binding is not addressed and the determination of such binding would not be a trivial task. The ICHS6 guidance is being considered for revision by ICH.

Since the responsibility for the review of biologics (e.g. replacement proteins and antibodies) has been moved from CBER to CDER, many pharmaceutical companies have encountered challenges when interacting with reviewers that have little, if any, experience in developing biologics. These challenges have included duration of chronic non-rodent toxicology studies (which tend to be six months in duration for a biologic per ICH S6 but may be 9 or 12 months for an NCE), the desire for the toxicity of the biologic to be determined in two animal species (even if only one species shows a pharmacological response and if both species that show a pharmacological response are rodents), the need for tissue distribution studies in order to determine potential target organs of toxicity (this is addressed by other means), the need for metabolism studies, and the need for certain *in vitro* studies (i.e. hERG binding and CYP inhibition).

The FDA guidance, "Points to Consider in the Manufacture and Testing of Monoclonal Antibody Products for Human Use" that was issued in 1997 raises the point that "For immunoconjugates containing radionuclides, animal biodistribution data may be used for initial human dose estimation" (US FDA, 1997b).

It further states that "There should be complete accounting of the metabolism of the total dose of administered radioactivity and an adequate number of time points to determine early and late elimination phases. The expected biodistribution and routes of clearance of the administered radiolabeled antibody dose fractions in tissues/organs should be defined. The expected biodistribution and routes of clearance that might occur in the presence of diseases in organs that are critical in radioimmunoconjugate metabolism or excretion should be described. The expected biodistribution and routes of clearance that might occur in the presence of immune responses (e.g., HAMA, HAHA, HARA) should be described. With reference to the radioactive fractions of the administered radiolabeled antibody dose and the patterns of biodistribution, the following issues should be addressed:

- From the biodistribution estimation, the expected residence time of the radiolabeled antibody fractions in the target tissues/organs and non-target tissues/organs should be determined.
- Based on the estimated residence times in each organ, the radiation exposure for each tissue/organ should be estimated.
- Based on the radiation exposure for each tissue/organ, the potential toxicity should be described.
- Based on the potential radiation toxicity to tissues/organs, toxicity monitoring protocols should be developed and incorporated into the clinical trial. If the study has increasing doses of radioactive materials (e.g., a study of the maximum tolerated dose), the radiation exposure for tissues/organs and the associated potential toxicities should be estimated for each radiation dose level."

In January 2007, the EMEA issued the following guideline: "Guideline on the Clinical Investigation of the Pharmacokinetics of Therapeutic Proteins" (EMEA, 2007e). To date, this is the only guidance on this topic, although it should be noted that a book was published in 2006 entitled *Clinical Pharmacology of Therapeutic Proteins* (Mahmood, 2006). The editor of the book was Iftekhar Mahmood of the FDA. All of the authors were employed by the FDA.

The EMEA guideline emphasized the importance of bioanalytical methods by stating:

One of the key elements of a pharmacokinetic study is the analytical method and its capability to detect and follow the time course of a given analyte (the parent compound and/or metabolites) in a complex biological matrix that contains many other proteins and with satisfactory specificity, sensitivity and a range of quantification with adequate accuracy and precision. The ability to distinguish the therapeutically applied protein from endogenously produced equivalents is an important criteria in selecting the analytical method, although it is recognized that developing an assay that distinguishes between the therapeutic agent and the endogenous molecule may not always be technically feasible.

The guideline again reiterated the message form ICHS6 regarding the catabolism of protein drugs by stating:

The main elimination pathway should be identified. However, for therapeutic proteins this could be predicted, to a large extent, from the molecular size and specific studies may not be necessary. Catabolism of proteins occurs, usually, by proteolysis. Small proteins of MW < 50,000 Da are eliminated through renal filtration (renal filtration becomes increasingly important the lower the molecular weight) followed by tubular re-absorption and subsequent metabolic catabolism. For larger protein molecules, elimination in other tissues and/or in target cells through for example receptor-mediated endocytosis followed by catabolism is more important relative to renal filtration. Mass-balance studies are not useful for determining the excretion pattern of the drug and drug-related material. Excreted proteins are not necessarily recovered in urine or faeces as intact substance, but are instead metabolised and reabsorbed as amino acids and incorporated in the general protein synthesis.

However, the EMEA guideline went one step further and made several additional points about the metabolism of therapeutic proteins and the consequences of their metabolism. For example, the guideline stated: "The need for, and the feasibility of, specific studies of the route of elimination and metabolism (e.g. microsomal, whole cell or tissue homogenate studies) and identification of metabolites in vitro should be considered and discussed on a case-by-case basis." The guideline did define metabolite as encompassing *in vivo* degradation products and other truncated forms of the protein. The subject of metabolites was further addressed in the following statement:

Metabolites that have pharmacodynamic activity should preferably be measured, e.g. through chromatographic separation, collection and further *in vivo* bioassay quantification. The metabolites may have different pharmacokinetic profiles compared with the parent compound. However, in cases where measurement of separate active metabolites or peptide fragments is not technically feasible, pharmacokinetics of the active moiety could be determined. Also, measurement of complexes between the protein and other components present in plasma should be considered (Section 2.2). The activity of a therapeutic protein may not only be related to the unbound fraction in plasma but also to bound fractions and to the binding kinetics. Thus, when interpreting the data it is important to understand what fraction is detected in the bioanalysis. It may even be important to analyse the bound fraction per se. Bioassays should be considered, especially if selective immunoassays for metabolites are lacking. Lack of relevant methods should be justified.

The guidance also states:

It is advantageous if both immunoassay and bioassay are used. If the activity is generated by several species (e.g. metabolites, isoforms), each with different activity, their relative content might change with the degree of renal function due to different degrees of renal clearance. If the different species have similar affinity in the immunoassay, a bioassay allows a more relevant interpretation of the data regarding total activity.

Hepatic impairment: Reduced hepatic function may decrease the elimination of a protein for which hepatic degradation is an important elimination pathway. The lack of studies should be justified by the Applicant (see guideline CPMP/EWP/2339/02).

There are several implications of the statements that have been made in this guidance. These include:

1. Therapeutic proteins may be metabolized to active metabolites and these metabolites should be quantitated.

2. Enhanced bioanalytical techniques may be required since ELISA assays cannot distinguish between parent drug and metabolites of the drug.

3. Methods to measure both drug (and metabolite) concentrations as well as total biological activity (bioassay) are required.

4. The protein therapeutic should be evaluated in similar manner in both humans and animals.

5. Bioanalytical methods may be required that can distinguish between parent drug and truncated versions of the parent drug (present in the drug that is administered) in biological samples.

6. A more expanded set of clinical pharmacology studies may be required (e.g. renal and hepatic impairment). At present, such studies are not routinely conducted.

7. Sponsors may be required to understand how antibodies are catabolized (which organs, which enzyme systems).

8. The focus on hepatic microsomal metabolism studies may lead to a need to assess the potential for these antibodies to inhibit DMEs.

9. The discussion of the need to discuss whether certain studies are required implies that the sponsor will request scientific advice on this topic. The scientific advice process is a long, drawn-out and expensive process.

10. Determination of the binding of the protein therapeutic agent to plasma (or other) proteins should be addressed. Again, this is not routinely done.

The impact of the molecular, biologic, and pharmacokinetic parameters of monoclonal antibodies on early clinical development of mAbs has recently been reviewed by Mascelli *et al.* (2007).

Interestingly, this guideline does not address the possibility that protein drugs, or the consequences of their administration, can affect the regulation of DMEs and hence the pharmacokinetics of the substrates of these enzymes. For example, cytokines have been shown to affect the expression of certain CYP isoforms (Renton, 2001; Aitken, Richardson and Morgan, 2006; Charles *et al.*, 2006). The administration of some protein therapeutics may cause a temporary increase in cytokine levels, resulting in a change in expression of CYP isoforms. However, this phenomenon is described in the FDA draft guidance "Drug Interaction Studies—Study Design, Data Analysis, and Implications for Dosing and Labeling" which was issued in 2006. This draft guidance states: "certain protein therapeutics modify the metabolism of drugs that are metabolized by the P450 enzymes. Type I interferons, for example, inhibit CYP1A2 production at the transcriptional and post-translational levels, inhibiting clearance of theophylline. The increased clinical use of therapeutic proteins may raise concerns regarding the potential for their impacts on drug metabolism." The selection of theophylline as an example is a good one since increased theophylline toxicity was observed in children in King County, Washington, during the 1980 Influenza B outbreak (Kraemer *et al.*, 1982). The role of viral infections on the induction of adverse drug interactions has been reviewed by Levy (1997). Data on the impact of vaccination on the metabolism of drugs is equivocal (Stults and Hashisaki, 1983; Meredith *et al.*, 1985; Hamdy *et al.*, 1995). The effect of disease on hepatic CYP expression has also been observed in human and in a rat model of inflammatory pain (Projean *et al.*, 2005; Charles *et al.*, 2006). Additionally, decreased intestinal CYP3A and p-glycoprotein activities have been observed in rats with adjuvant-induced arthritis (Uno *et al.*, 2007). The observation that the disease can affect the PK of a drug that is being administered to treat that disease raises the interesting possibility that the PK of the drug will change as the drug works to cure the disease. If this were to occur, it would appear that the animal (or patient) is becoming resistant to the drug.

The impact of interferons on CYP-mediated drug metabolism has been evaluated for some marketed drugs, including PegIntron (Mahmood and Green, 2007). The following statement is taken from the drug label for Pegintron (Schering-Plough Research Institute, 2006):

> The pharmacokinetics of representative drugs metabolized by CYP1A2 (caffeine), CYP2C8/9 (tolbutamide), CYP2D6 (dextromethorphan), CYP3A4 (midazolam), and N-acetyltransferase (dapsone) were studied in 22 patients with chronic hepatitis C who received PegIntron™ (1.5μg/kg) once weekly for 4 weeks. PegIntron™ treatment resulted in a 28% (mean) increase in a measure of CYP2C8/9 activity. PegIntron™ treatment also resulted in a 66% (mean) increase in a measure of CYP2D6 activity; however, the effect was variable as 13 patients had an increase, 5 patients had a decrease, and 4 patients had no significant change.
>
> No significant effect was observed on the pharmacokinetics of representative drugs metabolized by CYP1A2, CYP3A4, or N-acetyltransferase. The effects of PegIntron™ on CYP2C19 activity were not assessed.

It should be noted that the impact of interferons on the clearance of drugs will likely extend beyond their effects on CYP-mediated metabolism. The impact of interferon-alpha on p-glycoprotein activity in rats has been shown by Ben Reguiga *et al.* and it is not unreasonable to expect that such effects will also be observed in humans (Ben Reguiga *et al.*, 2005; Ben Reguiga, Bonhomme-Faivre and Farinotti, 2007).

DDI studies for therapeutic proteins or monoclonal antibodies have recently been reviewed by Mahmood and Green of the FDA (Mahmood and Green, 2007). They have noted that few DDI studies have been conducted for these classes of drugs, mainly because most macromolecules are not metabolized by CYPs and their mechanism of elimination is complex. The review does describe some DDI studies for monoclonal antibodies. These studies include adalimumab-methotrexate, etanercept-paclitaxel, etanercept-digoxin, etanercept-warfarin, etanercept-methotrexate, trastuzumab-paclitaxel, cetuximab-irinotecan, and panitumumab-irinotecan.

However, this review did not include the study of the effect of Avonex (interferon beta 1-a) on the pharmacokinetics of Tysabri (natalizumab) in MS patients. There was an increase in the C_{max} and half-life of natalizumab by 20% and 70%, respectively, following the administration of Avonex. The clearance of natalizumab decreased by 35% (US FDA, 2004f). However, based on the similarity of the side-effect profile of Tysabri either with or without coadministration of Avonex, no dose adjustment was recommended. This DDI was noted in the original label for Tysabri but is not present in the latest version of the label (Biogen-Idec, 2004, 2006). The reason for this change is unknown.

Over the past decade, the types of materials that may be classed as biologics have expanded from being mainly replacement proteins and antibodies to include antisense RNA, RNAi, and aptamers. The regulatory guidances do not specifically cover these agents with the exception of a reflection paper that was generated by the CHMP SWP. The reflection paper was issued in 2005 and was titled "CHMP SWP Reflection Paper on the Assessment of the Genotoxic Potential of Antisense Oligodeoxynucleotides"

(EMEA, 2005a). ICHS6 does mention these novel agents in passing by stating "The principles outlined in this guidance may also be applicable to recombinant DNA protein vaccines, chemically synthesised peptides, plasma derived products, endogenous proteins extracted from human tissue, and oligonucleotide drugs," although no details are provided as to specific testing that may be required for these agents (ICH, 1996b).

As can be seen, this reflection paper addresses only one aspect of the development of this new class of agents. The reflection paper states: "With regard to oligodeoxynucleotides at least two issues need to be assessed which may indicate a cause for concern, i.e. (a) degradation products of the phosphorothioate oligodeoxynucleotides being nucleotide analogues might lead to mispairing and thus induction of point mutation when integrated into newly synthesized DNA and (b) site-specific mutations might be induced by triplex formation of the oligodeoxynucleotides with the DNA fiber." In this instance, the word "degradation" can be taken to mean both chemical degradation and biological degradation (i.e. metabolism). Therefore, this reflection paper implies that metabolites of oligonucleotide drugs may cause toxicity.

The reflection paper goes on to state: "Although the major metabolic pathways eliminate nucleotides through catabolism, it is possible that phosporothioate mononucleotides may be substrates for various kinases, and thus be incorporated in the triphosphate nucleotide pool. As part of this pool, nucleotide thiotrisposphates could be incorporated into newly synthesized DNA with the consequence that binding to complementary nucleotides occurs with reduced fidelity thus leading to mispairing and eventually point mutations."

Again, the reflection paper alerts the reader to the importance of metabolism in the evaluation of the safety of this type of agent. ICHS6 clearly implies that if you have a biologic drug that degrades into molecules that exist in nature, there will be a low possibility of toxicity. However, many of the oligonucleotide drugs may have been chemically modified and therefore will not be metabolized into natural products. Additionally, oligonucleotide drugs will be metabolized by nucleases which will gradually remove nucleotides from the structure. It is feasible that a metabolite of a parent oligonucleotide drug will simply consist of the parent drug minus one nucleotide residue. This metabolite will likely still retain pharmacological activity and could contribute to overall efficacy.

Similar to the chemically modified oligonucleotide drugs, there exists a class of drugs called "immunoconjugates." These drugs may consist of antibodies to which a toxic agent is conjugated. Mylotarg is an example of an immunoconjugate. Often, the antibody part of the immunoconjugate must be chemically modified before the toxin can be conjugated to the antibody. This is often achieved by attaching a chemical linker to the antibody. Therefore, the protein is no longer a "pure" biotechnology-derived agent and may no longer degrade to compounds that would be regarded as being endogenous substances. Additionally, it is very difficult to control the number of toxin molecules that are conjugated to the protein and the immunoconjugates may consist of a mixture of different conjugates, each containing a different number of toxin molecules per antibody molecule. The toxin molecules will be released in a sequential, albeit random, manner (i.e. a conjugate containing four toxin molecules will be deconjugated to a conjugate containing three toxin molecules). Since the pharmacological activity of the immunoconjugate may depend upon the number of toxin molecules, and that the rate of degradation of the immunoconjugate in the blood will affect the safety profile of the immunoconjugate, it is clear that metabolism plays a hitherto unexplored and critical role in the safety and efficacy of immunoconjugates. Until recently, the existing bioanalytical technology has not been available to determine the metabolism of immunoconjugates. ELISA assays are traditionally used to quantitate protein-based drugs and are not capable of distinguishing between immunoconjugates that contain one toxin and an immunoconjugate that contains four toxins. However, they can be configured to be able to discriminate between immunoconjugates that contain a toxin moiety and those that do not. Recent advances in MALDI/TOF mass spectrometry has offered the possibility of being able to determine the relative levels of the immunoconjugates (i.e. levels of immunoconjugates containing three or four toxins). However, until technology is developed that allows for the generation of pure immunoconjugate standards (i.e. a conjugate that contains only one toxin molecule), this technology will not be able to readily provide absolute concentration data.

From the examples given, it can be seen that there exists an area in which metabolism plays an important role but for which there is little, if any, regulatory guidance.

27.7 MISCELLANEOUS TOPICS

There are several guidances that mention drug metabolism in passing that do not fit in the previous subsections of this chapter.

27.7.1 CMC-Related

For example, metabolites are mentioned in several CMC guidances, including "Impurities in New Drug Substances Q3A(R)" that states: "Impurities that are also significant metabolites present in animal and/or human studies are generally considered qualified" and "Impurities in New Drug Products Q3B(R2)" that states that "Degradation products that are also significant metabolites present in animal and/or human studies are generally considered qualified" (ICH, 2006a, b). It is clear that an early understanding of the metabolism profile of a drug can aid the qualification of impurities and eliminate the potential for unnecessary toxicology studies to qualify certain impurities or degradation products.

In December 2006, the EMEA has issued a guideline called "Guideline on the Environmental Risk Assessment of Medicinal Products for Human Use" (EMEA, 2006e). This guideline indicates that the environmental impact assessment will require an understanding of the metabolic fate of the drug.

27.7.2 Nanotechnology

The FDA's Nanotechnology Task Force issued a report in July 2007 that indicated the importance of an understanding of the ADME properties on nanoscale materials (US FDA, 2007q). The report stated that, for regulated products:

There may be a fundamental difference in the kind of uncertainty associated with nanoscale materials compared to conventional chemicals, both with respect to knowledge about them and the way that testing is performed. For conventional chemicals, there is a relatively long history of exploring, and a correspondingly relatively robust understanding, of interactions of molecular classes (such as compounds with particular structures or functional groups) with biological systems. In some cases, screening test methods are used to define what additional tests may need to be performed to gain sufficient knowledge about safety and/or effectiveness. For example, there are screening tests available to help identify whether DNA damage is a possible outcome from exposure to a certain chemical. Other tests can tell how the chemical is distributed in the body and in what forms it is present in various tissues. FDA has an expectation relevant to molecular forms of materials used in products that FDA regulates that if the molecule does not cause DNA damage during *in vitro* testing, or if it is metabolized quickly and does not reach sensitive organs, or if it is not absorbed, then it is less likely to present certain kinds of health hazards. This expectation is based on long experience with, and consequent understanding of, basic biological interactions of molecular forms of chemicals and of how these interactions correlate with the results of current testing methods.

27.7.3 Liposomes

Another guidance that provides some statements on how to determine the metabolism of a drug is the draft guidance issued by the FDA in 2002 called "Liposome Drug Products Chemistry, Manufacturing, and Controls; Human Pharmacokinetics and Bio-availability; and Labeling Documentation" (US FDA, 2002). Liposomes can be regarded as a nanoengineered product since their average diameter will be 60–100 nm. This draft guidance states that:

> The Agency recommends a comparative mass-balance study be performed to define and assess the differences in systemic exposure and pharmacokinetic measures or parameters between liposome and nonliposome drug products when (1) the two products have the same active moiety, (2) the two products are given by the same route of administration, and (3) one of the products is already approved for marketing. The disposition and pathways of elimination (including metabolism and excretion) and several important pharmacokinetic measures (C_{max}, AUC) and parameters (clearance, volume, half-life) of a liposomal formulation administered intravenously can be different from that of a nonliposomal formulation given by the same route of administration. Although no examples currently exist, absorption could also be altered for liposome drug product when given via nonintravenous routes. For these reasons, if satisfactory mass balance information is already available for the approved drug product, a limited mass balance study can be undertaken for the proposed drug product. In such a study, the quantity of the drug substance excreted via the major route should be compared in sufficient subjects by giving the liposomal and the nonliposomal formulations, using a crossover or a parallel study design. Comparison of the absorption, distribution, metabolism, and excretion (ADME) of the liposome and nonliposome drug product form should be made, using a crossover or noncrossover study design that employs an appropriate number of subjects. Depending on the drug substance under investigation, the dose of the liposome and nonliposome drug product may be different. The mass balance study should be based on drug substance tagged with a radioactive label (e.g., ^{14}C, ^{3}H) before its incorporation into liposomes to allow for sensitive monitoring of radioactive label after administration. Blood (plasma or serum as appropriate), urine, and fecal samples should be collected and assayed for radioactive label.

> Other routes of elimination should be monitored as appropriate. Both parent drug substances and any metabolites present should be quantitated. If feasible, mass balance studies can use nonlabeled drug moieties and ingredients. However, CDER recommends that a applicant contact the appropriate review division before conducting studies using nonlabeled drug substance. … Rarely, historical pharmacokinetic data for comparative purposes can be considered on a case-by-case basis in lieu of formal comparative mass balance and/or pharmacokinetic study, taking into account the following factors: (1) when and how the historical data was obtained, (2) similarities of study populations (e.g., disease

condition), (3) analytical procedures, and (4) data analysis. The appropriate CDER review division should be consulted to determine whether historical data can be relied upon.

27.7.4 Racemates

The topic of the development of stereoisomeric drugs is one that is very rarely discussed at the present, although this was a topic of concern in the past. This historical perspective is mentioned in the "FDA's Policy Statement for the Development of New Stereoisomeric Drugs" that was issued in 1992 (US FDA, 1992):

> When stereoisomers are biologically distinguishable, they might seem to be different drugs, yet it has been past practice to develop racemates (i.e., compound with 50:50 proportion of enantiomers). The properties of the individual enantiomers have not generally been well studied or characterized. Whether separated enantiomers should be developed was largely an academic question because commercial separation of racemates was difficult. Now that technological advances (large scale chiral separation procedures or asymmetric syntheses) permit production of many single enantiomers on a commercial scale, it is appropriate to consider what FDA's policy with respect to stereoisomeric mixtures should be. Development of racemates raises issues of acceptable manufacturing control of synthesis and impurities, adequate pharmacologic and toxicologic assessment, proper characterization of metabolism and distribution, and appropriate clinical evaluation. It should be appreciated that toxicity or unusual pharmacologic properties might reside not in the parent isomer, but in an isomer-specific metabolite.

Over the past decade or so, the vast majority of drugs that have entered clinical development have been single enantiomers. However, it should be recognized that the fact that the parent drug is a single enantiomer does not mean that the issue of the relative ADME, pharmacological or toxicological properties of the enantiomers is of no concern. For example, the parent drug can undergo enantiomeric conversion (e.g. ibuprofen) and that the parent drug could be metabolized to different enantiomers. This point is described in the Health Canada guidance (Stereochemical Issues in Chiral Drug Development) (Health Canada, 2000):

> The in vivo stability of the enantiomer must be established. If the antipode is formed in vivo, it should be considered to be a metabolite, and addressed as such during the drug development process. The metabolism and disposition of the enantiomer should be followed using enantioselective methods in each species used in preclinical studies, and in the Phase I studies in humans. If it is established that racemization or inversion does not occur, enantioselective methods may not be needed in all subsequent studies.

The 1994 EU Directive "Investigation of Chiral Active Substances. Legislative basis Directive 75/318/EEC as amended" covers very similar ground as the FDA and Health Canada guidances (European Union, 1994). This Directive describes several scenarios related to the development of enantiomers or racemates. For the development of a single enantiomer as a new active substance, the Directive states:

> An application for marketing authorisation of a single enantiomer as a new active substance should be considered and documented in the same way as any application for a new active substance. Studies should be carried out with the single enantiomer. However, where development studies commence with a racemate, the studies conducted up to the decision to develop the enantiomer may be taken into account to determine the necessity for further

studies. The possibility of the formation of the other enantiomer "in vivo" should be considered in relation to the chemical structure at an early stage in order to justify the need for any enantiospecific bioanalysis. If the other enantiomer is formed "in vivo" it should be evaluated with its biotransformation products in the same way as for other biotransformation products. In the case of endogenous human chiral compounds, it is considered that data on disposition (ADME) should not necessarily be based upon enantiospecific methods.

The enantiomeric purity of the active substance used in pre-clinical studies should be defined.

With respect to the development of a racemate as a new active substance, the EU Directive states that

An application for marketing authorization of a new racemic active substance should be considered and documented in the same way as any application for a new active substance.

The choice of the racemate instead of a single enantiomer should be explained. In practice one of the following two situations may occur:

Rapid interconversion "in vivo"

- if the interconversion rate "in vivo" is appreciably higher than the apparent distribution and elimination rates of the enantiomers, only the racemate should be studied as the active substance.

No or slow interconversion "in vivo," allowing for separate enantiomer effects on and fate in the organism in most situations, the following studies, in principle, will be needed for evaluation.

- Pharmacodynamics. The profile of the effects related to the therapeutic use should be provided for the racemate and each enantiomer. The racemate data related to the general pharmacodynamic properties should be extended with enantiomer data if necessary from the point of view of safety.
- Pharmacokinetics. The effective exposure to the enantiomers after administration of the racemate in pivotal studies should be measured by enantiospecific analytical methods allowing extrapolation to the human exposure.
- Toxicology. It is ordinarily sufficient to carry out toxicity studies on the racemate. If toxicity other than that predicted from the pharmacological properties of the medicinal product occurs at relatively low multiples of the exposure planned for clinical trials, relevant toxicity studies should be repeated with the individual enantiomers when possible.

The EU Directive clearly implies that the sponsor should develop a bioanalytical method that can distinguish between enantiomers and use the assay to evaluate the potential for chiral inversion *in vivo*. Interestingly, the FDA's guidance "Bioanalytical Method Validation" does not appear to address enantiomers and does not seem to require the sponsor to demonstrate the selectivity of the assay with respect to enantiomers (US FDA, 2001b).

The subject of the development of single enantiomers and racemates has been reviewed by several authors (Andersson, 2004; Mansfield, Henry and Tonkin, 2004; Brocks, 2006; Agrawal *et al.*, 2007).

27.8 PRODUCT LABELING

When a drug is approved for marketing, the available safety and efficacy information is summarized in materials that are intended to provide the basis of information for health professionals on how to use the medicinal product safely and effectively. They do not give general advice on the treatment of particular medical conditions. This information is included in either the product label (FDA terminology) or SPC (Summary of Product Characteristics; EU terminology). There are several guidances regarding the level of detail and the format of the information that will be included in the product label or SPC. The FDA has created a set of guidances (US FDA, 2008a) that address selected elements of the product label (e.g. Adverse Reaction sections; Warnings and Precautions, Contraindications, and Boxed Warning Sections; and Clinical Studies sections) or product labeling for selected classes of drugs (e.g. oral contraceptives). The EU has created a high-level guideline that describes the format and content of the SPC (European Commission, 2005). Additionally, the EU has guidelines for the SPC for a range of different classes of drugs (particularly blood products).

Information related to drug metabolism of metabolites can appear in many different places in the product label or SPC. These sections include (but are not limited to):

Posology and method of administration (SPC, Section 4.2)
This section will contain information regarding dosage adjustments for special populations such as hepatic/renal impairment or pediatrics.

Contraindications (SPC, Section 4.3)
This section will include information on situations where the medicinal product must not be given for safety reasons. These reasons could include demographic factors (e.g. age and gender) or predispositions (e.g. metabolic factors).

Interaction with other medicinal products and other forms of interaction (SPC). This section will also include information specific to a special age group.

Special warnings and precautions for use (SPC, Section 4.4)
This section will contain information regarding dosage adjustments for special populations such as hepatic/renal impairment or pediatrics.

Interactions with other medicinal products and other forms of interaction (SPC, Section 4.5)
This is one of two sections of the SPC that will contain the greatest extent of drug metabolism-related information. The EU guideline states:

> This section should provide information on the potential for clinically relevant interactions based on the pharmacodynamic properties and *in vivo* pharmacokinetic studies of the medicinal product, with a particular emphasis on the interactions, which result in a recommendation regarding the use of this medicinal product. This includes *in vivo* interaction results which are important for extrapolating an effect on a marker ('probe') substance to other medicinal products having the same pharmacokinetic property as the marker. Interactions affecting the use of this medicinal product should be given first, followed by those interactions resulting in clinically relevant changes on the use of others.

The following information should be given for each clinically relevant interaction:

   a. Recommendations: these might be

- contraindications of concomitant use (cross-refer to section 4.3),
- concomitant use not recommended (cross-refer to section 4.4), and
- precautions including dose adjustment (cross-refer to sections 4.2 and 4.4), mentioning specific situations where these may be required; for the actual dose recommendation, cross-refer to section 4.2.

   b. Any clinical manifestations and effects on plasma levels and AUC of parent compounds or active metabolites and/or on laboratory parameters.

   c. Mechanism, if known.

Interactions not studied *in vivo* but predicted from *in vitro* studies or deducible from other situations or studies should be described if they result in a change in the use of the medicinal product, crossreferring to sections 4.2 or 4.4.

This section should mention the duration of interaction when a medicinal product with clinically important interaction (e.g., enzyme inhibitor or inducer) is discontinued. Adjustment of dosing may be required as a result. The implication for the need for a washout period when using medicines consecutively should also be mentioned. Information on other relevant interactions such as with herbal medicinal products, food or, pharmacologically active substances not used for medical purpose, should also be given. With regard to pharmacodynamic effects where there is a possibility of a clinically relevant potentiation or a harmful additive effect, this should be stated. Results demonstrating an absence of interaction should only be mentioned here if this is of likely major interest to the prescriber. If no interaction studies have been performed, this should be clearly stated.

This section will contain additional information for special populations such as hepatic/renal impairment, elderly or pediatrics in which the impact of the interaction is more severe, or the magnitude of the interaction is expected to be larger.

Pregnancy and lactation (SPC, section 4.6)
The guideline includes the following statements regarding lactation:

"If available, clinical data should be mentioned including the conclusions of the studies on the transfer of the active substance and/or its metabolite(s) into human milk (positive/negative excretion, milk/serum ratio)." "Conclusion on animal studies on the transfer of the active substance and/or its metabolite(s) into milk should be given only if no human data are available."

Pharmacokinetic Properties (SPC, section 5.2)
This section will contain an extensive amount of information regarding the ADME properties of the drug. The guideline states:

Pharmacokinetic properties of the active substance(s) relevant for the advised dose, strength and the pharmaceutical formulation marketed should be given in this section. If these are not available, results obtained with other administration routes, other pharmaceutical forms or doses can be given as alternative.

Basic primary pharmacokinetic parameters, for instance bioavailability, clearance and half-life, should be given as mean values with a measure of variability.

Pharmacokinetics items, which could be included in this section when relevant, are given below.

a. General introduction, information about whether the medicinal product is a pro-drug or whether there are active metabolites, chirality, solubility etc.

b. General characteristics of the active substance(s) after administration of the medicinal product formulation to be marketed.

- Absorption: complete or incomplete absorption; absolute and/or relative bioavailability; first pass effect; T_{max}; the influence of food; in case of locally applied medicinal product the systemic bioavailability.

- Distribution: plasma protein binding; volume of distribution; tissue and/or plasma concentrations; pronounced multi-compartment behaviour.

- Biotransformation: degree of metabolism; which metabolites; activity of metabolites; enzymes involved in metabolism; site of metabolism; results from in vitro interaction studies that indicate whether the new compound can induce/inhibit metabolic enzymes.

- Elimination: elimination half-lives, the total clearance; inter and/or intra-subject variability in total clearance; excretion routes of the unchanged substance and the metabolites.

- Linearity/non-linearity: linearity/non-linearity of the pharmacokinetics of the new compound with respect to dose and/or time; if the pharmacokinetics are nonlinear with respect to dose and/or time, the underlying reason for the non-linearity should be presented.

Additional relevant information should be included here.

c. Characteristics in patients

- Variations with respect to factors such as age, gender, smoking status, polymorphic metabolism and concomitant pathological situations such as renal failure, hepatic insufficiency, including degree of impairment. If this influence on the pharmacokinetics is considered to be clinically relevant, it should be described here in quantitative terms (cross-referral to 4.2 when applicable).

d. Pharmacokinetic/pharmacodynamic relationship(s)

- Relationship between dose/concentration/pharmacokinetic parameter and effect (either true endpoint, validated surrogate endpoint or a side effect).

- Contribution (if any) of metabolite(s) to the effect.

The Annex to this guideline provides some generic statements that can be applied to these sections. It should be noted that the annex is currently under revision.

The FDA has provided guidance related to the product label in a different manner than has the EU. As mentioned previously, the FDA has created several guidances that relate to specific sections of the product label. In addition, labeling requirements are listed in 21 CFR Part 201 (US FDA, 2007b).

For example, the FDA issued a draft guidance, "Warnings and Precautions, Contraindications, and Boxed Warning Sections of Labeling for Human Prescription Drug and Biological Products—Content and Format" in January 2006 (US FDA, 2006j). This draft guidance provides information on when to contraindicate the administration of a particular drug. Several examples are provided, including "Likely Clinical Situations." The draft guidance states:

A contraindication usually involves one or more of the following clinical situations: Comorbid condition or coexistent physiological state (e.g., existing hepatic disease, renal disease, congenital long QT syndrome, hypokalemia, pregnancy or childbearing potential, CYP 2D6 poor metabolizer).

Coadministered drug where the combination is dangerous (e.g., MAO inhibitor and sympathomimetic drug, a drug known to prolong the QT interval and a drug known to interfere with the metabolism of that drug).

In 2007, the FDA issued a draft guidance entitled "Dosage and Administration Section of Labeling for Human Prescription Drug and Biological Products—Content and Format" (US FDA, 2007n). It should be noted that the title listed in the index of guidances indicates that the title is "Content and Format of the Dosage and Administration Section of Labeling for Human Prescription Drug and Biological Products." This guidance states that

> When it is important to maintain specific therapeutic blood levels of a drug or its metabolites, whether for effectiveness or safety reasons, the section must identify desirable levels (§ 201.57(c)(3)(i)(J)). The section should describe the monitoring needed to assess levels and how to adjust dose based on observed levels.

> Conversely, if a drug interaction is suspected based on a shared metabolic pathway, but there is not enough information to support a specific dosage adjustment recommendation, the interaction should ordinarily not be discussed in the DOSAGE AND ADMINISTRATION section.

Certain elements of drug labeling can be found in other guidances that are not solely devoted to labeling. Such guidances include "Drug Metabolism/Drug Interaction Studies in the Drug Development Process: Studies In Vitro (1997)" which states that

> Each year, large numbers of new drug-drug interactions are discovered, precluding the possibility that any prescriber could memorize them all. Based on the increasing amount of valuable information that is available, it is now possible to label for class effects for various enzymes, and the ability to extrapolate from partial data is growing. Standardized approaches to labeling are likely to emerge and be helpful, in a manner analogous to the class labeling used for certain categories of drugs. For example, certain powerful inhibitors (quinidine for CYP450 2D6, ketoconazole for CYP450 3A4) are likely to affect all drugs metabolized by these pathways. For this reason, if a new drug is found to be a substrate for certain CYP450 enzymes, then certain interactions may be anticipated, even though specific data are lacking. This understanding relies on knowledge about the activity of the drug and its metabolites.

> Similarly, it would be helpful to know what metabolic pathways are not involved in the elimination of a drug. When generalizations are made from studies *in vitro*, the conditions of extrapolation should be explicitly stated. Thus, conclusions based on data gained from *in vitro* studies that are extrapolated to the clinical situation should be identified and distinguished from conclusions based on clinical observations *in vivo*. Under these circumstances, the best advice available at any given time may be provided, and class effects may be updated as new information is obtained.

The guidance provides an example of class labeling based on studies *in vitro*:

> Although clinical studies have not been conducted, on the basis of this drug's metabolism by CYP450 3A4, ketoconazole, itraconazole, erythromycin, and grapefruit juice are likely to inhibit its metabolism. Furthermore, rifampin, dexamethasone, and certain anticonvulsants (phenytoin, phenobarbital, carbamazepine) may induce this drug's metabolism. Thus, if a patient has been titrated to a stable dosage on this drug, and then begins a course of treatment with one of these inducers or inhibitors, it's reasonable to expect that a dose adjustment may be necessary to prevent toxicity or therapeutic failure.

The guidance also provides an example of where the class effects would be inserted and also where information on the drug's inhibitory effects would be stated:

This drug is metabolized by CYP450 3A4 ⟨insert current statement⟩. At clinical doses, the drug itself does not inhibit the metabolism of other 3A4 substrates, but does inhibit the metabolism of substrates metabolized via the CYP450 2D6 pathway.

The guidance makes the point that:

Given the tendency to include many potential interactions, it is sometimes unclear if anything is noninteracting. In such a circumstance, labeling statements that denote both positive and negative expectations may be helpful.

For example:

This drug is a substrate for CYP450 1A2. Although inhibition of its metabolism by ciprofloxacin is observed, quinidine, erythromycin, ketoconazole, and itraconazole are not inhibitors.

In 1999, the FDA issued a complementary guidance (In Vivo Drug Metabolism/ Drug Interaction Studies—Study Design, Data Analysis, and Recommendations for Dosing and Labeling) to address *in vivo* studies.
This guidance states that:

All relevant information on the metabolic pathways and metabolites and pharmacokinetic interaction should be included in the CLINICAL PHARMACOLOGY section of the labeling.

The consequences of metabolism and interactions should be placed in PRECAUTIONS/ WARNINGS, CONTRAINDICATIONS, and DOSAGE AND ADMINISTRATION sections, as appropriate.

It further states that:

Relevant in vitro and in vivo metabolic drug-drug interaction data describing the drug's effects on substrates and the effects of inhibitors and inducers on the drug should be presented in the DRUG-DRUG INTERACTIONS section of the labeling in the CLINICAL PHARMACOLOGY section, including both positive and important negative findings. The types of studies on which statements are based should be identified briefly in the labeling. If findings indicate a known or potential interaction of clinical significance, or lack of an important interaction that might have been expected, these should be mentioned briefly in the clinical pharmacology interactions section and described more fully in the interaction section under PRECAUTIONS, with advice on how to adjust treatment placed in WARNINGS/PRECAUTIONS, DOSAGE AND ADMINISTRATION, and CONTRAINDICATIONS, as appropriate. In certain cases, information based on clinical studies not using the labeled drug under investigation can be described with an explanation that similar results may be expected for the labeled drug. For example, a strong inhibitor of CYP3A4 does not need to be tested with all 3A4 substrates to warn against an interaction.

Examples of appropriate labeling language are provided in this guidance for various situations. For example, when *in vivo* metabolic drug–drug interaction studies indicate little or no pharmacokinetic effect, the following phraseology should be used:

Data from a drug-drug interaction study involving (drug) and (probe drug) in _____ patients/healthy individuals indicate that the PK disposition of (probe drug) is not altered when the drugs are coadministered. This indicates that (drug) does not inhibit CYP3A4 and will not alter the metabolism of drugs metabolized by this enzyme.

When *in vivo* metabolic drug–drug interaction studies indicate a clinically significant pharmacokinetic interaction:

The effect of (drug) on the pharmacokinetics of (probe drug) has been studied in _____ patients/healthy subjects. The C_{max}, AUC, half-life and clearances of (probe drug) increased/decreased by ____ % (90% Confidence Interval: ____ to ____ %) in the presence of (drug). This indicates that (drug) can inhibit the metabolism of drugs metabolized by CYP3A4 and can increase blood concentrations of such drugs. (See PRECAUTIONS, WARNINGS, DOSAGE AND ADMINISTRATION, or CONTRAINDICATIONS sections.)

When specific enzymes have been identified as metabolizing the test drug, but no *in vivo* or *in vitro* drug interaction studies have been conducted:

In vitro drug metabolism studies reveal that (drug) is a substrate of the CYP _____ enzyme. No in vitro or clinical drug interaction studies have been performed to evaluate interactions. However, based on the in vitro data, blood concentrations of (drug) are expected to increase in the presence of inhibitors of _____ such as _____, _____, or_____.

When neither *in vivo* nor *in vitro* drug–drug interaction studies have been conducted and there is no significant metabolism of the drug:

In vivo or in vitro drug-drug interaction studies have not been conducted. The drug interaction potential resulting in changes of PK of (drug) is expected to be low because approximately 90% of the recovered dose of (drug) is excreted in the urine as unchanged drug. However, the role of other pathways of drug elimination, including drug transport systems, is not known. In addition, whether (drug) can inhibit or induce metabolic enzymes is not known. There is potential for drug interactions mediated via modulation of various CYP enzymes.

When *in vitro* interaction has been studied but no *in vivo* studies have been conducted to confirm or refute a finding:

In vitro interaction demonstrated: "In vitro drug interaction studies reveal that the metabolism of (drug) is by CYP3A4 and can be inhibited by the CYP3A4 inhibitor ketoconazole. No clinical studies have been performed to evaluate this finding. Based on the in vitro findings, it is likely that ketoconazole, itraconazole, ritonavir, and other 3A4 inhibitors may lead to substantial increase of (drug) blood concentrations. Refer to PRECAUTIONS, as appropriate.

In vitro interaction demonstrated and the substrate drug has substantial first-pass elimination:

In vitro drug interaction studies reveal that the metabolism of (drug) is by CYP3A4 and can be inhibited by the CYP3A4 inhibitor ketoconazole. No clinical studies have been performed to evaluate this finding. Based on the in vitro findings, it is likely that ketoconazole, itraconazole, ritonavir, grapefruit juice, and other 3A4 inhibitors may lead to substantial increase of (drug) blood concentrations. Refer to PRECAUTIONS, as appropriate.

In vitro interaction not demonstrated:

In vitro drug interaction studies reveal no inhibition of the metabolism of (drug) by the CYP3A4 inhibitor ketoconazole. No clinical studies have been performed to evaluate this finding. However, based on the in vitro findings, a metabolic interaction with ketoconazole, grapefruit juice, and other 3A4 inhibitors is not anticipated. Refer to PRECAUTIONS, as appropriate.

Information regarding DDIs may be included in PRECAUTIONS and/or WARNINGS when an interacting drug causes increased concentrations of the substrate but the administration of both drugs may continue with appropriate dosage adjustment. Results of the studies are described in CLINICAL PHARMACOLOGY, DRUG–DRUG INTERACTIONS, PRECAUTIONS and/or WARNINGS and may state:

Drug _____ /class of drug causes significant increases in concentrations of _____ when coadministered, so that dose of _____ must be adjusted (see DOSAGE AND ADMINISTRATION). If there is an important interaction, information for patients should point this out also.

When an interacting drug causes increased risk because of increased concentrations of the substrate and the interacting drug should not be used with the substrate. After describing the interaction in the CLINICAL PHARMACOLOGY section, there should be a CONTRAINDICATIONS section and possibly a boxed warning if the risk is serious.

Drug_____ /class of drug can cause significant increases in concentrations of drug_____ when coadministered. The two drugs should not be used together.

When no *in vitro* or *in vivo* drug interactions were conducted. These are described in CLINICAL PHARMACOLOGY, DRUG–DRUG INTERACTIONS, PRECAUTIONS and/or WARNINGS and may state:

There is potential for drug interactions mediated via modulation of various CYP enzymes.

Additional statement related to DDIs can be found in the section, DOSAGE AND ADMINISTRATION. When an interacting drug causes increased risk because of increased concentrations of the substrate, but the administration for both drugs may continue with suitable monitoring:

Drug _____ /class of drug leads to significant increases in blood concentrations of _____ by _____%. The dose of _____ should be decreased by _____% when the patient is also taking _____. Patients should be closely monitored when taking both drugs.

In the CONTRAINDICATIONS section, DDIs are addressed when an interacting drug causes increased risk because of increased concentrations of the substrate and should not be coadministered:

Drug _____ /class of drug leads to significant increases in blood concentrations of _____, with potentially serious adverse events. Administration of _____ to patients on drug _____ /class of drug is contraindicated.

As mentioned previously, in 2006 the FDA issued a draft guidance (Drug Interaction Studies—Study Design, Data Analysis, and Implications for Dosing and Labeling) that updates certain aspects of DDI studies.

This draft guidance states that:

It is important that all relevant information on the metabolic pathways and metabolites and pharmacokinetic interactions be included in the PHARMACOKINETICS subsection of the CLINICAL PHARMACOLOGY section of the labeling. The clinical consequences of metabolism and interactions should be placed in DRUG INTERACTIONS, WARNINGS AND PRECAUTIONS, BOXED WARNINGS, CONTRAINDICATIONS, or DOSAGE AND ADMINISTRATION sections, as appropriate. Information related to clinical consequences should not be included in detail in more than one section, but rather referenced from one section to other sections, as appropriate. When the metabolic pathway or interaction data results in recommendations for dosage adjustments, contraindications, or warnings (e.g., co-administration should be avoided) that are included in the BOXED WARNINGS, CONTRAINDICATIONS, WARNINGS AND PRECAUTIONS, or DOSAGE AND ADMINISTRATION sections, these recommendations should also be included in HIGHLIGHTS.

The guidance then states: "Refer to the guidance for industry on Labeling for Human Prescription Drug and Biological Products – Implementing the New Content and Format Requirements, and Clinical Pharmacology and Drug Interaction Labeling for more information on presenting drug interaction information in labeling."

However, the latter guidance does not seem to be located on the FDA's guidance page.

Finally, this draft guidance addresses the subject of class effects:

In certain cases, information based on clinical studies not using the labeled drug can be described, with an explanation that similar results may be expected for that drug. For example, if a drug has been determined to be a strong inhibitor of CYP3A, it does not need to be tested with all CYP3A substrates to warn about an interaction with sensitive CYP3A substrates and CYP3A substrates with narrow therapeutic range. An actual test involving a single substrate would lead to labeling concerning use with all sensitive and NTR substrates.

The draft guidance provides examples of sensitive CYP3A substrates and CYP3A substrates with narrow therapeutic range and examples of strong, moderate, and weak CYP3A inhibitors.

The guidance indicates that:

If a drug has been determined to be a sensitive CYP3A substrate or a CYP3A substrate with a narrow therapeutic range, it does not need to be tested with all strong or moderate inhibitors of CYP3A to warn about an interaction with strong or moderate CYP3A inhibitors, and it might be labeled in the absence of any actual study if its metabolism is predominantly by the CYP3A route. Similarly, if a drug has been determined to be a sensitive CYP3A substrate or a CYP3A substrate with a narrow therapeutic range, it does not need to be tested with all CYP3A inducers to warn about an interaction with CYP3A inducers.

Examples of CYP3A inducers include rifampin, rifabutin, rifapentin, dexamethasone, phenytoin, carbamazepine, phenobarbital, and St. John's wort.

A similar classification system can be used for inhibitors of other CYP enzymes. (US FDA, 2006d, 2007f)

Labeling requirements for special populations are also addressed in selected FDA guidances.

In 2003, the FDA issued a guidance called "Pharmacokinetics in Patients with Impaired Hepatic Function: Study Design, Data Analysis, and Impact on Dosing and Labeling" (US FDA, 2003d).

The guidance notes that "A general approach in developing dosage recommendations is appropriately based on the following considerations:

- If the effect of hepatic impairment on the PK of the drug is obvious (e.g., two-fold or greater increase in AUC), dosage adjustments should be recommended in labeling. It should be noted that for prodrugs (i.e., drugs with activity predominantly due to hepatically generated metabolite), it is possible that the dose would be increased, or the dosing interval shortened, in hepatically impaired patients.
- A conclusion that there is no effect (really, no clinically important effect) of hepatic impairment on the drug's PK, would usually be supported by the establishment of one of the following: (1) delineation of no effect boundaries, prior to the conduct of the studies, based on information available for the investigational drug (e.g., dose- and/or concentration-response studies), or (2) in the absence of other information to determine a different equivalence interval, the employment of a standard 90 percent confidence interval of 80–125 percent for AUC and C_{max}. FDA recognizes that documentation that a PK parameter remains within an 80–125 percent no effect boundary would be very difficult given the small numbers of subjects usually entered into hepatic impairment studies. If a wider boundary can be supported clinically, however, it may be possible to conclude that there is no need for dose adjustment."

Statements regarding the effect of hepatic impairment on the pharmacokinetics of the drug can be found throughout the product label, including the Clinical Pharmacology, Pharmacokinetics Section. The guidance states that "Information in this section of the labeling should include:

- The mechanism of hepatic elimination (e.g., enzyme pathways, glucuronidation, biliary excretion)
- The percent of drug that is eliminated by these mechanisms (e.g., metabolism, biliary excretion)
- The disposition of active metabolites in patients with impaired hepatic function, if applicable
- The effects of hepatic impairment on protein binding of parent drug and metabolites, if applicable"

These statements raise an important question, namely, "What is the appropriate route for the determination of the mass balance/*in vivo* metabolism of a drug in humans?" Studies to address the enzyme pathways involved in the hepatic elimination of the drug may be performed *in vitro*. An evaluation of the contribution of biliary elimination to hepatic elimination and the percent of drug that is eliminated by metabolism or biliary excretion may require the collection of bile samples if the drug is administered orally or the collection of bile and/or fecal samples if the drug is administered intravenously. For an orally administered drug, determination of the hepatic elimination of the drug is not a trivial task.

The FDA provides examples of labeling for different outcomes of the study in hepatically impaired patients. If studies show an effect of altered hepatic function, the guidance states:

> For drugs in which PK or PD is influenced by hepatic impairment, the following statement can be modified as appropriate and in accordance with what is known about the drug (e.g., racemate with different activity of stereoisomers, active or toxic metabolite) and from the studies performed in accordance with this guidance.
>
> *The disposition of _____ was compared in patients with hepatic impairment and subjects with normal hepatic function. Total body clearance of [unbound, if applicable] _____ / metabolite was reduced by _____ % in patients with moderate (as indicated by the Child-Pugh method) hepatic impairment. The half-life of _____ /metabolite is prolonged by ____ in patients with moderate hepatic impairment. Protein binding of _____ /metabolite [is/is not] affected by impaired hepatic function. The drug/metabolite accumulates to the extent of _____ in patients with impaired hepatic function on chronic administration. The dosage should be reduced in patients with mild and moderate hepatic impairment receiving _____. _____ should be [contraindicated/used with great caution] in severe hepatic impairment (see WARNINGS/PRECAUTIONS, CONTRAINDICATION and DOSAGE AND ADMINISTRATION).*

For the situation of "unknown hepatic elimination," the guidance states that the sponsor "consider the compound as extensively metabolized and use the above format."

The FDA guidance, "Pharmacokinetics in Patients with Impaired Renal Function— Study Design, Data Analysis, and Impact on Dosing and Labeling" provides clear information on the labeling of drugs for use in this patient population (US FDA, 1998b). It should be noted that the FDA's guidance page lists the title of this guidance as being "Pharmacokinetics in Patients with Impaired Renal Function." The guidance provides some general thoughts, including: "Specific dosing recommendations should be constructed based on the study results using the aforementioned model for the relationships between renal function and relevant PK parameters. Typically the dose is adjusted to produce a comparable range of unbound plasma concentrations of drug or active metabolites in both normal patients and patients with impaired renal function. Simulations are encouraged as a means to identify doses and dosing intervals that achieve that goal for patients with different levels of renal function." It is interesting to note that this is one of the few guidances that clearly encourages the use of simulations, a topic that is gaining importance at the FDA.

In addition to general thoughts, the guidance makes some specific recommendations. "The labeling should reflect the data pertaining to the effect of renal function on the pharmacokinetics and pharmacodynamics (if known) obtained from studies conducted. The various permutations of intrinsic drug characteristics and the effect of renal impairment on drug performance preclude precise specification of how such drugs should be labeled."

The guidance provides general suggestions on which sections of the labeling should include standardized information and how such information should be structured. Such information includes:

"The pharmacokinetics subsection should include information on the:

- Mechanism of renal elimination (e.g., filtration, secretion, active reabsorption)
- Percentage of drug that is eliminated by renal excretion and whether it is eliminated unchanged or as metabolites;
- Disposition of metabolites in patients with impaired renal function (if applicable);
- Effects of renal impairment on protein binding of parent drug and metabolites (if applicable);
- Effects of changes in urinary pH or other special situations that should be mentioned (e.g., tubular secretion inhibited by probenecid);
- If applicable, the effects of impaired renal function on stereospecific disposition of enantiomers of a racemic drug product should be described if there is evidence of differential stereoisomeric activity or toxicity."

The Special Populations subsection should "recapitulate, in brief, the pharmacokinetic changes found in various degrees of renal impairment and, if necessary, dosing adjustments for patients with varying degrees of renal impairment. This information should be based on the studies performed as described in this guidance. Reference should be made to the PRECAUTIONS/WARNINGS and the DOSAGE AND ADMINISTRATION sections." The guidance provides examples of appropriate wording for these sections.

The simplest situation involves drugs for which impaired renal function has little or no effect on PK:

Impaired renal function has little or no influence on _____ pharmacokinetics and no dosing adjustment is required.

Similarly, for drugs whose PK is influenced by renal impairment, the following statement may be modified as appropriate and in accordance with what is known about the drug (e.g., racemate with different activity of stereoisomers, active or toxic metabolite) and from the studies performed in accordance with this guidance:

The disposition of _____ was studied in patients with varying degrees of renal function. Elimination of the drug (and metabolite, if applicable) is significantly correlated with the creatinine clearance. Total body clearance of (unbound, if applicable) _____ /metabolite was reduced in patients with impaired renal function by _____ % in mild (CLcr = _____– _____ mL/min), _____ % in moderate (CLcr = _____–_____ mL/min) and _____ % in severe renal impairment (CLcr = _____–_____ mL/min), and _____ % in patients under dialysis compared to normal subjects (CLcr >_____mL/min). The terminal half-life of _____/metabolite is prolonged by _____, _____ , and _____ fold in mild, moderate, and severe renal impairment, respectively.

[Alternatively, the relationship between renal function and the PK parameters may be described in terms of equations, e.g., a linear equation relating unbound clearance and CLcr.]

Protein binding of _____/metabolite is/is not affected by decreasing renal function. The drug/metabolite accumulates in patients with impaired renal function on chronic administra-

*tion. The pharmacologic response is/is not affected by renal function. Approximately _____
% of the drug/metabolite in the body was cleared from the body during a standard 4-hour
hemodialysis procedure. The dosage should be reduced in patients with impaired renal function receiving _____ and supplemental doses should/should not be given to patients after
dialysis. (See DOSAGE AND ADMINISTRATION).*

Information on the effect of renal impairment on PK should also be included in the
Precautions/Warnings section. The Dosage and Administration section will also include
statements (as appropriate), including a statement on the relationship between drug
clearance and endogenous creatinine clearance; the need for dosage adjustment and
special considerations for combination drug products. Additionally, information may
be included in the Overdosage section. The guidance states:

Although the primary objective of a hemodialysis study is to evaluate the need for dosing
adjustments in ESRD (end stage renal disease), additional information regarding the value
of hemodialysis in overdose situations may reasonably be garnered from such studies (if
performed). In situations in which this information is known, the following wording may
be adapted as appropriate:

_____ *is not eliminated to a therapeutically significant degree by
hemodialysis.*

or

*Standard hemodialysis procedures result in significant clearance of _____ and
should be considered in cases of life-threatening overdose.*

A guidance called "Content and Format for Geriatric Labeling" was issued by the
FDA in 2001. However, this guidance does not provide any novel points related to drug
metabolism and pharmacokinetics that are not already covered in other guidances.

In 2005, the FDA issued a draft guidance entitled "Clinical Lactation Studies – Study
Design, Data Analysis, and Recommendations for Labeling" (US FDA, 2005g). This
draft guidance made the following statements regarding labeling:

Clinical Pharmacology, Pharmacokinetics Subsection

This section would include information pertinent to lactation on the:

- Disposition of parent drug and metabolites, if applicable
- Effects of lactation on protein binding, if applicable.

*Similarly, for drugs whose PK is influenced by lactation, the following statement can be
modified in accordance with what is known about the drug (e.g., active or toxic metabolite)
and from the studies performed in accordance with this guidance:*

*The disposition of [Drug X] was studied in [number of] lactating women from [a through
b months postpartum]. Elimination of the drug (and metabolite, if applicable) is significantly
changed during lactation. Total body clearance of (unbound, if applicable) [Drug X]/
metabolite was [reduced/increased] in lactating women compared to non-lactating women.
The terminal half-life of [Drug X]/metabolite is [prolonged/decreased] by [Y-fold]. (See
DOSAGE AND ADMINISTRATION.)*

Information should also be included in the Precautions/Nursing Mothers section.
"In addition to standard labeling for use in lactation, if studies performed during lactation demonstrate clinically important changes, the Agency recommends that such infor-

mation be included in the PRECAUTIONS/NURSING MOTHERS section with cross-reference to DOSAGE AND ADMINISTRATION and CLINICAL PHARMA-COLOGY sections. It is recommended that labeling contain information, to the extent possible, based on the lactation study conducted, including:

- PK/PD in lactation
- The effect of drug on milk production (e.g., quality and quantity of milk including milk production and composition)
- The presence of drug or metabolite in milk, including the limitation of the assay used if drug/metabolites are not detected in milk
- The amount of drug or metabolite in breast milk over a 24-hour period
- The amount of drug or metabolite consumed daily by the breast-fed infant
- The percent of maternal dose delivered via breast milk and consumed daily by the breast-fed infant (i.e., daily dose in human milk compared to the usual adult dose, or pediatric dose, if known)
- Possible ways to minimize exposure in the breast-fed child to drug via breast milk taking into account drug kinetics such as half-life in milk (e.g., timing of maternal dose relative to breast-feeding, the duration to discard breast milk relative to maternal dose, and how long to wait until resuming breast-feeding relative to maternal dose)
- Effects of drug exposure via breast milk in the breast-fed infant
- PK of drug in the breast-fed infant."

In 2004, the FDA issued a guidance called "Pharmacokinetics in Pregnancy—Study Design, Data Analysis, and Impact on Dosing and Labeling" (US FDA, 2004c). This guidance included recommendations related to labeling of the use of the drug in pregnancy. For the Clinical Pharmacology section (Pharmacokinetics subsection) the guidance stated that:

It is recommended that this section include information pertinent to pregnancy such as:

- Disposition of parent drug and metabolites, if applicable
- Effects of pregnancy on protein binding of parent drug and metabolites, if applicable
- Effects of changes in urinary pH or other special situations (e.g., tubular secretion inhibited by probenecid).

Similarly, for drugs whose PK is influenced by pregnancy, the statement similar to the following can be modified as appropriate and in accordance with what is known about the drug (e.g., active or toxic metabolite) and from the studies performed in accordance with this guidance:

The disposition of [Drug X] was studied in [number of] pregnant patients [in y trimester or from a through b weeks gestation]. Elimination of the drug (and metabolite, if applicable) is significantly changed during pregnancy. Total body clearance of (unbound, if applicable) [Drug X]/metabolite was reduced/increased in pregnant patients compared to [healthy post-partum women, the same women prior to pregnancy or c weeks postpartum]. The terminal half-life of [Drug X]/metabolite is [prolonged/decreased] by Y-, and Z- fold in second and third trimesters, respectively. Protein binding of [Drug X]/metabolite [is/is not] affected by pregnancy. The [drug/metabolite accumulates/does not accumulate] in pregnant patients on chronic administration resulting in increased/decreased plasma levels of drug/metabolite. The pharmacologic response [is/is not] affected by pregnancy. The dosage/dosing interval should be [decreased/increased] in pregnant patients receiving [Drug X] (see DOSAGE AND ADMINISTRATION).

The EMEA issued a draft guideline (Guideline on Risk Assessment of Medicinal Products on Human Reproduction and Lactation: From Data to Labelling) in 2006 that covers similar topics to the two FDA guidances described earlier (EMEA, 2006c). The CHMP work plan indicates that this guideline will be finalized in 3Q2007/1Q2008.

The FDA issued a guidance called "Pharmacogenomic Data Submissions" in 2007 (US FDA, 2005f). This guidance states that "The pharmacogenomic data and resulting test or tests may be intended to be included in the drug labeling to choose a dose and dose schedule, to identify patients at risk, or to identify patient responders. Inclusion of a pharmacogenomic test in the labeling would be contingent upon its performance characteristics. For example:

- Patients will be tested for drug metabolism genotype and dosed according to the test results.
- Patients will be selected as potential responders for an efficacy trial (or deselected because of a high risk) based on genotype (e.g., of either the patient or the patient's tumor) or gene expression profile.
- Patients will be excluded from a clinical trial based on genotype or gene expression profile (e.g., biomarker for risk of an adverse event)."

In all of these cases, FDA recommends co-development of the drug and the pharmacogenomic tests, if they are not currently available, and submission of complete information on the test/drug combination to the Agency. The FDA plans to issue further guidance on co-development of pharmacogenomic tests and drugs.

In October 2007, the EMEA issued a draft guideline called "Guideline on the Clinical Development of Medicinal Products for the Treatment of HIV Infection" (EMEA, 2007f). Annex A (Presentation of pharmacokinetic interaction data in the SPC) to this draft guideline provides recommendations that refer to antiretroviral drugs with high propensity for pharmacokinetic (PK) interactions (EMEA, 2007a). The draft guideline states that "The principles guiding data presentation should take the following into account:

The SPC is a tool to be used by the clinicians. For compounds with complex interaction potential, the most user-friendly way to present data is by therapeutic areas.

- The aim should be to provide clear recommendations as regards use/non-use and, for essential drugs, which dose to be used.
- If major interactions with a specific compound have been identified, there may be alternatives within the same therapeutic area without this interaction propensity. Therefore absence of PK interactions is informative and should be provided for therapeutic areas where (potentially) problematic interactions have been identified.
- For some compounds (e.g. substrates of CYP3A), the number of possible combinations of interacting compounds might be high in clinical practice. For such compounds therapeutic drug monitoring (TDM) might be useful. Information as regards target concentrations may be put forward in sections 5.2 and/or 4.2,

depending on the robustness of data and foreseen need for TDM in clinical practice."

Although all relevant ADME information required to aid the physician in appropriately prescribing the drug should be included in the product label at the time of marketing approval, this is not always the case. However, not all of the information for the latter may be available as is often the case with accelerated approvals. The FDA may give approval for the lawful marketing of the drug but with certain caveats. These caveats may be statements in the label that no information is available regarding the administration of a given drug with the marketed drug or in the form of a letter to the sponsor indicating a list of additional studies that will be required in the post-marketing (phase IV) phase.

Such phase IV commitments are shown in the approval letter issued by the FDA to Pfizer regarding the marketing approval for maraviroc (a CCR5 co-receptor antagonist that is used for the treatment of HIV infections) (US FDA, 2007r). The FDA approved maraviroc based on its risk–benefit analysis and the unmet medical need. In this letter, the FDA requested that Pfizer conduct the following studies:

- Conduct a study to evaluate the effect of renal impairment on the pharmacokinetics of maraviroc.
  a) at a dose of 150 mg when combined with a boosted protease inhibitor (e.g. saquinavir/ritonavir) in subjects with mild and moderate renal impairment and subjects with End-Stage Renal Disease (ESRD) that require dialysis.
  b) at a dose of 300 mg alone in subjects with severe renal impairment and subjects with end stage renal disease who require dialysis.

Protocol Submission: December 30, 2007

Final Report Submission: December 30, 2008

- Conduct a study to evaluate the potential for maraviroc metabolite(s) to inhibit CYP2D6 enzymes at a maraviroc dose of 600 mg.

Protocol Submission: December 30, 2007

Final Report Submission: June 30, 2008

- Conduct a study to evaluate the potential of maraviroc to inhibit P-gp.

Protocol Submission: December 30, 2007

Final Report Submission: June 30, 2008

- Conduct a study to investigate the potential for maraviroc to induce CP1A2 [sic].

Protocol Submission: December 30, 2007

Final Report Submission: June 30, 2008

- Conduct and submit a clinical study to evaluate the potential for pharmacodynamic interaction between maraviroc and inhibitors of phosphodiesterase type 5 (PDE5).

Protocol Submission: December 2007

Final Report Submission: June 2008

The FDA clearly found that there was a need for maraviroc in the treatment of HIV and that the information that would be generated by the post-marketing commitment studies would not presently preclude physicians from prescribing the drug, but that the information from these studies would enhance their ability to safely prescribe the drug.

The FDA has expressed concern regarding the number of post-marketing (phase IV) commitments that have not been completed. As of September 30, 2006 only 11% of the commitments for NDAs/ANDAs have been completed. Of the remainder, 71% are pending, 15% are ongoing, 3% are delayed, and <1% are terminated (US FDA, 2007s). Of the studies that concluded in the period October 1, 2005 to September 30, 2006, 83% of the studies met the commitment, 5% did not meet the commitment, and for 12% of the studies, the study was no longer required or was not feasible.

27.9 PROCEDURAL GUIDANCES

Since knowledge of the ADME characteristics of a drug is an integral part of the risk–benefit analysis for that drug, it is not surprising to find drug metabolism mentioned, albeit often in passing or by implication, in procedural guidances. Such guidances mainly relate to the format and content of submissions, rather than providing input into how the ADME characteristics of the drug should be determined.

The FDA guidance called "Content and Format of Investigational New Drug Applications (INDs) for Phase 1 Studies of Drugs, Including Well-Characterized, Therapeutic, Biotechnology-derived Products" was issued in 1995 and is an example of the lack of visibility for ADME in regulatory guidances (US FDA, 1995). This guidance provides a clear road map for the creation of an IND for submission to the US FDA. Section G of this guidance is titled "Pharmacology and Toxicology Information [21 CFR 312.23(a)(8)]." There is no mention of ADME in the title of this section and there is no section in this guidance that solely deals with ADME-related topics. This section also states that the following pharmacology and toxicology guidance is applicable to all phases of IND development of products covered by this guidance. Although ADME is not specifically mentioned in the title of section G, it is mentioned in subsections of this section. Subsection 1 is titled "Pharmacology and Drug Distribution [21 CFR 312.23(a)(8)(i)]" and states that this section should contain, if known:

1) a description of the pharmacologic effects and mechanism(s) of actions of the drug in animals, and
2) information on the absorption, distribution, metabolism, and excretions of the drug. This latter sentence is the sole mention of ADME in this guidance.

This subsection further states:

The regulations do not further describe the presentation of these data, in contrast to the more detailed description of how to submit toxicologic data. A summary report, without individual animal records or individual study results, usually suffices. In most circumstances, five pages or less should suffice for this summary. If this information is not known, it should simply be so stated.

To the extent that such studies may be important to address safety issues, or to assist in evaluation of toxicology data, they may be necessary; however, lack of this potential effectiveness information should not generally be a reason for a Phase 1 IND to be placed on clinical hold.

It is of interest to note that pharmacology and drug distribution have been linked together, giving the impression that these two items should be summarized together in the nonclinical section of the IND. This interpretation is enforced by the fact that subsection 2 refers to "Toxicology: Integrated Summary [21 CFR 312.23(a)(8)(ii)(a)]." This subsection also describes the submission of individual study reports as part of the IND, and raised the possibility of submitting either "final fully quality assured" individual study reports or earlier draft toxicological reports of the completed study(ies). It should be noted that the regulation does not specifically require individual toxicology study reports to be submitted. The regulations state that an integrated summary of the toxicological findings should be submitted along with a full tabulation of data from each study in order to allow for a detailed review. In practice, full study reports are submitted since the integrated assessment is often based on the findings of individual studies and therefore a draft or final report is available for submission to the IND. Although the guidance only refers to the submission of toxicology study reports, it should be borne in mind that the Pharm/Tox reviewer at the FDA will be reviewing the pharmacology, toxicology, and ADME data. It is reasonable to expect that the reviewer would have similar expectations regarding the level of documentation for the three components of the nonclinical section. Since the objective of the sponsor is to have the IND to be opened on schedule. The "approval" of an IND is essentially a passive process. If the FDA does not raise objections within 30 days, the IND will be viewed as being open and clinical trials can begin. If the FDA raises questions, these will need to be answered within the 30-day window or the IND will have to be withdrawn or the IND be placed on clinical hold by the FDA. The latter should be avoided at all costs. Therefore, the standard practice of sponsors is to provide the Pharm/Tox reviewer with all of the information that they may reasonably feel would be required to adequately review the application. To this end, the sponsor will not only provide an integrated summary of the pharmacology and ADME information but will also provide the full study reports for the studies that are summarized in the integrated summary. Similarly, although the guidance implies that it would be acceptable to submit no information on the ADME properties on the IND candidate if no such information was known, the practice is to provide a reasonable level of characterization of the ADME properties of the candidate drug. The guidance does not contain any details regarding what kind of information would be required to be summarized. If this guidance was to be written in today's regulatory environment, it would be reasonable to expect that the guidance would contain a far greater emphasis on the characterization of the ADME properties of the IND candidate and would not prescribe exactly what information should be included but would simply say that the level of characterization should be addressed on a "case-by-case" basis.

The subject of the level of documentation that should be submitted to the IND, as well as when the information should be submitted, has long been a topic of great debate and confusion. The level of documentation has been discussed earlier. Related to the timing of the submission of the documentation, the guidance states that full individual toxicology study reports should be available to the FDA, upon request, and individual study reports should be available to FDA, upon request, as final, fully quality-assured documents within 120 days of the start of the human study for which the animal study formed part of the safety conclusion basis. These final reports should contain in the introduction any changes from those reported in the integrated summary. If there are no changes, that should be so stated clearly at the beginning of the final, fully quality-assured report.

If the integrated summary is based upon unaudited draft reports, sponsors should submit an update to their integrated summary 120 days after the start of the human study(ies), identifying any differences found in the preparation of the final fully quality-assured study reports and the information submitted in the initial integrated summary. If there were no differences found, that should be stated in the integrated summary update.

In addition, any new finding discovered during the preparation of the final, fully quality-assured individual study reports that could affect subject safety must be reported to the FDA under 21 CFR 312.32.

The timing of the start of the 120-day clock has caused much debate. The guidance states in one place "within 120 days of the start of the human trial" and in another place states "120 days after the start of the human study(ies)." Additionally, the guidance states that the report should be available "upon request" which implies that the FDA could request the final report without notice. The sponsor would be obligated to submit the report or risk having the IND candidate placed on clinical hold. Since there is no mechanism for the FDA to determine when the clinical trial has started, the time frame suggested in the guidance is unenforceable.

This discrepancy was recognized by the FDA and a Q&A document for this guidance was issued in October 2000 (US FDA, 2000c). In this document, the FDA clarified that the 120-day period is measured based on the date of receipt stamped on the IND submission. In reality, this means that the finalized reports (and updated integrated summary) should be available to the FDA (either submitted to the IND or available upon request) 90 days after approval of the IND (i.e. after as 30-day review period).

Despite this clarification, debate exists to date on the definition of the "120-day rule" since the Q&A document cannot be found on the main FDA guidance page (US FDA, 2008g). The subsection of this web page that relates to investigational new drugs only lists the original guidance on format and content. The Q&A document can be found by searching the FDA web site for "120-day." The search will yield the location of the guidance (US FDA, 2000c). The existence of this document is not widely known in the pharmaceutical industry.

The creation of ADME-related reports for submission to the IND is a topic of debate between discovery and development scientists. If knowledge related to the ADME properties of a drug has been generated and was used by the company in decision making, it is most prudent to ensure that these studies have been written up and are submitted to the IND. This is critically important if the sponsor intends to utilize any of this information as a basis for the design or interpretation of future studies and for inclusion in the investigator's brochure. However, the studies that are conducted in the discovery phase (i.e. as part of a compound nomination package) are often conducted with a small number of animals or replicates. Therefore, consideration should be given to repeating these studies for inclusion in the IND, if appropriate. Some sponsors chose not to repeat these studies until later in the development timeline, once efficacy and safety has been confirmed in the clinic. However, if the studies are not repeated until late in development, it may be difficult to explain discrepancies between the early, discovery-mode studies and the later development studies.

The Pharm/Tox reviewer will review the integrated summary of nonclinical information, the more extensive and detailed summary of ADME information, and potentially individual study reports. All of these documents should contain a level of detail that is sufficient to allow the reviewer to come to the same conclusions that the author did.

In practice, this means that in addition to the appropriate mean data tables, the author should include appropriate individual data points. The intent of these summaries is to clearly and concisely convey the main points that the author is trying to make and is not to show how verbose or clever the author can be. To this end, graphics should be used wherever possible in order to convey a concept. This point is emphasized in the Format and Content guidance. Additionally, the author should appreciate that although the Pharm/Tox will have a good understanding of the science of ADME, this is unlikely to be their main area of expertise. Therefore, a simple and clear writing style is encouraged. Similarly, the author should acknowledge that English may not be the first language of the reviewer and hence the report or summary should be written with the bare minimum, if any, use of jargon or confusing words or phrases. Along the same lines, English may not be the language of the author of the scientific reports or summaries that form part of the regulatory submissions. Most, if not all, companies employ scientific writing groups to either write or edit these documents. It should be noted that peer-reviewed literature can be submitted to the IND as required and may replace the need for the creation of study reports. This is an interesting loophole, given the difference in the extent of detail that is given in a peer-reviewed journal article and a study report that contains all individual data. The FDA has also issued "Guidance for Reviewers: Pharm/Tox Review Format" (US FDA, 2001c). This guidance does not indicate which specific ADME studies should be conducted, but does imply that ADME information should be available. This guidance contains tables and checklists to help facilitate the review for the Pharm/Tox reviewer.

The format and content for an exploratory IND are very similar to those of a traditional IND and have been described in depth earlier in this chapter.

The process for the approval of clinical trials is somewhat different in the EU than it is in the United States. In the United States, an IND application is made that contains the investigator's brochure; a general investigational plan; the clinical trial protocol for the initial clinical trial; a summary of CMC, nonclinical and clinical data (as applicable) and individual study reports (primarily toxicology reports as noted earlier). For subsequent clinical trials, only the clinical protocol and investigator's brochure need to be submitted. For each clinical trial in Europe, the applicant will provide the investigator's brochure, the clinical trial protocol, and a summary of CMC, nonclinical and clinical information (the investigational medicinal product dossier or IMPD). Individual study reports are not submitted.

In many other guidances, ADME may be mentioned in passing. For example, in the EMEA guidance, "Detailed guidance for the request for authorization of a clinical trial on a medicinal product for human use to the competent authorities, notification of substantial amendments and declaration of the end of the trial" that was issued in October 2005, the only mention of metabolism is that there should be a subsection that describes "Pharmacokinetics of active metabolites" (European Commission, 2003). The Health Canada draft guidance, "Quality (Chemistry and Manufacturing) Guidance: New Drug Submissions (NDSs) and Abbreviated New Drug Submissions (ANDSs)" simply states that "It should also be indicated if the impurity is a metabolite of the drug substance" (Health Canada, 2001).

In 1984, the 505(b)(2) application route for a New Drug Application was created by Congress to allow applicants to create innovative medicines using currently available products without performing a full complement of safety and efficacy studies. This article reviews the history of the approach, provides examples, and considers some of the scientific and technical challenges associated with documenting safety and efficacy relative to the proposed change. The approach does not appear to have been used extensively in almost 18 years since it was created. The explanation for this is not fully apparent, but may relate to the limited exclusivity, usually three years, allowed for a 505(b)(2) application (Johnston and Williams, 2002). In 1999, the FDA issued a draft guidance, "505(b)(2)" (US FDA, 1999c). The guidance states that "In some cases a new molecular entity may have been studied by parties other than the applicant and published information may be pertinent to the new application. This is particularly likely if the NME is the prodrug of an approved drug or the active metabolite of an approved drug. In some cases, data on a drug with similar pharmacologic effects could be considered critical to approval."

Recently (March 2007), the FDA issued a draft guidance on "Target Product Profile—A Strategic Development Process Tool" (US FDA, 2007h). The Target Product Profile (TPP) is a tool that has been used for many years by pharmaceutical companies as a strategic communication tool that allows for continual dialog between different groups and allows for a common understanding of the desired characteristics of the drug under development and the steps that will be taken to achieve these goals. The FDA is interested in using this tool for the purposes listed earlier, with the purpose of improving the communications between the FDA and the sponsor. The TPP could potentially be a very useful tool that summarizes what is known about a compound, the development plan for the project, and how those goals will be achieved. Although the former is summarized in the investigator's brochure, the latter two topics are not presently gathered in an "evergreen" document. The guidance states: "Describe clinically significant pharmacokinetics of a drug or active metabolites (i.e., pertinent absorption, distribution, metabolism, and excretion parameters). Include results of pharmacokinetic studies that establish the absence of an effect, including pertinent human studies and in vitro data." The guidance also states: "Include a concise factual summary of the clinical pharmacology and actions of the drug in humans. Data that describe the drug's pharmacologic activity can be included in this section, including biochemical or physiological mechanism of action, pharmacokinetic information, degree of absorption, pathway for biotransformation, percent dose unchanged, metabolites, rate of half-lives including elimination concentration in body fluids at therapeutic and toxic levels, degree of binding to plasma, degree of uptake by a particular organ or fetus, and passage across the blood-brain barrier."

In an attempt to standardize the documents that would be submitted for marketing approval throughout the world, the ICH has issued several guidances, including "The Common Technical Document for the Registration of Pharmaceuticals for Human Use: Safety—M4S" (ICH, 2000d).

This guidance describes the sections that should be included in the CTD. These include:

"2.6.4 Pharmacokinetics Written Summary

The sequence of the Pharmacokinetics Written Summary should be as follows:

- Brief Summary
- Methods of Analysis
- Absorption
- Distribution
- Metabolism
- Excretion

- Pharmacokinetic Drug Interactions
- Other Pharmacokinetic Studies
- Discussion and Conclusions
- Tables and Figures (either here or included in text)

2.6.4.5 Metabolism (interspecies comparison)
The following data should be summarised in this section:

- Chemical structures and quantities of metabolites in biological samples
- Possible metabolic pathways
- Pre-systemic metabolism (GI/hepatic first-pass effects)
- In vitro metabolism including P450 studies
- Enzyme induction and inhibition.

2.6.5   Pharmacokinetics
2.6.5.1   Pharmacokinetics: Overview
2.6.5.2   Analytical Methods and Validation Reports*
2.6.5.3   Pharmacokinetics: Absorption after a Single Dose
2.6.5.4   Pharmacokinetics: Absorption after Repeated Doses
2.6.5.5   Pharmacokinetics: Organ Distribution
2.6.5.6   Pharmacokinetics: Plasma Protein Binding
2.6.5.7   Pharmacokinetics: Study in Pregnant or Nursing Animals
2.6.5.8   Pharmacokinetics: Other Distribution Study
2.6.5.9   Pharmacokinetics: Metabolism In Vivo
2.6.5.10 Pharmacokinetics: Metabolism In Vitro
2.6.5.11 Pharmacokinetics: Possible Metabolic Pathways
2.6.5.12 Pharmacokinetics: Induction/Inhibition of Drug-Metabolizing Enzymes
2.6.5.13 Pharmacokinetics: Excretion
2.6.5.14 Pharmacokinetics: Excretion into Bile
2.6.5.15 Pharmacokinetics: Drug-Drug Interactions
2.6.5.16 Pharmacokinetics: Other"

The FDA has created a web page for small businesses (US FDA, 2008l). The information provided includes a "Frequently Asked Questions" document on "Drug Development and Investigational New Drug Applications" (US FDA, 2001a). This document includes the FAQ, "What are the FDA requirements for pre-clinical studies?" The FDA's answer to this question is:

> Under FDA requirements, a sponsor must first submit data showing that the drug is reasonably safe for use in initial, small-scale clinical studies. During preclinical drug development, a sponsor evaluates the drug's toxic and pharmacologic effects through *in vitro* and *in vivo* laboratory animal testing. Genotoxicity screening is performed, as well as investigations on drug absorption and metabolism, the toxicity of the drug's metabolites, and the speed with which the drug and its metabolites are excreted from the body. At the preclinical stage, the FDA will generally ask, at a minimum, that sponsors: (1) develop a pharmacological profile of the drug; (2) determine the acute toxicity of the drug in at least two species of animals, and (3) conduct short-term toxicity studies ranging from 2 weeks to 3 months, depending on the proposed duration of use of the substance in the proposed clinical studies.

It is interesting to note that ADME studies are not mentioned in this latter list. The small business assistance web page also includes a document on "Frequently Asked Questions on the Pre-Investigational New Drug (IND) Meeting" (US FDA, 2005a).

It is of interest to note that even though drug metabolism and metabolites are mentioned, albeit often in passing, in many guidances, there is very little, if any, information included in these guidances regarding how to determine the metabolism of a drug, particularly *in vivo*. Occasionally, guidances or guidelines will provide some tips or insights. For example, the European guideline "Pharmacokinetics and Metabolic Studies in the Safety Evaluation of New Medicinal Products in Animals, Legislative basis Directive 75/318/EEC as amended" states that:

> When a labelled substance is used the position of the label in the molecule and the specific activity of the material must be stated. Consideration should be given when selecting the position of the label to its likely metabolic fate. When using labelled substances attention must be given to the fact that the measured label in body fluids may not correspond to that of the unmodified substance, but may include labelled metabolites and conjugates. Attention should be given to the possibility of isotope exchange with endogenous compounds. (EMEA, 1994b)

This may appear to be an obvious point, but one that is often overlooked. An example of this phenomenon is the exchange of tritium between a radiolabeled pharmaceutical and water.

27.10   INITIATIVES FROM REGULATORY AGENCIES

Guidances or guidelines are not the only tool that is used by regulatory agencies to communicate their expectations to sponsors. These expectations, or concerns, are also communicated by the participation of representatives from the regulatory agencies at scientific conferences; the publication of articles in peer-reviewed journals; the publication of book chapters and the creation of initiatives.

One such example of a regulatory initiative is the FDA's "Critical Path Initiative (Innovation or stagnation)" that was issued in 2004 (US FDA, 2004a). The original document and subsequent updates and related documents can be found on the FDA's web page on the Critical Path Initiative (US FDA, 2008d). In May 2007, the FDA expanded the scope of the Critical Path Initiative and created a document on Critical Path Opportunities for Generic Drugs (US FDA, 2007e).

The Critical Path Initiative document states that:

> Many of FDA's recent targeted efforts have involved working with the scientific community to define more reliable methods to predict and detect significant safety problems. For example, in the past, failure to predict unfavorable human metabolism of candidate drugs has led to costly failures in the clinic as well as multiple drug market withdrawals. FDA recommendations on the use of human cell lines to characterize drug metabolic pathways provide a straightforward in vitro method for prediction of human metabolism, allowing developers to eliminate early on compounds with unfavorable metabolic profiles (e.g., drug-drug interaction potential). Failures in the clinic due to drug interaction problems are now far less likely.

The FDA's Critical Path Opportunities List and Critical Path Opportunities Report that were issued in May 2007 identify several areas of interest to the drug metabolism scientist. (US FDA, 2006a, b). The Opportunities list includes Biomarker Qualification, Modernizing Predictive Toxicology, Noninvasive Therapeutic Monitoring, Improving Extrapolation from Animal Data to Human Experience, Identifying Safety Effects of Excipients, Characterizing and Qualifying Nanotechnologies, and Drug Metabolism and Therapeutic Response for Pediatrics. The Opportunities list represents an initial list that will be updated as necessary. There are 76 specific scientific opportunities, or priority topics, in the Opportunities report and list.

The FDA also issued a report called "Critical Path Opportunities Initiated During 2006" (US FDA, 2006c). This document lists more than 40 Critical Path collaborations and research activities that currently are under way with FDA participation. It can be seen that there are a large number of initiatives that have been created that will lead to the creation of a concept paper, followed by the creation of a draft guidance, and ultimately a final guidance. Therefore, there will be an increased amount of documentation that the drug development scientists will need to become familiar with. However, it should be noted that the Critical Path Initiative enables the drug development scientists to be part of the process that will ultimately lead to new, or updated, regulatory guidances. This is an opportunity that has not really existed until now.

The Critical Path Opportunities for Generics includes several topics of interest including Model Development and In Vitro-In Vivo Correlations, bioequivalence for novel delivery technologies, bioequivalence for pharmacokinetic profiles with multiple peaks, bioequivalence for transdermal products, bioequivalence for highly variable drugs, bioequivalence for locally acting and targeted delivery drugs, and bioequivalence of liposome-based formulations (US FDA, 2007e; Lionberger, 2008).

The Innovative Medicines Initiative (IMI) issued a report entitled "Strategic Research Agenda Creating Biomedical R&D Leadership for Europe to Benefit Patients and Society" in 2004 (EFPIA, 2004). This report noted that:

> the regulatory bodies hold data on the pharmacokinetics of a large number of drugs. Collective analysis of the data for all substrates of a particular metabolising enzyme e.g. cytochrome 2D6 or 3A4, should provide information not only on the inherent functional variability of these enzymes within the patient population, but also allow one to determine quantitatively the contribution of such factors as age, gender, disease, and inhibitors of these enzymes, to the variability. Armed with this generic information, one should be able to predict a priori the likely variability of the pharmacokinetics of a new drug within the patient population, under a variety of situations, thereby facilitating future design of clinical studies and subsequent product labeling, and also improving the cost-efficiency of such studies. This proposal will require not only inter-company collaboration but also the agreement of EMEA and national bodies to release these data.

If such data were to become available, the regulatory agencies may be able to help facilitate drug development. Such a close partnership would be relatively novel but would greatly benefit the patient in the long run. In 2007, the FDA announced that the formation of the Reagan-Udall foundation which is a private and independent non-profit organization will advance FDA's mission to modernize medical, veterinary, food, food ingredient, and cosmetic product development, accelerate innovation, and enhance product safety. The 14-member board consists of four representatives from the general pharmaceutical, device, food, cosmetic, and biotechnology industries; three from academic research organizations; two from patient or consumer advocacy groups; one representing health-care providers; and four at-large representatives with expertise or experience relevant to the foundation's purpose. Hopefully, close interaction between industry and regulators will expedite the development of safe medicines, although concern has been expressed that such a foundation will enable the pharmaceutical industry to exert influence over the FDA.

27.11 SUMMARY

Drug metabolism is playing an ever-increasing role in the drug development process and this is being reflected in the number of guidances that include mention of drug metabolism and the increased role of drug metabolism in clinical trials and the product label. The importance of drug metabolism will continue to increase as we learn more about the role of drug metabolism in special populations, the role of drug metabolism in sources of inter- and intra-individual variability, and the role of drug metabolism in certain toxicities. It is therefore of critical importance that the drug metabolism scientist become aware of the relevant regulatory guidances and take an active role in applying the existing guidances and the creation of new guidances. This is not an easy task, but is a task that will reap great rewards for the pharmaceutical industry in the long term.

REFERENCES

Abbott Laboratories (2007). *Kaletra Package Insert*, http://www.rxabbott.com/pdf/kaletrapi.pdf (accessed 7 March 2008).

Agrawal, Y.K. *et al.* (2007). Chirality—a new era of therapeutics. *Mini Reviews in Medicinal Chemistry*, 7, 451–460.

Agundez, J.A.G. (2004). Cytochrome P450 gene polymorphism and cancer. *Current Drug Metabolism*, 5, 211–224.

Aitken, A.E., Richardson, T.A., Morgan, E.T. (2006). Regulation of drug-metabolizing enzymes and transporters in inflammation. *Annual Review of Pharmacology and Toxicology*, 46 (1), 123–149.

Alexandre, J. *et al.* (2007). Relationship between cytochrome 3A activity, inflammatory status and the risk of docetaxel-induced febrile neutropenia: a prospective study. *Annals of Oncology*, 18 (1), 168–172.

Amgen (2005). *Comments on Draft Guidance for Industry on Safety Testing of Metabolites*, 3 August 2005, http://www.fda.gov/ohrms/dockets/dockets/05d0203/05D-0203-EC3-Attach-1.pdf (accessed 3 March 2008).

Andersson, T. (2004). Single-isomer drugs: true therapeutic advances. *Clinical Pharmacokinetics*, 43 (5), 279–285.

Ando, Y., Fuse, E., Figg, W.D. (2002). Thalidomide metabolism by the CYP2C subfamily. *Clinical Cancer Research*, 8 (6), 1964–1973.

Anon (2006). Slow start to phase 0 as researchers debate value. *Journal of the National Cancer Institute*, 98 (12), 804–805.

Ariyoshi, N. *et al.* (2001). Comparison of the levels of enzymes involved in drug metabolism between transgenic or gene-knockout and the parental mice. *Toxicologic Pathology*, 29 (Suppl.), 161–172.

Arlen, R.R., Wells, P.G. (1996). Inhibition of thalidomide teratogenicity by acetylsalicylic acid: evidence for prostaglandin H synthase-catalyzed bioactivation of thalidomide to a teratogenic reactive intermediate. *The Journal of Pharmacology and Experimental Therapeutics*, 277 (3), 1649–1658.

Association of the British Pharmaceutical Industry (2007). Guidelines for Phase I Clinical Trials. 2007 Edition.

Baillie, T.A. *et al.* (2002). Drug metabolites in safety testing. *Toxicology and Applied Pharmacology*, *182* (3), 188–196.

Baillie, T.A. *et al.* (2003). Reply. *Toxicology and Applied Pharmacology*, *190* (1), 93–94.

Baldrick, P. (2001). Preclinical considerations for regulatory submissions. *Drug Information Journal*, *35*, 99–105.

Bauer, R.J. *et al.* (1996). Alteration of the pharmacokinetics of small proteins by iodination. *Biopharmaceutics and Drug Disposition*, *17* (9), 761–774.

Ben Reguiga, M. *et al.* (2005). Modification of the P-glycoprotein dependent pharmacokinetics of digoxin in rats by human recombinant interferon-alpha. *Pharmaceutical Research*, *22* (11), 1829–1836.

Ben Reguiga, M., Bonhomme-Faivre, L., Farinotti, R. (2007). Bioavailability and tissular distribution of docetaxel, a P-glycoprotein substrate, are modified by interferon-alpha in rats. *Journal of Pharmacy and Pharmacology*, *59* (3), 401–408.

Bertino, J.S., Greenberg, Jr, H.E., Reed, M.D. (2007a). American college of clinical pharmacology position statement on the use of microdosing in the drug development process. *Journal of Clinical Pharmacol*, *47* (4), 418–422.

Bertino, J.S., Greenberg, Jr, H.E., Reed, M.D. (2007b). Response to comments on ACCP position statement on microdosing. *Journal of Clinical Pharmacol*, *47* (12), 1597–1598.

Beumer, J.H., Beijnen, J.H., Schellens, J.H.M. (2006). Mass balance studies, with a focus on anti-cancer drugs. *Clinical Pharmacokinetics*, *45* (1), 33–58.

Beumer, J.H. *et al.* (2007). Human mass balance study of the novel anticancer agent ixabepilone using accelerator mass spectrometry. *Investigational New Drugs*, *25*, 327–334.

Biogen-Idec (2004). *Tysabri. Prescribing Information*, http://www.fda.gov/cder/foi/label/2006/125104s015LBL.pdf (accessed 8 March 2008).

Biogen-Idec (2006). *Tysabri. Prescribing Information*, http://www.fda.gov/cder/foi/label/2004/125104lbl.pdf (accessed 8 March 2008).

Bjornsson, T.D. *et al.* (2003). The conduct of *in vitro* and *in vivo* drug–drug interaction studies: a pharmaceutical research and manufacturers of America (PhRMA) perspective. *Drug Metabolism and Disposition*, *31* (7), 815–832.

Bonate, P.L., Howard, D.R. (2004). *Pharmacokinetics in Drug Development: Regulatory and Development Paradigms*, AAPS Press.

Bowe, C. (2007). Wyeth Says Regulator Creating Monopolies. *Financial Times*, London. http://www.ft.com/cms/s/0/4aa84080-8b10-11dc-95f7-0000779fd2ac.html?nclick_check=1.

Boyd, R.A., Lalonde, R.L. (2007). Nontraditional approaches to first-in-human studies to increase efficiency of drug development: will microdose studies make a significant impact? *Clinical Pharmacol Therapy*, *81* (1), 24–26.

Brent, R.L. (2004). Utilization of animal studies to determine the effects and human risks of environmental toxicants (drugs, chemicals, and physical agents). *Pediatrics*, *113* (4), 984–995.

Brocks, D.R. (2006). Drug disposition in three dimensions: an update on stereoselectivity in pharmacokinetics. *Biopharmaceutics and Drug Disposition*, *27* (8), 387–406.

Brunner, M., Langer, O. (2006). Microdialysis versus other techniques for the clinical assessment of *in vivo* tissue drug distribution. *The AAPS Journal*, *8* (2), E263–E271.

Buehler, G.J. (2007). Letter to Sponsor of Pending Abbreviated New Drug Application(s) (ANDAs).

Cao, Y.J. *et al.* (2007). Effect of semen sampling frequency on seminal antiretroviral drug concentration. *Clinical Pharmacol Therapy*, *83*, 848–856.

Chainuvati, S. *et al.* (2003). Combined phenotypic assessment of cytochrome P450 1A2, 2C9, 2C19, 2D6, and 3A, N-acetyltransferase-2, and xanthine oxidase activities with the [ldquo]Cooperstown 5+1 cocktail [rdquo] [ast]. *Clinical Pharmacol Therapy*, 74 (5), 437–447.

Charles, K.A. *et al.* (2006). Transcriptional repression of hepatic cytochrome P450 3A4 gene in the presence of cancer. *Clinical Cancer Research*, 12 (24), 7492–7497.

Chaurand, P., Corentt, D., Caprioli, R. (2006). Molecular imaging of thin mammalian tissue sections by mass spectrometry. *Current Opinion in Biotechnology*, 17, 431–436.

Cheung, B. *et al.* (2005). The application of sample pooling methods for determining AUC, AUMC and mean residence times in pharmacokinetic studies. *Fundamental and Clinical Pharmacology*, 19, 347–354.

Chung, N.S., Wasan, K.M. (2004). Potential role of the low-density lipoprotein receptor family as mediators of cellular drug uptake. *Advanced Drug Delivery Reviews*, 56 (9), 1315–1334.

Correia, M.A. (2003). Hepatic cytochrome P450 degradation: mechanistic diversity of the cellular sanitation brigade. *Drug Metabolism Reviews*, 35 (2), 107–143.

Correia, M.A., Sadeghi, S., Mundo-Paredes, E. (2005). Cytochrome P450 ubiquitination: branding for the proteolytic slaughter? *Annual Review of Pharmacology and Toxicology*, 45 (1), 439–464.

Dain, J., Collins, J., Robinson, W. (1994). A regulatory and industrial perspective of the use of carbon-14 and tritium isotopes in human ADME studies. *Pharmaceutical Research*, 11 (6), 925–928.

Danko, I., Chaschin, N. (2005). Association of CYP2E1 gene polymorphism with predisposition to cancer development. *Experimental Oncology*, 27 (4), 248–256.

Dannan, H. (2003). Applying good laboratory practice regulations, in *Pharmaceutical Technology Europe*, Advanstar Publications.

Darpo, B., Nebout, T., Sager, P.T. (2006). Clinical evaluation of QT/QTc prolongation and proarrhythmic potential for nonantiarrhythmic drugs: the international conference on harmonization of technical requirements for registration of pharmaceuticals for human use E14 guideline. *Journal of Clinical Pharmacol*, 46 (5), 498–507.

Davis-Bruno, K.L., Atrakchi, A. (2006). A regulatory perspective on issues and approaches in characterizing human metabolites. *Chemical Research in Toxicology*, 19 (12), 1561–1563.

Davit, B. *et al.* (2008). Highly variable drugs: observations from bioequivalence data submitted to the FDA for new generic drug applications. *The AAPS Journal*, 10 (1), 148–156.

DeGeorge, J.J. *et al.* (1997). Regulatory considerations for preclinical development of anticancer drugs. *Cancer Chemotherapy and Pharmacology*, 41 (3), 173–185.

DeGregorio, M. *et al.* (2006). Accelerator mass spectrometry allows for cellular quantification of doxorubicin at femtomolar concentrations. *Cancer Chemotherapy and Pharmacology*, 57 (3), 335–342.

Dodds, P.F. (1991). Incorporation of xenobiotic carboxylic acids into lipids. *Life Sciences*, 49 (9), 629–649.

Dodds, P.F. (1995). Xenobiotic lipids: the inclusion of xenobiotic compounds in pathways of lipid biosynthesis. *Progress in Lipid Research*, 34 (3), 219–247.

Dressman, J. *et al.* (1998). Dissolution testing as a prognostic tool for oral drug absorption: immediate release dosage forms. *Pharmaceutical Research*, 15 (1), 11–22.

Dumez, H. *et al.* (2004). The relevance of therapeutic drug monitoring in plasma and erythrocytes in anti-cancer drug treatment. *Clinical Chemistry and Laboratory Medicine*, 42 (11), 147–173.

European Commission (2001). Directive 2001/20/EC of the European Parliament and of the Council of 4 April 2001 on the Approximation of the Laws, Regulations and Administrative Provisions of the Member States Relating to the Implementation of Good Clinical Practice in the Conduct of Clinical Trials on Medicinal Products for Human Use.

European Commission (2003). Detailed Guidance for the Request for Authorisation of a Clinical Trial on a Medicinal Product for Human use to the Competent Authorities, Notification of Substantial Amendments and Declaration of the End of the Trial.

European Commission (2005). A Guideline on Summary of Product Characteristics Rev 1.

European Commission (2008). *EUDRALEX. Volume 1—Pharmaceutical Legislation: Medicinal Products for Human Use*, http://ec.europa.eu/enterprise/pharmaceuticals/eudralex/homev1. htm (accessed 4 March 2008).

European Economic Community (1987). Pharmacokinetic Studies in Man. Legislative Basis Directive 75/318/EEC as Amended.

European Federation of Pharmaceutical Industries and Associations (EFPIA) (2004). Strategic Research Agenda Creating biomedical R&D Leadership for Europe to Benefit Patients and Society.

European Medicines Agency (EMEA) (1994a). Note for Guidance on Studies in Support of Special Populations: Geriatrics" (CPMP/ICH/379/95).

European Medicines Agency (EMEA) (1994b). Pharmacokinetics and Metabolic Studies in the Safety Evaluation of New Medicinal Products in Animals.

European Medicines Agency (EMEA) (1997). Note for Guidance on the Investigations of Drug Interactions (CPMP/EWP/560/95).

European Medicines Agency (EMEA) (1998). Note for Guidance on the Pre-Clinical Evaluation of Anticancer Medicinal Products (CPMP/SWP/997/96).

European Medicines Agency (EMEA) (1999). Note for Guidance on Modified Release Oral and Transdermal Dosage Forms: Section II (Pharmacokinetic and Clinical Evaluation) (CPMP/ EWP/280/96).

European Medicines Agency (EMEA) (2001). Note for Guidance on the Investigation of Bio-availability and Bioequivalence (CPMP/EWP/QWP/1401/98).

European Medicines Agency (EMEA) (2003). Note for Guidance on Evaluation of Anticancer Medicinal Products in Man. Addendum on Paediatric Oncology (EMEA/CPMP/EWP/ 569/02).

European Medicines Agency (EMEA) (2004). Position Paper on Non-Clinical Safety Studies to Support Clinical Trials with a Single Microdose (CPMP/SWP/2599/02/Rev 1).

European Medicines Agency (EMEA) (2005a). CHMP SWP Reflection Paper on the Assessment of the Genotoxic Potential of Antisense Oligodeoxynucleotides (EMEA/CHMP/SWP/199726/ 2004).

European Medicines Agency (EMEA) (2005b). Guideline on Risk Management Systems for Medicinal Products for Human Use (EMEA/CHMP/96268/2005).

European Medicines Agency (EMEA) (2005c). Guideline on the Evaluation of Control Samples in Nonclinical Safety Studies: Checking for Contamination With the Test Substance (CPMP/ SWP/1094/04).

European Medicines Agency (EMEA) (2005d). Guideline on the Evaluation of the Pharmaco-kinetics of Medicinal Products in Patients With Impaired Hepatic Function (CPMP/ EWP/2339/02).

European Medicines Agency (EMEA) (2005e). Guideline on the Need for Non-Clinical Testing in Juvenile Animals on Human Pharmaceuticals for Paediatric Indications (Draft) (EMEA/ CHMP/SWP/169215/2005).

European Medicines Agency (EMEA) (2005f). Guideline on the Non-Clinical Development of Fixed Combinations of Medicinal Products (Draft) (CHMP/EMEA/CHMP/SWP/ 258498/2005).

European Medicines Agency (EMEA) (2005g). Guideline on the Non-Clinical Documentation for Mixed Marketing Authorisation Applications (CPMP/SWP/799/95).

European Medicines Agency (EMEA) (2006a). Adequacy of Guidance on the Elderly Regarding Medicinal Products for Human Use (EMEA/498920/2006).

European Medicines Agency (EMEA) (2006b). Concept Paper on the Development of a CHMP Guideline on the Non-Clinical Requirements to Support Early Phase I Clinical Trials With Pharmaceutical Compounds (EMEA/CHMP/SWP/91850/2006).

European Medicines Agency (EMEA) (2006c). Guideline on Risk Assessment of Medicinal Products on Human Reproduction and Lactation: from Data to Labelling (Draft) (EMEA/CHMP/203927/2005).

European Medicines Agency (EMEA) (2006d). Guideline on the Development of a CHMP Guideline on the Non-Clinical Requirements to Support Early Phase I Clinical Trials With Pharmaceutical Compounds (EMEA/CHMP/SWP/91850/2006).

European Medicines Agency (EMEA) (2006e). Guideline on the Environmental Risk Assessment of Medicinal Products for Human Use (EMEA/CHMP/SWP/4447/00).

European Medicines Agency (EMEA) (2006f). Guideline on the Evaluation of Anticancer Medicinal Products in Man (CPMP/EWP/205/95/Rev.3/Corr.2).

European Medicines Agency (EMEA) (2006g). Guideline on the Non-Clinical Investigation of the Dependence Potential of Medicinal Products (EMEA/CHMP/SWP/94227/2004).

European Medicines Agency (EMEA) (2006h). Guideline on the Role of Pharmacokinetics in the Development of Medicinal Products in the Paediatric Population (EMEA/CHMP/EWP/147013/2004/Corr).

European Medicines Agency (EMEA) (2006i). Nexavar. European Public Assessment Report: Scientific Discussion.

European Medicines Agency (EMEA) (2006j). Questions & Answers on the Bioavailability and Bioequivalence Guideline (EMEA/CHMP/EWP/40326/2006).

European Medicines Agency (EMEA) (2006k). Reflection Paper: Formulations of Choice for the Paediatric Population (EMEA/CHMP/PEG/194810/2005).

European Medicines Agency (EMEA) (2007a). Annex a to Guideline on Clinical Development of Medicinal Products for Treatment of HIV Infection. Presentation of Pharmacokinetic Interaction Data in the SPC (Draft).

European Medicines Agency (EMEA) (2007b). Concept Paper on BCS-Based Biowaiver (EMEA/CHMP/EWP/213035/2007).

European Medicines Agency (EMEA) (2007c). EMEA Guidance for Companies Requesting Scientific Advice or Protocol Assistance (EMEA-H-4260-01-Rev. 4).

European Medicines Agency (EMEA) (2007d). Guideline on Strategies to Identify and Mitigate Risks for First-In Human Clinical Trials With Investigational Medicinal Products (EMEA/CHMP/SWP/28367/07).

European Medicines Agency (EMEA) (2007e). Guideline on the Clinical Investigation of the Pharmacokinetics of Therapeutic Proteins (CHMP/EWP/89249/2004).

European Medicines Agency (EMEA) (2007f). Guideline on the Clinical Development of Medicinal Products for the Treatment of HIV Infection (Draft) (EMEA/CPMP/EWP/633/02 Rev. 2).

European Medicines Agency (EMEA) (2007g). Note for Guidance on Definitions for Genomic Biomarkers, Pharmacogenomics, Pharmacogenetics, Genomic Data and Sample Coding Categories (EMEA/CHMP/ICH/437986/2006).

European Medicines Agency (EMEA) (2007h). Recommendation on the Need for Revision of "Note for Guidance on the Investigation of Bioavailability and Bioequivalence (CPMP/EWP/QWP/1401/98).

European Medicines Agency (EMEA) (2007i). Reflection Paper on the Use of Pharmacogenetics in the Pharmacokinetic Evaluation of Medicinal Products (EMEA/128517/2006).

European Medicines Agency (EMEA) (2007j). Reflection Paper on the Use of Genomics in Cardiovascular Clinical Intervention Trials (EMEA/CHMP/PGxWP/278789/2006).

European Medicines Agency (EMEA) (2007k). The Investigation of Medicinal Products in the Term and Preterm Neonate (Draft) (EMEA/267484/2007).

European Medicines Agency (EMEA) (2008a). *EPARs for Authorised Medicinal Products for Human Use*, http://www.emea.europa.eu/htms/human/epar/a.htm (accessed 3 March 2008).

European Medicines Agency (EMEA) (2008b). Guideline on Fixed Combination Medicinal Products (Draft) (CPMP/EWP/240/95 Rev 1).

European Medicines Agency (EMEA) (2008c). Guideline on Repeated Dose Toxicity (Draft) (EMEA/CHMP/SWP/488313/2007).

European Medicines Agency (EMEA) (2008d). Guideline on the Need for Non-Clinical Testing in Juvenile Animals of Pharmaceuticals for Paediatric Indications (EMEA/CHMP/SWP/169215/2005).

European Medicines Agency (EMEA) (2008e). Guideline on the Non-Clinical Development of Fixed Combinations of Medicinal Products (EMEA/CHMP/SWP/258498/2005).

European Medicines Agency (EMEA) (2008f). *Marketing Authorisation Withdrawals and Suspensions*, http://www.emea.europa.eu/htms/human/withdraw/withdraw.htm (accessed 3 March 2008).

European Medicines Agency (EMEA) (2008g). Non-Clinical Guideline on Drug-Induced Hepatotoxicity (Draft) (EMEA/CHMP/SWP/150115/2006).

European Medicines Agency (EMEA) (2008h). *Scientific Guidelines for Human Medicinal Products*, http://www.emea.europa.eu/htms/human/humanguidelines/background.htm (accessed 3 March 2008).

European Medicines Agency (EMEA) (2008i). *Work Plan for the Safety Working Party (SWP) 2008–2009*, 20 September 2007, http://www.emea.europa.eu/pdfs/human/swp/swpworkprogramme.pdf (accessed 4 March 2008).

European Parliament (2004). *Directive 2004/10/EC of the European Parliament and of the Council of 11 February 2004 on the Harmonisation of Laws, Regulations and Administrative Provisions Relating to the Application of the Principles of Good Laboratory Practice and the Verification of Their Applications for Tests on Chemical Substances (Codified Version) (Text with EEA Relevance)*, 20 February 2004, http://europa.eu.int/eur-lex/lex/LexUriServ/LexUriServ.do?uri=CELEX:32004L0010:EN:HTML (accessed 4 March 2008).

European Union (1994). Investigation of Chiral Active Substances. Legislative Basis Directive 75/318/EEC as Amended.

Expert Group on Phase One Clinical Trials (Chairman: Professor Gordon W. Duff) (2006). Expert Group on Phase One Clinical Trials: Final Report.

Farrell, A.T. *et al.* (2003). How oncology drug development differs from other fields, in *Handbook of Anticancer Drug Development* (eds D.R. Budman, A.H. Calvert and E.K. Rowinsky), Lippincott Williams & Wilkins, pp. 3–12.

Frueh, F.W. *et al.* (2006). Experience with voluntary and required genomic data submissions to the FDA: summary report from track 1 of the third FDA-DIA-PWG-PhRMA-BIO pharmacogenomics workshop. *The Pharmacogenomics Journal*, 6 (5), 296–300.

Fura, A. (2006). Role of pharmacologically active metabolites in drug discovery and development. *Drug Discovery Today*, 11 (3–4), 133–142.

G.D. Searle LLC, a Division of Pfizer Inc. (2008). *Inspra. Eplerenone Tablets*, http://www.fda.gov/cder/foi/label/2008/021437s005lbl.pdf (accessed 8 March 2008).

Galteau, M.M., Shamsa, F. (2003). Urinary 6ß-hydroxycortisol: a validated test for evaluating drug induction or drug inhibition mediated through CYP3A in humans and in animals. *European Journal of Clinical Pharmacology*, 59 (10), 713–733.

Goodsaid, F., Frueh, F. (2007). Implementing the U.S. FDA guidance on pharmacogenomic data submissions. *Environmental and Molecular Mutagenesis*, 48 (5), 354–358.

Haddad, A., Davis, M., Langman, R. (2007). The pharmacological importance of cytochrome CYP3A4 in the palliation of symptoms: review and recommendations for avoiding adverse drug interactions. *Support Care Cancer*, *15*, 251–257.

Haidar, S.H. *et al.* (2008). Bioequivalence approaches for highly variable drugs and drug products. *Pharmaceutical Research*, *25* (1), 237–241.

Hamdy, R. *et al.* (1995). Influenza vaccine may enhance theophylline toxicity. A case report and review of the literature. *Journal of the Tennessee Medical Association*, *88* (12), 463–464.

Han, H.-K. *et al.* (2006). An efficient approach for the rapid assessment of oral rat exposures for new chemical entities in drug discovery. *Journal of Pharmaceutical Sciences*, *95* (8), 1864–1692.

Hastings, K.L. *et al.* (2003). Letter to the editor. *Toxicology and Applied Pharmacology*, *190* (1), 91–92.

Health Canada (2000). Stereochemical Issues in Chiral Drug Development.

Health Canada (2001). Draft Guidance for Industry: Quality (Chemistry and Manufacturing) Guidance: New Drug Submissions (NDSs) and Abbreviated New Drug Submissions (ANDSs).

Health Canada (2003). Discussion Paper. Bioequivalence Requirements: Highly Variable Drugs and Highly Variable Drug Products: Issues and Options.

Health Canada (2007a). Clinical Assessment of Abuse Liability for Drugs with Central Nervous System Activity.

Health Canada (2007b). Withdrawal of Market Authorization for Prexige.

Hewitt, N.J., de Kanter, R., LeCluyse, E. (2007). Induction of drug metabolizing enzymes: a survey of *in vitro* methodologies and interpretations used in the pharmaceutical industry—do they comply with FDA recommendations? *Chemico-Biology Interactions*, *168* (1), 51–65.

Hill, T.P. (2007). Phase 0 trials: are they ethically challenged? *Clinical Cancer Research*, *13* (3), 783–784.

Hodge, L.S., Tracy, T.S. (2007). Alterations in drug disposition during pregnancy: implications for drug therapy. *Expert Opinion on Drug Metabolism and Toxicology*, *3* (4), 557.

Holmes, E., Nicholson, J. (2005). Variation in gut microbiota strongly influences individual rodent phenotypes. *Toxicological Sciences*, *87* (1), 1–2.

Hop, C. *et al.* (1998). Plasma-pooling methods to increase throughput for *in vivo* pharmacokinetic screening. *Journal of Pharmaceutical Sciences*, *87* (7), 901–903.

Hu, Z. *et al.* (2005). Herb-drug interactions: a literature review. *Drugs*, *65* (9), 1239–1282.

Huang, S.-M. *et al.* (2008). New era in drug interaction evaluation: US food and drug administration update on CYP enzymes, transporters, and the guidance process. *Journal of Clinical Pharmacol*, *48* (6), 662–670.

Humphreys, W.G., Unger, S.E. (2006). Safety assessment of drug metabolites: characterization of chemically stable metabolites. *Chemical Research in Toxicology*, *19* (12), 1564–1569.

Hurwitz, A. *et al.* (2003). Gastric function in the elderly: effects on absorption of ketoconazole. *Journal of Clinical Pharmacol*, *43* (9), 996–1002.

International Conference on Harmonisation of Technical Requirements for Registration of Pharmaceuticals for Human Use (ICH) (1994a). *ICH Harmonised Tripartite Guideline. Pharmacokinetics: Guidance for Repeated Dose Tissue Distribution Studies. S3B*, http://www.ich.org/LOB/media/MEDIA496.pdf (accessed 6 March 2008).

International Conference on Harmonisation of Technical Requirements for Registration of Pharmaceuticals for Human Use (ICH) (1994b). *ICH Harmonised Tripartite Guideline. Note for Guidance on Toxicokinetics: The Assessment of Systemic Exposure in Toxicity Studies. S3A*, 27 October 1994, http://www.ich.org/LOB/media/MEDIA495.pdf (accessed 4 March 2008).

International Conference on Harmonisation of Technical Requirements for Registration of Pharmaceuticals for Human Use (ICH) (1994c). *ICH Harmonised Tripartite Guideline. Studies in*

Support of Special Populations: Geriatrics. E7, http://www.ich.org/LOB/media/MEDIA483.pdf (accessed 8 March 2008).

International Conference on Harmonisation of Technical Requirements for Registration of Pharmaceuticals for Human Use (ICH) (1995). *ICH Harmonised Tripartite Guideline. Guidance on Specific Aspects of Regulatory Genotoxicity Tests for Pharmaceuticals. S2A*, http://www.ich.org/LOB/media/MEDIA493.pdf (accessed 6 March 2008).

International Conference on Harmonisation of Technical Requirements for Registration of Pharmaceuticals for Human Use (ICH) (1996a). *ICH Harmonised Tripartite Guideline. Good Clinical Practice: Consolidated Guideline. E6(R1)*, 10 June 1996, http://www.ich.org/LOB/media/MEDIA482.pdf (accessed 6 March 2008).

International Conference on Harmonisation of Technical Requirements for Registration of Pharmaceuticals for Human Use (ICH) (1996b). *ICH Harmonised Tripartite Guideline. Preclinical Safety Evaluation of Biotechnology-Derived Pharmaceuticals. S6*, http://www.ich.org/LOB/media/MEDIA503.pdf (accessed 8 March 2008).

International Conference on Harmonisation of Technical Requirements for Registration of Pharmaceuticals for Human Use (ICH) (1997a). *ICH Harmonised Tripartite Guideline. General Considerations for Clinical Trials. E8*, 14 July 1997, http://www.ich.org/LOB/media/MEDIA484.pdf (accessed 6 March 2008).

International Conference on Harmonisation of Technical Requirements for Registration of Pharmaceuticals for Human Use (ICH) (1997b). *ICH Harmonised Tripartite Guideline. Impurities: Guideline for Residual Solvents. Q3C(R3)*, http://www.ich.org/LOB/media/MEDIA423.pdf (accessed 7 March 2008).

International Conference on Harmonisation of Technical Requirements for Registration of Pharmaceuticals for Human Use (ICH) (1998). *ICH Harmonised Tripartite Guideline. Ethnic Factors in the Acceptability of Foreign Clinical Data. E5(R1)*, http://www.ich.org/LOB/media/MEDIA481.pdf (accessed 7 March 2008).

International Conference on Harmonisation of Technical Requirements for Registration of Pharmaceuticals for Human Use (ICH) (2000a). *ICH Harmonised Tripartite Guideline. Maintenance of the ICH Guideline on Non-Clinical Safety Studies for the Conduct of Human Clinical Trials for Pharmaceuticals. M3(R1)*, 9 November 2000, http://www.ich.org/LOB/media/MEDIA506.pdf (accessed 4 March 2008).

International Conference on Harmonisation of Technical Requirements for Registration of Pharmaceuticals for Human Use (ICH) (2000b). *ICH Harmonised Tripartite Guideline. Safety Pharmacology Studies for Human Pharmaceuticals. S7A*, 8 November 2000, http://www.ich.org/LOB/media/MEDIA504.pdf (accessed 5 March 2008).

International Conference on Harmonisation of Technical Requirements for Registration of Pharmaceuticals for Human Use (ICH) (2000c). *ICH Harmonised Tripartite Guideline. Clinical Investigation of Medicinal Products in the Pediatric Population. E11*.

International Conference on Harmonisation of Technical Requirements for Registration of Pharmaceuticals for Human Use (ICH) (2000d). *ICH Harmonised Tripartite Guideline. The Common Technical Document for the Registration of Pharmaceuticals for Human Use: Safety— M4S(R2). Nonclinical Overview and Nonclinical Summaries of Module 2 Organisation of Module 4*, http://www.ich.org/LOB/media/MEDIA559.pdf (accessed 10 March 2008).

International Conference on Harmonisation of Technical Requirements for Registration of Pharmaceuticals for Human Use (ICH) (2005a). *ICH Harmonised Tripartite Guideline. The Nonclinical Evaluation of the Potential for Delayed Ventricular Repolarization (QT Interval Prolongation) By Human Pharmaceuticals. S7B*, 12 May 2005, http://www.ich.org/LOB/media/MEDIA2192.pdf (accessed 5 March 2008).

International Conference on Harmonisation of Technical Requirements for Registration of Pharmaceuticals for Human Use (ICH) (2005b). *ICH Harmonised Tripartite Guideline. Detection of Toxicity to Reproduction for Medicinal Products & Toxicity to Male Fertility. S5(R2)*, November 2005, http://www.ich.org/LOB/media/MEDIA498.pdf (accessed 5 March 2008).

International Conference on Harmonisation of Technical Requirements for Registration of Pharmaceuticals for Human Use (ICH) (2005c). *ICH Harmonised Tripartite Guideline. Dose Selection for Carcinogenicity Studies of Pharmaceuticals & Limit Dose. S1C(R1)*, http://www.ich.org/LOB/media/MEDIA491.pdf (accessed 6 March 2008).

International Conference on Harmonisation of Technical Requirements for Registration of Pharmaceuticals for Human Use (ICH) (2005d). *ICH Harmonised Tripartite Guideline. Immunotoxicology Studies for Human Pharmaceuticals. S8*, http://www.ich.org/LOB/media/MEDIA1706.pdf (accessed 6 March 2008).

International Conference on Harmonisation of Technical Requirements for Registration of Pharmaceuticals for Human Use (ICH) (2005e). *ICH Harmonised Tripartite Guideline. The Clinical Evaluation of QT/QTc Interval Prolongation and Proarrhythmic Potential for Non-Antiarrhythmic Drugs. E14*, http://www.ich.org/LOB/media/MEDIA1476.pdf (accessed 7 March 2008).

International Conference on Harmonisation of Technical Requirements for Registration of Pharmaceuticals for Human Use (ICH) (2006a). *ICH Harmonised Tripartite Guideline. Impurities in New Drug Substances. Q3A(R2)*, 25 October 2006, http://www.ich.org/LOB/media/MEDIA422.pdf (accessed 4 March 2008).

International Conference on Harmonisation of Technical Requirements for Registration of Pharmaceuticals for Human Use (ICH) (2006b). *ICH Harmonised Tripartite Guideline. Impurities in New Drug Products. Q3B(R2)*, 2 June 2006, http://www.ich.org/LOB/media/MEDIA421.pdf (accessed 4 March 2008).

International Conference on Harmonisation of Technical Requirements for Registration of Pharmaceuticals for Human Use (ICH) (2006c). *Final Concept Paper: M3(R2): Revision of ICH M3(R1): Maintenance of the ICH Guideline on Non-Clinical Safety Studies for the Conduct of Human Clinical Trials for Pharmaceuticals*, http://www.ich.org/LOB/media/MEDIA3303.pdf (accessed 7 March 2008).

International Conference on Harmonisation of Technical Requirements for Registration of Pharmaceuticals for Human Use (ICH) (2006d). *Final Concept Paper S2(R1): Guidance on Genotoxicity Testing and Data Interpretation for Pharmaceuticals Intended for Human Use*, http://www.ich.org/LOB/media/MEDIA3304.pdf (accessed 18 March 2008).

International Conference on Harmonisation of Technical Requirements for Registration of Pharmaceuticals for Human Use (ICH) (2006e). *E5: Ethnic Factors in the Acceptability of Foreign Clinical Data. Questions & Answers (R1)*, http://www.ich.org/LOB/media/MEDIA1194.pdf (accessed 7 March 2008).

International Conference on Harmonisation of Technical Requirements for Registration of Pharmaceuticals for Human Use (ICH) (2007). *Final Concept Paper. S9: Preclinical Guideline on Oncology Therapeutic Development*, http://www.ich.org/LOB/media/MEDIA3922.pdf (accessed 6 March 2008).

International Conference on Harmonisation of Technical Requirements for Registration of Pharmaceuticals for Human Use (ICH) (2008). www.ich.org (accessed 4 March 2008).

Jacobson-Kram, D. (2007). Overview of exploratory IND (phase 0 trial): differences from the traditional IND, 2007 AAPS National Biotechnology Conference, San Diego, CA.

Japan Pharmaceutical Manufacturers Association (2007). *Pharmaceutical Administration and Regulations in Japan*, March 2007, http://www.nihs.go.jp/mhlw/jouhou/yakuji/yakuji-e0703.pdf (accessed 4 March 2008).

Johnston, G., Williams, R.L. (2002). 505(b)(2) applications: history, science, and experience. *Drug Information Journal, 36*, 319–323.

de Jonge, M.E. *et al.* (2005). Individualized cancer chemotherapy: strategies and performance of prospective studies on therapeutic drug monitoring with dose adaptation. *Clinical Pharmacokinetics, 44* (2), 147–173.

Joukhadar, C., Müller, M. (2005). Microdialysis: current applications in clinical pharmacokinetic studies and its potential role in the future. *Clinical Pharmacokinetics*, *44* (9), 895–913.

Kamimura, H. (2006). Genetic polymorphism of cytochrome P450s in beagles: possible influence of CYP1A2 deficiency on toxicological evaluations. *Archives Toxicology*, *80* (11), 732–736.

Kennedy, T. (1997). Managing the drug discovery/development interface. *Drug Discovery Today*, *2* (10), 436–444.

Kola, I., Landis, J. (2004). Can the pharmaceutical industry reduce attrition rates? *Nature Reviews. Drug Discovery*, *3* (8), 711–716.

Kraemer, M.J. *et al.* (1982). Altered theophylline clearance during an influenza B outbreak. *Pediatrics*, *69* (4), 476–480.

Kummar, S. *et al.* (2007). Inhibition of poly (ADP-ribose) polymerase (PARP) by ABT-888 in patients with advanced malignancies: results of a phase 0 trial. *Journal of Clinical Oncology*, *25* (18S), 3518.

Lacko, A.G. *et al.* (2007). Prospects and challenges of the development of lipoprotein-based formulations for anti-cancer drugs. *Expert Opinion on Drug Delivery*, *4* (6), 665–675.

Lange, D. *et al.* (1997). Effect of a cola beverage on the bioavailability of itraconazole in the presence of H2 blockers. *Journal of Clinical Pharmacol*, *37* (6), 535–540.

Lantz, R.J. *et al.* (2003). Metabolism, excretion, and pharmacokinetics of duloxetine in healthy human subjects. *Drug Metabolism and Disposition*, *31* (9), 1142–1150.

Lappin, G. *et al.* (2006). Use of microdosing to predict pharmacokinetics at the therapeutic dose: experience with 5 drugs. *Clinical Pharmacol Therapy*, *80* (3), 203–215.

Lappin, G., Temple, S. (2006). *Radiotracers in Drug Development*, CRC Press, Taylor & Francis Group.

Larson, J.L. (2005). Toxicity evaluations: ICH guidances and current practice, in *Preclinical Drug Development* (eds M.C. Rogge and D.R. Taft), Taylor and Francis, pp. 349–414.

Leblanc, B. *et al.* (1998). Binding of drugs to eye melanin is not predictive of ocular toxicity. *Regulatory Toxicology and Pharmacology*, *28* (2), 124–132.

Lesko, L.J. *et al.* (2003). Pharmacogenetics and pharmacogenomics in drug development and regulatory decision making: report of the first FDA-PWG-PhRMA-DruSafe workshop. *Journal of Clinical Pharmacol*, *43* (4), 342–358.

Levy, M. (1997). Role of viral infections in the induction of adverse drug reactions. *Drug Safety*, *16* (1), 1–8.

Li, J. *et al.* (2006). CYP3A phenotyping approach to predict systemic exposure to EGFR tyrosine kinase inhibitors. *Journal of the National Cancer Institute*, *98* (23), 1714–1723.

Li, M. *et al.* (2008). Symbiotic gut microbes modulate human metabolic phenotypes. *Proceedings of the National Academy of Sciences of the United States of America*, *105* (6), 2117–2122.

Lionberger, R. (2008). FDA critical path initiatives: opportunities for generic drug development. *The AAPS Journal*, *10* (1), 103–109.

Lytton, S.D. *et al.* (2002). Autoantibodies against cytochrome P450s in sera of children treated with immunosuppressive drugs. *Clinical and Experimental Immunology*, *127* (2), 293–302.

Mahmood, I. (2005). The correction factors do help in improving the prediction of human clearance from animal data. *Journal of Pharmaceutical Sciences*, *94* (5), 940–945.

Mahmood, I. (2006). *Clinical Pharmacology of Therapeutic Proteins*, Pine House Publishers.

Mahmood, I., Green, M.D. (2007). Drug interaction studies of therapeutic proteins or monoclonal antibodies. *Journal of Clinical Pharmacol*, *47* (12), 1540–1554.

Mahmood, I., Green, M.D., Fisher, J.E. (2003). Selection of the first-time dose in humans: comparison of different approaches based on interspecies scaling of clearance. *Journal of Clinical Pharmacol*, *43* (7), 692–697.

Malhotra, S., Dixit, R.K., Garg, S.K. (2002). Effect of an acidic beverage (Coca-Cola) on the pharmacokinetics of carbamazepine in healthy volunteers. *Methods and Findings in Experimental Clinical Pharmacology*, *24* (1), 31–33.

Mallet, L. (2008). *Drug Problems in the Elderly*, http://www.med.mcgill.ca/geriatrics/education/undergrad/ttag_pom_site/pharmacology.htm (accessed 8 March 2008).

Mallet, L., Spinewine, A., Huang, A. (2007). The challenge of managing drug interactions in elderly people. *Lancet*, *370* (9582), 185–191.

Mann, B.S. *et al.* (2007). FDA approval summary: Vorinostat for treatment of advanced primary cutaneous T-cell lymphoma. *Oncologist*, *12* (10), 1247–1252.

Mansfield, P., Henry, D., Tonkin, A. (2004). Single-enantiomer drugs: elegant science, disappointing effects. *Clinical Pharmacokinetics*, *43* (5), 287–290.

Marchetti, S., Schellens, J. (2007). The impact of FDA and EMEA guidelines on drug development in relation to Phase 0 trials. *British Journal of Cancer*, *97* (5), 577–581.

Marchetti, S. *et al.* (2007). Concise review: clinical relevance of drug–drug and herb–drug interactions mediated by the ABC transporter ABCB1 (MDR1, P-glycoprotein). *Oncologist*, *12* (8), 927–941.

Mascelli, M.A., Zhou, H., *et al.* (2007). Molecular, biologic, and pharmacokinetic properties of monoclonal antibodies: impact of these parameters on early clinical development. *J. Clin. Pharmacol.*, *47* (5), 553–65.

Meijerman, I., Beijnen, J.H., Schellens, J.H.M. (2006). Herb-drug interactions in oncology: focus on mechanisms of induction. *Oncologist*, *11* (7), 742–752.

Merck/Schering-Plough Pharmaceuticals (2008). *Zetia. Labelling information*, 7 February 2008, http://www.fda.gov/cder/foi/label/2008/021445s019lbl.pdf (accessed 4 March 2008).

Meredith, C. *et al.* (1985). Effects of influenza virus vaccine on hepatic drug metabolism. *Clinical Pharmacology and Therapeutics*, *37* (4), 396–401.

Micuda, S. *et al.* (2007). Diurnal variation of 6beta-hydroxycortisol in cardiac patients. *Physiological Research*, *57* (3), 307–313.

Ministry of Health and Welfare, Pharmaceutical Affairs Bureau (1995). *Japanese Guidelines for Nonclinical Studies of Drugs Manual 1995*, Yakuji Nippo, Ltd, Tokyo.

Miyakawa, H. *et al.* (2000). Immunoreactivity to various human cytochrome P450 proteins of sera from patients with autoimmune hepatitis, chronic hepatitis B, and chronic hepatitis C. *Autoimmunity*, *33* (1), 23–32.

Monahan, B.P. *et al.* (1990). Torsades de pointes occurring in association with terfenadine use. *The Journal of the American Medical Association*, *264* (21), 2788–2790.

Morita, K. *et al.* (1998). Strain differences in CYP3A-mediated C-8 hydroxylation (1,3,7-trimethyluric acid formation) of caffeine in wistar and dark agouti rats: rapid metabolism of caffeine in debrisoquine poor metabolizer model rats. *Biochemical Pharmacology*, *55* (9), 1405–1411.

Murray, M. (2006). Altered CYP expression and function in response to dietary factors: potential roles in disease pathogenesis. *Current Drug Metabolism*, *7*, 67–81.

Nagilla, R., Ward, K.W. (2004). A comprehensive analysis of the role of correction factors in the allometric predictivity of clearance from rat, dog, and monkey to humans. *Journal of Pharmaceutical Sciences*, *93* (10), 2522–2534.

Nagilla, R., Ward, K.W. (2005a). Erratum: a comprehensive analysis of the role of correction factors in the allometric predictivity of clearance from rat, dog, and monkey to humans. *Journal of Pharmaceutical Sciences*, *94* (1), 231–232.

Nagilla, R., Ward, K.W. (2005b). Correspondence. *Journal of Pharmaceutical Sciences*, *94* (5), 946–947.

Naito, S. *et al.* (2007). Current opinion: safety evaluation of drug metabolites in development of pharmaceuticals. *The Journal of Toxicological Sciences*, *32* (4), 329–341.

Nakai, K. *et al.* (2008). Decreased expression of cytochrome P450s 1A2, 2E1, and 3A4; and drug transporters Na+-taurocholate cotransporting polypeptide, organic cation transporter 1, and organic anion-transporting peptide-C correlates with the progression of liver fibrosis in chronic hepatitis C patients. *Drug Metabolism and Disposition*, *36* (9), 1786–1793.

Nanosphere Inc. (2008). *Verigene® Warfarin Metabolism Nucleic Acid Test (IVD)*, http://www.nanosphere-inc.com/VerigeneWarfarinMetabolismNucleicAcidTest_4472.aspx (accessed 7 March 2008).

National Cancer Institute (2007). *Phase 0 Trials in Oncologic Drug Development*, National Cancer Institute, Bethesda, MD.

Neuvonen, P., Kivisto, K. (1994). Enhancement of drug absorption by antacids: an unrecognised drug interaction. *Clinical Pharmacokinetics*, *27* (2), 120–128.

Nicholson, J.K., Holmes, E., Wilson, I.D. (2005). Gut microorganisms, mammalian metabolism and personalized health care. *Nature Reviews. Microbiology*, *3* (5), 431–438.

Novartis Australia Pty Limited (2007). *Novartis Withdraws Prexige® (lumiracoxib) in Australia in Response to Decision from Therapeutic Goods Administration (TGA)*, http://www.novartis.com.au/Prexige%20press%20release%2011%20August.pdf (accessed 3 March 2008).

Ohno, M. *et al.* (2000). Circadian variation of the urinary 6beta-hydroxycortisol to cortisol ratio that would reflect hepatic CYP3A activity. *European Journal of Clinical Pharmacology*, *55* (11–12), 861–865.

Organization for Economic Co-Operation and Development (OECD) (2007). *OECD Series on Principles of Good Laboratory Practice and Compliance Monitoring*, http://www.oecd.org/document/63/0,2340,en_2649_34381_2346175_1_1_1_37465,00.html (accessed 4 March 2008).

Organization for Economic Co-Operation and Development (OECD) (2008). *Good Laboratory Practice*, http://www.oecd.org/topic/0,2686,en_2649_34381_1_1_1_1_37465,00.html (accessed 4 March 2008).

Paulson, S.K. *et al.* (1999). Evidence for polymorphism in the canine metabolism of the cyclooxygenase 2 inhibitor, celecoxib. *Drug Metabolism and Disposition*, *27* (10), 1133–1142.

Pichini, S., Zuccaro, P., Pacifici, R. (1994). Drugs in semen. *Clinical Pharmacokinetics*, *26* (5), 356–373.

Projean, D. *et al.* (2005). Selective downregulation of hepatic cytochrome P450 expression and activity in a rat model of inflammatory pain. *Pharmaceutical Research*, *22* (1), 62–70.

Propper, D.J. *et al.* (2001). Phase I and pharmacokinetic study of PKC412, an inhibitor of protein kinase C. *Journal of Clinical Oncology*, *19* (5), 1485–1492.

Prueksaritanont, T., Lin, J.H., Baillie, T.A. (2006). Complicating factors in safety testing of drug metabolites: kinetic differences between generated and preformed metabolites. *Toxicology and Applied Pharmacology*, *217* (2), 143–152.

Qin, X. (2008). Gut microbiota: a new aspect that should be taken into more consideration when assessing the toxicity of chemicals or the adverse effects and efficacy of drugs. *Regulatory Toxicology and Pharmacology*, *51* (2), 251–251.

Renton, K.W. (2001). Alteration of drug biotransformation and elimination during infection and inflammation. *Pharmacology and Therapeutics*, *92*, 147–163.

Riechelmann, R., Saad, E. (2006). A systematic review on drug interactions in oncology. *Cancer Investigation*, *24*, 704–712.

Riechelmann, R. *et al.* (2005). Potential for drug interactions in hospitalized cancer patients. *Cancer Chemotherapy and Pharmacology*, *56*, 286–290.

Riechelmann, R.P. *et al.* (2007). Potential drug interactions and duplicate prescriptions among cancer patients. *Journal of the National Cancer Institute*, *99* (8), 592–600.

Robosky, L.C. *et al.* (2005). Metabonomic identification of two distinct phenotypes in Sprague-Dawley (Crl:CD(SD)) rats. *Toxicological Sciences*, *87* (1), 277–284.

Robosky, L.C. *et al.* (2006). Communication regarding metabonomic identification of two distinct phenotypes in Sprague-Dawley (Crl:CD(SD)) rats. *Toxicological Sciences*, *91* (1), 309.

Rocci, M.L. Jr, Shah, V.P., Rose, M.J., Sailstad, J.M. (2006). Bioanalytical method validation and implementation: best practices for chromatographic and ligand binding assays. *The AAPS Journal*, http://www.aapsj.org/theme_issues/theme_issue21.asp (accessed 4 March 2008).

Roche Diagnostics. (2008). *Amplichip*, http://www.amplichip.us/ (accessed 27 June 2008).

Rodriguez-Antona, C., Ingelman-Sundberg, M. (2006). Cytochrome P450 pharmacogenetics and cancer. *Oncogene*, *25* (11), 1679–1691.

Roffey, S.J. *et al.* (2007). What is the objective of the mass balance study? A retrospective analysis of data in animal and human excretion studies employing radiolabeled drugs. *Drug Metabolism Reviews*, *39* (1), 17–43.

Rogge, M.C., Taft, D.R. (2005). *Preclinical Drug Development. Drugs and the Pharmaceutical Sciences*, Vol. *152*, Taylor and Francis.

Rohner, T.C., Staab, D., Stoeckli, M. (2005). MALDI mass spectrometric imaging of biological tissue sections. *Mechanisms of Ageing and Development*, *126* (1), 177–185.

Rowland, M. (2007). Commentary on ACCP position statement on the use of microdosing in the drug development process. *Journal of Clinical Pharmacol*, *47* (12), 1595–1596.

Ryu, J.Y. *et al.* (2007). Development of the inje cocktail for high-throughput evaluation of five human cytochrome P450 isoforms *in vivo*. *Clinical Pharmacol Therapy*, *82* (5), 531–540.

Sadowski, D. (1994). Drug interactions with antacids. Mechanisms and clinical significance. *Drug Safety*, *11* (6), 395–407.

Salazar-Bookaman, M., Wainer, I., Patil, P. (1994). Relevance of drug-melanin interactions to ocular pharmacology and toxicology. *Journal of Ocular Pharmacology*, *10* (1), 217–239.

Sanders, J.M. *et al.* (2001). Comparative xenobiotic metabolism between Tg.AC and p53+/– genetically altered mice and their respective wild types. *Toxicological Sciences*, *61* (1), 54–61.

Sanderson, J., Naisbitt, D., Park, B. (2006). The role of bioactivation in drug-induced hypersensitivity reactions. *The AAPS Journal*, *8* (1), E55–E64.

Sarapa, N. (2007). Exploratory IND: a new regulatory strategy for early clinical drug development in the United States. *Ernst Schering Research Foundation Workshop*, *59*, 151–163.

Scharberg, D. (2007). *Critical Recommendations from the Third AAPS/FDA Bioanalytical Workshop*, http://www.pharmoutsource.com/publications/bp/articles/scharberg/critical_recommendations.pdf (accessed 4 March 2008).

Scharberg, D. (2008). *Four Lessons from MDS Montreal*, http://www.pharmoutsource.com/publications/bp/articles/scharberg/MDS_Best_Practice.pdf (accessed 4 March 2008).

Schering-Plough Research Institute (2006). *Product Information: PegIntron*, http://www.fda.gov/cder/foi/label/2006/103949s5124LBL.pdf (accessed 8 March 2008).

Schuster, D., Laggner, C., Langer, T. (2005). Why drugs fail—a study on side effects in new chemical entities. *Current Pharmaceutical Design*, *11*, 3545–3559.

Seeff, L. (2007). Herbal hepatotoxicity. *Clinics in Liver Disease*, *11* (3), 577–596.

Seitz, K., Zhou, H. (2007). Pharmacokinetic drug–drug interaction potentials for therapeutic monoclonal antibodies: reality check. *Journal of Clinical Pharmacol*, *47* (9), 1104–1118.

Senior, J. (2007). Drug hepatotoxicity from a regulatory perspective. *Clinics in Liver Disease*, *11* (3), 507–524.

Shah, V.P. *et al.* (1992). Analytical methods validation: bioavailability, bioequivalence, and pharmacokinetic studies. *Journal of Pharmaceutical Sciences*, *81* (3), 309–312.

Shah, V.P. *et al.* (2000). Bioanalytical method validation—a revisit with a decade of progress. *Pharmaceutical Research*, *17* (12), 1551–1557.

Skalli, S., Zaid, A., Soulaymani, R. (2007). Drug interactions with herbal medicines. *Therapeutic Drug Monitoring*, *29* (6), 679–686.

Smith, D.A., Obach, R.S. (2005). Seeing through the mist: abundance versus percentage. Commentary on metabolites in safety testing. *Drug Metabolism and Disposition*, *33* (10), 1409–1417.

Smith, D.A., Obach, R.S. (2006). Metabolites and safety: what are the concerns, and how should we address them? *Chemical Research in Toxicology*, *19* (12), 1570–1579.

Solon, E., Kraus, L. (2001). Quantitative whole-body autoradiography in the pharmaceutical industry. Survey results on study design, methods, and regulatory compliance. *Journal of Pharmacological and Toxicological Methods*, *46* (2), 73–81.

Solon, E.G., Lee, F. (2002). Methods determining phosphor imaging limits of quantitation in whole-body autoradiography rodent tissue distribution studies affect predictions of 14C human dosimetry. *Journal of Pharmacological and Toxicological Methods*, *46*, 83–91.

Sparreboom, A. *et al.* (2004). Herbal remedies in the united states: potential adverse interactions with anticancer agents. *Journal of Clinical Oncol*, *22* (12), 2489–2503.

Stephens, T., Brynner, R. (2001). *Dark Remedy: The impact of Thalidomide and Its Revival as a Vital Medicine*, Perseus Publishing, Cambridge, MA.

Stults, B.M., Hashisaki, P.A. (1983). Influenza vaccination and theophylline pharmacokinetics in patients with chronic obstructive lung disease. *Clinical Investigation*, *139* (5), 651–654.

Stumpf, W.E. (2005). Drug localization and targeting with receptor microscopic autoradiography. *Journal of Pharmacological and Toxicological Methods*, *51* (1), 25–40.

Tang, W. (2007). Drug metabolite profiling and elucidation of drug-induced hepatotoxicity. *Expert Opinion on Drug Metabolism and Toxicology*, *3* (3), 407–420.

Tchambaz, L. *et al.* (2006). Dose adaptation of antineoplastic drugs in patients with liver disease. *Drug Safety*, *29* (6), 509–522.

The Council of the European Union (1996). Council Directive 96/29/EURATOM of 13 May 1996 Laying Down Basic Safety Standards for the Protection of the Health of Workers and the General Public Against the Dangers Arising from Ionizing Radiation.

The Council of the European Union (1997). Council Directive 97/43/EURATOM of 30 June 1997 on Health Protection of Individuals Against the Dangers of Ionizing Radiation in Relation to Medical Exposure, And Repealing Directive 84/466/EURATOM.

Third Wave Technologies (2008). *Third Wave Technologies*, http://www.twt.com/ (accessed 7 March 2008).

U.S. Environmental Protection Agency (2008). *Good Laboratory Practices Standards*, http://www.epa.gov/Compliance/monitoring/programs/fifra/glp.html (accessed 4 March 2008).

U.S. Food and Drug Administration (US FDA) (1992). FDA's Policy Statement for the Development of New Stereoisomeric Drugs.

U.S. Food and Drug Administration (US FDA) (1995). Guidance for Industry: Content and Format of Investigational New Drug Applications (INDs) for Phase 1 Studies of Drugs, Including Well-Characterized, Therapeutic, Biotechnology-derived Products.

U.S. Food and Drug Administration (US FDA) (1996). Guidance for Industry: Single Dose Acute Toxicity Testing for Pharmaceuticals.

U.S. Food and Drug Administration (US FDA) (1997a). Guidance for Industry: Drug Metabolism/Drug Interaction Studies in the Drug Development Process: Studies In Vitro.

U.S. Food and Drug Administration (US FDA) (1997b). Points to Consider in the Manufacture and Testing of Monoclonal Antibody Products for Human Use.

U.S. Food and Drug Administration (US FDA) (1998a). *FDA Talk Paper. Roche Laboratories Announces Withdrawal of Posicor from the Market*, http://www.fda.gov/bbs/topics/ANSWERS/ANS00876.html (accessed 4 March 2008)

U.S. Food and Drug Administration (US FDA) (1998b). Guidance for Industry: Pharmacokinetics in Patients with Impaired Renal Function—Study Design, Data Analysis, and Impact on Dosing and Labeling.

U.S. Food and Drug Administration (US FDA) (1998c). Guidance for Industry: General Considerations for Pediatric Pharmacokinetic Studies for Drugs and Biological Products (Draft).

U.S. Food and Drug Administration (US FDA) (1999a). Guidance for Industry: in vivo Drug Metabolism/Drug Interaction Studies—Study Design, Data Analysis, and Recommendations for Dosing and Labeling.

U.S. Food and Drug Administration (US FDA) (1999b). Guidance for Industry: Development Programs for Drugs, Devices, and Biological Products for the Treatment of Rheumatoid Arthritis (RA).

U.S. Food and Drug Administration (US FDA) (1999c). Guidance for Industry: Applications Covered By Section 505(b)(2) (Draft).

U.S. Food and Drug Administration (US FDA) (2000a). Guidance for Industry: FDA guidance for industry on Formal Meetings with Sponsors and Applicants for PDUFA Products.

U.S. Food and Drug Administration (US FDA) (2000b). Guidance for Industry: Waiver of *in vivo* Bioavailability and Bioequivalence Studies for Immediate-Release Solid Oral Dosage Forms Based on a Biopharmaceutics Classification System.

U.S. Food and Drug Administration (US FDA) (2000c). Guidance for Industry: Q & A. Content and Format of INDs for Phase 1 Studies of Drugs, Including Well-Characterized, Therapeutic, Biotechnology-Derived Products.

U.S. Food and Drug Administration (US FDA) (2001a). *Frequently Asked Questions on Drug Development and Investigational New Drug Applications*, http://www.fda.gov/cder/about/smallbiz/faq.htm (accessed 10 March 2008).

U.S. Food and Drug Administration (US FDA) (2001b). Guidance for Industry: Bioanalytical Method Validation.

U.S. Food and Drug Administration (US FDA) (2001c). Guidance for Reviewers: Pharm/Tox Review Format.

U.S. Food and Drug Administration (US FDA) (2001d). Manual of Policies and Procedures. INDs: Screening INDs (MAPP 6030.4), Center for Drug Evaluation and Research (Offices of Review Management and Pharmaceutical Sciences) Editor.

U.S. Food and Drug Administration (US FDA) (2001e). Reviewer Guidance: Integration of Study Results to Assess Concerns about Human Reproductive and Developmental Toxicities (Draft).

U.S. Food and Drug Administration (US FDA) (2002a). Guidance for Industry: Carcinogenicity Study Protocol Submissions.

U.S. Food and Drug Administration (US FDA) (2002b). Guidance for Industry: Food-Effect Bioavailability and Fed Bioequivalence Studies.

U.S. Food and Drug Administration (US FDA) (2002c). Guidance for Industry: Liposome Drug Products Chemistry, Manufacturing, and Controls; Human Pharmacokinetics and Bioavailability; and Labeling Documentation (Draft).

U.S. Food and Drug Administration (US FDA) (2002d). Inspra (Epleronone) Tablets: Clinical Pharmacology Biopharmaceutics Review, pp. 70–72.

U.S. Food and Drug Administration (US FDA) (2002e). Zetia—Summary Basis of Approval—Pharmacology Review.

U.S. Food and Drug Administration (US FDA) (2003a). Guidance for Industry: Photosafety Testing.

U.S. Food and Drug Administration (US FDA) (2003b). Guidance for Industry: Exposure-Response Relationships—Study Design, Data Analysis, and Regulatory Applications.

U.S. Food and Drug Administration (US FDA) (2003c). Guidance for Industry: Bioavailability and Bioequivalence Studies for Orally Administered Drug Products—General Considerations.

U.S. Food and Drug Administration (US FDA) (2003d). Guidance for Industry: Pharmacokinetics in Patients with Impaired Hepatic Function: Study Design, Data Analysis, and Impact on Dosing and Labeling.

U.S. Food and Drug Administration (US FDA) (2004a). Challenges and Opportunity on the Critical Path to New Medical Products.

U.S. Food and Drug Administration (US FDA) (2004b). Guidance for Industry: IND Exemptions for Studies of Lawfully Marketed Drug or Biological Products for the Treatment of Cancer.

U.S. Food and Drug Administration (US FDA) (2004c). Guidance for Industry: Pharmacokinetics in Pregnancy—Study Design, Data Analysis, and Impact on Dosing and Labeling (Draft).

U.S. Food and Drug Administration (US FDA) (2004d). Letter from Joanne Rhoads to Gilbert Godin on 26 Apr 2004.

U.S. Food and Drug Administration (US FDA) (2004e). Letter from Joanne Rhoads to Gilbert Godin on 21 Dec 2004.

U.S. Food and Drug Administration (US FDA) (2004f). Natalizumab: Clinical Pharmacology and Biopharmaceutics Review(s).

U.S. Food and Drug Administration (US FDA) (2005a). *Frequently Asked Questions on the Pre-Investigational New Drug (IND) Meeting*, http://www.fda.gov/cder/about/smallbiz/pre_IND_qa.htm (accessed 3 March 2008).

U.S. Food and Drug Administration (US FDA) (2005b). Guidance for Industry, Investigators, and Reviewers: Exploratory IND Studies (Draft).

U.S. Food and Drug Administration (US FDA) (2005c). Guidance for Industry: Safety Testing of Drug Metabolites (Draft).

U.S. Food and Drug Administration (US FDA) (2005d). Guidance for Industry: Nonclinical Studies for the Safety Evaluation of Pharmaceutical Excipients.

U.S. Food and Drug Administration (US FDA) (2005e). Guidance for Industry: Estimating the Maximum Safe Starting Dose in Initial Clinical Trials for Therapeutics in Adult Healthy Volunteers.

U.S. Food and Drug Administration (US FDA) (2005f). Guidance for Industry: Pharmacogenomic Data Submissions.

U.S. Food and Drug Administration (US FDA) (2005g). Guidance for Industry: Clinical Lactation Studies—Study Design, Data Analysis, and Recommendations for Labeling (Draft).

U.S. Food and Drug Administration (US FDA) (2005h). Nexavar. Summary Basis of Approval: Pharmacology Review.

U.S. Food and Drug Administration (US FDA) (2006a). Critical Path Opportunities Report.

U.S. Food and Drug Administration (US FDA) (2006b). Critical Path Opportunities List.

U.S. Food and Drug Administration (US FDA) (2006c). Critical Path Opportunities Initiated During 2006.

U.S. Food and Drug Administration (US FDA) (2006d). *Drug Development and Drug Interactions*, http://www.fda.gov/cder/drug/drugInteractions/ (accessed 8 March 2008).

U.S. Food and Drug Administration (US FDA) (2006e). *Food and Drug Administration. Center for Drug Evaluation and Research. Advisory Committee for Pharmaceutical Science (ACPS) October 6, 2006. Highly Variable Drugs—Bioequivalence Issues*, http://www.fda.gov/ohrms/dockets/ac/cder06.html#PharmScience (accessed 7 March 2008).

U.S. Food and Drug Administration (US FDA) (2006f). Guidance for Industry, Investigators and Reviewers: Exploratory IND Studies.

U.S. Food and Drug Administration (US FDA) (2006g). Guidance for Industry: Drug Interaction Studies—Study Design, Data Analysis, and Implications for Dosing and Labeling (Draft).

U.S. Food and Drug Administration (US FDA) (2006h). Guidance for Industry: Nonclinical Safety Evaluation of Drug or Biologic Combinations.

U.S. Food and Drug Administration (US FDA) (2006i). Guidance for Industry: Nonclinical Safety Evaluation of Pediatric Drug Products.

U.S. Food and Drug Administration (US FDA) (2006j). Guidance for Industry: Warnings and Precautions, Contraindications, and Boxed Warning Sections of Labeling for Human Prescription Drug and Biological Products—Content and Format (Draft).

U.S. Food and Drug Administration (US FDA) (2006k). Letter from Joseph Salewski to Stephen DeFalco on 31 Aug 2006.

U.S. Food and Drug Administration (US FDA) (2006l). Sutent: Summary Basis of Approval. Clinical Pharmacology and Biopharmaceutics Review.

U.S. Food and Drug Administration (US FDA) (2007a). Code of Federal Regulations, 21 CFR Part 58. Good Laboratory Practice for Nonclinical Laboratory Studies.

U.S. Food and Drug Administration (US FDA) (2007b). Code of Federal Regulations, 21 CFR part 201. Product Labeling.

U.S. Food and Drug Administration (US FDA) (2007c). Code of Federal Regulations, 21 CFR Part 320. Bioavailability and Bioequivalence Requirements.

U.S. Food and Drug Administration (US FDA) (2007d). Code of Federal Regulations, 21 CFR Part 361.1. Radioactive Drugs for Certain Research Uses.

U.S. Food and Drug Administration (US FDA) (2007e). Critical Path Opportunities for Generic Drugs.

U.S. Food and Drug Administration (US FDA) (2007f). *Drug Development and Drug Interactions: Presentations*, http://www.fda.gov/CDER/drug/drugInteractions/presentations.htm (accessed 8 March 2008).

U.S. Food and Drug Administration (US FDA) (2007g). FDA Clears Genetic Lab Test for Warfarin Sensitivity.

U.S. Food and Drug Administration (US FDA) (2007h). Guidance for Industry and Review Staff: Target Product Profile—A Strategic Development Process Tool (Draft).

U.S. Food and Drug Administration (US FDA) (2007i). Guidance for Industry: Clinical Trial Endpoints for the Approval of Cancer Drugs and Biologics.

U.S. Food and Drug Administration (US FDA) (2007j). Guidance for Industry: Good Laboratory Practices Questions and Answers.

U.S. Food and Drug Administration (US FDA) (2007k). Guidance for Industry: Role of HIV Resistance Testing in Antiretroviral Drug Development.

U.S. Food and Drug Administration (US FDA) (2007l). Guidance for Industry: Drug-Induced Liver Injury: Premarketing Clinical Evaluation (Draft).

U.S. Food and Drug Administration (US FDA) (2007m). Guidance for Industry: Pharmacogenomic Data Submissions—Companion Guidance (Draft).

U.S. Food and Drug Administration (US FDA) (2007n). Guidance for Industry: Dosage and Administration Section of Labeling for Human Prescription Drug and Biological Products—Content and Format (Draft).

U.S. Food and Drug Administration (US FDA) (2007o). *Investigational New Drug (IND) Application Process*, http://www.fda.gov/cder/regulatory/applications/ind_page_1.htm (accessed 3 March 2008).

U.S. Food and Drug Administration (US FDA) (2007p). Janumet—Summary Basis of Approval—Pharmacology Review.

U.S. Food and Drug Administration (US FDA) (2007q). Nanotechnology. A Report of the U.S. Food and Drug Administration Nanotechnology Task Force.

U.S. Food and Drug Administration (US FDA) (2007r). NDA Approval (Maraviroc). Letter from Edward Cox (FDA) to Pfizer, Inc.

U.S. Food and Drug Administration (US FDA) (2007s). Report on the performance of drug and biologics firms in conducting postmarketing commitment studies availability. *Federal Register*, *72* (22), 5069–5070.

U.S. Food and Drug Administration (US FDA) (2008a). Center for Drug Evaluation and Research. List of Guidance Documents.

U.S. Food and Drug Administration (US FDA) (2008b). *Drug-Induced Liver Toxicity*, http://www.fda.gov/cder/livertox/default.htm (accessed 6 March 2008).

U.S. Food and Drug Administration (US FDA) (2008c). *FDA Approved Drug Products*, http://www.accessdata.fda.gov/scripts/cder/drugsatfda/index.cfm (accessed 3 March 2008).

U.S. Food and Drug Administration (US FDA) (2008d). *FDA's Critical Path Initiative*, http://www.fda.gov/oc/initiatives/criticalpath/ (accessed 7 March 2008).

U.S. Food and Drug Administration (US FDA) (2008e). *Genomics at FDA*, http://www.fda.gov/cder/genomics/ (accessed 7 March 2008).

U.S. Food and Drug Administration (US FDA) (2008f). Guidance Agenda: Guidances CDER is Planning to Develop During Calendar Year 2008.

U.S. Food and Drug Administration (US FDA) (2008g). *Guidance Documents*, http://www.fda.gov/cder/guidance/index.htm (accessed 8 March 2008).

U.S. Food and Drug Administration (US FDA) (2008h). Guidance for Industry and Review Staff: Nonclinical Safety Evaluation of Reformulated Drug Products Intended for Administration by an Alternate Route (Draft).

U.S. Food and Drug Administration (US FDA) (2008i). Guidance for Industry: Diabetes Mellitus: Developing Drugs and Therapeutic Biologics for Treatment and Prevention (Draft).

U.S. Food and Drug Administration (US FDA) (2008j). Guidance for Industry: Safety Testing of Drug Metabolites.

U.S. Food and Drug Administration (US FDA) (2008k). *Guidances We Develop and Enforcement Information*, http://www.fda.gov/cder/regulatory/default.htm#Legislation (accessed 18 March 2008).

U.S. Food and Drug Administration (US FDA) (2008l). *Small Business Assistance*, http://www.fda.gov/cder/about/smallbiz/default.htm (accessed 10 March 2008).

U.S. Food and Drug Administration, Center for Drug Evaluation and Research (US FDA CDER) (2006). Oncologic Drugs Advisory Committee Transcript: Pre-Clinical Requirements and Phase 1 Trial Design Issues for the Development of Oncologic Products.

U.S. National Institutes of Health (2008). *PKC412 With or Without Itraconazole in Treating Patients With Acute Myeloid Leukemia or Myelodysplastic Syndrome*, http://clinicaltrials.gov/ct2/show/NCT00045578?intr=%22Itraconazole%22&rank=18 (accessed 7 March 2008).

Uno, S. *et al.* (2007). Decreased intestinal CYP3A and P-glycoprotein activities in rats with adjuvant arthritis. *Drug Metabolism and Pharmacokinetics*, *22* (4), 313–321.

Vickers, A.E.M. *et al.* (1999). Multiple cytochrome P-450s involved in the metabolism of terbinafine suggest a limited potential for drug–drug interactions. *Drug Metabolism and Disposition*, *27* (9), 1029–1038.

Villeneuve, J., Pichette, V. (2004). Cytochrome P450 and liver disease. *Current Drug Metabolism*, *5* (3), 273–282.

Viswanathan, C.T. *et al.* (2007). Workshop/conference report—quantitative bioanalytical methods validation and implementation: best practices for chromatographic and ligand binding assays. *The AAPS Journal*, *9* (1), E30–E42.

Vuong, L.T. *et al.* (2007). Use of accelerator mass spectrometry to measure the pharmacokinetics and peripheral blood mononuclear cell concentrations of zidovudine. *Journal of Pharmaceutical Sciences*, *97* (7), 2833–2843.

Ward, K.W., Smith, B.R. (2004). A comprehensive quantitative and qualitative evaluation of extrapolation of intravenous pharmacokinetic parameters from rat, dog, and monkey to humans. I. clearance. *Drug Metabolism and Disposition*, *32* (6), 603–611.

Ward, K.W., Smith, B.R. (2004). A comprehensive quantitative and qualitative evaluation of extrapolation of intravenous pharmacokinetic parameters from rat, dog, and monkey to humans. II. volume of distribution and mean residence time. *Drug Metabolism and Disposition*, *32* (6), 612–619.

Wasan, K.M., Cassidy, S.M. (1998). Role of plasma lipoproteins in modifying the biological activity of hydrophobic drugs. *Journal of Pharmaceutical Sciences*, *87* (4), 411–424.

Wasan, K.M. *et al.* (2008). Impact of lipoproteins on the biological activity and disposition of hydrophobic drugs: implications for drug discovery. *Nature Reviews. Drug Discovery*, *7* (1), 84–99.

Weiner, M., Kotkoskte, L. (2007). *Excipient Toxicity and Safety. Drugs and the Pharmaceutical Sciences*, Vol. 103, Informa Healthcare, New York.

Wikipedia (2008a). *TeGenero*, 31 May 2007, http://en.wikipedia.org/wiki/TeGenero (accessed 3 March 2008).

Wikipedia (2008b). *Paracelsus*, 2 Mar 2008, http://en.wikipedia.org/wiki/Paracelsus (accessed 4 Mar 2008).

Wildiers, H. *et al.* (2003). Pharmacology of anticancer drugs in the elderly population. *Clinical Pharmacokinetics*, *42* (14), 1213–1242.

Williams, J.A. *et al.* (2008). PhRMA white paper on ADME pharmacogenomics. *Journal of Clinical Pharmacology*, *48* (7), 849–889.

Zhou, H., Tong, Z., McLeod, J.F. (2004). "Cocktail" approaches and strategies in drug development: valuable tool or flawed science? *Journal of Clinical Pharmacol*, *44* (2), 120–134.

Zhou, S. *et al.* (2007a). Metabolic activation of herbal and dietary constituents and its clinical and toxicological implications: an update. *Current Drug Metabolism*, *8* (6), 526–553.

Zhou, S. *et al.* (2007b). Identification of drugs that interact with herbs in drug development. *Drug Discovery Today*, *12* (15–16), 664–673.

Drug Metabolism Handbook: Concepts and Applications, Edited by Ala F. Nassar,
Paul F. Hollenberg, and JoAnn Scatina
Copyright © 2009 by John Wiley & Sons, Inc.